STUDENT SOLUTIONS MANUAL

DANIEL S. MILLER

Niagara County Community College

3RD EDITION

INTRODUCTORY & INTERMEDIATE ALGEBRA

for College Students

BLITZER

PEARSON

Prentice Hall

Upper Saddle River, NJ 07458

Editorial Director: Christine Hoag
Editor-in-Chief: Paul Murphy
Editorial Project Manager: Dawn Nuttall
Assistant Editor: Christine Whitlock
Senior Managing Editor: Linda Mihatov Behrens
Associate Managing Editor: Bayani Mendoza de Leon
Project Manager: Barbara Mack
Art Director: Heather Scott
Supplement Cover Manager: Paul Gourhan
Supplement Cover Designer: Victoria Colotta
Operations Specialist: Ilene Kahn
Senior Operations Supervisor: Diane Peirano

© 2009 Pearson Education, Inc.
Pearson Prentice Hall
Pearson Education, Inc.
Upper Saddle River, NJ 07458

The author and publisher of this book have used their best efforts in preparing this book. These efforts include the development, research, and testing of the theories and programs to determine their effectiveness. The author and publisher make no warranty of any kind, expressed or implied, with regard to these programs or the documentation contained in this book. The author and publisher shall not be liable in any event for incidental or consequential damages in connection with, or arising out of, the furnishing, performance, or use of these programs.

Printed in the United States of America

10 9 8 7 6 5 4 3 2

ISBN-13:   978-0-13-603166-6   Standalone
ISBN-10:   0-13-603166-8   Standalone
ISBN-13:   978-0-13-603167-3   Component
ISBN-10:   0-13-603167-6   Component

Pearson Education Ltd., London
Pearson Education Singapore, Pte. Ltd.
Pearson Education Canada, Inc.
Pearson Education—Japan
Pearson Education Australia PTY, Limited
Pearson Education North Asia, Ltd., Hong Kong
Pearson Educación de Mexico, S.A. de C.V.
Pearson Education Malaysia, Pte. Ltd.
Pearson Education Upper Saddle River, New Jersey

TABLE OF CONTENTS for STUDENT SOLUTIONS

INTRODUCTORY AND INTERMEDIATE ALGEBRA FOR COLLEGE STUDENTS 3E

Chapter 1   Variables, Real Numbers, and Mathematical Models 1

Chapter 2   Linear Equations and Inequalities in One Variable 38

Chapter 3   Linear Equations in Two Variables 100

Chapter 4   Systems of Linear Equations 145

Chapter 5   Exponents and Polynomials 215

Chapter 6   Factoring Polynomials 270

Chapter 7   Rational Expressions 313

                Mid-Textbook Check Point 410

Chapter 8   Basics of Functions 415

Chapter 9   Inequalities and Problem Solving 444

Chapter 10   Radicals, Radical Functions, and Rational Exponents 496

Chapter 11   Quadratic Equations and Functions 561

Chapter 12   Exponential and Logarithmic Functions 674

Chapter 13   Conic Sections and Systems of Nonlinear Equations 734

Chapter 14   Sequences, Series, and the Binomial Theorem 811

Appendices   872

Chapter 1
Variables, Real Numbers, and Mathematical Models

1.1 Check Points

1. **a.** $6 + 2x = 6 + 2(10) = 26$

   b. $2(x + 6) = 2(10 + 6) = 32$

2. **a.** $7x + 2y = 7 \cdot 3 + 2 \cdot 8 = 21 + 16 = 37$

   b. $\dfrac{6x - y}{2y - x - 8} = \dfrac{6 \cdot 3 - 8}{2 \cdot 8 - 3 - 8} = \dfrac{10}{5} = 2$

3. **a.** $6x$

   b. $4 + x$

   c. $3x + 5$

   d. $12 - 2x$

   e. $\dfrac{15}{x}$

4. **a.** $9x - 3 = 42$
   $9(6) - 3 = 42$
   $54 - 3 = 42$
   $\quad\quad 51 = 42, \;$ false
   6 is not a solution.

   b. $2(y + 3) = 5y - 3$
   $2(3 + 3) = 5(3) - 3$
   $\quad 2(6) = 15 - 3$
   $\quad\quad 12 = 12, \;$ true
   3 is a solution.

5. **a.** $\dfrac{x}{6} = 5$

   b. $7 - 2x = 1$

6. **a.** $d = 4n + 5$
   $d = 4(15) + 5 = 65$

   65% of marriages end in divorce after 15 years when the wife is under 18 at the time of marriage.

   b. According to the line graph, 60% of marriages end in divorce after 15 years when the wife is under 18 at the time of marriage.

   c. The mathematical model overestimates the actual percentage shown in the graph by 5%.

1.1 Exercise Set

1. $x + 8 = 4 + 8 = 12$

3. $12 - x = 12 - 4 = 8$

5. $5x = 5 \cdot 4 = 20$

7. $\dfrac{28}{x} = \dfrac{28}{4} = 7$

9. $5 + 3x = 5 + 3 \cdot 4 = 5 + 12 = 17$

11. $2(x + 5) = 2(4 + 5) = 2(9) = 18$

13. $\dfrac{12x - 8}{2x} = \dfrac{12 \cdot 4 - 8}{2 \cdot 4} = \dfrac{48 - 8}{8} = \dfrac{40}{8} = 5$

15. $2x + y = 2 \cdot 7 + 5 = 14 + 5 = 19$

17. $2(x + y) = 2(7 + 5) = 2(12) = 24$

19. $4x - 3y = 4 \cdot 7 - 3 \cdot 5 = 28 - 15 = 13$

21. $\dfrac{21}{x} + \dfrac{35}{y} = \dfrac{21}{7} + \dfrac{35}{5} = 3 + 7 = 10$

23. $\dfrac{2x - y + 6}{2y - x} = \dfrac{2 \cdot 7 - 5 + 6}{2 \cdot 5 - 7} = \dfrac{14 - 5 + 6}{10 - 7} = \dfrac{15}{3} = 5$

25. $x + 4$

27. $x - 4$

29. $x + 4$

31. $x - 9$

33. $9 - x$

35. $3x - 5$

37. $12x - 1$

39. $\dfrac{10}{x} + \dfrac{x}{10}$

41. $\dfrac{x}{30} + 6$

43. $x + 14 = 20$

$6 + 14 = 20$

$20 = 20,$ true

The number is a solution.

45. $30 - y = 10$

$30 - 20 = 10$

$10 = 10,$ true

The number is a solution.

47. $4z = 20$

$4(10) = 20$

$40 = 20,$ false

The number is not a solution.

49. $\dfrac{r}{6} = 8$

$\dfrac{48}{6} = 8$

$8 = 8,$ true

The number is a solution.

51. $4m + 3 = 23$

$4(6) + 3 = 23$

$24 + 3 = 23$

$27 = 23,$ false

The number is not a solution.

53. $5a - 4 = 2a + 5$

$5(3) - 4 = 2(3) + 5$

$15 - 4 = 6 + 5$

$11 = 11,$ true

The number is a solution.

55. $6(p - 4) = 3p$

$6(8 - 4) = 3(8)$

$6(4) = 24$

$24 = 24,$ true

The number is a solution.

57. $2(w + 1) = 3(w - 1)$

$2(7 + 1) = 3(7 - 1)$

$2(8) = 3(6)$

$16 = 18,$ false

The number is not a solution.

59. $4x = 28$

61. $\dfrac{14}{x} = \dfrac{1}{2}$

63. $20 - x = 5$

65. $2x + 6 = 16$

67. $3x - 5 = 7$

69. $4x + 5 = 33$

71. $4(x + 5) = 33$

73. $5x = 24 - x$

75. First find x.

$x = 7y + 2$

$x = 7(5) + 2 = 37$

Evaluate the expression.

$\dfrac{x - y}{4} = \dfrac{37 - 5}{4} = \dfrac{32}{4} = 8$

77. First find x.

$x = \dfrac{y}{4} - 1$

$x = \dfrac{12}{4} - 1 = 3 - 1 = 2$

Evaluate the expression.

$4x + 3(y + 5) = 4(2) + 3(12 + 5)$

$= 8 + 3(17)$

$= 8 + 51$

$= 59$

79. a. $2(x + 3y) = 2(4 + 3 \cdot 1) = 2(7) = 14$

b. $5z - 30 = 40$

$5(14) - 30 = 40$

$70 - 30 = 40$

$40 = 40,$ true

Yes, it is a solution.

81. a. $6x - 2y = 6 \cdot 3 - 2 \cdot 6 = 18 - 12 = 6$

b. $7w = 45 - 2w$

$7(6) = 45 - 2(6)$

$42 = 45 - 12$

$42 = 33,$ false

No, it is not a solution.

83. a. $p = 16 - n$

$p = 16 - 5$

$= 11$

The percentage of African American players was 11% in 2002.
The model overestimates the actual percentage shown in the bar graph by 1%.

b. $p = 16 - n$

$p = 16 - 7$

$= 9$

The percentage of African American players was 9% in 2004.
The model gives the actual percentage shown in the bar graph in 2004.

85. a. $p = 7.4h + 6$

$p = 7.4(4) + 6$

$= 35.6$

The percentage of U.S. adults that got 8 or more hours of sleep was 35.6% in 1998.
The model overestimates the actual percentage shown in the bar graph by 0.6%.

b. $p = 3.7h + 15$

$p = 3.7(4) + 15$

$= 29.8$

The percentage of U.S. adults that got 8 or more hours of sleep was 29.8% in 2005.
The model overestimates the actual percentage shown in the bar graph by 3.8%.

87. a. $H = 0.8(200 - A)$

$H = 0.8(200 - 145)$

$= 44$

A bowler with an average score of 145 will have a handicap of 44.

b. The bowler's final score will be $120 + 44$, or 164.

89. – 97. Answers will vary.

99. makes sense

101. makes sense

103. true

105. true

107. Choices of variables may vary.
Let h = hours worked.
Let s = salary.

$s = 20h$

109. a. O'Donnell: $p = \dfrac{s}{w} = \dfrac{2}{3.1} \approx \$0.65$

Vieira: $p = \dfrac{s}{w} = \dfrac{10}{5.6} \approx \$1.79$

Couric: $p = \dfrac{s}{w} = \dfrac{15}{7.3} \approx \$2.05$

b. Using the magazine's criterion, O'Donnell was the best buy.

110. $\dfrac{3}{7} \cdot \dfrac{2}{5} = \dfrac{3 \cdot 2}{7 \cdot 5} = \dfrac{6}{35}$

111. $\dfrac{2}{3} \div \dfrac{7}{5} = \dfrac{2}{3} \cdot \dfrac{5}{7} = \dfrac{2 \cdot 5}{3 \cdot 7} = \dfrac{10}{21}$

112. $\dfrac{9}{17} - \dfrac{5}{17} = \dfrac{9 - 5}{17} = \dfrac{4}{17}$

1.2 Check Points

1. $2\dfrac{5}{8} = \dfrac{2 \cdot 8 + 5}{8} = \dfrac{16 + 5}{8} = \dfrac{21}{8}$

2. 5 divided by 3 is 1 with a remainder of 2, so $\dfrac{5}{3} = 1\dfrac{2}{3}$.

3. Begin by selecting any two numbers whose product is 36.
Here is one possibility: $36 = 4 \cdot 9$
Because the factors 4 and 9 are not prime, factor each of these composite numbers.

$36 = 4 \cdot 9$

$= 2 \cdot 2 \cdot 3 \cdot 3$

Notice that 2 and 3 are both prime. The prime factorization of 36 is $2 \cdot 2 \cdot 3 \cdot 3$.

4. a. $\dfrac{10}{15} = \dfrac{2 \cdot \cancel{5}}{3 \cdot \cancel{5}} = \dfrac{2}{3}$

b. $\dfrac{42}{24} = \dfrac{\cancel{2} \cdot \cancel{3} \cdot 7}{2 \cdot 2 \cdot \cancel{2} \cdot \cancel{3}} = \dfrac{7}{4}$

When reducing fractions, it may not be necessary to write prime factorizations. We can use the greatest common factor to reduce this fraction.

$\dfrac{42}{24} = \dfrac{7 \cdot \cancel{6}}{4 \cdot \cancel{6}} = \dfrac{7}{4}$

c. $\dfrac{13}{15}$; Because 13 and 15 share no common factors (other than 1), $\dfrac{13}{15}$ is already reduced to its lowest terms.

d. $\dfrac{9}{45} = \dfrac{1 \cdot \cancel{9}}{5 \cdot \cancel{9}} = \dfrac{1}{5}$

5. a. $\dfrac{4}{11} \cdot \dfrac{2}{3} = \dfrac{4 \cdot 2}{11 \cdot 3} = \dfrac{8}{33}$

b. $6 \cdot \dfrac{3}{5} = \dfrac{6}{1} \cdot \dfrac{3}{5} = \dfrac{18}{5} = 3\dfrac{3}{5}$

c. $\dfrac{3}{7} \cdot \dfrac{2}{3} = \dfrac{3 \cdot 2}{7 \cdot 3} = \dfrac{6}{21} = \dfrac{2 \cdot \cancel{3}}{7 \cdot \cancel{3}} = \dfrac{2}{7}$

Remember that you can divide numerators and denominators by common factors before performing multiplication.

$\dfrac{3}{7} \cdot \dfrac{2}{3} = \dfrac{\cancel{3}}{7} \cdot \dfrac{2}{\cancel{3}} = \dfrac{2}{7}$

d. $\left(3\dfrac{2}{5}\right)\left(1\dfrac{1}{2}\right) = \dfrac{17}{5} \cdot \dfrac{3}{2} = \dfrac{51}{10} = 5\dfrac{1}{10}$

6. a. $\dfrac{5}{4} \div \dfrac{3}{8} = \dfrac{5}{4} \cdot \dfrac{8}{3} = \dfrac{5}{\cancel{4}} \cdot \dfrac{\cancel{4} \cdot 2}{3} = \dfrac{10}{3} = 3\dfrac{1}{3}$

b. $\dfrac{2}{3} \div 3 = \dfrac{2}{3} \div \dfrac{3}{1} = \dfrac{2}{3} \cdot \dfrac{1}{3} = \dfrac{2}{9}$

c. $3\dfrac{3}{8} \div 2\dfrac{1}{4} = \dfrac{27}{8} \div \dfrac{9}{4} = \dfrac{27}{8} \cdot \dfrac{4}{9}$

$\qquad = \dfrac{\cancel{9} \cdot 3}{\cancel{4} \cdot 2} \cdot \dfrac{\cancel{4}}{\cancel{9}} = \dfrac{3}{2} = 1\dfrac{1}{2}$

7. a. $\dfrac{2}{11} + \dfrac{3}{11} = \dfrac{5}{11}$

b. $\dfrac{5}{6} - \dfrac{1}{6} = \dfrac{4}{6} = \dfrac{2}{3}$

c. $3\dfrac{3}{8} - 1\dfrac{1}{8} = \dfrac{27}{8} - \dfrac{9}{8} = \dfrac{18}{8} = \dfrac{9}{4} = 2\dfrac{1}{4}$

8. $\dfrac{2}{3} = \dfrac{2 \cdot 7}{3 \cdot 7} = \dfrac{14}{21}$

9. a. $\dfrac{1}{2} + \dfrac{3}{5} = \dfrac{1 \cdot 5}{2 \cdot 5} + \dfrac{3 \cdot 2}{5 \cdot 2} = \dfrac{5}{10} + \dfrac{6}{10} = \dfrac{11}{10}$

b. $\dfrac{4}{3} - \dfrac{3}{4} = \dfrac{4 \cdot 4}{3 \cdot 4} - \dfrac{3 \cdot 3}{4 \cdot 3} = \dfrac{16}{12} - \dfrac{9}{12} = \dfrac{7}{12}$

c. $3\dfrac{1}{6} - 1\dfrac{11}{12} = \dfrac{19}{6} - \dfrac{23}{12} = \dfrac{19 \cdot 2}{6 \cdot 2} - \dfrac{23}{12}$

$\qquad = \dfrac{38}{12} - \dfrac{23}{12} = \dfrac{15}{12}$

$\qquad = \dfrac{5}{4} = 1\dfrac{1}{4}$

10. $10 = 2 \cdot 5$

$12 = 2 \cdot 2 \cdot 3$

$\text{LCD} = 2 \cdot 2 \cdot 3 \cdot 5 = 60$

$\dfrac{3}{10} + \dfrac{7}{12} = \dfrac{3 \cdot 6}{10 \cdot 6} + \dfrac{7 \cdot 5}{12 \cdot 5} = \dfrac{18}{60} + \dfrac{35}{60} = \dfrac{53}{60}$

11. a.
$$x - \frac{2}{9}x = 1$$
$$1\frac{2}{7} - \frac{2}{9}\left(1\frac{2}{7}\right) = 1$$
$$\frac{9}{7} - \frac{2}{9}\left(\frac{9}{7}\right) = 1$$
$$\frac{9}{7} - \frac{2}{7} = 1$$
$$\frac{7}{7} = 1$$
$$1 = 1, \text{ true}$$

The given fraction is a solution.

b.
$$\frac{1}{5} - w = \frac{1}{3}w$$
$$\frac{1}{5} - \frac{3}{20} = \frac{1}{3}\left(\frac{3}{20}\right)$$
$$\frac{4}{20} - \frac{3}{20} = \frac{1}{20}$$
$$\frac{1}{20} = \frac{1}{20}, \text{ true}$$

The given fraction is a solution.

12. a. $\dfrac{2}{3}(x-6)$

b. $\dfrac{3}{4}x - 2 = \dfrac{1}{5}x$

13. $C = \dfrac{5}{9}(F - 32)$

$C = \dfrac{5}{9}(77 - 32) = \dfrac{5}{9}(45) = 25$

77°F is equivalent to 25°C.

1.2 Exercise Set

1. $2\dfrac{3}{8} = \dfrac{2 \cdot 8 + 3}{8} = \dfrac{16 + 3}{8} = \dfrac{19}{8}$

3. $7\dfrac{3}{5} = \dfrac{7 \cdot 5 + 3}{5} = \dfrac{35 + 3}{5} = \dfrac{38}{5}$

5. $8\dfrac{7}{16} = \dfrac{8 \cdot 16 + 7}{16} = \dfrac{128 + 7}{16} = \dfrac{135}{16}$

7. 23 divided by 5 is 4 with a remainder of 3, so $\dfrac{23}{5} = 4\dfrac{3}{5}$.

9. 76 divided by 9 is 8 with a remainder of 4, so $\dfrac{76}{9} = 8\dfrac{4}{9}$.

11. 711 divided by 20 is 35 with a remainder of 11, so $\dfrac{711}{20} = 35\dfrac{11}{20}$.

13. composite; $22 = 2 \cdot 11$

15. composite; $20 = 4 \cdot 5 = 2 \cdot 2 \cdot 5$

17. 37 has no factors other than 1 and 37, so 37 is prime.

19. composite; $36 = 4 \cdot 9 = 2 \cdot 2 \cdot 3 \cdot 3$

21. composite; $140 = 10 \cdot 14 = 2 \cdot 5 \cdot 2 \cdot 7$
$$= 2 \cdot 2 \cdot 5 \cdot 7$$

23. 79 has no factors other than 1 and 79, so 79 is prime.

25. composite; $81 = 9 \cdot 9 = 3 \cdot 3 \cdot 3 \cdot 3$

27. composite; $240 = 10 \cdot 24$
$$= 2 \cdot 5 \cdot 2 \cdot 12$$
$$= 2 \cdot 5 \cdot 2 \cdot 3 \cdot 4$$
$$= 2 \cdot 5 \cdot 2 \cdot 3 \cdot 2 \cdot 2$$
$$= 2 \cdot 2 \cdot 2 \cdot 2 \cdot 3 \cdot 5$$

29. $\dfrac{10}{16} = \dfrac{\cancel{2} \cdot 5}{\cancel{2} \cdot 8} = \dfrac{5}{8}$

31. $\dfrac{15}{18} = \dfrac{\cancel{3} \cdot 5}{\cancel{3} \cdot 6} = \dfrac{5}{6}$

33. $\dfrac{35}{50} = \dfrac{\cancel{5} \cdot 7}{\cancel{5} \cdot 10} = \dfrac{7}{10}$

35. $\dfrac{32}{80} = \dfrac{\cancel{16} \cdot 2}{\cancel{16} \cdot 5} = \dfrac{2}{5}$

37. $\dfrac{44}{50} = \dfrac{\cancel{2} \cdot 22}{\cancel{2} \cdot 25} = \dfrac{22}{25}$

39. $\dfrac{120}{86} = \dfrac{\cancel{2} \cdot 60}{\cancel{2} \cdot 43} = \dfrac{60}{43}$

41. $\dfrac{2}{5} \cdot \dfrac{1}{3} = \dfrac{2 \cdot 1}{5 \cdot 3} = \dfrac{2}{15}$

43. $\dfrac{3}{8} \cdot \dfrac{7}{11} = \dfrac{3 \cdot 7 \cdot}{8 \cdot 11} = \dfrac{21}{88}$

45. $9 \cdot \dfrac{4}{7} = \dfrac{9}{1} \cdot \dfrac{4}{7} = \dfrac{9 \cdot 4}{1 \cdot 7} = \dfrac{36}{7}$ or $5\dfrac{1}{7}$

47. $\dfrac{1}{10} \cdot \dfrac{5}{6} = \dfrac{1 \cdot 5}{10 \cdot 6} = \dfrac{5}{60} = \dfrac{5 \cdot 1}{5 \cdot 12} = \dfrac{1}{12}$

49. $\dfrac{5}{4} \cdot \dfrac{6}{7} = \dfrac{5 \cdot 6}{4 \cdot 7} = \dfrac{30}{28} = \dfrac{2 \cdot 15}{2 \cdot 14} = \dfrac{15}{14}$ or $1\dfrac{1}{14}$

51. $\left(3\dfrac{3}{4}\right)\left(1\dfrac{3}{5}\right) = \dfrac{15}{4} \cdot \dfrac{8}{5} = \dfrac{120}{20} = \dfrac{20 \cdot 6}{20 \cdot 1} = 6$

53. $\dfrac{5}{4} \div \dfrac{4}{3} = \dfrac{5}{4} \cdot \dfrac{3}{4} = \dfrac{5 \cdot 3}{4 \cdot 4} = \dfrac{15}{16}$

55. $\dfrac{18}{5} \div 2 = \dfrac{18}{5} \cdot \dfrac{1}{2}$

$= \dfrac{18 \cdot 1}{5 \cdot 2} = \dfrac{18}{10} = \dfrac{2 \cdot 9}{2 \cdot 5} = \dfrac{9}{5}$ or $1\dfrac{4}{5}$

57. $2 \div \dfrac{18}{5} = \dfrac{2}{1} \cdot \dfrac{5}{18} = \dfrac{10}{18} = \dfrac{2 \cdot 5}{2 \cdot 9} = \dfrac{5}{9}$

59. $\dfrac{3}{4} \div \dfrac{1}{4} = \dfrac{3}{4} \cdot \dfrac{4}{1} = \dfrac{3 \cdot 4}{4 \cdot 1} = \dfrac{12}{4} = 3$

61. $\dfrac{7}{6} \div \dfrac{5}{3} = \dfrac{7}{6} \cdot \dfrac{3}{5} = \dfrac{7 \cdot 3}{6 \cdot 5} = \dfrac{21}{30} = \dfrac{3 \cdot 7}{3 \cdot 10} = \dfrac{7}{10}$

63. $\dfrac{1}{14} \div \dfrac{1}{7} = \dfrac{1}{14} \cdot \dfrac{7}{1} = \dfrac{7}{14} = \dfrac{7 \cdot 1}{7 \cdot 2} = \dfrac{1}{2}$

65. $6\dfrac{3}{5} \div 1\dfrac{1}{10} = \dfrac{33}{5} \div \dfrac{11}{10}$

$= \dfrac{33}{5} \cdot \dfrac{10}{11} = \dfrac{11 \cdot 3}{5} \cdot \dfrac{5 \cdot 2}{11} = \dfrac{6}{1} = 6$

67. $\dfrac{2}{11} + \dfrac{4}{11} = \dfrac{2 + 4}{11} = \dfrac{6}{11}$

69. $\dfrac{7}{12} + \dfrac{1}{12} = \dfrac{8}{12} = \dfrac{4 \cdot 2}{4 \cdot 3} = \dfrac{2}{3}$

71. $\dfrac{5}{8} + \dfrac{5}{8} = \dfrac{10}{8} = \dfrac{2 \cdot 5}{2 \cdot 4} = \dfrac{5}{4}$ or $1\dfrac{1}{4}$

73. $\dfrac{7}{12} - \dfrac{5}{12} = \dfrac{2}{12} = \dfrac{2 \cdot 1}{2 \cdot 6} = \dfrac{1}{6}$

75. $\dfrac{16}{7} - \dfrac{2}{7} = \dfrac{14}{7} = \dfrac{7 \cdot 2}{7 \cdot 1} = 2$

77. $\dfrac{1}{2} + \dfrac{1}{5} = \dfrac{1}{2} \cdot \dfrac{5}{5} + \dfrac{1}{5} \cdot \dfrac{2}{2}$

$= \dfrac{5}{10} + \dfrac{2}{10} = \dfrac{5 + 2}{10} = \dfrac{7}{10}$

79. $\dfrac{3}{4} + \dfrac{3}{20} = \dfrac{3}{4} \cdot \dfrac{5}{5} + \dfrac{3}{20}$

$= \dfrac{15}{20} + \dfrac{3}{20}$

$= \dfrac{18}{20} = \dfrac{2 \cdot 9}{2 \cdot 10} = \dfrac{9}{10}$

81. $\dfrac{3}{8} + \dfrac{5}{12} = \dfrac{3}{8} \cdot \dfrac{3}{3} + \dfrac{5}{12} \cdot \dfrac{2}{2}$

$= \dfrac{9}{24} + \dfrac{10}{24} = \dfrac{19}{24}$

83. $\dfrac{11}{18} - \dfrac{2}{9} = \dfrac{11}{18} - \dfrac{2}{9} \cdot \dfrac{2}{2} = \dfrac{11}{18} - \dfrac{4}{18} = \dfrac{7}{18}$

85. $\dfrac{4}{3} - \dfrac{3}{4} = \dfrac{4}{3} \cdot \dfrac{4}{4} - \dfrac{3}{4} \cdot \dfrac{3}{3}$

$= \dfrac{16}{12} - \dfrac{9}{12} = \dfrac{7}{12}$

87. $\dfrac{7}{10} - \dfrac{3}{16} = \dfrac{7}{10} \cdot \dfrac{8}{8} - \dfrac{3}{16} \cdot \dfrac{5}{5}$

$= \dfrac{56}{80} - \dfrac{15}{80} = \dfrac{41}{80}$

89. $3\dfrac{3}{4} - 2\dfrac{1}{3} = \dfrac{15}{4} - \dfrac{7}{3}$

$= \dfrac{15}{4} \cdot \dfrac{3}{3} - \dfrac{7}{3} \cdot \dfrac{4}{4}$

$= \dfrac{45}{12} - \dfrac{28}{12} = \dfrac{17}{12}$ or $1\dfrac{5}{12}$

91.

$$\frac{7}{2}x = 28$$

$$\frac{7}{2} \cdot 8 = 28$$

$$\frac{7}{2} \cdot \frac{8}{1} = 28$$

$$\frac{7}{\cancel{2}} \cdot \frac{4 \cdot \cancel{2}}{1} = 28$$

$$28 = 28, \text{ true}$$

The given number is a solution.

93.

$$w - \frac{2}{3} = \frac{3}{4}$$

$$1\frac{5}{12} - \frac{2}{3} = \frac{3}{4}$$

$$\frac{17}{12} - \frac{2}{3} = \frac{3}{4}$$

$$\frac{17}{12} - \frac{8}{12} = \frac{3}{4}$$

$$\frac{9}{12} = \frac{3}{4}$$

$$\frac{3}{4} = \frac{3}{4}, \text{ true}$$

The given number is a solution.

95.

$$20 - \frac{1}{3}z = \frac{1}{2}z$$

$$20 - \frac{1}{3} \cdot 12 = \frac{1}{2} \cdot 12$$

$$20 - \frac{1}{3} \cdot \frac{12}{1} = \frac{1}{2} \cdot \frac{12}{1}$$

$$20 - 4 = 12$$

$$16 = 12, \text{ false}$$

The given number is not a solution.

97.

$$\frac{2}{9}y + \frac{1}{3}y = \frac{3}{7}$$

$$\frac{2}{9} \cdot \frac{27}{35} + \frac{1}{3} \cdot \frac{27}{35} = \frac{3}{7}$$

$$\frac{2}{\cancel{9}} \cdot \frac{\cancel{9} \cdot 3}{35} + \frac{1}{\cancel{3}} \cdot \frac{\cancel{3} \cdot 9}{35} = \frac{3}{7}$$

$$\frac{6}{35} + \frac{9}{35} = \frac{3}{7}$$

$$\frac{15}{35} = \frac{3}{7}$$

$$\frac{3 \cdot \cancel{5}}{7 \cdot \cancel{5}} = \frac{3}{7}$$

$$\frac{3}{7} = \frac{3}{7}, \text{ true}$$

The given number is a solution.

99.

$$\frac{1}{3}(x - 2) = \frac{1}{5}(x + 4)$$

$$\frac{1}{3}(26 - 2) = \frac{1}{5}(26 + 4)$$

$$\frac{1}{3}(24) = \frac{1}{5}(30)$$

$$8 = 6, \text{ false}$$

The given number is not a solution.

101.

$$(y \div 6) + \frac{2}{3} = (y \div 2) - \frac{7}{9}$$

$$\left(4\frac{1}{3} \div 6\right) + \frac{2}{3} = \left(4\frac{1}{3} \div 2\right) - \frac{7}{9}$$

$$\left(\frac{13}{3} \div \frac{6}{1}\right) + \frac{2}{3} = \left(\frac{13}{3} \div \frac{2}{1}\right) - \frac{7}{9}$$

$$\left(\frac{13}{3} \cdot \frac{1}{6}\right) + \frac{2}{3} = \left(\frac{13}{3} \cdot \frac{1}{2}\right) - \frac{7}{9}$$

$$\frac{13}{18} + \frac{2}{3} = \frac{13}{6} - \frac{7}{9}$$

$$\frac{13}{18} + \frac{12}{18} = \frac{39}{18} - \frac{14}{18}$$

$$\frac{25}{18} = \frac{25}{18}, \text{ true}$$

The given number is a solution.

103. $\dfrac{1}{5}x$

105. $x - \dfrac{1}{4}x$

107. $x - \dfrac{1}{4} = \dfrac{1}{2}x$

109. $\dfrac{1}{7}x + \dfrac{1}{8}x = 12$

111. $\dfrac{2}{3}(x+6)$

113. $\dfrac{2}{3}x + 6 = x - 3$

115. $\dfrac{3}{4} \cdot \dfrac{a}{5} = \dfrac{3 \cdot a}{4 \cdot 5} = \dfrac{3a}{20}$

117. $\dfrac{11}{x} + \dfrac{9}{x} = \dfrac{11+9}{x} = \dfrac{20}{x}$

119. $\left(\dfrac{1}{2} - \dfrac{1}{3}\right) \div \dfrac{5}{8} = \left(\dfrac{3}{6} - \dfrac{2}{6}\right) \div \dfrac{5}{8}$

$\qquad = \dfrac{1}{6} \div \dfrac{5}{8}$

$\qquad = \dfrac{1}{6} \cdot \dfrac{8}{5} = \dfrac{8}{30} = \dfrac{\cancel{2} \cdot 4}{\cancel{2} \cdot 15} = \dfrac{4}{15}$

121. $\dfrac{1}{5}(x+2) = \dfrac{1}{2}\left(x - \dfrac{1}{5}\right)$

$\quad \dfrac{1}{5}\left(\dfrac{5}{8} + 2\right) = \dfrac{1}{2}\left(\dfrac{5}{8} - \dfrac{1}{5}\right)$

$\quad \dfrac{1}{5}\left(\dfrac{5}{8} + \dfrac{2}{1}\right) = \dfrac{1}{2}\left(\dfrac{5}{8} - \dfrac{1}{5}\right)$

$\quad \dfrac{1}{5}\left(\dfrac{5}{8} + \dfrac{16}{8}\right) = \dfrac{1}{2}\left(\dfrac{25}{40} - \dfrac{8}{40}\right)$

$\quad \dfrac{1}{5}\left(\dfrac{21}{8}\right) = \dfrac{1}{2}\left(\dfrac{17}{40}\right)$

$\qquad \dfrac{21}{40} = \dfrac{17}{80}$

$\qquad \dfrac{42}{80} = \dfrac{17}{80}$, false

The given number is not a solution.

123. $C = \dfrac{5}{9}(F - 32)$

$C = \dfrac{5}{9}(68 - 32) = \dfrac{5}{9}(36) = 20$

$68°F$ is equivalent to $20°C$.

125. a. $H = \dfrac{7}{10}(220 - a)$

$\qquad H = \dfrac{7}{10}(220 - 20)$

$\qquad = \dfrac{7}{10}(200)$

$\qquad = 140$

The lower limit of the heart rate for a 20-year-old with this exercise goal is 140 beats per minute.

b. $H = \dfrac{4}{5}(220 - a)$

$\qquad H = \dfrac{4}{5}(220 - 20)$

$\qquad = \dfrac{4}{5}(200)$

$\qquad = 160$

The upper limit of the heart rate for a 20-year-old with this exercise goal is 160 beats per minute.

127. a. $H = \dfrac{9}{10}(220 - a)$

b. $H = \dfrac{9}{10}(220 - a)$

$\qquad H = \dfrac{9}{10}(220 - 40)$

$\qquad = \dfrac{9}{10}(180)$

$\qquad = 162$

The heart rate for a 40-year-old with this exercise goal is 162 beats per minute.

129. 1995 is 15 years after 1980.

$c = 615 - \dfrac{47}{5}n$

$c = 615 - \dfrac{47}{5} \cdot 15$

$\quad = 615 - 141$

$\quad = 474$

The model estimates that 474 billion cigarettes were smoked in 1995.

This underestimates the actual total shown in the bar graph by 2 billion.

131. – 149. Answers will vary.

141. makes sense

143. makes sense

145. false; Changes to make the statement true will vary.
A sample change is: $\frac{1}{2} + \frac{1}{5} = \frac{5}{10} + \frac{2}{10} = \frac{7}{10}$.

147. true

149.

150. 5

151. $2\frac{1}{2}$ or $\frac{5}{2}$

152. −4

1.3 Check Points

1. **a.** −500

　　　b. −282

2. **a.**

(a)

$$\begin{array}{c} +\!\!-\!\!\bullet\!\!-\!\!+\!\!-\!\!+\!\!-\!\!+\!\!-\!\!+\!\!-\!\!+\!\!-\!\!+\!\!-\!\!+\!\!-\!\!+\!\!-\!\!+\!\!\rightarrow \\ -5\ -4\ -3\ -2\ -1\ \ 0\ \ 1\ \ 2\ \ 3\ \ 4\ \ 5 \end{array}$$

　　　b.

(b)

$$\begin{array}{c} +\!\!-\!\!+\!\!-\!\!+\!\!-\!\!+\!\!-\!\!+\!\!-\!\!\bullet\!\!-\!\!+\!\!-\!\!+\!\!-\!\!+\!\!-\!\!+\!\!-\!\!+\!\!\rightarrow \\ -5\ -4\ -3\ -2\ -1\ \ 0\ \ 1\ \ 2\ \ 3\ \ 4\ \ 5 \end{array}$$

　　　c.

(c)

$$\begin{array}{c} +\!\!-\!\!+\!\!-\!\!+\!\!-\!\!+\!\!-\!\!+\!\!-\!\!+\!\!-\!\!+\!\!-\!\!+\!\!-\!\!\bullet\!\!-\!\!+\!\!-\!\!+\!\!\rightarrow \\ -5\ -4\ -3\ -2\ -1\ \ 0\ \ 1\ \ 2\ \ 3\ \ 4\ \ 5 \end{array}$$

3. **a.**

(a)

$$\begin{array}{c} +\!\!-\!\!+\!\!-\!\!+\!\!-\!\!+\!\!-\!\!+\!\!-\!\!+\!\!-\!\!+\!\!-\!\!+\!\!-\!\!+\!\!-\!\!\bullet\!\!-\!\!+\!\!\rightarrow \\ -5\ -4\ -3\ -2\ -1\ \ 0\ \ 1\ \ 2\ \ 3\ \ 4\ \ 5 \end{array}$$

　　　b.

(b)

$$\begin{array}{c} +\!\!-\!\!+\!\!-\!\!+\!\!-\!\!+\!\!-\!\!\bullet\!\!-\!\!+\!\!-\!\!+\!\!-\!\!+\!\!-\!\!+\!\!-\!\!+\!\!-\!\!+\!\!\rightarrow \\ -5\ -4\ -3\ -2\ -1\ \ 0\ \ 1\ \ 2\ \ 3\ \ 4\ \ 5 \end{array}$$

4. **a.**

$$\begin{array}{r} 0.375 \\ 8\overline{)3.000} \\ \underline{24} \\ 60 \\ \underline{56} \\ 40 \\ \underline{40} \\ 0 \end{array}$$

$$\frac{3}{8} = 0.375$$

b.

$$\begin{array}{r} 0.454... \\ 11\overline{)5.000...} \\ \underline{44} \\ 60 \\ \underline{55} \\ 50 \\ \underline{44} \\ 60 \end{array}$$

$$\frac{5}{11} = 0.\overline{45}$$

5. **a.** $\sqrt{9}$

　　　b. $0, \sqrt{9}$

　　　c. $-9, 0, \sqrt{9}$

　　　d. $-9, -1.3, 0, 0.\overline{3}, \sqrt{9}$

　　　e. $\frac{\pi}{2}, \sqrt{10}$

　　　f. $-9, -1.3, 0, 0.\overline{3}, \frac{\pi}{2}, \sqrt{9}, \sqrt{10}$

6. **a.** $14 > 5$ since 14 is to the right of 5 on the number line.

　　　b. $-5.4 < 2.3$ since -5.4 is to the left of 2.3 on the number line.

　　　c. $-19 < -6$ since -19 is to the left of -6 on the number line.

　　　d. $\frac{1}{4} < \frac{1}{2}$ since $\frac{1}{4}$ is to the left of $\frac{1}{2}$ on the number line.

7. **a.** $-2 \le 3$ is true because $-2 < 3$ is true.

　　　b. $-2 \ge -2$ is true because $-2 = -2$ is true.

　　　c. $-4 \ge 1$ is false because neither $-4 > 1$ nor $-4 = 1$ is true.

8. **a.** $|-4| = 4$

　　　b. $|6| = 6$

　　　c. $\left|-\sqrt{2}\right| = \sqrt{2}$

1.3 Exercise Set

1. −20

3. 8

5. −3000

7. −4 billion

9. 2 is shown as a dot on the number line.

11. −5 is shown as a dot on the number line.

13. $3\frac{1}{2}$ is shown as a dot on the number line.

15. $\frac{11}{3}$ is shown as a dot on the number line.

17. −1.8 is shown as a dot on the number line.

19. $-\frac{16}{5}$ is shown as a dot on the number line.

21.
$$
\begin{array}{r}
0.75 \\
4\overline{)3.00} \\
\underline{28} \\
20 \\
\underline{20} \\
0
\end{array}
$$

$\frac{3}{4} = 0.75$

23.
$$
\begin{array}{r}
0.35 \\
20\overline{)7.00} \\
\underline{60} \\
100 \\
\underline{100} \\
0
\end{array}
$$

$\frac{7}{20} = 0.35$

25.
$$
\begin{array}{r}
0.875 \\
8\overline{)7.000} \\
\underline{64} \\
60 \\
\underline{56} \\
40 \\
\underline{40} \\
0
\end{array}
$$

$\frac{7}{8} = 0.875$

27.
$$
\begin{array}{r}
0.818... \\
11\overline{)9.000...} \\
\underline{88} \\
20 \\
\underline{11} \\
90 \\
\underline{88} \\
20
\end{array}
$$

$\frac{9}{11} = 0.\overline{81}$

29.
$$
\begin{array}{r}
0.5 \\
2\overline{)1.0} \\
\underline{1.0} \\
0
\end{array}
$$

$-\frac{1}{2} = -0.5$

31.
$$
\begin{array}{r}
0.833... \\
6\overline{)5.000...} \\
\underline{48} \\
20 \\
\underline{18} \\
20 \\
\underline{18} \\
20
\end{array}
$$

$\frac{5}{6} = 0.8\overline{3}$

33. a. $\sqrt{100}$ (=10)

b. $0, \sqrt{100}$

c. $-9, 0, \sqrt{100}$

d. $-9, -\dfrac{4}{5}, 0, 0.25, 9.2, \sqrt{100}$

e. $\sqrt{3}$

f. $-9, -\dfrac{4}{5}, 0, 0.25, \sqrt{3}, 9.2, \sqrt{100}$

35. a. $\sqrt{64}$ (= 8)

b. $0, \sqrt{64}$

c. $-11, 0, \sqrt{64}$

d. $-11, -\dfrac{5}{6}, 0, 0.75, \sqrt{64}$

e. $\sqrt{5}, \pi$

f. $-11, -\dfrac{5}{6}, 0, 0.75, \sqrt{5}, \pi, \sqrt{64}$

37. The only whole number that is not a natural number is 0.

39. Answers will vary. As an example, one rational number that is not an integer is $\dfrac{1}{2}$.

41. Answers will vary. As an example, 6 is a number that is an integer, a whole number, and a natural number.

43. Answers will vary. As an example, one number that is an irrational number and a real number is π.

45. $\dfrac{1}{2} < 2$ since $\dfrac{1}{2}$ is to the left of 2 on the number line.

47. $3 > -\dfrac{5}{2}$ since 3 is to the right of $-\dfrac{5}{2} = -2\dfrac{1}{2}$.

49. $-4 > -6$ since -4 is to the right of -6.

51. $-2.5 < 1.5$ since -2.5 is to the left of 1.5.

53. $-\dfrac{3}{4} > -\dfrac{5}{4}$ since $-\dfrac{3}{4}$ is to the right of $-\dfrac{5}{4}$.

55. $-4.5 < 3$ since -4.5 is to the left of 3.

57. $\sqrt{2} < 1.5$ since $\sqrt{2} \approx 1.414$ is to the left of 1.5.

59. $0.\bar{3} > 0.3$ since $0.\bar{3} = 0.333...$ is to the right of 0.3.

61. $-\pi > -3.5$ since $-\pi \approx -3.14$ is to the right of -3.5.

63. $-5 \geq -13$ is true because $-5 > -13$ is true.

65. $-9 \geq -9$ is true because $-9 = -9$ is true.

67. $0 \geq -6$ is true because $0 > -6$ is true.

69. $-17 \geq 6$ is false because neither $-17 > 6$ nor $-17 = 6$ is true.

71. $|6| = 6$ because the distance between 6 and 0 on the number line is 6 units.

73. $|-7| = 7$ because the distance between -7 and 0 on the number line is 7 units.

75. $\left|\dfrac{5}{6}\right| = \dfrac{5}{6}$ because the distance between $\dfrac{5}{6}$ and 0 on the number line is $\dfrac{5}{6}$ units.

77. $|-\sqrt{11}| = \sqrt{11}$ because the distance between $-\sqrt{11}$ and 0 on the number line is $\sqrt{11}$ units.

79. $|-6| \ \square\ |-3|$

$6 \ \square\ 3$

$6 > 3$

Since $6 > 3$, $|-6| > |-3|$.

81. $\left|\dfrac{3}{5}\right| \ \square\ |-0.6|$

$|0.6| \ \square\ |-0.6|$

$0.6 \ \square\ 0.6$

$0.6 = 0.6$

Since $0.6 = 0.6$, $\left|\dfrac{3}{5}\right| = |-0.6|$.

83.

$$\frac{30}{40}-\frac{3}{4} \;\square\; \frac{14}{15}\cdot\frac{15}{14}$$

$$\frac{30}{40}-\frac{30}{40} \;\square\; \frac{\cancel{14}}{\cancel{15}}\cdot\frac{\cancel{15}}{\cancel{14}}$$

$$0 \;\square\; 1$$

$$0 < 1$$

Since $0<1$, $\dfrac{30}{40}-\dfrac{3}{4} < \dfrac{14}{15}\cdot\dfrac{15}{14}$.

85.

$$\frac{8}{13}\div\frac{8}{13} \;\square\; \left|-1\right|$$

$$\frac{8}{13}\cdot\frac{13}{8} \;\square\; 1$$

$$1 \;\square\; 1$$

$$1 = 1$$

Since $1=1$, $\dfrac{8}{13}\div\dfrac{8}{13}=\left|-1\right|$.

87. rational numbers

89. integers

91. all real numbers

93. whole numbers

95. a.

b. Rhode Island, Georgia, Louisiana, Florida, Hawaii

97. – 107. Answers will vary

109. does not make sense; Explanations will vary. Sample explanation: The Bismarck's resting place is lower because it is further below sea level.

111. makes sense

113. false; Changes to make the statement true will vary. A sample change is: All whole numbers are integers.

115. false; Changes to make the statement true will vary. A sample change is: Irrational numbers can be negative.

117. false; Changes to make the statement true will vary. A sample change is: All integers are rational numbers.

119. $-\dfrac{1}{2}d$

121. $-\sqrt{12}\approx-3.464$ and should be graphed between -4 and -3.

123. $2-\sqrt{5}\approx-0.236$ and should be graphed between -1 and 0.

124. $3(x+5)=3(4+5)=3(9)=27$
and
$3x+15=3(4)+15=12+15=27$
Both expressions have the same value.

125. $3x+5x=3(4)+5(4)=12+20=32$
and
$8x=8(4)=32$
Both expressions have the same value.

126. $9x-2x=9(4)-2(4)=36-8=28$
and
$7x=7(4)=28$
Both expressions have the same value.

1.4 Check Points

1. a. 3 terms

   b. 6

   c. 11

   d. $6x$ and $2x$

2. a. $x+14=14+x$

   b. $7y=y7$

3. a. $5x+17=17+5x$

   b. $5x+17=x5+17$

4. a. $8+(12+x)=(8+12)+x$
   $\qquad\qquad\quad=20+x$ or $x+20$

   b. $6(5x)=(6\cdot5)x=30x$

5. $8+(x+4)=8+(4+x)$
   $\qquad\qquad=(8+4)+x$
   $\qquad\qquad=12+x$ or $x+12$

6. $5(x+3) = 5 \cdot x + 5 \cdot 3$
$\qquad = 5x + 15$

7. $6(4y+7) = 6 \cdot 4y + 6 \cdot 7$
$\qquad = 24y + 42$

8. a. $7x + 3x = (7+3)x = 10x$

$\quad$ **b.** $9a - 4a = (9-4)a = 5a$

9. a. $8x + 7 + 10x + 3 = (8x + 10x) + (7+3)$
$\qquad\qquad\qquad\quad = 18x + 10$

$\quad$ **b.** $9x + 6y + 5x + 2y = (9x + 5x) + (6y + 2y)$
$\qquad\qquad\qquad\qquad = 14x + 8y$

10. $7(2x+3) + 11x = 7 \cdot 2x + 7 \cdot 3 + 11x$
$\qquad\qquad\qquad = 14x + 21 + 11x$
$\qquad\qquad\qquad = (14x + 11x) + 21$
$\qquad\qquad\qquad = 25x + 21$

11. $7(4x+3y) + 2(5x+y) = 7 \cdot 4x + 7 \cdot 3y + 2 \cdot 5x + 2 \cdot y$
$\qquad\qquad\qquad\qquad\quad = 28x + 21y + 10x + 2y$
$\qquad\qquad\qquad\qquad\quad = (28x + 10x) + (21y + 2y)$
$\qquad\qquad\qquad\qquad\quad = 38x + 23y$

1.4 Exercise Set

1. $3x + 5$

$\quad$ **a.** 2 terms
$\quad$ **b.** 3
$\quad$ **c.** 5
$\quad$ **d.** no like terms

3. $x + 2 + 5x$

$\quad$ **a.** 3 terms
$\quad$ **b.** 1
$\quad$ **c.** 2
$\quad$ **d.** x and $5x$ are like terms.

5. $4y + 1 + 3$

$\quad$ **a.** 3 terms
$\quad$ **b.** 4
$\quad$ **c.** 1
$\quad$ **d.** no like terms

7. $y + 4 = 4 + y$

9. $5 + 3x = 3x + 5$

11. $4x + 5y = 5y + 4x$

13. $5(x+3) = 5(3+x)$

15. $9x = x \cdot 9$ or $x9$

17. $x + y6 = x + 6y$

19. $7x + 23 = x7 + 23$

21. $5(x+3) = (x+3)5$

23. $7 + (5+x) = (7+5) + x = 12 + x$

25. $7(4x) = (7 \cdot 4)x = 28x$

27. $3(x+5) = 3(x) + 3(5) = 3x + 15$

29. $8(2x+3) = 8(2x) + 8(3) = 16x + 24$

31. $\dfrac{1}{3}(12+6r) = \dfrac{1}{3}(12) + \dfrac{1}{3}(12) + \dfrac{1}{3}(6r)$
$\qquad\qquad\quad = 4 + 2r$

33. $5(x+y) = 5x + 5y$

35. $3(x-2) = 3(x) - 3(2) = 3x - 6$

37. $2(4x-5) = 2(4x) - 2(5) = 8x - 10$

39. $\dfrac{1}{2}(5x-12) = \dfrac{1}{2}(5x) + \dfrac{1}{2}(-12)$
$\qquad\qquad\quad = \dfrac{5}{2}x - 6$

41. $(2x+7)4 = 2x(4) + 7(4) = 8x + 28$

43. $6(x+3+2y) = 6(x) + 6(3) + 6(2y)$
$\qquad\qquad\qquad = 6x + 18 + 12y$

45. $5(3x-2+4y) = 5(3x) - 5(2) + 5(4y)$
$\qquad\qquad\qquad\quad = 15x - 10 + 20y$

47. $7x + 10x = (7+10)x = 17x$

49. $11a - 3a = (11-3)a = 8a$

51. $3 + (x+11) = (3+11) + x = 14 + x$

53. $5y + 3 + 6y = (5y + 6y) + 3 = 11y + 3$

55. $2x + 5 + 7x - 4 = (2x + 7x) + (5 - 4)$
$$= 9x + 1$$

57. $11a + 12 + 3a + 2 = (11a + 3a) + (12 + 2)$
$$= 14a + 14$$

59. $5(3x + 2) - 4 = 15x + 10 - 4 = 15x + 6$

61. $12 + 5(3x - 2) = 12 + 15x - 10$
$$= 15x + 12 - 10 = 15x + 2$$

63. $7(3a + 2b) + 5(4a + 2b)$
$$= 21a + 14b + 20a + 10b$$
$$= 21a + 20a + 14b + 10b$$
$$= 41a + 24b$$

65. $7 + 2(x + 9)$
$\quad = 7 + (2x + 18)$ Distributive Property
$\quad = 7 + (18 + 2x)$ Commutative Property of Addition
$\quad = (7 + 18) + 2x$ Associative Property of Addition
$\quad = 25 + 2x$
$\quad = 2x + 25$ Commutative Property of Addition

67. $7x + 2x$
$\quad 7x + 2x = 9x$

69. $12x - 3x$
$\quad 12x - 3x = 9x$

71. $6(4x)$
$\quad 6(4x) = 24x$

73. $6(4 + x)$
$\quad 6(4 + x) = 24 + 6x$

75. $8 + 5(x - 1)$
$\quad 8 + 5(x - 1) = 8 + 5x - 5 = 5x + 3$

77. a. $M = 2(2n + 25) + 0.5(n + 2)$
$$= 4n + 50 + 0.5n + 1$$
$$= 4.5n + 51$$

   b. $M = 4.5n + 51$
$\quad M = 4.5 \cdot 5 + 51$
$\quad = 22.5 + 51$
$\quad = 73.5$
The percentage of U.S. men who used the Internet in 2005 was 73.5%.
This underestimates the actual percentage shown in the bar graph by 1.5%.

79. – 87. Answers will vary.

89. does not make sense; Explanations will vary.
Sample explanation: Subtraction does not have a commutative property.

91. makes sense

93. false; Changes to make the statement true will vary.
A sample change is: $(24 \div 6) \div 2 \neq 24 \div (6 \div 2)$.

95. false; Changes to make the statement true will vary.
A sample change is: Addition cannot be distributed over multiplication.

97. 60 because $150 - 90 = 60$

98. -40 because $-30 - 10 = -40$

99. -5 because $30 - 35 = -5$

Chapter 1 Mid-Chapter Check Point

1. $2 + 10x = 2 + 10(6)$
$$= 2 + 60$$
$$= 62$$

2. $10x - 4 = 10\left(\dfrac{3}{5}\right) - 4$
$$= 6 - 4$$
$$= 2$$

3. $\dfrac{xy}{2} + 4(y-x) = \dfrac{3 \cdot 10}{2} + 4(10-3)$

$\qquad\qquad\qquad = \dfrac{30}{2} + 4(7)$

$\qquad\qquad\qquad = 15 + 28$

$\qquad\qquad\qquad = 43$

4. $\dfrac{1}{4}x - 2$

5. $\dfrac{x}{6} + 5 = 19$

6. $3(x+2) = 4x - 1$

$\quad 3(6+2) = 4 \cdot 6 - 1$

$\qquad 3(8) = 24 - 1$

$\qquad\; 24 = 23, \;\; \text{false}$

The number is not a solution.

7. $\quad 8y = 12\left(y - \dfrac{1}{2}\right)$

$\quad 8 \cdot \dfrac{3}{4} = 12\left(\dfrac{3}{4} - \dfrac{1}{2}\right)$

$\qquad 6 = 12\left(\dfrac{3}{4} - \dfrac{2}{4}\right)$

$\qquad 6 = 12\left(\dfrac{1}{4}\right)$

$\qquad 6 = 3, \;\; \text{false}$

The number is not a solution.

8. a. $V = 747 - 21n$

$\quad V = 747 - 21 \cdot 10 = 537$

The number of violent crimes in 2000 was 537 per 100,000.
This overestimates the actual number shown in the bar graph by 30 per 100,000.

b. $V = 747 - 21n$

$\quad V = 747 - 21 \cdot 20 = 327$

If trends continue, the number of violent crimes in 2010 will be 327 per 100,000.

9. $\dfrac{7}{10} - \dfrac{8}{15} = \dfrac{7}{10} \cdot \dfrac{3}{3} - \dfrac{8}{15} \cdot \dfrac{2}{2}$

$\qquad\quad = \dfrac{21}{30} - \dfrac{16}{30} = \dfrac{5}{30} = \dfrac{1}{6}$

10. $\dfrac{2}{3} \cdot \dfrac{3}{4} = \dfrac{2}{\cancel{3}} \cdot \dfrac{\cancel{3}}{4} = \dfrac{2}{4} = \dfrac{1}{2}$

11. $\dfrac{5}{22} + \dfrac{5}{33} = \dfrac{5}{22} \cdot \dfrac{3}{3} + \dfrac{5}{33} \cdot \dfrac{2}{2} = \dfrac{15}{66} + \dfrac{10}{66} = \dfrac{25}{66}$

12. $\dfrac{3}{5} \div \dfrac{9}{10} = \dfrac{3}{5} \cdot \dfrac{10}{9} = \dfrac{3}{5} \cdot \dfrac{2 \cdot 5}{3 \cdot 3} = \dfrac{\cancel{3}}{\cancel{5}} \cdot \dfrac{2 \cdot \cancel{5}}{\cancel{3} \cdot 3} = \dfrac{2}{3}$

13. $\dfrac{23}{105} - \dfrac{2}{105} = \dfrac{21}{105} = \dfrac{\cancel{3} \cdot \cancel{7}}{\cancel{3} \cdot 5 \cdot \cancel{7}} = \dfrac{1}{5}$

14. $2\dfrac{7}{9} \div 3 = \dfrac{25}{9} \div \dfrac{3}{1} = \dfrac{25}{9} \cdot \dfrac{1}{3} = \dfrac{25}{27}$

15. $5\dfrac{2}{9} - 3\dfrac{1}{6} = \dfrac{47}{9} - \dfrac{19}{6}$

$\qquad\qquad = \dfrac{47}{9} \cdot \dfrac{2}{2} - \dfrac{19}{6} \cdot \dfrac{3}{3}$

$\qquad\qquad = \dfrac{94}{18} - \dfrac{57}{18} = \dfrac{37}{18} \;\text{ or }\; 2\dfrac{1}{18}$

16. $C = \dfrac{5}{9}(F - 32)$

$\quad C = \dfrac{5}{9}(50 - 32) = \dfrac{5}{9}(18) = 10$

$50°F$ is equivalent to $10°C$.

17. $-8000 < -8\dfrac{1}{4}$

18. $\dfrac{1}{11} = 0.\overline{09}$

19. $|-19.3| = 19.3$

20. $-11, \; -\dfrac{3}{7}, \; 0, \; 0.45, \text{ and } \sqrt{25}$ are rational numbers.

21. $5(x+3) = (x+3)5$

22. $5(x+3) = 5(3+x)$

23. $5(x+3) = 5x + 15$

24. $7(9x+3) + \dfrac{1}{3}(6x) = 63x + 21 + 2x$

$\qquad\qquad\qquad\quad = 65x + 21$

25. $2(3x+5y) + 4(x+6y) = 6x + 10y + 4x + 24y$

$\qquad\qquad\qquad\qquad\quad = 10x + 34y$

1.5 Check Points

1. $4+(-7)=-3$
   Start at 4 and move 7 units to the left.

   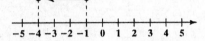

2. **a.** $-1+(-3)=-4$
   Start at -1 and move 3 units to the left.

   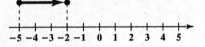

   b. $-5+3=-2$
   Start at -5 and move 3 units to the right.

3. **a.** $-10+(-25)=-35$

   b. $-0.3+(-1.2)=-1.5$

   c. $-\dfrac{2}{3}+\left(-\dfrac{1}{6}\right)=-\dfrac{4}{6}+\left(-\dfrac{1}{6}\right)=-\dfrac{5}{6}$

4. **a.** $-15+2=-13$

   b. $-0.4+1.6=1.2$

   c. $-\dfrac{2}{3}+\dfrac{1}{6}=-\dfrac{4}{6}+\dfrac{1}{6}=-\dfrac{3}{6}=-\dfrac{1}{2}$

5. **a.** $-20x+3x=(-20+3)x=-17x$

   b. $3y+(-10z)+(-10y)+16z$
   $=3y+(-10y)+(-10z)+16z$
   $=[3+(-10)]y+[(-10)+16]z$
   $=-7y+6z$

   c. $5(2x+3)+(-30x)=10x+15+(-30x)$
   $\qquad\qquad\qquad\quad=10x+(-30x)+15$
   $\qquad\qquad\qquad\quad=[10+(-30)]x+15$
   $\qquad\qquad\qquad\quad=-20x+15$

6. $2+(-4)+1+(-5)+3=(2+1+3)+[(-4)+(-5)]$
   $\qquad\qquad\qquad\qquad\qquad=6+(-9)$
   $\qquad\qquad\qquad\qquad\qquad=-3$
   At the end of 5 months the water level was down 3 feet.

1.5 Exercise Set

1. $7+(-3)=4$

   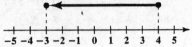

3. $-2+(-5)=-7$

   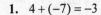

5. $-6+2=-4$

7. $3+(-3)=0$

9. $-7+0=-7$

11. $30+(-30)=0$

13. $-30+(-30)=-60$

15. $-8+(-10)=-18$

17. $-0.4+(-0.9)=-1.3$

19. $-\dfrac{7}{10}+\left(-\dfrac{3}{10}\right)=-\dfrac{10}{10}=-1$

21. $-9+4=-5$

23. $12+(-8)=4$

25. $6+(-9)=-3$

27. $-3.6+2.1=-1.5$

29. $-3.6+(-2.1)=-5.7$

31. $\dfrac{9}{10}+\left(-\dfrac{3}{5}\right)=\dfrac{9}{10}+\left(-\dfrac{6}{10}\right)=\dfrac{3}{10}$

33. $-\dfrac{5}{8}+\dfrac{3}{4}=-\dfrac{5}{8}+\dfrac{6}{8}=\dfrac{1}{8}$

35. $-\dfrac{3}{7}+\left(-\dfrac{4}{5}\right)=-\dfrac{15}{35}+\left(-\dfrac{28}{35}\right)=-\dfrac{43}{35}$

37. $\begin{aligned} 4+(-7)+(-5) &=\left[4+(-7)\right]+(-5) \\ &=-3+(-5) \\ &=-8 \end{aligned}$

39. $\begin{aligned} 85+&(-15)+(-20)+12 \\ &=\left[85+(-15)\right]+(-20)+12 \\ &=70+(-20)+12 \\ &=\left[70+(-20)\right]+12 \\ &=50+12 \\ &=62 \end{aligned}$

41. $\begin{aligned} 17+&(-4)+2+3+(-10) \\ &=13+2+3+(-10) \\ &=15+3+(-10) \\ &=18+(-10) \\ &=8 \end{aligned}$

43. $\begin{aligned} -45+&\left(-\dfrac{3}{7}\right)+25+\left(-\dfrac{4}{7}\right) \\ &=(-45+25)+\left[-\dfrac{3}{7}+\left(-\dfrac{4}{7}\right)\right] \\ &=-20+\left(-\dfrac{7}{7}\right) \\ &=-20+(-1) \\ &=-21 \end{aligned}$

45. $\begin{aligned} 3.5+&(-45)+(-8.4)+72 \\ &=\left[3.5+(-8.4)\right]+(-45+72) \\ &=-4.9+27 \\ &=22.1 \end{aligned}$

47. $-10x+2x=(-10+2)x=-8x$

49. $25y+(-12y)=\left[25+(-12)\right]y=13y$

51. $\begin{aligned} -8a+(-15a) &=\left[-8+(-15)\right]a \\ &=-23a \end{aligned}$

53. $\begin{aligned} 4y+&(-13z)+(-10y)+17z \\ &=4y+(-10y)+(-13z)+17z \\ &=-6y+4z \end{aligned}$

55. $\begin{aligned} -7b+&10+(-b)+(-6) \\ &=-7b+(-b)+10+(-6) \\ &=-8b+4 \end{aligned}$

57. $\begin{aligned} 7x+&(-5y)+(-9x)+19y \\ &=7x+(-9x)+(-5y)+19y \\ &=-2x+14y \end{aligned}$

59. $\begin{aligned} 8(4y+&3)+(-35y) \\ &=32y+24+(-35y) \\ &=32y+(-35y)+24 \\ &=-3y+24 \end{aligned}$

61. $\begin{aligned} \left|-3+(-5)\right|+\left|2+(-6)\right| &=\left|-8\right|+\left|-4\right| \\ &=8+4 \\ &=12 \end{aligned}$

63. $\begin{aligned} -20+&\left[-\left|15+(-25)\right|\right] \\ &=-20+\left[-\left|-10\right|\right] \\ &=-20+\left[-10\right] \\ &=-30 \end{aligned}$

65. $\begin{aligned} 6+\left[2+(-13)\right]&\square-3+\left[4+(-8)\right] \\ 6+\left[-11\right]&\square-3+\left[-4\right] \\ -5&\square-7 \\ -5&>-7 \end{aligned}$

67. $\begin{aligned} &-6x+(-13x) \\ &-6x+(-13x)=-19x \end{aligned}$

69. $\dfrac{-20}{x}+\dfrac{3}{x}$

$\dfrac{-20}{x}+\dfrac{3}{x}=\dfrac{-17}{x}$

71. $-56+100=44$

The high temperature was 44°F.

73. $-1312+712=-600$

The elevation of the person is 600 feet below sea level.

75. $-7+15-5=3$

The temperature at 4:00 P.M. was 3°F.

77. $27+4-2+8-12$

$=(27+4+8)+(-2-12)$

$=34-14$

$=25$

The location of the football at the end of the fourth play is at the 25-yard line.

79. a. $1880+(-2293)=-413$

The deficit in 2004 was –$413 billion.

b. $2154+(-2472)=-318$

The deficit in 2005 was –$318 billion.

c. $-413+(-318)=-731$

The combined deficit in 2004 and 2005 was –$731 billion.

81. – 87. Answers will vary.

89. makes sense

91. makes sense

93. true

95. false; Changes to make the statement true will vary. A sample change is: The sum of a positive number and a negative number is sometimes negative.

97. The sum is negative. When finding the sum of numbers with different signs, use the sign of the number with the greater absolute value as the sign of the sum. Since a is further from 0 than c, we use a negative sign.

99. Though the sum inside the absolute value is negative, the absolute value of this sum is positive.

101. The calculator verifies your results.

102. a. $\sqrt{4}\,(=2)$

b. $0, \sqrt{4}$

c. $-6, 0, \sqrt{4}$

d. $-6, 0, 0.\overline{7}, \sqrt{4}$

e. $-\pi, \sqrt{3}$

f. $-6, -\pi, 0, 0.\overline{7}, \sqrt{3}, \sqrt{4}$

103. $19 \geq -18$ is true because 19 is to the right of –18 on the number line.

104. $16=2(x-1)-x$

$16=2(18-1)-18$

$16=2(17)-18$

$16=34-18$

$16=16,\ \text{true}$

3 is a solution.

105. $7-10=7+(-10)=-3$

106. $-8-13=-8+(-13)=-21$

107. $-8-(-13)=-8+13=5$

1.6 Check Points

1. a. $3-11=3+(-11)=-8$

b. $4-(-5)=4+5=9$

c. $-7-(-2)=-7+2=-5$

2. a. $-3.4-(-12.6)=-3.4+12.6=9.2$

b. $-\dfrac{3}{5}-\dfrac{1}{3}=-\dfrac{3}{5}+\left(-\dfrac{1}{3}\right)=-\dfrac{9}{15}+\left(-\dfrac{5}{15}\right)=-\dfrac{14}{15}$

c. $5\pi-(-2\pi)=5\pi+2\pi=7\pi$

3. $10-(-12)-4-(-3)-6$

$=10+12+(-4)+3+(-6)$

$=(10+12+3)+\left[(-4)+(-6)\right]$

$=25+(-10)$

$=15$

4. $-6+4a-7ab$ has terms of -6, $4a$, and $-7ab$.

5. a. $4+2x-9x=4+(2-9)x$

$=4+\left[2+(-9)\right]x$

$=4-7x$

b. $-3x-10y-6x+14y=-3x-6x-10y+14y$

$=(-3-6)x+(-10+14)y$

$=-9x+4y$

6. $8848-(-10,915)=8848+10,915=19,763$

The difference in elevation between the peak of Mount Everest and the Marianas Trench is 19,763 meters.

1.6 Exercise Set

1. **a.** -12
   b. $5 - 12 = 5 + (-12)$

3. **a.** 7
   b. $5 - (-7) = 5 + 7$

5. $14 - 8 = 14 + (-8) = 6$

7. $8 - 14 = 8 + (-14) = -6$

9. $3 - (-20) = 3 + 20 = 23$

11. $-7 - (-18) = -7 + 18 = 11$

13. $-13 - (-2) = -13 + 2 = -11$

15. $-21 - 17 = -21 + (-17) = -38$

17. $-45 - (-45) = -45 + 45 = 0$

19. $23 - 23 = 23 + (-23) = 0$

21. $13 - (-13) = 13 + 13 = 26$

23. $0 - 13 = 0 + (-13) = -13$

25. $0 - (-13) = 0 + 13 = 13$

27. $\dfrac{3}{7} - \dfrac{5}{7} = \dfrac{3}{7} + \left(-\dfrac{5}{7}\right) = -\dfrac{2}{7}$

29. $\dfrac{1}{5} - \left(-\dfrac{3}{5}\right) = \dfrac{1}{5} + \dfrac{3}{5} = \dfrac{4}{5}$

31. $-\dfrac{4}{5} - \dfrac{1}{5} = -\dfrac{4}{5} + \left(-\dfrac{1}{5}\right) = -\dfrac{5}{5} = -1$

33. $-\dfrac{4}{5} - \left(-\dfrac{1}{5}\right) = -\dfrac{4}{5} + \dfrac{1}{5} = -\dfrac{3}{5}$

35. $\dfrac{1}{2} - \left(-\dfrac{1}{4}\right) = \dfrac{1}{2} + \dfrac{1}{4} = \dfrac{2}{4} + \dfrac{1}{4} = \dfrac{3}{4}$

37. $\dfrac{1}{2} - \dfrac{1}{4} = \dfrac{1}{2} + \left(-\dfrac{1}{4}\right) = \dfrac{2}{4} + \left(-\dfrac{1}{4}\right) = \dfrac{1}{4}$

39. $9.8 - 2.2 = 9.8 + (-2.2) = 7.6$

41. $-3.1 - (-1.1) = -3.1 + 1.1 = -2$

43. $1.3 - (-1.3) = 1.3 + 1.3 = 2.6$

45. $-2.06 - (-2.06) = -2.06 + 2.06 = 0$

47. $5\pi - 2\pi = 5\pi + (-2\pi) = 3\pi$

49. $3\pi - (-10\pi) = 3\pi + 10\pi = 13\pi$

51. $13 - 2 - (-8) = 13 + (-2) + 8$
$\qquad = (13 + 8) + (-2)$
$\qquad = 21 + (-2)$
$\qquad = 19$

53. $9 - 8 + 3 - 7 = 9 + (-8) + 3 + (-7)$
$\qquad = (9 + 3) + \left[(-8) + (-7)\right]$
$\qquad = 12 + (-15)$
$\qquad = -3$

55. $-6 - 2 + 3 - 10$
$\qquad = -6 + (-2) + 3 + (-10)$
$\qquad = \left[(-6) + (-2) + (-10)\right] + 3$
$\qquad = -18 + 3$
$\qquad = -15$

57. $-10 - (-5) + 7 - 2$
$\qquad = -10 + 5 + 7 + (-2)$
$\qquad = \left[(-10) + (-2)\right] + (5 + 7)$
$\qquad = -12 + 12$
$\qquad = 0$

59. $-23 - 11 - (-7) + (-25)$
$\qquad = (-23) + (-11) + 7 + (-25)$
$\qquad = \left[(-23) + (-11) + (-25)\right] + 7$
$\qquad = -59 + 7$
$\qquad = -52$

61. $-823 - 146 - 50 - (-832)$

$= -823 + (-146) + (-50) + 832$

$= \left[(-823) + (-146) + (-50)\right] + 832$

$= -1019 + 832$

$= -187$

63. $1 - \dfrac{2}{3} - \left(-\dfrac{5}{6}\right) = 1 + \left(-\dfrac{2}{3}\right) + \dfrac{5}{6}$

$= \left(1 + \dfrac{5}{6}\right) + \left(-\dfrac{2}{3}\right)$

$= \left(\dfrac{6}{6} + \dfrac{5}{6}\right) + \left(-\dfrac{2}{3}\right)$

$= \dfrac{11}{6} + \left(-\dfrac{2}{3} \cdot \dfrac{2}{2}\right)$

$= \dfrac{11}{6} + \left(-\dfrac{4}{6}\right)$

$= \dfrac{7}{6}$ or $1\dfrac{1}{6}$

65. $-0.16 - 5.2 - (-0.87)$

$= -0.16 + (-5.2) + 0.87$

$= \left[(-0.16) + (-5.2)\right] + 0.87$

$= -5.36 + 0.87$

$= -4.49$

67. $-\dfrac{3}{4} - \dfrac{1}{4} - \left(-\dfrac{5}{8}\right) = -\dfrac{3}{4} + \left(-\dfrac{1}{4}\right) + \dfrac{5}{8}$

$= -\dfrac{4}{4} + \dfrac{5}{8}$

$= -\dfrac{8}{8} + \dfrac{5}{8} = -\dfrac{3}{8}$

69. $-3x - 8y = -3x + (-8y)$

The terms are $-3x$ and $-8y$.

71. $12x - 5xy - 4 = 12x + (-5xy) + (-4)$

The terms are $12x$, $-5xy$, and -4.

73. $3x - 9x = 3x + (-9x)$

$= \left[3 + (-9)\right]x = -6x$

75. $4 + 7y - 17y = 4 + 7y + (-17y)$

$= 4 + \left[7 + (-17)\right]y$

$= 4 - 10y$

77. $2a + 5 - 9a = 2a + 5 + (-9a)$

$= 2a + (-9a) + 5$

$= \left[2 + (-9)\right]a + 5$

$= -7a + 5$ or $5 - 7a$

79. $4 - 6b - 8 - 3b$

$= 4 + (-6b) + (-8) + (-3b)$

$= 4 + (-8) + (-6b) + (-3b)$

$= 4 + (-8) + \left[-6 + (-3)\right]b$

$= -4 - 9b$

81. $13 - (-7x) + 4x - (-11)$

$= 13 + 7x + 4x + 11$

$= 13 + 11 + 7x + 4x$

$= 24 + 11x$

83. $-5x - 10y - 3x + 13y$

$= -5x + (-10y) + (-3x) + 13y$

$= -5x + (-3x) + (-10y) + 13y$

$= \left[-5 + (-3)\right]x + (-10 + 13)y$

$= -8x + 3y$ or $3y - 8x$

85. $-\left|-9 - (-6)\right| - (-12) = -\left|-9 + 6\right| + 12$

$= -\left|-3\right| + 12$

$= -3 + 12$

$= 9$

87. $\dfrac{5}{8} - \left(\dfrac{1}{2} - \dfrac{3}{4}\right) = \dfrac{5}{8} - \left(\dfrac{1}{2} \cdot \dfrac{2}{2} - \dfrac{3}{4}\right)$

$= \dfrac{5}{8} - \left(\dfrac{2}{4} + \left(-\dfrac{3}{4}\right)\right)$

$= \dfrac{5}{8} - \left(-\dfrac{1}{4}\right)$

$= \dfrac{5}{8} + \dfrac{1}{4}$

$= \dfrac{5}{8} + \dfrac{1}{4} \cdot \dfrac{2}{2}$

$= \dfrac{5}{8} + \dfrac{2}{8}$

$= \dfrac{7}{8}$

89. $\left| -9 - (-3+7) \right| - \left| -17 - (-2) \right|$

$= \left| -9 - 4 \right| - \left| -17 + 2 \right|$

$= \left| -9 + (-4) \right| - \left| -17 + 2 \right|$

$= \left| -13 \right| - \left| -15 \right|$

$= 13 + (-15)$

$= -2$

91. $6x - (-5x)$

$6x - (-5x) = 6x + 5x = 11x$

93. $\dfrac{-5}{x} - \left(\dfrac{-2}{x} \right)$

$\dfrac{-5}{x} - \left(\dfrac{-2}{x} \right) = \dfrac{-5}{x} + \dfrac{2}{x} = \dfrac{-3}{x}$

95. Elevation of Mount Kilimanjaro – elevation of Qattara Depression

$= 19,321 - (-436) = 19,757$

The difference in elevation between the two geographic locations is 19,757 feet.

97. $2 - (-19) = 2 + 19 = 21$

The difference between the average daily low temperature for March and February is 21°F.

99. $-19 - (-22) = -19 + 22 = 3$

February's average low temperature is 3°F warmer than January's.

101. a. $1.0 billion was exported in 2002.

b. $1.4 billion was imported in 2002.

c. $1.0 - 1.4 = 1.0 + (-1.4) = -0.4$

The trade deficit was –$0.4 billion in 2002.

103. a. $E = 0.04n + 0.7$

$E = 0.04(8) + 0.7$

$= 1.02$

$1.02 billion was exported in 2002.

b. $I = 0.09n + 0.8$

$I = 0.09(8) + 0.8$

$= 1.52$

$1.52 billion was imported in 2002.

c. $1.02 - 1.52 = 1.02 + (-1.52) = -0.5$

The trade deficit was –$0.5 billion in 2002.

d. $-0.4 - (-0.5) = -0.4 + 0.5 = 0.1$

The difference in the trade deficit for 2002 found in Exercise 101(c) and the trade deficit given by the models is $0.1 billion.

105. – 109. Answers will vary.

111. makes sense

113. makes sense

115. false; Changes to make the statement true will vary. A sample change is: $7 - (-2) = 7 + 2 = 9$

117. true

119. $a - b = \overbrace{(\text{negative number})}^{a} - \overbrace{(\text{negative number})}^{b}$

$\underbrace{}_{\substack{a \text{ has the greater} \\ \text{absolute value}}}$

$= (\text{negative number}) + (\text{positive number})$

$= \text{negative number}$

121. $0 - b = 0 - \overbrace{(\text{negative number})}^{b}$

$= 0 + (\text{positive number})$

$= \text{positive number}$

123. The calculator verifies your results.

125. $13x + 3 = 3(5x - 1)$

$13 \cdot 2 + 3 = 3(5 \cdot 2 - 1)$

$26 + 3 = 3(10 - 1)$

$29 = 3(9)$

$29 = 27, \ \text{false}$

The number is not a solution.

126. $5(3x + 2y) + 6(5y) = 15x + 10y + 30y$

$= 15x + 40y$

127. Answers will vary. -17 is an example of an integer that is not a natural number.

128. $4(-3) = (-3) + (-3) + (-3) + (-3) = -12$

129. $3(-3) = (-3) + (-3) + (-3) = -9$

130.
$$2(-3) = -6$$
$$1(-3) = -3$$
$$0(-3) = 0$$
$$-1(-3) = 3$$
$$-2(-3) = 6$$
$$-3(-3) = 9$$
$$-4(-3) = \boxed{12}$$

1.7 Check Points

1. a. $8(-5) = -40$

 b. $-\dfrac{1}{3} \cdot \dfrac{4}{7} = -\dfrac{4}{21}$

 c. $(-12)(-3) = 36$

 d. $(-1.1)(-5) = 5.5$

 e. $(-543)(0) = 0$

2. a. $(-2)(3)(-1)(4) = 24$

 b. $(-1)(-3)(2)(-1)(5) = -30$

3. a. The multiplicative inverse of 7 is $\dfrac{1}{7}$ because

 $7 \cdot \dfrac{1}{7} = 1.$

 b. The multiplicative inverse of $\dfrac{1}{8}$ is 8 because

 $\dfrac{1}{8} \cdot 8 = 1.$

 c. The multiplicative inverse of -6 is $-\dfrac{1}{6}$ because

 $(-6)\left(-\dfrac{1}{6}\right) = 1.$

 d. The multiplicative inverse of $-\dfrac{7}{13}$ is $-\dfrac{13}{7}$

 because $\left(-\dfrac{7}{13}\right)\left(-\dfrac{13}{7}\right) = 1.$

4. a. $-28 \div 7 = -28 \cdot \dfrac{1}{7} = -4$

 b. $\dfrac{-16}{-2} = -16 \cdot \left(-\dfrac{1}{2}\right) = 8$

5. a. $\dfrac{-32}{-4} = 8$

 b. $-\dfrac{2}{3} \div \dfrac{5}{4} = -\dfrac{2}{3} \cdot \dfrac{4}{5} = -\dfrac{8}{15}$

 c. $\dfrac{21.9}{-3} = -7.3$

 d. $\dfrac{0}{-5} = 0$

6. a. $-4(5x) = (-4 \cdot 5)x = -20x$

 b. $9x + x = 9x + 1x = (9 + 1)x = 10x$

 c. $13b - 14b = (13 - 14)b = -1b = -b$

 d. $-7(3x - 4) = -7(3x) - 7(-4) = -21x + 28$

 e. $-(7y - 6) = -(7y) - (-6) = -7y + 6$

7. $4(3y - 7) - (13y - 2) = 12y - 28 - 13y + 2$
$$= 12y - 13y - 28 + 2$$
$$= -1y - 26$$
$$= -y - 26$$

8. $\quad 2x - 5 = 8x + 7$
$$2(-3) - 5 = 8(-3) + 7$$
$$-6 - 5 = -24 + 7$$
$$-11 = -17, \text{ false}$$

The number is not a solution.

9. $M = -0.6n + 64.4$
$$M = -0.6(25) + 64.4$$
$$= -15 + 64.4$$
$$= 49.4$$

According to this model, 49.4% of doctorate degrees will be awarded to men in 2014. This overestimates the actual value shown in the bar graph by 0.4%.

1.7 Exercises Set

1. $5(-9) = -(5 \cdot 9) = -45$

3. $(-8)(-3) = +(8 \cdot 3) = 24$

5. $(-3)(7) = -21$

7. $(-19)(-1) = 19$

9. $0(-19) = 0$

11. $\dfrac{1}{2}(-24) = -12$

13. $\left(-\dfrac{3}{4}\right)(-12) = \dfrac{3 \cdot 12}{4 \cdot 1} = 9$

15. $-\dfrac{3}{5} \cdot \left(-\dfrac{4}{7}\right) = \dfrac{3 \cdot 4}{5 \cdot 7} = \dfrac{12}{35}$

17. $-\dfrac{7}{9} \cdot \dfrac{2}{3} = -\dfrac{7 \cdot 2}{9 \cdot 3} = -\dfrac{14}{27}$

19. $3(-1.2) = -3.6$

21. $-0.2(-0.6) = 0.12$

23. $(-5)(-2)(3) = 30$

25. $(-4)(-3)(-1)(6) = -72$

27. $-2(-3)(-4)(-1) = 24$

29. $(-3)(-3)(-3) = 9(-3) = -27$

31. $5(-3)(-1)(2)(3) = 90$

33. $(-8)(-4)(0)(-17)(-6) = 0$

35. The multiplicative inverse of 4 is $\dfrac{1}{4}$.

37. The multiplicative inverse of $\dfrac{1}{5}$ is 5.

39. The multiplicative inverse of -10 is $-\dfrac{1}{10}$.

41. The multiplicative inverse of $-\dfrac{2}{5}$ is $-\dfrac{5}{2}$.

43. a. $-32 \div 4 = -32 \cdot \dfrac{1}{4}$

b. $-32 \cdot \dfrac{1}{4} = -8$

45. a. $\dfrac{-60}{-5} = -60 \cdot \left(-\dfrac{1}{5}\right)$

b. $-60 \cdot \left(-\dfrac{1}{5}\right) = 12$

47. $\dfrac{12}{-4} = 12 \cdot \left(-\dfrac{1}{4}\right) = -3$

49. $\dfrac{-21}{3} = -21 \cdot \dfrac{1}{3} = -7$

51. $\dfrac{-90}{-3} = -90 \cdot \left(-\dfrac{1}{3}\right) = 30$

53. $\dfrac{0}{-7} = 0$

55. $\dfrac{-7}{0}$ is undefined.

57. $-15 \div 3 = -15 \cdot \dfrac{1}{3} = -5$

59. $12 \div (-10) = 120 \cdot \left(-\dfrac{1}{10}\right) = -12$

61. $(-180) \div (-30) = -180 \cdot \left(-\dfrac{1}{30}\right) = 6$

63. $0 \div (-4) = 0$

65. $-4 \div 0$ is undefined.

67. $\dfrac{-12.9}{3} = -12.9 \cdot \dfrac{1}{3} = -4.3$

69. $-\dfrac{1}{2} \div \left(-\dfrac{3}{5}\right) = -\dfrac{1}{2} \cdot \left(-\dfrac{5}{3}\right) = \dfrac{5}{6}$

71. $-\dfrac{14}{9} \div \dfrac{7}{8} = -\dfrac{14}{9} \cdot \dfrac{8}{7}$

$\qquad = -\dfrac{112}{63} = \dfrac{\not 7 \cdot 16}{\not 7 \cdot 9} = -\dfrac{16}{9}$

73. $\dfrac{1}{3} \div \left(-\dfrac{1}{3}\right) = \dfrac{1}{3} \cdot (-3) = -1$

75. $6 \div \left(-\dfrac{2}{5}\right) = 6 \cdot \left(-\dfrac{5}{2}\right) = -\dfrac{30}{2} = -15$

77. $-5(2x) = (-5 \cdot 2)x = -10x$

79. $-4\left(-\dfrac{3}{4}y\right) = \left[-4 \cdot \left(-\dfrac{3}{4}\right)\right]y = 3y$

81. $8x + x = 8x + 1x = (8+1)x = 9x$

83. $-5x + x = -5x + 1x = (-5+1)x = -4x$

85. $6b - 7b = (6-7)b = -1b = -b$

87. $-y + 4y = -1y + 4y = (-1+4)y = 3y$

89. $-4(2x-3) = -4(2x) - 4(-3) = -8x + 12$

91. $-3(-2x+4) = -3(-2x) - 3(4) = 6x - 12$

93. $-(2y-5) = -2y + 5$

95. $4(2y-3) - (7y+2)$

$\quad = 4(2y) + 4(-3) - 7y - 2$

$\quad = 8y - 12 + 7y - 2$

$\quad = 8y - 7y - 12 - 2$

$\quad = y - 14$

97. $\qquad 4x = 2x - 10$

$\qquad 4(-5) = 2(-5) - 10$

$\qquad -20 = -10 - 10$

$\qquad -20 = -20, \text{ true}$

The number is a solution.

99. $\qquad -7y + 18 = -10y + 6$

$\qquad -7(-4) + 18 = -10(-4) + 6$

$\qquad 28 + 18 = 40 + 6$

$\qquad 46 = 46, \text{ true}$

The number is a solution.

101. $\qquad 5(w+3) = 2w - 21$

$\qquad 5(-10+3) = 2(-10) - 21$

$\qquad 5(-7) = -20 - 21$

$\qquad -35 = -41, \text{ false}$

The number is not a solution.

103. $\qquad 4(6-z) + 7z = 0$

$\qquad 4(6-(-8)) + 7(-8) = 0$

$\qquad 4(6+8) - 56 = 0$

$\qquad 4(14) - 56 = 0$

$\qquad 56 - 56 = 0$

$\qquad 0 = 0, \text{ true}$

The number is a solution.

105. $\qquad 14 - 2x = -4x + 7$

$\qquad 14 - 2\left(-2\dfrac{1}{2}\right) = -4\left(-2\dfrac{1}{2}\right) + 7$

$\qquad 14 - 2\left(-\dfrac{5}{2}\right) = -4\left(-\dfrac{5}{2}\right) + 7$

$\qquad 14 + 5 = 10 + 7$

$\qquad 19 = 17, \text{ false}$

The number is not a solution.

107. $\qquad \dfrac{5m-1}{6} = \dfrac{3m-2}{4}$

$\qquad \dfrac{5(-4)-1}{6} = \dfrac{3(-4)-2}{4}$

$\qquad \dfrac{-20-1}{6} = \dfrac{-12-2}{4}$

$\qquad \dfrac{-21}{6} = \dfrac{-14}{4}$

$\qquad \dfrac{-7}{2} = \dfrac{-7}{2}, \text{ true}$

The number is a solution.

109. $4(-10) + 8 = -40 + 8 = -32$

111. $(-9)(-3) - (-2) = 27 + 2 = 29$

113. $\dfrac{-18}{-15+12} = \dfrac{-18}{-3} = 6$

115. $-6 - \left(\dfrac{12}{-4}\right) = -6 - (-3) = -6 + 3 = -3$

117. $C = \frac{5}{9}(F - 32)$

$C = \frac{5}{9}(-22 - 32) = \frac{5}{9}(-54) = -30$

$-22°F$ is equivalent to $-30°C$.

119. a. According to the line graph, 50% of seniors used alcohol in 2000.

   b. $A = -n + 70$

      $A = -(20) + 70 = 50$

      According to the formula, 50% of seniors used alcohol in 2000.
The model provides the same value as the line graph.

   c. According to the line graph, 22% of seniors used marijuana in 2000.

   d. $M = -0.5n + 28$

      $M = -0.5(20) + 28 = -10 + 28 = 18$

      According to the formula, 18% of seniors used marijuana in 2000.
The formula estimates a value that is lower than the value in the line graph.

   e. $R = \frac{M}{A} = \frac{-0.5n + 28}{-n + 70}$

121. a. The ratio is $\frac{11}{44} = \frac{1}{4}$.

      For each \$10,000 collected in individual income taxes, $\$10,000 \cdot \frac{1}{4} = \$2500$ is collected from corporate taxes.

   b. $\frac{1}{4} = 0.25$

   c. $D = \frac{-0.27n + 23}{44}$

      $D = \frac{-0.27(45) + 23}{44}$

      $= \frac{-12.15 + 23}{44}$

      $= \frac{10.85}{44}$

   d. $\frac{10.85}{44} \approx 0.25$

      This rounded value matches the value in part (b) extremely well.

123. – 129. Answers will vary.

131. makes sense

133. does not make sense; Explanations will vary. Sample explanation: The numerator and denominator of this formula are both positive.

135. false; Changes to make the statement true will vary. A sample change is: If the number is multiplied by 0, which is nonnegative, the result is 0.

137. true

139. $50x$

141. $\frac{40}{x + 40}$

143. The calculator verifies your results.

145. Answers will vary.

146. $-6 + (-3) = -9$

147. $-6 - (-3) = -6 + 3 = -3$

148. $-6 \div (-3) = -6\left(-\frac{1}{3}\right) = 2$

149. $(-6)^2 = (-6)(-6) = 36$

150. $(-5)^3 = (-5)(-5)(-5) = -125$

151. $(-2)^4 = (-2)(-2)(-2)(-2) = 16$

1.8 Check Points

1. a. $6^2 = 6 \cdot 6 = 36$

   b. $(-4)^3 = (-4)(-4)(-4) = -64$

   c. $(-1)^4 = (-1)(-1)(-1)(-1) = 1$

   d. $-1^4 = -(1 \cdot 1 \cdot 1 \cdot 1) = -1$

2. a. $16x^2 + 5x^2 = (16 + 5)x^2 = 21x^2$

   b. $7x^3 + x^3 = 7x^3 + 1x^3 = (7 + 1)x^3 = 8x^3$

   c. $10x^2 + 8x^3$ cannot be simplified.

3. $20 + 4 \cdot 3 - 17 = 20 + 12 - 17$
$$= 20 + 12 - 17$$
$$= 15$$

4. $7^2 - 48 \div 4^2 \cdot 5 - 2 = 49 - 48 \div 16 \cdot 5 - 2$
$$= 49 - 3 \cdot 5 - 2$$
$$= 49 - 15 - 2$$
$$= 34 - 2$$
$$= 32$$

5. a. $(3 \cdot 2)^2 = 6^2 = 36$

　b. $3 \cdot 2^2 = 3 \cdot 4 = 12$

6. $\left(-\dfrac{1}{2}\right)^2 - \left(\dfrac{7}{10} - \dfrac{8}{15}\right)^2 (-18) = \left(-\dfrac{1}{2}\right)^2 - \left(\dfrac{1}{6}\right)^2 (-18)$
$$= \dfrac{1}{4} - \dfrac{1}{36} \cdot \dfrac{-18}{1}$$
$$= \dfrac{1}{4} + \dfrac{1}{2}$$
$$= \dfrac{3}{4}$$

7. $4[3(6-11)+5] = 4[3(-5)+5]$
$$= 4[-15+5]$$
$$= 4[-10]$$
$$= -40$$

8. $25 \div 5 + 3[4 + 2(7-9)^3] = 25 \div 5 + 3[4 + 2(-2)^3]$
$$= 25 \div 5 + 3[4 + 2(-8)]$$
$$= 25 \div 5 + 3[4 - 16]$$
$$= 25 \div 5 + 3[-12]$$
$$= 5 + (-36)$$
$$= -31$$

9. $\dfrac{5(4-9)+10 \cdot 3}{2^3 - 1} = \dfrac{5(-5)+10 \cdot 3}{8-1}$
$$= \dfrac{-25+30}{7}$$
$$= \dfrac{5}{7}$$

10. $-x^2 - 4x = -(-5)^2 - 4(-5)$
$$= -25 + 20$$
$$= -5$$

11. $14x^2 + 5 - [7(x^2 - 2) + 4]$
$$= 14x^2 + 5 - [7x^2 - 14 + 4]$$
$$= 14x^2 + 5 - [7x^2 - 10]$$
$$= 14x^2 + 5 - 7x^2 + 10$$
$$= 14x^2 - 7x^2 + 5 + 10$$
$$= 7x^2 + 15$$

12. $P = -0.05x^2 + 3.6x - 15$
$$P = -0.05(20)^2 + 3.6(20) - 15$$
$$= -0.05(400) + 72 - 15$$
$$= -20 + 72 - 15$$
$$= -20 + 72 - 15$$
$$= 37$$
37% of 20-year-olds have been tested for HIV.

13. a. $\overline{C} = \dfrac{30x + 300,000}{x}$
$$\overline{C} = \dfrac{30(1000) + 300,000}{1000} = \dfrac{30,000 + 300,000}{1000}$$
$$= \dfrac{330,000}{1000}$$
$$= 330$$
The average cost is \$330.

　b. $\overline{C} = \dfrac{30x + 300,000}{x}$
$$\overline{C} = \dfrac{30(10,000) + 300,000}{10,000}$$
$$= \dfrac{300,000 + 300,000}{10,000}$$
$$= \dfrac{600,000}{10,000}$$
$$= 60$$
The average cost is \$60.

　c. $\overline{C} = \dfrac{30x + 300,000}{x}$
$$\overline{C} = \dfrac{30(100,000) + 300,000}{100,000}$$
$$= \dfrac{3,000,000 + 300,000}{100,000}$$
$$= \dfrac{3,300,000}{100,000}$$
$$= 33$$
The average cost is \$33.

1.8 Exercise Set

1. $9^2 = 9 \cdot 9 = 81$

3. $4^3 = 4 \cdot 4 \cdot 4 = 64$

5. $(-4)^2 = (-4)(-4) = 16$

7. $(-4)^3 = (-4)(-4)(-4) = -64$

9. $(-5)^4 = (-5)(-5)(-5)(-5) = 625$

11. $-5^4 = -5 \cdot 5 \cdot 5 \cdot 5 = -625$

13. $-10^2 = -10 \cdot 10 = -100$

15. $7x^2 + 12x^2 = (7+12)x^2 = 19x^2$

17. $10x^3 + 5x^3 = (10+5)x^3 = 15x^3$

19. $8x^4 + x^4 = 8x^4 + 1x^4 = (8+1)x^4 = 9x^4$

21. $26x^2 - 27x^2 = 26x^2 + \left(-27x^2\right)$
$\qquad = \left[26 + (-27)\right]x^2$
$\qquad = -1x^2 = -x^2$

23. $27x^3 - 26x^3 = 27x^3 + \left(-26x^2\right)$
$\qquad = 1x^3 = x^3$

25. $5x^2 + 5x^3$ cannot be simplified. The terms $5x^2$ and $5x^3$ are not like terms because they have different variable factors, namely, x^2 and x^3.

27. $16x^2 - 16x^2 = 16x^2 + \left(-16x^2\right)$
$\qquad = \left[16 + (-16)\right]x^2$
$\qquad = 0x^2 = 0$

29. $7 + 6 \cdot 3 = 7 + 18 = 25$

31. $45 \div 5 \cdot 3 = 9 + 18 = 27$

33. $6 \cdot 8 \div 4 = 48 \div 4 = 12$

35. $14 - 2 \cdot 6 + 3 = 14 - 12 + 3 = 2 + 3 = 5$

37. $8^2 - 16 \div 2^2 \cdot 4 - 3 = 64 - 16 \div 4 \cdot 4 - 3$
$\qquad = 64 - 4 \cdot 4 - 3$
$\qquad = 64 - 16 - 3$
$\qquad = 48 - 3$
$\qquad = 45$

39. $3(-2)^2 - 4(-3)^2 = 3 \cdot 4 - 4 \cdot 9$
$\qquad = 12 - 36$
$\qquad = 12 + (-36)$
$\qquad = -24$

41. $(4 \cdot 5)^2 - 4 \cdot 5^2 = 20^2 - 4 \cdot 25$
$\qquad = 400 - 100$
$\qquad = 300$

43. $(2-6)^2 - (3-7)^2 = (-4)^2 - (-4)^2$
$\qquad = 16 - 16$
$\qquad = 0$

45. $6(3-5)^3 - 2(1-3)^3$
$\qquad = 6(-2)^3 - 2(-2)^3$
$\qquad = 6(-8) - 2(-8)$
$\qquad = -48 + 16$
$\qquad = -32$

47. $\left[2(6-2)\right]^2 = (2 \cdot 4)^2 = 8^2 = 64$

49. $2\left[5 + 2(9-4)\right] = 2\left[5 + 2(5)\right]$
$\qquad = 2(5+10)$
$\qquad = 2 \cdot 15 = 30$

51. $\left[7 + 3\left(2^3 - 1\right)\right] + 21 = \left[7 + 3(8-1)\right] \div 21$
$\qquad = (7 + 3 \cdot 7) \div 21$
$\qquad = (7 + 21) \div 21$
$\qquad = 28 \div 21$
$\qquad = \dfrac{28}{21} = \dfrac{\cancel{7} \cdot 4}{\cancel{7} \cdot 3}$
$\qquad = \dfrac{4}{3}$

53. $\dfrac{10+8}{5^2 - 4^2} = \dfrac{18}{25-16} = \dfrac{18}{9} = 2$

55. $\dfrac{37+15\div(-3)}{2^4}=\dfrac{37+(-5)}{16}=\dfrac{32}{16}=2$

57. $\dfrac{(-11)(-4)+2(-7)}{7-(-3)}=\dfrac{44+(-14)}{7+3}$
$=\dfrac{30}{10}=3$

59. $4\,|10-(8-20)|=4\,|10-(-12)|=4\,|10+12|$
$=4\,|22|=4\cdot 22$
$=88$

61. $8(-10)+|4(-5)|=-80+|-20|$
$=-80+20=-60$

63. $-2^2+4\big[16+(3-5)\big]$
$=-4+4\big[16+(-2)\big]$
$=-4+4(-8)=-4-32=-36$

65. $24\div\dfrac{3^2}{8-5}-(-6)=24\div\dfrac{9}{3}-(-6)$
$=24\div 3-(-6)$
$=8+6=14$

67. $\dfrac{\frac{1}{4}-\frac{1}{2}}{\frac{1}{3}}=\dfrac{\frac{1}{4}-\frac{2}{4}}{\frac{1}{3}}=\dfrac{-\frac{1}{4}}{\frac{1}{3}}=-\dfrac{1}{4}\cdot\dfrac{3}{1}=-\dfrac{3}{4}$

69. $-\dfrac{9}{4}\left(\dfrac{1}{2}\right)+\dfrac{3}{4}\div\dfrac{5}{6}=-\dfrac{9}{4}\left(\dfrac{1}{2}\right)+\dfrac{3}{4}\cdot\dfrac{6}{5}$
$=-\dfrac{9}{8}+\dfrac{18}{20}$
$=-\dfrac{45}{40}+\dfrac{36}{40}=-\dfrac{9}{40}$

71. $\dfrac{\frac{7}{9}-3}{\frac{5}{6}}\div\dfrac{3}{2}+\dfrac{3}{4}=\dfrac{\frac{7}{9}-\frac{27}{9}}{\frac{5}{6}}\cdot\dfrac{3}{2}+\dfrac{3}{4}$
$=\dfrac{-\frac{20}{9}}{\frac{5}{6}}\cdot\dfrac{3}{2}+\dfrac{3}{4}$
$=-\dfrac{20}{9}\cdot\dfrac{6}{5}\cdot\dfrac{2}{3}+\dfrac{3}{4}$
$=-\dfrac{240}{135}+\dfrac{3}{4}$
$=-\dfrac{15\cdot 16}{15\cdot 9}+\dfrac{3}{4}$
$=-\dfrac{16}{9}+\dfrac{3}{4}$
$=-\dfrac{64}{36}+\dfrac{27}{36}$
$=-\dfrac{37}{36}\ \text{ or }\ -1\dfrac{1}{36}$

73. $x^2+5x;\ x=3$
$x^2+5x=3^2+5\cdot 3$
$=9+5\cdot 3=9+15=24$

75. $3x^2-8x;\ x=-2$
$3x^2-8x=3(-2)^2-9(-2)$
$=3\cdot 4-8(-2)=12+16=28$

77. $-x^2-10x;\ x=-1$
$-x^2-10x=-(-1)^2-10(-1)$
$=-1+10=9$

79. $\dfrac{6y-4y^2}{y^2-15};\ y=5$

$\dfrac{6y-4y^2}{y^2-15}=\dfrac{6(5)-4\left(5^2\right)}{5^2-15}$
$=\dfrac{6(5)-4(25)}{25-15}$
$=\dfrac{30-100}{25-15}=\dfrac{-70}{10}=-7$

81. $3\big[5(x-2)+1\big] = 3(5x-10+1)$
$$= 3(5x-9)$$
$$= 15x-27$$

83. $3\big[6-(y+1)\big] = 3(6-y-1)$
$$= 3(5-y)$$
$$= 15-3y$$

85. $7-4\big[3-(4y-5)\big]$
$$= 7-4(3-4y+5)$$
$$= 7-12+16y-20$$
$$= -25+16y \text{ or } 16y-25$$

87. $2(3x^2-5)-\big[4(2x^2-1)+3\big] = 6x^2-10-(8x^2-4+3)$
$$= 6x^2-10-(8x^2-1)$$
$$= 6x^2-10-8x^2-1$$
$$= -2x^2-9$$

89. $-10-(-2)^3 = -10-(-8) = -10+8 = -2$

91. $\big[2(7-10)\big]^2 = \big[2(-3)\big]^2 = [-6]^2 = 36$

93. $x-(5x+8) = x-5x-8 = -4x-8$

95. $5(x^3-4) = 5x^3-20$

97. $I = -0.09n^2+53n+315$

$I = -0.09(10)^2+53(10)+315 = 836$

In 1990 there were 836 thousand inmates.
This overestimates the value given in the bar graph by 62 thousand.

99. a. According to the line graph, about 92% of people at the college plus level of education used the Internet in 2005.

   b. $C = 0.006n^2+3.3n+75$

      $C = 0.006(5)^2+3.3(5)+75 = 91.65$

      According to the formula, about 91.65% of people at the college plus level of education used the Internet in 2005. This is very close to the value obtained from the line graph.

101. a.

Year	Percentage
2000	34%
2002	45%
2004	52%
2005	62%

b. Substitute the values of *n* into each model.

Year	n	Percentage from line graph	Model #1 $H = 0.41n^2 + 2.1n + 17$	Model #2 $H = 0.26n^2 + 3.9n + 34.5$
2000	0	34%	17%	34.5%
2002	2	45%	22.84%	43.34%
2004	4	52%	31.96%	54.26%
2005	5	62%	37.75%	60.5%

Model 2 better describes the data in the line graph.

103. a. $C = \dfrac{200x}{100 - x}$

$C = \dfrac{200(60)}{100 - 60} = 300$

It will cost $300 tens of thousands, or $3,000,000 to remove 60% of the contamination.

b. $C = \dfrac{200x}{100 - x}$

$C = \dfrac{200(90)}{100 - 90} = 1800$

It will cost $1800 tens of thousands, or $18,000,000 to remove 90% of the contamination.

c. The cost of cleanup increases as the percentage of contaminant removed increases.

105. Answers will vary.

107. makes sense

109. makes sense

111. false; Changes to make the statement true will vary. A sample change is: $-3(-3) - 9 = 9 - 9 = 0$

113. true

115. $\dfrac{1}{4} - 6(2+8) \div \left(-\dfrac{1}{3}\right)\left(-\dfrac{1}{9}\right)$

$= \dfrac{1}{4} - 6(10) \div \left(-\dfrac{1}{3}\right)\left(-\dfrac{1}{9}\right)$

$= \dfrac{1}{4} - 60 \div \left(-\dfrac{1}{3}\right)\left(-\dfrac{1}{9}\right)$

$= \dfrac{1}{4} - 60(-3)\left(-\dfrac{1}{9}\right)$

$= \dfrac{1}{4} + 180\left(-\dfrac{1}{9}\right)$

$= \dfrac{1}{4} - 20 = \dfrac{1}{4} - \dfrac{80}{4} = -\dfrac{79}{4}$

117. $\left(2 \cdot 5 - \dfrac{1}{2} \cdot 10\right) \cdot 9 = (10-5) \cdot 9 = 5 \cdot 9 = 45$

118. $-8 - 2 - (-5) + 11$

$= -8 + (-2) + 5 + 11 = \left[(-8) + (-2)\right] + (5+11)$

$= -10 + 16 = 6$

119. $-4(-1)(-3)(2) = -24$

120. Answers will vary. One example is 5.

121. $-\dfrac{1}{2} = x - \dfrac{2}{3}$

$-\dfrac{1}{2} = \dfrac{1}{6} - \dfrac{2}{3}$

$-\dfrac{1}{2} = \dfrac{1}{6} - \dfrac{4}{6}$

$-\dfrac{1}{2} = -\dfrac{3}{6}$

$-\dfrac{1}{2} = -\dfrac{1}{2}$, true

The number is a solution.

122. $5y + 3 - 4y - 8 = 15$

$5(20) + 3 - 4(20) - 8 = 15$

$100 + 3 - 80 - 8 = 15$

$15 = 15$, true

The number is a solution.

123. $4x + 2 = 3(x-6) + 8$

$4(-11) + 2 = 3(-11-6) + 8$

$-44 + 2 = 3(-17) + 8$

$-42 = -51 + 8$

$-42 = -43$, false

The number is not a solution.

Chapter 1 Review Exercises

1. $10 + 5x = 10 + 5(6) = 10 + 30 = 40$

2. $8(x-2) + 3x = 8(6-2) + 3(6)$

$= 8(4) + 18$

$= 32 + 18$

$= 50$

3. $\dfrac{40}{x} - \dfrac{y}{5} = \dfrac{40}{8} - \dfrac{10}{5}$

$= \dfrac{200}{40} - \dfrac{80}{40}$

$= \dfrac{120}{40}$

$= 3$

4. $3(2y + x) = 3(2(10) + 8)$

$= 3(20 + 8)$

$= 3(28)$

$= 84$

5. $7x - 6$

6. $\dfrac{x}{5} - 2 = 18$

7. $9 - 2x = 14$

8. $3(x + 7)$

9. $4x + 5 = 13$

$4(3) + 5 = 13$

$12 + 5 = 13$

$17 = 13$, false

The number is not a solution.

10. $2y + 7 = 4y - 5$

$2(6) + 7 = 4(6) - 5$

$12 + 7 = 24 - 5$

$19 = 19$, true

The number is a solution.

11. $3(w + 1) + 11 = 2(w + 8)$

$3(2+1) + 11 = 2(2 + 8)$

$3(3) + 11 = 2(10)$

$9 + 11 = 20$

$20 = 20$, true

The number is a solution.

12. According to the line graph, the average number of Latin words that the class remembered after 5 days was about 11.

13. $L = \dfrac{5n + 30}{n}$

$L = \dfrac{5(5) + 30}{5} = \dfrac{25 + 30}{5} = \dfrac{55}{5} = 11$

According to the mathematical model, the average number of Latin words that the class remembered after 5 days was 11. This is the same value given in the line graph.

14. a. $p = 1.7n + 42$

$p = 1.7(10) + 42 = 17 + 42 = 59$

59% of sophomores expected to earn a college degree in 1990.
This underestimates the actual value shown in the bar graph by 1%.

b. $p = 1.7n + 42$

$p = 1.7(30) + 42 = 51 + 42 = 93$

93% of sophomores will expect to earn a college degree in 2010.

15. $3\dfrac{2}{7} = \dfrac{3 \cdot 7 + 2}{7} = \dfrac{21 + 2}{7} = \dfrac{23}{7}$

16. $5\dfrac{9}{11} = \dfrac{5 \cdot 11 + 9}{11} = \dfrac{55 + 9}{11} = \dfrac{64}{11}$

17. 17 divided by 9 is 1 with a remainder of 8, so $\dfrac{17}{9} = 1\dfrac{8}{9}$.

18. 27 divided by 5 is 5 with a remainder of 2, so $\dfrac{27}{5} = 5\dfrac{2}{5}$.

19. Composite
$60 = 6 \cdot 10 = 2 \cdot 3 \cdot 2 \cdot 5 = 2 \cdot 2 \cdot 3 \cdot 5$

20. Composite
$63 = 7 \cdot 9 = 7 \cdot 3 \cdot 3 = 3 \cdot 3 \cdot 7$

21. 67 is a prime number.

22. $\dfrac{15}{33} = \dfrac{\cancel{3} \cdot 5}{\cancel{3} \cdot 11} = \dfrac{5}{11}$

23. $\dfrac{40}{75} = \dfrac{\cancel{5} \cdot 8}{\cancel{5} \cdot 15} = \dfrac{8}{15}$

24. $\dfrac{3}{5} \cdot \dfrac{7}{10} = \dfrac{3 \cdot 7}{5 \cdot 10} = \dfrac{21}{50}$

25. $\dfrac{4}{5} \div \dfrac{3}{10} = \dfrac{4}{5} \cdot \dfrac{10}{3} = \dfrac{40}{15} = \dfrac{\cancel{5} \cdot 8}{\cancel{5} \cdot 3} = \dfrac{8}{3}$

26. $1\dfrac{2}{3} \div 6\dfrac{2}{3} = \dfrac{5}{3} \div \dfrac{20}{3}$

$= \dfrac{5}{\cancel{3}} \cdot \dfrac{\cancel{3}}{20} = \dfrac{5}{20} = \dfrac{1 \cdot \cancel{5}}{4 \cdot \cancel{5}} = \dfrac{1}{4}$

27. $\dfrac{2}{9} + \dfrac{4}{9} = \dfrac{2 + 4}{9} = \dfrac{6}{9} = \dfrac{2 \cdot \cancel{3}}{3 \cdot \cancel{3}} = \dfrac{2}{3}$

28. $\dfrac{5}{6} + \dfrac{7}{9} = \dfrac{5}{6} \cdot \dfrac{3}{3} + \dfrac{7}{9} \cdot \dfrac{2}{2}$

$= \dfrac{15}{18} + \dfrac{14}{18} = \dfrac{29}{18}$ or $1\dfrac{11}{18}$

29. $\dfrac{3}{4} - \dfrac{2}{15} = \dfrac{3}{4} \cdot \dfrac{15}{15} - \dfrac{2}{15} \cdot \dfrac{4}{4} = \dfrac{45}{60} - \dfrac{8}{60} = \dfrac{37}{60}$

30. $x - \dfrac{3}{4} = \dfrac{7}{4}$

$2\dfrac{1}{2} - \dfrac{3}{4} = \dfrac{7}{4}$

$\dfrac{5}{2} - \dfrac{3}{4} = \dfrac{7}{4}$

$\dfrac{10}{4} - \dfrac{3}{4} = \dfrac{7}{4}$

$\dfrac{7}{4} = \dfrac{7}{4}$, true

The number is a solution.

31. $\dfrac{2}{3}w = \dfrac{1}{15}w + \dfrac{3}{5}$

$\dfrac{2}{3} \cdot 2 = \dfrac{1}{15} \cdot 2 + \dfrac{3}{5}$

$\dfrac{4}{3} = \dfrac{2}{15} + \dfrac{3}{5}$

$\dfrac{4}{3} = \dfrac{2}{15} + \dfrac{9}{15}$

$\dfrac{4}{3} = \dfrac{11}{15}$, false

The number is not a solution.

32. $2 - \dfrac{1}{2}x = \dfrac{1}{4}x$

33. $\dfrac{3}{5}(x+6)$

34. $H = \dfrac{4}{5}(220-a)$

$H = \dfrac{4}{5}(220-30) = \dfrac{4}{5}(190) = 152$

The target heart rate of a 30-year-old is 152 beats per minute.

35. -2.5

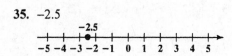

36. $4\dfrac{3}{4}$

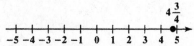

37.
$$
\begin{array}{r}
0.625 \\
8\overline{)5.000} \\
\underline{48} \\
20 \\
\underline{16} \\
40 \\
\underline{40} \\
0
\end{array}
$$

$\dfrac{5}{8} = 0.625$

38.
$$
\begin{array}{r}
0.2727... \\
11\overline{)3.0000...} \\
\underline{22} \\
80 \\
\underline{77} \\
30 \\
\underline{27} \\
30 \\
\underline{22} \\
8 \\
\vdots
\end{array}
$$

$\dfrac{3}{11} = 0.\overline{27}$

39. a. $\sqrt{81}\;(=9)$

b. $0, \sqrt{81}$

c. $-17, 0, \sqrt{81}$

d. $-17, -\dfrac{9}{13}, 0, 0.75, \sqrt{81}$

e. $\sqrt{2}, \pi$

f. $-17, -\dfrac{9}{13}, 0, 0.75, \sqrt{2}, \pi, \sqrt{81}$

40. Answers will vary. One example of an integer that is not a natural number is -7.

41. Answers will vary. One example of a rational number that is not an integer is $\dfrac{3}{4}$.

42. Answers will vary. One example of a real number that is not a rational number is π.

43. $-93 < 17$; -93 is to the left of 17, so $-93 < 17$.

44. $-2 > -200$; -2 is to the right of -200, so $-2 > -200$.

45. $0 > -\dfrac{1}{3}$; 0 is to the right of $-\dfrac{1}{3}$, so $0 > -\dfrac{1}{3}$.

46.

$-\dfrac{1}{4} < -\dfrac{1}{5}$; $-\dfrac{1}{4} = -0.25$ is to the left of

$-\dfrac{1}{5} = -0.2$, so $-\dfrac{1}{4} < -\dfrac{1}{5}$.

47. $-13 \geq -11$ is false because neither $-13 > -11$ nor $-13 = -11$ is true.

48. $-126 \leq -126$ is true because $-126 = -126$.

49. $|-58| = 58$ because the distance between -58 and 0 on the number line is 58.

50. $|2.75| = 2.75$ because the distance between 2.75 and 0 on the number line is 2.75.

51. $7 + 13y = 13y + 7$

52. $9(x+7) = (x+7)9$

53. $6 + (4+y) = (6+4) + y = 10 + y$

54. $7(10x) = (7 \cdot 10)x = 70x$

55. $6(4x - 2 + 5y) = 6(4x) + 6(-2) + 6(5y)$
$$= 24x - 12 + 30y$$

56. $4a + 9 + 3a - 7 = 4a + 3a + 9 - 7$
$$= (4 + 3)a + (9 - 7)$$
$$= 7a + 2$$

57. $6(3x + 4) + 5(2x - 1)$
$$= 6(3x) + 6(4) + 5(2x) + 5(-1)$$
$$= 18x + 24 + 10x - 5$$
$$= 18x + 10x + 24 - 5$$
$$= (18 + 10)x + [24 + (-5)]$$
$$= 28x + 19$$

58. $-6 + 8 = +2$ or 2.
Start at -6. Move 8 units to the right because 8 is positive.

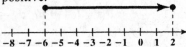

59. $8 + (-11) = -3$

60. $-\dfrac{3}{4} + \dfrac{1}{5} = -\dfrac{3}{4} \cdot \dfrac{5}{5} + \dfrac{1}{5} \cdot \dfrac{4}{4}$
$$= -\dfrac{15}{20} + \dfrac{4}{20} = -\dfrac{11}{20}$$

61. $7 + (-5) + (-13) + 4$
$$= [7 + (-5)] + (-13) + 4$$
$$= 2 + (-13) + 4$$
$$= [2 + (-13)] + 4 = -11 + 4 = -7$$

62. $8x + (-6y) + (-12x) + 11y$
$$= 8x + (-12x) + (-6y) + 11y$$
$$= [8 + (-12)]x + (-6 + 11)y$$
$$= -4x + 5y \text{ or } 5y - 4x$$

63. $10(3y + 4) + (-40y) = 30y + 40 + (-40y)$
$$= 30y + (-40y) + 40$$
$$= -10y + 40$$

64. $-1312 + 512 = -800$
The person's elevation is 800 feet below sea level.

65. $25 - 3 + 2 + 1 - 4 + 2$
$$= 25 + (-3) + 2 + 1 + (-4) + 2$$
$$= 23$$
The reservoir's water level at the end of five months is 23 feet.

66. $9 - 13 = 9 + (-13)$

67. $-9 - (-13) = -9 + 13 = 4$

68. $-\dfrac{7}{10} - \dfrac{1}{2} = -\dfrac{7}{10} - \dfrac{1}{2} \cdot \dfrac{5}{5}$
$$= -\dfrac{7}{10} - \dfrac{5}{10} = -\dfrac{12}{10} = -\dfrac{6}{5}$$

69. $-3.6 - (-2.1) = -3.6 + 2.1 = -1.5$

70. $-7 - (-5) + 11 - 16$
$$= -7 + 5 + 11 + (-16)$$
$$= [(-7) + (-16)] + (5 + 11)$$
$$= -23 + 16$$
$$= -7$$

71. $-25 - 4 - (-10) + 16$
$$= -25 + (-4) + 10 + 16$$
$$= [(-25) + (-4)] + (10 + 16)$$
$$= -29 + 26$$
$$= -3$$

72. $3 - 6a - 8 - 2a = 3 - 8 - 6a - 2a$
$$= [3 + (-8)] + [-6a - 2a]$$
$$= -5 + (-6 - 2)a$$
$$= -5 - 8a$$

73. $26,000 - (-650) = 26,500 + 650$
$$= 27,150$$
The difference in elevation is 27,150 feet.

74. $(-7)(-12) = 84$

75. $\dfrac{3}{5}\left(-\dfrac{5}{11}\right) = -\dfrac{3 \cdot \cancel{5}}{\cancel{5} \cdot 11} = -\dfrac{3}{11}$

76. $5(-3)(-2)(-4) = -120$

77. $\dfrac{45}{-5} = 45\left(-\dfrac{1}{5}\right) = -9$

78. $-17 \div 0$ is undefined.

79. $-\dfrac{4}{5} \div \left(-\dfrac{2}{5}\right) = -\dfrac{4}{5}\left(-\dfrac{5}{2}\right) = \dfrac{20}{10} = 2$

80. $-4\left(-\dfrac{3}{4}x\right)=\left[-4\left(-\dfrac{3}{4}\right)\right]x=3x$

81. $-3(2x-1)-(4-5x)$

$=-3(2x)+(-3)(-1)-4+5x$

$=-6x+3-4+5x$

$=-6x+5x+3-4$

$=(-6+5)x+\left[3+(-4)\right]$

$=-1x-1$

$=-x-1$

82. $\quad 5x+16=-8-x$

$\quad 5(-6)+16=-8-(-6)$

$\quad -30+16=-8+6$

$\quad\quad\quad -14=-2,\ \text{false}$

The number is not a solution.

83. $\quad 2(x+3)-18=5x$

$\quad 2(-4+3)-18=5(-4)$

$\quad\quad 2(-1)-18=-20$

$\quad\quad\quad -2-18=-20$

$\quad\quad\quad\quad -20=-20,\ \text{true}$

The number is a solution.

84. a. $\ I=-232n+3430$

$\quad I=-232(3)+3430=-696+3430=2734$

There were 2734 inmates younger than 18 in 2003.

This underestimates the actual total shown in the bar graph by 7.

b. $\ I=-232n+3430$

$\quad I=-232(10)+3430=-2320+3430=1110$

There will be 1110 inmates younger than 18 in 2010.

85. $(-6)^2=(-6)(-6)=36$

86. $-6^2=-6\cdot6=-36$

87. $(-2)^5=(-2)(-2)(-2)(-2)(-2)=-32$

88. $4x^3+2x^3=(4+2)x^3=6x^3$

89. $4x^3+4x^2$ cannot be simplified. The terms $4x^3$ and $4x^2$ are not like terms because they have different variable factors.

90. $-40\div5\cdot2=-8\cdot2=-16$

91. $-6+(-2)\cdot5=-6+(-10)=-16$

92. $6-5(-3+2)=6-4(-1)=6+4=10$

93. $28\div\left(2-4^2\right)=28\div(2-16)$

$\qquad\qquad\qquad=28\div\left[2+(-16)\right]$

$\qquad\qquad\qquad=28\div(-14)$

$\qquad\qquad\qquad=-2$

94. $36-24\div4\cdot3-1=36-6\cdot3-1$

$\qquad\qquad\qquad\quad=36-18-1$

$\qquad\qquad\qquad\quad=18-1$

$\qquad\qquad\qquad\quad=17$

95. $-8\left[-4-5(-3)\right]=-8(-4+15)$

$\qquad\qquad\qquad\quad=-8(11)=-88$

96. $\dfrac{6(-10+3)}{2(-15)-9(-3)}=\dfrac{6(-7)}{-30+27}$

$\qquad\qquad\qquad\quad=\dfrac{-42}{-3}=14$

97. $\left(\dfrac{1}{2}+\dfrac{1}{3}\right)\div\left(\dfrac{1}{4}-\dfrac{3}{8}\right)$

$=\left(\dfrac{3}{6}+\dfrac{2}{6}\right)\div\left(\dfrac{2}{8}-\dfrac{3}{8}\right)$

$=\dfrac{5}{6}\div\left(-\dfrac{1}{8}\right)=\dfrac{5}{6}\cdot\left(-\dfrac{8}{1}\right)=-\dfrac{40}{6}=-\dfrac{20}{3}$

98. $\dfrac{1}{2}-\dfrac{2}{3}\div\dfrac{5}{9}+\dfrac{3}{10}$

$=\dfrac{1}{2}-\dfrac{2}{\cancel{3}_1}\cdot\dfrac{\cancel{9}^3}{5}+\dfrac{3}{10}$

$=\dfrac{1}{2}-\dfrac{6}{5}+\dfrac{3}{10}$

$=\dfrac{5}{10}-\dfrac{12}{10}+\dfrac{3}{10}=-\dfrac{4}{10}=-\dfrac{2}{5}$

99. $x^2 - 2x + 3; \ x = -1$

$$x^2 - 2x + 3 = (-1)^2 - 2(-1) + 3$$
$$= 1 + 2 + 3$$
$$= 6$$

100. $-x^2 - 7x; \ x = -2$

$$-x^2 - 7x = -(-2)^2 - 7(-2)$$
$$= -4 + 14$$
$$= 10$$

101. $4[7(a-1) + 2] = 4(7a - 7 + 2)$
$$= 4(7a - 5)$$
$$= 4(7a) + 4(-5)$$
$$= 28a - 20$$

102. $-6[4 - (y+2)] = -6(4 - y - 2)$
$$= -6(2 - y)$$
$$= -6(2) + (-6)(-y)$$
$$= -12 + 6y \text{ or } 6y - 12$$

103. $V = -1.2n^2 + 10n + 45$

$V = -1.2(5)^2 + 10(5) + 45 = 65$

The number of Americans who volunteered at least once in 2005 was 65%.
This underestimates the actual value shown in the bar graph by 0.4%

Chapter 1 Test

1. $1.4 - (-2.6) = 1.4 + 2.6 = 4$

$-9 + 3 + (-11) + 6$

2. $= [-9 + (-11)] + (3 + 6)$
$$= -20 + 9 = -11$$

3. $3(-17) = -51$

4. $\left(-\dfrac{3}{7}\right) \div \left(-\dfrac{15}{7}\right) = \left(-\dfrac{3}{7}\right)\left(-\dfrac{7}{15}\right)$

$$= \dfrac{21}{105} = \dfrac{\cancel{21} \cdot 1}{\cancel{21} \cdot 5} = \dfrac{1}{5}$$

5. $\left(3\dfrac{1}{3}\right)\left(-1\dfrac{3}{4}\right) = \left(\dfrac{10}{3}\right)\left(-\dfrac{7}{4}\right)$

$$= -\dfrac{10 \cdot 7}{3 \cdot 4} = -\dfrac{70}{12}$$

$$= -\dfrac{\cancel{2} \cdot 35}{\cancel{2} \cdot 6}$$

$$= -\dfrac{35}{6} \text{ or } -5\dfrac{5}{6}$$

6. $-50 \div 10 = -50\left(\dfrac{1}{10}\right) = -5$

7. $-6 - (5 - 12) = -6 - (-7) = -6 + 7 = 1$

8. $(-3)(-4) \div (7 - 10)$
$$= (-3)(-4) \div [7 + (-10)]$$
$$= (-3)(-4) \div (-3)$$
$$= 12 \div (-3)$$
$$= -4$$

9. $(6-8)^2(5-7)^3 = (-2)^2(-2)^3$
$$= 4(-8) = -32$$

10. $\dfrac{3(-2) - 2(2)}{-2(8-3)} = \dfrac{-6 - 4}{-2(5)}$

$$= \dfrac{-6 + (-4)}{-2(5)} = \dfrac{-10}{-10} = 1$$

11. $11x - (7x - 4) = 11x - 7x + 4$
$$= 11x + (-7x) + 4$$
$$= [11 + (-7)]x + 4$$
$$= 4x + 4$$

12. $5(3x - 4y) - (2x - y)$
$$= 5(3x) - 5(4y) - 2x + y$$
$$= 15x - 20y - 2x + y$$
$$= 15x - 2x - 20y + y$$
$$= 13x - 19y$$

13. $6 - 2[3(x+1) - 5] = 6 - 2[3x + 3 - 5]$
$$= 6 - 2(3x - 2)$$
$$= 6 - 6x + 4$$
$$= 10 - 6x$$

14. Rational numbers can be written as the quotient of two integers.

$$-7 = -\frac{7}{1}, \, -\frac{4}{5} = \frac{-4}{5}, \, 0 = \frac{0}{1}, \, 0.25 = \frac{1}{4},$$

$$\sqrt{4} = 2 = \frac{2}{1}, \text{ and } \frac{22}{7} = \frac{22}{7}.$$

Thus, $-7, \, -\frac{4}{5}, \, 0, \, 0.25, \, \sqrt{4}, \text{ and } \frac{22}{7}$ are the rational numbers of the set.

15. $-1 > -100$; -1 is to the right of -100 on the number line, so -1 is greater than -100.

16. $|-12.8| = 12.8$ because the distance between 12.8 and 0 on the number line is 12.8

17. $5(x-7); \, x = 4$

$$5(x-7) = 5(4-7) = 5(-3) = -15$$

18. $x^2 - 5x; \, x = -10$

$$x^2 - 5x = (-10)^2 - 5(-10)$$
$$= 100 + 50 = 150$$

19. $2(x+3) = 2(3+x)$

20. $-6(4x) = (-6 \cdot 4)x = -24x$

21. $7(5x - 1 + 2y) = 7(5x) - 7(1) + 7(2y)$
$$= 35x - 7 + 14y$$

22. $16,200 - (-830) = 17,030$

The difference in elevations is 17,030 feet.

23. $\frac{1}{5}(x+2) = \frac{1}{10}x + \frac{3}{5}$

$$\frac{1}{5}(3+2) = \frac{1}{10} \cdot 3 + \frac{3}{5}$$

$$\frac{1}{5}(5) = \frac{3}{10} + \frac{3}{5}$$

$$1 = \frac{3}{10} + \frac{6}{10}$$

$$1 = \frac{9}{10}, \text{ false}$$

The number is not a solution.

24. $3(x+2) - 15 = 4x$

$$3(-9+2) - 15 = 4(-9)$$
$$3(-7) - 15 = -36$$
$$-21 - 15 = -36$$
$$-36 = -36, \text{ true}$$

The number is a solution.

25. $\frac{1}{4}x - 5 = 32$

26. $5(x+4) - 7$

27. $V = -n^2 + 8n - 4$

$$V = -(3)^2 + 8(3) - 4 = -9 + 24 - 4 = 11$$

According to the formula, there were 11 million viewers for the opening episode of season 3. This underestimates the actual number shown in the bar graph by 0.3 million.

28. According to the line graph, the target heart rate for a 40-year-old taking a stress test is about 144 beats per minute.

29. $H = \frac{4}{5}(220 - a)$

$$H = \frac{4}{5}(220 - 40) = \frac{4}{5}(180) = 144$$

According to the formula, the target heart rate for a 40-year-old taking a stress test is about 144 beats per minute.
This value is the same as the value estimated from the line graph.

Chapter 2
Linear Equations and Inequalities in One Variable

2.1 Check Points

1.
$$x - 5 = 12$$
$$x - 5 + 5 = 12 + 5$$
$$x + 0 = 17$$
$$x = 17$$
Check:
$$x - 5 = 12$$
$$17 - 5 = 12$$
$$12 = 12$$
The solution set is $\{17\}$.

2.
$$z + 2.8 = 5.09$$
$$z + 2.8 - 2.8 = 5.09 - 2.8$$
$$z + 0 = 2.29$$
$$z = 2.29$$
Check:
$$z + 2.8 = 5.09$$
$$2.29 + 2.8 = 5.09$$
$$5.09 = 5.09$$
The solution set is $\{2.29\}$.

3.
$$-\frac{1}{2} = x - \frac{3}{4}$$
$$-\frac{1}{2} + \frac{3}{4} = x - \frac{3}{4} + \frac{3}{4}$$
$$-\frac{2}{4} + \frac{3}{4} = x$$
$$\frac{1}{4} = x$$
Check:
$$-\frac{1}{2} = x - \frac{3}{4}$$
$$-\frac{1}{2} = \frac{1}{4} - \frac{3}{4}$$
$$-\frac{1}{2} = -\frac{2}{4}$$
$$-\frac{1}{2} = -\frac{1}{2}$$
The solution set is $\left\{\frac{1}{4}\right\}$.

4.
$$8y + 7 - 7y - 10 = 6 + 4$$
$$y - 3 = 10$$
$$y - 3 + 3 = 10 + 3$$
$$y = 13$$
Check:
$$8y + 7 - 7y - 10 = 6 + 4$$
$$8(13) + 7 - 7(13) - 10 = 6 + 4$$
$$104 + 7 - 91 - 10 = 10$$
$$111 - 101 = 10$$
$$10 = 10$$
The solution set is $\{13\}$.

5.
$$7x = 12 + 6x$$
$$7x - 6x = 12 + 6x - 6x$$
$$x = 12$$
Check:
$$7(12) = 12 + 6(12)$$
$$84 = 12 + 72$$
$$84 = 84$$
The solution set is $\{12\}$.

6.
$$3x - 6 = 2x + 5$$
$$3x - 2x - 6 = 2x - 2x + 5$$
$$x - 6 = 5$$
$$x - 6 + 6 = 5 + 6$$
$$x = 11$$
Check:
$$3x - 6 = 2x + 5$$
$$3(11) - 6 = 2(11) + 5$$
$$33 - 6 = 22 + 5$$
$$27 = 27$$
The solution set is $\{11\}$.

7.
$$V + 900 = 60A$$
$$V + 900 = 60(50)$$
$$V + 900 = 3000$$
$$V + 900 - 900 = 3000 - 900$$
$$V = 2100$$
At 50 months, a child will have a vocabulary of 2100 words.

2.1 Exercise Set

1. linear

3. not linear

5. not linear

7. linear

9. not linear

11.
$$x - 4 = 19$$
$$x - 4 + 4 = 19 + 4$$
$$x + 0 = 23$$
$$x = 23$$
Check:
$$x - 4 = 19$$
$$23 - 4 = 19$$
$$19 = 19$$
The solution set is $\{23\}$.

13.
$$z + 8 = -12$$
$$z + 8 - 8 = -12 - 8$$
$$z + 0 = -20$$
$$z = -20$$
Check:
$$z + 8 = -12$$
$$-20 + 8 = -12$$
$$-12 = -12$$
The solution set is $\{-20\}$.

15.
$$-2 = x + 14$$
$$-2 - 14 = x + 14 - 14$$
$$-16 = x$$
Check:
$$-2 = -16 + 14$$
$$-2 = -2$$
The solution set is $\{-16\}$.

17.
$$-17 = y - 5$$
$$-17 + 5 = y - 5 + 5$$
$$-12 = y$$
Check:
$$-17 = -12 - 5$$
$$-17 = -17$$
The solution set is $\{-12\}$.

19.
$$7 + z = 11$$
$$z = 11 - 7$$
$$z = 4$$
Check:
$$7 + 4 = 11$$
$$11 = 11$$
The solution set is $\{4\}$.

21.
$$-6 + y = -17$$
$$y = -17 + 6$$
$$y = -11$$
Check:
$$-6 - 11 = -17$$
$$-17 = -17$$
The solution set is $\{-11\}$.

23.
$$x + \frac{1}{3} = \frac{7}{3}$$
$$x = \frac{7}{3} - \frac{1}{3}$$
$$x = 2$$
Check:
$$2 + \frac{1}{3} = \frac{7}{3}$$
$$\frac{6}{3} + \frac{1}{3} = \frac{7}{3}$$
$$\frac{7}{3} = \frac{7}{3}$$
The solution set is $\{2\}$.

25.
$$t + \frac{5}{6} = -\frac{7}{12}$$
$$t = -\frac{7}{12} - \frac{5}{6}$$
$$t = -\frac{7}{12} - \frac{10}{12} = -\frac{17}{12}$$
Check:
$$-\frac{17}{15} + \frac{5}{6} = -\frac{7}{12}$$
$$-\frac{17}{12} + \frac{10}{12} = -\frac{7}{12}$$
$$-\frac{7}{12} = -\frac{7}{12}$$
The solution set is $\left\{-\frac{17}{12}\right\}$.

27.

$$x - \frac{3}{4} = \frac{9}{2}$$

$$x - \frac{3}{4} + \frac{3}{4} = \frac{9}{2} + \frac{3}{4}$$

$$x = \frac{21}{4}$$

Check:

$$\frac{21}{4} - \frac{3}{4} = \frac{9}{2}$$

$$\frac{18}{4} = \frac{9}{2}$$

$$\frac{9}{2} = \frac{9}{2}$$

The solution set is $\left\{ \frac{21}{4} \right\}$.

29.

$$-\frac{1}{5} + y = -\frac{3}{4}$$

$$y = -\frac{3}{4} + \frac{1}{5}$$

$$y = -\frac{15}{20} + \frac{4}{20} = -\frac{11}{20}$$

Check:

$$-\frac{1}{5} + \left(-\frac{11}{20} \right) = -\frac{3}{4}$$

$$-\frac{4}{20} - \frac{11}{20} = -\frac{3}{4}$$

$$-\frac{15}{20} = -\frac{3}{4}$$

$$-\frac{3}{4} = -\frac{3}{4}$$

The solution set is $\left\{ -\frac{11}{20} \right\}$.

31.

$$3.2 + x = 7.5$$

$$3.2 + x - 3.2 = 7.5 - 3.2$$

$$x = 4.3$$

Check:

$$3.2 + 4.3 = 7.5$$

$$7.5 = 7.5$$

The solution set is $\{4.3\}$.

33.

$$x + \frac{3}{4} = -\frac{9}{2}$$

$$x + \frac{3}{4} - \frac{3}{4} = -\frac{9}{2} - \frac{3}{4}$$

$$x = -\frac{21}{4}$$

Check:

$$-\frac{21}{4} + \frac{3}{4} = -\frac{9}{2}$$

$$-\frac{18}{4} = -\frac{9}{2}$$

$$-\frac{9}{2} = -\frac{9}{2}$$

The solution set is $\left\{ -\frac{21}{4} \right\}$.

35.

$$5 = -13 + y$$

$$5 + 13 = y$$

$$18 = y$$

Check:

$$5 = -13 + 18$$

$$5 = 5$$

The solution set is $\{18\}$.

37.

$$-\frac{3}{5} = -\frac{3}{2} + s$$

$$-\frac{3}{5} + \frac{3}{2} = s$$

$$-\frac{6}{10} + \frac{15}{10} = s$$

$$\frac{9}{10} = s$$

Check:

$$-\frac{3}{5} = -\frac{3}{2} + \frac{9}{10}$$

$$-\frac{6}{10} = -\frac{15}{10} + \frac{9}{10}$$

$$-\frac{6}{10} = -\frac{6}{10}$$

The solution set is $\left\{ \frac{9}{10} \right\}$.

39. $830 + y = 520$

$$y = 520 - 830$$
$$y = -310$$

Check:
$$830 - 310 = 520$$
$$520 = 520$$

The solution set is $\{-310\}$.

41. $r + 3.7 = 8$

$$r = 8 - 3.7$$
$$r = 4.3$$

Check:
$$4.3 + 3.7 = 8$$
$$8 = 8$$

The solution set is $\{4.3\}$.

43. $-3.7 + m = -3.7$

$$m = -3.7 + 3.7$$
$$m = 0$$

Check:
$$-3.7 + 0 = -3.7$$
$$-3.7 = -3.7$$

The solution set is $\{0\}$.

45. $6y + 3 - 5y = 14$

$$y + 3 = 14$$
$$y = 14 - 3$$
$$y = 11$$

Check:
$$6(11) + 3 - 5(11) = 14$$
$$66 + 3 - 55 = 14$$
$$14 = 14$$

The solution set is $\{11\}$.

47. $7 - 5x + 8 + 2x + 4x - 3 = 2 + 3 \cdot 5$

$$x + 12 = 2 + 15$$
$$x = 17 - 12$$
$$x = 5$$

Check:
$$7 - 5(5) + 8 + 2(5) + 4(5) - 3 = 2 + 3 \cdot 5$$
$$7 - 25 + 8 + 10 + 20 - 3 = 2 + 15$$
$$45 - 28 = 17$$
$$17 = 17$$

The solution set is $\{5\}$.

49. $7y + 4 = 6y - 9$

$$7y - 6y + 4 = -9$$
$$y = -9 - 4$$
$$y = -13$$

Check:
$$7(-13) + 4 = 6(-13) - 9$$
$$-91 + 4 = -78 - 9$$
$$-87 = -87$$

The solution set is $\{-13\}$.

51. $12 - 6x = 18 - 7x$

$$12 + x = 18$$
$$x = 6$$

Check:
$$12 - 6(6) = 18 - 7(6)$$
$$12 - 36 = 18 - 42$$
$$-24 = -24$$

The solution set is $\{6\}$.

53. $4x + 2 = 3(x - 6) + 8$

$$4x + 2 = 3x - 18 + 8$$
$$4x + 2 = 3x - 10$$
$$4x - 3x + 2 = -10$$
$$x + 2 = -10$$
$$x = -10 - 2$$
$$x = -12$$

Check:
$$4(-12) + 2 = 3(-12 - 6) + 8$$
$$-48 + 2 = 3(-18) + 8$$
$$-46 = -54 + 8$$
$$-46 = -46$$

The solution set is $\{-12\}$.

55. $x - \square = \triangle$

$$x - \square + \square = \triangle + \square$$
$$x = \triangle + \square$$

57. $2x + \triangle = 3x + \square$

$$\triangle = 3x - 2x + \square$$
$$\triangle = x + \square$$
$$\triangle - \square = x + \square - \square$$
$$\triangle - \square = x$$

59. $x - 12 = -2$

$\qquad x = -2 + 12$

$\qquad x = 10$

The number is 10.

61. $\dfrac{2}{5}x - 8 = \dfrac{7}{5}x$

$\qquad -8 = \dfrac{7}{5}x - \dfrac{2}{5}x$

$\qquad -8 = \dfrac{5}{5}x$

$\qquad -8 = x$

The number is -8.

63. $S = 1850, \; M = 150$

$\quad C + M = S$

$\quad C + 150 = 1850$

$\qquad C = 1850 - 150$

$\qquad C = 1700$

The cost of the computer is $1700.

65. $\quad C - 4.6x = 25$

$\quad C - 4.6(15) = 25$

$\qquad C - 69 = 25$

$\; C - 69 + 69 = 25 + 69$

$\qquad\qquad C = 94$

According to the formula, 94 channels were received by the average U.S. home in 2005. This underestimates the value given in the bar graph by 2.

67. a. According to the line graph, about 47% of U.S. workers were satisfied with their jobs in 2006.

b. 2006 is 19 years after 1987.

$\qquad S + 0.8x = 62$

$\qquad S + 0.8(19) = 62$

$\qquad S + 15.2 = 62$

$\quad S + 15.2 - 15.2 = 62 - 15.2$

$\qquad\qquad S = 46.8$

According to the formula, 46.8% of U.S. workers were satisfied with their jobs in 2006. This matches the line graph very well.

69. – 71. Answers will vary.

73. does not make sense; Explanations will vary. Sample explanation: It does not matter whether the number is added beside or below, as long as it is added to both sides of the equation.

75. makes sense

77. false; Changes to make the statement true will vary. A sample change is: If $y - a = -b$, then $y = a - b$.

79. true

81. Answers will vary. An example is: $x - 100 = -101$

83. $\qquad 6.9825 = 4.2296 + y$

$\quad 6.9825 - 4.2296 = y$

$\qquad\qquad 2.7529 = y$

The solution set is $\{2.7529\}$.

84. $\dfrac{9}{x} - 4x$

85. $-16 - 8 \div 4 \cdot (-2) = -16 - 2 \cdot (-2)$

$\qquad\qquad\qquad\qquad = -16 + (-2)(-2)$

$\qquad\qquad\qquad\qquad = -16 + 4$

$\qquad\qquad\qquad\qquad = -12$

87. $5 \cdot \dfrac{x}{5} = \dfrac{5}{1} \cdot \dfrac{x}{5} = x$

88. $\dfrac{-7y}{-7} = y$

89. $\quad 3x - 14 = -2x + 6$

$\quad 3(4) - 14 = -2(4) + 6$

$\qquad 12 - 14 = -8 + 6$

$\qquad\quad -2 = -2, \text{ true}$

Yes, 4 is a solution of the equation.

2.2 Check Points

1. $\dfrac{x}{3} = 12$

$3 \cdot \dfrac{x}{3} = 12 \cdot 3$

$1x = 36$

$x = 36$

Check:

$\dfrac{x}{3} = 12$

$\dfrac{36}{3} = 12$

$12 = 12$

The solution set is $\{36\}$.

2. a. $4x = 84$

$\dfrac{4x}{4} = \dfrac{84}{4}$

$1x = 21$

$x = 21$

The solution set is $\{21\}$.

b. $-11y = 44$

$\dfrac{-11y}{-11} = \dfrac{44}{-11}$

$1x = -4$

$x = -4$

The solution set is $\{-4\}$.

c. $-15.5 = 5z$

$\dfrac{-15.5}{5} = \dfrac{5z}{5}$

$-3.1 = 1z$

$-3.1 = z$

The solution set is $\{-3.1\}$.

3. a. $\dfrac{2}{3}y = 16$

$\dfrac{3}{2}\left(\dfrac{2}{3}y\right) = \dfrac{3}{2} \cdot 16$

$1y = 24$

$y = 24$

The solution set is $\{24\}$.

b. $28 = -\dfrac{7}{4}x$

$-\dfrac{4}{7} \cdot 28 = -\dfrac{4}{7}\left(-\dfrac{7}{4}x\right)$

$-16 = 1x$

$-16 = x$

The solution set is $\{-16\}$.

4. a. $-x = 5$

$-1x = 5$

$(-1)(-1x) = (-1)5$

$1x = -5$

$x = -5$

The solution set is $\{-5\}$.

b. $-x = -3$

$-1x = -3$

$(-1)(-1x) = (-1)(-3)$

$1x = 3$

$x = 3$

The solution set is $\{3\}$.

5. $4x + 3 = 27$

$4x + 3 - 3 = 27 - 3$

$4x = 24$

$\dfrac{4x}{4} = \dfrac{24}{4}$

$x = 6$

The solution set is $\{6\}$.

6. $-4y - 15 = 25$

$-4y - 15 + 15 = 25 + 15$

$-4y = 40$

$\dfrac{-4y}{-4} = \dfrac{40}{-4}$

$y = -10$

The solution set is $\{-10\}$.

7.
$$2x - 15 = -4x + 21$$
$$2x + 4x - 15 = -4x + 4x + 21$$
$$6x - 15 = 21$$
$$6x - 15 + 15 = 21 + 15$$
$$6x = 36$$
$$\frac{6x}{6} = \frac{36}{6}$$
$$x = 6$$

The solution set is $\{6\}$.

8. a. The bar graph indicates that there were 1.3 million pay phones in 2004. Since 2004 is 4 years after 2000, substitute 4 into the formula for *n*.
$$P = -0.18n + 2.1$$
$$P = -0.18(4) + 2.1$$
$$P = -0.72 + 2.1$$
$$P = 1.38$$
The formula indicates that there were 1.38 million pay phones in 2004.
The formula overestimates by 0.08 million.

b.
$$P = -0.18n + 2.1$$
$$0.3 = -0.18n + 2.1$$
$$0.3 - 2.1 = -0.18n + 2.1 - 2.1$$
$$-1.8 = -0.18n$$
$$\frac{-1.8}{-0.18} = \frac{-0.18n}{-0.18}$$
$$10 = n$$
The formula estimates that there will be 0.3 million pay phones 10 years after 2000, or in 2010.

2.2 Exercise Set

1.
$$\frac{x}{6} = 5$$
$$6 \cdot \frac{x}{6} = 6 \cdot 5$$
$$1x = 30$$
$$x = 30$$
Check:
$$\frac{30}{6} = 5$$
$$5 = 5$$
The solution set is $\{30\}$.

3.
$$\frac{x}{-3} = 11$$
$$-3 \cdot \frac{x}{-3} = -3(11)$$
$$1x = -33$$
$$x = -33$$
Check:
$$\frac{-33}{-3} = 11$$
$$11 = 11$$
The solution set is $\{-33\}$.

5.
$$5y = 35$$
$$\frac{5y}{5} = \frac{35}{5}$$
$$y = 7$$
Check:
$$5(7) = 35$$
$$35 = 35$$
The solution set is $\{7\}$.

7.
$$-7y = 63$$
$$\frac{-7y}{-7} = \frac{63}{-7}$$
$$y = -9$$
Check:
$$-7(-9) = 63$$
$$63 = 63$$
The solution set is $\{-9\}$.

9.
$$-28 = 8z$$
$$\frac{-28}{8} = \frac{8z}{8}$$
$$-\frac{7}{2} = z$$
Check:
$$-28 = 8\left(-\frac{7}{2}\right)$$
$$-28 = -\frac{56}{2}$$
$$-28 = -28$$
The solution set is $\left\{-\frac{7}{2}\right\}$. or $\left\{-3\frac{1}{2}\right\}$.

11. $-18 = -3z$

$$\frac{-18}{-3} = \frac{-3z}{-3}$$

$$6 = z$$

Check:

$$-18 = -3(6)$$

$$-18 = -18$$

The solution set is $\{6\}$.

13. $-8x = 6$

$$\frac{-8x}{-8} = \frac{6}{-8}$$

$$x = -\frac{6}{8} = -\frac{3}{4}$$

Check:

$$-8\left(-\frac{3}{4}\right) = 6$$

$$\frac{24}{4} = 6$$

$$6 = 6$$

The solution set is $\left\{-\frac{3}{4}\right\}$.

15. $17y = 0$

$$\frac{17y}{17} = \frac{0}{17}$$

$$y = 0$$

Check:

$$17(0) = 0$$

$$0 = 0$$

The solution set is $\{0\}$.

17. $\frac{2}{3}y = 12$

$$\frac{3}{2}\left(\frac{2}{3}y\right) = \frac{3}{2}(12)$$

$$1y = \frac{3}{2} \cdot \frac{12}{1} = \frac{36}{2}$$

$$y = 18$$

Check:

$$\frac{2}{3}(18) = 12$$

$$\frac{36}{3} = 12$$

$$12 = 12$$

The solution set is $\{18\}$.

19. $28 = -\frac{7}{2}x$

$$-\frac{2}{7}(28) = -\frac{2}{7}\left(-\frac{7}{2}x\right)$$

$$-\frac{56}{7} = 1x$$

$$-8 = x$$

Check:

$$28 = -\frac{7}{2}(-8)$$

$$28 = \frac{56}{2}$$

$$28 = 28$$

The solution set is $\{-8\}$.

21. $-x = 17$

$$-1x = 17$$

$$-1(-1x) = -1(17)$$

$$x = -17$$

Check:

$$-(-17) = 17$$

$$17 = 17$$

The solution set is $\{-17\}$.

23. $-47 = -y$

$$47 = y$$

Check:

$$-47 = -47$$

The solution set is $\{47\}$.

25. $-\frac{x}{5} = -9$

$$5\left(-\frac{x}{5}\right) = 5(-9)$$

$$-x = -45$$

$$x = 45$$

Check:

$$-\frac{45}{5} = -9$$

$$-9 = -9$$

The solution set is $\{45\}$.

27.

$$2x - 12x = 50$$
$$(2 - 12)x = 50$$
$$-10x = 50$$
$$\frac{-10x}{-10} = \frac{50}{-10}$$
$$x = -5$$

Check:
$$2(-5) - 12(-5) = 50$$
$$-10 + 60 = 50$$
$$50 = 50$$

The solution set is $\{-5\}$.

29.

$$2x + 1 = 11$$
$$2x + 1 - 1 = 11 - 1$$
$$2x = 10$$
$$\frac{2x}{2} = \frac{10}{2}$$
$$x = 5$$

Check:
$$2(5) + 1 = 11$$
$$10 + 1 = 11$$
$$11 = 11$$

The solution set is $\{5\}$.

31.

$$2x - 3 = 9$$
$$2x - 3 + 3 = 9 + 3$$
$$2x = 12$$
$$\frac{2x}{2} = \frac{12}{2}$$
$$x = 6$$

Check:
$$2(6) - 3 = 9$$
$$12 - 3 = 9$$
$$9 = 9$$

The solution set is $\{6\}$.

33.

$$-2y + 5 = 7$$
$$-2y + 5 - 5 = 7 - 5$$
$$-2y = 2$$
$$\frac{-2y}{2} = \frac{2}{-2}$$
$$y = -1$$

Check:
$$-2(-1) + 5 = 7$$
$$2 + 5 = 7$$
$$7 = 7$$

The solution set is $\{-1\}$.

35.

$$-3y - 7 = -1$$
$$-3y - 7 + 7 = -1 + 7$$
$$-3y = 6$$
$$\frac{-3y}{-3} = \frac{6}{-3}$$
$$y = -2$$

Check:
$$-3(-2) - 7 = -1$$
$$6 - 7 = -1$$
$$-1 = -1$$

The solution set is $\{-2\}$.

37.

$$12 = 4z + 3$$
$$12 - 3 = 4z + 3 - 3$$
$$9 = 4z$$
$$\frac{9}{4} = \frac{4z}{4}$$
$$\frac{9}{4} = z$$

Check:
$$12 = 4\left(\frac{9}{4}\right) + 3$$
$$12 = 9 + 3$$
$$12 = 12$$

The solution set is $\left\{\frac{9}{4}\right\}$.

39.
$$-x-3=3$$
$$-x-3+3=3+3$$
$$-x=6$$
$$x=-6$$
Check:
$$-(-6)-3=3$$
$$6-3=3$$
$$3=3$$
The solution set is $\{-6\}$.

41.
$$6y=2y-12$$
$$6y+12=2y-12+12$$
$$6y+12=2y$$
$$6y+12-6y=2y-6y$$
$$12=-4y$$
$$\frac{12}{-4}=\frac{-4y}{-4}$$
$$-3=y$$
Check:
$$6(-3)=2(-3)-12$$
$$-18=-6-12$$
$$-18=-18$$
The solution set is $\{-3\}$.

43.
$$3z=-2z-15$$
$$3z+2z=-2z-15+2z$$
$$5z=-15$$
$$\frac{5z}{5}=\frac{-15}{3}$$
$$z=-3$$
Check:
$$3(-3)=-2(-3)-15$$
$$-9=6-15$$
$$-9=-9$$
The solution set is $\{-3\}$.

45.
$$-5x=-2x-12$$
$$-5x+2x=-2x-12+2x$$
$$-3x=-12$$
$$\frac{-3x}{3}=\frac{-12}{-3}$$
$$x=4$$
Check:
$$-5(4)=2(4)-12$$
$$-20=-8-12$$
$$-20=-20$$
The solution set is $\{4\}$.

47.
$$8y+4=2y-5$$
$$8y+4-2y=2y-5-2y$$
$$6y+4=-5$$
$$6y+4-4=-5-4$$
$$6y=-9$$
$$\frac{6y}{6}=\frac{-9}{6}$$
$$y=-\frac{3}{2}$$
Check:
$$8\left(-\frac{3}{2}\right)+4=2\left(-\frac{3}{2}\right)-5$$
$$-12+4=-3-5$$
$$-8=-8$$
The solution set is $\left\{-\frac{3}{2}\right\}$.

49.
$$6z-5=z+5$$
$$6z-5-z=z+5-z$$
$$5z-5=5$$
$$5z-5+5=5+5$$
$$5z=10$$
$$\frac{5z}{5}=\frac{10}{5}$$
$$z=2$$
Check:
$$6(2)-5=2+5$$
$$12-5=2+5$$
$$7=7$$
The solution set is $\{2\}$.

51.
$$6x + 14 = 2x - 2$$
$$6x - 2x + 14 = -2$$
$$4x = -2 - 14$$
$$4x = -16$$
$$x = -4$$
Check:
$$6(-4) + 14 = 2(-4) - 2$$
$$-24 + 14 = -8 - 2$$
$$-10 = -10$$
The solution set is $\{-4\}$.

53.
$$-3y - 1 = 5 - 2y$$
$$-3y + 2y - 1 = 5$$
$$-y = 5 + 1$$
$$-y = 6$$
$$y = -6$$
Check:
$$-3(-6) - 1 = 5 - 2(-6)$$
$$18 - 1 = 5 + 12$$
$$17 = 17$$
The solution set is $\{-6\}$.

55.
$$\frac{x}{\square} = \triangle$$
$$\square \cdot \frac{x}{\square} = \triangle \cdot \square$$
$$x = \triangle \square$$

57.
$$\triangle = -x$$
$$\triangle(-1) = -x(-1)$$
$$-\triangle = x$$

59.
$$6x = 10$$
$$\frac{6x}{6} = \frac{10}{6}$$
$$x = \frac{10}{6} = \frac{5}{3}$$
The number is $\frac{5}{3}$.

61.
$$\frac{x}{-9} = 5$$
$$\frac{x}{-9}(-9) = 5(-9)$$
$$x = -45$$
The number is -45.

63.
$$4x - 8 = 56$$
$$4x - 8 + 8 = 56 + 8$$
$$4x = 64$$
$$\frac{4x}{4} = \frac{64}{4}$$
$$x = 16$$
The number is 16.

65.
$$-3x + 15 = -6$$
$$-3x + 15 - 15 = -6 - 15$$
$$-3x = -21$$
$$\frac{-3x}{-3} = \frac{-21}{-3}$$
$$x = 7$$
The number is 7.

67.
$$M = \frac{n}{5}$$
$$2 = \frac{n}{5}$$
$$5(2) = 5\left(\frac{n}{5}\right)$$
$$10 = n$$
If you are 2 miles away from the lightning flash, it will take 10 seconds for the sound of thunder to reach you.

69.
$$M = \frac{A}{740}$$
$$2.03 = \frac{A}{740}$$
$$740(2.03) = 740 \cdot \frac{A}{740}$$
$$1502.2 = A$$
The speed of the Concorde is 1502.2 miles per hour.

71. a. The bar graph indicates that the debt limit was $30,200 per citizen in 2006.
Since 2006 is 6 years after 2000, substitute 6 into the formula for n.

$$D = 1914n + 19,371$$

$$D = 1914(6) + 19,371$$

$$D = 11,484 + 19,371$$

$$D = 30,855$$

The formula indicates that the debt limit was $30,855 per citizen in 2006.
The formula overestimates by $655.

b.

$$D = 1914n + 19,371$$

$$40,425 = 1914n + 19,371$$

$$40,425 - 19,371 = 1914n + 19,371 - 19,371$$

$$21054 = 1914n$$

$$\frac{21054}{1914} = \frac{1914n}{1914}$$

$$11 = n$$

The formula estimates that the debt limit will be $40,425 per citizen 11 years after 2000, or in 2011.

73. – 75. Answers will vary.

77. does not make sense; Explanations will vary.
Sample explanation: When you subtract 12 from $12 - 3x$, you should obtain $-3x$, not positive $3x$.

79. does not make sense; Explanations will vary.
Sample explanation: To determine the number of pay phones in 2006, substitute 6 in for n and simplify.

81. false; Changes to make the statement true will vary.
A sample change is: If $3x - 4 = 16$, then $3x = 20$.

83. true

85. Answers will vary. As an example, start with an integer solution, such as 10, and set it equal to x.
That is, we have $x = 10$. The solution was obtained by multiplying both sides by $\dfrac{4}{5}$. To undo this, we multiply both sides of our equation by the reciprocal, $\dfrac{5}{4}$. This gives, $\dfrac{5}{4}x = \dfrac{5}{4}(10)$

$$\frac{5}{4}x = \frac{25}{2}$$

Therefore, an example equation would be $\dfrac{5}{4}x = \dfrac{25}{2}$.

87.

$$-72.8y - 14.6 = -455.43 - 4.98y$$

$$-72.8y - 14.6 + 4.98y =$$
$$-455.43 - 4.98y + 4.98y$$

$$-67.82y - 14.6 = -455.43$$

$$-67.82y - 14.6 + 14.6 = -455.43 + 14.6$$

$$-67.82y = -440.83$$

$$\frac{-67.82y}{-67.82} = \frac{-440.83}{-67.82}$$

$$y = 6.5$$

The solution set is $\{6.5\}$.

88. $(-10)^2 = (-10)(-10) = 100$

89. $-10^2 = -1 \cdot 10^2 = -1(10)(10) = -100$

90. $x^3 - 4x = (-1)^3 - 4(-1)$
$$= -1 + 4$$
$$= 3$$

91. $13 - 3(x + 2) = 13 - 3x - 6$
$$= -3x + 7$$

92. $2(x - 3) - 17 = 13 - 3(x + 2)$
$$2(6 - 3) - 17 = 13 - 3(6 + 2)$$
$$2(3) - 17 = 13 - 3(8)$$
$$6 - 17 = 13 - 24$$
$$-11 = -11, \text{ true}$$

Yes, 6 is a solution of the equation.

93. $10\left(\dfrac{x}{5} - \dfrac{39}{5}\right) = 10 \cdot \dfrac{x}{5} - 10 \cdot \dfrac{39}{5}$
$$= 2x - 78$$

2.3 Check Points

1. Simplify the algebraic expression on each side.
$$-7x + 25 + 3x = 16 - 2x - 3$$
$$-4x + 25 = 13 - 2x$$
Collect variable terms on one side and constant terms on the other side.
$$-4x + 25 = 13 - 2x$$
$$-4x + 25 + 2x = 13 - 2x + 2x$$
$$-2x + 25 = 13$$
$$-2x + 25 - 25 = 13 - 25$$
$$-2x = -12$$
Isolate the variable and solve.
$$\frac{-2x}{-2} = \frac{-12}{-2}$$
$$x = 6$$
The solution set is $\{6\}$.

2. Simplify the algebraic expression on each side.
$$8x = 2(x + 6)$$
$$8x = 2x + 12$$
Collect variable terms on one side and constant terms on the other side.
$$8x - 2x = 2x - 2x + 12$$
$$6x = 12$$
Isolate the variable and solve.
$$\frac{6x}{6} = \frac{12}{6}$$
$$x = 2$$
The solution set is $\{2\}$.

3. Simplify the algebraic expression on each side.
$$4(2x + 1) - 29 = 3(2x - 5)$$
$$8x + 4 - 29 = 6x - 15$$
$$8x - 25 = 6x - 15$$
Collect variable terms on one side and constant terms on the other side.
$$8x - 6x - 25 = 6x - 6x - 15$$
$$2x - 25 = -15$$
$$2x - 25 + 25 = -15 + 25$$
$$2x = 10$$
Isolate the variable and solve.
$$\frac{2x}{2} = \frac{10}{2}$$
$$x = 5$$
The solution set is $\{5\}$.

4. Begin by multiplying both sides of the equation by 12, the least common denominator.
$$\frac{x}{4} = \frac{2x}{3} + \frac{5}{6}$$
$$12 \cdot \frac{x}{4} = 12\left(\frac{2x}{3} + \frac{5}{6}\right)$$
$$12 \cdot \frac{x}{4} = 12 \cdot \frac{2x}{3} + 12 \cdot \frac{5}{6}$$
$$3x = 8x + 10$$
$$3x - 8x = 8x - 8x + 10$$
$$-5x = 10$$
$$\frac{-5x}{-5} = \frac{10}{-5}$$
$$x = -2$$
The solution set is $\{-2\}$.

5.
$$3x + 7 = 3(x + 1)$$
$$3x + 7 = 3x + 3$$
$$3x - 3x + 7 = 3x - 3x + 3$$
$$7 = 3$$
The original equation is equivalent to the false statement $7 = 3$.
The equation has no solution. The solution set is $\{\ \}$.

6.
$$3(x - 1) + 9 = 8x + 6 - 5x$$
$$3x - 3 + 9 = 3x + 6$$
$$3x + 6 = 3x + 6$$
$$3x - 3x + 6 = 3x - 3x + 6$$
$$6 = 6$$
The original equation is equivalent to $6 = 6$, which is true for every value of x.
The equation's solution is all real numbers or $\{x | x \text{ is a real number}\}$.

7.

$$D = \frac{10}{9}x + \frac{53}{9}$$

$$10 = \frac{10}{9}x + \frac{53}{9}$$

$$9 \cdot 10 = 9\left(\frac{10}{9}x + \frac{53}{9}\right)$$

$$90 = 10x + 53$$

$$90 - 53 = 10x + 53 - 53$$

$$37 = 10x$$

$$\frac{37}{10} = \frac{10x}{10}$$

$$3.7 = x$$

$$x = 3.7$$

The formula indicates that if the low-humor group averages a level of depression of 10 in response to a negative life event, the intensity of that event is 3.7. This is shown as the point whose corresponding value on the vertical axis is 10 and whose value on the horizontal axis is 3.7.

2.3 Exercise Set

1.

$$5x + 3x - 4x = 10 + 2$$

$$8x - 4x = 12$$

$$4x = 12$$

$$\frac{4x}{4} = \frac{12}{4}$$

$$x = 3$$

The solution set is $\{3\}$.

3.

$$4x - 9x + 22 = 3x + 30$$

$$-5x + 22 = 3x + 30$$

$$-5x - 3x + 22 = 30$$

$$-8x + 22 = 30$$

$$-8x = 30 - 22$$

$$-8x = 8$$

$$\frac{-8x}{-8} = \frac{8}{-8}$$

$$x = -1$$

The solution set is $\{-1\}$.

5.

$$3x + 6 - x = 8 + 3x - 6$$

$$2x + 6 = 2 + 3x$$

$$2x + 6 - 2 = 2 + 3x - 2$$

$$2x + 4 = 3x$$

$$2x + 4 - 2x = 3x - 2x$$

$$4 = x$$

The solution set is $\{4\}$.

7.

$$4(x+1) = 20$$

$$4x + 4 = 20$$

$$4x = 20 - 4$$

$$4x = 16$$

$$\frac{4x}{4} = \frac{16}{4}$$

$$x = 4$$

The solution set is $\{4\}$.

9.

$$7(2x-1) = 42$$

$$14x - 7 = 42$$

$$14x - 7 + 7 = 42 + 7$$

$$14x = 49$$

$$\frac{14x}{14} = \frac{49}{14}$$

$$x = \frac{7}{2}$$

The solution set is $\left\{\frac{7}{2}\right\}$.

11.

$$38 = 30 - 2(x-1)$$

$$38 = 30 - 2x + 2$$

$$38 = 32 - 2x$$

$$38 - 32 = -2x$$

$$6 = -2x$$

$$\frac{6}{-2} = \frac{-2x}{-2}$$

$$-3 = x$$

The solution set is $\{-3\}$.

13. $2(4z+3)-8=46$

$8z+6-8=46$

$8z-2=46$

$8z-2+2=46+2$

$8z=48$

$\dfrac{8z}{3}=\dfrac{48}{8}$

$z=6$

The solution set is $\{6\}$.

15. $6x-(3x+10)=14$

$6x-3x-10=14$

$3x-10=14$

$3x-10+10=14+10$

$3x=24$

$\dfrac{3x}{3}=\dfrac{24}{3}$

$x=8$

The solution set is $\{8\}$.

17. $5(2x+1)=12x-3$

$10x+5=12x-3$

$10x-10x+5=12x-10x-3$

$5=2x-3$

$5+3=2x-3+3$

$8=2x$

$\dfrac{8}{2}=\dfrac{2x}{2}$

$x=4$

The solution set is $\{4\}$.

19. $3(5-x)=4(2x+1)$

$15-3x=8x+4$

$15-3x-8x=8x+4-8x$

$15-11x=4$

$15-11x-15=4-15$

$-11x=-11$

$\dfrac{-11x}{-11}=\dfrac{-11}{-11}$

$x=1$

The solution set is $\{1\}$.

21. $8(y+2)=2(3y+4)$

$8y+16=6y+8$

$8y+16-16=6y+8-16$

$8y=6y-8$

$8y-6y=6y-8-6y$

$2y=-8$

$y=-4$

The solution set is $\{-4\}$.

23. $3x+3=7x-14-3$

$3x+3=7x-17$

$3x+3-3=7x-17-3$

$3x=7x-20$

$3x-7x=7x-20-7x$

$-4x=-20$

$\dfrac{-4x}{-4}=\dfrac{-20}{-4}$

$x=5$

The solution set is $\{5\}$.

25. $5(2x-8)-2=5(x-3)+3$

$10x-40-2=5x-15+3$

$10x-42=5x-12$

$10x-42+42=5x-12+42$

$10x=5x+30$

$10x=5x+30-5x$

$5x=30$

$\dfrac{5x}{5}=\dfrac{30}{5}$

$x=6$

The solution set is $\{6\}$.

27. $6=-4(1-x)+3(x+1)$

$6=-4+4x+3x+3$

$6=-1+7x$

$6+1=-1+7x+1$

$7=7x$

$\dfrac{7}{7}=\dfrac{7x}{7}$

$1=x$

The solution set is $\{1\}$.

29. $10(z+4)-4(z-2)=3(z-1)+2(z-3)$

$$10z+40-4z+8=3z-3+2z-6$$
$$6z+48=5z-9$$
$$6z+48-48=5z-9-48$$
$$6z-5z=5z-57-5z$$
$$z=-57$$

The solution set is $\{-57\}$.

31. $\dfrac{x}{5}-4=-6$

To clear the equation of fractions, multiply both sides by the least common denominator (LCD), which is 5.

$$5\left(\frac{x}{5}-4\right)=5(-6)$$
$$5\cdot\frac{x}{5}-5\cdot4=-30$$
$$x-20=-30$$
$$x-20+20=-30+20$$
$$x=-10$$

The solution set is $\{-10\}$.

33. $\dfrac{2x}{3}-5=7$

To clear the equation of fractions, multiply both sides by the least common denominator (LCD), which is 3.

$$3\left(\tfrac{2}{3}x-5\right)=3(7)$$
$$3\cdot\tfrac{2}{3}x-3\cdot5=21$$
$$2x-15=21$$
$$2x-15+15=21+15$$
$$2x=36$$
$$\frac{2x}{2}=\frac{36}{2}$$
$$x=18$$

The solution set is $\{18\}$.

35. $\dfrac{2y}{3}-\dfrac{3}{4}=\dfrac{5}{12}$

To clear the equation of fractions, multiply both sides by the least common denominator (LCD), which is 12.

$$12\left(\frac{2y}{3}-\frac{3}{4}\right)=12\left(\frac{5}{12}\right)$$
$$12\left(\frac{2y}{3}\right)-12\left(\frac{3}{4}\right)=5$$
$$8y-9=5$$
$$8y-9+9=5+9$$
$$8y=14$$
$$\frac{8y}{8}=\frac{14}{8}$$
$$y=\frac{14}{8}=\frac{7}{4}$$

The solution set is $\left\{\dfrac{7}{4}\right\}$.

37. $\dfrac{x}{3}+\dfrac{x}{2}=\dfrac{5}{6}$

To clear the equation of fractions, multiply both sides by the least common denominator (LCD), which is 6.

$$6\left(\frac{x}{3}+\frac{x}{2}\right)=6\left(\frac{5}{6}\right)$$
$$2x+3x=5$$
$$5x=5$$
$$\frac{5x}{5}=\frac{5}{5}$$
$$x=1$$

The solution set is $\{1\}$.

39. $20 - \dfrac{z}{3} = \dfrac{z}{2}$

To clear the equation of fractions, multiply both sides by the least common denominator (LCD), which is 6.

$$6\left(20 - \frac{z}{3}\right) = 6\left(\frac{z}{2}\right)$$

$$120 - 2z = 3z$$

$$120 - 2z + 2z = 3z + 2z$$

$$120 = 5z$$

$$\frac{120}{5} = \frac{5z}{5}$$

$$24 = z$$

The solution set is $\{24\}$.

41. $\dfrac{y}{3} + \dfrac{2}{5} = \dfrac{y}{5} - \dfrac{2}{5}$

To clear the equation of fractions, multiply both sides by the least common denominator (LCD), which is 15.

$$15\left(\frac{y}{3} + \frac{2}{5}\right) = 15\left(\frac{y}{5} + \frac{2}{5}\right)$$

$$15\left(\frac{y}{3}\right) + 15\left(\frac{2}{5}\right) = 15\left(\frac{y}{5}\right) + 15\left(-\frac{2}{5}\right)$$

$$5y + 6 = 3y - 6$$

$$5y + 6 - 3y = 3y - 6 - 3y$$

$$2y + 6 = -6$$

$$2y + 6 - 6 = -6 - 6$$

$$2y = -12$$

$$\frac{2y}{2} = \frac{-12}{2}$$

$$y = -6$$

The solution set is $\{-6\}$.

43. $\dfrac{3x}{4} - 3 = \dfrac{x}{2} + 2$

To clear the equation of fractions, multiply both sides by the least common denominator (LCD), which is 8.

$$8\left(\frac{3x}{4} - 3\right) = 8\left(\frac{x}{2} + 2\right)$$

$$8\left(\frac{3x}{4}\right) - 8 \cdot 3 = 8\left(\frac{x}{2}\right) + 8 \cdot 2$$

$$6x - 24 = 4x + 16$$

$$6x - 24 - 4x = 4x + 16 - 4x$$

$$2x - 24 = 16$$

$$2x - 24 + 24 = 16 + 24$$

$$2x = 40$$

$$\frac{2x}{2} = \frac{40}{2}$$

$$x = 20$$

The solution set is $\{20\}$.

45. $\dfrac{x-3}{5} - 1 = \dfrac{x-5}{4}$

To clear the equation of fractions, multiply both sides by the least common denominator (LCD), which is 20.

$$20\left(\frac{x-3}{5} - 1\right) = 20\left(\frac{x-5}{4}\right)$$

$$4(x-3) - 20 = 5(x-5)$$

$$4x - 12 - 20 = 5x - 25$$

$$4x - 5x - 32 = 5x - 5x - 25$$

$$-x - 32 = -25$$

$$-x - 32 + 32 = -25 + 32$$

$$-x = 7$$

$$-1(-x) = -1(7)$$

$$x = -7$$

The solution set is $\{-7\}$.

47. $3x - 7 = 3(x+1)$

$$3x - 7 = 3x + 3$$

$$3x - 7 - 3x = 3x + 3 - 3x$$

$$-7 = 3$$

The original equation is equivalent to the false statement $-7 = 3$, so the equation is inconsistent and has no solution. The solution set is $\{\ \}$.

49.
$$2(x+4) = 4x+5-2x+3$$
$$2x+8 = 2x+8$$
$$2x-8-2x = 2x+8-2x$$
$$8 = 8$$

The original equation is equivalent to the true statement $8 = 8$, so the equation is an identity and the solution set is all real numbers $\{x \mid x \text{ is a real number}\}$.

51.
$$7+2(3x-5) = 8-3(2x+1)$$
$$7+6x-10 = 8-6x-3$$
$$6x-3 = 5-6x$$
$$6x+6x-3 = 5-6x+6x$$
$$12x-3 = 5$$
$$12x-3+3 = 5+3$$
$$12x = 8$$
$$\frac{12x}{12} = \frac{8}{12}$$
$$x = \frac{2}{3}$$

The solution set is $\left\{\dfrac{2}{3}\right\}$.

53.
$$4x+1-5x = 5-(x+4)$$
$$-x+1 = 5-x-4$$
$$-x+1 = 1-x$$
$$-x+1+x = 1-x+x$$
$$1 = 1$$

The original equation is equivalent to the true statement $1 = 1$, so the equation is an identity and the solution set is all real numbers $\{x \mid x \text{ is a real number}\}$.

55.
$$4(x+2)+1 = 7x-3(x-2)$$
$$4x+8+1 = 7x-3x+6$$
$$4x+9 = 4x+6$$
$$4x-4x+9 = 4x-4x+6$$
$$9 = 6$$

Since $9 = 6$ is a false statement, the original equation is inconsistent and has no solution. The solution set is $\{\ \}$.

57.
$$3-x = 2x+3$$
$$3-x+x = 2x+x+3$$
$$3 = 3x+3$$
$$3-3 = 3x+3-3$$
$$0 = 3x$$
$$\frac{0}{3} = \frac{3x}{3}$$
$$0 = x$$

The solution set is $\{0\}$.

59. $\dfrac{x}{3}+2 = \dfrac{x}{3}$

Multiply by the LCD, which is 3.
$$3\left(\frac{x}{3}+2\right) = 3\left(\frac{x}{3}\right)$$
$$x+6 = x$$
$$x-x+6 = x-x$$
$$6 = 0$$

Since $6 = 0$ is a false statement, the original equation has no solution. The solution set is $\{\ \}$.

61. $\dfrac{x}{2}-\dfrac{x}{4}+4 = x+4$

Multiply by the LCD, which is 4.
$$4\left(\frac{x}{2}-\frac{x}{4}+4\right) = 4(x+4)$$
$$4\left(\frac{x}{2}\right)-4\left(\frac{x}{4}\right)+16 = 4x+16$$
$$2x-x+16 = 4x+16$$
$$x+16 = 4x+16$$
$$x-x+16 = 4x-x+16$$
$$16 = 3x+16$$
$$16-16 = 3x+16-16$$
$$0 = 3x$$
$$\frac{0}{3} = \frac{3x}{3}$$
$$0 = x$$

The solution set is $\{0\}$.

63. $\dfrac{2}{3}x = 2 - \dfrac{5}{6}x$

Multiply both sides by the LCD which is 6.

$6\left(\dfrac{2}{3}x\right) = 6(2) - 6\left(\dfrac{5}{6}x\right)$

$2(2x) = 12 - 5x$

$4x = 12 - 5x$

$4x + 5x = 12 - 5x + 5x$

$9x = 12$

$\dfrac{9x}{9} = \dfrac{12}{9}$

$x = \dfrac{12}{9} = \dfrac{4}{3}$

The solution set is $\left\{\dfrac{4}{3}\right\}$.

65. $\dfrac{x}{\square} + \triangle = \$$

$\dfrac{x}{\square} + \triangle - \triangle = \$ - \triangle$

$\dfrac{x}{\square} = \$ - \triangle$

$\square\left(\dfrac{x}{\square}\right) = \square(\$ - \triangle)$

$x = \square\$ - \square\triangle$

67. First solve the equation for x.

$\dfrac{x}{5} - 2 = \dfrac{x}{3}$

$\dfrac{x}{5} - \dfrac{x}{5} - 2 = \dfrac{x}{3} - \dfrac{x}{5}$

$-2 = \dfrac{5x}{15} - \dfrac{3x}{15}$

$-2 = \dfrac{2x}{15}$

$15(-2) = 15\left(\dfrac{2x}{15}\right)$

$-30 = 2x$

$\dfrac{-30}{2} = \dfrac{2x}{2}$

$-15 = x$

Now evaluate the expression $x^2 - x$ for $x = -15$.

$x^2 - x = (-15)^2 - (-15)$

$= 225 + 15$

$= 240$

69. $\dfrac{1}{3}x + \dfrac{1}{5}x = 16$

$\text{LCD} = 15$

$15\left(\dfrac{1}{3}x\right) + 15\left(\dfrac{1}{5}x\right) = 15(16)$

$5x + 3x = 240$

$8x = 240$

$\dfrac{8x}{8} = \dfrac{240}{8}$

$x = 30$

The number is 30.

71. $\dfrac{3}{4}x - 3 = \dfrac{1}{2}x$

$4\left(\dfrac{3}{4}x\right) - 4(3) = 4\left(\dfrac{1}{2}x\right)$

$3x - 12 = 2x$

$3x - 2x - 12 = 2x - 2x$

$x - 12 = 0$

$x - 12 + 12 = 0 + 12$

$x = 12$

The number is 12.

73. $F = 10(x - 65) + 50$

$250 = 10(x - 65) + 50$

$250 - 50 = 10(x - 65) + 50 - 50$

$200 = 10x - 650$

$200 + 650 = 10x - 650 + 650$

$850 = 10x$

$\dfrac{850}{10} = \dfrac{10x}{10}$

$85 = x$

A person receiving a $250 fine was driving 85 miles per hour.

75.
$$\frac{W}{2} - 3H = 53$$
$$\frac{W}{2} - 3(6) = 53$$
$$\frac{W}{2} - 18 = 53$$
$$\frac{W}{2} - 18 + 18 = 53 + 18$$
$$\frac{W}{2} = 71$$
$$2 \cdot \frac{W}{2} = 2 \cdot 71$$
$$W = 142$$

According to the formula, the healthy weight of a person of height 5'6" is 142 pounds. This is 13 pounds below the upper end of the range shown in the bar graph.

77.
$$p = 15 + \frac{5d}{11}$$
$$201 = 15 + \frac{5d}{11}$$
$$201 - 15 = 15 + \frac{5d}{11} - 15$$
$$186 = \frac{5d}{11}$$
$$11(186) = 11\left(\frac{5d}{11}\right)$$
$$2046 = 5d$$
$$\frac{2046}{5} = d$$
$$409.2 = d$$
He descended to a depth of 409.2 feet below the surface.

79. – 81. Answers will vary.

83. makes sense

85. does not make sense; Explanations will vary. Sample explanation: Though 5 is a solution, the complete solution is all real numbers.

87. false; Changes to make the statement true will vary. A sample change is: The solution of the equation is all real numbers.

89. true

91.
$$f = 0.432h - 10.44$$
$$16 = 0.432h - 10.44$$
$$16 + 10.44 = 0.432h - 10.44 + 10.44$$
$$26.44 = 0.432h$$
$$\frac{26.44}{0.432} = \frac{0.432h}{0.432}$$
$$61.2 \approx h$$
The woman's height was about 61 inches or 5 feet 1 inch, so the partial skeleton could be that of the missing woman.

93.
$$2(3x + 4) = 3x + 2\left[3(x - 1) + 2\right]$$
$$6x + 8 = 3x + 2(3x - 3 + 2)$$
$$6x + 8 = 3x + 2(3x - 1)$$
$$6x + 8 = 3x + 6x - 2$$
$$6x + 8 = 9x - 2$$
$$6x + 8 - 9x = 9x - 2 - 9x$$
$$-3x + 8 = -2$$
$$-3x + 8 - 8 = -2 - 8$$
$$-3x = -10$$
$$\frac{-3x}{-3} = \frac{-10}{3}$$
$$x = \frac{10}{3}$$
The solution set is $\left\{\frac{10}{3}\right\}$.

95.
$$4.8y + 32.5 = 124.8 - 9.4y$$
$$4.8y + 32.5 + 9.4y = 124.8 - 9.4y + 9.4y$$
$$14.2y + 32.5 = 124.8$$
$$14.2y + 32.5 - 32.5 = 124.8 - 32.5$$
$$14.2y = 92.3$$
$$\frac{14.2y}{14.2} = \frac{92.3}{14.2}$$
$$y = 6.5$$
The solution set is $\{6.5\}$.

96. $-24 < -20$ because -24 lies further to the left on a number line.

97. $-\frac{1}{3} < -\frac{1}{5}$ because $-\frac{1}{3}$ lies further to the left on a number line.

98. $-9-11+7-(-3) = -9-11+7+3$
$$= -20+10$$
$$= -10$$

99. a. $T = D + pm$
$$T - D = pm$$

b. $T - D = pm$
$$\frac{T-D}{p} = \frac{pm}{p}$$
$$\frac{T-D}{p} = m$$

100. $4 = 0.25B$
$$\frac{4}{0.25} = \frac{0.25B}{0.25}$$
$$16 = B$$
The solution set is $\{16\}$.

101. $1.3 = P \cdot 26$
$$\frac{1.3}{26} = \frac{P \cdot 26}{26}$$
$$0.05 = P$$
The solution set is $\{0.05\}$.

2.4 Check Points

1. $A = lw$
$$\frac{A}{w} = \frac{lw}{w}$$
$$\frac{A}{w} = l$$

2. $2l + 2w = P$
$$2l + 2w - 2w = P - 2w$$
$$2l = P - 2w$$
$$\frac{2l}{2} = \frac{P-2w}{2}$$
$$l = \frac{P-2w}{2}$$

3. $T = D + pm$
$$T - D = pm$$
$$\frac{T-D}{p} = \frac{pm}{p}$$
$$\frac{T-D}{p} = m$$
$$m = \frac{T-D}{p}$$

4. $\frac{x}{3} - 4y = 5$
$$3\left(\frac{x}{3} - 4y\right) = 3 \cdot 5$$
$$3 \cdot \frac{x}{3} - 3 \cdot 4y = 3 \cdot 5$$
$$x - 12y = 15$$
$$x - 12y + 12y = 15 + 12y$$
$$x = 15 + 12y$$

5. To change a percent to a decimal number, move the decimal point two places to the left and remove the percent sign.

a. $67\% = 0.67$

b. $250\% = 2.50$ or 2.5

6. To change a decimal number to a percent, move the decimal point two places to the right and add a percent sign.
$0.023 = 2.3\%$

7. Use the formula $A = PB$: A is P percent of B.

$$\boxed{\text{What}}\ \boxed{\text{is}}\ \boxed{9\%}\ \boxed{\text{of}}\ \boxed{50?}$$
$$A\quad =\ 0.09\ \cdot\quad 50$$
$$A = 4.5$$

8. Use the formula $A = PB$: A is P percent of B.

$$\boxed{9}\ \boxed{\text{is}}\ \boxed{60\%}\ \boxed{\text{of}}\ \boxed{\text{what?}}$$
$$9\ =\ 0.60\ \cdot\quad B$$
$$\frac{9}{0.60} = \frac{0.60B}{0.60}$$
$$15 = B$$

9. Use the formula $A = PB$: A is P percent of B.

$$\boxed{18} \; \boxed{\text{is}} \; \boxed{\text{what percent}} \; \boxed{\text{of}} \; \boxed{50?}$$
$$\overbrace{18} \; = \quad \overbrace{P} \quad \cdot \quad \overbrace{50}$$

$18 = P \cdot 50$

$$\frac{18}{50} = \frac{50P}{50}$$

$0.36 = P$

To change 0.36 to a percent, move the decimal point two places to the right and add a percent sign.
$0.36 = 36\%$

10. Use the formula $A = PB$: A is P percent of B.

Find the price decrease: $\$940 - \$611 = \$329$

$$\boxed{\begin{array}{c}\text{The price}\\\text{decrease}\end{array}} \; \boxed{\text{is}} \; \boxed{\begin{array}{c}\text{what}\\\text{percent}\end{array}} \; \boxed{\text{of}} \; \boxed{\begin{array}{c}\text{the original}\\\text{price?}\end{array}}$$
$$\overbrace{329} \quad = \quad \overbrace{P} \quad \cdot \quad \overbrace{940}$$

$329 = P \cdot 940$

$$\frac{329}{940} = \frac{940P}{940}$$

$0.35 = P$

To change 0.35 to a percent, move the decimal point two places to the right and add a percent sign.
$0.35 = 35\%$

11. a.

Year	Tax Paid the Year Before	increase/decrease	Taxes Paid This Year
1	$1200	20% decrease : $0.20 \cdot \$1200 = \$240$	$1200 - $240 = $960
2	$960	20% increase : $0.20 \cdot \$960 = \$192$	$960 + $192 = $1152

The taxes for year 2 will be $1152.

b. The taxes for year 2 are less than those originally paid.

Find the tax decrease: $\$1200 - \$1152 = \$48$

$$\begin{array}{ccccc}\text{The tax} & \text{what} & & \text{the original} \\ \text{decrease} & \text{is} & \text{percent} & \text{of} & \text{tax?}\end{array}$$
$$\overbrace{48} \quad = \quad \overbrace{P} \quad \cdot \quad \overbrace{1200}$$

$48 = P \cdot 1200$

$$\frac{48}{1200} = \frac{1200P}{1200}$$

$0.04 = P$

To change 0.04 to a percent, move the decimal point two places to the right and add a percent sign.
$0.04 = 4\%$

The overall tax decrease is 4%.

2.4 Exercise Set

1. $d = rt$ for r

$$\frac{d}{t} = \frac{rt}{t}$$

$$\frac{d}{t} = r \text{ or } r = \frac{d}{t}$$

This is the distance traveled formula: distance = rate · time.

3. $I = Prt$ for P

$$\frac{I}{rt} = \frac{Pr\,t}{rt}$$

$$\frac{I}{rt} = P \text{ or } P = \frac{I}{rt}$$

This is the formula for simple interest: interest = principal · rate · time.

5. $C = 2\pi r$ for r

$$\frac{C}{2\pi} = \frac{2\pi r}{2\pi}$$

$$\frac{C}{2\pi} = r \text{ or } r = \frac{C}{2\pi}$$

This is the formula for finding the circumference of a circle if you know its radius.

7. $E = mc^2$

$$\frac{E}{c^2} = \frac{mc^2}{c^2}$$

$$\frac{E}{c^2} = m \text{ or } m = \frac{E}{c^2}$$

This is Einstein's formula relating energy, mass, and the speed of light.

9. $y = mx + b$ for m

$$y - b = mx$$

$$\frac{y-b}{x} = \frac{mx}{x}$$

$$\frac{y-b}{x} = m \text{ or } m = \frac{y-b}{x}$$

This is the slope-intercept formula for the equation of a line.

11. $T = D + pm$ for D

$$T - pm = D + pm - pm$$

$$T - pm = D$$

$$D = T - pm$$

13. $A = \frac{1}{2}bh$ for b

$$2A = 2\left(\frac{1}{2}bh\right)$$

$$2A = bh$$

$$\frac{2A}{h} = \frac{bh}{h}$$

$$\frac{2A}{h} = b \text{ or } b = \frac{2A}{h}$$

This is the formula for the area of a triangle: area = $\frac{1}{2}$ · base · height.

15. $M = \frac{n}{5}$ for n

$$5M = 5\left(\frac{n}{5}\right)$$

$$5M = n \text{ or } n = 5M$$

17. $\frac{c}{2} + 80 = 2F$ for c

$$\frac{c}{2} + 80 - 80 = 2F - 80$$

$$\frac{c}{2} = 2F - 80$$

$$2\left(\frac{c}{2}\right) = 2(2F - 80)$$

$$c = 4F - 160$$

19. $A = \frac{1}{2}(a + b)$ for a

$$2A = 2\left[\frac{1}{2}(a+b)\right]$$

$$2A = a + b$$

$$2A - b = a + b - b$$

$$2A - b = a \text{ or } a = 2A - b$$

This is the formula for finding the average of two numbers.

21. $S = P + Prt$ for r

$S - P = P + Prt - P$

$S - P = Prt$

$\dfrac{S-P}{Pt} = \dfrac{Prt}{Pt}$

$\dfrac{S-P}{Pt} = r$ or $r = \dfrac{S-P}{Pt}$

This is the formula for finding the sum of principle and interest for simple interest problems.

23. $A = \dfrac{1}{2}h(a+b)$ for b

$2A = 2\left[\dfrac{1}{2}h(a+b)\right]$

$2A = h(a+b)$

$2A = ha + hb$

$2A - ha = ha + hb - ha$

$2A - ha = hb$

$\dfrac{2A-ha}{h} = \dfrac{hb}{h}$

$\dfrac{2A-ha}{h} = b$ or $b = \dfrac{2A}{h} - a$

This is the formula for the area of a trapezoid.

25. $Ax + By = C$ for x

$Ax + By - By = C - By$

$Ax = C - By$

$\dfrac{Ax}{A} = \dfrac{C-By}{A}$

$x = \dfrac{C-By}{A}$

This is the standard form of the equation of a line.

27. To change a percent to a decimal number, move the decimal point two places to the left and remove the percent sign.

$27\% = 0.27$

29. $63.4\% = 0.634$

31. $170\% = 1.7$

33. $3\% = 0.03$

35. $\dfrac{1}{2}\% = 0.5\% = 0.005$

37. To change a decimal number to a percent, move the decimal point two places to the right and add a percent sign.

$0.89 = 89\%$

39. $0.002 = 0.2\%$

41. $4.78 = 478\%$

43. $100 = 10,000\%$

45. $A = PB$; $P = 3\% = 0.03$, $B = 200$

$A = PB$

$A = 0.03 \cdot 200$

$A = 6$

3% of 200 is 6.

47. $A = PB$; $P = 18\% = 0.18$, $B = 40$

$A = PB$

$A = 0.18 \cdot 40$

$A = 7.2$

18% of 40 is 7.2.

49. $A = PB$; $A = 3$, $P = 60\% = 0.6$

$A = PB$

$3 = 0.6 \cdot B$

$\dfrac{3}{0.6} = \dfrac{0.6B}{0.6}$

$5 = B$

3 is 60% of 5.

51. $A = PB$; $A = 40.8$, $P = 24\% = 0.24$

$A = PB$

$40.8 = 0.24 \cdot B$

$\dfrac{40.8}{0.24} = \dfrac{0.24B}{0.24}$

$170 = B$

24% of 170 is 40.8.

53. $A = PB; \ A = 3, \ B = 15$

$A = PB$

$3 = P \cdot 15$

$\dfrac{3}{15} = \dfrac{P \cdot 15}{15}$

$0.2 = P$

$0.2 = 20\%$

3 is 20% of 15.

55. $A = PB; \ A = 0.3, \ B = 2.5$

$A = PB$

$0.3 = P \cdot 2.5$

$\dfrac{0.3}{2.5} = \dfrac{P \cdot 2.5}{2.5}$

$0.12 = P$

$0.12 = 12\%$

0.3 is 12% of 2.5

57. The increase is $8 - 5 = 3$.

$A = PB$

$3 = P \cdot 5$

$\dfrac{3}{5} = \dfrac{P \cdot 5}{5}$

$0.60 = P$

This is a 60% increase.

59. The decrease is $4 - 1 = 3$.

$A = PB$

$3 = P \cdot 4$

$\dfrac{3}{4} = \dfrac{4P}{4}$

$0.75 = P$

This is a 75% decrease.

61. $y = (a + b)x$

$\dfrac{y}{(a+b)} = \dfrac{(a+b)x}{(a+b)}$

$\dfrac{y}{a+b} = x \ $ or $ \ x = \dfrac{y}{a+b}$

63. $y = (a - b)x + 5$

$y - 5 = (a - b)x + 5 - 5$

$y - 5 = (a - b)x$

$\dfrac{y-5}{a-b} = \dfrac{(a-b)x}{a-b}$

$\dfrac{y-5}{a-b} = x \ $ or $ \ x = \dfrac{y-5}{a-b}$

65. $y = cx + dx$

$y = (c + d)x$

$\dfrac{y}{c+d} = \dfrac{(c+d)x}{c+d}$

$\dfrac{y}{c+d} = x \ $ or $ \ x = \dfrac{y}{c+d}$

67. $y = Ax - Bx - C$

$y = (A - B)x - C$

$y + C = (A - B)x - C + C$

$y + C = (A - B)x$

$\dfrac{y+C}{A-B} = \dfrac{(A-B)x}{A-B}$

$\dfrac{y+C}{A-B} = x \ $ or $ \ x = \dfrac{y+C}{A-B}$

69. a. $A = \dfrac{x + y + z}{3}$ for z

$3A = 3\left(\dfrac{x+y+z}{3}\right)$

$3A = x + y + z$

$3A - x - y = x + y + z - x - y$

$3A - x - y = z$

b. $A = 90, x = 86, y = 88$

$z = 3A - x - y$

$z = 3(90) - 86 - 88 = 96$

You need to get 96% on the third exam to have an average of 90%

71. a. $d = rt$ for t

$\dfrac{d}{r} = \dfrac{rt}{r}$

$\dfrac{d}{r} = t$

b. $t = \dfrac{d}{r}; \ d = 100, r = 40$

$t = \dfrac{100}{40} = 2.5$

You would travel for 2.5 $\left(\text{or } 2\dfrac{1}{2}\right)$ hours.

73. $0.34 \cdot 1200 = 408$

408 single women would marry someone other than the perfect mate.

75. This is the equivalent of asking: 175 is 35% of what?

$$A = P \cdot B$$
$$175 = 0.35 \cdot B$$
$$\frac{175}{0.35} = \frac{0.35B}{0.35}$$
$$500 = B$$

Americans throw away 500 billion pounds of trash each year.

77. a. The total number of countries in 1974 was $41 + 48 + 63 = 152$.

$$A = P \cdot B$$
$$41 = P \cdot 152$$
$$\frac{41}{152} = \frac{152B}{152}$$
$$0.27 \approx B$$

About 27% of countries were free in 1974.

b. The total number of countries in 2004 was $89 + 54 + 49 = 192$.

$$A = P \cdot B$$
$$89 = P \cdot 192$$
$$\frac{89}{192} = \frac{192B}{192}$$
$$0.46 \approx B$$

About 46% of countries were free in 2004.

c. The increase is $89 - 41 = 48$.

$$A = P \cdot B$$
$$48 = P \cdot 41$$
$$\frac{48}{41} = \frac{41B}{41}$$
$$1.17 \approx B$$

There was approximately a 117% increase in the number of free countries from 1974 to 2004.

79. $A = PB;\ A = 7500, B = 60,000$

$$A = PB$$
$$7500 = P \cdot 60,000$$
$$\frac{7500}{60,000} = \frac{P \cdot 60,000}{60,000}$$
$$0.125 = P$$

The charity has raised $0.125 = 12.5\%$ of its goal.

81. $A = PB;\ p = 15\% = 0.15, B = 60$

$$A = 0.15 \cdot 60 = 09$$

The tip was $9.

83. a. The sales tax is 6% of $16,800.

$$0.06(16,800) = 1008$$

The sales tax due on the car is $1008.

b. The total cost is the sum of the price of the car and the sales tax.

$$\$16,800 + \$1008 = \$17,808$$

The car's total cost is $17,808.

85. a. The discount is 12% of $860.

$$0.12(860) = 103.20$$

The discount amount is $103.20.

b. The sale price is the regular price minus the discount amount:

$$\$860 - \$103.20 = \$756.80$$

87. The decrease is $\$840 - \$714 = \$126$.

$$A = P \cdot B$$
$$126 = P \cdot 840$$
$$\frac{126}{840} = \frac{P \cdot 840}{840}$$
$$0.15 = P$$

This is a $0.15 = 15\%$ decrease.

89. Investment dollars decreased in year 1 are $0.30 \cdot \$10,000 = \$3000$. This means that $\$10,000 - \$3000 = \$7000$ remains. Investment dollars increased in year 2 are $0.40 \cdot \$7000 = \$2800$. This means that $\$7000 + \$2800 = \$9800$ of the original investment remains. This is an overall loss of $200 over the two years.

$$A = P \cdot B$$
$$200 = P \cdot 10,000$$
$$\frac{200}{10,000} = \frac{P \cdot 10,000}{10,000}$$
$$0.02 = P$$

The financial advisor is not using percentages properly. Instead of a 10% gain, this is a $0.02 = 2\%$ loss.

91. – 95. Answers will vary.

97. does not make sense; Explanations will vary. Sample explanation: Sometimes you will solve for one variable in terms of other variables.

99. does not make sense; Explanations will vary. Sample explanation: Since the sale price cannot be negative, the percent decrease cannot be more than 100%.

101. false; Changes to make the statement true will vary.

A sample change is: If $A = lw$, then $w = \dfrac{A}{l}$.

103. true

105.
$$5x + 20 = 8x - 16$$
$$5x + 20 - 8x = 8x - 16 - 8x$$
$$-3x + 20 = -16$$
$$-3x + 20 - 20 = -16 - 20$$
$$-3x = -36$$
$$\frac{-3x}{-3} = \frac{-36}{-3}$$
$$x = 12$$

Check:
$$5(12) + 20 = 8(12) - 16$$
$$60 + 20 = 96 - 16$$
$$80 = 80$$

The solution set is $\{12\}$.

106.
$$5(2y - 3) - 1 = 4(6 + 2y)$$
$$10y - 15 - 1 = 24 + 8y$$
$$10y - 16 = 24 + 8y$$
$$10y - 16 - 8y = 24 + 8y - 8y$$
$$2y - 16 = 24$$
$$2y - 16 + 16 = 24 + 16$$
$$2y = 40$$
$$\frac{2y}{2} = \frac{40}{2}$$
$$y = 20$$

Check:
$$5(2 \cdot 20 - 3) - 1 = 4(6 + 2 \cdot 20)$$
$$5(40 - 3) - 1 = 4(6 + 40)$$
$$5(37) - 1 = 4(46)$$
$$185 - 1 = 184$$
$$184 = 184$$

The solution set is $\{20\}$.

107. $x - 0.3x = 1x - 0.3x = (1 - 0.3)x = 0.7x$

108. $\dfrac{13}{x} - 7x$

109. $8(x + 14)$

110. $9(x - 5)$

Chapter 2 Mid-Chapter Check Point

1. Begin by multiplying both sides of the equation by 4, the least common denominator.
$$\frac{x}{2} = 12 - \frac{x}{4}$$
$$4\left(\frac{x}{2}\right) = 4(12) - 4\left(\frac{x}{4}\right)$$
$$2x = 48 - x$$
$$2x + x = 48 - x + x$$
$$3x = 48$$
$$\frac{3x}{3} = \frac{48}{3}$$
$$x = 16$$

The solution set is $\{16\}$.

2.
$$5x - 42 = -57$$
$$5x - 42 + 42 = -57 + 42$$
$$5x = -15$$
$$\frac{5x}{5} = \frac{-15}{5}$$
$$x = -3$$

The solution set is $\{-3\}$.

3.
$$H = \frac{EC}{825}$$
$$H \cdot 825 = \frac{EC}{825} \cdot 825$$
$$825H = EC$$
$$\frac{825H}{E} = \frac{EC}{E}$$
$$\frac{825H}{E} = C$$

4. $A = P \cdot B$
$$A = 0.06 \cdot 140$$
$$A = 8.4$$
8.4 is 6% of 140.

5.
$$\frac{-x}{10} = -3$$
$$10\left(\frac{-x}{10}\right) = 10(-3)$$
$$-x = -30$$
$$-1(-x) = -1(-30)$$
$$x = 30$$

The solution set is $\{30\}$.

6.
$$1 - 3(y - 5) = 4(2 - 3y)$$
$$1 - 3y + 15 = 8 - 12y$$
$$-3y + 16 = 8 - 12y$$
$$-3y + 12y + 16 = 8 - 12y + 12y$$
$$9y + 16 = 8$$
$$9y + 16 - 16 = 8 - 16$$
$$9y = -8$$
$$\frac{9y}{9} = \frac{-8}{9}$$
$$y = -\frac{8}{9}$$
The solution set is $\left\{-\dfrac{8}{9}\right\}$.

7.
$$S = 2\pi rh$$
$$\frac{S}{2\pi h} = \frac{2\pi rh}{2\pi h}$$
$$\frac{S}{2\pi h} = r$$

8.
$$A = P \cdot B$$
$$12 = 0.30 \cdot B$$
$$\frac{12}{0.30} = \frac{0.30 \cdot B}{0.30}$$
$$40 = B$$
12 is 30% of 40.

9. $\dfrac{3y}{5} + \dfrac{y}{2} = \dfrac{5y}{4} - 3$

To clear fractions, multiply both sides by the LCD, 20.
$$20\left(\frac{3y}{5}\right) + 20\left(\frac{y}{2}\right) = 20\left(\frac{5y}{4}\right) - 20(3)$$
$$4(3y) + 10y = 5(5y) - 60$$
$$12y + 10y = 25y - 60$$
$$22y = 25y - 60$$
$$22y - 25y = 25y - 25y - 60$$
$$-3y = -60$$
$$\frac{-3y}{-3} = \frac{-60}{-3}$$
$$y = 20$$
The solution set is $\{20\}$.

10.
$$5z + 7 = 6(z - 2) - 4(2z - 3)$$
$$5z + 7 = 6z - 12 - 8z + 12$$
$$5z + 7 = -2z$$
$$5z - 5z + 7 = -2z - 5z$$
$$7 = -7z$$
$$\frac{7}{-7} = \frac{-7z}{-7}$$
$$-1 = z$$
The solution set is $\{-1\}$.

11.
$$Ax - By = C$$
$$Ax - By + By = C + By$$
$$Ax = C + By$$
$$\frac{Ax}{A} = \frac{C + By}{A}$$
$$x = \frac{C + By}{A} \text{ or } \frac{By + C}{A}$$

12. $6y + 7 + 3y = 3(3y - 1)$
$$9y + 7 = 9y - 3$$
$$9y - 9y + 7 = 9y - 9y - 3$$
$$7 = -3$$
Since this is a false statement, there is no solution or $\{\ \}$.

13.
$$10\left(\frac{1}{2}x + 3\right) = 10\left(\frac{3}{5}x - 1\right)$$
$$10\left(\frac{1}{2}x\right) + 10(3) = 10\left(\frac{3}{5}x\right) - 10(1)$$
$$5x + 30 = 6x - 10$$
$$5x - 5x + 30 = 6x - 5x - 10$$
$$30 = x - 10$$
$$30 + 10 = x - 10 + 10$$
$$40 = x$$
The solution set is $\{40\}$.

14.
$$A = P \cdot B$$
$$50 = P \cdot 400$$
$$\frac{50}{400} = \frac{P \cdot 400}{400}$$
$$0.125 = P$$
50 is 0.125 = 12.5% of 400.

15.
$$\frac{3(m+2)}{4} = 2m+3$$
$$4 \cdot \frac{3(m+2)}{4} = 4(2m+3)$$
$$3(m+2) = 4(2m+3)$$
$$3m+6 = 8m+12$$
$$3m-3m+6 = 8m-3m+12$$
$$6 = 5m+12$$
$$6-12 = 5m+12-12$$
$$-6 = 5m$$
$$\frac{-6}{5} = \frac{5m}{5}$$
$$-\frac{6}{5} = m$$

The solution set is $\left\{ -\dfrac{6}{5} \right\}$.

16. The increase is $50 - 40 = 10$.
$$A = P \cdot B$$
$$10 = P \cdot 40$$
$$\frac{10}{40} = \frac{P \cdot 40}{40}$$
$$0.25 = P$$
This is a 0.25 = 25% increase.

17.
$$12w - 4 + 8w - 4 = 4(5w - 2)$$
$$20w - 8 = 20w - 8$$
$$20w - 20w - 8 = 20w - 20w - 8$$
$$-8 = -8$$
Since $-8 = -8$ is a true statement, the solution is all real numbers or $\left\{ x \mid x \text{ is a real number} \right\}$.

18. a.
$$G = -\frac{1}{2}n + 47$$
$$G = -\frac{1}{2}(10) + 47$$
$$= -5 + 47$$
$$= 42$$
According to the formula, 42% of Americans had guns in their homes in 1990.
This underestimates the actual percentage shown in the gar graph by 0.7%

b.
$$G = -\frac{1}{2}n + 47$$
$$30 = -\frac{1}{2}n + 47$$
$$2 \cdot 30 = 2\left(-\frac{1}{2}n + 47 \right)$$
$$60 = -n + 94$$
$$60 - 94 = -n + 94 - 94$$
$$-34 = -n$$
$$n = 34$$
According to the formula, 30% of Americans will have guns in their homes 34 years after 1980, or 2014.

2.5 Check Points

1. Let x = the number.
$$6x - 4 = 68$$
$$6x - 4 + 4 = 68 + 4$$
$$6x = 72$$
$$x = 12$$
The number is 12.

2. Let x = the average salary for elementary school teachers.
Let $x + 54,890$ = the average salary for lawyers
$$x + (x + 54,890) = 142,970$$
$$x + x + 54,890 = 142,970$$
$$2x + 54,890 = 142,970$$
$$2x + 54,890 - 54,890 = 142,970 - 54,890$$
$$2x = 88,080$$
$$x = 44,040$$
The average salary for elementary school teachers is $44,040 and the average salary for lawyers is $44,040 + $54,890 = $98,930.

3. Let $x =$ the page number of the first facing page.
Let $x + 1 =$ the page number of the second facing page.

$$x + (x + 1) = 145$$
$$x + x + 1 = 145$$
$$2x + 1 = 145$$
$$2x + 1 - 1 = 145 - 1$$
$$2x = 144$$
$$x = 72$$
$$x + 1 = 73$$

The page numbers are 72 and 73.

4. Let $x =$ the number of eighths of a mile traveled.

$$2 + 0.25x = 10$$
$$2 - 2 + 0.25x = 10 - 2$$
$$0.25x = 8$$
$$\frac{0.25x}{0.25} = \frac{8}{0.25}$$
$$x = 32$$

You can go 32 eighths of a mile. That is equivalent to $\frac{32}{8} = 4$ miles.

5. Let $x =$ the width of the swimming pool.
Let $3x =$ the length of the swimming pool.

$$P = 2l + 2w$$
$$320 = 2 \cdot 3x + 2 \cdot x$$
$$320 = 6x + 2x$$
$$320 = 8x$$
$$\frac{320}{8} = \frac{8x}{8}$$
$$40 = x$$
$$x = 40$$
$$3x = 120$$

The pool is 40 feet wide and 120 feet long.

6. Let $x =$ the original price.

Original price	minus	the reduction (40% of original price)	is	the reduced price, \$564
x	$-$	$0.4x$	$=$	564

$$x - 0.4x = 564$$
$$0.6x = 564$$
$$\frac{0.6x}{0.6} = \frac{564}{0.6}$$
$$x = 940$$

The original price was \$940.

2.5 Exercise Set

1.
$$x + 60 = 410$$
$$x + 60 - 60 = 410 - 60$$
$$x = 350$$
The number is 350.

3.
$$x - 23 = 214$$
$$x - 23 + 23 = 214 + 23$$
$$x = 237$$
The number is 237.

5.
$$7x = 126$$
$$\frac{7x}{7} = \frac{126}{7}$$
$$x = 18$$
The number is 18.

7.
$$\frac{x}{19} = 5$$
$$19\left(\frac{x}{19}\right) = 19(5)$$
$$x = 95$$
The number is 95.

9.
$$4 + 2x = 56$$
$$4 - 4 + 2x = 56 - 4$$
$$2x = 52$$
$$\frac{2x}{2} = \frac{52}{2}$$
$$x = 26$$
The number is 26.

11.
$$5x - 7 = 178$$
$$5x - 7 + 7 = 178 + 7$$
$$5x = 185$$
$$\frac{5x}{5} = \frac{185}{5}$$
$$x = 37$$
The number is 37.

13.
$$x + 5 = 2x$$
$$x + 5 - x = 2x - x$$
$$5 = x$$
The number is 5.

15. $2(x+4) = 36$

$2x + 8 = 36$

$2x = 28$

$x = 14$

The number is 14.

17. $9x = 30 + 3x$

$6x = 30$

$x = 5$

The number is 5.

19. $\dfrac{3x}{5} + 4 = 34$

$\dfrac{3x}{5} = 30$

$3x = 150$

$x = 50$

The number is 50.

21. Let $x =$ the time spent listening to radio.
Let $x + 581 =$ the time spent watching TV.

$x + (x + 581) = 2529$

$x + x + 581 = 2529$

$2x + 581 - 581 = 2529 - 581$

$2x = 1948$

$x = 974$

$x + 581 = 1555$

Americans spent 974 hours listening to radio and 1555 hours watching TV.

23. Let $x =$ the average salary for carpenters.
Let $2x - 7740 =$ the average salary for computer programmers.

$x + (2x - 7740) = 99,000$

$x + 2x - 7740 = 99,000$

$3x - 7740 = 99,000$

$3x - 7740 + 7740 = 99,000 + 7740$

$3x = 106,740$

$x = 35,580$

$2x - 7740 = 63,420$

The average salary for carpenters is \$35,580 and the average salary for computer programmers is \$63,420.

25. Let $x =$ the number of the left-hand page.
Let $x + 1 =$ the number of the right-hand page.

$x + (x + 1) = 629$

$x + x + 1 = 629$

$2x + 1 = 629$

$2x + 1 - 1 = 629 - 1$

$2x = 628$

$\dfrac{2x}{2} = \dfrac{628}{2}$

$x = 314$

The pages are 314 and 315.

27. Let $x =$ the first consecutive odd integer (Babe Ruth).
Let $x + 2 =$ the second consecutive odd integer (Roger Maris).

$x + (x + 2) = 120$

$x + x + 2 = 120$

$2x + 2 = 120$

$2x = 118$

$x = 59$

$x + 2 = 61$

Babe Ruth had 59 home runs and Roger Maris had 61.

29. Let $x =$ the number of miles you can travel in one week for \$320.

$200 + 0.15x = 320$

$200 + 0.15x - 200 = 320 - 200$

$0.15x = 120$

$\dfrac{0.15x}{0.15} = \dfrac{120}{0.15}$

$x = 800$

You can travel 800 miles in one week for \$320.
This checks because \$200 + 0.15(\$800) = \$320.

31. Let $x =$ the number of years after 2004.

$960 + 11.40x = 1074$

$960 - 960 + 11.40x = 1074 - 960$

$11.40x = 114$

$\dfrac{11.40x}{11.40} = \dfrac{114}{11.40}$

$x = 10$

Mortgage payments will average \$1074 ten years after 2004, or 2014.

33. Let x = the width of the field.
Let $4x$ = the length of the field.
$$P = 2l + 2w$$
$$500 = 2 \cdot 4x + 2 \cdot x$$
$$500 = 8x + 2x$$
$$500 = 10x$$
$$\frac{500}{10} = \frac{10x}{10}$$
$$50 = x$$
$$x = 50$$
$$4x = 200$$
The field is 50 yards wide and 200 yards long.

35. Let x = the width of a football field.
Let $x + 200$ = the length of a football field.
$$P = 2l + 2w$$
$$1040 = 2(x + 200) + 2 \cdot x$$
$$1040 = 2x + 400 + 2x$$
$$1040 = 4x + 400$$
$$640 = 4x$$
$$160 = x$$
$$x = 160$$
$$x + 200 = 360$$
A football field is 160 feet wide and 360 feet long.

37. As shown in the diagram,
let x = the height and $3x$ = the length.
To construct the bookcase, 3 heights and 4 lengths are needed.
Since 60 feet of lumber is available,
$$3x + 4(3x) = 60$$
$$3x + 12x = 60$$
$$15x = 60$$
$$x = 4$$
$$3x = 12$$
The bookcase is 12 feet long and 4 feet high.

39. Let x = the price before the reduction.
$$x - 0.20x = 320$$
$$0.80x = 320$$
$$\frac{0.80x}{0.80} = \frac{320}{0.80}$$
$$x = 400$$
The price before the reduction was $400.

41. Let x = the last year's salary.
$$x + 0.08x = 50,220$$
$$1.08x = 50,220$$
$$\frac{1.08x}{1.08} = \frac{50,220}{1.08}$$
$$x = 46,500$$
Last year's salary was $46,500.

43. Let x = the price of the car without tax.
$$x + 0.06x = 23,850$$
$$1.06x = 23,850$$
$$\frac{1.06x}{1.06} = \frac{23,850}{1.06}$$
$$x = 22,500$$
The price of the car without sales tax was $14,500.

45. Let x = the number of hours of labor.
$$63 + 35x = 448$$
$$63 + 35x - 63 = 448 - 63$$
$$35x = 385$$
$$\frac{35x}{35} = \frac{385}{35}$$
$$x = 11$$
It took 11 hours of labor to repair the car.

47. – 59. Answers will vary.

51. does not make sense; Explanations will vary.
Sample explanation: You should practice setting up and solving word problems.

53. makes sense

55. false; Changes to make the statement true will vary.
A sample change is: This should be modeled by $x - 10 = 160$.

57. true

59. Let x = the number of inches over 5 feet.
$$W = 100 + 5x$$
$$135 = 100 + 5x$$
$$135 - 100 = 100 - 100 + 5x$$
$$35 = 5x$$
$$\frac{35}{5} = \frac{5x}{5}$$
$$7 = x$$
The height 5' 7" corresponds to 135 pounds.

61. Let x = the woman's age.
Let $3x$ = the "uncle's" age.
$$3x + 20 = 2(x + 20)$$
$$3x + 20 = 2x + 40$$
$$3x - 2x + 20 = 2x - 2x + 40$$
$$x + 20 = 40$$
$$x + 20 - 20 = 40 - 20$$
$$x = 20$$
The woman is 20 years old and the "uncle" is $3x = 3(20) = 60$ years old.

63. $\dfrac{4}{5}x = -16$

$$\frac{5}{4}\left(\frac{4}{5}x\right) = \frac{5}{4}(-16)$$
$$x = -20$$

Check:
$$\frac{4}{5}(-20) = -16$$
$$\frac{4}{5} \cdot \frac{-20}{1} = -16$$
$$\frac{-80}{5} = -16$$
$$-16 = -16$$
The solution set is $\{-20\}$.

64. $6(y - 1) + 7 = 9y - y + 1$
$$6y - 6 + 7 = 9y - y + 1$$
$$6y + 1 = 8y + 1$$
$$6y + 1 - 1 = 8y + 1 - 1$$
$$6y = 8y$$
$$6y - 8y = 8y - 8y$$
$$-2y = 0$$
$$y = 0$$
Check:
$$6(0 - 1) + 7 = 9(0) - 0 + 1$$
$$6 - 10 + 7 = 0 - 0 + 1$$
$$1 = 1$$
The solution set is $\{0\}$.

65. $V = \dfrac{1}{3}lwh$ for w

$$V = \frac{1}{3}lwh$$
$$3V = 3\left(\frac{1}{3}lwh\right)$$
$$3V = lwh$$
$$\frac{3V}{lh} = \frac{lwh}{lh}$$
$$\frac{3V}{lh} = w \quad \text{or} \quad w = \frac{3V}{lh}$$

66. $A = \dfrac{1}{2}bh$

$$30 = \frac{1}{2}(12)h$$
$$\frac{30}{6} = \frac{6h}{6}$$
$$5 = h$$

67. $A = \dfrac{1}{2}h(a + b)$

$$A = \frac{1}{2} \cdot 7 \cdot (10 + 16) = \frac{1}{2} \cdot 7 \cdot (26) = 91$$

68. $x = 4(90 - x) - 40$
$$x = 360 - 4x - 40$$
$$x = 320 - 4x$$
$$x + 4x = 320 - 4x + 4x$$
$$5x = 320$$
$$\frac{5x}{5} = \frac{320}{5}$$
$$x = 64$$
The solution set is $\{64\}$.

2.6 Check Points

1. $A = 24, b = 4$

$$A = \frac{1}{2}bh$$
$$24 = \frac{1}{2} \cdot 4 \cdot h$$
$$24 = 2h$$
$$12 = h$$
The height of the sail is 12 ft.

2. Use the formulas for the area and circumference of a circle. The radius is 20 ft.

$A = \pi r^2$

$A = \pi(20)^2$

$= 400\pi$

≈ 1256 or 1257

The area is 400π ft^2 or approximately 1256 ft^2 or 1257 ft^2.

$C = 2\pi r$

$C = 2\pi(20)$

$= 40\pi$

≈ 126

The circumference is 40π ft or approximately 126 ft.

3. The radius of the large pizza is 9 inches, and the radius of the medium pizza is 7 inches.
large pizza:

$A = \pi r^2 = \pi(9 \text{ in.})^2 = 81\pi \text{ in.}^2 \approx 254 \text{ in.}^2$

medium pizza:

$A = \pi r^2 = \pi(7 \text{ in.})^2 = 49\pi \text{ in.}^2 \approx 154 \text{ in.}^2$

For each pizza, find the price per inch by dividing the price by the area.
Price per square inch for the large pizza

$= \dfrac{\$20.00}{81\pi \text{ in.}^2} \approx \dfrac{\$20.00}{254 \text{ in.}^2} \approx \dfrac{\$0.08}{\text{in.}^2}$

Price per square inch for the medium pizza

$= \dfrac{\$14.00}{49\pi \text{ in.}^2} \approx \dfrac{\$14.00}{154 \text{ in.}^2} \approx \dfrac{\$0.09}{\text{in.}^2}.$

The large pizza is the better buy.

4. Smaller cylinder: $r = 3$ in., $h = 5$ in.

$V = \pi r^2 h$

$V = \pi(3)^2 \cdot 5$

$= 45\pi$

The volume of the smaller cylinder is 45π in.^{3}.
Larger cylinder: $r = 3$ in., $h = 10$ in.

$V = \pi r^2 h$

$V = \pi(3)^2 \cdot 10$

$= 90\pi$

The volume of the smaller cylinder is 90π in.^{3}.
The ratio of the volumes of the two cylinders is

$\dfrac{V_{\text{larger}}}{V_{\text{smaller}}} = \dfrac{90\pi \text{ in.}^3}{45\pi \text{ in.}^3} = \dfrac{2}{1}.$

So, the volume of the larger cylinder is 2 times the volume of the smaller cylinder.

5. Use the formula for the volume of a sphere. The radius is 4.5 in.

$V = \dfrac{4}{3}\pi r^3$

$V = \dfrac{4}{3}\pi(4.5)^3$

$= 121.5\pi$

≈ 382

The volume is approximately 382 in.^{3}. Thus the 350 cubic inches will not be enough to fill the ball.

6. Let $3x = $ the measure of the first angle.
Let $x = $ the measure of the second angle.
Let $x - 20 = $ the measure of the third angle.

$3x + x + (x - 20) = 180$

$5x - 20 = 180$

$5x = 200$

$x = 40$

$3x = 120$

$x - 20 = 20$

The three angle measures are 120°, 40°, and 20°.

7. *Step 1* Let $x = $ the measure of the angle.

Step 2 Let $90 - x = $ the measure of its complement.

Step 3 The angle's measure is twice that of its complement, so the equation is
$x = 2 \cdot (90 - x)$.

Step 4 Solve this equation

$x = 2 \cdot (90 - x)$

$x = 180 - 2x$

$x + 2x = 180 - 2x + 2x$

$3x = 180$

$x = 60$

The measure of the angle is 60°.

Step 5 The complement of the angle is
$90° - 60° = 30°,$ and 60° is indeed twice 30°.

2.6 Exercise Set

1. Use the formulas for the perimeter and area of a rectangle. The length is 6 m and the width is 3 m.

$P = 2l + 2w$

$\quad = 2(6) + 2(3) = 12 + 6 = 18$

$A = lw = 6 \cdot 3 = 18$

The perimeter is 18 meters, and the area is 18 square meters.

3. Use the formula for the area of a triangle. The base is 14 in and the height is 8 in.

$A = \dfrac{1}{2}bh = \dfrac{1}{2}(14)(8) = 56$

The area is 56 square inches.

5. Use the formula for the area of a trapezoid. The bases are 16 m and 10 m and the height is 7 m.

$A = \dfrac{1}{2}h(a + b)$

$\quad = \dfrac{1}{2}(7)(16 + 10) = \dfrac{1}{2} \cdot 7 \cdot 26 = 91$

The area is 91 square meters.

7. $A = 1250, \ w = 25$

$A = lw$

$1250 = l \cdot 25$

$50 = l$

The length of the swimming pool is 50 feet.

9. $A = 20, b = 5$

$A = \dfrac{1}{2}bh$

$20 = \dfrac{1}{2} \cdot 5 \cdot h$

$20 = \dfrac{5}{2}h$

$\dfrac{2}{5}(20) = \dfrac{2}{5}\left(\dfrac{5}{2}h\right)$

$8 = h$

The height of the triangle is 8 feet.

11. $P = 188, \ w = 44$

$188 = 2l + 2(44)$

$188 = 2l + 88$

$100 = 2l$

$50 = l$

The length of the rectangle is 50 cm.

13. Use the formulas for the area and circumference of a circle. The radius is 4 cm.

$A = \pi r^2$

$A = \pi(4)^2$

$\quad = 16\pi$

$\quad \approx 50$

The area is 16π cm^2 or approximately 50 cm^2.

$C = 2\pi r$

$C = 2\pi(4)$

$\quad = 8\pi$

$\quad \approx 25$

The circumference is 8π cm or approximately 25 cm.

15. Since the diameter is 12 yd, the radius is $\dfrac{12}{2} = 6$ yd.

$A = \pi r^2$

$A = \pi(6)^2$

$\quad = 36\pi$

$\quad \approx 113$

The area is 36π yd^2 or approximately 113 yd^2.

$C = 2\pi r$

$C = 2\pi \cdot 6$

$\quad = 12\pi$

$\quad \approx 38$

The circumference is 12π yd or approximately 38 yd.

17. $C = 2\pi r$

$14\pi = 2\pi r$

$\dfrac{14\pi}{2\pi} = \dfrac{2\pi r}{2\pi}$

$7 = r$

The radius is 7 in. and the diameter is 2(7 in) = 14 in.

19. Use the formula for the volume of a rectangular solid. The length and width are each 3 inches and the height is 4 inches.

$V = lwh$

$V = 3 \cdot 3 \cdot 4$

$\quad = 36$

The volume is 36 in.^{3}.

21. Use the formula for the volume of a cylinder. The radius is 5 cm and the height is 6 cm.

$$V = \pi r^2 h$$
$$V = \pi (5)^2 6$$
$$= \pi (25) 6$$
$$= 150\pi$$
$$\approx 471$$

The volume of the cylinder is 150π cm^3 or approximately 471 cm^3.

23. Use the formula for the volume of a sphere. The diameter is 18 cm, so the radius is 9 cm.

$$V = \frac{4}{3}\pi r^3$$
$$V = \frac{4}{3}\pi (9)^3$$
$$= 972\pi$$
$$\approx 3052$$

The volume is 972π cm^3 or approximately 3052 cm^3.

25. Use the formula for the volume of a cone. The radius is 4 m and the height is 9 m.

$$V = \frac{1}{3}\pi r^2 h$$
$$V = \frac{1}{3}\pi (4)^2 \cdot 9$$
$$= 48\pi$$
$$\approx 151$$

The volume is 48π m^3 or approximately 151 m^3.

27.
$$\frac{V}{\pi r^2} = \frac{\pi r^2 h}{\pi r^2}$$
$$\frac{V}{\pi r^2} = h$$

29. Smaller cylinder: $r = 3$ in, $h = 4$ in.
$$V = \pi r^2 h = \pi (3)^2 \cdot 4 = 36\pi$$

The volume of the smaller cylinder is $36\pi \, in^3$.

Larger cylinder: $r = 3(3 \text{ in}) = 9$ in, $h = 4$ in.
$$V = \pi r^2 h = \pi (9)^2 \cdot 4 = 324\pi$$

The volume of the larger cylinder is 324π in.^{3}.
The ratio of the volumes of the two cylinders is
$$\frac{V_{larger}}{V_{smaller}} = \frac{324\pi}{36\pi} = \frac{9}{1}.$$

So, the volume of the larger cylinder is 9 times the volume of the smaller cylinder.

31. The sum of the measures of the three angles of any triangle is $180°$.
$$x + x + (x + 30) = 180$$
$$3x + 30 = 180$$
$$3x = 150$$
$$x = 50$$
$$x + 30 = 80$$

The three angle measures are $50°, 50°,$ and $80°$.

33.
$$4x + (3x + 4) + (2x + 5) = 180$$
$$9x + 9 = 180$$
$$9x = 171$$
$$x = 19$$
$$3x + 4 = 61$$
$$2x + 5 = 43$$

The three angle measures are $76°, 61°,$ and $43°$.

35. Let x = the measure of the smallest angle.
Let $2x$ = the measure of the second angle.
Let $x + 20$ = the measure of the third angle.
$$x + 2x + (x + 20) = 180$$
$$4x + 20 = 180$$
$$4x = 160$$
$$x = 40$$
$$2x = 80$$
$$x + 20 = 60$$

The three angle measures are $40°, 80°,$ and $60°$.

37. If the measure of an angle is $58°$, the measure of its complement is $90° - 58° = 32°$.

39. If the measure of an angle is $88°$, the measure of its complement is $2°$.

41. If the measure of an angle is $132°$, the measure of its supplement is $180° - 132° = 48°$.

43. If the measure of an angle is $90°$, the measure of its supplement is $180° - 90° = 90°$.

45. *Step 1* Let x = the measure of the angle.

Step 2 Let $90 - x$ = the measure of its complement.

Step 3 The angle's measure is $60°$ more than that of its complement, so the equation is $x = (90 - x) + 60$.

Step 4 Solve this equation

$$x = 90 - x + 60$$
$$x = 150 - x$$
$$2x = 150$$
$$x = 75$$

The measure of the angle is $75°$.

Step 5 The complement of the angle is $90° - 75° = 15°$, and $75°$ is $60°$ more than $15°$.

47. *Step 1* Let x = the measure of the angle.

Step 2 Then $180 - x$ = the measure of its supplement.

Step 3 The angle's measure is three times that of its supplement, so the equation is $x = 3(180 - x)$.

Step 4 Solve this equation

$$x = 3(180 - x)$$
$$x = 540 - 3x$$
$$4x = 540$$
$$x = 135$$

The measure of the angle is $135°$.

Step 5 The measure of its supplement is $180° - 135° = 45°$, and $135° = 3(45°)$, so the proposed solution checks.

49. *Step 1* Let x = the measure of the angle.

Step 2 Let $180 - x$ = the measure of its supplement, and, $90 - x$ = the measure of its complement.

Step 3 The measure of the angle's supplement is $10°$ more than three times that of its complement, so the equation is $180 - x = 3(90 - x) + 10$.

Step 4 Solve this equation

$$180 - x = 3(90 - x) + 10$$
$$180 - x = 270 - 3x + 10$$
$$180 - x = 280 - 3x$$
$$2x = 100$$
$$x = 50$$

The measure of the angle is $50°$.

Step 5 The measure of its supplement is $130°$ and the measure of its complement is $40°$. Since $130° = 3(40°) + 10°$, the proposed solution checks.

51. Divide the shape into two rectangles.

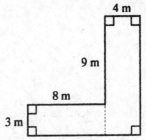

$A_{\text{entire figure}} = A_{\text{bottom rectangle}} + A_{\text{side rectangle}}$

$$A_{\text{entire figure}} = 3 \cdot 8 + 4(9 + 3)$$
$$= 24 + 4(12)$$
$$= 24 + 48$$
$$= 72$$

The area of the figure is 72 square meters.

53. Divide the shape into a rectangle and a triangle.

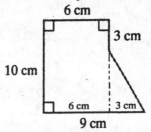

$A_{\text{entire figure}} = A_{\text{rectangle}} + A_{\text{triangle}}$

$$A_{\text{entire figure}} = lw + \frac{1}{2}bh$$
$$= 10(6) + \frac{1}{2}(3)(10 - 3)$$
$$= 60 + \frac{1}{2}(3)(7)$$
$$= 60 + 10.5 = 70.5$$

The area of the figure is 70.5 cm^2.

55. Subtract the volume of the three hollow portions from the volume of the whole rectangular solid.

$$V_{\text{cement block}} = V_{\text{rectangular solid}} - 3 \cdot V_{\text{hollow}}$$
$$= LWH - 3 \cdot lwh$$
$$= (8)(8)(16) - 3 \cdot (4)(6)(8)$$
$$= 1024 - 576$$
$$= 448$$

The volume of the cement block is 448 cubic inches.

57. The area of the office is $(20 \text{ ft})(16 \text{ ft}) = 320 \text{ ft}^2$. Use a proportion to determine how much of the yearly electric bill is deductible.

Let x = the amount of the electric bill that is deductible.

$$\frac{320}{2200} = \frac{x}{4800}$$
$$2200x = (320)(4800)$$
$$2200x = 1,536,000$$
$$\frac{2200x}{2200} = \frac{1,546,000}{2200}$$
$$x \approx 698.18$$

$698.18 of the yearly electric bill is deductible.

59. The radius of the large pizza is $\frac{1}{2} \cdot 14 = 7$ inches, and the radius of the medium pizza is $\frac{1}{2} \cdot 7$ inches $= 3.5$ inches.

large pizza:
$$A = \pi r^2 = \pi(7 \text{ in.})^2$$
$$= 49\pi \text{ in.}^2 \approx 154 \text{ in.}^2$$

medium pizza:
$$A = \pi r^2 = \pi(3.5 \text{ in.})^2$$
$$= 12.25 \text{ in.}^2 \approx 38.465 \text{ in.}^2$$

For each pizza, find the price per inch by dividing the price by the area.

Price per square inch for the large pizza $= \dfrac{\$12.00}{154 \text{ in.}^2} \approx \dfrac{\$0.08}{\text{in.}^2}$ and the price per square inch for the medium pizza

$= \dfrac{\$5.00}{28.465 \text{ in.}^2} \approx \dfrac{\$0.13}{\text{in.}^2}$.

The large pizza is the better buy.

61. The area of the larger circle is $A = \pi r^2 = \pi \cdot 50^2 = 2500\pi \text{ ft}^2$.

The area of the smaller circle is $A = \pi r^2 = \pi \cdot 40^2 = 1600\pi \text{ ft}^2$.

The area of the circular road is the difference between the area of the larger circle and the area of the smaller circle.

$$A = 2500\pi \text{ ft}^2 - 1600\pi \text{ ft}^2 = 900\pi \text{ ft}^2$$

The cost to pave the circular road is $\$0.80(900\pi) \approx \$2262$.

63. To find the perimeter of the entire window, first find the perimeter of the lower rectangular portion. This is the bottom and two sides of the window, which is 3 ft + 6 ft + 6 ft = 15 ft. Next, find the perimeter or circumference of the semicircular portion of the window. The radius of the semicircle is $\frac{1}{2} \cdot 3$ ft $= 1.5$ ft, so the circumference is

$$\frac{1}{2} \cdot 2\pi r \approx 3.14(1.5) = 4.7 \text{ ft}.$$

So, approximately 15 ft + 4.7 ft = 19.7 ft of stripping would be needed to frame the window.

65. First, find the volume of water when the reservoir was full.
$V = lwh = 50 \cdot 0 \cdot 20 = 30,000$
The volume was 30,000 yd^3.
Next, find the volume when the height of the water was 6 yards.
$V = 50 \cdot 30 \cdot 6 = 9000$
The volume was 9000 yd^3. The amount of water used in the three-month period was 30,000 yd^3 – 9000 yd^3 = 21,000 yd^3.

67. For the first can, the diameter is 6 in. so the radius is 3 in. and $V = \pi r^2 h = \pi(3)^2 \cdot 5 = 45\pi \approx 141.3$.
The volume of the first can is 141.3 in^3. For the second can, the diameter is 5 in., so the radius is 2.5 in. and
$V = \pi r^2 h = \pi(2.5)^2 \cdot 6 = 37.5\pi \approx 117.75$.
The volume of the second can is 117.75 in^2. Since the cans are the same price, the can with the greater volume is the better buy. Choose the can with the diameter of 6 inches and height of 5 inches.

69. Find the volume of a cylinder with radius 3 feet and height 2 feet 4 inches.
$$2 \text{ ft } 4 \text{ in } = 2\frac{1}{3} \text{ feet} = \frac{7}{3} \text{ feet}$$
$$V = \pi r^2 h$$
$$= \pi(3)^2 \left(\frac{7}{3}\right) = \pi \cdot 9 \cdot \frac{7}{3} = 21\pi \approx 65.94$$

The volume of the tank is approximately 65.94 ft^3. This is a little over 1 ft^3 smaller than 67 ft^3 so it is too small to hold 500 gallons of water. Yes, you should be able to win your case.

71. – 77. Answers will vary.

79. does not make sense; Explanations will vary. Sample explanation: Though the heights of the books are proportional to the data, the widths are also changing. This cause the larger values to be visually exaggerated.

81. does not make sense; Explanations will vary. Sample explanation: If the radius is doubled, the area is multiplied by 4.
$$A_{\text{radius } x} = \pi r^2 \qquad A_{\text{radius } 2x} = \pi r^2$$
$$= \pi(x)^2 \qquad\qquad = \pi(2x)^2$$
$$= \pi x^2 \qquad\qquad = 4\pi x^2$$

83. true

85. false; Changes to make the statement true will vary. A sample change is: 90° does not have a complement.

87. Area of smaller deck $= (8 \text{ ft})(10) = 80 \text{ ft}^2$.

Area of larger deck $= (12 \text{ ft})(15) = 180 \text{ ft}^2$.

Find the ratio of the areas.

$$\frac{A_{\text{larger}}}{A_{\text{smaller}}} = \frac{180 \text{ ft}^2}{80 \text{ ft}^2} = \frac{2.25}{1} \text{ or } 2.25:1$$

The cost will increase 2.25 times.

89. Let $x =$ the radius of the original sphere.
Let $2x =$ the radius of the larger sphere.
Find the ratio of the volumes of the two spheres.

$$\frac{A_{\text{larger}}}{A_{\text{original}}} = \frac{\dfrac{4}{3}\pi(2x)^3}{\dfrac{4}{3}\pi x^3} = \frac{8x^3}{x^3} = \frac{8}{1} \text{ or } 8:1$$

If the radius of a sphere is doubled, the volume increases 8 times.

91. The angles marked $2x$ and $2x + 40$ in the figure are supplementary, so their sum is $180°$.

$$2x + (2x + 40) = 180$$
$$2x + 2x + 40 = 180$$
$$4x + 40 = 180$$
$$4x = 10$$
$$x = 35$$

The angle of inclination is $35°$.

92. $P = 2s + b$ for s

$$P - b = 2s$$
$$\frac{P-b}{2} = \frac{2s}{2}$$
$$\frac{P-b}{2} = s \text{ or } s = \frac{P-b}{2}$$

93. $\dfrac{x}{2} + 7 = 13 - \dfrac{x}{4}$

Multiply both sides by the LCD, 4.

$$4\left(\frac{x}{2} + 7\right) = 4\left(13 - \frac{x}{4}\right)$$
$$2x + 28 = 52 - x$$
$$2x + 28 + x = 52 - x + x$$
$$3x + 28 = 52$$
$$3x + 28 - 28 = 52 - 28$$
$$3x = 24$$
$$\frac{3x}{3} = \frac{24}{3}$$
$$x = 8$$

The solution set is $\{8\}$.

94.
$$\left[3\left(12 \div 2^2 - 3\right)^2\right]^2$$
$$= \left[3\left(12 \div 4 - 3\right)^2\right]^2$$
$$= \left[3\left(3 - 3\right)^2\right]^2 = \left(3 \cdot 0^2\right)^2 = 0^2 = 0$$

95. $x + 3 < 8$

$$2 + 3 < 8$$
$$5 < 8, \text{ true}$$

2 is a solution to the inequality.

96. $4y - 7 \geq 5$

$$4(6) - 7 \geq 5$$
$$24 - 7 \geq 5$$
$$17 \geq 5, \text{ true}$$

6 is a solution to the inequality.

97. $2(x - 3) + 5x = 8(x - 1)$

$$2x - 6 + 5x = 8x - 8$$
$$7x - 6 = 8x - 8$$
$$7x - 8x - 6 = 8x - 8x - 8$$
$$-x - 6 + 6 = -8 + 6$$
$$-x = -2$$
$$x = 2$$

The solution set is $\{2\}$.

2.7 Check Points

1. a.

b.

c.

2. a. $[0, \infty)$

b. $(-\infty, 5)$

3. $x + 6 < 9$

$x + 6 - 6 < 9 - 6$

$x < 3$

The solution set is $(-\infty, 3)$ or $\{x | x < 3\}$.

4. $8x - 2 \geq 7x - 4$

$8x - 7x - 2 \geq 7x - 7x - 4$

$x - 2 \geq -4$

$x - 2 + 2 \geq -4 + 2$

$x \geq -2$

The solution set is $[-2, \infty)$ or $\{x | x \geq -2\}$.

5. a. $\dfrac{1}{4}x < 2$

$4 \cdot \dfrac{1}{4}x < 4 \cdot 2$

$x < 8$

The solution set is $(-\infty, 8)$ or $\{x | x < 8\}$.

b. $-6x < 18$

$\dfrac{-6x}{-6} > \dfrac{18}{-6}$

$x > -3$

The solution set is $(-3, \infty)$ or $\{x | x > -3\}$.

6. $5y - 3 \geq 17$

$5y - 3 + 3 \geq 17 + 3$

$5y \geq 20$

$\dfrac{5y}{5} \geq \dfrac{20}{5}$

$y \geq 4$

The solution set is $[4, \infty)$ or $\{y | y \geq 4\}$.

7. $6 - 3x \leq 5x - 2$

$6 - 3x - 5x \leq 5x - 5x - 2$

$6 - 8x \leq -2$

$6 - 6 - 8x \leq -2 - 6$

$-8x \leq -8$

$\dfrac{-8x}{-8} \geq \dfrac{-8}{-8}$

$x \geq 1$

The solution set is $[1, \infty)$ or $\{x | x \geq 1\}$.

8. $2(x - 3) - 1 \leq 3(x + 2) - 14$

$2x - 6 - 1 \leq 3x + 6 - 14$

$2x - 7 \leq 3x - 8$

$2x - 3x - 7 \leq 3x - 3x - 8$

$-x - 7 \leq -8$

$-x - 7 + 7 \leq -8 + 7$

$-x \leq -1$

$\dfrac{-x}{-1} \geq \dfrac{-1}{-1}$

$x \geq 1$

The solution set is $[1, \infty)$ or $\{x | x \geq 1\}$.

9. $4(x + 2) > 4x + 15$

$4x + 8 > 4x + 15$

$4x - 4x + 8 > 4x - 4x + 15$

$8 > 15, \text{ false}$

There is no solution or $\{ \ \}$.

10. $3(x + 1) \geq 2x + 1 + x$

$3x + 3 \geq 3x + 1$

$3x - 3x + 3 \geq 3x - 3x + 1$

$3 \geq 1, \text{ true}$

The solution is $(-\infty, \infty)$ or $\{x | x \text{ is a real number}\}$.

11. Let x = your grade on the final examination.

$$\frac{82 + 74 + 78 + x + x}{5} \geq 80$$

$$\frac{234 + 2x}{5} \geq 80$$

$$5\left(\frac{234 + 2x}{5}\right) \geq 5 \cdot 80$$

$$234 + 2x \geq 400$$

$$234 - 234 + 2x \geq 400 - 234$$

$$2x \geq 166$$

$$x \geq 83$$

To earn a B you must get at least an 83% on the final examination.

2.7 Exercise Set

1. $x > 5$

3. $x < -2$

5. $x \geq -4$

7. $x \leq 4.5$

9. $-2 < x \leq 6$

11. $-1 < x < 3$

13. $(-\infty, 3]$

15. $\left(\frac{5}{2}, \infty\right)$

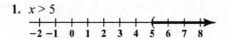

17. $(-\infty, 0]$

19. $(-\infty, 4)$

21. $x - 3 > 4$

$x - 3 + 3 > 4 + 3$

$x > 7$

$(7, \infty)$

23. $x + 4 \leq 10$

$x + 4 - 4 \leq 10 - 4$

$x \leq 6$

$(-\infty, 6]$

25. $y - 2 < 0$

$y - 2 + 2 < 0 + 2$

$y < 2$

$(-\infty, 2)$

27. $3x + 4 \leq 2x + 7$

$3x - 2x \leq 7 - 4$

$x \leq 3$

$(-\infty, 3]$

29. $5x - 9 < 4x + 7$

$5x - 4x < 7 + 9$

$x < 16$

$(-\infty, 16)$

31. $7x - 7 > 6x - 3$

$7x - 6x > -3 + 7$

$x > 4$

$(4, \infty)$

33.
$$x - \frac{2}{3} > \frac{1}{2}$$
$$x - \frac{2}{3} + \frac{2}{3} > \frac{1}{2} + \frac{2}{3}$$
$$x > \frac{3}{6} + \frac{4}{6}$$
$$x > \frac{7}{6}$$
$$\left(\frac{7}{6}, \infty\right)$$

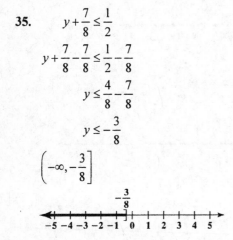

35.
$$y + \frac{7}{8} \le \frac{1}{2}$$
$$y + \frac{7}{8} - \frac{7}{8} \le \frac{1}{2} - \frac{7}{8}$$
$$y \le \frac{4}{8} - \frac{7}{8}$$
$$y \le -\frac{3}{8}$$
$$\left(-\infty, -\frac{3}{8}\right]$$

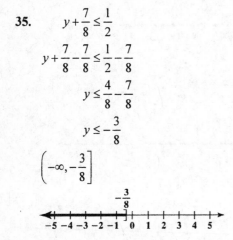

37.
$$-15y + 13 > 13 - 16y$$
$$-15y + 13 + 16y > 13 - 16y + 16y$$
$$y + 13 > 13$$
$$y + 13 - 13 > 13 - 13$$
$$y > 0$$
$$(0, \infty)$$

39.
$$\frac{1}{2}x < 4$$
$$2\left(\frac{1}{2}x\right) < 2(4)$$
$$1x < 8$$
$$x < 8$$
$$(-\infty, 8)$$

41.
$$\frac{x}{3} > -2$$
$$3\left(\frac{x}{3}\right) > 3(-2)$$
$$x > -6$$
$$(-6, \infty)$$

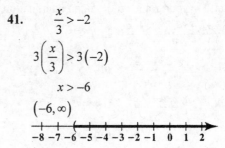

43.
$$4x < 20$$
$$\frac{4x}{4} < 20$$
$$x < 5$$
$$(-\infty, 5)$$

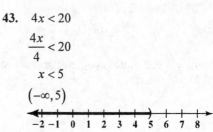

45.
$$3x \ge -21$$
$$\frac{3x}{3} \ge \frac{-21}{3}$$
$$x \ge -7$$
$$[-7, \infty)$$

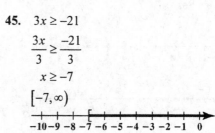

47.
$$-3x < 15$$
$$\frac{-3x}{-3} > \frac{15}{-3}$$
$$x > -5$$
$$(-5, \infty)$$

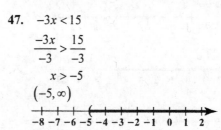

49.
$$-3x \ge 15$$
$$\frac{-3x}{-3} \le \frac{15}{-3}$$
$$x \le -5$$
$$(-\infty, -5]$$

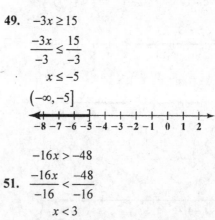

$$-16x > -48$$

51.
$$\frac{-16x}{-16} < \frac{-48}{-16}$$
$$x < 3$$
$$(-\infty, 3)$$

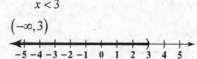

53. $-4y \leq \dfrac{1}{2}$

$2(-4y) \leq 2\left(\dfrac{1}{2}\right)$

$-8y \leq 1$

$\dfrac{-8y}{-8} \geq \dfrac{1}{-8}$

$y \geq -\dfrac{1}{8}$

$\left[-\dfrac{1}{8}, \infty\right)$

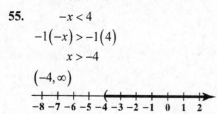

55. $-x < 4$

$-1(-x) > -1(4)$

$x > -4$

$(-4, \infty)$

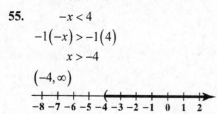

57. $2x - 3 > 7$

$2x - 3 + 3 > 7 + 3$

$2x > 10$

$\dfrac{2x}{2} > \dfrac{10}{2}$

$x > 5$

$(5, \infty)$

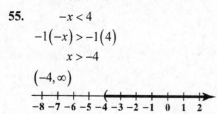

59. $3x + 3 < 18$

$3x + 3 - 3 < 18 - 3$

$3x < 15$

$\dfrac{3x}{3} < \dfrac{15}{3}$

$x < 5$

$(-\infty, 5)$

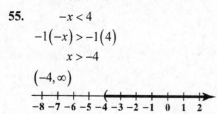

61. $3 - 7x \leq 17$

$3 - 7x - 3 \leq 17 - 3$

$-7x \leq 14$

$\dfrac{-7x}{-7} \geq \dfrac{14}{-7}$

$x \geq -2$

$[-2, \infty)$

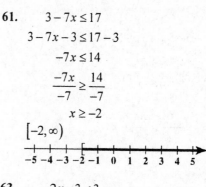

63. $-2x - 3 < 3$

$-2x - 3 + 3 < 3 + 3$

$-2x < 6$

$\dfrac{-2x}{-2} > \dfrac{6}{-2}$

$x > -3$

$(-3, \infty)$

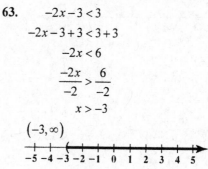

65. $5 - x \leq 1$

$5 - x - 5 \leq 1 - 5$

$-x \leq -4$

$-1(-x) \geq -1(-4)$

$x \geq 4$

$[4, \infty)$

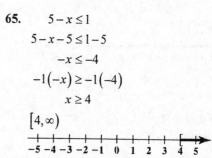

67. $2x - 5 > -x + 6$

$2x - 5 + x > -x + 6 + x$

$3x - 5 > 6$

$3x - 5 + 5 > 6 + 5$

$3x > 11$

$\dfrac{3x}{3} > \dfrac{11}{3}$

$x > \dfrac{11}{3}$

$\left(\dfrac{11}{3}, \infty\right)$

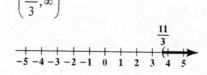

69.
$$2y - 5 < 5y - 11$$
$$2y - 5 - 5y < 5y - 11 - 5y$$
$$-3y - 5 < -11$$
$$-3y - 5 + 5 < -11 + 5$$
$$-3y < -6$$
$$\frac{-3y}{-3} > \frac{-6}{-3}$$
$$y > 2$$
$$(2, \infty)$$

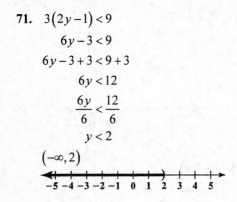

71.
$$3(2y - 1) < 9$$
$$6y - 3 < 9$$
$$6y - 3 + 3 < 9 + 3$$
$$6y < 12$$
$$\frac{6y}{6} < \frac{12}{6}$$
$$y < 2$$
$$(-\infty, 2)$$

73.
$$3(x + 1) - 5 < 2x + 1$$
$$3x + 3 - 5 < 2x + 1$$
$$3x - 2 < 2x + 1$$
$$3x - 2 - 2x < 2x + 1 - 2x$$
$$x - 2 < 1$$
$$x - 2 + 2 < 1 + 2$$
$$x < 3$$
$$(-\infty, 3)$$

75.
$$8x + 3 > 3(2x + 1) - x + 5$$
$$8x + 3 > 6x + 3 - x + 5$$
$$8x + 3 > 5x + 8$$
$$8x + 3 - 5x > 5x + 8 - 5x$$
$$3x + 3 > 8$$
$$3x + 3 - 3 > 8 - 3$$
$$3x > 5$$
$$x > \frac{5}{3}$$
$$\left(\frac{5}{3}, \infty\right)$$

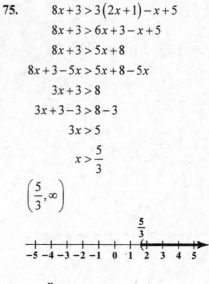

77.
$$\frac{x}{3} - 2 \geq 1$$
$$\frac{x}{3} - 2 + 2 \geq 1 + 2$$
$$\frac{x}{3} \geq 3$$
$$3\left(\frac{x}{3}\right) \geq 3(3)$$
$$x \geq 9$$
$$[9, \infty)$$

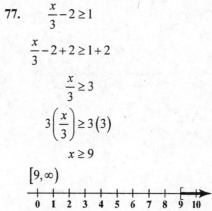

79.
$$1 - \frac{x}{2} - 1 > 4 - 1$$
$$-\frac{x}{2} > 3$$
$$2\left(-\frac{x}{2}\right) > 2(3)$$
$$-x > 6$$
$$-1(-x) < -1(6)$$
$$x < -6$$
$$(-\infty, -6)$$

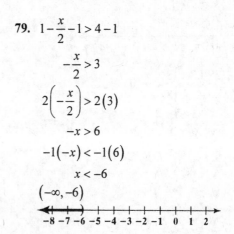

81.
$$4x - 4 < 4(x - 5)$$
$$4x - 4 < 4x - 20$$
$$4x - 4 + 4 < 4x - 20 + 4$$
$$4x < 4x - 16$$
$$4x - 4x < 4x - 16 - 4x$$
$$0 < -16$$

The original inequality is equivalent to the false statement $0 < -16$, so the inequality has no solution. The solution set is $\{\ \}$.

83.
$$x + 3 < x + 7$$
$$x + 3 - x < x + 7 - x$$
$$3 < 7$$

The original inequality is equivalent to the true statement $3 < 7$.
The solution is the set of all real numbers, written $\{x | x \text{ is a real number}\}$ or $(-\infty, \infty)$.

85.
$$7x \leq 7(x - 2)$$
$$7x \leq 7x - 14$$
$$7x - 7x \leq 7x - 14 - 7x$$
$$0 \leq -14$$

Since $0 \leq -14$ is a false statement, the original inequality has no solution.
The solution set is $\{\ \}$.

87.
$$2(x + 3) > 2x + 1$$
$$2x + 6 > 2x + 1$$
$$2x + 6 - 2x > 2x + 1 - 2x$$
$$6 > 1$$

Since $6 > 1$ is a true statement, the original inequality is true for all real numbers the solution set is $\{x | x \text{ is a real number}\}$ or $(-\infty, \infty)$.

89.
$$5x - 4 \leq 4(x - 1)$$
$$5x - 4 \leq 4x - 4$$
$$5x - 4 + 4 \leq 4x - 4 + 4$$
$$5x \leq 4x$$
$$5x - 4x \leq 4x - 4x$$
$$x \leq 0$$
$$(-\infty, 0]$$

91.
$$3x + a > b$$
$$3x > b - a$$
$$\frac{3x}{3} > \frac{b - a}{3}$$
$$x > \frac{b - a}{3}$$

93.
$$y \leq mx + b$$
$$y - b \leq mx$$
$$\frac{y - b}{m} \geq \frac{mx}{m}$$
$$\frac{y - b}{m} \geq x \text{ or } x \leq \frac{y - b}{m}$$

95. x is between -2 and 2, so $|x| < 2$.

97. x is less than -2 or greater than 2, so $|x| > 2$.

99. Denmark, Netherlands, and Norway

101. Japan and Mexico

103. Netherlands, Norway, Canada, and U.S.

105. $N = 550 - 9x; N < 370$
$$550 - 9x < 370$$
$$550 - 9x - 550 < 370 - 550$$
$$-9x < -180$$
$$\frac{-9x}{-9} > \frac{-180}{-9}$$
$$x > 20$$

According to the model, there will be 370 billion cigarettes consumed in $1988 + 20 = 2008$ and less than 370 billion after 2008 (from 2009 onward).

107. a. Let x = your grade on the final exam.
$$\frac{86 + 88 + x}{3} \geq 90$$
$$3\left(\frac{86 + 88 + x}{3}\right) \geq 3(90)$$
$$86 + 88 + x \geq 270$$
$$174 + x \geq 270$$
$$174 + x - 174 \geq 270 - 174$$
$$x \geq 96$$

You must get at least a 96% on the final exam to earn an A in the course.

b. $\dfrac{86+88+x}{3} < 80$

$$3\left(\dfrac{86+88+x}{3}\right) < 3(80)$$

$$86+88+x < 240$$

$$174+x < 240$$

$$174+x-174 < 240-174$$

$$x < 66$$

If you get less than a 66 on the final exam, your grade will be below a B.

109. Let x = number of miles driven.

$$80+0.25x \le 400$$

$$80+0.25x-80 \le 400-80$$

$$0.25x \le 320$$

$$\dfrac{0.25x}{0.25} \le \dfrac{320}{0.25}$$

$$x \le 1280$$

You can drive up to 1280 miles.

111. Let x = number of cement bags.

$$245+95x \le 3000$$

$$245+95x-245 \le 3000-245$$

$$95x \le 2755$$

$$\dfrac{95x}{95} \le \dfrac{2755}{95}$$

$$x \le 29$$

Up to 29 bags of cement can safely be listed on the elevator in one trip.

113. – 115. Answers will vary.

117. makes sense

119. makes sense

121. false; Changes to make the statement true will vary. A sample change is: The inequality $x-3 > 0$ is equivalent to $x > 3$.

123. false; Changes to make the statement true will vary. A sample change is: The inequality $-4x < -20$ is equivalent to $x > 5$.

125. Let x = number of miles driven.
Weekly cost for Basic Rental: $260.
Weekly cost for Continental: $80 + 0.25x$
The cost for Basic Rental is a better deal if $80 + 0.25x > 260$.
Solve this inequality.

$$80+0.25x-80 > 260-80$$

$$0.25x > 180$$

$$\dfrac{0.25x}{0.25} > \dfrac{180}{0.25}$$

$$x > 720$$

Basic Car Rental is a better deal if you drive more than 720 miles in a week.

127. $\qquad 1.45-7.23x > -1.442$

$$1.45-7.23x-1.45 > -1.442-1.45$$

$$-7.23x > -2.892$$

$$\dfrac{-7.23x}{-7.23} < \dfrac{-2.892}{-7.23}$$

$$x < 0.4$$

$$(-\infty, 0.4)$$

129. $A = PB$, $A = 8$, $P = 40\% = 0.4$

$$A = PB$$

$$8 = 0.4B$$

$$\dfrac{8}{0.4} = \dfrac{0.4B}{0.4}$$

$$20 = B$$

8 is 40% of 20.

130. Let x = the width of the rectangle.
Let $x + 5$ = the length of the rectangle.

$$P = 2l + 2w$$

$$34 = 2(x+5) + 2 \cdot x$$

$$34 = 2x+10+2x$$

$$34 = 4x+10$$

$$34-10 = 4x+10-10$$

$$24 = 4x$$

$$6 = x$$

$$x = 6$$

$$x + 5 = 11$$

The width is 6 inches and the length is 11 inches.

131.
$$5x+16=3(x+8)$$
$$5x+16=3x+24$$
$$5x+16-3x=3x+24-3x$$
$$2x+16=24$$
$$2x+16-16=24-16$$
$$2x=8$$
$$\frac{2x}{2}=\frac{8}{2}$$
$$x=4$$
Check: $5(4)+16=3(4+8)$
$$20+16=3(12)$$
$$36=36, \text{ true}$$
The solution is set is $\{4\}$.

132.
$$x-4y=14$$
$$2-4(-3)=14$$
$$2+12=14$$
$$14=14, \text{ true}$$
Yes, the values make it a true statement.

133.
$$x-4y=14$$
$$12-4(1)=14$$
$$12-4=14$$
$$8=14, \text{ false}$$
No, the values make it a false statement.

134.
$$y=\frac{2}{3}x+1$$
$$y=\frac{2}{3}(-6)+1$$
$$y=-4+1$$
$$y=-3$$

Chapter 2 Review Exercises

1.
$$x-10=22$$
$$x-10+10=22+10$$
$$x=32$$
The solution is set is $\{32\}$.

2.
$$-14=y+8$$
$$-14-8=y+8-8$$
$$-22=y$$
The solution is set is $\{-22\}$.

3.
$$7z-3=6z+9$$
$$7z-3-6z=6z+9-6z$$
$$z-3=9$$
$$z-3+3=9+3$$
$$z=12$$
The solution is set is $\{12\}$.

4.
$$4(x+3)=3x-10$$
$$4x+12=3x-10$$
$$4x+12-3x=3x-10-3x$$
$$x+12=-10$$
$$x+12-12=-10-12$$
$$x=-22$$
The solution is set is $\{-22\}$.

5.
$$6x-3x-9+1=-5x+7x-3$$
$$3x-8=2x-3$$
$$3x-8-2x=2x-3-2x$$
$$x-8=-3$$
$$x-8+8=-3+8$$
$$x=5$$
The solution is set is $\{5\}$.

6.
$$\frac{x}{8}=10$$
$$8\left(\frac{x}{8}\right)=8(10)$$
$$x=80$$
The solution is set is $\{80\}$.

7.
$$\frac{y}{-8}=7$$
$$-8\left(\frac{y}{-8}\right)=-8(7)$$
$$y=-56$$
The solution is set is $\{-56\}$.

8.
$$7z=77$$
$$\frac{7z}{7}=\frac{77}{7}$$
$$z=11$$
The solution is set is $\{11\}$.

9. $-36 = -9y$

$$\frac{-36}{-9} = \frac{-9y}{-9}$$

$$4 = y$$

The solution is set is $\{4\}$.

10. $\frac{3}{5}x = -9$

$$\frac{5}{3}\left(\frac{3}{5}x\right) = \frac{5}{3}(-9)$$

$$1x = -15$$

$$x = -15$$

The solution is set is $\{-15\}$.

11. $30 = -\frac{5}{2}y$

$$-\frac{2}{5}(30) = -\frac{2}{5}\left(-\frac{5}{2}y\right)$$

$$-12 = y$$

The solution is set is $\{-12\}$.

12. $-x = 25$

$$-1(-x) = -1(25)$$

$$x = -25$$

The solution is set is $\{-25\}$.

13. $\frac{-x}{10} = -1$

$$10\left(\frac{-x}{10}\right) = 10(-1)$$

$$-x = -10$$

$$-1(-x) = -1(-10)$$

$$x = 10$$

The solution is set is $\{10\}$.

14. $4x + 9 = 33$

$$4x + 9 - 9 = 33 - 9$$

$$4x = 24$$

$$\frac{4x}{4} = \frac{24}{4}$$

$$x = 6$$

The solution is set is $\{6\}$.

15. $-3y - 2 = 13$

$$-3y - 2 + 2 = 13 + 2$$

$$-3y = 15$$

$$\frac{-3y}{-3} = \frac{15}{-3}$$

$$y = -5$$

The solution is set is $\{-5\}$.

16. $5z + 20 = 3z$

$$5z + 20 - 3z = 3z - 3z$$

$$2z + 20 = 0$$

$$2z + 20 - 20 = 0 - 20$$

$$2z = -20$$

$$\frac{2z}{2} = \frac{-20}{2}$$

$$z = -10$$

The solution is set is $\{-10\}$.

17. $5x - 3 = x + 5$

$$5x - 3 - x = x + 5 - x$$

$$4x - 3 = 5$$

$$4x - 3 + 3 = 5 + 3$$

$$4x = 8$$

$$\frac{4x}{4} = \frac{8}{4}$$

$$x = 2$$

The solution is set is $\{2\}$.

18. $3 - 2x = 9 - 8x$

$$3 - 2x + 8x = 9 - 8x + 8x$$

$$3 + 6x = 9$$

$$3 + 6x - 3 = 9 - 3$$

$$6x = 6$$

$$\frac{6x}{6} = \frac{6}{6}$$

$$x = 1$$

The solution is set is $\{1\}$.

19. a. 2005 is 20 years after 1985.

$$I = 12n + 151$$

$$I = 12(20) + 151 = 240 + 151 = 391$$

According to the formula, the U.S. imported 391 million barrels of oil per month in 2005. The formula underestimated the actual value given in the bar graph by 8 million barrels.

b.
$$I = 12n + 151$$
$$511 = 12n + 151$$
$$511 - 151 = 12n + 151 - 151$$
$$360 = 12n$$
$$\frac{360}{12} = \frac{12n}{12}$$
$$30 = n$$

If trends continue, the U.S. will import an average of 511 million barrels of oil per month 30 years after 1985, or 2015.

20. $\quad 5x + 9 - 7x + 6 = x + 18$
$$-2x + 15 = x + 18$$
$$-2x + 15 - x = x + 18 - x$$
$$-3x + 15 = 18$$
$$-3x + 15 - 15 = 18 - 15$$
$$-3x = 3$$
$$\frac{-3x}{-3} = \frac{3}{-3}$$
$$x = -1$$

The solution is set is $\{-1\}$.

21. $\quad 3(x + 4) = 5x - 12$
$$3x + 12 = 5x - 12$$
$$3x + 12 - 5x = 5x - 12 - 5x$$
$$-2x + 12 = -12$$
$$-2x + 12 - 12 = -12 - 12$$
$$-2x = -24$$
$$\frac{-2x}{-2} = \frac{-24}{-2}$$
$$x = 12$$

The solution is set is $\{12\}$.

22. $\quad 1 - 2(6 - y) = 3y + 2$
$$1 - 12 + 2y = 3y + 2$$
$$2y - 11 = 3y + 2$$
$$2y - 11 - 3y = 3y + 2 - 3y$$
$$-y - 11 = 2$$
$$-y - 11 + 11 = 2 + 11$$
$$-y = 13$$
$$y = -13$$

The solution is set is $\{-13\}$.

23. $\quad 2x - 8 + 3x + 15 = 2x - 2$
$$5x + 7 = 2x - 2$$
$$5x + 7 - 2x = 2x - 2 - 2x$$
$$3x + 7 = -2$$
$$3x + 7 - 7 = -2 - 7$$
$$3x = -9$$
$$\frac{3x}{3} = \frac{-9}{3}$$
$$x = -3$$

The solution is set is $\{-3\}$.

24. $\quad -2(y - 4) - (3y - 2) = -2 - (6y - 2)$
$$-2y + 8 - 3y + 2 = -2 - 6y + 2$$
$$-5y + 10 = -6y$$
$$-5y + 10 + 6y = -6y + 6y$$
$$10 + y = 0$$
$$10 + y - 10 = 0 - 10$$
$$y = -10$$

The solution is set is $\{-10\}$.

25. $\quad \dfrac{2x}{3} = \dfrac{x}{6} + 1$

To clear fractions, multiply both sides by the LCD, which is 6.

$$6\left(\frac{2x}{3}\right) = 6\left(\frac{x}{6} + 1\right)$$
$$6\left(\frac{2x}{3}\right) = 6\left(\frac{x}{6}\right) + 6(1)$$
$$4x = x + 6$$
$$4x - x = x + 6 - x$$
$$3x = 6$$
$$\frac{3x}{3} = \frac{6}{3}$$
$$x = 2$$

The solution is set is $\{2\}$.

26. $\dfrac{x}{2} - \dfrac{1}{10} = \dfrac{x}{5} + \dfrac{1}{2}$

Multiply both sides by the LCD, which is 10.

$$10\left(\dfrac{x}{2} - \dfrac{1}{10}\right) = 10\left(\dfrac{x}{5} + \dfrac{1}{2}\right)$$

$$10\left(\dfrac{x}{2}\right) - 10\left(\dfrac{1}{10}\right) = 10\left(\dfrac{x}{5}\right) + 10\left(\dfrac{1}{2}\right)$$

$$5x - 1 = 2x + 5$$

$$5x - 1 - 2x = 2x + 5 - 2x$$

$$3x - 1 = 5$$

$$3x - 1 + 1 = 5 + 1$$

$$3x = 6$$

$$\dfrac{3x}{3} = \dfrac{6}{3}$$

$$x = 2$$

The solution is set is $\{2\}$.

27. $3(8x - 1) = 6(5 + 4x)$

$$24x - 3 = 30 + 24x$$

$$24x - 3 - 24x = 30 + 24x - 24x$$

$$-3 = 30$$

Since $-3 = 30$ is a false statement, the original equation is inconsistent and has no solution or $\{\ \}$.

28. $4(2x - 3) + 4 = 8x - 8$

$$8x - 12 + 4 = 8x - 8$$

$$8x - 8 = 8x - 8$$

$$8x - 8 - 8x = 8x - 8 - 8x$$

$$-8 = -8$$

Since $-8 = -8$ is a true statement, so the solution is the set of all real numbers, written $\{x | x \text{ is a real number}\}$.

29. $H = 0.7(220 - a)$

$$133 = 0.7(220 - a)$$

$$133 - 154 = 154 - 154 - 0.7a$$

$$-21 = -0.7a$$

$$\dfrac{-21}{-0.7} = \dfrac{-0.7a}{-0.7}$$

$$30 = a$$

If the optimal heart rate is 133 beats per minute, the person is 30 years old.

30. $I = Pr$ for r

$$\dfrac{I}{P} = \dfrac{Pr}{P}$$

$$\dfrac{I}{P} = r \text{ or } r = \dfrac{I}{P}$$

31. $V = \dfrac{1}{3}Bh$ for h

$$3V = 3\left(\dfrac{1}{3}Bh\right)$$

$$3V = Bh$$

$$\dfrac{3V}{B} = \dfrac{Bh}{B}$$

$$\dfrac{3V}{B} = h \text{ or } h = \dfrac{3V}{B}$$

32. $P = 2l + 2w$ for w

$$P - 2l = 2l + 2w - 2l$$

$$P - 2l = 2w$$

$$\dfrac{P - 2l}{2} = \dfrac{2w}{2}$$

$$\dfrac{P - 2l}{2} = w \text{ or } w = \dfrac{P - 2l}{2}$$

33. $A = \dfrac{B + C}{2}$ for B

$$2A = 2\left(\dfrac{B + C}{2}\right)$$

$$2A = B + C$$

$$2A - C = B + C - C$$

$$2A - C = B \text{ or } B = 2A - C$$

34. $T = D + pm$ for m

$$T - D = D + pm - D$$

$$T - D = pm$$

$$\dfrac{T - D}{p} = \dfrac{pm}{p}$$

$$\dfrac{T - D}{p} = m \text{ or } m = \dfrac{T - D}{p}$$

35. $65\% = 0.65$

36. $150\% = 1.50$

37. $3\% = 0.03$

38. $0.72 = 72\%$

39. $0.0035 = 0.35\%$

40. $A = PB$; $P = 8\% = 0.08$, $B = 120$

$A = 0.08 \cdot 120$

$A = 9.6$

8% of 120 is 9.6

41. $A = PB$; $A = 90$, $P = 45\% = 0.45$

$90 = 0.45B$

$\dfrac{90}{0.45} = \dfrac{0.45B}{0.45}$

$200 = B$

90 is 45% of 200.

42. $A = PB$; $A = 36$, $B = 75$

$36 = P \cdot 75$

$\dfrac{36}{75} = \dfrac{P \cdot 75}{75}$

$0.48 = P$

36 is 48% of 75.

43. Increase = Percent · Original

First, find the increase: $12 - 6 = 6$

$6 = P \cdot 6$

$\dfrac{6}{6} = \dfrac{P \cdot 6}{6}$

$1 = P$

The percent increase is 100%.

44. Decrease = Percent · Original

First, find the decrease: $5 - 3 = 2$

$2 = P \cdot 5$

$\dfrac{2}{5} = \dfrac{P \cdot 5}{5}$

$0.4 = P$

The percent decrease is 40%.

45. Increase = Percent · Original

First, find the increase: $45 - 40 = 5$

$5 = P \cdot 40$

$\dfrac{5}{40} = \dfrac{P \cdot 40}{40}$

$0.125 = P$

The percent increase is 12.5%.

46. Investment dollars lost last year were $0.10 \cdot \$10,000 = \$1000$. This means that $\$10,000 - \$1000 = \$9000$ remains. Investment dollars gained this year are $0.10 \cdot \$9000 = \$900$. This means that $\$9000 + \$900 = \$9900$ of the original investment remains. This is an overall loss of $\$100$.

decrease = percent · original

$100 = P \cdot 10,000$

$\dfrac{100}{10,000} = \dfrac{P \cdot 10,000}{10,000}$

$0.01 = P$

The statement is not true. Instead of recouping losses, there is an overall 1% decrease in the portfolio.

47. a. $r = \dfrac{h}{7}$

$7r = 7\left(\dfrac{h}{7}\right)$

$7r = h$ or $h = 7r$

b. $h = 7r$; $r = 9$

$h = 7(9) = 63$

The woman's height is 63 inches or 5 feet, 3 inches.

48. $A = P \cdot B$

$91 = 0.26 \cdot B$

$\dfrac{91}{0.26} = \dfrac{0.26 \cdot B}{0.26}$

$350 = B$

The average U.S. household uses 350 gallons of water per day.

49. Let x = the unknown number.

$6x - 20 = 4x$

$6x - 20 - 4x = 4x - 4x$

$2x - 20 = 0$

$2x - 20 + 20 = 0 + 20$

$2x = 20$

$x = 10$

The number is 10.

50. Let x = the average amount spent on cat food.
Let $x + 2$ = the average amount spent on dog food.

$$x + (x + 29) = 405$$
$$x + x + 29 = 405$$
$$2x + 29 = 405$$
$$2x + 29 - 29 = 405 - 29$$
$$2x = 376$$
$$x = 188$$
$$x + 29 = 217$$

$188 was spent on cats and $217 was spent on dogs.

51. Let x = the smaller page number.
Let $x + 1$ = the larger page number.

$$x + (x + 1) = 93$$
$$2x + 1 = 93$$
$$2x = 92$$
$$x = 46$$

The page numbers are 46 and 47.

52. Let x = the percentage of females.
Let $x + 2$ = the percentage of males.

$$x + (x + 2) = 100$$
$$x + x + 2 = 100$$
$$2x + 2 = 100$$
$$2x + 2 - 2 = 100 - 2$$
$$2x = 98$$
$$x = 49$$
$$x + 2 = 51$$

For Americans under 20, 49% are female and 51% are male.

53. Let x = number of years after 2003.

$$612 + 15x = 747$$
$$612 + 15x - 612 = 747 - 612$$
$$15x = 135$$
$$\frac{15x}{15} = \frac{135}{15}$$
$$x = 9$$

According to this model, the average weekly salary will reach $747 in 9 years after 2003, or 2012.

54. Let x = the number of checks written.

$$6 + 0.05x = 6.90$$
$$6 + 0.05x - 6 = 6.90 - 6$$
$$0.05x = 0.90$$
$$\frac{0.05x}{0.05} = \frac{0.90}{0.05}$$
$$x = 18$$

You wrote 18 checks that month.

55. Let x = the width of the field.
Let $3x$ = the length of the field.

$$P = 2l + 2w$$
$$400 = 2 \cdot 3x + 2 \cdot x$$
$$400 = 6x + 2x$$
$$400 = 8x$$
$$\frac{400}{8} = \frac{8x}{8}$$
$$50 = x$$
$$x = 50$$
$$3x = 150$$

The field is 50 yards wide and 150 yards long.

56. Let x = the original price of the table.

$$x - 0.25x = 180$$
$$0.75x = 180$$
$$\frac{0.75x}{0.75} = \frac{180}{0.75}$$
$$x = 240$$

The table's price before the reduction was $240.

57. Find the area of a rectangle with length 6.5 ft and width 5 ft.

$$A = lw = (6.5)(5) = 32.5$$

The area is 32.5 ft^2.

58. Find the area of a triangle with base 20 cm and height 5 cm.

$$A = \frac{1}{2}bh = \frac{1}{2}(20)(5) = 50$$

The area is 50 cm^2.

59. Find the area of a trapezoid with bases 22 yd and 5 yd and height 10 yd.

$$A = \frac{1}{2}h(a+b)$$

$$= \frac{1}{2}(10)(22+5)$$

$$= \frac{1}{2}\cdot 10\cdot 27 = 135$$

The area is 135 yd^2.

60. Notice that the height of the middle rectangle is $64 - 12 - 12 = 40$ m.

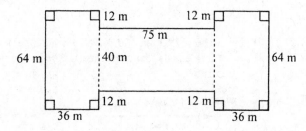

Using $A = lw$ we must find the sum of areas of the middle rectangle and the two side rectangles.

$$A = (40)(75) + 2\cdot(64)(36)$$

$$= 3000 + 2\cdot 2304$$

$$= 3000 + 4608$$

$$= 7608$$

The area is 7608 m^2.

61. Since the diameter is 20 m, the radius is $\frac{20}{2} = 10$ m.

$$C = 2\pi = 2\pi(10) = 20\pi \approx 63$$

$$A = \pi r^2 = \pi(10)^2 = 100\pi \approx 314$$

The circumference is 20π m or approximately 63 m; the area is 100π m^2 or approximately 314 m^2.

62. $A = 42, b = 14$

$$A = \frac{1}{2}bh$$

$$42 = \frac{1}{2}\cdot 14\cdot h$$

$$42 = 7h$$

$$6 = h$$

The height of the sail is 6 ft.

63. Area of floor:

$$A = bh = (12\,\text{ft})(15\,\text{ft}) = 180\,\text{ft}^2$$

Area of base of stove:

$$A = bh = (3\,\text{ft})(4\,\text{ft}) = 12\,\text{ft}^2$$

Area of bottom of refrigerator:

$$A = bh = (3\,\text{ft})(4\,\text{ft}) = 12\,\text{ft}^2$$

The area to be covered with floor tile is

$$180\,\text{ft}^2 - 12\,\text{ft}^2 - 12\,\text{ft}^2 = 156\,\text{ft}^2.$$

64. First, find the area of a trapezoid with bases 80 ft and 100 ft and height 60 ft.

$$A = \frac{1}{2}h(a+b)$$

$$= \frac{1}{2}(60)(80+100) = 5400$$

The area of the yard is 5400 ft^2. The cost is $0.35(5400) = $1890.

65. The radius of the medium pizza is

$\frac{1}{2}\cdot 14$ inches $= 7$ inches, and the radius of each

small pizza is $\frac{1}{2}\cdot 8$ inches $= 4$ inches.

Medium pizza:

$$A = \pi r^2 = \pi(7\ \text{in.})^2$$

$$= 49\pi\ \text{in.}^2 \approx 154\ \text{in.}^2$$

Small pizza:

$$A = \pi r^2 = \pi(4\,\text{in.})^2$$

$$= 16\pi\ \text{in.}^2 \approx 50\ \text{in.}^2$$

The area of one medium pizza is approximately 154 in.^{2} and the area of two small pizzas is approximately $2(50) = 100$ in.^{2}. Since the price of one medium pizza is the same as the price of two small pizzas and the medium pizza has the greater area, the medium pizza is the better buy. (Because the prices are the same, it is not necessary to find price per square inch in this case.)

66. Find the volume of a rectangular solid with length 5 cm, width 3 cm, and height 4 cm.

$$A = lwh = 5\cdot 3\cdot 4 = 60$$

The volume is 60 cm^3.

67. Find the volume of a cylinder with radius 4 yd and height 8 yd.

$$V = \pi r^2 h$$

$$= \pi(4)^2 \cdot 8 = 128\pi \approx 402$$

The volume is 128π yd$^3 \approx 402$ yd^3.

68. Find the volume of a sphere with radius 6 m.

$$V = \frac{4}{3}\pi r^3$$
$$= \frac{4}{3}\pi(6)^3 = \frac{4}{3}\cdot\pi\cdot 216$$
$$= 288\pi \approx 905$$

The volume is 288π m$^3 \approx 905$ m^3.

69. Find the volume of each box.

$$V = lwh = (8\text{m})(4\text{m})(3\text{m}) = 96\text{m}^3$$

The space required for 50 containers is

$$50(96\text{ m}^3) = 4800\text{ m}^3.$$

70. Since the diameter of the fish tank 6 ft, the radius is 3 ft.

$$V = \pi r^2 h = \pi(3)^2\cdot 3 = 27\pi \approx 84.82$$

The volume of the tank is approximately 85 ft^3. Divide by 5 to determine how many fish can be put in the tank.

$$\frac{84.82}{5} \approx 16.96$$

There is enough water in the tank for 16 fish. Round down to 16, since 0.96 of a fish cannot be purchased.

71. The sum of the measures of the angles of any triangle is $180°$, so $x + 3x + 2x = 180$.

$$x + 3x + 2x = 180$$
$$6x = 180$$
$$x = 30$$

If $x = 30$, then $3x = 90$ and $2x = 60$, so the angles measure $30°$, $60°$, and $90°$.

72. Let x = the measure of the second angle.
Let $2x + 15$ = the measure of the first angle.
Let $x + 25$ = the measure of the third angle.

$$x + (2x + 15) + (x + 25) = 180$$
$$4x + 40 = 180$$
$$4x = 140$$
$$x = 35$$

If $x = 35$, then $2x + 15 = 2(35) + 15 = 85$ and $x + 25 = 35 + 25 = 60$. The angles measure $85°$, $35°$, and $60°$.

73. If the measure of an angle is $57°$, the measure of its complement is $90° - 57° = 33°$

74. If the measure of an angle is $75°$, the measure of its supplement is $180° - 75° = 105°$.

75. Let x = the measure of the angle.
Let $90 - x$ = the measure of its complement.

$$x = (90 - x) + 25$$
$$x = 115 - x$$
$$2x = 115$$
$$x = 57.5$$

The measure of the angle is $57.5°$.

76. Let x = the measure of the angle.
Let $180 - x$ = the measure of its supplement.

$$180 - x = 4x - 45$$
$$180 - 5x = -45$$
$$-5x = -225$$
$$x = 45$$

If $x = 45$, then $180 - x = 135$. The measure of the angle is $45°$ and the measure of its supplement is $135°$.

77. $x < -1$

78. $-2 < x \le 4$

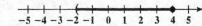

79. $\left[\dfrac{3}{2}, \infty\right)$

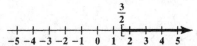

80. $(-\infty, 0)$

81.
$$2x - 5 < 3$$
$$2x - 5 + 5 < 3 + 5$$
$$2x < 8$$
$$\frac{2x}{2} < \frac{8}{2}$$
$$x < 4$$

$$(-\infty, 4)$$

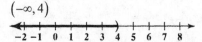

82. $\dfrac{x}{2} > -4$

$2\left(\dfrac{x}{2}\right) > 2(-4)$

$x > -8$

$(-8, \infty)$

83. $3 - 5x \le 18$

$3 - 5x - 3 \le 18 - 3$

$-5x \le 15$

$\dfrac{-5x}{-5} \ge \dfrac{15}{-5}$

$x \ge -3$

$[-3, \infty)$

84. $4x + 6 < 5x$

$4x + 6 - 5x < 5x - 5x$

$-x + 6 < 0$

$-x + 6 - 6 < 0 - 6$

$-x < -6$

$-1(-x) > -1(-6)$

$x > 6$

$(6, \infty)$

85. $6x - 10 \ge 2(x + 3)$

$6x - 10 \ge 2x + 6$

$6x - 10 - 2x \ge 2x + 6 - 2x$

$4x - 10 \ge 6$

$4x - 10 + 10 \ge 6 + 10$

$4x \ge 16$

$\dfrac{4x}{4} \ge \dfrac{16}{4}$

$x \ge 4$

$[4, \infty)$

86. $4x + 3(2x - 7) \le x - 3$

$4x + 6x - 21 \le x - 3$

$10x - 21 \le x - 3$

$10x - 21 - x \le x - 3 - x$

$9x - 21 \le -3$

$9x - 21 + 21 \le -3 + 21$

$9x \le 18$

$\dfrac{9x}{9} \le \dfrac{18}{9}$

$x \le 2$

$(-\infty, 2]$

87. $2(2x + 4) > 4(x + 2) - 6$

$4x + 8 > 4x + 8 - 6$

$4x + 8 > 4x + 2$

$4x + 8 - 4x > 4x + 2 - 4x$

$8 > 2$

Since $8 > 2$ is a true statement, the original inequality is true for all real numbers, and the solution set is $\{x | x \text{ is a real number}\}$.

88. $-2(x - 4) \le 3x + 1 - 5x$

$-2x + 8 \le -2x + 1$

$-2x + 8 + 2x \le -2x + 1 + 2x$

$8 \le 1$

Since $8 \le 1$ is a false statement, the original inequality has no solution. The solution set is $\{ \ \}$.

89. Let x = the student's score on the third test.

$\dfrac{42 + 74 + x}{3} \ge 60$

$3\left(\dfrac{42 + 74 + x}{3}\right) \ge 3(60)$

$42 + 74 + x \ge 180$

$116 + x \ge 180$

$116 + x - 116 \ge 180 - 116$

$x \ge 64$

The student must score at least 64 on the third test to pass the course.

90. $C = 10 + 5(x-1)$; $C \le 500$

$$10 + 5(x-1) \le 500$$
$$10 + 5x - 5 \le 500$$
$$5x + 5 \le 500$$
$$5x + 5 - 5 \le 500 - 5$$
$$5x \le 495$$
$$\frac{5x}{5} \le \frac{495}{5}$$
$$x \le 99$$

You can talk no more than 99 minutes.

Chapter 2 Test

1.
$$4x - 5 = 13$$
$$4x + 5 + 5 = 13 + 5$$
$$4x = 18$$
$$\frac{4x}{4} = \frac{18}{4} = \frac{9}{2}$$
$$x = \frac{9}{2}$$

The solution set is $\left\{ \dfrac{9}{2} \right\}$.

2.
$$12x + 4 = 7x - 21$$
$$12x + 4 - 7x = 7x - 21 - 7x$$
$$5x + 4 = -21$$
$$5x + 4 - 4 = -21 - 4$$
$$5x = -25$$
$$\frac{5x}{5} = \frac{-25}{5}$$
$$x = -5$$

The solution set is $\{-5\}$.

3.
$$8 - 5(x-2) = x + 26$$
$$8 - 5x + 10 = x + 26$$
$$18 - 5x = x + 26$$
$$18 - 5x - x = x + 26 - x$$
$$18 - 6x = 26$$
$$18 - 6x - 18 = 26 - 18$$
$$-6x = 8$$
$$\frac{-6x}{-6} = \frac{8}{-6}$$
$$x = -\frac{8}{6} = -\frac{4}{3}$$

The solution set is $\left\{ -\dfrac{4}{3} \right\}$.

4.
$$3(2y-4) = 9 - 3(y+1)$$
$$6y - 12 = 9 - 3y - 3$$
$$6y - 12 = 6 - 3y$$
$$6y - 12 + 3y = 6 - 3y + 3y$$
$$9y - 12 = 6$$
$$9y - 12 + 12 = 6 + 12$$
$$9y = 18$$
$$\frac{9y}{9} = \frac{18}{9}$$
$$y = 2$$

The solution set is $\{2\}$.

5.
$$\frac{3}{4}x = -15$$
$$\frac{4}{3}\left(\frac{3}{4}x \right) = \frac{4}{3}(-15)$$
$$x = -20$$

The solution set is $\{-20\}$.

6. $\dfrac{x}{10} + \dfrac{1}{3} = \dfrac{x}{5} + \dfrac{1}{2}$

Multiply both sides by the LCD, 30.

$$30\left(\dfrac{x}{10} + \dfrac{1}{3}\right) = 30\left(\dfrac{x}{5} + \dfrac{1}{2}\right)$$

$$30\left(\dfrac{x}{10}\right) + 30\left(\dfrac{1}{3}\right) = 30\left(\dfrac{x}{5}\right) + 30\left(\dfrac{1}{2}\right)$$

$$3x + 10 = 6x + 15$$

$$3x + 10 - 6x = 6x + 15 - 6x$$

$$-3x + 10 = 15$$

$$-3x + 10 - 10 = 15 - 10$$

$$-3x = 5$$

$$\dfrac{-3x}{-3} = \dfrac{5}{-3}$$

$$x = -\dfrac{5}{3}$$

The solution set is $\left\{-\dfrac{5}{3}\right\}$.

7. $N = 2.4x + 180;\; N = 324$

$$2.4x + 180 = 324$$

$$2.4x + 180 - 180 = 324 - 180$$

$$2.4x = 144$$

$$\dfrac{2.4x}{2.4} = \dfrac{144}{2.4}$$

$$x = 60$$

The US population is expected to reach 324 million 60 years after 1960, in the year 2020.

8. $V = \pi r^2 h$ for h

$$\dfrac{V}{\pi r^2} = \dfrac{\pi r^2 h}{\pi r^2}$$

$$\dfrac{V}{\pi r^2} = h \text{ or } h = \dfrac{V}{\pi r^2}$$

9. $l = \dfrac{P - 2w}{2}$ for w

$$2l = 2\left(\dfrac{P - 2w}{2}\right)$$

$$2l = P - 2w$$

$$2l - P = P - 2w - P$$

$$2l - P = -2w$$

$$\dfrac{2l - P}{-2} = \dfrac{-2w}{-2}$$

$$\dfrac{2l - P}{-2} = w \text{ or } w = \dfrac{P - 2l}{2}$$

10. $A = PB;\; P = 6\% = 0.06,\; B = 140$

$$A = 0.06(140)$$

$$A = 8.4$$

6% of 140 is 8.4.

11. $A = PB;\; A = 120,\; P = 80\% = 0.80$

$$120 = 0.80B$$

$$\dfrac{120}{0.80} = \dfrac{0.80B}{0.80}$$

$$150 = B$$

120 is 80% of 150.

12. $A = PB;\; A = 12,\; B = 240$

$$12 = P \cdot 240$$

$$\dfrac{12}{240} = \dfrac{P \cdot 240}{240}$$

$$0.05 = P$$

12 is 5% of 240.

13. Let x = the unknown number.

$$5x - 9 = 306$$

$$5x - 9 + 9 = 306 + 9$$

$$5x = 315$$

$$\dfrac{5x}{5} = \dfrac{315}{5}$$

$$x = 63$$

The number is 63.

14. Let x = the number of people, in millions, with an income of $50,000 - $74,999.

Let $x + 2.6$ = the number of people, in millions, with an income of $75,000 or more.

$$x + (x + 2.6) = 14.2$$

$$x + x + 2.6 = 14.2$$

$$2x + 2.6 = 14.2$$

$$2x + 2.6 - 2.6 = 14.2 - 2.6$$

$$2x = 11.6$$

$$x = 5.8$$

$$x + 2.6 = 8.4$$

5.8 million people with an income of $50,000 - $74,999 and 8.4 million people with an income of $75,000 or more had at least one major depressive episode.

15. Let x = number of minutes of long distance calls.

$$15 + 0.05x = 45$$

$$0.05x = 30$$

$$x = \frac{30}{0.05}$$

$$x = 600$$

You can talk long distance for 600 minutes.

16. Let x = the width of the field.
Let $2x$ = the length of the field.

$$P = 2l + 2w$$

$$450 = 2 \cdot 2x + 2 \cdot x$$

$$450 = 4x + 2x$$

$$450 = 6x$$

$$\frac{450}{6} = \frac{6x}{6}$$

$$75 = x$$

$$x = 75$$

$$2x = 150$$

The field is 75 yards wide and 150 yards long.

17. Let x = the book's original price.

$$x - 0.20x = 28$$

$$0.80x = 28$$

$$x = \frac{28}{0.80}$$

$$x = 35$$

The price of the book before the reduction was \$35.

18. Find the area of a triangle with base 47 meters and height 22 meters.

$$A = \frac{1}{2}bh = \frac{1}{2}(47)(22) = 517$$

The area of the triangle is 517 m^2.

19. Find the area of a trapezoid with height 15 in, lower base 40 in and upper base 30 in.

$$A = \frac{1}{2}h(a+b)$$

$$= \frac{1}{2}(15)(40+30)$$

$$= \frac{1}{2} \cdot 15 \cdot 70 = 525$$

The area is 525 in^2.

20. Notice that the height of the side rectangle is $6 + 3 = 9$ ft.

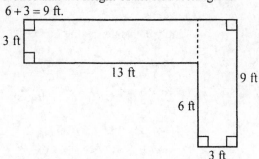

Using $A = lw$ we must find the sum of areas of the upper rectangle and the side rectangle.

$$A = (3)(13) + (3)(9)$$

$$= 39 + 27$$

$$= 66$$

The area is 66 ft^2.

21. Find the volume of a rectangular solid with length 3 in, width 2 in, and height 3 in.

$$V = lwh = 3 \cdot 2 \cdot 3 = 18$$

The volume is 18 in^3.

22. Find the volume of a cylinder with radius 5 cm and height 7 cm.

$$V = \pi r^2 h$$

$$= \pi(5)^2 \cdot 7 = \pi \cdot 25 \cdot 7$$

$$= 175\pi \approx 550$$

The volume is $175\pi \text{ cm}^3$ or approximately 550 cm^3.

23. The area of the floor is $A = (40 \text{ ft})(50 \text{ ft}) = 2000 \text{ ft}^2$.

The area of each tile is $A = (2 \text{ ft})(2 \text{ ft}) = 4 \text{ ft}^2$.

The number of tiles needed is $\dfrac{2000 \text{ ft}^2}{4 \text{ ft}^2} = 500$.

Since there are 10 tiles in a package, the number of packages needed is $\dfrac{500}{10} = 50$.

Since each package costs \$13, the cost for enough tiles to cover the floor is $50(\$13) = \$650$.

24. $A = 56, b = 8$

$$A = \frac{1}{2}bh$$

$$56 = \frac{1}{2} \cdot 8 \cdot h$$

$$56 = 4h$$

$$14 = h$$

The height of the sail is 14 feet.

25. Let x = the measure of the second angle.
Let $3x$ = the measure of the first angle.
Let $x - 30$ = the measure of the third angle.
$$x + 3x + (x - 30) = 180$$
$$5x - 30 = 180$$
$$5x = 210$$
$$x = 42$$
The measure of the first angle: $3x = 3(42°) = 126°$.
The measure of the second angle: $x = 42°$.
The measure of the third angle: $x - 30 = 42° - 30°$
$= 12°$.

26. Let x = the measure of the angle.
Let $90 - x$ = the measure of its complement.
$$x = (90 - x) + 16$$
$$x = 106 - x$$
$$2x = 106$$
$$x = 53$$
The measure of the angle is $53°$.

27. $(-2, \infty)$

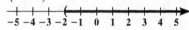

28. $(-\infty, 3]$

29. $\dfrac{x}{2} < -3$
$$2\left(\dfrac{x}{2}\right) < 2(-3)$$
$$x < -6$$
$(-\infty, -6)$

30. $6 - 9x \geq 33$
$$6 - 9x - 6 \geq 33 - 6$$
$$-9x \geq 27$$
$$\dfrac{-9x}{-9} \leq \dfrac{27}{-9}$$
$$x \leq -3$$
$(-\infty, -3]$

31.
$$4x - 2 > 2(x + 6)$$
$$4x - 2 > 2x + 12$$
$$4x - 2 - 2x > 2x + 12 - 2x$$
$$2x - 2 > 12$$
$$2x > 14$$
$$x > 7$$
$(7, \infty)$

32. Let x = the student's score on the fourth exam.
$$\dfrac{76 + 80 + 72 + x}{4} \geq 80$$
$$4\left(\dfrac{76 + 80 + 72 + x}{4}\right) \geq 4(80)$$
$$76 + 80 + 72 + x \geq 320$$
$$228 + x \geq 320$$
$$x \geq 92$$
The student must score at least 92 on the fourth exam to have an average of at least 80.

33. Let x = the width of the rectangle.
$$2(20) + 2x > 56$$
$$40 + 2x > 56$$
$$40 - 40 + 2x > 56 - 40$$
$$2x > 16$$
$$x > 8$$
The perimeter is greater than 56 inches when the width is greater than 8 inches.

Cumulative Review Exercises (Chapters 1-2)

1. $-8 - (12 - 16) = -8 - (-4) = -8 + 4 = -4$

2. $(-3)(-2) + (-2)(4) = 6 + (-8) = -2$

3. $(8 - 10)^3 (7 - 11)^2 = (-2)^3 (-4)^2$
$$= -8(16) = -128$$

4. $2 - 5[x + 3(x + 7)]$
$$= 2 - 5(x + 3x + 21)$$
$$= 2 - 5(4x + 21)$$
$$= 2 - 20x - 105$$
$$= -103 - 20x$$

5. The rational numbers are

$$-4, -\frac{1}{3}, 0, \sqrt{4}\,(=2), \text{ and } 1063.$$

6. $\dfrac{5}{x} - (x+2)$

7. $-10,000 < -2$ since $-10,000$ is to the left of -2 on the number line.

8. $6(4x - 1 - 5y) = 6(4x) - 6(1) - 6(5y)$
$$= 24x - 6 - 30y$$

9. $T = 5.3n + 9.5$
$T = 5.3(4) + 9.5$
$T = 21.2 + 9.5$
$T = 30.7$
According to the formula, 30.7% of consumers looked for trans fats on food labels in 2007. This underestimates the actual value shown in the bar graph by 0.3%

10.
$$T = 5.3n + 9.5$$
$$89 = 5.3n + 9.5$$
$$89 - 9.5 = 5.3n + 9.5 - 9.5$$
$$79.5 = 5.3n$$
$$\frac{79.5}{5.3} = \frac{5.3n}{5.3}$$
$$15 = n$$
If trends continue, 89% of consumers will look for trans fats on food labels 15 years after 2003, or 2018.

11. $5 - 6(x + 2) = x - 14$
$5 - 6x - 12 = x - 14$
$-7 - 6x = x - 14$
$-7 - 6x - x = x - 14 - x$
$-7 - 7x = -14$
$-7 - 7x + 7 = -14 + 7$
$-7x = -7$
$\dfrac{-7x}{-7} = \dfrac{-7}{-7}$
$x = 1$
The solution set is $\{1\}$.

12. $\dfrac{x}{5} - 2 = \dfrac{x}{3}$
Multiply both sides by the LCD, 15.
$$15\left(\frac{x}{5} - 2\right) = 15\left(\frac{x}{3}\right)$$
$$15\left(\frac{x}{5}\right) - 15(2) = 15\left(\frac{x}{3}\right)$$
$$3x - 30 = 5x$$
$$3x - 30 - 3x = 5x - 3x$$
$$-30 = 2x$$
$$\frac{-30}{2} = \frac{2x}{2}$$
$$-15 = x$$
The solution set is $\{-15\}$.

13. $V = \dfrac{1}{3}Ah$ for A
$$V = \frac{1}{3}Ah$$
$$3V = 3\left(\frac{1}{3}Ah\right)$$
$$3V = Ah$$
$$\frac{3V}{h} = \frac{Ah}{h}$$
$$\frac{3V}{h} = A \text{ or } A = \frac{3V}{h}$$

14. $A = PB; A = 48, P = 30\% = 0.30$
$$48 = 0.30B$$
$$\frac{48}{0.30} = \frac{0.30B}{0.30}$$
$$160 = B$$
48 is 30% of 160.

15. Let x = the width of the parking lot.
Let $2x - 10 =$ the length of the parking lot.

$$P = 2l + 2w$$
$$400 = 2(2x - 10) + 2 \cdot x$$
$$400 = 4x - 20 + 2x$$
$$400 = 6x - 20$$
$$400 + 20 = 6x - 20 + 20$$
$$420 = 6x$$
$$\frac{420}{6} = \frac{6x}{6}$$
$$70 = x$$
$$x = 70$$
$$2x - 10 = 130$$

The parking lot is 70 yards wide and 130 yards long.

16. Let x = number of gallons of gasoline.

$$0.40x = 30,000$$
$$\frac{0.40x}{0.40} = \frac{30,000}{0.40}$$
$$x = 75,000$$

75,000 gallons of gasoline must be sold

17. $\left(-\infty, \frac{1}{2} \right]$

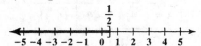

18. $$3 - 3x > 12$$
$$3 - 3x - 3 > 12 - 3$$
$$-3x > 9$$
$$\frac{-3x}{-3} < \frac{9}{-3}$$
$$x < -3$$

$(-\infty, -3)$

19. $$5 - 2(3 - x) \le 2(2x + 5) + 1$$
$$5 - 6 + 2x \le 4x + 10 + 1$$
$$2x - 1 \le 4x + 11$$
$$2x - 1 - 4x \le 4x + 11 - 4x$$
$$-2x - 1 \le 11$$
$$-2x - 1 + 1 \le 11 + 1$$
$$-2x \le 12$$
$$\frac{-2x}{-2} \ge \frac{12}{-2}$$
$$x \ge -6$$

$[-6, \infty)$

20. Let x = value of medical supplies sold.

$$600 + 0.04x > 2500$$
$$600 + 0.04x - 600 > 2500 - 600$$
$$0.04x > 1900$$
$$\frac{0.04x}{0.04} > \frac{1900}{0.04}$$
$$x > 47,500$$

You must sell more than \$47,500 worth of medical supplies.

Chapter 3
Linear Equations in Two Variables

3.1 Check Points

1.

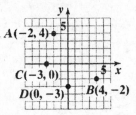

2. $E(-4,-2)$ $F(-2,0)$ $G(6,0)$

3. a. $x - 3y = 9$

$3 - 3(-2) = 9$

$3 + 6 = 9$

$9 = 9,$ true

$(3, -2)$ is a solution.

b. $x - 3y = 9$

$-2 - 3(3) = 9$

$-2 - 9 = 9$

$-11 = 9,$ false

$(-2, 3)$ is not a solution.

4.

x	$y = 3x + 2$	(x, y)
-2	$y = 3(-2) + 2 = -4$	$(-2, -4)$
-1	$y = 3(-1) + 2 = -1$	$(-1, -1)$
0	$y = 3(0) + 2 = 2$	$(0, 2)$
1	$y = 3(1) + 2 = 5$	$(1, 5)$
2	$y = 3(2) + 2 = 8$	$(2, 8)$

5.

x	$y = 2x$	(x, y)
-2	$y = 2(-2) = -4$	$(-2, -4)$
-1	$y = 2(-1) = -2$	$(-1, -2)$
0	$y = 2(0) = 0$	$(0, 0)$
1	$y = 2(1) = 2$	$(1, 2)$
2	$y = 2(2) = 4$	$(2, 4)$

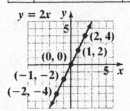

6.

x	$y = 2x - 2$	(x, y)
-2	$y = 2(-2) - 2 = -6$	$(-2, -6)$
-1	$y = 2(-1) - 2 = -4$	$(-1, -4)$
0	$y = 2(0) - 2 = -2$	$(0, -2)$
1	$y = 2(1) - 2 = 0$	$(1, 0)$
2	$y = 2(2) - 2 = 2$	$(2, 2)$

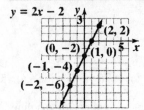

7.

x	$y = \frac{1}{2}x + 2$	(x, y)
-4	$y = \frac{1}{2}(-4) + 2 = 0$	$(-4, 0)$
-2	$y = \frac{1}{2}(-2) + 2 = 1$	$(-2, 1)$
0	$y = \frac{1}{2}(0) + 2 = 2$	$(0, 2)$
2	$y = \frac{1}{2}(2) + 2 = 3$	$(2, 3)$
4	$y = \frac{1}{2}(4) + 2 = 4$	$(4, 4)$

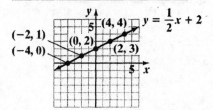

8. a.

n	$D = 1.4n + 1$	(n, D)
0	$D = 1.4(0) + 1 = 1$	$(0, 1)$
5	$D = 1.4(5) + 1 = 8$	$(5, 8)$
10	$D = 1.4(10) + 1 = 15$	$(10, 15)$
15	$D = 1.4(15) + 1 = 22$	$(15, 22)$

b. Graph formula:

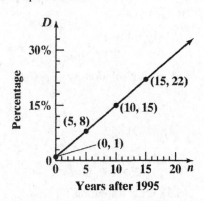

c. According to the graph, about 29% of consumers will pay primarily with debit cards in 2015.

d. $D = 1.4n + 1$

   $D = 1.4(20) + 1$

   $\quad = 29$

   According to the formula, about 29% of consumers will pay primarily with debit cards in 2015.

3.1 Exercise Set

1. Quadrant I

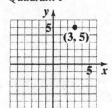

3. Quadrant II

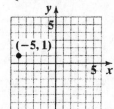

5. Quadrant III

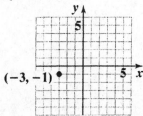

7. Quadrant IV

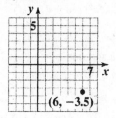

9. – 23.

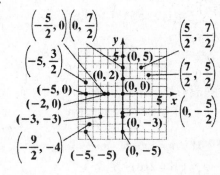

25. $A\,(5,2)$

27. $C\,(-6,5)$

29. $E(-2,-3)$

31. $G(5,-3)$

33. The y-coordinates are positive in Quadrants I and II.

35. The x- and y-coordinates have the same sign in Quadrants I and III.

37. $y = 3x$

   $y = 3x$

   $3 = 3(2)$

   $3 = 6,\ \text{false}$

   $(2,3)$ is not a solution.

   $y = 3x$

   $2 = 3(3)$

   $2 = 9,\ \text{false}$

   $(3,2)$ is not a solution.

   $y = 3x$

   $-12 = 3(-4)$

   $-12 = -12,\ \text{true}$

   $(-4,-12)$ is a solution.

39. $y = -4x$

$-20 = -4(-5)$

$-20 = 20,$ false

$(-5, -20)$ is not a solution.

$y = -4x$

$0 = -4(0)$

$0 = 0,$ true

$(0, 0)$ is a solution.

$y = -4x$

$-36 = -4(9)$

$-36 = -36,$ true

$(9, -36)$ is a solution.

41. $y = 2x + 6$

$6 = 2(0) + 6$

$6 = 6,$ true

$(0, 6)$ is a solution.

$y = 2x + 6$

$0 = 2(-3) + 6$

$0 = 0,$ true

$(-3, 0)$ is a solution.

$y = 2x + 6$

$-2 = 2(2) + 6$

$-2 = 10,$ false

$(2, -2)$ is not a solution.

43. $3x + 5y = 15$

$3(-5) + 5(6) = 15$

$-15 + 30 = 15$

$15 = 15,$ true

$(-5, 6)$ is a solution.

$3x + 5y = 15$

$3(0) + 5(5) = 15$

$0 + 25 = 15$

$25 = 15,$ false

$(0, 5)$ is not a solution.

$3x + 5y = 15$

$3(10) + 5(-3) = 15$

$30 - 15 = 15$

$15 = 15,$ true

$(10, -3)$ is a solution.

45. $x + 3y = 0$

$0 + 3(0) = 0$

$0 = 0,$ true

$(0, 0)$ is a solution.

$x + 3y = 0$

$1 + 3\left(\dfrac{1}{3}\right) = 0$

$1 + 1 = 0$

$2 = 0,$ false

$\left(1, \dfrac{1}{3}\right)$ is not a solution.

$x + 3y = 0$

$2 + 3\left(-\dfrac{2}{3}\right) = 0$

$2 - 2 = 0$

$0 = 0,$ true

$\left(2, -\dfrac{2}{3}\right)$ is a solution.

47. $x - 4 = 0$

$4 - 4 = 0$

$0 = 0$, true

$(4, 7)$ is a solution.

$x - 4 = 0$

$3 - 4 = 0$

$-1 = 0$, false

$(3, 4)$ is not a solution.

$x - 4 = 0$

$0 - 4 = 0$

$-4 = 0$, false

$(0, -4)$ is not a solution.

49.

x	$y = 12x$	(x, y)
-2	$y = 12(-2) = -24$	$(-2, -24)$
-1	$y = 12(-1) = -12$	$(-1, -12)$
0	$y = 12(0) = 0$	$(0, 0)$
1	$y = 12(1) = 12$	$(1, 12)$
2	$y = 12(2) = 24$	$(2, 24)$

51.

x	$y = -10x$	(x, y)
-2	$y = -10(-2) = 20$	$(-2, 20)$
-1	$y = -10(-1) = 10$	$(-1, 10)$
0	$y = -10(0) = 0$	$(0, 0)$
1	$y = -10(1) = -10$	$(1, -10)$
2	$y = -10(2) = -20$	$(2, -20)$

53.

x	$y = 8x - 5$	(x, y)
-2	$y = 8(-2) - 5 = -21$	$(-2, -21)$
-1	$y = 8(-1) - 5 = -13$	$(-1, -13)$
0	$y = 8(0) - 5 = -5$	$(0, -5)$
1	$y = 8(1) - 5 = 3$	$(1, 3)$
2	$y = 8(2) - 5 = 11$	$(2, 11)$

55.

x	$y = -3x + 7$	(x, y)
-2	$y = -3(-2) + 7 = 13$	$(-2, 13)$
-1	$y = -3(-1) + 7 = 10$	$(-1, 10)$
0	$y = -3(0) + 7 = 7$	$(0, 7)$
1	$y = -3(1) + 7 = 4$	$(1, 4)$
2	$y = -3(2) + 7 = 1$	$(2, 1)$

57.

x	$y = x$	(x, y)
-2	$y = -2$	$(-2, -2)$
-1	$y = -1$	$(-1, -1)$
0	$y = 0$	$(0, 0)$
1	$y = 1$	$(1, 1)$
2	$y = 2$	$(2, 2)$

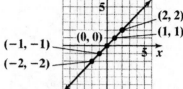

59.

x	$y = x - 1$	(x, y)
-2	$y = -2 - 1 = -3$	$(-2, -3)$
-1	$y = -1 - 1 = -2$	$(-1, -2)$
0	$y = 0 - 1 = -1$	$(0, -1)$
1	$y = 1 - 1 = 0$	$(1, 0)$
2	$y = 2 - 1 = 1$	$(2, 1)$

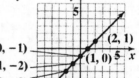

61.

x	$y = 2x + 1$	(x, y)
-2	$y = 2(-2) + 1 = -3$	$(-2, -3)$
-1	$y = 2(-1) + 1 = -1$	$(-1, -1)$
0	$y = 2(0) + 1 = 1$	$(0, 1)$
1	$y = 2(1) + 1 = 3$	$(1, 3)$
2	$y = 2(2) + 1 = 5$	$(2, 5)$

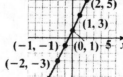

63.

x	$y = -x + 2$	(x, y)
-2	$y = -(-2) + 2 = 4$	$(-2, 4)$
-1	$y = -(-1) + 2 = 3$	$(-1, 3)$
0	$y = -0 + 2 = 2$	$(0, 2)$
1	$y = -1 + 2 = 1$	$(1, 1)$
2	$y = -2 + 2 = 0$	$(2, 0)$

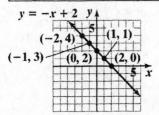

65.

x	$y = -3x - 1$	(x, y)
-2	$y = -3(-2) - 1 = 5$	$(-2, 5)$
-1	$y = -3(-1) - 1 = 2$	$(-1, 2)$
0	$y = -3(0) - 1 = -1$	$(0, -1)$
1	$y = -3(1) - 1 = -4$	$(1, -4)$
2	$y = -3(2) - 1 = -7$	$(2, -7)$

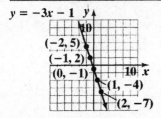

67.

x	$y = \frac{1}{2}x$	(x, y)
-4	$y = \frac{1}{2}(-4) = -2$	$(-4, -2)$
-2	$y = \frac{1}{2}(-2) = -1$	$(-2, -1)$
0	$y = \frac{1}{2}(0) = 0$	$(0, 0)$
2	$y = \frac{1}{2}(2) = 1$	$(2, 1)$
4	$y = \frac{1}{2}(4) = 2$	$(4, 2)$

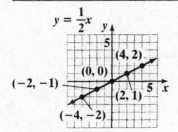

69.

x	$y = -\frac{1}{4}x$	(x, y)
-8	$y = -\frac{1}{4}(-8) = 2$	$(-8, 2)$
-4	$y = -\frac{1}{4}(-4) = 1$	$(-4, 1)$
0	$y = -\frac{1}{4}(0) = 0$	$(0, 0)$
4	$y = -\frac{1}{4}(4) = -1$	$(4, -1)$
8	$y = -\frac{1}{4}(8) = -2$	$(8, -2)$

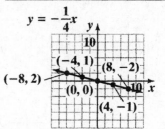

71.

x	$y = \frac{1}{3}x + 1$	(x, y)
-6	$y = \frac{1}{3}(-6) + 1 = -1$	$(-6, -1)$
-3	$y = \frac{1}{3}(-3) + 1 = 0$	$(-3, 0)$
0	$y = \frac{1}{3}(0) + 1 = 1$	$(0, -1)$
3	$y = \frac{1}{3}(3) + 1 = 2$	$(3, 2)$
6	$y = \frac{1}{3}(6) + 1 = 3$	$(6, 3)$

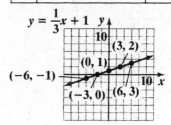

73.

x	$y=-\frac{3}{2}x+1$	(x,y)
-4	$y=-\frac{3}{2}(-4)+1=7$	$(-4,7)$
-2	$y=-\frac{3}{2}(-2)+1=4$	$(-2,4)$
0	$y=-\frac{3}{2}(0)+1=1$	$(0,1)$
2	$y=-\frac{3}{2}(2)+1=-2$	$(2,-2)$
4	$y=-\frac{3}{2}(4)+1=-5$	$(4,-5)$

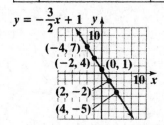

75.

x	$y=-\frac{5}{2}x-1$	(x,y)
-4	$y=-\frac{5}{2}(-4)-1=9$	$(-4,9)$
-2	$y=-\frac{5}{2}(-2)-1=4$	$(-2,4)$
0	$y=-\frac{5}{2}(0)-1=-1$	$(0,-1)$
2	$y=-\frac{5}{2}(2)-1=-6$	$(2,-6)$
4	$y=-\frac{5}{2}(4)-1=-11$	$(4,-11)$

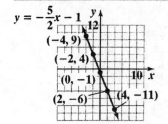

77.

x	$y=x+\frac{1}{2}$	(x,y)
-4	$y=-4+\frac{1}{2}=-3.5$	$(-4,-3.5)$
-2	$y=-2+\frac{1}{2}=-1.5$	$(-2,-1.5)$
0	$y=0+\frac{1}{2}=0.5$	$(0,0.5)$
2	$y=2+\frac{1}{2}=2.5$	$(2,2.5)$
4	$y=4+\frac{1}{2}=4.5$	$(4,4.5)$

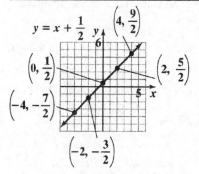

79.

x	$y=0x+4$	(x,y)
-6	$y=0(-6)+4=4$	$(-6,4)$
-3	$y=0(-3)+4=4$	$(-3,4)$
0	$y=0(0)+4=4$	$(0,4)$
3	$y=0(3)+4=4$	$(3,4)$
6	$y=0(6)+4=4$	$(6,4)$

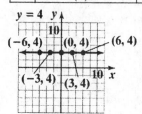

81. $y=x+3$

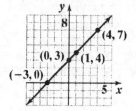

83. $y=2x+5$

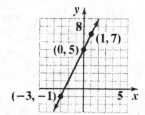

85. a. $8x + 6y = 14.50$

    b. $8x + 6(0.75) = 14.50$
$$8x + 4.50 = 14.50$$
$$8x = 10.00$$
$$x = 1.25$$
One pen costs $1.25.

87. The coordinates of point *A* are $(2,7)$. When the football is 2 yards from the quarterback, its height is 7 feet.

89. The coordinates of point *C* are approximately $(6, 9.25)$.

91. The football's maximum height is 12 feet. It reaches this height when it is 15 yards from the quarterback.

93. a.

n	$C = 1.3n + 28$	(n, C)
0	$C = 1.3(0) + 28 = 28$	$(0, 28)$
10	$C = 1.3(10) + 28 = 41$	$(10, 41)$
20	$C = 1.3(20) + 28 = 54$	$(20, 54)$
30	$C = 1.3(30) + 28 = 67$	$(30, 67)$
40	$C = 1.3(40) + 28 = 80$	$(40, 80)$

    b. Graph formula:

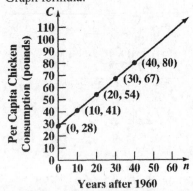

    c. According to the graph, per capita chicken consumption will be about 99 pounds in 2015.

    d. $C = 1.3n + 28$
$$C = 1.3(55) + 28$$
$$= 99.5$$
According to the formula, per capita chicken consumption will be about 99.5 pounds in 2015.

95. – 101. Answers will vary.

103. makes sense

105. does not make sense; Explanations will vary. Sample explanation: These points do not lie along a line.

107. false; Changes to make the statement true will vary. A sample change is: The lines are not parallel as they both contain the point $(-1, -2)$.

109. false; Changes to make the statement true will vary. A sample change is: The point $(4, 3)$ satisfies the equation.

111. a. $\left(1, \dfrac{1}{2}\right)$, $(2, 1)$, $\left(3, \dfrac{3}{2}\right)$, $(4, 2)$

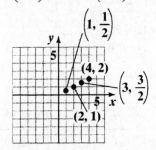

    b. In order for the resulting graph to be a mirror-image reflection about the *y*-axis of the graph in part (a), the sign of each *x*-coordinate should be changed: $\left(-1, \dfrac{1}{2}\right)$, $(-2, 1)$, $\left(-3, \dfrac{3}{2}\right)$, $(-4, 2)$

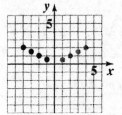

    c. In order for the resulting graph to be a mirror-image reflection about the *x*-axis of the graph in part (a), the sign of each *y*-coordinate should be changed: $\left(1, -\dfrac{1}{2}\right)$, $(2, -1)$, $\left(3, -\dfrac{3}{2}\right)$, $(4, -2)$

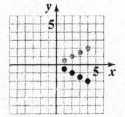

d. In order for the resulting graph to be a straight-line extension of the graph in part (a), the signs of both coordinates of each ordered pair should be changed:

$$\left(-1,-\frac{1}{2}\right), \ (-2,-1), \ \left(-3,-\frac{3}{2}\right), \ (-4,-2)$$

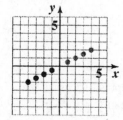

113. Answers will vary depending upon the points chosen. One example is shown here.

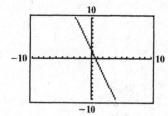

115. Answers will vary depending upon the points chosen. One example is shown here.

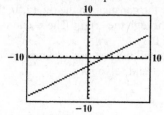

117.
$$3x+5 = 4(2x-3)+7$$
$$3x+5 = 8x-12+7$$
$$3x+5 = 8x-5$$
$$3x+5-8x = 8x-5-8x$$
$$-5x+5 = -5$$
$$-5x+5-5 = -5-5$$
$$-5x = -10$$
$$\frac{-5x}{-5} = \frac{-10}{-5}$$
$$x = 2$$
The solution set is $\{2\}$.

118. $3(1-2\cdot5)-(-28) = 3(1-10)+28$
$$= 3(-9)+28$$
$$= -27+28 = 1$$

119. $V = \frac{1}{3}Ah$ for h
$$V = \frac{1}{3}Ah$$
$$3V = 3\left(\frac{1}{3}Ah\right)$$
$$3V = Ah$$
$$\frac{3V}{A} = \frac{Ah}{A}$$
$$\frac{3V}{A} = h \text{ or } h = \frac{3V}{A}$$

120.
$$3x-4y = 24$$
$$3x-4(0) = 24$$
$$3x = 24$$
$$x = 8$$
The equation is satisfied by the ordered pair $(8,0)$.

121.
$$3x-4y = 24$$
$$3(0)-4y = 24$$
$$-4y = 24$$
$$y = -6$$
The equation is satisfied by the ordered pair $(0,-6)$.

122. $x+2y = 0$
$$0+2y = 0$$
$$2y = 0$$
$$y = 0$$
The equation is satisfied by the ordered pair $(0,0)$.

3.2 Check Points

1. a. The graph crosses the x-axis at $(-3,0)$. Thus, the x-intercept is –3.
The graph crosses the y-axis at $(0,5)$. Thus, the y-intercept is 5.

b. The graph does not cross the x-axis. Thus, there is no x-intercept.
The graph crosses the y-axis at $(0,4)$. Thus, the y-intercept is 4.

c. The graph crosses the x-axis at $(0,0)$. Thus, the x-intercept is 0.
The graph crosses the y-axis at $(0,0)$. Thus, the y-intercept is 0.

2. To find the *x*-intercept, let $y = 0$ and solve for *x*.
$$4x - 3y = 12$$
$$4x - 3(0) = 12$$
$$4x = 12$$
$$x = 3$$
The *x*-intercept is 3.

3. To find the *y*-intercept, let $x = 0$ and solve for *y*.
$$4x - 3y = 12$$
$$4(0) - 3y = 12$$
$$-3y = 12$$
$$y = -4$$
The *y*-intercept is –4.

4. Find the *x*-intercept. Let $y = 0$ and solve for *x*.
$$2x + 3y = 6$$
$$2x + 3(0) = 6$$
$$2x = 6$$
$$x = 3$$
The *x*-intercept is 3.
Find the *y*- intercept. Let $x = 0$ and solve for *y*.
$$2x + 3y = 6$$
$$2(0) + 3y = 6$$
$$3y = 6$$
$$y = 2$$
The *y*-intercept is 2.
Find a checkpoint. For example, let $x = 1$ and solve for *y*.
$$2x + 3y = 6$$
$$2(1) + 3y = 6$$
$$2 + 3y = 6$$
$$3y = 4$$
$$y = \frac{4}{3} \text{ or } 1\frac{1}{3}$$

 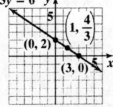

5. Find the *x*-intercept. Let $y = 0$ and solve for *x*.
$$x - 2y = 4$$
$$x - 2(0) = 4$$
$$x = 4$$
The *x*-intercept is 4.
Find the *y*- intercept. Let $x = 0$ and solve for *y*.
$$x - 2y = 4$$
$$0 - 2y = 4$$
$$-2y = 4$$
$$y = -2$$
The *y*-intercept is –2.
Find a checkpoint. For example, let $x = 2$ and solve for *y*.
$$x - 2y = 4$$
$$2 - 2y = 4$$
$$-2y = 2$$
$$y = -1$$

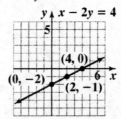

6. Because the constant on the right is 0, the graph passes through the origin. The *x*- and *y*-intercepts are both 0.
Thus we will need to find two more points.
Let $y = -1$ and solve for *x*.
$$x + 3y = 0$$
$$x + 3(-1) = 0$$
$$x - 3 = 0$$
$$x = 3$$
Let $y = 1$ and solve for *x*.
$$x + 3y = 0$$
$$x + 3(1) = 0$$
$$x + 3 = 0$$
$$x = -3$$
Use these three solutions of (0,0), (3,–1), and (–3,1).

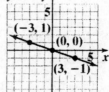

7. As demonstrated in the table below, all ordered pairs that are solutions of $y = 3$ have a value of y that is always 3.

x	$y = 3$	(x, y)
-2	3	$(-2, 3)$
0	3	$(0, 3)$
1	3	$(1, 3)$

Thus the line is horizontal.

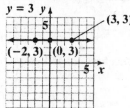

8. As demonstrated in the table below, all ordered pairs that are solutions of $x = -2$ have a value of x that is always -2.

$x = -2$	y	(x, y)
-2	-3	$(-2, -3)$
-2	0	$(-2, 0)$
-2	2	$(-2, 2)$

Thus the line is vertical.

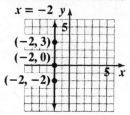

3.2 Exercise Set

1. a. The graph crosses the *x*-axis at (3,0). Thus, the *x*-intercept is 3.

b. The graph crosses the *y*-axis at (0,4). Thus, the *y*-intercept is 4.

3. a. The graph crosses the *x*-axis at (−4,0). Thus, the *x*-intercept is −4.

b. The graph crosses the *y*-axis at (0,−2). Thus, the *y*-intercept is −2.

5. a. The graph crosses the *x*-axis at (0,0) (the origin). Thus, the *x*-intercept is 0.

b. The graph also crosses the *y*-axis at (0,0). Thus, the *y*-intercept is 0.

7. a. The graph does not cross the *x*-axis. Thus, there is no *x*-intercept.

b. The graph crosses the *y*-axis at (0,−2). Thus the *y*-intercept is −2.

9. To find the *x*-intercept, let $y = 0$ and solve for *x*.
$$2x + 5y = 20$$
$$2x + 5(0) = 20$$
$$2x = 20$$
$$x = 10$$
The *x*-intercept is 10.

To find the *y*-intercept, let $x = 0$ and solve for *y*.
$$2x + 5y = 20$$
$$2(0) + 5y = 20$$
$$5y = 20$$
$$y = 4$$
The *y*-intercept is 4.

11. To find the *x*-intercept, let $y = 0$ and solve for *x*.
$$2x - 3y = 15$$
$$2x - 3(0) = 15$$
$$2x = 15$$
$$x = \frac{15}{2}$$
The *x*-intercept is $\frac{15}{2}$.

To find the *y*-intercept, let $x = 0$ and solve for *y*.
$$2x - 3y = 15$$
$$2(0) - 3y = 15$$
$$-3y = 15$$
$$y = -5$$
The *y*-intercept is −5.

13. To find the *x*-intercept, let *y* = 0 and solve for *x*.
$$-x + 3y = -8$$
$$-x + 3(0) = -8$$
$$-x = -8$$
$$x = 8$$
The *x*-intercept is 8.

To find the *y*-intercept, let *x* = 0 and solve for *y*.
$$-x + 3y = -8$$
$$-0 + 3y = -8$$
$$3y = -8$$
$$y = -\frac{8}{3}$$
The *y*-intercept is $-\frac{8}{3}$.

15. To find the *x*-intercept, let *y* = 0 and solve for *x*.
$$7x - 9y = 0$$
$$7x - 9(0) = 0$$
$$7x = 0$$
$$x = 0$$
The *x*-intercept is 0.

To find the *y*-intercept, let *x* = 0 and solve for *y*.
$$7x - 9y = 0$$
$$7(0) - 9y = 0$$
$$-9y = 0$$
$$y = 0$$
The *y*-intercept is 0.

17. To find the *x*-intercept, let *y* = 0 and solve for *x*.
$$2x = 3y - 11$$
$$2x = 3(0) - 11$$
$$2x = -11$$
$$x = -\frac{11}{2}$$
The *x*-intercept is $-\frac{11}{2}$.

To find the *y*-intercept, let *x* = 0 and solve for *y*.
$$2x = 3y - 11$$
$$2(0) = 3y - 11$$
$$0 = 3y - 11$$
$$11 = 3y$$
$$\frac{11}{3} = y$$
The *y*-intercept is $\frac{11}{3}$.

19. $x + y = 5$
x-intercept: 5
y-intercept: 5
checkpoint: (2,3)
Draw a line through (5,0), (0,5), and (2,3).

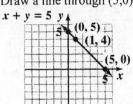

In Exercises 21-41, checkpoints will vary.

21. $x + 3y = 6$
x-intercept: 6
y-intercept: 2
checkpoint: (3,1)
Draw a line through (6,0), (0,2), and (3,1).

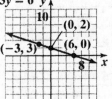

23. $6x - 9y = 18$
x-intercept: 3
y-intercept: −2
checkpoint: $\left(1, -\frac{4}{3}\right)$

Draw a line through (3,0), (0,−2), and $\left(1, -\frac{4}{3}\right)$.

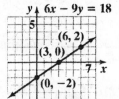

25. $-x + 4y = 6$

x-intercept: -6

y-intercept: $\dfrac{3}{2}$

checkpoint: $(2,2)$

Draw a line through $(-6,0)$, $\left(0, \dfrac{3}{2}\right)$, and $(2,2)$.

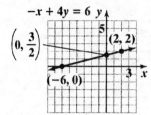

27. $2x - y = 7$

x-intercept: $\dfrac{7}{2}$

y-intercept: -7

checkpoint: $(1,-5)$

Draw a line through $\left(\dfrac{7}{2}, 0\right)$, $(0,7)$, and $(1,-5)$.

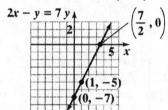

29. $3x = 5y - 15$

x-intercept: -5

y-intercept: 3

checkpoint: $\left(-\dfrac{10}{3}, 1\right)$

Draw a line through $(-5,0)$, $(0,3)$, and $\left(-\dfrac{10}{3}, 1\right)$.

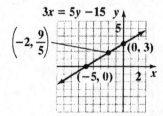

31. $25y = 100 - 50x$

x-intercept: 2

y-intercept: 4

checkpoint: $(1, 2)$

Draw a line through $(2,0)$, $(0,4)$, and $(1, 2)$.

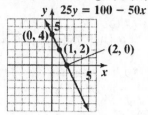

33. $2x - 8y = 12$

x-intercept: 6

y-intercept: $-\dfrac{3}{2}$

checkpoint: $(2, -1)$

Draw a line through $(6,0)$, $\left(0, -\dfrac{3}{2}\right)$, and $(2, -1)$.

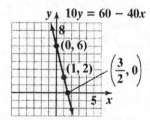

35. $x + 2y = 0$

x-intercept: 0

y-intercept: 0

Since the line goes through the origin, find two additional points.

checkpoint: $(2, -1)$

checkpoint: $(4, -2)$

Draw a line through $(0,0)$, $(2, -1)$, and $(4, -2)$.

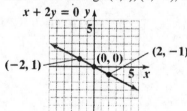

37. $y - 3x = 0$

x-intercept: 0

y-intercept: 0

Since the line goes through the origin, find two additional points.

checkpoint: $(1, 3)$

checkpoint: $(2, 6)$

Draw a line through $(0,0)$, $(1, 3)$, and $(2, 6)$.

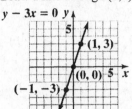

39. $2x - 3y = -11$

x-intercept: $-\dfrac{11}{2}$

y-intercept: $\dfrac{11}{3}$

checkpoint: $(-1, 3)$

Draw a line through $\left(-\dfrac{11}{2}, 0\right)$, $\left(0, \dfrac{11}{3}\right)$ and $(-1, 3)$.

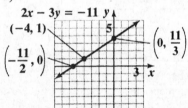

41. The equation for this horizontal line is $y = 3$.

43. The equation for this vertical line is $x = -3$.

45. The equation for this horizontal line, which is the x-axis is $y = 0$.

47. $y = 4$

All ordered pairs that are solutions will have a value of y that is 4. Any value can be used for x. Three ordered pairs that are solutions are $(-2,4)$, $(0,4)$, and $(3,4)$.

Plot these points and draw the line through them. The graph is a horizontal line.

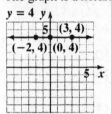

49. $y = -2$

Three ordered pairs are $(-3,-2)$, $(0,-2)$, and $(4,-2)$. The graph is a horizontal line.

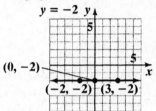

51. $x = 2$

All ordered pairs that are solutions will have a value of x that is 2. Any value can be used for y. Three ordered pairs that are solutions are $(2, -3)$, $(2,0)$, and $(2,2)$.

The graph is a vertical line.

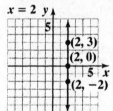

53. $x + 1 = 0$

$x = -1$

Three ordered pairs are $(-1,-3)$, $(-1,0)$, and $(-1,3)$. The graph is a vertical line.

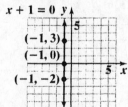

55. $y - 3.5 = 0$

$y = 3.5$

Three ordered pairs are $(-2, 3.5)$, $(0, 3.5)$, and $(3.5, 3.5)$. The graph is a horizontal line.

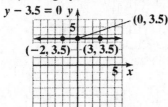

57. $x = 0$

Three ordered pairs are $(0,-2)$, $(0,0)$, and $(0,4)$. The graph is a vertical line, the y-axis.

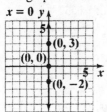

59. $3y = 9$

$y = 3$

Three ordered pairs are $(-3,3)$, $(0,3)$, and $(3,3)$. The graph is a horizontal line.

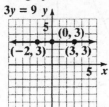

61. $12 - 3x = 0$

$-3x = -12$

$x = 4$

Three ordered pairs are $(4,-2)$, $(4,1)$, and $(4,3)$. The graph is a vertical line.

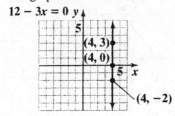

63. Using intercepts, we see that $3x + 2y = -6$ corresponds to Exercise 4.

x-intercept: $3x + 2y = -6$

$3x + 2(0) = -6$

$3x = -6$

$x = -2$

y-intercept: $3x + 2y = -6$

$3(0) + 2y = -6$

$2y = -6$

$y = -3$

65. Since $y = -2$ is a horizontal line at -2, it corresponds to Exercise 7.

67. Using intercepts, we see that $4x + 3y = 12$ corresponds to Exercise 1.

x-intercept: $4x + 3y = 12$

$4x + 3(0) = 12$

$4x = 12$

$x = 3$

y-intercept: $4x + 3y = 12$

$4(0) + 3y = 12$

$3y = 12$

$y = 4$

69. a. Let $x + 5 + 5 = x + 10 =$ the length.
Let $y + 8 =$ width.
Using the formula for the perimeter of a rectangle, we have $2l + 2w = P$

$$2(x+10) + 2(y+8) = 58$$
$$2x + 20 + 2y + 16 = 58$$
$$2x + 2y + 36 = 58$$
$$2x + 2y = 22$$
$$x + y = 11$$

b. x and y must be non-negative because they are dimensions.

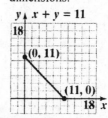

71. The eagle's height is decreasing from 3 seconds to 12 seconds.

73. The *y*-intercept is 45. This means that the eagle's height was 45 meters at the beginning of the observation.

75. Five *x*-intercepts of the graph are 12, 13, 14, 15, and 16. During these times (12-16 minutes), the eagle was on the ground.

77. a.
$$y = -5000x + 45,000$$
$$0 = -5000x + 45,000$$
$$5000x = 45,000$$
$$x = 9$$
After 9 years, the car is worth nothing.

b.
$$y = -5000x + 45,000$$
$$y = -5000(0) + 45,000$$
$$y = 45,000$$
The new car is worth $45,000.

c. *x* and *y* must be non-negative because they represent time and the car's value.

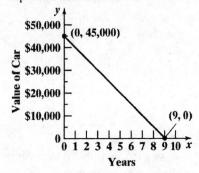

d. According to the graph, the car's value will be about $20,000 after five years. Estimates will vary.

79. – 85. Answers will vary.

87. makes sense

89. makes sense

91. Since the *x*-intercept is 5, *y* = 0 when *x* = 5.
$$\boxed{?}x + \boxed{}y = 10$$
$$\boxed{?}(5) + \boxed{}(0) = 10$$
$$\boxed{?}(5) = 10$$
$$\boxed{?} = 2$$
So, the coefficient of *x* is 2.
Similarly, since the *y*-intercept is 2, *x* = 0 when *y* = 2.
$$2x + \boxed{?}y = 10$$
$$2(0) + \boxed{?}(2) = 10$$
$$\boxed{?}(2) = 10$$
$$\boxed{?} = 5$$
So, the coefficient of *y* is 5.
The equation of the line is $2x + 5y = 10$.

93. Answers will vary.

95.
$$3x - y = 9$$
$$-y = -3x + 9$$
$$(-1)(-y) = -1(3x + 9)$$
$$y = 3x - 9$$

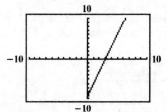

The *y*-intercept is –9.
The *x*-intercept is 3.

97.
$$4x - 2y = -40$$
$$-2y = -4x - 40$$
$$\frac{-2y}{-2} = \frac{-4x - 40}{-2}$$
$$y = 2x + 20$$

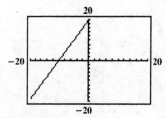

The *y*-intercept is 20.
The *x*-intercept is –10.

98. $|-13.4| = 13.4$

99. $7x - (3x - 5) = 7x - 3x + 5 = 4x + 5$

100. $8(x - 2) - 2(x - 3) \le 8x$

$8x - 16 - 2x + 6 \le 8x$

$6x - 10 \le 8x$

$6x - 8x - 10 \le 8x - 8x$

$-2x - 10 \le 0$

$-2x \le 10$

$\dfrac{-2x}{-2} \ge \dfrac{10}{-2}$

$x \ge -5$

The solution set is $[-5, \infty)$.

101. $\dfrac{y_2 - y_1}{x_2 - x_1} = \dfrac{13 - 3}{6 - 1} = \dfrac{10}{5} = 2$

102. $\dfrac{y_2 - y_1}{x_2 - x_1} = \dfrac{-4 - (-2)}{6 - 4} = \dfrac{-2}{2} = -1$

103. $\dfrac{y_2 - y_1}{x_2 - x_1} = \dfrac{4 - 4}{5 - 3} = \dfrac{0}{2} = 0$

3.3 Check Points

1. a. Let $(x_1, y_1) = (-3, 4)$ and $(x_2, y_2) = (-4, -2)$.

$m = \dfrac{\text{Change in } y}{\text{Change in } x} = \dfrac{y_2 - y_1}{x_2 - x_1} = \dfrac{-2 - 4}{-4 - (-3)} = \dfrac{-6}{-1} = 6$

The slope is 6. Since the slope is positive, the line rises from left to right.

b. Let $(x_1, y_1) = (4, -2)$ and $(x_2, y_2) = (-1, 5)$.

$m = \dfrac{\text{Change in } y}{\text{Change in } x} = \dfrac{y_2 - y_1}{x_2 - x_1} = \dfrac{5 - (-2)}{-1 - 4} = \dfrac{7}{-5} = -\dfrac{7}{5}$

The slope is $-\dfrac{7}{5}$. Since the slope is negative, the line falls from left to right.

2. a. Let $(x_1, y_1) = (6, 5)$ and $(x_2, y_2) = (2, 5)$.

$m = \dfrac{\text{Change in } y}{\text{Change in } x} = \dfrac{y_2 - y_1}{x_2 - x_1} = \dfrac{5 - 5}{2 - 6} = \dfrac{0}{-4} = 0$

Since the slope is 0, the line is horizontal.

b. Let $(x_1, y_1) = (1, 6)$ and $(x_2, y_2) = (1, 4)$.

$m = \dfrac{\text{Change in } y}{\text{Change in } x} = \dfrac{y_2 - y_1}{x_2 - x_1} = \dfrac{4 - 6}{1 - 1} = \dfrac{-2}{0}$

Because division by 0 is undefined the slope is undefined. Since the slope is undefined, the line is vertical.

3. Line through (4,2) and (6,6):

$m = \dfrac{\text{Change in } y}{\text{Change in } x} = \dfrac{6 - 2}{6 - 4} = \dfrac{4}{2} = 2$

Line through (0,−2) and (1,0):

$m = \dfrac{\text{Change in } y}{\text{Change in } x} = \dfrac{0 - (-2)}{1 - 0} = \dfrac{2}{1} = 2$

Since their slopes are equal, the lines are parallel.

4. Line through (−1,4) and (3,2):

$m = \dfrac{\text{Change in } y}{\text{Change in } x} = \dfrac{2 - 4}{3 - (-1)} = \dfrac{-2}{4} = -\dfrac{1}{2}$

Line through (−2,−1) and (2,7):

$m = \dfrac{\text{Change in } y}{\text{Change in } x} = \dfrac{7 - (-1)}{2 - (-2)} = \dfrac{8}{4} = 2$

Since the product of their slopes is $-\dfrac{1}{2}(2) = -1$, the lines are perpendicular.

5. Let

$(x_1, y_1) = (1990, 9.0)$ and $(x_2, y_2) = (2005, 12.7)$.

$m = \dfrac{\text{Change in } y}{\text{Change in } x} = \dfrac{y_2 - y_1}{x_2 - x_1} = \dfrac{12.7 - 9.0}{2005 - 1990} = \dfrac{3.7}{15} \approx 0.25$

The number of men living alone increased at a rate of 0.25 million per year. The rate of change is 0.25 million men per year.

3.3 Exercise Set

1. Let $(x_1, y_1) = (4, 7)$ and $(x_2, y_2) = (8, 10)$.

$m = \dfrac{\text{Change in } y}{\text{Change in } x} = \dfrac{y_2 - y_1}{x_2 - x_1} = \dfrac{10 - 7}{8 - 4} = \dfrac{3}{4}$

Since the slope is positive, the line rises from left to right.

3. Let $(x_1, y_1) = (-2, 1)$ and $(x_2, y_2) = (2, 2)$.

$m = \dfrac{\text{Change in } y}{\text{Change in } x} = \dfrac{y_2 - y_1}{x_2 - x_1} = \dfrac{2 - 1}{2 - (-2)} = \dfrac{1}{4}$

Since the slope is positive, the line rises from left to right.

5. Let $(x_1, y_1) = (4, -2)$ and $(x_2, y_2) = (3, -2)$.

$$m = \frac{\text{Change in } y}{\text{Change in } x} = \frac{y_2 - y_1}{x_2 - x_1} = \frac{-2 - (-2)}{3 - 4} = \frac{0}{-1} = 0$$

Since the slope is zero, the line is horizontal.

7. Let $(x_1, y_1) = (-2, 4)$ and $(x_2, y_2) = (-1, -1)$.

$$m = \frac{\text{Change in } y}{\text{Change in } x} = \frac{y_2 - y_1}{x_2 - x_1} = \frac{-1 - 4}{-1 - (-2)} = \frac{-5}{1} = -5$$

Since the slope is negative, the line falls from left to right.

9. Let $(x_1, y_1) = (5, 3)$ and $(x_2, y_2) = (5, -2)$.

$$m = \frac{\text{Change in } y}{\text{Change in } x} = \frac{y_2 - y_1}{x_2 - x_1} = \frac{-2 - 3}{5 - 5} = \frac{-5}{0}$$

Since the slope is undefined, the line is vertical.

11. Line through $(-2, 2)$ and $(2, 4)$:

$$m = \frac{4 - 2}{2 - (-2)} = \frac{2}{4} = \frac{1}{2}$$

13. Line through $(-3, 4)$ and $(3, 2)$:

$$m = \frac{2 - 4}{3 - (-3)} = \frac{-2}{6} = -\frac{1}{3}$$

15. Line through $(-2, 1)$, $(0, 0)$, and $(2, -1)$

Use any two of these points to find the slope.

$$m = \frac{0 - 1}{0 - (-2)} = \frac{-1}{2} = -\frac{1}{2}$$

17. Line through $(0, 2)$ and $(3, 0)$:

$$m = \frac{0 - 2}{3 - 0} = -\frac{2}{3}$$

19. Line through $(-2, 1)$ and $(4, 1)$:

$$m = \frac{1 - 1}{4 - (-2)} = \frac{0}{6} = 0$$

(Since the line is horizontal, it is not necessary to do this computation. The slope of every horizontal line is 0.)

21. Line through $(-3, 4)$ and $(-3, -2)$:

$$m = \frac{-2 - 4}{-3 - (-3)} = \frac{-6}{0}; \text{ undefined}$$

(Since the line is vertical, it is not necessary to do this computation. The slope of every vertical line is undefined.)

23. Line through $(-2, 0)$ and $(0, 6)$:

$$m = \frac{6 - 0}{0 - (-2)} = 3$$

Line through $(1, 8)$ and $(0, 5)$:

$$m = \frac{5 - 8}{0 - 1} = \frac{-3}{-1} = 3$$

Since their slopes are equal, the lines are parallel.

25. Line through $(0, 3)$ and $(1, 5)$:

$$m = \frac{5 - 3}{1 - 0} = \frac{2}{1} = 2$$

Line through $(-1, 7)$ and $(1, 10)$:

$$m = \frac{10 - 7}{1 - (-1)} = \frac{3}{2}$$

Since their slopes are not equal, the lines are not parallel.

27. Line through $(1, 5)$ and $(0, 3)$:

$$m = \frac{3 - 5}{0 - 1} = 2$$

Line through $(-2, 8)$ and $(2, 6)$:

$$m = \frac{6 - 8}{2 - (-2)} = -\frac{1}{2}$$

Since the product of their slopes is $2\left(-\frac{1}{2}\right) = -1$, the lines are perpendicular.

29. Line through $(-1, -6)$ and $(2, 9)$:

$$m = \frac{9 - (-6)}{2 - (-1)} = 5$$

Line through $(-15, -1)$ and $(5, 3)$:

$$m = \frac{3 - (-1)}{5 - (-15)} = \frac{1}{5}$$

Since the product of their slopes is $5\left(\frac{1}{5}\right) = 1 \neq -1$, the lines are not perpendicular.

31. Line through $(-2, -5)$ and $(3, 10)$:

$$m = \frac{10 - (-5)}{3 - (-2)} = 3$$

Line through $(-1, -9)$ and $(4, 6)$:

$$m = \frac{6 - (-9)}{4 - (-1)} = 3$$

Since their slopes are equal the lines are parallel.

33. Line through $(-4,-12)$ and $(0,-4)$:

$$m = \frac{-4-(-12)}{0-(-4)} = 2$$

Line through $(0,-5)$ and $(2,-4)$:

$$m = \frac{-4-(-5)}{2-0} = \frac{1}{2}$$

Since their slopes are not equal, nor are the slopes negative reciprocals, the lines are neither parallel nor perpendicular.

35. Line through $(-5,-1)$ and $(0,2)$:

$$m = \frac{2-(-1)}{0-(-5)} = \frac{3}{5}$$

Line through $(-6,9)$ and $(3,-6)$:

$$m = \frac{-6-9}{3-(-6)} = -\frac{5}{3}$$

Since the product of their slopes is $\frac{3}{5}\left(-\frac{5}{3}\right) = -1$,

the lines are perpendicular.

37.

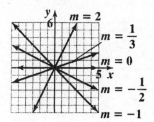

39. $m = \dfrac{y_2 - y_1}{x_2 - x_1} = \dfrac{-3-1}{-3-0} = \dfrac{-4}{-3} = \dfrac{4}{3}$

$m = \dfrac{y_2 - y_1}{x_2 - x_1} = \dfrac{-1-1}{5-0} = \dfrac{-2}{5} = -\dfrac{2}{5}$

$m = \dfrac{y_2 - y_1}{x_2 - x_1} = \dfrac{-5-(-1)}{2-5} = \dfrac{-4}{-3} = \dfrac{4}{3}$

$m = \dfrac{y_2 - y_1}{x_2 - x_1} = \dfrac{-3-(-5)}{-3-2} = \dfrac{2}{-5} = -\dfrac{2}{5}$

Slopes of opposite sides are equal, so the figure is a parallelogram.

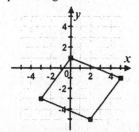

41. First find the slope of the line passing through $(2, 3)$ and $(-2, 1)$.

$$m = \frac{y_2 - y_1}{x_2 - x_1} = \frac{3-1}{2-(-2)} = \frac{2}{4} = \frac{1}{2}$$

Now, use the slope formula, the slope and the points $(5, y)$ and $(1, 0)$ to find y.

$$\frac{1}{2} = \frac{y-0}{5-1}$$

$$\frac{1}{2} = \frac{y}{4}$$

$$4\left(\frac{1}{2}\right) = 4\left(\frac{y}{4}\right)$$

$$2 = y$$

43. Find the slope of the line passing through $(-1, y)$ and $(1, 0)$.

$$m = \frac{y_2 - y_1}{x_2 - x_1} = \frac{0-y}{1-(-1)} = \frac{-y}{2}$$

Find the slope of the line passing through $(2, 3)$ and $(-2, 1)$.

$$m = \frac{y_2 - y_1}{x_2 - x_1} = \frac{1-3}{-2-2} = \frac{1}{2}$$

Since the lines are perpendicular, the product of their slopes is -1..

$$\left(\frac{-y}{2}\right)\left(\frac{1}{2}\right) = -1$$

$$\frac{-y}{4} = -1$$

$$\frac{y}{4} = 1$$

$$y = 4$$

45. a. $m = \dfrac{\text{Change in } y}{\text{Change in } x} = \dfrac{42-26}{180-80} = \dfrac{16}{100} = 0.16$

b. For each minute of brisk walking, the percentage of patients with depression in remission increased by 0.16%. The rate of change is 0.16% per minute of brisk walking.

47. $m = \dfrac{\text{Change in } y}{\text{Change in } x} = \dfrac{6}{18} = \dfrac{1}{3}$

The pitch of the roof is $\dfrac{1}{3}$.

49. The grade of an access ramp is

$$\frac{1 \text{ foot}}{12 \text{ feet}} = \frac{1}{12} \approx 0.083 = 8.3\%.$$

51. – 55. Answers will vary.

57. does not make sense; Explanations will vary.
Sample explanation: Either point can be considered
(x_1, y_1) or (x_2, y_2).

59. makes sense

61. false; Changes to make the statement true will vary.
A sample change is: Slope is rise divided by run.

63. false; Changes to make the statement true will vary.
A sample change is: A line with slope 3 can not be
parallel to a line with slope –3.

65. The positive slopes are m_1 and m_2, and of these,
the line with slope m_1 is the steeper one so has the
larger slope. The negative slopes are m_2 and m_4,
and of these, the line with slope m_4 is the steeper
one, so has the slope with the largest absolute value,
which is the smaller slope. Therefore, in decreasing
size, the slopes are m_1, m_3, m_2 and m_4.

67. $y = 2x + 4$

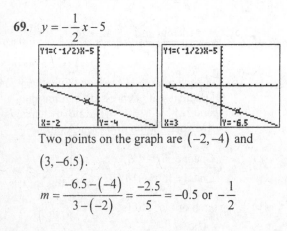

Two points on the graph are $(-2.5, -1)$, and

$(1.5, 7)$.

$$m = \frac{7 - (-1)}{1.5 - (-2.5)} = \frac{8}{4} = 2$$

69. $y = -\dfrac{1}{2}x - 5$

Two points on the graph are $(-2, -4)$ and

$(3, -6.5)$.

$$m = \frac{-6.5 - (-4)}{3 - (-2)} = \frac{-2.5}{5} = -0.5 \text{ or } -\frac{1}{2}$$

71. The slope is always the coefficient of x.

72. Let $x =$ length of shorter piece (in inches).
Let $2x =$ length of longer piece.
$$x + 2x = 36$$
$$3x = 36$$
$$x = 12$$
The pieces are 12 inches and 24 inches.

73.
$$-10 + 16 \div 2(-4) = -10 + 8(-4)$$
$$= -10 - 32$$
$$= -10 + (-32) = -42$$

74. $2x - 3 \le 5$
$$2x \le 8$$
$$x \le 4$$
$$(-\infty, 4]$$

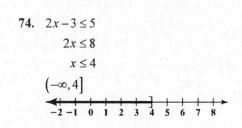

75. $(0 \overset{\text{1 unit right}}{+ 1}, -3 \overset{\text{4 units up}}{+ 4}) = (1, 1)$

76. $(0 \overset{\text{3 units right}}{+ 3}, 1 \overset{\text{2 units down}}{- 2}) = (3, -1)$

77.
$$2x + 5y = 0$$
$$2x - 2x + 5y = 0 - 2x$$
$$5y = -2x$$
$$y = \frac{-2x}{5}$$
$$y = -\frac{2}{5}x$$

3.4 Check Points

1. a. $y = 5x - 3$

The slope is the x-coefficient, which is $m = 5$.
The y-intercept is the constant term, which is –3.

b. $y = \dfrac{2}{3}x + 4$

The slope is the x-coefficient, which is $m = \dfrac{2}{3}$.
The y-intercept is the constant term, which is 4.

c. $7x + y = 6 \rightarrow y = -7x + 6$

The slope is the x-coefficient, which is $m = -7$.
The y-intercept is the constant term, which is 6.

2. $y = 3x - 2$

The y-intercept is –2, so plot the point $(0, -2)$.

The slope is $m = 3$ or $m = \frac{3}{1}$. Find another point

by going up 3 units and to the right 1 unit.
Use a straightedge to draw a line through the two
points.

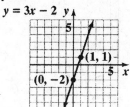

3. $y = \frac{3}{5}x + 1$

The y-intercept is 1, so plot the point $(0, 1)$.

The slope is $m = \frac{3}{5}$. Find another point by going up

3 units and to the right 5 units.
Use a straightedge to draw a line through the two
points.

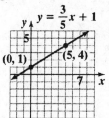

4. $3x + 4y = 0$

$$4y = -3x$$

$$y = -\frac{3}{4}x$$

The y-intercept is 0, so plot the point $(0, 0)$.

The slope is $m = \frac{-3}{4}$. Find another point by going

down 3 units and to the right 4 units.
Use a straightedge to draw a line through the two
points.

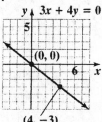

5. a. The y-intercept is 3.9 and the slope is
$$m = \frac{\text{Change in } y}{\text{Change in } x} = \frac{7.7 - 3.9}{30 - 0} = \frac{3.8}{30} \approx 0.13$$
The equation is $y = 0.13x + 3.9$.

b. $y = 0.13x + 3.9 = 0.13(40) + 3.9 = 9.1$
The model projects that 9.1% of Hispanics will
be divorced in 2010.

3.4 Exercise Set

1. $y = 3x + 2$

The slope is the x-coefficient, which is 3. The y-intercept is the constant term, which is 2.

3. $y = 3x - 5$
$y = 3x + (-5)$
$m = 3$; $y - \text{intercept} = -5$

5. $y = -\frac{1}{2}x + 5$

$m = -\frac{1}{2}$; $y\text{-intercept} = 5$

7. $y = 7x$
$y = 7x + 0$
$m = 7$; $y\text{-intercept} = 0$

9. $y = 10$
$y = 0x + 10$
$m = 0$; $y\text{-intercept} = 10$

11. $y = 4 - x$
$y = -x + 4 = -1x + 4$
$m = -1$; $y\text{-intercept} = 4$

13. $-5x + y = 7$
$-5x + y + 5x = 5x + 7$
$y = 5x + 7$
$m = 5$; $y\text{-intercept} = 7$

15. $x + y = 6$
$y = -x + 6 = -1x + 6$
$m = -1$; $y\text{-intercept} = 6$

17. $6x + y = 0$

$$y = -6x = -6x + 0$$

$m = -6$; y-intercept $= 0$

19. $3y = 6x$

$$y = 2x$$

$m = 2$; y-intercept $= 0$

21. $2x + 7y = 0$

$$7y = -2x$$

$$y = -\frac{2}{7}x$$

$m = -\frac{2}{7}$; y-intercept $= 0$

23. $3x + 2y = 3$

$$2y = -3x + 3$$

$$y = -\frac{3}{2}x + \frac{3}{2}$$

$m = -\frac{3}{2}$; y-intercept $= \frac{3}{2}$

25. $3x - 4y = 12$

$$-4y = -3x + 12$$

$$y = \frac{3}{4}x - 3$$

$m = \frac{3}{4}$; y-intercept $= -3$

27. $y = 2x + 4$

Step 1. Plot (0,4) on the y-axis.

Step 2. $m = \frac{2}{1} = \frac{\text{rise}}{\text{run}}$

Start at (0,4). Using the slope, move 2 units *up* (the rise) and 1 unit to the *right* (the run) to reach the point (1,6).

Step 3. Draw a line through (0,4) and (1,6).

$y = 2x + 4$

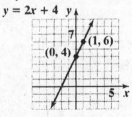

29. $y = -3x + 5$

Slope $= -3 = \frac{-3}{1}$; y-intercept $= 5$.

Plot (0,5) on the y-axis. From this point, move 3 units *down* (because -3 is negative) and 1 unit to the *right* to reach the point (1,2). Draw a line through (0,5) and (1,2).

$y = -3x + 5$

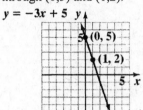

31. $y = \frac{1}{2}x + 1$

Slope $= \frac{1}{2}$; y-intercept $= 1$

Plot (0,1). From this point, move 1 unit *up* and 2 units to the *right* to reach the point (2,2). Draw a line through (0,1) and (2,2).

$y = \frac{1}{2}x + 1$

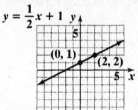

33. $y = \frac{2}{3}x - 5$

Slope $= \frac{2}{3}$; y-intercept $= -5$

Plot (0,−5). From this point move 2 units *up* and 3 units to the *right* to reach the point (3,−3). Draw a line through (0,−5) and (3,−3).

$y = \frac{2}{3}x - 5$

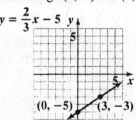

35.　$y = -\dfrac{3}{4}x + 2$

Slope $= -\dfrac{3}{4} = \dfrac{-3}{4}$; y-intercept $= 2$

Plot $(0,2)$. From this point move 3 units *down* and 4 units to the *right* to reach the point $(4,-1)$.
Draw a line through $(0,2)$ and $(4,-1)$

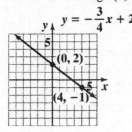

37.　$y = -\dfrac{5}{3}x$

Slope $= -\dfrac{5}{3} = \dfrac{-5}{3}$; y-intercept $= 0$

Plot $(0,0)$. From this point, move 5 units *down* and 3 units to the *right* to reach the point $(3,-5)$. Draw a line through $(0,0)$ and $(3,-5)$.

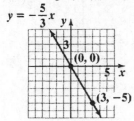

39.　a.　$3x + y = 0$

　　　　　$y = -3x$

b.　$m = -3$; y-intercept $= 0$

c.　Plot $(0,0)$. Since $m = -3 = -\dfrac{3}{1}$, move 3 units *down* and 1 units to the *right* to reach the point $(1,-3)$. Draw a line through $(0,0)$ and $(1,-3)$.

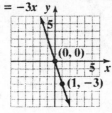

41.　a.　$3y = 4x$

　　　　　$y = \dfrac{4}{3}x$

b.　$m = \dfrac{4}{3}$; y-intercept $= 0$

c.　Plot $(0,0)$. Move 4 units *up* and 3 units to the *right* to reach the point $(3,4)$.
　　Draw a line through $(0,0)$ and $(3,4)$.

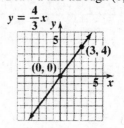

43.　a.　$2x + y = 3$

　　　　　$y = -2x + 3$

b.　$m = -2$; y-intercept $= 3$

c.　Plot $(0,3)$. Since $m = -2 = -\dfrac{2}{1}$, move 2 units *down* and 1 units to the *right* to reach the point $(1,1)$.
Draw a line through $(0,3)$ and $(1,1)$.

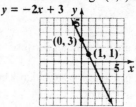

45.　a.　$7x + 2y = 14$

　　　　　$2y = -7x + 14$

　　　　　$\dfrac{2y}{2} = \dfrac{-7x + 14}{2}$

　　　　　$y = -\dfrac{7}{2}x + 7$

b.　$m = -\dfrac{7}{2}$; y-intercept $= 7$

c. Plot (0,7). Since $m = -\dfrac{7}{2} = -\dfrac{7}{2}$, move 7 units *down* and 2 units to the *right* to reach the point (2,0).

Draw a line through (0,7) and (2,0).

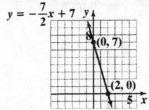

47. $y = 3x + 1$:

$m = 3$; y-intercept $= 1$

$y = 3x - 3$:

$m = 3$; y-intercept $= -3$

The lines are parallel because their slopes are equal.

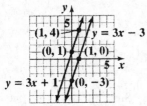

49. $y = -3x + 2$:

$m = -3$; y-intercept $= 2$

$y = 3x + 2$:

$m = 3$; y-intercept $= 2$

The lines are not parallel because their slopes are not equal.

The lines are not perpendicular because the product of their slopes is not -1.

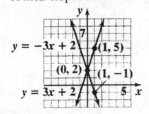

51. $y = x + 3$

$m = 1$; y-intercept $= 3$

$y = -x + 1$

$m = -1$; y-intercept $= 1$

The lines are perpendicular because the product of their slopes is -1.

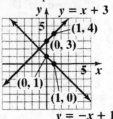

53. $x - 2y = 2 \;\rightarrow\; y = \dfrac{1}{2}x - 1$

$2x - 4y = 3 \;\rightarrow\; y = \dfrac{1}{2}x - \dfrac{3}{4}$

The lines are parallel because their slopes are equal.

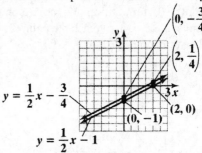

55. $2x - y = -1 \;\rightarrow\; y = 2x + 1$

$x + 2y = -6 \;\rightarrow\; y = -\dfrac{1}{2}x - 3$

The lines are perpendicular because the product of their slopes is $2\left(-\dfrac{1}{2}\right) = -1$.

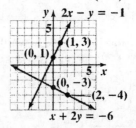

57. Find the slope of the parallel line. $3x + y = 6$

$$y = -3x + 6$$

The slope is -3. We are given that the y-intercept is 5, so using slope-intercept form, we have $y = -3x + 5$.

59. The slope of the line $y = 5x - 1$ is 5. The negative reciprocal of 5 is $-\dfrac{1}{5}$.

We are given that the y-intercept is 6, so using the slope-intercept form, we have $y = -\dfrac{1}{5}x + 6$.

61. Find the y-intercept of the line.
$$16y = 8x + 32$$
$$\frac{16}{16}y = \frac{8}{16}x + \frac{32}{16}$$
$$y = \frac{1}{2}x + 2$$
The y-intercept is 2.
Find the slope of the parallel line.
$$3x + 3y = 9$$
$$3y = -3x + 9$$
$$y = -x + 3$$
The slope is -1. Using slope-intercept form, we have $y = -x + 2$.

63. If the line rises from left to right, it has a positive slope. It passes through the origin, (0, 0) and a second point with equal x- and y-coordinates. The point (2, 2) is one example.
Use the two points to find the slope.
$$m = \frac{0-2}{0-2} = \frac{-2}{-2} = 1$$
The slope is 1. The y-intercept is 0. Using slope-intercept form, we have $y = 1x + 0$ or $y = x$.

65. a. The y-intercept is 68 and the slope is
$$m = \frac{\text{Change in } y}{\text{Change in } x} = \frac{51-68}{47-0} = \frac{-17}{47} \approx -0.36$$
The equation is $y = -0.36x + 68$.

b. $y = -0.36x + 68 = -0.36(100) + 68 = 32$
The model projects that 32% of the U.S. population will be white in 2103.

67. – 69. Answers will vary.

71. does not make sense; Explanations will vary.
Sample explanation: The slope can be determined if the equation is solved for y.

73. does not make sense; Explanations will vary.
Sample explanation: Under these circumstances, you can either find a different point to plot or change the scale on your graph paper.

75. false; Changes to make the statement true will vary. A sample change is: Vertical lines cannot be expressed in slope-intercept form.

77. false; Changes to make the statement true will vary. A sample change is: By solving the equation for y, you can determine that the y-intercept is $\dfrac{7}{2}$.

79. $\dfrac{x}{2} + 7 = 13 - \dfrac{x}{4}$
Multiply by the LCD, which is 4.
$$4\left(\frac{x}{2} + 7\right) = 4\left(13 - \frac{x}{4}\right)$$
$$2x + 28 = 52 - x$$
$$3x + 28 = 52$$
$$3x = 24$$
$$x = 8$$
The solution is {8}.

80. $3\left(12 \div 2^2 - 3\right)^2$
$$= 3\left(12 \div 4 - 3\right)^2$$
$$= 3\left(3 - 3\right)^2$$
$$= 3 \cdot 0^2 = 3 \cdot 0 = 0$$

81. $A = 14, P = 25\% = 0.25$
$$A = PB$$
$$14 = 0.25 \cdot B$$
$$\frac{14}{0.25} = \frac{0.25B}{0.25}$$
$$56 = B$$
14 is 25% of 56.

82. $y - 3 = 4(x + 1)$
$$y - 3 = 4x + 4$$
$$y = 4x + 7$$

83. $y + 3 = -\dfrac{3}{2}(x - 4)$
$$y + 3 = -\frac{3}{2}x + 6$$
$$y = -\frac{3}{2}x + 3$$

84. $y - 30.0 = 0.265(x - 10)$
$$y - 30.0 = 0.265x - 2.65$$
$$y = 0.265x + 27.35$$

Chapter 3 Mid-Chapter Check Point

1. **a.** The *x*-intercept is 4.

    b. The *y*-intercept is 2.

    c. The points (4, 0) and (0, 2) lie on the line.
    $$m = \frac{2-0}{0-4} = \frac{2}{-4} = -\frac{1}{2}$$

2. **a.** The *x*-intercept is −5.

    b. There is no *y*-intercept.

    c. It is a vertical line, so the slope is undefined.

3. **a.** The *x*-intercept is 0.

    b. The *y*-intercept is 0.

    c. The points (0, 0) and (5, 3) lie on the line.
    $$m = \frac{3-0}{5-0} = \frac{3}{5}$$

4. $y = -2x$

    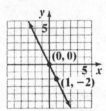

5. $y = -2$

    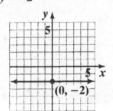

6. $x + y = -2$
    $$y = -x - 2$$

    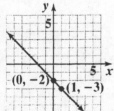

7. $y = \frac{1}{3}x - 2$

    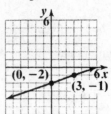

8. $x = 3.5$

    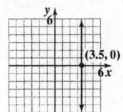

9. $4x - 2y = 8$
    $$-2y = -4x + 8$$
    $$\frac{-2y}{-2} = \frac{-4x}{-2} + \frac{8}{-2}$$
    $$y = 2x - 4$$

    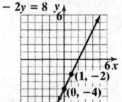

10. $y = 3x + 2$

    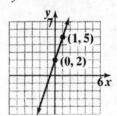

11. $3x + y = 0$
    $$y = -3x$$

    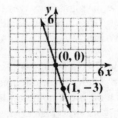

12. $y = -x + 4$

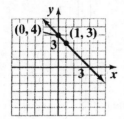

13. $y = x - 4$

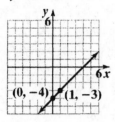

14. $5y = -3x$

$$y = -\frac{3}{5}x$$

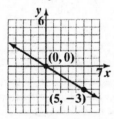

15. $5y = 20$

$$y = 4$$

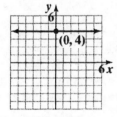

16. $5x - 2y = 10$

$$-2y = -5x + 10$$

$$y = \frac{5}{2}x - 5$$

The slope is $\dfrac{5}{2}$ and the y-intercept is -5.

17. Line through $(-5,-3)$ and $(0,-4)$:

$$m = \frac{\text{Change in } y}{\text{Change in } x} = \frac{-4-(-3)}{0-(-5)} = \frac{-1}{5} = -\frac{1}{5}$$

Line through $(-2,-8)$ and $(1,7)$:

$$m = \frac{\text{Change in } y}{\text{Change in } x} = \frac{7-(-8)}{1-(-2)} = \frac{15}{3} = 5$$

Since the product of their slopes is $-\dfrac{1}{5}(5) = -1$, the lines are perpendicular.

18. Line through $(-4,1)$ and $(2,7)$:

$$m = \frac{\text{Change in } y}{\text{Change in } x} = \frac{7-1}{2-(-4)} = \frac{6}{6} = 1$$

Line through $(-5,13)$ and $(4,-5)$:

$$m = \frac{\text{Change in } y}{\text{Change in } x} = \frac{-5-13}{4-(-5)} = \frac{-18}{9} = -2$$

Since their slopes are not equal, the lines are not parallel.
Since the product of their slopes is
$1(-2) = -2 \neq -1$, the lines are not perpendicular.

19. Line through $(2,-4)$ and $(7,0)$:

$$m = \frac{\text{Change in } y}{\text{Change in } x} = \frac{0-(-4)}{7-2} = \frac{4}{5}$$

Line through $(-4,2)$ and $(1,6)$:

$$m = \frac{\text{Change in } y}{\text{Change in } x} = \frac{6-2}{1-(-4)} = \frac{4}{5}$$

Since their slopes are equal, the lines are parallel.

20. a. The y-intercept is 1098 and the slope is

$$m = \frac{\text{Change in } y}{\text{Change in } x} = \frac{2277-1098}{9-0} = \frac{1179}{9} = 131$$

The equation is $y = 131x + 1098$.

b. The average annual household gasoline bill increased at an average of $131 per year.

c. 2013 is 16 years after 1997.
$y = 131x + 1098 = 131(16) + 1098 = 3194$
The model projects that in 2013 the average annual household gasoline bill will be $3194.

3.5 Check Points

1. Begin with the point-slope equation of a line.
$$y - y_1 = m(x - x_1)$$
$$y - (-5) = 6(x - 2)$$
$$y + 5 = 6(x - 2)$$
Now solve this equation for y to write the equation in slope-intercept form.
$$y + 5 = 6(x - 2)$$
$$y + 5 = 6x - 12$$
$$y = 6x - 17$$

2. **a.** Begin by finding the slope:
$$m = \frac{-6 - (-1)}{-1 - (-2)} = \frac{-5}{1} = -5$$
Using the slope and either point, find the point-slope equation of a line.

$$y - y_1 = m(x - x_1) \qquad \text{or} \qquad y - y_1 = m(x - x_1)$$
$$y - (-1) = -5(x - (-2)) \qquad\qquad y - (-6) = -5(x - (-1))$$
$$y + 1 = -5(x + 2) \qquad\qquad\qquad y + 6 = -5(x + 1)$$

   b. Solve the above equation for y:

$$y + 1 = -5(x + 2) \quad \text{or} \quad y + 6 = -5(x + 1)$$
$$y + 1 = -5x - 10 \qquad\qquad y + 6 = -5x - 5$$
$$y = -5x - 11 \qquad\qquad\quad y = -5x - 11$$

3. Since the line is parallel to $y = 3x + 1$, we know it will have slope $m = 3$. We are given that it passes through $(-2, 5)$. We use the slope and point to write the equation in point-slope form.
$$y - y_1 = m(x - x_1)$$
$$y - 5 = 3(x - (-2))$$
$$y - 5 = 3(x + 2)$$
Solve for y to obtain slope-intercept form.
$$y - 5 = 3(x + 2)$$
$$y - 5 = 3x + 6$$
$$y = 3x + 11$$

4. **a.** Solve the given equation for y to obtain slope-intercept form.
$$x + 3y = 12$$
$$3y = -x + 12$$
$$y = -\frac{1}{3}x + 4$$

   Since the slope of the given line is $-\frac{1}{3}$, the slope of any line perpendicular to the given line is 3.

   b. We use the slope of 3 and the point $(-2, -6)$ to write the equation in point-slope form.
$$y - y_1 = m(x - x_1)$$
$$y - (-6) = 3(x - (-2))$$
$$y + 6 = 3(x + 2)$$
Solve for y to obtain slope-intercept form.
$$y + 6 = 3(x + 2)$$
$$y + 6 = 3x + 6$$
$$y = 3x$$

5. Find slope: $m = \dfrac{32.8 - 30.0}{20 - 10} = \dfrac{2.8}{10} = 0.28$
Use the point-slope form to write the equation. Then solve for y to obtain slope-intercept form.
$$y - y_1 = m(x - x_1)$$
$$y - 30.0 = 0.28(x - 10)$$
$$y - 30.0 = 0.28x - 2.8$$
$$y = 0.28x + 27.2$$
Because 2020 is 50 years after 1970, substitute 50 for x and compute y.
$$y = 0.28x + 27.2 = 0.28(50) + 27.2 = 14 + 27.2 = 41.2$$

The model predicts that 41.2 will be the median age in 2020.

3.5 Exercise Set

1. Begin with the point-slope equation of a line.
$$y - y_1 = m(x - x_1)$$
$$y - 5 = 3(x - 2)$$
Now solve this equation for y to write the equation in slope-intercept form.
$$y - 5 = 3x - 6$$
$$y = 3x - 1$$

3. Begin with the point-slope equation of a line.
$$y - y_1 = m(x - x_1)$$
$$y - 6 = 5(x - (-2))$$
$$y - 6 = 5(x + 2)$$
Now solve this equation for y to write the equation in slope-intercept form.
$$y - 6 = 5(x + 2)$$
$$y - 6 = 5x + 10$$
$$y = 5x + 16$$

5. Begin with the point-slope equation of a line.
$$y - y_1 = m(x - x_1)$$
$$y - (-2) = -8(x - (-3))$$
$$y + 2 = -8(x + 3)$$
Now solve this equation for y to write the equation in slope-intercept form.
$$y + 2 = -8(x + 3)$$
$$y + 2 = -8x - 24$$
$$y = -8x - 26$$

7. Begin with the point-slope equation of a line.
$$y - y_1 = m(x - x_1)$$
$$y - 0 = -12(x - (-8))$$
$$y = -12(x + 8)$$
$$y = -12x - 96$$
Now solve this equation for y to write the equation in slope-intercept form.
$$y = -12(x + 8)$$
$$y = -12x - 96$$

9. Begin with the point-slope equation of a line.
$$y - y_1 = m(x - x_1)$$
$$y - (-2) = -1\left(x - \left(-\frac{1}{2}\right)\right)$$
$$y + 2 = -1\left(x + \frac{1}{2}\right)$$
Now solve this equation for y to write the equation in slope-intercept form.
$$y + 2 = -1\left(x + \frac{1}{2}\right)$$
$$y + 2 = -x - \frac{1}{2}$$
$$y = -x - \frac{5}{2}$$

11. Begin with the point-slope equation of a line.
$$y - y_1 = m(x - x_1)$$
$$y - 0 = \frac{1}{2}(x - 0)$$
Now solve this equation for y to write the equation in slope-intercept form.
$$y - 0 = \frac{1}{2}(x - 0)$$
$$y = \frac{1}{2}x$$

13. Begin with the point-slope equation of a line.
$$y - y_1 = m(x - x_1)$$
$$y - (-2) = -\frac{2}{3}(x - 6)$$
$$y + 2 = -\frac{2}{3}(x - 6)$$
Now solve this equation for y to write the equation in slope-intercept form.
$$y + 2 = -\frac{2}{3}(x - 6)$$
$$y + 2 = -\frac{2}{3}x + 4$$
$$y = -\frac{2}{3}x + 2$$

15. slope $= \dfrac{10 - 2}{5 - 1} = \dfrac{8}{4} = 2$
Using the slope and either point, find the point-slope equation of a line.
$$y - y_1 = m(x - x_1)$$
$$y - 2 = 2(x - 1) \text{ or } y - 10 = 2(x - 5)$$
Now solve this equation for y to write the equation in slope-intercept form.
$$y - 2 = 2(x - 1)$$
$$y - 2 = 2x - 2$$
$$y = 2x$$

17. slope $= \dfrac{3 - 0}{0 + 3} = \dfrac{3}{3} = 1$
Using the slope and either point, find the point-slope equation of a line.
$$y - y_1 = m(x - x_1)$$
$$y - 0 = 1(x + 3) \text{ or } y - 3 = 1(x - 0)$$
Now solve this equation for y to write the equation in slope-intercept form.
$$y - 0 = 1(x + 3)$$
$$y = x + 3$$

19. slope $= \dfrac{4+1}{2+3} = \dfrac{5}{5} = 1$

Using the slope and either point, find the point-slope equation of a line.

$$y - y_1 = m(x - x_1)$$
$$y - (-1) = 1(x - (-3))$$
$$y + 1 = 1(x + 3) \text{ or } y - 4 = 1(x - 2)$$

Now solve this equation for y to write the equation in slope-intercept form.

$$y + 1 = 1(x + 3)$$
$$y + 1 = x + 3$$
$$y = x + 2$$

21. slope $= \dfrac{4 - (-1)}{3 - (-4)} = \dfrac{5}{7}$

Using the slope and either point, find the point-slope equation of a line.

$$y - y_1 = m(x - x_1)$$
$$y - 4 = \dfrac{5}{7}(x - 3) \text{ or } y + 1 = \dfrac{5}{7}(x + 4)$$

Now solve this equation for y to write the equation in slope-intercept form.

$$y - 4 = \dfrac{5}{7}(x - 3)$$
$$y - 4 = \dfrac{5}{7}x - \dfrac{15}{7}$$
$$y = \dfrac{5}{7}x + \dfrac{13}{7}$$

23. slope $= \dfrac{-1 + 1}{4 + 3} = \dfrac{0}{7} = 0$

Using the slope and either point, find the point-slope equation of a line.

$$y - y_1 = m(x - x_1)$$
$$y - (-1) = 0(x - (-3))$$
$$y + 1 = 0(x + 3) \text{ or } y + 1 = 0(x - 4)$$

Now solve this equation for y to write the equation in slope-intercept form.

$$y + 1 = 0(x + 3)$$
$$y + 1 = 0$$
$$y = -1$$

25. Use the points $(2,4)$ and $(-2,0)$ to find the slope.

slope $= \dfrac{0 + 4}{-2 - 2} = \dfrac{-4}{-4} = 1$

Find the point-slope equation of a line.

$$y - y_1 = m(x - x_1)$$
$$y - 4 = 1(x - 2)$$

Now solve this equation for y to write the equation in slope-intercept form.

$$y - 4 = 1(x - 2)$$
$$y - 4 = x - 2$$
$$y = x + 2$$

27. Use the points $\left(-\dfrac{1}{2}, 0\right)$ and $(0,4)$ to find the slope.

slope $= \dfrac{4 - 0}{0 + \dfrac{1}{2}} = \dfrac{4}{\dfrac{1}{2}} = 8$

Find the point-slope equation of a line.

$$y - y_1 = m(x - x_1)$$
$$y - 0 = 8\left(x - \left(-\dfrac{1}{2}\right)\right)$$
$$y - 0 = 8\left(x + \dfrac{1}{2}\right) \text{ or } y - 4 = 8(x - 0)$$

Now solve this equation for y to write the equation in slope-intercept form.

$$y - 4 = 8(x - 0)$$
$$y - 4 = 8x$$
$$y = 8x + 4$$

29. For $y = 5x$, $m = 5$.

a. A line parallel to this line would have the same slope, $m = 5$.

b. A line perpendicular to it would have slope $m = -\dfrac{1}{5}$.

31. For $y = -7x$, $m = -7$.

a. A line parallel to this line would have the same slope, $m = -7$.

b. A line perpendicular to it would have slope $m = \dfrac{1}{7}$.

33. For $y = \dfrac{1}{2}x + 3$, $m = \dfrac{1}{2}$.

    a. A line parallel to this line would have the same slope, $m = \dfrac{1}{2}$.

    b. A line perpendicular to it would have slope $m = -2$.

35. For $y = -\dfrac{2}{5}x - 1$, $m = -\dfrac{2}{5}$.

    a. A line parallel to this line would have the same slope, $m = -\dfrac{2}{5}$.

    b. A line perpendicular to it would have slope $m = \dfrac{5}{2}$.

37. To find the slope, we rewrite the equation in slope-intercept form.
$$4x + y = 7$$
$$y = -4x + 7$$
So, $m = -4$.

    a. A line parallel to this line would have the same slope, $m = -4$.

    b. A line perpendicular to it would have slope $m = \dfrac{1}{4}$.

39. To find the slope, we rewrite the equation in slope-intercept form.
$$2x + 4y = 8$$
$$4y = -2x + 8$$
$$y = -\dfrac{1}{2}x + 2$$
So, $m = -\dfrac{1}{2}$.

    a. A line parallel to this line would have the same slope, $m = -\dfrac{1}{2}$.

    b. A line perpendicular to it would have slope $m = 2$.

41. To find the slope, we rewrite the equation in slope-intercept form.
$$2x - 3y = 5$$
$$-3y = -2x + 5$$
$$y = \dfrac{2}{3}x - \dfrac{5}{3}$$
So, $m = \dfrac{2}{3}$.

    a. A line parallel to this line would have the same slope, $m = \dfrac{2}{3}$.

    b. A line perpendicular to it would have slope $m = -\dfrac{3}{2}$.

43. We know that $x = 6$ is a vertical line with undefined slope.

    a. A line parallel to it would also be vertical with undefined slope.

    b. A line perpendicular to it would be horizontal with slope $m = 0$.

45. Since L is parallel to $y = 2x$, we know it will have slope $m = 2$. We are given that it passes through $(4, 2)$. We use the slope and point to write the equation in point-slope form.
$$y - y_1 = m(x - x_1)$$
$$y - 2 = 2(x - 4)$$
Solve for y to obtain slope-intercept form.
$$y - 2 = 2(x - 4)$$
$$y - 2 = 2x - 8$$
$$y = 2x - 6$$

47. Since L is perpendicular to $y = 2x$, we know it will have slope $m = -\dfrac{1}{2}$. We are given that it passes through

(2, 4). We use the slope and point to write the equation in point-slope form.

$$y - y_1 = m(x - x_1)$$

$$y - 4 = -\frac{1}{2}(x - 2)$$

Solve for y to obtain slope-intercept form.

$$y - 4 = -\frac{1}{2}(x - 2)$$

$$y - 4 = -\frac{1}{2}x + 1$$

$$y = -\frac{1}{2}x + 5$$

49. Since the line is parallel to $y = -4x + 3$, we know it will have slope $m = -4$. We are given that it passes through

(−8, −10). We use the slope and point to write the equation in point-slope form.

$$y - y_1 = m(x - x_1)$$

$$y - (-10) = -4(x - (-8))$$

$$y + 10 = -4(x + 8)$$

Solve for y to obtain slope-intercept form.

$$y + 10 = -4(x + 8)$$

$$y + 10 = -4x - 32$$

$$y = -4x - 42$$

51. Since the line is perpendicular to $y = \dfrac{1}{5}x + 6$, we know it will have slope $m = -5$. We are given that it passes through (2, −3). We use the slope and point to write the equation in point-slope form.

$$y - y_1 = m(x - x_1)$$

$$y - (-3) = -5(x - 2)$$

$$y + 3 = -5(x - 2)$$

Solve for y to obtain slope-intercept form.

$$y + 3 = -5(x - 2)$$

$$y + 3 = -5x + 10$$

$$y = -5x + 7$$

53. To find the slope, we rewrite the equation in slope-intercept form.

$$2x - 3y = 7$$

$$-3y = -2x + 7$$

$$y = \frac{2}{3}x - \frac{7}{3}$$

Since the line is parallel to $y = \dfrac{2}{3}x - \dfrac{7}{3}$, we know it

will have slope $m = \dfrac{2}{3}$. We are given that it passes

through (−2, 2). We use the slope and point to write the equation in point-slope form.

$$y - y_1 = m(x - x_1)$$

$$y - 2 = \frac{2}{3}(x - (-2))$$

$$y - 2 = \frac{2}{3}(x + 2)$$

Solve for y to obtain slope-intercept form.

$$y - 2 = \frac{2}{3}(x + 2)$$

$$y - 2 = \frac{2}{3}x + \frac{4}{3}$$

$$y = \frac{2}{3}x + \frac{10}{3}$$

55. To find the slope, we rewrite the equation in slope-intercept form.

$$x - 2y = 3$$

$$-2y = -x + 3$$

$$y = \frac{1}{2}x - \frac{3}{2}$$

Since the line is perpendicular to $y = \dfrac{1}{2}x - \dfrac{3}{2}$, we

know it will have slope $m = -2$. We are given that it passes through (4, −7). We use the slope and point to write the equation in point-slope form.

$$y - y_1 = m(x - x_1)$$

$$y - (-7) = -2(x - 4)$$

$$y + 7 = -2(x - 4)$$

Solve for y to obtain slope-intercept form.

$$y + 7 = -2(x - 4)$$

$$y + 7 = -2x + 8$$

$$y = -2x + 1$$

57. Through $(2, 4)$ and same y-intercept as $x - 4y = 8$.
Solve the equation to obtain the y-intercept.
$$x - 4y = 8$$
$$-4y = -x + 8$$
$$y = \frac{1}{4}x - 2$$
Now, use the two points to find the slope.
$$m = \frac{4 - (-2)}{2 - 0} = \frac{6}{2} = 3$$
Now use the slope and one of the points to find the equation of the line.
$$y - 4 = 3(x - 2)$$
$$y - 4 = 3x - 6$$
$$y = 3x - 2$$

59. x-intercept at -4 and parallel to the line containing $(3, 1)$ and $(2, 6)$
First, find the slope of the line going through the points $(3, 1)$ and $(2, 6)$.
$$m = \frac{6 - 1}{2 - 3} = \frac{5}{-1} = -5$$
The slope of the line is -5. Since this line is parallel to the line we are writing the equation for, its slope is also -5. Since the x-intercept is -4, the line goes through the point $(-4, 0)$. Use the point and the slope to find the equation of the line.
$$y - 0 = -5(x - (-4))$$
$$y = -5(x + 4)$$
$$y = -5x - 20$$

61. Since the line is perpendicular to $x = 6$ which is a vertical line, we know the graph is a horizontal line with 0 slope. The graph passes through $(-1, 5)$, so the equation is $y = 5$.

63. First we need to find the equation of the line with x – intercept of 2 and y – intercept of -4. This line will pass through $(2, 0)$ and $(0, -4)$. We use these points to find the slope.
$$m = \frac{-4 - 0}{0 - 2} = \frac{-4}{-2} = 2$$
Since the graph is perpendicular to this line, it will have slope $m = -\frac{1}{2}$.

Use the point $(-6, 4)$ and the slope $-\frac{1}{2}$ to find the equation of the line.
$$y - y_1 = m(x - x_1)$$
$$y - 4 = -\frac{1}{2}(x - (-6))$$
$$y - 4 = -\frac{1}{2}(x + 6)$$
$$y - 4 = -\frac{1}{2}x - 3$$
$$y = -\frac{1}{2}x + 1$$

65. First put the equation $3x - 2y = 4$ in slope-intercept form.
$$3x - 2y = 4$$
$$-2y = -3x + 4$$
$$y = \frac{3}{2}x - 2$$
The equation will have slope $-\frac{2}{3}$ since it is perpendicular to the line above and has the same y – intercept, -2.
So the equation is $y = -\frac{2}{3}x - 2$.

67. To find the slope of the line whose equation is $Ax + By = C$, put this equation in slope-intercept form by solving for y.
$$Ax + By = C$$
$$By = -Ax + C$$
$$y = -\frac{A}{B}x + \frac{C}{B}$$
The slope of this line is $m = -\frac{A}{B}$ so the slope of the line that is parallel to it is the same, $-\frac{A}{B}$.

69. a. Find slope: $m = \frac{71.4 - 64.6}{15 - 5} = \frac{6.8}{10} = 0.68$
$$y - y_1 = m(x - x_1)$$
$$y - 64.6 = 0.68(x - 5)$$
$$y - 64.6 = 0.68x - 3.4$$
$$y = 0.68x + 61.2$$

b. $y = 0.68x + 61.2 = 0.68(35) + 61.2 = 85$
According to the model, women will earn 85% of men in 2015.

71. a. Scatter plot:

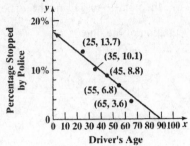

b. Find slope: $m = \dfrac{6.8 - 8.8}{55 - 45} = \dfrac{-2}{10} = -0.2$

$$y - y_1 = m(x - x_1)$$
$$y - 8.8 = -0.2(x - 45)$$
$$y - 8.8 = -0.2x + 9$$
$$y = -0.2x + 17.8$$

c. $y = -0.2x + 17.8 = -0.2(35) + 17.8 = 10.8$

According to the model, 10.8% of 35-year-old drivers are stopped by police.
This overestimates the actual percentage shown in the bar graph by 0.7%.

73. – 75. Answers will vary.

77. does not make sense; Explanations will vary.
Sample explanation: Typically, we go from point-slope form to slope-intercept form. It is not common to go from slope-intercept form to point-slope form.

79. makes sense

81. false; Changes to make the statement true will vary. A sample change is: This line passes through $(-3,-4)$.

83. false; Changes to make the statement true will vary. A sample change is: Solving this line for y shows that its slope is -3.

85. Answers will vary.

87. Let $x =$ the number of sheets of paper.
$$4 + 2x \le 29$$
$$2x \le 25$$
$$x \le \frac{25}{2} \text{ or } 12\frac{1}{2}$$

Since the number of sheets of paper must be a whole number, at most 12 sheets of paper can be put in the envelope.

88. The only natural numbers in the given set are 1 and $\sqrt{4}\,(=2)$

89. $3x - 5y = 15$
x-intercept:
$$3x - 5(0) = 15$$
$$3x = 15$$
$$x = 5$$
y-intercept:
$$3(0) - 5y = 15$$
$$-5y = 15$$
$$y = -3$$

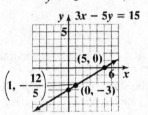

90. $x + 2y = 2$
$$4 + 2(-1) = 2$$
$$4 - 2 = 2$$
$$2 = 2, \text{ true}$$

$$x - 2y = 6$$
$$4 - 2(-1) = 6$$
$$4 + 2 = 6$$
$$6 = 6, \text{ true}$$

Yes, the values are a solution of both equations.

91. $x + 2y = 2$
$$-4 + 2(3) = 2$$
$$-4 + 6 = 2$$
$$2 = 2, \text{ true}$$

$$x - 2y = 6$$
$$-4 - 2(3) = 6$$
$$-4 - 6 = 6$$
$$-10 = 6, \text{ false}$$

No, the values are not a solution of both equations.

92. Graph $2x + 3y = 6$ by finding intercepts.

Find the *x*-intercept.

$$2x + 3y = 6$$
$$2x + 3(0) = 6$$
$$2x = 6$$
$$x = 3$$

Find the *y*-intercept.

$$2x + 3y = 6$$
$$2(0) + 3y = 6$$
$$3y = 6$$
$$y = 2$$

Graph $2x + y = -2$ by finding intercepts.

Find the *x*-intercept.

$$2x + y = -2$$
$$2x + 0 = -2$$
$$2x = -2$$
$$x = -1$$

Find the *y*-intercept.

$$2x + y = -2$$
$$2(0) + y = -2$$
$$y = -2$$

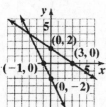

The point of intersection is $(-3, 4)$.

Chapter 3 Review Exercises

1. Quadrant IV

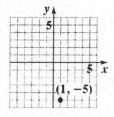

2. Quadrant IV

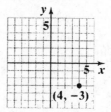

3. Quadrant I

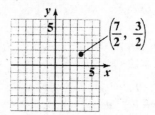

4. Quadrant II

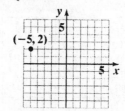

5. $A(5, 6)$ $B(-3, 0)$ $C(-5, 2)$
$D(-4, -2)$ $E(0, -5)$ $F(3, -1)$

6. $y = 3x + 6$

$$3 = 3(-3) + 6$$
$$3 = -6 + 9$$
$$3 = -3, \text{ false}$$
$$(-3, 3) \text{ is not a solution.}$$

$y = 3x + 6$

$$6 = 3(0) + 6$$
$$6 = 6, \text{ true}$$
$$(0, 6) \text{ is a solution.}$$

$y = 3x + 6$

$$9 = 3(1) + 6$$
$$9 = 9, \text{ true}$$
$$(1, 9) \text{ is a solution.}$$

7. $3x - y = 12$

$3(0) - 4 = 12$

$-4 = 12$, false

$(0, 4)$ is not a solution.

$3x - y = 12$

$3(4) - 0 = 12$

$12 = 12$, true

$(4, 0)$ is a solution.

$3x - y = 12$

$3(-1) - 15 = 12$

$-3 - 15 = 12$

$-18 = 12$, false

$(-1, 15)$ is not a solution.

8. a.

x	$y = 2x - 3$	(x, y)
-2	$y = 2(-2) - 3 = -7$	$(-2, -7)$
-1	$y = 2(-1) - 3 = -5$	$(-1, -5)$
0	$y = 2(0) - 3 = -3$	$(0, -3)$
1	$y = 2(1) - 3 = -1$	$(1, -1)$
2	$y = 2(2) - 3 = 1$	$(2, 1)$

b. $y = 2x - 3$

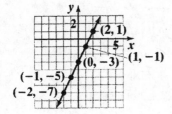

9. a.

x	$y = \frac{1}{2}x + 1$	(x, y)
-2	$y = \frac{1}{2}(-2) + 1 = 0$	$(-2, 0)$
-1	$y = \frac{1}{2}(-1) + 1 = \frac{1}{2}$	$\left(-1, \frac{1}{2}\right)$
0	$y = \frac{1}{2}(0) + 1 = 1$	$(0, 1)$
1	$y = \frac{1}{2}(1) + 1 = \frac{3}{2}$	$\left(1, \frac{3}{2}\right)$
2	$y = \frac{1}{2}(2) + 1 = 2$	$(2, 2)$

b. $y = \frac{1}{2}x + 1$

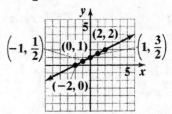

10. a. The graph crosses the x-axis at $(-2, 0)$, so the x-intercept is -2.

b. The graph crosses the y-axis at $(0, -4)$, so the y-intercept is -4.

11. a. The graph does not cross the x-axis, so there is no x-intercept.

b. The graph crosses the y-axis at $(0, 2)$, so the y-intercept is 2.

12. a. The graph crosses the x-axis at $(0, 0)$ (the origin), so the x-intercept is 0.

b. The graph also crosses the y-axis at $(0, 0)$, so the y-intercept is 0.

13. Find the *x*-intercept. Let $y = 0$ and solve for *x*.

$2x + y = 4$

$2x + 0 = 4$

$2x = 4$

$x = 2$

The *x*-intercept is 2.

Find the *y*- intercept. Let $x = 0$ and solve for *y*.

$2x + y = 4$

$2(0) + y = 4$

$y = 4$

The *y*-intercept is 4.

Find a checkpoint. For example, let $x = 1$ and solve for *y*.

$2x + y = 4$

$2(1) + y = 4$

$2 + y = 4$

$y = 2$

The checkpoint is (1,2).

Draw the line through these three points.

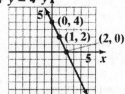

14. Find the *x*-intercept. Let $y = 0$ and solve for *x*.

$3x - 2y = 12$

$3x - 2(0) = 12$

$3x = 12$

$x = 4$

The *x*-intercept is 4.

Find the *y*- intercept. Let $x = 0$ and solve for *y*.

$3x - 2y = 12$

$3(0) - 2y = 12$

$-2y = 12$

$y = -6$

The *y*-intercept is –6.

Find a checkpoint. For example, let $x = 2$ and solve for *y*.

$3x - 2y = 12$

$3(2) - 2y = 12$

$6 - 2y = 12$

$-2y = 6$

$y = -3$

The checkpoint is (2,–3).

Draw the line through these three points.

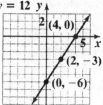

15. Find the *x*-intercept. Let $y = 0$ and solve for *x*.

$3x = 6 - 2y$

$3x = 6 - 2(0)$

$3x = 6$

$x = 2$

The *x*-intercept is 2.

Find the *y*- intercept. Let $x = 0$ and solve for *y*.

$3x = 6 - 2y$

$3(0) = 6 - 2y$

$0 = 6 - 2y$

$2y = 6$

$y = 3$

The *y*-intercept is 3.

Find a checkpoint. For example, let $x = 4$ and solve for *y*.

$3x = 6 - 2y$

$3(4) = 6 - 2y$

$12 = 6 - 2y$

$6 = -2y$

$-3 = y$

The checkpoint is (4,–3).

Draw the line through these three points.

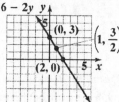

16. Because the constant on the right is 0, the graph passes through the origin. The x- and y-intercepts are both 0.

Thus we will need to find two more points.

Let $x = 1$ and solve for y.

$$3x - y = 0$$
$$3(1) - y = 0$$
$$3 - y = 0$$
$$3 = y$$

This gives the point $(1,3)$.

Let $x = -1$ and solve for y.

$$3x - y = 0$$
$$3(-1) - y = 0$$
$$-3 - y = 0$$
$$-3 = y$$

This gives the point $(-1,-3)$.

Draw the line through these three points.

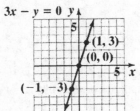

17. $x = 3$

Three ordered pairs are $(3, -2)$, $(3,0)$, and $(3,2)$. The graph is a vertical line.

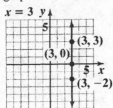

18. $y = -5$

Three ordered pairs are $(-2, -5)$, $(0, -5)$, and $(2, -5)$. The graph is a horizontal line.

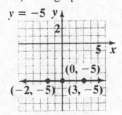

19. $y + 3 = 5$
$$y = 2$$

Three ordered pairs are $(-2,2)$, $(0,2)$, and $(2,2)$. The graph is a horizontal line.

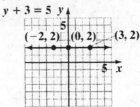

20. $2x = -8$
$$x = -4$$

Three ordered pairs $(-4, -2)$, $(-4,0)$, and $(-4,2)$. The graph is a vertical line.

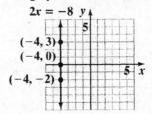

21. a. The minimum temperature occurred at 5 P.M. and was $-4°$F.

b. The maximum temperature occurred at 8 P.M. and was at $16°$F.

c. The x-intercepts are 4 and 6. This indicates that 4 P.M. and 6 P.M., the temperature was $0°$F.

d. The y-intercept is 12. This indicates that at noon the temperature was $12°$F.

e. This indicates that the temperature stayed the same, at $12°$F, from 9 P.M. until midnight.

22. $m = \dfrac{y_2 - y_1}{x_2 - x_1} = \dfrac{1 - 2}{5 - 3} = -\dfrac{1}{2}$

The slope is $-\dfrac{1}{2}$. Since the slope is negative, the line falls from left to right.

23. $m = \dfrac{-4 - 2}{-3 - (-1)} = \dfrac{-6}{-2} = 3$

Since the slope is positive, the line rises from left to right.

24. $m = \dfrac{4 - 4}{6 - (-3)} = \dfrac{0}{9} = 0$

Since the slope is 0, the line is horizontal.

25. $m = \dfrac{-3-3}{5-5} = \dfrac{-6}{0}$; undefined

Since the slope is undefined, the line is vertical.

26. $m = \dfrac{1-(-2)}{2-(-3)} = \dfrac{3}{5}$

27. The line is vertical, so its slope is undefined.

28. $m = \dfrac{-3-(-1)}{2-(-4)} = \dfrac{-2}{6} = -\dfrac{1}{3}$

29. The line is horizontal, so its slope is 0.

30. Line through $(-1, -3)$ and $(2, -8)$:

$m = \dfrac{-8-(-3)}{2-(-1)} = \dfrac{-5}{3} = -\dfrac{5}{3}$

Line through $(8, -7)$ and $(9,10)$:

$m = \dfrac{10-(-7)}{9-8} = \dfrac{17}{1} = 17$

Since their slopes are not equal, the lines are not parallel.
Since the product of their slopes is not -1, the lines are not perpendicular.

31. Line through $(0,-4)$ and $(5, -1)$:

$m = \dfrac{-1-(-4)}{5-0} = \dfrac{3}{5}$

Line through $(-6,8)$ and $(3,-7)$:

$m = \dfrac{-7-8}{3-(-6)} = \dfrac{-15}{9} = \dfrac{-5}{3}$

Since the product of their slopes is

$\dfrac{3}{5}\left(-\dfrac{5}{3}\right) = -1$, the lines are perpendicular.

32. Line through $(5,4)$ and $(9,7)$:

$m = \dfrac{7-4}{9-5} = \dfrac{3}{4}$

Line through $(-6,0)$ and $(-2,3)$:

$m = \dfrac{3-0}{-2-(-6)} = \dfrac{3}{4}$

Since their slopes are equal, the lines are parallel.

33. a. $m = \dfrac{47-7}{2005-1965} = \dfrac{40}{40} = 1$

b. For each year from 1965 through 2005, the percentage of female medical school graduates increased by 1%. The rate of change was 1% per year.

34. $y = 5x - 7$

$y = 5x + (-7)$

The slope is the x-coefficient, which is 5. The y-intercept is the constant term, which is -7.

35. $y = 6 - 4x$

$y = -4x + 6$

$m = -4$; y-intercept $= 6$

36. $y = 3$

$m = 0$; y-intercept $= 3$

37. $2x + 3y = 6$

$3y = -2x + 6$

$y = \dfrac{-2x+6}{3}$

$y = -\dfrac{2}{3}x + 2$

$m = -\dfrac{2}{3}$; y-intercept $= 2$

38. $y = 2x - 4$

slope $= 2 = \dfrac{2}{1}$; y-intercept $= -4$

Plot $(0, -4)$ on the y-axis. From this point, move 2 units *up* (because 2 is positive) and 1 unit to the *right* to reach the point $(1, -2)$. Draw a line through $(0, -4)$ and $(1, -2)$.

$y = 2x - 4$
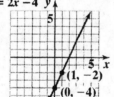

39. $y = \dfrac{1}{2}x - 1$

slope $= \dfrac{1}{2}$; y-intercept $= -1$

Plot $(0, -1)$. From the point, move 1 unit *up* and 2 units to the *right* to reach the point $(2,0)$. Draw a line through $(0, -1)$ and $(2,0)$.

$y = \dfrac{1}{2}x - 1$
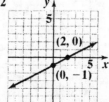

40. $y = -\frac{2}{3}x + 5$

slope $= -\frac{2}{3} = \frac{-2}{3}$; y-intercept $= 5$

Plot (0,5). Move 2 units *down* (because −2 is negative) and 3 units to the *right* to reach the point (3,3). Draw a line through (0,5) and (3,3).

$y = -\frac{2}{3}x + 5$

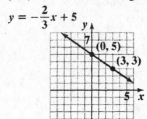

41. $y - 2x = 0$

$y = 2x$

slope $= 2 = \frac{2}{1}$; y-intercept $= 0$

Plot (0,0) (the origin). Move 2 units *up* and 1 unit to the *right* to reach the point (1,2). Draw a line through (0,0) and (1,2).

$y = 2x$

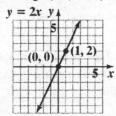

42. $\frac{1}{3}x + y = 2$

$y = -\frac{1}{3}x + 2$

slope $= -\frac{1}{3} = \frac{-1}{3}$; y-intercept $= 2$

Plot (0,2). Move 1 unit *down* and 3 units to the *right* to reach the point (3,1). Draw line through (0,2) and (3,1).

$y = -\frac{1}{3}x + 2$

43. $y = -\frac{1}{2}x + 4$

slope $= -\frac{1}{2} = \frac{-1}{2}$

y-intercept $= 4$

$y = -\frac{1}{2}x - 1$

slope $= -\frac{1}{2} = \frac{-1}{2}$

y-intercept $= -1$

Graph each line using its slope and y-intercept.

$y = -\frac{1}{2}x + 4$

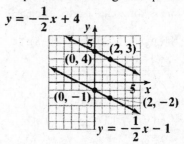

$y = -\frac{1}{2}x - 1$

Yes, they are parallel since both lines have a slope of $-\frac{1}{2}$ and different y-intercepts.

44. a. $m = \frac{51.1 - 12.6}{10 - 0} = \frac{38.5}{10} = 3.85$

The y-intercept is 12.6 as shown in the graph.
$y = mx + b$
$y = 3.85x + 12.6$

b. $y = 3.85x + 12.6$

$y = 3.85(20) + 12.6 = 89.6$

The model predicts that 89.6% of tax returns will be e-filed in 2015.

45. Slope = 6, passing through (−4,7)
point-slope form:
$y - y_1 = m(x - x_1)$
$y - 7 = 6[x - (-4)]$
$y - 7 = 6(x + 4)$

slope-intercept form:
$y - 7 = 6(x + 4)$
$y - 7 = 6x + 24$
$y = 6x + 31$

46. Passing through (3,4) and (2,1)
First, find the slope.
$$m = \frac{1-4}{2-3} = \frac{-3}{-1} = 3$$
Next, use the slope and one of the points to write the equation of the line in point-slope form.
$$y - y_1 = m(x - x_1)$$
$$y - 4 = 3(x - 3)$$
Solve for y to obtain slope-intercept form.
$$y - 4 = 3x - 9$$
$$y = 3x - 5$$

47. Rewrite $3x + y - 9 = 0$ in slope-intercept form.
$$3x + y - 9 = 0$$
$$y = -3x + 9$$
Since the line we are concerned with is parallel to this line, we know it will have slope $m = -3$. We are given that it passes through $(4, -7)$. We use the slope and point to write the equation in point-slope form.
$$y - y_1 = m(x - x_1)$$
$$y - (-7) = -3(x - 4)$$
$$y + 7 = -3(x - 4)$$
Solve for y to obtain slope-intercept form.
$$y + 7 = -3(x - 4)$$
$$y + 7 = -3x + 12$$
$$y = -3x + 5$$

48. The line is perpendicular to $y = \frac{1}{3}x + 4$, so the slope is –3. We are given that it passes through $(-2, 6)$. We use the slope and point to write the equation in point-slope form.
$$y - y_1 = m(x - x_1)$$
$$y - 6 = -3(x - (-2))$$
$$y - 6 = -3(x + 2)$$
Solve for y to obtain slope-intercept form.
$$y - 6 = -3(x + 2)$$
$$y - 6 = -3x - 6$$
$$y = -3x$$

49. a. First, find the slope.
$$m = \frac{5.3 - 3.7}{40 - 20} = \frac{1.6}{20} = 0.08$$
Next, use the slope and one of the points to write the equation of the line in point-slope form.
$$y - y_1 = m(x - x_1)$$
$$y - 3.7 = 0.08(x - 20)$$
Solve for y to obtain slope-intercept form.
$$y - 3.7 = 0.08(x - 20)$$
$$y - 3.7 = 0.08x - 1.6$$
$$y = 0.08x + 2.1$$

b. $y = 0.08x + 2.1$
$$y = 0.08(75) + 2.1 = 8.1$$
The model projects a world population of 8.1 billion in 2025.

Chapter 3 Test

1.
$$4x - 2y = 10$$
$$4(0) - 2(-5) = 10$$
$$0 + 10 = 10$$
$$10 = 10, \text{ true}$$
$(0, -5)$ is a solution.

$$4x - 2y = 10$$
$$4(-2) - 2(1) = 10$$
$$-8 - 2 = 10$$
$$-10 = 10, \text{ false}$$
$(-2, 1)$ is not a solution.

$$4x - 2y = 10$$
$$4(4) - 2(3) = 10$$
$$16 - 6 = 10$$
$$10 = 10, \text{ true}$$
$(4, 3)$ is a solution.

2.

x	$y = 3x + 1$	(x, y)
-2	$y = 3(-2) + 1 = -5$	$(-2, -5)$
-1	$y = 3(-1) + 1 = -2$	$(-1, -2)$
0	$y = 3(0) + 1 = 1$	$(0, 1)$
1	$y = 3(1) + 1 = 4$	$(1, 4)$
2	$y = 3(2) + 1 = 7$	$(2, 7)$

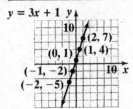

3. a. The graph crosses the x-axis at $(2,0)$, so the x-intercept is 2.

b. The graph crosses the y-axis at $(0, -3)$, so the y-intercept is -3.

4. Find the x-intercept. Let $y = 0$ and solve for x.
$$4x - 2y = -8$$
$$4x - 2(0) = -8$$
$$4x = -8$$
$$x = -2$$
The x-intercept is -2.
Find the y- intercept. Let $x = 0$ and solve for y.
$$4x - 2y = -8$$
$$4(0) - 2y = -8$$
$$-2y = -8$$
$$y = 4$$
The y-intercept is 4.
Find a checkpoint. For example, let $x = -1$ and solve for y.
$$4x - 2y = -8$$
$$4(-1) - 2y = -8$$
$$-4 - 2y = -8$$
$$-2y = -4$$
$$y = 2$$
The checkpoint is $(-1, 2)$.
Draw the line through these three points.

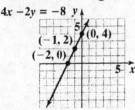

5. $y = 4$
The graph is a horizontal line.

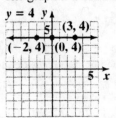

6. $m = \dfrac{-2 - 4}{-5 - (-3)} = \dfrac{-6}{-2} = 3$

The slope is 3. Since the slope is positive, the line rises from left to right.

7. $m = \dfrac{3 - (-1)}{6 - 6} = \dfrac{4}{0}$; undefined

Since the slope is undefined, the line is vertical.

8. Use the points $(-1, -2)$ and $(1, 1)$.
$$m = \frac{1 - (-2)}{1 - (-1)} = \frac{3}{2}$$

9. Line through $(-2, 10)$ and $(0, 2)$:
$$m = \frac{2 - 10}{0 - (-2)} = \frac{-8}{2} = -4$$
Line through $(-8, -7)$ and $(24, 1)$:
$$m = \frac{1 - (-7)}{24 - (-8)} = \frac{8}{32} = \frac{1}{4}$$

Since the product of their slopes is $-4\left(\dfrac{1}{4}\right) = -1$, the lines are perpendicular.

10. Line through $(2, 4)$ and $(6, 1)$:
$$m = \frac{1 - 4}{6 - 2} = \frac{-3}{4} = -\frac{3}{4}$$
Line through $(-3, 1)$ and $(1, -2)$:
$$m = \frac{-2 - 1}{1 - (-3)} = \frac{-3}{4} = -\frac{3}{4}$$
Since the slopes are equal, the lines are parallel.

11. $y = -x + 10$
$$y = -1x + 10$$
The slope is the coefficient of x, which is -1. The y-intercept is the constant term, which is 10.

12. $2x + y = 6$

$$y = -2x + 6$$

$m = -2$; y-intercept $= 6$

13. $y = \dfrac{2}{3}x - 1$

slope $= \dfrac{2}{3}$; y-intercept $= -1$

Plot $(0, -1)$. From this point, move 2 units *up* and 3 units to the *right* to reach the point $(3, 1)$.
Draw a line through $(0, -1)$ and $(3, 1)$.

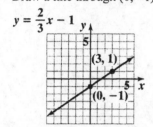

14. $y = -2x + 3$

slope $= -2 = \dfrac{-2}{1}$; y-intercept $= 3$

Plot $(0, 3)$. Move 2 units *down* and 1 unit to the right to reach the point $(1, 1)$. Draw a line through $(0, 3)$ and $(1, 1)$.

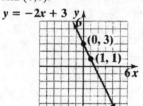

15. point-slope form:

$$y - y_1 = m(x - x_1)$$

$$y - 4 = -2\big[x - (-1)\big]$$

$$y - 4 = -2(x + 1)$$

slope-intercept form:

$$y - 4 = -2(x + 1)$$

$$y - 4 = -2x - 2$$

$$y = -2x + 2$$

16. Passing through $(2, 1)$ and $(-1, -8)$
First, find the slope.

$$m = \frac{-8 - 1}{-1 - 2} = \frac{-9}{-3} = 3$$

Next, use the slope and one of the points to write the equation of the line in point-slope form.

$$y - y_1 = m(x - x_1)$$

$$y - 1 = 3(x - 2)$$

Solve for y to obtain slope-intercept form.

$$y - 1 = 3(x - 2)$$

$$y - 1 = 3x - 6$$

$$y = 3x - 5$$

17. The line is perpendicular to $y = -\dfrac{1}{2}x - 4$, so the slope is 2. We are given that it passes through $(-2, 3)$. We use the slope and point to write the equation in point-slope form.

$$y - y_1 = m(x - x_1)$$

$$y - 3 = 2\big(x - (-2)\big)$$

$$y - 3 = 2(x + 2)$$

Solve for y to obtain slope-intercept form.

$$y - 3 = 2(x + 2)$$

$$y - 3 = 2x + 4$$

$$y = 2x + 7$$

18. The line is parallel to $x + 2y = 5$.

Put this equation in slope-intercept form by solving for y.

$$x + 2y = 5$$
$$2y = -x + 5$$
$$y = -\frac{1}{2}x + \frac{5}{2}$$

Therefore the slopes are the same; $m = -\frac{1}{2}$.

We are given that it passes through $(6, -4)$.

We use the slope and point to write the equation in point-slope form.

$$y - y_1 = m(x - x_1)$$
$$y - (-4) = -\frac{1}{2}(x - 6)$$
$$y + 4 = -\frac{1}{2}(x - 6)$$

Solve for y to obtain slope-intercept form.

$$y + 4 = -\frac{1}{2}(x - 6)$$
$$y + 4 = -\frac{1}{2}x + 3$$
$$y = -\frac{1}{2}x - 1$$

19. a. $m = \dfrac{1616 - 886}{2006 - 2002} = \dfrac{730}{4} = 182.5$

  b. For the period shown, corporate profits increased each year by approximately <u>$182.5 billion</u>. The rate of change was <u>$182.5 billion per year</u>.

Cumulative Review Exercises (Chapters 1-3)

1. $\dfrac{10 - (-6)}{3^2 - (4 - 3)} = \dfrac{10 + 6}{9 - 1} = \dfrac{16}{8} = 2$

2. $6 - 2\big[3(x - 1) + 4\big]$
$= 6 - 2(3x - 3 + 4) = 6 - 2(3x + 1)$
$= 6 - 6x - 2 = 4 - 6x$

3. The only irrational number in the given set is $\sqrt{5}$.

4. $6(2x - 1) - 6 = 11x + 7$
$$12x - 6 - 6 = 11x + 7$$
$$12x - 12 = 11x + 7$$
$$x - 12 = 7$$
$$x = 19$$
The solution set is $\{19\}$.

5. $x - \dfrac{3}{4} = \dfrac{1}{2}$
$$x - \frac{3}{4} + \frac{3}{4} = \frac{1}{2} + \frac{3}{4}$$
$$x = \frac{2}{4} + \frac{3}{4} = \frac{5}{4}$$
The solution set is $\left\{\dfrac{5}{4}\right\}$.

6. $y = mx + b$
$$y - b = mx + b - b$$
$$y - b = mx$$
$$\frac{y - b}{m} = \frac{mx}{m}$$
$$\frac{y - b}{m} = x \text{ or } x = \frac{y - b}{m}$$

7. $A = 120; P = 15\% = 0.15$
$$A = PB$$
$$120 = 0.15 \cdot B$$
$$\frac{120}{0.15} = \frac{0.15B}{0.15}$$
$$800 = B$$
120 is 15% of 800.

8. $y = 4.5x - 46.7$
$$133.3 = 4.5x - 46.7$$
$$133.3 + 46.7 = 4.5x - 46.7 + 46.7$$
$$180 = 4.5x$$
$$\frac{180}{4.5} = \frac{4.5x}{4.5}$$
$$40 = x$$
The car is traveling 40 miles per hour.

9.
$$2 - 6x \geq 2(5 - x)$$
$$2 - 6x \geq 10 - 2x$$
$$2 - 6x + 2x \geq 10 - 2x + 2x$$
$$2 - 4x \geq 10$$
$$2 - 4x - 2 \geq 10 - 2$$
$$-4x \geq 8$$
$$\frac{-4x}{-4} \leq \frac{8}{-4}$$
$$x \leq -2$$
$$(-\infty, -2]$$

10.
$$6(2 - x) > 12$$
$$12 - 6x > 12$$
$$12 - 6x - 12 > 12 - 12$$
$$-6x > 0$$
$$\frac{-6x}{-6} < \frac{0}{-6}$$
$$x < 0$$
$$(-\infty, 0)$$

11. Let x = the number of hours the plumber worked.
$$18 + 35x = 228$$
$$35x = 210$$
$$x = 6$$
The plumber worked 6 hours.

12. Let x = the width of the rectangle.
Let $2x + 14$ = the length of the rectangle.
$$P = 2l + 2w$$
$$346 = 2(2x + 14) + 2(x)$$
$$346 = 4x + 28 + 2x$$
$$346 = 6x + 28$$
$$318 = 6x$$
$$x = 53$$
$$2x + 14 = 120$$
The width is 53 meters and the length 120 meters.

13. Let x = the weight before the loss
Let $0.10x$ = the weight lost
$$x - 0.10x = 180$$
$$0.90x = 180$$
$$x = \frac{180}{0.90}$$
$$x = 200$$
The weight before the loss was 200 pounds.

14. Let x = the measure of the first angle.
Let $x + 20$ = the measure of the second angle.
Let $2x$ = the measure of third angle.
$$x + (x + 20) + 2x = 180$$
$$4x + 20 = 180$$
$$4x = 160$$
$$x = 40$$
The angles measure $x = 40°$, $x + 20 = 60°$, and $2x = 80°$.

15.
$$x^2 - 10x = (-3)^2 - 10(-3)$$
$$= 9 + 30$$
$$= 39$$

16. $-2000 < -3$

17. *x*-intercept:

$$2x - y = 4$$
$$2x - 0 = 4$$
$$2x = 4$$
$$x = 2$$

y-intercept :

$$2x - y = 4$$
$$2(0) - y = 4$$
$$-y = 4$$
$$y = -4$$

checkpoint:

$$2x - y = 4$$
$$2(1) - y = 4$$
$$2 - y = 4$$
$$-y = 2$$
$$y = -2$$

Draw a line through (2,0), (0, −4), and (1,−2.)

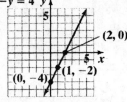

18. $x = -5$

The graph is a vertical line.

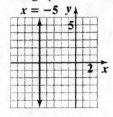

19. $y = -4x + 3$

slope $= -4 = \dfrac{-4}{1}$; *y*-intercept = 3

Plot (0,3). Move 4 units *down* and 1 unit to the *right* to reach the point
(1, −1). Draw a line through (0,3) and (1, −1).

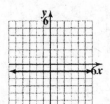

20. $y = -1$

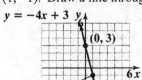

Chapter 4
Systems of Linear Equations

4.1 Check Points

1. **a.** To determine if $(1,2)$ is a solution to the system, replace x with 1 and y with 2 in both equations.
$$2x - 3y = -4$$
$$2(1) - 3(2) = -4$$
$$2 - 6 = -4$$
$$-4 = -4, \text{ true}$$
$$2x + y = 4$$
$$2(1) + 2 = 4$$
$$2 + 2 = 4$$
$$4 = 4, \text{ true}$$
The ordered pair satisfies both equations, so it is a solution to the system.

b. To determine if $(7,6)$ is a solution to the system, replace x with 7 and y with 6 in both equations.
$$2x - 3y = -4$$
$$2(7) - 3(6) = -4$$
$$14 - 18 = -4$$
$$-4 = -4, \text{ true}$$
$$2x + y = 4$$
$$2(7) + 6 = 4$$
$$14 + 6 = 4$$
$$20 = 4, \text{ false}$$
The ordered pair does not satisfy both equations, so it is not a solution to the system.

2. Graph $2x + y = 6$ by using intercepts.
x-intercept (Set $y = 0$.)
$$2x + y = 6$$
$$2x + 0 = 6$$
$$2x = 6$$
$$x = 3$$
y-intercept (Set $x = 0$.)
$$2x + y = 6$$
$$2(0) + y = 6$$
$$0 + y = 6$$
$$y = 6$$

Graph $2x - y = -2$ by using intercepts.
x-intercept (Set $y = 0$.)
$$2x - y = -2$$
$$2x - 0 = -2$$
$$2x = -2$$
$$x = -1$$
y-intercept (Set $x = 0$.)
$$2x - y = -2$$
$$2(0) - y = -2$$
$$-y = -2$$
$$y = 2$$

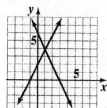

The lines intersect at $(1,4)$.
The solution set is $\{(1,4)\}$.

3. Graph $y = -x + 6$ by using the y-intercept of 6 and the slope of -1.
Graph $y = 3x - 6$ by using the y-intercept of -6 and the slope of 3.

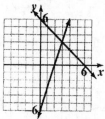

The lines intersect at $(3,3)$.
The solution set is $\{(3,3)\}$.

4. Graph $y = 3x - 2$ by using the y-intercept of -2 and the slope of 3.
Graph $y = 3x + 1$ by using the y-intercept of 1 and the slope of 3.

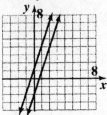

Because both equations have the same slope, 3, but different y-intercepts, the lines are parallel. Thus, the system is inconsistent and has no solution. The solution set is the empty set, $\{\ \}$.

5. Graph $x + y = 3$ by using intercepts.
x-intercept (Set $y = 0$.)
$$x + y = 3$$
$$x + 0 = 3$$
$$x = 3$$
y-intercept (Set $x = 0$.)
$$x + y = 3$$
$$0 + y = 3$$
$$y = 3$$
Graph $2x + 2y = 6$ by using intercepts.
x-intercept (Set $y = 0$.)
$$2x + 2y = 6$$
$$2x + 2(0) = 6$$
$$2x = 6$$
$$x = 3$$
y-intercept (Set $x = 0$.)
$$2x + 2y = 6$$
$$2(0) + 2y = 6$$
$$2y = 6$$
$$y = 3$$
Both lines have the same x-intercept and the same y-intercept. Thus, the graphs of the two equations in the system are the same line.

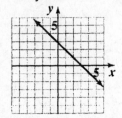

Any ordered pair that is a solution to one equation is a solution to the other, and, consequently, a solution of the system. The system has an infinite number of solutions, namely all points that are solutions of either line.
The solution set is $\{(x, y) \,|\, x + y = 3\}$.

6. **a.** Graph $y = 2x$ by using the y-intercept of 0 and the slope of 2.
Graph $y = x + 10$ by using the y-intercept of 10 and the slope of 1.

The solution is the ordered pair (10,20).

b. If the bridge is used 10 times in a month, the total monthly cost without the coupon book is the same as the monthly cost with the coupon book, namely $20.

4.1 Exercise Set

1. To determine if $(2, -3)$ is a solution to the system, replace x with 2 and y with -3 in both equations.
$$2x + 3y = -5$$
$$2(2) + 3(-3) = -5$$
$$4 + (-9) = -5$$
$$-5 = -5, \text{ true}$$

$$7x - 3y = 23$$
$$7(2) - 3(-3) = 23$$
$$14 + 9 = 23$$
$$23 = 23, \text{ true}$$
The ordered pair satisfies both equations, so it is a solution to the system.

3. $x + 3y = 1$

$\dfrac{2}{3} + 3\left(\dfrac{1}{9}\right) = 1$

$\dfrac{2}{3} + \dfrac{1}{3} = 1$

$1 = 1,$ true

$4x + 3y = 3$

$4\left(\dfrac{2}{3}\right) + 3\left(\dfrac{1}{9}\right) = 3$

$\dfrac{8}{3} + \dfrac{1}{3} = 3$

$\dfrac{9}{3} = 3$

$3 = 3,$ true

The ordered pair satisfies both equations, so it is a solution to the system.

5. $5x + 3y = 2$

$5(-5) + 3(9) = 2$

$-25 + 27 = 2$

$2 = 2,$ true

$x + 4y = 14$

$-5 + 4(9) = 14$

$-5 + 36 = 14$

$31 = 14,$ false

The ordered pair does not satisfy both equations, so it is not a solution to the system.

7. $x - 2y = 500$

$1400 - 2(450) = 500$

$1400 - 900 = 500$

$500 = 500,$ true

$0.03x + 0.02y = 51$

$0.03(1400) + 0.02(450) = 51$

$42 + 9 = 51$

$51 = 51,$ true

The ordered pair satisfies both equations, so the ordered pair is a solution to the system.

9. $5x - 4y = 20$

$5(8) - 4(5) = 20$

$40 - 20 = 20$

$20 = 20,$ true

$3y = 2x + 1$

$3(5) = 2(8) + 1$

$15 = 16 + 1$

$15 = 17,$ false

The ordered pair does not satisfy both equations, so it is not a solution to the system.

11. Graph both equations on the same axes.
$x + y = 6:$

x-intercept = 6; y-intercept = 6
$x - y = 2:$

x-intercept = 2; y-intercept = −2

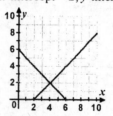

The solution set is $\{(4, 2)\}$.

13. Graph both equations on the same axes.
$x + y = 1:$

x-intercept = 1; y-intercept = 1
$y - x = 3:$

x-intercept = −3; y-intercept = 3

The solution set is $\{(-1, 2)\}$.

15. Graph both equations.

$2x - 3y = 6$:

x-intercept = 3: y-intercept = -2
$4x + 3y = 12$:

x-intercept = 3: y-intercept = 4

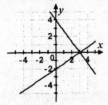

The solution set is $\{(3,0)\}$.

17. Graph both equations.

$4x + y = 4$:

x-intercept = 1: y-intercept = 4
$3x - y = 3$:

x-intercept = 1: y-intercept = -3

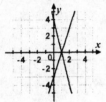

The solution set is $\{(1,0)\}$.

19. Graph both equations.

$y = x + 5$:

Slope = 1; y-intercept = 5
$y = -x + 3$:

Slope = -1; y-intercept = 3

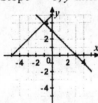

The solution set is $\{(-1,4)\}$.

21. Graph both equations.

$y = 2x$:

slope = 2; y-intercept = 0
$y = -x + 6$:

slope = -1; y-intercept = 6

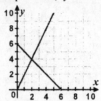

The solution set is $\{(2,4)\}$.

23. Graph both equations.

$y = -2x + 3$:

slope = -2; y-intercept = 3
$y = -x + 1$:

slope = -1; y-intercept = 1

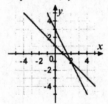

The solution set is $\{(2,-1)\}$.

25. Graph both equations.

$y = 2x - 1$:

Slope = 2; y-intercept = -1
$y = 2x + 1$:

Slope = 2; y-intercept = 1

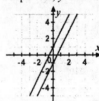

The lines are parallel, so the solution set is $\{\ \}$.

27. Graph each equation.

$x + y = 4$:

x-intercept $= 4$; y-intercept $= 4$

$x = -2$:

vertical line with x-intercept -2

The solution set is $\{(-2, 6)\}$.

29. Graph each equation.

$x - 2y = 4$:

x-intercept $= 4$; y-intercept $= -2$

$2x - 4y = 8$:

x-intercept $= 4$; y-intercept $= -2$

The graph of the two equations are the same line. (Note that they have the same slope and same y-intercept.) Because the lines coincide, the system has an infinite number of solutions. The solution set is $\{(x, y) \mid x - 2y = 4\}$ or $\{(x, y) \mid 2x - 4y = 8\}$.

31. Graph both lines.

$y = 2x - 1$:

slope $= 2$; y-intercept $= -1$

$x - 2y = -4$:

x-intercept $= -4$; y-intercept $= 2$

The solution set is $\{(2, 3)\}$.

33. Graph both lines.

$x + y = 5$:

x-intercept $= 5$; y-intercept $= 5$

$2x + 2y = 12$:

x-intercept $= 6$; y-intercept $= 6$

The lines are parallel, so the solution set is $\{\ \}$.

35. $x - y = 0$

$y = x$

Because the lines coincide, the system has an infinite number of solutions. The solution set is $\{(x, y) \mid x - y = 0\}$ or $\{(x, y) \mid y = x\}$.

37. $x = 2$

$y = 4$

The solution set is $\{(2, 4)\}$.

39. $x = 2$

$x = -1$

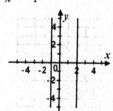

The two vertical lines are parallel, so the solution set is $\{\ \}$.

41. $y = 0$
$y = 4$

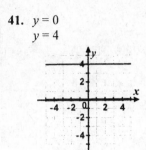

The two horizontal lines are parallel, so the solution set is $\{\ \}$.

43. $y = \frac{1}{2}x - 3$:

slope = $\frac{1}{2}$, y-intercept = -3

$y = \frac{1}{2}x - 5$:

slope = $\frac{1}{2}$, y-intercept = -5

Since the slopes are the same, but the y-intercepts are different, the lines are parallel and there is no solution.

45. $y = -\frac{1}{2}x + 4$

slope = $-\frac{1}{2}$, y-intercept = 4
$3x - y = -4$
$-y = -3x - 4$
$y = 3x + 4$
slope = 3, y-intercept = 4
Since the lines have different slopes, there will be one solution.

47. $3x - y = 6$
$-y = -3x + 6$
$y = 3x - 6$
slope = 3, y-intercept = -6
$x = \frac{y}{3} + 2$
$3x = y + 6$
$3x - 6 = y$
$y = 3x - 6$
slope = 3, y-intercept = -6
Since the lines have the same slopes and y-intercepts, the graphs will coincide and there are an infinite number of solutions.

49. $3x + y = 0$
$y = -3x$
slope = -3, y-intercept = 0
$y = -3x + 1$
slope = -3, y-intercept = 1
Since the slopes are the same, but the y-intercepts are different, the lines are parallel and there is no solution.

51. a. The x-coordinate of the intersection point is 40. Both companies charge the same for 40 miles driven.

b. The y-coordinate of the intersection point is about 55.

c. $y = 0.35x + 40$
$y = 0.35(40) + 40 = 54$

$y = 0.45x + 36$
$y = 0.45(40) + 36 = 54$
Both companies charge $54 for 40 miles driven.

53. a. The solution is the ordered pair (5,20).

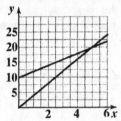

b. Nonmembers and members pay the same amount per month for taking 5 classes, namely $20.

55. – 61. Answers will vary.

63. does not make sense; Explanations will vary. Sample explanation: Some linear systems have no solutions or one solution.

65. makes sense

67. false; Changes to make the statement true will vary. A sample change is: The system's lines could have different slopes.

69. false; Changes to make the statement true will vary. A sample change is: The two lines of a linear system that has one solution must have different slopes.

71. Answers will vary.

73. $y = 2x + 2$

$y = -2x + 6$

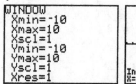

The solution set is $\{(1, 4)\}$.

75. $x + 2y = 2$

$x - y = 2$

In order to enter the equations into a graphing calculator, each of them must be solved for y.

$x + 2y = 2$

$2y = -x + 2$

$\dfrac{2y}{2} = \dfrac{-x + 2}{2}$

$y = -\dfrac{1}{2}x + 1$

$x - y = 2$

$-y = -x + 2$

$y = x - 2$

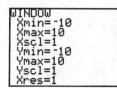

The solution set is $\{(2, 0)\}$.

77. $3x - y = 5$

$-5x + 2y = -10$

Solve each equation for y.

$3x - y = 5$

$-y = -3x + 5$

$y = 3x - 5$

$-5x + 2y = -10$

$2y = 5x - 10$

$y = \dfrac{5}{2}x - 5$

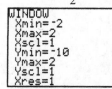

The solution set is $\{(0, -5)\}$.

79. $y = \dfrac{1}{3}x + \dfrac{2}{3}$

$y = \dfrac{5}{7}x - 2$

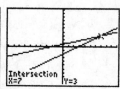

The solution set is $\{(7, 3)\}$.

81. $-3 + (-9) = -3 - 9 = -12$

82. $-3 - (-9) = -3 + 9 = 6$

83. $-3(-9) = 27$

84. $4x - 3(-x - 1) = 24$

$4x + 3x + 3 = 24$

$7x + 3 = 24$

$7x = 21$

$x = 3$

The solution set is $\{3\}$.

85. $5(2y - 3) - 4y = 9$

$10y - 15 - 4y = 9$

$6y - 15 = 9$

$6y = 24$

$y = 4$

The solution set is $\{4\}$.

86. $(5x - 1) + 1 = 5x + 5$

$5x - 1 + 1 = 5x + 5$

$5x = 5x + 5$

$0 = 5$

The solution set is $\{\ \}$.

4.2 Check Points

1. $y = 5x - 13$

$2x + 3y = 12$

Substitute $5x - 13$ for y in the second equation.

$2x + 3(5x - 13) = 12$

$2x + 15x - 39 = 12$

$17x - 39 = 12$

$17x = 51$

$x = 3$

Back substitute 3 for x into the first equation.

$y = 5x - 13$

$y = 5(3) - 13 = 2$

The solution set is $\{(3, 2)\}$.

2. $3x + 2y = -1$

$x - y = 3$

Solve the second equation for x.

$x - y = 3$

$x = y + 3$

Substitute $y + 3$ for x in the first equation.

$3x + 2y = -1$

$3(y + 3) + 2y = -1$

$3y + 9 + 2y = -1$

$5y + 9 = -1$

$5y = -10$

$y = -2$

Back substitute -2 for y into $x = y + 3$.

$x = y + 3$

$x = -2 + 3 = 1$

The solution set is $\{(1, -2)\}$.

3. $3x + y = -5$

$y = -3x + 3$

Substitute $-3x + 3$ for y in the first equation.

$3x + y = -5$

$$3x + \overbrace{(-3x + 3)}^{y} = -5$$

$3x - 3x + 3 = -5$

$3 = -5$, false

The false statement indicates that the system is inconsistent and has no solution.

The solution set is $\{\ \}$.

4. $y = 3x - 4$

$9x - 3y = 12$

Substitute $3x - 4$ for y in the second equation.

$9x - 3y = 12$

$$9x - 3(\overbrace{3x - 4}^{y}) = 12$$

$9x - 3(3x - 4) = 12$

$9x - 9x + 12 = 12$

$12 = 12$, true

The true statement indicates that the system contains dependent equations and has infinitely many solutions.

The solution set is $\{(x, y) \mid y = 3x - 4\}$.

5. a. $p = -30x + 1800$

$p = 30x$

Substitute $30x$ for p in the first equation.

$p = -30x + 1800$

$$\overbrace{30x}^{p} = -30x + 1800$$

$60x = 1800$

$x = 30$

Back-substitute to find p.

$p = 30x$

$p = 30(30) = 900$

The solution set is $\{(30, 900)\}$.

Equilibrium quantity: 30,000

Equilibrium price: $900

b. When rents are $\underline{\$900}$ per month, consumers will demand $\underline{30,000}$ apartments and suppliers will offer $\underline{30,000}$ apartments for rent.

4.2 Exercise Set

1. $x + y = 4$

    $y = 3x$

    Substitute $3x$ for y in the first equation.

    $x + y = 4$

    $x + (3x) = 4$

    Solve this equation for x.

    $4x = 4$

    $x = 1$

    Back substitute 1 for x into the second equation.

    $y = 3x$

    $y = 3(1) = 3$

    The solution set is $\{(1, 3)\}$.

3. $x + 3y = 8$

    $y = 2x - 9$

    Substitute $2x - 9$ for y in the first equation and solve for x.

    $x + 3y = 8$

    $x + 3(2x - 9) = 8$

    $x + 6x - 27 = 8$

    $7x - 27 = 8$

    $7x = 35$

    $x = 5$

    Back-substitute 5 for x into the second equation and solve for y.

    $y = 2x - 9$

    $y = 2(5) - 9 = 1$

    The solution set is $\{5, 1\}$.

5. $x + 3y = 5$

    $4x + 5y = 13$

    Solve the first equation for x.

    $x + 3y = 5$

    $x = 5 - 3y$

    Substitute $5 - 3y$ for x in the second equation and solve for y.

    $4x + 5y = 13$

    $4(5 - 3y) + 5y = 13$

    $20 - 12y + 5y = 13$

    $20 - 7y = 13$

    $-7y = -7$

    $y = 1$

Back-substitute 1 for y in the equation $x = 5 - 3y$ and solve for x.

$x = 5 - 3y$

$x = 5 - 3(1) = 2$

The solution set is $\{(2, 1)\}$.

7. $2x - y = -5$

    $x + 5y = 14$

    Solve the second equation for x.

    $x + 5y = 14$

    $x = 14 - 5y$

    Substitute $14 - 5y$ for x in the first equation.

    $2(14 - 5y) - y = -5$

    $28 - 10y - y = -5$

    $28 - 11y = -5$

    $-11y = -33$

    $y = 3$

    Back-substitute.

    $x = 14 - 5y$

    $x = 14 - 5(3) = 14 - 15 = -1$

    The solution set is $\{(-1, 3)\}$.

9. $2x - y = 3$

    $5x - 2y = 10$

    Solve the first equation for y.

    $2x - y = 3$

    $-y = -2x + 3$

    $y = 2x - 3$

    Substitute $2x - 3$ for y in the second equation.

    $5x - 2(2x - 3) = 10$

    $5x - 4x + 6 = 10$

    $x + 6 = 10$

    $x = 4$

    Back-substitute.

    $y = 2x - 3$

    $y = 2(4) - 3 = 8 - 3 = 5$

    The solution set is $\{(4, 5)\}$.

11. $-3x + y = -1$

$x - 2y = 4$

Solve the second equation for x.

$x - 2y = 4$

$x = 2y + 4$

Substitute $2y + 4$ for x in the first equation.

$-3x + y = -1$

$-3(2y + 4) + y = -1$

$-6y - 12 + y = -1$

$-5y - 12 = -1$

$-5y = 11$

$$y = -\frac{11}{5}$$

Back-substitute.

$x = 2y + 4$

$x = 2\left(-\frac{11}{5}\right) + 4$

$x = -\frac{22}{5} + 4$

$x = -\frac{22}{5} + \frac{20}{5}$

$x = -\frac{2}{5}$

The solution set is $\left\{\left(-\frac{2}{5}, -\frac{11}{5}\right)\right\}$.

13. $x = 9 - 2y$

$x + 2y = 13$

The first equation is already solved for x.
Substitute $9 - 2y$ for x in the second equation.

$x + 2y = 13$

$(9 - 2y) + 2y = 13$

$9 = 13$, false

The false statement 9=13 indicates that the system is inconsistent and has no solution.
The solution set is $\{ \ \}$.

15. $y = 3x - 5$

$21x - 35 = 7y$

Substitute $3x-5$ for y in the second equation.

$21x - 35 = 7y$

$21x - 35 = 7(3x - 5)$

$21x - 35 = 21x - 35$

$-35 = -35$, true

The true statement $-35 = -35$ indicates that the system contains dependent equations and has an infinite number of solutions.
The solution set is $\{(x, y) | y = 3x - 5\}$.

17. $5x + 2y = 0$

$x - 3y = 0$

Solve the second equation for x.

$x - 3y = 0$

$x = 3y$

Substitute $3y$ for x in the first equation.

$5x + 2y = 0$

$5(3y) + 2y = 0$

$15y + 2y = 0$

$17y = 0$

$y = 0$

Back-substitute to find x.

$x = 3y$

$x = 3(0) = 0$

The solution set is $\{(0, 0)\}$.

19. $2x - y = 6$

$3x + 2y = 5$

Solve the first equation for y.

$2x - y = 6$

$-y = -2x + 6$

$y = 2x - 6$

Substitute $2x - 6$ for y in the second equation.

$3x + 2y = 5$

$3x + 2(2x - 6) = 5$

$3x + 4x - 12 = 5$

$7x - 12 = 5$

$7x = 17$

$x = \dfrac{17}{7}$

Back-substitute to find y.

$y = 2x - 6 = 2\left(\dfrac{17}{7}\right) - 6 = -\dfrac{8}{7}$

The solution set is $\left\{\left(\dfrac{17}{7}, -\dfrac{8}{7}\right)\right\}$.

21. $2(x - 1) - y = -3$

$y = 2x + 3$

Substitute $2x + 3$ for y in the first equation.

$2(x - 1) - (2x + 3) = -3$

$2x - 2 - 2x - 3 = -3$

$-5 = -3, \ \text{false}$

The false statement $-5 = -5$ indicates that the system has no solution.

The solution set is $\{ \ \}$.

23. $x = 2y + 9$

$x = 7y + 10$

Substitute $7y + 10$ for x in the first equation.

$x = 2y + 9$

$7y + 10 = 2y + 9$

$5y + 10 = 9$

$5y = -1$

$y = -\dfrac{1}{5}$

Back-substitute to find x.

$x = 2y + 9 = 2\left(-\dfrac{1}{5}\right) + 9$

$= -\dfrac{2}{5} + 9 = -\dfrac{2}{5} + \dfrac{45}{5} = \dfrac{43}{5}$

The solution set is $\left\{\left(\dfrac{43}{5}, -\dfrac{1}{5}\right)\right\}$.

25. $4x - y = 100$

$0.05x - 0.06y = -32$

Solve the first equation for y.

$4x - y = 100$

$-y = -4x + 100$

$y = 4x - 100$

Substitute $4x - 100$ for y in the second equation.

$0.05x - 0.06y = -32$

$0.05x - 0.06(4x - 100) = -32$

$0.05x - 0.24x + 6 = -32$

$-0.19x + 6 = -32$

$-0.19x = -38$

$x = 200$

Back-substitute to find y.

$y = 4x - 100$

$= 4(200) - 100$

$= 800 - 100 = 700$

$\left(-\dfrac{44}{3}, -\dfrac{7}{3}\right)$

The solution set is $\{(200, 700)\}$.

27. $y = \frac{1}{3}x + \frac{2}{3}$

$y = \frac{5}{7}x - 2$

First, clear both equations of fractions. Multiply the first equation by the LCD, 3.

$$3y = 3\left(\frac{1}{3}x + \frac{2}{3}\right)$$

$$3y = 3x + 2$$

Multiply the second equation by the LCD, 7.

$$7y = 7\left(\frac{5}{7}x - 2\right)$$

$$7y = 5x - 14$$

Now solve the new system

$3y = x + 2$

$7y = 5x - 14$

Solve the first of these equations for x.

$3y = x + 2$

$3y - 2 = x$

Substitute $3y - 2$ for x in the second equation of the new system.

$7y = 5x - 14$

$7y = 5(3y - 2) - 14$

$7y = 15y - 10 - 14$

$7y = 15y - 24$

$-8y = -24$

$y = 3$

Back-substitute to find x.

$x = 3y - 2$

$x = 3(3) - 2 = 9 - 2 = 7$

The solution set is $\{(7, 3)\}$.

29. $\frac{x}{6} - \frac{y}{2} = \frac{1}{3}$

$x + 2y = -3$

Clear the first equation of fractions by multiplying 6.

$$6\left(\frac{x}{6} - \frac{y}{2}\right) = 6\left(\frac{1}{3}\right)$$

$$x - 3y = 2$$

Solve this equation for x.

$x = 3y + 2$

Substitute $3y + 2$ for x in the second equation of the system.

$(3y + 2) + 2y = -3$

$5y + 2 = -3$

$5y = -5$

$y = -1$

Back-substitute to find x.

$x = 3y + 2 = 3(-1) + 2 = -1$

The solution set is $\{(-1, -1)\}$.

31. $2x - 3y = 8 - 2x$

$3x + 4y = x + 3y + 14$

Simplify the first equation.

$2x - 3y = 8 - 2x$

$2x - 3y + 2x = 8 - 2x + 2x$

$4x - 3y = 8$

Simplify the second equation.

$3x + 4y = x + 3y + 14$

$3x + 4y - x - 3y = x + 3y + 14 - x - 3y$

$2x + y = 14$

Solve the last equation for y.

$2x + y = 14$

$y = 14 - 2x$

Substitute $14 - 2x$ for y in the equation $4x - 3y = 8$.

$4x - 3y = 8$

$4x - 3(14 - 2x) = 8$

$4x - 42 + 6x = 8$

$10x - 42 = 8$

$10x = 50$

$x = 5$

Back-substitute to find y.

$y = 14 - 2x$

$y = 14 - 2(5) = 4$

The solution set is $\{(5, 4)\}$.

33. $x + y = 81$

$x = y + 41$

Substitute $y + 41$ for x in the first equation.

$x + y = 81$

$(y + 41) + y = 81$

$y + 41 + y = 81$

$2y + 41 = 81$

$2y = 40$

$y = 20$

Back-substitute.

$x = y + 41 = 20 + 41 = 61$

The numbers are 20 and 61.

35. $x - y = 5$

$4x = 6y$

Solve the first equation for x.

$x - y = 5$

$x = y + 5$

Substitute $y + 5$ for x in the second equation.

$4x = 6y$

$4(y + 5) = 6y$

$4y + 20 = 6y$

$20 = 2y$

$10 = y$

Back-substitute.

$x = y + 5 = 10 + 5 = 15$

The numbers are 10 and 15.

37. $x - y = 1$

$x + 2y = 7$

Solve the first equation for x.

$x - y = 1$

$x = y + 1$

Substitute $y + 1$ for x in the second equation.

$x + 2y = 7$

$(y + 1) + 2y = 7$

$y + 1 + 2y = 7$

$3y + 1 = 7$

$3y = 6$

$y = 2$

Back-substitute.

$x = y + 1 = 2 + 1 = 3$

The numbers are 2 and 3.

39. $0.7x - 0.1y = 0.6$

$0.8x - 0.3y = -0.8$

Multiply both sides of both equations by 10.

$7x - y = 6$

$8x - 3y = -8$

Solve the first equation for y.

$7x - y = 6$

$7x = 6 + y$

$7x - 6 = y$

Substitute $7x - 6$ for y in the second equation.

$8x - 3y = -8$

$8x - 3(7x - 6) = -8$

$8x - 21x + 18 = -8$

$-13x + 18 = -8$

$-13x = -26$

$x = 2$

Back-substitute.

$y = 7x - 6 = 7(2) - 6 = 14 - 6 = 8$

The solution set is $\{(2, 8)\}$.

41. a. Substitute $0.375x + 3$ for p in the first equation.

$p = -0.325x + 5.8$

$\overbrace{0.375x + 3}^{p} = -0.325x + 5.8$

$0.375x + 3 = -0.325x + 5.8$

$0.375x + 0.325x + 3 = -0.325x + 0.325x + 5.8$

$0.7x + 3 = 5.8$

$0.7x + 3 - 3 = 5.8 - 3$

$0.7x = 2.8$

$\dfrac{0.7x}{0.7} = \dfrac{2.8}{0.7}$

$x = 4$

Back-substitute to find p.

$p = -0.325x + 5.8$

$p = -0.325(4) + 5.8 = 4.5$

The ordered pair is (4, 4.5).
Equilibrium number of workers: 4 million
Equilibrium hourly wage: $4.50

b. If workers are paid $\underline{\$4.50}$ per hour, there will be $\underline{4}$ million available workers and $\underline{4}$ million workers will be hired. In this state of market equilibrium, there is no unemployment.

c.
$$p = -0.325x + 5.8$$
$$5.15 = -0.325x + 5.8$$
$$0.65 = -0.325x$$
$$\frac{-0.65}{-0.325} = \frac{-0.325x}{-0.325}$$
$$2 = x$$
At $5.15 per hour, 2 million workers will be hired.

d.
$$p = 0.375x + 3$$
$$5.15 = 0.375x + 3$$
$$2.15 = 0.375x$$
$$\frac{2.15}{0.375} = \frac{0.375x}{0.375}$$
$$x \approx 5.7$$
At $5.15 per hour, there will be about 5.7 million available workers.

e. $5.7 - 2 = 3.7$
At $5.15 per hour, there will be about 3.7 million more people looking for work than employers are willing to hire.

43. – 47. Answers will vary.

49. does not make sense; Explanations will vary. Sample explanation: Solving for x in the second equation will allow us to avoid fractions.

51. does not make sense; Explanations will vary. Sample explanation: Equilibrium is the point at which demand is equal to supply.

53. true

55. false; Changes to make the statement true will vary. A sample change is: Replace y in the second equation with $2x - 5$.

57. $y = mx + 3$

$5x - 2y = 7$

Start by writing the second equation in slope-intercept form.
$$5x - 2y = 7$$
$$-2y = -5x + 7$$
$$y = \frac{5}{2}x - \frac{7}{2}$$
The system will be inconsistent if the graphs of the two equations have the same slope and different y-intercepts. The y-intercepts are different. Therefore, the system will be inconsistent if $m = \frac{5}{2}$.

58. $4x + 6y = 12$

x-intercept:
$$4x + 6y = 12$$
$$4x + 6(0) = 12$$
$$4x = 12$$
$$x = 3$$
y-intercept:
$$4x + 6y = 12$$
$$4(0) + 6y = 12$$
$$6y = 12$$
$$y = 2$$
Checkpoint:
$$4x + 6y = 12$$
$$4(-3) + 6y = 12$$
$$-12 + 6y = 12$$
$$6y = 24$$
$$y = 4$$
Draw a line through $(3,0)$, $(0, 2)$, and $(-3, 4)$.

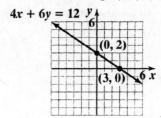

4x + 6y = 12

59. $4(x + 1) = 25 + 3(x - 3)$
$$4x + 4 = 25 + 3x - 9$$
$$x + 4 = 16$$
$$x = 12$$
The solution set is $\{12\}$.

60. The integers in the given set are -73, 0, and $\frac{3}{1} = 3$.

61.
$$3x + 2y = 48$$
$$3x + 2(12) = 48$$
$$3x + 24 = 48$$
$$3x = 24$$
$$x = 8$$

$$9x - 8y = -24$$
$$9x - 8(12) = -24$$
$$9x - 96 = -24$$
$$9x = 72$$
$$x = 8$$

The same value of x is obtained in both equations, so $(8,12)$ is the solution.

62. $-14y = -168$
$$\frac{-14y}{-14} = \frac{-168}{-14}$$
$$y = 12$$
The solution set is $\{12\}$.

63.
$$x - 5y = 3$$
$$-4(x - 5y) = -4(3)$$
$$-4x + 20y = -12$$

4.3 Check Points

1. $x + y = 5$
$x - y = 9$
Add the equations to eliminate the y-terms.
$$x + y = 5$$
$$\underline{x - y = 9}$$
$$2x \quad = 14$$
Now solve for x.
$$2x = 14$$
$$x = 7$$
Back-substitute into either of the original equations to solve for y.
$$x + y = 5$$
$$7 + y = 5$$
$$y = -2$$
The solution set is $\{(7, -2)\}$.

2. $4x - y = 22$
$3x + 4y = 26$
Multiply each term of the first equation by 4 and add the equations to eliminate y.
$$16x - 4y = 88$$
$$\underline{3x + 4y = 26}$$
$$19x \quad = 114$$
$$x = 6$$
Back-substitute into either of the original equations to solve for y.
$$4x - y = 22$$
$$4(6) - y = 22$$
$$24 - y = 22$$
$$-y = -2$$
$$y = 2$$
The solution set is $\{(6, 2)\}$.

3. $4x + 5y = 3$
$2x - 3y = 7$
Multiply each term of the second equation by -2 and add the equations to eliminate x.
$$4x + 5y = 3$$
$$\underline{-4x + 6y = -14}$$
$$11y = -11$$
$$y = -1$$
Back-substitute into either of the original equations to solve for x.
$$2x - 3y = 7$$
$$2x - 3(-1) = 7$$
$$2x + 3 = 7$$
$$2x = 4$$
$$x = 2$$
The solution set is $\{(2, -1)\}$.

4. $2x = 9 + 3y$

$4y = 8 - 3x$

Rewrite each equation in the form $Ax + By = C$.

$2x - 3y = 9$

$3x + 4y = 8$

Multiply the top equation by 4 and multiply the bottom equation by 3.

$8x - 12y = 36$

$\underline{9x + 12y = 24}$

$17x \qquad = 60$

$x = \dfrac{60}{17}$

Back-substitution of $\dfrac{60}{17}$ to find y would cause cumbersome arithmetic.

Instead, use the system that is in the form $Ax + By = C$ to eliminate x and find y.

$2x - 3y = 9$

$3x + 4y = 8$

Multiply the top equation by -3 and multiply the bottom equation by 2.

$-6x + 9y = -27$

$\underline{6x + 8y = \; 16}$

$17y = -11$

$y = \dfrac{-11}{17}$

The solution set is $\left\{ \left(\dfrac{60}{17}, -\dfrac{11}{17} \right) \right\}$.

5. $x + 2y = 4$

$3x + 6y = 13$

Multiply the first equation by -3.

$-3x - 6y = -12$

$\underline{3x + 6y = \; 13}$

$0 = \; 1, \; \text{false}$

The false statement indicates that the system is inconsistent and has no solution.

The solution set is $\{ \; \}$.

6. $x - 5y = 7$

$3x - 15y = 21$

Multiply the first equation by -3.

$-3x + 15y = -21$

$\underline{3x - 15y = \; 21}$

$0 = \; 0, \; \text{true}$

The true statement indicates that the system has infinitely many solutions.

The solution set is $\left\{ (x, y) \mid x - 5y = 7 \right\}$ or

$\left\{ (x, y) \mid 3x - 15y = 21 \right\}$.

4.3 Exercise Set

1. $x + y = -3$

$x - y = 11$

Add the equations to eliminate the y-terms.

$x + y = -3$

$\underline{x - y = 11}$

$2x \qquad = 8$

Now solve for x.

$2x = 8$

$x = 4$

Back-substitute into either of the original equations to solve for y.

$x + y = -3$

$4 + y = -3$

$y = -7$

The solution set is $\left\{ (4, -7) \right\}$.

3. $2x + 3y = 6$

$\underline{2x - 3y = 6}$

$4x = 12$

$x = 3$

Back-substitute into either of the original equations to solve for y.

$2x + 3y = 6$

$2(3) + 3y = 6$

$3y = 0$

$y = 0$

The solution set is $\left\{ (3, 0) \right\}$.

5. $x + 2y = 7$

  $\underline{-x + 3y = 18}$

  $\quad 5y = 25$

  $\quad\; y = 5$

Back-substitute into either of the original equations to solve for x.

  $x + 2y = 7$

  $x + 2(5) = 7$

  $\quad x + 10 = 7$

  $\quad\quad x = -3$

The solution set is $\{(-3, 5)\}$.

7. $5x - y = 14$

  $\underline{-5x + 2y = -13}$

  $\quad\quad y = 1$

Back-substitute into either of the original equations to solve for x.

  $5x - (1) = 14$

  $\quad 5x = 15$

  $\quad\; x = 3$

The solution set is $\{(3, 1)\}$.

9. $3x + y = 7$

  $2x - 5y = -1$

Multiply each term of the first equation by 5 and add the equations to eliminate y.

  $15x + 5y = 35$

  $\underline{2x - 5y = -1}$

  $17x \quad\;\; = 34$

  $\quad x \quad\;\; = 2$

Back-substitute into either of the original equations to solve for y.

  $3x + y = 7$

  $3(2) + y = 7$

  $\quad 6 + y = 7$

  $\quad\quad y = 1$

The solution set is $\{(2, 1)\}$.

11. $x + 3y = 4$

  $4x + 5y = 2$

Multiply each term of the first equation by -4 and add the equations to eliminate x.

  $-4x - 12y = -16$

  $\underline{4x + 5y = \quad 2}$

  $\quad -7y = -14$

  $\quad\; y = \quad 2$

Back-substitute into either of the original equations to solve for x.

  $x + 3y = 4$

  $x + 3(2) = 4$

  $\quad x + 6 = 4$

  $\quad\quad x = -2$

The solution set is $\{(-2, 2)\}$.

13. $-3x + 7y = \quad 14$

  $2x - y = -13$

Multiply each term of the second equation by 7 and add the equations to eliminate y.

  $-3x + 7y = \quad 14$

  $\underline{14x - 7y = -91}$

  $11x \quad\quad = -77$

  $\quad x \quad\quad = -7$

Back-substitute into either of the original equations to solve for y.

  $2x - y = -13$

  $2(-7) - y = -13$

  $\quad -14 - y = -13$

  $\quad\quad -y = 1$

  $\quad\quad\; y = -1$

The solution set is $\{(-7, -1)\}$.

15. $3x - 14y = 6$

$5x + 7y = 10$

Multiply each term of the second equation by 2 and add the equations to eliminate y.

$3x - 14y = 6$

$\underline{10x + 14y = 20}$

$13x \qquad = 26$

$x \qquad = 2$

Back-substitute into either of the original equations to solve for y.

$5x + 7y = 10$

$5(2) + 7y = 10$

$10 + 7y = 10$

$7y = 0$

$y = 0$

The solution set is $\{(2, 0)\}$.

17. $3x - 4y = 11$

$2x + 3y = -4$

Multiply the first equation by 3, and the second equation by 4.

$9x - 12y = 33$

$\underline{8x + 12y = -16}$

$17x \qquad = 17$

$x \qquad = 1$

Back-substitute into either of the original equations to solve for y.

$2x + 3y = -4$

$2(1) + 3y = -4$

$3y = -6$

$y = -2$

The solution set is $\{(1, -2)\}$.

19. $3x + 2y = -1$

$-2x + 7y = 9$

Multiply the first equation by 2 and the second equation by 3.

$6x + 4y = -2$

$\underline{-6x + 21y = 27}$

$25y = 25$

$y = 1$

Back-substitute into either of the original equations to solve for x.

$3x + 2(1) = -1$

$3x = -3$

$x = -1$

The solution set is $\{(-1, 1)\}$.

21. $3x = 2y + 7$

$5x = 2y + 13$

Rewrite each equation in the form $Ax + By = C$.

$3x - 2y = 7$

$5x - 2y = 13$

Multiply the first equation by -1 and add the equations to eliminate y.

$-3x + 2y = -7$

$\underline{5x - 2y = 13}$

$2x = 6$

$x = 3$

Back-substitute into either of the original equations to solve for y.

$3x = 2y + 7$

$3(3) = 2y + 7$

$2 = 2y$

$1 = y$

The solution set is $\{(3, 1)\}$.

23. $2x = 3y - 4$

$-6x + 12y = 6$

Rewrite the first equation in the form $Ax + By = C$.

$2x - 3y = -4$

Multiply the first equation by 3 and add to eliminate x.

$6x - 9y = -12$

$\underline{-6x + 12y = 6}$

$3y = -6$

$y = -2$

Back-substitute into either of the original equations to solve for x.

$2x = 3(-2) - 4$

$2x = -6 - 4$

$2x = -10$

$x = -5$

The solution set is $\{(-5, -2)\}$.

25. $2x - y = 3$

$4x + 4y = -1$

Multiply the first equation by 4 and add to eliminate y.

$8x - 4y = 12$

$\underline{4x + 4y = -1}$

$12x = 11$

$x = \dfrac{11}{12}$

Instead of back-substituting $\dfrac{11}{12}$ and working with fractions, go back to the original system. Multiply the first equation by -2 and add the equations to eliminate x.

$-4x + 2y = -6$

$\underline{4x + 4y = -1}$

$6y = -7$

$y = -\dfrac{7}{6}$

The solution set is $\left\{\left(\dfrac{11}{12}, -\dfrac{7}{6}\right)\right\}$.

27. $4x = 5 + 2y$

$2x + 3y = 4$

Rewrite the first equation in the form $Ax + By = C$, and multiply the second equation by -2.

$4x - 2y = 5$

$\underline{-4x - 6y = -8}$

$-8y = -3$

$y = \dfrac{3}{8}$

Instead of back-substituting $\dfrac{3}{8}$ and working with fractions, go back to the original system. Use the rewritten form of the first equation, and multiply by -3. Solve by addition.

$-12x + 6y = -15$

$\underline{{-4x} - 6y = -8}$

$-16x = -23$

$x = \dfrac{23}{16}$

The solution set is $\left\{\left(\dfrac{23}{16}, \dfrac{3}{8}\right)\right\}$.

29. $3x - y = 1$

$3x - y = 2$

Multiply the first equation by -1.

$-3x + y = -1$

$\underline{3x - y = 2}$

$0 = 1, \text{ false}$

The false statement indicates that the system is inconsistent and has no solution.

The solution set is $\{\ \}$.

31. $x + 3y = 2$

$3x + 9y = 6$

Multiply the first equation by -3.

$-3x - 9y = -6$

$\underline{3x + 9y = 6}$

$0 = 0, \text{ true}$

The true statement indicates that the system has infinitely many solutions.

The solution set is $\{(x, y) \mid x + 3y = 2\}$ or

$\{(x, y) \mid 3x + 9y = 6\}$.

33. $7x - 3y = 4$

$-14x + 6y = -7$

Multiply the first equation by 2.

$14x - 6y = 8$

$\underline{-14x + 6y = -7}$

$ 0 = 1, \text{ false}$

The false statement indicates that the system is inconsistent and has no solution.

The solution set is $\{\ \}$.

35. $5x + y = 2$

$3x + y = 1$

Multiply the second equation by -1.

$5x + y = 2$

$\underline{-3x - y = -1}$

$ 2x = 1$

$ x = \dfrac{1}{2}$

Back-substitute $\dfrac{1}{2}$ for x and solve for y.

$3\left(\dfrac{1}{2}\right) + y = 1$

$ y = -\dfrac{1}{2}$

The solution set is $\left\{\left(\dfrac{1}{2}, -\dfrac{1}{2}\right)\right\}$.

37. $x = 5 - 3y$

$2x + 6y = 10$

Rewrite the first equation in the form $Ax + By = C$, and multiply the second equation by -2.

$-2x - 6y = -10$

$\underline{2x + 6y = 10}$

$ 0 = 0, \text{ true}$

The true statement indicates that the system has infinitely many solutions.

The solution set is $\{(x, y) | x = 5 - 3y\}$ or

$\{(x, y) | 2x + 6y = 10\}$.

39. $4(3x - y) = 0$

$3(x + 3) = 10y$

Rewrite both equations.

$12x - 4y = 0$

$3x - 10y = -9$

Multiply the second equation by -4 and add the equations to eliminate x.

$12x - 4y = 0$

$\underline{-12x + 40y = 36}$

$ 36y = 36$

$ y = 1$

Back-substitute 1 for y in one of the original equations and solve for x.

$12x - 4y = 0$

$12x - 4(1) = 0$

$ 12x = 4$

$ x = \dfrac{1}{3}$

The solution set is $\left\{\left(\dfrac{1}{3}, 1\right)\right\}$.

41. $x + y = 11$

$\dfrac{x}{5} + \dfrac{y}{7} = 1$

Multiply the second equation by the LCD, 35, to clear fractions.

$35\left(\dfrac{x}{5} + \dfrac{y}{7}\right) = 35(1)$

$ 7x + 5y = 35$

Now solve the system.

$x + y = 11$

$7x + 5y = 35$

Multiply the top equation by -5 and add the result to the second equation.

$-5x - 5y = -55$

$\underline{7x + 5y = 35}$

$ 2x = -20$

$ x = -10$

Back-substitute to find y.

$-10 + y = 11$

$ y = 21$

The solution set is $\{(-10, 21)\}$.

43. $\dfrac{4}{5}x - y = -1$

$\dfrac{2}{5}x + y = 1$

Multiply both equations by 5 to clear fractions.

$4x - 5y = -5$

$\underline{2x + 5y = 5}$

$6x = 0$

$x = 0$

Back-substitute 0 for x and solve for y.

$\dfrac{2}{5}(0) + y = 1$

$y = 1$

The solution set is $\{(0,1)\}$.

45. $3x - 2y = 8$

$x = -2y$

The substitution method is a good choice because the second equation is already solved for x.
Substitute $-2y$ for x in the first equation

$3x - 2y = 8$

$3(-2y) - 2y = 8$

$-6y - 2y = 8$

$-8y = 8$

$y = -1$

Back-substitute -1 for y in the second equation.

$x = -2y = -2(-1) = 2$

The solution set is $\{(2,-1)\}$.

47. $3x + 2y = -3$

$2x - 5y = 17$

The addition method is a good choice because both equations are written in the form $Ax + By = C$.

Multiply the first equation by 2 and the second equation by -3.

$6x + 4y = -6$

$\underline{-6x + 15y = -51}$

$19y = -57$

$y = -3$

Back-substitute -3 for y and solve for x.

$3x + 2(-3) = -3$

$3x - 6 = -3$

$3x = 3$

$x = 1$

The solution set is $\{(1,-3)\}$.

49. $3x - 2y = 6$

$y = 3$

The substitution method is a good choice because the second equation is already solved for y.
Substitute 3 for y in the first equation.

$3x - 2y = 6$

$3x - 2(3) = 6$

$3x - 6 = 6$

$3x = 12$

$x = 4$

It is not necessary to back-substitute to find the value of y because $y = 3$ is one of the equations of the given system.

The solution set is $\{(4,3)\}$.

51. $y = 2x + 1$

$y = 2x - 3$

The substitution method is a good choice, because both equations are already solved for y. Substitute $2x + 1$ for y in the second equation.

$y = 2x - 3$

$2x + 1 = 2x - 3$

$2x + 1 - 2x = 2x - 3 - 2x$

$1 = -3, \quad \text{false}$

The false statement indicates that the system has no solution.

The solution set is $\{\ \}$.

53. $2(x + 2y) = 6$

$3(x + 2y - 3) = 0$

The addition method is a good choice since the equations can easily be simplified to give equations of the form $Ax + By = C$.

$2x + 4y = 6 \qquad\qquad 3x + 6y - 9 = 0$

$ 3x + 6y = 9$

Solve the resulting system.

$2x + 4y = 6$

$3x + 6y = 9$

Multiply the first equation by -3 and the second by 2 and solve by addition.

$-6x - 12y = -18$

$\underline{6x + 12y = 18}$

$0 = 0, \text{ true}$

The true statement indicates that the system has infinitely many solutions.

The solution set is $\{(x,y) | 2(x + 2y) = 6\}$ or

$\{(x,y) | 3(x + 2y - 3) = 0\}$.

55.
$$3y = 2x$$
$$2x + 9y = 24$$

The substitution method is a good choice because the first equation can easily be solved for one of the variables. Solve this equation for y.

$$3y = 2x$$
$$y = \frac{2}{3}x$$

Substitute $\frac{2}{3}x$ for y in the second equation.

$$2x + 9y = 24$$
$$2x + 9\left(\frac{2}{3}x\right) = 24$$
$$2x + 6x = 24$$
$$8x = 24$$
$$x = 3$$

Back-substitute 3 for x in the equation, $y = \frac{2}{3}x$.

$$y = \frac{2}{3}x = \frac{2}{3}(3) = 2$$

The solution set is $\{(3,2)\}$.

57.
$$\frac{3x}{5} + \frac{4y}{5} = 1$$
$$\frac{x}{4} - \frac{3y}{8} = -1$$

Multiply the first equation by 5 and the second equation by 8 to clear fractions.

$$\frac{3x}{5} + \frac{4y}{5} = 1 \qquad \frac{x}{4} - \frac{3y}{8} = -1$$
$$3x + 4y = 5 \qquad 2x - 3y = -8$$

The addition method is a good choice since both equations are of the form $Ax + By = C$.

$$3x + 4y = 5$$
$$2x - 3y = -8$$

Multiply the first equation by 3 and the second equation by 4.

$$9x + 12y = 15$$
$$\underline{8x - 12y = -32}$$
$$17x = -17$$
$$x = -1$$

Back-substitute -1 for x in the equation and solve for y.

$$3x + 4y = 5$$
$$3(-1) + 4y = 5$$
$$-3 + 4y = 5$$
$$4y = 8$$
$$y = 2$$

The solution set is $\{(-1,2)\}$.

59.
$$5(x+1) = 7(y+1) - 7$$
$$6(x+1) + 5 = 5(y+1)$$

Simplify both equations.

$$5(x+1) = 7(y+1) - 7$$
$$5x + 5 = 7y + 7 - 7$$
$$5x + 5 = 7y$$
$$5x - 7y + 5 = 0$$
$$5x - 7y = -5$$

$$6(x+1) + 5 = 5(y+1)$$
$$6x + 6 + 5 = 5y + 5$$
$$6x + 11 = 5y + 5$$
$$6x - 5y + 11 = 5$$
$$6x - 5y = -6$$

Solve the rewritten system.
$$5x - 7y = -5$$
$$6x - 5y = -6$$

Multiply the first equation by -6 and the second equation by 5, and solve by addition.

$$-30x + 42y = 30$$
$$\underline{30x - 25y = -30}$$
$$17y = 0$$
$$y = 0$$

Back-substitute 0 for y to find x.
$$5x - 7y = -5$$
$$5x - 7(0) = -5$$
$$5x - 0 = -5$$
$$5x = -5$$
$$x = -1$$

The solution set is $\{(-1,0)\}$.

61. $0.4x + y = 2.2$
$0.5x - 1.2y = 0.3$
Multiply the first equation by 1.2 and solve by addition.

$$0.48x + 1.2y = 2.64$$
$$\underline{0.50x - 1.2y = 0.30}$$
$$0.98x \quad\quad = 2.94$$
$$x \quad\quad = 3$$

Back-substitute 3 for x to find y.
$$0.4x + y = 2.2$$
$$0.4(3) + y = 2.2$$
$$1.2 + y = 2.2$$
$$y = 1$$

The solution set is $\{(3,1)\}$.

63. $\dfrac{x}{2} = \dfrac{y+8}{3}$

$\dfrac{x+2}{2} = \dfrac{y+11}{3}$

Simplify the first equation.
$$\frac{x}{2} = \frac{y+8}{3}$$
$$3x = 2(y+8)$$
$$3x = 2y + 16$$
$$3x - 2y = 16$$

Simplify the second equation.
$$\frac{x+2}{2} = \frac{y+11}{3}$$
$$3(x+2) = 2(y+11)$$
$$3x + 6 = 2y + 22$$
$$3x - 2y + 6 = 22$$
$$3x - 2y = 16$$

When simplified, the equations are the same. This means that the system is dependent and there are an infinite number of solutions.

The solution set is $\left\{(x,y) \,\middle|\, \dfrac{x}{2} = \dfrac{y+8}{3}\right\}$ or

$\left\{(x,y) \,\middle|\, \dfrac{x+2}{2} = \dfrac{y+11}{3}\right\}$.

65. $x + 2y = 50$
$-x + 2y = 24$
Add the equations.
$$x + 2y = 50$$
$$\underline{-x + 2y = 24}$$
$$4y = 74$$
$$y = 18.5$$

Back-substitute 18.5 for y and solve for x.
$$x + 2y = 50$$
$$x + 2(18.5) = 50$$
$$x + 37 = 50$$
$$x = 13$$

The solution is $(13, 18.5)$. This means that 13 years after 1996, or 2009, the percentage who will be pro-life and pro-choice will be the same at 18.5%.

67. – 71. Answers will vary.

73. does not make sense; Explanations will vary. Sample explanation: The addition method does not involve graphing.

75. does not make sense; Explanations will vary. Sample explanation: When one of equations has a variable on one side by itself, it is typically best to use the substitution method.

77. false; Changes to make the statement true will vary. A sample change is: If $(2,-2)$ satisfies $Ax + 2y = 2$ and $2x + By = 10$, then A and B can be found by substitution.
Find A.
$$Ax + 2y = 2$$
$$A(2) + 2(-2) = 2$$
$$2A - 4 = 2$$
$$2A = 6$$
$$A = 3$$
Find B.
$$2x + By = 10$$
$$2(2) + B(-2) = 10$$
$$4 - 2B = 10$$
$$-2B = 6$$
$$B = -3$$

79. false; Changes to make the statement true will vary. A sample change is: After these multiplications, the coefficients of x will be the same. Thus, adding them will not eliminate them.

81.
$$Ax - 3y = 16$$
$$A(5) - 3(-2) = 16$$
$$5A + 6 = 16$$
$$5A = 10$$
$$A = 2$$

$$3x + By = 7$$
$$3(5) + B(-2) = 7$$
$$15 - 2B = 7$$
$$-2B = -8$$
$$B = 4$$

82. Let x = the unknown number.
$$5x = x + 40$$
$$4x = 40$$
$$x = 10$$
The number is 10.

83. Because the x-coordinate is negative and the y-coordinate is positive, $\left(-\dfrac{3}{2}, 15\right)$ is located in quadrant II.

84.
$$29,700 + 150x = 5000 + 1100x$$
$$29,700 - 950x = 5000$$
$$-950x = -24,700$$
$$x = 26$$
The solution set is $\{26\}$.

85. a. $x + y = 28$
$x - y = 6$

b.
$$x + y = 28$$
$$\underline{x - y = 6}$$
$$2x = 34$$
$$x = 17$$
Substitute 17 for x to find y.
$$x + y = 28$$
$$17 + y = 28$$
$$y = 11$$
The numbers are 17 and 11.

86. $3x + 2y$

87. a. $\$20 + \$0.05(200) = \$30$

  b. $y = 20 + 0.05x$

Chapter 4 Mid-Chapter Check Points

1. $3x + 2y = 6$
$2x - \ y = 4$

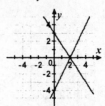

The lines intersect at $(2, 0)$.
The solution set is $\{(2,0)\}$.

2. $y = 2x - 1$
$y = 3x - 2$

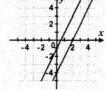

The lines intersect at $(1, 1)$.
The solution set is $\{(1,1)\}$.

3. $y = 2x - 1$
$6x - 3y = 12$

The lines are parallel, so the system is inconsistent and there is no solution.
The solution set is $\{\ \}$.

4. $5x - 3y = 1$

   $y = 3x - 7$

Substitute $3x - 7$ for y in the first equation and solve for x.

    $5x - 3y = 1$

  $5x - 3(3x - 7) = 1$

   $5x - 9x + 21 = 1$

    $-4x + 21 = 1$

     $-4x = -20$

      $x = 5$

Back-substitute to find y.

$y = 3x - 7 = 3(5) - 7 = 15 - 7 = 8$

The solution set is $\{(5, 8)\}$.

5. $6x + 5y = 7$

   $3x - 7y = 13$

Multiply the second equation by -2 and add to eliminate x.

   $6x + 5y = 7$

  $\underline{-6x + 14y = -26}$

     $19y = -19$

      $y = -1$

Back-substitute -1 for y and solve for x.

   $6x + 5y = 7$

  $6x + 5(-1) = 7$

   $6x - 5 = 7$

    $6x = 12$

     $x = 2$

The solution set is $\{(2, -1)\}$.

6. $x = \dfrac{y}{3} - 1$

   $6x + y = 21$

Substitute $\dfrac{y}{3} - 1$ for x in the second equation and solve for y.

    $6x + y = 21$

  $6\left(\dfrac{y}{3} - 1\right) + y = 21$

   $2y - 6 + y = 21$

    $3y - 6 = 21$

     $3y = 27$

      $y = 9$

Back-substitute to find x.

$x = \dfrac{y}{3} - 1 = \dfrac{9}{3} - 1 = 3 - 1 = 2$

The solution set is $\{(2, 9)\}$.

7. $3x - 4y = 6$

   $5x - 6y = 8$

Multiply the first equation by -5, the second equation by 3 and add to eliminate x.

  $-15x + 20y = -30$

  $\underline{15x - 18y = 24}$

    $2y = -6$

     $y = -3$

Back-substitute -3 for y and solve for x.

   $3x - 4y = 6$

  $3x - 4(-3) = 6$

   $3x + 12 = 6$

    $3x = -6$

     $x = -2$

The solution set is $\{(-2, -3)\}$.

8. $3x - 2y = 32$

   $\dfrac{x}{5} + 3y = -1$

Multiply the second equation by 5 to clear the fraction.

  $5\left(\dfrac{x}{5} + 3y\right) = 5(-1)$

   $x + 15y = -5$

The system is as follows.

$3x - 2y = 32$

$x + 15y = -5$

Multiply the second equation by -3 and solve by addition.

   $3x - 2y = 32$

  $\underline{-3x - 45y = 15}$

    $-47y = 47$

     $y = -1$

Back-substitute -1 for y and solve for x.

   $x + 15y = -5$

  $x + 15(-1) = -5$

   $x - 15 = -5$

    $x = 10$

The solution set is $\{(10, -1)\}$.

9. $x - y = 3$

$2x = 4 + 2y$

Solve the first equation for x.

$x - y = 3$

$x = y + 3$

Substitute $y + 3$ for x in the second equation and solve for y.

$2x = 4 + 2y$

$2(y + 3) = 4 + 2y$

$2y + 6 = 4 + 2y$

$6 = 4$

Notice that y has also been eliminated. The false statement $6 = 4$ indicates that the system is inconsistent and has no solution. The solution set is $\{\ \}$.

10. $x = 2(y - 5)$

$4x + 40 = y - 7$

Substitute $2(y - 5)$ for x in the second equation and solve for y.

$4x + 40 = y - 7$

$4(2(y - 5)) + 40 = y - 7$

$4(2y - 10) + 40 = y - 7$

$8y - 40 + 40 = y - 7$

$8y = y - 7$

$7y = -7$

$y = -1$

Back-substitute -1 for y and solve for x.

$x = 2(y - 5)$

$= 2(-1 - 5) = 2(-6) = -12$

The solution set is $\{(-12, -1)\}$.

11. $y = 3x - 2$

$y = 2x - 9$

Substitute $3x - 2$ for y in the second equation and solve for x.

$y = 2x - 9$

$3x - 2 = 2x - 9$

$x - 2 = -9$

$x = -7$

Back-substitute -7 for x and solve for y.

$y = 3(-7) - 2 = -21 - 2 = -23$

The solution set is $\{(-7, -23)\}$.

12. $2x - 3y = 4$

$3x + 4y = 0$

Multiply the first equation by 4, the second equation by 3 and add to eliminate y.

$8x - 12y = 16$

$\underline{9x + 12y = 0}$

$17x \qquad = 16$

$x \qquad = \dfrac{16}{17}$

Back-substitution of $\dfrac{16}{17}$ to find y would cause cumbersome arithmetic.

Instead, use the addition to eliminate x and find y.

$2x - 3y = 4$

$3x + 4y = 0$

Multiply the top equation by -3 and multiply the bottom equation by 2.

$-6x + 9y = -12$

$\underline{6x + 8y = \ 0}$

$17y = -12$

$y = \dfrac{-12}{17}$

The solution set is $\left\{\left(\dfrac{16}{17}, -\dfrac{12}{17}\right)\right\}$.

13. $y - 2x = 7$

$4x = 2y - 14$

Solve the first equation for y.

$y - 2x = 7$

$y = 2x + 7$

Substitute $2x + 7$ for y and solve for x.

$4x = 2y - 14$

$4x = 2(2x + 7) - 14$

$4x = 4x + 14 - 14$

$0 = 14 - 14$

$0 = 0$

Notice that x has also been eliminated. The true statement $0 = 0$ indicates that the system is dependent and has infinitely many solutions. The solution set is $\{(x, y) | y - 2x = 7\}$ or $\{(x, y) | 4x = 2y - 14\}$.

14. $4(x+3) = 3y+7$

$2(y-5) = x+5$

First, rewrite both equations in the form $Ax + By = C$.

$$4(x+3) = 3y+7$$
$$4x+12 = 3y+7$$
$$4x-3y+12 = 7$$
$$4x-3y = -5$$

$$2(y-5) = x+5$$
$$2y-10 = x+5$$
$$-x+2y-10 = 5$$
$$-x+2y = 15$$

The system is as follows.

$4x-3y = -5$

$-x+2y = 15$

Multiply the second equation by 4 and add to eliminate x.

$$4x-3y = -5$$
$$\underline{-4x+8y = 60}$$
$$5y = 55$$
$$y = 11$$

Back-substitute 11 for y and solve for x.

$$-x+2y = 15$$
$$-x+2(11) = 15$$
$$-x+22 = 15$$
$$-x = -7$$
$$x = 7$$

The solution set is $\{(7,11)\}$.

15. $\dfrac{x}{2} - \dfrac{y}{5} = 1$

$y - \dfrac{x}{3} = 8$

Multiply the first equation by 10 and the second equation by 3 to clear fractions.

$$10\left(\frac{x}{2}\right) - 10\left(\frac{y}{5}\right) = 10(1)$$
$$5x-2y = 10$$

$$3y-3\left(\frac{x}{3}\right) = 3(8)$$
$$3y-x = 24$$
$$-x+3y = 24$$

The system is as follows.

$5x-2y = 10$

$-x+3y = 24$

Multiply the second equation by 5 and solve by addition.

$$5x- 2y = 10$$
$$\underline{-5x+15y = 120}$$
$$13y = 130$$
$$y = 10$$

Back-substitute and solve for x.

$$-x+3y = 24$$
$$-x+3(10) = 24$$
$$-x+30 = 24$$
$$-x = -6$$
$$x = 6$$

The solution set is $\{(6,10)\}$.

4.4 Check Points

1. Let x = average time per day women spend socializing.
   Let y = average time per day men spend socializing.

   $$x + y = 138$$
   $$\underline{x - y = 8}$$
   $$2x = 146$$
   $$x = 73$$

   Back-substitute 73 for x to find y.

   $$x + y = 138$$
   $$73 + y = 138$$
   $$y = 65$$

   Men average 65 minutes per day socializing and women average 73 minutes.

2. Let x = the number of calories in a Quarter Pounder.
   Let y = the number of calories in a Whopper with cheese.

   $$2x + 3y = 2607$$
   $$x + y = 1000 + 9$$

   Solve the second equation for x.

   $$x + y = 1000 + 9$$
   $$x = -y + 1009$$

   Substitute $-y + 1009$ for x to find y.

   $$2x + 3y = 2607$$
   $$2(\overbrace{-y + 1009}^{x}) + 3y = 2607$$
   $$2(-y + 1009) + 3y = 2607$$
   $$-2y + 2018 + 3y = 2607$$
   $$y + 2018 = 2607$$
   $$y = 589$$

   Back substitute to find x.

   $$x = -y + 1009 = -589 + 1009 = 420$$

   There are 420 calories in a Quarter Pounder and 589 calories in a Whopper with cheese.

3. Let x = the length of the lot.
   Let y = the width of the lot.
   Use the formula for the perimeter of a rectangle to write the first equation.

   $$P = 2l + 2w$$
   $$360 = 2x + 2y$$

   Use the other information in the problem to write the second equation.

   $$20x + 8 \cdot 2y = 3280$$

   The two equations form the system.

   $$2x + 2y = 360$$
   $$20x + 16y = 3280$$

   Multiply the first equation by -8 and add the result to the second equation.

   $$-16x - 16y = -2880$$
   $$\underline{20x + 16y = 3280}$$
   $$4x = 400$$
   $$x = 100$$

   Back-substitute to find y.

   $$2x + 2y = 360$$
   $$2(100) + 2y = 360$$
   $$200 + 2y = 360$$
   $$2y = 160$$
   $$y = 80$$

   The length is 100 feet and the width is 80 feet.

4. Let x = the number of years the heating system is used.
   Let y = the total cost of the heating system.

   $$y = 5000 + 1100x$$
   $$y = 12,000 + 700x$$

   Solve by the substitution method.

   $$y = 12,000 + 700x$$
   $$\overbrace{5000 + 1100x}^{y} = 12,000 + 700x$$
   $$5000 + 1100x = 12,000 + 700x$$
   $$5000 + 400x = 12,000$$
   $$400x = 7000$$
   $$x = 17.5$$

   Back-substitute to find y.

   $$y = 5000 + 1100(17.5)$$
   $$y = 24,250$$

   After 17.5 years, the total costs for the two systems will be the same, namely $24,250.

5. Let x = the amount invested at 9%.
Let y = the amount invested at 11%.
$$x + y = 5000$$
$$0.09x + 0.11y = 487$$

This system can be solved by substitution.
Solve for y in terms of x.
$$x + y = 5000$$
$$y = -x + 5000$$

Substitute this value into the other equation.
$$0.09x + 0.11y = 487$$

$$0.09x + 0.11(\overset{y}{\overbrace{-x + 5000}}) = 487$$
$$0.09x - 0.11x + 550 = 487$$
$$-0.02x + 550 = 487$$
$$-0.02x = -63$$
$$x = 3150$$

Back-substitute to find y.
$$y = -x + 5000$$
$$y = -(3150) + 5000$$
$$y = 1850$$

There was $3150 invested at 9% and $1850 invested at 11%.

6. Let x = the number of ounces of 12% acid solution.
Let y = the number of ounces of 20% acid solution.
$$x + y = 160$$
$$0.12x + 0.20y = 0.15(160)$$

This system can be solved by substitution.
Solve for y in terms of x.
$$x + y = 160$$
$$y = -x + 160$$

Substitute this value into the other equation.
$$0.12x + 0.20y = 0.15(160)$$
$$0.12x + 0.20y = 24$$

$$0.12x + 0.20(\overset{y}{\overbrace{-x + 160}}) = 24$$
$$-0.08x + 32 = 24$$
$$-0.08x = -8$$
$$x = 100$$

Back-substitute to find y.
$$y = -x + 160$$
$$y = -(100) + 160$$
$$y = 60$$

The chemist should mix 100 ounces of the 12% acid solution and 60 ounces of the 20% acid solution.

7. Let x = the rate of the motorboat in still water.
Let y = the rate of the current.

	Rate	×	Time	=	Distance
Trip with the Current	$x + y$		2		$2(x+y)$
Trip against the Current	$x - y$		3		$3(x-y)$

This gives,
$$2(x + y) = 84$$
$$3(x - y) = 84$$

This system simplifies to:
$$x + y = 42$$
$$x - y = 28$$

This system can be solved by addition.
$$x + y = 42$$
$$\underline{x - y = 28}$$
$$2x\quad\ = 70$$
$$x = 35$$

Back-substitute to find y.
$$x + y = 42$$
$$35 + y = 42$$
$$y = 7$$

The rate of the motorboat in still water is 35 miles per hour and the rate of the current is 7 miles per hour.

4.4 Exercise Set

1. Let x = one number.
Let y = the other number.
$$x + y = 17$$
$$\underline{x - y = -3}$$
$$2x\quad\ = 14$$
$$x\quad\ = 7$$

Back-substitute 7 for x to find y.
$$x + y = 17$$
$$7 + y = 17$$
$$y = 10$$

The numbers are 7 and 10.

3. Let x = one number.
Let y = the other number.
$$3x - y = -1$$
$$x + 2y = 23$$
Solve the second equation for x.
$$x + 2y = 23$$
$$x = -2y + 23$$
Substitute $-2y + 23$ for x to find y.
$$3x - y = -1$$
$$3(-2y + 23) - y = -1$$
$$-6y + 69 - y = -1$$
$$-7y + 69 = -1$$
$$-7y = -70$$
$$y = 10$$
Back substitute 10 for y to find x.
$$x = -2y + 23 = -2(10) + 23 = 3$$
The numbers are 3 and 10.

5. Let x = the average number of minutes 20- to 24-year-old women spend grooming.
Let y = the average number of minutes 20- to 24-year-old men spend grooming.
$$x + y = 86$$
$$x - y = 12$$
Add the equations to eliminate y.
$$x + y = 86$$
$$\underline{x - y = 12}$$
$$2x \quad = 98$$
$$x = 49$$
Back-substitute to find y.
$$x + y = 86$$
$$49 + y = 86$$
$$y = 37$$
20- to 24-year-old women averaged 49 minutes grooming and men averaged 37 minutes.

7. Let x = the number of calories in a Mr. Goodbar.
Let y = the number of calories in a Mounds bar.
$$x + 2y = 780$$
$$2x + y = 786$$
Multiply the bottom equation by -2 and then add the equations to eliminate y.
$$x + 2y = \quad 780$$
$$\underline{-4x - 2y = -1572}$$
$$-3x \quad = -792$$
$$x = 264$$

Back-substitute to find y.
$$x + 2y = 780$$
$$264 + 2y = 780$$
$$2y = 516$$
$$y = 258$$
There are 264 calories in a Mr. Goodbar and 258 calories in a Mounds bar.

9. Let x = the number of Mr. Goodbars.
Let y = the number of Mounds bars.
$$x + y = 5$$
$$16.3x + 14.1y - 70 = 7.1$$
Solve the first equation for y in terms of x.
$$x + y = 5$$
$$y = -x + 5$$
Substitute $-x + 5$ for y in the second equation.
$$16.3x + 14.1y - 70 = 7.1$$
$$16.3x + 14.1\overbrace{(-x + 5)}^{y} - 70 = 7.1$$
$$16.3x + 14.1(-x + 5) - 70 = 7.1$$
$$16.3x - 14.1x + 70.5 - 70 = 7.1$$
$$2.2x + 0.5 = 7.1$$
$$2.2x = 6.6$$
$$x = 3$$
Back-substitute to find y.
$$x + y = 5$$
$$3 + y = 5$$
$$y = 2$$
There are 3 Mr. Goodbars and 2 Mounds bars.

11. Let x = the price of one sweater.
Let y = the price of one shirt.
$$x + 3y = 42$$
$$3x + 2y = 56$$
Multiply the first equation by -3 and add the result to the second equation.
$$-3x - 9y = -126$$
$$\underline{3x + 2y = \quad 56}$$
$$-7y = -70$$
$$y = \quad 10$$
Back-substitute 10 for y and solve for x.
$$x + 3y = 42$$
$$x + 3(10) = 42$$
$$x + 30 = 42$$
$$x = 12$$
The price of one sweater is \$12 and the price of one shirt is \$10.

13. Let x = the length of a badminton court.
Let y = the width of a badminton court.
Use the formula for the perimeter of a rectangle to write the first equation.

$P = 2l + 2w$

$128 = 2x + 2y$

Use the other information in the problem to write the second equation.

$6x + 9y = 444$

The two equations form the system.

$2x + 2y = 128$

$6x + 9y = 444$

Multiply the first equation by -3 and add the result to the second equation.

$-6x - 6y = -384$

$\underline{6x + 9y = \ \ 444}$

$3y = 60$

$y = 20$

Back-substitute 20 for y and solve for x.

$2x + 2y = 128$

$2x + 2(20) = 128$

$2x + 40 = 128$

$2x = 88$

$x = 44$

The length is 44 feet and the width is 20 feet, so the dimensions of a standard badminton court are 44 feet by 20 feet.

15. Let x = the length of the lot.
Let y = the width of the lot.
Use the formula for the perimeter of a rectangle to write the first equation.

$2x + 2y = 320$

Use the other information in the problem to write the second equation.

$16x + 5(2y) = 2140$

These two equations form the system.

$2x + 2y = 320$

$16x + 10y = 2140$

Multiply the first equation by -5 and add the result to the second equation.

$-10x - 10y = -1600$

$\underline{16x + 10y = \ \ 2140}$

$6x \quad = \quad 540$

$x \quad = \quad 90$

Back-substitute 90 for x to find y.

$2(90) + 2y = 320$

$180 + 2y = 320$

$2y = 140$

$y = 70$

The length is 90 feet and the width is 70 feet, so the dimensions of the lot are 90 feet by 70 feet.

17. a. Let x = the number of minutes of long distance calls.
Let y = the monthly cost of a telephone plan.
Plan A: $y = 20 + 0.05x$

Plan B: $y = 5 + 0.10x$

Solve by substitution. Substitute $5 + 0.10x$ for y in the first equation.

$5 + 0.10x = 20 + 0.05x$

$5 + 0.05x = 20$

$0.05x = 15$

$x = 300$

Back-substitute 300 for x.

$y = 20 + 0.05x = 20 + 0.05(300) = 35$

The costs for the two plans will be equal for 300 minutes of long-distance calls per month. The cost for each plan will be $35.

b. $x = 10(20) = 200$

Plan A: $y = 20 + 0.05(200) = 30$

Plan B: $y = 5 + 0.10(200) = 25$

The monthly cost would be $30 for Plan A and $25 for Plan B, so Plan B should be selected to get the lower cost.

19. Let x = the number of dollars of merchandise purchased in a year.
Let y = the total cost for a year.
Plan A: $y = 100 + 0.80x$

Plan B: $y = 40 + 0.90x$

Substitute $40 = 0.90x$ for y in the first equation and solve for x.

$40 + 0.90x = 100 + 0.80x$

$40 + 0.10x = 100$

$0.10x = 60$

$x = 600$

Back-substitute 600 for x to find y.

$y = 100 + 0.80(600) = 580.$

If you purchase $600 worth of merchandise, you will pay the $580 under both plans.

21. Let x = the number of adult tickets sold.
Let y = the number of student tickets sold.

$$x + y = 301$$
$$3x + y = 487$$

Multiply the first equation by -1 and add it to the second equation.

$$-x - y = -301$$
$$\underline{3x + y = \ \ 487}$$
$$2x \quad \ \ = 186$$
$$x = \ \ 93$$

Back-substitute to find y.

$$x + y = 301$$
$$93 + y = 301$$
$$y = 208$$

There were 93 adult tickets sold and 208 student tickets sold.

23. Let x = the cost of an item from column A.
Let y = the cost of an item from column B.

$$x + y = 5.49$$
$$x + 2y = 6.99$$

Multiply the first equation by -1 and add it to the second equation.

$$-x - \ y = -5.49$$
$$\underline{x + 2y = \ \ 6.99}$$
$$y = 1.50$$

Back-substitute to find x.

$$x + y = 5.49$$
$$x + 1.50 = 5.49$$
$$x = 3.99$$

The cost of an item from column A is \$3.99 and the cost of an item from column B is \$1.50.

25. Let x = the number of servings of macaroni.
Let y = the number of servings of broccoli.

$$3x + 2y = 14$$
$$16x + 4y = 48$$

Multiply the first equation by -2 and add to second equation.

$$-6x - 4y = -28$$
$$\underline{16x + 4y = \ \ 48}$$
$$10x \quad \ \ = 20$$
$$x \quad \ \ = 2$$

Back-substitute 2 for x to find y.

$$3(2) + 2y = 14$$
$$2y = 8$$
$$y = 4$$

It would take 2 servings of macaroni and 4 servings of broccoli to get 14 grams of protein and 48 grams of carbohydrate.

27. The sum of the measures of the three angles of any triangle is $180°$, so

$$(x + 8y - 1) + (3y + 4) + (7x + 5) = 180.$$

Simplify this equation.

$$8x + 11y + 8 = 180$$
$$8x + 11y = 172$$

The base angles of an isosceles triangle have equal measures, so

$$3y + 4 = 7x + 5$$

Rewrite this equation in the form $Ax + By = C$.

$$7x + 5 = 3y + 4$$
$$7x - 3y = -1$$

Use the addition method to solve the system.

$$8x + 11y = 172$$
$$7x - 3y = -1$$

Multiply the first equation by 3 and the second equation by 11; then add the results.

$$24x + 33y = 516$$
$$\underline{77x - 33y = -11}$$
$$101x \quad \quad = 505$$
$$x \quad \quad = \ \ 5$$

Back-substitute 5 for x to find y.

$$7(5) - 3y = -1$$
$$35 - 3y = -1$$
$$-3y = -36$$
$$y = 12$$

Use the values of x and y to find the angle measures.

Angle A: $(x + 8y - 1)° = (5 + 8 \cdot 12 - 1)° = 100°$

Angle B: $(3y + 4)° = (3 \cdot 12 + 4)° = 40°$

Angle C: $(7x + 5)° = (7 \cdot 5 + 5)° = 40°$

29. Let x = the amount invested at 6%.
Let y = the amount invested at 8%.
$$x + y = 7000$$
$$0.06x + 0.08y = 520$$
Solve the first equation for x.
$$x = 7000 - y$$
Substitute this result for x in the second equation.
$$0.06(7000 - y) + 0.08y = 520$$
$$420 - 0.06y + 0.08y = 520$$
$$0.02y = 100$$
$$y = \frac{100}{0.02} = 5000$$
Back-substitute to solve for x.
$$x + y = 7000$$
$$x + 5000 = 7000$$
$$x = 2000$$
$2000 was invested at 6% and $5000 was invested at 8%.

31. Let x = the amount in the first fund.
Let y = the amount in the second fund.
$$0.09x + 0.03y = 900$$
$$0.10x + 0.01y = 860$$
Multiply the second equation by -3 and add the two equations.
$$0.09x + 0.03y = 900$$
$$\underline{-0.30x - 0.03y = -2580}$$
$$-0.21x = -1680$$
$$x = 8000$$
Back-substitute to solve for y.
$$0.10x + 0.01y = 860$$
$$0.10(8000) + 0.01y = 860$$
$$800 + 0.01y = 860$$
$$0.01y = 60$$
$$y = \frac{60}{0.01} = 6000$$
$8000 was invested in the first fund and $6000 was invested in the second fund.

33. Let x = amount invested with 12% return.
Let y = amount invested with the 5% loss.
$$x + y = 20,000$$
$$0.12x - 0.05y = 1890$$
Multiply the first equation by 0.05 and add the two equations.
$$0.05x + 0.05y = 1000$$
$$\underline{0.12x - 0.05y = 1890}$$
$$0.17x = 2890$$
$$x = 17,000$$
Back-substitute to solve for y.
$$x + y = 20,000$$
$$17,000 + y = 20,000$$
$$y = 3,000$$
$17,000 was invested at 12% interest and $3000 was invested at a 5% loss.

35. Let x = gallons of 5% wine.
Let y = gallons of 9% wine.
$$x + y = 200$$
$$0.05x + 0.09y = 0.07(200)$$
or
$$x + y = 200$$
$$0.05x + 0.09y = 14$$
Solve the first equation for x.
$$x + y = 200$$
$$x = 200 - y$$
Substitute this expression for x in the second equation.
$$0.05(200 - y) + 0.09y = 14$$
$$10 - 0.05y + 0.09y = 14$$
$$0.04y = 4$$
$$y = 100$$
Back-substitute and solve for x.
$$x = 200 - y$$
$$= 200 - 100$$
$$= 100$$
The wine company should mix 100 gallons of the 5% California wine with 100 gallons of the 9% French wine.

37. Let x = grams of 18-karat gold.
Let y = grams of 12-karat gold.
$$x + y = 300$$
$$0.75x + 0.5y = 0.58(300)$$
or
$$x + y = 300$$
$$0.75x + 0.5y = 174$$
Solve the first equation for x.
$$x = 300 - y$$
Substitute this result for x into the second equation and solve for y.
$$0.75(300 - y) + 0.5y = 174$$
$$225 - 0.75 + 0.5y = 174$$
$$-0.25y = -51$$
$$y = 204$$
Back-substitute to solve for x.
$$x = 300 - y = 300 - 204 = 96$$
You would need 96 grams of 18-karat gold and 204 grams of 12-karat gold.

39. Let x = pounds of cheaper candy.
Let y = pounds of more expensive candy.
$$x + y = 75$$
$$1.6x + 2.1y = 1.9(75)$$
or
$$x + y = 75$$
$$1.6x + 2.1y = 142.5$$
Multiply the first equation by -1.6 and add the two equations.
$$-1.6x - 1.6y = -120$$
$$\underline{1.6x + 2.1y = 142.5}$$
$$0.5y = 22.5$$
$$y = 45$$
Back-substitute to solve for x.
$$x + 45 = 75$$
$$x = 30$$
The manager should mix 30 pounds of the cheaper candy and 45 pounds of the more expensive candy.

41. Let n = the number of nickels.
Let d = the number of dimes.
$$n + d = 15$$
$$0.05n + 0.1d = 1.10$$
Solve the first equation for n.
$$n = 15 - d$$
$$0.05(15 - d) + 0.1d = 1.10$$
$$0.75 - 0.05d + 0.1d = 1.10$$
$$0.05d = 0.35$$
$$d = 7$$
Back-substitute to solve for n.
$$n + 7 = 15$$
$$n = 8$$
The purse has 8 nickels and 7 dimes.

43. Let x = the speed of the plane in still air.
Let y = the speed of the wind.

	Rate	×	Time	=	Distance
Trip with the Wind	$x + y$		5		$5(x + y)$
Trip against the Wind	$x - y$		8		$8(x - y)$

$$5(x + y) = 800$$
$$8(x - y) = 800$$
$$5x + 5y = 800$$
$$8x - 8y = 800$$
Multiply the first equation by 8 and the second equation by 5.
$$40x + 40y = 6400$$
$$40x - 40y = 4000$$
$$80x = 10400$$
$$x = 130$$
Back-substitute to find y.
$$5x + 5y = 800$$
$$5(130) + 5y = 800$$
$$650 + 5y = 800$$
$$5y = 150$$
$$y = 30$$
The speed of the plane in still air is 130 miles per hour and the speed of the wind is 30 miles per hour.

45. Let x = the crew's rowing rate.
Let y = the rate of the current.

	Rate	$\times$ Time	= Distance
Trip with the Current	$x + y$	2	$2(x + y)$
Trip against the Current	$x - y$	4	$4(x - y)$

$2(x + y) = 16$

$4(x - y) = 16$

Rewrite the system in $Ax + By = C$ form.

$2x + 2y = 16$

$4x - 4y = 16$

Multiply the first equation by –2.

$-4x - 4y = -32$

$\underline{4x - 4y = 16}$

$-8y = -16$

$y = 2$

Back-substitute to find x.

$2x + 2(2) = 16$

$2x + 4 = 16$

$2x = 12$

$x = 6$

The crew's rowing rate is 6 kilometers per hour and the rate of the current is 2 kilometers per hour.

47. Let x = the speed in still water.
Let y = the speed of the current.

	Rate	$\times$ Time	= Distance
Trip with the Current	$x + y$	4	$4(x + y)$
Trip against the Current	$x - y$	6	$6(x - y)$

$4(x + y) = 24$

$6(x - y) = \dfrac{3}{4}(24)$

Rewrite the system in $Ax + By = C$ form.

$4x + 4y = 24$

$6x - 6y = 18$

Multiply the first equation by –3 and the second equation by 2.

$-12x - 12y = -72$

$\underline{12x - 12y = 36}$

$-24y = -36$

$y = 1.5$

Back-substitute to find x.

$4x + 4y = 24$

$4x + 4(1.5) = 24$

$4x + 6 = 24$

$4x = 18$

$x = 4.5$

The speed in still water is 4.5 miles per hour and the speed of the current is 1.5 miles per hour.

49. – 51. Answers will vary.

53. does not make sense; Explanations will vary.
Sample explanation: The model is $y = 4 + 0.07x$.

55. does not make sense; Explanations will vary.
Sample explanation: The model is
$(4x - 2y + 4) + (12x + 6y + 12) = 180$.

57. Let x = the number of birds.
Let y = the number of lions.
Since each bird has one head and each lion has one head, $x + y = 30$.
Since each bird has two feet and each lion has four feet, $2x + 4y = 100$.
Solve the first equation for y. $y = 30 - x$
Substitute $30-x$ for y in the second equation.
$$2x + 4(30 - x) = 100$$
$$2x + 120 - 4x = 100$$
$$-2x + 120 = 100$$
$$-2x = -20$$
$$x = 10$$
Back-substitute 10 for x to find y.
$$10 + y = 30$$
$$y = 20$$
There were 10 birds and 20 lions in the zoo.

59. Let x = the number of people in the downstairs apartment.
Let y = the number of people in the upstairs apartment.
If one of the people in the upstairs apartment goes downstairs, there will be the same number of people in both apartments, so $y - 1 = x + 1$.

If one of the people in the downstairs apartment goes upstairs, there will be twice as many people upstairs as downstairs, so $y + 1 = 2(x - 1)$.

Solve the first equation for y.
$$y = x + 2$$
Also solve the second equation for y.
$$y + 1 = 2x - 2$$
$$y = 2x - 3$$
Substitute $x + 2$ for y in the last equation.
$$x + 2 = 2x - 3$$
$$-x + 2 = -3$$
$$-x = -5$$
$$x = 5$$
Back-substitute to find y.
$$y = 5 + 2 = 7$$
There are 5 people downstairs and 7 people upstairs.

61. Answers will vary.

62.
$$2(x + 3) = 24 - 2(x + 4)$$
$$2x + 6 = 24 - 2x - 8$$
$$2x + 6 = 16 - 2x$$
$$4x = 10$$
$$x = \frac{10}{4}$$
$$x = \frac{5}{2}$$
The solution set is $\left\{\dfrac{5}{2}\right\}$.

63.
$$5 + 6(x + 1) = 5 + 6x + 6$$
$$= 6x + 11$$

64. Find slope: $m = \dfrac{y_2 - y_1}{x_2 - x_1} = \dfrac{-10 - 6}{3 - (-5)} = \dfrac{-16}{8} = -2$
Use either point in the point-slope form and then solve for y.
$$y - y_1 = m(x - x_1)$$
$$y - (-10) = -2(x - 3)$$
$$y + 10 = -2x + 6$$
$$y = -2x - 4$$

65.
$$2x - y + 4z = -8$$
$$2(3) - (2) + 4(-3) = -8$$
$$-8 = -8, \text{ true}$$
Yes, the ordered triple satisfies the equation.

66.
$$5x - 2y - 4z = 3$$
$$3x + 3y + 2z = -3$$
Multiply Equation 2 by 2.
$$5x - 2y - 4z = 3$$
$$6x + 6y + 4z = -6$$
Then add to eliminate z.
$$5x - 2y - 4z = 3$$
$$\underline{6x + 6y + 4z = -6}$$
$$11x + 4y \qquad = -3$$

67.
$$ax^2 + bx + c = y$$
$$a(4)^2 + b(4) + c = 1682$$
$$16a + 4b + c = 1682$$

4.5 Check Points

1. Test the ordered triple in each equation.
$$x - 2y + 3z = 22$$
$$(-1) - 2(-4) + 3(5) = 22$$
$$22 = 22, \text{ true}$$
$$2x - 3y - z = 5$$
$$2(-1) - 3(-4) - (5) = 5$$
$$5 = 5, \text{ true}$$
$$3x + y - 5z = -32$$
$$3(-1) + (-4) - 5(5) = -32$$
$$-32 = -32, \text{ true}$$
The ordered triple $(-1, -4, 5)$ makes all three equations true, so it is a solution to the system.

2. $\quad x + 4y - z = 20$
$$3x + 2y + z = 8$$
$$2x - 3y + 2z = -16$$
Add the first two equations to eliminate z.
$$x + 4y - z = 20$$
$$\underline{3x + 2y + z = 8}$$
$$4x + 6y = 28$$
Multiply the first equation by 2 and add to the third equation to eliminate z again.
$$2x + 8y - 2z = 40$$
$$\underline{2x - 3y + 2z = -16}$$
$$4x + 5y = 24$$
Solve the system of two equations in two variables.
$$4x + 6y = 28$$
$$4x + 5y = 24$$
Multiply the second equation by -1 and add the equations.
$$4x + 6y = 28$$
$$\underline{-4x - 5y = -24}$$
$$y = 4$$
Back-substitute 4 for y to find x.
$$4x + 6y = 28$$
$$4x + 6(4) = 28$$
$$4x + 24 = 28$$
$$4x = 4$$
$$x = 1$$

Back-substitute into an original equation.
$$3x + 2y + z = 8$$
$$3(1) + 2(4) + z = 8$$
$$11 + z = 8$$
$$z = -3$$
The solution is $(1, 4, -3)$ and the solution set is $\{(1, 4, -3)\}$.

3. $\qquad 2y - z = 7$
$$x + 2y + z = 17$$
$$2x - 3y + 2z = -1$$
Since the first equation already has only two variables, use the second and third equations to eliminate x.
Multiply the second equation by -2 and add to the third equation.
$$-2x - 4y - 2z = -34$$
$$\underline{2x - 3y + 2z = -1}$$
$$-7y = -35$$
$$\frac{-7y}{-7} = \frac{-35}{-7}$$
$$y = 5$$
Back-substitute 5 for y to find z.
$$2y - z = 7$$
$$2(5) - z = 7$$
$$10 - z = 7$$
$$-z = -3$$
$$z = 3$$
Back-substitute into an original equation to find x.
$$x + 2y + z = 17$$
$$x + 2(5) + (3) = 17$$
$$x + 13 = 17$$
$$x = 4$$
The solution is $(4, 5, 3)$ and the solution set is $\{(4, 5, 3)\}$.

4. Use each ordered pair to write an equation.

$$(x, y) = (1, 4)$$

$$y = ax^2 + bx + c$$

$$4 = a(1)^2 + b(1) + c$$

$$4 = a + b + c$$

$$(x, y) = (2, 1)$$

$$y = ax^2 + bx + c$$

$$1 = a(2)^2 + b(2) + c$$

$$1 = 4a + 2b + c$$

$$(x, y) = (3, 4)$$

$$y = ax^2 + bx + c$$

$$4 = a(3)^2 + b(3) + c$$

$$4 = 9a + 3b + c$$

The system of three equations in three variables is:

$$a + b + c = 4$$

$$4a + 2b + c = 1$$

$$9a + 3b + c = 4$$

Multiplying the first equation by -1 and adding it to the second gives $3a + b = -3$.
Multiplying the first equation by -1 and adding it to the third gives $8a + 2b = 0$.
Solve this system of two equations in two variables.

$$3a + b = -3$$

$$8a + 2b = 0$$

Multiply the first equation by -2 and add to the second equation.

$$-6a - 2b = 6$$

$$\underline{8a + 2b = 0}$$

$$2a = 6$$

$$a = 3$$

Back-substitute to find b.

$$3a + b = -3$$

$$3(3) + b = -3$$

$$9 + b = -3$$

$$b = -12$$

Back-substitute into an original equation to find c.

$$a + b + c = 4$$

$$(3) + (-12) + c = 4$$

$$-9 + c = 4$$

$$c = 13$$

The quadratic equation is $y = 3x^2 - 12x + 13$

4.5 Exercise Set

1. Test the ordered triple in each equation.

$$x + y + z = 4 \qquad\qquad x - 2y - z = 1 \qquad\qquad 2x - y - z = -1$$
$$2 - 1 + 3 = 4 \qquad\qquad 2 - 2(-1) - 3 = 1 \qquad\qquad 2(2) - (-1) - 3 = -1$$
$$4 = 4, \text{ true} \qquad\qquad 2 + 2 - 3 = 1 \qquad\qquad 4 + 1 - 3 = -1$$
$$1 = 1, \text{ true} \qquad\qquad 2 = -1, \text{ false}$$

The ordered triple $(2, -1, 3)$ does not make all three equations true, so it is not a solution.

3. Test the ordered triple in each equation.

$$x - 2y = 2 \qquad\qquad 2x + 3y = 11 \qquad\qquad y - 4z = -7$$
$$4 - 2(1) = 2 \qquad\qquad 2(4) + 3(1) = 11 \qquad\qquad 1 - 4(2) = -7$$
$$4 - 2 = 2 \qquad\qquad 8 + 3 = 11 \qquad\qquad 1 - 8 = -7$$
$$2 = 2, \text{ true} \qquad\qquad 11 = 11, \text{ true} \qquad\qquad -7 = -7, \text{ true}$$

The ordered triple makes all three equations true, so it is a solution.

5. $x + y + 2z = 11$
   $x + y + 3z = 14$
   $x + 2y - z = 5$

   Multiply the second equation by -1 and add to the first equation.

   $$\begin{array}{r} x + y + 2z = 11 \\ -x - y - 3z = -14 \\ \hline -z = -3 \\ z = 3 \end{array}$$

   Back-substitute 3 for z in the first and third equations.

   $$\begin{array}{ll} x + y + 2z = 11 & x + 2y - z = 5 \\ x + y + 2(3) = 11 & x + 2y - 3 = 5 \\ x + y + 6 = 11 & x + 2y = 8 \\ x + y = 5 & \end{array}$$

   We now have two equations in two variables.
   $x + y = 5$
   $x + 2y = 8$

   Multiply the first equation by -1 and solve by addition.

   $$\begin{array}{r} -x - y = -5 \\ x + 2y = 8 \\ \hline y = 3 \end{array}$$

   Back-substitute 3 for y into one of the equations in two variables.
   $x + y = 5$
   $x + 3 = 5$
   $x = 2$

   The solution is $(2, 3, 3)$ and the solution set is $\{(2, 3, 3)\}$.

7. $4x - y + 2z = 11$
$\quad x + 2y - z = -1$
$\quad 2x + 2y - 3z = -1$

Multiply the second equation by -4 and add to the first equation.

$\quad 4x - y + 2z = 11$
$\underline{-4x - 8y + 4z = 4}$
$\qquad -9y + 6z = 15$

Multiply the second equation by -2 and add it to the third equation.

$\quad -2x - 4y + 2z = 2$
$\underline{\quad 2x + 2y - 3z = -1}$
$\qquad -2y - z = 1$

We now have two equations in two variables.
$-9y + 6z = 15$
$-2y - z = 1$

Multiply the second equation by 6 and solve by addition.

$\quad -9y + 6z = 15$
$\underline{-12y - 6z = 6}$
$-21y \qquad = 21$
$\quad y \qquad = -1$

Back-substitute -1 for y in one of the equations in two variables.

$\quad -2y - z = 1$
$-2(-1) - z = 1$
$\quad 2 - z = 1$
$\qquad -z = -1$
$\qquad z = 1$

Back-substitute -1 for y and 1 for z in one of the original equations in three variables.

$\quad x + 2y - z = -1$
$x + 2(-1) - 1 = -1$
$\quad x - 2 - 1 = -1$
$\qquad x - 3 = -1$
$\qquad x = 2$

The solution is $(2, -1, 1)$ and the solution set is $\{(2, -1, 1)\}$.

9. $3x + 2y - 3z = -2$
$\quad 2x - 5y + 2z = -2$
$\quad 4x - 3y + 4z = 10$

Multiply the second equation by -2 and add to the third equation.

$\quad -4x + 10y - 4z = 4$
$\underline{\quad 4x - 3y + 4z = 10}$
$\qquad 7y \qquad = 14$
$\qquad y \qquad = 2$

Back-substitute 2 for y in the first and third equations to obtain two equations in two unknowns.

$\quad 3x + 2y - 3z = -2$
$3x + 2(2) - 3z = -2$
$\quad 3x + 4 - 3z = -2$
$\qquad 3x - 3z = -6$
$\quad 4x - 3y + 4z = 10$
$4x - 3(2) + 4z = 10$
$\quad 4x - 6 + 4z = 10$
$\qquad 4x + 4z = 16$

The system of two equations in two variables becomes:

$3x - 3z = -6$
$4x + 4z = 16$

Multiply the first equation by -4 and the second equation by 3.

$\quad -12x + 12z = 24$
$\underline{\quad 12x + 12z = 48}$
$\qquad 24z = 72$
$\qquad z = 3$

Back-substitute 3 for z to find x.

$\quad 3x - 3z = -6$
$3x - 3(3) = -6$
$\quad 3x - 9 = -6$
$\qquad 3x = 3$
$\qquad x = 1$

The solution is $(1, 2, 3)$ and the solution set is $\{(1, 2, 3)\}$.

11. $2x - 4y + 3z = 17$

$x + 2y - z = 0$

$4x - y - z = 6$

Multiply the second equation by -1 and add it to the third equation.

$-x - 2y + z = 0$

$\underline{4x - y - z = 6}$

$3x - 3y \quad = 6$

Multiply the second equation by 3 and add it to the first equation.

$2x - 4y + 3z = 17$

$\underline{3x + 6y - 3z = 0}$

$5x + 2y \quad = 17$

The system in two variables becomes:

$3x - 3y = 6$

$5x + 2y = 17$

Multiply the first equation by 2 and the second equation by 3 and solve by addition.

$6x - 6y = 12$

$\underline{15x + 6y = 51}$

$21x \quad = 63$

$x \quad = 3$

Back-substitute 3 for x in one of the equations in two variables.

$3x - 3y = 6$

$3(3) - 3y = 6$

$9 - 3y = 6$

$-3y = -3$

$y = 1$

Back-substitute 3 for x and 1 for y in one of the original equations in three variables.

$x + 2y - z = 0$

$3 + 2(1) - z = 0$

$3 + 2 - z = 0$

$5 - z = 0$

$5 = z$

The solution is $(3, 1, 5)$ and the solution set is

$\{(3, 1, 5)\}$.

13. $2x + y \quad = 2$

$x + y - z = 4$

$3x + 2y + z = 0$

Add the second and third equations together to obtain an equation in two variables.

$x + y - z = 4$

$\underline{3x + 2y + z = 0}$

$4x + 3y \quad = 4$

Use this equation and the first equation in the original system to write two equations in two variables.

$2x + y = 2$

$4x + 3y = 4$

Multiply the first equation by -2 and solve by addition.

$-4x - 2y = -4$

$\underline{4x + 3y = \quad 4}$

$y = \quad 0$

Back-substitute 0 for y in one of the equations in two unknowns.

$2x + y = 2$

$2x + 0 = 2$

$2x = 2$

$x = 1$

Back-substitute 1 for x and 0 for y in one of the equations in three unknowns.

$x + y - z = 4$

$1 + 0 - z = 4$

$1 - z = 4$

$-z = 3$

$z = -3$

The solution is $(1, 0, -3)$ and the solution set is

$\{(1, 0, -3)\}$.

15. $x + y \quad = -4$

$\quad\quad y - z = \quad 1$

$2x + y + 3z = -21$

Multiply the first equation by –1 and add to the second equation.

$-x - y \quad = 4$

$\underline{\quad\quad y - z = 1}$

$-x \quad - z = 5$

Multiply the second equation by –1 and add to the third equation.

$\quad -y + \ z = -1$

$\underline{2x + y + 3z = -21}$

$2x \quad + 4z = -22$

The system of two equations in two variables becomes.

$-x - \ z = \quad 5$

$2x + 4z = -22$

Multiply the first equation by 2 and add to the second equation.

$-2x - 2z = \quad 10$

$\underline{2x + 4z = -22}$

$\quad\quad 2z = -12$

$\quad\quad z = -6$

Back-substitute –6 for z in one of the equations in two variables.

$-x - z = 5$

$-x - (-6) = 5$

$\quad -x + 6 = 5$

$\quad\quad -x = -1$

$\quad\quad\quad x = 1$

Back-substitute 1 for x in the first equation of the original system.

$x + y = -4$

$1 + y = -4$

$\quad y = -5$

The solution is $(1, -5, -6)$ and the solution set is $\{(1, -5, -6)\}$.

17. $2x + \ y + 2z = 1$

$3x - \ y + \ z = 2$

$\ x - 2y - \ z = 0$

Add the first and second equations to eliminate y.

$2x + y + 2z = 1$

$\underline{3x - y + \ z = 2}$

$5x \quad + 3z = 3$

Multiply the second equation by –2 and add to the third equation.

$-6x + 2y - 2z = -4$

$\underline{\ x - 2y - \ z = \ 0}$

$-5x \quad\quad - 3z = -4$

We obtain two equations in two variables.

$5x + 3z = \quad 3$

$-5x - 3z = -4$

Adding the two equations, we obtain:

$5x + 3z = \quad 3$

$\underline{-5x - 3z = -4}$

$\quad\quad 0 = -1$

The system is inconsistent. There are no values of x, y, and z for which $0 = -1$. The solution set is $\varnothing$ or $\{\ \}$.

19. $5x - 2y - \ 5z = 1$

$10x - 4y - 10z = 2$

$15x - 6y - 15z = 3$

Multiply the first equation by –2 and add to the second equation.

$-10x + 4y + 10z = -2$

$\underline{10x - 4y - 10z = \ 2}$

$\quad\quad\quad 0 = \ 0$

The system is dependent and has infinitely many solutions.

21. $3(2x + y) + 5z = -1$

$2(x - 3y + 4z) = -9$

$\quad 4(1 + x) = -3(z - 3y)$

Rewrite each equation and obtain the system of three equations in three variables.

$6x + 3y + 5z = -1$

$2x - 6y + 8z = -9$

$4x - 9y + 3z = -4$

Multiply the second equation by –3 and add to the first equation.

$6x + \ 3y + \ 5z = -1$

$\underline{-6x + 18y - 24z = 27}$

$\quad\quad 21y - 19z = 26$

Multiply the second equation by –2 and add to the third equation.

$-4x + 12y - 16z = \ 18$

$\underline{\ 4x - \ 9y + \ 3z = -4}$

$\quad\quad 3y - 13z = 14$

The system of two variables in two equations is:

$21y - 19z = 26$

$3y - 13z = 14$

Multiply the second equation by -7 and add to the third equation.

$$21y - 19z = 26$$
$$\underline{-21y + 91z = -98}$$
$$72z = -72$$
$$z = -1$$

Back-substitute -1 for z in one of the equations in two variables to find y.

$$3y - 13z = 14$$
$$3y - 13(-1) = 14$$
$$3y + 13 = 14$$
$$3y = 1$$
$$y = \frac{1}{3}$$

Back-substitute -1 for z and $\frac{1}{3}$ for y in one of the original equations in three variables.

$$6x + 3y + 5z = -1$$
$$6x + 1 - 5 = -1$$
$$6x - 4 = -1$$
$$6x = 3$$
$$x = \frac{1}{2}$$

The solution is $\left(\frac{1}{2}, \frac{1}{3}, -1\right)$ and the solution set is

$\left\{\left(\frac{1}{2}, \frac{1}{3}, -1\right)\right\}$.

23. Use each ordered pair to write an equation.

$$(x, y) = (-1, 6)$$

$$y = ax^2 + bx + c$$

$$6 = a(-1)^2 + b(-1) + c$$

$$6 = a - b + c$$

$$(x, y) = (1, 4)$$

$$y = ax^2 + bx + c$$

$$4 = a(1)^2 + b(1) + c$$

$$4 = a + b + c$$

$$(x, y) = (2, 9)$$

$$y = ax^2 + bx + c$$

$$9 = a(2)^2 + b(2) + c$$

$$9 = a(4) + 2b + c$$

$$9 = 4a + 2b + c$$

The system of three equations in three variables is:

$$a - b + c = 6$$
$$a + b + c = 4$$
$$4a + 2b + c = 9$$

Add the first and second equations.

$$a - b + c = 6$$
$$\underline{a + b + c = 4}$$
$$2a + 2c = 10$$

Multiply the first equation by 2 and add to the third equation.

$$2a - 2b + 2c = 12$$
$$\underline{4a + 2b + c = 9}$$
$$6a + 3c = 21$$

The system of two equations in two variables becomes:

$$2a + 2c = 10$$
$$6a + 3c = 21$$

Multiply the first equation by -3 and add to the second equation.

$$-6a - 6c = -30$$
$$\underline{6a + 3c = 21}$$
$$-3c = -9$$
$$c = 3$$

Back-substitute 3 for c in one of the equations in two variables.

$$2a + 2c = 10$$
$$2a + 2(3) = 10$$
$$2a + 6 = 10$$
$$2a = 4$$
$$a = 2$$

Back-substitute 3 for c and 2 for a in one of the equations in three variables.

$$a + b + c = 4$$
$$2 + b + 3 = 4$$
$$b + 5 = 4$$
$$b = -1$$

The quadratic equation is $y = 2x^2 - x + 3$.

25. Use each ordered pair to write an equation.

$$(x, y) = (-1, -4)$$

$$y = ax^2 + bx + c$$

$$-4 = a(-1)^2 + b(-1) + c$$

$$-4 = a - b + c$$

$$(x, y) = (1, -2)$$

$$y = ax^2 + bx + c$$

$$-2 = a(1)^2 + b(1) + c$$

$$-2 = a + b + c$$

$$(x, y) = (2, 5)$$

$$y = ax^2 + bx + c$$

$$5 = a(2)^2 + b(2) + c$$

$$5 = a(4) + 2b + c$$

$$5 = 4a + 2b + c$$

The system of three equations in three variables is:

$$a - b + c = -4$$

$$a + b + c = -2$$

$$4a + 2b + c = 5$$

Multiply the second equation by -1 and add to the first equation.

$$a - b + c = -4$$

$$\underline{-a - b - c = 2}$$

$$-2b = -2$$

$$b = 1$$

Back-substitute 4 for b in first and third equations to obtain two equations in two variables.

$$a - b + c = -4 \qquad 4a + 2b + c = 5$$

$$a - 1 + c = -4 \qquad 4a + 2(1) + c = 5$$

$$a + c = -3 \qquad 4a + 2 + c = 5$$

$$4a + c = 3$$

The system of two equations in two variables becomes:

$$a + c = -3$$

$$4a + c = 3$$

Multiply the first equation by -1 and add to the second equation.

$$-a - c = 3$$

$$\underline{4a + c = 3}$$

$$3a = 6$$

$$a = 2$$

Back-substitute 2 for a and 1 for b in one of the equations in three variables.

$$a - b + c = -4$$

$$2 - 1 + c = -4$$

$$1 + c = -4$$

$$c = -5$$

The quadratic equation is $y = 2x^2 + x - 5$.

27. Let x = the first number.
Let y = the second number.
Let z = the third number.

$$x + y + z = 16$$

$$2x + 3y + 4z = 46$$

$$5x - y = 31$$

Multiply the first equation by -4 and add to the second equation.

$$-4x - 4y - 4z = -64$$

$$\underline{2x + 3y + 4z = 46}$$

$$-2x - y = -18$$

The system of two equations in two variables becomes:

$$5x - y = 31$$

$$-2x - y = -18$$

Multiply the first equation by -1 and add to the second equation.

$$-5x + y = -31$$

$$\underline{-2x - y = -18}$$

$$-7x = -49$$

$$x = 7$$

Back-substitute 7 for x in one of the equations in two variables.

$$5x - y = 31$$

$$5(7) - y = 31$$

$$35 - y = 31$$

$$-y = -4$$

$$y = 4$$

Back-substitute 7 for x and 4 for y in one of the equations in two variables.

$$x + y + z = 16$$

$$7 + 4 + z = 16$$

$$11 + z = 16$$

$$z = 5$$

The numbers are 7, 4 and 5.

29. Simplify each equation.
$$\frac{x+2}{6} - \frac{y+4}{3} + \frac{z}{2} = 0$$
$$6\left(\frac{x+2}{6} - \frac{y+4}{3} + \frac{z}{2}\right) = 6(0)$$
$$(x+2) - 2(y+4) + 3z = 0$$
$$x + 2 - 2y - 8 + 3z = 0$$
$$x - 2y + 3z = 6$$

$$\frac{x+1}{2} + \frac{y-1}{2} - \frac{z}{4} = \frac{9}{2}$$
$$4\left(\frac{x+1}{2} + \frac{y-1}{2} - \frac{z}{4}\right) = 4\left(\frac{9}{2}\right)$$
$$2(x+1) + 2(y-1) - z = 18$$
$$2x + 2 + 2y - 2 - z = 18$$
$$2x + 2y - z = 18$$

$$\frac{x-5}{4} + \frac{y+1}{3} + \frac{z-2}{2} = \frac{19}{4}$$
$$12\left(\frac{x-5}{4} + \frac{y+1}{3} + \frac{z-2}{2}\right) = 12\left(\frac{19}{4}\right)$$
$$3(x-5) + 4(y+1) + 6(z-2) = 57$$
$$3x - 15 + 4y + 4 + 6z - 12 = 57$$
$$3x + 4y + 6z = 80$$

Now solve the equivalent system.
$$x - 2y + 3z = 6$$
$$2x + 2y - z = 18$$
$$3x + 4y + 6z = 80$$

Add the first two equations together.
$$x - 2y + 3z = 6$$
$$\underline{2x + 2y - z = 18}$$
$$3x + 2z = 24$$

Multiply the second equation by -2 and add it to the third equation.
$$-4x - 4y + 2z = -36$$
$$\underline{3x + 4y + 6z = 80}$$
$$-x + 8z = 44$$

Using the two reduced equations, we solve the system.
$$3x + 2z = 24$$
$$-x + 8z = 44$$

Multiply the second equation by 3 and add the equations.
$$3x + 2z = 24$$
$$\underline{-3x + 24z = 132}$$
$$26z = 156$$
$$z = 6$$

Back-substitute to find x.
$$-x + 8(6) = 44$$
$$-x + 48 = 44$$
$$-x = -4$$
$$x = 4$$

Back substitute to find y.
$$x - 2y + 3z = 6$$
$$4 - 2y + 3(6) = 6$$
$$-2y = -16$$
$$y = 8$$

The solution is $(4, 8, 6)$ and the solution set is $\{(4, 8, 6)\}$.

31. Selected points may vary, but the equation will be the same.
$$y = ax^2 + bx + c$$
Use the points $(2, -2)$, $(4, 1)$, and $(6, -2)$ to get the system
$$4a + 2b + c = -2$$
$$16a + 4b + c = 1$$
$$36a + 6b + c = -2$$
Multiply the first equation by -1 and add to the second equation.
$$-4a - 2b - c = 2$$
$$\underline{16a + 4b + c = 1}$$
$$12a + 2b = 3$$
Multiply the first equation by -1 and add to the third equation.
$$-4a - 2b - c = 2$$
$$\underline{36a + 6b + c = -2}$$
$$32a + 4b = 0$$
Using the two reduced equations, we get the system
$$12a + 2b = 3$$
$$32a + 4b = 0$$

Multiply the first equation by -2 and add to the second equation.

$$-24a - 4b = -6$$
$$\underline{32a + 4b = 0}$$
$$8a = -6$$
$$a = -\frac{3}{4}$$

Back-substitute to solve for b.

$$12a + 2b = 3$$
$$12\left(-\frac{3}{4}\right) + 2b = 3$$
$$-9 + 2b = 3$$
$$2b = 12$$
$$b = 6$$

Back-substitute to solve for c.

$$4a + 2b + c = -2$$
$$4\left(-\frac{3}{4}\right) + 2(6) + c = -2$$
$$-3 + 12 + c = -2$$
$$c = -11$$

The equation is $y = -\frac{3}{4}x^2 + 6x - 11$.

33. $ax - by - 2cz = 21$
$ax + by + cz = 0$
$2ax - by + cz = 14$

Add the first two equations.

$$ax - by - 2cz = 21$$
$$\underline{ax + by + cz = 0}$$
$$2ax - cz = 21$$

Multiply the first equation by -1 and add to the third equation.

$$-ax + by + 2cz = -21$$
$$\underline{2ax - by + cz = 14}$$
$$ax + 3cz = -7$$

Use the two reduced equations to get the following system:

$2ax - cz = 21$

$ax + 3cz = -7$

Multiply the second equation by -2 and add the equations.

$$2ax - cz = 21$$
$$\underline{-2ax - 6cz = 14}$$
$$-7cz = 35$$
$$z = -\frac{5}{c}$$

Back-substitute to solve for x.

$$ax + 3cz = -7$$
$$ax + 3c\left(-\frac{5}{c}\right) = -7$$
$$ax - 15 = -7$$
$$ax = 8$$
$$x = \frac{8}{a}$$

Back-substitute to solve for y.

$$ax + by + cz = 0$$
$$a\left(\frac{8}{a}\right) + by + c\left(-\frac{5}{c}\right) = 0$$
$$8 + by - 5 = 0$$
$$by = -3$$
$$y = -\frac{3}{b}$$

The solution is $\left(\frac{8}{a}, -\frac{3}{b}, -\frac{5}{c}\right)$ and the solution set is

$$\left\{\left(\frac{8}{a}, -\frac{3}{b}, -\frac{5}{c}\right)\right\}.$$

35. a. $(0, 1.8), (5, 0.8), (13, 1.9)$

b. Substituting each ordered pair gives:

$$y = ax^2 + bx + c$$
$$1.8 = a(0)^2 + b(0) + c$$
$$0.8 = a(5)^2 + b(5) + c$$
$$1.9 = a(13)^2 + b(13) + c$$

Simplifying gives the following system:

$$0a + 0b + c = 1.8$$
$$25a + 5b + c = 0.8$$
$$169a + 13b + c = 1.9$$

37. a. Using the three ordered pairs, $(1, 224)$, $(3, 176)$, and $(4, 104)$, we get the following system:

$$a + b + c = 224$$
$$9a + 3b + c = 176$$
$$16a + 4b + c = 104$$

Multiply the first equation by -1 and add to the second equation.

$$-a - b - c = -224$$
$$\underline{9a + 3b + c = 176}$$
$$8a + 2b = -48$$

Multiply the first equation by -1 and add to the third equation.

$$-a - b - c = -224$$
$$\underline{16a + 4b + c = 104}$$
$$15a + 3b = -120$$

Using the two reduced equations, we get the following system:

$$8a + 2b = -48$$
$$15a + 3b = -120$$

Multiply the first equation by -3 and multiply the second equation by 2, then add to the equations.

$$-24a - 6b = 144$$
$$\underline{30a + 6b = -240}$$
$$6a = -96$$
$$a = -16$$

Back-substitute to solve for b.

$$8a + 2b = -48$$
$$8(-16) + 2b = -48$$
$$-128 + 2b = -48$$
$$2b = 80$$
$$b = 40$$

Back-substitute to solve for c.

$$a + b + c = 224$$
$$-16 + 40 + c = 224$$
$$c = 200$$

The equation is $y = -16x^2 + 40x + 200$.

b. When $x = 5$, we get

$$y = -16(5)^2 + 40(5) + 200$$
$$= -16(25) + 200 + 200$$
$$= -400 + 400$$
$$= 0$$

After 5 seconds, the ball hits the ground.

39. Let x = average annual U.S. household spending on rent.
Let y = average annual U.S. household spending on cars.
Let z = average annual U.S. household spending on books.

$$x + y + z = 12{,}691$$
$$x - y = 6204$$
$$y - z = 2573$$

Solve the second equation for x.

$$x - y = 6204$$
$$x = y + 6204$$

Solve the third equation for z.

$$y - z = 2573$$
$$-z = -y + 2573$$
$$z = y - 2573$$

Substitute the expressions for x and z into the first equation and solve for y.

$$x + y + z = 12{,}691$$
$$(\overset{x}{\overbrace{y + 6204}}) + y + (\overset{z}{\overbrace{y - 2573}}) = 12{,}691$$
$$y + 6204 + y + y - 2573 = 12{,}691$$
$$3y + 3631 = 12{,}691$$
$$3y = 9060$$
$$y = 3020$$

Back-substitute to solve for x and z.

$$x = y + 6204$$
$$= 3020 + 6204$$
$$= 9224$$

$$z = y - 2573$$
$$= 3020 - 2573$$
$$= 447$$

The average annual U.S. household spending on rent is \$9224, on cars is \$3020, and on books is \$447.

41. Let x = the amount invested at 8%.
Let y = the amount invested at 10%.
Let z = the amount invested at 12%.

$x + y + z = 6700$

$0.08x + 0.10y + 0.12z = 716$

$z - x - y = 300$

Rewrite the system in $Ax + By + Cz = D$ form.

$x + y + z = 6700$

$0.08x + 0.10y + 0.12z = 716$

$-x - y + z = 300$

Add the first and third equations to find z.

$x + y + z = 6700$

$\underline{-x - y + z = 300}$

$2z = 7000$

$z = 3500$

Back-substitute 3500 for z to obtain two equations in two variables.

$x + y + z = 6700$

$x + y + 3500 = 6700$

$x + y = 3200$

$0.08x + 0.10y + 0.12(3500) = 716$

$0.08x + 0.10y + 420 = 716$

$0.08x + 0.10y = 296$

The system of two equations in two variables becomes:

$x + y = 3200$

$0.08x + 0.10y = 296$

Multiply the second equation by -10 and add it to the first equation.

$x + y = 3200$

$\underline{-0.8x + -y = -2960}$

$0.2x = 240$

$x = 1200$

Back-substitute 1200 for x in one of the equations in two variables.

$x + y = 3200$

$1200 + y = 3200$

$y = 2000$

$1200 was invested at 8%, $2000 was invested at 10%, and $3500 was invested at 12%.

43. Let x = the number of $8 tickets.
Let y = the number of $10 tickets.
Let z = the number of $12 tickets.

$x + y + z = 400$

$8x + 10y + 12z = 3700$

$x + y = 7z$

Rewrite the system in $Ax + By + Cz = D$ form.

$x + y + z = 400$

$8x + 10y + 12z = 3700$

$x + y - 7z = 0$

Multiply the first equation by -1 and add to the third equation.

$-x - y - z = -400$

$\underline{x + y - 7z = 0}$

$-8z = -400$

$z = 50$

Back-substitute 50 for z in two of the original equations to obtain two of equations in two variables.

$x + y + z = 400$

$x + y + 50 = 400$

$x + y = 350$

$8x + 10y + 12z = 3700$

$8x + 10y + 12(50) = 3700$

$8x + 10y + 600 = 3700$

$8x + 10y = 3100$

The system of two equations in two variables becomes:

$x + y = 350$

$8x + 10y = 3100$

Multiply the first equation by -8 and add to the second equation.

$-8x - 8y = -2800$

$\underline{8x + 10y = 3100}$

$2y = 300$

$y = 150$

Back-substitute 50 for z and 150 for y in one of the original equations in three variables.

$x + y + z = 400$

$x + 150 + 50 = 400$

$x + 200 = 400$

$x = 200$

There were 200 $8 tickets, 150 $10 tickets, and 50 $12 tickets sold.

45. Let A = the number of servings of A.
Let B = the number of servings of B.
Let C = the number of servings of C.

$$40A + 200B + 400C = 660$$
$$5A + 2B + 4C = 25$$
$$30A + 10B + 300C = 425$$

Multiply the second equation by −8 and add to the first equation to obtain an equation in two variables.

$$40A + 200B + 400C = 660$$
$$\underline{-40A - 16B - 32C = -200}$$
$$184B + 368C = 460$$

Multiply the second equation by −6 and add to the third equation to obtain an equation in two variables.

$$-30A - 12B - 24C = -150$$
$$\underline{30A + 10B + 300C = 425}$$
$$-2B + 276C = 275$$

The system of two equations in two variables becomes:

$$184B + 368C = 460$$
$$-2B + 276C = 275$$

Multiply the second equation by 92 and eliminate B.

$$184B + 368C = 460$$
$$\underline{-184B + 25392C = 25300}$$
$$25760C = 25760$$
$$C = 1$$

Back-substitute 1 for C in one of the equations in two variables.

$$-2B + 276C = 275$$
$$-2B + 276(1) = 275$$
$$-2B + 276 = 275$$
$$-2B = -1$$
$$B = \frac{1}{2}$$

Back-substitute 1 for C and $\frac{1}{2}$ for B in one of the original equations in three variables.

$$5A + 2B + 4C = 25$$
$$5A + 2\left(\frac{1}{2}\right) + 4(1) = 25$$
$$5A + 1 + 4 = 25$$
$$5A + 5 = 25$$
$$5A = 20$$
$$A = 4$$

To meet the requirements, 4 ounces of Food A, $\frac{1}{2}$ ounce of Food B, and 1 ounce of Food C should be used.

47. – 53. Answers will vary.

55. does not make sense; Explanations will vary. Sample explanation: The third variable could possibly have the same variable as one of the other two.

57. makes sense

59. false; Changes to make the statement true will vary. A sample change is: The given ordered triple is one solution to the equation, but there are an infinite number of other ordered triples which satisfy the equation.

61. true

63.
$$x + y + z = 180$$
$$(2x + 5) + y = 180$$
$$(2x - 5) + z = 180$$

Rewrite the system in standard form as
$$x + y + z = 180$$
$$2x + y = 175$$
$$2x + z = 185$$

Multiply the first equation by −1 and add to the second equation to obtain an equation with two variables.

$$-x - y - z = -180$$
$$\underline{2x + y = 175}$$
$$x - z = -5$$

Combine this equation with the third equation to make a system of two equations.

$$x - z = -5$$
$$\underline{2x + z = 185}$$
$$3x = 180$$
$$x = 60$$

Back-substitute to find z.
$$x - z = -5$$
$$60 - z = -5$$
$$-z = -65$$
$$z = 65$$

Back-substitute to find y.
$$x + y + z = 180$$
$$60 + y + 65 = 180$$
$$y = 55$$

The angles measure $55°$, $60°$, and $65°$.

65. Let x = height of the table.
Let y = length of the wood blocks.
Let z = width of the wood blocks.

From the problem, we have the following two equations.
$x + y - z = 32$
$x - y + z = 28$
Add the two equations.

$$\begin{array}{r} x + y - z = 32 \\ \underline{x - y + z = 28} \\ 2x \qquad = 60 \\ x = 30 \end{array}$$

The height of the table is 30 centimeters.

66. $y = -\dfrac{3}{4}x + 3$

Use the slope and the y–intercept to graph the line.

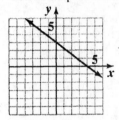

67. $-2x + y = 6$

Rewrite the equation in slope-intercept form.
$-2x + y = 6$
$\qquad y = 2x + 6$
Use the slope and the y–intercept to graph the line.

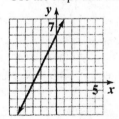

68. $y = -5$ is a horizontal line.

69. $5x^3 + 12x^3 = (5 + 12)x^3 = 17x^3$

70. $-8x^2 + 6x^2 = (-8 + 6)x^2 = -2x^2$

71. $-9y^4 - (-2y^4) = -9y^4 + 2y^4$
$\qquad\qquad\qquad = (-9 + 2)y^4$
$\qquad\qquad\qquad = -7y^4$

Chapter 4 Review Exercises

1. $\qquad 4x - y = 9$
$\quad 4(1) - (-5) = 9$
$\qquad\quad 4 + 5 = 9$
$\qquad\qquad 9 = 9,\ \text{true}$

$\qquad 2x + 3y = -13$
$2(1) + 3(-5) = -13$
$\qquad 2 - 15 = -13$
$\qquad -13 = -13,\ \text{true}$

Since the ordered pair $(1, -5)$ satisfies both equations, it is a solution of the given system.

2. $\qquad 2x + 3y = -4$
$2(-5) + 3(2) = -4$
$\qquad -10 + 6 = -4$
$\qquad\quad -4 = -4,\ \text{true}$

$\qquad x - 4y = -10$
$-5 - 4(-2) = -10$
$\qquad -5 + 8 = -10$
$\qquad\quad 3 = -10,\ \text{false}$

Since $(-5, 2)$ fails to satisfy *both* equations, it is not a solution of the given system.

3. $x + y = 2$
$-1 + 3 = 2$
$\quad 2 = 2,\ \text{true}$

$\quad 2x + y = -5$
$2(-1) + 3 = -5$
$\quad -2 + 3 = -5$
$\qquad 1 = -5,\ \text{false}$

Since $(-1, 3)$ fails to satisfy *both* equations, it is not a solution of the given system. Also, the second equation in the system, which can be rewritten as $y = -2x - 5$, is a line with slope -2 and y-intercept -5, while the graph shows a line with slope 2 and y-intercept 5.

4. $x + y = 2$

$x - y = 6$

Graph both lines on the same axes.

$x + y = 2$: x-intercept = 2; y-intercept 2

$x - y = 6$: x-intercept = 6, y-intercept = -6

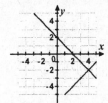

The solution set is $\{(4, -2)\}$.

5. $2x - 3y = 12$

$-2x + y = -8$

Graph both equations.

$2x - 3y = 12$: x-intercept = 6; y-intercept = -4

$-2x + y = -8$: x-intercept = 4; y-intercept = -8

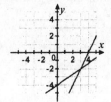

The solution set is $\{(3, -2)\}$.

6. $3x + 2y = 6$

$3x - 2y = 6$

Graph both equations.

$3x + 2y = 6$: x-intercept = 2; y-intercept = 3

$3x - 2y = 6$: x-intercept = 2; y-intercept = -3

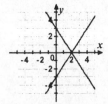

The solution set is $\{(2, 0)\}$.

7. $y = \frac{1}{2}x$

$y = 2x - 3$

Graph both equations.

$y = \frac{1}{2}x$: slope = $\frac{1}{2}$; y-intercept = 0

$y = 2x - 3$: slope = 2; y-intercept = -3

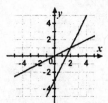

The solution set is $\{(2, 1)\}$.

8. $x + 2y = 2$

$y = x - 5$

Graph both equations.

$x + 2y = 2$: x-intercept = 2; y-intercept = 1

$y = x - 5$: slope = 1; y-intercept = -5

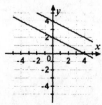

The solution set is $\{(4, -1)\}$.

9. $x + 2y = 8$

$3x + 6y = 12$

Graph both equations.

$x + 2y = 8$: x-intercept = 8; y-intercept = 4

$3x + 6y = 12$: x-intercept = 4; y-intercept = 2

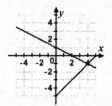

The lines are parallel. The system is inconsistent and has no solution. The solution set is $\{\ \}$.

10. $2x - 4y = 8$

$x - 2y = 4$

Graph both equations.

$2x - 4y = 8$: x-intercept = 4; y-intercept = -2

$x - 2y = 4$: x-intercept = 4; y-intercept = -2

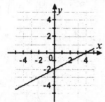

The graphs of the two equations are the same line. The system is dependent and has infinitely many solutions.

The solution set is $\{(x, y) | 2x - 4y = 8\}$ or

$\{(x, y) | x - 2y = 4\}$.

11. $y = 3x - 1$

$y = 3x + 2$

Graph both equations.

$y = 3x - 1$: slope = 3; y-intercept = -1

$y = 3x + 2$: slope = 3; y-intercept = 2

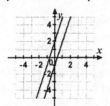

The lines are parallel, so the system is inconsistent and has no solution. The solution set is $\{\ \}$.

12. $x - y = 4$

$x = -2$

Graph both equations:

$x - y = 4$: x-intercept = 4; y-intercept = -4

$x = 2$: vertical line with x-intercept = -2

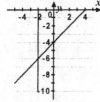

The solution set is $\{(-2, -6)\}$.

13. $x = 2$

$y = 5$

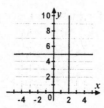

The solution set is $\{(2, 5)\}$.

14. $x = 2$

$x = 5$

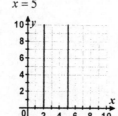

The lines are parallel, so the system inconsistent and has no solution.

The solution set is $\{\ \}$.

15. $2x - 3y = 7$

$y = 3x - 7$

Substitute $3x - 7$ for y in the first equation.

$2x - 3y = 7$

$2x - 3(3x - 7) = 7$

$2x - 9x + 21 = 7$

$-7x + 21 = 7$

$-7x = -14$

$x = 2$

Back-substitute 7 for x into the second equation and solve for y.

$y = 3x - 7$

$y = 3(2) - 7 = -1$

The solution set is $\{(2, -1)\}$.

16. $2x - y = 6$

$\qquad x = 13 - 2y$

Substitute $13 - 2y$ for x into the first equation.

$\qquad 2x - y = 6$

$\qquad 2(13 - 2y) - y = 6$

$\qquad 26 - 4y - y = 6$

$\qquad 26 - 5y = 6$

$\qquad -5y = -20$

$\qquad y = 4$

Back-substitute 4 for y in the second equation.

$x = 13 - 2y$

$x = 13 - 2(4) = 5$

The solution set is $\{(5, 4)\}$.

17. $2x - 5y = 1$

$\qquad 3x + y = -7$

Solve the second equation for y.

$3x + y = -7$

$\qquad y = -3x - 7$

Substitute $-3x - 7$ in the first equation.

$\qquad 2x - 5y = 1$

$\qquad 2x - 5(-3x - 7) = 1$

$\qquad 2x + 15x + 35 = 1$

$\qquad 17x + 35 = 1$

$\qquad 17x = -34$

$\qquad x = -2$

Back-substitute in the equation $y = -3x - 7$.

$y = -3x - 7$

$y = -3(-2) - 7 = -1$

The solution set is $\{(-2, -1)\}$.

18. $3x + 4y = -13$

$\qquad 5y - x = -21$

Solve the second equation for x.

$5y - x = -21$

$\qquad -x = -5y - 21$

$\qquad x = 5y + 21$

Substitute $5y + 21$ for x in the first equation.

$\qquad 3x + 4y = -13$

$\qquad 3(5y + 21) + 4y = -13$

$\qquad 15y + 63 + 4y = -13$

$\qquad 19y + 63 = -13$

$\qquad 19y = -76$

$\qquad y = -4$

Back-substitute.

$\qquad 3x + 4y = -13$

$\qquad 3x + 4(-4) = -13$

$\qquad 3x - 16 = -13$

$\qquad 3x = 3$

$\qquad x = 1$

The solution set is $\{(1, -4)\}$.

19. $y = 39 - 3x$

$\qquad y = 2x - 61$

Substitute $2x - 61$ for y in the first equation.

$2x - 61 = 39 - 3x$

$5x - 61 = 39$

$\qquad 5x = 100$

$\qquad x = 20$

Back-substitute.

$y = 2x - 61 = 2(20) - 61 = -21$

The solution set is $\{(20, -21)\}$.

20. $\qquad 4x + y = 5$

$\qquad 12x + 3y = 15$

Solve the first equation for y.

$4x + y = 5$

$\qquad y = -4x + 5$

Substitute $-4x + 5$ for y in the second equation.

$\qquad 12x + 3y = 15$

$\qquad 12x + 3(-4x + 5) = 15$

$\qquad 12x - 12x + 15 = 15$

$\qquad 15 = 15, \text{ true}$

The true statement indicates that the given system has infinitely many solutions.

The solution set is $\{(x, y) | 4x + y = 5\}$ or

$\{(x, y) | 12x + 3y = 15\}$.

21. $4x - 2y = 10$

$\qquad y = 2x + 3$

Substitute $2x + 3$ for y in the first equation.

$\qquad 4x - 2y = 10$

$\qquad 4x - 2(2x + 3) = 10$

$\qquad 4x - 4x - 6 = 10$

$\qquad -6 = 10, \text{ false}$

The false statement $-6 = 10$ indicates that the system is inconsistent and has no solution.

The solution set is $\{\ \}$.

22. $x - 4 = 0$

$9x - 2y = 0$

Solve the first equation for x.

$x - 4 = 0$

$x = 4$

Substitute 4 for x in the second equation.

$9x - 2y = 0$

$9(4) - 2y = 0$

$36 - 2y = 0$

$-2y = -36$

$y = 18$

The solution set is $\{(4, 18)\}$.

23. $8y = 4x$

$7x + 2y = -8$

Solve the first equation for y.

$8y = 4x$

$y = \frac{1}{2}x$

Substitute $\frac{1}{2}x$ for y in the second equation.

$7x + 2y = -8$

$7x + 2\left(\frac{1}{2}x\right) = -8$

$7x + x = -8$

$8x = -8$

$x = -1$

Back-substitute.

$y = \frac{1}{2}x = \frac{1}{2}(-1) = -\frac{1}{2}$

The solution set is $\left\{\left(-1, -\frac{1}{2}\right)\right\}$.

24. a. Demand model: $p = -50x + 2000$

Supply model: $p = 50x$

Use the substitution method.

$p = -50x + 2000$

$\overbrace{50x}^{p} = -50x + 2000$

$50x = -50x + 2000$

$100x = 2000$

$x = 20$

Back-substitute 20 for x and find p.

$p = 50x$

$p = 50(20) = 1000$

The solution set is $\{(20, 1000)\}$.

The equilibrium quantity is 20,000 and the equilibrium price is $1000.

b. When rents are $\underline{\$1000}$ per month, consumers will demand $\underline{20,000}$ apartments and suppliers will offer $\underline{20,000}$ apartments for rent.

25. $x + y = 6$

$2x + y = 8$

Multiply the first equation by -1 and add the result to the second equation to eliminate the y-terms.

$-x - y = -6$

$\underline{2x + y = 8}$

$x = 2$

Back-substitute into either of the original equations to solve for y.

$x + y = 6$

$2 + y = 6$

$y = 4$

The solution set is $\{(2, 4)\}$.

26. $3x - 4y = 1$

$12x - y = -11$

Multiply the first equation by -4 and add the result to the second equation.

$-12x + 16y = -4$

$\underline{12x - y = -11}$

$15y = -15$

$y = -1$

Back-substitute.

$3x - 4y = 1$

$3x - 4(-1) = 1$

$3x + 4 = 1$

$3x = -3$

$x = -1$

The solution set is $\{(-1, -1)\}$.

27. $3x - 7y = 13$

$6x + 5y = 7$

Multiply the first equation by -2.

$-6x + 14y = -26$

$\underline{6x + 5y = 7}$

$19y = -19$

$y = -1$

Back-substitute.

$3x - 7y = 13$

$3x - 7(-1) = 13$

$3x + 7 = 13$

$3x = 6$

$x = 2$

The solution set is $\{(2, -1)\}$.

28. $8x - 4y = 16$

$4x + 5y = 22$

Multiply the second equation by -2.

$8x - 4y = 16$

$\underline{-8x - 10y = -44}$

$-14y = -28$

$y = 2$

Back-substitute.

$8x - 4y = 16$

$8x - 4(2) = 16$

$8x - 8 = 16$

$8x = 24$

$x = 3$

The solution set is $\{(3, 2)\}$.

29. $5x - 2y = 8$

$3x - 5y = 1$

Multiply the first equation by 3.
Multiply the second equation by -5.

$15x - 6y = 24$

$\underline{-15x + 25y = -5}$

$19y = 19$

$y = 1$

Back-substitute.

$5x - 2y = 8$

$5x - 2(1) = 8$

$5x - 2 = 8$

$5x = 10$

$x = 2$

The solution set is $\{(2, 1)\}$.

30. $2x + 7y = 0$

$7x + 2y = 0$

Multiply the first equation by 7.
Multiply the second equation by -2.

$14x + 49y = 0$

$\underline{-14x - 4y = 0}$

$45y = 0$

$y = 0$

Back-substitute.

$2x + 7y = 0$

$2x + 7(0) = 0$

$2x = 0$

$x = 0$

The solution set is $\{(0, 0)\}$.

31. $x + 3y = -4$

$3x + 2y = 3$

Multiply the first equation by −3.

$-3x - 9y = 12$

$\underline{3x + 2y = 3}$

$-7y = 15$

$y = -\dfrac{15}{7}$

Instead of back-substituting $-\dfrac{15}{7}$ and working with

fractions, go back to the original system. Multiply the first equation by 2 and the second equation by −3.

$2x + 6y = -8$

$\underline{-9x - 6y = -9}$

$-7x = -17$

$x = \dfrac{17}{7}$

The solution set is $\left\{\left(\dfrac{17}{7}, -\dfrac{15}{7}\right)\right\}$.

32. $2x + y = 5$

$2x + y = 7$

Multiply the first equation by −1.

$-2x - y = -5$

$\underline{2x + y = 7}$

$0 = 2, \text{ false}$

The false statement indicates that the system has no solution. The solution set is $\{\ \}$.

33. $3x - 4y = -1$

$-6x + 8y = 2$

Multiply the first equation by 2.

$6x - 8y = -2$

$\underline{-6x + 8y = 2}$

$0 = 0, \text{ true}$

The true statement indicates that the system is dependent and has infinitely many solutions.

The solution set is $\{(x, y) | 3x - 4y = -1\}$ or

$\{(x, y) | -6x + 8y = 2\}$.

34. $2x = 8y + 24$

$3x + 5y = 2$

Rewrite the first equation in the form $Ax + By = C$.

$2x - 8y = 24$

Multiply this equation by 3.
Multiply the second equation by −2.

$6x - 24y = 72$

$\underline{-6y - 10y = -4}$

$-34y = 68$

$y = -2$

Back-substitute.

$3x + 5y = 2$

$3x + 5(-2) = 2$

$3x - 10 = 2$

$3x = 12$

$x = 4$

The solution set is $\{(4, -2)\}$.

35. $5x - 7y = 2$

$3x = 4y$

Rewrite the second equation in the form $Ax + By = C$.

$3x - 4y = 0$

Multiply this equation by −5.
Multiply the first equation by 3.

$15x - 21y = 6$

$\underline{-15x + 20y = 0}$

$-y = 6$

$y = -6$

Back-substitute.

$3x - 4y = 0$

$3x - 4(-6) = 0$

$3x + 24 = 0$

$3x = -24$

$x = -8$

The solution set is $\{(-8, -6)\}$.

36. $3x + 4y = -8$

$2x + 3y = -5$

Multiply the first equation by 2.

Multiply the second equation by −3.

$6x + 8y = -16$

$\underline{-6x - 9y = 15}$

$-y = -1$

$y = 1$

Back-substitute.

$3x + 4y = -8$

$3x + 4(1) = -8$

$3x + 4 = -8$

$3x = -12$

$x = -4$

The solution set is $\{(-4, 1)\}$.

37. $6x + 8y = 39$

$y = 2x - 2$

Substitute $2x - 2$ for y in the first equation.

$6x + 8y = 39$

$6x + 8(2x - 2) = 39$

$6x + 16x - 16 = 39$

$22x - 16 = 39$

$22x = 55$

$x = \dfrac{55}{22} = \dfrac{5}{2}$

Back-substitute $\dfrac{5}{2}$ for x into the second equation of

the system.

$y = 2x - 2$

$y = 2\left(\dfrac{5}{2}\right) - 2 = 5 - 2 = 3$

The solution set is $\left\{\left(\dfrac{5}{2}, 3\right)\right\}$.

38. $x + 2y = 7$

$2x + y = 8$

Multiply the first equation by −2.

$-2x - 4y = -14$

$\underline{2x + y = 8}$

$-3y = -6$

$y = 2$

Back-substitute.

$x + 2y = 7$

$x + 2(2) = 7$

$x + 4 = 7$

$x = 3$

The solution set is $\{(3, 2)\}$.

39. $y = 2x - 3$

$y = -2x - 1$

Substitute $-2x - 1$ for y in the first equation.

$-2x - 1 = 2x - 3$

$-4x - 1 = -3$

$-4x = -2$

$x = \dfrac{1}{2}$

Back-substitute.

$y = 2x - 3$

$y = 2\left(\dfrac{1}{2}\right) - 3 = -2$

The solution set is $\left\{\left(\dfrac{1}{2}, -2\right)\right\}$.

40. $3x - 6y = 7$

$3x = 6y$

Solve the second equation for x.

$3x = 6y$

$x = 2y$

Substitute $2y$ for x in the first equation.

$3x - 6y = 7$

$3(2y) - 6y = 7$

$6y - 6y = 7$

$0 = 7$

The false statement indicates that the system has no solution. The solution set is $\{\ \}$.

41.
$$y - 7 = 0$$
$$7x - 3y = 0$$
Solve the first equation for y.
$$y - 7 = 0$$
$$y = 7$$
Substitute 7 for y in the second equation.
$$7x - 3y = 0$$
$$7x - 3(7) = 0$$
$$7x - 21 = 0$$
$$7x = 21$$
$$x = 3$$
The solution set is $\{(3,7)\}$.

42. Let $x =$ the selling price for Klint's work.
Let $y =$ the selling price for Picasso's work.
$$x + y = 239$$
$$x - y = 31$$
Add the equations to eliminate y and solve for x.
$$x + y = 239$$
$$\underline{x - y = 31}$$
$$2x \quad = 270$$
$$x = 135$$
Back-substitute to find y.
$$x + y = 239$$
$$135 + y = 239$$
$$y = 104$$
Klint's work sold for $135 million and Picasso's work sold for $104 million.

43. Let $x =$ the cholesterol content of one ounce of shrimp (in milligrams).
Let $y =$ the cholesterol content in one ounce of scallops.
$$3x + 2y = 156$$
$$5x + 3y = 300 - 45$$
Simplify the second equation.
$$5x + 3y = 255$$
Multiply this equation by 2.
Multiply the first equation by -3.
$$-9x - 6y = -468$$
$$\underline{10x + 6y = 510}$$
$$x \quad = 42$$
Back-substitute to find y.

$$3x + 2y = 156$$
$$3(42) + 2y = 156$$
$$126 + 2y = 156$$
$$2y = 30$$
$$y = 15$$
There are 42 mg of cholesterol in an ounce of shrimp and 15 mg in an ounce of scallops.

44. Let $x =$ the length of a tennis table top.
Let $y =$ the width.
Use the formula for perimeter of a rectangle to write the first equation and the other information in the problem to write the second equation.
$$2x + 2y = 28$$
$$4x - 3y = 21$$
Multiply the first equation by -2.
$$-4x + 4y = -56$$
$$\underline{4x - 3y = 21}$$
$$-7y = -35$$
$$y = 5$$
Back-substitute to find x.
$$2x + 2(5) = 28$$
$$2x + 10 = 28$$
$$2x = 18$$
$$x = 9$$
The length is 9 feet and the width is 5 feet, so the dimensions of the table are 9 feet by 5 feet.

45. Let $x =$ the length of the garden.
Let $y =$ the width of the garden.
The perimeter of the garden is 24 yards, so $2x + 2y = 24$.
Since there are two lengths and two widths to be fenced, the information about the cost of fencing leads to the equation $3(2x) + 2(2y) = 62$.
Simplify the second equation.
$6x + 4y = 62$.
Multiply the first equation by -2.
$$-4x - 4y = -48$$
$$\underline{6x + 4y = 62}$$
$$2x \quad = 14$$
$$x \quad = 7$$
Back-substitute to find y.
$$2(7) + 2y = 24$$
$$14 + 2y = 24$$
$$2y = 10$$
$$y = 5$$
The length of the garden is 7 yards and the width is 5 yards.

46. Let x = daily cost for room.
Let y = daily cost for car.
First plan: $3x + 2y = 360$

Second plan: $4x + 3y = 500$

Multiply the first equation by 3.
Multiply the second equation by −2.

$$9x + 6y = 1080$$
$$\underline{-8x - 6y = -1000}$$
$$x = 80$$

Back-substitute to find y.

$$3(80) + 2y = 360$$
$$240 + 2y = 360$$
$$2y = 120$$
$$y = 60$$

The cost per day is \$80 for the room and \$60 for the car.

47. Let x = the number of minutes of long-distance calls.
Let y = the monthly cost of a telephone plan.
Plan A: $y = 15 + 0.05x$

Plan B: $y = 10 + 0.075x$

To determine the amount of calling time that will result in the same cost for both plans, solve this system by the substitution method. Substitute $15 + 0.05x$ for y in the first equation.

$$15 + 0.05x = 10 + 0.075x$$
$$15 - 0.025x = 10$$
$$-0.025x = -5$$
$$\frac{-0.025x}{-0.025} = \frac{-5}{-0.025}$$
$$x = 200$$

Back-substitute to find y.

$$y = 15 + 0.05(200) = 25$$

The costs for the two plans will be equal for 200 minutes of long-distance calls per month. The cost of each plan will be \$25.

48. Let x = the number orchestra tickets.
Let y = the number balcony tickets.

$$x + y = 9$$
$$90x + 60y = 720$$

Solve the first equation for y.

$$x + y = 9$$
$$y = -x + 9$$

Use the substitution method.

$$90x + 60y = 720$$

$$90x + 60(\overbrace{-x + 9}^{y}) = 720$$
$$90x + 60(-x + 9) = 720$$
$$90x - 60x + 540 = 720$$
$$30x + 540 = 720$$
$$30x = 180$$
$$x = 6$$

Back-substitute to find y.

$$y = -x + 9$$
$$y = -6 + 9 = 3$$

You purchased 6 orchestra tickets and 3 balcony tickets.

49. Let x = the amount invested at 4%.
Let y = the amount invested at 7%.

$$x + y = 9000$$
$$0.04x + 0.07y = 555$$

Multiply the first equation by −0.04 and add.

$$-0.04x - 0.04y = -360$$
$$\underline{0.04x + 0.07y = 555}$$
$$0.03y = 195$$
$$y = 6500$$

Back-substitute 6500 for y in one of the original equations to find x.

$$x + y = 9000$$
$$x + 6500 = 9000$$
$$x = 2500$$

There was \$2500 invested at 4% and \$6500 invested at 7%.

50. Let $x =$ the amount of the 34% solution.
Let $y =$ the amount of the 4% solution.
$$x + y = 100$$
$$0.34x + 0.04y = 0.07(100)$$

Simplified, the system becomes
$$x + y = 100$$
$$0.34x + 0.04y = 7$$

Multiply the first equation by –0.34 and add to the second equation.
$$-0.34x - 0.34y = -34$$
$$\underline{0.34x + 0.04y = 7}$$
$$-0.30 = -27$$
$$y = 90$$

Back-substitute 90 for y to find x.
$$x + y = 100$$
$$x + 90 = 100$$
$$x = 10$$

10 ml of the 34% solution and 90 ml of the 4% solution must be used.

51. Let $r =$ the speed of the plane in still air.
Let $w =$ the speed of the wind.

	Rate	× Time	= Distance
Trip with the Wind	$r + w$	3	$3(r + w)$
Trip against the Wind	$r - w$	4	$4(r - w)$

$$3(r + w) = 2160$$
$$4(r - w) = 2160$$
Simplified, the system becomes
$$3r + 3w = 2160$$
$$4r - 4w = 2160$$
Multiply the first equation by 4, the second equation by 3, and solve by addition.
$$12r + 12w = 8640$$
$$\underline{12r - 12w = 6480}$$
$$24r = 15120$$
$$r = 630$$
Back-substitute 630 for r to find w.
$$3r + 3w = 2160$$
$$3(630) + 3w = 2160$$
$$1890 + 3w = 2160$$
$$3w = 270$$
$$w = 90$$
The speed of the plane in still air is 630 miles per hour and the speed of the wind is 90 miles per hour.

52.
$$x + y + z = 0$$
$$-3 + (-2) + 5 = 0$$
$$-5 + 5 = 0$$
$$0 = 0, \text{ true}$$

$$2x - 3y + z = 5$$
$$2(-3) - 3(-2) + 5 = 5$$
$$-6 + 6 + 5 = 5$$
$$5 = 5, \text{ true}$$

$$4x + 2y + 4z = 3$$
$$4(-3) + 2(-2) + 4(5) = 3$$
$$-12 - 4 + 20 = 3$$
$$4 = 3, \text{ false}$$

The ordered triple (–3, –2, 5) does not satisfy all three equations, so it is not a solution.

53.
$$2x - y + z = 1$$
$$3x - 3y + 4z = 5$$
$$4x - 2y + 3z = 4$$

Multiply the first equation by –2 and add to the third.

$$-4x + 2y - 2z = -2$$
$$\underline{4x - 2y + 3z = 4}$$
$$z = 2$$

Back-substitute 2 for z in two of the original equations to obtain a system of two equations in two variables.

$$2x - y + z = 1 \qquad 3x - 3y + 4z = 5$$
$$2x - y + 2 = 1 \qquad 3x - 3y + 4(2) = 5$$
$$2x - y = -1 \qquad 3x - 3y + 8 = 5$$
$$3x - 3y = -3$$

The system of two equations in two variables becomes:
$$2x - y = -1$$
$$3x - 3y = -3$$

Multiply the first equation by –3 and solve by addition.

$$-6x + 3y = 3$$
$$\underline{3x - 3y = -3}$$
$$-3x = 0$$
$$x = 0$$

Back-substitute 0 for x to find y.
$$2x - y = -1$$
$$2(0) - y = -1$$
$$-y = -1$$
$$y = 1$$

The solution set is $\{(0, 1, 2)\}$.

54.
$$x + 2y - z = 5$$
$$2x - y + 3z = 0$$
$$2y + z = 1$$

Multiply the first equation by –2 and add to the second equation.
$$-2x - 4y + 2z = -10$$
$$\underline{2x - y + 3z = 0}$$
$$-5y + 5z = -10$$

We now have two equations in two variables.
$$2y + z = 1$$
$$-5y + 5z = -10$$

Multiply the first equation by –5 and solve by addition.
$$-10y - 5z = -5$$
$$\underline{-5y + 5z = -10}$$
$$-15y = -15$$
$$y = 1$$

Back-substitute 1 for y to find z.
$$2(1) + z = 1$$
$$2 + z = 1$$
$$z = -1$$

Back-substitute 1 for y and –1 for z to find x.
$$x + 2y - z = 5$$
$$x + 2(1) - (-1) = 5$$
$$x + 2 + 1 = 5$$
$$x + 3 = 5$$
$$x = 2$$

The solution set is $\{(2, 1, -1)\}$.

55. $3x - 4y + 4z = 7$
$x - y - 2z = 2$
$2x - 3y + 6z = 5$

Multiply the second equation by –3 and add to the third equation.
$-3x + 3y + 6z = -6$
$\underline{2x - 3y + 6z = 5}$
$-x + 12z = -1$

Multiply the second equation by –4 and add to the first equation.
$3x - 4y + 4z = 7$
$\underline{-4x + 4y + 8z = -8}$
$-x + 12z = -1$

The system of two equations in two variables becomes:
$-x + 12z = -1$
$-x + 12z = -1$

The two equations in two variables are identical. The system is dependent. There are an infinite number of solutions to the system.

56. Use each ordered pair to write an equation as follows:
$(x, y) = (1, 4)$
$y = ax^2 + bx + c$
$4 = a(1)^2 + b(1) + c$
$4 = a + b + c$

$(x, y) = (3, 20)$
$y = ax^2 + bx + c$
$20 = a(3)^2 + b(3) + c$
$20 = a(9) + 3b + c$
$20 = 9a + 3b + c$

$(x, y) = (-2, 25)$
$y = ax^2 + bx + c$
$25 = a(-2)^2 + b(-2) + c$
$25 = a(4) - 2b + c$
$25 = 4a - 2b + c$

The system of three equations in three variables is:

$a + b + c = 4$
$9a + 3b + c = 20$
$4a - 2b + c = 25$

Multiply the first equation by –1 and add to the second equation.
$-a - b - c = -4$
$\underline{9a + 3b + c = 20}$
$8a + 2b = 16$

Multiply the first equation by –1 and add to the third equation.
$-a - b - c = -4$
$\underline{4a - 2b + c = 25}$
$3a - 3b = 21$

The system of two equations in two variables becomes:
$8a + 2b = 16$
$3a - 3b = 21$

Multiply the first equation by 3, the second equation by 2 and solve by addition.
$24a + 6b = 48$
$\underline{6a - 6b = 42}$
$30a = 90$
$a = 3$

Back-substitute 3 for a to find b.
$3(3) - 3b = 21$
$9 - 3b = 21$
$-3b = 12$
$b = -4$

Back-substitute 3 for a and –4 for b to find c.
$a + b + c = 4$
$3 + (-4) + c = 4$
$-1 + c = 4$
$c = 5$

The quadratic equation is $y = 3x^2 - 4x + 5$.

57. Let x = average debt for the $18 - 29$ age group in the U.S.

Let y = average debt for the $30 - 39$ age group in the U.S.

Let z = average debt for the $40 - 49$ age group in the U.S.

$$x + y + z = 44,200$$
$$y - x = 8100$$
$$z - y = 3100$$

Solve the second equation for x.

$$y - x = 8100$$
$$-x = -y + 8100$$
$$x = y - 8100$$

Solve the third equation for z.

$$z - y = 3100$$
$$z = y + 3100$$

Substitute the expressions for x and z into the first equation and solve for y.

$$x + y + z = 44,200$$
$$(\overbrace{y - 8100}^{x}) + y + (\overbrace{y + 3100}^{z}) = 44,200$$
$$y - 8100 + y + y + 3100 = 44,200$$
$$3y - 5000 = 44,200$$
$$3y = 49,200$$
$$y = 16,400$$

Back-substitute to solve for x and z.

$$x = y - 8100$$
$$= 16,400 - 8100$$
$$= 8300$$

$$z = y + 3100$$
$$= 16,400 + 3100$$
$$= 19,500$$

The average debt for the $18 - 29$ age group in the U.S. is \$8300, for the $30 - 39$ age group is \$16,400, and for the $40 - 49$ age group is \$19,500.

Chapter 4 Test

1.
$$2x + y = 5$$
$$2(5) + (-5) = 5$$
$$10 + (-5) = 5$$
$$5 = 5, \text{ true}$$

$$x + 3y = -10$$
$$5 + 3(-5) = -10$$
$$5 + (-15) = -10$$
$$-10 = -10, \text{ true}$$

Since the ordered pair $(5, -5)$ satisfies both equations, it is a solution of the given system.

2.
$$x + 5y = 7$$
$$-3 + 5(2) = 7$$
$$-3 + 10 = 7$$
$$7 = 7, \text{ true}$$

$$3x - 4y = 1$$
$$3(-3) - 4(2) = 1$$
$$-9 - 8 = 1$$
$$-17 = 1, \text{ false}$$

Since the ordered pair $(-3, 2)$ fails to satisfy *both* equations, it is not a solution of the given system.

3.
$$x + y = 6$$
$$4x - y = 4$$

Graph both lines on the same axes.

$x + y = 6$: x-intercept = 6; y-intercept = 6

$4x - y = 4$: x-intercept: 1; y-intercept = -4

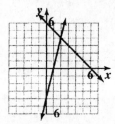

The solution set is $\{(2, 4)\}$.

4. $2x + y = 8$

   $y = 3x - 2$

Graph both lines on the same axes.

$2x + y = 8$: x-intercept = 4; y-intercept = 8

$y = 3x - 2$: slope = 3; y-intercept = -2

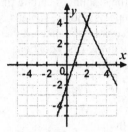

The solution set is $\{(2, 4)\}$.

5. $x = y + 4$

   $3x + 7y = -18$

Substitute $y + 4$ for x in the second equation.

   $3x + 7y = -18$

   $3(y + 4) + 7y = -18$

   $3y + 12 + 7y = -18$

   $10y + 12 = -18$

   $10y = -30$

   $y = -3$

Back-substitute -3 for y in the first equation.

$x = y + 4$

$x = -3 + 4 = 1$

The solution set is $\{(1, -3)\}$.

6. $2x - y = 7$

   $3x + 2y = 0$

Solve the first equation for y.

$2x - y = 7$

   $-y = -2x + 7$

   $y = 2x - 7$

Substitute $2x - 7$ for y in the second equation.

   $3x + 2y = 0$

$3x + 2(2x - 7) = 0$

   $3x + 4x - 14 = 0$

   $7x - 14 = 0$

   $7x = 14$

   $x = 2$

Back-substitute 2 for x in the equation $3x + 2y = 0$.

   $3x + 2y = 0$

   $3(2) + 2y = 0$

   $6 + 2y = 0$

   $2y = -6$

   $y = -3$

The solution set is $\{(2, -3)\}$.

7. $2x - 4y = 3$

   $x = 2y + 4$

Substitute $2y + 4$ for x in the first equation.

   $2x - 4y = 3$

   $2(2y + 4) - 4y = 3$

   $4y + 8 - 4y = 3$

   $8 = 3$, false

The false statement indicates that the system has no solution. The solution set is $\{\ \}$.

8. $2x + y = 2$

   $\underline{4x - y = -8}$

   $6x\quad = -6$

   $x = -1$

Back-substitute to find y.

   $2x + y = 2$

   $2(-1) + y = 2$

   $-2 + y = 2$

   $y = 4$

The solution set is $\{(-1, 4)\}$.

9. $2x + 3y = 1$

   $3x + 2y = -6$

Multiply the first equation by 3.

Multiply the second equation by -2.

   $6x + 9y = 3$

   $\underline{-6x - 4y = 12}$

   $5y = 15$

   $y = 3$

Back-substitute to find x.

   $2x + 3y = 1$

   $2x + 3(3) = 1$

   $2x + 9 = 1$

   $2x = -8$

   $x = -4$

The solution set is $\{(-4, 3)\}$.

10. $3x - 2y = 2$

    $-9x + 6y = -6$

    Multiply the first equation by 3.

    $9x - 6y = 6$

    $-9x + 6y = -6$

    _____

    $0 = 0$, true

    The true statement $0 = 0$ indicates that the system is dependent and the equation has infinitely many solutions.

    The solution set is $\{(x, y) | 3x - 2y = 2\}$ or

    $\{(x, y) | -9x + 6y = -6\}$.

11. Let $x =$ the percentage of females named Mary.
    Let $y =$ the percentage of females named Patricia.
    The system is $x + y = 3.7$

    $x - y = 1.5$

    Add the two equations:

    $x + y = 3.7$

    $x - y = 1.5$

    _____

    $2x \quad = 5.2$

    $x \quad = 2.6$

    Back substitute to find y.

    $2.6 + y = 3.7$

    $y = 1.1$

    2.6% of females are named Mary and 1.1% are named Patricia.

12. Let $x =$ the length of the garden.
    Let $y =$ the width of the garden.
    The perimeter of the garden is 34 yards so
    $2x + 2y = 34$.

    Since there are two lengths and two widths to be fenced, the information about the cost of fencing leads to the equation $2(2x) + 1(2y) = 58$.

    Simplify the second equation.

    $4x + 2y = 58$

    Multiply this equation by -1 and add the result to the first equation.

    $2x + 2y = 34$

    $-4x - 2y = -58$

    _____

    $-2x \qquad = -24$

    $x \qquad = \quad 12$

    Back substitute to find y.

    $2(12) + 2y = 34$

    $24 + 2y = 34$

    $2y = 10$

    $y = 5$

    The length of the garden is 12 yards and the width is 5 yards.

13. Let $x =$ the number of minutes of long-distance calls.
    Let $y =$ the monthly cost of a telephone plan.
    Plan A: $y = 15 + 0.05x$

    Plan B: $y = 5 + 0.07x$

    To determine the amount of calling time that will result in the same cost for both plans, solve this system by the substitution method. Substitute $5 + 0.07x$ for y in the first equation.

    $5 + 0.07x = 15 + 0.05x$

    $5 + 0.02x = 15$

    $0.02x = 10$

    $\dfrac{0.02x}{0.02} = \dfrac{10}{0.02}$

    $x = 500$

    If $x = 500$, $y = 15 + 0.05(500) = 40$.

    The cost of the two plans will be equal for 500 minutes per month. The cost of each plan will be $40.

14. Let $x =$ the amount invested at 6%
    Let $y =$ the amount invested at 7%

    $x + \qquad y = 9000$

    $0.06x + 0.07y = 610$

    Multiply the first equation by -0.06 and add to the second equation.

    $-0.06x - 0.06y = -540$

    $0.06x + 0.07y = 610$

    _____

    $0.01y = 70$

    $y = 7000$

    Back-substitute 7000 for y to find x.

    $x + y = 9000$

    $x + 7000 = 9000$

    $x = 2000$

    There is $2000 invested at 6% and $7000 invested at 7%.

15. Let x = the number of ounces of 6% solution
Let y = the number of ounces of 9% solution
$$x + y = 36$$
$$0.06x + 0.09y = 0.08(36)$$

Rewrite the system in standard form.
$$x + y = 36$$
$$0.06x + 0.09y = 2.88$$

Multiply the first equation by –0.06 and add to the second equation.
$$-0.06x - 0.06y = -2.16$$
$$\underline{0.06x + 0.09y = 2.88}$$
$$0.03y = 0.72$$
$$y = 24$$

Back-substitute 24 for y to find x.
$$x + y = 36$$
$$x + 24 = 36$$
$$x = 12$$

12 ounces of 6% peroxide solution and 24 ounces of 9% peroxide solution must be used.

16. Let r = the speed of the paddleboat in still water.
Let c = the speed of the current.

	Rate $\times$	Time $=$	Distance
Trip with the Current	$r + c$	3	$3(r + c)$
Trip against the Current	$r - c$	4	$4(r - c)$

$$3(r + c) = 48$$
$$4(r - c) = 48$$

Simplified, the system becomes
$$3r + 3c = 48$$
$$4r - 4c = 48$$

Multiply the first equation by 4, the second equation by 3, and solve by addition.
$$12r + 12c = 192$$
$$\underline{12r - 12c = 144}$$
$$24r = 336$$
$$r = 14$$

Back-substitute 14 for r to find c.
$$3r + 3c = 48$$
$$3(14) + 3c = 48$$
$$42 + 3c = 48$$
$$3c = 6$$
$$c = 2$$

The speed of the paddleboat in still water is 14 miles per hour and the speed of the current is 2 miles per hour.

17.
$$x + y + z = 6$$
$$3x + 4y - 7z = 1$$
$$2x - y + 3z = 5$$

Multiply the first equation by 7 and add to the second equation.
$$7x + 7y + 7z = 42$$
$$\underline{3x + 4y - 7z = 1}$$
$$10x + 11y = 43$$

Multiply the first equation by −3 and add to the third equation.
$$-3x - 3y - 3z = -18$$
$$\underline{2x - y + 3z = 5}$$
$$-x - 4y = -13$$

The system of two equations in two variables.
$$10x + 11y = 43$$
$$-x - 4y = -13$$

Multiply the second equation by 10 and solve by addition.
$$10x + 11y = 43$$
$$\underline{-10x - 40y = -130}$$
$$-29y = -87$$
$$y = 3$$

Back-substitute 3 for y to find x.
$$-x - 4y = -13$$
$$-x - 4(3) = -13$$
$$-x - 12 = -13$$
$$-x = -1$$
$$x = 1$$

Back-substitute 1 for x and 3 for y to find z.
$$x + y + z = 6$$
$$1 + 3 + z = 6$$
$$4 + z = 6$$
$$z = 2$$

The solution set is $\{(1, 3, 2)\}$.

Cumulative Review Exercises (Chapters 1-4)

1. $-14 - \left[18 - (6 - 10)\right]$
$$= -14 - \left[18 - (-4)\right]$$
$$= -14 - \left[18 + 4\right]$$
$$= -14 - 22$$
$$= -14 + (-22)$$
$$= -36$$

2. $6(3x - 2) - (x - 1) = 18x - 12 - x + 1$
$$= 17x - 11$$

3. $17(x + 3) = 13 + 4(x - 10)$
$$17x + 51 = 13 + 4x - 40$$
$$17x + 51 = 4x - 27$$
$$13x = -78$$
$$x = -6$$
The solution set is $\{-6\}$.

4. $\dfrac{x}{4} - 1 = \dfrac{x}{5}$

To clear fractions, multiply both sides by 20.
$$20\left(\frac{x}{4} - 1\right) = 20\left(\frac{x}{5}\right)$$
$$5x - 20 = 4x$$
$$x - 20 = 0$$
$$x = 20$$
The solution set is $\{20\}$.

5. $A = P + Prt$
$$A - P = Prt$$
$$\frac{A - P}{Pr} = \frac{Prt}{Pr}$$
$$\frac{A - P}{Pr} = t \quad \text{or} \quad t = \frac{A - P}{Pr}$$

6. $2x - 5 < 5x - 11$
$$-3x - 5 < -11$$
$$-3x < -6$$
$$\frac{-3x}{-3} > \frac{-6}{-3}$$
$$x > 2$$
The solution set is $\{x \mid x > 2\}$.

7. $x - 3y = 6$

x-intercept:

$x - 3y = 6$

$x - 3(0) = 6$

$x = 6$

y-intercept:

$x - 3y = 6$

$0 - 3y = 6$

$-3y = 6$

$y = -2$

Check point:

$x - 3y = 6$

$3 - 3y = 6$

$-3y = 3$

$y = -1$

A check point is $(3, -1)$.

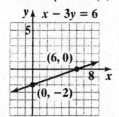

8.

x	$y = 0x + 4$	(x, y)
-6	$y = 0(-6) + 4 = 4$	$(-6, 4)$
-3	$y = 0(-3) + 4 = 4$	$(-3, 4)$
0	$y = 0(0) + 4 = 4$	$(0, 4)$
3	$y = 0(3) + 4 = 4$	$(3, 4)$
6	$y = 0(6) + 4 = 4$	$(6, 4)$

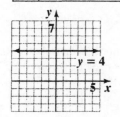

9. $y = -\dfrac{3}{5}x + 2$

slope $= -\dfrac{3}{5} = \dfrac{-3}{5}$; y-intercept $= 2$

Plot the point (0,2). From this point, move 3 units down (because -3 is negative) and 5 units to the right to reach the point $(5, -1)$. Draw a line through $(0,2)$ and $(5, -1)$.

$y = -\dfrac{3}{5}x + 2$

10. $3x - 4y = 8$

$4x + 5y = -10$

To solve this system by the addition method, multiply the first equation by 4 and the second equation by -3. Then add the results.

$12x - 16y = 32$

$\underline{-12x - 15y = 30}$

$-31y = 62$

$y = -2$

Back-substitute to find x.

$3x - 4y = 8$

$3x - 4(-2) = 8$

$3x + 8 = 8$

$3x = 0$

$x = 0$

The solution set is $\{(0, -2)\}$.

11. $2x - 3y = 9$

$y = 4x - 8$

To solve this system by the substitution method, substitute $4x - 8$ for y in the first equation.

$2x - 3y = 9$

$2x - 3(4x - 8) = 9$

$2x - 12x + 24 = 9$

$-10x + 24 = 9$

$-10x = -15$

$x = \dfrac{3}{2}$

Back-substitute $\dfrac{3}{2}$ for x in the second equation.

$$y = 4x - 8 = 4\left(\dfrac{3}{2}\right) - 8 = -2$$

The solution set is $\left\{\left(\dfrac{3}{2}, -2\right)\right\}$.

12. $m = \dfrac{y_2 - y_1}{x_2 - x_1} = \dfrac{-5 - (-6)}{6 - 5} = \dfrac{1}{1} = 1$

13. $y - y_1 = m(x - x_1)$

$y - 6 = -4[x - (-1)]$

$y - 6 = -4(x + 1)$

slope-intercept form:

$y - 6 = -4x - 4$

$y = -4x + 2$

14. Use the formula for the area of a triangle.

$A = \dfrac{1}{2}bh$

$80 = \dfrac{1}{2} \cdot 16 \cdot h$

$80 = 8h$

$10 = h$

The height is 10 feet.

15. Let x = the cost of one pen.
Let y = the cost of one pad.

$10x + 15y = 26$

$5x + 10y = 16$

Multiply the second equation by -2, and add the result to the first equation.

$\quad 10x + 15y = 26$

$\underline{-10x - 20y = -32}$

$\qquad -5y = -6$

$y = \dfrac{6}{5} = 1.20$

Back-substitute 1.20 for y and solve for x.

$10x + 15(1.20) = 26$

$10x + 18 = 26$

$10x = 8$

$x = \dfrac{8}{10} = 0.8$

One pen costs $0.80 and one pad costs $1.20.

16. The integers in the given set are -93, 0, $\dfrac{7}{1}$ $(=7)$ and $\sqrt{100}$ $(=10)$.

17. The skim milk line has a positive slope. This means that skim milk consumption, in gallons per person, increased from 1945 through 2003.

18. The whole milk line has a negative slope. This means that whole milk consumption, in gallons per person, decreased from 1945 through 2003.

19. If y is whole milk consumption, in gallons per person and x is the number of years after 2003, then this can be modeled by the equation $y = -0.57x + 7.6$.

To find when no whole milk will be consumed, let $y = 0$ and solve for x.

$y = -0.57x + 7.6$

$0 = -0.57x + 7.6$

$0.57x = 7.6$

$x = \dfrac{7.6}{0.57}$

$x \approx 13$

According to the model no whole milk will be consumed 13 years after 2003, or 2016.

20. To find when skim milk consumption will be 17.6 gallons per person, let $y = 15.6$ and solve for x.

$y = 0.17x + 13.9$

$15.6 = 0.17x + 13.9$

$1.7 = 0.17x$

$\dfrac{1.7}{0.17} = x$

$x = 10$

According to the model skim milk consumption will be 17.6 gallons per person 10 years after 2003, or 2013.

Chapter 5
Exponents and Polynomials

5.1 Check Points

1. $(-11x^3 + 7x^2 - 11x - 5) + (16x^3 - 3x^2 + 3x - 15)$

    $= -11x^3 + 7x^2 - 11x - 5 + 16x^3 - 3x^2 + 3x - 15$

    $= -11x^3 + 16x^3 + 7x^2 - 3x^2 - 11x + 3x - 5 - 15$

    $= 5x^3 + 4x^2 - 8x - 20$

2. $\begin{array}{r} -11x^3 + 7x^2 - 11x - 5 \\ +16x^3 - 3x^2 + 3x - 15 \\ \hline 5x^3 + 4x^2 - 8x - 20 \end{array}$

3. $(9x^2 + 7x - 2) - (2x^2 - 4x - 6)$

    $= 9x^2 + 7x - 2 - 2x^2 + 4x + 6$

    $= 9x^2 - 2x^2 + 7x + 4x - 2 + 6$

    $= 7x^2 + 11x + 4$

4. $(10x^3 - 5x^2 + 7x - 2) - (3x^3 - 8x^2 - 5x + 6)$

    $= 10x^3 - 5x^2 + 7x - 2 - 3x^3 + 8x^2 + 5x - 6$

    $= 10x^3 - 3x^3 - 5x^2 + 8x^2 + 7x + 5x - 2 - 6$

    $= 7x^3 + 3x^2 + 12x - 8$

5. $\begin{array}{r} 8y^3 - 10y^2 - 14y - 2 \\ -\left(5y^3 \qquad\quad - 3y + 6\right) \\ \hline \end{array}$

    To subtract, add the opposite of the polynomial being subtracted.

    $\begin{array}{r} 8y^3 - 10y^2 - 14y - 2 \\ -5y^3 \qquad\quad + 3y - 6 \\ \hline 3y^3 - 10y^2 - 11y - 8 \end{array}$

6. Make a table of values.

x	$y = x^2 - 1$	(x, y)
-3	$y = (-3)^2 - 1 = 8$	$(-3, 8)$
-2	$y = (-2)^2 - 1 = 3$	$(-2, 3)$
-1	$y = (-1)^2 - 1 = 0$	$(-1, 0)$
0	$y = (0)^2 - 1 = -1$	$(0, -1)$
1	$y = (1)^2 - 1 = 0$	$(1, 0)$
2	$y = (2)^2 - 1 = 3$	$(2, 3)$
3	$y = (3)^2 - 1 = 8$	$(3, 8)$

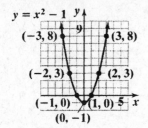

5.1 Exercise Set

1. $3x + 7$ is a binomial of degree 1.

3. $x^2 - 2x$ is a binomial of degree 3.

5. $8x^2$ is a monomial of degree 2.

7. 5 is a monomial. Because it is a nonzero constant, its degree is 0.

9. $x^2 - 3x + 4$ is a trinomial of degree 2.

11. $7y^2 - 9y^4 + 5$ is a trinomial of degree 4.

13. $15x - 7x^3$ is a binomial of degree 3.

15. $-9y^{23}$ is a monomial of degree 23.

17. $(9x + 8) + (-17x + 5)$
$= 9x + 8 + (-17)x + 5$
$= 9x + 8 - 17x + 5$
$= (9x - 17x) + (8 + 5)$
$= -8x + 13$

19. $\left(4x^2+6x-7\right)+\left(8x^2+9x-2\right)$

$= 4x^2+6x-7+8x^2+9x-2$

$= \left(4x^2+8x^2\right)+\left(6x+9x\right)+\left(-7-2\right)$

$= 12x^2+15x-9$

21. $\left(7x^2-11x\right)+\left(3x^2-x\right)$

$= 7x^2-11x+3x^2-x$

$= \left(7x^2+3x^2\right)+\left(-11x-x\right)$

$= 10x^2-12x$

23. $\left(4x^2-6x+12\right)+\left(x^2+3x+1\right)$

$= 4x^2-6x+12+x^2+3x+1$

$= \left(4x^2+x^2\right)+\left(-6x+3x\right)+\left(12+1\right)$

$= 5x^2-3x+13$

25. $\left(4y^3+7y-5\right)+\left(10y^2-6y+3\right)$

$= \left(4y^3+7y-5\right)+\left(10y^2-6y+3\right)$

$= 4y^3+7y-5+10y^2-6y+3$

$= 4y^3+10y^2+\left(7y-6y\right)+\left(-5+3\right)$

$= 4y^3+10y^2+y-2$

27. $= \left(2x^2-6x+7\right)+\left(3x^3-3x\right)$

$= 2x^2-6x+7+3x^3-3x$

$= 3x^3+2x^2+\left(-6x-3x\right)+7$

$= 3x^3+2x^2-9x+7$

29. $\left(4y^2+8y+11\right)+\left(-2y^3+5y+2\right)$

$= 4y^2+8y+11+\left(-2\right)y^3+5y+2$

$= -2y^3+4y^2+\left(8y+5y\right)+\left(11+2\right)$

$= -2y^3+4y^2+13y+13$

31. $\left(-2y^6+3y^4-y^2\right)+\left(-y^6+5y^4+2y^2\right) = -2y^6+3y^4-y^2-y^6+5y^4+2y^2$

$= \left(-2y^6-y^6\right)+\left(3y^4+5y^4\right)+\left(-y^2+2y^2\right)$

$= -3y^6+8y^4+y^2$

33. $\left(9x^3 - x^2 - x - \dfrac{1}{3}\right) + \left(x^3 + x^2 + x + \dfrac{4}{3}\right) = \left(9x^3 + x^3\right) + \left(-x^2 + x^2\right) + \left(-x + x\right) + \left(-\dfrac{1}{3} + \dfrac{4}{3}\right)$

$$= 10x^3 + \dfrac{3}{3}$$

$$= 10x^3 + 1$$

35. $\left(\dfrac{1}{5}x^4 + \dfrac{1}{3}x^3 + \dfrac{3}{8}x^2 + 6\right) + \left(-\dfrac{3}{5}x^4 + \dfrac{2}{3}x^3 - \dfrac{1}{2}x^2 - 6\right)$

$= \left[\dfrac{1}{5}x^4 + \left(-\dfrac{3}{5}x^4\right)\right] + \left(\dfrac{1}{3}x^2 + \dfrac{2}{3}x^3\right) + \left[\dfrac{3}{8}x^2 + \left(-\dfrac{1}{2}x^2\right)\right] + \left[6 + (-6)\right]$

$= -\dfrac{2}{5}x^4 + x^3 - \dfrac{1}{8}x^2$

37. $\left(0.03x^5 - 0.1x^3 + x + 0.03\right) + \left(-0.02x^5 + x^4 - 0.7x + 0.3\right) = \left(0.03x^5 - 0.02x^5\right) + x^4 - 0.1x^3 + \left(x - 0.07x\right) + \left(0.03 + 0.3\right)$

$$= 0.01x^5 + x^4 - 0.1x^3 + 0.3x + 0.33$$

39. $5y^3 - 7y^2$

$\dfrac{6y^3 + 4y^2}{11y^3 - 3y^2}$

41. $3x^2 - 7x + 4$

$\dfrac{-5x^2 + 6x - 3}{-2x^2 - x + 1}$

43. $\dfrac{1}{4}x^4 - \dfrac{2}{3}x^3 - 5$

$-\dfrac{1}{2}x^4 + \dfrac{1}{5}x^3 + 4.7$

To add, rewrite using common denominators for common terms.

$\dfrac{1}{4}x^4 - \dfrac{10}{15}x^3 - 5.0$

$-\dfrac{1}{2}x^4 + \dfrac{3}{15}x^3 + 4.7$

$-\dfrac{1}{4}x^4 - \dfrac{7}{15}x^3 - 0.3$

45. $y^3 + 5y^2 - 7y - 3$

$\dfrac{-2y^3 + 3y^2 + 4y - 11}{-y^3 + 8y^2 - 3y - 14}$

47.
$$4x^3 - 6x^2 + 5x - 7$$
$$\underline{-9x^3 \qquad - 4x + 3}$$
$$-5x^3 - 6x^2 + x - 4$$

49.
$$7x^4 - 3x^3 + x^2$$
$$\underline{\qquad x^3 - x^2 + 4x - 2}$$
$$7x^4 - 2x^3 \qquad + 4x - 2$$

51.
$$7x^2 - 9x + 3$$
$$4x^2 + 11x - 2$$
$$\underline{-3x^2 + 5x - 6}$$
$$8x^2 + 7x - 5$$

53.
$$1.2x^3 - 3x^2 + 9.1$$
$$7.8x^3 - 3.1x^2 + 8$$
$$\underline{\qquad 1.2x^2 - 6}$$
$$9x^3 - 4.9x^2 + 11.1$$

55. $(x - 8) - (3x + 2) = (x - 8) + (-3x - 2)$
$$= (x - 3x) + (-8 - 2)$$
$$= -2x - 10$$

57. $\left(x^2 - 5x - 3\right) - \left(6x^2 - 4x + 9\right)$
$$= \left(x^2 - 5x - 3\right) + \left(-6x^2 + 4x - 9\right)$$
$$= \left(x^2 - 6x^2\right) + (-5x - 4x) + (-3 - 9)$$
$$= -5x^2 - 9x - 12$$

59. $\left(x^2 - 5x\right) - \left(6x^2 - 4x\right)$
$$= \left(x^2 - 5x\right) + \left(-6x^2 + 4x\right)$$
$$= \left(x^2 - 6x^2\right) + (-5x + 4x)$$
$$= -5x^2 - x$$

61. $\left(x^2 - 8x - 9\right) - \left(5x^2 - 4x - 3\right)$
$$= \left(x^2 - 8x - 9\right) + \left(-5x^2 + 4x + 3\right)$$
$$= -4x^2 - 4x - 6$$

63. $(y - 8) - (3y - 2) = (y - 8) + (-3y + 2)$
$$= -2y - 6$$

65. $\left(6y^3 + 2y^2 - y - 11\right) - \left(y^2 - 8y + 9\right)$
$$= \left(6y^3 + 2y^2 - y - 11\right) + \left(-y^2 + 8y - 9\right)$$
$$= 6y^3 + y^2 + 7y - 20$$

67. $\left(7n^3 - n^7 - 8\right) - \left(6n^3 - n^7 - 10\right)$
$$= \left(7n^3 - n^7 - 8\right) + \left(-6n^3 + n^7 + 10\right)$$
$$= \left(7n^3 - 6n^3\right) + \left(-n^7 + n^7\right) + (-8 + 10)$$
$$= n^3 + 2$$

69. $\left(y^6 - y^3\right) - \left(y^2 - y\right)$
$$= \left(y^6 - y^3\right) + \left(-y^2 + y\right)$$
$$= y^6 - y^3 - y^2 + y$$

71. $\left(7x^4 + 4x^2 + 5x\right) - \left(-19x^4 - 5x^2 - x\right)$
$$= \left(7x^4 + 4x^2 + 5x\right) + \left(19x^4 + 5x^2 + x\right)$$
$$= 26x^4 + 9x^2 + 6x$$

73. $\left(\dfrac{3}{7}x^3 - \dfrac{1}{5}x - \dfrac{1}{3}\right) - \left(-\dfrac{2}{7}x^3 + \dfrac{1}{4}x - \dfrac{1}{3}\right)$
$$= \left(\dfrac{3}{7}x^3 - \dfrac{1}{5}x - \dfrac{1}{3}\right) + \left(\dfrac{2}{7}x^3 - \dfrac{1}{4}x + \dfrac{1}{3}\right)$$
$$= \left(\dfrac{3}{7}x^3 + \dfrac{2}{7}x^3\right) + \left(-\dfrac{1}{5}x - \dfrac{1}{4}x\right) + \left(-\dfrac{1}{3} + \dfrac{1}{3}\right)$$
$$= \left(\dfrac{3}{7}x^3 + \dfrac{2}{7}x^3\right) + \left(-\dfrac{4}{20}x - \dfrac{5}{20}x\right)$$
$$= \dfrac{5}{7}x^3 - \dfrac{9}{20}x$$

75.
$$7x + 1$$
$$\underline{-(3x - 5)}$$

To subtract, add the opposite of the polynomial being subtracted.
$$7x + 1$$
$$\underline{-3x + 5}$$
$$4x + 6$$

77.
$$7x^2 - 3$$
$$-\left(-3x^2 + 4\right)$$

To subtract, add the opposite of the polynomial being subtracted.

$$7x^2 - 3$$
$$3x^2 - 4$$
$$\overline{10x^2 - 7}$$

79.
$$7y^2 - 5y + 2$$
$$-\left(11y^2 + 2y - 3\right)$$

To subtract, add the opposite of the polynomial being subtracted.

$$7y^2 - 5y + 2$$
$$+ -11y^2 - 2y + 3$$
$$\overline{-4y^2 - 7y + 5}$$

81.
$$7x^3 + 5x^2 - 3$$
$$-\left(-2x^3 - 6x^2 + 5\right)$$

To subtract, add the opposite of the polynomial being subtracted.

$$7x^3 + 5x^2 - 3$$
$$2x^3 + 6x^2 - 5$$
$$\overline{9x^3 + 11x^2 - 8}$$

83.
$$5y^3 + 6y^2 - 3y + 10$$
$$-\left(6y^3 - 2y^2 - 4y - 4\right)$$

To subtract, add the opposite of the polynomial being subtracted.

$$5y^3 + 6y^2 - 3y + 10$$
$$+ -6y^3 + 2y^2 + 4y + 4$$
$$\overline{-y^3 + 8y^2 + y + 14}$$

85.
$$7x^4 - 3x^3 + 2x^2$$
$$-\left(\quad - x^3 - x^2 + x - 2\right)$$

To subtract, add the opposite of the polynomial being subtracted.

$$7x^4 - 3x^3 + 2x^2$$
$$+ \quad x^3 + x^2 - x + 2$$
$$\overline{7x^4 - 2x^3 + 3x^2 - x + 2}$$

87.
$$0.07x^3 - 0.01x^2 + 0.02x$$
$$-\left(0.02x^3 - 0.03x^2 - \quad x\right)$$

To subtract, add the opposite of the polynomial being subtracted.

$$0.07x^3 - 0.01x^2 + 0.02x$$
$$-0.02x^3 + 0.03x^2 + \quad x$$
$$\overline{0.05x^3 + 0.02x^2 + 1.02x}$$

89.

x	$y = x^2$	(x, y)
-3	$y = (-3)^2 = 9$	$(-3, 9)$
-2	$y = (-2)^2 = 4$	$(-2, 4)$
-1	$y = (-1)^2 = 1$	$(-1, 1)$
0	$y = (0)^2 = 0$	$(0, 0)$
1	$y = (1)^2 = 1$	$(1, 1)$
2	$y = (2)^2 = 4$	$(2, 4)$
3	$y = (3)^2 = 9$	$(3, 9)$

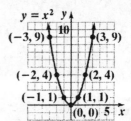

91.

x	$y = x^2 + 1$	(x, y)
-3	$y = (-3)^2 + 1 = 10$	$(-3, 10)$
-2	$y = (-2)^2 + 1 = 5$	$(-2, 5)$
-1	$y = (-1)^2 + 1 = 2$	$(-1, 2)$
0	$y = (0)^2 + 1 = 1$	$(0, 1)$
1	$y = (1)^2 + 1 = 2$	$(1, 2)$
2	$y = (2)^2 + 1 = 5$	$(2, 5)$
3	$y = (3)^2 + 1 = 10$	$(3, 10)$

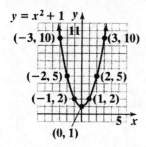

93.

x	$y = 4 - x^2$	(x, y)
-3	$y = 4 - (-3)^2 = -5$	$(-3, -5)$
-2	$y = 4 - (-2)^2 = 0$	$(-2, 0)$
-1	$y = 4 - (-1)^2 = 3$	$(-1, 3)$
0	$y = 4 - (0)^2 = 4$	$(0, 4)$
1	$y = 4 - (1)^2 = 3$	$(1, 3)$
2	$y = 4 - (2)^2 = 0$	$(2, 0)$
3	$y = 4 - (3)^2 = -5$	$(3, -5)$

95. $\left[\left(4x^2 + 7x - 5 \right) - \left(2x^2 - 10x + 3 \right) \right] - \left(x^2 + 5x - 8 \right)$

$= \left[2x^2 + 17x - 8 \right] - x^2 - 5x + 8$

$= x^2 + 12x$

97. $\left[\left(4y^2 - 3y + 8\right) - \left(5y^2 + 7y - 4\right)\right] - \left[\left(8y^2 + 5y - 7\right) + \left(-10y^2 + 4y + 3\right)\right]$

$= \left[-y^2 - 10y + 12\right] - \left[-2y^2 + 9y - 4\right]$

$= y^2 - 19y + 16$

99. $\left[\left(4x^3 + x^2\right) + \left(-x^3 + 7x - 3\right)\right] - \left(x^3 - 2x^2 + 2\right)$

$= \left[3x^3 + x^2 + 7x - 3\right] + \left(-x^3 + 2x^2 - 2\right)$

$= 2x^3 + 3x^2 + 7x - 5$

101. $\left[\left(-5 + y^2 + 4y^3\right) - \left(-8 - y + 7y^3\right)\right] - \left(-y^2 + 7y^3\right)$

$= \left[-3y^3 + y^2 + y + 3\right] + \left(y^2 - 7y^3\right)$

$= -10y^3 + 2y^2 + y + 3$

103. a. $M - W = (-18x^3 + 923x^2 - 9603x + 48,446) - (17x^3 - 450x^2 + 6392x - 14,764)$

$M - W = -18x^3 + 923x^2 - 9603x + 48,446 - 17x^3 + 450x^2 - 6392x + 14,764$

$M - W = -18x^3 - 17x^3 + 923x^2 + 450x^2 - 9603x - 6392x + 48,446 + 14,764$

$M - W = -35x^3 + 1373x^2 - 15,995x + 63,210$

b. $M - W = -35x^3 + 1373x^2 - 15,995x + 63,210$

$M - W = -35(14)^3 + 1373(14)^2 - 15,995(14) + 63,210 = 12,348$

The difference in the median income between men and women with 14 years experience is $12,348.

c. $44,404 - 33,481 = 10,923$

The actual difference displayed in the graph in the median income between men and women with 14 years experience is $10,923.
The model overestimates this difference by $12,348 - \$10,923 = \$1425$.

105. a. $M = 177x^2 + 288x + 7075$

$M = 177(16)^2 + 288(16) + 7075 = 56,995$

The model estimates the median annual income for a man with 16 years of education to be $56,995.
The model underestimates the actual value of $57,220 shown in the bar graph by $225.

b. The solution found in part (a) is represented by the point (16, 56,995) on the graph for men.

c. According to the graph in part (b) the median annual income for a woman with 16 years of education is about $42,000 (although answers will vary).

107. – 113. Answers will vary.

115. does not make sense; Explanations will vary. Sample explanation: The term of highest degree of a polynomial is not necessarily the first term. For example, the degree of $8x^2 - 7x + 9x^3 + 5$ is 3.

117. does not make sense; Explanations will vary. Sample explanation: $y = x^2 - 4$ is not a linear equation. Using two points and a checkpoint is an appropriate method for graphing a linear equation.

119. false; Changes to make the statement true will vary. A sample change is: The expression $\dfrac{1}{5x^2}+\dfrac{1}{3x}$ is not a polynomial expression (variables must not be in the denominator).

121. false; Changes to make the statement true will vary. A sample change is: The coefficient of x is -5.

123. $\left(5t-3t^2+t^3\right)-\left(t-t^2+\dfrac{1}{3}t^3\right)$

$=\left(5t-3t^2+t^3\right)+\left(-t+t^2-\dfrac{1}{3}t^3\right)$

$=\left(5t-t\right)+\left(-3t^2+t^2\right)+\left(t^3-\dfrac{1}{3}t^3\right)$

$=4t-2t^2+\dfrac{2}{3}t^3$

$=\dfrac{2}{3}t^3-2t^2+4t$

125. $(-10)(-7)\div(1-8)=(-10)(-7)\div(-7)$
$=70\div(-7)=-10$

126. $-4.6-(-10.2)=-4.6+10.2=5.6$

127. $3(x-2)=9(x+2)$
$3x-6=9x+18$
$3x-6-9x=9x+18-9x$
$-6x-6=18$
$-6x-6+6=18+6$
$-6x=24$
$\dfrac{-6x}{-6}=\dfrac{24}{-6}$
$x=-4$
The solution set is $\{-4\}$.

128. $x^3\cdot x^4=(x\cdot x\cdot x)\cdot(x\cdot x\cdot x\cdot x)=x^7$

129. $3x(x+5)=3x\cdot x+3x\cdot 5=3x^2+15x$

130. $x(x+2)+3(x+2)=x^2+2x+3x+6=x^2+5x+6$

5.2 Check Points

1. a. $2^2\cdot 2^4=2^{2+4}=2^6$ or 64

b. $x^6\cdot x^4=x^{6+4}=x^{10}$

c. $y\cdot y^7=y^{1+7}=y^8$

d. $y^4\cdot y^3\cdot y^2=y^{4+3+2}=y^9$

2. a. $\left(3^4\right)^5=3^{4\cdot 5}=3^{20}$

b. $\left(x^9\right)^{10}=x^{9\cdot 10}=x^{90}$

c. $\left[(-5)^7\right]^3=(-5)^{7\cdot 3}=(-5)^{21}$

3. a. $(2x)^4=2^4x^4=16x^4$

b. $\left(-4y^2\right)^3=(-4)^3\left(y^2\right)^3=(-4)^3y^{2\cdot 3}=-64y^6$

4. a. $(7x^2)(10x)=(7\cdot 10)(x^2\cdot x)=70x^3$

b. $(-5x^4)(4x^5)=(-5\cdot 4)(x^4\cdot x^5)=-20x^9$

5. a. $3x(x+5)=3x\cdot x+3x\cdot 5=3x^2+15x$

b. $6x^2(5x^3-2x+3)$
$=6x^2\cdot 5x^3-6x^2\cdot 2x+6x^2\cdot 3$
$=30x^5-12x^3+18x^2$

6. a. $(x+4)(x+5)=x^2+5x+4x+20$
$=x^2+9x+20$

b. $(5x+3)(2x-7)=10x^2-35x+6x-21$
$=10x^2-29x-21$

7. $(5x+2)(x^2-4x+3)$
$=5x\cdot x^2-5x\cdot 4x+5x\cdot 3+2\cdot x^2-2\cdot 4x+2\cdot 3$
$=5x^3-20x^2+15x+2x^2-8x+6$
$=5x^3-20x^2+2x^2+15x-8x+6$
$=5x^3-18x^2+7x+6$

8.

$$2x^3 - 5x^2 + 4x$$
$$3x^2 - 2x$$
$$\overline{-4x^4 + 10x^3 - 8x^2}$$
$$\underline{6x^5 - 15x^4 + 12x^3}$$
$$6x^5 - 19x^4 + 22x^3 - 8x^2$$

5.2 Exercise Set

1. $x^{15} \cdot x^3 = x^{15+3} = x^{18}$

3. $y \cdot y^{11} = y^1 \cdot y^{11} = y^{1+11} = y^{12}$

5. $x^2 \cdot x^6 \cdot x^3 = x^{2+6+3} = x^{11}$

7. $7^9 \cdot 7^{10} = 7^{9+10} = 7^{19}$

9. $\left(6^9\right)^{10} = 6^{9 \cdot 10} = 6^{90}$

11. $\left(x^{15}\right)^3 = x^{15 \cdot 3} = x^{45}$

13. $\left[(-20)^3\right]^3 = (-20)^{3 \cdot 3} = (-20)^9$

15. $(2x)^3 = 2^3 \cdot x^3 = 8x^3$

17. $(-5x)^2 = (-5)^2 x^2 = 25x^2$

19. $\left(4x^3\right)^2 = 4^2 \left(x^3\right)^2 = 16x^6$

21. $\left(-2y^6\right)^4 = (-2)^4 \left(y^6\right)^4 = 16y^{24}$

23. $\left(-2x^7\right)^5 = (-2)^5 \left(x^7\right)^5 = -32x^{35}$

25. $(7x)(2x) = (7 \cdot 2)(x \cdot x) = 14x^2$

27. $(6x)\left(4x^2\right) = (6 \cdot 4)\left(x \cdot x^2\right) = 24x^3$

29. $\left(-5y^4\right)\left(3y^3\right) = (-5 \cdot 3)\left(y^4 \cdot y^3\right) = -15y^7$

31. $\left(-\dfrac{1}{2}a^3\right)\left(-\dfrac{1}{4}a^2\right) = \left(-\dfrac{1}{2} \cdot -\dfrac{1}{4}\right)\left(a^3 \cdot a^2\right)$
$$= \dfrac{1}{8}a^5$$

33. $\left(2x^2\right)(-3x)\left(8x^4\right)$
$$= (2 \cdot -3 \cdot 8)\left(x^2 \cdot x \cdot x^4\right) = -48x^7$$

35. $4x(x+3) = 4x \cdot x + 4x \cdot 3$
$$= 4x^2 + 12x$$

37. $x(x-3) = x \cdot x - x \cdot 3$
$$= x^2 - 3x$$

39. $2x(x-6) = 2x \cdot x - 2x \cdot 6$
$$= 2x^2 - 12x$$

41. $-4y(3y+5) = -4y \cdot 3y - 4y \cdot 5$
$$= -12y^2 - 20y$$

43. $4x^2(x+2) = 4x^2 \cdot x + 4x^2 \cdot 2$
$$= 4x^3 + 8x^2$$

45. $2y^2\left(y^2+3y\right) = 2y^2 \cdot y^2 + 2y^2 \cdot 3y$
$$= 2y^4 + 6y^3$$

47. $2y^2\left(3y^2 - 4y + 7\right)$
$$= 2y^2\left(3y^2\right) + 2y^2(-4y) + 2y^2(7)$$
$$= 6y^4 - 8y^3 + 14y^2$$

49. $\left(3x^3 + 4x^2\right)(2x) = 3x^3 \cdot 2x + 4x^2 \cdot 2x$
$$= 6x^4 + 8x^3$$

51. $\left(x^2 + 5x - 3\right)(-2x)$
$$= x^2(-2x) + 5x(-2x) - 3(-2x)$$
$$= -2x^3 - 10x^2 + 6x$$

53. $-3x^2\left(-4x^2+x-5\right)$

$= -3x^2\left(-4x^2\right)-3x^2\left(x\right)-3x^2\left(-5\right)$

$= 12x^4-3x^3+15x^2$

55. $\left(x+3\right)\left(x+5\right)$

$= x\left(x+5\right)+3\left(x+5\right)$

$= x\cdot x+x\cdot5+3\cdot x+3\cdot5$

$= x^2+5x+3x+15$

$= x^2+8x+15$

57. $\left(2x+1\right)\left(x+4\right)$

$= 2x\left(x+4\right)+1\left(x+4\right)$

$= 2x^2+8x+x+4$

$= 2x^2+9x+4$

59. $\left(x+3\right)\left(x-5\right)=x\left(x-5\right)+3\left(x-5\right)$

$= x^2-5x+3x-15$

$= x^2-2x-15$

61. $\left(x-11\right)\left(x+9\right)=x\left(x+9\right)-11\left(x+9\right)$

$= x^2+9x-11x-99$

$= x^2-2x-99$

63. $\left(2x-5\right)\left(x+4\right)$

$= 2x\left(x+4\right)-5\left(x+4\right)$

$= 2x^2+8x-5x-20$

$= 2x^2+3x-20$

65. $\left(\dfrac{1}{4}x+4\right)\left(\dfrac{3}{4}x-1\right)$

$= \dfrac{1}{4}x\left(\dfrac{3}{4}x-1\right)+4\left(\dfrac{3}{4}x-1\right)$

$= \dfrac{1}{4}x\cdot\dfrac{3}{4}x+\dfrac{1}{4}x\left(-1\right)$

$\quad +4\left(\dfrac{3}{4}x\right)+4\left(-1\right)$

$= \dfrac{3}{16}x^2-\dfrac{1}{4}x+\dfrac{12}{4}x-4$

$= \dfrac{3}{16}x^2+\dfrac{11}{4}x-4$

67. $\left(x+1\right)\left(x^2+2x+3\right)$

$= x\left(x^2+2x+3\right)+1\left(x^2+2x+3\right)$

$= x^3+2x^2+3x+x^2+2x+3$

$= x^3+3x^2+5x+3$

69. $\left(y-3\right)\left(y^2-3y+4\right)$

$= y\left(y^2-3y+4\right)-3\left(y^2-3y+4\right)$

$= y^3-3y^2+4y-3y^2+9y-12$

$= y^3-6y^2+13y-12$

71. $\left(2a-3\right)\left(a^2-3a+5\right)$

$= 2a\left(a^2-3a+5\right)-3\left(a^2-3a+5\right)$

$= 2a^3-6a^2+10a-3a^2+9a-15$

$= 2a^3-9a^2+19a-15$

73. $\left(x+1\right)\left(x^3+2x^2+3x+4\right)$

$= x\left(x^3+2x^2+3x+4\right)+1\left(x^3+2x^2+3x+4\right)$

$= x^4+2x^3+3x^2+4x+x^3+2x^2+3x+4$

$= x^4+\left(2x^3+x^3\right)+\left(3x^2+2x^2\right)+\left(4x+3x\right)+4$

$= x^4+3x^3+5x^2+7x+4$

75. $\left(x-\dfrac{1}{2}\right)\left(4x^3-2x^2+5x-6\right)$

$= x\left(4x^3-2x^2+5x-6\right)-\dfrac{1}{2}\left(4x^3-2x^2+5x-6\right)$

$= 4x^4-2x^3+5x^2-6x-2x^3+x^2-\dfrac{5}{2}x+3$

$= 4x^4-4x^3+6x^2-\dfrac{17}{2}x+3$

77. $\left(x^2+2x+1\right)\left(x^2-x+2\right)$

$= x^2\left(x^2-x+2\right)+2x\left(x^2-x+2\right)+1\left(x^2-x+2\right)$

$= x^4-x^3+2x^2+2x^3-2x^2+4x+x^2-x+2$

$= x^4+x^3+x^2+3x+2$

79.
$$\begin{array}{r} x^2 - 5x + 3 \\ x + 8 \\ \hline 8x^2 - 40x + 24 \\ x^3 - 5x^2 + 3x \\ \hline x^3 + 3x^2 - 37x + 24 \end{array}$$

81.
$$\begin{array}{r} x^2 - 3x + 9 \\ 2x - 3 \\ \hline -3x^2 + 9x - 27 \\ 2x^3 - 6x^2 + 18x \\ \hline 2x^3 - 9x^2 + 27x - 27 \end{array}$$

83.
$$\begin{array}{r} 2x^3 + x^2 + 2x + 3 \\ x + 4 \\ \hline 8x^3 + 4x^2 + 8x + 12 \\ 2x^4 + x^3 + 2x^2 + 3x \\ \hline 2x^4 + 9x^3 + 6x^2 + 11x + 12 \end{array}$$

85.
$$\begin{array}{r} 4z^3 - 2z^2 + 5z - 4 \\ 3z - 2 \\ \hline -8z^3 + 4z^2 - 10z + 8 \\ 12z^4 - 5z^3 + 15z^2 - 12z \\ \hline 12z^4 - 14z^3 + 19z^2 - 22z + 8 \end{array}$$

87.
$$\begin{array}{r} 7x^3 - 5x^2 + 6x \\ 3x^2 - 4x \\ \hline -28x^4 + 20x^3 - 24x^2 \\ 21x^5 - 15x^4 + 18x^3 \\ \hline 21x^5 - 43x^4 + 38x^3 - 24x^2 \end{array}$$

89.
$$\begin{array}{r} 2y^5 \qquad -3y^3 + y^2 - 2y + 3 \\ 2y - 1 \\ \hline -2y^5 \qquad + 3y^3 - y^2 + 2y - 3 \\ 4y^6 - \qquad 6y^4 + 2y^3 - 4y^2 + 6y \\ \hline 4y^6 - 2y^5 - 6y^4 + 5y^3 - 5y^2 + 8y - 3 \end{array}$$

91.
$$\begin{array}{r} x^2 + 7x - 3 \\ x^2 - x - 1 \\ \hline -x^2 - 7x + 3 \\ -x^3 - 7x^2 + 3x \\ x^4 + 7x^3 - 3x^2 \\ \hline x^4 + 6x^3 - 11x^2 - 4x + 3 \end{array}$$

93. $(x+4)(x-5) - (x+3)(x-6)$
$$= \left(x^2 - x - 20\right) - \left(x^2 - 3x - 18\right)$$
$$= \left(x^2 - x - 20\right) + \left(-x^2 + 3x + 18\right)$$
$$= 2x - 2$$

95. $4x^2\left(5x^3 + 3x - 2\right) - 5x^3\left(x^2 - 6\right)$
$$= \left(20x^5 + 12x^3 - 8x^2\right) + \left(-5x^5 + 30x^3\right)$$
$$= 15x^5 + 42x^3 - 8x^2$$

97. $(y+1)\left(y^2 - y + 1\right) + (y-1)\left(y^2 + y + 1\right)$
$$= y\left(y^2 - y + 1\right) + 1\left(y^2 - y + 1\right)$$
$$\quad + y\left(y^2 + y + 1\right) - 1\left(y^2 + y + 1\right)$$
$$= y^3 - y^2 + y + y^2 - y + 1$$
$$\quad + y^3 + y^2 + y - y^2 - y - 1$$
$$= 2y^3$$

99. $(y+6)^2 - (y-2)^2$
$$= (y+6)(y+6) - (y-2)(y-2)$$
$$= \left(y^2 + 12y + 36\right) - \left(y^2 - 4y + 4\right)$$
$$= 16y + 32$$

101. Use the formula for the area of a rectangle.
$$A = l \cdot w$$
$$A = (x+5)(2x-3)$$
$$= x(2x-3) + 5(2x-3)$$
$$= 2x^2 - 3x + 10x - 15$$
$$= 2x^2 + 7x - 15$$

The area of the rug is $2x^2 + 7x - 15$ square feet.

103. a. $(x+2)(2x+1)$

b. $x \cdot 2x + 2 \cdot 2x + x \cdot 1 + 2 \cdot 1$
$$= 2x^2 + 4x + x + 2$$
$$= 2x^2 + 5x + 2$$

c. $(x+2)(2x+1) = x(2x+1) + 2(2x+1)$
$$= 2x^2 + x + 4x + 2$$
$$= 2x^2 + 5x + 2$$

105. When multiplying numbers with the same base, keep the base and add the exponents.

Example: $2^3 \cdot 2^5 = 2^{3+5} = 2^8$

107. – 111. Answers will vary.

113. makes sense

115. makes sense

117. false; Changes to make the statement true will vary. A sample change is: $4x^3 \cdot 3x^4 = 4 \cdot 3x^{3+4} = 12x^7$

119. true

121. The area of the outer square is
$$(x+4)(x+4) = x(x+4) + 4(x+4)$$
$$= x^2 + 4x + 4x + 16$$
$$= x^2 + 8x + 16$$

The area of the inner square is x^2. The area of the shaded region is the difference between the areas of the two squares, which is
$$\left(x^2 + 8x + 16\right) - x^2 = 8x + 16$$

123. $\left(-8x^4\right)\left(-\dfrac{1}{4}xy^3\right) = 2x^5y^3$, so the missing factor is $-8x^4$.

124.
$$4x - 7 > 9x - 2$$
$$4x - 7 - 9x > 9x - 2 - 9x$$
$$-5x - 7 > -2$$
$$-5x - 7 + 7 > -2 + 7$$
$$-5x > 5$$
$$\frac{-5x}{-5} < \frac{5}{-5}$$
$$x < -1$$
Solution: $(-\infty, -1)$

125. $3x - 2y = 6$
x-intercept:
$$3x - 2y = 6$$
$$3x - 2(0) = 6$$
$$3x = 6$$
$$x = 2$$
y-intercept:
$$3x - 2y = 6$$
$$3(0) - 2y = 6$$
$$-2y = 6$$
$$y = -3$$
checkpoint:
$$3x - 2y = 6$$
$$3(4) - 2y = 6$$
$$12 - 2y = 6$$
$$-2y = -6$$
$$y = 3$$
A checkpoint is (4,3).

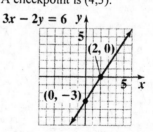

$3x - 2y = 6$

126. $m = \dfrac{y_2 - y_1}{x_2 - x_2} = \dfrac{6 - 8}{1 - (-2)} = -\dfrac{2}{3}$

127. a. $(x+3)(x+4) = x^2 + 4x + 3x + 12$
$$= x^2 + 7x + 12$$

b. $(x+5)(x+20) = x^2 + 5x + 20x + 100$
$$= x^2 + 25x + 100$$

A fast method is $(x + a)(x + b) = x^2 + (a + b)x + ab$.

128. **a.** $(x+3)(x-3) = x^2 + 3x - 3x - 9$

$\qquad\qquad = x^2 - 9$

b. $(x+5)(x-5) = x^2 + 5x - 5x - 25$

$\qquad\qquad = x^2 - 25$

A fast method is $(x+a)(x-a) = x^2 - a^2$.

129. **a.** $(x+3)^2 = (x+3)(x+3)$

$\qquad\qquad = x^2 + 3x + 3x + 9$

$\qquad\qquad = x^2 + 6x + 9$

b. $(x+5)^2 = (x+5)(x+5)$

$\qquad\qquad = x^2 + 5x + 5x + 9$

$\qquad\qquad = x^2 + 10x + 25$

A fast method is $(x+a)^2 = x^2 + 2ax + a^2$.

5.3 Check Points

1. $(x+5)(x+6) = x^2 + 6x + 5x + 30$

$\qquad\qquad = x^2 + 11x + 30$

2. $(7x+5)(4x-3) = 28x^2 - 21x + 20x - 15$

$\qquad\qquad = 28x^2 - x - 15$

3. $(4-2x)(5-3x) = 20 - 12x - 10x + 6x^2$

$\qquad\qquad = 20 - 22x + 6x^2$

$\qquad\qquad = 6x^2 - 22x + 20$

4. **a.** $(7y+8)(7y-8) = \overset{\substack{\text{first term}\\\text{squared}}}{\overbrace{(7y)^2}} - \overset{\substack{\text{second term}\\\text{squared}}}{\overbrace{8^2}}$

$\qquad\qquad = 49y^2 - 64$

b. $(4x-5)(4x+5) = \overset{\substack{\text{first term}\\\text{squared}}}{\overbrace{(4x)^2}} - \overset{\substack{\text{second term}\\\text{squared}}}{\overbrace{5^2}}$

$\qquad\qquad = 16x^2 - 25$

c. $(2a^3+3)(2a^3-3) = \overset{\substack{\text{first term}\\\text{squared}}}{\overbrace{(2a^3)^2}} - \overset{\substack{\text{second term}\\\text{squared}}}{\overbrace{3^2}}$

$\qquad\qquad = 4a^6 - 9$

5. **a.** $(x+10)^2 = \overset{\substack{\text{first term}\\\text{squared}}}{\overbrace{x^2}} + \overset{\substack{2\cdot\text{product}\\\text{of the terms}}}{\overbrace{2\cdot 10x}} + \overset{\substack{\text{last term}\\\text{squared}}}{\overbrace{10^2}}$

$\qquad\qquad = x^2 + 20x + 100$

b. $(5x+4)^2 = \overset{\substack{\text{first term}\\\text{squared}}}{\overbrace{(5x)^2}} + \overset{\substack{2\cdot\text{product}\\\text{of the terms}}}{\overbrace{2\cdot 20x}} + \overset{\substack{\text{last term}\\\text{squared}}}{\overbrace{4^2}}$

$\qquad\qquad = 25x^2 + 40x + 16$

6. **a.** $(x-9)^2 = \overset{\substack{\text{first term}\\\text{squared}}}{\overbrace{x^2}} - \overset{\substack{2\cdot\text{product}\\\text{of the terms}}}{\overbrace{2\cdot 9x}} + \overset{\substack{\text{last term}\\\text{squared}}}{\overbrace{9^2}}$

$\qquad\qquad = x^2 - 18x + 81$

b. $(7x-3)^2 = \overset{\substack{\text{first term}\\\text{squared}}}{\overbrace{(7x)^2}} - \overset{\substack{2\cdot\text{product}\\\text{of the terms}}}{\overbrace{2\cdot 21x}} + \overset{\substack{\text{last term}\\\text{squared}}}{\overbrace{3^2}}$

$\qquad\qquad = 49x^2 - 42x + 9$

5.3 Exercise Set

1. $(x+4)(x+6) = x^2 + 6x + 4x + 24$

$\qquad\qquad = x^2 + 10x + 24$

3. $(y-7)(y+3) = y^2 + 3y - 7y - 21$

$\qquad\qquad = y^2 - 4y - 21$

5. $(2x-3)(x+5) = 2x^2 + 10x - 3x - 15$

$\qquad\qquad = 2x^2 + 7x - 15$

7. $(4y+3)(y-1) = 4y^2 - 4y + 3y - 3$

$\qquad\qquad = 4y^2 - y - 3$

9. $(2x-3)(5x+3) = 10x^2 + 6x - 15x - 9$

$\qquad\qquad = 10x^2 - 9x - 9$

11. $(3y-7)(4y-5) = 12y^2 - 15y - 28y + 35$

$\qquad\qquad = 12y^2 - 43y + 35$

13. $(7+3x)(1-5x) = 7 - 35x + 3x - 15x^2$
$$= -15x^2 - 32x + 7$$

15. $(5-3y)(6-2y) = 30 - 10y - 18y + 6y^2$
$$= 30 - 28y + 6y^2$$
$$= 6y^2 - 28y + 30$$

17. $(5x^2 - 4)(3x^2 - 7)$
$$= (5x^2)(3x^2) + (5x^2)(-7) + (-4)(3x^2) + (-4)(-7)$$
$$= 15x^4 - 35x^2 - 12x^2 + 28$$
$$= 15x^4 - 47x^2 + 28$$

19. $(6x-5)(2-x) = 12x - 6x^2 - 10 + 5x$
$$= -6x^2 + 17x - 10$$

21. $(x+5)(x^2 + 3) = x^3 + 3x + 5x^2 + 15$
$$= x^3 + 5x^2 + 3x + 15$$

23. $(8x^3 + 3)(x^2 + 5) = 8x^5 + 40x^3 + 3x^2 + 15$

25. $(x+3)(x-3) = x^2 - 3^2 = x^2 - 9$

27. $(3x+2)(3x-2) = (3x)^2 - 2^2 = 9x^2 - 4$

29. $(3r-4)(3r+4) = (3r)^2 - 4^2$
$$= 9r^2 - 16$$

31. $(3+r)(3-r) = 3^2 - r^2 = 9 - r^2$

33. $(5-7x)(5+7x) = 5^2 - (7x^2) = 25 - 49x^2$

35. $\left(2x + \frac{1}{2}\right)\left(2x - \frac{1}{2}\right) = (2x)^2 - \left(\frac{1}{2}\right)^2$
$$= 4x^2 - \frac{1}{4}$$

37. $(y^2 + 1)(y^2 - 1) = (y^2)^2 - 1^2 = y^4 - 1$

39. $(r^3 + 2)(r^3 - 2) = (r^3)^2 - 2^2 = r^6 - 4$

41. $(1-y^4)(1+y^4) = 1^2 - (y^4)^2 = 1 - y^8$

43. $(x^{10} + 5)(x^{10} - 5) = (x^{10})^2 - 5^2$
$$= x^{20} - 25$$

45. $(x+2)^2 = x^2 + 2(2x) + 2^2$
$$= x^2 + 4x + 4$$

47. $(2x+5)^2 = (2x)^2 + 2(2x)(5) + 5^2$
$$= 4x^2 + 20x + 25$$

49. $(x-3)^2 = x^2 - 2(3x) + 3^2$
$$= x^2 - 6x + 9$$

51. $(3y-4)^2 = (3y)^2 - 2(3y)(4) + 4^2$
$$= 9y^2 - 24y + 16$$

53. $(4x^2 - 1)^2 = (4x^2)^2 - 2(4x^2)(1) + 1^2$
$$= 16x^4 - 8x^2 + 1$$

55. $(7-2x)^2 = 7^2 - 2(7)(2x) + (2x)^2$
$$= 49 - 28x + 4x^2$$

57. $\left(2x + \frac{1}{2}\right)^2 = 4x^2 + 2(2x)\left(\frac{1}{2}\right) + \left(\frac{1}{2}\right)^2$
$$= 4x^2 + 2x + \frac{1}{4}$$

59. $\left(4y - \frac{1}{4}\right)^2 = (4y)^2 - 2(4y)\left(\frac{1}{4}\right) + \left(\frac{1}{4}\right)^2$
$$= 16y^2 - 2y + \frac{1}{16}$$

61. $(x^8 + 3)^2 = (x^8)^2 + 2(x^8)(3) + 3^2$
$$= x^{16} + 6x^8 + 9$$

63. $(x-1)(x^2+x+1)$

$\quad = x(x^2+x+1)-1(x^2+x+1)$

$\quad = x^3+x^2+x-x^2-x-1$

$\quad = x^3-1$

65. $(x-1)^2 = x^2-2(x)(1)+1^2$

$\quad\quad = x^2-2x+1$

67. $(3y+7)(3y-7) = (3y^2)-7^2$

$\quad\quad\quad\quad = 9y^2-49$

69. $3x^2(4x^2+x+9)$

$\quad = 3x^2(4x^2)+3x^2(x)+3x^2(9)$

$\quad = 12x^4+3x^3+27x^2$

71. $(7y+3)(10y-4)$

$\quad = 70y^2-28y+30y-12$

$\quad = 70y^2+2y-12$

73. $(x^2+1)^2 = (x^2)^2+2(x^2)(1)+1^2$

$\quad\quad\quad = x^4+2x^2+1$

75. $(x^2+1)(x^2+2)$

$\quad = x^2\cdot x^2+x^2\cdot 2+1\cdot x^2+1\cdot 2$

$\quad = x^4+3x^2+2$

77. $(x^2+4)(x^2-4) = (x^2)^2-4^2$

$\quad\quad\quad\quad = x^4-16$

79. $(2-3x^5)^2 = 2^2-2(2)(3x^5)+(3x^5)^2$

$\quad\quad\quad = 4-12x^5+9x^{10}$

81. $\left(\dfrac{1}{4}x^2+12\right)\left(\dfrac{3}{4}x^2-8\right)$

$\quad = \dfrac{1}{4}x^2\left(\dfrac{3}{4}x^2\right)+\dfrac{1}{4}x^2(-8)+12\left(\dfrac{3}{4}x^2\right)+12(-8)$

$\quad = \dfrac{3}{16}x^4-2x^2+9x^2-96$

$\quad = \dfrac{3}{16}x^2+7x^2-96$

83. $A=(x+1)^2 = x^2+2x+1$

85. $A=(2x-3)(2x+3) = (2x)^2-3^2$

$\quad\quad = 4x^2-9$

87. Area of outer rectangle:

$\quad (x+9)(x+3) = x^2+12x+27$

Area of inner rectangle:

$\quad (x+5)(x+1) = x^2+6x+5$

Area of shaded region:

$\quad (x^2+12x+27)-(x^2+6x+5)=6x+22$

89. $\left[(2x+3)(2x-3)\right]^2$

$\quad = \left[4x^2-9\right]^2$

$\quad = 16x^4-72x^2+81$

91. $(4x^2+1)\left[(2x+1)(2x-1)\right]$

$\quad = (4x^2+1)\left[4x^2-1\right]$

$\quad = 16x^4-1$

93. $(x+2)^3$

$\quad = (x+2)(x+2)^2$

$\quad = (x+2)(x^2+4x+4)$

$\quad = x(x^2+4x+4)+2(x^2+4x+4)$

$\quad = x^3+4x^2+4x+2x^2+8x+8$

$\quad = x^3+6x^2+12x+8$

95. $\left[(x+3)-y\right]\left[(x+3)+y\right]$

$\quad = (x+3)^2-y^2$

$\quad = x^2+6x+9-y^2$

97. $(x+2)(x+1)$

The area of the larger garden is given by $(x+2)(x+1)$ square yards.

99. If the original garden measures $x = 6$ yards on a side, the area of the larger garden would be

$(6)^2 + 3(6) + 2 = 56$ square yards.

This corresponds to the point $(6, 56)$ on the graph.

101. $A_{\text{total}} = (x+2)^2 = x^2 + 2 \cdot (2x) + 2^2 = x^2 + 4x + 4$

The total area is $(x^2 + 4x + 4)$ square inches.

103. – 107. Answers will vary.

109. makes sense

111. makes sense

113. true

115. false; Changes to make the statement true will vary.
A sample change is: $(x-5)^2 = x^2 - 10x + 25$.

117. $V = l \cdot w \cdot h$

$= (10 - 2x)(8 - 2x)(x)$

$= (80 - 20x - 16x + 4x^2)(x)$

$= (80 - 36x + 4x^2)(x)$

$= 80x - 36x^2 + 4x^3$

$= 4x^3 - 36x^2 + 80x$

The volume of the box is $(4x^3 - 36x^2 + 80x)$ cubic units.

119. Let $y_1 = (x+1)^2$ and $y_2 = x^2 + 1$.

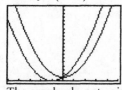

The graphs do not coincide so the multiplication is not correct.

To correct, let $y_2 = x^2 + 2x + 1$

121. Let $y_1 = (x+1)(x-1)$ and $y_2 = x^2 - 1$.

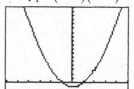

The graphs coincide so the multiplication is correct.

$(x+1)(x-1) = x^2 - x + x - 1 = x^2 - 1$

123. $2x + 3y = 1$

$y = 3x - 7$

The substitution method is a good choice because the second equation is already solved for y.
Substitute $3x - 7$ for y into the first equation.

$2x + 3y = 1$

$2x + 3(3x - 7) = 1$

$2x + 9x - 21 = 1$

$11x - 21 = 1$

$11x = 22$

$x = 2$

Back-substitute to find y.

$y = 3x - 7$

$y = 3(2) - 7 = 6 + 7 = -1$

The solution set is $\{(2, -1)\}$.

124. $3x + 4y = 7$

$2x + 7y = 9$

The addition method is a good choice because both equations are written in the form $Ax + By = C$. To eliminate x, multiply the first equation by 2 and the second equation by -3. Then add the results.

$6x + 8y = 14$

$\underline{-6x - 21y = -27}$

$-13y = -13$

$y = 1$

Back-substitute 1 for y in either equation of the original system.

$3x + 4y = 7$

$3x + 4(1) = 7$

$3x + 4 = 7$

$3x = 3$

$x = 1$

The solution set is $\{(1, 1)\}$.

125. $y = \dfrac{1}{3}x$

Graph $y = \dfrac{1}{3}x$ by using its slope, $\dfrac{1}{3}$, and y-intercept, 0.

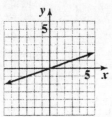

126.
$$
\begin{aligned}
x^3 y + 2xy^2 + 5x - 2 &= (-2)^3(3) + 2(-2)(3)^2 + 5(-2) - 2 \\
&= (-8)(3) + 2(-2)(9) + 5(-2) - 2 \\
&= -24 - 36 - 10 - 2 \\
&= -72
\end{aligned}
$$

127. $5xy + 6xy = (5 + 6)xy = 11xy$

128.
$$
\begin{aligned}
(x + 2y)(3x + 5y) &= x \cdot 3x + x \cdot 5y + 2y \cdot 3x + 2y \cdot 5y \\
&= 3x^2 + 5xy + 6xy + 10y^2 \\
&= 3x^2 + 11xy + 10y^2
\end{aligned}
$$

5.4 Check Points

1.
$$
\begin{aligned}
3x^3 y + xy^2 + 5y + 6 &= 3(-1)^3(5) + (-1)(5)^2 + 5(5) + 6 \\
&= 3(-1)(5) + (-1)(25) + 5(5) + 6 \\
&= -15 - 25 + 25 + 6 \\
&= -9
\end{aligned}
$$

2. $8x^4 y^5 - 7x^3 y^2 - x^2 y - 5x + 11$

Term	Coefficient	Degree
$8x^4 y^5$	8	$4 + 5 = 9$
$-7x^3 y^2$	-7	$3 + 2 = 5$
$-x^2 y$	-1	$2 + 1 = 3$
$-5x$	-5	1
11	11	0

The degree of the polynomial is the highest degree of all its terms, which is 9.

3. $(-8x^2 y - 3xy + 6) + (10x^2 y + 5xy - 10)$

$= -8x^2 y - 3xy + 6 + 10x^2 y + 5xy - 10$

$= -8x^2 y + 10x^2 y - 3xy + 5xy + 6 - 10$

$= 2x^2 y + 2xy - 4$

4. $(7x^3 - 10x^2 y + 2xy^2 - 5) - (4x^3 - 12x^2 y - 3xy^2 + 5)$

$= 7x^3 - 10x^2 y + 2xy^2 - 5 - 4x^3 + 12x^2 y + 3xy^2 - 5$

$= 7x^3 - 4x^3 - 10x^2 y + 12x^2 y + 2xy^2 + 3xy^2 - 5 - 5$

$= 3x^3 + 2x^2 y + 5xy^2 - 10$

5. $(6xy^3)(10x^4 y^2) = (6 \cdot 10)(x \cdot x^4)(y^3 \cdot y^2)$

$= 60x^{1+4} y^{3+2}$

$= 60x^5 y^5$

6. $6xy^2(10x^4 y^5 - 2x^2 y + 3)$

$= 6xy^2 \cdot 10x^4 y^5 - 6xy^2 \cdot 2x^2 y + 6xy^2 \cdot 3$

$= 60x^{1+4} y^{2+5} - 12x^{1+2} y^{2+1} + 18xy^2$

$= 60x^5 y^7 - 12x^3 y^3 + 18xy^2$

7. a. $(7x - 6y)(3x - y)$

$= \overbrace{(7x)(3x)}^{F} + \overbrace{(7x)(-y)}^{O} + \overbrace{(-6y)(3x)}^{I} + \overbrace{(-6y)(-y)}^{L}$

$= 21x^2 - 7xy - 18xy + 6y^2$

$= 21x^2 - 25xy + 6y^2$

b. $\overbrace{(2x + 4y)^2}^{(A+B)^2} = \overbrace{(2x)^2}^{A^2} + \overbrace{2(2x)(4y)}^{2 \cdot A \cdot B} + \overbrace{(4y)^2}^{B^2}$

$= 4x^2 + 16xy + 16y^2$

$= 4x^2 + 16xy + 16y^2$

8. a. $\overbrace{(6xy^2 + 5x)(6xy^2 - 5x)}^{(A+B) \cdot (A-B)} = \overbrace{(6xy^2)^2}^{A^2} - \overbrace{(5x)^2}^{B^2}$

$= 36x^2 y^4 - 25x^2$

b. $(x - y)(x^2 + xy + y^2) = x(x^2 + xy + y^2) - y(x^2 + xy + y^2)$

$= x \cdot x^2 + x \cdot xy + x \cdot y^2 - y \cdot x^2 - y \cdot xy - y \cdot y^2$

$= x^3 + x^2 y + xy^2 - x^2 y - xy^2 - y^3$

$= x^3 + x^2 y - x^2 y + xy^2 - xy^2 - y^3$

$= x^3 - y^3$

5.4 Exercise Set

1. $x^2 + 2xy + y^2 = 2^2 + 2(2)(-3) + (-3)^2$

$= 4 - 12 + 9 = 1$

3. $xy^3 - xy + 1 = 2(-3)^3 - 2(-3) + 1$
$$= 2(-27) + 6 + 1$$
$$= -54 + 6 + 1 = -47$$

5. $2x^2 y - 5y + 3 = 2(2^2)(-3) - 5(-3) + 3$
$$= 2(4)(-3) - 5(-3) + 3$$
$$= -24 + 15 + 3$$
$$= -6$$

7. $x^3 y^2 - 5x^2 y^7 + 6y^2 - 3$

Term	Coefficient	Degree
$x^3 y^2$	1	$3 + 2 = 5$
$-5x^2 y^7$	-5	$2 + 7 = 9$
$6y^2$	6	2
-3	-3	0

The degree of the polynomial is the highest degree of all its terms, which is 9.

9. $(5x^2 y - 3xy) + (2x^2 y - xy)$
$$= (5x^2 y + 2x^2 y) + (-3xy - xy)$$
$$= 7x^2 y - 4xy$$

11. $(4x^2 y + 8xy + 11) + (-2x^2 y + 5xy + 2)$
$$= (4x^2 y - 2x^2 y) + (8xy + 5xy) + (11 + 2)$$
$$= 2x^2 y + 13xy + 13$$

13. $(7x^4 y^2 - 5x^2 y^2 + 3xy) + (-18x^4 y^2 - 6x^2 y^2 - xy)$
$$= (7x^4 y^2 - 18x^4 y^2) + (-5x^2 y^2 - 6x^2 y^2) + (3xy - xy)$$
$$= -11x^4 y^2 - 11x^2 y^2 + 2xy$$

15. $(x^3 + 7xy - 5y^2) - (6x^3 - xy + 4y^2)$
$$= (x^3 + 7xy - 5y^2) + (-6x^3 + xy - 4y^2)$$
$$= (x^3 - 6x^3) + (7xy + xy) + (-5y^2 - 4y^2)$$
$$= -5x^3 + 8xy - 9y^2$$

17. $\left(3x^4y^2 + 5x^3y - 3y\right) - \left(2x^4y^2 - 3x^3y - 4y + 6x\right)$

$= \left(3x^4y^2 + 5x^3y - 3y\right) + \left(-2x^4y^2 + 3x^3y + 4y - 6x\right)$

$= \left(3x^4y^2 - 2x^4y^2\right) + \left(5x^3y + 3x^3y\right) + \left(-3y + 4y\right) + \left(-6x\right)$

$= x^4y^2 + 8x^3y + y - 6x$

19. $\left(x^3 - y^3\right) - \left(-4x^3 - x^2y + xy^2 + 3y^3\right)$

$= \left(x^3 - y^3\right) + \left(4x^3 + x^2y - xy^2 - 3y^3\right)$

$= \left(x^3 + 4x^3\right) + \left(-y^3 - 3y^3\right) + x^2y - xy^2$

$= 5x^3 - 4y^3 + x^2y - xy^2$

$= 5x^3 + x^2y - xy^2 - 4y^3$

21. $5x^2y^2 - 4xy^2 + 6y^2$

$\underline{-8x^2y^2 + 5xy^2 - y^2}$

$-3x^2y^2 + xy^2 + 5y^2$

23. $3a^2b^4 - 5ab^2 + 7ab$

$-\left(\underline{-5a^2b^4 - 8ab^2 - ab}\right)$

To subtract, add the opposite of the polynomial being subtracted.

$3a^2b^4 - 5ab^2 + 7ab$

$\underline{+5a^2b^4 + 8ab^2 + ab}$

$8a^2b^4 + 3ab^2 + 8ab$

25. $\left[\left(7x + 13y\right) + \left(-26x + 19y\right)\right] - \left(11x - 5y\right) = 7x + 13y - 26x + 19y - 11x + 5y$

$= 7x - 26x - 11x + 13y + 19y + 5y$

$= -30x + 37y$

27. $\left(5x^2y\right)\left(8xy\right) = \left(5 \cdot 8\right)\left(x^2 \cdot x\right)\left(y \cdot y\right)$

$= 40x^3y^2$

29. $\left(-8x^3y^4\right)\left(3x^2y^5\right) = \left(-8 \cdot 3\right)\left(x^3 \cdot x^2\right)\left(y^4 \cdot y^5\right)$

$= -24x^5y^9$

31. $9xy\left(5x + 2y\right) = 9xy\left(5x\right) + 9xy\left(2y\right)$

$= 45x^2y + 18xy^2$

33. $5xy^2 \left(10x^2 - 3y\right) = 5xy^2 \left(10x^2\right) - 5xy^2 \left(3y\right)$

$$= 50x^3y^2 - 15xy^3$$

35. $4ab^2 \left(7a^2b^3 + 2ab\right) = 4ab^2 \left(7a^2b^3\right) + 4ab^2 \left(2ab\right)$

$$= 28a^3b^5 + 8a^2b^3$$

37. $-b\left(a^2 - ab + b^2\right) = -b\left(a^2\right) - b\left(-ab\right) - b\left(b^2\right)$

$$= -a^2b + ab^2 - b^3$$

39. $(x + 5y)(7x + 3y)$

$$= x(7x) + x(3y) + 5y(7x) + 5y(3y)$$

$$= 7x^2 + 3xy + 35xy + 15y^2$$

$$= 7x^2 + 38xy + 15y^2$$

41. $(x - 3y)(2x + 7y)$

$$= x(2x) + x(7y) - 3y(2x) - 3y(7y)$$

$$= 2x^2 + 7xy - 6xy - 21y^2$$

$$= 2x^2 + xy - 21y^2$$

43. $(3xy - 1)(5xy + 2)$

$$= 3xy(5xy) + 3xy(2) - 1(5xy) - 1(2)$$

$$= 15x^2y^2 + 6xy - 5xy - 2$$

$$= 15x^2y^2 + xy - 2$$

45. $(2x + 3y)^2 = (2x)^2 + 2(2x)(3y) + (3y)^2$

$$= 4x^2 + 12xy + 9y^2$$

47. $(xy - 3)^2 = (xy)^2 - 2(xy)(3) + (-3)^2$

$$= x^2y^2 - 6xy + 9$$

49. $\left(x^2 + y^2\right)^2 = \left(x^2\right)^2 + 2\left(x^2\right)\left(y^2\right) + \left(y^2\right)^2$

$$= x^4 + 2x^2y^2 + y^4$$

51. $\left(x^2 - 2y^2\right)^2$

$$= \left(x^2\right) - 2\left(x^2\right)\left(2y^2\right) + \left(-2y^2\right)^2$$

$$= x^4 - 4x^2y^2 + 4y^4$$

53. $(3x + y)(3x - y) = (3x)^2 - y^2 = 9x^2 - y^2$

55. $(ab + 1)(ab - 1) = (ab)^2 - 1^2 = a^2b^2 - 1$

57. $\left(x + y^2\right)\left(x - y^2\right) = x^2 - \left(y^2\right)^2 = x^2 - y^4$

59. $\left(3a^2b + a\right)\left(3a^2b - a\right) = \left(3a^2b\right)^2 - a^2$

$$= 9a^4b^2 - a^2$$

61. $\left(3xy^2 - 4y\right)\left(3xy^2 + 4y\right) = \left(3xy^2\right)^2 - (4y)^2$

$$= 9x^2y^4 - 16y^2$$

63. $(a + b)\left(a^2 - b^2\right)$

$$= a\left(a^2\right) + a\left(-b^2\right) + b\left(a^2\right) + b\left(-b^2\right)$$

$$= a^3 - ab^2 + a^2b - b^3$$

65. $(x + y)\left(x^2 + 3xy + y^2\right)$

$$= x\left(x^2 + 3xy + y^2\right) + y\left(x^2 + 3xy + y^2\right)$$

$$= x^3 + 3x^2y + xy^2 + x^2y + 3xy^2 + y^3$$

$$= x^3 + 4x^2y + 4xy^2 + y^3$$

67. $(x - y)\left(x^2 - 3xy + y^2\right)$

$$= x\left(x^2 - 3xy + y^2\right) - y\left(x^2 - 3xy + y^2\right)$$

$$= x^3 - 3x^2y + xy^2 - x^2y + 3xy^2 - y^3$$

$$= x^3 - 4x^2y + 4xy^2 - y^3$$

69. $(xy + ab)(xy - ab) = (xy)^2 - (ab)^2$

$$= x^2y^2 - a^2b^2$$

71. $\left(x^2 + 1\right)\left(x^4y + x^2 + 1\right)$

$$= x^2\left(x^4y + x^2 + 1\right) + 1\left(x^4y + x^2 + 1\right)$$

$$= x^6y + x^4 + x^2 + x^4y + x^2 + 1$$

$$= x^6y + x^4y + x^4 + 2x^2 + 1$$

73. $\left(x^2y^2 - 3\right)^2$

$$= \left(x^2y^2\right)^2 - 2\left(x^2y^2\right)(3) + (-3)^2$$

$$= x^4y^4 - 6x^2y^2 + 9$$

75. $(x+y+1)(x+y-1)$

$= x(x+y-1)+y(x+y-1)+1(x+y-1)$

$= x^2+xy-x+yx+y^2-y+x+y-1$

$= x^2+2xy+y^2-1$

77. $A = (3x+5y)(x+y)$

$= 3x(x)+3x(y)+5y(x)+5y(y)$

$= 3x^2+3xy+5xy+5y^2$

$= 3x^2+8xy+5y^2$

The area of the shaded region is $3x^2+8xy+5y^2$ square units.

79. Area of shaded region $= \underbrace{(x+y)^2}_{\substack{\text{Area of}\\\text{larger square}}} - \underbrace{x^2}_{\substack{\text{Area of}\\\text{smaller square}}}$

$= (x^2+2xy+y^2)-x^2$

$= 2xy+y^2$

81. $\left[\left(x^3y^3+1\right)\left(x^3y^3-1\right)\right]^2$

$= \left[\left(x^3y^3\right)^2-(1)^2\right]^2$

$= \left[x^6y^6-1\right]^2$

$= \left(x^6y^6\right)^2-2\left(x^6y^6\right)(1)+(1)^2$

$= x^{12}y^{12}-2x^6y^6+1$

83. $(xy-3)^2(xy+3)^2$

$= \left[(xy-3)(xy+3)\right]^2$

$= \left[(xy)^2-3^2\right]^2$

$= \left[x^2y^2-9\right]^2$

$= \left(x^2y^2\right)^2-2\left(x^2y^2\right)(9)+(9)^2$

$x^4y^4-18x^2y^2+81$

85. $[x+y+z][x-(y+z)]$

$= [x+(y+z)][x-(y+z)]$

$= x^2-(y+z)^2$

$= x^2-(y^2+2yz+z^2)$

$= x^2-y^2-2yz-z^2$

87. $N = \frac{1}{4}x^2y-2xy+4y;\ x=10,\ y=16$

$N = \frac{1}{4}x^2y-2xy+4y$

$= \frac{1}{4}(10)^2(16)-2(10)(16)+4(16)$

$= \frac{1}{4}(100)(16)-2(10)(16)+4(16)$

$= 400-320+64$

$= 144$

Each tree provides 144 board feet of lumber, so 20 trees will provide $20(144)=2880$ board feet. This is not enough lumber to complete the job. Since $3000-2880=120$, the contractor will need 120 more board feet.

89. $v_0=80,\ s_0=96$ and $t=2$

$s = -16t^2+v_0t+s_0$

$s = -16t^2+80t+96$

$s = -16(2)^2+80(2)+96$

$= -16(4)+80(6)+96$

$= -64+160+96$

$= 192$

The ball will be 192 feet above the ground 2 seconds after being thrown.

91. $v_0=80,\ s_0=96$ and $t=6$

$s = -16t^2+v_0t+s_0$

$s = -16t^2+80t+96$

$s = -16(6)^2+80(6)+96$

$= -16(36)+80(6)+96$

$= -576+480+96=0$

The ball will be 0 feet above the ground after 6 seconds. This means that the ball hits the ground 6 seconds after being thrown.

93. The ball is falling from 2.5 seconds to 6 seconds. The graph is decreasing over this interval.

95. $(2,192)$

97. The ball reaches its maximum height 2.5 seconds after it is thrown.

$v_0 = 80,\ s_0 = 96$ and $t = 2.5$

$s = -16t^2 + v_0 t + s_0$

$s = -16t^2 + 80t + 96$

$s = -16(2.5)^2 + 80(2.5) + 96$

$\quad = -16(6.25) + 80(2.5) + 96$

$\quad = -100 + 200 + 96$

$\quad = 196$

The maximum height is 196 feet.

99. Answers will vary.

101. makes sense

103. does not make sense; Explanations will vary.
Sample explanation: FOIL is used to multiply a binomial by a binomial.

105. false; Changes to make the statement true will vary.
A sample change is: The term $-3x^{16}y^9$ has degree $16 + 9 = 25$.

107. true

109. Area of rectangle:

$(x + 10y)(x + 8y) = x^2 + 18xy + 80y^2$

Area of two corner squares: $x^2 + x^2 = 2x^2$
Area of shaded region:

$(x^2 + 18xy + 80y^2) - 2x^2 = -x^2 + 18xy + 80y^2$

111. Note that the shed consists of a rectangular solid with half a cylinder on top.
Radius of cylinder: x
Length of cylinder: y
Width of base: $2x$
Length of base: y
Height of base: x
Volume of rectangular solid: $l \cdot w \cdot h$
Volume of cylinder: $\pi r^2 \cdot l$

Volume of half-cylinder: $\dfrac{1}{2}\pi r^2 \cdot l$

$V_{\text{shed}} = V_{\text{base}} + V_{\text{top}}$

$\quad = l \cdot w \cdot h + \dfrac{1}{2}\pi r^2 \cdot l$

$\quad = y \cdot 2x \cdot x + \dfrac{1}{2}\pi \cdot (x)^2 \cdot y$

$\quad = 2x^2 y + \dfrac{1}{2}\pi x^2 y$

112. $R = \dfrac{L + 3W}{2}$; for W

$R = \dfrac{L + 3W}{2}$

$2R = 2\left(\dfrac{L + 3W}{2}\right)$

$2R = L + 3W$

$2R - L = L + 3W - L$

$2R - L = 3W$

$\dfrac{2R - L}{3} = \dfrac{3W}{3}$

$\dfrac{2R - L}{3} = W$ or $W = \dfrac{2R - L}{3}$

113. $-6.4 - (-10.2) = -6.4 + 10.2 = 3.8$

114. First, solve $3x - y = 9$ for y to find slope.

$3x - y = 9$

$\quad -y = -3x + 9$

$\quad y = 3x - 9$

The slope is 3.
Next find the point-slope form of the line.

$y - y_1 = m(x - x_1)$

$y - 5 = 3(x - (-2))$

$y - 5 = 3(x + 2)$

Now find slope-intercept form.

$y - 5 = 3(x + 2)$

$y - 5 = 3x + 6$

$\quad y = 3x + 11$

115. $\dfrac{x^7}{x^3} = \dfrac{\cancel{x}\cdot\cancel{x}\cdot\cancel{x}\cdot x \cdot x \cdot x \cdot x}{\cancel{x}\cdot\cancel{x}\cdot\cancel{x}} = x^4$

116. $\dfrac{(x^2)^3}{5^3} = \dfrac{x^6}{125}$

117. $\dfrac{(2a^3)^5}{(b^4)^5} = \dfrac{2^5(a^3)^5}{b^{20}} = \dfrac{32a^{15}}{b^{20}}$

Chapter 5 Mid-Chapter Check Point

1. $(11x^2y^3)(-5x^2y^3)$

$= -55x^{2+2}y^{3+3} = -55x^4y^6$

2. $11x^2y^3 - 5x^2y^3 = 6x^2y^3$

3. $(3x+5)(4x-7)$

$= 12x^2 - 21x + 20x - 35$

$= 12x^2 - x - 35$

4. $(3x+5)-(4x-7)$

$= (3x+5)+(-4x+7)$

$= -x + 12$

5. $(2x-5)(x^2-3x+1)$

$= 2x(x^2-3x+1)-5(x^2-3x+1)$

$= 2x^3 - 6x^2 + 2x - 5x^2 + 15x - 5$

$= 2x^3 - 11x^2 + 17x - 5$

6. $(2x-5)+(x^2-3x+1)$

$= x^2 - x - 4$

7. $(8x-3)^2$

$= (8x)^2 - 2(8x)(3) + 3^2$

$= 64x^2 - 48x + 9$

8. $(-10x^4)(-7x^5) = 70x^9$

9. $(x^2+2)(x^2-2)$

$= (x^2)^2 - 2^2$

$= x^4 - 4$

10. $(x^2+2)^2$

$= (x^2)^2 + 2(x^2)(2) + 2^2$

$= x^4 + 4x^2 + 4$

11. $(9a-10b)(2a+b)$

$= 18a^2 + 9ab - 20ba - 10b^2$

$= 18a^2 + 9ab - 20ab - 10b^2$

$= 18a^2 - 11ab - 10b^2$

12. $7x^2(10x^3 - 2x + 3)$

$= 70x^5 - 14x^3 + 21x^2$

13. $(3a^2b^3 - ab + 4b^2) - (-2a^2b^3 - 3ab + 5b^2)$

$= (3a^2b^3 - ab + 4b^2) + (2a^2b^3 + 3ab - 5b^2)$

$= 5a^2b^3 + 2ab - b^2$

14. $2(3y-5)(3y+5) = 2(9y^2 - 25)$

$\qquad\qquad\qquad = 18y^2 - 50$

15. $(-9x^3 + 5x^2 - 2x + 7)$

$\quad + (11x^3 - 6x^2 + 3x - 7)$

$= (-9x^3 + 11x^3) + (5x^2 - 6x^2)$

$\quad + (-2x + 3x) + (7 - 7)$

$= 2x^3 - x^2 + x$

16. $10x^2 - 8xy - 3(y^2 - xy)$

$= 10x^2 - 8xy - 3y^2 + 3xy$

$= 10x^2 - 5xy - 3y^2$

17. $(-2x^5 + x^4 - 3x + 10)$

$\quad - (2x^5 - 6x^4 + 7x - 13)$

$= (-2x^5 + x^4 - 3x + 10)$

$\quad + (-2x^5 + 6x^4 - 7x + 13)$

$= -4x^5 + 7x^4 - 10x + 23$

18. $(x+3y)\left(x^2-3xy+9y^2\right)$

$= x\left(x^2-3xy+9y^2\right)+3y\left(x^2-3xy+9y^2\right)$

$= x^3-3x^2y+9xy^2+3x^2y-9xy^2+27y^3$

$= x^3+27y^3$

19. $\left(5x^4+4\right)\left(2x^3-1\right)$

$= 10x^7-5x^4+8x^3-4$

20. $(y-6z)^2 = y^2-2(y)(6z)+(6z)^2$

$= y^2-12yz+36z^2$

21. $(2x+3)(2x-3)-(5x+4)(5x-4)$

$= \left(4x^2-9\right)-\left(25x^2-16\right)$

$= \left(4x^2-9\right)+\left(-25x^2+16\right)$

$= -21x^2+7$

22. Make a table of values.

x	$y=1-x^2$	(x,y)
-3	$y=1-(-3)^2=-8$	$(-3,-8)$
-2	$y=1-(-2)^2=-3$	$(-2,-3)$
-1	$y=1-(-1)^2=0$	$(-1,0)$
0	$y=1-(0)^2=1$	$(0,1)$
1	$y=1-(1)^2=0$	$(1,0)$
2	$y=1-(2)^2=-3$	$(2,-3)$
3	$y=1-(3)^2=-8$	$(3,-8)$

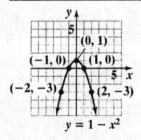

$y=1-x^2$

5.5 Check Points

1. a. $\dfrac{5^{12}}{5^4}=5^{12-4}=5^8$

   b. $\dfrac{x^9}{x^2}=x^{9-2}=x^7$

   c. $\dfrac{y^{20}}{y}=y^{20-1}=y^{19}$

2. a. $14^0=1$

   b. $(-10)^0=1$

   c. $-10^0=-1\cdot 10^0=-1\cdot 1=-1$

   d. $20x^0=20\cdot 1=20$

   e. $(20x)^0=1$

3. a. $\left(\dfrac{x}{5}\right)^2=\dfrac{x^2}{5^2}=\dfrac{x^2}{25}$

   b. $\left(\dfrac{x^4}{2}\right)^3=\dfrac{x^{4\cdot 3}}{2^3}=\dfrac{x^{12}}{8}$

   c. $\left(\dfrac{2a^{10}}{b^3}\right)^4=\dfrac{2^4(a^{10})^4}{(b^3)^4}=\dfrac{16a^{40}}{b^{12}}$

4. a. $\dfrac{-20x^{12}}{10x^4}=\dfrac{-20}{10}x^{12-4}=-2x^8$

   b. $\dfrac{3x^4}{15x^4}=\dfrac{3}{15}x^{4-4}=\dfrac{1}{5}x^0=\dfrac{1}{5}$

   c. $\dfrac{9x^6y^5}{3xy^2}=\dfrac{9}{3}\cdot x^{6-1}y^{5-2}=3x^5y^3$

5. $\dfrac{-15x^9+6x^5-9x^3}{3x^2}=\dfrac{-15x^9}{3x^2}+\dfrac{6x^5}{3x^2}-\dfrac{9x^3}{3x^2}$

$= -5x^7+2x^3-3x$

6. $\dfrac{25x^9 - 7x^4 + 10x^3}{5x^3} = \dfrac{25x^9}{5x^3} - \dfrac{7x^4}{5x^3} + \dfrac{10x^3}{5x^3}$

$\qquad\qquad = 5x^6 - \dfrac{7}{5}x + 2$

7. $\dfrac{18x^7 y^6 - 6x^2 y^3 + 60xy^2}{6xy^2}$

$= \dfrac{18x^7 y^6}{6xy^2} - \dfrac{6x^2 y^3}{6xy^2} + \dfrac{60xy^2}{6xy^2}$

$= 3x^6 y^4 - xy + 10$

5.5 Exercise Set

1. $\dfrac{3^{20}}{3^5} = 3^{20-5} = 3^{15}$

3. $\dfrac{x^6}{x^2} = x^{6-2} = x^4$

5. $\dfrac{y^{13}}{y^5} = y^{13-5} = y^8$

7. $\dfrac{5^6 \cdot 2^8}{5^3 \cdot 2^4} = 5^{6-3} \cdot 2^{8-4} = 5^3 \cdot 2^4$

9. $\dfrac{x^{100} y^{50}}{x^{25} y^{10}} = x^{100-25} y^{50-10} = x^{75} y^{40}$

11. $2^0 = 1$

13. $(-2)^0 = 1$

15. $-2^0 = -\left(2^0\right) = -(1) = -1$

17. $100 y^0 = 100 \cdot 1 = 100$

19. $(100y)^0 = 1$

21. $-5^0 + (-5)^0 = -1 + 1 = 0$

23. $-\pi^0 - (-\pi)^0 = -1 - 1 = -2$

25. $\left(\dfrac{x}{3}\right)^2 = \dfrac{x^2}{3^2} = \dfrac{x^2}{9}$

27. $\left(\dfrac{x^2}{4}\right)^3 = \dfrac{\left(x^2\right)^3}{4^3} = \dfrac{x^{2 \cdot 3}}{4^3} = \dfrac{x^6}{64}$

29. $\left(\dfrac{2x^3}{5}\right)^2 = \dfrac{2^2 \left(x^3\right)^2}{5^2} = \dfrac{4x^6}{25}$

31. $\left(\dfrac{-4}{3a^3}\right)^3 = \dfrac{(-4)^3}{3^3 \left(a^3\right)^3} = \dfrac{-64}{27a^9} = -\dfrac{64}{27a^9}$

33. $\left(\dfrac{-2a^7}{b^4}\right)^5 = \dfrac{\left(-2a^7\right)^5}{\left(b^4\right)^5} = \dfrac{(-2)^5 \left(a^7\right)^5}{\left(b^4\right)^5}$

$= \dfrac{-32a^{35}}{b^{20}} = -\dfrac{32a^{35}}{b^{20}}$

35. $\left(\dfrac{x^2 y^3}{2z}\right)^4 = \dfrac{\left(x^2\right)^4 \left(y^3\right)^4}{2^4 z^4} = \dfrac{x^8 y^{12}}{16z^4}$

37. $\dfrac{30x^{10}}{10x^5} = \dfrac{30}{10} x^{10-5} = 3x^5$

39. $\dfrac{-8x^{22}}{4x^2} = \dfrac{-8}{4} x^{22-2} = -2x^{20}$

41. $\dfrac{-9y^8}{18y^5} = \dfrac{-9}{18} y^{8-5} = -\dfrac{1}{2} y^3$

43. $\dfrac{7y^{17}}{5y^5} = \dfrac{7}{5} y^{17-5} = \dfrac{7}{5} y^{12}$

45. $\dfrac{30x^7 y^5}{5x^2 y} = \dfrac{30}{5} x^{7-2} y^{5-1} = 6x^5 y^4$

47. $\dfrac{-18x^{14} y^2}{36x^2 y^2} = \dfrac{-18}{36} x^{14-2} y^{2-2}$

$= -\dfrac{1}{2} x^{12} y^0 = -\dfrac{1}{2} x^{12} \cdot 1$

$= -\dfrac{1}{2} x^{12}$

49. $\dfrac{9x^{20}y^{20}}{7x^{20}y^{20}} = \dfrac{9}{7}x^{20-20}y^{20-20}$

$\qquad\qquad = \dfrac{9}{7}x^0y^0 = \dfrac{9}{7}\cdot 1\cdot 1 = \dfrac{9}{7}$

51. $\dfrac{-5x^{10}y^{12}z^6}{50x^2y^3z^2} = -\dfrac{1}{10}x^{10-2}y^{12-3}z^{6-2}$

$\qquad\qquad = -\dfrac{1}{10}x^8y^9z^4$

53. $\dfrac{10x^4+2x^3}{2} = \dfrac{10x^4}{2} + \dfrac{2x^3}{2} = 5x^4 + x^3$

55. $\dfrac{14x^4-7x^3}{7x} = \dfrac{14x^4}{7x} - \dfrac{7x^3}{7x}$

$\qquad\qquad = 2x^{4-1} - x^{3-1} = 2x^3 - x^2$

57. $\dfrac{y^7-9y^2+y}{y} = \dfrac{y^7}{y} - \dfrac{9y^2}{y} + \dfrac{y}{y}$

$\qquad\qquad = y^{7-1} - 9y^{2-1} + y^{1-1}$

$\qquad\qquad = y^6 - 9y^1 + y^0$

$\qquad\qquad = y^6 - 9y + 1$

59. $\dfrac{24x^3-15x^2}{-3x} = \dfrac{24x^3}{-3x} + \dfrac{-15x^2}{-3x}$

$\qquad\qquad = -8x^{3-1} + 5x^{2-1} = -8x^2 + 5x$

61. $\dfrac{18x^5+6x^4+9x^3}{3x^2} = \dfrac{18x^5}{3x^2} + \dfrac{6x^4}{3x^2} + \dfrac{9x^3}{3x^2}$

$\qquad\qquad = 6x^3 + 2x^2 + 3x$

63. $\dfrac{12x^4-8x^3+40x^2}{4x} = \dfrac{12x^4}{4x} - \dfrac{8x^3}{4x} + \dfrac{40x^2}{4x}$

$\qquad\qquad = 3x^3 - 2x^2 + 10x$

65. $\left(4x^2-6x\right)\div x = \dfrac{4x^2-6x}{x} = \dfrac{4x^2}{x} - \dfrac{6x}{x}$

$\qquad\qquad = 4x - 6$

67. $\dfrac{30z^3+10z^2}{-5z} = \dfrac{30z^3}{-5z} + \dfrac{10z^2}{-5z} = -6z^2 - 2z$

69. $\dfrac{8x^3+6x^2-2x}{2x} = \dfrac{8x^3}{2x} + \dfrac{6x^2}{2x} - \dfrac{2x}{2x}$

$\qquad\qquad = 4x^2 + 3x - 1$

71. $\dfrac{25x^7-15x^5-5x^4}{5x^3} = \dfrac{25x^7}{5x^3} - \dfrac{15x^5}{5x^3} - \dfrac{5x^4}{5x^3}$

$\qquad\qquad = 5x^4 - 3x^2 - x$

73. $\dfrac{18x^7-9x^6+20x^5-10x^4}{-2x^4}$

$\qquad = \dfrac{18x^7}{-2x^4} - \dfrac{9x^6}{-2x^4} + \dfrac{20x^5}{-2x^4} - \dfrac{10x^4}{-2x^4}$

$\qquad = -9x^3 + \dfrac{9}{2}x^2 - 10x + 5$

75. $\dfrac{12x^2y^2+6x^2y-15xy^2}{3xy}$

$\qquad = \dfrac{12x^2y^2}{3xy} + \dfrac{6x^2y}{3xy} - \dfrac{15xy^2}{3xy}$

$\qquad = 4xy + 2x - 5y$

77. $\dfrac{20x^7y^4-15x^3y^2-10x^2y}{-5x^2y}$

$\qquad = \dfrac{20x^7y^4}{-5x^2y} + \dfrac{-15x^3y^2}{-5x^2y} + \dfrac{-10x^2y}{-5x^2y}$

$\qquad = -4x^5y^3 + 3xy + 2$

79. $\dfrac{2x^3\left(4x+2\right)-3x^2\left(2x-4\right)}{2x^2}$

$\qquad = \dfrac{8x^4+4x^3-6x^3+12x^2}{2x^2}$

$\qquad = \dfrac{8x^4-2x^3+12x^2}{2x^2}$

$\qquad = \dfrac{8x^4}{2x^2} - \dfrac{2x^3}{2x^2} + \dfrac{12x^2}{2x^2}$

$\qquad = 4x^2 - x + 6$

81. $\left(\dfrac{18x^2y^4}{9xy^2}\right) - \left(\dfrac{15x^5y^6}{5x^4y^4}\right) = 2xy^2 - 3xy^2$

$\qquad\qquad\qquad\qquad\qquad = -xy^2$

83. $\dfrac{(y+5)^2+(y+5)(y-5)}{2y}$

$=\dfrac{(y^2+10y+25)+(y^2-25)}{2y}$

$=\dfrac{2y^2+10y}{2y}=\dfrac{2y^2}{2y}+\dfrac{10y}{2y}=y+5$

85. $\dfrac{12x^{15n}-24x^{12n}+8x^{3n}}{4x^{3n}}$

$=\dfrac{12x^{15n}}{4x^{3n}}-\dfrac{24x^{12n}}{4x^{3n}}+\dfrac{8x^{3n}}{4x^{3n}}$

$=3x^{12n}-6x^{9n}+2$

87. a. $\text{Average}=\dfrac{\text{Receipts}}{\text{Admissions}}=\dfrac{7661}{1421}=5.39$

The average admission charge for a film in 2000 was $5.39.

b. $\text{Average}=\dfrac{\text{Receipts}}{\text{Admissions}}=\dfrac{3.6x^2+158x+2790}{-0.2x^2+21x+1015}$

c.

2000 is 20 years after 1980.

$\dfrac{3.6x^2+158x+2790}{-0.2x^2+21x+1015}=\dfrac{3.6(20)^2+158(20)+2790}{-0.2(20)^2+21(20)+1015}$

$=\dfrac{7390}{1355}$

$=5.45$

According to the model, the average admission charge for a film in 2000 was $5.45.
This overestimates the actual value by $0.06.

d. Polynomial division cannot be performed using the methods in this section because the divisor is not a monomial.

89. – 95. Answers will vary.

97. does not make sense; Explanations will vary.
Sample explanation: The quotient rule involves subtracting exponents.

99. does not make sense; Explanations will vary.
Sample explanation: Divide each term of the polynomial by the monomial.

101. false; Changes to make the statement true will vary.
A sample change is:
$\dfrac{12x^3-6x}{2x}=\dfrac{12x^3}{2x}-\dfrac{6x}{2x}=6x^2-3$

103. true

105. $\dfrac{?x^8-?x^6}{3x^?}=3x^5-4x^3$

To get 3 as the coefficient of the first term of the quotient, the coefficient of the first term of the dividend must be 9. For the exponent in the first term of the quotient to be 5, the exponent in the divisor must be 3. Since now know that the divisor is $3x^3$, the coefficient of the second term in the dividend must be 12. Therefore,

$\dfrac{?x^8-?x^6}{3x^?}=\dfrac{9x^8-12x^6}{3x^3}$

107. $|-20.3|=20.3$

108.
$$\begin{array}{r}0.875\\8\overline{)7.000}\\\underline{64}\\60\\\underline{56}\\40\\\underline{40}\\0\end{array}$$

$\dfrac{7}{8}=0.875$

109. $y=\dfrac{1}{3}x+2$

slope $=\dfrac{1}{3}$; y-intercept $=2$

Plot (0,2). From this point move 1 unit *up* and 3 units to the *right* to reach the point (3,3). Draw a line through (0,2) and (3,3).

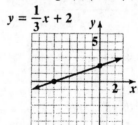

110.
$$\begin{array}{r}26\\19\overline{)494}\\\underline{38}\\114\\\underline{114}\\0\end{array}$$

The quotient is 26 and the remainder is 0.

111.

$$\begin{array}{r} 123 \\ 24\overline{)2958} \\ \underline{24} \\ 55 \\ \underline{48} \\ 78 \\ \underline{72} \\ 6 \end{array}$$

The quotient is 123 and the remainder is 6.

112.

$$\begin{array}{r} 257 \\ 98\overline{)25187} \\ \underline{196} \\ 558 \\ \underline{490} \\ 687 \\ \underline{686} \\ 1 \end{array}$$

The quotient is 257 and the remainder is 1.

5.6 Check Points

1.

$$\begin{array}{r} x+5 \\ x+9\overline{)x^2+14x+45} \\ \underline{x^2+9x} \\ 5x+45 \\ \underline{5x+45} \\ 0 \end{array}$$

$$\frac{x^2+14x+45}{x+9}=x+5$$

2. $\dfrac{6x+8x^2-12}{2x+3}=\dfrac{8x^2+6x-12}{2x+3}$

$$\begin{array}{r} 4x-3 \\ 2x+3\overline{)8x^2+6x-12} \\ \underline{8x^2+12x} \\ -6x-12 \\ \underline{-6x-9} \\ -3 \end{array}$$

$$\frac{6x+8x^2-12}{2x+3}=4x-3-\frac{3}{2x+3}$$

3. Rewriting x^3-1 using coefficients of 0 on the missing terms gives x^3+0x^2+0x-1.

$$\begin{array}{r} x^2+x+1 \\ x-1\overline{)x^3+0x^2+0x-1} \\ \underline{x^3-x^2} \\ x^2+0x \\ \underline{x^2-x} \\ x-1 \\ \underline{x-1} \\ 0 \end{array}$$

$$\frac{x^3-1}{x-1}=x^2+x+1$$

4. Rewrite the dividend with the missing power of x and divide.

$$\begin{array}{r} 2x^2+7x+14 \\ x^2-2x\overline{)2x^4+3x^3+0x^2-7x-10} \\ \underline{2x^4-4x^3} \\ 7x^3-0x^2 \\ \underline{7x^3-14x^2} \\ 14x^2-7x \\ \underline{14x^2-28x} \\ 21x-10 \end{array}$$

Thus,

$$\frac{2x^4+3x^3-7x-10}{x^2-2x}=2x^2+7x+14+\frac{21x-10}{x^2-2x}$$

5. $\left(x^3-7x-6\right)\div\left(x+2\right)=x^2-2x-3$

$$\begin{array}{r|rrrr} -2 & 1 & 0 & -7 & -6 \\ & & -2 & 4 & 6 \\ \hline & 1 & -2 & -3 & 0 \end{array}$$

5.6 Exercise Set

1.

$$\begin{array}{r} x+4 \\ x+2\overline{)x^2+6x+8} \\ \underline{x^2+2x} \\ 4x+8 \\ \underline{4x+8} \\ 0 \end{array}$$

$$\frac{x^2+6x+8}{x+2}=x+4$$

3.
$$
\begin{array}{r}
2x+5 \\
x-2\overline{\smash{)}2x^2+x-10} \\
\underline{2x^2-4x} \\
5x-10 \\
\underline{5x-10} \\
0
\end{array}
$$

$$\frac{2x^2+x-10}{x-2}=2x+5$$

5.
$$
\begin{array}{r}
x-2 \\
x-3\overline{\smash{)}x^2-5x+6} \\
\underline{x^2-3x} \\
-2x+6 \\
\underline{-2x+6} \\
0
\end{array}
$$

$$\frac{x^2-5x+6}{x-3}=x-2$$

7.
$$
\begin{array}{r}
2y+1 \\
y+2\overline{\smash{)}2y^2+5y+2} \\
\underline{2y^2+4y} \\
y+2 \\
\underline{y+2} \\
0
\end{array}
$$

$$\frac{2y^2+5y+2}{y+2}=2y+1$$

9.
$$
\begin{array}{r}
x-5 \\
x+2\overline{\smash{)}x^2-3x+4} \\
\underline{x^2+2x} \\
-5x+4 \\
\underline{-5x-10} \\
14
\end{array}
$$

$$\frac{x^2-3x+4}{x+2}=x-5+\frac{14}{x+2}$$

11.
$$
\begin{array}{r}
y+3 \\
y+2\overline{\smash{)}y^2+5y+10} \\
\underline{y^2+2y} \\
3y+10 \\
\underline{3y+6} \\
4
\end{array}
$$

$$\frac{5y+10+y^2}{y+2}=\frac{y^2+5y+10}{y+2}=y+3+\frac{4}{y+2}$$

13.
$$
\begin{array}{r}
x^2-5x+2 \\
x-1\overline{\smash{)}x^3-6x^2+7x-2} \\
\underline{x^3-x^2} \\
-5x^2+7x \\
\underline{-5x^2+5x} \\
2x-2 \\
\underline{2x-2} \\
0
\end{array}
$$

$$\frac{x^3-6x^2+7x-2}{x-1}=x^2-5x+2$$

15.
$$
\begin{array}{r}
6y-1 \\
2y-3\overline{\smash{)}12y^2-20y+3} \\
\underline{12y^2-18y} \\
-2y+3 \\
\underline{-2y+3} \\
0
\end{array}
$$

$$\frac{12y^2-20y+3}{2y-3}=6y-1$$

17.
$$
\begin{array}{r}
2a+3 \\
2a-1\overline{\smash{)}4a^2+4a-3} \\
\underline{4a^2-2a} \\
6a-3 \\
\underline{6a-3} \\
0
\end{array}
$$

$$\frac{4a^2+4a-3}{2a-1}=2a+3$$

19.

$$2y+1 \overline{\smash{\big)}\ 2y^3 - y^2 + 3y + 2} \quad \overset{\textstyle y^2 - y + 2}{}$$

$$\underline{2y^3 + y^2}$$
$$-2y^2 + 3y$$
$$\underline{-2y^2 - y}$$
$$4y + 2$$
$$\underline{4y + 2}$$
$$0$$

$$\frac{3y - y^2 + 2y^3 + 2}{2y + 1}$$

$$= \frac{2y^3 - y^2 + 3y + 2}{2y + 1} = y^2 - y + 2$$

21.

$$2x-5 \overline{\smash{\big)}\ 6x^2 - 5x - 30} \quad \overset{\textstyle 3x + 5}{}$$

$$\underline{6x^2 - 15x}$$
$$10x - 30$$
$$\underline{10x - 25}$$
$$-5$$

$$\frac{6x^2 - 5x - 30}{2x - 5} = 3x + 5 - \frac{5}{2x - 5}$$

23.

$$x-2 \overline{\smash{\big)}\ x^3 + 0x^2 + 4x - 3} \quad \overset{\textstyle x^2 + 2x + 8}{}$$

$$\underline{x^3 - 2x^2}$$
$$2x^2 + 4x$$
$$\underline{2x^2 - 4x}$$
$$8x - 3$$
$$\underline{8x - 16}$$
$$13$$

$$\frac{x^2 + 4x - 3}{x - 2} = x^2 + 2x + 8 + \frac{13}{x - 2}$$

25.

$$2y+3 \overline{\smash{\big)}\ 4y^3 + 8y^2 + 5y + 9} \quad \overset{\textstyle 2y^2 + y + 1}{}$$

$$\underline{4y^3 + 6y^2}$$
$$2y^2 + 5y$$
$$\underline{2y^2 + 3y}$$
$$2y + 9$$
$$\underline{2y + 3}$$
$$6$$

$$\frac{4y^3 + 8y^2 + 5y + 9}{2y + 3} = 2y^2 + y + 1 + \frac{6}{2y + 3}$$

27.

$$3y+2 \overline{\smash{\big)}\ 6y^3 - 5y^2 + 0y + 5} \quad \overset{\textstyle 2y^2 - 3y + 2}{}$$

$$\underline{6y^3 + 4y^2}$$
$$-9y^2 + 0y$$
$$\underline{-9y^2 - 6y}$$
$$6y + 5$$
$$\underline{6y + 4}$$
$$1$$

$$\frac{6y^3 - 5y^2 + 5}{3y + 2} = 2y^2 - 3y + 2 + \frac{1}{3y + 2}$$

29.

$$3x-1 \overline{\smash{\big)}\ 27x^3 + 0x^2 + 0x - 1} \quad \overset{\textstyle 9x^2 + 3x + 1}{}$$

$$\underline{27x^3 - 9x^2}$$
$$9x^2 + 0x$$
$$\underline{9x^2 - 3x}$$
$$3x - 1$$
$$\underline{3x - 1}$$
$$0$$

$$\frac{27x^3 - 1}{3x - 1} = 9x^2 + 3x + 1$$

31.

$$y-3 \overline{)y^4 - 12y^3 + 54y^2 - 108y + 81} \quad \frac{y^3 - 9y^2 + 27y - 27}{}$$

$$\underline{y^4 - 3y^3}$$
$$-9y^3 + 54y^2$$
$$\underline{-9y^3 + 27y^2}$$
$$27y^2 - 108y$$
$$\underline{27y^2 - 81y}$$
$$-27y + 81$$
$$\underline{-27y + 81}$$
$$0$$

$$\frac{81 - 12y^3 + 54y^2 + y^4 - 108y}{y - 3}$$

$$= \frac{y^4 - 12y^3 + 54y^2 - 108y + 81}{y - 3}$$

$$= y^3 - 9y^2 + 27y - 27$$

33.

$$2y - 1 \overline{)4y^2 + 6y + 0} \quad \frac{2y + 4}{}$$

$$\underline{4y^2 - 2y}$$
$$8y + 0$$
$$\underline{8y - 4}$$
$$4$$

$$\frac{4y^2 + 6y}{2y - 1} = 2y + 4 + \frac{4}{2y - 1}$$

35.

$$y - 1 \overline{)y^4 + 0y^3 - 2y^2 + 0y + 5} \quad \frac{y^3 + y^2 - y - 1}{}$$

$$\underline{y^4 - y^3}$$
$$y^3 - 2y^2$$
$$\underline{y^3 - y^2}$$
$$-y^2 + 0y$$
$$\underline{-y^2 + y}$$
$$-y + 5$$
$$\underline{-y + 1}$$
$$4$$

$$\frac{y^4 - 2y^2 + 5}{y - 1} = y^3 + y^2 - y - 1 + \frac{4}{y - 1}$$

37.

$$x^2 + 0x + 3 \overline{)4x^4 + 3x^3 + 4x^2 + 9x - 6} \quad \frac{4x^2 + 3x - 8}{}$$

$$\underline{4x^4 + 0x^3 + 12x^2}$$
$$3x^3 - 8x^2 + 9x$$
$$\underline{3x^3 + 0x^2 + 9x}$$
$$-8x^2 + 0x - 6$$
$$\underline{-8x^2 + 0x - 24}$$
$$18$$

$$\frac{4x^4 + 3x^3 + 4x^2 + 9x - 6}{x^2 + 3}$$

$$= 4x^2 + 3x - 8 + \frac{18}{x^2 + 3}$$

39.

$$3x^2 + 0x - 1 \overline{)15x^4 + 3x^3 + 4x^2 + 0x + 4} \quad \frac{5x^2 + x + 3}{}$$

$$\underline{15x^4 + 0x^3 - 5x^2}$$
$$3x^3 + 9x^2 + 0x$$
$$\underline{3x^3 + 0x^2 - x}$$
$$9x^2 + x + 4$$
$$\underline{9x^2 + 0x - 3}$$
$$x + 7$$

$$\frac{15x^4 + 3x^3 + 4x^2 + 4}{3x^2 - 1}$$

$$= 5x^2 + x + 3 + \frac{x + 7}{3x^2 - 1}$$

41. $\left(2x^2 + x - 10\right) \div \left(x - 2\right)$

$$\underline{2|} \quad \begin{array}{ccc} 2 & 1 & -10 \\ & 4 & 10 \\ \hline 2 & 5 & 0 \end{array}$$

$$\left(2x^2 + x - 10\right) \div \left(x - 2\right) = 2x + 5$$

43. $\left(3x^2 + 7x - 20\right) \div \left(x + 5\right)$

$$\underline{-5|} \quad \begin{array}{ccc} 3 & 7 & -20 \\ & -15 & 40 \\ \hline 3 & -8 & 20 \end{array}$$

$$\left(3x^2 + 7x - 20\right) \div \left(x + 5\right) = 3x - 8 + \frac{20}{x + 5}$$

45. $\left(4x^3 - 3x^2 + 3x - 1\right) \div \left(x - 1\right)$

$$\begin{array}{r|rrrr} 1 & 4 & -3 & 3 & -1 \\ & & 4 & 1 & 4 \\ \hline & 4 & 1 & 4 & 3 \end{array}$$

$\left(4x^3 - 3x^2 + 3x - 1\right) \div \left(x - 1\right)$

$= 4x^2 + x + 4 + \dfrac{3}{x-1}$

47. $\left(6x^5 - 2x^3 + 4x^2 - 3x + 1\right) \div \left(x - 2\right)$

$$\begin{array}{r|rrrrrr} 2 & 6 & 0 & -2 & 4 & -3 & 1 \\ & & 12 & 24 & 44 & 96 & 186 \\ \hline & 6 & 12 & 22 & 48 & 93 & 187 \end{array}$$

$\left(6x^5 - 2x^3 + 4x^3 - 3x + 1\right) \div \left(x - 2\right)$

$= 6x^4 + 12x^3 + 22x^2 + 48x + 93 + \dfrac{187}{x-2}$

49. $\left(x^2 - 5x - 5x^3 + x^4\right) \div \left(5 + x\right)$

Rewrite the polynomials in descending order.

$\left(x^4 - 5x^3 + x^2 - 5x\right) \div \left(x + 5\right)$

$$\begin{array}{r|rrrrr} -5 & 1 & -5 & 1 & -5 & 0 \\ & & -5 & 50 & -255 & 1300 \\ \hline & 1 & -10 & 51 & -260 & 1300 \end{array}$$

$\left(x^2 - 5x - 5x^3 + x^4\right) \div \left(5 + x\right)$

$= x^3 - 10x^2 + 51x - 260 + \dfrac{1300}{5+x}$

51. $\left(3x^3 + 2x^2 - 4x + 1\right) \div \left(x - \dfrac{1}{3}\right)$

$$\begin{array}{r|rrrr} \frac{1}{3} & 3 & 2 & -4 & 1 \\ & & 1 & 1 & -1 \\ \hline & 3 & 3 & -3 & 0 \end{array}$$

$\left(3x^3 + 2x^2 - 4x - 1\right) \div \left(x - \dfrac{1}{3}\right) = 3x^2 + 3x - 3$

53. $\dfrac{x^5 + x^3 - 2}{x - 1}$

$$\begin{array}{r|rrrrrr} 1 & 1 & 0 & 1 & 0 & 0 & -2 \\ & & 1 & 1 & 2 & 2 & 2 \\ \hline & 1 & 1 & 2 & 2 & 2 & 0 \end{array}$$

$\dfrac{x^5 + x^3 - 2}{x - 1} = x^4 + x^3 + 2x^2 + 2x + 2$

55. $\dfrac{x^4 - 256}{x - 4}$

$$\begin{array}{r|rrrrr} 4 & 1 & 0 & 0 & 0 & -256 \\ & & 4 & 16 & 64 & 256 \\ \hline & 1 & 4 & 16 & 64 & 0 \end{array}$$

$\dfrac{x^4 - 256}{x - 4} = x^3 + 4x^2 + 16x + 64$

57. $\dfrac{2x^5 - 3x^4 + x^3 - x^2 + 2x - 1}{x + 2}$

$$\begin{array}{r|rrrrrr} -2 & 2 & -3 & 1 & -1 & 2 & -1 \\ & & -4 & 14 & -30 & 62 & -128 \\ \hline & 2 & -7 & 15 & -31 & 64 & -129 \end{array}$$

$\dfrac{2x^5 - 3x^4 + x^3 - x^2 + 2x - 1}{x + 2}$

$= 2x^4 - 7x^3 + 15x^2 - 31x + 64 - \dfrac{129}{x+2}$

59.

$$\begin{array}{r}
x^3 - x^2y + xy^2 - y^3 \\
x+y \overline{\smash{\big)}\, x^4 + 0x^3 + 0x^2 + 0x + y^4} \\
\underline{x^4 + x^3y} \\
-x^3y + 0x^2 \\
\underline{-x^3y - x^2y^2} \\
x^2y^2 + 0x \\
\underline{x^2y^2 + xy^3} \\
-xy^3 + y^4 \\
\underline{-xy^3 - y^4} \\
2y^4
\end{array}$$

$\dfrac{x^4 + y^4}{x + y} = x^3 - x^2y + xy^2 - y^3 + \dfrac{2y^4}{x+y}$

61.

$$
\begin{array}{r}
3x^2 + 2x - 1 \\
x^2 + x + 2 \overline{\smash{\big)}\, 3x^4 + 5x^3 + 7x^2 + 3x - 2} \\
\underline{3x^4 + 3x^3 + 6x^2} \\
2x^3 + x^2 + 3x \\
\underline{2x^3 + 2x^2 + 4x} \\
-x^2 - x - 2 \\
\underline{-x^2 - x - 2} \\
0
\end{array}
$$

$$\frac{3x^4 + 5x^3 + 7x^2 + 3x - 2}{x^2 + x + 2} = 3x^2 + 2x - 1$$

63.

$$
\begin{array}{r}
4x - 7 \\
x^2 + x + 1 \overline{\smash{\big)}\, 4x^3 - 3x^2 + x + 1} \\
\underline{4x^3 + 4x^2 + 4x} \\
-7x^2 - 3x + 1 \\
\underline{-7x^2 - 7x - 7} \\
4x + 8
\end{array}
$$

$$\frac{4x^3 - 3x^2 + x + 1}{x^2 + x + 1} = 4x - 7 + \frac{4x + 8}{x^2 + x + 1}$$

65.

$$
\begin{array}{r}
x^3 + x^2 - x - 3 \\
x^2 - x + 2 \overline{\smash{\big)}\, x^5 + 0x^4 + 0x^3 + 0x^2 + 0x - 1} \\
\underline{x^5 - x^4 + 2x^3} \\
x^4 - 2x^3 + 0x^2 \\
\underline{x^4 - x^3 + 2x^2} \\
-x^3 - 2x^2 + 0x \\
\underline{-x^3 + x^2 - 2x} \\
-3x^2 + 2x - 1 \\
\underline{-3x^2 + 3x - 6} \\
-x + 5
\end{array}
$$

$$\frac{x^5 - 1}{x^2 - x + 2} = x^3 + x^2 - x - 3 + \frac{-x + 5}{x^2 - x + 2}$$

67.

$$
\begin{array}{r}
4x^2 + 5xy - y^2 \\
x - 3y \overline{\smash{\big)}\, 4x^3 - 7x^2 y - 16xy^2 + 3y^3} \\
\underline{4x^3 - 12x^2 y} \\
5x^2 y - 16xy^2 \\
\underline{5x^2 y - 15xy^2} \\
-xy^2 + 3y^3 \\
\underline{-xy^2 + 3y^3} \\
0
\end{array}
$$

$$\frac{4x^3 - 7x^2 y - 16xy^2 + 3y^3}{x - 3y}$$
$$= 4x^2 + 5xy - y^2$$

69. First, compute the difference:

$$\left(4x^3 + x^2 - 2x + 7\right) - \left(3x^3 - 2x^2 - 7x + 4\right)$$

$$= x^3 + 3x^2 + 5x + 3$$

Now, complete the division:

$$\frac{x^3 + 3x^2 + 5x + 3}{x + 1}$$

$$
\begin{array}{r}
x^2 + 2x + 3 \\
x + 1 \overline{\smash{\big)}\, x^3 + 3x^2 + 5x + 3} \\
\underline{x^3 + x^2} \\
2x^2 + 5x \\
\underline{2x^2 + 2x} \\
3x + 3 \\
\underline{3x + 3} \\
0
\end{array}
$$

$$\frac{x^3 + 3x^2 + 5x + 3}{x + 1} = x^2 + 2x + 3$$

71. $A = l \cdot w$ so $l = \dfrac{A}{w} = \dfrac{x^3 + 3x^2 + 5x + 3}{x+1}$

$$
\begin{array}{r}
x^2 + 2x + 3 \\
x+1\overline{)x^3 + 3x^2 + 5x + 3} \\
\underline{x^3 + x^2} \\
2x^2 + 5x \\
\underline{2x^2 + 2x} \\
3x + 3 \\
\underline{3x + 3} \\
0
\end{array}
$$

The base of the parallelogram is $(x^2 + 2x + 3)$ units.

73. a. Substitute $n = 3$ into the formula:
$$\frac{30{,}000x^3 - 30{,}000}{x-1}$$

b. Factor out 30,000 from the numerator:
$$\frac{30{,}000x^3 - 30{,}000}{x-1} = 30{,}000\,\frac{x^3 - 1}{x-1}$$

Now use the formula from #39 with $a = 1$:
$$\frac{x^3 - a^3}{x-a} = x^2 + ax + a^2$$
$$\frac{x^3 - 1}{x-1} = \frac{x^3 - 1^3}{x-1}$$
$$= x^2 + 1 \cdot x + 1^2 = x^2 + x + 1$$

So $\dfrac{30{,}000x^3 - 30{,}000}{x-1} = 30{,}000 \cdot \dfrac{x^3 - 1}{x-1}$

$= 30{,}000\left(x^2 + x + 1\right)$

$= 30{,}000x^2 + 30{,}000x + 30{,}000$

c. Substitute in $x = 1.05$ into your formulas from parts (a) and (b) above:
$$\frac{30{,}000x^3 - 30{,}000}{x-1}$$
$$= \frac{30{,}000(1.05)^3 - 30{,}000}{(1.05) - 1} = 94{,}575$$

$30{,}000x^2 + 30{,}000x + 30{,}000$

$= 30{,}000(1.05)^2 + 30{,}000(1.05)$

$ + 30{,}000$

$= 94{,}575$

Total salary over three years is \$94,575.

75. – 77. Answers will vary.

79. makes sense

81. does not make sense; Explanations will vary. Sample explanation: The correct answer is $x^2 - x + 1$.

83. false; Changes to make the statement true will vary. A sample change is: The remainder is –9.

85. true

87. We can find the polynomial by multiplying the divisor by the quotient and adding the remainder.
$(x-3)(2x+4) + 17$

$= 2x^2 + 4x - 6x - 12 + 17$

$= 2x^2 - 2x + 5$

89. Answers will vary. The quotient starts with x to a power that is one less than the power in the dividend. It is made up of terms that are all powers of x down to 1, but with alternating signs. The remainder is always -2. Following this pattern,
$$\frac{x^7 - 1}{x+1} = x^6 - x^5 + x^4 - x^3 + x^2 - x + 1 - \frac{2}{x+1}$$

$$
\begin{array}{r}
x^6 - x^5 + x^4 - x^3 + x^2 - x + 1 \\
x+1\overline{)x^7 + 0x^6 + 0x^5 + 0x^4 + 0x^3 + 0x^2 + 0x - 1} \\
\underline{x^7 + x^6} \\
-x^6 + 0x^5 \\
\underline{-x^6 - x^5} \\
x^5 + 0x^4 \\
\underline{x^5 + x^4} \\
-x^4 + 0x^3 \\
\underline{-x^4 - x^3} \\
x^3 + 0x^2 \\
\underline{x^3 + x^2} \\
-x^2 + 0x \\
\underline{-x^2 - x} \\
x - 1 \\
\underline{x + 1} \\
-2
\end{array}
$$

Long division yields the same result.

91. Let $y_1 = \dfrac{x^2 - 25}{x - 5}$ and $y_2 = x - 5$.

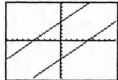

The graphs do not coincide so the division is incorrect.

$$x - 5 \overline{\smash{\big)}\ x^2 + 0x - 25} \atop \quad x + 5$$

$$\underline{x^2 - 5x}$$
$$5x - 25$$
$$\underline{5x - 25}$$
$$0$$

The right side should be $x + 5$.

93. Let $y_1 = \dfrac{6x^2 + 16x + 8}{3x + 2}$ and $y_2 = 2x - 4$.

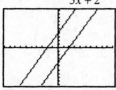

The graphs do not coincide so the division is incorrect.

$$3x + 2 \overline{\smash{\big)}\ 6x^2 + 16x + 8} \atop \quad 2x + 4$$

$$\underline{6x^2 + 4x}$$
$$12x + 8$$
$$\underline{12x + 8}$$
$$0$$

The right side should be $2x + 4$.

95. $7x - 6y = 17$

$3x + y = 18$

The addition method is a good choice because both equations are written in the form $Ax + By = C$. To eliminate y, multiply second equation by 6, and add the result to the first equation.

$$7x - 6y = 17$$
$$\underline{18x + 6y = 108}$$
$$25x = 125$$
$$x = 5$$

Back-substitute 5 for x in either equation of the original system. We choose the original second equation:

$$3x + y = 18$$
$$3(5) + y = 18$$
$$15 + y = 18$$
$$y = 3$$

Solution: $\{(5,3)\}$

96. $P = 6\% = 0.06,\ B = 20$

$A = PB$

$A = (0.06)(20)$

$A = 1.2$

1.2 is 6% of 20.

97. $\dfrac{x}{3} + \dfrac{2}{5} = \dfrac{x}{5} - \dfrac{2}{5}$

To clear fractions, multiply by the LCD, 15.

$$15\left(\frac{x}{3} + \frac{2}{5}\right) = 15\left(\frac{x}{5} - \frac{2}{5}\right)$$

$$15\left(\frac{x}{3}\right) + 15\left(\frac{2}{5}\right) = 15\left(\frac{x}{5}\right) - 15\left(\frac{2}{5}\right)$$

$$5x + 6 = 3x - 6$$
$$2x + 6 = -6$$
$$2x = -12$$
$$x = -6$$

The solution set is $\{-6\}$.

98. a. $\dfrac{7^3}{7^5} = \dfrac{\cancel{7} \cdot \cancel{7} \cdot \cancel{7}}{\cancel{7} \cdot \cancel{7} \cdot \cancel{7} \cdot 7 \cdot 7} = \dfrac{1}{7^2}$

$$\frac{7^3}{7^5} = 7^{3-5} = 7^{-2}$$

b. $\dfrac{1}{7^2} = 7^{-2}$

99. $\dfrac{(2x^3)^4}{x^{10}} = \dfrac{2^4(x^3)^4}{x^{10}} = \dfrac{16x^{12}}{x^{10}} = 16x^{12-10} = 16x^2$

100. $\left(\dfrac{x^5}{x^2}\right)^3 = \left(x^{5-2}\right)^3 = \left(x^3\right)^3 = x^9$

5.7 Check Points

1. a. $6^{-2} = \dfrac{1}{6^2} = \dfrac{1}{36}$

b. $5^{-3} = \dfrac{1}{5^3} = \dfrac{1}{125}$

c. $(-3)^{-4} = \dfrac{1}{(-3)^4} = \dfrac{1}{81}$

d. $-3^{-4} = -\dfrac{1}{3^4} = -\dfrac{1}{81}$

e. $8^{-1} = \dfrac{1}{8^1} = \dfrac{1}{8}$

2. a. $\dfrac{2^{-3}}{7^{-2}} = \dfrac{7^2}{2^3} = \dfrac{49}{8}$

b. $\left(\dfrac{4}{5}\right)^{-2} = \dfrac{5^2}{4^2} = \dfrac{25}{16}$

c. $\dfrac{1}{7y^{-2}} = \dfrac{y^2}{7}$

d. $\dfrac{x^{-1}}{y^{-8}} = \dfrac{y^8}{x^1} = \dfrac{y^8}{x}$

3. $x^{-12} \cdot x^2 = x^{-12+2} = x^{-10} = \dfrac{1}{x^{10}}$

4. a. $\dfrac{x^2}{x^{10}} = x^{2-10} = x^{-8} = \dfrac{1}{x^8}$

b. $\dfrac{75x^3}{5x^9} = \dfrac{75}{5} \cdot \dfrac{x^3}{x^9} = 15x^{3-9} = 15x^{-6} = \dfrac{15}{x^6}$

c. $\dfrac{50y^8}{-25y^{14}} = \dfrac{50}{-25} \cdot \dfrac{y^8}{y^{14}} = -2y^{8-14} = -2y^{-6} = -\dfrac{2}{y^6}$

5. $\dfrac{(6x^4)^2}{x^{11}} = \dfrac{6^2(x^4)^2}{x^{11}} = \dfrac{36x^{4 \cdot 2}}{x^{11}} = \dfrac{36x^8}{x^{11}}$

$= 36x^{8-11} = 36x^{-3} = \dfrac{36}{x^3}$

6. $\left(\dfrac{x^8}{x^4}\right)^{-5} = \left(x^4\right)^{-5} = x^{-20} = \dfrac{1}{x^{20}}$

7. a. $7.4 \times 10^9 = 7,400,000,000$

b. $3.017 \times 10^{-6} = 0.000003017$

8. a. $7,410,000,000 = 7.41 \times 10^9$

b. $0.000000092 = 9.2 \times 10^{-8}$

9. a. $(3 \times 10^8)(2 \times 10^2) = (3 \times 2) \times (10^8 \times 10^2)$

$= 6 \times 10^{8+2}$

$= 6 \times 10^{10}$

b. $\dfrac{8.4 \times 10^7}{4 \times 10^{-4}} = \dfrac{8.4}{4} \cdot \dfrac{10^7}{10^{-4}}$

$= 2.1 \times 10^{7-(-4)}$

$= 2.1 \times 10^{11}$

c. $(4 \times 10^{-2})^3 = 4^3 \times (10^{-2})^3$

$= 64 \times 10^{-6}$

$= 6.4 \times 10^{-5}$

10. $\dfrac{13 \times 10^9}{5.1 \times 10^6} = \dfrac{13}{5.1} \times \dfrac{10^9}{10^6} \approx 2.5 \times 10^3 = 2500$

The average grant was about $2500.

5.7 Exercise Set

1. $8^{-2} = \dfrac{1}{8^2} = \dfrac{1}{64}$

3. $5^{-3} = \dfrac{1}{5^3} = \dfrac{1}{125}$

5. $(-6)^{-2} = \dfrac{1}{(-6)^2} = \dfrac{1}{36}$

7. $-6^{-2} = -\dfrac{1}{6^2} = -\dfrac{1}{36}$

9. $4^{-1} = \dfrac{1}{4^1} = \dfrac{1}{4}$

11. $2^{-1} + 3^{-1} = \dfrac{1}{2^1} + \dfrac{1}{3^1} = \dfrac{1}{2} + \dfrac{1}{3}$

$\qquad = \dfrac{3}{6} + \dfrac{2}{6} = \dfrac{5}{6}$

13. $\dfrac{1}{3^{-2}} = 3^2 = 9$

15. $\dfrac{1}{(-3)^{-2}} = (-3)^2 = 9$

17. $\dfrac{2^{-3}}{8^{-2}} = \dfrac{8^2}{2^3} = \dfrac{64}{8} = 8$

19. $\left(\dfrac{1}{4}\right)^{-2} = \dfrac{1^{-2}}{4^{-2}} = \dfrac{4^2}{1^2} = \dfrac{16}{1} = 16$

21. $\left(\dfrac{3}{5}\right)^{-3} = \dfrac{3^{-3}}{5^{-3}} = \dfrac{5^3}{3^3} = \dfrac{125}{27}$

23. $\dfrac{1}{6x^{-5}} = \dfrac{1 \cdot x^5}{6} = \dfrac{x^5}{6}$

25. $\dfrac{x^{-8}}{y^{-1}} = \dfrac{y^1}{x^8} = \dfrac{y}{x^8}$

27. $\dfrac{3}{(-5)^{-3}} = 3 \cdot (-5)^3 = 5(-125) = -375$

29. $x^{-8} \cdot x^3 = x^{-8+3} = x^{-5} = \dfrac{1}{x^5}$

31. $\left(4x^{-5}\right)\left(2x^2\right) = 8x^{-5+2} = 8x^{-3} = \dfrac{8}{x^3}$

33. $\dfrac{x^3}{x^9} = x^{3-9} = x^{-6} = \dfrac{1}{x^6}$

35. $\dfrac{y}{y^{100}} = \dfrac{y^1}{y^{100}} = y^{1-100} = y^{-99} = \dfrac{1}{y^{99}}$

37. $\dfrac{30z^5}{10z^{10}} = \dfrac{30}{10} \cdot \dfrac{z^5}{z^{10}} = 3z^{5-10}$

$\qquad = -3z^{-5} = \dfrac{3}{z^5}$

39. $\dfrac{-8x^3}{2x^7} = \dfrac{-8}{2} \cdot \dfrac{x^3}{x^7} = -4x^{-4} = -\dfrac{4}{x^4}$

41. $\dfrac{-9a^5}{27a^8} = \dfrac{-9}{27} \cdot \dfrac{a^5}{a^8} = -\dfrac{1}{3}a^{-3} = -\dfrac{1}{3a^3}$

43. $\dfrac{7w^5}{5w^{13}} = \dfrac{7}{5} \cdot \dfrac{w^5}{w^{13}} = \dfrac{7}{5}w^{-8} = \dfrac{7}{5w^8}$

45. $\dfrac{x^3}{\left(x^4\right)^2} = \dfrac{x^3}{x^{4 \cdot 2}} = \dfrac{x^3}{x^8} = x^{-5} = \dfrac{1}{x^5}$

47. $\dfrac{y^{-3}}{\left(y^4\right)^2} = \dfrac{y^{-3}}{y^8} = y^{-3-8} = y^{-11} = \dfrac{1}{y^{11}}$

49. $\dfrac{\left(4x^3\right)^2}{x^8} = \dfrac{4^2 x^6}{x^8} = 16x^{-2} = \dfrac{16}{x^2}$

51. $\dfrac{\left(6y^4\right)^3}{y^{-5}} = \dfrac{6^3 y^{12}}{y^{-5}} = 216y^{12-(-5)} = 216y^{17}$

53. $\left(\dfrac{x^4}{x^2}\right)^{-3} = \left(x^2\right)^{-3} = x^{-6} = \dfrac{1}{x^6}$

55. $\left(\dfrac{4x^5}{2x^2}\right)^{-4} = \left(2x^3\right)^{-4} = 2^{-4}x^{-12} = \dfrac{1}{2^4 x^{12}}$

$\qquad = \dfrac{1}{16x^{12}}$

57. $\left(3x^{-1}\right)^{-2} = 3^{-2}\left(x^{-1}\right)^{-2} = 3^{-2}x^2$

$\qquad = \dfrac{x^2}{3^2} = \dfrac{x^2}{9}$

59. $\left(-2y^{-1}\right)^{-3} = (-2)^{-3}\left(y^{-1}\right)^{-3} = \dfrac{y^3}{(-2)^3}$

$= \dfrac{y^3}{-8} = -\dfrac{y^3}{8}$

61. $\dfrac{2x^5 \cdot 3x^7}{15x^6} = \dfrac{6x^{12}}{15x^6} = \dfrac{6}{15} \cdot \dfrac{x^{12}}{x^6}$

$= \dfrac{2}{5} \cdot x^6 = \dfrac{2x^6}{5}$

63. $\left(x^3\right)^5 \cdot x^{-7} = x^{15} \cdot x^{-7} = x^{15+(-7)} = x^8$

65. $\left(2y^3\right)^4 y^{-6} = 2^4\left(y^3\right)^4 y^{-6} = 16y^{12}y^{-6}$

$= 16y^6$

67. $\dfrac{\left(y^3\right)^4}{\left(y^2\right)^7} = \dfrac{y^{12}}{y^{14}} = y^{-2} = \dfrac{1}{y^2}$

69. $\left(y^{10}\right)^{-5} = y^{(10)(-5)} = y^{-50} = \dfrac{1}{y^{50}}$

71. $\left(a^4 b^5\right)^{-3} = \left(a^4\right)^{-3}\left(b^5\right)^{-3} = a^{-12}b^{-15}$

$= \dfrac{1}{a^{12}b^{15}}$

73. $\left(a^{-2}b^6\right)^{-4} = a^8 b^{-24} = \dfrac{a^8}{b^{24}}$

75. $\left(\dfrac{x^2}{2}\right)^{-2} = \dfrac{x^{-4}}{2^{-2}} = \dfrac{2^2}{x^4} = \dfrac{4}{x^4}$

77. $\left(\dfrac{x^2}{y^3}\right)^{-3} = \dfrac{\left(x^2\right)^{-3}}{\left(y^3\right)^{-3}} = \dfrac{x^{-6}}{y^{-9}} = \dfrac{y^9}{x^6}$

79. $8.7 \times 10^2 = 870$ (Move decimal point 2 places to the right.)

81. $9.23 \times 10^5 = 923,000$ (Move right 5.)

83. $3.4 \times 10^0 = 3.4$ (Don't move decimal point.)

85. $7.9 \times 10^{-1} = 0.79$ (Move left 1.)

87. $2.15 \times 10^{-2} = 0.0215$ (Move left 2.)

89. $7.86 \times 10^{-4} = 0.000786$ (Move left 4.)

91. $32,400 = 3.24 \times 10^4$

93. $220,000,000 = 2.2 \times 10^8$

95. $713 = 7.13 \times 10^2$

97. $6751 = 6.751 \times 10^3$

99. $0.0027 = 2.7 \times 10^{-3}$

101. $0.000020 = 2.02 \times 10^{-5}$

103. $0.005 = 5 \times 10^{-3}$

105. $3.14159 = 3.14159 \times 10^0$

107. $\left(2 \times 10^3\right)\left(3 \times 10^2\right) = 6 \times 10^{3+2} = 6 \times 10^5$

109. $\left(2 \times 10^5\right)\left(8 \times 10^3\right) = 16 \times 10^{5+3} = 16 \times 10^8$

$= 1.6 \times 10^9$

111. $\dfrac{12 \times 10^6}{4 \times 10^2} = 3 \times 10^{6-2} = 3 \times 10^4$

113. $\dfrac{15 \times 10^4}{5 \times 10^{-2}} = 3 \times 10^{4+2} = 3 \times 10^6$

115. $\dfrac{15 \times 10^{-4}}{5 \times 10^2} = 3 \times 10^{-4-2} = 3 \times 10^{-6}$

117. $\dfrac{180 \times 10^6}{2 \times 10^3} = 90 \times 10^{6-3} = 90 \times 10^3$

$= 9 \times 10^4$

119. $\dfrac{3 \times 10^4}{12 \times 10^{-3}} = 0.25 \times 10^{4+3} = 0.25 \times 10^7$

$= 2.5 \times 10^6$

121. $\left(5\times10^2\right)^3 = 5^3\times10^{2(3)} = 125\times10^6$

$\qquad\qquad = 1.25\times10^8$

123. $\left(3\times10^{-2}\right)^4 = 3^4\times10^{-2(4)} = 81\times10^{-8}$

$\qquad\qquad = 8.1\times10^{-7}$

125. $\left(4\times10^6\right)^{-1} = 4^{-1}\times10^{6(-1)} = 0.25\times10^{-6}$

$\qquad\qquad = 2.5\times10^{-7}$

127. $\dfrac{\left(x^{-2}y\right)^{-3}}{\left(x^2y^{-1}\right)^3} = \dfrac{x^6y^{-3}}{x^6y^{-3}}$

$\qquad\qquad = x^{6-6}y^{-3-(-3)} = x^0y^0 = 1$

129. $\left(2x^{-3}yz^{-6}\right)\left(2x\right)^{-5} = 2x^{-3}yz^{-6}\cdot2^{-5}x^{-5}$

$\qquad = 2^{-4}x^{-8}yz^{-6} = \dfrac{y}{2^4x^8z^6} = \dfrac{y}{16x^8z^6}$

131. $\left(\dfrac{x^3y^4z^5}{x^{-3}y^{-4}z^{-5}}\right)^{-2} = \left(x^6y^8z^{10}\right)^{-2}$

$\qquad\qquad = x^{-12}y^{-16}z^{-20} = \dfrac{1}{x^{12}y^{16}z^{20}}$

133. $\dfrac{\left(2^{-1}x^{-2}y^{-1}\right)^{-2}\left(2x^{-4}y^3\right)^{-2}\left(16x^{-3}y^3\right)^0}{\left(2x^{-3}y^{-5}\right)^2}$

$\qquad = \dfrac{\left(2^2x^2y^2\right)\left(2^{-2}x^8y^{-6}\right)(1)}{\left(2^2x^{-6}y^{-10}\right)}$

$\qquad = \dfrac{x^{18}y^6}{4}$

135. $\dfrac{\left(5\times10^3\right)\left(1.2\times10^{-4}\right)}{\left(2.4\times10^2\right)} = 2.5\times10^{-3}$

137. $\dfrac{\left(1.6\times10^4\right)\left(7.2\times10^{-3}\right)}{\left(3.6\times10^8\right)\left(4\times10^{-3}\right)} = 0.8\times10^{-4}$

$\qquad\qquad = 8\times10^{-5}$

139. $9200 = 9.2\times10^3$

141. $0.00000000000000025 = 2.5\times10^{-16}$

143. $90.5\times10^6 = 9.05\times10^1\times10^6 = 9.05\times10^7$

145. $465\times10^9 = 4.65\times10^2\times10^9 = 4.65\times10^{11}$

147. **a.** 2.27×10^{12}

b. $298\times10^6 = 2.98\times10^2\times10^6 = 2.98\times10^8$

c. $\dfrac{2.27\times10^{12}}{2.98\times10^8} = \dfrac{2.27}{2.98}\cdot\dfrac{10^{12}}{10^8} \approx 0.7617\times10^4 = 7617$

The amount per citizen would be $7617.

149. $120(3.1\times10^8) = 372\times10^8$

$\qquad\qquad = 3.72\times10^2\times10^8 = 3.72\times10^{10}$

The total annual spending in the United States on ice cream is about 3.72×10^{10}.

151. $\dfrac{2.325\times10^5}{1.86\times10^5} = 1.25$ seconds

153. – 157. Answers will vary.

159. does not make sense; Explanations will vary.
Sample explanation: $36(x^3)^9 = 36x^{27} \neq 36x^{12}$

161. does not make sense; Explanations will vary.
Sample explanation: This value is much greater than the world population.

163. false; Changes to make the statement true will vary.
A sample change is: $4^{-2} > 4^{-3}$

165. false; Changes to make the statement true will vary.
A sample change is: $(-2)^4 = 16$ and

$2^{-4} = \dfrac{1}{2^4} = \dfrac{1}{16}$.

167. false; Changes to make the statement true will vary.
A sample change is: $534.7 = 5.347\times10^2$

169. false; Changes to make the statement true will vary. A sample change is:

$$\left(7 \times 10^5\right) + \left(2 \times 10^{-3}\right)$$

$$= \left(7 \times 10^5\right) + \left(0.00000002 \times 10^5\right)$$

$$= \left(7 + 0.00000002\right) \times 10^5$$

$$= 7.00000002 \times 10^5$$

171. $2^{-1} + 2^{-2} = \dfrac{1}{2^1} + \dfrac{1}{2^2}$

$$= \dfrac{1}{2} + \dfrac{1}{4}$$

$$= \dfrac{2}{4} + \dfrac{1}{4} = \dfrac{3}{4}$$

Dr. Frankenstein has gathered enough bits and pieces for $\dfrac{3}{4}$ of his creature. He still needs parts for

$$1 - \dfrac{3}{4} = \dfrac{1}{4}.$$

173. The calculator verifies your results.

175. The calculator verifies your results.

176. $\quad 8 - 6x > 4x - 12$

$$-6x - 4x > -12 - 8$$

$$-10x > -20$$

$$\dfrac{-10x}{-10} < \dfrac{-20}{-10}$$

$$x < 2$$

$$(-\infty, 2)$$

177. $24 \div 8 \cdot 3 + 28 \div (-7) = 3 \cdot 3 + 28 \div (-7)$

$$= 9 + (-4) = 5$$

178. The whole numbers in the given set are 0 and $\sqrt{16}$. Note that $\sqrt{16} = 4$.

179. $4x^3(4x^2 - 3x + 1) = 4x^3 \cdot 4x^2 - 4x^3 \cdot 3x + 4x^3 \cdot 1$

$$= 16x^5 - 12x^4 + 4x^3$$

180. $9xy(3xy^2 - y + 9) = 9xy \cdot 3xy^2 - 9xy \cdot y + 9xy \cdot 9$

$$= 27x^2y^3 - 9xy^2 + 81xy$$

181. $(x + 3)(x^2 + 5) = (x)(x^2) + (x)(5) + (3)(x^2) + (3)(5)$

$$= x^3 + 5x + 3x^2 + 15$$

$$= x^3 + 3x^2 + 5x + 15$$

Chapter 5 Review Exercises

1. $7x^4 + 9x$ is a binomial of degree 4.

2. $3x + 5x^2 - 2$ is a trinomial of degree 2.

3. $16x$ is a monomial of degree 1.

4. $\left(-6x^3 + 7x^2 - 9x + 3\right) + \left(14x^3 + 3x^2 - 11x - 7\right)$

$$= \left(-6x^3 + 14x^3\right) + \left(7x^2 + 3x^2\right) + \left(-9x - 11x\right) + \left(3 - 7\right)$$

$$= 8x^3 + 10x^2 - 20x - 4$$

5. $\left(9y^3 - 7y^2 + 5\right) + \left(4y^3 - y^2 + 7y - 10\right)$

$$= \left(9y^3 + 4y^3\right) + \left(-7y^2 - y^2\right) + 7y + \left(5 - 10\right)$$

$$= 13y^3 - 8y^2 + 7y - 5$$

6. $\left(5y^2 - y - 8\right) - \left(-6y^2 + 3y - 4\right)$

$$= \left(5y^2 - y - 8\right) + \left(6y^2 - 3y + 4\right)$$

$$= \left(5y^2 + 6y^2\right) + \left(-y - 3y\right) + \left(-8 + 4\right)$$

$$= 11y^2 - 4y - 4$$

7. $\left(13x^4 - 8x^3 + 2x^2\right) - \left(5x^4 - 3x^3 + 2x^2 - 6\right)$

$$= \left(13x^4 - 8x^3 + 2x^2\right)$$

$$\quad + \left(-5x^4 + 3x^3 - 2x^2 + 6\right)$$

$$= \left(13x^4 - 5x^4\right) + \left(-8x^3 + 3x^3\right)$$

$$\quad + \left(2x^2 - 2x^2\right) + 6$$

$$= 8x^4 - 5x^3 + 6$$

8. $\left(-13x^4 - 6x^2 + 5x\right) - \left(x^4 + 7x^2 - 11x\right)$

$= \left(-13x^4 - 6x^2 + 5x\right) + \left(-x^4 - 7x^2 + 11x\right)$

$= \left(-13x^4 - x^4\right) + \left(-6x^2 - 7x^2\right) + \left(5x + 11x\right)$

$= -14x^4 - 13x^2 + 16x$

9. $7y^4 - 6y^3 + 4y^2 - 4y$

$\underline{y^3 - y^2 + 3y - 4}$

$7y^4 - 5y^3 + 3y^2 - y - 4$

10. $7x^2 - 9x + 2$

$\underline{-\left(4x^2 - 2x - 7\right)}$

To subtract, add the opposite of the polynomial being subtracted.

$7x^2 - 9x + 2$

$\underline{-4x^2 + 2x + 7}$

$3x^2 - 7x + 9$

11. $5x^3 - 6x^2 - 9x + 14$

$\underline{-\left(-5x^3 + 3x^2 - 11x + 3\right)}$

To subtract, add the opposite of the polynomial being subtracted.

$5x^3 - 6x^2 - 9x + 14$

$\underline{5x^3 - 3x^2 + 11x - 3}$

$10x^3 - 9x^2 + 2x + 11$

12.

x	$y = x^2 + 3$	(x, y)
-3	$y = (-3)^2 + 3 = 12$	$(-3, 12)$
-2	$y = (-2)^2 + 3 = 7$	$(-2, 7)$
-1	$y = (-1)^2 + 3 = 4$	$(-1, 4)$
0	$y = (0)^2 + 3 = 3$	$(0, 3)$
1	$y = (1)^2 + 3 = 4$	$(1, 4)$
2	$y = (2)^2 + 3 = 7$	$(2, 7)$
3	$y = (3)^2 + 3 = 12$	$(3, 12)$

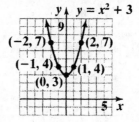

13.

x	$y = 1 - x^2$	(x, y)
-3	$y = 1 - (-3)^2 = -8$	$(-3, -8)$
-2	$y = 1 - (-2)^2 = -3$	$(-2, -3)$
-1	$y = 1 - (-1)^2 = 0$	$(-1, 0)$
0	$y = 1 - (0)^2 = 1$	$(0, 1)$
1	$y = 1 - (1)^2 = 0$	$(1, 0)$
2	$y = 1 - (2)^2 = -3$	$(2, -3)$
3	$y = 1 - (3)^2 = -8$	$(3, -8)$

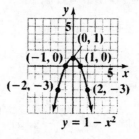

14. $x^{20} \cdot x^3 = x^{20+3} = x^{23}$

15. $y \cdot y^5 \cdot y^8 = y^1 \cdot y^5 \cdot y^8 = y^{1+5+8} = y^{14}$

16. $\left(x^{20}\right)^5 = x^{20 \cdot 5} = x^{100}$

17. $\left(10y\right)^2 = 10^2 y^2 = 100y^2$

18. $\left(-4x^{10}\right)^3 = (-4)^3 \left(x^{10}\right)^3 = -64x^{30}$

19. $(5x)\left(10x^3\right) = (5 \cdot 10)\left(x^1 \cdot x^3\right) = 50x^4$

20. $\left(-12y^7\right)\left(3y^4\right) = -36y^{11}$

21. $\left(-2x^5\right)\left(-3x^4\right)\left(5x^3\right) = 30x^{12}$

22. $7x\left(3x^2 + 9\right) = 7x\left(3x^2\right) + (7x)(9)$
$ = 21x^3 + 63x$

23. $5x^3\left(4x^2 - 11x\right) = 5x^3\left(4x^2\right) - 5x^3\left(11x\right)$
$ = 20x^5 - 55x^4$

24. $3y^2\left(-7y^2 + 3y - 6\right)$
$ = 3y^2\left(-7y^2\right) + 3y^2\left(3y\right) + 3y^2\left(-6\right)$
$ = -21y^4 + 9y^3 - 18y^2$

25. $2y^5\left(8y^3 - 10y^2 + 1\right)$
$ = 2y^5\left(8y^3\right) + 2y^5\left(-10y^2\right) + 2y^5\left(1\right)$
$ = 16y^8 - 20y^7 + 2y^5$

26. $(x+3)\left(x^2 - 5x + 2\right)$
$ = x\left(x^2 - 5x + 2\right) + 3\left(x^2 - 5x + 2\right)$
$ = x^3 - 5x^2 + 2x + 3x^2 - 15x + 6$
$ = x^3 - 2x^2 - 13x + 6$

27. $(3y-2)\left(4y^2 + 3y - 5\right)$
$ = 3y\left(4y^2 + 3y - 5\right) - 2\left(4y^2 + 3y - 5\right)$
$ = 12y^3 + 9y^2 - 15y - 8y^2 - 6y + 10$
$ = 12y^3 + y^2 - 21y + 10$

28. $\quad y^2 - 4y + 7$
$\quad\quad\underline{\ 3y - 5}$
$\quad -5y^2 + 20y - 35$
$\quad\underline{3y^3 - 12y^2 + 21y}$
$\quad 3y^3 - 17y^2 + 41y - 35$

29. $\quad 4x^3 - 2x^2 - 6x - 1$
$\quad\quad\underline{\ 2x + 3}$
$\quad 12x^3 - 6x^2 - 18x - 3$
$\quad\underline{8x^4 - 4x^3 - 12x^2 - 2x}$
$\quad 8x^4 + 8x^3 - 18x^2 - 20x - 3$

30. $(x+6)(x+2)$
$ = x \cdot x + x \cdot 2 + 6 \cdot x + 6 \cdot 2$
$ = x^2 + 2x + 6x + 12$
$ = x^2 + 8x + 12$

31. $(3y-5)(2y+1) = 6y^2 + 3y - 10y - 5$
$ = 6y^2 - 7y - 5$

32. $\left(4x^2 - 2\right)\left(x^2 - 3\right)$
$ = 4x^2 \cdot x^2 + 4x^2(-3) - 2 \cdot x^2 - 2(-3)$
$ = 4x^4 - 12x^2 - 2x^2 + 6$
$ = 4x^4 - 14x^2 + 6$

33. $(5x+4)(5x-4) = (5x)^2 - 4^2$
$ = 25x^2 - 16$

34. $(7-2y)(7+2y) = 7^2 - (2y)^2$
$ = 49 - 4y^2$

35. $\left(y^2 + 1\right)\left(y^2 - 1\right) = \left(y^2\right)^2 - 1^2 = y^4 - 1$

36. $(x+3)^2 = x^2 + 2(x)(3) + 3^2$
$ = x^2 + 6x + 9$

37. $(3y+4)^2 = (3y)^2 + 2(3y)(4) + 16$
$ = 9y^2 + 24y + 16$

38. $(y-1)^2 = y^2 - 2y + 1$

39. $(5y-2)^2 = (5y)^2 - 2(5y)(2) + 2^2$
$$= 25y^2 - 20y + 4$$

40. $(x^2+4)^2 = (x^2)^2 + 2(x^2)(4) + 4^2$
$$= x^4 + 8x^2 + 16$$

41. $(x^2+4)(x^2-4) = (x^2)^2 - 4^2 = x^4 - 16$

42. $(x^2+4)(x^2-5) = (x^2)^2 - 5x^2 + 4x^2 - 20$
$$= x^4 - x^2 - 20$$

43. $A = (x+3)(x+4)$
$$= x^2 + 4x + 3x + 12$$
$$= x^2 + 7x + 12$$

44. $A = (x+30)(x+20)$
$$= x^2 + 20x + 30x + 600$$
$$= x^2 + 50x + 600$$

The area of the expanded garage is $\left(x^2 + 50x + 600\right)$ yards^2.

45. $2x^3 y - 4xy^2 + 5y + 6$
$$= 2(-1)^3(2) - 4(-1)(2)^2 + 5(2) + 6$$
$$= 2(-1)(2) + 4(1)(4) + 5(2) + 6$$
$$= -4 + 16 + 10 + 6 = 28$$

46. $4x^2 y + 9x^3 y^2 - 17x^4 - 12$

Term	Coefficient	Degree
$4x^2 y$	4	$2+1=3$
$9x^3 y^2$	9	$3+2=5$
$-17x^4$	-17	4
-12	-12	0

The degree of the polynomial is the highest degree of all its terms, which is 5.

47. $\left(7x^2 - 8xy + y^2\right) + \left(-8x^2 - 9xy + 4y^2\right)$
$$= \left(7x^2 - 8x^2\right) + (-8xy - 9xy) + \left(y^2 + 4y^2\right)$$
$$= -x^2 - 17xy + 5y^2$$

48. $\left(13x^3y^2 - 5x^2y - 9x^2\right) - \left(11x^3y^2 - 6x^2y - 3x^2 + 4\right)$

$= \left(13x^3y^2 - 5x^2y - 9x^2\right) + \left(-11x^3y^2 + 6x^2y + 3x^2 - 4\right)$

$= \left(13x^3y^2 - 11x^3y^2\right) + \left(-5x^2y + 6x^2y\right) + \left(-9x^2 + 3x^2\right) - 4$

$= 2x^3y^2 + x^2y - 6x^2 - 4$

49. $\left(-7x^2y^3\right)\left(5x^4y^6\right) = (-7)(-5)x^{2+4}y^{3+6}$

$\qquad\qquad\qquad\qquad = -35x^6y^9$

50. $5ab^2\left(3a^2b^3 - 4ab\right)$

$= 5ab^2\left(3a^2b^3\right) + 5ab^2\left(-4ab\right)$

$= 15a^3b^5 - 20a^2b^3$

51. $(x + 7y)(3x - 5y)$

$= x(3x) + x(-5y) + 7y(3x) + 7y(-5y)$

$= 3x^2 - 5xy + 21xy - 35y^2$

$= 3x^2 + 16xy - 35y^2$

52. $(4xy - 3)(9xy - 1)$

$= 4xy(9xy) + 4xy(-1) - 3(9xy) - 3(-1)$

$= 36x^2y^2 - 4xy - 27xy + 3$

$= 36x^2y^2 - 31xy + 3$

53. $(3x + 5y)^2 = (3x)^2 + 2(3x)(5y) + (5y)^2$

$\qquad\qquad\quad = 9x^2 + 30xy + 25y^2$

54. $(xy - 7)^2 = (xy)^2 - 2(xy)(7) + 7^2$

$\qquad\qquad\quad = x^2y^2 - 14xy + 49$

55. $(7x + 4y)(7x - 4y) = (7x)^2 - (4y)^2$

$\qquad\qquad\qquad\qquad = 49x^2 - 16y^2$

56. $(a - b)\left(a^2 + ab + b^2\right)$

$= a\left(a^2 + ab + b^2\right) - b\left(a^2 + ab + b^2\right)$

$= a^3 + a^2b + ab^2 - a^2b - ab^2 - b^3$

$= a^3 + \left(a^2b - a^2b\right) + \left(ab^2 - ab^2\right) - b^3$

$= a^3 - b^3$

57. $\dfrac{6^{40}}{6^{10}} = 6^{40-10} = 6^{30}$

58. $\dfrac{x^{18}}{x^3} = x^{18-3} = x^{15}$

59. $(-10)^0 = 1$

60. $-10^0 = -(1) = -1$

61. $400x^0 = 400 \cdot 1 = 400$

62. $\left(\dfrac{x^4}{2}\right)^3 = \dfrac{\left(x^4\right)^3}{2^3} = \dfrac{x^{4 \cdot 3}}{8} = \dfrac{x^{12}}{8}$

63. $\left(\dfrac{-3}{2y^6}\right)^4 = \dfrac{(-3)^4}{\left(2y^6\right)^4} = \dfrac{81}{\left(2^4 y^6\right)^4} = \dfrac{81}{16y^{24}}$

64. $\dfrac{-15y^8}{3y^2} = \dfrac{-15}{3} \cdot \dfrac{y^8}{y^2} = -5y^6$

65. $\dfrac{40x^8 y^6}{5xy^3} = \dfrac{40}{5} \cdot \dfrac{x^8}{x^1} \cdot \dfrac{y^6}{y^3} = 8x^7 y^3$

66. $\dfrac{18x^4 - 12x^2 + 36x}{6x} = \dfrac{18x^4}{6x} - \dfrac{12x^2}{6x} + \dfrac{36x}{6x}$

$\qquad = 3x^3 - 2x + 6$

67. $\dfrac{30x^8 - 25x^7 - 40x^5}{-5x^3}$

$= \dfrac{30x^8}{-5x^3} - \dfrac{25x^7}{-5x^3} - \dfrac{40x^5}{-5x^3}$

$= -6x^5 + 5x^4 + 8x^2$

68. $\dfrac{27x^3 y^2 - 9x^2 y - 18xy^2}{3xy}$

$= \dfrac{27x^3 y^2}{3xy} - \dfrac{9x^2 y}{3xy} - \dfrac{18xy^2}{3xy}$

$= 9x^2 y - 3x - 6y$

69.

$$
\begin{array}{r}
2x + 7 \\
x-2 \overline{)\,2x^2 + 3x - 14\,} \\
\underline{2x^2 - 4x} \\
7x - 14 \\
\underline{7x - 14} \\
0
\end{array}
$$

$\dfrac{2x^2 + 3x - 14}{x - 2} = 2x + 7$

70.

$$
\begin{array}{r}
x^2 - 3x + 5 \\
2x+1 \overline{)\,2x^3 - 5x^2 + 7x + 5\,} \\
\underline{2x^3 + x^2} \\
-6x^2 + 7x \\
\underline{-6x^2 - 3x} \\
10x + 5 \\
\underline{10x + 5} \\
0
\end{array}
$$

$\dfrac{2x^3 - 5x^2 + 7x + 5}{2x + 1} = x^2 - 3x + 5$

71.

$$
\begin{array}{r}
x^2 + 5x + 2 \\
x-7 \overline{)\,x^3 - 2x^2 - 33x - 7\,} \\
\underline{x^3 - 7x^2} \\
5x^2 - 33x \\
\underline{5x^2 - 35x} \\
2x - 7 \\
\underline{2x - 14} \\
7
\end{array}
$$

$\dfrac{x^3 - 2x^2 - 33x - 7}{x - 7} = x^2 + 5x + 2 + \dfrac{7}{x - 7}$

72.

$$
\begin{array}{r}
y^2+3y+9 \\
y-3{\overline{\smash{\big)}\,y^3+0y^2+0y-27}} \\
\underline{y^3-3y^2} \\
3y^2+0y \\
\underline{3y^2-9y} \\
9y-27 \\
\underline{9y-27} \\
0
\end{array}
$$

$$\frac{y^2-27}{y-3}=y^2+3y+9$$

73.

$$
\begin{array}{r}
2x^2+3x-1 \\
2x^2+0x+1{\overline{\smash{\big)}\,4x^4+6x^3+0x^2+3x-1}} \\
\underline{4x^4+0x^3+2x^2} \\
6x^3-2x^2+3x \\
\underline{6x^3+0x^2+3x} \\
-2x^2+0x-1 \\
\underline{-2x^2+0x-1} \\
0
\end{array}
$$

$$\frac{4x^4+6x^3+3x-1}{2x^2+1}=2x^2+3x-1$$

74. $\left(4x^3-3x^2-2x+1\right)\div(x+1)$

$$
\begin{array}{r|rrrr}
-1 & 4 & -3 & -2 & 1 \\
& & -4 & 7 & -5 \\
\hline
& 4 & -7 & 5 & -4
\end{array}
$$

$$\left(4x^3-3x^2-2x+1\right)\div(x+1)$$

$$=4x^2-7x+5-\frac{4}{x+1}$$

75. $\left(3x^4-2x^2-10x-20\right)\div(x-2)$

$$
\begin{array}{r|rrrrr}
2 & 3 & 0 & -2 & -10 & -20 \\
& & 6 & 12 & 20 & 20 \\
\hline
& 3 & 6 & 10 & 10 & 0
\end{array}
$$

$$\left(3x^4-2x^2-10x-20\right)\div(x-2)$$

$$=3x^3+6x^2+10x+10$$

76. $\left(x^4+16\right)\div(x+4)$

$$
\begin{array}{r|rrrrr}
-4 & 1 & 0 & 0 & 0 & 16 \\
& & -4 & 16 & -64 & 256 \\
\hline
& 1 & -4 & 16 & -64 & 272
\end{array}
$$

$$\left(x^4+16\right)\div(x+4)$$

$$=x^3-4x^2+16x-64+\frac{272}{x+4}$$

77. $7^{-2}=\dfrac{1}{7^2}=\dfrac{1}{49}$

78. $(-4)^{-3}=\dfrac{1}{(-4)^3}=\dfrac{1}{-64}=-\dfrac{1}{64}$

79. $2^{-1}+4^{-1}=\dfrac{1}{2}+\dfrac{1}{4}=\dfrac{3}{4}$

80. $\dfrac{1}{5^{-2}}=5^2=25$

81. $\left(\dfrac{2}{5}\right)^{-3}=\dfrac{2^{-3}}{5^{-3}}=\dfrac{5^3}{2^3}=\dfrac{125}{8}$

82. $\dfrac{x^3}{x^9}=x^{3-9}=x^{-6}=\dfrac{1}{x^6}$

83. $\dfrac{30y^6}{5y^8}=\dfrac{30}{5}\cdot\dfrac{y^6}{y^8}=6y^{-2}=\dfrac{6}{y^2}$

84. $\left(5x^{-7}\right)\left(6x^2\right)=(5\cdot6)\left(x^{-7+2}\right)$

$$=30x^{-5}=\dfrac{30}{x^5}$$

85. $\dfrac{x^4\cdot x^{-2}}{x^{-6}}=\dfrac{x^{4+(-2)}}{x^{-6}}=\dfrac{x^2}{x^{-6}}$

$$=x^{2-(-6)}=x^8$$

86. $\dfrac{\left(3y^3\right)^4}{y^{10}}=\dfrac{3^4y^{3(4)}}{y^{10}}=\dfrac{81y^{12}}{y^{10}}$

$$=81y^{12-10}=81y^2$$

87. $\dfrac{y^{-7}}{\left(y^4\right)^3} = \dfrac{y^{-7}}{y^{12}} = y^{-7-12} = y^{-19} = \dfrac{1}{y^{19}}$

88. $\left(2x^{-1}\right)^{-3} = 2^{-3}\left(x^{-1}\right)^{-3} = 2^{-3}x^3$

$\qquad = \dfrac{x^3}{2^3} = \dfrac{x^3}{8}$

89. $\left(\dfrac{x^7}{x^4}\right)^{-2} = \left(x^3\right)^{-2} = x^{-6} = \dfrac{1}{x^6}$

90. $\dfrac{\left(y^3\right)^4}{\left(y^{-2}\right)^4} = \dfrac{y^{12}}{y^{-8}} = y^{12-(-8)} = y^{20}$

91. $2.3 \times 10^4 = 23,000$

92. $1.76 \times 10^{-3} = 0.00176$

93. $9 \times 10^{-1} = 0.9$

94. $73,900,000 = 7.39 \times 10^7$

95. $0.00062 = 6.2 \times 10^{-4}$

96. $0.38 = 3.8 \times 10^{-1}$

97. $3.8 = 3.8 \times 10^0$

98. $\left(6 \times 10^{-3}\right)\left(1.5 \times 10^6\right) = 6(1.5) \times 10^{-3+6}$

$\qquad = 9 \times 10^3$

99. $\dfrac{2 \times 10^2}{4 \times 10^{-3}} = 0.5 \cdot 10^{2+3} = 0.5 \times 10^5$

$\qquad = 5 \times 10^{-1} \times 10^5$

$\qquad = 5.0 \times 10^4$

100. $\left(4 \times 10^{-2}\right)^2 = 4^2 \times 10^{-2(2)} = 16 \times 10^{-4}$

$\qquad = 1.6 \times 10^1 \times 10^{-4} = 1.6 \times 10^{1-4}$

$\qquad = 1.6 \times 10^{-3}$

101. $257 \times 10^9 = 2.57 \times 10^2 \times 10^9 = 2.57 \times 10^{11}$

102. $175 \times 10^6 = 1.75 \times 10^2 \times 10^6 = 1.75 \times 10^8$

103. $\dfrac{2.57 \times 10^{11}}{1.75 \times 10^8} = \dfrac{2.57}{1.75} \cdot \dfrac{10^{11}}{10^8} \approx 1.469 \times 10^3 = 1469$

Chapter 5 Test

1. $9x + 6x^2 - 4$ is a trinomial of degree 2.

2. $\left(7x^3 + 3x^2 - 5x - 11\right) + \left(6x^3 - 2x^2 + 4x - 13\right)$

$= \left(7x^3 + 6x^3\right) + \left(3x^2 - 2x^2\right) + \left(-5x + 4x\right) + \left(-11 - 13\right)$

$= 13x^3 + x^2 - x - 24$

3. $\left(9x^3 - 6x^2 - 11x - 4\right) - \left(4x^3 - 8x^2 - 13x + 5\right)$

$= \left(9x^3 - 6x^2 - 11x - 4\right) + \left(-4x^3 + 8x^2 + 13x - 5\right)$

$= \left(9x^3 - 4x^3\right) + \left(-6x^2 + 8x^2\right) + \left(-11x + 13x\right) + \left(-4 - 5\right)$

$= 5x^3 + 2x^2 + 2x - 9$

4.

x	$y = x^2 - 3$	(x, y)
-3	$y = (-3)^2 - 3 = 6$	$(-3, 6)$
-2	$y = (-2)^2 - 3 = 1$	$(-2, 1)$
-1	$y = (-1)^2 - 3 = -2$	$(-1, -2)$
0	$y = (0)^2 - 3 = -3$	$(0, -3)$
1	$y = (1)^2 - 3 = -2$	$(1, -2)$
2	$y = (2)^2 - 3 = 1$	$(2, 1)$
3	$y = (3)^2 - 3 = 6$	$(3, 6)$

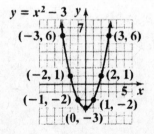

5. $\left(-7x^3\right)\left(5x^8\right) = (-7 \cdot 5)\left(x^{3+8}\right) = -35x^{11}$

6. $6x^2\left(8x^3 - 5x - 2\right)$

$= 6x^2\left(8x^3\right) + 6x^2\left(-5x\right) + 6x^2\left(-2\right)$

$= 48x^5 - 30x^3 - 12x^2$

7. $\left(3x + 2\right)\left(x^2 - 4x - 3\right)$

$= 3x\left(x^2 - 4x - 3\right) + 2\left(x^2 - 4x - 3\right)$

$= 3x^3 - 12x^2 - 9x + 2x^2 - 8x - 6$

$= 3x^3 - 10x^2 - 17x - 6$

8. $\left(3y + 7\right)\left(2y - 9\right)$

$= 6y^2 + 14y - 27y - 63$

$= 6y^2 - 13y - 63$

9. $\left(7x + 5\right)\left(7x - 5\right) = \left(7x\right)^2 - 5^2$

$= 49x^2 - 25$

10. $\left(x^2 + 3\right)^2 = \left(x^2\right)^2 + 2\left(x^2\right)\left(3\right) + 3^2$

$= x^4 + 6x^2 + 9$

11. $\left(5x - 3\right)^2 = \left(5x\right)^2 - 2\left(5x\right)\left(3\right) + 3^2$

$= 25x^2 - 30x + 9$

12. $4x^2 y + 5xy - 6x$

$= 4\left(-2\right)^2\left(3\right) + 5\left(-2\right)\left(3\right) - 6\left(-2\right)$

$= 4\left(4\right)\left(3\right) + 5\left(-2\right)\left(3\right) - 6\left(-2\right)$

$= 48 - 30 + 12 = 30$

13. $\left(8x^2 y^3 - xy + 2y^2\right) - \left(6x^2 y^3 - 4xy - 10y^2\right)$

$= \left(8x^2 y^3 - xy + 2y^2\right) + \left(-6x^2 y^3 + 4xy + 10y^2\right)$

$= \left(8x^2 y^3 - 6x^2 y^3\right) + \left(-xy + 4xy\right) + \left(2y^2 + 10y^2\right)$

$= 2x^2 y^3 + 3xy + 12y^2$

14. $\left(3a - 7b\right)\left(4a + 5b\right)$

$= \left(3a\right)\left(4a\right) + \left(3a\right)\left(5b\right) - \left(7b\right)\left(4a\right) - \left(7b\right)\left(5b\right)$

$= 12a^2 + 15ab - 28ab - 35b^2$

$= 12a^2 - 13ab - 35b^2$

15. $\left(2x + 3y\right)^2 = \left(2x\right)^2 + 2\left(2x\right)\left(3y\right) + \left(3y\right)^2$

$= 4x^2 + 12xy + 9y^2$

16. $\dfrac{-25x^{16}}{5x^4} = \dfrac{-25}{5} \cdot \dfrac{x^{16}}{x^4} = -5x^{16-4}$

$= -5x^{12}$

Check by multiplication:

$5x^4\left(-5x^{12}\right) = -25x^{4+12} = -25x^{16}$

17. $\dfrac{15x^4 - 10x^3 + 25x^2}{5x}$

$= \dfrac{15x^4}{5x} - \dfrac{10x^3}{5x} + \dfrac{25x^2}{5x}$

$= 3x^3 - 2x^2 + 5x$

Check by multiplication:

$5x\left(3x^3 - 2x^2 + 5x\right)$

$= 5x\left(3x^3\right) + 5x\left(-2x^2\right) + 5x\left(5x\right)$

$= 15x^4 - 10x^3 + 25x^2$

18.

$$\begin{array}{r} x^2 - 2x + 3 \\ 2x+1{\overline{\smash{\big)}\,2x^3 - 3x^2 + 4x + 4}} \\ \underline{2x^3 + \ x^2\phantom{{}+4x+4}} \\ -4x^2 + 4x \\ \underline{-4x^2 - 2x} \\ 6x + 4 \\ \underline{6x + 3} \\ 1 \end{array}$$

$\dfrac{2x^3 - 3x^2 + 4x + 4}{2x + 1} = x^2 - 2x + 3 + \dfrac{1}{2x + 1}$

Check by multiplication:

$\left(2x + 1\right)\left(x^2 - 2x + 3\right) + 1$

$= \left[2x\left(x^2 - 2x + 3\right) + 1\left(x^2 - 2x + 3\right)\right] + 1$

$= \left(2x^3 - 4x^2 + 6x + x^2 - 2x + 3\right) + 1$

$= \left(2x^3 - 3x^2 + 4x + 3\right) + 1$

$= 2x^3 - 3x^2 + 4x + 4$

19. $\left(3x^4 + 11x^3 - 20x^2 + 7x + 35\right) \div (x+5)$

$$\underline{-5|} \quad 3 \quad 11 \quad -20 \quad 7 \quad 35$$
$$\quad\quad\quad -15 \quad 20 \quad 0 \quad -35$$
$$\overline{\quad 3 \quad -4 \quad\; 0 \quad 7 \quad\; 0}$$

$\left(3x^4 + 11x^3 - 20x^2 + 7x + 35\right) \div (x+5)$

$= 3x^3 - 4x^2 + 7$

20. $10^{-2} = \dfrac{1}{10^2} = \dfrac{1}{100}$

21. $\dfrac{1}{4^{-3}} = 1 \cdot 4^3 = 4^3 = 64$

22. $\left(-3x^2\right)^3 = (-3)^3 \left(x^2\right)^3 = -27x^6$

23. $\dfrac{20x^3}{5x^8} = \dfrac{20}{5} \cdot \dfrac{x^3}{x^8} = 4x^{3-8} = 4x^{-5} = \dfrac{4}{x^5}$

24. $\left(-7x^{-8}\right)\left(3x^2\right) = -21x^{-8+2} = -\dfrac{21}{x^6}$

25. $\dfrac{\left(2y^3\right)^4}{y^8} = \dfrac{2^4\left(y^3\right)^4}{y^8} = \dfrac{16y^{12}}{y^8} = 16y^4$

26. $\left(5x^{-4}\right)^{-2} = 5^{-2}\left(x^{-4}\right)^{-2} = 5^{-2}x^8$

$= \dfrac{x^8}{5^2} = \dfrac{x^8}{25}$

27. $\left(\dfrac{x^{10}}{x^5}\right)^{-3} = \left(x^{10-5}\right)^{-3} = \left(x^5\right)^{-3}$

$= x^{-15} = \dfrac{1}{x^{15}}$

28. $3.7 \times 10^{-4} = 0.00037$

29. $7,600,000 = 7.6 \times 10^6$

30. $\left(4.1 \times 10^2\right)\left(3 \times 10^{-5}\right)$

$= (4.1 \cdot 3)\left(10^2 \cdot 10^{-5}\right)$

$= 12.3 \times 10^{-3}$

$= 1.23 \times 10^{-2}$

31. $\dfrac{8.4 \times 10^6}{4 \times 10^{-2}} = \dfrac{8.4}{4} \times \dfrac{10^6}{10^{-2}}$

$= 2.1 \times 10^{6-(-2)}$

$= 2.1 \times 10^8$

32. $A = (x+8)(x+2)$

$= x^2 + 2x + 8x + 16$

$= x^2 + 10x + 16$

Cumulative Review Exercises (Chapters 1-5)

1. $(-7)(-5) \div (12-3) = (-7)(-5) \div 9$

$= 35 \div 9 = \dfrac{35}{9}$

2. $(3-7)^2(9-11)^3 = (-4)^2(-2)^3$

$= 16(-8) = -128$

3. $14,300 - (-750) = 14,300 + 750$

$= 15,050$

The difference in elevation between the plane and the submarine is 15,050 feet.

4. $2(x+3) + 2x = x + 4$

$2x + 6 + 2x = x + 4$

$4x + 6 = x + 4$

$3x + 6 = 4$

$3x = -2$

$x = -\dfrac{2}{3}$

The solution set is $\left\{-\dfrac{2}{3}\right\}$.

5. $\dfrac{x}{5} - \dfrac{1}{3} = \dfrac{x}{10} - \dfrac{1}{2}$

To clear fractions, multiply by the LCD = 30.

$$30\left(\dfrac{x}{5} - \dfrac{1}{3}\right) = 30\left(\dfrac{x}{10} - \dfrac{1}{2}\right)$$

$$30\left(\dfrac{x}{5}\right) - 30\left(\dfrac{1}{3}\right) = 30\left(\dfrac{x}{10}\right) - 30\left(\dfrac{1}{2}\right)$$

$$6x - 10 = 3x - 15$$

$$3x - 10 = -15$$

$$3x = -5$$

$$x = -\dfrac{5}{3}$$

The solution set is $\left\{-\dfrac{5}{3}\right\}$.

6. Let x = width of sign.
Then $3x - 2$ = length of sign.

$$2x + 2(3x - 2) = 28$$

$$2x + 6x - 4 = 28$$

$$8x - 4 = 28$$

$$8x = 32$$

$$x = 4$$

$$3x - 2 = 3(4) - 2 = 10$$

The length of the sign is 10 feet and the width is 4 feet, so the dimensions are 10 feet by 4 feet.

7. $\qquad 7 - 8x \le -6x - 5$

$$7 - 8x + 6x \le -6x - 5 + 6x$$

$$-2x + 7 \le -5$$

$$-2x + 7 - 7 \le -5 - 7$$

$$-2x \le -12$$

$$\dfrac{-2x}{-2} \ge \dfrac{-12}{-2}$$

$$x \ge 6$$

$[6, \infty)$

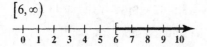

8. Let x = amount invested at 12%
$\qquad y$ = amount invested at 14%

	Amount invested	Interest rate	Interest earned
12%	x	0.12	$0.12x$
14%	y	0.14	$0.14y$

Form a system of equations based on the amount invested and on the amount of interest earned:

$x + y = 6000$

$0.12x + 0.14y = 772$

Solve the first equation for y and substitute the result into the second equation:

$$x + y = 6000$$

$$y = 6000 - x$$

$$0.12x + 0.14y = 772$$

$$0.12x + 0.14(6000 - x) = 772$$

$$0.12x + 840 - 0.14x = 772$$

$$-0.02x + 840 = 772$$

$$-0.02x = -68$$

$$x = 3400$$

Back-substitute 3400 for x into either of the original equations. We choose the first equation:

$$x + y = 6000$$

$$3400 + y = 6000$$

$$y = 2600$$

Thus, $3400 was invested at 12%, and $2600 was invested at 14%.

9. Let x = number of liters of 70% antifreeze.
$\qquad y$ = number of liters of 30% antifreeze.

	No. of liters	Percent antifreeze	Amount of antifreeze
70%	x	70% = 0.7	$0.7x$
30%	y	30% = 0.3	$0.3y$
60%	20	60% = 0.6	$0.6(20) = 12$

Form a system of equations based on the number of liters of mixture and on the amount of antifreeze:

$x + y = 20$

$0.7x + 0.3y = 12$

Solve the first equation for y and substitute the result into the second equation:

$$x + y = 20$$

$$y = 20 - x$$

$$0.7x + 0.3y = 12$$
$$0.7x + 0.3(20 - x) = 12$$
$$0.7x + 6 - 0.3x = 12$$
$$0.4x + 6 = 12$$
$$0.4x = 6$$
$$x = 15$$

Back-substitute 15 for x into one of the original equations. We choose the first equation:
$$x + y = 20$$
$$15 + y = 20$$
$$y = 5$$

Thus, 15 liters of 70% antifreeze and 5 liters of 30% antifreeze must be used.

10. $y = -\dfrac{2}{5}x + 2$

slope $= -\dfrac{2}{5} = \dfrac{-2}{5}$; y-intercept $= 2$

Plot $(0,2)$. Move 2 units *down* (since -2 is negative) and 5 units to the *right* to reach the point $(5,0)$. Draw a line through $(0,2)$ and $(5,0)$.

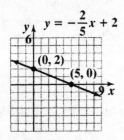

11. $x - 2y = 4$

x-intercept: 4
y-intercept: -2
checkpoint: $(-2, -3)$

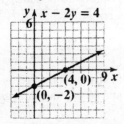

12. $m = \dfrac{y_2 - y_1}{x_2 - x_1} = \dfrac{-4 - 2}{2 - (-3)} = \dfrac{-6}{5} = -\dfrac{6}{5}$

Because the slope is negative, the line is falling.

13. $y - (-1) = -2(x - 3)$
$\quad y + 1 = -2(x - 3)$ point-slope form
$\quad y + 1 = -2x + 6$
$\quad\quad y = -2x + 5$ slope-intercept form

14. $3x + 2y = 10$
$\quad 4x - 3y = -15$

Multiply the first equation by 3 and the second equation by 2:
$$9x + 6y = \ \ 30$$
$$\underline{8x - 6y = -30}$$
$$17x \quad\ \ = \ \ 0$$
$$x = \ \ 0$$

Back-substitute $x = 0$ to find y:
$$3(0) + 2y = 10$$
$$2y = 10$$
$$y = 5$$

The solution set is $\{(0,5)\}$.

15. $2x + 3y = -6$
$\quad y = 3x - 13$

Substitute the second equation in for y in the first equation:
$$2x + 3y = -6$$
$$2x + 3(3x - 13) = -6$$
$$2x + 9x - 39 = -6$$
$$11x - 39 = -6$$
$$11x = 33$$
$$x = 3$$

Back-substitute $x = 3$ to find y:
$$y = 3x - 13 = 3(3) - 13 = -4$$

The solution set is $\{(3, -4)\}$.

16. Let y = total charge.

Let x = # of minutes.

Plan A: $y = 0.05x + 15$

Plan B: $y = 0.07x + 5$

To find when the plans are the same, substitute the second equation into the first equation:

$0.07x + 5 = 0.05x + 15$

$0.02x + 5 = 15$

$\quad 0.02x = 10$

$\qquad x = 500$

Back-substitute to find y:

$y = 0.07x + 5 = 0.07(500) + 5 = 40$

The plans will be the same for 500 minutes at $40 a plan.

17. $0.0024 = 2.4 \times 10^{-3}$

18. $\left(9x^5 - 3x^3 + 2x - 7\right) - \left(6x^5 + 3x^3 - 7x - 9\right)$

$= \left(9x^5 - 3x^3 + 2x - 7\right) + \left(-6x^5 - 3x^3 + 7x + 9\right)$

$= \left(9x^5 - 6x^5\right) + \left(-3x^3 - 3x^3\right) + \left(2x + 7x\right) + \left(-7 + 9\right)$

$= 3x^5 - 6x^3 + 9x + 2$

19.

$$
\begin{array}{r}
x^2 + 2x + 3 \\
x+1 \overline{\smash{\big)}\ x^3 + 3x^2 + 5x + 3} \\
\underline{x^3 +\ \ x^2} \\
2x^2 + 5x \\
\underline{2x^2 + 2x} \\
3x + 3 \\
\underline{3x + 3} \\
0
\end{array}
$$

$\dfrac{x^3 + 3x^2 + 5x + 3}{x + 1} = x^2 + 2x + 3$

20. $\dfrac{\left(3x^2\right)^4}{x^{10}} = \dfrac{3^4\left(x^2\right)^4}{x^{10}} = \dfrac{81x^8}{x^{10}}$

$\qquad = 81x^{8-10} = 81x^{-2} = \dfrac{81}{x^2}$

Chapter 6
Factoring Polynomials

6.1 Check Points

1. **a.** $18x^3 = 3x^2 \cdot 6x$

    $15x^2 = 3x^2 \cdot 5$

    The GCF is $3x^2$.

    b. $-20x^2 = 4x^2 \cdot (-5)$

    $12x^4 = 4x^2 \cdot 3x^2$

    $40x^3 = 4x^2 \cdot 10x$

    The GCF is $4x^2$.

    c. $x^4 y = x^2 y \cdot x^2$

    $x^3 y^2 = x^2 y \cdot xy$

    $x^2 y = x^2 y$

    The GCF is $x^2 y$.

2. $6x^2 + 18 = 6 \cdot x^2 + 6 \cdot 3 = 6(x^2 + 3)$

3. $25x^2 + 35x^3 = 5x^2 \cdot 5 + 5x^2 \cdot 7x = 5x^2(5 + 7x)$

4. $15x^5 + 12x^4 - 27x^3 = 3x^3 \cdot 5x^2 + 3x^3 \cdot 4x - 3x^3 \cdot 9$

    $\qquad\qquad\qquad = 3x^3(5x^2 + 4x - 9)$

5. $8x^3 y^2 - 14x^2 y + 2xy = 2xy \cdot 4x^2 y - 2xy \cdot 7x + 2xy \cdot 1$

    $\qquad\qquad\qquad = 2xy \cdot (4x^2 y - 7x + 1)$

6. **a.** $x^2 \overset{\text{GCF}}{(x+1)} + 7 \overset{\text{GCF}}{(x+1)} = (x+1)(x^2 + 7)$

    b. $x \overset{\text{GCF}}{(y+4)} - 7 \overset{\text{GCF}}{(y+4)} = (y+4)(x-7)$

7. $x^3 + 5x^2 + 2x + 10 = (x^3 + 5x^2) + (2x + 10)$

    $\qquad\qquad\qquad = x^2(x+5) + 2(x+5)$

    $\qquad\qquad\qquad = (x+5)(x^2 + 2)$

8. $xy + 3x - 5y - 15 = x(y+3) - 5(y+3)$

    $\qquad\qquad\qquad = (y+3)(x-5)$

6.1 Exercise Set

1. The GCF of 4 and $8x$ is 4.

3. The GCF of $12x^2$ and $8x$ is $4x$.

5. The GCF of $-2x^4$ and $6x^3$ is $2x^3$.

7. The GCF of $9y^5, 18y^2$, and $-3y$ is $3y$.

9. The GCF of xy, xy^2, and xy^3 is xy.

11. The GCF of $16x^5y^4, 8x^6y^3$, and $20x^4y^5$ is $4x^4y^3$.

13. $8x + 8 = 8 \cdot x + 8 \cdot 1$
$$= 8(x+1)$$

15. $4y - 4 = 4 \cdot y - 4 \cdot 1$
$$= 4(y-1)$$

17. $5x + 30 = 5 \cdot x + 5 \cdot 6$
$$= 5(x+6)$$

19. $30x - 12 = 6 \cdot 5x - 6 \cdot 2$
$$= 6(5x-2)$$

21. $x^2 + 5x = x \cdot x + x \cdot 5$
$$= x(x+5)$$

23. $18y^2 + 12 = 6 \cdot 3y^2 + 6 \cdot 2$
$$= 6(3y^2+2)$$

25. $14x^3 + 21x^2 = 7x^2 \cdot 2x + 7x^2 \cdot 3$
$$= 7x^2(2x+3)$$

27. $13y^2 - 25y = y \cdot 13y - y \cdot 25$
$$= y(13y-25)$$

29. $9y^4 + 27y^6 = 9y^4 \cdot 1 + 9y^4 \cdot 3y^2$
$$= 9y^4(1+3y^2)$$

31. $8x^2 - 4x^4 = 4x^2(2) - 4x^2(x^2)$
$$= 4x^2(2-x^2)$$

33. $12y^2 + 16y - 8 = 4\left(3y^2\right) + 4\left(4y\right) - 4\left(2\right)$

$$= 4\left(3y^2 + 4y - 2\right)$$

35. $9x^4 + 18x^3 + 6x^2$

$$= 3x^2\left(3x^2\right) + 3x^2\left(6x\right) + 3x^2\left(2\right)$$

$$= 3x^2\left(3x^2 + 6x + 2\right)$$

37. $100y^5 - 50y^3 + 100y^2$

$$= 50y^2\left(2y^3\right) - 50y^2\left(y\right) + 50y^2\left(2\right)$$

$$= 50y^2\left(2y^3 - y + 2\right)$$

39. $10x - 20x^2 + 5x^3$

$$= 5x\left(2\right) - 5x\left(4x\right) + 5x\left(x^2\right)$$

$$= 5x\left(2 - 4x + x^2\right)$$

41. $11x^2 - 23$ cannot be factored because the two terms have no common factor other than 1.

43. $6x^3y^2 + 9xy = 3xy\left(2x^2 + y\right) + 3xy\left(3\right)$

$$= 3xy\left(2x^2y + 3\right)$$

45. $30x^2y^2 - 10xy^2 + 20xy$

$$= 10xy\left(3xy^2\right) - 10xy\left(y\right) + 10xy\left(2\right)$$

$$= 10xy\left(3xy^2 - y + 2\right)$$

47. $32x^3y^2 - 24x^3y - 16x^2y$

$$= 8x^2y\left(4xy\right) - 8x^2y\left(3x\right) - 8x^2y\left(2\right)$$

$$= 8x^2y\left(4xy - 3x - 2\right)$$

49. $x\left(x+5\right) + 3\left(x+5\right) = \left(x+5\right)\left(x+3\right)$

51. $x\left(x+2\right) - 4\left(x+2\right) = \left(x+2\right)\left(x-4\right)$

53. $x\left(y+6\right) - 7\left(y+6\right) = \left(y+6\right)\left(x-7\right)$

55. $3x\left(x+y\right) - \left(x+y\right)$

$$= 3x\left(x+y\right) - 1\left(x+y\right)$$

$$= \left(x+y\right)\left(3x-1\right)$$

57. $4x(3x+1)+3x+1$

$\quad = 4x(3x+1)+1(3x+1)$

$\quad = (3x+1)(4x+1)$

59. $7x^2(5x+4)+5x+4$

$\quad = 7x^2(5x+4)+1(5x+4)$

$\quad = (5x+4)(7x^2+1)$

61. $x^2+2x+4x+8 = (x^2+2x)+(4x+8)$

$\qquad\qquad\qquad\ = x(x+2)+4(x+2)$

$\qquad\qquad\qquad\ = (x+2)(x+4)$

63. $x^2+3x-5x-15 = (x^2+3x)+(-5x-15)$

$\qquad\qquad\qquad\quad\ = x(x+3)-5(x+3)$

$\qquad\qquad\qquad\quad\ = (x+3)(x-5)$

65. $x^3-2x^2+5x-10$

$\quad = (x^3-2x^2)+(5x-10)$

$\quad = x^2(x-2)+5(x-2)$

$\quad = (x-2)(x^2+5)$

67. $x^3-x^2+2x-2 = x^2(x-1)+2(x-1)$

$\qquad\qquad\qquad\ = (x-1)(x^2+2)$

69. $xy+5x+9y+45 = x(y+5)+9(y+5)$

$\qquad\qquad\qquad\quad = (y+5)(x+9)$

71. $xy-x+5y-5 = x(y-1)+5(y-1)$

$\qquad\qquad\qquad\ = (y-1)(x+5)$

73. $3x^2-6xy+5xy-10y^2$

$\quad = 3x(x-2y)+5y(x-2y)$

$\quad = (x-2y)(3x+5y)$

75. $3x^3-2x^2-6x+4$

$\quad = x^2(3x-2)-2(3x-2)$

$\quad = (3x-2)(x^2-2)$

77. $x^2 - ax - bx + ab = x(x-a) - b(x-a)$
$$= (x-a)(x-b)$$

79. $24x^3y^3z^3 + 30x^2y^2z + 18x^2yz^2$
$$= 6x^2yz(4xy^2z^2) + 6x^2yz(5y) + 6x^2yz(3z)$$
$$= 6x^2yz(4xy^2z^2 + 5y + 3z)$$

81. $x^3 - 4 + 3x^3y - 12y = 1(x^3 - 4) + 3y(x^3 - 4)$
$$= (x^3 - 4)(1 + 3y)$$

83. $4x^5(x+1) - 6x^3(x+1) - 8x^2(x+1)$
$$= 2x^2(x+1) \cdot 2x^3 - 2x^2(x+1) \cdot 3x - 2x^2(x+1) \cdot 4$$
$$= 2x^2(x+1)(2x^3 - 3x - 4)$$

85. $3x^5 - 3x^4 + x^3 - x^2 + 5x - 5$
$$= (3x^5 - 3x^4) + (x^3 - x^2) + (5x - 5)$$
$$= 3x^4(x-1) + x^2(x-1) + 5(x-1)$$
$$= (x-1)(3x^4 + x^2 + 5)$$

87. The area of the square is $6x \cdot 6x = 36x^2$. The area of the circle is $\pi(2x)^2 = \pi \cdot 4x^2 = 4\pi x^2$. So the shaded area is the area of the square minus the area of the circle, which is $36x^2 - 4\pi x^2 = 4x^2(9 - \pi)$.

89. a. Use the formula, $64x - 16x^2$, for the height of the debris above the ground. Substitute 3 for x.
$$64x - 16x^2 = 64(3) - 16(3)^2$$
$$= 192 - 16(9) = 192 - 144 = 48$$
Therefore, the height of the debris after 3 seconds is 48 feet.

b. $64x - 16x^2 = 16x(4 - x)$

c. Substitute 3 for x in the factored polynomial.
$16 \cdot 3(4 - 3) = 48(1) = 48$
You do get the same answer as in part (a) but this does not prove your factorization is correct.

91. Use the formula for the area of a rectangle, $A = l \cdot w$. Substitute $5x^4 - 10x$ for A and $5x$ for w.
$$A = l \cdot w$$
$$5x^4 - 10x = l(5x)$$
$$\frac{5x^4 - 10x}{5x} = \frac{l(5x)}{5x}$$
$$\frac{5x(x^3 - 2)}{5x} = l$$
$$x^3 - 2 = l$$
The length, l, is $(x^3 - 2)$ units.

93. – 97. Answers will vary.

99. makes sense

101. does not make sense; Explanations will vary. Sample explanation: The power must be the greatest that is *common* for this variable.

103. false; Changes to make the statement true will vary. A sample change is: Since $\frac{3x}{3x} = 1$, it is necessary to write the 1.

105. false; Changes to make the statement true will vary. A sample change is: $a^2 + b^2$ is not factorable.

107. The polynomial will be
$$x + (x + 100) + (x + 200) + (x + 300)$$
$$= 4x + 600$$
$$= 4(x + 150)$$

109. Answers will vary. One example is $5x^2 + 10x - 4x - 8$.

111. The graphs do not coincide.

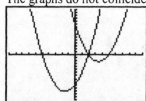

Factor by grouping.
$$x^2 - 2x + 5x - 10 = x(x - 2) + 5(x - 2)$$
$$= (x - 2)(x + 5)$$
Change the expression on the right side to $(x - 2)(x + 5)$.

113. $(x+7)(x+10) = x^2 + 10x + 7x + 70$
$$= x^2 + 17x + 70$$

114. $2x - y = -4$

$x - 3y = 3$

Graph both equations on the same axes.

$2x - y = -4$: x-intercept: -2; y-intercept: 4

$x - 3y = 3$: x-intercept: 3; y-intercept: -1

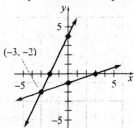

The lines intersect as $(-3, -2)$.

The solution set is $\{(-3, -2)\}$.

115. First, find the slope $m = \dfrac{5-2}{-4-(-7)} = \dfrac{3}{3} = 1$

Write the point-slope equation using
$m = 1$ and $(x_1, y_1) = (-7, 2)$.

$y - y_1 = m(x - x_1)$

$y - 2 = 1\left[x - (-7)\right]$

$y - 2 = 1(x + 7)$

Now rewrite this equation in slope-intercept form.

$y - 2 = x + 7$

$y = x + 9$

Note: If $(-4, 5)$ is used as $(x_1 y_1)$, the point-slope
equation will be

$y - 5 = 1\left[x - (-4)\right]$

$y - 5 = x + 4$

This also leads to the slope-intercept equation
$y = x + 9$.

116. $2 \times 4 = 8$ and $2 + 4 = 6$

117. $(-3)(-2) = 6$ and $(-3) + (-2) = -5$

118. $(-5)(7) = -35$ and $(-5) + 7 = 2$

6.2 Check Points

1. $x^2 + 5x + 6$

Factors of 6	6,1	−6,−1	2,3	−2,−3
Sum of Factors	7	−7	5	−5

The factors of 6 whose sum is 5 are 2 and 3.

Thus, $x^2 + 5x + 6 = (x + 2)(x + 3)$.

Check:

$(x + 2)(x + 3) = x^2 + 3x + 2x + 6$
$$= x^2 + 5x + 6$$

2. $x^2 - 6x + 8$

Factors of 8	8,1	−8,−1	2,4	−2,−4
Sum of Factors	9	−9	6	−6

The factors of 8 whose sum is −6 are −2 and −4.

Thus, $x^2 - 6x + 8 = (x - 2)(x - 4)$.

Check:

$(x - 2)(x - 4) = x^2 - 4x - 2x + 8$
$$= x^2 - 6x + 8$$

3. $x^2 + 3x - 10$

Factors of −10	−10,1	10,−1	−5,2	5,−2
Sum of Factors	−9	9	−3	3

The factors of −10 whose sum is 3 are 5 and −2.

Thus, $x^2 + 3x - 10 = (x + 5)(x - 2)$.

Check:

$(x + 5)(x - 2) = x^2 - 2x + 5x - 10$
$$= x^2 + 3x - 10$$

4. The factors of −27 whose sum is −6 are −9 and 3.

Thus, $y^2 - 6y - 27 = (y - 9)(y + 3)$.

5. No factor pair of −7 has a sum of 1.

Thus, $x^2 + x - 7$ is prime.

6. The factors of 3 whose sum is −4 are −3 and −1.

Thus, $x^2 - 4xy + 3y^2 = (x - 3y)(x - y)$.

7. First factor out the common factor of $2x$.

$2x^3 + 6x^2 - 56x = 2x(x^2 + 6x - 28)$

Continue by factoring the trinomial.

$2x^3 + 6x^2 - 56x = 2x(x^2 + 6x - 28)$
$$= 2x(x - 4)(x + 7)$$

6.2 Exercise Set

1. $x^2 + 7x + 6$

Factors of 6	6,1	−6,−1	3,2	−3,−2
Sum of Factors	7	−7	5	−5

The factors of 6 whose sum is 7 are 6 and 1.

Thus, $x^2 + 7x + 6 = (x+6)(x+1)$.

Check:

$$(x+6)(x+1) = x^2 + 1x + 6x + 6$$
$$= x^2 + 7x + 6$$

3. $x^2 + 7x + 10 = (x+5)(x+2)$

$5(2) = 10;\ 5+2 = 7$

5. $x^2 + 11x + 10 = (x+10)(x+1)$

$10(1) = 10;\ 10+1 = 11$

7. $x^2 - 7x + 12 = (x-4)(x-3)$

$-4(-3) = 12;\ -4 + -3 = -7$

9. $x^2 - 12x + 36 = (x-6)(x-6)$

$-6(-6) = 36;\ -6 + -6 = -12$

11. $y^2 - 8y + 15 = (y-5)(y-3)$

$-5(-3) = 15;\ -5 + -3 = -8$

13. $x^2 + 3x - 10 = (x+5)(x-2)$

$5(-2) = -10;\ 5 + -2 = 3$

15. $y^2 + 10y - 39 = (y+13)(y-3)$

$(13)(-3) = -39;\ 13 + -3 = 10$

17. $x^2 - 2x - 15 = (x-5)(x+3)$

$(-5)(3) = -15;\ -5 + 3 = -2$

19. $x^2 - 2x - 8 = (x-4)(x+2)$

$(-4)(2) = -8;\ -4 + 2 = -2$

21. $x^2 + 4x + 12$ is prime because there is no pair of integers whose product is 12 and whose sum is 4.

23. $y^2 - 16y + 48 = (y-4)(y-12)$

$(-4)(-12) = 48;\ -4 + -12 = -16$

25. $x^2 - 3x + 6$ is prime because there is no pair of integers whose product is 6 and whose sum is −3.

27. $w^2 - 30w - 64 = (w-32)(w+2)$

$(-32)(2) = -64;\ -32 + 2 = -30$

29. $y^2 - 18y + 65 = (y-5)(y-13)$

$(-5)(-13) = 65;\ -5 + -13 = -18$

31. $r^2 + 12r + 27 = (r+3)(r+9)$

$(3)(9) = 27;\ 3+9 = 12$

33. $y^2 - 7y + 5$ is prime because there is no pair of integers whose product is 5 and whose sum is −7.

35. $x^2 + 7xy + 6y^2 = (x+6y)(x+y)$

$(6)(1) = 6;\ 6+1 = 7$

37. $x^2 - 8xy + 15y^2 = (x-3y)(x-5y)$

$(-3)(-5) = 15;\ -3 + -5 = -8$

39. $x^2 - 3xy - 18y^2 = (x-6y)(x+3y)$

$(-6)(3) = -18;\ -6 + 3 = -3$

41. $a^2 - 18ab + 45b^2 = (a-15b)(a-3b)$

$(-15)(-3) = 45;\ -15 + -3 = -18$

43. $3x^2 + 15x + 18$

First factor out the GCF, 3. Then factor the resulting binomial.

$$3x^2 + 15x + 18 = 3(x^2 + 5x + 6)$$
$$= 3(x+2)(x+3)$$

45. $4y^2 - 4y - 8 = 4(y^2 - y - 2)$
$$= 4(y-2)(y+1)$$

47. $10x^2 - 40x - 600 = 10(x^2 - 4x - 60)$
$$= 10(x-10)(x+6)$$

49. $3x^2 - 33x + 54 = 3(x^2 - 11x + 18)$
$$3(x-2)(x-9)$$

51. $2r^3 + 6r^2 + 4r = 2r\left(r^2 + 3r + 2\right)$
$$= 2r(r+2)(r+1)$$

53. $4x^3 + 12x^2 - 72x = 4x\left(x^2 + 3x - 18\right)$
$$= 4x(x+6)(x-3)$$

55. $2r^3 + 8r^2 - 64r = 2r\left(r^2 + 4r - 32\right)$
$$= 2r(r+8)(r-4)$$

57. $y^4 + 2y^3 - 80y^2 = y^2\left(y^2 + 2y - 80\right)$
$$= y^2(y+10)(y-8)$$

59. $x^4 - 3x^3 - 10x^2 = x^2\left(x^2 - 3x - 10\right)$
$$= x^2(x-5)(x+2)$$

61. $2w^4 - 26w^3 - 96w^2$
$$= 2w^2\left(w^2 - 13w - 48\right)$$
$$= 2w^2(w-16)(w+3)$$

63. $15xy^2 + 45xy - 60x = 15x\left(y^2 + 3y - 4\right)$
$$= 15x(y+4)(y-1)$$

65. $x^5 + 3x^4y - 4x^3y^2 = x^3\left(x^2 + 3xy - 4y^2\right)$
$$= x^3(x+4y)(x-y)$$

67. $2x^2y^2 - 32x^2yz + 30x^2z^2$
$$= 2x^2\left(y^2 - 16yz + 15z^2\right)$$
$$= 2x^2(y-15z)(y-z)$$

69. $(a+b)x^2 + (a+b)x - 20(a+b)$
$$= (a+b)\left(x^2 + x - 20\right)$$
$$= (a+b)(x+5)(x-4)$$

71. $x^2 + 0.5x + 0.06 = (x+0.2)(x+0.3)$
$$0.2(0.3) = 0.06; \quad 0.2 + 0.3 = 0.5$$

73. $x^2 - \dfrac{2}{5}x + \dfrac{1}{25} = \left(x - \dfrac{1}{5}\right)\left(x - \dfrac{1}{5}\right)$
$$\frac{1}{5}\left(\frac{1}{5}\right) = \frac{1}{25}; \quad -\frac{1}{5} + -\frac{1}{5} = -\frac{2}{5}$$

75. $-x^2 - 3x + 40 = -\left(x^2 + 3x - 40\right)$
$$= -(x+8)(x-5)$$
$$8(-5) = -40; \quad 8 + -5 = 3$$

77. a. $-16t^2 + 16t + 32 = -16\left(t^2 - t - 2\right)$
$$= -16(t-2)(t+1)$$

b. Substitute 2 for t in the original polynomial:
$$-16t^2 + 16t + 32 = -16(2)^2 + 16(2) + 32$$
$$= -16(4) + 32 + 32$$
$$= -64 + 64$$
$$= 0$$
Substitute 2 for t in the factored polynomial:
$$-16(t-2)(t+1) = -16(2-2)(2+1)$$
$$= -16(0)(3)$$
$$= 0$$
The answers are the same.
This answer means that after 2 seconds you hit the water.

79. – 81. Answers will vary.

83. does not make sense; Explanations will vary. Sample explanation: If the order of the terms is switched, the factors of an expression will not change.

85. does not make sense; Explanations will vary. Sample explanation: $x^2 + x + 1$ is prime.

87. false; Changes to make the statement true will vary. A sample change is: $x^2 + x + 20$ is prime.

89. true

91. In order for $x^2 + bx + 15$ to be factorable, b must be the sum of two integers that are positive factors of 15. The only positive factor pairs for 15 are $3 \cdot 5$ and $1 \cdot 15$. Since $3 + 5 = 8, 1 + 15 = 16$, the possible values of b are 8 and 16.

93. Answers will vary. An example is $x^2 + 14x + 2$.

95. $x^3 + 3x^2 + 2x = x\left(x^2 + 3x + 2\right)$

$\qquad = x(x+1)(x+2)$

The trinomial represents the product of three consecutive integers.

97. The graphs coincide.

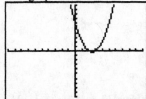

This verifies the factorization
$x^2 - 5x + 6 = (x-2)(x-3).$

99. The graphs do not coincide.

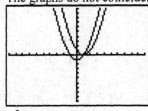

$x^2 - 2x + 1 = (x-1)(x-1)$

Change the polynomial on the right to $(x-1)(x-1).$

101. $4(x-2) = 3x + 5$

$\qquad 4x - 8 = 3x + 5$

$\qquad x - 8 = 5$

$\qquad x = 13$

The solution set is $\{13\}$.

102. $6x - 5y = 30$

Find the y-intercept

$\qquad 6x - 5y = 30$

$\qquad 6(0) - 5y = 30$

$\qquad -5y = 30$

$\qquad y = -6$

Find the x-intercept

$\qquad 6x - 5y = 30$

$\qquad 6x - 5(0) = 30$

$\qquad 6x = 30$

$\qquad x = 5$

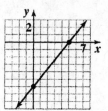

103 $y = -\dfrac{1}{2}x + 2$

The y-intercept is 2. Find an additional point by using the slope. From the y-intercept, move down one unit and to the right 2 units.
Draw the line through these points.

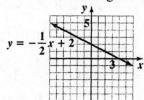

104. $(2x+3)(x-2) = 2x^2 - 4x + 3x - 6$

$\qquad = 2x^2 - x - 6$

105. $(3x+4)(3x+1) = 9x^2 + 3x + 12x + 4$

$\qquad = 9x^2 + 15x + 4$

106. $8x^2 - 2x - 20x + 5 = 2x(4x-1) - 5(4x-1)$

$\qquad = (4x-1)(2x-5)$

6.3 Check Points

1. Factor $5x^2 - 14x + 8$ by trial and error.

Step 1 $5x^2 - 14x + 8 = (5x \quad)(x \quad)$

Step 2 The number 8 has pairs of factors that are either both positive or both negative. Because the middle term, $-14x$, is negative, both factors must be negative.

Step 3

Possible Factors of $5x^2 - 14x + 8$	Sum of Outside and Inside Products
	required middle term
$(5x-4)(x-2)$	$-10x - 4x = -14x$
$(5x-2)(x-4)$	$-20x - 2x = -22x$
$(5x-1)(x-8)$	$-40x - x = -41x$
$(5x-8)(x-1)$	$-5x - 8x = -13x$

Check:

$$(5x-4)(x-2) = 5x^2 - 10x - 4x + 8$$
$$= 5x^2 - 14x + 8$$

Thus, $5x^2 - 14x + 8 = (5x-4)(x-2)$.

2. Factor $6x^2 + 19x - 7$ by trial and error.

Step 1 Find two First terms whose product is $6x^2$.

$$6x^2 + 19x - 7 = (6x \quad)(x \quad)$$

$$6x^2 + 19x - 7 = (3x \quad)(2x \quad)$$

Step 2 The last term, -7, has possible factorizations of $1(-7)$ and $-1(7)$.

Step 3

Possible Factors of $6x^2 + 19x - 7$	Sum of Outside and Inside Products
$(6x+1)(x-7)$	$-42x + x = -41x$
$(6x-7)(x+1)$	$6x - 7x = -x$
$(6x-1)(x+7)$	$42x - x = 41x$
$(6x+7)(x-1)$	$-6x + 7x = x$
$(3x+1)(2x-7)$	$-21x + 2x = -19x$
$(3x-7)(2x+1)$	$3x - 14x = -11x$
	required middle term
$(3x-1)(2x+7)$	$21x - 2x = 19x$
$(3x+7)(2x-1)$	$-3x + 14x = 11x$

Check:

$$(3x-1)(2x+7) = 6x^2 + 21x - 2x - 7$$
$$= 6x^2 + 19x - 7$$

Thus, $6x^2 + 19x - 7 = (3x-1)(2x+7)$

3. Factor $3x^2 - 13xy + 4y^2$ by trial and error.

Step 1 Find two First terms whose product is $2x^2$.

$$3x^2 - 13xy + 4y^2 = (3x \quad)(x \quad)$$

Step 2 The last term, $4y^2$, has pairs of factors that are either both positive or both negative. Because the middle term, $-13xy$, is negative, both factors must be negative. Thus the last term has possible factorizations of $-2y(-2y)$ or $-y(-4y)$.

Step 3

Possible Factors of $3x^2 - 13xy + 4y^2$	Sum of Outside and Inside Products
$(3x-4y)(x-y)$	$-3xy - 4xy = -7xy$
	required middle term
$(3x-y)(x-4y)$	$-12xy - xy = -13xy$
$(3x-2y)(x-2y)$	$-6xy - 2xy = -8xy$

Check:

$$(3x - y)(x - 4y) = 3x^2 - 12xy - xy + 4y^2$$
$$= 3x^2 - 13xy + 4y^2$$

Thus, $3x^2 - 13xy + 4y^2 = (3x - y)(x - 4y)$

4. Factor $3x^2 - x - 10$ by grouping.

$a = 3$ and $c = -10$, so $ac = 3(-10) = -30$.

The factors of -30 whose sum is -1 are 5 and -6.

$3x^2 - x - 10 = 3x^2 + 5x - 6x - 10$
$$= x(3x + 5) - 2(3x + 5)$$
$$= (3x + 5)(x - 2)$$

5. Factor $8x^2 - 10x + 3$ by grouping.

$a = 8$ and $c = 3$, so $ac = 8(3) = 24$.

The factors of 24 whose sum is -10 are -6 and -4.

$8x^2 - 10x + 3 = 8x^2 - 4x - 6x + 3$
$$= 4x(2x - 1) - 3(2x - 1)$$
$$= (2x - 1)(4x - 3)$$

6. First factor out the GCF.

$5y^4 + 13y^3 + 6y^2 = y^2(5y^2 + 13y + 6)$

Then factor the resulting trinomial.

$5y^4 + 13y^3 + 6y^2 = y^2(5y^2 + 13y + 6)$
$$= y^2(5y + 3)(y + 2)$$

6.3 Exercise Set

1. Factor $2x^2 + 5x + 3$ by trial and error.

   Step 1 $2x^2 + 5x + 3 = (2x\quad)(x\quad)$

   Step 2 The number 3 has pairs of factors that are either both positive or both negative. Because the middle term, $5x$, is positive, both factors must be positive. The only positive factorization is $(1)(3)$.

   Step 3

Possible Factors of $2x^2 + 5x + 3$	Sum of Outside and Inside Products
$(2x+1)(x+3)$	$6x + x = 7x$
$(2x+3)(x+1)$	$2x + 3x = 5x$

   Check:

   $$(2x+3)(x+1) = 2x^2 + 2x + 3x + 3$$
   $$= 2x^2 + 5x + 3$$

   Thus, $2x^2 + 5x + 3 = (2x+3)(x+1)$.

3. Factor $3x^2 + 13x + 4$ by trial and error. The only possibility for the first terms is
   $$(3x)(x) = 3x^2.$$
   Because the middle term is positive and the last term is also positive, the possible factorizations of 4 are $(1)(4)$ and $(2)(2)$.

Possible Factors of $3x^2 + 13x + 4$	Sum of Outside and Inside Products
$(3x+1)(x+4)$	$12x + x = 13x$
$(3x+4)(x+1)$	$3x + 4x = 7x$
$(3x+2)(x+2)$	$6x + 2x = 8x$

   Thus, $3x^2 + 13x + 4 = (3x+1)(x+4)$.

5. Factor $2x^2 + 11x + 12$ by grouping.
   $a = 2$ and $c = 12$, so $ac = 2(12) = 24$.
   The factors of 24 whose sum is 11 are 8 and 3.
   $$2x^2 + 11x + 12 = 2x^2 + 8x + 3x + 12$$
   $$= 2x(x+4) + 3(x+4)$$
   $$= (x+4)(2x+3)$$

7. Factor $5y^2 - 16y + 3$ by trial and error. The first terms must be $5y$ and y. Because the middle term is negative, the factors of 3 must be -3 and -1.

Possible Factors of $5y^2 - 16y + 3$	Sum of Outside and Inside Products
$(5y-1)(y-3)$	$-15y - y = -16y$
$(5y-3)(y-1)$	$-5y - 3y = -8y$

   Thus, $y^2 - 16y + 3 = (5y-1)(y-3)$.

9. Factor $3y^2 + y - 4$ by trial and error.
   $$(3y+1)(y-4) = 3y^2 - 11y - 4$$
   $$(3y-1)(y+4) = 3y^2 + 11y - 4$$
   $$(3y+4)(y-1) = 3y^2 + y - 4$$
   $$(3y-4)(y+1) = 3y^2 - y - 4$$
   $$(3y+2)(y-2) = 3y^2 - 4y - 4$$
   $$(3y-2)(y+2) = 3y^2 + 4y - 4$$
   Thus, $3y^2 + y - 4 = (3y+4)(y-1)$.

11. Factor $3x^2 + 13x - 10$ by grouping.
    $a = 3$ and $c = -10$, so $ac = -30$.
    The factors of -30 whose sum is 13 are 15 and -2.
    $$3x^2 + 13x - 10 = 3x^2 + 15x - 2x - 10$$
    $$= 3x(x+5) - 2(x+5)$$
    $$= (x+5)(3x-2)$$

13. Factor $3x^2 - 22x + 7$ by trial and error.
    $$(3x-7)(x-1) = 3x^2 - 10x + 7$$
    $$(3x-1)(x-7) = 3x^2 - 22x + 7$$
    Thus, $3x^2 - 22x + 7 = (3x-1)(x-7)$.

15. Factor $5y^2 - 16y + 3$ by trial and error.
    $$(5y-3)(y-1) = 5y^2 - 8y + 3$$
    $$(5y-1)(y-3) = 5y^2 - 16y + 3$$
    Thus, $5y^2 - 16y + 3 = (5y-1)(y-3)$.

17. Factor $3x^2 - 17x + 10$ by grouping.

$a = 3$ and $c = 10$, so $ac = 30$.

The factors of 30 whose sum is -17 are -15 and -2.

$3x^2 - 17x + 10 = 3x^2 - 15x - 2x + 10$
$$= 3x(x-5) - 2(x-5)$$
$$= (x-5)(3x-2)$$

19. Factor $6w^2 - 11w + 4$ by grouping.

$a = 6$ and $c = 4$, so $ac = 24$.

The factors of 24 whose sum is -11 are -3 and -8.

$6w^2 - 11w + 4 = 6w^2 - 3w - 8w + 4$
$$= 3w(2w-1) - 4(2w-1)$$
$$= (2w-1)(3w-4)$$

21. Factor $8x^2 + 33x + 4$ by grouping.

$a = 8$ and $c = 4$, so $ac = 32$.

The factors of 32 whose sum is 33 are 32 and 1.

$8x^2 + 33x + 4 = 8x^2 + 32x + x + 4$
$$= 8x(x+4) + 1(x+4)$$
$$= (x+4)(8x+1)$$

23. Factor $5x^2 + 33x - 14$ by trial and error.

$(5x-7)(x+2) = 5x^2 + 3x - 14$

$(5x+7)(x-2) = 5x^2 - 3x - 14$

$(5x-2)(x+7) = 5x^2 + 33x - 14$

Because the correct factorization has been found, there is no need to try additional possibilities.

Thus, $5x^2 + 33x - 14 = (5x-2)(x+7)$.

25. Factor $14y^2 + 15y - 9$ by trial and error. The sign in one factor must be positive and the other negative.

$(7y+9)(2y-1) = 14y^2 + 11y - 9$

$(7y+1)(2y-9) = 14y^2 - 61y - 9$

$(7y+3)(2y-3) = 14y^2 - 15y - 9$

$(7y-3)(2y+3) = 14y^2 + 15y - 9$

Thus, $14y^2 + 15y - 9 = (7y-3)(2y+3)$.

27. Factor $6x^2 - 7x + 3$ by trial and error. List all the possibilities in which both signs are negative.

$(6x-1)(x-3) = 6x^2 - 19x + 3$

$(6x-3)(x-1) = 6x^2 - 9x + 3$

$(3x-1)(2x-3) = 6x^2 - 11x + 3$

$(3x-3)(2x-1) = 6x^2 - 9x + 3$

None of these possibilities gives the required middle term, $-7x$, and there are no more possibilities to try, so $6x^2 - 7x + 3$ is prime.

29. Factor $25z^2 - 30z + 9$ by trial and error until the correct factorization is obtained. The signs in both factors must be negative.

$(5z-1)(5z-9) = 25z^2 - 50z + 9$

$(5z-3)(5z-3) = 25z^2 - 30z + 9$

Thus, $25z^2 - 30z + 9 = (5z-3)(5z-3)$.

31. Factor $15y^2 - y - 2$ by grouping.

$a = 15$ and $c = -2$, so $ac = -30$.

The factors of -30 whose sum is -1 are -6 and 5.

$15y^2 - y - 2 = 15y^2 - 6y + 5y - 2$
$$= 3y(5y-2) + 1(5y-2)$$
$$= (5y-2)(3y+1)$$

33. Factor $5x^2 + 2x + 9$ by trial and error. The signs in both factors must be positive.

$(5x+3)(x+3) = 5x^2 + 18x + 9$

$(5x+9)(x+1) = 5x^2 + 14x + 9$

$(5x+1)(x+9) = 5x^2 + 46x + 9$

None of these possibilities gives the required middle term, $2x$, and there are no more possibilities to try, so $5x^2 + 2x + 9$ is prime.

35. Factor $10y^2 + 43y - 9$ by grouping.

$a = 10$ and $c = -9$ so $ac = -90$.

The factors of -90 whose sum is 43 are 45 and -2.

$10y^2 + 43y - 9 = 10y^2 + 45y - 2y - 9$
$$= 5y(2y+9) - 1(2y+9)$$
$$= (2y+9)(5y-1)$$

37. Factor $8x^2 - 2x - 1$ by trial and error until the correct factorization is obtained. The sign must be negative in one factor and positive in the other.

$(4x - 1)(2x + 1) = 8x^2 + 2x - 1$

$(4x + 1)(2x - 1) = 8x^2 - 2x - 1$

Thus, $8x^2 - 2x - 1 = (4x + 1)(2x - 1)$.

39. Factor $9y^2 - 9y + 2$ by grouping.

$a = 9$ and $c = 2$, so $ac = 18$.

The factors of 18 whose sum is -9 are -3 and -6.

$9y^2 - 9y + 2 = 9y^2 - 3y - 6y + 2$
$= 3y(3y - 1) - 2(3y - 1)$
$= (3y - 1)(3y - 2)$

41. Factor $20x^2 + 27x - 8$ by grouping.

$a = 20$ and $c = -8$, so $ac = -160$.

The factors of -160 whose sum is 27 are -5 and 32.

$20x^2 + 27x - 8 = 20x^2 - 5x + 32x - 8$
$= 5x(4x - 1) + 8(4x - 1)$
$= (4x - 1)(5x + 8)$

43. $2x^2 + 3xy + y^2 = (2x + y)(x + y)$

45. Factor $3x^2 + 5xy + 2y^2$ by trial and error.

$(3x + y)(x + 2y) = 3x^2 + 7xy + 2y^2$

$(3x + 2y)(x + y) = 3x^2 + 5xy + 2y^2$

Thus, $3x^2 + 5xy + 2y^2 = (3x + 2y)(x + y)$.

47. Factor $2x^2 - 9xy + 9y^2$ by trial and error until the correct factorization is obtained. The signs in both factors must be negative.

$(2x - 9y)(x - y) = 2x^2 - 11xy + 9y^2$

$(2x + 9y)(x + y) = 2x^2 + 11xy + 9y^2$

$(2x - 3y)(x - 3y) = 2x^2 - 9xy + 9y^2$

Thus, $2x^2 - 9xy + 9y^2 = (2x - 3y)(x - 3y)$.

49. Factor $6x^2 - 5xy - 6y^2$ by grouping.

$a = 6$ and $c = -6$, so $ac = -36$.

The factors of -36 whose sum is -5 are -9 and 4.

$6x^2 - 5xy - 6y^2$
$= 6x^2 - 9xy + 4xy - 6y^2$
$= 3x(2x - 3y) + 2y(2x - 3y)$
$= (2x - 3y)(3x + 2y)$

51. Factor $15x^2 + 11xy - 14y^2$ by grouping.

$a = 15$ and $c = -14$, so $ac = -210$.

The factors of -210 whose sum is 11 are 21 and -10.

$15x^2 + 11xy - 14y^2$
$= 15x^2 + 21xy - 10xy - 14y^2$
$= 3x(5x + 7y) - 2y(5x + 7y)$
$= (5x + 7y)(3x - 2y)$

53. Factor $2a^2 + 7ab + 5b^2$ by trial and error.

$(2a + 5b)(a + b) = 2a^2 + 7ab + 5b^2$

$(2a + b)(a + 5b) = 2a^2 + 11ab + 5b^2$

Thus, $2a^2 + 7ab + 5b^2 = (2a + 5b)(a + b)$.

55. Factor $15a^2 - ab - 6b^2$ by grouping.

$a = 15$ and $c = -6$, so $ac = -90$.

The factors of -90 whose sum is -1 are 9 and -10.

$15a^2 - ab - 6b^2 = 15a^2 + 9ab - 10ab - 6b^2$
$= 3a(5a + 3b) - 2b(5a + 3b)$
$= (5a + 3b)(3a - 2b)$

57. Factor $12x^2 - 25xy + 12y^2$ by grouping.

$a = 12$ and $c = 12$, so $ac = 144$.

The factors of 144 whose sum is -25 are -9 and -16.

$12x^2 - 25xy + 12y^2 = 12x^2 - 9xy - 16xy + 12y^2$
$= 3x(4x - 3y) - 4y(4x - 3y)$
$= (4x - 3y)(3x - 4y)$

59. $4x^2 + 26x + 30$

First factor out the GCF, 2. Then factor the resulting trinomial by trial and error or grouping.

$$4x^2 + 26x + 30 = 2\left(2x^2 + 13x + 15\right)$$
$$= 2\left(2x + 3\right)\left(x + 5\right)$$

61. $9x^2 - 6x - 24$

First factor out the GCF, 3. Then factor the resulting trinomial by trial and error or grouping.

$$9x^2 - 6x - 24 = 3\left(3x^2 - 2x - 8\right)$$
$$= 3\left(3x + 4\right)\left(x - 2\right)$$

63. $4y^2 + 2y - 30 = 2\left(2y^2 + y - 15\right)$
$$= 2\left(2y - 5\right)\left(y + 3\right)$$

65. $9y^2 + 33y - 60 = 3\left(3y^2 + 11y - 20\right)$
$$= 3\left(3y - 4\right)\left(y + 5\right)$$

67. $3x^3 + 4x^2 + x$

First factor out the GCF, x. Then factor the resulting trinomial by trial and error or grouping.

$$3x^3 + 4x^2 + x = x\left(3x^2 + 4x + 1\right)$$
$$= x\left(3x + 1\right)\left(x + 1\right)$$

69. $2x^3 - 3x^2 - 5x = x\left(2x^2 - 3x - 5\right)$
$$= x\left(2x - 5\right)\left(x + 1\right)$$

71. $9y^3 - 39y^2 + 12y$

First factor out the GCF, $3y$. Then factor the resulting trinomial by trial and error or grouping.

$$9y^3 - 39y^2 + 12y = 3y\left(3y^2 - 13y + 4\right)$$
$$= 3y\left(3y - 1\right)\left(y - 4\right)$$

73. $60z^3 + 40z^2 + 5z = 5z\left(12z^2 + 8z + 1\right)$
$$= 5z\left(6z + 1\right)\left(2z + 1\right)$$

75. $15x^4 - 39x^3 + 18x^2 = 3x^2\left(5x^2 - 13x + 6\right)$
$$= 3x^2\left(5x - 3\right)\left(x - 2\right)$$

77. $10x^5 - 17x^4 + 3x^3 = x^3\left(10x^2 - 17x + 3\right)$
$$= x^3\left(2x - 3\right)\left(5x - 1\right)$$

79. $6x^2 - 3xy - 18y^2 = 3\left(2x^2 - xy - 6y^2\right)$
$$= 3\left(2x + 3y\right)\left(x - 2y\right)$$

81. $12x^2 + 10xy - 8y^2 = 2\left(6x^2 + 5xy - 4y^2\right)$
$$= 2\left(2x - y\right)\left(3x + 4y\right)$$

83. $8x^2y + 34xy - 84y = 2y\left(4x^2 + 17x - 42\right)$
$$= 2y\left(4x - 7\right)\left(x + 6\right)$$

85. $12a^2b - 46ab^2 + 14b^3 = 2b\left(6a^2 - 23ab + 7b^2\right)$
$$= 2b\left(2a - 7b\right)\left(3a - b\right)$$

87. $30\left(y + 1\right)x^2 + 10\left(y + 1\right)x - 20\left(y + 1\right)$
$$= 10\left(y + 1\right)\left(3x^2 + x - 2\right)$$
$$= 10\left(y + 1\right)\left(3x - 2\right)\left(x + 1\right)$$

89. $-32x^2y^4 + 20xy^4 + 12y^4$
$$= -4y^4\left(8x^2 - 5x - 3\right)$$
$$= -4y^4\left(8x + 3\right)\left(x - 1\right)$$

91. a. $2x^2 - 5x - 3 = \left(2x + 1\right)\left(x - 3\right)$

b. $2\left(y + 1\right)^2 - 5\left(y + 1\right) - 3$
$$= \left[2\left(y + 1\right) + 1\right]\left[\left(y + 1\right) - 3\right]$$
$$= \left[2y + 2 + 1\right]\left[y + 1 - 3\right]$$
$$= \left(2y + 3\right)\left(y - 2\right)$$

93.

$$
\require{enclose}
\begin{array}{r}
3x^2 - 5x + 2 \\
x-2 \enclose{longdiv}{3x^3 - 11x^2 + 12x - 4} \\
\underline{3x^3 - 6x^2} \\
-5x^2 + 12x \\
\underline{-5x^2 + 10x} \\
2x - 4 \\
\underline{2x - 4} \\
0
\end{array}
$$

The quotient $3x^2 - 5x + 2$ factors into $\left(3x - 2\right)\left(x - 1\right)$.

Thus, $3x^3 - 11x^2 + 12x - 4 = \left(x - 2\right)\left(3x - 2\right)\left(x - 1\right)$.

95. a. $x^2 + 3x + 2$

b. $(x+2)(x+1)$

c. Yes, the pieces are the same in both figures: one large square, three long rectangles, and two small squares. This geometric model illustrates the factorization:
$$x^2 + 3x + 2 = (x+2)(x+1).$$

97. – 99. Answers will vary.

101. makes sense

103. does not make sense; Explanations will vary. Sample explanation: If a polynomial has a greatest common factor then it is not prime.

105. true

107. false; Changes to make the statement true will vary. A sample change is: The factorization of $4y^2 - 11y - 3$ is $(y-3)(4y+1)$.

109. $3x^2 + bx + 2$

The possible factorizations that will give $3x^2$ as the first term and 2 as the last term are:
$(3x+1)(x+2) = 3x^2 + 7x + 2$
$(3x+2)(x+1) = 3x^2 + 5x + 2$
$(3x-1)(x-2) = 3x^2 - 7x + 2$
$(3x-2)(x-1) = 3x^2 - 5x + 2$
The possible middle terms are $5x, 7x, -5x$ and $-7x$, so $3x^2 + bx + 2$ can be factored if b is 5, 7, −5, or −7.

111. $3x^{10} - 4x^5 - 15$

Since $\left(x^5\right)^2 = x^{10}$, the first terms of the factors must be $3x^5$ and x^5 so the middle term will contain x^5. Use trial and error or grouping to obtain the correct factorization.
$3x^{10} - 4x^5 - 15 = \left(3x^5 + 5\right)\left(x^5 - 3\right)$

113. $4x - y = 105$
$x + 7y = -10$
Multiply the second equation by −4 and then add the equations.
$$\begin{array}{r} 4x - y = 105 \\ -4x - 28y = 40 \\ \hline -29y = 145 \end{array}$$
$\dfrac{-29y}{-29} = \dfrac{145}{-29}$
$y = -5$
Back-substitute to find x.
$x + 7y = -10$
$x + 7(-5) = -10$
$x - 35 = -10$
$x = 25$
The solution set is $\{(25, -5)\}$.

114. $0.00086 = 8.6 \times 10^{-4}$

115. $8x - \dfrac{x}{6} = \dfrac{1}{6} - 8$
Multiply both sides by the LCD of 6.
$6\left(8x - \dfrac{x}{6}\right) = 6\left(\dfrac{1}{6} - 8\right)$
$48x - x = 1 - 48$
$47x = -47$
$x = -1$
The solution set is $\{-1\}$.

116. $(9x+10)(9x-10) = (9x)^2 - 10^2$
$= 81x^2 - 100$

117. $(4x+5y)^2 = (4x)^2 + 2(4x)(5y) + (5y)^2$
$= 16x^2 + 40xy + 25y^2$

118. $(x+2)\left(x^2 - 2x + 4\right)$
$= x\left(x^2 - 2x + 4\right) + 2\left(x^2 - 2x + 4\right)$
$= x^3 - 2x^2 + 4x + 2x^2 - 4x + 8$
$= x^3 + 8$

Chapter 6 Mid-Chapter Check Point

1. The GCF is x^4.
$$x^5 + x^4 = x^4(x+1)$$

2. $x^2 + 7x - 18 = (x-2)(x+9)$

3. The GCF is $x^2 y$. The polynomial in the parentheses is prime because there are no factors of 1 whose sum is -1.
$$x^2 y^3 - x^2 y^2 + x^2 y = x^2 y(y^2 - y + 1)$$

4. $x^2 - 2x + 4$ is prime because there are no factors of 4 whose sum is -2.

5. Factor $7x^2 - 22x + 3$ by grouping.
$a = 7$ and $c = 3$, so $ac = 21$.
The only factors of 21 whose sum is -22 are -21 and -1.
$$7x^2 - 22x + 3 = 7x^2 - 21x - x + 3$$
$$= 7x(x-3) - 1(x-3)$$
$$= (x-3)(7x-1)$$

6. Factor $x^3 + 5x^2 + 3x + 15$ by grouping.
$$x^3 + 5x^2 + 3x + 15 = (x^3 + 5x^2) + (3x + 15)$$
$$= x^2(x+5) + 3(x+5)$$
$$= (x+5)(x^2 + 3)$$

7. The GCF is x.
$$2x^3 - 11x^2 + 5x = x(2x^2 - 11x + 5)$$
$$= x(2x-1)(x-5)$$

8. Factor $xy - 7x - 4y + 28$ by grouping.
$$xy - 7x - 4y + 28 = (xy - 7x) + (-4y + 28)$$
$$= x(y-7) - 4(y-7)$$
$$= (y-7)(x-4)$$

9. Factor $x^2 - 17xy + 30y^2$ by trial and error. The only factors of 30 whose sum is -17 are -15 and -2.
$$x^2 - 17xy + 30y^2 = (x - 15y)(x - 2y)$$

10. Factor $25x^2 - 25x - 14$ by trial and error.
$$(5x-2)(5x+7) = 25x^2 + 25x - 14$$
$$(5x+2)(5x-7) = 25x^2 - 25x - 14$$
Because the correct factorization has been found, there is no need to try additional possibilities.
$$25x^2 - 25x - 14 = (5x+2)(5x-7).$$

11. The GCF is 2.
$$16x^2 - 70x + 24 = 2(8x^2 - 35x + 12)$$
Factor the polynomial in parentheses by grouping.
$a = 8$ and $c = 12$, so $ac = 96$.
The only factors of 96 whose sum is -35 are -32 and -3.
$$16x^2 - 70x + 24 = 2(8x^2 - 35x + 12)$$
$$= 2(8x^2 - 32x - 3x + 12)$$
$$= 2\left[8x(x-4) - 3(x-4)\right]$$
$$= 2(x-4)(8x-3)$$

12. Factor $3x^2 + 10xy + 7y^2$ by grouping.
$a = 3$ and $c = 7$, so $ac = 21$.
The only factors of 21 whose sum is 10 are 3 and 7.
$$3x^2 + 10xy + 7y^2 = 3x^2 + 3xy + 7xy + 7y^2$$
$$= 3x(x+y) + 7y(x+y)$$
$$= (x+y)(3x+7y)$$

6.4 Check Points

1. a. $x^2 - 81 = (x+9)(x-9)$

   b. $36x^2 - 25 = (6x+5)(6x-5)$

2. a. $25 - 4x^{10} = (5 + 2x^5)(5 - 2x^5)$

   b. $100x^2 - 9y^2 = (10x + 3y)(10x - 3y)$

3. a. $18x^3 - 2x = 2x(9x^2 - 1)$
$$= 2x(3x+1)(3x-1)$$

   b. $72 - 18x^2 = 18(4 - x^2)$
$$= 18(2 + x)(2 - x)$$

4. $81x^4 - 16 = (9x^2 + 4)(9x^2 - 4)$

$\qquad = (9x^2 + 4)(3x + 2)(3x - 2)$

5. a. Notice that the trinomial fits the form

$A^2 + 2AB + B^2$.

Thus it factors as $(A + B)^2$.

$x^2 + 14x + 49 = (x + 7)^2$

b. Notice that the trinomial fits the form

$A^2 - 2AB + B^2$.

Thus it factors as $(A - B)^2$.

$x^2 - 6x + 9 = (x - 3)^2$

c. Notice that the trinomial fits the form

$A^2 - 2AB + B^2$.

Thus it factors as $(A - B)^2$.

$16x^2 - 56x + 49 = (4x - 7)^2$

6. Notice that the trinomial fits the form

$A^2 + 2AB + B^2$.

Thus it factors as $(A + B)^2$.

$4x^2 + 12xy + 9y^2 = (2x + 3y)^2$

7. Notice that the polynomial fits the form $A^3 + B^3$.

Thus it factors as $(A + B)(A^2 - AB + B^2)$.

$x^3 + 27 = x^3 + 3^3$

$\qquad = (x + 3)(x^2 - 3x + 3^2)$

$\qquad = (x + 3)(x^2 - 3x + 9)$

8. Notice that the polynomial fits the form $A^3 - B^3$.

Thus it factors as $(A - B)(A^2 + AB + B^2)$.

$1 - y^3 = 1^3 - y^3$

$\qquad = (1 - y)(1^2 + 1 \cdot y + y^2)$

$\qquad = (1 - y)(1 + y + y^2)$

9. Notice that the polynomial fits the form $A^3 + B^3$.

Thus it factors as $(A + B)(A^2 - AB + B^2)$.

$125x^3 + 8 = (5x)^3 + 2^3$

$\qquad = (5x + 2)\left[(5x)^2 - (5x)(2) + 2^2\right]$

$\qquad = (5x + 2)(25x^2 - 10x + 4)$

6.4 Exercise Set

1. $x^2 - 25 = x^2 - 5^2 = (x + 5)(x - 5)$

3. $y^2 - 1 = y^2 - 1^2 = (y + 1)(y - 1)$

5. $4x^2 - 9 = (2x)^2 - 3^2 = (2x + 3)(2x - 3)$

7. $25 - x^2 = 5^2 - x^2 = (5 + x)(5 - x)$

9. $1 - 49x^2 = 1^2 - (7x)^2 = (1 + 7x)(1 - 7x)$

11. $9 - 25y^2 = 3^2 - (5y)^2 = (3 + 5y)(3 - 5y)$

13. $x^4 - 9 = \left(x^2\right)^2 - 3^2 = \left(x^2 + 3\right)\left(x^2 - 3\right)$

15. $49y^4 - 16 = \left(7y^2\right)^2 - 4^2$

$\qquad = \left(7y^2 + 4\right)\left(7y^2 - 4\right)$

17. $x^{10} - 9 = \left(x^5\right)^2 - 3^2 = \left(x^5 + 3\right)\left(x^5 - 3\right)$

19. $25x^2 - 16y^2 = (5x)^2 - (4y)^2$

$\qquad = (5x + 4y)(5x - 4y)$

21. $x^4 - y^{10} = \left(x^2\right)^2 - \left(y^5\right)^2$

$\qquad = \left(x^2 + y^5\right)\left(x^2 - y^5\right)$

23. $x^4 - 16 = \left(x^2\right)^2 - 4^2 = \left(x^2 + 4\right)\left(x^2 - 4\right)$

Because $x^2 - 4$ is also the difference of two squares, the factorization must be continued.

$x^4 - 16 = \left(x^2 + 4\right)\left(x^2 - 4\right)$

$\qquad = \left(x^2 + 4\right)\left(x^2 - 2^2\right)$

$\qquad = \left(x^2 + 4\right)(x + 2)(x - 2).$

25. $16x^4 - 81 = \left(4x^2\right)^2 - 9^2$

$$= \left(4x^2 + 9\right)\left(4x^2 - 9\right)$$

$$= \left(4x^2 + 9\right)\left[\left(2x\right)^2 - 3^2\right]$$

$$= \left(4x^2 + 9\right)\left(2x + 3\right)\left(2x - 3\right)$$

27. $2x^2 - 18 = 2\left(x^2 - 9\right) = 2\left(x + 3\right)\left(x - 3\right)$

29. $2x^3 - 72x = 2x\left(x^2 - 36\right)$

$$= 2x\left(x + 6\right)\left(x - 6\right)$$

31. $x^2 + 36$ is prime because it is the sum of two squares with no common factor other than 1.

33. $3x^3 + 27x = 3x\left(x^2 + 9\right)$

35. $18 - 2y^2 = 2\left(9 - y^2\right) = 2\left(3 + y\right)\left(3 - y\right)$

37. $3y^3 - 48y = 3y\left(y^2 - 16\right)$

$$= 3y\left(y + 4\right)\left(y - 4\right)$$

39. $18x^3 - 2x = 2x\left(9x^2 - 1\right)$

$$= 2x\left(3x + 1\right)\left(3x - 1\right)$$

41. $x^2 + 2x + 1 = x^2 + 2\left(1x\right) + 1^2$

$$= \left(x + 1\right)^2$$

43. $x^2 - 14x + 49 = x^2 - 2\left(7x\right) + 7^2$

$$= \left(x - 7\right)^2$$

45. $x^2 - 2x + 1 = x^2 - 2\left(1x\right) + 1^2$

$$= \left(x - 1\right)^2$$

47. $x^2 + 22x + 121 = x^2 + 2\left(11x\right) + 11^2$

$$= \left(x + 11\right)^2$$

49. $4x^2 + 4x + 1 = \left(2x\right)^2 + 2\left(2x\right) + 1^2$

$$= \left(2x + 1\right)^2$$

51. $25y^2 - 10y + 1 = \left(5y\right)^2 - 2\left(5y\right) + 1^2$

$$= \left(5y - 1\right)^2$$

53. $x^2 - 10x + 100$ is prime.
To be a perfect square trinomial, the middle term would have to be $2\left(-10x\right) = -20x$ rather than $-10x$.

55. $x^2 + 14xy + 49y^2 = x^2 + 2\left(7xy\right) + \left(7y\right)^2$

$$= \left(x + 7y\right)^2$$

57. $x^2 - 12xy + 36y^2 = x^2 - 2\left(6xy\right) + \left(6y\right)^2$

$$= \left(x - 6y\right)^2$$

59. $x^2 - 8xy + 64y^2$ is prime.
To be a perfect square trinomial, the middle term would have to be $2\left(-8xy\right) = -16xy$ rather than $-8xy$.

61. $16x^2 - 40xy + 25y^2$

$$= \left(4x\right)^2 - 2\left(4x \cdot 5y\right) + \left(5y\right)^2$$

$$= \left(4x - 5y\right)^2$$

63. $12x^2 - 12x + 3 = 3\left(4x^2 - 4x + 1\right)$

$$= 3\left[\left(2x\right)^2 - 2\left(2x\right) + 1^2\right]$$

$$= 3\left(2x - 1\right)^2$$

65. $9x^3 + 6x^2 + x$

$$= x\left(9x^2 + 6x + 1\right)$$

$$= x\left[\left(3x\right)^2 + 2\left(3x\right) + 1^2\right]$$

$$= x\left(3x + 1\right)^2$$

67. $2y^2 - 4y + 2 = 2\left(y^2 - 2y + 1\right)$

$$= 2\left(y - 1\right)^2$$

69. $2y^3 + 28y^2 + 98y = 2y\left(y^2 + 14y + 49\right)$

$$= 2y\left(y + 7\right)^2$$

71. $x^3 + 1 = x^3 + 1^3$

$\qquad = (x+1)(x^2 - x \cdot 1 + 1^2)$

$\qquad = (x+1)(x^2 - x + 1)$

73. $x^3 - 27 = x^3 - 3^3$

$\qquad = (x-3)(x^2 + x \cdot 3 + 3^2)$

$\qquad = (x-3)(x^2 + 3x + 9)$

75. $8y^3 - 1 = (2y)^3 - 1^3$

$\qquad = (2y-1)\left[(2y)^2 + 2y \cdot 1 + 1\right]$

$\qquad = (2y-1)(4y^2 + 2y + 1)$

76. $27y^3 - 1 = (3y)^3 - 1^3$

$\qquad = (3y-1)\left[(3y)^2 + 3y \cdot 1 + 1^2\right]$

$\qquad = (3y-1)(9y^2 + 3y + 1)$

77. $27x^3 + 8 = (3x)^3 + 2^3$

$\qquad = (3x+2)\left[(3x)^2 - 3x \cdot 2 + 2^2\right]$

$\qquad = (3x+2)(9x^2 - 6x + 4)$

79. $x^3 y^3 - 64 = (xy)^3 - 4^3$

$\qquad = (xy-4)\left[(xy)^2 + xy \cdot 4 + 4^2\right]$

$\qquad = (xy-4)(x^2 y^2 + 4xy + 16)$

81. $27y^4 + 8y = y(27y^3 + 8)$

$\qquad = y\left[(3y)^3 + 2^3\right]$

$\qquad = y(3y+2)\left[(3y)^2 - 3y \cdot 2 + 2^2\right]$

$\qquad = y(3y+2)(9y^2 - 6y + 4)$

83. $54 - 16y^3 = 2(27 - 8y^3)$

$\qquad = 2\left[3^3 - (2y)^3\right]$

$\qquad = 2(3 - 2y)\left[3^2 + 3 \cdot 2y + (2y)^2\right]$

$\qquad = 2(3 - 2y)(9 + 6y + 4y^2)$

85. $64x^3 + 27y^3 = (4x)^3 + (3y)^3$

$\qquad = (4x+3y)\left[(4x)^2 - 4x \cdot 3y + (3y)^2\right]$

$\qquad = (4x+3y)(16x^2 - 12xy + 9y^2)$

87. $125x^3 - 64y^3 = (5x)^3 - (4y)^3$

$\qquad = (5x-4y)\left[(5x)^2 + 5x \cdot 4y + (4y)^2\right]$

$\qquad = (5x-4y)(25x^2 + 20xy + 16y^2)$

89. $25x^2 - \dfrac{4}{49} = (5x)^2 - \left(\dfrac{2}{7}\right)^2$

$\qquad = \left(5x + \dfrac{2}{7}\right)\left(5x - \dfrac{2}{7}\right)$

91. $y^4 - \dfrac{y}{1000} = y\left(y^3 - \dfrac{1}{1000}\right)$

$\qquad = y\left[y^3 - \left(\dfrac{1}{10}\right)^3\right]$

$\qquad = y\left(y - \dfrac{1}{10}\right)\left[y^2 + y \cdot \dfrac{1}{10} + \left(\dfrac{1}{10}\right)^2\right]$

$\qquad = y\left(y - \dfrac{1}{10}\right)\left(y^2 + \dfrac{y}{10} + \dfrac{1}{100}\right)$

93. $0.25x - x^3 = x(0.25 - x^2)$

$\qquad = x\left[(0.5)^2 - x^2\right]$

$\qquad = x(0.5 + x)(0.5 - x)$

95. $(x+1)^2 - 25 = (x+1)^2 - 5^2$

$\qquad = \left[(x+1) + 5\right]\left[(x+1) - 5\right]$

$\qquad = (x+6)(x-4)$

97.
$$\begin{array}{r} x^2 + 2x + 1 \\ x-3\overline{\smash{\big)}\,x^3 - x^2 - 5x - 3} \\ \underline{x^3 - 3x^2} \\ 2x^2 - 5x \\ \underline{2x^2 - 6x} \\ x - 3 \\ \underline{x - 3} \end{array}$$

The quotient $x^2 + 2x + 1$ factors further.
$x^2 + 2x + 1 = (x+1)^2$.
Thus, $x^3 - x^2 - 5x - 3 = (x-3)(x+1)^2$.

99. Area of outer square $= x^2$
Area of inner square $= 5^2 = 25$
Area of shaded region $= x^2 - 25 = (x+5)(x-5)$

101. Area of large square $= x^2$
Area of each small corner squares $= 2^2 = 4$
Area of four corner squares $= 4 \cdot 4 = 16$
Area of shaded region $= x^2 - 16$
$$= (x+4)(x-4)$$

103. – 105. Answers will vary.

107. does not make sense; Explanations will vary.
Sample explanation: The original expression has a common factor of 9.
$9x^2 - 36 = 9(x^2 - 4) = 9(x+2)(x-2)$

109. does not make sense; Explanations will vary.
Sample explanation: The second factor involves a subtraction. The commutative property does not apply to subtraction.

111. false; Changes to make the statement true will vary.
A sample change is: $x^2 + 25$ is prime.

113. false; Changes to make the statement true will vary.
A sample change is: $4x^2 + 36$ is a polynomial that is the sum of two squares. $4x^2 + 36$ factors into $4(x^2 + 9)$.

115. The error in the proof is in step 7 where you are asked to divide by $a - b = 0$.
Division by 0 is not permitted because division by zero is undefined.

117. $x^{2n} - 25y^{2n} = (x^n)^2 - (5y^n)^2$
$\qquad = (x^n + 5y^n)(x^n - 5y^n)$

119. $(x+3)^2 - 2(x+3) + 1$
$\qquad = [(x+3)-1]^2$
$\qquad = (x+2)^2$

121. $64x^2 - 16x + k$
Let r be the number such that $r^2 = k$.
Then, $64x^2 - 16x + k = (8x)^2 - 2 \cdot 8x \cdot r + r^2$.
Comparing the middle terms, we see that
$-2 \cdot 8x \cdot r = -16x$
$\qquad -16xr = -16x$
$\qquad\qquad r = 1$.
Therefore, $k = r^2 = 1^2 = 1$.

123. The graphs coincide.

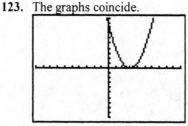

This verifies that $x^2 - 6x + 9 = (x-3)^2$.

125. The graphs do not coincide.

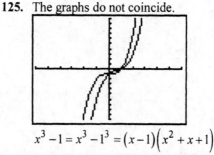

$x^3 - 1 = x^3 - 1^3 = (x-1)(x^2 + x + 1)$

The polynomial on the right side should be changed to $(x-1)(x^2 + x + 1)$.

126. $\left(2x^2y^3\right)^4\left(5xy^2\right)$

$= \left[2^4\left(x^2\right)^4\left(y^3\right)^4\right]\cdot\left(5xy^2\right)$

$= \left(16x^8y^{12}\right)\left(5xy^2\right)$

$= (16\cdot5)\left(x^8\cdot x^1\right)\left(y^{12}\cdot y^2\right)$

$= 80x^9y^{14}$

127. $\left(10x^2-5x+2\right)-\left(14x^2-5x-1\right)$

$= \left(10x^2-5x+2\right)+\left(-14x^2+5x+1\right)$

$= \left(10x^2-14x^2\right)+(-5x+5x)+(2+1)$

$= -4x^2+3$

128.
$$\begin{array}{r} 2x+5 \\ 3x-2\overline{\smash{)}6x^2+11x-10} \\ \underline{6x^2-\ 4x} \\ 15x-10 \\ \underline{15x-10} \\ 0 \end{array}$$

$\dfrac{6x^2+11x-10}{3x-2}=2x+5$

129. $3x^3-75x=3x(x^2-25)=3x(x+5)(x-5)$

130. $2x^2-20x+50=2(x^2-10x+25)=2(x-5)^2$

131. $x^3-2x^2-x+2=x^2(x-2)-1(x-2)$

$= (x-2)(x^2-1)$

$= (x-2)(x+1)(x-1)$

6.5 Check Points

1. $5x^4-45x^2=5x^2(x^2-9)$

$= 5x^2(x+3)(x-3)$

2. $4x^2-16x-48=4(x^2-4x-12)$

$= 4(x-6)(x+2)$

3. $4x^5-64x=4x(x^4-16)$

$= 4x(x^2+4)(x^2-4)$

$= 4x(x^2+4)(x+2)(x-2)$

4. $x^3-4x^2-9x+36=x^2(x-4)-9(x-4)$

$= (x-4)(x^2-9)$

$= (x-4)(x+3)(x-3)$

5. $3x^3-30x^2+75x=3x(x^2-10x+25)$

$= 3x(x-5)^2$

6. $2x^5+54x^2=2x^2(x^3+27)$

$= 2x^2(x+3)(x^2-3x+9)$

7. $3x^4y-48y^5=3y(x^4-16y^4)$

$= 3y(x^2+4y^2)(x^2-4y^2)$

$= 3y(x^2+4y^2)(x+2y)(x-2y)$

8. $12x^3+36x^2y+27xy^2=3x(4x^2+12xy+9y^2)$

$= 3x(2x+3y)^2$

6.5 Exercise Set

1. $5x^3-20x=5x\left(x^2-4\right)$

$= 5x\left(x+2\right)\left(x-2\right)$

3. $7x^3+7x=7x\left(x^2+1\right)$

5. $5x^2-5x-30=5\left(x^2-x-6\right)$

$= 5\left(x+2\right)\left(x-3\right)$

7. $2x^4-162=2\left(x^4-81\right)$

$= 2\left(x^2+9\right)\left(x^2-9\right)$

$= 2\left(x^2+9\right)\left(x+3\right)\left(x-3\right)$

9. $x^3 + 2x^2 - 9x - 18 = \left(x^3 + 2x^2\right) + (-9x - 18)$

$\qquad\qquad = x^2(x+2) - 9(x+2)$

$\qquad\qquad = (x+2)\left(x^2 - 9\right)$

$\qquad\qquad = (x+2)(x+3)(x-3)$

11. $3x^3 - 24x^2 + 48x = 3x\left(x^2 - 8x + 16\right)$

$\qquad\qquad = 3x(x-4)^2$

13. $2x^5 + 2x^2 = 2x^2\left(x^3 + 1\right)$

$\qquad\qquad = 2x^2(x+1)\left(x^2 - x + 1\right)$

15. $6x^2 + 8x = 2x(3x+4)$

17. $2y^2 - 2y - 112 = 2\left(y^2 - y - 56\right)$

$\qquad\qquad = 2(y-8)(y+7)$

19. $7y^4 + 14y^3 + 7y^2 = 7y^2\left(y^2 + 2y + 1\right)$

$\qquad\qquad = 7y^2(y+1)^2$

21. $y^2 + 8y - 16$ is prime because there are no two integers whose product is -16 and whose sum is 8.

23. $16y^2 - 4y - 2 = 2\left(8y^2 - 2y - 1\right)$

$\qquad\qquad = 2(4y+1)(2y-1)$

25. $r^2 - 25r = r(r-25)$

27. $4w^2 + 8w - 5 = (2w+5)(2w-1)$

29. $x^3 - 4x = x\left(x^2 - 4\right) = x(x+2)(x-2)$

31. $x^2 + 64$ is prime because it is the sum of two squares with no common factor other than 1.

33. $9y^2 + 13y + 4 = (9y+4)(y+1)$

35. $y^3 + 2y^2 - 4y - 8 = \left(y^3 + 2y^2\right) + (-4y - 8)$

$\qquad\qquad = y^2(y+2) - 4(y+2)$

$\qquad\qquad = (y+2)\left(y^2 - 4\right)$

$\qquad\qquad = (y+2)(y+2)(y-2)$

$\qquad\qquad = (y+2)^2(y-2)$

37. $16y^2 + 24y + 9 = (4y)^2 + 2(4y \cdot 3) + 3^2$

$\qquad\qquad = (4y+3)^2$

39. $4y^3 - 28y^2 + 40y = 4y\left(y^2 - 7y + 10\right)$

$\qquad\qquad = 4y(y-5)(y-2)$

41. $y^5 - 81y = y\left(y^4 - 81\right)$

$\qquad\qquad = y\left(y^2 + 9\right)\left(y^2 - 9\right)$

$\qquad\qquad = y\left(y^2 + 9\right)(y+3)(y-3)$

43. $20a^4 - 45a^2 = 5a^2\left(4a^2 - 9\right)$

$\qquad\qquad = 5a^2(2a+3)(2a-3)$

45. $9x^4 + 18x^3 + 6x^2 = 3x^2(3x^2 + 6x + 2)$

47. $12y^2 - 11y + 2 = (4y-1)(3y-2)$

49. $9y^2 - 64 = (3y)^2 - 8^2$

$\qquad\qquad = (3y+8)(3y-8)$

51. $9y^2 + 64$ is prime because it is the sum of two squares with no common factor other than 1.

53. $2y^3 + 3y^2 - 50y - 75 = \left(2y^3 + 3y^2\right) + (-50y - 75)$

$\qquad\qquad = y^2(2y+3) - 25(2y+3)$

$\qquad\qquad = (2y+3)\left(y^2 - 25\right)$

$\qquad\qquad = (2y+3)(y+5)(y-5)$

55. $2r^3 + 30r^2 - 68r = 2r\left(r^2 + 15r - 34\right)$

$\qquad\qquad = 2r(r+17)(r-2)$

57. $8x^5 - 2x^3 = 2x^3\left(4x^2 - 1\right)$

$$= 2x^3\left[(2x)^2 - 1^2\right]$$

$$= 2x^3\left(2x + 1\right)\left(2x - 1\right)$$

59. $3x^2 + 243 = 3\left(x^2 + 81\right)$

61. $x^4 + 8x = x\left(x^3 + 8\right)$

$$= x\left(x^3 + 2^3\right)$$

$$= x(x + 2)\left(x^2 - 2x + 4\right)$$

63. $2y^5 - 2y^2 = 2y^2\left(y^3 - 1\right)$

$$= 2y^2\left(y - 1\right)\left(y^2 + y + 1\right)$$

65. $6x^2 + 8xy = 2x\left(3x + 4y\right)$

67. $xy - 7x + 3y - 21 = \left(xy - 7x\right) + \left(3y - 21\right)$

$$= x\left(y - 7\right) + 3\left(y - 7\right)$$

$$= \left(y - 7\right)\left(x + 3\right)$$

69. $x^2 - 3xy - 4y^2 = \left(x - 4y\right)\left(x + y\right)$

71. $72a^3b^2 + 12a^2 - 24a^4b^2$

$$= 12a^2\left(6ab^2 + 1 - 2a^2b^2\right)$$

73. $3a^2 + 27ab + 54b^2 = 3\left(a^2 + 9ab + 18b^2\right)$

$$= 3\left(a + 6b\right)\left(a + 3b\right)$$

75. $48x^4y - 3x^2y = 3x^2y\left(16x^2 - 1\right)$

$$= 3x^2y\left(4x + 1\right)\left(4x - 1\right)$$

77. $6a^2b + ab - 2b = b\left(6a^2 + a - 2\right)$

$$= b\left(3a + 2\right)\left(2a - 1\right)$$

79. $7x^5y - 7xy^5 = 7xy\left(x^4 - y^4\right)$

$$= 7xy\left(x^2 + y^2\right)\left(x^2 - y^2\right)$$

$$= 7xy\left(x^2 + y^2\right)\left(x + y\right)\left(x - y\right)$$

81. $10x^3y - 14x^2y^2 + 4xy^3 = 2xy\left(5x^2 - 7xy + 2y^2\right)$

$$= 2xy\left(5x - 2y\right)\left(x - y\right)$$

83. $2bx^2 + 44bx + 242b = 2b\left(x^2 + 22x + 121\right)$

$$= 2b\left(x^2 + 2(11x) + 11^2\right)$$

$$= 2b\left(x + 11\right)^2$$

85. $15a^2 + 11ab - 14b^2 = \left(5a + 7b\right)\left(3a - 2b\right)$

87. $36x^3y - 62x^2y^2 + 12xy^3$

$$= 2xy\left(18x^2 - 31xy + 6y^2\right)$$

$$= 2xy\left(9x - 2y\right)\left(2x - 3y\right)$$

89. $a^2y - b^2y - a^2x + b^2x$

$$= \left(a^2y - b^2y\right) + \left(-a^2x + b^2x\right)$$

$$= y\left(a^2 - b^2\right) - x\left(a^2 - b^2\right)$$

$$= \left(a^2 - b^2\right)\left(y - x\right)$$

$$= \left(a + b\right)\left(a - b\right)\left(y - x\right)$$

91. $9ax^3 + 15ax^2 - 14ax = ax\left(9x^2 + 15x - 14\right)$

$$= ax\left(3x + 7\right)\left(3x - 2\right)$$

93. $2x^4 + 6x^3y + 2x^2y^2 = 2x^2\left(x^2 + 3xy + y^2\right)$

95. $81x^4y - y^5 = y\left(81x^4 - y^4\right)$

$$= y\left(9x^2 + y^2\right)\left(9x^2 - y^2\right)$$

$$= y\left(9x^2 + y^2\right)\left(3x + y\right)\left(3x - y\right)$$

97. $10x^2\left(x + 1\right) - 7x\left(x + 1\right) - 6\left(x + 1\right)$

$$= \left(x + 1\right)\left(10x^2 - 7x - 6\right)$$

$$= \left(x + 1\right)\left(5x - 6\right)\left(2x + 1\right)$$

99. $6x^4 + 35x^2 - 6 = \left(x^2 + 6\right)\left(6x^2 - 1\right)$

101. $(x-7)^2 - 4a^2 = (x-7)^2 - (2a)^2$

$\qquad = \left[(x-7)+2a\right]\left[(x-7)-2a\right]$

$\qquad = (x-7+2a)(x-7-2a)$

103. $x^2 + 8x + 16 - 25a^2$

$\qquad = \left(x^2 + 8x + 16\right) - (5a)^2$

$\qquad = (x+4)^2 - (5a)^2$

$\qquad = \left[(x+4)+5a\right]\left[(x+4)-5a\right]$

$\qquad = (x+4+5a)(x+4-5a)$

105. $y^7 + y = y\left(y^6 + 1\right) = y\left[\left(y^2\right)^3 + 1^3\right]$

$\qquad = y\left(y^2+1\right)\left(y^4 - y^2 + 1\right)$

107. $256 - 16t^2 = 16\left(16 - t^2\right)$

$\qquad = 16(4+t)(4-t)$

109. Area of outer circle $= \pi b^2$

Area of inner circle $= \pi a^2$

Area of shaded ring $= \pi b^2 - \pi a^2$

$\pi b^2 - \pi a^2 = \pi\left(b^2 - a^2\right)$

$\qquad = \pi(b+a)(b-a)$

111. Answers will vary.

113. makes sense

115. makes sense

117. false; Changes to make the statement true will vary. A sample change is: $x^2 - 9 = (x+3)(x-3)$ for any real number x.

119. false; Changes to make the statement true will vary. A sample change is: Some polynomials are completely factored after one step.

121. $3x^5 - 21x^3 - 54x = 3x\left(x^4 - 7x^2 - 18\right)$

$\qquad = 3x\left(x^2 + 2\right)\left(x^2 - 9\right)$

$\qquad = 3x\left(x^2 + 2\right)(x+3)(x-3)$

123. $4x^4 - 9x^2 + 5 = \left(4x^2 - 5\right)\left(x^2 - 1\right)$

$\qquad = \left(4x^2 - 5\right)(x+1)(x-1)$

125. $3x^{2n} - 27y^{2n} = 3\left(x^{2n} - 9y^{2n}\right)$

$\qquad = 3\left[\left(x^n\right)^2 - \left(3y^n\right)^2\right]$

$\qquad = 3\left(x^n + 3y^n\right)\left(x^n - 3y^n\right)$

127. The graphs do not coincide.

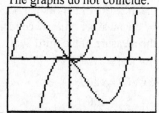

$3x^3 - 12x^2 - 15x = 3x\left(x^2 - 4x - 5\right)$

$\qquad = 3x(x-5)(x+1)$

Change the polynomial on the right side to $3x(x-5)(x+1)$.

129. The graphs coincide.

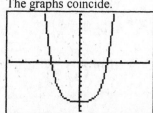

This verifies that the factorization

$x^4 - 16 = \left(x^2 + 4\right)(x+2)(x-2)$ is correct.

131. $9x^2 - 16 = (3x)^2 - 4^2$

$\qquad = (3x+4)(3x-4)$

132. $5x - 2y = 10$

To find the x-intercept, let $y = 0$.

$5x - 2(0) = 10$

$\qquad 5x = 10$

$\qquad x = 2$

To find the y-intercept, let $x = 0$.

$5(0) - 2y = 10$

$\qquad -2y = 10$

$\qquad y = -5$

Checkpoint: $(4,5)$

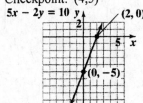

133. Let $x =$ the measure of the first angle.

Then $3x =$ the measure of the second angle,

$x + 80 =$ the measure of the third angle.

$x + 3x + (x + 80) = 180$

$\qquad 5x + 80 = 180$

$\qquad 5x = 100$

$\qquad x = 20$

Measure of first angle $= x = 20°$

Measure of second angle $= 3x = 60°$

Measure of third angle $= x + 80 = 100°$

134. $(3x - 1)(x + 2) = \left(3 \cdot \dfrac{1}{3} - 1\right)\left(\dfrac{1}{3} + 2\right)$

$\qquad = (1 - 1)\left(2\dfrac{1}{3}\right)$

$\qquad = (0)\left(2\dfrac{1}{3}\right)$

$\qquad = 0$

135. $2x^2 + 7x - 4 = 2\left(\dfrac{1}{2}\right)^2 + 7\left(\dfrac{1}{2}\right) - 4$

$\qquad = \dfrac{1}{2} + \dfrac{7}{2} - 4$

$\qquad = \dfrac{8}{2} - 4$

$\qquad = 4 - 4$

$\qquad = 0$

136. $(x - 2)(x + 3) - 6 = x^2 + 3x - 2x - 6 - 6$

$\qquad\qquad\qquad\qquad = x^2 + x - 12$

$\qquad\qquad\qquad\qquad = (x + 4)(x - 3)$

6.6 Check Points

1. $(2x + 1)(x - 4) = 0$

$2x + 1 = 0 \qquad$ or $\qquad x - 4 = 0$

$\quad 2x = -1 \qquad\qquad\qquad x = 4$

$\quad x = -\dfrac{1}{2}$

The solution set is $\left\{-\dfrac{1}{2}, 4\right\}$.

2. $\quad x^2 - 6x + 5 = 0$

$(x - 1)(x - 5) = 0$

$x - 1 = 0 \quad$ or $\quad x - 5 = 0$

$\quad x = 1 \qquad\qquad x = 5$

The solution set is $\{1, 5\}$.

3. $\qquad 4x^2 = 2x$

$\quad 4x^2 - 2x = 0$

$2x(2x - 1) = 0$

$2x = 0 \quad$ or $\quad 2x - 1 = 0$

$\quad x = 0 \qquad\qquad 2x = 1$

$\qquad\qquad\qquad\qquad x = \dfrac{1}{2}$

The solution set is $\left\{0. \dfrac{1}{2}\right\}$.

4. $\qquad x^2 = 10x - 25$

$x^2 - 10x + 25 = 0$

$\qquad (x - 5)^2 = 0$

Because both factors are the same, it is only necessary to set one of them equal to zero.

$x - 5 = 0$

$\quad x = 5$

The solution set is $\{5\}$.

5.
$$16x^2 = 25$$
$$16x^2 - 25 = 0$$
$$(4x+5)(4x-5) = 0$$
$$4x+5 = 0 \quad \text{or} \quad 4x-5 = 0$$
$$4x = -5 \qquad\qquad 4x = 5$$
$$x = -\frac{5}{4} \qquad\qquad x = \frac{5}{4}$$
The solution set is $\left\{-\dfrac{5}{4}, \dfrac{5}{4}\right\}$.

6. $(x-5)(x-2) = 28$
$$x^2 - 7x + 10 = 28$$
$$x^2 - 7x - 18 = 0$$
$$(x-9)(x+2) = 0$$
$$x-9 = 0 \quad \text{or} \quad x+2 = 0$$
$$x = 9 \qquad\qquad x = -2$$
The solution set is $\{-2, 9\}$.

7. $h = -16t^2 + 48t + 160$
$$192 = -16t^2 + 48t + 160$$
$$0 = -16t^2 + 48t - 32$$
$$0 = -16(t^2 - 3t + 2)$$
$$0 = -16(t-1)(t-2)$$
$$x-1 = 0 \quad \text{or} \quad x-2 = 0$$
$$x = 1 \qquad\qquad x = 2$$
The ball's height will be 192 feet at 1 second and 2 seconds after it is thrown.
This is represented on the graph by the points (1,192) and (2,192).

8. Let $x =$ the width of the sign.
Let $x + 3 =$ the length of the sign.
The area of 54 square units can be represented as follows.
$$A = l \cdot w$$
$$54 = (x+3) \cdot x$$
$$54 = x^2 + 3x$$
$$0 = x^2 + 3x - 54$$
$$0 = (x-6)(x+9)$$
$$x-6 = 0 \quad \text{or} \quad x+9 = 0$$
$$x = 6 \qquad\qquad x = -9$$
Reject -9 because the width cannot be negative.
The width of the sign is 6 units and the length is 9 units.

6.6 Exercise Set

1. $x(x+7) = 0$
$$x = 0 \quad \text{or} \quad x+7 = 0$$
$$x = -7$$
The solution set is $\{-7, 0\}$.

3. $(x-6)(x+4) = 0$
$$x-6 = 0 \quad \text{or} \quad x+4 = 0$$
$$x = 6 \qquad\qquad x = -4$$
The solution set is $\{-4, 6\}$.

5. $(x-9)(5x+4) = 0$
$$x-9 = 0 \quad \text{or} \quad 5x+4 = 0$$
$$x = 9 \qquad\qquad 5x = -4$$
$$x = -\frac{4}{5}$$
The solution set is $\left\{-\dfrac{4}{5}, 9\right\}$.

7. $10(x-4)(2x+9) = 0$
$$x-4 = 0 \quad \text{or} \quad 2x+9 = 0$$
$$x = 4 \qquad\qquad 2x = -9$$
$$x = -\frac{9}{2}$$
The solution set is $\left\{-\dfrac{9}{2}, 4\right\}$.

9. $x^2 + 8x + 15 = 0$
$$(x+5)(x+3) = 0$$
$$x+5 = 0 \quad \text{or} \quad x+3 = 0$$
$$x = -5 \qquad\qquad x = -3$$
Check -5:
$$x^2 + 8x + 15 = 0$$
$$(-5)^2 + 8(-5) + 15 = 0$$
$$25 - 40 + 15 = 0$$
$$0 = 0 \text{ true}$$
Check -3:
$$x^2 + 8x + 15 = 0$$
$$(-3)^2 + 8(-3) + 15 = 0$$
$$9 - 24 + 15 = 0$$
$$0 = 0 \text{ true}$$
The solution set is $\{-5, -3\}$.

11. $x^2 - 2x - 15 = 0$

$(x+3)(x-5) = 0$

$x+3 = 0$ or $x-5 = 0$

$x = -3$ $\qquad x = 5$

The solution set is $\{-3, 5\}$.

13. $x^2 - 4x = 21$

$x^2 - 4x - 21 = 0$

$(x+3)(x-7) = 0$

$x+3 = 0$ or $x-7 = 0$

$x = -3$ $\qquad x = 7$

The solution set is $\{-3, 7\}$.

15. $x^2 + 9x = -8$

$x^2 + 9x + 8 = 0$

$(x+8)(x+1) = 0$

$x+8 = 0$ or $x+1 = 0$

$x = -8$ $\qquad x = -1$

The solution set is $\{-8, -1\}$.

17. $x^2 + 4x = 0$

$x(x+4) = 0$

$x = 0$ or $x+4 = 0$

$\qquad\qquad x = -4$

The solution set is $\{-4, 0\}$.

19. $x^2 - 5x = 0$

$x(x-5) = 0$

$x = 0$ or $x-5 = 0$

$\qquad\qquad x = 5$

The solution set is $\{0, 5\}$.

21. $x^2 = 4x$

$x^2 - 4x = 0$

$x(x-4) = 0$

$x = 0$ or $x-4 = 0$

$\qquad\qquad x = 4$

The solution set is $\{0, 4\}$.

23. $2x^2 = 5x$

$2x^2 - 5x = 0$

$x(2x-5) = 0$

$x = 0$ or $2x-5 = 0$

$\qquad\qquad 2x = 5$

$\qquad\qquad x = \dfrac{5}{2}$

The solution set is $\left\{0, \dfrac{5}{2}\right\}$.

25. $3x^2 = -5x$

$3x^2 + 5x = 0$

$x(3x+5) = 0$

$x = 0$ or $3x+5 = 0$

$\qquad\qquad 3x = -5$

$\qquad\qquad x = -\dfrac{5}{3}$

The solution set is $\left\{-\dfrac{5}{3}, 0\right\}$.

27. $x^2 + 4x + 4 = 0$

$(x-2)^2 = 0$

$x+2 = 0$

$x = -2$

The solution set is $\{-2\}$.

29. $x^2 = 12x - 36$

$x^2 - 12x + 36 = 0$

$(x-6)^2 = 0$

$x-6 = 0$

$x = 6$

The solution set is $\{6\}$.

31.
$$4x^2 = 12x - 9$$
$$4x^2 - 12x + 9 = 0$$
$$(2x - 3)^2 = 0$$
$$2x - 3 = 0$$
$$2x = 3$$
$$x = \frac{3}{2}$$
The solution set is $\left\{\frac{3}{2}\right\}$.

33.
$$2x^2 = 7x + 4$$
$$2x^2 - 7x - 4 = 0$$
$$(2x + 1)(x - 4) = 0$$
$$2x + 1 = 0 \quad \text{or} \quad x - 4 = 0$$
$$2x = -1 \qquad x = 4$$
$$x = -\frac{1}{2}$$
The solution set is $\left\{-\frac{1}{2}, 4\right\}$.

35.
$$5x^2 = 18 - x$$
$$5x^2 + x - 18 = 0$$
$$(5x - 9)(x + 2) = 0$$
$$5x - 9 = 0 \quad \text{or} \quad x + 2 = 0$$
$$5x = 9 \qquad x = -2$$
$$x = \frac{9}{5}$$
The solution set is $\left\{-2, \frac{9}{5}\right\}$.

37.
$$x^2 - 49 = 0$$
$$(x + 7)(x - 7) = 0$$
$$x + 7 = 0 \quad \text{or} \quad x - 7 = 0$$
$$x = -7 \qquad x = 7$$
The solution set is $\{-7, 7\}$.

39.
$$4x^2 - 25 = 0$$
$$(2x + 5)(2x - 5) = 0$$
$$2x + 5 = 0 \quad \text{or} \quad 2x - 5 = 0$$
$$2x = -5 \qquad 2x = 5$$
$$x = -\frac{5}{2} \qquad x = \frac{5}{2}$$
The solution set is $\left\{-\frac{5}{2}, \frac{5}{2}\right\}$.

41.
$$81x^2 = 25$$
$$81x^2 - 25 = 0$$
$$(9x + 5)(9x - 5) = 0$$
$$9x + 5 = 0 \quad \text{or} \quad 9x - 5 = 0$$
$$9x = -5 \qquad 9x = 5$$
$$x = -\frac{5}{9} \qquad x = \frac{5}{9}$$
The solution set is $\left\{-\frac{5}{9}, \frac{5}{9}\right\}$.

43.
$$x(x - 4) = 21$$
$$x^2 - 4x = 21$$
$$x^2 - 4x - 21 = 0$$
$$(x + 3)(x - 7) = 0$$
$$x + 3 = 0 \quad \text{or} \quad x - 7 = 0$$
$$x = -3 \qquad x = 7$$
The solution set is $\{-3, 7\}$.

45.
$$4x(x + 1) = 15$$
$$4x^2 + 4x = 15$$
$$4x^2 + 4x - 15 = 0$$
$$(2x + 5)(2x - 3) = 0$$
$$2x + 5 = 0 \quad \text{or} \quad 2x - 3 = 0$$
$$2x = -5 \qquad 2x = 3$$
$$x = -\frac{5}{2} \qquad x = \frac{3}{2}$$
The solution set is $\left\{-\frac{5}{2}, \frac{3}{2}\right\}$.

47. $(x-1)(x+4) = 14$

$\qquad x^2 + 3x - 4 = 14$

$\qquad x^2 + 3x - 18 = 0$

$\qquad (x+6)(x-3) = 0$

$\qquad x+6 = 0 \qquad x-3 = 0$

$\qquad\qquad\qquad \text{or}$

$\qquad x = -6 \qquad\qquad x = 3$

The solution set is $\{-6, 3\}$.

49. $(x+1)(2x+5) = -1$

$\qquad 2x^2 + 7x + 5 = -1$

$\qquad 2x^2 + 7x + 6 = 0$

$\qquad (2x+3)(x+2) = 0$

$\qquad 2x+3 = 0 \qquad \text{or} \qquad x+2 = 0$

$\qquad 2x = -3 \qquad\qquad\qquad x = -2$

$\qquad x = -\dfrac{3}{2}$

The solution set is $\left\{-2, -\dfrac{3}{2}\right\}$.

51. $\qquad y(y+8) = 16(y-1)$

$\qquad\quad y^2 + 8y = 16y - 16$

$\qquad y^2 - 8y + 16 = 0$

$\qquad\quad (y-4)^2 = 0$

$\qquad\qquad y - 4 = 0$

$\qquad\qquad\qquad y = 4$

The solution set is $\{4\}$.

53. $4y^2 + 20y + 25 = 0$

$\qquad (2y+5)^2 = 0$

$\qquad\quad 2y+5 = 0$

$\qquad\qquad 2y = -5$

$\qquad\qquad\quad y = -\dfrac{5}{2}$

The solution set is $\left\{-\dfrac{5}{2}\right\}$.

55. $\qquad 64w^2 = 48w - 9$

$\qquad 64w^2 - 48w + 9 = 0$

$\qquad\quad (8w-3)^2 = 0$

$\qquad\qquad 8w - 3 = 0$

$\qquad\qquad\quad 8w = 3$

$\qquad\qquad\quad\; w = \dfrac{3}{8}$

The solution set is $\left\{\dfrac{3}{8}\right\}$.

57. $(x-4)(x^2 + 5x + 6) = 0$

$\qquad (x-4)(x+3)(x+2) = 0$

$\qquad x-4 = 0 \;\text{ or }\; x+3 = 0 \quad \text{ or }\; x+2 = 0$

$\qquad\quad x = 4 \qquad\quad x = -3 \qquad\qquad x = -2$

The solution set is $\{-3, -2, 4\}$.

59. $\qquad x^3 - 36x = 0$

$\qquad\quad x(x^2 - 36) = 0$

$\qquad x(x+6)(x-6) = 0$

$\qquad x = 0 \;\text{ or }\; x+6 = 0 \;\text{ or }\; x-6 = 0$

$\qquad\qquad\qquad\quad x = -6 \qquad\quad x = 6$

The solution set is $\{-6, 0, 6\}$.

61. $\qquad y^3 + 3y^2 + 2y = 0$

$\qquad\quad y(y^2 + 3y + 2) = 0$

$\qquad y(y+2)(y+1) = 0$

$\qquad y = 0 \;\text{ or }\; y+2 = 0 \;\text{ or }\; y+1 = 0$

$\qquad\qquad\qquad\quad y = -2 \qquad\quad y = -1$

The solution set is $\{-2, -1, 0\}$.

63. $\qquad 2(x-4)^2 + x^2 = x(x+50) - 46x$

$\qquad 2(x^2 - 8x + 16) + x^2 = x^2 + 50x - 46x$

$\qquad 2x^2 - 16x + 32 + x^2 = x^2 + 4x$

$\qquad\qquad 3x^2 - 16x + 32 = x^2 + 4x$

$\qquad\qquad 2x^2 - 20x + 32 = 0$

$\qquad\qquad 2(x^2 - 10x + 16) = 0$

$\qquad\qquad 2(x-8)(x-2) = 0$

$\qquad x-8 = 0 \;\text{ or }\; x-2 = 0$

$\qquad\quad x = 8 \qquad\qquad x = 2$

The solution set is $\{2, 8\}$.

65. $\quad (x-2)^2 - 5(x-2) + 6 = 0$

$$[(x-2)-3][(x-2)-2] = 0$$

$$(x-5)(x-4) = 0$$

$$x-5 = 0 \quad \text{or} \quad x-4 = 0$$

$$x = 5 \qquad\qquad x = 4$$

The solution set is $\{4, 5\}$.

67. $\quad h = -16t^2 + 20t + 300$

Substitute 0 for h and solve for t.

$$0 = -16t^2 + 20t + 300$$

$$-16t^2 + 20t + 300 = 0$$

$$-4t\left(4t^2 - 5t - 75\right) = 0$$

$$-4t(4t + 15)(t - 5) = 0$$

$$-4t = 0 \text{ or } 4t + 15 = 0 \quad \text{or} \quad t - 5 = 0$$

$$t = 0 \qquad 4t = -15 \qquad t = 5$$

$$t = -\frac{15}{4} = -3.75$$

Reject $t = 0$ since this represents the time before the ball was thrown. Also discard $t = -3.75$ since time cannot be negative. The only solution that makes sense is 5. So it will take 5 seconds for the ball to hit the ground. Each tick mark represents one second.

69. Substitute 276 for h and solve for t.

$$276 = -16t^2 + 20t + 300$$

$$16t^2 - 20t - 24 = 0$$

$$4\left(4t^2 - 5t - 6\right) = 0$$

$$4(4t + 3)(t - 2) = 0$$

$$4t + 3 = 0 \qquad \text{or} \qquad t - 2 = 0$$

$$4t = -3 \qquad\qquad t = 2$$

$$t = -\frac{3}{4}$$

Reject $t = -\dfrac{3}{4}$ since time cannot be negative. The ball's height will be 276 feet 2 seconds after it is thrown. This corresponds to the point $(2, 276)$ on the graph.

71. $\quad h = -16t^2 + 72t$

Substitute 32 for h and solve for t.

$$32 = -16t^2 + 72t$$

$$16t^2 - 72t + 32 = 0$$

$$8\left(2t^2 - 9t + 4\right) = 0$$

$$8(2t - 1)(t - 4) = 0$$

$$2t - 1 = 0 \quad \text{or} \quad t - 4 = 0$$

$$t = \frac{1}{2} \qquad\qquad t = 4$$

The debris will be 32 feet above the ground $\dfrac{1}{2}$ second after the explosion and 4 seconds after the explosion.

73. $\quad S = 2x^2 - 12x + 82$

Substitute 72 for S and solve for x.

$$S = 2x^2 - 12x + 82$$

$$66 = 2x^2 - 12x + 82$$

$$0 = 2x^2 - 12x + 16$$

$$0 = 2(x^2 - 6x + 8)$$

$$0 = x^2 - 6x + 8$$

$$0 = (x - 2)(x - 4)$$

$$x - 2 = 0 \qquad \text{or} \qquad x - 4 = 0$$

$$x = 2 \qquad\qquad x = 4$$

International travelers spent \$66 billion 2 years and 4 years after 2000, or 2002 and 2004.

75. International travelers spent \$66 billion 2 years and 4 years after 2000. This corresponds to the point $(2, 66)$ and the point $(4, 66)$ on the graph.

77. $\quad P = -10x^2 + 475x + 3500$

Substitute 7250 for P and solve for x.

$$7250 = -10x^2 + 475x + 3500$$

$$10x^2 - 475x + 3750 = 0$$

$$5\left(2x^2 - 95x + 750\right) = 0$$

$$5(x - 10)(2x - 75) = 0$$

$$x - 10 = 0 \quad \text{or} \quad 2x - 75 = 0$$

$$x = 10 \qquad\qquad 2x = 75$$

$$x = \frac{75}{2} \text{ or } 37.5$$

The alligator population will have increased to 7250 after 10 years. (Discard 37.5 because this value is outside of $0 \le x \le 12$.)

79. The solution in Exercise 77 corresponds to the point $(10, 7250)$ on the graph.

81. $N = \dfrac{t^2 - t}{2}$

Substitute 45 for N and solve for t.

$$45 = \frac{t^2 - t}{2}$$

$$2 \cdot 45 = 2\left(\frac{t^2 - t}{2}\right)$$

$$90 = t^2 - t$$

$$0 = t^2 - t - 90$$

$$0 = (t - 10)(t + 9)$$

$$t - 10 = 0 \quad \text{or} \quad t + 9 = 0$$

$$t = 10 \qquad\qquad t = -9$$

Reject $t = -9$ since the number of teams cannot be negative. If 45 games are scheduled, there are 10 teams in the league.

83. Let x = the width of the parking lot.
Then $x + 3$ = the length.

$$l \cdot w = A$$

$$(x + 3)(x) = 180$$

$$x^2 + 3x = 180$$

$$x^2 + 3x - 180 = 0$$

$$(x + 15)(x - 12) = 0$$

$$x + 15 = 0 \quad \text{or} \quad x - 12 = 0$$

$$x = -15 \qquad\qquad x = 12$$

Reject $x = -15$ since the width cannot be negative. Then $x = 12$ and $x + 3 = 15$, so the length is 15 yards and the width is 12 yards.

85. Use the formula for the area of a triangle where x is the base and $x + 1$ is the height.

$$\frac{1}{2}bh = A$$

$$\frac{1}{2}x(x + 1) = 15$$

$$2\left[\frac{1}{2}x(x + 1)\right] = 2 \cdot 15$$

$$x(x + 1) = 30$$

$$x^2 + x = 30$$

$$x^2 + x - 30 = 0$$

$$(x + 6)(x - 5) = 0$$

$$x + 6 = 0 \quad \text{or} \quad x - 5 = 0$$

$$x = -6 \qquad\qquad x = 5$$

Reject $x = -6$ since the length of the base cannot be negative. Then $x = 5$ and $x + 1 = 6$, so the base is 5 centimeters and the height is 6 centimeters.

87. a. Area of border

$$= \overbrace{(2x + 12)(2x + 10)}^{\text{Area of a large rectangle}} - \overbrace{10 \cdot 12}^{\substack{\text{Area of} \\ \text{flower bed}}}$$

$$= 4x^2 + 20x + 24x + 120 - 120$$

$$= 4x^2 + 44x$$

b. Find the width of the border for which the area of the border would be 168 square feet.

$$4x^2 + 44x = 168$$

$$4x^2 + 44x - 168 = 0$$

$$4\left(x^2 + 11x - 42\right) = 0$$

$$4(x + 14)(x - 3) = 0$$

$$x + 14 = 0 \quad \text{or} \quad x - 3 = 0$$

$$x = -14 \qquad\qquad x = 3$$

Reject $x = -14$ since the width of the border cannot be negative. You should prepare a strip that is 3 feet wide for the border.

89. Answers will vary.

91. does not make sense; Explanations will vary. Sample explanation: Though 4 cannot equal 0 and can therefore be ignored, $4x$ will equal 0 when $x = 0$ and so it cannot be ignored.

93. makes sense

95. false; Changes to make the statement true will vary. A sample change is: If $(x+3)(x-4)=0$, then $x+3=0$ or $x-4=0$.

97. false; Changes to make the statement true will vary. A sample change is: Some equations solved by factoring have more than 2 solutions and some have fewer than 2 solutions

99. If -3 and 5 are solutions of the quadratic equation, then $x-(-3)=x+3$ and $x-5$ must be factors of the polynomial on the left side when the quadratic equation is written in standard form.
$$(x+3)(x-5)=0$$
$$x^2-5x+3x-15=0$$
$$x^2-2x-15=0$$
Thus, $x^2-2x-15=0$ is a quadratic equation in standard form whose solutions are -3 and 5.

101. $3^{x^2-9x+20}=1$

Because $3^0=1$ (and there is no other power of 3 that is equal to 1), $x^2-9x+20=0$. Solve this equation.
$$x^2-9x+20=0$$
$$(x-4)(x-5)=0$$
$$x-4=0 \quad \text{or} \quad x-5=0$$
$$x=4 \qquad\qquad x=5$$
The solution set is $\{4,5\}$.

103. $y=x^2-x-2$

To match this equation with its graph, find the intercepts.
To find the y-intercept, let $x=0$ and solve for y.
$$y=0^2-0-2=-2$$
The y-intercept is -2.
To find the x intercepts (if any), let $y=0$ and solve for x.
$$0=x^2-x-2$$
$$0=(x+1)(x-2)$$
The x intercepts are -1 and 2.
The only graph with y-intercept -2 and x-intercepts -1 and 2 is graph, c, so this is the graph of $y=x^2-x-2$

105. $y=x^2-4$

To match this equation with its graph, find the intercepts.
To find the y-intercepts, let $x=0$ and solve for y.
$$y=0^2-4=-4$$
The y-intercept is -4.
To find the x-intercepts (if any), let $y=0$ and solve for x.
$$0=x^2-4$$
$$0=(x+2)(x-2)$$
$$x+2=0 \quad \text{or} \quad x-2=0$$
$$x=-2 \qquad\qquad x=2$$
The x-intercepts are -2 and 2.
The only graph with y-intercept -4 and x-intercepts -2 and 2 is graph d, so this is the graph of $y=x^2-4$.

107. $y=x^2+3x-4$
$$x^2+3x-4=0$$
Graph $y=x^2+3x-4$ and use the graph to find the x-intercepts.

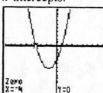 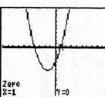

The x-intercepts of $y=x^2+3x-4$ are -4 and 1, so the solution set of $x^2+3x-4=0$ is $\{-4,1\}$.
Check -4:
$$x^2+3x-4=0$$
$$(-4)^2+3(-4)-4=0$$
$$16-12-4=0$$
$$4-4=0$$
$$0=0, \text{ true}$$
Check 1:
$$x^2+3x-4=0$$
$$1^2+3(1)-4=0$$
$$1+3-4=0$$
$$4-4=0$$
$$0=0, \text{ true}$$

109. $y = (x-2)(x+3) - 6$

$(x-2)(x+3) - 6 = 0$

Graph $y = (x-2)(x-3) - 6$ and use the graph to find the x-intercepts.

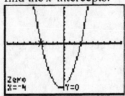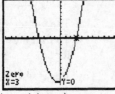

The x-intercepts of $y = (x-2)(x+3) - 6$ are -4

and 3, so the solution set of $(x-2)(x+3) - 6 = 0$ is

$\{-4, 3\}$.

Check -4:

$(x-2)(x+3) - 6 = 0$

$(-4-2)(-4+3) - 6 = 0$

$(-6)(-1) - 6 = 0$

$6 - 6 = 0$

$= 0$, true

Check 3:

$(x-2)(x+3) - 6 = 0$

$(3-1)(3+3) - 6 = 0$

$(1)(6) - 6 = 0$

$6 - 6 = 0$

$0 = 0$, true

111. Answers will vary depending on the exercises chosen.

113. Graph $y = -\dfrac{2}{3}x + 1$ using the slope $-\dfrac{2}{3} = \dfrac{-2}{3}$ and y-intercept 1.

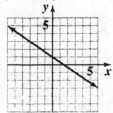

114. $\left(\dfrac{8x^4}{4x^7}\right)^2 = \left(\dfrac{8}{4} \cdot x^{4-7}\right) = \left(2x^{-3}\right)^2$

$= 2^2 \cdot \left(x^{-3}\right)^2 = 4x^{-6} = \dfrac{4}{x^6}$

115.
$$5x + 28 = 6 - 6x$$
$$5x + 6x + 28 = 6 - 6x + 6x$$
$$11x + 28 = 6$$
$$11x + 28 - 28 = 6 - 28$$
$$11x = -22$$
$$\dfrac{11x}{11} = \dfrac{-22}{11}$$
$$x = -2$$

The solution set is $\{-2\}$.

116. $\dfrac{250x}{100-x} = \dfrac{250(60)}{100-60} = \dfrac{15000}{40} = 375$

117. When x is replaced with 4, the denominator is 0. Division by zero is undefined.

$7x - 28 = 7(4) - 28 = 28 - 28 = 0$

118. $\dfrac{x^2 + 6x + 5}{x^2 - 25} = \dfrac{(x+5)(x+1)}{(x+5)(x-5)}$

$= \dfrac{\cancel{(x+5)}(x+1)}{\cancel{(x+5)}(x-5)}$

$= \dfrac{x+1}{x-5}$

Chapter 6 Review Exercises

1. $30x - 45 = 15(2x - 3)$

2. $12x^3 + 16x^2 - 400x = 4x\left(3x^2 + 4x - 100\right)$

3. $30x^4 y + 15x^3 y + 5x^2 y = 5x^2 y\left(6x^2 + 3x + 1\right)$

4. $7(x+3) - 2(x+3) = (x+3)(7-2)$
$= (x+3) \cdot 5$ or $5(x+3)$

5. $7x^2(x+y) - (x+y) = 7x^2(x+y) - 1(x+y)$
$= (x+y)(7x^2 - 1)$

6. $x^3 + 3x^2 + 2x + 6 = \left(x^3 + 3x^2\right) + (2x+6)$
$= x^2(x+3) + 2(x+3)$
$= (x+3)\left(x^2 + 2\right)$

7. $xy + y + 4x + 4 = (xy + y) + (4x + 4)$
    $= y(x + 1) + 4(x + 1)$
    $= (x + 1)(y + 4)$

8. $x^3 + 5x + x^2 + 5 = (x^3 + 5x) + (x^2 + 5)$
    $= x(x^2 + 5) + 1(x^2 + 5)$
    $= (x^2 + 5)(x + 1)$

9. $xy + 4x - 2y - 8 = (xy + 4x) + (-2y - 8)$
    $= x(y + 4) - 2(y + 4)$
    $= (y + 4)(x - 2)$

10. $x^2 - 3x + 2 = (x - 2)(x - 1)$

11. $x^2 - x - 20 = (x - 5)(x + 4)$

12. $x^2 + 19x + 48 = (x + 3)(x + 16)$

13. $x^2 - 6xy + 8y^2 = (x - 4y)(x - 2y)$

14. $x^2 + 5x - 9$ is prime because there is no pair of integers whose product is -9 and whose sum is 5.

15. $x^2 + 16xy - 17y^2 = (x + 17y)(x - y)$

16. $3x^2 + 6x - 24 = 3(x^2 + 2x - 8)$
    $= 3(x + 4)(x - 2)$

17. $3x^3 - 36x^2 + 33x = 3x(x^2 - 12x + 11)$
    $= 3x(x - 11)(x - 1)$

18. Factor $3x^2 + 17x + 10$ by trial and error or by grouping. To factor by grouping, find two integers whose product is $ac = 3 \cdot 10 = 30$ and whose sum is $b = 17$. These integers are 15 and 2.
    $3x^2 + 17x + 10 = 3x^2 + 15x + 2x + 10$
    $= 3x(x + 5) + 2(x + 5)$
    $= (x + 5)(3x + 2)$

19. Factor $5y^2 - 17y + 6$ by trial and error or by grouping. To factor by trial and error, start with the First terms, which must be $5y$ and y. Because the middle term is negative, the factors of 6 must both be negative. Try various combinations until the correct middle term is obtained.
    $(5y - 1)(y - 6) = 5y^2 - 31y + 6$
    $(5y - 6)(y - 1) = 5y^2 - 11y + 6$
    $(5y - 3)(y - 2) = 5y^2 - 13y + 6$
    $(5y - 2)(y - 3) = 5y^2 - 17y + 6$
    Thus, $5y^2 - 17y + 6 = (5y - 2)(y - 3)$.

20. $4x^2 + 4x - 15 = (2x + 5)(2x - 3)$

21. Factor $5y^2 + 11y + 4$ by trial and error. The first terms must be $5y$ and y. Because the middle term is positive, the factors of 4 must both be positive. Try all the combinations.
    $(5y + 2)(y + 2) = 5y^2 + 12y + 4$
    $(5y + 4)(y + 1) = 5y^2 + 9y + 4$
    $(5y + 1)(y + 4) = 5y^2 + 21y + 4$
    None of these possibilities gives the required middle term, $11x$, and there are no more possibilities to try, so $5y^2 + 11y + 4$ is prime.

22. First factor out the GCF, 2. Then factor the resulting trinomial by trial and error or by grouping.
    $8x^2 + 8x - 6 = 2(4x^2 + 4x - 3)$
    $= 2(2x + 3)(2x - 1)$

23. $2x^3 + 7x^2 - 72x = x(2x^2 + 7x - 72)$
    $= x(2x - 9)(x + 8)$

24. $12y^3 + 28y^2 + 8y = 4y(3y^2 + 7y + 2)$
    $= 4y(3y + 1)(y + 2)$

25. $2x^2 - 7xy + 3y^2 = (2x - y)(x - 3y)$

26. $5x^2 - 6xy - 8y^2 = (5x + 4y)(x - 2y)$

27. $4x^2 - 1 = (2x)^2 - 1^2 = (2x + 1)(2x - 1)$

28. $81 - 100y^2 = 9^2 - (10y)^2$
$= (9 + 10y)(9 - 10y)$

29. $25a^2 - 49b^2 = (5a)^2 - (7b)^2$
$= (5a + 7b)(5a - 7b)$

30. $z^4 - 16 = \left(z^2\right)^2 - 4^2$
$= \left(z^2 + 4\right)\left(z^2 - 4\right)$
$= \left(z^2 + 4\right)(z + 2)(z - 2)$

31. $2x^2 - 18 = 2\left(x^2 - 9\right) = 2(x + 3)(x - 3)$

32. $x^2 + 1$ is prime because it is the sum of two squares with no common factor other than 1.

33. $9x^3 - x = x\left(9x^2 - 1\right) = x(3x + 1)(3x - 1)$

34. $18xy^2 - 8x = 2x\left(9y^2 - 4\right)$
$= 2x(3y + 2)(3y - 2)$

35. $x^2 + 22x + 121 = x^2 + 2(11x) + 11^2$
$= (x + 11)^2$

36. $x^2 - 16x + 64 = x^2 - 2(8 \cdot x) + 8^2$
$= (x - 8)^2$

37. $9y^2 + 48y + 64 = (3y)^2 + 2(3y \cdot 8) + 8^2$
$= (3y + 8)^2$

38. $16x^2 - 40x + 25 = (4x)^2 - 2(4x \cdot 5) + 5^2$
$= (4x - 5)^2$

39. $25x^2 + 15x + 9$ is prime.
Note that to be a perfect square trinomial, the middle term would have to be $2(5x \cdot 3) = 30x$.

40. $36x^2 + 60xy + 25y^2$
$= (6x)^2 + 2(6x \cdot 5y) + (5y)^2$
$= (6x + 5y)^2$

41. $25x^2 - 40xy + 16y^2$
$= (5x)^2 - 2(5x \cdot 4y) + (4y)^2$
$= (5x - 4y)^2$

42. $x^3 - 27 = x^3 - 3^2 = (x - 3)\left(x^2 + 3x + 9\right)$

43. $64x^3 + 1 = (4x)^3 + 1^3$
$= (4x + 1)\left[(4x)^2 - 4x \cdot 1 + 1^2\right]$
$= (4x + 1)\left(16x^2 - 4x + 1\right)$

44. $54x^3 - 16y^3$
$= 2\left(27x^3 - 8y^3\right)$
$= 2\left[(3x)^3 - (2y)^3\right]$
$= 2(3x - 2y)\left[(3x)^2 + 3x \cdot 2y + (2y)^2\right]$
$= 2(3x - 2y)\left(9x^2 + 6xy + 4y^2\right)$

45. $27x^3y + 8y = y\left(27x^3 + 8\right)$
$= y\left[(3x)^3 + 2^3\right]$
$= y(3x + 2)\left[(3x)^2 - 3x \cdot 2 + 2^2\right]$
$= y(3x + 2)\left(9x^2 - 6x + 4\right)$

46. Area of outer square $= a^2$
Area of inner square $= 3^2 = 9$
Area of shaded region $= a^2 - 9$
$= (a + 3)(a - 3)$

47. Area of large square $= a^2$
Area of each small corner square $= b^2$
Area of four corner squares $= 4b^2$
Area of shaded region $= a^2 - 4b^2$
$= (a + 2b)(a - 2b)$

48. Area on the left:

Area of large square = A^2
Area of each rectangle: $A \cdot 1 = A$
Area of two rectangles = $2A$
Area of small square = $1^2 = 1$
Area on the right:

Area of square = $(A+1)^2$

This geometric model illustrates the

factorization $A^2 + 2A - 1 = (A+1)^2$

49. $x^3 - 8x^2 + 7x = x(x^2 - 8x + 7) = x(x-7)(x-1)$

50. $10y^2 + 9y + 2 = (5y+2)(2y+1)$

51. $128 - 2y^2 = 2(64 - y^2) = 2(8+y)(8-y)$

52. $9x^2 + 6x + 1 = (3x)^2 + 2(3x) + 1^2 = (3x+1)^2$

53. $20x^7 - 36x^3 = 4x^3(5x^4 - 9)$

54. $x^3 - 3x^2 - 9x + 27$

$= (x^3 - 3x^2) + (-9x + 27)$

$= x^2(x-3) - 9(x-3)$

$= (x-3)(x^2 - 9)$

$= (x-3)(x+3)(x-3)$

or $(x-3)^2(x+3)$

55. $y^2 + 16$ is prime because it is the sum of two squares with no common factor other than 1.

56. $2x^3 + 19x^2 + 35x = x(2x^2 + 19x + 35)$

$= x(2x+5)(x+7)$

57. $3x^3 - 30x^2 + 75x = 3x(x^2 - 10x + 25)$

$= 3x(x-5)^2$

58. $3x^5 - 24x^2 = 3x^2(x^3 - 8)$

$= 3x^2(x^3 - 2^3)$

$= 3x^2(x-2)(x^2 + 2x + 4)$

59. $4y^4 - 36y^2 = 4y^2(y^2 - 9)$

$= 4y^2(y+3)(y-3)$

60. $5x^2 + 20x - 105 = 5(x^2 + 4x - 21)$

$= 5(x+7)(x-3)$

61. $9x^2 + 8x - 3$ is prime because there are no two integers whose product is $ac = -27$ and whose sum is 8.

62. $10x^5 - 44x^4 + 16x^3 = 2x^3(5x^2 - 22x + 8)$

$= 2x^3(5x-2)(x-4)$

63. $100y^2 - 49 = (10y)^2 - 7^2$

$= (10y+7)(10y-7)$

64. $9x^5 - 18x^4 = 9x^4(x-2)$

65. $x^4 - 1 = (x^2)^2 - 1^2$

$= (x^2 + 1)(x^2 - 1)$

$= (x^2 + 1)(x+1)(x-1)$

66. $2y^3 - 16 = 2(y^3 - 8)$

$= 2(y^3 - 2^3)$

$= 2(y-2)(y^2 + 2y + 2^2)$

$= 2(y-2)(y^2 + 2y + 4)$

67. $x^3 + 64 = x^3 + 4^3$

$= (x+4)(x^2 - 4x + 4^2)$

$= (x+4)(x^2 - 4x + 16)$

68. $6x^2 + 11x - 10 = (3x-2)(2x+5)$

69. $3x^4 - 12x^2 = 3x^2(x^2 - 4)$

$= 3x^2(x+2)(x-2)$

70. $x^2 - x - 90 = (x-10)(x+9)$

71. $25x^2 + 25xy + 6y^2 = (5x+2y)(5x+3y)$

72. $x^4 + 125x = x(x^3 + 125)$

$\qquad = x(x^3 + 5^3)$

$\qquad = x(x+5)(x^2 - 5x + 5^2)$

$\qquad = x(x+5)(x^2 - 5x + 25)$

73. $32y^3 + 32y^2 + 6y = 2y(16y^2 + 16y + 3)$

$\qquad\qquad = 2y(4y+3)(4y+1)$

74. $2y^2 - 16y + 32 = 2(y^2 - 8y + 16)$

$\qquad\qquad = 2(y-4)^2$

75. $x^2 - 2xy - 35y^2 = (x+5y)(x-7y)$

76. $x^2 + 7x + xy + 7y = x(x+7) + y(x+7)$

$\qquad\qquad = (x+7)(x+y)$

77. $9x^2 + 24xy + 16y^2$

$\qquad = (3x)^2 + 2(3x \cdot 4y) + (4y)^2$

$\qquad = (3x+4y)^2$

78. $2x^4 y - 2x^2 y = 2x^2 y(x^2 - 1)$

$\qquad\qquad = 2x^2 y(x+1)(x-1)$

79. $100y^2 - 49z^2 = (10y)^2 - (7z)^2$

$\qquad\qquad = (10y+7z)(10y-7z)$

80. $x^2 + xy + y^2$ is prime.

Note that to be a perfect square trinomial, the middle term would have to be $2xy$.

81. $3x^4 y^2 - 12x^2 y^4 = 3x^2 y^2(x^2 - 4y^2)$

$\qquad\qquad = 3x^2 y^2(x+2y)(x-2y)$

82. $x(x-12) = 0$

$x = 0 \quad \text{or} \quad x - 12 = 0$

$\qquad\qquad x = 12$

The solution set is $\{0, 12\}$.

83. $3(x-7)(4x+9) = 0$

$x - 7 = 0 \quad \text{or} \quad 4x + 9 = 0$

$x = 7 \qquad\qquad 4x = -9$

$\qquad\qquad\qquad x = -\dfrac{9}{4}$

The solution set is $\left\{-\dfrac{9}{4}, 7\right\}$.

84. $x^2 + 5x - 14 = 0$

$(x+7)(x-2) = 0$

$x + 7 = 0 \quad \text{or} \quad x - 2 = 0$

$x = -7 \qquad\qquad x = 2$

The solution set is $\{-7, 2\}$.

85. $5x^2 + 20x = 0$

$5x(x+4) = 0$

$5x = 0 \quad \text{or} \quad x + 4 = 0$

$x = 0 \qquad\qquad x = -4$

The solution set is $\{-4, 0\}$.

86. $\qquad 2x^2 + 15x = 8$

$2x^2 + 15x - 8 = 0$

$(2x-1)(x+8) = 0$

$2x - 1 = 0 \quad \text{or} \quad x + 8 = 0$

$2x = 1 \qquad\qquad x = -8$

$x = \dfrac{1}{2}$

The solution set is $\left\{-8, \dfrac{1}{2}\right\}$.

87. $\qquad x(x-4) = 32$

$x^2 - 4x = 32$

$x^2 - 4x - 32 = 0$

$(x+4)(x-8) = 0$

$x + 4 = 0 \quad \text{or} \quad x - 8 = 0$

$x = -4 \qquad\qquad x = 8$

The solution set is $\{-4, 8\}$.

88. $(x+3)(x-2)=50$

$x^2+x-6=50$

$x^2+x-56=0$

$(x+8)(x-7)=0$

$x+8=0$ or $x-7=0$

$x=-8$ $x=7$

The solution set is $\{-8,7\}$.

89. $x^2=14x-49$

$x^2-14x+49=0$

$(x-7)^2=0$

$x-7=0$

$x=7$

The solution set is $\{7\}$.

90. $9x^2=100$

$9x^2-100=0$

$(3x+10)(3x-10)=0$

$3x+10=0$ or $3x-10=0$

$3x=-10$ $3x=10$

$x=-\dfrac{10}{3}$ $x=\dfrac{10}{3}$

The solution set is $\left\{-\dfrac{10}{3},\dfrac{10}{3}\right\}$.

91. $3x^2+21x+30=0$

$3\left(x^2+7x+10\right)=0$

$3(x+5)(x+2)=0$

$x+5=0$ or $x+2=0$

$x=-5$ $x=-2$

The solution set is $\{-5,-2\}$.

92. $3x^2=22x-7$

$3x^2-22x+7=0$

$(3x-1)(x-7)=0$

$3x-1=0$ or $x-7=0$

$3x=1$ $x=7$

$x=\dfrac{1}{3}$

The solution set is $\left\{\dfrac{1}{3},7\right\}$.

93. $h=-16t^2+16t+32$

Substitute 0 for h and solve for t.

$0=-16t^2+16t+32$

$16t^2-16t-32=0$

$16\left(t^2-t-2\right)=0$

$16(t+1)(t-2)=0$

$t+1=0$ or $t-2=0$

$t=-1$ $t=2$

Because time cannot be negative, reject the solution $t=-1$.

It will take you 2 seconds to hit the water.

94. Let x = the width of the sign.

Then $x+3$ = the length of the sign.

Use the formula for the area of a rectangle.

$l\cdot w=A$

$(x+3)(x)=40$

$x^2+3x=40$

$x^2+3x-40=0$

$(x+8)(x-5)=0$

$x+8=0$ or $x-5=0$

$x=-8$ $x=5$

A rectangle cannot have a negative length. Thus $x=5$, and $x+3=8$.

The length of the sign is 8 feet and the width is 5 feet.

95. Area of garden $=x(x-3)=88$

$x(x-3)=88$

$x^2-3x=88$

$x^2-3x-88=0$

$(x-11)(x+8)=0$

$x-11=0$ or $x+8=0$

$x=11$ $x=-8$

Because a length cannot be negative, reject $x=-8$. Each side of the square lot is 11 meters, that is, the dimensions of the square lot are 11 meters by 11 meters.

Chapter 6 Test

1. $x^2 - 9x + 18 = (x-3)(x-6)$

2. $x^2 - 14x + 49 = x^2 - 2(x \cdot 7) + 7^2$
   $$= (x-7)^2$$

3. $15y^4 - 35y^3 + 10y^2 = 5y^2\left(3y^2 - 7y + 2\right)$
   $$= 5y^2(3y-1)(y-2)$$

4. $x^3 + 2x^2 + 3x + 6 = \left(x^3 + 2x^2\right) + (3x+6)$
   $$= x^2(x+2) + 3(x+2)$$
   $$= (x+2)\left(x^2+3\right)$$

5. $x^2 - 9x = x(x-9)$

6. $x^3 + 6x^2 - 7x = x\left(x^2 + 6x - 7\right)$
   $$= x(x+7)(x-1)$$

7. $14x^2 + 64x - 30 = 2\left(7x^2 + 32x - 15\right)$
   $$= 2(7x-3)(x+5)$$

8. $25x^2 - 9 = (5x)^2 - 3^2$
   $$= (5x+3)(5x-3)$$

9. $x^3 + 8 = x^3 + 2^3 = (x+2)\left(x^2 - 2x + 2^2\right)$
   $$= (x+2)\left(x^2 - 2x + 4\right)$$

10. $x^2 - 4x - 21 = (x+3)(x-7)$

11. $x^2 + 4$ is prime.

12. $6y^3 + 9y^2 + 3y = 3y\left(2y^2 + 3y + 1\right)$
    $$= 3y(2y+1)(y+1)$$

13. $4y^2 - 36 = 4\left(y^2 - 9\right) = 4(y+3)(y-3)$

14. $16x^2 + 48x + 36$
    $$= 4\left(4x^2 + 12x + 9\right)$$
    $$= 4\left[(2x)^2 + 2(2x \cdot 3) + 3^2\right]$$
    $$= 4(2x+3)^2$$

15. $2x^4 - 32 = 2\left(x^4 - 16\right)$
    $$= 2\left(x^2 + 4\right)\left(x^2 - 4\right)$$
    $$= 2\left(x^2 + 4\right)(x+2)(x-2)$$

16. $36x^2 - 84x + 49 = (6x)^2 - 2(6x \cdot 7) + 7^2$
    $$= (6x-7)^2$$

17. $7x^2 - 50x + 7 = (7x-1)(x-7)$

18. $x^3 + 2x^2 - 5x - 10$
    $$= \left(x^3 + 2x^2\right) + (-5x - 10)$$
    $$= x^2(x+2) - 5(x+2)$$
    $$= (x+2)\left(x^2 - 5\right)$$

19. $12y^3 - 12y^2 - 45y = 3y\left(4y^2 - 4y - 15\right)$
    $$= 3y(2y+3)(2y-5)$$

20. $y^3 - 125 = y^3 - 5^3$
    $$= (y-5)\left(y^2 + 5y + 5^2\right)$$
    $$= (y-5)\left(y^2 + 5y + 25\right)$$

21. $5x^2 - 5xy - 30y^2 = 5\left(x^2 - xy - 6y^2\right)$
    $$= 5(x-3y)(x+2y)$$

22. $x^2 + 2x - 24 = 0$
    $(x+6)(x-4) = 0$
    $x + 6 = 0$ or $x - 4 = 0$
    $x = -6 \qquad x = 4$
    The solution set is $\{-6, 4\}$.

23.
$$3x^2 - 5x = 2$$
$$3x^2 - 5x - 2 = 0$$
$$(3x+1)(x-2) = 0$$
$$3x+1 = 0 \quad \text{or} \quad x-2 = 0$$
$$3x = -1 \qquad\qquad x = 2$$
$$x = -\frac{1}{3}$$

The solution set is $\left\{-\dfrac{1}{3}, 2\right\}$.

24.
$$x(x-6) = 16$$
$$x^2 - 6x = 16$$
$$x^2 - 6x - 16 = 0$$
$$(x+2)(x-8) = 0$$
$$x+2 = 0 \quad \text{or} \quad x-8 = 0$$
$$x = -2 \qquad\qquad x = 8$$

The solution set is $\{-2, 8\}$.

25.
$$6x^2 = 21x$$
$$6x^2 - 21x = 0$$
$$3x(2x-7) = 0$$
$$3x = 0 \quad \text{or} \quad 2x-7 = 0$$
$$x = 0 \qquad\qquad 2x = 7$$
$$x = \frac{7}{2}$$

The solution set is $\left\{0, \dfrac{7}{2}\right\}$.

26.
$$16x^2 = 81$$
$$16x^2 - 81 = 0$$
$$(4x+9)(4x-9) = 0$$
$$4x+9 = 0 \quad \text{or} \quad 4x-9 = 0$$
$$4x = -9 \qquad\qquad 4x = 9$$
$$x = -\frac{9}{4} \qquad\qquad x = \frac{9}{4}$$

The solution set is $\left\{-\dfrac{9}{4}, \dfrac{9}{4}\right\}$.

27.
$$(5x+4)(x-1) = 2$$
$$5x^2 - x - 4 = 2$$
$$5x^2 - x - 6 = 0$$
$$(5x-6)(x+1) = 0$$
$$5x-6 = 0 \quad \text{or} \quad x+1 = 0$$
$$5x = 6 \qquad\qquad x = -1$$
$$x = \frac{6}{5}$$

The solution set is $\left\{-1, \dfrac{6}{5}\right\}$.

28. Area of large square $= x^2$
Area of each small (corner) square $= 1^2 = 1$
Area of four corner squares $= 4 \cdot 1 = 4$
Area of shaded region $= x^2 - 4$
$$= (x+2)(x-2)$$

29. $h = -16t^2 + 80t + 96$
Substitute 0 for h and solve for t.
$$0 = -16t^2 + 80t + 96$$
$$16t^2 - 80t - 96 = 0$$
$$16\left(t^2 - 5t - 6\right) = 0$$
$$16(t-6)(t+1) = 0$$
$$t-6 = 0 \quad \text{or} \quad t+1 = 0$$
$$t = 6 \qquad\qquad t = -1$$

Since time cannot be negative, reject $t = -1$. The rocket will reach the ground after 6 seconds.

30. Let x = the width of the garden.
Then $x+6$ = the length of the garden.
$$(x+6)(x) = 55$$
$$x^2 + 6x = 55$$
$$x^2 + 6x - 55 = 0$$
$$(x+11)(x-5) = 0$$
$$x+11 = 0 \quad \text{or} \quad x-5 = 0$$
$$x = -11 \qquad\qquad x = 5$$

Since the width cannot be negative, reject $x = -11$.
Then $x = 5$ and $x+6 = 11$, so the width is 5 feet and the length is 11 feet.

Cumulative Review Exercises (Chapters 1-6)

1. $6\left[5+2(3-8)-3\right] = 6\left[5+2(-5)-3\right]$
$$= 6\left[5-10-3\right]$$
$$= 6(-8) = -48$$

2. $4(x-2) = 2(x-4)+3x$
$$4x-8 = 2x-8+3x$$
$$4x-8 = 5x-8$$
$$-x = 0$$
$$x = 0$$
The solution set is $\{0\}$.

3. $\dfrac{x}{2}-1 = \dfrac{x}{3}+1$
$$6\left(\dfrac{x}{2}-1\right) = 6\left(\dfrac{x}{3}+1\right)$$
$$3x-6 = 2x+6$$
$$x = 12$$
The solution set is $\{12\}$.

The solution is 12.

4. $5-5x > 2(5-x)+1$
$$5-5x > 10-2x+1$$
$$5-5x > 11-2x$$
$$5-5x+2x > 11-2x+2x$$
$$5-3x > 11$$
$$5-3x-5 > 11-5$$
$$-3x > 6$$
$$\dfrac{-3x}{-3} < \dfrac{6}{-3}$$
$$x < -2$$
$(-\infty, -2)$

5. Let x = the measure of each of the two base angles. Then $3x-10$ = the measure of the third angle.
$$x+x+(3x-10) = 180$$
$$5x-10 = 180$$
$$5x = 190$$
$$x = 38$$
$$3x-10 = 104$$
The measures of the three angles of the triangle are $38°$, $38°$, and $104°$.

6. Let x = the cost of the dinner before tax.
$$x+0.06x = 159$$
$$1.06x = 159$$
$$\dfrac{1.06x}{1.06} = \dfrac{159}{1.06}$$
$$x \approx 150$$
The cost of the dinner before tax was \$150.

7. $y = -\dfrac{3}{5}x+3$

slope = $-\dfrac{3}{5} = \dfrac{-3}{5}$; y-intercept = 3

Plot $(0,3)$. From this point, move 3 units *down* (because -3 is negative) and 5 units to the *right* to reach the point $(5,0)$. Draw a line through $(0,3)$ and $(5,0)$.

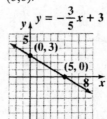

8. First, find slope $m = \dfrac{1-(-4)}{3-2} = \dfrac{5}{1} = 5$.

Use the point $(2, -4)$ in the point-slope equation.
$$y-y_1 = m(x-x_1)$$
$$y-(-4) = 5(x-2)$$
$$y+4 = 5(x-2)$$
Rewrite this equation in slope-intercept form.
$$y+4 = 5x-10$$
$$y = 5x-14$$

9. $x - 2y + 2z = 4$

$3x - y + 4z = 4$

$2x + y - 3z = 5$

Multiply the third equation by 2 and add to the first equation.

$x - 2y + 2z = 4$

$\underline{4x + 2y - 6z = 10}$

$5x \qquad - 4z = 14$

Add the second and third equations.

$3x - y + 4z = 4$

$\underline{2x + y - 3z = 5}$

$5x \qquad + z = 9$

Multiply this equation by -1 and add the result to the other equation in two variables.

$5x - 4z = 14$

$\underline{-5x - z = -9}$

$\qquad -5z = 5$

$\qquad z = -1$

Back-substitute -1 for z in one of the equations in two variables to find x.

$5x + z = 9$

$5x + (-1) = 9$

$5x = 10$

$x = 2$

Back-substitute 2 for x and -1 for z in one of the original equations in three variables to find y.

$2x + y - 3z = 5$

$2(2) + y - 3(-1) = 5$

$4 + y + 3 = 5$

$y + 7 = 5$

$y = -2$

The solution is $(2, -2, -1)$, and the solution set is $\{(2, -2, -1)\}$.

10. $5x + 2y = 13$

$y = 2x - 7$

The substitution method is a good choice for solving this system because the second equation is already solved for y.

Substitute $2x - 7$ for y in the first equation.

$5x + 2y = 13$

$5x + 2(2x - 7) = 13$

$5x + 4x - 14 = 13$

$9x - 14 = 13$

$9x = 27$

$x = 3$

Back-substitute into the second given equation.

$y = 2x - 7$

$y = 2(3) - 7 = -1$

The solution set is $\{(3, -1)\}$.

11. $2x + 3y = 5$

$3x - 2y = -4$

The addition method is a good choice for solving this system because both equations are written in the form $Ax + By = C$.

Multiply the first equation by 2 and the second equation by 3; then add the results.

$4x + 6y = \ \ 10$

$\underline{9x - 6y = -12}$

$13x \qquad = -2$

$x = -\dfrac{2}{13}$

Instead of back-substituting $-\dfrac{2}{13}$ and working with fractions, go back to the original system and eliminate x. Multiply the first equation by 3 and the second equation by -2; then add the results.

$6x + 9y = 15$

$\underline{-6x + 4y = \ \ 8}$

$13y = 23$

$y = \dfrac{23}{13}$

The solution set is $\left\{\left(-\dfrac{2}{13}, \dfrac{23}{13}\right)\right\}$.

12. $\dfrac{4}{5} - \dfrac{9}{8} = \dfrac{4}{5} \cdot \dfrac{8}{5} - \dfrac{9}{8} \cdot \dfrac{5}{5}$

$= \dfrac{32}{40} - \dfrac{45}{40} = -\dfrac{13}{40}$

13. $$\frac{6x^5 - 3x^4 + 9x^2 + 27x}{3x}$$

$$= \frac{6x^5}{3x} - \frac{3x^4}{3x} + \frac{9x^2}{3x} + \frac{27x}{3x}$$

$$= 2x^4 - x^3 + 3x + 9$$

14. $(3x - 5y)(2x + 9y)$

$= 6x^2 + 27xy - 10xy - 45y^2$

$= 6x^2 + 17xy - 45y^2$

15.

$$\begin{array}{r} 2x^2 + 5x - 3 \\ 3x - 5 \overline{)6x^3 + 5x^2 - 34x + 13} \\ \underline{6x^3 - 10x^2} \\ 15x^2 - 34x \\ \underline{15x^2 - 25x} \\ -9x + 13 \\ \underline{-9x + 15} \\ -2 \end{array}$$

$$\frac{6x^3 + 5x^2 - 34x + 13}{3x - 5} = 2x^2 + 5x - 3 + \frac{-2}{3x - 5}$$

$$\text{or } 2x^2 + 5x - 3 - \frac{2}{3x - 5}$$

16. $0.0071 = 7.1 \times 10^{-3}$

To write 0.0071 in scientific notation, move the decimal point 3 places to the right. Because the given number is between 0 and 1, the exponent will be negative.

17. Factor $3x^2 + 11x + 6$ by trial and error or by grouping. To Factor by grouping, find two integers whose product is $ac = 3 \cdot 6 = 18$ and whose sum is $b = 11$. These integers are 9 and 2.

$3x^2 + 11x + 6 = 3x^2 + 9x + 2x + 6$

$\qquad = 3x(x + 3) + 2(x + 3)$

$\qquad = (x + 3)(3x + 2)$

18. $y^5 - 16y = y\left(y^4 - 16\right)$

$\qquad = y\left(y^2 + 4\right)\left(y^2 - 4\right)$

$\qquad = y\left(y^2 + 4\right)(y + 2)(y - 2)$

19. $4x^2 + 12x + 9 = (2x)^2 + 2(2x \cdot 3) + 3x^2$

$\qquad = (2x + 3)^2$

20. Let x = the width of the rectangle.
Then $x + 2$ = the length of the rectangle.
Use the formula for the area of a rectangle.

$$l \cdot w = A$$

$$(x + 2)(x) = 24$$

$$x^2 + 2x = 24$$

$$x^2 + 2x - 24 = 0$$

$$(x + 6)(x - 4) = 0$$

$x + 6 = 0 \qquad$ or $\qquad x - 4 = 0$

$\qquad x = -6 \qquad\qquad\qquad x = 4$

Reject −6 because the width cannot be negative.
$x = 4$ and $x + 2 = 6$.
The dimensions of the rectangle are 6 feet by 4 feet.

Chapter 7
Rational Expressions

7.1 Check Points

1. **a.** $\dfrac{7x-28}{8x-40}$

   Set the denominator equal to 0 and solve for x.
   $$8x-40=0$$
   $$8x=40$$
   $$x=5$$
   The rational expression is undefined for $x=5$.

   b. $\dfrac{8x-40}{x^2+3x-28}$

   Set the denominator equal to 0 and solve for x.
   $$x^2+3x-28=0$$
   $$(x+7)(x-4)=0$$
   $$x+7=0 \quad \text{or} \quad x-4=0$$
   $$x=-7 \qquad \qquad x=4$$
   The rational expression is undefined for $x=-7$ and $x=4$.

2. $\dfrac{7x+28}{21x}=\dfrac{7(x+4)}{7\cdot 3x}=\dfrac{x+4}{3x}$

3. $\dfrac{x^3-x^2}{7x-7}=\dfrac{x^2(x-1)}{7(x-1)}=\dfrac{x^2}{7}$

4. $\dfrac{x^2-1}{x^2+2x+1}=\dfrac{(x+1)(x-1)}{(x+1)^2}=\dfrac{\cancel{(x+1)}(x-1)}{\cancel{(x+1)}(x+1)}=\dfrac{x-1}{x+1}$

5. $\dfrac{9x^2-49}{28-12x}=\dfrac{(3x+7)(3x-7)}{4(7-3x)}$

   $$=\dfrac{(3x+7)\overset{-1}{\cancel{(3x-7)}}}{4\cancel{(7-3x)}}$$

   $$=\dfrac{-(3x+7)}{4} \quad \text{or} \quad -\dfrac{3x+7}{4} \quad \text{or} \quad \dfrac{-3x-7}{4}$$

7.1 Exercise Set

1. $\dfrac{5}{2x}$

   Set the denominator equal to 0 and solve for x.
   $$2x=0$$
   $$x=0$$
   The rational expression is undefined for $x=0$.

3. $\dfrac{x}{x-8}$

   Set the denominator equal to 0 and solve for x.
   $$x-8=0$$
   $$x=8$$
   The rational expression is undefined for $x=8$.

5. $\dfrac{13}{5x-20}$

   Set the denominator equal to 0 and solve for x.
   $$5x-20=0$$
   $$5x=20$$
   $$x=4$$
   The rational expression is undefined for $x=4$.

7. $\dfrac{x+3}{(x+9)(x-2)}$

   Set the denominator equal to 0 and solve for x.
   $$(x+9)(x-2)=0$$
   $$x+9=0 \quad \text{or} \quad x-2=0$$
   $$x=-9 \qquad \qquad x=2$$
   The rational expression is undefined for $x=-9$ and $x=2$.

9. $\dfrac{4x}{(3x-17)(x+3)}$

   Set the denominator equal to 0 and solve for x.
   $$(3x-17)(x+3)=0$$
   $$3x-17=0 \quad \text{or} \quad x+3=0$$
   $$3x=17 \qquad \qquad x=-3$$
   $$x=\dfrac{17}{3}$$

   The rational expression is undefined for $x=\dfrac{17}{3}$ and $x=-3$.

11. $\dfrac{x+5}{x^2+x-12}$

Set the denominator equal to 0 and solve for x.

$$x^2+x-12=0$$
$$(x+4)(x-3)=0$$
$$x+4=0 \quad \text{or} \quad x-3=0$$
$$x=-4 \qquad x=3$$

The rational expression is undefined for $x=-4$ and $x=3$.

13. $\dfrac{x+5}{5}$

Because the denominator, 5, is not zero for any value of x, the rational expression is defined for all real numbers.

15. $\dfrac{y+3}{4y^2+y-3}$

Set the denominator equal to 0 and solve for x.

$$4y^2+y-3=0$$
$$(y+1)(4y-3)=0$$
$$y+1=0 \quad \text{or} \quad 4y-3=0$$
$$y=-1 \qquad 4y=3$$
$$y=\frac{3}{4}$$

The rational expression is undefined for $y=-1$ and $y=\dfrac{3}{4}$.

17. $\dfrac{y+5}{y^2-25}$

Set the denominator equal to 0 and solve for x.

$$y^2-25=0$$
$$(y+5)(y-5)=0$$
$$y+5=0 \quad \text{or} \quad y-5=0$$
$$y=-5 \qquad y=5$$

The rational expression is undefined for $y=-5$ and $y=5$.

19. $\dfrac{5}{x^2+1}$

The smallest possible value of x^2 is 0, so $x^2+1\geq 1$ for all real numbers of x. This means that there is no real number x for which $x^2+1=0$. Thus, the rational expression is defined for all real numbers.

21. $\dfrac{14x^2}{7x}=\dfrac{2\cdot 7\cdot x\cdot x}{7\cdot x}=\dfrac{2x}{1}=2x$

23. $\dfrac{5x-15}{25}=\dfrac{5(x-3)}{5\cdot 5}=\dfrac{x-3}{5}$

25. $\dfrac{2x-8}{4x}=\dfrac{2(x-4)}{2\cdot 2x}=\dfrac{x-4}{2x}$

27. $\dfrac{3}{3x-9}=\dfrac{3}{3(x-3)}=\dfrac{1}{x-3}$

29. $\dfrac{-15}{3x-9}=\dfrac{-15}{3(x-3)}=\dfrac{-5}{x-3}$ or $-\dfrac{5}{x-3}$

31. $\dfrac{3x+9}{x+3}=\dfrac{3(x+3)}{x+3}=\dfrac{3}{1}=3$

33. $\dfrac{x+5}{x^2-25}=\dfrac{x+5}{(x+5)(x-5)}=\dfrac{1}{x-5}$

35. $\dfrac{2y-10}{3y-15}=\dfrac{2(y-5)}{3(y-5)}=\dfrac{2}{3}$

37. $\dfrac{x+1}{x^2-2x-3}=\dfrac{x+1}{(x+1)(x-3)}=\dfrac{1}{x-3}$

39. $\dfrac{4x-8}{x^2-4x+4}=\dfrac{4(x-2)}{(x-2)(x-2)}=\dfrac{4}{x-2}$

41. $\dfrac{y^2-3y+2}{y^2+7y-18}=\dfrac{(y-1)(y-2)}{(y+9)(y-2)}=\dfrac{y-1}{y+9}$

43. $\dfrac{2y^2-7y+3}{2y^2-5y+2}=\dfrac{(2y-1)(y-3)}{(2y-1)(y-2)}=\dfrac{y-3}{y-2}$

45. $\dfrac{2x+3}{2x+5}$

The numerator and denominator have no common factor (other than 1), so this rational expression cannot be simplified.

47. $\dfrac{x^2+12x+36}{x^2-36}=\dfrac{(x+6)(x+6)}{(x+6)(x-6)}=\dfrac{x+6}{x-6}$

49.
$$\frac{x^3 - 2x^2 + x - 2}{x - 2} = \frac{x^2(x-2) + 1(x-2)}{x-2}$$
$$= \frac{(x-2)(x^2+1)}{x-2}$$
$$= x^2 + 1$$

51.
$$\frac{x^3 - 8}{x - 2} = \frac{(x-2)(x^2 + 2x + 4)}{x-2}$$
$$= x^2 + 2x + 4$$

53.
$$\frac{(x-4)^2}{x^2 - 16} = \frac{(x-4)(x-4)}{(x+4)(x-4)} = \frac{x-4}{x+4}$$

55. $\dfrac{x}{x+1}$; The numerator and denominator have no common factor (other than 1), so this rational expression cannot be simplified.

57. $\dfrac{x+4}{x^2+16}$; The numerator and denominator are both prime polynomials. They have no common factor (other than 1), so this rational expression cannot be simplified.

59.
$$\frac{x-5}{5-x} = \frac{-1(5-x)}{5-x} = -1$$

Notice that the numerator and denominator of the given rational expression are additive inverses.

61. The numerator and denominator of this rational expression are additive inverses, so $\dfrac{2x-3}{3-2x} = -1$.

63. $\dfrac{x-5}{x+5}$; The numerator and denominator have no common factor and they are not additive inverses, so this rational expression cannot be simplified.

65.
$$\frac{4x-6}{3-2x} = \frac{2(2x-3)}{3-2x} = \frac{-2(3-2x)}{3-2x} = -2$$

67.
$$\frac{4-6x}{3x^2-2x} = \frac{2(2-3x)}{x(3x-2)}$$
$$= \frac{-2(3x-2)}{x(3x-2)}$$
$$= -\frac{2}{x}$$

69.
$$\frac{x^2-1}{1-x} = \frac{(x+1)(x-1)}{1-x}$$
$$= \frac{(x+1) \cdot -1(1-x)}{1-x}$$
$$= -1(x+1) = -x-1$$

71.
$$\frac{y^2 - y - 12}{4-y} = \frac{(y-4)(y+3)}{4-y}$$
$$= \frac{-1(4-y)(y+3)}{4-y}$$
$$= -1(y+3) = -y-3$$

73.
$$\frac{x^2 y - x^2}{x^3 - x^3 y} = \frac{x^2(y-1)}{x^3(1-y)}$$
$$= \frac{x^2 \cdot -1(1-y)}{x^3(1-y)}$$
$$= -\frac{1}{x}$$

75.
$$\frac{x^2 + 2xy - 3y^2}{2x^2 + 5xy - 3y^2} = \frac{(x-y)(x+3y)}{(2x-y)(x+3y)}$$
$$= \frac{x-y}{2x-y}$$

77.
$$\frac{x^2 - 9x + 18}{x^3 - 27} = \frac{(x-3)(x-6)}{(x-3)(x^2+3x+9)}$$
$$= \frac{x-6}{x^2+3x+9}$$

79.
$$\frac{9-y^2}{y^2 - 3(2y-3)} = \frac{(3+y)(3-y)}{y^2 - 6y + 9}$$
$$= \frac{(3+y)(3-y)}{(y-3)(y-3)} = \frac{(3+y) \cdot -1(y-3)}{(y-3)(y-3)}$$
$$= \frac{-1(3+y)}{y-3} \text{ or } \frac{3+y}{-1(y-3)} = \frac{3+y}{3-y}$$

81.
$$\frac{xy + 2y + 3x + 6}{x^2 + 5x + 6} = \frac{y(x+2) + 3(x+2)}{(x+2)(x+3)}$$
$$= \frac{(x+2)(y+3)}{(x+2)(x+3)} = \frac{y+3}{x+3}$$

83.
$$\frac{8x^2 + 4x + 2}{1 - 8x^3} = \frac{2\left(4x^2 + 2x + 1\right)}{\left(1 - 2x\right)\left(1 + 2x + 4x^2\right)}$$
$$= \frac{2}{1 - 2x}$$

85. $\dfrac{130x}{100 - x}$

a. $x = 40$:
$$\frac{130x}{100 - x} = \frac{130(40)}{100 - 40}$$
$$= \frac{5200}{60}$$
$$\approx 86.67$$

This means it costs about $86.67 million to inoculate 40% of the population.

$x = 80$:
$$\frac{130x}{100 - x} = \frac{130(80)}{100 - 80}$$
$$= \frac{10,400}{20}$$
$$= 520$$

This means it costs $520 million to inoculate 80% of the population.

$x = 90$:
$$\frac{130x}{100 - x} = \frac{130(90)}{100 - 90}$$
$$= \frac{11,700}{10}$$
$$= 1170$$

This means it costs $1170 million ($1,170,000,000) to inoculate 90% of the population.

b. Set the denominator equal to 0 and solve for x.
$$100 - x = 0$$
$$100 = x$$
The rational expression is undefined for $x = 100$.

c. The cost keeps rising as x approaches 100. No amount of money will be enough to inoculate 100% of the population.

87. $D = 1000, A = 8$
$$\frac{DA}{A + 12} = \frac{1000 \cdot 8}{8 + 12}$$
$$= \frac{8000}{20} = 400$$
The correct dosage for an 8-year old is 400 milligrams.

89. $C = \dfrac{100x + 100,000}{x}$

a. $x = 500$
$$C = \frac{100(500) + 100,000}{500}$$
$$= \frac{150,000}{500} = 300$$
The cost per bicycle when manufacturing 500 bicycles is $300.

b. $x = 4000$
$$C = \frac{100(4000) + 100,000}{4000}$$
$$= \frac{400,000 + 100,000}{4000}$$
$$= \frac{500,000}{4000} = 125$$
The cost per bicycle when manufacturing 4000 bicycles is $125.

c. The cost per bicycle decreases as more bicycles are manufactured. One possible reason for this is that there could be fixed costs for equipment, so the more the equipment is used, the lower the cost per bicycle.

91. $y = \dfrac{5x}{x^2 + 1}; x = 3$
$$y = \frac{5 \cdot 3}{3^2 + 1} + \frac{15}{10} = 1.5$$
The equation indicates that the drug's concentration after 3 hours is 1.5 milligram per liter. This is represented on the graph by the point $\left(3, 1.5\right)$.

93. a. Substitute 4 for x in the model.

$$W = -82x^2 + 654x + 620$$

$$W = -82(4)^2 + 654(4) + 620$$

$$W = 1924$$

According to the model, women between the ages of 19 and 30 with this lifestyle need 1924 calories per day. This underestimates the actual value shown in the bar graph by 76 calories.

b. Substitute 4 for x in the model.

$$M = -96x^2 + 802x + 660$$

$$M = -96(4)^2 + 802(4) + 660$$

$$M = 2332$$

According to the model, men between the ages of 19 and 30 with this lifestyle need 2332 calories per day. This underestimates the actual value shown in the bar graph by 68 calories.

c. $\dfrac{W}{M} = \dfrac{-82x^2 + 654x + 620}{-96x^2 + 802x + 660}$

$$= \dfrac{2\left(-41x^2 + 327x + 310\right)}{2\left(-48x^2 + 401x + 330\right)}$$

$$= \dfrac{-41x^2 + 327x + 310}{-48x^2 + 401x + 330}$$

95. – 99. Answers will vary.

101. makes sense

103. does not make sense; Explanations will vary.
Sample explanation: 1 makes the denominator equal to 0 and thus the expression is undefined at 1.

105. false; Changes to make the statement true will vary.
A sample change is: 3 is not a factor of $x^2 + 3$.

107. false; Changes to make the statement true will vary.
A sample change is: x is not a factor of the numerator or the denominator.

109. Answers will vary. The denominator should be $x + 4$ or contain a factor of $x + 4$.

111. The graphs coincide.

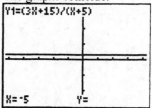

This verifies that the simplification $\dfrac{3x + 15}{x + 5} = 3, x \neq -5$, is correct. Notice the screen shows no y – value for $x = -5$.

113. The graphs do not coincide.

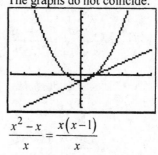

$$\dfrac{x^2 - x}{x} = \dfrac{x(x-1)}{x}$$

$$= x - 1, x \neq 0$$

Change the expression on the right from $x^2 - 1$ to $x - 1$.

115. $\dfrac{5}{6} \cdot \dfrac{9}{25} = \dfrac{\cancel{5}^{1}}{\cancel{6}_{2}} \cdot \dfrac{\cancel{9}^{3}}{\cancel{25}_{5}} = \dfrac{3}{10}$

116. $\dfrac{2}{3} \div 4 = \dfrac{2}{3} \cdot \dfrac{1}{4} = \dfrac{2}{12} = \dfrac{2 \cdot 1}{2 \cdot 6} = \dfrac{1}{6}$

117. $2x - 5y = -2$

$3x + 4y = 20$

Multiply the first equation by 3 and the second equation by -2; then add the results.

$6x - 15y = -6$

$\underline{-6x - 8y = -40}$

${-23y = -46}$

$y = 2$

Back-substitute to find x.

$2x - 5y = -2$

$2x - 5(2) = -2$

$2x - 10 = -2$

$2x = 8$

$x = 4$

The solution set is $\{(4, 2)\}$.

118. $\dfrac{2}{5} \cdot \dfrac{3}{7} = \dfrac{6}{35}$

119. $\dfrac{3}{4} \div \dfrac{1}{2} = \dfrac{3}{4} \cdot \dfrac{2}{1} = \dfrac{6}{4} = \dfrac{2 \cdot 3}{2 \cdot 2} = \dfrac{3}{2}$

120. $\dfrac{5}{4} \div \dfrac{15}{8} = \dfrac{5}{4} \cdot \dfrac{8}{15} = \dfrac{\overset{1}{\cancel{5}}}{\cancel{4}} \cdot \dfrac{\overset{2}{\cancel{8}}}{\cancel{15}} = \dfrac{2}{3}$

7.2 Check Points

1. $\dfrac{9}{x+4} \cdot \dfrac{x-5}{2} = \dfrac{9(x-5)}{(x+4)2} = \dfrac{9x-45}{2x+8}$

2. $\dfrac{x+4}{x-7} \cdot \dfrac{3x-21}{8x+32} = \dfrac{x+4}{x-7} \cdot \dfrac{3(x-7)}{8(x+4)}$

$\qquad = \dfrac{\cancel{x+4}}{\cancel{x-7}} \cdot \dfrac{3\,\cancel{(x-7)}}{8\,\cancel{(x+4)}}$

$\qquad = \dfrac{3}{8}$

3. $\dfrac{x-5}{x-2} \cdot \dfrac{x^2-4}{9x-45} = \dfrac{x-5}{x-2} \cdot \dfrac{(x+2)(x-2)}{9(x-5)}$

$\qquad = \dfrac{\cancel{x-5}}{\cancel{x-2}} \cdot \dfrac{(x+2)\,\cancel{(x-2)}}{9\,\cancel{(x-5)}}$

$\qquad = \dfrac{x+2}{9}$

4. $\dfrac{5x+5}{7x-7x^2} \cdot \dfrac{2x^2+x-3}{4x^2-9}$

$\qquad = \dfrac{5(x+1)}{7x(1-x)} \cdot \dfrac{(2x+3)(x-1)}{(2x+3)(2x-3)}$

$\qquad = \dfrac{5(x+1)}{7x\,\cancel{(1-x)}} \cdot \dfrac{\cancel{(2x+3)}\,\overset{-1}{\cancel{(x-1)}}}{\cancel{(2x+3)}(2x-3)}$

$\qquad = \dfrac{-5(x+1)}{7x(2x-3)}$ or $-\dfrac{5(x+1)}{7x(2x-3)}$

5. $(x+3) \div \dfrac{x-4}{x+7} = \dfrac{x+3}{1} \cdot \dfrac{x+7}{x-4}$

$\qquad = \dfrac{(x+3)(x+7)}{x-4}$

6. $\dfrac{x^2+5x+6}{x^2-25} \div \dfrac{x+2}{x+5} = \dfrac{x^2+5x+6}{x^2-25} \cdot \dfrac{x+5}{x+2}$

$\qquad = \dfrac{(x+3)(x+2)}{(x+5)(x-5)} \cdot \dfrac{x+5}{x+2}$

$\qquad = \dfrac{(x+3)\,\cancel{(x+2)}}{\cancel{(x+5)}(x-5)} \cdot \dfrac{\cancel{x+5}}{\cancel{x+2}}$

$\qquad = \dfrac{x+3}{x-5}$

7. $\dfrac{y^2+3y+2}{y^2+1} \div \left(5y^2+10y\right)$

$\qquad = \dfrac{y^2+3y+2}{y^2+1} \div \dfrac{5y^2+10y}{1}$

$\qquad = \dfrac{y^2+3y+2}{y^2+1} \cdot \dfrac{1}{5y^2+10y}$

$\qquad = \dfrac{(y+2)(y+1)}{y^2+1} \cdot \dfrac{1}{5y(y+2)}$

$\qquad = \dfrac{\cancel{(y+2)}(y+1)}{y^2+1} \cdot \dfrac{1}{5y\,\cancel{(y+2)}}$

$\qquad = \dfrac{y+1}{5y(y^2+1)}$

7.2 Exercise Set

1. $\dfrac{4}{x+3} \cdot \dfrac{x-5}{9} = \dfrac{4(x-5)}{(x+3)9} = \dfrac{4x-20}{9x+27}$

3. $\dfrac{x}{3} \cdot \dfrac{12}{x+5} = \dfrac{3 \cdot 4x}{3(x+5)} = \dfrac{4x}{x+5}$

5. $\dfrac{3}{x} \cdot \dfrac{4x}{15} = \dfrac{3 \cdot 4x}{3 \cdot 5x} = \dfrac{4}{5}$

7. $\dfrac{x-3}{x+5} \cdot \dfrac{4x+20}{9x-27} = \dfrac{x-3}{x+5} \cdot \dfrac{4(x+5)}{9(x-3)} = \dfrac{4}{9}$

9. $\dfrac{x^2+9x+14}{x+7} \cdot \dfrac{1}{x+2} = \dfrac{(x+7)(x+2) \cdot 1}{(x+7)(x+2)} = 1$

11. $\dfrac{x^2-25}{x^2-3x-10} \cdot \dfrac{x+2}{x} = \dfrac{(x+5)(x-5)}{(x+2)(x-5)} \cdot \dfrac{(x+2)}{x}$

$\qquad = \dfrac{x+5}{x}$

13. $\dfrac{4y+30}{y^2-3y} \cdot \dfrac{y-3}{2y+15} = \dfrac{2(2y+15)}{y(y-3)} \cdot \dfrac{(y-3)}{(2y+15)}$

$\qquad\qquad = \dfrac{2}{y}$

15. $\dfrac{y^2-7y-30}{y^2-6y-40} \cdot \dfrac{2y^2+5y+2}{2y^2+7y+3}$

$\qquad = \dfrac{(y+3)(y-10)}{(y+4)(y-10)} \cdot \dfrac{(2y+1)(y+2)}{(2y+1)(y+3)}$

$\qquad = \dfrac{y+2}{y+4}$

17. $\left(y^2-9\right) \cdot \dfrac{4}{y-3} = \dfrac{y^2-9}{1} \cdot \dfrac{4}{y-3}$

$\qquad\qquad = \dfrac{(y+3)(y-3)}{1} \cdot \dfrac{4}{y-3}$

$\qquad\qquad = 4(y+3) \text{ or } 4y+12$

19. $\dfrac{x^2-5x+6}{x^2-2x-3} \cdot \dfrac{x^2-1}{x^2-4}$

$\qquad = \dfrac{(x-2)(x-3)}{(x+1)(x-3)} \cdot \dfrac{(x+1)(x-1)}{(x+2)(x-2)}$

$\qquad = \dfrac{x-1}{x+2}$

21. $\dfrac{x^3-8}{x^2-4} \cdot \dfrac{x+2}{3x}$

$\qquad = \dfrac{(x-2)\left(x^2+2x+4\right)}{(x+2)(x-2)} \cdot \dfrac{(x+2)}{3x}$

$\qquad = \dfrac{x^2+2x+4}{3x}$

23. $\dfrac{(x-2)^3}{(x-1)^3} \cdot \dfrac{x^2-2x+1}{x^2-4x+4}$

$\qquad = \dfrac{(x-2)^3}{(x-1)^3} \cdot \dfrac{(x-1)^2}{(x-2)^2}$

$\qquad = \dfrac{x-2}{x-1}$

25. $\dfrac{6x+2}{x^2-1} \cdot \dfrac{1-x}{3x^2+x}$

$\qquad = \dfrac{2(3x+1)}{(x+1)(x-1)} \cdot \dfrac{(1-x)}{x(3x+1)}$

$\qquad = \dfrac{2(3x+1)}{(x+1)(x-1)} \cdot \dfrac{-1(x-1)}{x(3x+1)}$

$\qquad = \dfrac{-2}{x(x+1)} \text{ or } -\dfrac{2}{x(x+1)}$

27. $\dfrac{25-y^2}{y^2-2y-35} \cdot \dfrac{y^2-8y-20}{y^2-3y-10}$

$\qquad = \dfrac{(5+y)(5-y)}{(y+5)(y-7)} \cdot \dfrac{(y-10)(y+2)}{(y-5)(y+2)}$

$\qquad = \dfrac{-(y-10)}{y-7} \text{ or } -\dfrac{y-10}{y-7}$

29. $\dfrac{x^2-y^2}{x} \cdot \dfrac{x^2+xy}{x+y}$

$\qquad = \dfrac{(x+y)(x-y)}{x} \cdot \dfrac{x(x+y)}{(x+y)}$

$\qquad = (x-y)(x+y) \text{ or } x^2-y^2$

31. $\dfrac{x^2+2xy+y^2}{x^2-2xy+y^2} \cdot \dfrac{4x-4y}{3x+3y}$

$\qquad = \dfrac{(x+y)(x+y)}{(x-y)(x-y)} \cdot \dfrac{4(x-y)}{3(x+y)}$

$\qquad = \dfrac{4(x+y)}{3(x-y)}$

33. $\dfrac{x}{7} \div \dfrac{5}{3} = \dfrac{x}{7} \cdot \dfrac{3}{5} = \dfrac{3x}{35}$

35. $\dfrac{3}{x} \div \dfrac{12}{x} = \dfrac{3}{x} \cdot \dfrac{x}{12} = \dfrac{1}{4}$

37. $\dfrac{15}{x} \div \dfrac{3}{2x} = \dfrac{15}{x} \cdot \dfrac{2x}{3} = 10$

39. $\dfrac{x+1}{3} \div \dfrac{3x+3}{7} = \dfrac{x+1}{3} \cdot \dfrac{7}{3x+3}$

$\qquad\qquad = \dfrac{x+1}{3} \cdot \dfrac{7}{3(x+1)}$

$\qquad\qquad = \dfrac{7}{9}$

41. $\dfrac{7}{x-5} \div \dfrac{28}{3x-15} = \dfrac{7}{x-5} \cdot \dfrac{3x-15}{28}$

$\qquad = \dfrac{7}{(x-5)} \cdot \dfrac{3(x-5)}{7 \cdot 4}$

$\qquad = \dfrac{3}{4}$

43. $\dfrac{x^2-4}{x} \div \dfrac{x+2}{x-2} = \dfrac{x^2-4}{x} \cdot \dfrac{x-2}{x+2}$

$\qquad = \dfrac{(x+2)(x-2)}{x} \cdot \dfrac{x-2}{x+2}$

$\qquad = \dfrac{(x-2)^2}{x}$

45. $\left(y^2-16\right) \div \dfrac{y^2+3y-4}{y^2+4}$

$\qquad = \dfrac{y^2-16}{1} \cdot \dfrac{y^2+4}{y^2+3y-4}$

$\qquad = \dfrac{(y+4)(y-4)}{1} \cdot \dfrac{y^2+4}{(y+4)(y-1)}$

$\qquad = \dfrac{(y-4)\left(y^2+4\right)}{y-1}$

47. $\dfrac{y^2-y}{15} \div \dfrac{y-1}{5} = \dfrac{y^2-y}{15} \cdot \dfrac{5}{y-1}$

$\qquad = \dfrac{y(y-1)}{15} \cdot \dfrac{5}{(y-1)}$

$\qquad = \dfrac{y}{3}$

49. $\dfrac{4x^2+10}{x-3} \div \dfrac{6x^2+15}{x^2-9}$

$\qquad = \dfrac{4x^2+10}{x-3} \cdot \dfrac{x^2-9}{6x^2+15}$

$\qquad = \dfrac{2\left(2x^2+5\right)}{(x-3)} \cdot \dfrac{(x+3)(x-3)}{3\left(2x^2+5\right)}$

$\qquad = \dfrac{2(x+3)}{3} \quad$ or $\quad \dfrac{2x+6}{3}$

51. $\dfrac{x^2-25}{2x-2} \div \dfrac{x^2+10x+25}{x^2+4x-5}$

$\qquad = \dfrac{x^2-25}{2x-2} \cdot \dfrac{x^2+4x-5}{x^2+10x+25}$

$\qquad = \dfrac{(x+5)(x-5)}{2(x-1)} \cdot \dfrac{(x+5)(x-1)}{(x+5)(x+5)}$

$\qquad = \dfrac{x-5}{2}$

53. $\dfrac{y^3+y}{y^2-y} \div \dfrac{y^3-y^2}{y^2-2y+1}$

$\qquad = \dfrac{y^3+y}{y^2-y} \cdot \dfrac{y^2-2y+1}{y^3-y^2}$

$\qquad = \dfrac{y\left(y^2+1\right)}{y(y-1)} \cdot \dfrac{(y-1)(y-1)}{y^2(y-1)}$

$\qquad = \dfrac{y^2+1}{y^2}$

55. $\dfrac{y^2+5y+4}{y^2+12y+32} \div \dfrac{y^2-12y+35}{y^2+3y-40}$

$\qquad = \dfrac{y^2+5y+4}{y^2+12y+32} \cdot \dfrac{y^2+3y-40}{y^2-12y+35}$

$\qquad = \dfrac{(y+4)(y+1)}{(y+4)(y+8)} \cdot \dfrac{(y+8)(y-5)}{(y-7)(y-5)}$

$\qquad = \dfrac{y+1}{y-7}$

57. $\dfrac{2y^2-128}{y^2+16y+64} \div \dfrac{y^2-6y-16}{3y^2+30y+48}$

$\qquad = \dfrac{2y^2-128}{y^2+16y+64} \cdot \dfrac{3y^2+30y+48}{y^2-6y-16}$

$\qquad = \dfrac{2\left(y^2-64\right)}{(y+8)(y+8)} \cdot \dfrac{3\left(y^2+10y+16\right)}{(y+2)(y-8)}$

$\qquad = \dfrac{2(y+8)(y-8)}{(y+8)(y+8)} \cdot \dfrac{3(y+2)(y+8)}{(y+2)(y-8)}$

$\qquad = 6$

59. $\dfrac{2x+2y}{3} \div \dfrac{x^2-y^2}{x-y}$

$= \dfrac{2x+2y}{3} \cdot \dfrac{x-y}{x^2-y^2}$

$= \dfrac{2(x+y)}{3} \cdot \dfrac{x-y}{(x+y)(x-y)}$

$= \dfrac{2}{3}$

61. $\dfrac{x^2-y^2}{8x^2-16xy+8y^2} \div \dfrac{4x-4y}{x+y}$

$= \dfrac{x^2-y^2}{8x^2-16xy+8y^2} \cdot \dfrac{x+y}{4x-4y}$

$= \dfrac{(x+y)(x-y)}{8\left(x^2-2xy+y^2\right)} \cdot \dfrac{x+y}{4(x-y)}$

$= \dfrac{(x+y)(x-y)}{8(x-y)(x-y)} \cdot \dfrac{x+y}{4(x-y)}$

$= \dfrac{(x+y)^2}{32(x-y)^2}$

63. $\dfrac{xy-y^2}{x^2+2x+1} \div \dfrac{2x^2+xy-3y^2}{2x^2+5xy+3y^2}$

$= \dfrac{xy-y^2}{x^2+2x+1} \cdot \dfrac{2x^2+5xy+3y^2}{2x^2+xy-3y^2}$

$= \dfrac{y(x-y)}{(x+1)(x+1)} \cdot \dfrac{(2x+3y)(x+y)}{(2x+3y)(x-y)}$

$= \dfrac{y(x+y)}{(x+1)^2}$

65. $\left(\dfrac{y-2}{y^2-9y+18} \cdot \dfrac{y^2-4y-12}{y+2} \right) \div \dfrac{y^2-4}{y^2+5y+6} = \left(\dfrac{y-2}{y^2-9y+18} \cdot \dfrac{y^2-4y-12}{y+2} \right) \cdot \dfrac{y^2+5y+6}{y^2-4}$

$= \left(\dfrac{y-2}{(y-6)(y-3)} \cdot \dfrac{(y-6)(y+2)}{y+2} \right) \cdot \dfrac{(y+2)(y+3)}{(y+2)(y-2)} = \left(\dfrac{y-2}{y-3} \right) \cdot \dfrac{y+3}{y-2} = \dfrac{y+3}{y-3}$

67. $\dfrac{3x^2+3x-60}{2x-8} \div \left(\dfrac{30x^2}{x^2-7x+10} \cdot \dfrac{x^3+3x^2-10x}{25x^3} \right)$

$= \dfrac{3x^2+3x-60}{2x-8} \div \left(\dfrac{30x^2}{(x-2)(x-5)} \cdot \dfrac{x\left(x^2+3x-10\right)}{25x^3} \right)$

$= \dfrac{3\left(x^2+x-20\right)}{2x-8} \div \left(\dfrac{30x^2}{(x-2)(x-5)} \cdot \dfrac{x(x+5)(x-2)}{25x^3} \right)$

$= \dfrac{3(x+5)(x-4)}{2(x-4)} \div \dfrac{6(x+5)}{5(x-5)} = \dfrac{3(x+5)(x-4)}{2(x-4)} \cdot \dfrac{5(x-5)}{6(x+5)} = \dfrac{5(x-5)}{4}$

69. $\dfrac{x^2+xz+xy+yz}{x-y} \div \dfrac{x+z}{x+y} = \dfrac{x(x+z)+y(x+z)}{x-y} \cdot \dfrac{x+y}{x+z} = \dfrac{(x+z)(x+y)}{x-y} \cdot \dfrac{x+y}{x+z} = \dfrac{(x+y)^2}{x-y}$

71. $\dfrac{3xy+ay+3xb+ab}{9x^2-a^2} \div \dfrac{y^3+b^3}{6x-2a} = \dfrac{3xy+ay+3xb+ab}{9x^2-a^2} \cdot \dfrac{6x-2a}{y^3+b^3}$

$= \dfrac{y(3x+a)+b(3x+a)}{(3x+a)(3x-a)} \cdot \dfrac{2(3x-a)}{(y+b)\left(y^2-by+b^2\right)}$

$= \dfrac{(3x+a)(y+b)}{(3x+a)(3x-a)} \cdot \dfrac{2(3x-a)}{(y+b)\left(y^2-by+b^2\right)}$

$= \dfrac{2}{y^2-by+b^2}$

73. $A = l \cdot w$

$A = \dfrac{x+1}{x^2+2x} \cdot \dfrac{x^2-4}{x^2-1}$

$= \dfrac{x+1}{x(x+2)} \cdot \dfrac{(x+2)(x-2)}{(x+1)(x-1)}$

$= \dfrac{x+1}{x(x+2)} \cdot \dfrac{(x+2)(x-2)}{(x+1)(x-1)}$

$= \dfrac{x-2}{x(x-1)} \text{ in.}^2$

75. $A = \dfrac{1}{2} \cdot b \cdot h$

$A = \dfrac{1}{2} \cdot \dfrac{18x}{x^2 + 3x + 2} \cdot \dfrac{x+1}{3}$

$= \dfrac{1}{2} \cdot \dfrac{2 \cdot 3 \cdot 3x}{(x+2)(x+1)} \cdot \dfrac{x+1}{3}$

$= \dfrac{1}{\cancel{2}} \cdot \dfrac{\cancel{2} \cdot \cancel{3} \cdot 3x}{(x+2)\cancel{(x+1)}} \cdot \dfrac{\cancel{x+1}}{\cancel{3}}$

$= \dfrac{3x}{x+2}$ in.^{2}

77. – 79. Answers will vary.

81. makes sense

83. makes sense

85. true

87. false; Changes to make the statement true will vary. A sample change is: The quotient of two rational expressions can be found by inverting the divisor and then multiplying .

89. $-\dfrac{1}{2x-3} \div \dfrac{?}{?} = \dfrac{1}{3}$

The numerator of the unknown rational expression must contain a factor of -3 . The denominator of the unknown rational expression must contain a factor of $(2x-3)$. Therefore, the simplest pair of polynomials that will work are -3 in the numerator and $2x-3$ in the denominator, to give the rational expression $\dfrac{-3}{2x-3}$.

Check:

$-\dfrac{1}{2x-3} \div \dfrac{-3}{2x-3} = -\dfrac{1}{2x-3} \cdot \dfrac{2x-3}{-3} = \dfrac{1}{3}$

91. The graph coincides.

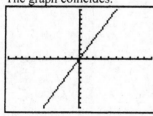

This verifies that $\dfrac{x^3 + x}{3x} \cdot \dfrac{6x}{x+1} = 2x$.

93. The graphs do not coincide.

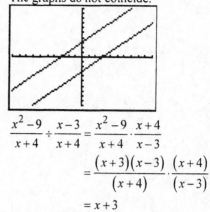

$\dfrac{x^2 - 9}{x+4} \div \dfrac{x-3}{x+4} = \dfrac{x^2-9}{x+4} \cdot \dfrac{x+4}{x-3}$

$= \dfrac{(x+3)(x-3)}{(x+4)} \cdot \dfrac{(x+4)}{(x-3)}$

$= x+3$

Change the expression on the right from $(x-3)$ to $(x+3)$.

95. $2x+3 < 3(x-5)$

$2x+3 < 3x-15$

$-x+3 < -15$

$-x < -18$

$x > 18$

$(18, \infty)$

96. $3x^2 - 15x - 42 = 3(x^2 - 5x - 14)$

$= 3(x-7)(x+2)$

97. $x(2x+9) = 5$

$2x^2 + 9x = 5$

$2x^2 + 9x - 5 = 0$

$(2x-1)(x+5) = 0$

$2x - 1 = 0$ or $x+5 = 0$

$2x = 1 \qquad\qquad x = -5$

$x = \dfrac{1}{2}$

The solution set is $\left\{ -5, \dfrac{1}{2} \right\}$.

98. $\dfrac{7}{9} - \dfrac{1}{9} = \dfrac{6}{9} = \dfrac{2}{3}$

99. $\dfrac{2x}{3} + \dfrac{x}{3} = \dfrac{3x}{3} = x$

100. $\dfrac{x^2 - 6x + 9}{x^2 - 9} = \dfrac{(x-3)^2}{(x+3)(x-3)} = \dfrac{(x-3)(x-3)}{(x+3)(x-3)} = \dfrac{x-3}{x+3}$

7.3 Check Points

1. $\dfrac{3x-2}{5} + \dfrac{2x+12}{5} = \dfrac{3x-2+2x+12}{5}$

$\qquad = \dfrac{5x+10}{5}$

$\qquad = \dfrac{\overset{1}{\cancel{5}}(x+2)}{\underset{1}{\cancel{5}}}$

$\qquad = x + 2$

2. $\dfrac{x^2}{x^2 - 25} + \dfrac{25 - 10x}{x^2 - 25} = \dfrac{x^2 - 10x + 25}{x^2 - 25}$

$\qquad = \dfrac{(x-5)^2}{(x+5)(x-5)}$

$\qquad = \dfrac{(x-5)\cancel{(x-5)}}{(x+5)\cancel{(x-5)}}$

$\qquad = \dfrac{x-5}{x+5}$

3. a. $\dfrac{4x+5}{x+7} - \dfrac{x}{x+7} = \dfrac{4x+5-x}{x+7} = \dfrac{3x+5}{x+7}$

b. $\dfrac{3x^2+4x}{x-1} - \dfrac{11x-4}{x-1} = \dfrac{3x^2+4x-(11x-4)}{x-1}$

$\qquad = \dfrac{3x^2+4x-11x+4}{x-1}$

$\qquad = \dfrac{3x^2-7x+4}{x-1}$

$\qquad = \dfrac{(3x-4)(x-1)}{x-1}$

$\qquad = 3x-4$

4. $\dfrac{y^2+3y-6}{y^2-5y+4} - \dfrac{4y-4-2y^2}{y^2-5y+4}$

$\qquad = \dfrac{y^2+3y-6-(4y-4-2y^2)}{y^2-5y+4}$

$\qquad = \dfrac{y^2+3y-6-4y+4+2y^2}{y^2-5y+4}$

$\qquad = \dfrac{3y^2-y-2}{y^2-5y+4}$

$\qquad = \dfrac{(3y+2)(y-1)}{(y-4)(y-1)}$

$\qquad = \dfrac{3y+2}{y-4}$

5. $\dfrac{x^2}{x-7} + \dfrac{4x+21}{7-x} = \dfrac{x^2}{x-7} + \dfrac{(-1)}{(-1)} \cdot \dfrac{4x+21}{7-x}$

$\qquad = \dfrac{x^2}{x-7} + \dfrac{-4x-21}{x-7}$

$\qquad = \dfrac{x^2-4x-21}{x-7}$

$\qquad = \dfrac{(x+3)(x-7)}{x-7}$

$\qquad = x+3$

6. $\dfrac{7x-x^2}{x^2-2x-9} - \dfrac{5x-3x^2}{9+2x-x^2}$

$\qquad = \dfrac{7x-x^2}{x^2-2x-9} - \dfrac{3x^2-5x}{x^2-2x-9}$

$\qquad = \dfrac{7x-x^2-(3x^2-5x)}{x^2-2x-9}$

$\qquad = \dfrac{7x-x^2-3x^2+5x}{x^2-2x-9}$

$\qquad = \dfrac{-4x^2+12x}{x^2-2x-9}$

7.3 Exercise Set

1. $\dfrac{7x}{13} + \dfrac{2x}{13} = \dfrac{9x}{13}$

3. $\dfrac{8x}{15} + \dfrac{x}{15} = \dfrac{9x}{15} = \dfrac{3x}{5}$

5. $\dfrac{x-3}{12} + \dfrac{5x+21}{12} = \dfrac{6x+18}{12}$

$\qquad\qquad = \dfrac{6(x+3)}{12}$

$\qquad\qquad = \dfrac{x+3}{2}$

7. $\dfrac{4}{x} + \dfrac{2}{x} = \dfrac{6}{x}$

9. $\dfrac{8}{9x} + \dfrac{13}{9x} = \dfrac{21}{9x} = \dfrac{7}{3x}$

11. $\dfrac{5}{x+3} + \dfrac{4}{x+3} = \dfrac{9}{x+3}$

13. $\dfrac{x}{x-3} + \dfrac{4x+5}{x-3} = \dfrac{5x+5}{x-3}$

15. $\dfrac{4x+1}{6x+5} + \dfrac{8x+9}{6x+5} = \dfrac{12x+10}{6x+5}$

$\qquad\qquad = \dfrac{2(6x+5)}{6x+5} = 2$

17. $\dfrac{y^2+7y}{y^2-5y} + \dfrac{y^2-4y}{y^2-5y} = \dfrac{y^2+7y+y^2-4y}{y^2-5y}$

$\qquad\qquad = \dfrac{2y^2+3y}{y^2-5y}$

$\qquad\qquad = \dfrac{y(2y+3)}{y(y-5)}$

$\qquad\qquad = \dfrac{2y+3}{y-5}$

19. $\dfrac{4y-1}{5y^2} + \dfrac{3y+1}{5y^2} = \dfrac{4y-1+3y+1}{5y^2}$

$\qquad\qquad = \dfrac{7y}{5y^2} = \dfrac{7}{5y}$

21. $\dfrac{x^2-2}{x^2+x-2} + \dfrac{2x-x^2}{x^2+x-2}$

$\qquad = \dfrac{x^2-2+2x-x^2}{x^2+x-2}$

$\qquad = \dfrac{2x-2}{x^2+x-2} = \dfrac{2(x-1)}{(x+2)(x-1)} = \dfrac{2}{x+2}$

23. $\dfrac{x^2-4x}{x^2-x-6} + \dfrac{4x-4}{x^2-x-6}$

$\qquad = \dfrac{x^2-4x+4x-4}{x^2-x-6}$

$\qquad = \dfrac{x^2-4}{x^2-x-6} = \dfrac{(x+2)(x-2)}{(x-3)(x+2)} = \dfrac{x-2}{x-3}$

25. $\dfrac{3x}{5x-4} - \dfrac{4}{5x-4} = \dfrac{3x-4}{5x-4}$

27. $\dfrac{4x}{4x-3} - \dfrac{3}{4x-3} = \dfrac{4x-3}{4x-3} = 1$

29. $\dfrac{14y}{7y+2} - \dfrac{7y-2}{7y+2} = \dfrac{14y-(7y-2)}{7y+2}$

$\qquad\qquad = \dfrac{14y-7y+2}{7y+2}$

$\qquad\qquad = \dfrac{7y+2}{7y+2} = 1$

31. $\dfrac{3x+1}{4x-2} - \dfrac{x+1}{4x-2} = \dfrac{(3x+1)-(x+1)}{4x-2}$

$\qquad\qquad = \dfrac{3x+1-x-1}{4x-2}$

$\qquad\qquad = \dfrac{2x}{4x-2}$

$\qquad\qquad = \dfrac{2x}{2(2x-1)}$

$\qquad\qquad = \dfrac{x}{2x-1}$

33. $\dfrac{3y^2-1}{3y^3} - \dfrac{6y^2-1}{3y^3}$

$\qquad = \dfrac{(3y^2-1)-(6y^2-1)}{3y^3}$

$\qquad = \dfrac{3y^2-1-6y^2+1}{3y^3} = \dfrac{-3y^2}{3y^3} = -\dfrac{1}{y}$

35. $\dfrac{4y^2+5}{9y^2-64} - \dfrac{y^2-y+29}{9y^2-64}$

$= \dfrac{\left(4y^2+5\right)-\left(y^2-y+29\right)}{9y^2-64}$

$= \dfrac{4y^2+5-y^2+y-29}{9y^2-64}$

$= \dfrac{3y^2+y-24}{9y^2-64}$

$= \dfrac{(3y-8)(y+3)}{(3y+8)(3y-8)} = \dfrac{y+3}{3y+8}$

37. $\dfrac{6y^2+y}{2y^2-9y+9} - \dfrac{2y+9}{2y^2-9y+9} - \dfrac{4y-3}{2y^2-9y+9}$

$= \dfrac{\left(6y^2+y\right)-(2y+9)-(4y-3)}{2y^2-9y+9}$

$= \dfrac{6y^2+y-2y-9-4y+3}{2y^2-9y+9}$

$= \dfrac{6y^2-5y-6}{2y^2-9y+9}$

$= \dfrac{(2y-3)(3y+2)}{(2y-3)(y-3)}$

$= \dfrac{3y+2}{y-3}$

39. $\dfrac{4}{x-3} + \dfrac{2}{3-x} = \dfrac{4}{x-3} + \dfrac{(-1)}{(-1)} \cdot \dfrac{2}{3-x}$

$= \dfrac{4}{x-3} + \dfrac{-2}{x-3}$

$= \dfrac{2}{x-3}$

41. $\dfrac{6x+7}{x-6} + \dfrac{3x}{6-x} = \dfrac{6x+7}{x-6} + \dfrac{(-1)}{(-1)} \cdot \dfrac{3x}{6-x}$

$= \dfrac{6x+7}{x-6} + \dfrac{-3x}{x-6}$

$= \dfrac{3x+7}{x-6}$

43. $\dfrac{5x-2}{3x-4} + \dfrac{2x-3}{4-3x} = \dfrac{5x-2}{3x-4} + \dfrac{(-1)}{(-1)} \cdot \dfrac{2x-3}{4-3x}$

$= \dfrac{5x-2}{3x-4} + \dfrac{-2x+3}{3x-4}$

$= \dfrac{5x-2-2x+3}{3x-4}$

$= \dfrac{3x+1}{3x-4}$

45. $\dfrac{x^2}{x-2} + \dfrac{4}{2-x} = \dfrac{x^2}{x-2} + \dfrac{(-1)}{(-1)} \cdot \dfrac{4}{2-x}$

$= \dfrac{x^2}{x-2} + \dfrac{-4}{x-2}$

$= \dfrac{x^2-4}{x-2}$

$= \dfrac{(x+2)(x-2)}{x-2}$

$= x+2$

47. $\dfrac{y-3}{y^2-25} + \dfrac{y-3}{25-y^2}$

$= \dfrac{y-3}{y^2-25} + \dfrac{(-1)}{(-1)} \cdot \dfrac{y-3}{25-y^2}$

$= \dfrac{y-3}{y^2-25} + \dfrac{-y+3}{y^2-25}$

$= \dfrac{y-3-y+3}{y^2-25} = \dfrac{0}{y^2-25} = 0$

49. $\dfrac{6}{x-1} - \dfrac{5}{1-x} = \dfrac{6}{x-1} - \dfrac{(-1)}{(-1)} \cdot \dfrac{5}{1-x}$

$= \dfrac{6}{x-1} - \dfrac{-5}{x-1}$

$= \dfrac{6+5}{x-1} = \dfrac{11}{x-1}$

51. $\dfrac{10}{x+3} - \dfrac{2}{-x-3} = \dfrac{10}{x+3} - \dfrac{(-1)}{(-1)} \cdot \dfrac{2}{-x-3}$

$= \dfrac{10}{x+3} - \dfrac{-2}{x+3}$

$= \dfrac{10+2}{x+3} = \dfrac{12}{x+3}$

53. $\dfrac{y}{y-1} - \dfrac{1}{1-y} = \dfrac{y}{y-1} - \dfrac{(-1)}{(-1)} \cdot \dfrac{1}{1-y}$

$\quad\quad = \dfrac{y}{y-1} - \dfrac{-1}{y-1}$

$\quad\quad = \dfrac{y+1}{y-1}$

55. $\dfrac{3-x}{x-7} - \dfrac{2x-5}{7-x} = \dfrac{3-x}{x-7} - \dfrac{(-1)}{(-1)} \cdot \dfrac{2x-5}{7-x}$

$\quad\quad = \dfrac{3-x}{x-7} - \dfrac{-2x+5}{x-7}$

$\quad\quad = \dfrac{(3-x)-(-2x+5)}{x-7}$

$\quad\quad = \dfrac{3-x+2x-5}{x-7}$

$\quad\quad = \dfrac{x-2}{x-7}$

57. $\dfrac{x-2}{x^2-25} - \dfrac{x-2}{25-x^2}$

$\quad = \dfrac{x-2}{x^2-25} - \dfrac{(-1)}{(-1)} \cdot \dfrac{x-2}{25-x^2}$

$\quad = \dfrac{x-2}{x^2-25} - \dfrac{-x+2}{x^2-25}$

$\quad = \dfrac{(x-2)-(-x+2)}{x^2-25}$

$\quad = \dfrac{x-2+x-2}{x^2-25} = \dfrac{2x-4}{x^2-25}$

59. $\dfrac{x}{x-y} + \dfrac{y}{y-x} = \dfrac{x}{x-y} + \dfrac{(-1)}{(-1)} \cdot \dfrac{y}{y-x}$

$\quad\quad = \dfrac{x}{x-y} + \dfrac{-y}{x-y}$

$\quad\quad = \dfrac{x-y}{x-y} = 1$

61. $\dfrac{2x}{x^2-y^2} + \dfrac{2y}{y^2-x^2}$

$\quad = \dfrac{2x}{x^2-y^2} + \dfrac{(-1)}{(-1)} \cdot \dfrac{2y}{y^2-x^2}$

$\quad = \dfrac{2x}{x^2-y^2} + \dfrac{-2y}{x^2-y^2}$

$\quad = \dfrac{2x-2y}{x^2-y^2} = \dfrac{2(x-y)}{(x+y)(x-y)} = \dfrac{2}{x+y}$

63. $\dfrac{x^2-2}{x^2+6x-7} + \dfrac{19-4x}{7-6x-x^2}$

$\quad = \dfrac{x^2-2}{x^2+6x-7} + \dfrac{(-1)}{(-1)} \cdot \dfrac{19-4x}{7-6x-x^2}$

$\quad = \dfrac{x^2-2}{x^2+6x-7} + \dfrac{-19+4x}{-7+6x+x^2}$

$\quad = \dfrac{x^2-2}{x^2+6x-7} + \dfrac{-19+4x}{x^2+6x-7}$

$\quad = \dfrac{x^2-2-19+4x}{x^2+6x-7}$

$\quad = \dfrac{x^2+4x-21}{x^2+6x-7}$

$\quad = \dfrac{(x+7)(x-3)}{(x+7)(x-1)} = \dfrac{x-3}{x-1}$

65. $\dfrac{6b^2-10b}{16b^2-48b+27} + \dfrac{7b^2-20b}{16b^2-48b+27} - \dfrac{6b-3b^2}{16b^2-48b+27}$

$\quad = \dfrac{6b^2-10b+7b^2-20b-6b+3b^2}{16b^2-48b+27}$

$\quad = \dfrac{16b^2-36b}{16b^2-48b+27} = \dfrac{4b(4b-9)}{(4b-9)(4b-3)}$

$\quad = \dfrac{4b}{4b-3}$

67. $\dfrac{2y}{y-5} - \left(\dfrac{2}{y-5} + \dfrac{y-2}{y-5} \right)$

$\quad = \dfrac{2y}{y-5} - \left(\dfrac{2+y-2}{y-5} \right) = \dfrac{2y}{y-5} - \dfrac{y}{y-5}$

$\quad = \dfrac{y}{y-5}$

69. $\dfrac{b}{ac+ad-bc-bd} - \dfrac{a}{ac+ad-bc-bd}$

$\quad = \dfrac{b-a}{ac+ad-bc-bd}$

$\quad = \dfrac{b-a}{a(c+d)-b(c+d)}$

$\quad = \dfrac{b-a}{(c+d)(a-b)} = \dfrac{(-1)}{(-1)} \cdot \dfrac{b-a}{(c+d)(a-b)}$

$\quad = \dfrac{a-b}{-(c+d)(a-b)} = \dfrac{-1}{c+d}$ or $-\dfrac{1}{c+d}$

71. $\dfrac{(y-3)(y+2)}{(y+1)(y-4)} - \dfrac{(y+2)(y+3)}{(y+1)(4-y)} - \dfrac{(y+5)(y-1)}{(y+1)(4-y)} = \dfrac{y^2-y-6}{(y+1)(y-4)} - \dfrac{y^2+5y+6}{(y+1)(4-y)} - \dfrac{y^2+4y-5}{(y+1)(4-y)}$

$= \dfrac{y^2-y-6}{(y+1)(y-4)} - \dfrac{(-1)}{(-1)} \cdot \dfrac{y^2+5y+6}{(y+1)(4-y)} - \dfrac{(-1)}{(-1)} \cdot \dfrac{y^2+4y-5}{(y+1)(4-y)}$

$= \dfrac{y^2-y-6}{(y+1)(y-4)} + \dfrac{y^2+5y+6}{(y+1)(y-4)} + \dfrac{y^2+4y-5}{(y+1)(y-4)} = \dfrac{y^2-y-6+y^2+5y+6+y^2+4y-5}{(y+1)(y-4)}$

$= \dfrac{3y^2+8y-5}{(y+1)(y-4)}$

73. a. $\dfrac{L+60W}{L} - \dfrac{L-40W}{L}$

$= \dfrac{(L+60W)-(L-40W)}{L}$

$= \dfrac{L+60W-L+40W}{L}$

$= \dfrac{100W}{L}$

b. $\dfrac{100W}{L}$; $W=5, L=6$

$\dfrac{100W}{L} = \dfrac{100 \cdot 5}{6} \approx 83.3$

Since this value is over 80, the skull is round.

75. $P = 2L + 2W$

$= 2\left(\dfrac{5x+10}{x+3}\right) + 2\left(\dfrac{5}{x+3}\right)$

$= \dfrac{10x+20}{x+3} + \dfrac{10}{x+3}$

$= \dfrac{10x+30}{x+3} = \dfrac{10(x+3)}{x+3} = 10$

The perimeter is 10 meters.

77. – 79. Answers will vary.

81. does not make sense; Explanations will vary. Sample explanation: $\dfrac{3x+1}{4} + \dfrac{x+2}{4} = \dfrac{4x+3}{4}$ and $4x+3$ is not divisible by 4.

83. makes sense

85. false; Changes to make the statement true will vary. A sample change is: You do not add the common denominators. You just keep the common denominator in your answer.

87. false; Changes to make the statement true will vary. A sample change is: Some such rational expressions cannot be simplified.

89. $\left(\dfrac{3x-1}{x^2+5x-6}-\dfrac{2x-7}{x^2+5x-6}\right)\div\dfrac{x+2}{x^2-1}$

$=\left(\dfrac{(3x-1)-(2x-7)}{x^2+5x-6}\right)\div\dfrac{x+2}{x^2-1}$

$=\dfrac{3x-1-2x+7}{x^2+5x-6}\div\dfrac{x+2}{x^2-1}$

$=\dfrac{x+6}{x^2+5x-6}\div\dfrac{x+2}{x^2-1}$

$=\dfrac{x+6}{x^2+5x-6}\cdot\dfrac{x^2-1}{x+2}$

$=\dfrac{(x+6)}{(x+6)(x-1)}\cdot\dfrac{(x+1)(x-1)}{(x+2)}$

$=\dfrac{x+1}{x+2}$

91. $\dfrac{2x}{x+3}+\dfrac{?}{x+3}=\dfrac{4x+1}{x+3}$

The sum of numerators on the left side must be $4x+1,$ so the missing expression is $2x+1.$
Check:

$\dfrac{2x}{x+3}+\dfrac{2x+1}{x+3}=\dfrac{2x+2x+1}{x+3}$

$=\dfrac{4x+1}{x+3}$

93. $\dfrac{6}{x-2}+\dfrac{?}{2-x}=\dfrac{13}{x-2}$

$\dfrac{6}{x-2}+\dfrac{(-1)}{(-1)}\cdot\dfrac{?}{2-x}=\dfrac{13}{x-2}$

$\dfrac{6}{x-2}+\dfrac{(-1)?}{x-2}=\dfrac{13}{x-2}$

Since $6+7=13,$ the opposite of the missing expression must be 7, so the missing expression is $-7.$
Check:

$\dfrac{6}{x-2}+\dfrac{-7}{2-x}=\dfrac{6}{x-2}+\dfrac{(-1)\cdot-7}{(-1)(2-x)}$

$=\dfrac{6}{x-2}+\dfrac{7}{x-2}=\dfrac{13}{x-2}$

95. $\dfrac{3x}{x-5}+\dfrac{?}{5-x}=\dfrac{7x+1}{x-5}$

$\dfrac{3x}{x-5}+\dfrac{(-1)}{(-1)}\cdot\dfrac{?}{5-x}=\dfrac{7x+1}{x-5}$

$\dfrac{3x}{x-5}+\dfrac{(-1)?}{x-5}=\dfrac{7x+1}{x-5}$

Since $3x+(4x+1)=7x+1,$ the opposite of the missing expression must be $4x+1,$ so the missing expression is $-4x-1.$
Check:

$\dfrac{3x}{x-5}+\dfrac{-4x-1}{5-x}=\dfrac{3x}{x-5}+\dfrac{4x+1}{x-5}$

$=\dfrac{7x+1}{x-5}$

97. The graphs do not coincide.

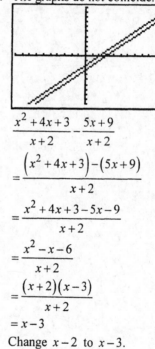

$\dfrac{x^2+4x+3}{x+2}-\dfrac{5x+9}{x+2}$

$=\dfrac{\left(x^2+4x+3\right)-(5x+9)}{x+2}$

$=\dfrac{x^2+4x+3-5x-9}{x+2}$

$=\dfrac{x^2-x-6}{x+2}$

$=\dfrac{(x+2)(x-3)}{x+2}$

$=x-3$

Change $x-2$ to $x-3.$

99. $\dfrac{13}{15}-\dfrac{8}{45}=\dfrac{13}{15}\cdot\dfrac{3}{3}-\dfrac{8}{45}$

$=\dfrac{39}{45}-\dfrac{8}{45}=\dfrac{31}{45}$

100. $81x^4-1=\left(9x^2+1\right)\left(9x^2-1\right)$

$=\left(9x^2+1\right)(3x+1)(3x-1)$

101.

$$x+3\overline{)3x^3+2x^2-26x-15}$$ with quotient $3x^2-7x-5$

$$\underline{3x^3+9x^2}$$
$$-7x^2-26x$$
$$\underline{-7x^2-21x}$$
$$-5x-15$$
$$\underline{-5x-15}$$
$$0$$

$$\frac{3x^3+2x^2-26x-15}{x+3}=3x^2-7x-5$$

102. $\dfrac{1}{2}+\dfrac{2}{3}=\dfrac{3}{6}+\dfrac{4}{6}=\dfrac{7}{6}$

103. $\dfrac{1}{8}-\dfrac{5}{6}=\dfrac{3}{24}-\dfrac{20}{24}=-\dfrac{17}{24}$

104.
$$\frac{(y+2)y-2\cdot 4}{4y(y+4)}=\frac{y^2+2y-8}{4y(y+4)}$$
$$=\frac{(y+4)(y-2)}{4y(y+4)}$$
$$=\frac{y-2}{4y}$$

7.4 Check Points

1. $\dfrac{3}{10x^2}$ and $\dfrac{7}{15x}$

$$10x^2=2\cdot 5x^2$$
$$15x=3\cdot 5x$$
$$\text{LCD}=2\cdot 3\cdot 5\cdot x^2=30x^2$$

2. $\dfrac{2}{x+3}$ and $\dfrac{4}{x-3}$

$$x+3=1(x+3)$$
$$x-3=1(x-3)$$
$$\text{LCD}=(x+3)(x-3)$$

3. $\dfrac{9}{7x^2+28x}$ and $\dfrac{11}{x^2+8x+16}$

$$7x^2+28x=7x(x+4)$$
$$x^2+8x+16=(x+4)^2$$
$$\text{LCD}=7x(x+4)^2$$

4. $\dfrac{3}{10x^2}+\dfrac{7}{15x}$

$$\text{LCD}=30x^2$$
$$\frac{3}{10x^2}+\frac{7}{15x}=\frac{3}{3}\cdot\frac{3}{10x^2}+\frac{2x}{2x}\cdot\frac{7}{15x}$$
$$=\frac{9}{30x^2}+\frac{14x}{30x^2}$$
$$=\frac{9+14x}{30x^2}$$

5. $\dfrac{2}{x+3}+\dfrac{4}{x-3}$

$$\text{LCD}=(x+3)(x-3)$$
$$\frac{2}{x+3}+\frac{4}{x-3}=\frac{x-3}{x-3}\cdot\frac{2}{x+3}+\frac{x+3}{x+3}\cdot\frac{4}{x-3}$$
$$=\frac{2x-6}{(x+3)(x-3)}+\frac{4x+12}{(x+3)(x-3)}$$
$$=\frac{2x-6+4x+12}{(x+3)(x-3)}$$
$$=\frac{6x+6}{(x+3)(x-3)}$$

6. $\dfrac{x}{x+5}-1$

$$\text{LCD}=x+5$$
$$\frac{x}{x+5}-1=\frac{x}{x+5}-\frac{x+5}{x+5}$$
$$=\frac{x-(x+5)}{x+5}$$
$$=\frac{x-x-5}{x+5}$$
$$=\frac{-5}{x+5}\text{ or }-\frac{5}{x+5}$$

7. $\dfrac{5}{y^2-5y}-\dfrac{y}{5y-25}=\dfrac{5}{y(y-5)}-\dfrac{y}{5(y-5)}$

$\quad$ LCD $=5y(y-5)$

$\quad \dfrac{5}{y^2-5y}-\dfrac{y}{5y-25}=\dfrac{5}{y(y-5)}-\dfrac{y}{5(y-5)}$

$\qquad\qquad = \dfrac{5}{5}\cdot\dfrac{5}{y(y-5)}-\dfrac{y}{y}\cdot\dfrac{y}{5(y-5)}$

$\qquad\qquad = \dfrac{25}{5y(y-5)}-\dfrac{y^2}{5y(y-5)}$

$\qquad\qquad = \dfrac{25-y^2}{5y(y-5)}$

$\qquad\qquad = \dfrac{(5+y)(5-y)}{5y(y-5)}$

$\qquad\qquad = \dfrac{(5+y)\,\overset{-1}{\cancel{(5-y)}}}{5y\,\cancel{(y-5)}_{1}}$

$\qquad\qquad = -\dfrac{5+y}{5y}$

8. $\dfrac{4x}{x^2-25}+\dfrac{3}{5-x}=\dfrac{4x}{(x+5)(x-5)}+\dfrac{3}{5-x}$

$\quad$ LCD $=(x+5)(x-5)$

$\quad \dfrac{4x}{x^2-25}+\dfrac{3}{5-x}=\dfrac{4x}{(x+5)(x-5)}+\dfrac{3}{5-x}$

$\qquad\qquad = \dfrac{4x}{(x+5)(x-5)}+\dfrac{-1(x+5)}{-1(x+5)}\cdot\dfrac{3}{5-x}$

$\qquad\qquad = \dfrac{4x}{(x+5)(x-5)}+\dfrac{-1(x+5)\cdot 3}{(x+5)(x-5)}$

$\qquad\qquad = \dfrac{4x}{(x+5)(x-5)}+\dfrac{-3x-15}{(x+5)(x-5)}$

$\qquad\qquad = \dfrac{4x-3x-15}{(x+5)(x-5)}$

$\qquad\qquad = \dfrac{x-15}{(x+5)(x-5)}$

7.4 Exercise Set

1. $\dfrac{7}{15x^2}$ and $\dfrac{13}{24x}$

$\quad 15x^2=3\cdot5x^2$

$\quad 24x=2^3\cdot3x$

$\quad$ LCD $=2^3\cdot3\cdot5x^2=120x^2$

3. $\dfrac{8}{15x^2}$ and $\dfrac{5}{6x^5}$

$\quad 15x^2=3\cdot5x^2$

$\quad 6x^5=2\cdot3x^5$

$\quad$ LCD $=2\cdot3\cdot5\cdot x^5=30x^5$

5. $\dfrac{4}{x-3}$ and $\dfrac{7}{x+1}$

$\quad$ LCD $=(x-3)(x+1)$

7. $\dfrac{5}{7(y+2)}$ and $\dfrac{10}{y}$

$\quad$ LCD $=7y(y+2)$

9. $\dfrac{17}{x+4}$ and $\dfrac{18}{x^2-16}$

$\quad x+4=1(x+4)$

$\quad x^2-16=(x+4)(x-4)$

$\quad$ LCD $=(x+4)(x-4)$

11. $\dfrac{8}{y^2-9}$ and $\dfrac{14}{y(y+3)}$

$\quad y^2-9=(y+3)(y-3)$

$\quad y(y+3)=y(y+3)$

$\quad$ LCD $=y(y+3)(y-3)$

13. $\dfrac{7}{y^2-1}$ and $\dfrac{y}{y^2-2y+1}$

$\quad y^2-1=(y+1)(y-1)$

$\quad y^2-2y+1=(y-1)(y-1)$

$\quad$ LCD $=(y+1)(y-1)(y-1)$

15. $\dfrac{3}{x^2-x-20}$ and $\dfrac{x}{2x^2+7x-4}$

$x^2-x-20 = (x-5)(x+4)$

$2x^2+7x-4 = (2x-1)(x+4)$

LCD $= (x-5)(x+4)(2x-1)$

17. $\dfrac{3}{x}+\dfrac{5}{x^2}$

LCD $= x^2$

$\dfrac{3}{x}+\dfrac{5}{x^2} = \dfrac{3}{x}\cdot\dfrac{x}{x}+\dfrac{5}{x^2} = \dfrac{3x+5}{x^2}$

19. $\dfrac{2}{9x}+\dfrac{11}{6x}$

LCD $= 18x$

$\dfrac{2}{9x}+\dfrac{11}{6x} = \dfrac{2}{9x}\cdot\dfrac{2}{2}+\dfrac{11}{6x}\cdot\dfrac{3}{3}$

$= \dfrac{4}{18x}+\dfrac{33}{18x} = \dfrac{37}{18x}$

21. $\dfrac{4}{x}+\dfrac{7}{2x^2}$

LCD $= 2x^2$

$\dfrac{4}{x}+\dfrac{7}{2x^2} = \dfrac{4}{x}\cdot\dfrac{2x}{2x}+\dfrac{7}{2x^2} = \dfrac{8x}{2x^2}+\dfrac{7}{2x^2}$

$= \dfrac{8x+7}{2x^2}$

23. $6+\dfrac{1}{x}$

LCD $= x$

$6+\dfrac{1}{x} = \dfrac{6}{1}\cdot\dfrac{x}{x}+\dfrac{1}{x} = \dfrac{6x}{x}+\dfrac{1}{x} = \dfrac{6x+1}{x}$

25. $\dfrac{2}{x}+9$

LCD $= x$

$\dfrac{2}{x}+9 = \dfrac{2}{x}+\dfrac{9}{1}\cdot\dfrac{x}{x} = \dfrac{2}{x}+\dfrac{9x}{x} = \dfrac{2+9x}{x}$

27. $\dfrac{x-1}{6}+\dfrac{x+2}{3}$

LCD $= 6$

$\dfrac{x-1}{6}+\dfrac{x+2}{3} = \dfrac{x-1}{6}+\dfrac{(x+2)}{3}\cdot\dfrac{2}{2}$

$= \dfrac{x-1}{6}+\dfrac{2x+4}{6} = \dfrac{3x+3}{6} = \dfrac{3(x+1)}{6}$

$= \dfrac{x+1}{2}$

29. $\dfrac{4}{x}+\dfrac{3}{x-5}$

LCD $= x(x-5)$

$\dfrac{4}{x}+\dfrac{3}{x-5} = \dfrac{4(x-5)}{x(x-5)}+\dfrac{3}{x-5}\cdot\dfrac{x}{x}$

$= \dfrac{4(x-5)}{x(x-5)}+\dfrac{3x}{x(x-5)}$

$= \dfrac{4x-20+3x}{x(x-5)}$

$= \dfrac{7x-20}{x(x-5)}$

31. $\dfrac{2}{x-1}+\dfrac{3}{x+2}$

LCD $= (x-1)(x+2)$

$\dfrac{2}{x-1}+\dfrac{3}{x+2}$

$= \dfrac{2(x+2)}{(x-1)(x+2)}+\dfrac{3(x-1)}{(x-1)(x+2)}$

$= \dfrac{2x+4+3x-3}{(x-1)(x+2)}$

$= \dfrac{5x+1}{(x-1)(x+2)}$

33. $\dfrac{2}{y+5}+\dfrac{3}{4y}$

LCD $=4y(y+5)$

$\dfrac{2}{y+5}+\dfrac{3}{4y}=\dfrac{2(4y)}{(4y)(y+5)}+\dfrac{3(y+5)}{4y(y+5)}$

$\qquad=\dfrac{2(4y)+3(y+5)}{4y(y+5)}$

$\qquad=\dfrac{8y+3y+15}{4y(y+5)}$

$\qquad=\dfrac{11y+15}{4y(y+5)}$

35. $\dfrac{x}{x+7}-1$

LCD $=x+7$

$\dfrac{x}{x+7}-1=\dfrac{x}{x+7}-\dfrac{x+7}{x+7}$

$\qquad=\dfrac{x-(x+7)}{x+7}$

$\qquad=\dfrac{x-x-7}{x+7}$

$\qquad=\dfrac{-7}{x+7}$ or $-\dfrac{7}{x+7}$

37. $\dfrac{7}{x+5}-\dfrac{4}{x-5}$

LCD $=(x+5)(x-5)$

$\dfrac{7}{x+5}-\dfrac{4}{x-5}$

$=\dfrac{7(x-5)}{(x+5)(x-5)}-\dfrac{4(x+5)}{(x+5)(x-5)}$

$=\dfrac{7(x-5)-4(x+5)}{(x+5)(x-5)}$

$=\dfrac{7x-35-4x-20}{(x+5)(x-5)}$

$=\dfrac{3x-55}{(x+5)(x-5)}$

39. $\dfrac{2x}{x^2-16}+\dfrac{x}{x-4}$

$x^2-16=(x+4)(x-4)$

$x-4=1(x-4)$

LCD $=(x+4)(x-4)$

$\dfrac{2x}{x^2-16}+\dfrac{x}{x-4}$

$=\dfrac{2x}{(x+4)(x-4)}+\dfrac{x}{x-4}$

$=\dfrac{2x}{(x+4)(x-4)}+\dfrac{x(x+4)}{(x+4)(x-4)}$

$=\dfrac{2x+x(x+4)}{(x+4)(x-4)}$

$=\dfrac{2x+x^2+4x}{(x+4)(x-4)}$

$=\dfrac{x^2+6x}{(x+4)(x-4)}$

41. $\dfrac{5y}{y^2-9}-\dfrac{4}{y+3}$

LCD $=(y+3)(y-3)$

$\dfrac{5y}{y^2-9}-\dfrac{4}{y+3}$

$=\dfrac{5y}{(y+3)(y-3)}-\dfrac{4}{y+3}$

$=\dfrac{5y}{(y+3)(y-3)}-\dfrac{4(y-3)}{(y+3)(y-3)}$

$=\dfrac{5y-4(y-3)}{(y+3)(y-3)}$

$=\dfrac{5y-4y+12}{(y+3)(y-3)}$

$=\dfrac{y+12}{(y+3)(y-3)}$

43. $\dfrac{7}{x-1} - \dfrac{3}{(x-1)(x-1)}$

$\text{LCD} = (x-1)(x-1)$

$\dfrac{7}{x-1} - \dfrac{3}{(x-1)(x-1)}$

$= \dfrac{7(x-1)}{(x-1)(x-1)} - \dfrac{3}{(x-1)(x-1)}$

$= \dfrac{7x-7-3}{(x-1)(x-1)}$

$= \dfrac{7x-10}{(x-1)(x-1)}$ or $\dfrac{7x-10}{(x-1)^2}$

45. $\dfrac{3y}{4y-20} + \dfrac{9y}{6y-30}$

$4y - 20 = 4(y-5)$

$6y - 30 = 6(y-5)$

$\text{LCD} = 12(y-5)$

$\dfrac{3y}{4y-20} + \dfrac{9y}{6y-30}$

$= \dfrac{4y}{4(y-5)} + \dfrac{9y}{6(y-5)}$

$= \dfrac{4y}{4(y-5)} \cdot \dfrac{3}{3} + \dfrac{9y}{6(y-5)} \cdot \dfrac{2}{2}$

$= \dfrac{12y}{12(y-5)} + \dfrac{18y}{12(y-5)}$

$= \dfrac{9y+18y}{12(y-5)} = \dfrac{27y}{12(y-5)}$

$= \dfrac{9y}{4(y-5)}$

47. $\dfrac{y+4}{y} - \dfrac{y}{y+4}$

$\text{LCD} = y(y+4)$

$\dfrac{y+4}{y} - \dfrac{y}{y+4}$

$= \dfrac{(y+4)(y+4)}{y(y+4)} - \dfrac{y \cdot y}{y(y+4)}$

$= \dfrac{y^2 + 8y + 16 - y^2}{y(y+4)}$

$= \dfrac{8y+16}{y(y+4)}$

49. $\dfrac{2x+9}{x^2-7x+12} = \dfrac{2}{x-3}$

$x^2 - 7x + 12 = (x-3)(x-4)$

$x - 3 = 1(x-3)$

$\text{LCD} = (x-3)(x-4)$

$\dfrac{2x+9}{x^2-7x+12} - \dfrac{2}{x-3}$

$= \dfrac{2x+9}{(x-3)(x-4)} - \dfrac{2}{x-3}$

$= \dfrac{2x+9}{(x-3)(x-4)} - \dfrac{2(x-4)}{(x-3)(x-4)}$

$= \dfrac{2x+9-2(x-4)}{(x-3)(x-4)}$

$= \dfrac{2x+9-2x+8}{(x-3)(x-4)}$

$= \dfrac{17}{(x-3)(x-4)}$

51. $\dfrac{3}{x^2-1} + \dfrac{4}{(x+1)^2}$

$x^2 - 1 = (x+1)(x-1)$

$(x+1)^2 = (x+1)(x+1)$

$\text{LCD} = (x+1)(x+1)(x-1)$

$\dfrac{3}{x^2-1} + \dfrac{4}{(x+1)^2}$

$= \dfrac{3}{(x+1)(x-1)} + \dfrac{4}{(x+1)(x+1)}$

$= \dfrac{3(x+1)}{(x+1)(x+1)(x-1)} + \dfrac{4(x-1)}{(x+1)(x+1)(x-1)}$

$= \dfrac{3(x+1)+4(x-1)}{(x+1)(x+1)(x-1)}$

$= \dfrac{3x+3+4x-4}{(x+1)(x+1)(x-1)}$

$= \dfrac{7x-1}{(x+1)(x+1)(x-1)}$

53. $\dfrac{3x}{x^2+3x-10}-\dfrac{2x}{x^2+x-6}$

$x^2+3x-10=(x-2)(x+5)$

$x^2+x-6=(x+3)(x-2)$

$\text{LCD}=(x+3)(x-2)(x+5)$

$\dfrac{3x}{x^2+3x-10}-\dfrac{2x}{x^2+x-6}$

$=\dfrac{3x}{(x-2)(x+5)}-\dfrac{2x}{(x+3)(x-2)}$

$=\dfrac{3x(x+3)}{(x+3)(x-2)(x+5)}-\dfrac{2x(x+5)}{(x+3)(x-2)(x+5)}$

$=\dfrac{3x(x+3)-2x(x+5)}{(x+3)(x-2)(x+5)}$

$=\dfrac{3x^2+9x-2x^2-10x}{(x+3)(x-2)(x+5)}$

$=\dfrac{x^2-x}{(x+3)(x-2)(x+5)}$

55. $\dfrac{y}{y^2+2y+1}+\dfrac{4}{y^2+5y+4}$

$y^2+2y+1=(y+1)(y+1)$

$y^2+5y+4=(y+4)(y+1)$

$\text{LCD}=(y+4)(y+1)(y+1)$

$\dfrac{y}{y^2+2y+1}+\dfrac{4}{y^2+5y+4}$

$=\dfrac{y}{(y+1)(y+1)}+\dfrac{4}{(y+4)(y+1)}$

$=\dfrac{y(y+4)}{(y+4)(y+1)(y+1)}+\dfrac{4(y+1)}{(y+4)(y+1)(y+1)}$

$=\dfrac{y(y+4)+4(y+1)}{(y+4)(y+1)(y+1)}$

$=\dfrac{y^2+4y+4y+4}{(y+4)(y+1)(y+1)}$

$=\dfrac{y^2+8y+4}{(y+4)(y+1)(y+1)}$

57. $\dfrac{x-5}{x+3}+\dfrac{x+3}{x-5}$

$\text{LCD}=(x+3)(x-5)$

$\dfrac{x-5}{x+3}+\dfrac{x+3}{x-5}$

$=\dfrac{(x-5)(x-5)}{(x+3)(x-5)}+\dfrac{(x+3)(x+3)}{(x-5)(x+3)}$

$=\dfrac{(x-5)(x-5)+(x+3)(x+3)}{(x+3)(x-5)}$

$=\dfrac{\left(x^2-10x+25\right)+\left(x^2+6x+9\right)}{(x+3)(x-5)}$

$=\dfrac{2x^2-4x+34}{(x+3)(x-5)}$

59. $\dfrac{5}{2y^2-2y}-\dfrac{3}{2y-2}$

$2y^2-2y=2y(y-1)$

$2y-2=2(y-1)$

$\text{LCD}=2y(y-1)$

$\dfrac{5}{2y^2-2y}-\dfrac{3}{2y-2}$

$=\dfrac{5}{2y(y-1)}-\dfrac{3}{2(y-1)}$

$=\dfrac{5}{2y(y-1)}-\dfrac{3\cdot y}{2y(y-1)}$

$=\dfrac{5-3y}{2y(y-1)}$

61. $\dfrac{4x+3}{x^2-9}-\dfrac{x+1}{x-3}$

LCD $=(x+3)(x-3)$

$\dfrac{4x+3}{x^2-9}-\dfrac{x+1}{x-3}$

$=\dfrac{4x+3}{(x+3)(x-3)}-\dfrac{(x+1)(x+3)}{(x+3)(x-3)}$

$=\dfrac{(4x+3)-(x+1)(x+3)}{(x+3)(x-3)}$

$=\dfrac{(4x+3)-\left(x^2+4x+3\right)}{(x+3)(x-3)}$

$=\dfrac{4x+3-x^2-4x-3}{(x+3)(x-3)}$

$=\dfrac{-x^2}{(x+3)(x-3)}=-\dfrac{x^2}{(x+3)(x-3)}$

63. $\dfrac{y^2-39}{y^2+3y-10}-\dfrac{y-7}{y-2}$

$y^2+3y-10=(y-2)(y+5)$

$\qquad y-2=1(y-2)$

LCD $=(y-2)(y+5)$

$\dfrac{y^2-39}{y^2+3y-10}-\dfrac{y-7}{y-2}$

$=\dfrac{y^2-39}{(y-2)(y+5)}-\dfrac{y-7}{y-2}$

$=\dfrac{y^2-39}{(y-2)(y+5)}-\dfrac{(y-7)(y+5)}{(y-2)(y+5)}$

$=\dfrac{\left(y^2-39\right)-(y-7)(y+5)}{(y-2)(y+5)}$

$=\dfrac{\left(y^2-39\right)-\left(y^2-2y-35\right)}{(y-2)(y+5)}$

$=\dfrac{y^2-39-y^2+2y+35}{(y-2)(y+5)}$

$=\dfrac{2y-4}{(y-2)(y+5)}=\dfrac{2(y-2)}{(y-2)(y+5)}$

$=\dfrac{2}{y+5}$

65. $4+\dfrac{1}{x-3}$

LCD $=x-3$

$4+\dfrac{1}{x-3}=\dfrac{4(x-3)}{x-3}+\dfrac{1}{x-3}$

$=\dfrac{4(x-3)+1}{x-3}$

$=\dfrac{4x-12+1}{x-3}$

$=\dfrac{4x-11}{x-3}$

67. $3-\dfrac{3y}{y+1}$

LCD $=y+1$

$3-\dfrac{3y}{y+1}=\dfrac{3(y+1)}{y+1}-\dfrac{3y}{y+1}$

$=\dfrac{3(y+1)-3y}{y+1}$

$=\dfrac{3y+3-3y}{y+1}$

$=\dfrac{3}{y+1}$

69. $\dfrac{9x+3}{x^2-x-6}+\dfrac{x}{3-x}$

$x^2-x-6=(x-3)(x+2)$

$\qquad 3-x=-1(x-3)$

LCD $=(x-3)(x+2)$

$\dfrac{9x+3}{x^2-x-6}+\dfrac{x}{3-x}$

$=\dfrac{9x+3}{(x-3)(x+2)}+\dfrac{(-1)}{(-1)}\cdot\dfrac{x}{3-x}$

$=\dfrac{9x+3}{(x-3)(x+2)}+\dfrac{-x}{x-3}$

$=\dfrac{9x+3}{(x-3)(x+2)}+\dfrac{-x(x+2)}{(x-3)(x+2)}$

$=\dfrac{9x+3-x(x+2)}{(x-3)(x+2)}$

$=\dfrac{9x+3-x^2-2x}{(x-3)(x+2)}$

$=\dfrac{-x^2+7x+3}{(x-3)(x+2)}$

71. $\dfrac{x+3}{x^2+x-2}-\dfrac{2}{x^2-1}$

$x^2+x-2=(x-1)(x+2)$

$x^2-1=(x+1)(x-1)$

LCD $=(x+1)(x-1)(x+2)$

$\dfrac{x+3}{x^2+x-2}-\dfrac{2}{x^2-1}$

$=\dfrac{x+3}{(x-1)(x+2)}-\dfrac{2}{(x+1)(x-1)}$

$=\dfrac{(x+3)(x+1)}{(x+1)(x-1)(x+2)}$

$\quad-\dfrac{2(x+2)}{(x+1)(x-1)(x+2)}$

$=\dfrac{(x+3)(x+1)-2(x+2)}{(x+1)(x-1)(x+2)}$

$=\dfrac{x^2+4x+3-2x-4}{(x+1)(x-1)(x+2)}$

$=\dfrac{x^2+2x-1}{(x+1)(x-1)(x+2)}$

73. $\dfrac{y+3}{5y^2}-\dfrac{y-5}{15y}$

LCD $=15y^2$

$\dfrac{y+3}{5y^2}-\dfrac{y-5}{15y}$

$=\dfrac{(y+3)(3)}{5y^2(3)}-\dfrac{(y-5)(y)}{15y(y)}$

$=\dfrac{(3y+9)-\left(y^2-5y\right)}{15y^2}$

$=\dfrac{3y+9-y^2+5y}{15y^2}$

$=\dfrac{-y^2+8y+9}{15y^2}$

75. $\dfrac{x+3}{3x+6}+\dfrac{x}{4-x^2}$

$3x+6=3(x+2)$

$4-x^2=(2+x)(2-x)$

Note that $-1(2-x)=x-2$

LCD $=3(x+2)(x-2)$

$\dfrac{x+3}{3x+6}+\dfrac{x}{4-x^2}$

$=\dfrac{x+3}{3(x+2)}+\dfrac{x}{(2+x)(2-x)}$

$=\dfrac{x+3}{3(x+2)}+\dfrac{(-1)}{(-1)}\cdot\dfrac{x}{(2+x)(2-x)}$

$=\dfrac{x+3}{3(x+2)}+\dfrac{-x}{(x+2)(x-2)}$

$=\dfrac{(x+3)(x-2)}{3(x+2)(x-2)}+\dfrac{-x(3)}{3(x+2)(x-2)}$

$=\dfrac{x^2+x-6-3x}{3(x+2)(x-2)}$

$=\dfrac{x^2-2x-6}{3(x+2)(x-2)}$

77. $\dfrac{y}{y^2-1}+\dfrac{2y}{y-y^2}$

$y^2-1=(y+1)(y-1)$

$y-y^2=y(1-y)$

Note that $-1(1-y)=y-1$

LCD $=y(y+1)(y-1)$

$\dfrac{y}{y^2-1}+\dfrac{2y}{y-y^2}$

$=\dfrac{y}{(y+1)(y-1)}+\dfrac{2y}{y(1-y)}$

$=\dfrac{y}{(y+1)(y-1)}+\dfrac{(-1)}{(-1)}\cdot\dfrac{2y}{y(1-y)}$

$=\dfrac{y}{(y+1)(y-1)}+\dfrac{-2y}{y(y-1)}$

$=\dfrac{y\cdot y}{y(y+1)(y-1)}+\dfrac{-2y(y+1)}{y(y+1)(y-1)}$

$=\dfrac{y^2-2y(y+1)}{y(y+1)(y-1)}=\dfrac{y^2-2y^2-2y}{y(y+1)(y-1)}$

$=\dfrac{-y^2-2y}{y(y+1)(y-1)}=\dfrac{-y(y+2)}{y(y+1)(y-1)}$

$=\dfrac{-1(y+2)}{(y+1)(y-1)}=\dfrac{-y-2}{(y+1)(y-1)}$

79. $\dfrac{x-1}{x}+\dfrac{y+1}{y}$

LCD $=xy$

$\dfrac{x-1}{x}+\dfrac{y+1}{y}$

$=\dfrac{(x-1)(y)}{xy}+\dfrac{(y+1)(x)}{xy}$

$=\dfrac{xy-y+xy+x}{xy}$

$=\dfrac{x+2xy-y}{xy}$

81. $\dfrac{3x}{x^2-y^2}-\dfrac{2}{y-x}$

$x^2-y^2=(x+y)(x-y)$

Note that $y-x=-1(x-y)$

LCD $=(x+y)(x-y)$

$\dfrac{3x}{x^2-y^2}-\dfrac{2}{y-x}$

$=\dfrac{3x}{(x+y)(x-y)}-\dfrac{(-1)}{(-1)}\cdot\dfrac{2}{y-x}$

$=\dfrac{3x}{(x+y)(x-y)}-\dfrac{-2}{x-y}$

$=\dfrac{3x}{(x+y)(x-y)}-\dfrac{-2(x+y)}{(x+y)(x-y)}$

$=\dfrac{3x+2(x+y)}{(x+y)(x-y)}=\dfrac{3x+2x+2y}{(x+y)(x-y)}$

$=\dfrac{5x+2y}{(x+y)(x-y)}$

83. $\dfrac{x+6}{x^2-4}-\dfrac{x+3}{x+2}+\dfrac{x-3}{x-2}$

LCD $=(x+2)(x-2)$

$\dfrac{x+6}{x^2-4}-\dfrac{x+3}{x+2}+\dfrac{x-3}{x-2}=\dfrac{x+6}{(x+2)(x-2)}-\dfrac{x+3}{x+2}+\dfrac{x-3}{x-2}$

$=\dfrac{x+6}{(x+2)(x-2)}-\dfrac{(x+3)(x-2)}{(x+2)(x-2)}+\dfrac{(x-3)(x+2)}{(x-2)(x+2)}$

$=\dfrac{x+6-(x+3)(x-2)+(x-3)(x+2)}{(x+2)(x-2)}=\dfrac{x+6-\left(x^2+x-6\right)+\left(x^2-x-6\right)}{(x+2)(x-2)}$

$=\dfrac{x+6-x^2-x+6+x^2-x-6}{(x+2)(x-2)}=\dfrac{-x+6}{(x+2)(x-2)}$

85. $\dfrac{5}{x^2-25}+\dfrac{4}{x^2-11x+30}-\dfrac{3}{x^2-x-30}$

$x^2-25=(x+5)(x-5)$

$x^2-11x+30=(x-6)(x-5)$

$x^2-x-30=(x-6)(x+5)$

$\text{LCD}=(x+5)(x-5)(x-6)$

$\dfrac{5}{x^2-25}+\dfrac{4}{x^2-11x+30}-\dfrac{3}{x^2-x-30}$

$=\dfrac{5}{(x+5)(x-5)}+\dfrac{4}{(x-6)(x-5)}-\dfrac{3}{(x-6)(x+5)}$

$=\dfrac{5(x-6)}{(x+5)(x-5)(x-6)}+\dfrac{4(x+5)}{(x-6)(x-5)(x+5)}-\dfrac{3(x-5)}{(x-6)(x+5)(x-5)}$

$=\dfrac{5(x-6)+4(x+5)-3(x-5)}{(x+5)(x-5)(x-6)}=\dfrac{5x-30+4x+20-3x+15}{(x+5)(x-5)(x-6)}$

$=\dfrac{6x+5}{(x+5)(x-5)(x-6)}$

87. $\dfrac{x+6}{x^3-27}-\dfrac{x}{x^3+3x^2+9x}$

$x^3-27=(x-3)\left(x^2+3x+9\right)$

$x^3+3x^2+9x=x\left(x^2+3x+9\right)$

$\text{LCD}=x(x-3)\left(x^2+3x+9\right)$

$\dfrac{x+6}{x^3-27}-\dfrac{x}{x^3+3x^2+9x}=\dfrac{x+6}{(x-3)\left(x^2+3x+9\right)}-\dfrac{x}{x\left(x^2+3x+9\right)}$

$=\dfrac{(x+6)x}{(x-3)\left(x^2+3x+9\right)x}-\dfrac{x(x-3)}{x\left(x^2+3x+9\right)(x-3)}$

$=\dfrac{x(x+6)-x(x-3)}{x(x-3)\left(x^2+3x+9\right)}=\dfrac{x^2+6x-x^2+3x}{x(x-3)\left(x^2+3x+9\right)}$

$=\dfrac{9x}{x(x-3)\left(x^2+3x+9\right)}=\dfrac{9}{(x-3)\left(x^2+3x+9\right)}$

89. $\dfrac{9y+3}{y^2-y-6}+\dfrac{y}{3-y}+\dfrac{y-1}{y+2}$

$y^2-y-6=(y-3)(y+2)$

$3-y=-1(y-3)$

$y+2=1(y+2)$

$\text{LCD}=(y-3)(y+2)$

$\dfrac{9y+3}{y^2-y-6}+\dfrac{y}{3-y}+\dfrac{y-1}{y+2}=\dfrac{9y+3}{(y-3)(y+2)}+\dfrac{y}{-1(y-3)}+\dfrac{y-1}{y+2}$

$=\dfrac{9y+3}{(y-3)(y+2)}+\dfrac{-y(y+2)}{(y-3)(y+2)}+\dfrac{(y-1)(y-3)}{(y+2)(y-3)}$

$=\dfrac{9y+3+-y(y+2)+(y-1)(y-3)}{(y-3)(y+2)}=\dfrac{9y+3-y^2-2y+y^2-4y+3}{(y-3)(y+2)}$

$=\dfrac{3y+6}{(y-3)(y+2)}=\dfrac{3(y+2)}{(y-3)(y+2)}=\dfrac{3}{y-3}$

91. $\dfrac{3}{x^2+4xy+3y^2}-\dfrac{5}{x^2-2xy-3y^2}+\dfrac{2}{x^2-9y^2}$

$x^2+4xy+3y^2=(x+y)(x+3y)$

$x^2-2xy-3y^2=(x-3y)(x+y)$

$x^2-9y^2=(x+3y)(x-3y)$

$\text{LCD}=(x+y)(x+3y)(x-3y)$

$\dfrac{3}{x^2+4xy+3y^2}-\dfrac{5}{x^2-2xy-3y^2}+\dfrac{2}{x^2-9y^2}$

$=\dfrac{3}{(x+y)(x+3y)}-\dfrac{5}{(x-3y)(x+y)}+\dfrac{2}{(x+3y)(x-3y)}$

$=\dfrac{3(x-3y)}{(x+y)(x+3y)(x-3y)}-\dfrac{5(x+3y)}{(x-3y)(x+y)(x+3y)}+\dfrac{2(x+y)}{(x+3y)(x-3y)(x+y)}$

$=\dfrac{3(x-3y)-5(x+3y)+2(x+y)}{(x+y)(x+3y)(x-3y)}=\dfrac{3x-9y-5x-15y+2x+2y}{(x+y)(x+3y)(x-3y)}$

$=\dfrac{-22y}{(x+y)(x+3y)(x-3y)}$

93. Young's Rule: $C = \dfrac{DA}{A+12}$

$A = 8;$

$C = \dfrac{D \cdot 8}{8+12} = \dfrac{8D}{20} = \dfrac{2D}{5}$

$A = 3;$

$C = \dfrac{D \cdot 3}{3+12} = \dfrac{3D}{15} = \dfrac{D}{5}$

Difference:

$\dfrac{2D}{5} - \dfrac{D}{5} = \dfrac{D}{5}$

The difference in dosages for an 8-year-old child and a 3-year-old child is $\dfrac{D}{5}$. This means that an 8-year-old should be given $\dfrac{1}{5}$ of the adult dosage more than a 3-year-old.

95. Young's Rule: $C = \dfrac{DA}{A+12}$

Cowling's Rule: $C = \dfrac{D(A+1)}{24}$

For $A = 12$, Young's Rule gives

$C = \dfrac{D \cdot 12}{12+12} = \dfrac{12D}{24} = \dfrac{D}{2}$ and Cowling's Rule gives

$C = \dfrac{D(12+1)}{24} = \dfrac{13D}{24}$.

The difference between the dosages given by Cowling's Rule and Young's Rule is

$\dfrac{13D}{24} - \dfrac{12D}{24} = \dfrac{D}{24}$.

This means that Cowling's Rule says to give a 12-year-old $\dfrac{1}{24}$ of the adult dose more than Young's Rule says the dosage should be.

97. No, because the graphs cross, neither formula gives a consistently smaller dosage.

99. The difference in dosage is greatest at 5 years. This is where the graphs are farthest apart.

101. $P = 2L + 2W$

$= 2\left(\dfrac{x}{x+3}\right) + 2\left(\dfrac{x}{x-4}\right)$

$= \dfrac{2x}{x+3} + \dfrac{2x}{x+4}$

$= \dfrac{2x(x+4)}{(x+3)(x+4)} + \dfrac{2x(x+3)}{(x+3)(x+4)}$

$= \dfrac{2x^2 + 8x + 2x^2 + 6x}{(x+3)(x+4)}$

$= \dfrac{4x^2 + 14x}{(x+3)(x+4)}$

103. Answers will vary.

105. Explanations will vary. The right side of the equation should be charged from $\dfrac{3}{x+5}$ to $\dfrac{5+2x}{5x}$.

107. Answers will vary.

109. makes sense

111. does not make sense; Explanations will vary. Sample explanation: It is acceptable to leave the numerator in factored form.

113. false; Changes to make the statement true will vary. A sample change is: The LCD is $x(x-1)$ or $x^2 - x$.

115. true

117. $\left(\dfrac{1}{x+h} - \dfrac{1}{x}\right) \div h$

$= \left(\dfrac{1(x)}{(x+h)(x)} - \dfrac{1(x+h)}{x(x+h)}\right) \div h$

$= \left(\dfrac{x - (x+h)}{x(x+h)}\right) \div h$

$= \left(\dfrac{x - x - h}{x(x+h)}\right) \div h = \dfrac{-h}{x(x+h)} \div h$

$= \dfrac{-h}{x(x+h)} \cdot \dfrac{1}{h} = \dfrac{-1}{x(x+h)}$

119. $\dfrac{4}{x-2} - \dfrac{?}{?} = \dfrac{2x+8}{(x-2)(x+1)}$

The missing rational expression must have $(x+1)$ as a factor in its denominator or as the complete denominator. Let $y =$ the numerator of the missing rational expression.

$$\dfrac{4}{x-2} - \dfrac{y}{x+1} = \dfrac{2x+8}{(x-2)(x+1)}$$

Then, $\dfrac{4(x+1) - y(x-2)}{(x-2)(x+1)} - \dfrac{2x+8}{(x-2)(x+1)}$

So $4x+4 - yx + 2y = 2x+8$, which implies that $4x - yx = 2x$ and $4 + 2y = 8$. Both of these equations give $y = 2$. Thus, the missing rational expression is $\dfrac{2}{x+1}$.

120. $(3x+5)(2x-7)$
$= 6x^2 - 21x + 10x - 35$
$= 6x^2 - 11x - 35$

121. $3x - y = 3$
Find the x-intercept.
$3x - y = 3$
$3x - 0 = 3$
$3x = 3$
$x = 1$
Find the y-intercept.
$3x - y = 3$
$3(0) - y = 3$
$-y = 3$
$y = -3$

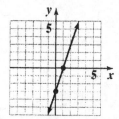

122. First find the slope $m = \dfrac{0-(-4)}{1-(-3)} = \dfrac{4}{4} = 1$

Use $m = 1$ and $(x_1, y_1) = (1, 0)$ in the point-slope form and simplify to find the slope-intercept form.
$y - y_1 = m(x - x_1)$
$\quad y - 0 = 1(x - 1)$
$\qquad y = x - 1$

123. a. $\dfrac{1}{3} + \dfrac{2}{5} = \dfrac{1 \cdot 5}{3 \cdot 5} + \dfrac{2 \cdot 3}{5 \cdot 3} = \dfrac{5}{15} + \dfrac{6}{15} = \dfrac{11}{15}$

b. $\dfrac{2}{5} - \dfrac{1}{3} = \dfrac{2 \cdot 3}{5 \cdot 3} - \dfrac{1 \cdot 5}{3 \cdot 5} = \dfrac{6}{15} - \dfrac{5}{15} = \dfrac{1}{15}$

c. $\left(\dfrac{1}{3} + \dfrac{2}{5}\right) \div \left(\dfrac{2}{5} - \dfrac{1}{3}\right) = \dfrac{11}{15} \div \dfrac{1}{15} = \dfrac{11}{15} \cdot \dfrac{15}{1} = 11$

124. a. $\dfrac{1}{x} + \dfrac{1}{y} = \dfrac{1 \cdot y}{x \cdot y} + \dfrac{1 \cdot x}{y \cdot x} = \dfrac{y}{xy} + \dfrac{x}{xy} = \dfrac{y+x}{xy}$

b. $\dfrac{1}{xy} \div \left(\dfrac{1}{x} + \dfrac{1}{y}\right) = \dfrac{1}{xy} \div \left(\dfrac{y+x}{xy}\right)$
$\qquad = \dfrac{1}{xy} \cdot \dfrac{xy}{y+x}$
$\qquad = \dfrac{1}{y+x}$

125. $xy\left(\dfrac{1}{x} + \dfrac{1}{y}\right) = \dfrac{xy}{x} + \dfrac{xy}{y} = y + x$

Chapter 7 Mid-Chapter Check Point

1. $\dfrac{x^2 - 4}{x^2 - 2x - 8}$
$\quad x^2 - 2x - 8 = 0$
$\quad (x-4)(x+2) = 0$
$\quad x - 4 = 0 \quad \text{or} \quad x + 2 = 0$
$\qquad x = 4 \qquad\qquad x = -2$
The rational expression is undefined for $x = 4$ and $x = -2$.

2. $\dfrac{3x^2 - 7x + 2}{6x^2 + x - 1} = \dfrac{(3x-1)(x-2)}{(3x-1)(2x+1)}$
$\qquad = \dfrac{x-2}{2x+1}$

3. $\dfrac{9-3y}{y^2-5y+6} = \dfrac{3(3-y)}{(y-3)(y-2)}$

$ = \dfrac{-3(y-3)}{(y-3)(y-2)} = \dfrac{-3}{y-2}$

4. $\dfrac{16w^3-24w^2}{8w^4-12w^3} = \dfrac{8w^2(2w-3)}{4w^3(2w-3)} = \dfrac{2}{w}$

5. $\dfrac{7x-3}{x^2+3x-4} - \dfrac{3x+1}{x^2+3x-4} = \dfrac{7x-3-3x-1}{(x-1)(x+4)}$

$ = \dfrac{4x-4}{(x-1)(x+4)} = \dfrac{4(x-1)}{(x-1)(x+4)}$

$ = \dfrac{4}{x+4}$

6. $\dfrac{x+2}{2x-4} \cdot \dfrac{8}{x^2-4}$

$ = \dfrac{x+2}{2(x-2)} \cdot \dfrac{8}{(x+2)(x-2)}$

$ = \dfrac{4}{(x-2)(x-2)} \text{ or } \dfrac{4}{(x-2)^2}$

7. $1 + \dfrac{7}{x-2} = \dfrac{x-2}{x-2} + \dfrac{7}{x-2} = \dfrac{x+5}{x-2}$

8. $\dfrac{2x^2+x-1}{2x^2-7x+3} \div \dfrac{x^2-3x-4}{x^2-x-6}$

$ = \dfrac{2x^2+x-1}{2x^2-7x+3} \cdot \dfrac{x^2-x-6}{x^2-3x-4}$

$ = \dfrac{(2x-1)(x+1)}{(2x-1)(x-3)} \cdot \dfrac{(x-3)(x+2)}{(x-4)(x+1)}$

$ = \dfrac{x+2}{x-4}$

9. $\dfrac{1}{x^2+2x-3}+\dfrac{1}{x^2+5x+6}$

$x^2+2x-3=(x+3)(x-1)$

$x^2+5x+6=(x+2)(x+3)$

LCD $=(x+3)(x-1)(x+2)$

$\dfrac{1}{x^2+2x-3}+\dfrac{1}{x^2+5x+6}=\dfrac{1}{(x+3)(x-1)}+\dfrac{1}{(x+2)(x+3)}$

$=\dfrac{1(x+2)}{(x+3)(x-1)(x+2)}+\dfrac{1(x-1)}{(x+2)(x+3)(x-1)}$

$=\dfrac{x+2+x-1}{(x+3)(x-1)(x+2)}=\dfrac{2x+1}{(x+3)(x-1)(x+2)}$

10. $\dfrac{17}{x-5}+\dfrac{x+8}{5-x}$

Note: $5-x=-1(x-5)$

LCD $=x-5$

$\dfrac{17}{x-5}+\dfrac{-1(x+8)}{-1(5-x)}=\dfrac{17}{x-5}+\dfrac{-x-8}{x-5}=\dfrac{17-x-8}{x-5}=\dfrac{9-x}{x-5}$

11. $\dfrac{4y^2-1}{9y-3y^2}\cdot\dfrac{y^2-7y+12}{2y^2-7y-4}=\dfrac{(2y+1)(2y-1)}{3y(3-y)}\cdot\dfrac{(y-4)(y-3)}{(2y+1)(y-4)}$

$=\dfrac{-1(2y+1)(2y-1)}{-1\cdot3y(3-y)}\cdot\dfrac{(y-4)(y-3)}{(2y+1)(y-4)}=\dfrac{-1(2y+1)(2y-1)}{3y(y-3)}\cdot\dfrac{(y-4)(y-3)}{(2y+1)(y-4)}$

$=\dfrac{-(2y-1)}{3y}=\dfrac{-2y+1}{3y}$

12. $\dfrac{y}{y+1}-\dfrac{2y}{y+2}$

LCD $=(y+1)(y+2)$

$\dfrac{y(y+2)}{(y+1)(y+2)}-\dfrac{2y(y+1)}{(y+2)(y+1)}=\dfrac{y^2+2y-2y^2-2y}{(y+1)(y+2)}=\dfrac{-y^2}{(y+1)(y+2)}$

13. $\dfrac{w^2+6w+5}{7w^2-63}\div\dfrac{w^2+10w+25}{7w+21}=\dfrac{w^2+6w+5}{7w^2-63}\cdot\dfrac{7w+21}{w^2+10w+25}$

$=\dfrac{(w+5)(w+1)}{7(w^2-9)}\cdot\dfrac{7(w+3)}{(w+5)(w+5)}=\dfrac{(w+5)(w+1)}{7(w+3)(w-3)}\cdot\dfrac{7(w+3)}{(w+5)(w+5)}=\dfrac{w+1}{(w-3)(w+5)}$

14. $\dfrac{2z}{z^2-9} - \dfrac{5}{z^2+4z+3}$

$z^2-9 = (z+3)(z-3)$

$z^2+4z+3 = (z+3)(z+1)$

$\text{LCD} = (z+3)(z-3)(z+1)$

$\dfrac{2z}{z^2-9} - \dfrac{5}{z^2+4z+3} = \dfrac{2z}{(z+3)(z-3)} - \dfrac{5}{(z+3)(z+1)}$

$= \dfrac{2z(z+1)}{(z+3)(z-3)(z+1)} - \dfrac{5(z-3)}{(z+3)(z+1)(z-3)} = \dfrac{2z^2+2z-5z+15}{(z+3)(z-3)(z+1)} = \dfrac{2z^2-3z+15}{(z+3)(z-3)(z+1)}$

15. $\dfrac{z+2}{3z-1} + \dfrac{5}{(3z-1)^2}$

$\text{LCD} = (3z-1)(3z-1)$

$\dfrac{(z+2)(3z-1)}{(3z-1)(3z-1)} + \dfrac{5}{(3z-1)^2} = \dfrac{3z^2+5z-2+5}{(3z-1)^2} = \dfrac{3z^2+5z+3}{(3z-1)^2}$

16. $\dfrac{8}{x^2+4x-21} + \dfrac{3}{x+7}$

$x^2+4x-21 = (x+7)(x-3)$

$x+7 = 1(x+7)$

$\text{LCD} = (x+7)(x-3)$

$\dfrac{8}{x^2+4x-21} + \dfrac{3}{x+7}$

$= \dfrac{8}{(x+7)(x-3)} + \dfrac{3}{x+7}$

$= \dfrac{8}{(x+7)(x-3)} + \dfrac{3(x-3)}{(x+7)(x-3)}$

$= \dfrac{8+3x-9}{(x+7)(x-3)} = \dfrac{3x-1}{(x+7)(x-3)}$

17. $\dfrac{x^4-27x}{x^2-9} \cdot \dfrac{x+3}{x^2+3x+9}$

$= \dfrac{x(x^3-27)}{(x+3)(x-3)} \cdot \dfrac{x+3}{x^2+3x+9}$

$= \dfrac{x(x-3)(x^2+3x+9)}{(x+3)(x-3)} \cdot \dfrac{x+3}{x^2+3x+9}$

$= \dfrac{x}{1} = x$

18. $\dfrac{x-1}{x^2-x-2} - \dfrac{x+2}{x^2+4x+3}$

$x^2-x-2 = (x-2)(x+1)$

$x^2+4x+3 = (x+3)(x+1)$

LCD $= (x-2)(x+1)(x+3)$

$\dfrac{x-1}{x^2-x-2} - \dfrac{x+2}{x^2+4x+3}$

$= \dfrac{x-1}{(x-2)(x+1)} - \dfrac{x+2}{(x+1)(x+3)}$

$= \dfrac{(x-1)(x+3)}{(x-2)(x+1)(x+3)} - \dfrac{(x+2)(x-2)}{(x+1)(x+3)(x-2)}$

$= \dfrac{x^2+2x-3-\left(x^2-4\right)}{(x-2)(x+1)(x+3)}$

$= \dfrac{x^2+2x-3-x^2+4}{(x-2)(x+1)(x+3)}$

$= \dfrac{2x+1}{(x-2)(x+1)(x+3)}$

19. $\dfrac{x^2-2xy+y^2}{x+y} \div \dfrac{x^2-xy}{5x+5y}$

$= \dfrac{x^2-2xy+y^2}{x+y} \cdot \dfrac{5x+5y}{x^2-xy}$

$= \dfrac{(x-y)(x-y)}{x+y} \cdot \dfrac{5(x+y)}{x(x-y)}$

$= \dfrac{5(x-y)}{x} = \dfrac{5x-5y}{x}$

20. $\dfrac{5}{x+5} + \dfrac{x}{x-4} - \dfrac{11x-8}{x^2+x-20}$

$x^2+x-20 = (x+5)(x-4)$

LCD $= (x+5)(x-4)$

$\dfrac{5}{x+5} + \dfrac{x}{x-4} - \dfrac{11x-8}{x^2+x-20}$

$= \dfrac{5}{x+5} + \dfrac{x}{x-4} - \dfrac{11x-8}{(x+5)(x-4)}$

$= \dfrac{5(x-4)}{(x+5)(x-4)} + \dfrac{x(x+5)}{(x-4)(x+5)}$

$\qquad - \dfrac{11x-8}{(x+5)(x-4)}$

$= \dfrac{5x-20+x^2+5x-11x+8}{(x+5)(x-4)}$

$= \dfrac{x^2-x-12}{(x+5)(x-4)} = \dfrac{(x+3)(x-4)}{(x+5)(x-4)}$

$= \dfrac{x+3}{x+5}$

7.5 Check Points

1. $\dfrac{\dfrac{1}{4}+\dfrac{2}{3}}{\dfrac{2}{3}-\dfrac{1}{4}}$

Add to get a single rational expression in the numerator.

$$\frac{1}{4}+\frac{2}{3}=\frac{3}{12}+\frac{8}{12}=\frac{11}{12}$$

Subtract to get a single rational expression in the denominator.

$$\frac{2}{3}-\frac{1}{4}=\frac{8}{12}-\frac{3}{12}=\frac{5}{12}$$

Perform the division indicated by the fraction bar. Invert and multiply.

$$\frac{\dfrac{1}{4}+\dfrac{2}{3}}{\dfrac{2}{3}-\dfrac{1}{4}}=\frac{\dfrac{11}{12}}{\dfrac{5}{12}}=\frac{11}{12}\cdot\frac{12}{5}=\frac{11}{5}$$

2. $\dfrac{2-\dfrac{1}{x}}{2+\dfrac{1}{x}}$

Subtract to get a single rational expression in the numerator.

$$2-\frac{1}{x}=\frac{2x}{x}-\frac{1}{x}=\frac{2x-1}{x}$$

Add to get a single rational expression in the denominator.

$$2+\frac{1}{x}=\frac{2x}{x}+\frac{1}{x}=\frac{2x+1}{x}$$

Perform the division indicated by the fraction bar. Invert and multiply.

$$\frac{2-\dfrac{1}{x}}{2+\dfrac{1}{x}}=\frac{2x-1}{2x+1}$$

3. $\dfrac{\dfrac{1}{x}-\dfrac{1}{y}}{\dfrac{1}{xy}}$

Subtract to get a single rational expression in the numerator.

$$\frac{1}{x}-\frac{1}{y}=\frac{y}{xy}-\frac{x}{xy}=\frac{y-x}{xy}$$

Perform the division indicated by the fraction bar. Invert and multiply.

$$\frac{\dfrac{y-x}{xy}}{\dfrac{1}{xy}}=\frac{y-x}{xy}\cdot\frac{xy}{1}=y-x$$

4. $\dfrac{\dfrac{1}{4}+\dfrac{2}{3}}{\dfrac{2}{3}-\dfrac{1}{4}}$

Multiply the numerator and the denominator by the LCD of 12.

$$\frac{\dfrac{1}{4}+\dfrac{2}{3}}{\dfrac{2}{3}-\dfrac{1}{4}}=\frac{12\left(\dfrac{1}{4}+\dfrac{2}{3}\right)}{12\left(\dfrac{2}{3}-\dfrac{1}{4}\right)}=\frac{12\cdot\dfrac{1}{4}+12\cdot\dfrac{2}{3}}{12\cdot\dfrac{2}{3}-12\cdot\dfrac{1}{4}}=\frac{3+8}{8-3}=\frac{11}{5}$$

5. $\dfrac{2-\dfrac{1}{x}}{2+\dfrac{1}{x}}$

Multiply the numerator and the denominator by the LCD of x.

$$\frac{2-\dfrac{1}{x}}{2+\dfrac{1}{x}}=\frac{x\left(2-\dfrac{1}{x}\right)}{x\left(2+\dfrac{1}{x}\right)}=\frac{x\cdot2-x\cdot\dfrac{1}{x}}{x\cdot2+x\cdot\dfrac{1}{x}}=\frac{2x-1}{2x+1}$$

6. $\dfrac{\dfrac{1}{x}-\dfrac{1}{y}}{\dfrac{1}{xy}}$

Multiply the numerator and the denominator by the LCD of xy.

$$\frac{\dfrac{1}{x}-\dfrac{1}{y}}{\dfrac{1}{xy}}=\frac{xy\left(\dfrac{1}{x}-\dfrac{1}{y}\right)}{xy\left(\dfrac{1}{xy}\right)}=\frac{xy\cdot\dfrac{1}{x}-xy\cdot\dfrac{1}{y}}{xy\cdot\dfrac{1}{xy}}=\frac{y-x}{1}=y-x$$

7.5 Exercise Set

1. $\dfrac{\dfrac{1}{2}+\dfrac{1}{4}}{\dfrac{1}{2}+\dfrac{1}{3}}$

Add to get a single rational expression in the numerator.

$$\frac{1}{2}+\frac{1}{4}=\frac{2}{4}+\frac{1}{4}=\frac{3}{4}$$

Add to get a single rational expression in the denominator.

$$\frac{1}{2}+\frac{1}{3}=\frac{3}{6}+\frac{2}{6}=\frac{5}{6}$$

Perform the division indicated by the fraction bar. Invert and multiply.

$$\frac{\dfrac{1}{2}+\dfrac{1}{4}}{\dfrac{1}{2}+\dfrac{1}{3}}=\frac{\dfrac{3}{4}}{\dfrac{5}{6}}=\frac{3}{4}\cdot\frac{6}{5}=\frac{9}{10}$$

3. $\dfrac{5+\dfrac{2}{5}}{7-\dfrac{1}{10}}=\dfrac{\dfrac{25}{5}+\dfrac{2}{5}}{\dfrac{70}{10}-\dfrac{1}{10}}$

$$=\frac{\dfrac{27}{5}}{\dfrac{69}{10}}=\frac{27}{5}\cdot\frac{10}{69}=\frac{9\cdot3\cdot2\cdot5}{5\cdot3\cdot23}=\frac{18}{23}$$

5. $\dfrac{\dfrac{2}{5}-\dfrac{1}{3}}{\dfrac{2}{3}-\dfrac{3}{4}}$

LCD = 60

$$\frac{\dfrac{2}{5}-\dfrac{1}{3}}{\dfrac{2}{3}-\dfrac{3}{4}}=\frac{60\cdot\left(\dfrac{2}{5}-\dfrac{1}{3}\right)}{60\cdot\left(\dfrac{2}{3}-\dfrac{3}{4}\right)}$$

$$=\frac{60\cdot\dfrac{2}{5}-60\cdot\dfrac{1}{3}}{60\cdot\dfrac{2}{3}-60\cdot\dfrac{3}{4}}$$

$$=\frac{24-20}{40-45}=\frac{4}{-5}=-\frac{4}{5}$$

7. $\dfrac{\dfrac{3}{4}-x}{\dfrac{3}{4}+x}=\dfrac{\dfrac{3}{4}-\dfrac{4x}{4}}{\dfrac{3}{4}+\dfrac{4x}{4}}$

$$=\frac{\dfrac{3-4x}{4}}{\dfrac{3+4x}{4}}$$

$$=\frac{3-4x}{4}\cdot\frac{4}{3+4x}=\frac{3-4x}{3+4x}$$

9. $\dfrac{7-\dfrac{2}{x}}{5+\dfrac{1}{x}}=\dfrac{\dfrac{7x-2}{x}}{\dfrac{5x+1}{x}}=\dfrac{7x-2}{x}\cdot\dfrac{x}{5x+1}=\dfrac{7x-2}{5x+1}$

11. $\dfrac{2+\dfrac{3}{y}}{1-\dfrac{7}{y}}=\dfrac{\dfrac{2y+3}{y}}{\dfrac{y-7}{y}}$

$$=\frac{2y+3}{y}\cdot\frac{y}{y-7}=\frac{2y+3}{y-7}$$

13. $\dfrac{\dfrac{1}{y}-\dfrac{3}{2}}{\dfrac{1}{y}+\dfrac{3}{4}}=\dfrac{\dfrac{2-3y}{2y}}{\dfrac{4+3y}{4y}}$

$$=\frac{2-3y}{2y}\cdot\frac{4y}{4+3y}$$

$$=\frac{2(2-3y)}{4+3y}=\frac{4-6y}{4+3y}$$

15. $\dfrac{\dfrac{x}{5}-\dfrac{5}{x}}{\dfrac{1}{5}+\dfrac{1}{x}}$

LCD = 5x

$$\frac{\dfrac{x}{5}-\dfrac{5}{x}}{\dfrac{1}{5}+\dfrac{1}{x}}=\frac{5x\cdot\left(\dfrac{x}{5}-\dfrac{5}{x}\right)}{5x\cdot\left(\dfrac{1}{5}+\dfrac{1}{x}\right)}$$

$$=\frac{5x\cdot\dfrac{x}{5}-5x\cdot\dfrac{5}{x}}{5x\cdot\dfrac{1}{5}+5x\cdot\dfrac{1}{x}}$$

$$=\frac{x^2-25}{x+5}$$

$$=\frac{(x+5)(x-5)}{x+5}=x-5$$

17. $\dfrac{1+\dfrac{1}{x}}{1-\dfrac{1}{x^2}} = \dfrac{\dfrac{x+1}{x}}{\dfrac{x^2-1}{x^2}}$

$= \dfrac{x+1}{x} \cdot \dfrac{x^2}{x^2-1}$

$= \dfrac{x+1}{x} \cdot \dfrac{x^2}{(x+1)(x-1)}$

$= \dfrac{x}{x-1}$

19. $\dfrac{\dfrac{1}{7}-\dfrac{1}{y}}{\dfrac{7-y}{7}}$

LCD = $7y$

$\dfrac{\dfrac{1}{7}-\dfrac{1}{y}}{\dfrac{7-y}{7}} = \dfrac{7y\left(\dfrac{1}{7}-\dfrac{1}{y}\right)}{7y\left(\dfrac{7-y}{7}\right)}$

$= \dfrac{7y\left(\dfrac{1}{7}\right)-7y\left(\dfrac{1}{y}\right)}{7y\left(\dfrac{7-y}{7}\right)}$

$= \dfrac{y-7}{y(7-y)}$

$= \dfrac{-1(7-y)}{y(7-y)} = -\dfrac{1}{y}$

21. $\dfrac{x+\dfrac{2}{y}}{\dfrac{x}{y}} = \dfrac{\dfrac{xy+2}{y}}{\dfrac{x}{y}} = \dfrac{xy+2}{y} \cdot \dfrac{y}{x} = \dfrac{xy+2}{x}$

23. $\dfrac{\dfrac{1}{x}+\dfrac{1}{y}}{xy}$

LCD = xy

$\dfrac{\dfrac{1}{x}+\dfrac{1}{y}}{xy} = \dfrac{xy\left(\dfrac{1}{x}+\dfrac{1}{y}\right)}{xy(xy)} = \dfrac{y+x}{x^2y^2}$

25. $\dfrac{\dfrac{x}{y}+\dfrac{1}{x}}{\dfrac{y}{x}+\dfrac{1}{x}} = \dfrac{\dfrac{x^2+y}{xy}}{\dfrac{y+1}{x}} = \dfrac{x^2+y}{xy} \cdot \dfrac{x}{y+1}$

$= \dfrac{x^2+y}{y(y+1)}$

27. $\dfrac{\dfrac{1}{y}+\dfrac{2}{y^2}}{\dfrac{2}{y}+1}$

LCD = y^2

$\dfrac{\dfrac{1}{y}+\dfrac{2}{y^2}}{\dfrac{2}{y}+1} = \dfrac{y^2\left(\dfrac{1}{y}+\dfrac{2}{y^2}\right)}{y^2\left(\dfrac{2}{y}+1\right)}$

$= \dfrac{y^2\left(\dfrac{1}{y}\right)+y^2\left(\dfrac{2}{y^2}\right)}{y^2\left(\dfrac{2}{y}\right)+y^2(1)}$

$= \dfrac{y+2}{2y+y^2}$

$= \dfrac{(y+2)}{y(2+y)} = \dfrac{1}{y}$

29. $\dfrac{\dfrac{12}{x^2}-\dfrac{3}{x}}{\dfrac{15}{x}-\dfrac{9}{x^2}} = \dfrac{\dfrac{12}{x^2}-\dfrac{3x}{x^2}}{\dfrac{15x}{x^2}-\dfrac{9}{x^2}} = \dfrac{\dfrac{12-3x}{x^2}}{\dfrac{15x-9}{x^2}}$

$= \dfrac{12-3x}{x^2} \cdot \dfrac{x^2}{15x-9} = \dfrac{12-3x}{15x-9}$

$= \dfrac{3(4-x)}{3(5x-3)} = \dfrac{4-x}{5x-3}$

31. $\dfrac{2+\dfrac{6}{y}}{1-\dfrac{9}{y^2}}$

LCD = y^2

$$\dfrac{2+\dfrac{6}{y}}{1-\dfrac{9}{y^2}} = \dfrac{y^2\left(2+\dfrac{6}{y}\right)}{y^2\left(1-\dfrac{9}{y^2}\right)}$$

$$= \dfrac{2y^2+6y}{y^2-9}$$

$$= \dfrac{2y(y+3)}{(y+3)(y-3)} = \dfrac{2y}{y-3}$$

33. $\dfrac{\dfrac{1}{x+2}}{1+\dfrac{1}{x+2}}$

LCD = $x+2$

$$\dfrac{\dfrac{1}{x+2}}{1+\dfrac{1}{x+2}} = \dfrac{(x+2)\left(\dfrac{1}{x+2}\right)}{(x+2)\left(1+\dfrac{1}{x+2}\right)}$$

$$= \dfrac{1}{x+2+1} = \dfrac{1}{x+3}$$

35. $\dfrac{x-5+\dfrac{3}{x}}{x-7+\dfrac{2}{x}}$

LCD = x

$$\dfrac{x-5+\dfrac{3}{x}}{x-7+\dfrac{2}{x}} = \dfrac{x\left(x-5+\dfrac{3}{x}\right)}{x\left(x-7+\dfrac{2}{x}\right)}$$

$$= \dfrac{x^2-5x+3}{x^2-7x+2}$$

37. $\dfrac{\dfrac{3}{xy^2}+\dfrac{2}{x^2y}}{\dfrac{1}{x^2y}+\dfrac{2}{xy^3}} = \dfrac{\dfrac{3x}{x^2y^2}+\dfrac{2y}{x^2y^2}}{\dfrac{y^2}{x^2y^3}+\dfrac{2x}{x^2y^3}}$

$$= \dfrac{\dfrac{3x+2y}{x^2y^2}}{\dfrac{y^2+2x}{x^2y^3}}$$

$$= \dfrac{3x+2y}{x^2y^2} \cdot \dfrac{x^2y^3}{y^2+2x}$$

$$= \dfrac{3x+2y}{x^2y^2} \cdot \dfrac{x^2y^3}{y^2+2x}$$

$$= \dfrac{(3x+2y)(y)}{y^2+2x}$$

$$= \dfrac{3xy+2y^2}{y^2+2x}$$

39. $\dfrac{\dfrac{3}{x+1}-\dfrac{3}{x-1}}{\dfrac{5}{x^2-1}}$

$$= \dfrac{\dfrac{3(x-1)-3(x+1)}{(x+1)(x-1)}}{\dfrac{5}{x^2-1}}$$

$$= \dfrac{\dfrac{3x-3-3x-3}{(x+1)(x-1)}}{\dfrac{5}{x^2-1}}$$

$$= \dfrac{\dfrac{-6}{(x+1)(x-1)}}{\dfrac{5}{x^2-1}}$$

$$= \dfrac{-6}{(x+1)(x-1)} \cdot \dfrac{x^2-1}{5}$$

$$= \dfrac{-6}{(x+1)(x-1)} \cdot \dfrac{(x+1)(x-1)}{5}$$

$$= -\dfrac{6}{5}$$

41.

$$\frac{\dfrac{6}{x^2+2x-15}-\dfrac{1}{x-3}}{\dfrac{1}{x+5}+1} = \frac{\dfrac{6}{(x+5)(x-3)}-\dfrac{1}{x-3}}{\dfrac{1}{x+5}+1}$$

LCD $= (x+5)(x-3)$

$$\frac{\dfrac{6}{(x+5)(x-3)}-\dfrac{1}{x-3}}{\dfrac{1}{x+5}+1} = \frac{(x+5)(x-3)\left[\dfrac{6}{(x+5)(x-3)}-\dfrac{1}{x-3}\right]}{(x+5)(x-3)\left[\dfrac{1}{x+5}+1\right]}$$

$$= \frac{6-(x+5)}{x-3+(x+5)(x-3)} = \frac{6-x-5}{x-3+x^2+2x-15} = \frac{-x+1}{x^2+3x-18} = \frac{1-x}{(x-3)(x+6)}$$

43.

$$\frac{y^{-1}-(y+5)^{-1}}{5} = \frac{\dfrac{1}{y}-\dfrac{1}{y+5}}{5}$$

LCD$= y(y+5)$

$$\frac{\dfrac{1}{y}-\dfrac{1}{y+5}}{5} = \frac{y(y+5)\left(\dfrac{1}{y}-\dfrac{1}{y+5}\right)}{y(y+5)(5)} = \frac{y+5-y}{5y(y+5)} = \frac{5}{5y(y+5)} = \frac{1}{y(y+5)}$$

45.

$$\frac{1}{1-\dfrac{1}{x}}-1 = \frac{x(1)}{x\left(1-\dfrac{1}{x}\right)}-1 = \frac{x}{x-1}-1 = \frac{x}{x-1}-\frac{x-1}{x-1} = \frac{x-x+1}{x-1} = \frac{1}{x-1}$$

47.

$$\frac{1}{1+\dfrac{1}{1+\dfrac{1}{x}}} = \frac{1}{1+\dfrac{x(1)}{x\left(1+\dfrac{1}{x}\right)}} = \frac{1}{1+\dfrac{x}{x+1}} = \frac{(x+1)(1)}{(x+1)\left(1+\dfrac{x}{x+1}\right)} = \frac{x+1}{x+1+x} = \frac{x+1}{2x+1}$$

49.

$$\frac{2d}{\dfrac{d}{r_1}+\dfrac{d}{r_2}}$$

LCD $= r_1 r_2$

$$\frac{2d}{\dfrac{d}{r_1}+\dfrac{d}{r_2}} = \frac{r_1 r_2(2d)}{r_1 r_2\left(\dfrac{d}{r_1}+\dfrac{d}{r_2}\right)}$$

$$= \frac{2r_1 r_2 d}{r_2 d+r_1 d}$$

$$= \frac{2r_1 r_2 d}{d(r_2+r_1)} = \frac{2r_1 r_2}{r_2+r_1}$$

If $r_1 = 40$ and $r_2 = 30$, the value of this expression will be $\dfrac{2\cdot40\cdot30}{30+40} = \dfrac{2400}{70} = 34\dfrac{2}{7}$.

Your average speed will be $34\dfrac{2}{7}$ miles per hour.

51. a.

$$\frac{\dfrac{11}{10}+\dfrac{4}{x}+\dfrac{104}{x^2}}{\dfrac{6}{5}+\dfrac{23}{x}+\dfrac{219}{x^2}}=\frac{10x^2\left(\dfrac{11}{10}+\dfrac{4}{x}+\dfrac{104}{x^2}\right)}{10x^2\left(\dfrac{6}{5}+\dfrac{23}{x}+\dfrac{219}{x^2}\right)}$$

$$=\frac{10x^2\cdot\dfrac{11}{10}+10x^2\cdot\dfrac{4}{x}+10x^2\cdot\dfrac{104}{x^2}}{10x^2\cdot\dfrac{6}{5}+10x^2\cdot\dfrac{23}{x}+10x^2\cdot\dfrac{219}{x^2}}$$

$$=\frac{11x^2+40x+1040}{12x^2+230x+2190}$$

b. According to the data in the bar graph, $\dfrac{1707.2}{2708.7}$ or about 63% was spent on human resources in 2006.

c. $\dfrac{11x^2+40x+1040}{12x^2+230x+2190}=\dfrac{11(36)^2+40(36)+1040}{12(36)^2+230(36)+2190}\approx 0.64$

According to the model about 64% was spent on human resources in 2006.
This overestimates the actual percent by 1%.

53. – 55. Answers will vary.

57. makes sense

59. does not make sense; Explanations will vary. Sample explanation: The expression simplifies to $\dfrac{x+y}{x-y}$.

61. true

63. false; Changes to make the statement true will vary. A sample change is: All complex rational expressions can simplified by both methods.

65.

$$\frac{2y}{2+\dfrac{2}{y}}+\frac{y}{1+\dfrac{1}{y}}=\frac{y}{y}\left(\frac{2y}{2+\dfrac{2}{y}}\right)+\frac{y}{y}\left(\frac{y}{1+\dfrac{1}{y}}\right)$$

$$=\frac{2y^2}{2y+2}+\frac{y^2}{y+1}=\frac{2y^2}{2(y+1)}+\frac{y^2}{y+1}$$

$$=\frac{y^2}{y+1}+\frac{y^2}{y+1}=\frac{2y^2}{y+1}$$

67. The graphs coincide.

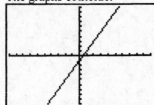

69. The graphs do not coincide.

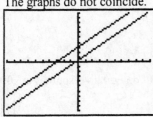

$$\frac{\dfrac{1}{x}+\dfrac{1}{3}}{\dfrac{1}{3x}}=\frac{3x\left(\dfrac{1}{x}+\dfrac{1}{3}\right)}{3x\left(\dfrac{1}{3x}\right)}=\frac{3+x}{1}=3+x$$

Therefore, $x+\dfrac{1}{3}$ should be $3+x$.

70. $2x^3-20x^2+50x$

$=2x\left(x^2-10x+25\right)$

$=2x\left(x-5\right)^2$

71. $2-3\left(x-2\right)=5\left(x+5\right)-1$

$2-3x+6=5x+25-1$

$8-3x=5x+24$

$8-3x-5x=5x+24-5x$

$8-8x=24$

$8-8x-8=24-8$

$-8x=16$

$\dfrac{-8x}{-8}=\dfrac{16}{-8}$

$x=-2$

The solution set is $\{-2\}$.

72. $\left(x+y\right)\left(x^2-xy+y^2\right)$

$=x\left(x^2-xy+y^2\right)+y\left(x^2-xy+y^2\right)$

$=x^3-x^2y+xy^2+x^2y-xy^2+y^3$

$=x^3+y^3$

73. $\dfrac{x}{3}+\dfrac{x}{2}=\dfrac{5}{6}$

$6\left(\dfrac{x}{3}+\dfrac{x}{2}\right)=6\left(\dfrac{5}{6}\right)$

$\dfrac{6x}{3}+\dfrac{6x}{2}=\dfrac{6\cdot5}{6}$

$2x+3x=5$

$5x=5$

$x=1$

The solution set is $\{1\}$.

74. $\dfrac{2x}{3}=\dfrac{14}{3}-\dfrac{x}{2}$

$6\left(\dfrac{2x}{3}\right)=6\left(\dfrac{14}{3}-\dfrac{x}{2}\right)$

$\dfrac{6\cdot2x}{3}=\dfrac{6\cdot14}{3}-\dfrac{6\cdot x}{2}$

$4x=28-3x$

$7x=28$

$x=4$

The solution set is $\{4\}$.

75. $2x^2+2=5x$

$2x^2-5x+2=0$

$(2x-1)(x-2)=0$

$2x-1=0 \quad$ or $\quad x-2=0$

$2x=1 \qquad\qquad x=2$

$x=\dfrac{1}{2}$

The solution set is $\left\{\dfrac{1}{2},2\right\}$.

7.6 Check Points

1. $\dfrac{x}{6} = \dfrac{1}{6} + \dfrac{x}{8}$

The LCD is 24.

$$\dfrac{x}{6} = \dfrac{1}{6} + \dfrac{x}{8}$$

$$24\left(\dfrac{x}{6}\right) = 24\left(\dfrac{1}{6} + \dfrac{x}{8}\right)$$

$$24 \cdot \dfrac{x}{6} = 24 \cdot \dfrac{1}{6} + 24 \cdot \dfrac{x}{8}$$

$$4x = 4 + 3x$$

$$x = 4$$

The solution set is $\{4\}$.

Check:

$$\dfrac{x}{6} = \dfrac{1}{6} + \dfrac{x}{8}$$

$$\dfrac{4}{6} = \dfrac{1}{6} + \dfrac{4}{8}$$

$$\dfrac{16}{24} \overset{?}{=} \dfrac{4}{24} + \dfrac{12}{24}$$

$$\dfrac{16}{24} = \dfrac{16}{24}$$

2. $\dfrac{5}{2x} = \dfrac{17}{18} - \dfrac{1}{3x}$

The restriction is $x \neq 0$.

The LCD is $18x$.

$$\dfrac{5}{2x} = \dfrac{17}{18} - \dfrac{1}{3x}$$

$$18x\left(\dfrac{5}{2x}\right) = 18x\left(\dfrac{17}{18} - \dfrac{1}{3x}\right)$$

$$18x \cdot \dfrac{5}{2x} = 18x \cdot \dfrac{17}{18} - 18x \cdot \dfrac{1}{3x}$$

$$45 = 17x - 6$$

$$51 = 17x$$

$$3 = x$$

The solution set is $\{3\}$.

3. $x + \dfrac{6}{x} = -5$

The restriction is $x \neq 0$.

The LCD is x.

$$x + \dfrac{6}{x} = -5$$

$$x\left(x + \dfrac{6}{x}\right) = x(-5)$$

$$x \cdot x + x \cdot \dfrac{6}{x} = -5x$$

$$x^2 + 6 = -5x$$

$$x^2 + 5x + 6 = 0$$

$$(x + 3)(x + 2) = 0$$

$$x + 3 = 0 \quad \text{or} \quad x + 2 = 0$$

$$x = -3 \qquad\qquad x = -2$$

The solution set is $\{-3, -2\}$.

4.
$$\frac{11}{x^2-25}+\frac{4}{x+5}=\frac{3}{x-5}$$

$$\frac{11}{(x+5)(x-5)}+\frac{4}{x+5}=\frac{3}{x-5}$$

The restrictions are $x \neq -5$ and $x \neq 5$.

The LCD is $(x+5)(x-5)$.

$$\frac{11}{(x+5)(x-5)}+\frac{4}{x+5}=\frac{3}{x-5}$$

$$(x+5)(x-5)\left(\frac{11}{(x+5)(x-5)}+\frac{4}{x+5}\right)=(x+5)(x-5)\left(\frac{3}{x-5}\right)$$

$$\frac{11(x+5)(x-5)}{(x+5)(x-5)}+\frac{4(x+5)(x-5)}{x+5}=\frac{3(x+5)(x-5)}{x-5}$$

$$11+4(x-5)=3(x+5)$$

$$11+4x-20=3x+15$$

$$4x-9=3x+15$$

$$x-9=15$$

$$x=24$$

The solution set is $\{24\}$.

5. $\dfrac{x}{x-3}=\dfrac{3}{x-3}+9$

The restriction is $x \neq 3$.

The LCD is $x-3$.

$$\frac{x}{x-3}=\frac{3}{x-3}+9$$

$$(x-3)\left(\frac{x}{x-3}\right)=(x-3)\left(\frac{3}{x-3}+9\right)$$

$$\frac{x(x-3)}{x-3}=\frac{3(x-3)}{x-3}+9(x-3)$$

$$x=3+9(x-3)$$

$$x=3+9x-27$$

$$x=9x-24$$

$$-8x=-24$$

$$x=3$$

The proposed solution, 3, is not a solution because of the restriction $x \neq 3$.

The solution set is $\{\ \}$.

6. $y = \dfrac{250x}{100 - x}$

The restriction is $x \neq 100$.

$$750 = \dfrac{250x}{100 - x}$$

$$(100 - x)(750) = (100 - x)\left(\dfrac{250x}{100 - x}\right)$$

$$75{,}000 - 750x = 250x$$

$$75{,}000 = 1000x$$

$$75 = x$$

If government funding is increased to \$750 million, then 75% of pollutants can be removed.

7.
$$\dfrac{1}{x} + \dfrac{1}{y} = \dfrac{1}{z}$$

$$xyz\left(\dfrac{1}{x} + \dfrac{1}{y}\right) = xyz\left(\dfrac{1}{z}\right)$$

$$xyz\left(\dfrac{1}{x}\right) + xyz\left(\dfrac{1}{y}\right) = xy$$

$$yz + xz = xy$$

$$xz - xy = -yz$$

$$x(z - y) = -yz$$

$$x = \dfrac{-yz}{z - y}$$

$$x = \dfrac{yz}{y - z}$$

7.6 Exercise Set

1. $\dfrac{x}{3} = \dfrac{x}{2} - 2$

There are no restrictions on the variable because the variable does not appear in any denominator.
The LCD is 6.

$$\dfrac{x}{3} = \dfrac{x}{2} - 2$$

$$6\left(\dfrac{x}{3}\right) = 6\left(\dfrac{x}{2} - 2\right)$$

$$6 \cdot \dfrac{x}{3} = 6 \cdot \dfrac{x}{2} - 6 \cdot 2$$

$$2x = 3x - 12$$

$$0 = x - 12$$

$$12 = x$$

The solution set is $\{12\}$.

3. $\dfrac{4x}{3} = \dfrac{x}{18} - \dfrac{x}{6}$

There are no restrictions.
The LCD is 18.

$$\dfrac{4x}{3} = \dfrac{x}{18} - \dfrac{x}{6}$$

$$18\left(\dfrac{4x}{3}\right) = 18\left(\dfrac{x}{18} - \dfrac{x}{6}\right)$$

$$18 \cdot \dfrac{4x}{3} = 18 \cdot \dfrac{x}{18} - 18 \cdot \dfrac{x}{6}$$

$$24x = x - 3x$$

$$24x = -2x$$

$$26x = 0$$

$$x = 0$$

The solution set is $\{0\}$.

5. $2 - \dfrac{8}{x} = 6$

The restriction is $x \neq 0$.
The LCD is x.

$$2 - \dfrac{8}{x} = 6$$

$$x\left(2 - \dfrac{8}{x}\right) = x \cdot 6$$

$$x \cdot 2 - x \cdot \dfrac{8}{x} = x \cdot 6$$

$$2x - 8 = 6x$$

$$-8 = 4x$$

$$-2 = x$$

The solution set is $\{-2\}$.

7. $\dfrac{2}{x} + \dfrac{1}{3} = \dfrac{4}{x}$

The restriction is $x \neq 0$.
The LCD is $3x$.

$$\dfrac{2}{x} + \dfrac{1}{3} = \dfrac{4}{x}$$

$$3x\left(\dfrac{2}{x} + \dfrac{1}{3}\right) = 3x\left(\dfrac{4}{x}\right)$$

$$3x \cdot \dfrac{2}{x} + 3x \cdot \dfrac{1}{3} = 3x \cdot \dfrac{4}{x}$$

$$6 + x = 12$$

$$x = 6$$

The solution set is $\{6\}$.

9. $\dfrac{2}{x} + 3 = \dfrac{5}{2x} + \dfrac{13}{4}$

The restriction is $x \neq 0$

The LCD is $4x$.

$$\dfrac{2}{x} + 3 = \dfrac{5}{2x} + \dfrac{13}{4}$$

$$4x\left(\dfrac{2}{x} + 3\right) = 4x\left(\dfrac{5}{2x} + \dfrac{13}{4}\right)$$

$$8 + 12x = 10 + 13x$$

$$8 = 10 + x$$

$$-2 = x$$

The solution set is $\{-2\}$.

11. $\dfrac{2}{3x} + \dfrac{1}{4} = \dfrac{11}{6x} - \dfrac{1}{3}$

The restriction is $x \neq 0$.

The LCD is $12x$.

$$\dfrac{2}{3x} + \dfrac{1}{4} = \dfrac{11}{6x} - \dfrac{1}{3}$$

$$12x\left(\dfrac{2}{3x} + \dfrac{1}{4}\right) = 12x\left(\dfrac{11}{6x} - \dfrac{1}{3}\right)$$

$$8 + 3x = 22 - 4x$$

$$8 + 7x = 22$$

$$7x = 14$$

$$x = 2$$

The solution set is $\{2\}$.

13. $\dfrac{6}{x+3} = \dfrac{4}{x-3}$

Restrictions: $x \neq -3, x \neq 3$

LCD $= (x+3)(x-3)$

$$\dfrac{6}{x+3} = \dfrac{4}{x-3}$$

$$(x+3)(x-3) \cdot \dfrac{6}{x+3} = (x+3)(x-3) \cdot \dfrac{4}{x-3}$$

$$(x-3) \cdot 6 = (x+3) \cdot 4$$

$$6x - 18 = 4x + 12$$

$$2x - 18 = 12$$

$$2x = 30$$

$$x = 15$$

The solution set is $\{15\}$.

15. $\dfrac{x-2}{2x} + 1 = \dfrac{x+1}{x}$

Restriction: $x \neq 0$

LCD $= 2x$.

$$\dfrac{x-2}{2x} + 1 = \dfrac{x+1}{x}$$

$$2x\left(\dfrac{x-2}{2x} + 1\right) = 2x\left(\dfrac{x+1}{x}\right)$$

$$x - 2 + 2x = 2(x+1)$$

$$3x - 2 = 2x + 2$$

$$x - 2 = 2$$

$$x = 4$$

The solution set is $\{4\}$.

17. $x + \dfrac{6}{x} = -7$

Restriction: $x \neq 0$

LCD $= x$

$$x + \dfrac{6}{x} = -7$$

$$x\left(x + \dfrac{6}{x}\right) = x(-7)$$

$$x^2 + 6 = -7x$$

$$x^2 + 7x + 6 = 0$$

$$(x+6)(x+1) = 0$$

$$x + 6 = 0 \quad \text{or} \quad x + 1 = 0$$

$$x = -6 \qquad\qquad x = -1$$

The solution set is $\{-6, -1\}$.

19. $\dfrac{x}{5} - \dfrac{5}{x} = 0$

Restriction: $x \neq 0$

LCD $= 5x$

$$\dfrac{x}{5} - \dfrac{5}{x} = 0$$

$$5x\left(\dfrac{x}{5} - \dfrac{5}{x}\right) = 5x \cdot 0$$

$$5x \cdot \dfrac{x}{5} - 5x \cdot \dfrac{5}{x} = 0$$

$$x^2 - 25 = 0$$

$$(x+5)(x-5) = 0$$

$$x + 5 = 0 \quad \text{or} \quad x - 5 = 0$$

$$x = -5 \qquad\qquad x = 5$$

The solution set is $\{-5, 5\}$.

21. $x + \dfrac{3}{x} = \dfrac{12}{x}$

Restriction: $x \neq 0$

LCD $= x$

$$x + \frac{3}{x} = \frac{12}{x}$$

$$x\left(x + \frac{3}{x}\right) = x\left(\frac{12}{x}\right)$$

$$x^2 + 3 = 12$$

$$x^2 - 9 = 0$$

$$(x+3)(x-3) = 0$$

$x + 3 = 0 \quad \text{or} \quad x - 3 = 0$

$x = -3 \qquad\qquad x = 3$

The solution set is $\{-3, 3\}$.

23. $\dfrac{4}{y} - \dfrac{y}{2} = \dfrac{7}{2}$

Restrictions: $y \neq 0$

LCD $= 2y$

$$\frac{4}{y} - \frac{y}{2} = \frac{7}{2}$$

$$2y\left(\frac{4}{y} - \frac{y}{2}\right) = 2y\left(\frac{7}{2}\right)$$

$$8 - y^2 = 7y$$

$$0 = y^2 + 7y - 8$$

$$0 = (y+8)(y-1)$$

$y + 8 = 0 \quad \text{or} \quad y - 1 = 0$

$y = -8 \qquad\qquad y = 1$

The solution set is $\{-8, 1\}$.

25. $\dfrac{x-4}{x} = \dfrac{15}{x+4}$

Restrictions: $x \neq 0, x \neq -4$

LCD $= x(x+4)$

$$\frac{x-4}{x} = \frac{15}{x+4}$$

$$x(x+4)\left(\frac{x-4}{x}\right) = x(x+4)\left(\frac{15}{x+4}\right)$$

$$(x+4)(x-4) = x \cdot 15$$

$$x^2 - 16 = 15x$$

$$x^2 - 15x - 16 = 0$$

$$(x+1)(x-16) = 0$$

$x + 1 = 0 \quad \text{or} \quad x - 16 = 0$

$x = -1 \qquad\qquad x = 16$

The solution set is $\{-1, 16\}$.

27. $\dfrac{2}{x^2 - 1} = \dfrac{4}{x+1}$

$$\frac{2}{(x+1)(x-1)} = \frac{4}{x+1}$$

Restrictions: $x \neq -1, x \neq 1$

LCD $= (x+1)(x-1)$

$$\frac{2}{(x+1)(x-1)} = \frac{4}{x+1}$$

$$(x+1)(x-1)\frac{2}{(x+1)(x-1)} = (x+1)(x-1)\frac{4}{x+1}$$

$$2 = 4(x-1)$$

$$2 = 4x - 4$$

$$6 = 4x$$

$$\frac{6}{4} = x$$

$$\frac{3}{2} = x$$

The solution set is $\left\{\dfrac{3}{2}\right\}$.

29. $\dfrac{1}{x-1}+5=\dfrac{11}{x-1}$

Restriction: $x\neq 1$

LCD $= x-1$

$$\dfrac{1}{x-1}+5=\dfrac{11}{x-1}$$

$$(x-1)\left(\dfrac{1}{x-1}+5\right)=(x-1)\left(\dfrac{11}{x-1}\right)$$

$$1+(x-1)\cdot 5=11$$

$$1+5x-5=11$$

$$5x-4=11$$

$$5x=15$$

$$x=3$$

The solution set is $\{3\}$.

31. $\dfrac{8y}{y+1}=4-\dfrac{8}{y+1}$

Restriction: $y\neq -1$

LCD $= y+1$

$$\dfrac{8y}{y+1}=4-\dfrac{8}{y+1}$$

$$(y+1)\left(\dfrac{8y}{y+1}\right)=(y+1)\left(4-\dfrac{8}{y+1}\right)$$

$$8y=(y+1)\cdot 4-8$$

$$8y=4y+4-8$$

$$8y=4y-4$$

$$4y=-4$$

$$y=-1$$

The proposed solution, -1, is *not* a solution because of the restriction $x\neq -1$. Notice that -1 makes two of the denominators zero in the original equation. Therefore, the equation has no solution. The solution set is $\{\ \}$.

33. $\dfrac{3}{x-1}+\dfrac{8}{x}=3$

Restrictions: $x\neq 1, x\neq 0$

LCD $= x(x-1)$

$$\dfrac{3}{x-1}+\dfrac{8}{x}=3$$

$$x(x-1)\left(\dfrac{3}{x-1}+\dfrac{8}{x}\right)=x(x-1)\cdot 3$$

$$x(x-1)\left(\dfrac{3}{x-1}+\dfrac{8}{x}\right)=3x(x-1)$$

$$3x+8(x-1)=3x^2-3x$$

$$3x+8x-8=3x^2-3x$$

$$11x-8=3x^2-3x$$

$$0=3x^2-14x+8$$

$$0=(3x-2)(x-4)$$

$$3x-2=0 \quad \text{or} \quad x-4=0$$

$$3x=2 \qquad\qquad x=4$$

$$x=\dfrac{2}{3}$$

The solution set is $\left\{\dfrac{2}{3},4\right\}$.

35. $\dfrac{3y}{y-4}-5=\dfrac{12}{y-4}$

Restriction: $y\neq 4$

LCD $= y-4$

$$\dfrac{3y}{y-4}-5=\dfrac{12}{y-4}$$

$$(y-4)\left(\dfrac{3y}{y-4}-5\right)=(y-4)\left(\dfrac{12}{y-4}\right)$$

$$3y-5(y-4)=12$$

$$3y-5y+20=12$$

$$-2y+20=12$$

$$-2y=-8$$

$$y=4$$

The proposed solution, 4, is *not* a solution because of the restriction $y\neq 4$. Therefore, this equation has no solution. The solution set is $\{\ \}$.

37. $\dfrac{1}{x} + \dfrac{1}{x-3} = \dfrac{x-2}{x-3}$

Restrictions: $x \neq 0, x \neq 3$

LCD $= x(x-3)$

$$\frac{1}{x} + \frac{1}{x-3} = \frac{x-2}{x-3}$$

$$x(x-3)\left(\frac{1}{x} + \frac{1}{x-3}\right) = x(x-3) \cdot \frac{x-2}{x-3}$$

$$x - 3 + x = x(x-2)$$

$$2x - 3 = x^2 - 2x$$

$$0 = x^2 - 4x + 3$$

$$0 = (x-3)(x-1)$$

$x - 3 = 0 \ $ or $ \ x - 1 = 0$

$\quad x = 3 \qquad \quad x = 1$

The proposed solution 3 is *not* a solution because of the restriction $x \neq 3$.

The solution set is $\{1\}$.

39. $\dfrac{x+1}{3x+9} + \dfrac{x}{2x+6} = \dfrac{2}{4x+12}$

To find any restrictions and the LCD, factor the denominators.

$$\frac{x+1}{3(x+3)} + \frac{x}{2(x+3)} = \frac{2}{4(x+3)}$$

Restriction: $x \neq -3$

LCD $= 12(x+3)$

$$12(x+3)\left[\frac{x+1}{3(x+3)} + \frac{x}{2(x+3)}\right] = 12(x+3)\left[\frac{2}{4(x+3)}\right]$$

$$4(x+1) + 6x = 6$$

$$4x + 4 + 6x = 6$$

$$10x + 4 = 6$$

$$10x = 2$$

$$x = \frac{2}{10}$$

$$x = \frac{1}{5}$$

The solution set is $\left\{\dfrac{1}{5}\right\}$.

41. $\dfrac{4y}{y^2-25}+\dfrac{2}{y-5}=\dfrac{1}{y+5}$

To find any restrictions and the LCD, factor the first denominator.

$\dfrac{4y}{(y+5)(y-5)}+\dfrac{2}{y-5}=\dfrac{1}{y+5}$

Restrictions: $y \neq -5, y \neq 5$

$\text{LCD}=(y+5)(y-5)$

$(y+5)(y-5)\left[\dfrac{4y}{(y+5)(y-5)}+\dfrac{2}{y-5}\right]=(y+5)(y-5)\cdot\dfrac{1}{y+5}$

$$4y+2(y+5)=y-5$$
$$4y+2y+10=y-5$$
$$6y+10=y-5$$
$$5y+10=-5$$
$$5y=-15$$
$$y=-3$$

The solution set is $\{-3\}$.

43. $\dfrac{1}{x-4}-\dfrac{5}{x+2}=\dfrac{6}{x^2-2x-8}$

Factor the last denominator.

$\dfrac{1}{x-4}-\dfrac{5}{x+2}=\dfrac{6}{(x-4)(x+2)}$

Restrictions: $x \neq 4, x \neq -2$

$\text{LCD}=(x-4)(x+2)$

$(x-4)(x+2)\left[\dfrac{1}{x-4}-\dfrac{5}{x+2}\right]=(x-4)(x+2)\left[\dfrac{6}{(x-4)(x+2)}\right]$

$$(x+2)\cdot 1-(x-4)\cdot 5=6$$
$$x+2-5x+20=6$$
$$-4x+22=6$$
$$-4x=-16$$
$$x=4$$

The proposed solution 4 is *not* a solution because of the restriction $x \neq 4$. Therefore, the given equation has no solution. The solution set is $\{\ \}$.

45. $\dfrac{2}{x+3} - \dfrac{2x+3}{x-1} = \dfrac{6x-5}{x^2+2x-3}$

Factor the denominators.

$\dfrac{2}{x+3} - \dfrac{2x+3}{x-1} = \dfrac{6x-5}{(x+3)(x-1)}$

Restrictions: $x \neq -3, x \neq 1$

LCD $= (x+3)(x-1)$

$(x+3)(x-1)\left[\dfrac{2}{x+3} - \dfrac{2x+3}{x-1}\right] = (x+3)(x-1)\left[\dfrac{6x-5}{(x+3)(x-1)}\right]$

$(x-1)\cdot 2 - (x+3)(2x+3) = 6x-5$

$2x - 2 - \left(2x^2 + 9x + 9\right) = 6x-5$

$2x - 2 - 2x^2 - 9x - 9 = 6x-5$

$-2x^2 - 7x - 11 = 6x-5$

$0 = 2x^2 + 13x + 6$

$0 = (x+6)(2x+1)$

$x + 6 = 0 \quad \text{or} \quad 2x+1 = 0$

$x = -6 \qquad\qquad 2x = -1$

$\qquad\qquad\qquad\qquad x = -\dfrac{1}{2}$

The solution set is $\left\{-6, -\dfrac{1}{2}.\right\}$.

47. $\dfrac{V_1}{V_2} = \dfrac{P_2}{P_1}$

$P_1 V_2 \left(\dfrac{V_1}{V_2}\right) = P_1 V_2 \left(\dfrac{P_2}{P_1}\right)$

$P_1\left(V_1\right) = V_2\left(P_2\right)$

$P_1 V_1 = P_2 V_2$

$\dfrac{P_1 V_1}{V_1} = \dfrac{P_2 V_2}{V_1}$

$P_1 = \dfrac{P_2 V_2}{V_1}$

49.

$$\frac{1}{p} + \frac{1}{q} = \frac{1}{f}$$

$$fpq\left(\frac{1}{p} + \frac{1}{q}\right) = fpq\left(\frac{1}{f}\right)$$

$$fpq\left(\frac{1}{p}\right) + fpq\left(\frac{1}{q}\right) = pq$$

$$fq + fp = pq$$

$$f(q + p) = pq$$

$$\frac{f(q + p)}{q + p} = \frac{pq}{q + p}$$

$$f = \frac{pq}{q + p}$$

51.

$$P = \frac{A}{1 + r}$$

$$(1 + r)(P) = (1 + r)\left(\frac{A}{1 + r}\right)$$

$$P + Pr = A$$

$$Pr = A - P$$

$$\frac{Pr}{P} = \frac{A - P}{P}$$

$$r = \frac{A - P}{P}$$

53.

$$F = \frac{Gm_1 m_2}{d^2}$$

$$d^2(F) = d^2\left(\frac{Gm_1 m_2}{d^2}\right)$$

$$d^2 F = Gm_1 m_2$$

$$\frac{d^2 F}{Gm_2} = \frac{Gm_1 m_2}{Gm_2}$$

$$m_1 = \frac{d^2 F}{Gm_2}$$

55.

$$z = \frac{x - \overline{x}}{s}$$

$$s(z) = s\left(\frac{x - \overline{x}}{s}\right)$$

$$zs = x - \overline{x}$$

$$x = \overline{x} + zs$$

57.
$$I = \frac{E}{R+r}$$

$$(R+r)(I) = (R+r)\left(\frac{E}{R+r}\right)$$

$$IR + Ir = E$$

$$IR = E - Ir$$

$$\frac{IR}{I} = \frac{E - Ir}{I}$$

$$R = \frac{E - Ir}{I}$$

59.
$$f = \frac{f_1 f_2}{f_1 + f_2}$$

$$(f_1 + f_2)(f) = (f_1 + f_2)\left(\frac{f_1 f_2}{f_1 + f_2}\right)$$

$$ff_1 + ff_2 = f_1 f_2$$

$$ff_2 = f_1 f_2 - ff_1$$

$$ff_2 = f_1 (f_2 - f)$$

$$\frac{ff_2}{f_2 - f} = \frac{f_1 (f_2 - f)}{f_2 - f}$$

$$f_1 = \frac{ff_2}{f_2 - f}$$

61. Solve $\dfrac{x^2 - 10}{x^2 - x - 20} = 1 + \dfrac{7}{x-5} - \dfrac{1}{2}$.

Factor the first denominator.

$$\frac{x^2 - 10}{(x-5)(x+4)} = 1 + \frac{7}{x-5} - \frac{1}{2}.$$

Restrictions: $x \neq 5, x \neq -4$

LCD $= (x-5)(x+4)$

$$(x-5)(x+4)\left(\frac{x^2 - 10}{(x-5)(x+4)}\right) = (x-5)(x+4)\cdot 1 + (x-5)(x+4)\left(\frac{7}{x-5}\right)$$

$$x^2 - 10 = (x-5)(x+4) + (x+4)\cdot 7$$

$$x^2 - 10 = x^2 - x - 20 + 7x + 28$$

$$x^2 - 10 = x^2 + 6x + 8$$

$$-10 = 6x + 8$$

$$-18 = 6x$$

$$-3 = x$$

The solution set is $\{-3\}$.

63. Simplify $\dfrac{x^2-10}{x^2-x-20}-1-\dfrac{7}{x-5}$.

Factor the first denominator.

$$\frac{x^2-10}{(x-5)(x+4)}-1-\frac{7}{x-5}$$

$$\text{LCD}=(x-5)(x+4)$$

$$\frac{x^2-10}{(x-5)(x+4)}-\frac{(x-5)(x+4)}{(x-5)(x+4)}-\frac{7(x+4)}{(x-5)(x+4)}=\frac{x^2-10-(x-5)(x+4)-7(x+4)}{(x-5)(x+4)}$$

$$=\frac{x^2-10-\left(x^2-x-20\right)-7x-28}{(x-5)(x+4)}$$

$$=\frac{x^2-10-x^2+x+20-7x-28}{(x-5)(x+4)}$$

$$=\frac{-6x-18}{(x-5)(x+4)}$$

65. Solve $5y^{-2}+1=6y^{-1}$

$$\frac{5}{y^2}+1=\frac{6}{y}$$

Restrictions: $y\neq 0$

$$\text{LCD}=y^2$$

$$y^2\left(\frac{5}{y^2}+1\right)=y^2\left(\frac{6}{y}\right)$$

$$y^2\cdot\frac{5}{y^2}+y^2\cdot 1=6y$$

$$5+y^2=6y$$

$$y^2-6y+5=0$$

$$(y-5)(y-1)=0$$

$$y-5=0 \quad\text{or}\quad y-1=0$$

$$y=5 \qquad\qquad y=1$$

The solution set is $\{1,5\}$.

67. Solve $\dfrac{3}{y+1} - \dfrac{1}{1-y} = \dfrac{10}{y^2-1}$.

Factor the denominators.

$$\dfrac{3}{y+1} - \dfrac{(-1)\cdot 1}{(-1)(1-y)} = \dfrac{10}{(y+1)(y-1)}$$

$$\dfrac{3}{y+1} - \dfrac{-1}{y-1} = \dfrac{10}{(y+1)(y-1)}$$

$$\dfrac{3}{y+1} + \dfrac{1}{y-1} = \dfrac{10}{(y+1)(y-1)}$$

Restrictions: $y \neq -1, y \neq 1$

$\text{LCD} = (y+1)(y-1)$

$$(y+1)(y-1)\left(\dfrac{3}{y+1} + \dfrac{1}{y-1}\right) = (y+1)(y-1)\left(\dfrac{10}{(y+1)(y-1)}\right)$$

$$(y-1)\cdot 3 + (y+1)\cdot 1 = 10$$

$$3y - 3 + y + 1 = 10$$

$$4y - 2 = 10$$

$$4y = 12$$

$$y = 3$$

The solution set is $\{3\}$.

69. $C = \dfrac{400x + 500,000}{x}; C = 450$

$$450 = \dfrac{400x + 500,000}{x}$$

$\text{LCD} = x$

$$x \cdot 450 = x\left(\dfrac{400x + 500,000}{x}\right)$$

$$450x = 400x + 500,000$$

$$50x = 500,000$$

$$x = 10,000$$

At an average cost of \$450 per wheelchair, 10,000 wheelchairs can be produced.

71. $C = \dfrac{2x}{100-x}; C = 2$

$$2 = \dfrac{2x}{100-x}$$

$\text{LCD} = 100 - x$

$$(100-x)\cdot 2 = (100-x)\cdot \dfrac{2x}{100-x}$$

$$200 - 2x = 2x$$

$$200 = 4x$$

$$50 = x$$

For \$2 million, 50% of the contaminants can be removed.

73. $C = \dfrac{DA}{A+12}; C = 300, D = 1000$

$300 = \dfrac{1000A}{A+12}$

$\text{LCD} = A+12$

$(A+12) \cdot 300 = (A+12)\left(\dfrac{1000A}{A+12}\right)$

$300A + 3600 = 1000A$

$3600 = 700A$

$\dfrac{3600}{700} = A$

$A = \dfrac{36}{7} \approx 5.14$

To the nearest year, the child is 5 years old.

75. $C = \dfrac{10,000}{x} + 3x; C = 350$

$350 = \dfrac{10,000}{x} + 3x$

$\text{LCD} = x$

$x \cdot 350 = x\left(\dfrac{10,000}{x} + 3x\right)$

$350x = 10,000 + 3x^2$

$0 = 3x^2 - 350x + 10,000$

$0 = (3x - 200)(x - 50)$

$3x - 200 = 0 \quad \text{or} \quad x - 50 = 0$

$3x = 200 \qquad\qquad x = 50$

$x = \dfrac{200}{3}$

$= 66\dfrac{2}{3} \approx 67$

For yearly inventory costs to be \$350, the owner should order either 50 or approximately 67 cases. These solutions correspond to the points $(50, 350)$ and $\left(66\dfrac{2}{3}, 350\right)$ on the graph.

77. – 81. Answers will vary.

83. makes sense

85. does not make sense; Explanations will vary. Sample explanation: If all potential solutions make any of the denominators of the rational equation equal to zero, then the equation will have no solution.

87. true

89. false; Changes to make the statement true will vary. A sample change is: You could begin by multiplying both sides by the LCD, $3x$.

91. $\left(\dfrac{x+1}{x+7}\right)^2 \div \left(\dfrac{x+1}{x+7}\right)^4 = 0$

$\left(\dfrac{x+1}{x+7}\right)^2 \cdot \left(\dfrac{x+7}{x+1}\right)^4 = 0$

$\dfrac{(x+1)^2}{(x+7)^2} \cdot \dfrac{(x+7)^4}{(x+1)^4} = 0$

Restrictions: $x \neq -7, x \neq -1$

$\dfrac{(x+7)^2}{(x+1)^2} = 0$

Multiply both sides by $(x+1)^2$.

$(x+1)^2 \left[\dfrac{(x+7)^2}{(x+1)^2}\right] = (x+1)^2 \cdot 0$

$(x+7)^2 = 0$

$x+7 = 0$

$x = -7$

The proposed solution, -7, is *not* a solution of the original equation because it is on the list of restrictions. Therefore, the given equation has no solution. The solution set is $\{\ \}$.

93. $\dfrac{x}{2} + \dfrac{x}{4} = 6$

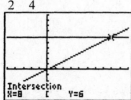

The solution set is $\{8\}$.

Check $x = 8$:

$\dfrac{8}{2} + \dfrac{8}{4} = 6$

$4 + 2 = 6$

$6 = 6$ true

94. $\dfrac{50}{x} = 2x$

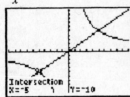

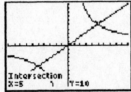

The solution set is $\{-5, 5\}$.

95. $x + \dfrac{6}{x} = -5$

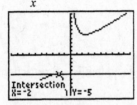

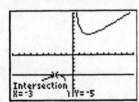

The solution set is $\{-3, -2\}$.

$$\text{Check } -3: \qquad \text{Check } -2:$$

$$x + \frac{6}{x} = -5 \qquad\quad x + \frac{6}{x} = -5$$

$$-3 + \frac{6}{-3} = -5 \qquad -2 + \frac{6}{-2} = -5$$

$$-3 + (-2) = -5 \qquad -2 + (-3) = -5$$

$$-5 = -5, \text{ true} \qquad -5 = -5, \text{ true}$$

96. $x^4 + 2x^3 - 3x - 6$

Factor by grouping.

$$x^4 + 2x^3 - 3x - 6$$

$$= \left(x^4 + 2x^3\right) + \left(-3x - 6\right)$$

$$= x^3 (x + 2) - 3(x + 2)$$

$$= (x + 2)\left(x^3 - 3\right)$$

97. $\left(3x^2\right)\left(-4x^{-10}\right)$

$= \left(3 \cdot -4\right)\left(x^2 \cdot x^{-10}\right) = -12x^{2+(-10)}$

$= -12x^{-8} = -\dfrac{12}{x^8}$

98. $-5\left[4(x-2)-3\right] = -5\left[4x-8-3\right]$

$= -5\left[4x-11\right]$

$= -20x+55$

99.
$$\frac{15}{8+x} = \frac{9}{8-x}$$

$$(8+x)(8-x)\frac{15}{8+x} = (8+x)(8-x)\frac{9}{8-x}$$

$$15(8-x) = 9(8+x)$$

$$120-15x = 72+9x$$

$$-24x = -48$$

$$x = 2$$

The solution set is $\{2\}$.

100. In 1 hour you can complete $\dfrac{1}{5}$ of the job.

In 3 hours you can complete $\dfrac{3}{5}$ of the job.

In x hours you can complete $\dfrac{x}{5}$ of the job.

101. $\dfrac{63}{x} = \dfrac{7}{5}$

7.7 Check Points

1. Let x = the rate of the current.
   Then $3 + x =$ the canoe's rate with the current.
   and $3 - x =$ the canoe's rate against the current.

	Distance	Rate	Time $= \dfrac{\text{Distance}}{\text{Rate}}$
With the current	10	$3 + x$	$\dfrac{10}{3+x}$
Against the current	2	$3 - x$	$\dfrac{2}{3-x}$

$$\frac{10}{3+x} = \frac{2}{3-x}$$

Use the cross-products principle to solve this equation.

$$\frac{10}{3+x} = \frac{2}{3-x}$$
$$10(3-x) = 2(3+x)$$
$$30 - 10x = 6 + 2x$$
$$30 - 12x = 6$$
$$-12x = -24$$
$$x = 2$$

The rate of the current is 2 miles per hour.

2. Let x = the number of hours for both people to paint a house together.

	Fractional part of job completed in 1 hour	Time working together	Fractional part of job completed in x hours
First person	$\dfrac{1}{8}$	x	$\dfrac{x}{8}$
Second person	$\dfrac{1}{4}$	x	$\dfrac{x}{4}$

Working together, the two people can complete the whole job, so $\dfrac{x}{8} + \dfrac{x}{4} = 1$.

Multiply both sides by the LCD, 8.

$$8\left(\frac{x}{8} + \frac{x}{4}\right) = 8 \cdot 1$$
$$x + 2x = 8$$
$$3x = 8$$
$$x = \frac{8}{3}$$
$$x = 2\frac{2}{3}$$

It will take $2\frac{2}{3}$ hours (or 2 hours 40 minutes) if the work together.

3. Let x = the property tax on the $420,000 house.

$$\frac{\text{Tax on \$250,000 house}}{\text{Assessed value (\$250,000)}} = \frac{\text{Tax on \$420,000 house}}{\text{Assessed value (\$420,000)}}$$

$$\frac{\$3500}{\$250,000} = \frac{\$x}{\$420,000}$$

$$\frac{3500}{250,000} = \frac{x}{420,000}$$

$$250,000x = (3500)(420,000)$$

$$250,000x = 1,470,000,000$$

$$\frac{250,000x}{250,000} = \frac{1,470,000,000}{250,000}$$

$$x = 5880$$

The property tax is $5880.

4. Let x = the total number of deer in the refuge.

$$\frac{120}{x} = \frac{25}{150}$$

$$25x = (120)(150)$$

$$25x = 18,000$$

$$\frac{25x}{25} = \frac{18,000}{25}$$

$$x = 720$$

There are about 720 deer in the refuge.

5. $\dfrac{3}{8} = \dfrac{12}{x}$

$$3x = 8 \cdot 12$$

$$3x = 96$$

$$x = 32$$

The missing length is 32 inches.

6. $\dfrac{h}{2} = \dfrac{56}{3.5}$

$$3.5h = 2 \cdot 56$$

$$3.5h = 112$$

$$h = \frac{112}{3.5}$$

$$h = 32$$

The height of the tower is 32 yards.

7.7 Exercise Set

1. The times are equal, so $\dfrac{10}{x} = \dfrac{15}{x+3}$

To solve this equation, multiply both sides by the LCD, $x(x+3)$.

$$x(x+3) \cdot \frac{10}{x} = x(x+3) \cdot \frac{15}{x+3}$$
$$10(x+3) = 15x$$
$$10x + 30 = 15x$$
$$30 = 5x$$
$$6 = x$$

If $x = 6, x + 3 = 9$.

Note: The equation $\dfrac{10}{x} = \dfrac{15}{x+3}$ is a proportion, so it can also be solved by using the cross-products principle.

$$10(x+3) = 15x$$

This allows you to skip the first step of the solution process shown above.
The walking rate is 6 miles per hour and the car's rate is 9 miles per hour.

3. Let $x =$ the jogger's rate running uphill.
Then $x + 4 =$ the jogger's rate running downhill.

	Distance	Rate	Time $= \dfrac{\text{Distance}}{\text{Rate}}$
Cycling	5	$x+4$	$\dfrac{5}{x+4}$
Walking	3	x	$\dfrac{3}{x}$

The times are equal, so $\dfrac{5}{x+4} = \dfrac{3}{x}$.
Use the cross-products principle to solve this equation.

$$5x = 3(x+4)$$
$$5x = 3x + 12$$
$$2x = 12$$
$$x = 6$$

If $x = 6, x + 4 = 10$.
The jogger runs 10 miles per hour downhill and 6 miles per hour uphill.

5. Let $x =$ the rate of the current.
Then $15 + x =$ the boat's rate with the current.
and $15 - x =$ the boat's rate against the current.

	Distance	Rate	Time $= \dfrac{\text{Distance}}{\text{Rate}}$
With the current	20	$15+x$	$\dfrac{20}{15+x}$
Against the current	10	$15-x$	$\dfrac{10}{15-x}$

$$\frac{20}{15+x} = \frac{10}{15-x}$$

Use the cross-products principle to solve this equation.

$$20(15-x)=10(15+x)$$
$$300-20x=150+10x$$
$$300=150+30x$$
$$150=30x$$
$$5=x$$

The rate of the current is 5 miles per hour.

7. Let $x =$ walking rate.
Then $2x =$ jogging rate.

	Distance	Rate	Time $= \dfrac{\text{Distance}}{\text{Rate}}$
Walking	2	x	$\dfrac{2}{x}$
Jogging	2	$2x$	$\dfrac{2}{2x}$

The total time is 1 hour, so $\dfrac{2}{x}+\dfrac{2}{2x}=1$

$$\dfrac{2}{x}+\dfrac{1}{x}=1.$$

To solve this equation, multiply both sides by the LCD, x.

$$x\left(\dfrac{2}{x}+\dfrac{1}{x}\right)=x\cdot 1$$
$$2+1=x$$
$$3=x$$

If $x = 3$, $2x = 6$.
The walking rate is 3 miles per hour and the jogging rate is 6 miles per hour.

9. Let $x =$ the boat's average rate in still water.
Then $x+2 =$ the boat's rate with the current (downstream).
and $x-2 =$ the boat's rate against the current (upstream).

	Distance	Rate	Time $= \dfrac{\text{Distance}}{\text{Rate}}$
Downstream	6	$x+2$	$\dfrac{6}{x+2}$
Upstream	4	$x-2$	$\dfrac{4}{x-2}$

The times are equal so solve the following equation.

$$\dfrac{6}{x+2}=\dfrac{4}{x-2}$$
$$6(x-2)=4(x+2)$$
$$6x-12=4x+8$$
$$2x-12=8$$
$$2x=20$$
$$x=10$$

The boat's average rate in still water is 10 miles per hour.

11. Let x = the time in minutes, for both people to shovel the driveway together.

	Fractional part of job completed in 1 minute	Time working together	Fractional part of job completed in x minutes
You	$\dfrac{1}{20}$	x	$\dfrac{x}{20}$
Your brother	$\dfrac{1}{15}$	x	$\dfrac{x}{15}$

Working together, you and your brother complete the whole job, so $\dfrac{x}{20}+\dfrac{x}{15}=1$.

Multiply both sides by the LCD, 60.

$$60\left(\frac{x}{20}+\frac{x}{15}\right)=60\cdot1$$
$$3x+4x=60$$
$$7x=60$$
$$x=\frac{60}{7}\approx8.6$$

It will take about 8.6 minutes, which is enough time.

13. Let x = the time, in hours, for both teams to clean the streets working together.

	Fractional part of job completed in 1 hour	Time working together	Fractional part of job completed in x hours
First team	$\dfrac{1}{400}$	x	$\dfrac{x}{400}$
Second team	$\dfrac{1}{300}$	x	$\dfrac{x}{300}$

Working together, the two teams complete one whole job, so $\dfrac{x}{400}+\dfrac{x}{300}=1$.

Multiply both sides by the LCD, 1200.

$$1200\left(\frac{x}{400}+\frac{x}{300}\right)=1200\cdot1$$
$$3x+4x=1200$$
$$7x=1200$$
$$x=\frac{1200}{7}\approx171.4$$

It will take about 171.4 hours for both teams to clean the streets working together. One week is $7\cdot24=168$ hours, so even if both crews work 24 hours a day, there is not enough time.

15. Let x = the time, in hours, for both pipes to fill in the pool.

	Fractional part of job completed in 1 hour	Time working together	Fractional part of job completed in x hours
First pipe	$\dfrac{1}{4}$	x	$\dfrac{x}{4}$
Second pipe	$\dfrac{1}{6}$	x	$\dfrac{x}{6}$

$$\frac{x}{4}+\frac{x}{6}=1$$
$$12\left(\frac{x}{4}+\frac{x}{6}\right)=12\cdot 1$$
$$3x+2x=12$$
$$5x=12$$
$$x=\frac{12}{5}=2.4$$

Using both pipes, it will take 2.4 hours or 2 hours 24 minutes to fill the pool.

17. Let x = the tax on a property with an assessed value of \$162,500.

$$\frac{\text{Tax on \$62,000 house}}{\text{Assessed value (\$62,000)}}=\frac{\text{Tax on \$162,500 house}}{\text{Assessed value (\$162,500)}}$$
$$\frac{\$720}{\$65,000}=\frac{\$x}{\$162,500}$$
$$\frac{720}{65,000}=\frac{x}{162,500}$$
$$65,000x=(720)(162,500)$$
$$65,000x=117,000,000$$
$$\frac{65,000x}{65,000}=\frac{117,000,000}{65,000}$$
$$x=1800$$

The tax on a property assessed at \$162,500 is \$1800.

19. Let x = the total number of fur seal pups in the rookery.

$$\frac{\text{Original \# tagged}}{\text{Total \# fur seal pups}}=\frac{\text{\# tagged in sample}}{\text{\# in sample}}$$
$$\frac{4963}{x}=\frac{218}{900}$$
$$218x=(4963)(900)$$
$$218x=4,466,700$$
$$\frac{218x}{218}=\frac{4,466,700}{218}$$
$$x\approx 20,489$$

There were approximately 20,489 fur seal pups in the rookery.

21. Let $x =$ the number of people, in billions, suffering from malnutrition (in 2006).

$$\frac{28}{200} = \frac{x}{6.5}$$

$$200x = (28)(6.5)$$

$$200x = 182$$

$$\frac{200x}{200} = \frac{182}{200}$$

$$x \approx 0.9$$

0.9 billion people suffered from malnutrition in 2006.

23. Let $x =$ the height of the critter.

$$\frac{\text{foot length person}}{\text{height of person}} = \frac{\text{foot length critter}}{\text{height of critter}}$$

$$\frac{10 \text{ inches}}{67 \text{ inches}} = \frac{23 \text{ inches}}{x}$$

$$\frac{10}{67} = \frac{23}{x}$$

$$10x = (67)(23)$$

$$10x = 1541$$

$$x = 154.1$$

The height of the critter is 154.1 in.

25.
$$\frac{18}{9} = \frac{10}{x}$$

$$18x = 9 \cdot 10$$

$$18x = 90$$

$$x = 5$$

The length of the side marked x is 5 inches.

27.
$$\frac{10}{30} = \frac{x}{18}$$

$$30x = 10 \cdot 18$$

$$30x = 180$$

$$x = 6$$

The length of the side marked x is 6 meters.

29.
$$\frac{20}{15} = \frac{x}{12}$$

$$15x = 12 \cdot 20$$

$$15x = 240$$

$$x = 16$$

The length of the side marked x is 16 inches.

31.　$\dfrac{8}{6} = \dfrac{x}{12}$

$6x = 8 \cdot 12$

$6x = 96$

$x = 16$

The tree is 16 feet tall.

33. – 41.　Answers will vary.

43.　does not make sense; Explanations will vary. Sample explanation: In the same amount of time, you will go further with the current than against it.

45.　makes sense

47.　Let x = the usual average rate.
Then $x - 15$ = the average rate, in miles per hour, of the bus in the snowstorm.

	Distance	Rate	Time $= \dfrac{\text{Distance}}{\text{Rate}}$
Usual conditions	60	x	$\dfrac{60}{x}$
Snowstorm conditions	60	$x - 15$	$\dfrac{60}{x-15}$

Since the time during the snowstorm is 2 hours longer than the usual time, solve the following equation.

$\dfrac{60}{x} + 2 = \dfrac{60}{x-15}$

Multiply both sides of the equation by the LCD, $x(x-15)$

$x(x-15)\left(\dfrac{60}{x} + 2\right) = x(x-15)\left(\dfrac{60}{x-15}\right)$

$(x-15) \cdot 60 + x(x-15) \cdot 2 = 60x$

$60x - 900 + 2x^2 - 30x = 60x$

$2x^2 + 30x - 900 = 60x$

$2x^2 - 30x - 900 = 0$

$x^2 - 15x - 450 = 0$

$(x-30)(x+15) = 0$

$x - 30 = 0$　or　$x + 15 = 0$

$\quad x = 30 \qquad\qquad x = -15$

Since the rate cannot be negative, the solution is 30. Therefore, the usual average rate of the bus is 30 miles per hour.

49. Let $x =$ the time, in hours, it takes to prepare one report working together.

	Fractional part of job completed in 1 hour	Time working together	Fractional part of job completed in x hours
Ben	$\dfrac{1}{3}$	x	$\dfrac{x}{3}$
Shane	$\dfrac{1}{4.2} = \dfrac{10}{42} = \dfrac{5}{21}$	x	$\dfrac{5x}{21}$

Working together, Ben and Shane prepare one report, so $\dfrac{x}{3} + \dfrac{5x}{21} = 1$.

To solve this equation, multiply both sides by the LCD, 21.

$$21 \cdot \left(\frac{x}{3} + \frac{5x}{21} \right) = 21 \cdot 1$$
$$7x + 5x = 21$$
$$12x = 21$$
$$x = \frac{21}{12} = 1.75$$

Ben and Shane can prepare one report in 1.75 hours. Multiply this by four to determine how many hours it takes to prepare four reports. $4(1.75) = 7$. Therefore, working together Ben and Shane can prepare four reports in 7 hours.

51. Let $x =$ time, in hours, to fill empty swimming pool.

	Fractional part of job completed in 1 hour	Time working together	Fractional part of job completed in x hours
Normal filling of pool	$\dfrac{1}{2}$	x	$\dfrac{x}{2}$
Leak	$-\dfrac{1}{10}$	x	$-\dfrac{x}{10}$

The situation can be modeled by the equation $\dfrac{x}{2} - \dfrac{x}{10} = 1$.

Multiply both sides of the equation by the LCD, 20.

$$20 \cdot \left(\frac{x}{2} - \frac{x}{10} \right) = 20 \cdot 1$$
$$10x - 2x = 20$$
$$8x = 20$$
$$x = 2.5$$

It will take 2.5 hours to fill the empty swimming pool.

53. $25x^2 - 81 = (5x)^2 - 9^2$
$$= (5x + 9)(5x - 9)$$

54. $x^2 - 12x + 36 = 0$
$$(x - 6)^2 = 0$$
$$x - 6 = 0$$
$$x = 6$$

The solution set is $\{6\}$.

55. $y = -\dfrac{2}{3}x + 4$

slope $= -\dfrac{2}{3} = \dfrac{-2}{3}$

y-intercept $= 4$

Plot (0,4). From this point, move 2 units *down* and 3 units to the *right* to reach the point (3,2). Draw a line through (0,4) and (3,2).

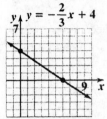

56. a. Substitute to find k.

$y = kx^2$

$y = kx^2$

$64 = k \cdot 2^2$

$64 = 4k$

$k = 16$

b. $y = kx^2$

$y = 16x^2$

c. $y = 16x^2$

$y = 16 \cdot 5^2$

$y = 400$

57. a. Substitute to find k.

$y = \dfrac{k}{x}$

$12 = \dfrac{k}{8}$

$k = 12 \cdot 8$

$k = 96$

b. $y = \dfrac{k}{x}$

$y = \dfrac{96}{x}$

c. $y = \dfrac{96}{x}$

$y = \dfrac{96}{3}$

$y = 32$

58. $S = \dfrac{kA}{P}$

$12,000 = \dfrac{k \cdot 60,000}{40}$

$12,000 = \dfrac{k \cdot 60,000}{40}$

$12,000 = 1500k$

$k = 8$

7.8 Check Points

1. Since W varies directly with t, we have $W = kt$. Use the given values to find k.

$W = kt$

$30 = k \cdot 5$

$\dfrac{30}{5} = \dfrac{k \cdot 5}{5}$

$6 = k$

The equation becomes $W = 6t$.

Find W when $t = 11$.

$W = 6t$

$W = 6 \cdot 11$

$W = 66$

An 11 minute shower will use 66 gallons of water.

2. Beginning with $y = \dfrac{k}{x}$, we will use l for the length of the string and f for the frequency.

Use the given values to find k.

$f = \dfrac{k}{l}$

$640 = \dfrac{k}{8}$

$8 \cdot 640 = 8 \cdot \dfrac{k}{8}$

$5120 = k$

The equation becomes $f = \dfrac{k}{l}$

$f = \dfrac{5120}{l}$

Find f when $l = 10$.

$f = \dfrac{5120}{l}$

$f = \dfrac{5120}{10}$

$f = 512$

A string length of 10 inches will vibrate at 512 cycles per second.

3. Let $m =$ the number of minutes needed to solve an exercise set.

   Let $p =$ the number of people working on the problems.

   Let $x =$ the number of problems in the exercise set.

   Use $m = \dfrac{kx}{p}$ to find k.

   $$m = \frac{kx}{p}$$

   $$32 = \frac{k16}{4}$$

   $$32 = 4k$$

   $$k = 8$$

   Thus, $m = \dfrac{8x}{p}$.

   Find m when $p = 8$ and $x = 24$.

   $$m = \frac{8 \cdot 24}{8}$$

   $$m = 24$$

   It will take 24 minutes for 8 people to solve 24 problems.

4. Find k: $V = khr^2$

   $$120\pi = k \cdot 10 \cdot 6^2$$

   $$120\pi = k \cdot 360$$

   $$\frac{120\pi}{360} = \frac{k \cdot 360}{360}$$

   $$\frac{\pi}{3} = k$$

   Thus, $V = \dfrac{\pi}{3} hr^2 = \dfrac{\pi hr^2}{3}$.

   $$V = \frac{\pi hr^2}{3}$$

   $$V = \frac{\pi \cdot 2 \cdot 12^2}{3} = 96\pi$$

   The volume of a cone having a radius of 12 feet and a height of 2 feet is 96π cubic feet.

7.8 Exercise Set

1. Since y varies directly with x, we have $y = kx$. Use the given values to find k.

   $$y = kx$$

   $$65 = k \cdot 5$$

   $$\frac{65}{5} = \frac{k \cdot 5}{5}$$

   $$13 = k$$

   The equation becomes $y = 13x$.

   When $x = 12$, $y = 13x = 13 \cdot 12 = 156$.

3. Since y varies inversely with x, we have $y = \dfrac{k}{x}$.

   Use the given values to find k.

   $$y = \frac{k}{x}$$

   $$12 = \frac{k}{5}$$

   $$5 \cdot 12 = 5 \cdot \frac{k}{5}$$

   $$60 = k$$

   The equation becomes $y = \dfrac{60}{x}$.

   When $x = 2$, $y = \dfrac{60}{2} = 30$.

5. Since y varies inversely as x and inversely as the square of z, we have $y = \dfrac{kx}{z^2}$.

   Use the given values to find k.

   $$y = \frac{kx}{z^2}$$

   $$20 = \frac{k(50)}{5^2}$$

   $$20 = \frac{k(50)}{25}$$

   $$20 = 2k$$

   $$10 = k$$

   The equation becomes $y = \dfrac{10x}{z^2}$.

   When $x = 3$ and $z = 6$,

   $$y = \frac{10x}{z^2} = \frac{10(3)}{6^2} = \frac{10(3)}{36} = \frac{30}{36} = \frac{5}{6}.$$

7. Since y varies jointly as x and y, we have $y = kxz$.
Use the given values to find k.
$$y = kxz$$
$$25 = k(2)(5)$$
$$25 = k(10)$$
$$\frac{25}{10} = \frac{k(10)}{10}$$
$$\frac{5}{2} = k$$

The equation becomes $y = \frac{5}{2}xz$.
When $x = 8$ and $z = 12$,
$$y = \frac{5}{2}(8)(12) = \frac{5}{2}(\overset{4}{\cancel{8}})(12) = 240.$$

9. Since y varies jointly as a and b and inversely as the square root of c, we have $y = \frac{kab}{\sqrt{c}}$.
Use the given values to find k.
$$y = \frac{kab}{\sqrt{c}}$$
$$12 = \frac{k(3)(2)}{\sqrt{25}}$$
$$12 = \frac{k(6)}{5}$$
$$12(5) = \frac{k(6)}{5}(5)$$
$$60 = 6k$$
$$\frac{60}{6} = \frac{6k}{6}$$
$$10 = k$$

The equation becomes $y = \frac{10ab}{\sqrt{c}}$.
When $a = 5$, $b = 3$, $c = 9$,
$$y = \frac{10ab}{\sqrt{c}} = \frac{10(5)(3)}{\sqrt{9}} = \frac{150}{3} = 50.$$

11. $x = kyz$;
Solving for y:
$$x = kyz$$
$$\frac{x}{kz} = \frac{kyz}{yz}.$$
$$y = \frac{x}{kz}$$

13. $x = \frac{kz^3}{y}$;
Solving for y:
$$x = \frac{kz^3}{y}$$
$$xy = y \cdot \frac{kz^3}{y}$$
$$xy = kz^3$$
$$\frac{xy}{x} = \frac{kz^3}{x}$$
$$y = \frac{kz^3}{x}$$

15. $x = \frac{kyz}{\sqrt{w}}$;
Solving for y:
$$x = \frac{kyz}{\sqrt{w}}$$
$$x(\sqrt{w}) = (\sqrt{w})\frac{kyz}{\sqrt{w}}$$
$$x\sqrt{w} = kyz$$
$$\frac{x\sqrt{w}}{kz} = \frac{kyz}{kz}$$
$$y = \frac{x\sqrt{w}}{kz}$$

17. $x = kz(y + w)$;
Solving for y:
$$x = kz(y + w)$$
$$x = kzy + kzw$$
$$x - kzw = kzy$$
$$\frac{x - kzw}{kz} = \frac{kzy}{kz}$$
$$y = \frac{x - kzw}{kz}$$

19. $x = \dfrac{kz}{y-w}$;

Solving for y:

$$x = \frac{kz}{y-w}$$

$$(y-w)x = (y-w)\frac{kz}{y-w}$$

$$xy - wx = kz$$

$$xy = kz + wx$$

$$\frac{xy}{x} = \frac{kz + wx}{x}$$

$$y = \frac{kz + wx}{x}$$

21. Since T varies directly as B, we have $T = kB$. Use the given values to find k.

$$T = kB$$

$$3.6 = k(4)$$

$$\frac{3.6}{4} = \frac{k(4)}{4}$$

$$0.9 = k$$

The equation becomes $T = 0.9B$.
When $B = 6$, $T = 0.9(6) = 5.4$. The tail length is 5.4 feet.

23. Since B varies directly as D, we have $B = kD$. Use the given values to find k.

$$B = kD$$

$$8.4 = k(12)$$

$$\frac{8.4}{12} = \frac{k(12)}{12}$$

$$k = \frac{8.4}{12} = 0.7$$

The equation becomes $B = 0.7D$.
When $B = 56$,

$$56 = 0.7D$$

$$\frac{56}{0.7} = \frac{0.7D}{0.7}$$

$$D = \frac{56}{0.7} = 80$$

It was dropped from 80 inches.

25. Since a man's weight varies directly as the cube of his height, we have $w = kh^3$.
Use the given values to find k.

$$w = kh^3$$

$$170 = k(70)^3$$

$$170 = k(343,000)$$

$$\frac{170}{343,000} = \frac{k(343,000)}{343,000}$$

$$0.000496 = k$$

The equation becomes $w = 0.000496h^3$.

When $h = 107$, $\quad w = 0.000496(107)^3$
$$= 0.000496(1,225,043) \approx 607.$$

Robert Wadlow's weight was approximately 607 pounds.

27. Since the banking angle varies inversely as the turning radius, we have $B = \dfrac{k}{r}$.
Use the given values to find k.

$$B = \frac{k}{r}$$

$$28 = \frac{k}{4}$$

$$28(4) = 28\left(\frac{k}{4}\right)$$

$$112 = k$$

The equation becomes $B = \dfrac{112}{r}$.

When $r = 3.5$, $B = \dfrac{112}{r} = \dfrac{112}{3.5} = 32$.

The banking angle is $32°$ when the turning radius is 3.5 feet.

29. a. Use $L = \dfrac{k}{R}$ to find k.

$$L = \frac{k}{R}$$

$$30 = \frac{k}{63}$$

$$63 \cdot 30 = 63 \cdot \frac{k}{63}$$

$$1890 = k$$

Thus, $L = \dfrac{1890}{R}$.

b. This is an approximate model.

c. $L = \dfrac{1890}{R}$

$$L = \frac{1890}{27} = 70$$

The average life span of an elephant is 70 years.

31. a. A mammal with a life span of 20 years will have a heart rate of about 90 beats per minute.

b. $L = \dfrac{1890}{R}$

$$20 = \frac{1890}{R}$$

$$20R = 1890$$

$$R = \frac{1890}{20}$$

$$R \approx 95$$

The model estimates a mammal with a life span of 20 years will have a heart rate of about 95 beats per minute.

c. The data for horses is represented on the graph by the point $(63, 30)$.

33. Since intensity varies inversely as the square of the distance, we have $I = \dfrac{k}{d^2}$.

Use the given values to find k.

$$I = \frac{k}{d^2}$$

$$62.5 = \frac{k}{3^2}$$

$$62.5 = \frac{k}{9}$$

$$9(62.5) = 9\left(\frac{k}{9}\right)$$

$$562.5 = k$$

The equation becomes $I = \dfrac{562.5}{d^2}$.

When $d = 2.5$, $I = \dfrac{562.5}{2.5^2} = \dfrac{562.5}{6.25} = 90$

The intensity is 90 milliroentgens per hour.

35. Since index varies directly as weight and inversely as the square of one's height, we have $I = \dfrac{kw}{h^2}$.

Use the given values to find k.

$$I = \frac{kw}{h^2}$$

$$35.15 = \frac{k(180)}{60^2}$$

$$35.15 = \frac{k(180)}{3600}$$

$$(3600)35.15 = \frac{k(180)}{3600}$$

$$126540 = k(180)$$

$$k = \frac{126540}{180} = 703$$

The equation becomes $I = \dfrac{703w}{h^2}$.

When $w = 170$ and $h = 70$, $I = \dfrac{703(170)}{(70)^2} \approx 24.4$.

This person has a BMI of 24.4 and is not overweight.

37. Since heat loss varies jointly as the area and temperature difference, we have $L = kAD$. Use the given values to find k.

$$L = kAD$$
$$1200 = k(3 \cdot 6)(20)$$
$$1200 = 360k$$
$$\frac{1200}{360} = \frac{360k}{360}$$
$$k = \frac{10}{3}$$

The equation becomes $L = \frac{10}{3}AD$

When $A = 6 \cdot 9 = 54$, $D = 10$,

$L = \frac{10}{3}(9 \cdot 6)(10) = 1800$.

The heat loss is 1800 Btu.

39. Since intensity varies inversely as the square of the distance from the sound source, we have $I = \frac{k}{d^2}$.

If you move to a seat twice as far, then $d = 2d$. So we have $I = \frac{k}{(2d)^2} = \frac{k}{4d^2} = \frac{1}{4} \cdot \frac{k}{d^2}$. The intensity will be multiplied by a factor of $\frac{1}{4}$. So the sound intensity is $\frac{1}{4}$ of what it was originally.

41. a. Since the average number of phone calls varies jointly as the product of the populations and inversely as the square of the distance, we have $C = \frac{kP_1 P_2}{d^2}$.

b. Use the given values to find k.

$$C = \frac{kP_1 P_2}{d^2}$$
$$326,000 = \frac{k(777,000)(3,695,000)}{(420)^2}$$
$$326,000 = \frac{k(2.87 \times 10^{12})}{176,400}$$
$$326,000 = 16,269,841.27k$$
$$0.02 \approx k$$

The equation becomes $C = \frac{0.02 P_1 P_2}{d^2}$.

c. $C = \frac{0.02(650,000)(490,000)}{(400)^2}$

$\approx 39,813$

The average number of calls is approximately 39,813 daily phone calls.

43. a.

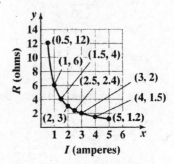

b. Current varies inversely as resistance. Answers will vary.

c. Since the current varies inversely as resistance we have $R = \frac{k}{I}$. Use one of the given ordered pairs to find k. $12 = \frac{k}{0.5}$ $12(0.5) = \frac{k}{0.5}(0.5)$ $k = 6$

The equation becomes $R = \frac{6}{I}$.

45. – 47. Answers will vary.

49. z varies directly as the square root of x and inversely as the square root of y.

51. Answers will vary.

53. does not make sense; Explanations will vary. Sample explanation: This would cause division by 0. Division by zero is undefined.

55. makes sense

57. Since wind pressure varies directly as the square of the wind velocity, we have $P = kv^2$. If the wind speed doubles then the value of v has been multiplied by two. In the formula, $P = k(2v)^2 = k(4v^2) = 4kv^2$. Then the wind pressure will be multiplied by a factor of 4. So if the wind speed doubles, the wind pressure is 4 times more destructive.

59. Since the brightness of a source point varies inversely as the square of its distance from an observer, we have $B = \dfrac{k}{d^2}$. We can now see things that are only $\dfrac{1}{50}$ as bright.

$$B = \frac{1}{50} \cdot \frac{k}{d^2} = \frac{k}{50d^2}$$

$$= \frac{k}{(7.07)^2 d^2} = \frac{k}{(7.07d)^2}$$

The distance that can be seen is about 7.07 times farther with the space telescope.

60. $8(2-x) = -5x$

$16 - 8x = -5x$

$16 = 3x$

$\dfrac{16}{3} = x$

The solution set is $\left\{ \dfrac{16}{3} \right\}$.

61.

$$\begin{array}{r} 9x^2 - 6x + 4 \\ 3x+2\overline{\smash{\big)}\,27x^3 + 0x^2 + 0x - 8} \\ \underline{27x^3 + 18x^2} \\ -18x^2 + 0x \\ \underline{-18x^2 - 12x} \\ 12x - 8 \\ \underline{12x + 8} \\ -16 \end{array}$$

$$\frac{27x^3 + 8}{3x+2} = 9x^2 - 6x + 4 - \frac{16}{3x+2}$$

62. $6x^3 - 6x^2 - 120x = 6x\left(x^2 - x - 20\right)$

$$= 6x(x-5)(x+4)$$

63. Set 1 has each x-coordinate paired with only one y-coordinate.

64. $r^3 - 2r^2 + 5 = (-5)^3 - 2(-5)^2 + 5$

$$= -125 - 50 + 5$$

$$= -170$$

65. $5x + 7 = 5(a+h) + 7$

$$= 5a + 5h + 7$$

Chapter 7 Review Exercises

1. $\dfrac{5x}{6x - 24}$

Set the denominator equal to 0 and solve for x.

$6x - 24 = 0$

$6x = 24$

$x = 4$

The rational expression is undefined for $x = 4$.

2. $\dfrac{x+3}{(x-2)(x+5)}$

Set the denominator equal to 0 and solve for x.

$(x-2)(x+5) = 0$

$x - 2 = 0 \quad \text{or} \quad x + 5 = 0$

$\quad x = 2 \qquad\qquad x = -5$

The rational expression is undefined for $x = 2$ and $x = -5$.

3. $\dfrac{x^2 + 3}{x^2 - 3x + 2}$

$x^2 - 3x + 2 = 0$

$(x-1)(x-2) = 0$

$x - 1 = 0 \quad \text{or} \quad x - 2 = 0$

$\quad x = 1 \qquad\qquad x = 2$

The rational expression is undefined for $x = 1$ and $x = 2$.

4. $\dfrac{7}{x^2 + 81}$

The smallest possible value of x^2 is 0, so $x^2 + 81 \geq 81$ for all real numbers x. This means that there is no real number for x for which $x^2 + 81 = 0$. Thus, the rational expression is defined for all real numbers.

5. $\dfrac{16x^2}{12x} = \dfrac{4 \cdot 4 \cdot x \cdot x}{4 \cdot 3 \cdot x} = \dfrac{4x}{3}$

6. $\dfrac{x^2 - 4}{x - 2} = \dfrac{(x+2)(x-2)}{(x-2)} = x + 2$

7. $\dfrac{x^3 + 2x^2}{x+2} = \dfrac{x^2(x+2)}{(x+2)} = x^2$

8. $\dfrac{x^2 + 3x - 18}{x^2 - 36} = \dfrac{(x+6)(x-3)}{(x+6)(x-6)}$

$\quad = \dfrac{x-3}{x-6}$

9. $\dfrac{x^2 - 4x - 5}{x^2 + 8x + 7} = \dfrac{(x+1)(x-5)}{(x+1)(x+7)}$

$\quad = \dfrac{x-5}{x+7}$

10. $\dfrac{y^2 + 2y}{y^2 + 4y + 4} = \dfrac{y(y+2)}{(y+2)(y+2)}$

$\quad = \dfrac{y}{y+2}$

11. $\dfrac{x^2}{x^2 + 4}$

The numerator and denominator have no common factor, so this rational expression cannot be simplified.

12. $\dfrac{2x^2 - 18y^2}{3y - x} = \dfrac{2(x^2 - 9y^2)}{3y - x}$

$\quad = \dfrac{2(x+3y)(x-3y)}{(3y-x)}$

$\quad = \dfrac{2(x+3y)(-1)(3y-x)}{(3y-x)}$

$\quad = -2(x+3y) \text{ or } -2x - 6y$

13. $\dfrac{x^2 - 4}{12x} \cdot \dfrac{3x}{x+2} = \dfrac{(x+2)(x-2)}{12x} \cdot \dfrac{3x}{(x+2)}$

$\quad = \dfrac{x-2}{4}$

14. $\dfrac{5x+5}{6} \cdot \dfrac{3x}{x^2 + x} = \dfrac{5(x+1)}{6} \cdot \dfrac{3x}{x(x+1)}$

$\quad = \dfrac{5}{2}$

15. $\dfrac{x^2 + 6x + 9}{x^2 - 4} \cdot \dfrac{x-2}{x+3}$

$\quad = \dfrac{(x+3)(x+3)}{(x+2)(x-2)} \cdot \dfrac{x-2}{x+3}$

$\quad = \dfrac{x+3}{x+2}$

16. $\dfrac{y^2 - 2y + 1}{y^2 - 1} \cdot \dfrac{2y^2 + y - 1}{5y - 5}$

$\quad = \dfrac{(y-1)(y-1)}{(y+1)(y-1)} \cdot \dfrac{(2y-1)(y+1)}{5(y-1)}$

$\quad = \dfrac{2y-1}{5}$

17. $\dfrac{2y^2 + y - 3}{4y^2 - 9} \cdot \dfrac{3y+3}{5y - 5y^2}$

$\quad = \dfrac{(2y+3)(y-1)}{(2y+3)(2y-3)} \cdot \dfrac{3(y+1)}{5y(1-y)}$

$\quad = \dfrac{-3(y+1)}{5y(2y-3)} \text{ or } -\dfrac{3(y+1)}{5y(2y-3)}$

18. $\dfrac{x^2 + x - 2}{10} \div \dfrac{2x+4}{5}$

$\quad = \dfrac{x^2 + x - 2}{10} \cdot \dfrac{5}{2x+4}$

$\quad = \dfrac{(x-1)(x+2)}{10} \cdot \dfrac{5}{2(x+2)}$

$\quad = \dfrac{x-1}{4}$

19. $\dfrac{6x+2}{x^2 - 1} \div \dfrac{3x^2 + x}{x-1}$

$\quad = \dfrac{6x+2}{x^2 - 1} \cdot \dfrac{x-1}{3x^2 + x}$

$\quad = \dfrac{2(3x+1)}{(x+1)(x-1)} \cdot \dfrac{(x-1)}{x(3x+1)}$

$\quad = \dfrac{2}{x(x+1)}$

20. $\dfrac{1}{y^2+8y+15} \div \dfrac{7}{y+5}$

$= \dfrac{1}{y^2+8y+15} \cdot \dfrac{y+5}{7}$

$= \dfrac{1}{(y+3)(y+5)} \cdot \dfrac{(y+5)}{7}$

$= \dfrac{1}{7(y+3)}$

21. $\dfrac{y^2+y-42}{y-3} \div \dfrac{y+7}{(y-3)^2}$

$= \dfrac{y^2+y-42}{y-3} \cdot \dfrac{(y-3)^2}{y+7}$

$= \dfrac{(y+7)(y-6)}{(y-3)} \cdot \dfrac{(y-3)(y-3)}{y+7}$

$= (y-6)(y-3)$ or $y^2-9y+18$

22. $\dfrac{8x+8y}{x^2} \div \dfrac{x^2-y^2}{x^2}$

$= \dfrac{8x+8y}{x^2} \cdot \dfrac{x^2}{x^2-y^2}$

$= \dfrac{8(x+y)}{x^2} \cdot \dfrac{x^2}{(x+y)(x-y)}$

$= \dfrac{8}{x-y}$

23. $\dfrac{4x}{x+5} + \dfrac{20}{x+5} = \dfrac{4x+20}{x+5} = \dfrac{4(x+5)}{x+5} = 4$

24. $\dfrac{8x-5}{3x-1} + \dfrac{4x+1}{3x-1} = \dfrac{8x-5+4x+1}{3x-1}$

$= \dfrac{12x-4}{3x-1}$

$= \dfrac{4(3x-1)}{3x-1} = 4$

25. $\dfrac{3x^2+2x}{x-1} - \dfrac{10x-5}{x-1}$

$= \dfrac{\left(3x^2+2x\right)-(10x-5)}{x-1}$

$= \dfrac{3x^2+2x-10x+5}{x-1}$

$= \dfrac{3x^2-8x+5}{x-1}$

$= \dfrac{(3x-5)(x-1)}{x-1}$

$= 3x-5$

26. $\dfrac{6y^2-4y}{2y-3} - \dfrac{12-3y}{2y-3}$

$= \dfrac{\left(6y^2-4y\right)-(12-3y)}{2y-3}$

$= \dfrac{6y^2-4y-12+3y}{2y-3}$

$= \dfrac{6y^2-y-12}{2y-3}$

$= \dfrac{(2y-3)(3y+4)}{2y-3}$

$= 3y+4$

27. $\dfrac{x}{x-2} + \dfrac{x-4}{2-x} = \dfrac{x}{x-2} + \dfrac{(-1)}{(-1)} \cdot \dfrac{x-4}{x-2}$

$= \dfrac{x}{x-2} + \dfrac{-x+4}{x-2}$

$= \dfrac{x-x+4}{x-2} = \dfrac{4}{x-2}$

28. $\dfrac{x+5}{x-3} - \dfrac{x}{3-x} = \dfrac{x+5}{x-3} - \dfrac{(-1)}{(-1)} \cdot \dfrac{x}{3-x}$

$= \dfrac{x+5}{x-3} + \dfrac{x}{x-3}$

$= \dfrac{x+5+x}{x-3} = \dfrac{2x+5}{x-5}$

29. $\dfrac{7}{9x^3}$ and $\dfrac{5}{12x}$

$9x^3 = 3^2 x^3$

$12x = 2^2 \cdot 3x$

$\text{LCD} = 2^2 \cdot 3^2 \cdot x^3 = 36x^3$

30. $\dfrac{3}{x^2(x-1)}$ and $\dfrac{11}{x(x-1)^2}$

LCD $= x^2(x-1)^2$

31. $\dfrac{x}{x^2+4x+3}$ and $\dfrac{17}{x^2+10x+21}$

$x^2+4x+3 = (x+3)(x+1)$

$x^2+10x+21 = (x+3)(x+7)$

LCD $= (x+3)(x+1)(x+7)$

32. $\dfrac{7}{3x}+\dfrac{5}{2x^2}$

LCD $= 6x^2$

$\dfrac{7}{3x}+\dfrac{6}{2x^2} = \dfrac{7}{3x}\cdot\dfrac{2x}{2x}+\dfrac{5}{2x^2}\cdot\dfrac{3}{3}$

$\qquad = \dfrac{14x+15}{6x^2}$

33. $\dfrac{5}{x+1}+\dfrac{2}{x}$

LCD $= x(x+1)$

$\dfrac{5}{x+1}+\dfrac{2}{x} = \dfrac{5x}{x(x+1)}+\dfrac{2(x+1)}{x(x+1)}$

$\qquad = \dfrac{5x+2(x+1)}{x(x+1)} = \dfrac{5x+2x+2}{x(x+1)}$

$\qquad = \dfrac{7x+2}{x(x+1)}$

34. $\dfrac{7}{x+3}+\dfrac{4}{(x+3)^2}$

LCD $= (x+3)^2$ or $(x+3)(x+3)$

$\dfrac{7}{x+3}+\dfrac{4}{(x+3)^2}$

$= \dfrac{7}{x+3}+\dfrac{4}{(x+3)(x+3)}$

$= \dfrac{7(x+3)}{(x+3)(x+3)}+\dfrac{4}{(x+3)(x+3)}$

$= \dfrac{7(x+3)+4}{(x+3)(x+3)} = \dfrac{7x+21+4}{(x+3)(x+3)}$

$= \dfrac{7x+25}{(x+3)(x+3)}$ or $\dfrac{7x+25}{(x+3)^2}$

35. $\dfrac{6y}{y^2-4}-\dfrac{3}{y+2}$

$y^2-4 = (y+2)(y-2)$

$y+2 = 1(y+2)$

LCD $= (y+2)(y-2)$

$\dfrac{6y}{y^2-4}-\dfrac{3}{y+2}$

$= \dfrac{6y}{(y+2)(y-2)}-\dfrac{3}{y+2}$

$= \dfrac{6y}{(y+2)(y-2)}-\dfrac{3(y-2)}{(y+2)(y-2)}$

$= \dfrac{6y-3(y-2)}{(y+2)(y-2)} = \dfrac{6y-3y+6}{(y+2)(y-2)}$

$= \dfrac{3y+6}{(y+2)(y-2)} = \dfrac{3(y+2)}{(y+2)(y-2)}$

$= \dfrac{3}{y-2}$

36. $\dfrac{y-1}{y^2-2y+1}-\dfrac{y+1}{y-1}$

$= \dfrac{y-1}{(y-1)(y-1)}-\dfrac{y+1}{y-1}$

$= \dfrac{1}{y-1}-\dfrac{y+1}{y-1}$

$= \dfrac{1-(y+1)}{y-1} = \dfrac{1-y-1}{y-1}$

$= \dfrac{-y}{y-1}$ or $-\dfrac{y}{y-1}$

37. $\dfrac{x+y}{y} - \dfrac{y-x}{x}$

LCD = xy

$$\dfrac{x+y}{y} - \dfrac{y-x}{x} = \dfrac{(x+y)}{y} \cdot \dfrac{x}{x} - \dfrac{(x-y)}{x} \cdot \dfrac{y}{y}$$

$$= \dfrac{x^2 + xy}{xy} - \dfrac{xy - y^2}{xy}$$

$$= \dfrac{\left(x^2 + xy\right) - \left(xy - y^2\right)}{xy}$$

$$= \dfrac{x^2 + xy - xy + y^2}{xy}$$

$$= \dfrac{x^2 + y^2}{xy}$$

38. $\dfrac{2x}{x^2 + 2x + 1} + \dfrac{x}{x^2 - 1}$

$x^2 + 2x + 1 = (x+1)(x+1)$

$x^2 - 1 = (x+1)(x-1)$

LCD = $(x+1)(x+1)(x-1)$

$$\dfrac{2x}{x^2 + 2x + 1} + \dfrac{x}{x^2 - 1}$$

$$= \dfrac{2x}{(x+1)(x+1)} + \dfrac{x}{(x+1)(x-1)}$$

$$= \dfrac{2x(x-1)}{(x+1)(x+1)(x-1)} + \dfrac{x(x+1)}{(x+1)(x-1)(x+1)}$$

$$= \dfrac{2x(x-1) + x(x+1)}{(x+1)(x+1)(x-1)}$$

$$= \dfrac{2x^2 - 2x + x^2 + x}{(x+1)(x+1)(x-1)}$$

$$= \dfrac{3x^2 - x}{(x+1)(x+1)(x-1)}$$

39. $\dfrac{5x}{x+1} - \dfrac{2x}{1-x^2}$

$x + 1 = 1(x+1)$

$1 - x^2 = -1\left(x^2 - 1\right) = -(x+1)(x-1)$

LCD = $(x+1)(x-1)$

$$\dfrac{5x}{x+1} - \dfrac{2x}{1-x^2}$$

$$= \dfrac{5x}{x+1} - \dfrac{(-1)}{(-1)} \cdot \dfrac{2x}{1-x^2} = \dfrac{5x}{x+1} - \dfrac{-2x}{x^2 - 1}$$

$$= \dfrac{5x(x-1)}{(x+1)(x-1)} - \dfrac{-2x}{(x+1)(x-1)}$$

$$= \dfrac{5x(x-1) + 2x}{(x+1)(x-1)} = \dfrac{5x^2 - 5x + 2x}{(x+1)(x-1)}$$

$$= \dfrac{5x^2 - 3x}{(x+1)(x-1)}$$

40. $\dfrac{4}{x^2 - x - 6} - \dfrac{4}{x^2 - 4}$

$x^2 - x - 6 = (x+2)(x-3)$

$x^2 - 4 = (x+2)(x-2)$

LCD = $(x+2)(x-3)(x-2)$

$$\dfrac{4}{x^2 - x - 6} - \dfrac{4}{x^2 - 4}$$

$$= \dfrac{4}{(x+2)(x-3)} - \dfrac{4}{(x+2)(x-2)}$$

$$= \dfrac{4(x-2)}{(x+2)(x-3)(x-2)} - \dfrac{4(x-3)}{(x+2)(x-3)(x-2)}$$

$$= \dfrac{4(x-2) - 4(x-3)}{(x+2)(x-3)(x-2)}$$

$$= \dfrac{4x - 8 - 4x + 12}{(x+2)(x-3)(x-2)}$$

$$= \dfrac{4}{(x+2)(x-3)(x-2)}$$

41. $\dfrac{7}{x+3}+2$

LCD $= x+3$

$\dfrac{7}{x+3}+2 = \dfrac{7}{x+3}+\dfrac{2(x+3)}{x+3}$

$\qquad = \dfrac{7+2(x+3)}{x+3}$

$\qquad = \dfrac{7+2x+6}{x+3}$

$\qquad = \dfrac{2x+13}{x+3}$

42. $\dfrac{2y-5}{6y+9}-\dfrac{4}{2y^2+3y}$

$6y+9 = 3(2y+3)$

$2y^2+3y = y(2y+3)$

LCD $= 3y(2y+3)$

$\dfrac{2y-5}{6y+9}-\dfrac{4}{2y^2+3y}$

$= \dfrac{2y-5}{3(2y+3)}-\dfrac{4}{y(2y+3)}$

$= \dfrac{(2y-5)(y)}{3(2y+3)(y)}-\dfrac{4(3)}{y(2y+3)(3)}$

$= \dfrac{2y^2-5y-12}{3y(2y+3)} = \dfrac{(2y+3)(y-4)}{3y(2y+3)}$

$= \dfrac{y-4}{3y}$

43. $\dfrac{\frac{1}{2}+\frac{3}{8}}{\frac{3}{4}-\frac{1}{2}} = \dfrac{\frac{4}{8}+\frac{3}{8}}{\frac{3}{4}-\frac{2}{4}} = \dfrac{\frac{7}{8}}{\frac{1}{4}} = \dfrac{7}{8}\cdot\dfrac{4}{1} = \dfrac{7}{2}$

44. $\dfrac{\frac{1}{x}}{1-\frac{1}{x}}$

LCD $= x$

$\dfrac{\frac{1}{x}}{1-\frac{1}{x}} = \dfrac{x}{x}\cdot\dfrac{\left(\frac{1}{x}\right)}{\left(1-\frac{1}{x}\right)}$

$= \dfrac{x\cdot\frac{1}{x}}{x\cdot 1-x\cdot\frac{1}{x}} = \dfrac{1}{x-1}$

45. $\dfrac{\frac{1}{x}+\frac{1}{y}}{\frac{1}{xy}}$

LCD $= xy$

$\dfrac{\frac{1}{x}+\frac{1}{y}}{\frac{1}{xy}} = \dfrac{xy}{xy}\cdot\dfrac{\left(\frac{1}{x}+\frac{1}{y}\right)}{\left(\frac{1}{xy}\right)}$

$= \dfrac{xy\cdot\frac{1}{x}+xy\cdot\frac{1}{y}}{xy\cdot\frac{1}{xy}}$

$= \dfrac{y+x}{1} = y+x \text{ or } x+y$

46. $\dfrac{\frac{1}{x}-\frac{1}{2}}{\frac{1}{3}-\frac{x}{6}} = \dfrac{\frac{2}{2x}-\frac{x}{2x}}{\frac{2}{6}-\frac{x}{6}} = \dfrac{\frac{2-x}{2x}}{\frac{2-x}{6}}$

$= \dfrac{2-x}{2x}\cdot\dfrac{6}{2-x} = \dfrac{3}{x}$

47. $\dfrac{3+\frac{12}{x}}{1-\frac{16}{x^2}}$

LCD $= x^2$

$\dfrac{3+\frac{12}{x}}{1-\frac{16}{x^2}} = \dfrac{x^2}{x^2}\cdot\dfrac{\left(3+\frac{12}{x}\right)}{\left(1-\frac{16}{x^2}\right)}$

$= \dfrac{x^2\cdot 3+x^2\cdot\frac{12}{x}}{x^2\cdot 1-x^2\cdot\frac{16}{x^2}} = \dfrac{3x^2+12x}{x^2-16}$

$= \dfrac{3x(x+4)}{(x+4)(x-4)} = \dfrac{3x}{x-4}$

48. $\dfrac{3}{x} - \dfrac{1}{6} = \dfrac{1}{x}$

The restriction is $x \neq 0$.

The LCD is $6x$.

$$\dfrac{3}{x} - \dfrac{1}{6} = \dfrac{1}{x}$$

$$6x\left(\dfrac{3}{x} - \dfrac{1}{6}\right) = 6x\left(\dfrac{1}{x}\right)$$

$$6x \cdot \dfrac{3}{x} - 6x \cdot \dfrac{1}{6} = 6x \cdot \dfrac{1}{x}$$

$$18 - x = 6$$

$$-x = -12$$

$$x = 12$$

The solution set is $\{12\}$.

49. $\dfrac{3}{4x} = \dfrac{1}{x} + \dfrac{1}{4}$

The restriction is $x \neq 0$.

The LCD is $4x$.

$$\dfrac{3}{4x} = \dfrac{1}{x} + \dfrac{1}{4}$$

$$4x\left(\dfrac{3}{4x}\right) = 4x\left(\dfrac{1}{x} + \dfrac{1}{4}\right)$$

$$3 = 4 + x$$

$$-1 = x$$

The solution set is $\{-1\}$.

50. $x + 5 = \dfrac{6}{x}$

The restriction is $x \neq 0$.

The LCD is x.

$$x + 5 = \dfrac{6}{x}$$

$$x(x + 5) = x\left(\dfrac{6}{x}\right)$$

$$x^2 + 5x = 6$$

$$x^2 + 5x - 6 = 0$$

$$(x + 6)(x - 1) = 0$$

$$x + 6 = 0 \quad \text{or} \quad x - 1 = 0$$

$$x = -6 \qquad x = 1$$

The solution set is $\{-6, 1\}$.

51. $4 - \dfrac{x}{x+5} = \dfrac{5}{x+5}$

The restriction is $x \neq -5$.

The LCD is $x + 5$.

$$(x+5)\left(4 - \frac{x}{x+5}\right) = (x+5)\left(\frac{5}{x+5}\right)$$

$$(x+5)\cdot 4 - (x+5)\left(\frac{x}{x+5}\right) = (x+5)\left(\frac{5}{x+5}\right)$$

$$4x + 20 - x = 5$$

$$3x + 20 = 5$$

$$3x = -15$$

$$x = -5$$

The only proposed solution, -5, is *not* a solution because of the restriction $x \neq -5$. Notice that -5 makes two of the denominators zero in the original equation. The solution set is $\{\ \}$.

52. $\dfrac{2}{x-3} = \dfrac{4}{x+3} + \dfrac{8}{x^2 - 9}$

To find any restrictions and the LCD, all denominators should be written in factored form.

$$\frac{2}{x-3} = \frac{4}{x+3} + \frac{8}{(x+3)(x-3)}$$

Restrictions: $x \neq 3, x \neq -3$

LCD $= (x+3)(x-3)$

$$(x+3)(x-3)\cdot\frac{2}{x-3} = (x+3)(x-3)\left(\frac{4}{x+3} + \frac{8}{(x+3)(x-3)}\right)$$

$$2(x+3) = 4(x-3) + 8$$

$$2x + 6 = 4x - 12 + 8$$

$$2x + 6 = 4x - 4$$

$$6 = 2x - 4$$

$$10 = 2x$$

$$5 = x$$

The solution set is $\{5\}$.

53. $\dfrac{2}{x} = \dfrac{2}{3} + \dfrac{x}{6}$

Restriction: $x \neq 0$

LCD $= 6x$

$$6x\left(\frac{2}{x}\right) = 6x\left(\frac{2}{3} + \frac{x}{6}\right)$$

$$12 = 4x + x^2$$

$$0 = x^2 + 4x - 12$$

$$0 = (x+6)(x-2)$$

$$x + 6 = 0 \quad \text{or} \quad x - 2 = 0$$

$$x = -6 \qquad\quad x = 2$$

The solution set is $\{-6, 2\}$.

54. $\dfrac{13}{y-1} - 3 = \dfrac{1}{y-1}$

Restriction: $y \neq 1$

LCD $= y-1$

$(y-1)\left(\dfrac{13}{y-1} - 3\right) = (y-1)\left(\dfrac{1}{y-1}\right)$

$13 - 3(y-1) = 1$

$13 - 3y + 3 = 1$

$16 - 3y = 1$

$-3y = -15$

$y = 5$

The solution set is $\{5\}$.

55. $\dfrac{1}{x+3} - \dfrac{1}{x-1} = \dfrac{x+1}{x^2 + 2x - 3}$

$\dfrac{1}{x+3} - \dfrac{1}{x-1} = \dfrac{x+1}{(x+3)(x-1)}$

Restrictions: $x \neq -3, x \neq 1$

LCD $= (x+3)(x-1)$

$(x+3)(x-1)\left[\dfrac{1}{x+3} - \dfrac{1}{x-1}\right] = (x+3)(x-1) \cdot \left[\dfrac{x+1}{(x+3)(x-1)}\right]$

$(x-1) - (x+3) = x+1$

$x - 1 - x - 3 = x + 1$

$-4 = x + 1$

$-5 = x$

The solution set is $\{-5\}$.

56. $P = \dfrac{250(3t+5)}{t+25}$

$125 = \dfrac{250(3t+5)}{t+25}$

$125(t+25) = \dfrac{250(3t+5)}{t+25} \cdot (t+25)$

$125t + 3125 = 250(3t+5)$

$125t + 3125 = 750t + 1250$

$3125 = 625t + 1250$

$1875 = 625t$

$3 = t$

It will take 3 years for the population to reach 125 elk.

57.
$$S = \frac{C}{1-r}$$
$$200 = \frac{140}{1-r}$$
$$200(1-r) = \frac{140}{1-r} \cdot 1 - r$$
$$200 - 200r = 140$$
$$-200r = -60$$
$$r = \frac{-60}{-200} = \frac{3}{10} = 30\%$$

The markup is 30%.

58.
$$P = \frac{R-C}{n}$$
$$n(P) = n\left(\frac{R-C}{n}\right)$$
$$nP = R - C$$
$$nP + C = R$$
$$C = R - nP$$

59.
$$\frac{P_1 V_1}{T_1} = \frac{P_2 V_2}{T_2}$$
$$T_1 T_2 \left(\frac{P_1 V_1}{T_1}\right) = T_1 T_2 \left(\frac{P_2 V_2}{T_2}\right)$$
$$P_1 T_2 V_1 = P_2 T_1 V_2$$

$$\frac{P_1 T_2 V_1}{P_2 V_2} = \frac{P_2 T_1 V_2}{P_2 V_2}$$
$$T_1 = \frac{P_1 T_2 V_1}{P_2 V_2}$$

60.
$$T = \frac{A-P}{Pr}$$
$$Pr(T) = Pr\left(\frac{A-P}{Pr}\right)$$
$$PrT = A - P$$
$$PrT + P = A$$
$$P(rT+1) = A$$
$$\frac{P(rT+1)}{rT+1} = \frac{A}{rT+1}$$
$$P = \frac{A}{rT+1}$$

61. $\dfrac{1}{R} = \dfrac{1}{R_1} + \dfrac{1}{R_2}$

$$RR_1R_2\left(\dfrac{1}{R}\right) = RR_1R_2\left(\dfrac{1}{R_1} + \dfrac{1}{R_2}\right)$$

$$R_1R_2 = RR_1R_2\left(\dfrac{1}{R_1}\right) + RR_1R_2\left(\dfrac{1}{R_2}\right)$$

$$R_1R_2 = RR_2 + RR_1$$

$$R_1R_2 = R\left(R_2 + R_1\right)$$

$$\dfrac{R_1R_2}{R_2 + R_1} = \dfrac{R\left(R_2 + R_1\right)}{R_2 + R_1}$$

$$R = \dfrac{R_1R_2}{R_2 + R_1}$$

62. $I = \dfrac{nE}{R + nr}$

$$\left(R + nr\right)\left(I\right) = \left(R + nr\right)\left(\dfrac{nE}{R + nr}\right)$$

$$IR + Inr = nE$$

$$IR = nE - Inr$$

$$IR = n\left(E - Ir\right)$$

$$\dfrac{IR}{E - Ir} = \dfrac{n\left(E - Ir\right)}{E - Ir}$$

$$n = \dfrac{IR}{E - Ir}$$

63. Let x = the rate of the current. Then $20 + x$ = the rate of the boat with the current and $20 - x$ = the rate of the boat against the current.

	Distance	Rate	Time = $\dfrac{\text{Distance}}{\text{Rate}}$
Downstream	72	$20 + x$	$\dfrac{72}{20 + x}$
Upstream	48	$20 - x$	$\dfrac{48}{20 - x}$

The times are equal, so $\dfrac{72}{20 + x} = \dfrac{48}{20 - x}$

$$72\left(20 - x\right) = 48\left(20 + x\right)$$

$$1440 - 72x = 960 + 48x$$

$$1440 = 960 + 120x$$

$$480 = 120x$$

$$4 = x$$

The rate of the current is 4 miles per hour

64. Let x = the rate of the slower car. Then $x+10$ = the rate of the faster car.

	Distance	Rate	Time $= \dfrac{\text{Distance}}{\text{Rate}}$
Slow car	60	x	$\dfrac{60}{x}$
Faster Car	90	$x+10$	$\dfrac{90}{x+10}$

$$\frac{60}{x} = \frac{90}{x+10}$$
$$60(x+10) = 90x$$
$$60x + 600 = 90x$$
$$600 = 30x$$
$$20 = x$$

If $x = 20$, $x + 10 = 30$.

The rate of the slower car is 20 miles per hour and the rate of the faster car is 30 miles per hour.

65. Let x = the time, in hours, for both people to paint the fence together.

	Fractional part of job completed in 1 hour	Time working together	Fractional part of job completed in x hours
Painter	$\dfrac{1}{6}$	x	$\dfrac{x}{6}$
Apprentice	$\dfrac{1}{12}$	x	$\dfrac{x}{12}$

Working together, the two people complete one whole job, so $\dfrac{x}{6} + \dfrac{x}{12} = 1$.

Multiply both sides by the LCD, 12.

$$12\left(\frac{x}{6} + \frac{x}{12}\right) = 12 \cdot 1$$
$$2x + x = 12$$
$$3x = 12$$
$$x = 4$$

It would take them 4 hours to paint the fence working together.

66. Let x = number of teachers needed for 5400 students.

$$\frac{3}{50} = \frac{x}{5400}$$
$$50x = 3 \cdot 5400$$
$$50x = 16,200$$
$$\frac{50x}{50} = \frac{16,200}{50}$$
$$x = 324$$

For an enrollment of 5400 students, 324 teachers are needed.

67. Let x = number of trout in the lake.

$$\frac{\text{Original Number Tagged Deer}}{\text{Total Number of Deer}} = \frac{\text{Number Tagged Deer in Sample}}{\text{Total Number Deer in Sample}}$$

$$\frac{112}{x} = \frac{32}{82}$$

$$32x = 112 \cdot 82$$

$$32x = 9184$$

$$\frac{32x}{32} = \frac{9184}{32}$$

$$x = 287$$

There are 287 trout in the lake.

68. $\dfrac{8}{4} = \dfrac{10}{x}$

$8x = 40$

$x = 5$

The length of the side marked with an x is 5 feet.

69. Write a proportion relating the corresponding sides of the large and small triangle. Notice that the length of the base of the larger triangle is 9 ft + 6ft = 15 ft.

$$\frac{x}{5} = \frac{15}{6}$$

$$6x = 5 \cdot 15$$

$$6x = 75$$

$$x = \frac{75}{6} = 12.5$$

The height of the lamppost is 12.5 feet.

70. Since the profit varies directly as the number of products sold, we have $p = kn$. Use the given values to find k.

$$p = kn.$$

$$1175 = k(25)$$

$$\frac{1175}{25} = \frac{k(25)}{25}$$

$$47 = k$$

The equation becomes $p = 47n$

When $n = 105$ products, $p = 47(105) = 4935$.

If 105 products are sold, the company's profit is $4935.

71. Since distance varies directly as the square of the time, we have $d = kt^2$.

Use the given values to find k.

$$d = kt^2$$
$$144 = k(3)^2$$
$$144 = k(9)$$
$$\frac{144}{9} = \frac{k(9)}{9}$$
$$16 = k$$

The equation becomes $d = 16t^2$. When $t = 10$,
$$d = 16(10)^2 = 16(100) = 1600.$$
A skydiver will fall 1600 feet in 10 seconds.

72. Since the pitch of a musical tone varies inversely as its wavelength, we have $p = \dfrac{k}{w}$.

Use the given values to find k.

$$p = \frac{k}{w}$$
$$660 = \frac{k}{1.6}$$
$$660(1.6) = 1.6\left(\frac{k}{1.6}\right)$$
$$1056 = k$$

The equation becomes $p = \dfrac{1056}{w}$.

When $w = 2.4$, $p = \dfrac{1056}{2.4} = 440$.

The tone's pitch is 440 vibrations per second.

73. Since loudness varies inversely as the square of the distance, we have $l = \dfrac{k}{d^2}$.

Use the given values to find k.

$$l = \frac{k}{d^2}$$
$$28 = \frac{k}{8^2}$$
$$28 = \frac{k}{64}$$
$$64(28) = 64\left(\frac{k}{64}\right)$$
$$1792 = k$$

The equation becomes $l = \dfrac{1792}{d^2}$.

When $d = 4$, $l = \dfrac{1792}{(4)^2} = \dfrac{1792}{16} = 112$.

At a distance of 4 feet, the loudness of the stereo is 112 decibels.

74. Since time varies directly as the number of computers and inversely as the number of workers, we have $t = \dfrac{kn}{w}$.

Use the given values to find k.

$$t = \frac{kn}{w}$$
$$10 = \frac{k(30)}{6}$$
$$10 = 5k$$
$$\frac{10}{5} = \frac{5k}{5}$$
$$2 = k$$

The equation becomes $t = \dfrac{2n}{w}$.

When $n = 40$ and $w = 5$, $t = \dfrac{2(40)}{5} = \dfrac{80}{5} = 16$.

It will take 16 hours for 5 workers to assemble 40 computers.

75. Since the volume varies jointly as height and the area of the base, we have $v = kha$.

Use the given values to find k.

$$175 = k(15)(35)$$

$$175 = k(525)$$

$$\frac{175}{525} = \frac{k(525)}{525}$$

$$\frac{1}{3} = k$$

The equation becomes $v = \frac{1}{3}ha$. When $h = 20$ feet

and $a = 120$ square feet, $v = \frac{1}{3}(20)(120) = 800$.

If the height is 20 feet and the area is 120 square feet, the volume will be 800 cubic feet.

Chapter 7 Test

1. $\dfrac{x+7}{x^2+5x-36}$

Set the denominator equal to 0 and solve for x.

$$x^2 + 5x - 36 = 0$$

$$(x+9)(x-4) = 0$$

$$x+9 = 0 \quad \text{or} \quad x-4 = 0$$

$$x = -9 \qquad x = 4$$

The rational expression is undefined for $x = -9$ and $x = 4$.

2. $\dfrac{x^2+2x-3}{x^2-3x+2} = \dfrac{(x-1)(x+3)}{(x-1)(x-2)} = \dfrac{x+3}{x-2}$

3. $\dfrac{4x^2-20x}{x^2-4x-5} = \dfrac{4x(x-5)}{(x+1)(x-5)} = \dfrac{4x}{x+1}$

4. $\dfrac{x^2-16}{10} \cdot \dfrac{5}{x+4} = \dfrac{(x+4)(x-4)}{10} \cdot \dfrac{5}{(x+4)}$

$$= \dfrac{x-4}{2}$$

5. $\dfrac{x^2-7x+12}{x^2-4x} \cdot \dfrac{x^2}{x^2-9}$

$$= \dfrac{(x-3)(x-4)}{x(x-4)} \cdot \dfrac{x^2}{(x+3)(x-3)}$$

$$= \dfrac{x}{x+3}$$

6. $\dfrac{2x+8}{x-3} \div \dfrac{x^2+5x+4}{x^2-9}$

$$= \dfrac{2x+8}{x-3} \cdot \dfrac{x^2-9}{x^2+5x+4}$$

$$= \dfrac{2(x+4)}{(x-3)} \cdot \dfrac{(x+3)(x-3)}{(x+4)(x+1)}$$

$$= \dfrac{2(x+3)}{x+1} = \dfrac{2x+6}{x+1}$$

7. $\dfrac{5y+5}{(y-3)^2} \div \dfrac{y^2-1}{y-3}$

$$= \dfrac{5y+5}{(y-3)^2} \cdot \dfrac{y-3}{y^2-1}$$

$$= \dfrac{5(y+1)}{(y-3)(y-3)} \cdot \dfrac{(y-3)}{(y+1)(y-1)}$$

$$= \dfrac{5}{(y-3)(y-1)}$$

8. $\dfrac{2y^2+5}{y+3} + \dfrac{6y-5}{y+3}$

$$= \dfrac{(2y^2+5)+(6y-5)}{y+3}$$

$$= \dfrac{2y^2+5+6y-5}{y+3}$$

$$= \dfrac{2y^2+6y}{y+3}$$

$$= \dfrac{2y(y+3)}{y+3} = 2y$$

9. $\dfrac{y^2-2y+3}{y^2+7y+12}-\dfrac{y^2-4y-5}{y^2+7y+12}$

$=\dfrac{\left(y^2-2y+3\right)-\left(y^2-4y-5\right)}{y^2+7y+12}$

$=\dfrac{y^2-2y+3-y^2+4y+5}{y^2+7y+12}$

$=\dfrac{2y+8}{y^2+7y+12}$

$=\dfrac{2(y+4)}{(y+3)(y+4)}$

$=\dfrac{2}{y+3}$

10. $\dfrac{x}{x+3}+\dfrac{5}{x-3}$

LCD $=(x+3)(x-3)$

$\dfrac{x}{x+3}+\dfrac{5}{x-3}$

$=\dfrac{x(x-3)}{(x+3)(x-3)}+\dfrac{5(x+3)}{(x+3)(x-3)}$

$=\dfrac{x(x-3)+5(x+3)}{(x+3)(x-3)}$

$=\dfrac{x^2-3x+5x+15}{(x+3)(x-3)}$

$=\dfrac{x^2+2x+15}{(x+3)(x-3)}$

11. $\dfrac{2}{x^2-4x+3}+\dfrac{6}{x^2+x-2}$

$x^2-4x+3=(x-1)(x-3)$

$x^2+x-2=(x-1)(x+2)$

LCD $=(x-1)(x-3)(x+2)$

$\dfrac{2}{x^2-4x+3}+\dfrac{6}{x^2+x-2}$

$=\dfrac{2}{(x-1)(x-3)}+\dfrac{6}{(x-1)(x+2)}$

$=\dfrac{2(x+2)}{(x-1)(x-3)(x+2)}$

$\quad+\dfrac{6(x-3)}{(x-1)(x-3)(x+2)}$

$=\dfrac{2(x+2)+6(x-3)}{(x-1)(x-3)(x+2)}$

$=\dfrac{2x+4+6x-18}{(x-1)(x-3)(x+2)}$

$=\dfrac{8x-14}{(x-1)(x-3)(x+2)}$

12. $\dfrac{4}{x-3}+\dfrac{x+5}{3-x}$

$3-x=-1(x-3)$

LCD $=x-3$

$\dfrac{4}{x-3}+\dfrac{x+5}{3-x}$

$=\dfrac{4}{x-3}+\dfrac{(-1)}{(-1)}\cdot\dfrac{(x+5)}{(3-x)}$

$=\dfrac{4}{x-3}+\dfrac{-x-5}{x-3}$

$=\dfrac{4-x-5}{x-3}=\dfrac{-x-1}{x-3}$

13. $1+\dfrac{3}{x-1}$

LCD $=x-1$

$1+\dfrac{3}{x-1}=\dfrac{1(x-1)}{x-1}+\dfrac{3}{x-1}$

$=\dfrac{x-1+3}{x-1}=\dfrac{x+2}{x-1}$

14. $\dfrac{2x+3}{x^2-7x+12}-\dfrac{2}{x-3}$

$x^2-7x+12=(x-3)(x-4)$

$x-3=1(x-3)$

$\text{LCD}=(x-3)(x-4)$

$\dfrac{2x+3}{x^2-7x+12}-\dfrac{2}{x-3}$

$=\dfrac{2x+3}{(x-3)(x-4)}-\dfrac{2(x-4)}{(x-3)(x-4)}$

$=\dfrac{2x+3-2(x-4)}{(x-3)(x-4)}$

$=\dfrac{2x+3-2x+8}{(x-3)(x-4)}$

$=\dfrac{11}{(x-3)(x-4)}$

15. $\dfrac{8y}{y^2-16}-\dfrac{4}{y-4}$

$y^2-16=(y+4)(y-4)$

$y-4=1(y-4)$

$\text{LCD}=(y+4)(y-4)$

$\dfrac{8y}{y^2-16}-\dfrac{4}{y-4}$

$=\dfrac{8y}{(y+4)(y-4)}-\dfrac{4}{y-4}$

$=\dfrac{8y}{(y+4)(y-4)}-\dfrac{4(y+4)}{(y+4)(y-4)}$

$=\dfrac{8y-4(y+4)}{(y+4)(y-4)}$

$=\dfrac{8y-4y-16}{(y+4)(y-4)}$

$=\dfrac{4y-16}{(y+4)(y-4)}$

$=\dfrac{4(y-4)}{(y+4)(y-4)}$

$=\dfrac{4}{y+4}$

16. $\dfrac{(x-y)^2}{x+y}\div\dfrac{x^2-xy}{3x+3y}$

$=\dfrac{(x-y)^2}{x+y}\cdot\dfrac{3x+3y}{x^2-xy}$

$=\dfrac{(x-y)(x-y)}{(x+y)}\cdot\dfrac{3(x+y)}{x(x-y)}$

$=\dfrac{3(x-y)}{x}=\dfrac{3x-3y}{x}$

17. $\dfrac{5+\dfrac{5}{x}}{2+\dfrac{1}{x}}=\dfrac{\dfrac{5x}{x}+\dfrac{5}{x}}{\dfrac{2x}{x}+\dfrac{1}{x}}=\dfrac{\dfrac{5x+5}{x}}{\dfrac{2x+1}{x}}$

$=\dfrac{5x+5}{x}\cdot\dfrac{x}{2x+1}=\dfrac{5x+5}{2x+1}$

18. $\dfrac{\dfrac{1}{x}-\dfrac{1}{y}}{\dfrac{1}{x}}$

$\text{LCD}=xy$

$\dfrac{\dfrac{1}{x}-\dfrac{1}{y}}{\dfrac{1}{x}}=\dfrac{xy}{xy}\cdot\dfrac{\left(\dfrac{1}{x}-\dfrac{1}{y}\right)}{\left(\dfrac{1}{x}\right)}$

$=\dfrac{xy\cdot\dfrac{1}{x}-xy\cdot\dfrac{1}{y}}{xy\cdot\dfrac{1}{x}}=\dfrac{y-x}{y}$

19. $\dfrac{5}{x}+\dfrac{2}{3}=2-\dfrac{2}{x}-\dfrac{1}{6}$

Restriction: $x\neq 0$

$\text{LCD}=6x$

$6x\left(\dfrac{5}{x}+\dfrac{2}{3}\right)=6x\left(2-\dfrac{2}{x}-\dfrac{1}{6}\right)$

$6x\cdot\dfrac{5}{x}+6x\cdot\dfrac{2}{3}=6x\cdot2-6x\cdot\dfrac{2}{x}-6x\cdot\dfrac{1}{6}$

$30+4x=12x-12-x$

$30+4x=11x-12$

$30=7x-12$

$42=7x$

$6=x$

The solution set is $\{6\}$.

20. $\dfrac{3}{y+5} - 1 = \dfrac{4-y}{2y+10}$

$\dfrac{3}{y+5} - 1 = \dfrac{4-y}{2(y+5)}$

Restriction: $y \neq -5$

LCD $= 2(y+5)$

$2(y+5)\left(\dfrac{3}{y+5} - 1\right) = 2(y+5)\left[\dfrac{4-y}{2(y+5)}\right]$

$6 - 2(y+5) = 4 - y$

$6 - 2y - 10 = 4 - y$

$-4 - 2y = 4 - y$

$-4 = 4 + y$

$-8 = y$

The solution set is $\{-8\}$.

21. $\dfrac{2}{x-1} = \dfrac{3}{x^2-1} + 1$

$\dfrac{2}{x-1} = \dfrac{3}{(x+1)(x-1)} + 1$

Restrictions: $x \neq 1, x \neq -1$

LCD $= (x+1)(x-1)$

$(x+1)(x-1)\left(\dfrac{2}{x-1}\right) = (x+1)(x-1)\left[\dfrac{3}{(x+1)(x-1)} + 1\right]$

$2(x+1) = 3 + (x+1)(x-1)$

$2x + 2 = 3 + x^2 - 1$

$2x + 2 = 2 + x^2$

$0 = x^2 - 2x$

$0 = x(x-2)$

$x = 0$ or $x - 2 = 0$

$x = 2$

The solution set is $\{0, 2\}$.

22. $R = \dfrac{as}{a+s}$

$\qquad (a+s)\, R = (a+s)\left(\dfrac{as}{a+s}\right)$

$\qquad aR + Rs = as$

$\qquad\quad aR = as - Rs$

$\qquad aR - as = -Rs$

$\qquad a(R-s) = -Rs$

$\qquad \dfrac{a(R-s)}{R-s} = -\dfrac{Rs}{R-s}$

$\qquad\qquad a = -\dfrac{Rs}{R-s} \ \text{ or } \ \dfrac{Rs}{s-R}$

23. Let $x =$ the rate of the current.

Then $30 + x =$ the rate of the boat with the current and $30 - x =$ the rate of the boat against the current.

	Distance	Rate	Time $= \dfrac{\text{Distance}}{\text{Rate}}$
Downstream	16	$30 + x$	$\dfrac{16}{30+x}$
Upstream	14	$30 - x$	$\dfrac{14}{30-x}$

$\qquad \dfrac{16}{30+x} = \dfrac{14}{30-x}$

$\quad 16(30-x) = 14(30+x)$

$\ \ 480 - 16x = 420 + 14x$

$\qquad\quad 480 = 420 + 30x$

$\qquad\quad\ 60 = 30x$

$\qquad\quad\ \ 2 = x$

The rate of the current is 2 miles per hour.

24. Let $x =$ the time (in minutes) for both pipes to fill the hot tub.

$\qquad \dfrac{x}{20} + \dfrac{x}{30} = 1$

$\qquad \text{LCD} = 60$

$\qquad 60\left(\dfrac{x}{20} + \dfrac{x}{30}\right) = 60 \cdot 1$

$\qquad\quad 3x + 2x = 60$

$\qquad\qquad\ \ 5x = 60$

$\qquad\qquad\ \ \ x = 12$

It will take 12 minutes for both pipes to fill the hot tub.

25. Let x = number of tule elk in the park.

$$\frac{200}{x} = \frac{5}{150}$$

$$5x = 30,000$$

$$x = 6000$$

There are 6000 tule elk in the park.

26. $\dfrac{10}{4} = \dfrac{8}{x}$

$$10x = 8 \cdot 4$$

$$10x = 32$$

$$x = 3.2$$

The length of the side marked with an x is 3.2 inches.

27. Let $\quad C$ = the current (in amperes).

Then R = the resistance (in ohms).

Step 1 $\quad C = \dfrac{k}{R}$

Step 2 To find k, substitute 42 for C and 5 for R.

$$42 = \frac{k}{5}$$

$$42 \cdot 5 = \frac{k}{5} \cdot 5$$

$$210 = k$$

Step 3 $\quad C = \dfrac{210}{R}$

Step 4 Substitute 4 for R and solve for C.

$$C = \frac{210}{4} = 52.5$$

When the resistance is 4 ohms, the current is 52.5 amperes.

Chapter 7 Cumulative Review Exercises (Chapters 1-7)

1. $2(x-3) + 5x = 8(x-1)$

$$2x - 6 + 5x = 8x - 8$$

$$7x - 6 = 8x - 8$$

$$-6 = x - 8$$

$$2 = x$$

The solution set is $\{2\}$.

2. $-3(2x-4) > 2(6x-12)$

$\quad\quad -6x+12 > 12x-24$

$\quad\quad -18x+12 > -24$

$\quad\quad\quad -18x > -36$

$\quad\quad\quad \dfrac{-18x}{-18} < \dfrac{-36}{-18}$

$\quad\quad\quad\quad x < 2$

$(-\infty, 2)$

3. $\quad\quad x^2 + 3x = 18$

$\quad\quad x^2 + 3x - 18 = 0$

$\quad\quad (x+6)(x-3) = 0$

$\quad\quad x+6 = 0 \quad \text{or} \quad x-3 = 0$

$\quad\quad\quad x = -6 \quad\quad\quad x = 3$

The solution set is $\{-6, 3\}$.

4. $\dfrac{2x}{x^2-4} + \dfrac{1}{x-2} = \dfrac{2}{x+2}$

$\quad\quad x^2 - 4 = (x+2)(x-2)$

Restrictions: $x \neq 2, x \neq -2$

LCD $= (x+2)(x-2)$

$(x+2)(x-2)\left[\dfrac{2x}{(x+2)(x-2)} + \dfrac{1}{x-2}\right] = (x+2)(x-2) \cdot \dfrac{2}{x+2}$

$\quad\quad\quad\quad 2x + (x+2) = 2(x-2)$

$\quad\quad\quad\quad\quad 3x + 2 = 2x - 4$

$\quad\quad\quad\quad\quad\quad x = -6$

The solution set is $\{-6\}$.

5. $y = 2x - 3$

$x + 2y = 9$

To solve this system by the substitution method, substitute $2x - 3$ for y in the second equation.

$\quad\quad x + 2y = 9$

$\quad x + 2(2x-3) = 9$

$\quad\quad x + 4x - 6 = 9$

$\quad\quad\quad 5x - 6 = 9$

$\quad\quad\quad\quad 5x = 15$

$\quad\quad\quad\quad x = 3$

Back-substitute 3 for x into the first equation.

$y = 2x - 3$

$y = 2 \cdot 3 - 3 = 3$

The solution set is $\{(3, 3)\}$.

6. $3x + 2y = -2$

$-4x + 5y = 18$

To solve this system by the addition method, multiply the first equation by 4 and the second equation by 3.

Then add the equations.

$12x + 8y = -8$

$\underline{-12x + 15y = 54}$

$23y = 46$

$y = 2$

Back-substitute 2 for y in the first equation of the original system.

$3x + 2y = -2$

$3x + 2(2) = -2$

$3x + 4 = -2$

$3x = -6$

$x = -2$

The solution set is $\{(-2, 2)\}$.

7. $3x - 2y = 6$

x-intercept: 2

y-intercept: -3

checkpoint: $(4, 3)$

Draw a line through $(2, 0)$, $(0, -3)$ and $(4, 3)$.

$3x - 2y = 6$

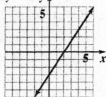

8. Graph $y = -2x + 3$ by using its slope, $-2 = \dfrac{-2}{1}$, and its y-intercept, 3.

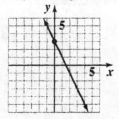

9. $y = -3$

The graph is a horizontal line with y-intercept -3.

$y = -3$

10. $-21 - 16 - 3(2 - 8) = -21 - 16 - 3(-6)$

$= -21 - 16 + 18$

$= -37 + 18 = -19$

11. $\left(\dfrac{4x^5}{2x^2}\right)^3 = \left(2x^3\right)^3 = 2^3 \cdot \left(x^3\right)^3 = 8x^9$

12. $\dfrac{\dfrac{1}{x} - 2}{4 - \dfrac{1}{x}}$

LCD $= x$

$\dfrac{\dfrac{1}{x} - 2}{4 - \dfrac{1}{x}} = \dfrac{x\left(\dfrac{1}{x} - 2\right)}{x\left(4 - \dfrac{1}{x}\right)}$

$= \dfrac{x \cdot \dfrac{1}{x} - x \cdot 2}{x \cdot 4 - x \cdot \dfrac{1}{x}}$

$= \dfrac{1 - 2x}{4x - 1}$

13. $4x^2 - 13x + 3$

Factor by trial and error. Try various combinations until the correct one is found.

$4x^2 - 13x + 3 = (4x - 1)(x - 3)$

14. $4x^2 - 20x + 25 = (2x)^2 - 2(2x \cdot 5) + 5^2$

$= (2x - 5)^2$

15. $3x^2 - 75 = 3(x^2 - 25)$

$= 3(x + 5)(x - 5)$

16. $\left(4x^2 - 3x + 2\right) - \left(5x^2 - 7x - 6\right)$

$= \left(4x^2 - 3x + 2\right) + \left(-5x^2 + 7x + 6\right)$

$= -x^2 + 4x + 8$

17. $\dfrac{-8x^6 + 12x^4 - 4x^2}{4x^2} = \dfrac{-8x^6}{4x^2} + \dfrac{12x^4}{4x^2} - \dfrac{4x^2}{4x^2}$

$\qquad\qquad\qquad = -2x^4 + 3x^2 - 1$

18. $\dfrac{x+6}{x-2} + \dfrac{2x+1}{x+3}$

$\text{LCD} = (x-2)(x+3)$

$\dfrac{x+6}{x-2} + \dfrac{2x+1}{x+3}$

$= \dfrac{(x+6)(x+3)}{(x-2)(x+3)} + \dfrac{(2x+1)(x-2)}{(x-2)(x+3)}$

$= \dfrac{(x+6)(x+3) + (2x+1)(x-2)}{(x-2)(x+3)}$

$= \dfrac{x^2 + 9x + 18 + 2x^2 - 3x - 2}{(x-2)(x+3)}$

$= \dfrac{3x^2 + 6x + 16}{(x-2)(x+3)}$

19. Let x = the amount invested at 5%.
Then $4000 - x$ = the amount invested at 9%.

$0.05x + 0.09(4000 - x) = 311$

$0.05x + 360 - 0.09x = 311$

$-0.04x + 360 = 311$

$-0.04x = -49$

$x = \dfrac{-49}{-0.04}$

$x = 1225$

If $x = 1225$, then $4000 - x = 2775$.
$1225 was invested at 5% and $2775 at 9%.

20. Let x = the length of the shorter piece.
Then $3x$ = the length of the larger piece.

$x + 3x = 68$

$4x = 68$

$x = 17$

If $x = 17$, then $3x = 51$.
The lengths of the pieces are 17 inches and 51 inches.

1. $2-4(x+2)=5-3(2x+1)$

$2-4x-8=5-6x-3$

$-4x-6=-6x+2$

$-4x-6+6x=-6x+2+6x$

$2x-6=2$

$2x-6+6=2+6$

$2x=8$

$\dfrac{2x}{2}=\dfrac{8}{2}$

$x=4$

The solution set it $\{4\}$.

2. $\dfrac{x}{2}-3=\dfrac{x}{5}$

Multiply both sides by the least common denominator of the fractions, 10:

$10\left(\dfrac{x}{2}-3\right)=10\left(\dfrac{x}{5}\right)$

$5x-30=2x$

$3x-30=0$

$3x=30$

$x=10$

The solution set it $\{10\}$.

3. $3x+9\geq 5(x-1)$

$3x+9\geq 5x-5$

$3x+9-5x\geq 5x-5-5x$

$-2x+9\geq -5$

$-2x+9-9\geq -5-9$

$-2x\geq -14$

$\dfrac{-2x}{-2}\leq \dfrac{-14}{-2}$

$x\leq 7$

The solution set is $\{x\mid x\leq 7\}$ or $(-\infty,7]$.

4. $2x+3y=6$

$x+2y=5$

Multiply the second equation by -2 and add the result to the first equation.

$2x+3y=6$

$\underline{-2x-4y=-10}$

$-y=-4$

$y=4$

Back-substitute 4 for y into either of the original equations to find x. We choose to use the second equation:

$x+2y=5$

$x+2(4)=5$

$x+8=5$

$x=-3$

The solution set is $\{(-3,4)\}$.

5. $3x-2y=1$

$y=10-2x$

Substitute the expression $10-2x$ for y into the first equation and solve for x.

$3x-2(10-2x)=1$

$3x-20+4x=1$

$7x-20=1$

$7x=21$

$x=3$

Back-substitute 3 for x into either of the original equations to find y. We choose to use the second equation:

$y=10-2x$

$y=10-2(3)$

$y=10-6$

$y=4$

The solution set is $\{(3,4)\}$.

6. $\dfrac{3}{x+5}-1=\dfrac{4-x}{2x+10}$

$\dfrac{3}{x+5}-1=\dfrac{4-x}{2(x+5)}$

Multiply both sides of the equation by the LCD, $2(x+5)$.

$2(x+5)\left(\dfrac{3}{x+5}-1\right)=2(x+5)\left(\dfrac{4-x}{2(x+5)}\right)$

$2(3)-2(x+5)(1)=4-x$

$6-2x-10=4-x$

$-2x-4=4-x$

$-x-4=4$

$-x=8$

$x=-8$

This proposed solution checks, so the solution set is $\{-8\}$.

7. $x+\dfrac{6}{x}=-5$

Multiply both sides of the equation by the LCD, x.

$x\left(x+\dfrac{6}{x}\right)=x(-5)$

$x^2+6=-5x$

$x^2+5x+6=0$

$(x+3)(x+2)=0$

$x+3=0 \quad$ or $\quad x+2=0$

$x=-3 \qquad\qquad x=-2$

Both proposed solutions check, so the solution set is $\{-3,-2\}$.

8. $\dfrac{12x^3}{3x^{12}}=\dfrac{12}{3}\cdot\dfrac{x^3}{x^{12}}=4x^{3-12}=4x^{-9}=\dfrac{4}{x^9}$

9. $4\cdot6\div2\cdot3+(-5)=24\div2\cdot3+(-5)$

$=12\cdot3+(-5)$

$=36+(-5)$

$=31$

10. $\left(6x^2-8x+3\right)-\left(-4x^2+x-1\right)$

$=6x^2-8x+3+4x^2-x+1$

$=10x^2-4x+4$

11. $(7x+4)(3x-5)$

$=7x(3x)-7x(5)+4(3x)-4(5)$

$=21x^2-35x+12x-20$

$=21x^2-23x-20$

12. $(5x-2)^2=(5x)^2-2(5x)(2)+2^2$

$=25x^2-20x+4$

13. $(x+y)\left(x^2-xy+y^2\right)$

$=x\left(x^2\right)-x(xy)+x\left(y^2\right)$

$\qquad +y\left(x^2\right)-y(xy)+y\left(y^2\right)$

$=x^3-x^2y+xy^2+x^2y-xy^2+y^3$

$=x^3+y^3$

14. $\dfrac{x^2+6x+8}{x^2}\div\left(3x^2+6x\right)$

$=\dfrac{x^2+6x+8}{x^2}\cdot\dfrac{1}{3x^2+6x}$

$=\dfrac{(x+4)(x+2)}{x^2}\cdot\dfrac{1}{3x(x+2)}$

$=\dfrac{(x+4)\cancel{(x+2)}}{x^2}\cdot\dfrac{1}{3x\cancel{(x+2)}}$

$=\dfrac{x+4}{3x^3}$

15. $\dfrac{x}{x^2+2x-3}-\dfrac{x}{x^2-5x+4}$

$=\dfrac{x}{(x+3)(x-1)}-\dfrac{x}{(x-4)(x-1)}$

The LCD is $(x+3)(x-1)(x-4)$

$\dfrac{x(x-4)}{(x+3)(x-1)(x-4)}-\dfrac{x(x+3)}{(x-4)(x-1)(x+3)}$

$=\dfrac{x(x-4)-x(x+3)}{(x+3)(x-1)(x-4)}$

$=\dfrac{x^2-4x-x^2-3x}{(x+3)(x-1)(x-4)}$

$=\dfrac{-7x}{(x+3)(x-1)(x-4)}$

16. $\dfrac{x-\dfrac{1}{5}}{5-\dfrac{1}{x}}$

Multiply the numerator and denominator by the LCD, $5x$.

$$\frac{5x}{5x} \cdot \frac{x-\dfrac{1}{5}}{5-\dfrac{1}{x}} = \frac{5x \cdot x - 5x \cdot \dfrac{1}{5}}{5x \cdot 5 - 5x \cdot \dfrac{1}{x}}$$

$$= \frac{5x^2 - x}{25x - 5} = \frac{x(5x-1)}{5(5x-1)} = \frac{x}{5}$$

17. $4x^2 - 49 = (2x)^2 - 7^2$
$$= (2x+7)(2x-7)$$

18. $x^3 + 3x^2 - x - 3 = x^2(x+3) - 1(x+3)$
$$= (x+3)(x^2-1)$$
$$= (x+3)(x+1)(x-1)$$

19. $2x^2 + 8x - 42 = 2(x^2 + 4x - 21)$
$$= 2(x-3)(x+7)$$

20. $x^5 - 16x = x(x^4 - 16)$
$$= x(x^2+4)(x^2-4)$$
$$= x(x^2+4)(x+2)(x-2)$$

21. $x^3 - 10x^2 + 25x = x(x^2 - 10x + 25)$
$$= x(x-5)^2$$

22. $x^3 - 8 = x^3 - 2^3$
$$= (x-2)(x^2 + x \cdot 2 + 2^2)$$
$$= (x-2)(x^2 + 2x + 4)$$

23. $y = \dfrac{1}{3}x - 1$

The slope is $\dfrac{1}{3}$, and the y-intercept is -1. Plot the point $(0,-1)$. We move up 1 unit and to the right 3 units to the point $(3,0)$. Find additional points as needed and connect with a line.

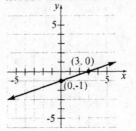

24. $3x + 2y = -6$

Let $x = 0$: $\quad 3(0) + 2y = -6$
$$2y = -6$$
$$y = -3$$

The y-intercept is -3, so the line passes through the point $(0,-3)$.

Let $y = 0$: $\quad 3x + 2(0) = -6$
$$3x = -6$$
$$x = -2$$

The x-intercept is -2, so the line passes through the point $(-2,0)$. Find additional points to check and connect them with a straight line.

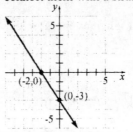

25. $y = -2$

The graph is a horizontal line that passes through the point $(0,-2)$.

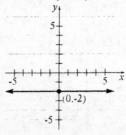

412

26. $m = \dfrac{y_2 - y_1}{x_2 - x_1} = \dfrac{-3 - 3}{2 - (-1)} = \dfrac{-6}{3} = -2$

27. $m = \dfrac{y_2 - y_1}{x_2 - x_1} = \dfrac{6 - 2}{3 - 1} = \dfrac{4}{2} = 2$

The point-slope equation is

$y - y_1 = m(x - x_1)$

$y - 2 = 2(x - 1)$ or $y - 6 = 2(x - 3)$

The slope-intercept equation is

$y - 2 = 2(x - 1)$

$y - 2 = 2x - 2$

$y = 2x$

28. Let x = the number.

$5x - 7 = 208$

$5x = 215$

$x = 43$

The number is 43.

29. Let x = the price before the reduction.

$x - 0.20x = 256$

$0.80x = 256$

$x = 320$

The price of the digital camera before the reduction is $320.

30. Let x = the width of the field.

$3x$ = the length of the field.

$2x + 2(3x) = 400$

$2x + 6x = 400$

$8x = 400$

$x = 50$

$3x = 150$

The width of the field is 50 yards, and the length of the field is 150 yards.

31. Let x = the amount invested at 7%.

y = the amount invested at 9%.

	Principal	Rate	Interest
7%	x	0.07	0.07x
9%	y	0.09	0.09y
	20,000		1550

$x + y = 20,000$

$0.07x + 0.09y = 1550$

Multiply the first equation by -0.07 and add the result to the second equation.

$\begin{array}{r} -0.07x - 0.07y = -1400 \\ \underline{0.07x + 0.09y = 1550} \\ 0.02y = 150 \\ y = 7500 \end{array}$

Back-substitute 7500 for y into one of the original equations. We use the first equation.

$x + 7500 = 20,000$

$x = 12,500$

$12,500 was invested at 7% and $7500 was invested at 9%.

32. Let x = the number of liters of 40% acid solution.

y = the number of liters of 70% acid solution.

	No. of liters	Percent of acid	Amount of acid
40% Sol	x	40% = 0.4	0.4x
70% Sol	y	70% = 0.7	0.7y
50% Mix	12	50%=0.5	0.5(12) = 6

$x + y = 12$

$0.4x + 0.7y = 6$

Multiply the first equation by -0.4 and add the result to the second equation.

$\begin{array}{r} -0.4x - 0.4y = -4.8 \\ \underline{0.4x + 0.7y = 6} \\ 0.3y = 1.2 \\ y = 4 \end{array}$

Back-substitute 4 for y into one of the original equations. We use the first equation.

$x + 4 = 12$

$x = 8$

The chemist should mix 8 liters of 40% acid solution with 4 liters of 70% acid solution.

33. Let x = the height.

$$\frac{1}{2} \cdot 15 \cdot x = 120$$

$$\frac{15x}{2} = 120$$

$$15x = 240$$

$$x = 16$$

The triangular sail is 16 feet high.

34. Let $\;\;x$ = the measure of the 2nd angle.
$x + 10$ = the measure of the 1st angle.
$4x + 20$ = the measure of the 3rd angle.

$$x + (x + 10) + (4x + 20) = 180$$

$$6x + 30 = 180$$

$$6x = 150$$

$$x = 25$$

$$x + 10 = 35$$

$$4x + 20 = 120$$

The 1st angle measures $35°$; the 2nd angle measures $25°$, and the 3rd angle measures $120°$.

35. Let x = the price of a TV.

y = the price of a stereo.

$$3x + 4y = 2530$$

$$4x + 3y = 2510$$

Multiply the first equation by 4, multiply the second equation by -3, and add the results.

$$12x + 16y = 10,120$$

$$\underline{-12x - 9y = -7530}$$

$$7y = 2590$$

$$y = 370$$

Back-substitute 370 for y into one of the original equations. We use the first equation.

$$3x + 4(370) = 2530$$

$$3x + 1480 = 2530$$

$$3x = 1050$$

$$x = 350$$

The price of the TV is $350, and the price of the stereo is $370.

36. Let x = the width of the rectangle.
$x + 6$ = the length of the rectangle.

$$x(x + 6) = 55$$

$$x^2 + 6x = 55$$

$$x^2 + 6x - 55 = 0$$

$$(x + 11)(x - 5) = 0$$

$$x + 11 = 0 \quad \text{of} \quad x - 5 = 0$$

$$x = -11 \qquad x = 5$$

Disregard -11 because the width of a rectangle cannot be negative. So $x = 5$ and $x + 6 = 11$. The width of the rectangle is 5 meters, and the length is 11 meters.

Chapter 8
Basics of Functions

8.1 Check Points

1. The domain is {0, 10, 20, 30, 34}.
   The range is {9.1, 6.7, 10.7, 13.2, 15.5}.

2. **a.** The relation is not a function because an element, 5, in the domain corresponds to two elements in the range.

   b. The relation is a function.

3. **a.** $f(x) = 4x + 5$
   $f(6) = 4(6) + 5$
   $f(6) = 29$

   b. $g(x) = 3x^2 - 10$
   $g(-5) = 3(-5)^2 - 10$
   $g(-5) = 65$

   c. $h(r) = r^2 - 7r + 2$
   $h(-4) = (-4)^2 - 7(-4) + 2$
   $h(-4) = 46$

   d. $F(x) = 6x + 9$
   $F(a+h) = 6(a+h) + 9$
   $F(a+h) = 6a + 6h + 9$

4. **a.** Every element in the domain corresponds to exactly one element in the range.

   b. The domain is {0, 1, 2, 3, 4}.
   The range is {3, 0, 1, 2}.

   c. $g(1) = 0$

   d. $g(3) = 2$

   e. $x = 0$ and $x = 4$.

8.1 Exercise Set

1. The relation is a function.
   The domain is {1, 3, 5}.
   The range is {2, 4, 5}.

3. The relation is not a function.
   The domain is {3, 4}.
   The range is {4, 5}.

5. The relation is a function.
   The domain is {-3, -2, -1, 0}.
   The range is {-3, -2, -1, 0}.

7. The relation is not a function.
   The domain is {1}.
   The range is {4, 5, 6}.

9. **a.** $f(0) = 0 + 1 = 1$

   b. $f(5) = 5 + 1 = 6$

   c. $f(-8) = -8 + 1 = -7$

   d. $f(2a) = 2a + 1$

   e. $f(a+2) = (a+2) + 1$
   $= a + 2 + 1 = a + 3$

11. **a.** $g(0) = 3(0) - 2 = 0 - 2 = -2$

   b. $g(-5) = 3(-5) - 2$
   $= -15 - 2 = -17$

   c. $g\left(\dfrac{2}{3}\right) = 3\left(\dfrac{2}{3}\right) - 2 = 2 - 2 = 0$

   d. $g(4b) = 3(4b) - 2 = 12b - 2$

   e. $g(b+4) = 3(b+4) - 2$
   $= 3b + 12 - 2 = 3b + 10$

13. a. $h(0) = 3(0)^2 + 5 = 3(0) + 5$

$= 0 + 5 = 5$

b. $h(-1) = 3(-1)^2 + 5 = 3(1) + 5$

$= 3 + 5 = 8$

c. $h(4) = 3(4)^2 + 5 = 3(16) + 5$

$= 48 + 5 = 53$

d. $h(-3) = 3(-3)^2 + 5 = 3(9) + 5$

$= 27 + 5 = 32$

e. $h(4b) = 3(4b)^2 + 5 = 3(16b^2) + 5$

$= 48b^2 + 5$

15. a. $f(0) = 2(0)^2 + 3(0) - 1$

$= 0 + 0 - 1 = -1$

b. $f(3) = 2(3)^2 + 3(3) - 1$

$= 2(9) + 9 - 1$

$= 18 + 9 - 1 = 26$

c. $f(-4) = 2(-4)^2 + 3(-4) - 1$

$= 2(16) - 12 - 1$

$= 32 - 12 - 1 = 19$

d. $f(b) = 2(b)^2 + 3(b) - 1$

$= 2b^2 + 3b - 1$

e. $f(5a) = 2(5a)^2 + 3(5a) - 1$

$= 2(25a^2) + 15a - 1$

$= 50a^2 + 15a - 1$

17. a. $f(0) = \dfrac{2(0) - 3}{(0) - 4} = \dfrac{0 - 3}{0 - 4}$

$= \dfrac{-3}{-4} = \dfrac{3}{4}$

b. $f(3) = \dfrac{2(3) - 3}{(3) - 4} = \dfrac{6 - 3}{3 - 4}$

$= \dfrac{3}{-1} = -3$

c. $f(-4) = \dfrac{2(-4) - 3}{(-4) - 4} = \dfrac{-8 - 3}{-8}$

$= \dfrac{-11}{-8} = \dfrac{11}{8}$

d. $f(-5) = \dfrac{2(-5) - 3}{(-5) - 4} = \dfrac{-10 - 3}{-9}$

$= \dfrac{-13}{-9} = \dfrac{13}{9}$

e. $f(a + h) = \dfrac{2(a + h) - 3}{(a + h) - 4}$

$= \dfrac{2a + 2h - 3}{a + h - 4}$

f. Four must be excluded from the domain, because four would make the denominator zero. Division by zero is undefined.

19. a. $f(-2) = 6$

b. $f(2) = 12$

c. $x = 0$

21. a. $h(-2) = 2$

b. $h(1) = 1$

c. $x = -1$ and $x = 1$

23. $g(1) = 3(1) - 5 = 3 - 5 = -2$

$f(g(1)) = f(-2) = (-2)^2 - (-2) + 4$

$= 4 + 2 + 4 = 10$

25. $\sqrt{3 - (-1)} - (-6)^2 + 6 \div -6 \cdot 4$

$= \sqrt{3 + 1} - 36 + -1 \cdot 4$

$= \sqrt{4} - 36 + -4 = 2 - 36 - 4 = -38$

27. $f(-x) - f(x)$

$= (-x)^3 + (-x) - 5 - \left[x^3 + x - 5 \right]$

$= -x^3 - x - 5 - x^3 - x + 5$

$= -2x^3 - 2x$

29. a. $f(-2) = 3(-2) + 5 = -6 + 5 = -1$

b. $f(0) = 4(0) + 7 = 0 + 7 = 7$

c. $f(3) = 4(3) + 7 = 12 + 7 = 19$

d. $f(-100) + f(100)$
$= 3(-100) + 5 + 4(100) + 7$
$= -300 + 5 + 400 + 7 = 112$

31. a. {(Extremely Liberal, 1%), (Liberal, 7%), (Slightly Liberal, 11%), (Moderate, 52%), (Slightly Conservative, 13%), (Conservative, 13%), (Extremely Conservative, 3%)}

b. Yes, the relation is a function because each ideology in the domain corresponds to exactly one percentage in the range.

c. {(1%, Extremely Liberal), (7%, Liberal), (11%, Slightly Liberal), (52%, Moderate), (13%, Slightly Conservative), (13%, Conservative), (3%, Extremely Conservative)}

d. No, the relation is not a function because 13% in the domain corresponds to two ideologies in the range, Slightly Conservative and Conservative.

33. – 35. Answers will vary.

37. makes sense

39. makes sense

41. false; Changes to make the statement true will vary. A sample change is: All functions are relations.

43. true

45. true

47. $f(a + h) = 3(a + h) + 7 = 3a + 3h + 7$
$f(a) = 3a + 7$
$\dfrac{f(a + h) - f(a)}{h}$
$= \dfrac{(3a + 3h + 7) - (3a + 7)}{h}$
$= \dfrac{3a + 3h + 7 - 3a - 7}{h} = \dfrac{3h}{h} = 3$

49. It is given that $f(x + y) = f(x) + f(y)$ and $f(1) = 3$.
To find $f(2)$, rewrite 2 as 1 + 1.
$f(2) = f(1 + 1) = f(1) + f(1)$
$\quad = 3 + 3 = 6$
Similarly:
$f(3) = f(2 + 1) = f(2) + f(1)$
$\quad = 6 + 3 = 9$
$f(4) = f(3 + 1) = f(3) + f(1)$
$\quad = 9 + 3 = 12$
While $f(x + y) = f(x) + f(y)$ is true for this function, it is not true for all functions. It is not true for $f(x) = x^2$, for example.

50. $24 \div 4 \left[2 - (5 - 2) \right]^2 - 6$
$= 24 \div 4 \left[2 - (3) \right]^2 - 6$
$= 24 \div 4 (-1)^2 - 6$
$= 24 \div 4 (1) - 6$
$= 6(1) - 6 = 6 - 6 = 0$

51. $\left(\dfrac{3x^2 y^{-2}}{y^3} \right)^{-2} = \left(\dfrac{3x^2}{y^5} \right)^{-2} = \left(\dfrac{y^5}{3x^2} \right)^2 = \dfrac{y^{10}}{9x^4}$

52. $\dfrac{x}{3} = \dfrac{3x}{5} + 4$
$15 \left(\dfrac{x}{3} \right) = 15 \left(\dfrac{3x}{5} + 4 \right)$
$15 \left(\dfrac{x}{3} \right) = 15 \left(\dfrac{3x}{5} \right) + 15(4)$
$5x = 3(3x) + 60$
$5x = 9x + 60$
$5x - 9x = 9x - 9x + 60$
$-4x = 60$
$\dfrac{-4x}{-4} = \dfrac{60}{-4}$
$x = -15$
The solution set is $\{-15\}$.

53. When the y-coordinate is 4, the x-coordinates are -3 and 3.

54. The x-coordinates are all real numbers.

55. The y-coordinates are all real numbers greater than or equal to 1.

8.2 Check Points

1. a. The graph represents a function. It passes the vertical line test.

    b. The graph represents a function. It passes the vertical line test.

    c. The graph does not represent a function. It fails the vertical line test.

2. a. $f(5) = 400$

    b. When x is 9, the function's value is 100. i.e. $f(9) = 100$

    c. The minimum T cell count during the asymptomatic stage is approximately 425.

3. a. $\{x \mid -2 \le x < 5\}$

    b. $\{x \mid 1 \le x \le 3.5\}$

    c. $\{x \mid x < -1\}$

4. a. The domain is $\{x \mid -2 \le x \le 1\}$ or $[-2,1]$.
        The range is $\{y \mid 0 \le y \le 3\}$ or $[0,3]$.

    b. The domain is $\{x \mid -2 < x \le 1\}$ or $(-2,1]$.
        The range is $\{y \mid -1 \le y < 2\}$ or $[-1,2)$.

    c. The domain is $\{x \mid -3 \le x < 0\}$ or $[-3,0)$.
        The range is $\{y \mid y = -3,-2,-1\}$.

8.2 Exercise Set

1. The graph represents a function. It passes the vertical line test.

3. The graph does not represent a function. It fails the vertical line test.

5. The graph represents a function. It passes the vertical line test.

7. The graph does not represent a function. It fails the vertical line test.

9. $f(-2) = -4$

11. $f(4) = 4$

13. $f(-3) = 0$

15. $g(-4) = 2$

17. $g(-10) = 2$

19. When $x = -2$, $g(x) = 1$.

21. $\{x \mid 1 < x \le 6\}$

23. $\{x \mid -5 \le x < 2\}$

25. $\{x \mid -3 \le x \le 1\}$

27. $\{x \mid x > 2\}$

29. $\{x \mid x \ge -3\}$

31. $\{x \mid x < 3\}$

33. $\{x \mid x < 5.5\}$

35. The domain is $\{x \mid 0 \le x < 5\}$ or $[0,5)$.
    The range is $\{y \mid -1 \le y < 5\}$ or $[-1,5)$.

37. The domain is $\{x \mid x \ge 0\}$ or $[0,\infty)$.
    The range is $\{y \mid y \ge 1\}$ or $[1,\infty)$.

39. The domain is $\{x \mid -2 \le x \le 6\}$ or $[-2,6]$.
    The range is $\{y \mid -2 \le y \le 6\}$ or $[-2,6]$.

41. The domain is $\{x \mid x \text{ is a real number}\}$ or (∞,∞).
    The range is $\{y \mid y \le -2\}$ or $(-\infty,-2]$.

43. The domain is $\{x \mid x = -5, -2, 0, 1, 3\}$.

   The range is $\{y \mid y = 2\}$.

45. a. The domain is

   $\{x \mid x \text{ is a real number}\}$ or $(-\infty, \infty)$.

   b. The range is $\{y \mid y \geq -4\}$ or $[-4, \infty)$.

   c. $f(-3) = 4$

   d. 2 and 6; i.e. $f(2) = f(6) = -2$

   e. f crosses the x-axis at $(1, 0)$ and $(7, 0)$.

   f. f crosses the y-axis at $(0, 4)$.

   g. $f(x) < 0$ for $\{x \mid 1 < x < 7\}$ or $(1, 7)$.

   h. $f(-8)$ is positive.

47. a. $f(x) = 0.013x + 56.46$

   $f(80) = 0.013(80) + 56.46$

   $\qquad = 57.5$

   The average global temperature in 1980 was 57.5°F.

   This is represented by the point $(80, 57.5)$.

   b. $f(80)$ overestimates the actual data shown in the bar graph by 0.1°F.

49. a. $f(x) = 1.24x + 305$

   $f(50) = 1.24(50) + 305$

   $\qquad = 367$

   The average carbon dioxide concentration in 2000 was 367 parts per million.

   This is represented by the point $(50, 367)$.

   b. $g(x) = 1.01x^2 + 0.6x + 310$

   $g(50) = 0.01(50)^2 + 0.6(50) + 310$

   $\qquad = 365$

   The average carbon dioxide concentration in 2000 was 365 parts per million.

   This is represented by the point $(50, 365)$.

   c. The linear function describes the actual value better.

51. $f(20) = 0.4(20)^2 - 36(20) + 1000$

   $\qquad = 0.4(400) - 720 + 1000$

   $\qquad = 160 - 720 + 1000$

   $\qquad = -560 + 1000 = 440$

   Twenty-year-old drivers have 440 accidents per 50 million miles driven.

   This is represented on the graph by point $(20, 440)$.

53. The graph reaches its lowest point at $x = 45$.

   $f(45) = 0.4(45)^2 - 36(45) + 1000$

   $\qquad = 0.4(2025) - 1620 + 1000$

   $\qquad = 810 - 1620 + 1000$

   $\qquad = -810 + 1000$

   $\qquad = 190$

   Drivers at age 45 have 190 accidents per 50 million miles driven. This is the least number of accidents for any driver between ages 16 and 74.

55. $f(60) \approx 3.1$

   In 1960, 3.1% of the U.S. population was made up of Jewish Americans.

57. In 1919 and 1964, $f(x) = 3$. This means that in 1919 and 1964, 3% of the U.S. population was made up of Jewish Americans.

59. The percentage of Jewish Americans in the U.S. population reached a maximum in 1940. Using the graph to estimate, approximately 3.7% of the U.S. population was Jewish American.

61. Each year is paired with exactly one percentage. This means that each member of the domain is paired with one member of the range.

63. $f(3) = 0.75$

   The cost of mailing a first-class letter weighing 3 ounces is $0.75.

65. The cost to mail a letter weighing 1.5 ounces is $0.58.

67. – 69. Answers will vary.

71. makes sense

73. does not make sense; Explanations will vary. Sample explanation: The domain is the set of number of years that people work for a company.

75. false; Changes to make the statement true will vary. A sample change is: The graph of a vertical line is not a function.

77. true

79. true

81. $\sqrt{f(-1.5) + f(-0.9)} - [f(\pi)]^2 + f(-3) \div f(1) \cdot f(-\pi)$

$= \sqrt{1+0} - [-4]^2 + 2 \div (-2)(3)$

$= \sqrt{1} - 16 + (-1)(3)$

$= 1 - 16 - 3$

$= -15 - 3$

$= -18$

83. The relation is a function. Every element in the domain corresponds to exactly one element in the range.

84. $12 - 2(3x + 1) = 4x - 5$

$12 - 6x - 2 = 4x - 5$

$10 - 6x = 4x - 5$

$-6x - 4x = -5 - 10$

$-10x = -15$

$\dfrac{-10x}{-10} = \dfrac{-15}{-10}$

$x = \dfrac{3}{2}$

The solution set is $\left\{\dfrac{3}{2}\right\}$.

85. Let $x =$ the width of the rectangle.
Let $3x + 8 =$ length of the rectangle.
$P = 2l + 2w$
$624 = 2(3x + 8) + 2x$
$624 = 6x + 16 + 2x$
$624 = 8x + 16$
$-8x = -608$
$x = 76$
$3x + 8 = 236$
The dimensions of the rectangle are 76 yards by 236 yards.

86. 3 must be excluded from the domain of f because it would cause the denominator, $x - 3$, to be equal to zero. Division by 0 is undefined.

87. $f(4) + g(4) = \overbrace{(4^2 + 4)}^{f(4)} + \overbrace{(4 - 5)}^{g(4)}$

$= 20 + (-1)$

$= 19$

88. $7.4x^2 - 15x + 4046 - \left(-3.5x^2 + 20x + 2405\right) = 7.4x^2 - 15x + 4046 + 3.5x^2 - 20x - 2405$

$= 10.9x^2 - 35x + 1641$

8.3 Check Points

1. **a.** Domain of $f = \{x \mid x \text{ is a real number}\}$ or $(-\infty, \infty)$

   b. Domain of $g = \{x \mid x \text{ is a real number and } x \neq -5\}$

2. **a.** $(f + g)(x) = \left(3x^2 + 4x - 1\right) + (2x + 7)$

   $$= 3x^2 + 6x + 6$$

   b. $(f + g)(4) = 3(4)^2 + 6(4) + 6$

   $$= 78$$

3. **a.** $(f - g)(x) = \dfrac{5}{x} - \dfrac{7}{x - 8}$

   b. Domain of $f - g = \{x \mid x \text{ is a real number and } x \neq 0 \text{ and } x \neq 8\}$.

4. **a.** $(f + g)(5) = f(5) + g(5) = [5^2 - 2 \cdot 5] + [5 + 3] = 23$

   b. $(f - g)(x) = f(x) - g(x) = [x^2 - 2x] - [x + 3] = x^2 - 3x - 3$

   $(f - g)(-1) = (-1)^2 - 3(-1) - 3 = 1$

   c. $(fg)(x) = f(x) \cdot g(x) = \left(x^2 - 2x\right)(x + 3) = x^3 + x^2 - 6x$

   $(fg)(-4) = f(-4) \cdot g(-4) = \left((-4)^2 - 2(-4)\right)\left((-4) + 3\right) = -24$

   d. $\left(\dfrac{f}{g}\right)(x) = \dfrac{f(x)}{g(x)} = \dfrac{x^2 - 2x}{x + 3}$

   $\left(\dfrac{f}{g}\right)(7) = \dfrac{(7)^2 - 2(7)}{(7) + 3} = \dfrac{35}{10} = \dfrac{7}{2}$

5. **a.** $(B + D)(x) = B(x) + D(x)$

   $$= (7.4x^2 - 15x + 4046) + (-3.5x^2 + 20x + 2405)$$

   $$= 7.4x^2 - 15x + 4046 - 3.5x^2 + 20x + 2405$$

   $$= 3.9x^2 + 5x + 6451$$

   b. $(B + D)(x) = 3.9x^2 + 5x + 6451$

   $(B + D)(5) = 3.9(5)^2 + 5(5) + 6451$

   $$= 6573.5$$

   The number of births and deaths in the U.S. in 2005 is 6573.5 thousand.

   c. $(B + D)(x)$ underestimates the actual number of births and deaths in 2005 by 1.5 thousand.

8.3 Exercise Set

1. Domain of $f = \{x \mid x \text{ is a real number}\}$ or $(-\infty, \infty)$

3. Domain of $g = \{x \mid x \text{ is a real number and } x \neq -4\}$.

5. Domain of $f = \{x \mid x \text{ is a real number and } x \neq 3\}$.

7. Domain of $g = \{x \mid x \text{ is a real number and } x \neq 5\}$.

9. Domain of $f = \{x \mid x \text{ is a real number and } x \neq -7 \text{ and } x \neq 9\}$.

11. $(f+g)(x) = (3x+1)+(2x-6)$
    $ = 3x+1+2x-6$
    $ = 5x-5$

    $(f+g)(5) = 5(5)-5$
    $ = 25-5 = 20$

13. $(f+g)(x) = (x-5)+\left(3x^2\right)$
    $ = x-5+3x^2$
    $ = 3x^2+x-5$

    $(f+g)(5) = 3(5)^2+5-5$
    $ = 3(25) = 75$

15. $(f+g)(x)$
    $= \left(2x^2-x-3\right)+(x+1)$
    $= 2x^2-x-3+x+1$
    $= 2x^2-2$

    $(f+g)(5) = 2(5)^2-2$
    $ = 2(25)-2$
    $ = 50-2 = 48$

17. Domain of $f+g = \{x \mid x \text{ is a real number}\}$ or $(-\infty, \infty)$

19. Domain of $f+g = \{x \mid x \text{ is a real number and } x \neq 5\}$

21. Domain of $f+g = \{x \mid x \text{ is a real number and } x \neq 0 \text{ and } x \neq 5\}$

23. Domain of $f+g = \{x \mid x \text{ is a real number and } x \neq 2 \text{ and } x \neq -3\}$

25. Domain of $f + g = \{x | x \text{ is a real number and } x \neq 2\}$

27. Domain of $f + g = \{x | x \text{ is a real number}\}$ or $(-\infty, \infty)$

29. $(f + g)(x) = f(x) + g(x)$

$\qquad = x^2 + 4x + 2 - x$

$\qquad = x^2 + 3x + 2$

$(f + g)(3) = (3)^2 + 3(3) + 2 = 20$

31. $f(-2) + g(-2) = \left((-2)^2 + 4(-2)\right) + \left(2 - (-2)\right) = -4 + 4 = 0$

33. $(f - g)(x) = f(x) - g(x)$

$\qquad = \left(x^2 + 4x\right) - (2 - x)$

$\qquad = x^2 + 4x - 2 + x$

$\qquad = x^2 + 5x - 2$

$(f - g)(5) = (5)^2 + 5(5) - 2$

$\qquad = 25 + 25 - 2 = 48$

35. $f(-2) - g(-2) = \left((-2)^2 + 4(-2)\right) - \left(2 - (-2)\right) = -4 - 4 = -8$

37. $(fg)(x) = f(x) \cdot g(x) = \left(x^2 + 4x\right) \cdot (2 - x) = -x^3 - 2x^2 + 8x$

$(fg)(2) = f(2) \cdot g(2) = \left(2^2 + 4 \cdot 2\right) \cdot (2 - 2) = (12)(0) = 0$

39. $(fg)(5) = f(5) \cdot g(5) = \left((5)^2 + 4(5)\right) \cdot (2 - (5)) = 45(-3) = -135$

41. $\left(\dfrac{f}{g}\right)(x) = \dfrac{f(x)}{g(x)} = \dfrac{x^2 + 4x}{2 - x}$

$\left(\dfrac{f}{g}\right)(1) = \dfrac{(1)^2 + 4(1)}{2 - (1)} = \dfrac{1 + 4}{1} = \dfrac{5}{1} = 5$

43. $\left(\dfrac{f}{g}\right)(x) = \dfrac{f(x)}{g(x)} = \dfrac{x^2 + 4x}{2 - x}$

$\left(\dfrac{f}{g}\right)(-1) = \dfrac{(-1)^2 + 4(-1)}{2 - (-1)}$

$\qquad = \dfrac{1 - 4}{3} = \dfrac{-3}{3} = -1$

45. Domain of $f + g = \{x | x \text{ is a real number}\}$ or $(-\infty, \infty)$

46. Domain of $f - g = \{x | x \text{ is a real number}\}$ or $(-\infty, \infty)$

48. $(fg)(x) = f(x) \cdot g(x) = (x^2 + 4x)(2-x)$

Domain of $fg = \{x | x \text{ is a real number}\}$ or $(-\infty, \infty)$

51. $(fg)(2) = f(2)g(2) = (-1)(1) = -1$

53. The domain of $f + g$ is $\{x | -4 \le x \le 3\}$ or $[-4, 3]$.

55. The graph of $f + g$

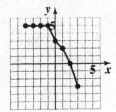

57. $(f+g)(1) - (g-f)(-1)$

$= f(1) + g(1) - [g(-1) - f(-1)]$

$= f(1) + g(1) - g(-1) + f(-1)$

$= -6 + -3 - (-2) + 3$

$= -6 + -3 + 2 + 3 = -4$

59. $(fg)(-2) - \left[\left(\dfrac{f}{g}\right)(1)\right]^2$

$= f(-2)g(-2) - \left[\dfrac{f(1)}{g(1)}\right]^2$

$= 5 \cdot 0 - \left[\dfrac{-6}{-3}\right]^2 = 0 - 2^2 = 0 - 4 = -4$

61. **a.** $(M+F)(x) = M(x) + F(x) = (1.58x + 114.4) + (1.48x + 120.6) = 3.06x + 235$

b. $(M+F)(x) = 3.06x + 235$

$(M+F)(20) = 3.06(20) + 235 = 296.2$

The total U.S. population in 2005 was 296.2 million.

c. The result in part (b) underestimates the actual total by 1.8 million.

63. **a.** $\left(\dfrac{M}{F}\right)(x) = \left(\dfrac{M(x)}{F(x)}\right) = \dfrac{1.58x + 114.4}{1.48x + 120.6}$

b. $\left(\dfrac{M}{F}\right)(x) = \dfrac{1.58x + 114.4}{1.48x + 120.6}$

$\left(\dfrac{M}{F}\right)(15) = \dfrac{1.58(15) + 114.4}{1.48(15) + 120.6} \approx 0.967$

In 2000 the ratio of men to women was 0.967.

c. The result in part (b) overestimates the actual ratio of $\dfrac{138}{143} \approx 0.965$ by about 0.002.

65. – 67. Answers will vary.

69. $y_1 = 2x + 3$ $y_2 = 2 - 2x$ $y_3 = y_1 + y_2$

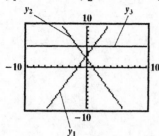

71. $y_1 = x$ $y_2 = x - 4$ $y_3 = y_1 \cdot y_2$

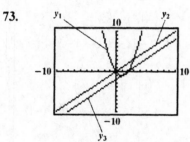

73.

No y-value is displayed because y_3 is undefined at $x = 0$.

75. makes sense

77. makes sense

79. true

81. false; Changes to make the statement true will vary. A sample change is: $f(a)$ or $f(b)$ is 0.

82.
$$R = 3(a + b)$$
$$R = 3a + 3b$$
$$R - 3a = 3b$$
$$b = \frac{R - 3a}{3} \text{ or } b = \frac{R}{3} - a$$

83. $3(6 - x) = 3 - 2(x - 4)$
$$18 - 3x = 3 - 2x + 8$$
$$18 - 3x = 11 - 2x$$
$$18 = 11 + x$$
$$7 = x$$
The solution set is $\{7\}$.

84. $f(b + 2) = 6(b + 2) - 4$
$$= 6b + 12 - 4 = 6b + 8$$

85. a. $f(x) = 3x - 4$
$$f(5) = 3(5) - 4 = 11$$

b. $g(x) = x^2 + 6$
$$g(f(5)) = g(11) = (11)^2 + 6 = 127$$

86. $3\left(\dfrac{x - 2}{3}\right) + 2 = x - 2 + 2 = x$

87.
$$x = 7y - 5$$
$$x + 5 = 7y$$
$$\frac{x + 5}{7} = y$$
$$y = \frac{x + 5}{7}$$

Chapter 8 Mid-Chapter Check Point

1. The relation is not a function.
The domain is $\{1, 2\}$.
The range is $\{-6, 4, 6\}$.

2. The relation is a function.
The domain is $\{0, 2, 3\}$.
The range is $\{1, 4\}$.

3. The relation is a function.
The domain is $\{x \mid -2 \le x < 2\}$ or $[-2, 2)$.
The range is $\{y \mid 0 \le y \le 3\}$ or $[0, 3]$.

4. The relation is not a function.
The domain is $\{x \mid -3 < x \le 4\}$ or $[-3, 4]$.
The range is $\{y \mid -1 \le y \le 2\}$ or $[-1, 2]$.

5. The relation is not a function.
   The domain is $\{-2,-1,0,1,2\}$.
   The range is $\{-2,-1,1,3\}$.

6. The relation is a function.
   The domain is $\{x \mid x \le 1\}$ or $(-\infty, 1]$.
   The range is $\{y \mid y \ge -1\}$ or $[-1, \infty)$.

7. The graph of f represents the graph of a function because every element in the domain is corresponds to exactly one element in the range. It passes the vertical line test.

8. $f(-4) = 3$

9. The function $f(x) = 4$ when $x = -2$.

10. The function $f(x) = 0$ when $x = 2$ and $x = -6$.

11. The domain of f is
    $\{x \mid x$ is a real number$\}$ or $(-\infty, \infty)$.

12. The range of f is $\{y \mid y \le 4\}$ or $(-\infty, 4]$.

13. The domain is $\{x \mid x$ is a real number$\}$ or $(-\infty, \infty)$.

14. The domain of g is
    $\{x \mid x$ is a real number and $x \ne -2$ and $x \ne 2\}$.

15. $f(0) = 0^2 - 3(0) + 8 = 8$
    $g(-10) = -2(-10) - 5 = 20 - 5 = 15$
    $f(0) + g(-10) = 8 + 15 = 23$

16. $f(-1) = (-1)^2 - 3(-1) + 8 = 1 + 3 + 8 = 12$
    $g(3) = -2(3) - 5 = -6 - 5 = -11$
    $f(-1) - g(3) = 12 - (-11) = 12 + 11 = 23$

17. $f(a) = a^2 - 3a + 8$
    $g(a+3) = -2(a+3) - 5$
    $\qquad = -2a - 6 - 5 = -2a - 11$
    $f(a) + g(a+3) = a^2 - 3a + 8 + -2a - 11$
    $\qquad = a^2 - 5a - 3$

18. $(f+g)(x) = x^2 - 3x + 8 + -2x - 5$
    $\qquad = x^2 - 5x + 3$
    $(f+g)(-2) = (-2)^2 - 5(-2) + 3$
    $\qquad = 4 + 10 + 3 = 17$

19. $(f-g)(x) = x^2 - 3x + 8 - (-2x - 5)$
    $\qquad = x^2 - 3x + 8 + 2x + 5$
    $\qquad = x^2 - x + 13$
    $(f-g)(5) = (5)^2 - 5 + 13$
    $\qquad = 25 - 5 + 13 = 33$

20. $(fg)(x) = f(x)g(x)$
    $\qquad = (x^2 - 3x + 8)(-2x - 5)$
    $\qquad = -2x^3 + x^2 - x - 40$
    $(fg)(-1) = f(-1)g(-1)$
    $\qquad = \left[(-1)^2 - 3(-1) + 8\right]\left[-2(-1) - 5\right]$
    $\qquad = 12(-3)$
    $\qquad = -36$

21. $\left(\dfrac{f}{g}\right)(x) = \dfrac{x^2 - 3x + 8}{-2x - 5}$
    $\left(\dfrac{f}{g}\right)(-4) = \dfrac{(-4)^2 - 3(-4) + 8}{-2(-4) - 5}$
    $\qquad = \dfrac{16 + 12 + 8}{8 - 5} = \dfrac{36}{3} = 12$

22. The domain of $\dfrac{f}{g}$ is $\left\{x \mid x \ne -\dfrac{5}{2}\right\}$.

8.4 Check Points

1. a. $(f \circ g)(x) = f\big(g(x)\big)$

$= f\big(x^2 - 1\big)$

$= 5\big(x^2 - 1\big) + 6$

$= 5x^2 - 5 + 6$

$= 5x^2 + 1$

b. $(g \circ f)(x) = g\big(f(x)\big)$

$= g\big(5x + 6\big)$

$= (5x + 6)^2 - 1$

$= 25x^2 + 60x + 36 - 1$

$= 25x^2 + 60x + 35$

2. $f\big(g(x)\big) = 7\left(\dfrac{x}{7}\right) = x$

$g\big(f(x)\big) = \dfrac{7x}{7} = x$

3. $f\big(g(x)\big) = 4\left(\dfrac{x+7}{4}\right) - 7 = x + 7 - 7 = x$

$g\big(f(x)\big) = \dfrac{(4x - 7) + 7}{4} = \dfrac{4x}{4} = x$

4. $f(x) = 2x + 7$

$y = 2x + 7$

Interchange x and y and solve for y.

$x = 2y + 7$

$x - 7 = 2y$

$\dfrac{x - 7}{2} = y$

$f^{-1}(x) = \dfrac{x - 7}{2}$

5. a. Since a horizontal line can be drawn that intersects the graph more than once, it fails the horizontal line test. Thus, this graph does not represent a function that has an inverse function.

b. Since a horizontal line cannot be drawn that intersects the graph more than once, it passes the horizontal line test. Thus, this graph represents a function that has an inverse function.

c. Since a horizontal line cannot be drawn that intersects the graph more than once, it passes the horizontal line test. Thus, this graph represents a function that has an inverse function.

6. Since f has a line segment from $(-2, -2)$ to $(-1, 0)$, then f^{-1} has a line segment from $(-2, -2)$ to $(0, -1)$.

Since f has a line segment from $(-1, 0)$ to $(1, 2)$, then f^{-1} has a line segment from $(0, -1)$ to $(2, 1)$.

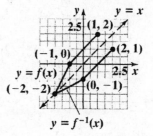

8.4 Exercise Set

1. a. $(f \circ g)(x) = f\big(g(x)\big)$

$= f\big(x + 7\big)$

$= 2(x + 7) = 2x + 14$

b. $(g \circ f)(x) = g\big(f(x)\big)$

$= g(2x) = 2x + 7$

c. $(f \circ g)(2) = 2(2) + 14$

$= 4 + 14 = 18$

3. a. $(f \circ g)(x) = f\big(g(x)\big)$

$= f\big(2x + 1\big)$

$= (2x + 1) + 4 = 2x + 5$

b. $(g \circ f)(x) = g\big(f(x)\big)$

$= g\big(x + 4\big)$

$= 2(x + 4) + 1$

$= 2x + 8 + 1 = 2x + 9$

c. $(f \circ g)(2) = 2(2) + 5 = 4 + 5 = 9$

5. a. $(f \circ g)(x) = f(g(x))$

$\qquad = f(5x^2 - 2)$

$\qquad = 4(5x^2 - 2) - 3$

$\qquad = 20x^2 - 8 - 3$

$\qquad = 20x^2 - 11$

b. $(g \circ f)(x) = g(f(x))$

$\qquad = g(4x - 3)$

$\qquad = 5(4x - 3)^2 - 2$

$\qquad = 5(16x^2 - 24x + 9) - 2$

$\qquad = 80x^2 - 120x + 45 - 2$

$\qquad = 80x^2 - 120x + 43$

c. $(f \circ g)(2) = 20(2)^2 - 11$

$\qquad = 20(4) - 11$

$\qquad = 80 - 11 = 69$

7. a. $(f \circ g)(x) = f(g(x))$

$\qquad = f(x^2 - 2)$

$\qquad = (x^2 - 2)^2 + 2$

$\qquad = x^4 - 4x^2 + 4 + 2$

$\qquad = x^4 - 4x^2 + 6$

b. $(g \circ f)(x) = g(f(x))$

$\qquad = g(x^2 + 2)$

$\qquad = (x^2 + 2)^2 - 2$

$\qquad = x^4 + 4x^2 + 4 - 2$

$\qquad = x^4 + 4x^2 + 2$

c. $(f \circ g)(2) = 2^4 - 4(2)^2 + 6$

$\qquad = 16 - 4(4) + 6$

$\qquad = 16 - 16 + 6 = 6$

9. a. $(f \circ g)(x) = f(g(x))$

$\qquad = f(x - 1) = \sqrt{x - 1}$

b. $(g \circ f)(x) = g(f(x))$

$\qquad = g(\sqrt{x}) = \sqrt{x} - 1$

c. $(f \circ g)(2) = \sqrt{2 - 1} = \sqrt{1} = 1$

11. a. $(f \circ g)(x) = f(g(x))$

$\qquad = f\left(\dfrac{x + 3}{2}\right)$

$\qquad = 2\left(\dfrac{x + 3}{2}\right) - 3$

$\qquad = x + 3 - 3 = x$

b. $(g \circ f)(x) = g(f(x))$

$\qquad = g(2x - 3)$

$\qquad = \dfrac{(2x - 3) + 3}{2}$

$\qquad = \dfrac{2x - 3 + 3}{2} = \dfrac{2x}{2} = x$

c. $(f \circ g)(2) = 2$

13. a. $(f \circ g)(x) = f(g(x))$

$\qquad = f\left(\dfrac{1}{x}\right)$

$\qquad = \dfrac{1}{\frac{1}{x}} = 1 \cdot \dfrac{x}{1} = x$

b. $(g \circ f)(x) = g(f(x))$

$\qquad = g\left(\dfrac{1}{x}\right)$

$\qquad = \dfrac{1}{\frac{1}{x}} = 1 \cdot \dfrac{x}{1} = x$

c. $(f \circ g)(2) = 2$

15. $f\big(g(x)\big) = f\left(\dfrac{x}{4}\right) = 4\left(\dfrac{x}{4}\right) = x$

$g\big(f(x)\big) = g(4x) = \dfrac{4x}{4} = x$

The functions are inverses.

17. $f\big(g(x)\big) = f\left(\dfrac{x-8}{3}\right)$

$= 3\left(\dfrac{x-8}{3}\right) + 8$

$= x - 8 + 8 = x$

$g\big(f(x)\big) = g(3x+8)$

$= \dfrac{(3x+8) - 8}{3}$

$= \dfrac{3x + 8 - 8}{3} = \dfrac{3x}{3} = x$

The functions are inverses.

19. $f\big(g(x)\big) = f\left(\dfrac{x+5}{9}\right)$

$= 5\left(\dfrac{x+5}{9}\right) - 9$

$= \dfrac{5x+25}{9} - \dfrac{81}{9}$

$= \dfrac{5x+25-81}{9} = \dfrac{5x-56}{9}$

$g\big(f(x)\big) = g(5x-9)$

$= \dfrac{(5x-9)+5}{9} = \dfrac{5x-4}{9}$

Since $f\big(g(x)\big) \neq g\big(f(x)\big) \neq x$, we conclude the functions are not inverses.

21. $f\big(g(x)\big) = f\left(\dfrac{3}{x} + 4\right)$

$= \dfrac{3}{\left(\dfrac{3}{x} + 4\right) - 4}$

$= \dfrac{3}{\dfrac{3}{x} + 4 - 4} = \dfrac{3}{\dfrac{3}{x}} = 3 \cdot \dfrac{x}{3} = x$

$g\big(f(x)\big) = g\left(\dfrac{3}{x-4}\right)$

$= \dfrac{3}{\dfrac{3}{x-4}} + 4$

$= 3 \cdot \dfrac{x-4}{3} + 4 = x - 4 + 4 = x$

The functions are inverses.

23. $f\big(g(x)\big) = f(-x) = -(-x) = x$

$g\big(f(x)\big) = g(-x) = -(-x) = x$

The functions are inverses.

25. a. $f(x) = x + 3$

$y = x + 3$

Interchange x and y and solve for y.

$x = y + 3$

$x - 3 = y$

$f^{-1}(x) = x - 3$

b. $f\big(f^{-1}(x)\big) = f(x-3)$

$= (x-3) + 3$

$= x - 3 + 3 = x$

$f^{-1}\big(f(x)\big) = f(x+3)$

$= (x+3) - 3$

$= x + 3 - 3 = x$

27. a. $f(x) = 2x$

$y = 2x$

Interchange x and y and solve for y.

$x = 2y$

$\dfrac{x}{2} = y$

$f^{-1}(x) = \dfrac{x}{2}$

b. $f\left(f^{-1}(x)\right) = f\left(\dfrac{x}{2}\right) = 2\left(\dfrac{x}{2}\right) = x$

$f^{-1}\left(f(x)\right) = f(2x) = \dfrac{2x}{2} = x$

29. a. $f(x) = 2x + 3$

$y = 2x + 3$

Interchange x and y and solve for y.

$x = 2y + 3$

$x - 3 = 2y$

$\dfrac{x-3}{2} = y$

$f^{-1}(x) = \dfrac{x-3}{2}$

b. $f\left(f^{-1}(x)\right) = f\left(\dfrac{x-3}{2}\right)$

$= 2\left(\dfrac{x-3}{2}\right) + 3$

$= x - 3 + 3 = x$

$f^{-1}\left(f(x)\right) = f^{-1}(2x + 3)$

$= \dfrac{(2x+3)-3}{2}$

$= \dfrac{2x+3-3}{2} = \dfrac{2x}{2} = x$

31. a. $f(x) = \dfrac{1}{x}$

$y = \dfrac{1}{x}$

Interchange x and y and solve for y.

$x = \dfrac{1}{y}$

$xy = 1$

$y = \dfrac{1}{x}$

$f^{-1}(x) = \dfrac{1}{x}$

b. $f\left(f^{-1}(x)\right) = f\left(\dfrac{1}{x}\right)$

$= \dfrac{1}{\dfrac{1}{x}} = 1 \cdot \dfrac{x}{1} = x$

$f^{-1}\left(f(x)\right) = f^{-1}\left(\dfrac{1}{x}\right)$

$= \dfrac{1}{\dfrac{1}{x}} = 1 \cdot \dfrac{x}{1} = x$

33. a. $f(x) = \dfrac{2x+1}{x-3}$

$y = \dfrac{2x+1}{x-3}$

Interchange x and y and solve for y.

$x = \dfrac{2y+1}{y-3}$

$x(y-3) = 2y + 1$

$xy - 3x = 2y + 1$

$xy - 2y = 3x + 1$

$(x-2)y = 3x + 1$

$y = \dfrac{3x+1}{x-2}$

$f^{-1}(x) = \dfrac{3x+1}{x-2}$

b. $f\left(f^{-1}(x)\right) = f\left(\dfrac{3x+1}{x-2}\right)$

$= \dfrac{2\left(\dfrac{3x+1}{x-2}\right)+1}{\left(\dfrac{3x+1}{x-2}\right)-3}$

$= \dfrac{x-2}{x-2} \cdot \dfrac{2\left(\dfrac{3x+1}{x-2}\right)+1}{\left(\dfrac{3x+1}{x-2}\right)-3}$

$= \dfrac{2(3x+1)+1(x-2)}{(3x+1)-3(x-2)}$

$= \dfrac{6x+2+x-2}{3x+1-3x+6}$

$= \dfrac{7x}{7}$

$= x$

$f^{-1}\left(f(x)\right) = f^{-1}\left(\dfrac{2x+1}{x-3}\right)$

$= \dfrac{3\left(\dfrac{2x+1}{x-3}\right)+1}{\left(\dfrac{2x+1}{x-3}\right)-2}$

$= \dfrac{x-3}{x-3} \cdot \dfrac{3\left(\dfrac{2x+1}{x-3}\right)+1}{\left(\dfrac{2x+1}{x-3}\right)-2}$

$= \dfrac{3(2x+1)+1(x-3)}{(2x+1)-2(x-3)}$

$= \dfrac{6x+3+x-3}{2x+1-2x+6}$

$= \dfrac{7x}{7}$

$= x$

35. The graph does not satisfy the horizontal line test so the function does not have an inverse.

37. The graph does not satisfy the horizontal line test so the function does not have an inverse.

39. The graph satisfies the horizontal line test so the function has an inverse.

41.

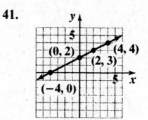

43.

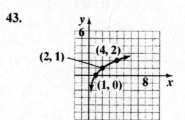

45. $f\left(g(1)\right) = f(1) = 5$

47. $(g \circ f)(-1) = g\left(f(-1)\right) = g(1) = 1$

49. $f^{-1}\left(g(10)\right) = f^{-1}(-1) = 2$, since $f(2) = -1$.

51. $(f \circ g)(-1) = f\left(g(-1)\right) = f(-3) = 1$

53. $(g \circ f)(0) = g\left(f(0)\right) = g(2) = -6$

55. $(f \circ g)(0) = f\left(g(0)\right)$

$= f(4 \cdot 0 - 1)$

$= f(-1) = 2(-1) - 5 = -7$

57. Let $f^{-1}(1) = x$. Then

$f(x) = 1$

$2x - 5 = 1$

$2x = 6$

$x = 3$

Thus, $f^{-1}(1) = 3$

59. $g\left(f[h(1)]\right) = g\left(f\left[1^2 + 1 + 2\right]\right)$

$= g\left(f(4)\right)$

$= g(2 \cdot 4 - 5)$

$= g(3)$

$= 4 \cdot 3 - 1 = 11$

61. a. f represents the price after a \$400 discount; g represents the price after a 25% discount (75% of the regular price).

b. $(f \circ g)(x) = f(g(x))$
$$= f(0.75x)$$
$$= 0.75x - 400$$

$f \circ g$ represents and additional \$400 discount on a price that has already been reduced by 25%.

c. $(g \circ f)(x) = g(f(x))$
$$= g(x - 400)$$
$$= 0.75(x - 400)$$
$$= 0.75x - 300$$

$g \circ f$ represents an additional 25% discount on a price that has already been reduced by \$400.

d. $0.75x - 400 < 0.75x - 300$, so $f \circ g$ models the greater discount. It has a savings of \$100 over $g \circ f$.

e. $f(x) = x - 400$
$$y = x - 400$$

Interchange x and y and solve for y.
$$x = y - 400$$
$$x + 400 = y$$
$$f^{-1}(x) = x + 400$$

f^{-1} represents the regular price of the computer, since the value of x here is the price after a \$400 discount.

63. a. f: {(Zambia, 4.2), (Columbia, 4.5), (Poland, 3.3), (Italy, 3.3), (U.S., 2.5)}

b. Inverse: {(4.2, Zambia), (4.5, Columbia), (3.3, Poland), (3.3, Italy), (2.5, U.S.)}; The inverse is not a function because the input 3.3 is associated with two different outputs: Poland and Italy.

65. a. We know that f has an inverse because no horizontal line intersects the graph of f in more than one point.

b. $f^{-1}(0.25)$, or approximately 15, represents the number of people who must be in a room so that the probability of two sharing a birthday would be 0.25; $f^{-1}(0.5)$, or approximately 23, represents the number of people who must be in a room so that the probability of 2 sharing a birthday would be 0.5;

$f^{-1}(0.7)$, or approximately 30, represents the number of people who must be in a room so that the probability of two sharing a birthday would be 0.70.

67. $f(g(x)) = f\left(\dfrac{5}{9}(x - 32)\right)$
$$= \dfrac{9}{5}\left[\dfrac{5}{9}(x - 32)\right] + 32$$
$$= (x - 32) + 32$$
$$= x$$

$$g(f(x)) = g\left(\dfrac{9}{5}x + 32\right)$$
$$= \dfrac{5}{9}\left[\left(\dfrac{9}{5}x + 32\right) - 32\right]$$
$$= \dfrac{5}{9}\left(\dfrac{9}{5}x\right)$$
$$= x$$

Since $f(g(x)) = x$ and $g(f(x)) = x$, then f and g are inverses.

69. – 73. Answers will vary

75. $f(x) = \sqrt[3]{2 - x}$

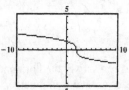

f has an inverse function because it passes the horizontal line test.

77. $f(x) = \dfrac{x^4}{4}$

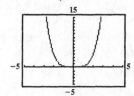

f does not have an inverse function because it does not pass the horizontal line test.

79. $f(x) = (x-1)^3$

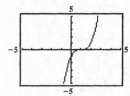

f has an inverse function because it passes the horizontal line test.

81. $f(x) = x^3 + x + 1$

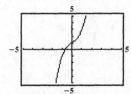

f has an inverse function because it passes the horizontal line test.

83. $f(x) = 4x + 4$

$g(x) = 0.25x - 1$

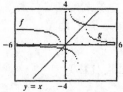

f and *g* are inverses because they are reflections of each other over the line *y = x*.

85. does not make sense; Explanations will vary. Sample explanation: The diagram illustrates $g(f(x))$.

87. does not make sense; Explanations will vary. Sample explanation: A quadratic function does not have an inverse because it fails the horizontal line test.

89. false; Changes to make the statement true will vary. A sample change is: The inverse is $\{(4,1),(7,2)\}$.

91. true

93. To find $(f \circ g)^{-1}(x)$, first find $(f \circ g)(x)$.

$$(f \circ g)(x) = f(g(x)) = f(x+5)$$
$$= 3(x+5) = 3x + 15$$

Now find $(f \circ g)^{-1}(x)$.

$$(f \circ g)(x) = 3x + 15$$
$$y = 3x + 15$$

Interchange *x* and *y* and solve for *y*.

$$x = 3y + 15$$
$$x - 15 = 3y$$
$$\frac{x-15}{3} = y$$

$$(f \circ g)^{-1}(x) = \frac{x-15}{3}$$

To find $(g^{-1} \circ f^{-1})(x)$, find $g^{-1}(x)$ and $f^{-1}(x)$.

$$g(x) = x + 5 \qquad f(x) = 3x$$
$$y = x + 5 \qquad y = 3x$$

Interchange *x* and *y* and solve for *y*.

$$x = y + 5 \qquad\qquad x = 3y$$
$$x - 5 = y \qquad\qquad \frac{x}{3} = y$$
$$g^{-1}(x) = x - 5 \qquad f^{-1}(x) = \frac{x}{3}$$

Now find $(g^{-1} \circ f^{-1})(x)$.

$$(g^{-1} \circ f^{-1})(x) = g^{-1}(f^{-1}(x))$$
$$= g^{-1}\left(\frac{x}{3}\right) = \frac{x}{3} - 5$$
$$= \frac{x}{3} - \frac{15}{3} = \frac{x-15}{3}$$

Notice that $(f \circ g)^{-1}(x) = (g^{-1} \circ f^{-1})(x)$.

95. $f(x) = m_1 x + b_1$

$g(x) = m_2 x + b_2$

First find $(f \circ g)(x)$.

$$(f \circ g)(x) = f(g(x)) = f(m_2 x + b_2)$$
$$= m_1(m_2 x + b_2) + b_1$$
$$= m_1 m_2 x + m_1 b_2 + b_1$$

The slope of the composite function is $m_1 m_2$. The slope of f is m_1 and the slope of g is m_2, thus the product $m_1 m_2$ is the same as the slope of the composite function.

96. $\dfrac{4.3 \times 10^5}{8.6 \times 10^{-4}} = \dfrac{4.3}{8.6} \times \dfrac{10^5}{10^{-4}} = 0.5 \times 10^9$

$$= 5 \times 10^{-1} \times 10^9 = 5 \times 10^8$$

97.

$$
\require{enclose}
\begin{array}{r}
x^2 + 9x + 16 \\[2pt]
x - 2 \enclose{longdiv}{x^3 - 7x^2 - 20x + 3} \\
\end{array}
$$

$$\underline{x^3 - 2x^2}$$
$$9x^2 - 2x$$
$$\underline{9x^2 - 18x}$$
$$16x + 3$$
$$\underline{16x - 32}$$
$$35$$

$$\frac{x^3 + 7x^2 - 20x + 3}{x - 2} = x^2 + 9x + 16 + \frac{35}{x - 2}$$

98. $3x + 2y = 6$

$8x - 3y = 1$

To solve this system by the addition method, multiply the first equation by 3 and then second equation by 2; then add the equations.

$9x + 6y = 18$

$\underline{16x - 6y = \; 2}$

$25x \qquad = 20$

$$x = \frac{20}{25} = \frac{4}{5}$$

Instead of substituting $\dfrac{4}{5}$ for x for working with fractions, go back to the original system and eliminate x.

To do this, multiply the first equation by 8 and the second equation by -3; then add.

$24x + 16y = 48$

$\underline{-24x + \; 9y = -3}$

$25y = 45$

$$y = \frac{45}{25} = \frac{9}{5}$$

The solution set is $\left\{ \left(\dfrac{4}{5}, \dfrac{9}{5} \right) \right\}$.

99. $2 - 12x = 7(x - 1)$

$2 - 12x = 7x - 7$

$-12x - 7x = -7 - 2$

$-19x = -9$

$$x = \frac{9}{19}$$

The solution set is $\left\{ \dfrac{9}{19} \right\}$.

100. $\dfrac{x+3}{4} = \dfrac{x-2}{3} + \dfrac{1}{4}$

$12 \cdot \dfrac{x+3}{4} = 12 \cdot \dfrac{x-2}{3} + 12 \cdot \dfrac{1}{4}$

$3(x + 3) = 4(x - 2) + 3$

$3x + 9 = 4x - 8 + 3$

$3x + 9 = 4x - 5$

$3x - 4x = -5 - 9$

$-x = -14$

$x = 14$

The solution set is $\{14\}$.

101. $600x - (500{,}000 + 400x) > 0$

$600x - 500{,}000 - 400x > 0$

$200x - 500{,}000 > 0$

$200x > 500{,}000$

$x > 2500$

$(2500, \infty)$

Chapter 8 Review Exercises

1. The relation is a function.
   Domain {3, 4, 5}
   Range {10}

2. The relation is a function.
   Domain {1, 2, 3, 4}
   Range {−6, π, 12, 100}

3. The relation is not a function.
   Domain {13, 15}
   Range {14, 16, 17}

4. **a.** $f(0) = 7(0) - 5 = 0 - 5 = -5$

   b. $f(3) = 7(3) - 5 = 21 - 5 = 16$

   c. $f(-10) = 7(-10) - 5 = -75$

   d. $f(2a) = 7(2a) - 5 = 14a - 5$

   e. $f(a+2) = 7(a+2) - 5$
   $\qquad = 7a + 14 - 5 = 7a + 9$

5. **a.** $g(0) = 3(0)^2 - 5(0) + 2 = 2$

   b. $g(5) = 3(5)^2 - 5(5) + 2$
   $\qquad = 3(25) - 25 + 2$
   $\qquad = 75 - 25 + 2 = 52$

   c. $g(-4) = 3(-4)^2 - 5(-4) + 2 = 70$

   d. $g(b) = 3(b)^2 - 5(b) + 2$
   $\qquad = 3b^2 - 5b + 2$

   e. $g(4a) = 3(4a)^2 - 5(4a) + 2$
   $\qquad = 3(16a^2) - 20a + 2$
   $\qquad = 48a^2 - 20a + 2$

6. The vertical line test shows that this is not the graph of a function.

7. The vertical line test shows that this is the graph of a function.

8. The vertical line test shows that this is the graph of a function.

9. The vertical line test shows that this is not the graph of a function.

10. The vertical line test shows that this is not the graph of a function.

11. The vertical line test shows that this is the graph of a function.

12. $\left\{ x \mid -2 < x \le 3 \right\}$

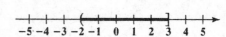

13. $\left\{ x \mid -1.5 \le x \le 2 \right\}$

14. $\left\{ x \mid x > -1 \right\}$

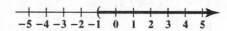

15. $f(-2) = -3$

16. $f(0) = -2$

17. When $x = 3$, $f(x) = -5$.

18. The domain of f is $\{x \mid -3 \le x < 5\}$.

19. The range of f is $\{y \mid -5 \le y \le 0\}$.

20. **a.** The eagle's height is a function of its time in flight because every time, t, is associated with at most one height.

    b. $f(15) = 0$
    At time $t = 15$ seconds, the eagle is at height zero. This means that after 15 seconds, the eagle is on the ground.

    c. The eagle's maximum height is 45 meters.

    d. For $x = 7$ and 22, $f(x) = 20$. This means that at times 7 seconds and 22 seconds, the eagle is at a height of 20 meters.

    e. The eagle began the flight at 45 meters and remained there for approximately 3 seconds. At that time, the eagle descended for 9 seconds. It landed on the ground and stayed there for 5 seconds. The eagle then began to climb back up to a height of 44 meters.

21. The domain of f is
$\{x \mid x \text{ is a real number}\}$ or $(-\infty, \infty)$.

22. The domain of
f is $\{x \mid x \text{ is a real number and } x \neq -8\}$.

23. The domain of
f is $\{x \mid x \text{ is a real number and } x \neq 5\}$.

24. a. $(f+g)(x) = (4x-5) + (2x+1)$
$$= 4x - 5 + 2x + 1$$
$$= 6x - 4$$

b. $(f+g)(3) = 6(3) - 4$
$$= 18 - 4 = 14$$

25. a. $(f+g)(x)$
$$= (5x^2 - x + 4) + (x - 3)$$
$$= 5x^2 - x + 4 + x - 3 = 5x^2 + 1$$

b. $(f+g)(3) = 5(3)^2 + 1 = 5(9) + 1$
$$= 45 + 1 = 46$$

26. The domain of $f + g$ is
$\{x \mid x \text{ is a real number and } x \neq 4\}$

27. The domain of $f + g$ is
$\{x \mid x \text{ is a real number and } x \neq -6 \text{ and } x \neq -1\}$.

28. $f(x) = x^2 - 2x, \quad g(x) = x - 5$
$(f+g)(x) = (x^2 - 2x) + (x - 5)$
$$= x^2 - 2x + x - 5$$
$$= x^2 - x - 5$$
$(f+g)(-2) = (-2)^2 - (-2) - 5$
$$= 4 + 2 - 5 = 1$$

29. From Exercise 27 we know
$(f+g)(x) = x^2 - x - 5$. We can use this to find
$f(3) + g(3)$.
$f(3) + g(3) = (f+g)(3)$
$$= (3)^2 - (3) - 5$$
$$= 9 - 3 - 5 = 1$$

30. $f(x) = x^2 - 2x, \quad g(x) = x - 5$
$(f-g)(x) = (x^2 - 2x) - (x - 5)$
$$= x^2 - 2x - x + 5$$
$$= x^2 - 3x + 5$$
$(f-g)(x) = x^2 - 3x + 5$
$(f-g)(1) = (1)^2 - 3(1) + 5$
$$= 1 - 3 + 5 = 3$$

31. From Exercise 29 we know
$(f-g)(x) = x^2 - 3x + 5$. We can use this to find
$f(4) - g(4)$.
$f(4) - g(4) = (f-g)(4)$
$$= (4)^2 - 3(4) + 5$$
$$= 16 - 12 + 5 = 9$$

32. $(fg)(x) = f(x) \cdot g(x)$
$$= (x^2 - 2x)(x - 5)$$
$$= x^3 - 7x^2 + 10x$$
$(fg)(-3) = (-3)^3 - 7(-3)^2 + 10(-3)$
$$= -27 - 63 - 30$$
$$= -120$$

33. $f(x) = x^2 - 2x, \quad g(x) = x - 5$
$\left(\dfrac{f}{g}\right)(x) = \dfrac{x^2 - 2x}{x - 5}$
$\left(\dfrac{f}{g}\right)(4) = \dfrac{(4)^2 - 2(4)}{4 - 5} = \dfrac{16 - 8}{-1}$
$$= \dfrac{8}{-1} = -8$$

34. $(f-g)(x) = x^2 - 3x + 5$
The domain of $f - g$ is $\{x \mid x \text{ is a real number}\}$.

436

35. $\left(\dfrac{f}{g}\right)(x) = \dfrac{x^2 - 2x}{x - 5}$

The domain of

$\dfrac{f}{g}$ is $\{x \mid x \text{ is a real number and } x \neq 5\}$.

36. a. $(f \circ g)(x) = f(g(x))$

$\qquad = f(4x - 1)$

$\qquad = (4x - 1)^2 + 3$

$\qquad = 16x^2 - 8x + 1 + 3$

$\qquad = 16x^2 - 8x + 4$

b. $(g \circ f)(x) = g(f(x))$

$\qquad = g(x^2 + 3)$

$\qquad = 4(x^2 + 3) - 1$

$\qquad = 4x^2 + 12 - 1$

$\qquad = 4x^2 + 11$

c. $(f \circ g)(3) = 16(3)^2 - 8(3) + 4$

$\qquad = 16(9) - 24 + 4$

$\qquad = 144 - 24 + 4$

$\qquad = 124$

37. a. $(f \circ g)(x) = f(g(x))$

$\qquad = f(x + 1) = \sqrt{x + 1}$

b. $(g \circ f)(x) = g(f(x))$

$\qquad = g(\sqrt{x}) = \sqrt{x} + 1$

c. $(f \circ g)(3) = \sqrt{3 + 1} = \sqrt{4} = 2$

38. $f(x) = \dfrac{3}{5}x + \dfrac{1}{2}$ and $g(x) = \dfrac{5}{3}x - 2$

$f(g(x)) = f\left(\dfrac{5}{3}x - 2\right)$

$\qquad = \dfrac{3}{5}\left(\dfrac{5}{3}x - 2\right) + \dfrac{1}{2}$

$\qquad = \dfrac{3}{5}\left(\dfrac{5}{3}x\right) - \left(\dfrac{3}{5}\right)2 + \dfrac{1}{2}$

$\qquad = x - \dfrac{6}{5} + \dfrac{1}{2}$

$\qquad = x - \dfrac{7}{10}$

$g(f(x)) = g\left(\dfrac{3}{5}x + \dfrac{1}{2}\right)$

$\qquad = \dfrac{5}{3}\left(\dfrac{3}{5}x + \dfrac{1}{2}\right) - 2$

$\qquad = \dfrac{5}{3}\left(\dfrac{3}{5}x\right) + \left(\dfrac{5}{3}\right)\dfrac{1}{2} - 2$

$\qquad = x + \dfrac{5}{6} - 2$

$\qquad = x - \dfrac{7}{6}$

The functions are not inverses.

39. $f(x) = 2 - 5x$ and $g(x) = \dfrac{2 - x}{5}$

$f(g(x)) = f\left(\dfrac{2 - x}{5}\right)$

$\qquad = 2 - 5\left(\dfrac{2 - x}{5}\right)$

$\qquad = 2 - (2 - x) = 2 - 2 + x = x$

$g(f(x)) = g(2 - 5x)$

$\qquad = \dfrac{2 - (2 - 5x)}{5}$

$\qquad = \dfrac{2 - 2 + 5x}{5} = \dfrac{5x}{5} = x$

The functions are inverses.

40. a. $f(x) = 4x - 3$

$y = 4x - 3$

Interchange x and y and solve for y.

$x = 4y - 3$

$x + 3 = 4y$

$\dfrac{x+3}{4} = y$

$f^{-1}(x) = \dfrac{x+3}{4}$

b. $f\left(f^{-1}(x)\right) = f\left(\dfrac{x+3}{4}\right)$

$\qquad = 4\left(\dfrac{x+3}{4}\right) - 3$

$\qquad = x + 3 - 3 = x$

$f^{-1}\left(f(x)\right) = f(4x - 3)$

$\qquad = \dfrac{(4x-3)+3}{4}$

$\qquad = \dfrac{4x-3+3}{4} = \dfrac{4x}{4} = x$

41. a. $f(x) = -\dfrac{1}{x}$

$y = -\dfrac{1}{x}$

Interchange x and y and solve for y.

$x = -\dfrac{1}{y}$

$y = -\dfrac{1}{x}$

$f^{-1}(x) = -\dfrac{1}{x}$

b. $f\left(f^{-1}(x)\right) = f\left(-\dfrac{1}{x}\right) = -\dfrac{1}{\left(-\dfrac{1}{x}\right)} = x$

$f^{-1}\left(f(x)\right) = f^{-1}\left(-\dfrac{1}{x}\right)$

$\qquad = -\dfrac{1}{\left(-\dfrac{1}{x}\right)} = x$

42. Since the graph satisfies the horizontal line test, it has an inverse function.

43. Since the graph does not satisfy the horizontal line test, it does not have an inverse function.

44. Since the graph satisfies the horizontal line test, it has an inverse function.

45. Since the graph does not satisfy the horizontal line test, it does not have an inverse function.

46. Since the points $(-3,-1), (0,0)$ and $(2,4)$ lie on the graph of the function, the points $(-1,-3)$, $(0,0)$ and $(4,2)$ lie on the inverse function.

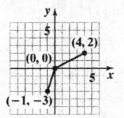

Chapter 8 Test

1. The relation is a function.
Domain $\{1, 3, 5, 6\}$
Range $\{2, 4, 6\}$

2. The relation is not a function.
Domain $\{2, 4, 6\}$
Range $\{1, 3, 5, 6\}$

3. $f(a+4) = 3(a+4) - 2$
$\qquad = 3a + 12 - 2 = 3a + 10$

4. $f(-2) = 4(-2)^2 - 3(-2) + 6$
$\qquad = 4(4) + 6 + 6 = 16 + 6 + 6 = 28$

5. The vertical line test shows that this is the graph of a function.

6. The vertical line test shows that this is not the graph of a function.

7. $f(6) = -3$

8. $f(x) = 0$ when $x = -2$ and $x = 3$.

9. The domain of f is
$\{x | x$ is a real number$\}$ or $(-\infty, \infty)$.

10. The range of f is $\{y | y \le 3\}$ or $(-\infty, 3]$.

11. The domain of f is
$\{x | x$ is a real number and $x \ne 10\}$.

12. $f(x) = x^2 + 4x$ and $g(x) = x + 2$
$(f+g)(x) = f(x) + g(x)$
$= (x^2 + 4x) + (x + 2)$
$= x^2 + 4x + x + 2$
$= x^2 + 5x + 2$
$(f+g)(3) = (3)^2 + 5(3) + 2$
$= 9 + 15 + 2 = 26$

13. $f(x) = x^2 + 4x$ and $g(x) = x + 2$
$(f-g)(x) = f(x) - g(x)$
$= (x^2 + 4x) - (x + 2)$
$= x^2 + 4x - x - 2$
$= x^2 + 3x - 2$
$(f-g)(-1) = (-1)^2 + 3(-1) - 2$
$= 1 - 3 - 2 = -4$

14. $(fg)(x) = f(x) \cdot g(x)$
$= (x^2 + 4x)(x + 2)$
$= x^3 + 6x^2 + 8x$
$(fg)(-5) = (-5)^3 + 6(-5)^2 + 8(-5)$
$= -125 + 150 - 40$
$= -15$

15. $f(x) = x^2 + 4x$ and $g(x) = x + 2$
$\left(\dfrac{f}{g}\right)(x) = \dfrac{x^2 + 4x}{x + 2}$
$\left(\dfrac{f}{g}\right)(2) = \dfrac{(2)^2 + 4(2)}{2 + 2} = \dfrac{4 + 8}{4} = \dfrac{12}{4} = 3$

16. Domain of $\dfrac{f}{g}$ is $\{x | x$ is a real number and $x \ne -2\}$.

17. $f(x) = x^2 + x$ and $g(x) = 3x - 1$
$(f \circ g)(x) = f(g(x)) = f(3x - 1)$
$= (3x - 1)^2 + (3x - 1)$
$= 9x^2 - 6x + 1 + 3x - 1$
$= 9x^2 - 3x$
$(g \circ f)(x) = g(f(x))$
$= g(x^2 + x)$
$= 3(x^2 + x) - 1$
$= 3x^2 + 3x - 1$

18. $f(x) = 5x - 7$
$y = 5x - 7$

Interchange x and y and solve for y.
$x = 5y - 7$
$x + 7 = 5y$
$\dfrac{x + 7}{5} = y$

$f^{-1}(x) = \dfrac{x + 7}{5}$

19. a. The function passes the horizontal line test so we know its inverse is a function.

b. $f(80) = 2000$

c. $f^{-1}(2000)$ represents the income, $80 thousand, of a family that gives $2000 to charity.

Cumulative Review Exercises (Chapters 1 – 8)

1. $2x + 3x - 5 + 7 = 10x + 3 - 6x - 4$

$5x + 2 = 4x - 1$

$x + 2 = -1$

$x = -3$

The solution set is $\{-3\}$.

2. $2x^2 + 5x = 12$

$2x^2 + 5x - 12 = 0$

$(2x - 3)(x + 4) = 0$

$2x - 3 = 0 \quad$ or $\quad x + 4 = 0$

$2x = 3 \qquad\qquad x = -4$

$x = \dfrac{3}{2}$

The solution set is $\left\{-4, \dfrac{3}{2}\right\}$.

3. $8x - 5y = -4$

$2x + 15y = -66$

Eliminate y by multiplying both sides of the first equation by 3 and adding the two equations.

$24x - 15y = -12$

$\underline{2x + 15y = -66}$

$26x \qquad = -78$

$x \qquad = -3$

Let $x = -3$ in the first equation and solve for y.

$8(-3) - 5y = -4$

$-24 - 5y = -4$

$-5y = 20$

$y = -4$

The solution is $(-3, -4)$ and the solution set is $\{(-3, -4)\}$.

4. $\dfrac{15}{x} - 4 = \dfrac{6}{x} + 3$

Multiply both sides of the equation by x to eliminate the fractions.

$x\left(\dfrac{15}{x} - 4\right) = x\left(\dfrac{6}{x} + 3\right)$

$15 - 4x = 6 + 3x$

$15 = 6 + 7x$

$9 = 7x$

$\dfrac{9}{7} = x$

The solution set is $\left\{\dfrac{9}{7}\right\}$.

5. $-3x - 7 = 8$

$-3x = 15$

$x = -5$

The solution set is $\{-5\}$.

6. $f(x) = 2x^2 - 5x + 2$; $g(x) = x^2 - 2x + 3$

$(f - g)(x)$

$= f(x) - g(x)$

$= \left(2x^2 - 5x + 2\right) - \left(x^2 - 2x + 3\right)$

$= 2x^2 - 5x + 2 - x^2 + 2x - 3$

$= x^2 - 3x - 1$

$(f - g)(3) = (3)^2 - 3(3) - 1$

$= 9 - 9 - 1$

$= -1$

7. $\dfrac{8x^3}{-4x^7} = \dfrac{8}{-4} \cdot \dfrac{x^3}{x^7} = -2x^{3-7}$

$= -2x^{-4} = -\dfrac{2}{x^4}$

8. $-8 - (-3) \cdot 4 = -8 - (-12)$

$= -8 + 12$

$= 4$

9. $\dfrac{\dfrac{1}{x}-\dfrac{1}{2}}{\dfrac{1}{3}-\dfrac{x}{6}} = \dfrac{\dfrac{2}{2x}-\dfrac{x}{2x}}{\dfrac{2}{6}-\dfrac{x}{6}} = \dfrac{\dfrac{2-x}{2x}}{\dfrac{2-x}{6}}$

$= \dfrac{2-x}{2x}\cdot\dfrac{6}{2-x} = \dfrac{6}{2x} = \dfrac{3}{x}$

10. $\dfrac{4-x^2}{3x^2-5x-2} = \dfrac{(2-x)(2+x)}{(3x+1)(x-2)}$

$= \dfrac{-1(x-2)(2+x)}{(3x+1)(x-2)}$

$= \dfrac{-(2+x)}{3x+1}$ or $-\dfrac{2+x}{3x+1}$

11. $-5-(-8)-(4-6)$

$=-5-(-8)-(-2)$

$=-5+8+2$

$=3+2$

$=5$

12. $x^2-18x+77$

We need two factors of 77 whose sum is -18.
Since the product is positive, the factors have the
same sign, and since the sum is negative, they are
both negative.

Since $-11\cdot-7=77$ and $-11+(-7)=-18$, we get

$x^2-18x+77=(x-11)(x-7)$

13. $x^3-25x = x\left(x^2-25\right)$

$= x(x-5)(x+5)$

14. $x-2\overline{)6x^3-19x^2+16x-4}$ with quotient $6x^2-7x+2$

$\underline{6x^3-12x^2}$

$-7x^2+16x$

$\underline{-7x^2+14x}$

$2x-4$

$\underline{2x-4}$

0

$\dfrac{6x^3-19x^2+16x-4}{x-2}=6x^2-7x+2$

15. $(2x-3)\left(4x^2+6x+9\right)$

$=(2x)\left(4x^2\right)+(2x)(6x)+(2x)(9)$

$-3\left(4x^2\right)-3(6x)-3(9)$

$=8x^3+12x^2+18x-12x^2-18x-27$

$=8x^3-27$

16. $\dfrac{3x}{x^2+x-2}-\dfrac{2}{x+2}$

$=\dfrac{3x}{(x+2)(x-1)}-\dfrac{2(x-1)}{(x+2)(x-1)}$

$=\dfrac{3x-2(x-1)}{(x+2)(x-1)}$

$=\dfrac{3x-2x+2}{(x+2)(x-1)}$

$=\dfrac{x+2}{(x+2)(x-1)}$

$=\dfrac{1}{x-1}$

17. $\dfrac{5x^2-6x+1}{x^2-1}\div\dfrac{16x^2-9}{4x^2+7x+3}$

$=\dfrac{5x^2-6x+1}{x^2-1}\cdot\dfrac{4x^2+7x+3}{16x^2-9}$

$=\dfrac{(5x-1)(x-1)}{(x+1)(x-1)}\cdot\dfrac{(4x+3)(x+1)}{(4x-3)(4x+3)}$

$=\dfrac{(5x-1)\cancel{(x-1)}}{\cancel{(x+1)}\cancel{(x-1)}}\cdot\dfrac{\cancel{(4x+3)}\cancel{(x+1)}}{(4x-3)\cancel{(4x+3)}}$

$=\dfrac{5x-1}{4x-3}$

18.
$$x + 3y - z = 5$$
$$-x + 2y + 3z = 13$$
$$2x - 5y - z = -8$$

Eliminate x from the second equation by adding the first two equations together.
$$x + 3y - z = 5$$
$$\underline{-x + 2y + 3z = 13}$$
$$5y + 2z = 18$$

Eliminate x from the third equation by multiplying both sides of the second equation by 2 and adding to the third equation.
$$-2x + 4y + 6z = 26$$
$$\underline{2x - 5y - z = -8}$$
$$-y + 5z = 18$$

Using the two reduced equations, we can form the following system of linear equations in two variables:
$$5y + 2z = 18$$
$$-y + 5z = 18$$

Multiply the second equation by 5 and add to the first equation.
$$5y + 2z = 18$$
$$\underline{-5y + 25z = 90}$$
$$27z = 108$$
$$z = 4$$

Back-substitute this value to solve for y.
$$5y + 2z = 18$$
$$5y + 2(4) = 18$$
$$5y + 8 = 18$$
$$5y = 10$$
$$y = 2$$

Back-substitute the values for y and z to solve for x.
$$x + 3y - z = 5$$
$$x + 3(2) - (4) = 5$$
$$x + 6 - 4 = 5$$
$$x + 2 = 5$$
$$x = 3$$

The solution is $(3, 2, 4)$ and the solution set is $\{(3, 2, 4)\}$.

19.
$$2x - y = 4$$
$$-y = -2x + 4$$
$$y = 2x - 4$$

The slope is $m = \dfrac{2}{1} = 2$ and the y-intercept is $b = -4$.

One point is $(0, -4)$ and using the slope, we can get a second point: $(1, -2)$

Let $x = -1$. $y = 2(-1) - 4 = -6$, so the point $(-1, -6)$ must be on the graph as well.

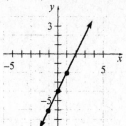

20. $y = -\dfrac{2}{3}x$

The slope is $m = -\dfrac{2}{3} = \dfrac{-2}{3}$ and the y-intercept is 0.

One point is $(0, 0)$ and using the slope we can get a second point: $(3, -2)$.

Let $x = -3$. $y = -\dfrac{2}{3}(-3) = 2$, so the point $(-3, 2)$ should be on the graph.

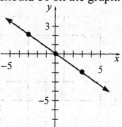

21. Since each element from the domain corresponds to exactly one element of the range, the relation is a function.

Domain: $\{1, 2, 3, 4, 6\}$

Range: $\{5\}$

22. $m = \dfrac{y_2 - y_1}{x_2 - x_1} = \dfrac{-3 - 5}{2 - (-1)} = \dfrac{-8}{3} = -\dfrac{8}{3}$

23.　$m = 5$; $(x_1, y_1) = (-2, -3)$

$$y - y_1 = m(x - x_1)$$

$$y - (-3) = 5(x - (-2))$$

$$y + 3 = 5(x + 2)$$

To obtain the slope-intercept form, solve this equation for y.

$$y + 3 = 5(x + 2)$$

$$y + 3 = 5x + 10$$

$$y = 5x + 7$$

24.　$(7 \times 10^{-8})(3 \times 10^2)$

$$= (7 \cdot 3) \times (10^{-8} \cdot 10^2)$$

$$= 21 \times 10^{-6}$$

$$= 2.1 \times 10^1 \times 10^{-6}$$

$$= 2.1 \times 10^{-5}$$

25.　$f(x) = \dfrac{1}{15 - x}$

This is a rational function so the domain is all real numbers except where the denominator equals 0.

$$15 - x = 0$$

$$15 = x$$

Therefore, the domain is

$\{x \mid x \text{ is a real number and } x \neq 15\}$.

Chapter 9
Inequalities and Problem Solving

9.1 Check Points

1. $4x - 3 > -23$
   $4x > -20$
   $x > -5$
   $(-5, \infty)$ or $\{x | x > -5\}$

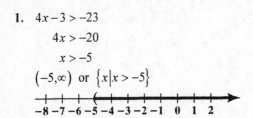

2. $3x + 1 > 7x - 15$
   $-4x > -16$
   $\dfrac{-4x}{-4} < \dfrac{-16}{-4}$
   $x < 4$
   $(-\infty, 4)$ or $\{x | x < 4\}$

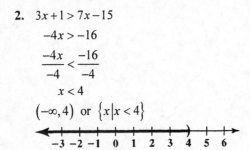

3. $\dfrac{x-4}{2} \geq \dfrac{x-2}{3} + \dfrac{5}{6}$
   $6\left(\dfrac{x-4}{2}\right) \geq 6\left(\dfrac{x-2}{3} + \dfrac{5}{6}\right)$
   $3(x-4) \geq 2(x-2) + 5$
   $3x - 12 \geq 2x - 4 + 5$
   $3x - 12 \geq 2x + 1$
   $x \geq 13$
   $[13, \infty)$ or $\{x | x \geq 13\}$

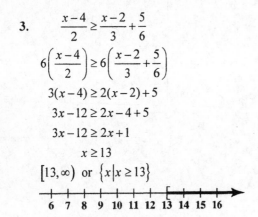

4. **a.** $P(x) = R(x) - C(x)$
   $P(x) = 200x - (160,000 + 75x)$
   $= 200x - 160,000 - 75x$
   $= 125x - 160,000$

   b. $P(x) > 0$
   $125x - 160,000 > 0$
   $125x > 160,000$
   $\dfrac{125x}{125} > \dfrac{160,000}{125}$
   $x > 1280$
   For the business to make money, more than 1280 units must be produced and sold.

5. **a.** $C(x) = 300,000 + 30x$

   b. $R(x) = 80x$

   c. $P(x) = R(x) - C(x)$
   $P(x) = 80x - (300,000 + 30x)$
   $= 80x - 300,000 - 30x$
   $= 50x - 300,000$

   d. $P(x) > 0$
   $50x - 300,000 > 0$
   $50x > 300,000$
   $\dfrac{50x}{50} > \dfrac{300,000}{50}$
   $x > 6000$
   For the business to make money, more than 6000 pairs must be produced and sold.

9.1 Exercise Set

1. $5x + 11 < 26$
   $5x < 15$
   $x < 3$
   The solution set is $\{x | x < 3\}$ or $(-\infty, 3)$.

2. $3x - 8 \geq 13$
   $3x \geq 21$
   $x \geq 7$
   The solution set is $\{x | x \geq 7\}$ or $[7, \infty)$.

5. $-9x \geq 36$
   $x \leq -4$
   The solution set is $\{x | x \leq -4\}$ or $(-\infty, -4]$.

7. $8x - 11 \le 3x - 13$

$\quad\quad 5x - 11 \le -13$

$\quad\quad\quad 5x \le -2$

$\quad\quad\quad\quad x \le -\dfrac{2}{5}$

The solution set is $\left\{ x \middle| x \le -\dfrac{2}{5} \right\}$ or $\left(-\infty, -\dfrac{2}{5} \right]$.

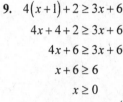

9. $4(x+1) + 2 \ge 3x + 6$

$\quad\quad 4x + 4 + 2 \ge 3x + 6$

$\quad\quad\quad 4x + 6 \ge 3x + 6$

$\quad\quad\quad\quad x + 6 \ge 6$

$\quad\quad\quad\quad\quad x \ge 0$

The solution set is $\left\{ x \middle| x \ge 0 \right\}$ or $[0, \infty)$.

11. $2x - 11 < -3(x + 2)$

$\quad\quad 2x - 11 < -3x - 6$

$\quad\quad 5x - 11 < -6$

$\quad\quad\quad 5x < 5$

$\quad\quad\quad\quad x < 1$

The solution set is $\left\{ x \middle| x < 1 \right\}$ or $(-\infty, 1)$.

13. $1 - (x + 3) \ge 4 - 2x$

$\quad\quad 1 - x - 3 \ge 4 - 2x$

$\quad\quad\quad -x - 2 \ge 4 - 2x$

$\quad\quad\quad\quad x - 2 \ge 4$

$\quad\quad\quad\quad\quad x \ge 6$

The solution set is $\left\{ x \middle| x \ge 6 \right\}$ or $[6, \infty)$.

15. $\quad\quad \dfrac{x}{4} - \dfrac{1}{2} \le \dfrac{x}{2} + 1$

$\quad 4\left(\dfrac{x}{4} \right) - 4\left(\dfrac{1}{2} \right) \le 4\left(\dfrac{x}{2} \right) + 4(1)$

$\quad\quad\quad x - 2 \le 2x + 4$

$\quad\quad\quad -x - 2 \le 4$

$\quad\quad\quad\quad -x \le 6$

$\quad\quad\quad\quad\quad x \ge -6$

The solution set is $\left\{ x \middle| x \ge -6 \right\}$ or $[-6, \infty)$.

17. $\quad\quad 1 - \dfrac{x}{2} > 4$

$\quad 2(1) - 2\left(\dfrac{x}{2} \right) > 2(4)$

$\quad\quad\quad 2 - x > 8$

$\quad\quad\quad\quad -x > 6$

$\quad\quad\quad\quad\quad x < -6$

The solution set is $\left\{ x \middle| x < -6 \right\}$ or $(-\infty, -6)$.

19. $\quad\quad \dfrac{x-4}{6} \ge \dfrac{x-2}{9} + \dfrac{5}{18}$

$\quad 18\left(\dfrac{x-4}{6} \right) \ge 18\left(\dfrac{x-2}{9} + \dfrac{5}{18} \right)$

$\quad\quad 3(x-4) \ge 2(x-2) + 5$

$\quad\quad 3x - 12 \ge 2x - 4 + 5$

$\quad\quad 3x - 12 \ge 2x + 1$

$\quad\quad\quad x \ge 13$

The solution set is $\left\{ x \middle| x \ge 13 \right\}$ or $[13, \infty)$.

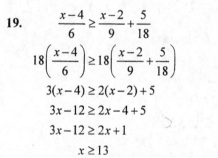

21. $7(x + 4) - 13 < 12 + 13(3 + x)$

$\quad 7x + 28 - 13 < 12 + 39 + 13x$

$\quad\quad 7x + 15 < 13x + 51$

$\quad\quad -6x + 15 < 51$

$\quad\quad\quad -6x < 36$

$\quad\quad\quad \dfrac{-6x}{-6} > \dfrac{36}{-6}$

$\quad\quad\quad\quad x > -6$

The solution set is $\{ x \mid x > -6 \}$ or $(-6, \infty)$.

23. $6 - \dfrac{2}{3}(3x - 12) \le \dfrac{2}{5}(10x + 50)$

$6 - 2x + 8 \le 4x + 20$

$-2x + 14 \le 4x + 20$

$-6x \le 6$

$\dfrac{-6x}{-6} \ge \dfrac{6}{-6}$

$x \ge -1$

The solution set is $\{x \mid x \ge -1\}$ or $[-1, \infty)$.

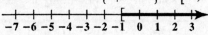

25. $3\big[3(x + 5) + 8x + 7\big] + 5\big[3(x - 6) - 2(3x - 5)\big] < 2(4x + 3)$

$3\big[3x + 15 + 8x + 7\big] + 5\big[3x - 18 - 6x + 10\big] < 8x + 6$

$3\big[11x + 22\big] + 5\big[-3x - 8\big] < 8x + 6$

$33x + 66 - 15x - 40 < 8x + 6$

$18x + 26 < 8x + 6$

$10x + 26 < 6$

$10x < -20$

$x < -2$

The solution set is $\{x \mid x < -2\}$ or $(-\infty, -2)$.

27. $f(x) > g(x)$

$3x + 2 > 5x - 8$

$-2x + 2 > -8$

$-2x > -10$

$\dfrac{-2x}{-2} < \dfrac{-10}{-2}$

$x < 5$

The solution set is $\{x \mid x < 5\}$ or $(-\infty, 5)$.

29. $f(x) \le g(x)$

$\dfrac{1}{4}(8 - 12x) \le \dfrac{2}{5}(10x + 15)$

$2 - 3x \le 4x + 6$

$2 - 7x \le 6$

$-7x \le 4$

$\dfrac{-7x}{-7} \ge \dfrac{4}{-7}$

$x \ge -\dfrac{4}{7}$

The solution set is $\left\{x \,\middle|\, x \ge -\dfrac{4}{7}\right\}$ or $\left[-\dfrac{4}{7}, \infty\right)$.

31. $1-(x+3)+2x \geq 4$

$1-x-3+2x \geq 4$

$-2+x \geq 4$

$x \geq 6$

The solution set is $\{x \mid x \geq 6\}$ or $[6,\infty)$.

33. a. $P(x) = R(x) - C(x)$

$= 32x - (25,500 + 15x)$

$= 32x - 25,500 - 15x$

$= 17x - 25,500$

b. $P(x) > 0$

$17x - 25,500 > 0$

$17x > 25,500$

$\dfrac{17x}{17} > \dfrac{25,500}{17}$

$x > 1500$

More than 1500 units must be produced and sold to have a profit.

35. a. $P(x) = R(x) - C(x)$

$= 245x - (105x + 70,000)$

$= 245x - 105x - 70,000$

$= 140x - 70,000$

b. $P(x) > 0$

$140x - 70,000 > 0$

$140x > 70,000$

$\dfrac{140x}{140} > \dfrac{70,000}{140}$

$x > 500$

More than 500 units must be produced and sold to have a profit.

37. $2(x+3) > 6 - \left\{4\left[x-(3x-4)-x\right]+4\right\}$

$2x+6 > 6 - \left\{4\left[x-3x+4-x\right]+4\right\}$

$2x+6 > 6 - \left\{4\left[-3x+4\right]+4\right\}$

$2x+6 > 6 - \left\{-12x+16+4\right\}$

$2x+6 > 6 - \left\{-12x+20\right\}$

$2x+6 > 6 + 12x - 20$

$2x+6 > 12x - 14$

$6 > 10x - 14$

$20 > 10x$

$2 > x$

$x < 2$

The solution set is $\{x \mid x < 2\}$ or $(-\infty, 2)$.

39. $ax + b > c, a < 0$

$ax + b - b > c - b$

$ax > c - b$

$\dfrac{ax}{a} < \dfrac{c-b}{a}, a < 0$

$x < \dfrac{c-b}{a}$

41. $\{x \mid x \leq -3\}$ or $(-\infty, -3]$

43. $\{x \mid x > -1.4\}$ or $(-1.4, \infty)$

45. $(0, 4)$

47. intimacy $\geq$ passion or passion $\leq$ intimacy

49. commitment $>$ passion or passion $<$ commitment

51. 9, after 3 years

53. $3.1x + 25.8 > 63$

$3.1x > 37.2$

$x > 12$

Since x is the number of years after 1994, we calculate $1994+12=2006$. 63% of voters will use electronic systems after 2006.

55. a. cost = fixed costs + variable cost

$$C(x) = 18,000 + 20x$$

b. revenue = price · quantity

$$R(x) = 80x$$

c. $P(x) = R(x) - C(x)$

$$= 80x - (18,000 + 20x)$$

$$= 80x - 18,000 - 20x$$

$$= 60x - 18,000$$

d. $P(x) > 0$

$$60x - 18,000 > 0$$

$$60x > 18,000$$

$$\frac{60x}{60} > \frac{18,000}{60}$$

$$x > 300$$

More than 300 canoes need to be produced and sold in order to make a profit.

57. a. cost = overhead + per show cost

$$C(x) = 30,000 + 2500x$$

b. revenue = (receipts) · (# of sell-outs)

$$R(x) = 3125x$$

c. $P(x) = R(x) - C(x)$

$$= 3125x - (30,000 + 2500x)$$

$$= 3125 - 30,000 - 2500x$$

$$= 625x - 30,000$$

d. $P(x) > 0$

$$625x - 30,000 > 0$$

$$625x > 30,000$$

$$\frac{625x}{625} > \frac{30,000}{625}$$

$$x > 48$$

More than 48 sold-out performances are needed to make a profit.

59. Let x = number of tapes produced and sold each week.

$$P(x) > 0$$

$$R(x) - C(x) > 0$$

$$2x - (10,000 + 0.40x) > 0$$

$$2x - 10,000 - 0.40x > 0$$

$$1.60x - 10,000 > 0$$

$$1.60x > 10,000$$

$$x > 6250$$

More than 6250 tapes need to be produced and sold each week to make a profit.

61. Let x = number of minutes of long-distance calls in a month.

Plan A will be a better deal if the cost of Plan A is less than the cost of Plan B.

$$\text{Cost}_A < \text{Cost}_B$$

$$15 + 0.08x < 3 + 0.12x$$

$$12 + 0.08x < 0.12x$$

$$12 < 0.04x$$

$$\frac{12}{0.04} < \frac{0.04x}{0.04}$$

$$300 < x \qquad \text{or} \quad x > 300$$

Plan A is a better deal if you have more than 300 minutes of long-distance calls.

63. – 67. Answers will vary.

69. $-2(x + 4) > 6x + 16$

Moving from left to right on the graphing calculator screen, we see that the graph of $-2(x + 4)$ is above the graph of $6x + 16$ from $-\infty$ to -3. The solution set is $\{x \mid x < -3\}$ or $(-\infty, -3)$.

71. a. Plan A: $4 + 0.10x$
Plan B: $2 + 0.15x$

b. Window: [0,50,1] by [0,10,1]

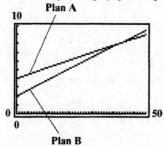

c. Plan A is better than Plan B for more than 40 checks per month.

d.
$$A < B$$
$$4 + 0.10x < 2 + 0.15x$$
$$4 < 2 + 0.05x$$
$$0.05x + 2 > 4$$
$$0.05x > 2$$
$$x > 40$$

73. does not make sense; Explanations will vary. Sample explanation: It is a good thing for a business if revenue is greater than cost.

75. makes sense

77. false; Changes to make the statement true will vary. A sample change is: 4 is not the smallest real number in the solution set of $2x > 6$. For example, 3.1 is a smaller real number in the solution set.

79. true

81. Since $x > y$, then $y - x < 0$. When multiplying both sides of the inequality by $(y - x)$, remember to *flip* the inequality.
$$2 > 1$$
$$2(y - x) < 1(y - x)$$
$$2y - 2x < y - x$$
$$y - 2x < -x$$
$$y < x$$

82. $f(x) = x^2 - 2x + 5$
$$f(-4) = (-4)^2 - 2(-4) + 5$$
$$= 16 + 8 + 5$$
$$= 29$$

83. $2x - y - z = -3$
$3x - 2y - 2z = -5$
$-x + y + 2z = 4$
Add the first and third equations to eliminate y.
$$2x - y - z = -3$$
$$\underline{-x + y + 2z = 4}$$
$$x + z = 1$$
Multiply the third equation by 2 and add to the second equation.
$$3x - 2y - 2z = -5$$
$$\underline{-2x + 2y + 4z = 8}$$
$$x + 2z = 3$$
The system of two equations in two variables becomes:
$$x + z = 1$$
$$x + 2z = 3$$
Multiply the second equation by -1 and solve for z.
$$x + z = 1$$
$$\underline{-x - 2z = -3}$$
$$-z = -2$$
$$z = 2$$
Back-substitute 2 for z to find x.
$$x + z = 1$$
$$x + 2 = 1$$
$$x = -1$$
Back-substitute 2 for z and -1 and x in one of the original equations in three variables to find y.
$$2x - y - z = -3$$
$$2(-1) - y - 2 = -3$$
$$-2 - y - 2 = -3$$
$$-y - 4 = -3$$
$$-y = 1$$
$$y = -1$$
The solution is $(-1, -1, 2)$ and the solution set is $\{(-1, -1, 2)\}$.

84. $25x^2 - 81 = (5x)^2 - (9)^2$
$$= (5x + 9)(5x - 9)$$

85. a. $\{3,4\}$

b. $\{1,2,3,4,5,6,7\}$

86. a. $x - 3 < 5$

$x < 8$

The solution set is $\{x \mid x < 8\}$ or $(-\infty, 8)$.

b. $2x + 4 < 14$

$2x < 10$

$x < 5$

The solution set is $\{x \mid x < 5\}$ or $(-\infty, 5)$.

c. Answers will vary. Any number less than 5.

d. Answers will vary. Any number in $[5, 8)$.

87. a. $2x - 6 \geq -4$

$2x \geq 2$

$x \geq 1$

The solution set is $\{x \mid x \geq 1\}$ or $[1, \infty)$.

b. $5x + 2 \geq 17$

$5x \geq 15$

$x \geq 3$

The solution set is $\{x \mid x \geq 3\}$ or $[3, \infty)$.

c. Answers will vary. Any number greater than or equal to 3.

d. Answers will vary. Any number in $[1, 3)$.

9.2 Check Points

1. $\{3,4,5,6,7\} \cap \{3,7,8,9\} = \{3,7\}$

2. Solve and graph each inequality, and graph the intersection.

$x + 2 < 5 \quad$ and $\quad 2x - 4 < -2$

$x < 3 \qquad\qquad 2x < 2$

$\qquad\qquad\qquad x < 1$

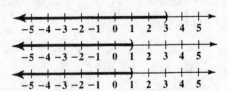

The solution set is $\{x \mid x < 1\}$ or $(-\infty, 1)$.

3. Solve and graph each inequality, and graph the intersection.

$4x - 5 > 7 \quad$ and $\quad 5x - 2 < 3$

$4x > 12 \qquad\qquad 5x < 5$

$x > 3 \qquad\qquad\quad x < 1$

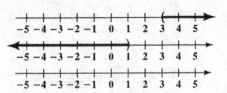

Since the two sets do not intersect, the solution set is $\varnothing$ or $\{\ \}$.

4. $\quad 1 \leq 2x + 3 < 11$

$1 - 3 \leq 2x + 3 - 3 < 11 - 3$

$-2 \leq 2x < 8$

$\dfrac{-2}{2} \leq \dfrac{2x}{2} < \dfrac{8}{2}$

$-1 \leq x < 4$

The solution set is $\{x \mid -1 \leq x < 4\}$ or $[-1, 4)$.

5. $\{3,4,5,6,7\} \cup \{3,7,8,9\} = \{3,4,5,6,7,8,9\}$

6. Solve and graph each inequality, and graph the intersection.

$3x - 5 \leq -2 \quad$ or $\quad 10 - 2x < 4$

$3x \leq 3 \qquad\qquad -2x < -6$

$x \leq 1 \qquad\qquad \dfrac{-2x}{-2} > \dfrac{-6}{-2}$

$\qquad\qquad\qquad\quad x > 3$

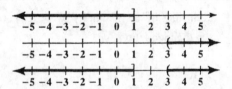

The solution set is $\{x \mid x \leq 1 \text{ or } x > 3\}$ or $(-\infty, 1] \cup (3, \infty)$.

7. Solve and graph each inequality, and graph the intersection.

$$2x + 5 \geq 3 \quad \text{or} \quad 2x + 3 < 3$$
$$2x \geq -2 \qquad\qquad 2x < 0$$
$$x \geq -1 \qquad\qquad x < 0$$

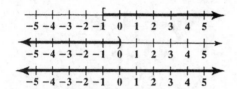

The solution set is $\{x \mid x \text{ is a real number}\}$ or $\mathbb{R}$ or $(-\infty, \infty)$.

9.2 Exercise Set

1. $\{1,2,3,4\} \cap \{2,4,5\} = \{2,4\}$

3. $\{1,3,5,7\} \cap \{2,4,6,8,10\} = \{\ \} \text{ or } \varnothing$

5. $\{a,b,c,d\} \cap \varnothing = \varnothing$

7. $x > 3$ and $x > 6$

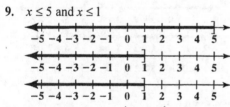

The solution set is $\{x \mid x > 6\}$ or $(6, \infty)$.

9. $x \leq 5$ and $x \leq 1$

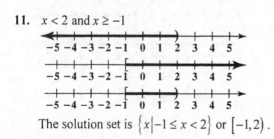

The solution set is $\{x \mid x \leq 1\}$ or $(-\infty, 1]$.

11. $x < 2$ and $x \geq -1$

The solution set is $\{x \mid -1 \leq x < 2\}$ or $[-1, 2)$.

13. $x > 2$ and $x < -1$

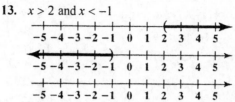

Since the two sets do not intersect, the solution set is $\varnothing$ or $\{\ \}$.

15. $5x < -20 \quad \text{and} \quad 3x > -18$
$$x < -4 \qquad\qquad x > -6$$

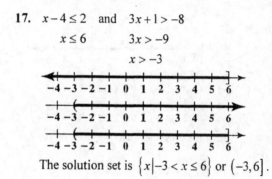

The solution set is $\{x \mid -6 < x < -4\}$ or $(-6, -4)$.

17. $x - 4 \leq 2 \quad \text{and} \quad 3x + 1 > -8$
$$x \leq 6 \qquad\qquad 3x > -9$$
$$\qquad\qquad\qquad x > -3$$

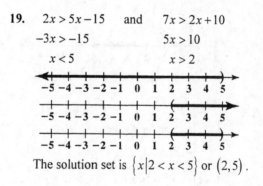

The solution set is $\{x \mid -3 < x \leq 6\}$ or $(-3, 6]$.

19. $2x > 5x - 15 \quad \text{and} \quad 7x > 2x + 10$
$$-3x > -15 \qquad\qquad 5x > 10$$
$$x < 5 \qquad\qquad\qquad x > 2$$

The solution set is $\{x \mid 2 < x < 5\}$ or $(2, 5)$.

21. $4(1-x) < -6$ and $\dfrac{x-7}{5} \le -2$

$\qquad 4 - 4x < -6$

$\qquad -4x < -10 \qquad 5\left(\dfrac{x-7}{5}\right) \le 5(-2)$

$\qquad x > \dfrac{5}{2} \qquad\qquad x - 7 \le -10$

$\qquad\qquad\qquad\qquad\quad x \le -3$

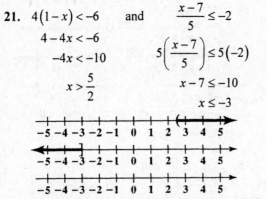

Since the two sets do not intersect, the solution set is $\varnothing$ or $\{\ \}$.

23. $x - 1 \le 7x - 1$ and $4x - 7 < 3 - x$

$\qquad -1 \le 6x - 1 \qquad\quad 5x - 7 < 3$

$\qquad\quad 0 \le 6x \qquad\qquad\quad 5x < 10$

$\qquad\quad 0 \le x \qquad\qquad\qquad x < 2$

$\qquad\quad x \ge 0$

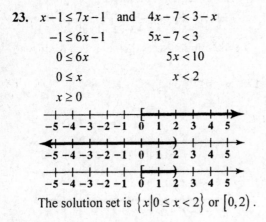

The solution set is $\{x \mid 0 \le x < 2\}$ or $[0, 2)$.

25. $\qquad 6 < x + 3 < 8$

$\qquad 6 - 3 < x + 3 - 3 < 8 - 3$

$\qquad\qquad 3 < x < 5$

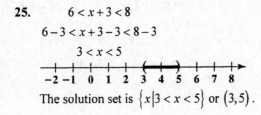

The solution set is $\{x \mid 3 < x < 5\}$ or $(3, 5)$.

27. $\qquad -3 \le x - 2 < 1$

$\qquad -3 + 2 \le x - 2 + 2 < 1 + 2$

$\qquad\qquad -1 \le x < 3$

The solution set is $\{x \mid -1 \le x < 3\}$ or $[-1, 3)$.

29. $\qquad -11 < 2x - 1 \le -5$

$\qquad -11 + 1 < 2x - 1 + 1 \le -5 + 1$

$\qquad\qquad -10 < 2x \le -4$

$\qquad\qquad -5 < x \le -2$

The solution set is $\{x \mid -5 < x \le -2\}$ or $(-5, -2]$.

31. $\qquad -3 \le \dfrac{2x}{3} - 5 < -1$

$\qquad -3 + 5 \le \dfrac{2x}{3} - 5 + 5 < -1 + 5$

$\qquad\qquad 2 \le \dfrac{2x}{3} < 4$

$\qquad 3(2) \le 3\left(\dfrac{2x}{3}\right) < 3(4)$

$\qquad\qquad 6 \le 2x < 12$

$\qquad\qquad 3 \le x < 6$

The solution set is $\{x \mid 3 \le x < 6\}$ or $[3, 6)$.

33. $\{1, 2, 3, 4\} \cup \{2, 4, 5\} = \{1, 2, 3, 4, 5\}$

35. $\{1, 3, 5, 7\} \cup \{2, 4, 6, 8, 10\}$
$\qquad = \{1, 2, 3, 4, 5, 6, 7, 8, 10\}$

37. $\{a, e, i, o, u\} \cup \varnothing = \{a, e, i, o, u\}$

39. $x > 3$ or $x > 6$

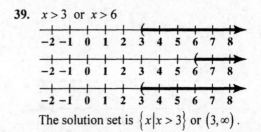

The solution set is $\{x \mid x > 3\}$ or $(3, \infty)$.

41. $x \le 5$ or $x \le 1$

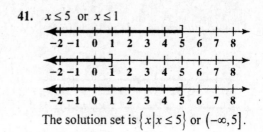

The solution set is $\{x \mid x \le 5\}$ or $(-\infty, 5]$.

43. $x < 2$ or $x \ge -1$

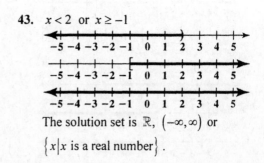

The solution set is $\mathbb{R}$, $(-\infty, \infty)$ or $\{x \mid x \text{ is a real number}\}$.

45. $x \geq 2$ or $x < -1$

The solution set is $\{x | x < -1$ or $x \geq 2\}$ or $(-\infty, -1) \cup [2, \infty)$.

47. $3x > 12$ or $2x < -6$
    $x > 4$ $x < -3$

The solution set is $\{x | x < -3$ or $x > 4\}$ or $(-\infty, -3) \cup (4, \infty)$.

49. $3x + 2 \leq 5$ or $5x - 7 \geq 8$
    $3x \leq 3$ $5x \geq 15$
    $x \leq 1$ $x \geq 3$

The solution set is $\{x | x \leq 1$ or $x \geq 3\}$ or $(-\infty, 1] \cup [3, \infty)$.

51. $4x + 3 < -1$ or $2x - 3 \geq -11$
    $4x < -4$ $2x \geq -8$
    $x < -1$ $x \geq -4$

The solution set is $\mathbb{R}$, $(-\infty, \infty)$ or $\{x | x$ is a real number$\}$.

53. $-2x + 5 > 7$ or $-3x + 10 > 2x$
    $-2x > 2$ $-5x + 10 > 0$
    $x < -1$ $-5x > -10$
                $x < 2$

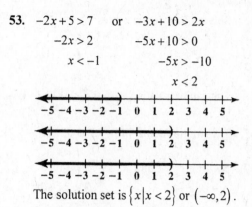

The solution set is $\{x | x < 2\}$ or $(-\infty, 2)$.

55. $2x + 3 \geq 5$ and $3x - 1 > 11$
    $2x \geq 2$ $3x > 12$
    $x \geq 1$ $x > 4$

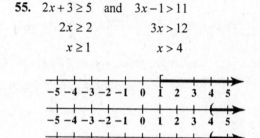

The solution set is $\{x | x > 4\}$ or $(4, \infty)$.

57. $3x - 1 < -1$ or $4 - x < -2$
    $3x < 0$ $4 < -2 + x$
    $x < 0$ $6 < x$
                $x > 6$

The solution set is $\{x | x < 0$ or $x > 6\}$ or $(-\infty, 0) \cup (6, \infty)$.

59. $a > 0$, $b > 0$, $c > 0$
    $-c < ax - b < c$
    $b - c < ax < b + c$
    $\dfrac{b - c}{a} < x < \dfrac{b + c}{a}$

61. $\{x | -1 \leq x \leq 3\}$ or $[-1, 3]$

63. Solving in separate pieces:

$x - 2 < 2x - 1$ and $2x - 1 < x + 2$

$\quad -2 < x - 1 \qquad\qquad x - 1 < 2$

$\quad -1 < x \qquad\qquad\quad\ x < 3$

The solution set is $\{x | -1 < x < 3\}$ or $(-1, 3)$.

65. The solution set is $\{x | -1 \leq x < 2\}$ or $[-1, 2)$.

67. $5 - 4x \geq 1 \qquad$ and $\qquad 3 - 7x < 31$

$\quad -4x \geq -4 \qquad\qquad\qquad 3 < 7x + 31$

$\qquad x \leq 1 \qquad\qquad\qquad\ -28 < 7x$

$\qquad\qquad\qquad\qquad\qquad\quad -4 < x$

The solution set is $\{x | -4 < x \leq 1\}$ or $(-4, 1]$. The set of negative integers that fall within this set is $\{-3, -2, -1\}$.

69. $\{x | x$ is a year for which women's weight $\geq 144\} \cap \{x | x$ is a year for which men's weight $\leq 182\}$

$\{1974, 1980, 1994, 2002\} \cap \{1962, 1974, 1980, 1994\}$

$\{1974, 1980, 1994\}$

71. $\{x | x$ is a year for which women's weight $\geq 144\} \cup \{x | x$ is a year for which men's weight $\leq 182\}$

$\{1974, 1980, 1994, 2002\} \cup \{1962, 1974, 1980, 1994\}$

$\{1962, 1974, 1980, 1994, 2002\}$

73. $\{x | x$ is a year for which women's weight $< 144\} \cup \{x | x$ is a year for which men's weight $> 182\}$

$\{1962\} \cup \{2002\}$

$\{1962, 2002\}$

75. $\{x | x$ is a year for which women's weight $< 144\} \cap \{x | x$ is a year for which men's weight $> 182\}$

$\{1962\} \cap \{2002\}$

$\varnothing$

77. $\{x | x$ is a year for which $166 \leq$ men's weight $< 174\}$

$\{1962, 1974\}$

79. $28 \leq 20 + 0.40(x - 60) \leq 40$

$\quad 28 \leq 20 + 0.40x - 24 \leq 40$

$\qquad 28 \leq 0.40x - 4 \leq 40$

$\qquad 32 \leq 0.40x \leq 44$

$\qquad 80 \leq x \leq 110$

Between 80 and 110 minutes, inclusive.

81. Let x = the score on the fifth exam.

$$80 \le \frac{70 + 75 + 87 + 92 + x}{5} < 90$$

$$80 \le \frac{324 + x}{5} < 90$$

$$5(80) \le 5\left(\frac{324 + x}{5}\right) < 5(90)$$

$$400 \le 324 + x < 450$$

$$400 - 324 \le 324 - 324 + x < 450 - 324$$

$$76 \le x < 126$$

A grade between 76 and 125 is needed on the fifth exam.

Because the inequality states the score must be less than 126, we say 125 is the highest possible score. In interval notation, we can use parentheses to exclude the maximum value. The range of scores can be expressed as $[76, 126)$.

If the highest grade is 100, the grade would need to be between 76 and 100, inclusive.

83. Let x = the number of times the bridge is crossed per three month period.

The cost with the 3-month pass is $C_3 = 7.50 + 0.50x$.

The cost with the 6-month pass is $C_6 = 30$.

Because we need to buy two 3-month passes per 6-month pass, we multiply the cost with the 3-month pass by 2.

$$2(7.50 + 0.50x) < 30$$

$$15 + x < 30$$

$$x < 15$$

We also must consider the cost without purchasing a pass. We need this cost to be less than the cost with a 3-month pass.

$$3x > 7.50 + 0.50x$$

$$2.50x > 7.50$$

$$x > 3$$

The 3-month pass is the best deal when making more than 3 but less than 15 crossings per 3-month period.

85. – 89. Answers will vary.

91. $1 < x + 3 < 9$

Using the intersection feature, find the range of the x-values of the points lying between the two constant functions.

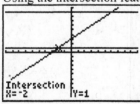

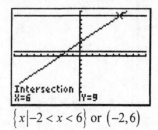

$\{x \mid -2 < x < 6\}$ or $(-2, 6)$

93. $1 \le 4x - 7 \le 3$

Using the intersection feature, find the range of the x-values of the points lying between the two constant functions.

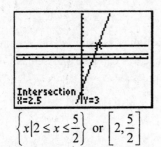

$\left\{ x \mid 2 \le x \le \dfrac{5}{2} \right\}$ or $\left[2, \dfrac{5}{2} \right]$

95. a.

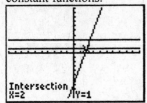

b.

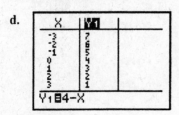

c.

d.

97. makes sense

99. makes sense

101. false; Changes to make the statement true will vary. A sample change is: $(-\infty, 3) \cup (-\infty, -2) = (-\infty, 3)$

103. false; Changes to make the statement true will vary. A sample change is: The solution set of $x < a$ and $x > a$ is $\varnothing$.

105. The domain of $f = \{ x \mid x \le 4 \}$ or $(-\infty, 4]$.

107. The domain of $f = (-\infty, 4]$, and the domain of $g = [-1, \infty)$. The domain of $f + g$ is the intersection of the domains of f and g. The domain of $f + g = \{ x \mid -1 \le x \le 4 \}$ or $[-1, 4]$.

109. Let n = number of nickels.
Let d = number of dimes.
Let q = number of quarters.
$320 \le 5n + 10d + 25q \le 545$
$320 \le 5n + 10(2n - 3) + 25(2n + 2) \le 545$
$320 \le 5n + 20n - 30 + 50n + 50 \le 545$
$320 \le 75n + 20 \le 545$
$30 \le 75n \le 525$
$4 \le n \le 7$

The least possible number of nickels was 4 and the greatest possible number of nickels was 7.

110. $(g - f)(x) = g(x) - f(x)$
$= (2x - 5) - (x^2 - 3x + 4)$
$= 2x - 5 - x^2 + 3x - 4$
$= -x^2 + 5x - 9$
$(g - f)(-1) = -(-1)^2 + 5(-1) - 9$
$= -1 - 5 - 9 = -15$

111. $4x - 2y = 8$

$-2y = -4x + 8$

$y = 2x - 4$

The slope of this line is 2. The slope of the line

perpendicular to this line is $-\dfrac{1}{2}$.

$y - 2 = -\dfrac{1}{2}(x - 4)$

$y - 2 = -\dfrac{1}{2}x + 2$

$y = -\dfrac{1}{2}x + 4$

$f(x) = -\dfrac{1}{2}x + 4$

112. $4 - \left[2(x - 4) - 5\right] = 4 - \left[2x - 8 - 5\right]$

$= 4 - \left[2x - 13\right]$

$= 4 - 2x + 13$

$= 17 - 2x$

113. $1 - 4x = 3$ or $1 - 4x = -3$

$-4x = 2$ $-4x = -4$

$x = -\dfrac{1}{2}$ $x = 1$

114. $3x - 1 = -(x + 5)$ or $3x - 1 = x + 5$

$3x - 1 = -x - 5$ $2x = 6$

$4x = -4$ $x = 3$

$x = -1$

115. a. $|2x + 3| \ge 5$

$|2(-5) + 3| \ge 5$

$|-10 + 3| \ge 5$

$|-7| \ge 5$

$7 \ge 5,$ true

-5 satisfies the inequality.   **b.**

$|2x + 3| \ge 5$

$|2(0) + 3| \ge 5$

$|0 + 3| \ge 5$

$|3| \ge 5$

$3 \ge 5,$ false

0 does not satisfy the inequality.

9.3 Check Points

1. $|2x - 1| = 5$

$2x - 1 = 5$ or $2x - 1 = -5$

$2x = 6$ $2x = -4$

$x = 3$ $x = -2$

The solution set is $\{-2, 3\}$.

2. $2|1 - 3x| - 28 = 0$

$2|1 - 3x| = 28$

$|1 - 3x| = 14$

$1 - 3x = 14$ or $1 - 3x = -14$

$-3x = 13$ $-3x = -15$

$x = -\dfrac{13}{3}$ $x = 5$

The solution set is $\left\{-\dfrac{13}{3}, 5\right\}$.

3. $|2x - 7| = |x + 3|$

$2x - 7 = x + 3$ or $2x - 7 = -(x + 3)$

$x = 10$ $2x - 7 = -x - 3$

$3x = 4$

$x = \dfrac{4}{3}$

The solution set is $\left\{\dfrac{4}{3}, 10\right\}$.

4. $|x-2| < 5$

Solve the related equation.

$|x-2| = 5$

$x - 2 = 5$ or $x - 2 = -5$

$x = 7$ $x = -3$

Interval	Test Value	Substitute	Conclusion
$(-\infty, -3)$	-10	$\begin{aligned}&|x-2| < 5\\ &|-10-2| < 5\\ &\quad 12 < 5,\ \text{false}\end{aligned}$	$(-\infty, -3)$ does not belong to the solution set
$(-3, 7)$	0	$\begin{aligned}&|x-2| < 5\\ &|0-2| < 5\\ &\quad 2 < 5,\ \text{true}\end{aligned}$	$(-3, 7)$ belongs to the solution set
$(7, \infty)$	10	$\begin{aligned}&|x-2| < 5\\ &|10-2| < 5\\ &\quad 8 < 5,\ \text{false}\end{aligned}$	$(7, \infty)$ does not belong to the solution set

The solution set is $(-3, 7)$ or $\{x \mid -3 < x < 7\}$.

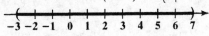

5. $|2x - 5| \geq 3$

Solve the related equation.

$|2x - 5| = 3$

$2x - 5 = 3$ or $2x - 5 = -3$

$2x = 8$ $2x = 2$

$x = 4$ $x = 1$

Interval	Test Value	Substitute	Conclusion
$(-\infty, 1)$	0	$\begin{aligned}&|2x-5| \geq 3\\ &|2(0)-5| \geq 3\\ &\quad 5 \geq 3,\ \text{true}\end{aligned}$	$(-\infty, 1)$ belongs to the solution set
$(1, 4)$	2	$\begin{aligned}&|2x-5| \geq 3\\ &|2(2)-5| \geq 3\\ &\quad 1 \geq 3,\ \text{false}\end{aligned}$	$(1, 4)$ does not belong to the solution set
$(4, \infty)$	5	$\begin{aligned}&|2x-5| \geq 3\\ &|2(5)-5| \geq 3\\ &\quad 5 \geq 3,\ \text{true}\end{aligned}$	$(4, \infty)$ belongs to the solution set

The solution set is $(-\infty, 1] \cup [4, \infty)$ or $\{x \mid x \leq 1 \text{ or } x \geq 4\}$.

6. a. $|x-2| < 5$

$$-5 < x-2 < 5$$

$$-5+2 < x-2+2 < 5+2$$

$$-3 < x < 7$$

The solution set is $(-3,7)$ or $\{x|-3 < x < 7\}$.

b. $|2x-5| \geq 3$

$$2x-5 \geq 3 \quad \text{or} \quad 2x-5 \leq -3$$

$$2x \geq 8 \qquad\qquad 2x \leq 2$$

$$x \geq 4 \qquad\qquad x \leq 1$$

The solution set is $(-\infty,1]\cup[4,\infty)$ or $\{x|x \leq 1 \text{ or } x \geq 4\}$.

7. $-3|5x-2| + 20 \geq -19$

$$-3|5x-2| \geq -39$$

$$\frac{-3|5x-2|}{-3} \leq \frac{-39}{-3}$$

$$|5x-2| \leq 13$$

$$-13 \leq 5x-2 \leq 13$$

$$-13+2 \leq 5x-2+2 \leq 13+2$$

$$-11 \leq 5x \leq 15$$

$$\frac{-11}{5} \leq \frac{5x}{5} \leq \frac{15}{5}$$

$$-\frac{11}{5} \leq x \leq 3$$

The solution set is $\left[\dfrac{-11}{5},3\right]$ or $\left\{x\left|-\dfrac{11}{5} \leq x \leq 3\right.\right\}$.

8. $|x-11| \leq 2.9$

$$-2.9 \leq x-11 \leq 2.9$$

$$-2.9+11 \leq x-11+11 \leq 2.9+11$$

$$8.1 \leq x \leq 13.9$$

The solution set is $\{x|8.1 \leq x \leq 13.9\}$ or $[8.1, 13.9]$.

The percentage of children in the population who think that not being able to do everything they want is a bad thing is between a low of 8.1% and a high of 13.9%.

9.3 Exercise Set

1. $|x| = 8$

$x = 8$ or $x = -8$

The solution set is $\{-8, 8\}$.

3. $|x - 2| = 7$

$x - 2 = 7$ or $x - 2 = -7$

$\quad x = 9 \qquad\qquad x = -5$

The solutions set is $\{-5, 9\}$.

5. $|2x - 1| = 7$

$2x - 1 = 7$ or $2x - 1 = -7$

$\quad 2x = 8 \qquad\qquad 2x = -6$

$\quad x = 4 \qquad\qquad x = -3$

The solutions set is $\{-3, 4\}$.

7. $\left|\dfrac{4x - 2}{3}\right| = 2$

$\dfrac{4x - 2}{3} = 2 \quad$ or $\quad \dfrac{4x - 2}{3} = -2$

$4x - 2 = 3(2) \qquad 4x - 2 = 3(-2)$

$4x - 2 = 6 \qquad\quad 4x - 2 = -6$

$\quad 4x = 8 \qquad\qquad 4x = -4$

$\quad x = 2 \qquad\qquad\quad x = -1$

The solutions set is $\{-1, 2\}$.

9. $|x| = -8$

The solution set is $\varnothing$ or $\{\ \}$. There are no values of x for which the absolute value of x is a negative number. By definition, absolute values are always zero or positive.

11. $|x + 3| = 0$

Since the absolute value of the expression equals zero, we set the expression equal to zero and solve.

$x + 3 = 0$

$\quad x = -3$

The solution set is $\{-3\}$.

13. $2|y + 6| = 10$

$|y + 6| = 5$

$y + 6 = 5 \quad$ or $\quad y + 6 = -5$

$\quad y = -1 \qquad\qquad y = -11$

The solutions set is $\{-11, -1\}$.

15. $3|2x - 1| = 21$

$|2x - 1| = 7$

$2x - 1 = 7 \quad$ or $\quad 2x - 1 = -7$

$\quad 2x = 8 \qquad\qquad 2x = -6$

$\quad x = 4 \qquad\qquad x = -3$

The solutions set is $\{-3, 4\}$.

17. $|6y - 2| + 4 = 32$

$|6y - 2| = 28$

$6y - 2 = 28 \quad$ or $\quad 6y - 2 = -28$

$\quad 6y = 30 \qquad\qquad 6y = -26$

$\quad y = 5 \qquad\qquad\quad y = -\dfrac{26}{6}$

$\qquad\qquad\qquad\qquad\quad y = -\dfrac{13}{3}$

The solutions set is $\left\{-\dfrac{13}{3}, 5\right\}$.

19. $7|5x| + 2 = 16$

$7|5x| = 14$

$|5x| = 2$

$5x = 2 \quad$ or $\quad 5x = -2$

$x = \dfrac{2}{5} \qquad\qquad x = -\dfrac{2}{5}$

The solutions set is $\left\{-\dfrac{2}{5}, \dfrac{2}{5}\right\}$.

21. $|x + 1| + 5 = 3$

$|x + 1| = -2$

The solution set is $\varnothing$ or $\{\ \}$. By definition, absolute values are always zero or positive.

23. $|4y + 1| + 10 = 4$

$|4y + 1| = -6$

The solution set is $\varnothing$ or $\{\ \}$. By definition, absolute values are always zero or positive.

25. $|2x-1|+3=3$

$|2x-1|=0$

Since the absolute value of the expression equals zero, we set the expression equal to zero and solve.

$2x-1=0$

$2x=1$

$x=\dfrac{1}{2}$

The solutions set is $\left\{\dfrac{1}{2}\right\}$.

27. $|5x-8|=|3x+2|$

$5x-8=3x+2$ or $5x-8=-3x-2$

$2x-8=2$ $8x-8=-2$

$2x=10$ $8x=6$

$x=5$ $x=\dfrac{6}{8}=\dfrac{3}{4}$

The solutions set is $\left\{\dfrac{3}{4},5\right\}$.

29. $|2x-4|=|x-1|$

$2x-4=x-1$ or $2x-4=-x+1$

$x-4=-1$ $3x-4=1$

$x=3$ $3x=5$

$x=\dfrac{5}{3}$

The solutions set is $\left\{\dfrac{5}{3},3\right\}$.

31. $|2x-5|=|2x+5|$

$2x-5=2x+5$ or $2x-5=-2x-5$

$-5\neq5$ $4x-5=-5$

$4x=0$

$x=0$

The solutions set is $\{0\}$.

33. $|x-3|=|5-x|$

$x-3=5-x$ or $x-3=-(5-x)$

$2x-3=5$ $x-3=-5+x$

$2x=8$ $-3\neq-5$

$x=4$

The solutions set is $\{4\}$.

35. $|2y-6|=|10-2y|$

$2y-6=10-2y$ or $2y-6=-10+2y$

$4y-6=10$ $-6\neq-10$

$4y=16$

$y=4$

The solutions set is $\{4\}$.

37. $\left|\dfrac{2x}{3}-2\right|=\left|\dfrac{x}{3}+3\right|$

$\dfrac{2x}{3}-2=\dfrac{x}{3}+3$ or

$3\left(\dfrac{2x}{3}\right)-3(2)=3\left(\dfrac{x}{3}\right)+3(3)$

$2x-6=x+9$

$x-6=9$

$x=15$

$\dfrac{2x}{3}-2=-\left(\dfrac{x}{3}+3\right)$

$\dfrac{2x}{3}-2=-\dfrac{x}{3}-3$

$3\left(\dfrac{2x}{3}\right)-3(2)=3\left(-\dfrac{x}{3}\right)-3(3)$

$2x-6=-x-9$

$3x-6=-9$

$3x=-3$

$x=-1$

The solutions set is $\{-1,15\}$.

39. $|x|<3$

$-3<x<3$

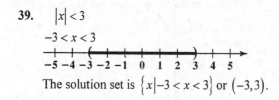

The solution set is $\{x|-3<x<3\}$ or $(-3,3)$.

41. $|x-2|<1$

$-1<x-2<1$

$-1+2<x-2+2<1+2$

$1<x<3$

The solution set is $\{x|1<x<3\}$ or $(1,3)$.

43.
$$|x+2| \le 1$$
$$-1 \le x+2 \le 1$$
$$-1-2 \le x+2-2 \le 1-2$$
$$-3 \le x \le -1$$

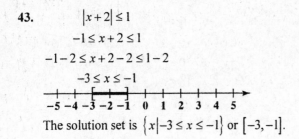

The solution set is $\{x | -3 \le x \le -1\}$ or $[-3, -1]$.

45.
$$|2x-6| < 8$$
$$-8 < 2x-6 < 8$$
$$-8+6 < 2x-6+6 < 8+6$$
$$-2 < 2x < 14$$
$$-1 < x < 7$$

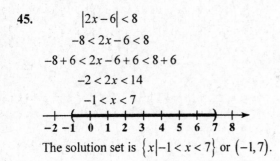

The solution set is $\{x | -1 < x < 7\}$ or $(-1, 7)$.

47. $|x| > 3$
$$x < -3 \quad \text{or} \quad x > 3$$

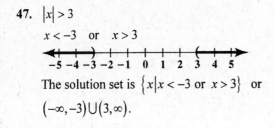

The solution set is $\{x | x < -3 \text{ or } x > 3\}$ or $(-\infty, -3) \cup (3, \infty)$.

49. $|x+3| > 1$
$$x+3 < -1 \quad \text{or} \quad x+3 > 1$$
$$x < -4 \qquad\qquad x > -2$$

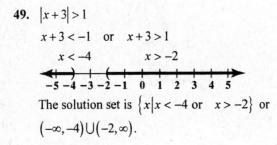

The solution set is $\{x | x < -4 \text{ or } x > -2\}$ or $(-\infty, -4) \cup (-2, \infty)$.

51. $|x-4| \ge 2$
$$x-4 \le -2 \quad \text{or} \quad x-4 \ge 2$$
$$x \le 2 \qquad\qquad x \ge 6$$

The solution set is $\{x | x \le 2 \text{ or } x \ge 6\}$ or $(-\infty, 2] \cup [6, \infty)$.

53. $|3x-8| > 7$
$$3x-8 < -7 \quad \text{or} \quad 3x-8 > 7$$
$$3x < 1 \qquad\qquad 3x > 15$$
$$x < \frac{1}{3} \qquad\qquad x > 5$$

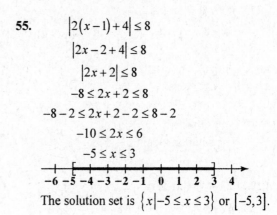

The solution set is $\left\{ x \middle| x < \frac{1}{3} \text{ or } x > 5 \right\}$ or

$\left(-\infty, \frac{1}{3}\right) \cup (5, \infty)$.

55.
$$|2(x-1)+4| \le 8$$
$$|2x-2+4| \le 8$$
$$|2x+2| \le 8$$
$$-8 \le 2x+2 \le 8$$
$$-8-2 \le 2x+2-2 \le 8-2$$
$$-10 \le 2x \le 6$$
$$-5 \le x \le 3$$

The solution set is $\{x | -5 \le x \le 3\}$ or $[-5, 3]$.

57. $\left| \dfrac{2y+6}{3} \right| < 2$
$$-2 < \frac{2y+6}{3} < 2$$
$$3(-2) < 3\left(\frac{2y+6}{3}\right) < 3(2)$$
$$-6 < 2y+6 < 6$$
$$-6-6 < 2y+6-6 < 6-6$$
$$-12 < 2y < 0$$
$$-6 < y < 0$$

The solution set is $\{x | -6 < x < 0\}$ or $(-6, 0)$.

59. $\left|\dfrac{2x+2}{4}\right| \geq 2$

$$\dfrac{2x+2}{4} \leq -2 \quad \text{or} \quad \dfrac{2x+2}{4} \geq 2$$

$$2x+2 \leq -8 \qquad\qquad 2x+2 \geq 8$$

$$2x \leq -10 \qquad\qquad 2x \geq 6$$

$$x \leq -5 \qquad\qquad x \geq 3$$

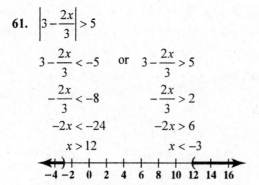

The solution set is $\{x | x \leq -5 \text{ or } x \geq 3\}$ or $(-\infty, -5] \cup [3, \infty)$.

61. $\left|3 - \dfrac{2x}{3}\right| > 5$

$$3 - \dfrac{2x}{3} < -5 \quad \text{or} \quad 3 - \dfrac{2x}{3} > 5$$

$$-\dfrac{2x}{3} < -8 \qquad\qquad -\dfrac{2x}{3} > 2$$

$$-2x < -24 \qquad\qquad -2x > 6$$

$$x > 12 \qquad\qquad x < -3$$

The solution set is $\{x | x < -3 \text{ or } x > 12\}$ or $(-\infty, -3) \cup (12, \infty)$.

63. $|x - 2| < -1$

The solution set is $\varnothing$ or $\{\ \}$. Since all absolute values are zero or positive, there are no values of x that will make the absolute value of the expression less than -1.

65. $|x + 6| > -10$

Since all absolute values are zero or positive, we know that when simplified, the left hand side will be a positive number. We also know that any positive number is greater than any negative number. This means that regardless of the value of x, the left hand side will be greater than the right hand side of the inequality. The solution set is $\{x | x \text{ is a real number}\}$, $\mathbb{R}$ or $(-\infty, \infty)$.

67. $|x + 2| + 9 \leq 16$

$$|x + 2| \leq 7$$

$$-7 \leq x + 2 \leq 7$$

$$-7 - 2 \leq x + 2 - 2 \leq 7 - 2$$

$$-9 \leq x \leq 5$$

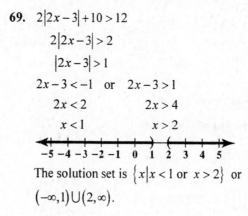

The solution set is $\{x | -9 \leq x \leq 5\}$ or $[-9, 5]$.

69. $2|2x - 3| + 10 > 12$

$$2|2x - 3| > 2$$

$$|2x - 3| > 1$$

$$2x - 3 < -1 \quad \text{or} \quad 2x - 3 > 1$$

$$2x < 2 \qquad\qquad 2x > 4$$

$$x < 1 \qquad\qquad x > 2$$

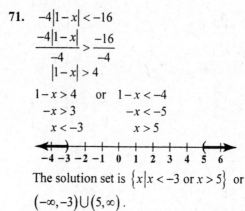

The solution set is $\{x | x < 1 \text{ or } x > 2\}$ or $(-\infty, 1) \cup (2, \infty)$.

71. $-4|1 - x| < -16$

$$\dfrac{-4|1 - x|}{-4} > \dfrac{-16}{-4}$$

$$|1 - x| > 4$$

$$1 - x > 4 \quad \text{or} \quad 1 - x < -4$$

$$-x > 3 \qquad\qquad -x < -5$$

$$x < -3 \qquad\qquad x > 5$$

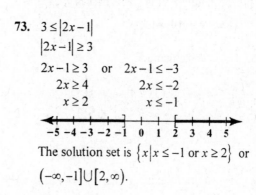

The solution set is $\{x | x < -3 \text{ or } x > 5\}$ or $(-\infty, -3) \cup (5, \infty)$.

73. $3 \leq |2x - 1|$

$$|2x - 1| \geq 3$$

$$2x - 1 \geq 3 \quad \text{or} \quad 2x - 1 \leq -3$$

$$2x \geq 4 \qquad\qquad 2x \leq -2$$

$$x \geq 2 \qquad\qquad x \leq -1$$

The solution set is $\{x | x \leq -1 \text{ or } x \geq 2\}$ or $(-\infty, -1] \cup [2, \infty)$.

75. $|5 - 4x| = 11$

$5 - 4x = 11$ or $5 - 4x = -11$

$-4x = 6$ $\qquad -4x = -16$

$x = -\dfrac{3}{2}$ $\qquad y = 4$

The solution set is $\left\{ -\dfrac{3}{2}, 4 \right\}$.

77. $|3 - x| = |3x + 11|$

$3 - x = 3x + 11$ or $3 - x = -(3x + 11)$

$-4x + 3 = 11$ $\qquad 3 - x = -3x - 11$

$-4x = 8$ $\qquad 2x + 3 = -11$

$x = -2$ $\qquad 2x = -14$

$\qquad\qquad x = -7$

The solution set is $\{-7, -2\}$.

79. $|-1 + 3(x + 1)| \le 5$

$-5 \le -1 + 3x + 3 \le 5$

$-5 \le 3x + 2 \le 5$

$-7 \le 3x \le 3$

$-\dfrac{7}{3} \le x \le 1$

The solution set is $\left\{ x \mid -\dfrac{7}{3} \le x \le 1 \right\}$ or $\left[-\dfrac{7}{3}, 1 \right]$.

81. $|2x - 3| + 1 > 6$

$|2x - 3| > 5$

$2x - 3 > 5$ or $2x - 3 < -5$

$2x > 8$ $\qquad 2x < -2$

$x > 4$ $\qquad x < -1$

The solution set is $\{x \mid x < -1 \text{ or } x > 4\}$ or

$(-\infty, -1) \cup (4, \infty)$.

83. Let x be the number.

$|4 - 3x| \ge 5$

$3x - 4 \ge 5$ or $3x - 4 \le -5$

$3x \ge 9$ $\qquad 3x \le -1$

$x \ge 3$ $\qquad x \le -\dfrac{1}{3}$

The solution set is $\left\{ x \mid x \le -\dfrac{1}{3} \text{ or } x \ge 3 \right\}$ or

$\left(-\infty, -\dfrac{1}{3} \right] \cup [3, \infty)$.

85. $|ax + b| < c$

When solving, we do not reverse the inequality symbol from "<" to ">" when dividing by a since $a > 0$.

$-c < ax + b < c$

$-c - b < ax < c - b$

$\dfrac{-c - b}{a} < x < \dfrac{c - b}{a}$

The solution set is $\left\{ x \mid \dfrac{-c - b}{a} < x < \dfrac{c - b}{a} \right\}$.

87. $|4 - x| = 1$

The graphs of $f(x) = |4 - x|$ and $y = 1$ intersect when $x = 3$ and when $x = 5$. Thus, the solution set is $\{3, 5\}$.

89. The solution set is $\{x \mid -2 \le x \le 1\}$ or $[-2, 1]$

91. $|x - 60.2| \le 1.6$

$-1.6 \le x - 60.2 \le 1.6$

$-1.6 + 60.2 \le x - 60.2 + 60.2 \le 1.6 + 60.2$

$58.6 \le x \le 61.8$

The percentage of the U.S. population that watched M*A*S*H is between 58.6% and 61.8%, inclusive. The margin of error is 1.6%.

93. $|T - 57| \le 7$

$-7 \le T - 57 \le 7$

$-7 + 57 \le T - 57 + 57 \le 7 + 57$

$50 \le T \le 64$

The monthly average temperature for San Francisco, California ranges from $50°F$ to $64°F$, inclusive.

95. $|x - 8.6| \le 0.01$

$-0.01 \le x - 8.6 \le 0.01$

$-0.01 + 8.6 \le x - 8.6 + 8.6 \le 0.01 + 8.6$

$8.59 \le x \le 8.61$

The length of the machine part must be between 8.59 and 8.61 centimeters, inclusive.

97. $\left|\dfrac{h-50}{5}\right| \geq 1.645$

$\dfrac{h-50}{5} \leq -1.645 \quad$ or $\quad \dfrac{h-50}{5} \geq 1.645$

$h - 50 \leq 5(-1.645) \qquad h - 50 \geq 5(1.645)$

$h - 50 \leq -8.225 \qquad h - 50 \geq 8.225$

$h \leq 41.775 \qquad h \geq 58.225$

The coin would be considered unfair if the tosses resulted in 41 or less heads, or 59 or more heads.

99. – 103. Answers will vary.

105. $|x+1| = 5$

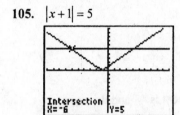

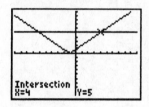

The solutions are –6 and 4 and the solution set is $\{-6, 4\}$.

107. $|2x - 3| = |9 - 4x|$

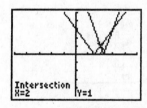

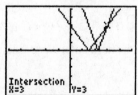

The solutions are 2 and 3 and the solution set is $\{2, 3\}$.

109. $\left|\dfrac{2x-1}{3}\right| < \dfrac{5}{3}$

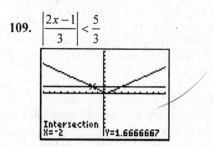

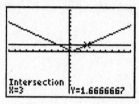

The solution set is $\{x \mid -2 < x < 3\}$ or $(-2, 3)$.

111. $|2x - 1| > 7$

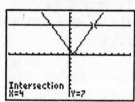

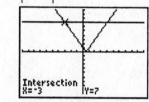

The solution set is $\{x \mid x < -3 \text{ or } x > 4\}$ or $(-\infty, -3) \cup (4, \infty)$.

113. $|x + 4| > -1$

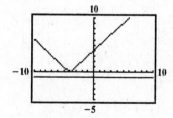

$\{x \mid x$ is a real number$\}$, $\mathbb{R}$, or $(-\infty, \infty)$

115. makes sense

117. does not make sense; Explanations will vary. Sample explanation: By adding 9 to both sides, we see that the absolute value expression must be less than 5. This is possible so the inequality does have a solution.

119. false; Changes to make the statement true will vary. A sample change is: Some absolute value equations, such as $|x-3| = 0$ have 1 solution. Some absolute value equations, such as $|x+5| = -2$ have no solution.

121. true

123. a. $|x-4| < 3$

b. $|x-4| \geq 3$

124. The proportion of refunds will be

$|p - 0.3\%| \leq 0.2\%$

$|p - 0.003| \leq 0.002$

$-0.002 \leq p - 0.003 \leq 0.002$

$0.001 \leq p \leq 0.005$

The number of refunds will be

$100,000(0.001) \leq 100,000p \leq 100,000(0.005)$

$100 \leq 100,000p \leq 500$

The cost of refunds will be

$5(100) \leq 5(100,000p) \leq 5(500)$

$500 \leq 5(100,000p) \leq 2500$

The cost for refunds will be between a low of $500 and a high of $2500.

125. $|2x+5| = 3x+4$

$2x+5 = 3x+4$ or $2x+5 = -(3x+4)$

$-x+5 = 4$ $2x+5 = -3x-4$

$-x = -1$ $5x+5 = -4$

$x = 1$ $5x = -9$

$x = -\dfrac{9}{5}$

Check:

$x = 1: \quad |2(1)+5| = 3(1)+4$

$\qquad\qquad |7| = 7, \quad \text{true}$

$x = -\dfrac{9}{5}: \quad \left|2\left(-\dfrac{9}{5}\right)+5\right| = 3\left(-\dfrac{9}{5}\right)+4$

$\qquad\qquad \left|-\dfrac{18}{5}+\dfrac{25}{5}\right| = -\dfrac{27}{5}+\dfrac{20}{5}$

$\qquad\qquad \left|\dfrac{7}{5}\right| = -\dfrac{7}{5}, \quad \text{false}$

Thus, 1 checks and $-\dfrac{9}{5}$ does not check, so the solution set is $\{1\}$.

126. $3x - 5y = 15$

$\qquad -5y = -3x+15$

$\qquad\quad y = \dfrac{3}{5}x - 3$

The y-intercept is -3 and the slope is $\dfrac{3}{5}$.

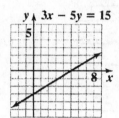

127. $f(x) = -\dfrac{2}{3}x$

The y-intercept is 0 and the slope is $-\dfrac{2}{3}$.

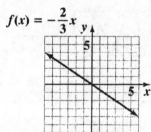

128. $f(x) = -2$ is the horizontal line positioned at $y = -2$.

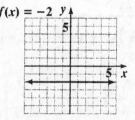

466

Chapter 9 Mid-Chapter Check Point

1. $4 - 3x \geq 12 - x$

$\qquad 4 \geq 12 + 2x$

$\qquad -8 \geq 2x$

$\qquad -4 \geq x$

$\qquad x \leq -4$

The solution set is $\{x \mid x \leq -4\}$ or $(-\infty, -4]$.

2. $5 \leq 2x - 1 < 9$

$\quad 6 \leq 2x < 10$

$\quad 3 \leq x < 5$

The solution set is $\{x \mid 3 \leq x < 5\}$ or $[3, 5)$.

3. $\mid 4x - 7 \mid = 5$

$\quad 4x - 7 = 5 \qquad$ or $\qquad 4x - 7 = -5$

$\qquad 4x = 12 \qquad\qquad\qquad 4x = 2$

$\qquad\; x = 3 \qquad\qquad\qquad\; x = \dfrac{1}{2}$

The solution set is $\left\{\dfrac{1}{2}, 3\right\}$.

4. $-10 - 3(2x + 1) > 8x + 1$

$\quad -10 - 6x - 3 > 8x + 1$

$\qquad -6x - 13 > 8x + 1$

$\qquad\quad -13 > 14x + 1$

$\qquad\quad -14 > 14x$

$\qquad\qquad -1 > x$

$\qquad\qquad\; x < -1$

The solution set is $\{x \mid x < -1\}$ or $(-\infty, -1)$.

5. $2x + 7 < -11 \quad$ or $\quad -3x - 2 < 13$

$\quad 2x < -18 \qquad\qquad -3x < 15$

$\quad\; x < -9 \qquad\qquad\quad x > -5$

The solution set is

$\{x \mid x < -9 \text{ or } x > -5\}$ or $(-\infty, -9) \cup (-5, \infty)$.

6. $\mid 3x - 2 \mid \leq 4$

$\quad -4 \leq 3x - 2 \leq 4$

$\quad -2 \leq 3x \leq 6$

$\quad -\dfrac{2}{3} \leq x \leq 2$

The solution set is $\left\{x \;\middle|\; -\dfrac{2}{3} \leq x \leq 2\right\}$ or $\left[-\dfrac{2}{3}, 2\right]$.

7. $\mid x + 5 \mid = \mid 5x - 8 \mid$

$\quad x + 5 = 5x - 8 \quad$ or $\quad x + 5 = -(5x - 8)$

$\; -4x + 5 = -8 \qquad\qquad x + 5 = -5x + 8$

$\quad -4x = -13 \qquad\qquad\quad 6x + 5 = 8$

$\qquad x = \dfrac{13}{4} \qquad\qquad\qquad\; 6x = 3$

$\qquad\qquad\qquad\qquad\qquad\qquad x = \dfrac{1}{2}$

The solution set is $\left\{\dfrac{1}{2}, \dfrac{13}{4}\right\}$.

8. $5 - 2x \geq 9 \qquad$ and $\; 5x + 3 > -17$

$\quad 5 \geq 2x + 9 \qquad\qquad 5x > -20$

$\quad -4 \geq 2x \qquad\qquad\quad x > -4$

$\quad -2 \geq x$

$\quad\; x \leq -2$

The solution set is $\{x \mid -4 < x \leq -2\}$ or $(-4, -2]$.

9. $3x - 2 > -8 \quad$ or $\; 2x + 1 < 9$

$\quad 3x > -6 \qquad\qquad 2x < 8$

$\quad\; x > -2 \qquad\qquad\; x < 4$

The union of these sets is the entire number line.

The solution set is $\{x \mid x \text{ is a real number}\}$ or $(-\infty, \infty)$.

10. $\qquad \dfrac{x}{2} + 3 \leq \dfrac{x}{3} + \dfrac{5}{2}$

$\quad 6\left(\dfrac{x}{2} + 3\right) \leq 6\left(\dfrac{x}{3} + \dfrac{5}{2}\right)$

$\qquad 3x + 18 \leq 2x + 15$

$\qquad\; x + 18 \leq 15$

$\qquad\qquad\; x \leq -3$

The solution set is $\{x \mid x \leq -3\}$ or $(-\infty, -3]$.

11. $\dfrac{2}{3}(6x - 9) + 4 > 5x + 1$

$\quad 4x - 6 + 4 > 5x + 1$

$\qquad 4x - 2 > 5x + 1$

$\qquad\quad -2 > x + 1$

$\qquad\quad -3 > x$

$\qquad\qquad x < -3$

The solution set is $\{x \mid x < -3\}$ or $(-\infty, -3)$.

12. $|5x+3| > 2$

$5x+3 > 2 \quad$ or $\quad 5x+3 < -2$

$5x > -1 \qquad\qquad 5x < -5$

$x > -\dfrac{1}{5} \qquad\qquad x < -1$

The solution set is $\left\{ x \mid x < -1 \text{ or } x > -\dfrac{1}{5} \right\}$ or

$(-\infty, -1) \cup \left(-\dfrac{1}{5}, \infty \right)$.

13. $7 - \left| \dfrac{x}{2} + 2 \right| \le 4$

$-\left| \dfrac{x}{2} + 2 \right| \le -3$

$\left| \dfrac{x}{2} + 2 \right| \ge 3$

$|x+4| \ge 6$

$x+4 \ge 6 \quad$ or $\quad x+4 \le -6$

$x \ge 2 \qquad\qquad x \le -10$

The solution set is $\{ x \mid x \le -10 \text{ or } x \ge 2 \}$ or

$(-\infty, -10] \cup [2, \infty)$.

14. $\dfrac{x+3}{4} < \dfrac{1}{3}$

$3x + 9 < 4$

$3x < -5$

$x < -\dfrac{5}{3}$

The solution set is $\left\{ x \mid x < -\dfrac{5}{3} \right\}$ or $\left(-\infty, -\dfrac{5}{3} \right)$.

15. $5x+1 \ge 4x-2 \quad$ and $\quad 2x-3 > 5$

$x+1 \ge -2 \qquad\qquad 2x > 8$

$x \ge -3 \qquad\qquad\quad x > 4$

The solution set is $\{ x \mid x > 4 \}$ or $(4, \infty)$.

16. $3 - |2x-5| = -6$

$-|2x-5| = -9$

$|2x-5| = 9$

$2x-5 = 9 \quad$ or $\quad 2x-5 = -9$

$2x = 14 \qquad\qquad 2x = -4$

$x = 7 \qquad\qquad\quad x = -2$

The solution set is $\{ -2, 7 \}$.

17. $3 + |2x-5| = -6$

$|2x-5| = -9$

Since absolute values cannot be negative, there are no solutions. The solution set is $\varnothing$.

18. a. cost = fixed costs + variable cost

$C(x) = 60,000 + 0.18x$

b. revenue = price $\cdot$ quantity

$R(x) = 0.30x$

c. profit = revenue $-$ cost

$\begin{aligned}
P(x) &= R(x) - C(x) \\
&= 0.30x - (60,000 + 0.18x) \\
&= 0.30x - 60,000 - 0.18x \\
&= 0.12x - 60,000
\end{aligned}$

d. Let x = number of compact discs.

We need

$0.30x - (60,000 + 0.18x) \ge 30,000$

$0.30x - 60,000 - 0.18x \ge 30,000$

$0.12x - 60,000 \ge 30,000$

$0.12x \ge 90,000$

$x \ge 750,000$

The company should produce and sell at least 750,000 compact discs.

19. Let x = number of miles.

$24 + 0.20x \le 40$

$0.20x \le 16$

$x \le 80$

No more than 80 miles per day.

20. Let x = grade on the fifth exam.

$80 \le \dfrac{95 + 79 + 91 + 86 + x}{5} < 90$

$80 \le \dfrac{351 + x}{5} < 90$

$400 \le x + 351 < 450$

$49 \le x < 99$

$[49, 99)$

21. Let x = amount invested.

$x(0.075) \ge 9000$

$x \ge 120,000$

The retiree should invest at least $120,000.

9.4 Check Points

1. $4x - 2y \geq 8$

    First, graph the equation $4x - 2y = 8$ with a solid line.

    Set $y = 0$ to find the x-intercept.
    $$4x - 2y = 8$$
    $$4x - 2(0) = 8$$
    $$4x = 8$$
    $$x = 2$$
    Set $x = 0$ to find the y-intercept.
    $$4x - 2y = 8$$
    $$4(0) - 2y = 8$$
    $$-2y = 8$$
    $$y = -4$$
    Next, use the origin as a test point.
    $$4x - 2y \geq 8$$
    $$4(0) - 2(0) \geq 8$$
    $$0 \geq 8, \ \text{false}$$

    Since the statement is false, shade the half-plane that does not contain the test point.

    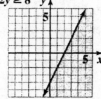

2. $y > -\dfrac{3}{4}x$

    First, graph the equation $y = -\dfrac{3}{4}x$ with a dashed line.

x	$y = -\dfrac{3}{4}x$	(x, y)
-4	$-\dfrac{3}{4}(-4) =$	$(-4, 3)$
0	$-\dfrac{3}{4}(0) =$	$(0, 0)$
4	$-\dfrac{3}{4}(4) =$	$(4, -3)$

    Next, use a test point such as $(4, 0)$.
    $$y > -\frac{3}{4}x$$
    $$4 > -\frac{3}{4}(0)$$
    $$4 > 0, \ \text{true}$$
    Since the statement is true, shade the half-plane that contains the test point.

    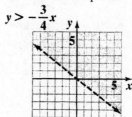

3. **a.** $y > 1$

    Graph the line $y = 1$ with a dashed line.
    Since the inequality is of the form $y > a$, shade the half-plane above the line.

    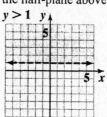

    b. $x \leq -2$

    Graph the line $x = -2$ with a solid line.
    Since the inequality is of the form $x \leq a$, shade the half-plane to the left of the line.

    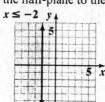

4. Point $B = (60, 20)$. Check this point in each of the three inequalities for grasslands.

$T \geq 35$	$5T - 7P \geq 70$	$3T - 35P \leq -140$
$60 \geq 35$, true	$5(60) - 7(20) \geq 70$	$3(60) - 35(20) \leq -140$
	$160 \geq 70$, true	$-520 \leq -140$, true

Since the three inequalities for grasslands are true, the point $B = (60, 20)$ does describe where grasslands occur.

5. $x - 3y < 6$

$2x + 3y \geq -6$

Graph the line $x - 3y = 6$ with a dashed line. Graph the line $2x + 3y = -6$ with a solid line.

For $x - 3y < 6$ use a test point such as (0, 0).

$x - 3y < 6$

$0 - 3(0) < 6$

$\quad 0 < 6$, true

Since the statement is true, shade the half-plane that contains the test point.

For $2x + 3y \geq -6$ use a test point such as (0, 0).

$2x + 3y \geq -6$

$2(0) + 3(0) \geq -6$

$\quad 0 \geq -6$, true

Since the statement is true, shade the half-plane that contains the test point.

The solution set of the system is the intersection (the overlap) of the two half-planes.

$x - 3y < 6$

$2x + 3y \geq -6$

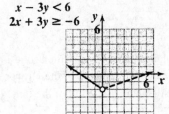

6. $x + y < 2$

$\quad -2 \leq x < 1$

$\quad\quad y > -3$

Graph the line $x + y = 2$ with a dashed line.

Graph the line $x = -2$ with a solid line.

Graph the line $x = 1$ with a dashed line.

Graph the line $y = -3$ with a dashed line.

For $x + y < 2$ use a test point such as (0, 0).

$x + y < 2$

$0 + 0 < 2$

$\quad 0 < 2$, true

Since the statement is true, shade the half-plane below the line, as it contains the test point.

For $-2 \leq x < 1$ use a test point such as (0, 0).

$-2 \leq x < 1$

$-2 \leq 0 < 1$, true

Since the statement is true, shade the region between the lines.

For $y > -3$ use a test point such as (0, 0).

$y > -3$

$0 > -3$, true

Since the statement is true, shade the half-plane above the line, as it contains the test point.

The solution set of the system is the intersection the shaded regions.

$x + y < 2$

$-2 \leq x < 1$

$\quad y > -3$

9.4 Exercise Set

1. $x + y \geq 3$

First, graph the equation $x + y = 3$. Rewrite the equation in slope-intercept form by solving for y.
$x + y = 3$

$\quad y = -x + 3$

y-intercept $= 3$
$\quad$ slope $= -1$

Next, use the origin as a test point.
$x + y \geq 3$

$0 + 0 \geq 3$

$\quad 0 \geq 3$

This is a false statement. This means that the point $(0,0)$ will not fall in the shaded half-plane.

$x + y \geq 3$ y

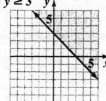

3. $x - y < 5$

First, graph the equation $x - y = 5$. Rewrite the equation in slope-intercept form by solving for y.
$x - y = 5$

$\quad -y = -x + 5$

$\quad y = x - 5$

y-intercept $= -5$
slope $= 1$
Next, use the origin as a test point.
$x - y < 5$

$0 - 0 < 5$

$\quad 0 < 5$

This is a true statement. This means that the point $(0,0)$ will fall in the shaded half-plane.

$x - y < 5$ $\quad y$

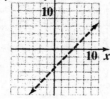

5. $x + 2y > 4$

First, graph the equation $x + 2y = 4$. Rewrite the equation in slope-intercept form by solving for y.
$x + 2y = 4$

$\quad 2y = -x + 4$

$\quad y = -\dfrac{1}{2}x + 2$

y-intercept $= 2$

slope $= -\dfrac{1}{2}$

Next, use the origin as a test point.
$0 + 2(0) > 4$

$\quad 0 + 0 > 4$

$\quad\quad 0 > 4$

This is a false statement. This means that the point $(0,0)$ will not fall in the shaded half-plane.

$x + 2y > 4$ $\quad y$

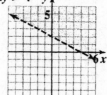

7. $3x - y \leq 6$

First, graph the equation $3x - y = 6$. Rewrite the equation in slope-intercept form by solving for y.
$3x - y = 6$

$\quad -y = -3x + 6$

$\quad\quad y = 3x - 6$

y-intercept $= -6$ $\quad\quad$ slope $= 3$
Next, use the origin as a test point.
$3(0) - 0 \leq 6$

$\quad 0 - 0 \leq 6$

$\quad\quad 0 \leq 6$

This is a true statement. This means that the point $(0,0)$ will fall in the shaded half-plane.

9. $\dfrac{x}{2} + \dfrac{y}{3} < 1$

First, graph the equation $\dfrac{x}{2} + \dfrac{y}{3} = 1$. Rewrite the

equation in slope-intercept form by solving for y.

$$\dfrac{x}{2} + \dfrac{y}{3} = 1$$

$$6\left(\dfrac{x}{2}\right) + 6\left(\dfrac{y}{3}\right) = 6(1)$$

$$3x + 2y = 6$$

$$2y = -3x + 6$$

$$y = -\dfrac{3}{2}x + 3$$

y-intercept $= 3$ slope $= -\dfrac{3}{2}$

Next, use the origin as a test point.

$$\dfrac{0}{2} + \dfrac{0}{3} < 1$$

$$0 + 0 < 1$$

$$0 < 1$$

This is a true statement. This means that the point $(0,0)$ will fall in the shaded half-plane.

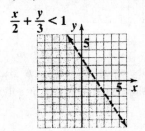

11. $y > \dfrac{1}{3}x$

Replacing the inequality symbol with an equal sign, we have $y = \dfrac{1}{3}x$. Since the equation is in slope-intercept form, use the slope and the intercept to graph the equation. The y-intercept is 0 and the slope is $\dfrac{1}{3}$.

Next, we need to find a test point. We cannot use the origin this time, because it lies on the line. Use $(1,1)$ as a test point.

$$1 > \dfrac{1}{3}(1)$$

$$1 > \dfrac{1}{3}$$

This is a true statement, so we know the point $(1,1)$ lies in the shaded half-plane.

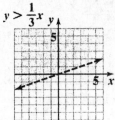

13. $y \le 3x + 2$

First, graph the equation $y = 3x + 2$. Since the equation is in slope-intercept form, use the slope and the intercept to graph the equation. The y-intercept is 2 and the slope is 3.

Next, use the origin as a test point.

$$0 \le 3(0) + 2$$

$$0 \le 2$$

This is a true statement. This means that the point $(0,0)$ will fall in the shaded half-plane.

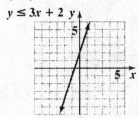

15. $y < -\dfrac{1}{4}x$

Replacing the inequality symbol with an equal sign, we have $y = -\dfrac{1}{4}x$. Since the equation is in slope-intercept form, use the slope and the intercept to graph the equation. The *y*–intercept is 0 and the slope is $-\dfrac{1}{4}$.

Next, we need to find a test point. We cannot use the origin this time, because it lies on the line. Use $(1,1)$ as a test point.

$$1 < -\frac{1}{4}(1)$$

$$1 < -\frac{1}{4}$$

This is a false statement, so we know the point $(1,1)$ does not lie in the shaded half-plane.

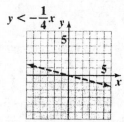

17. $x \le 2$

Replacing the inequality symbol with an equal sign, we have $x = 2$. We know that equations of the form $x = a$ are vertical lines with x-intercept $= a$. Next, use the origin as a test point.

$$x \le 2$$

$$0 \le 2$$

This is a true statement, so we know the point $(0,0)$ lies in the shaded half-plane.

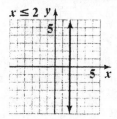

19. $y > -4$

Replacing the inequality symbol with an equal sign, we have $y = -4$. We know that equations of the form $y = b$ are horizontal lines with y-intercept $= b$. Next, use the origin as a test point.

$$y > -4$$

$$0 > -4$$

This is a true statement, so we know the point $(0,0)$ lies in the shaded half-plane.

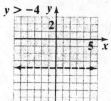

21. $y \ge 0$

Replacing the inequality symbol with an equal sign, we have $y = 0$. We know that equations of the form $y = b$ are horizontal lines with y-intercept $= b$. In this case, we have $y = 0$, the equation of the x-axis.

Next, we need to find a test point. We cannot use the origin, because it lies on the line. Use $(1,1)$ as a test point.

$$y \ge 0$$

$$1 \ge 0$$

This is a true statement, so we know the point $(1,1)$ lies in the shaded half-plane.

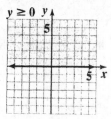

23. $3x + 6y \leq 6$

$2x + y \leq 8$

Graph the equations using the intercepts.

$3x + 6y = 6$ $2x + y = 8$

x – intercept = 2 x – intercept = 4

y – intercept = 1 y – intercept = 8

Use the origin as a test point to determine shading.

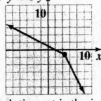

$3x + 6y \leq 6$

$2x + y \leq 8$

The solution set is the intersection of the shaded half-planes.

25. $2x - 5y \leq 10$

$3x - 2y > 6$

Graph the equations using the intercepts.

$2x - 5y = 10$ $3x - 2y = 6$

x – intercept = 5 x – intercept = 2

y – intercept = –2 y – intercept = –3

Use the origin as a test point to determine shading.

$2x - 5y \leq 10$

$3x - 2y > 6$

The solution set is the intersection of the shaded half-planes.

27. $y > 2x - 3$

$y < -x + 6$

Graph the equations using the intercepts.

$y = 2x - 3$ $y = -x + 6$

x – intercept = $\dfrac{3}{2}$ x – intercept = 6

             y – intercept = 6

y – intercept = –3

Use the origin as a test point to determine shading.

$y > 2x - 3$

$y < -x + 6$

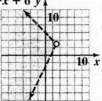

The solution set is the intersection of the shaded half-planes.

29. $x + 2y \leq 4$

$y \geq x - 3$

Graph the equations using the intercepts.

$x + 2y = 4$ $y = x - 3$

x – intercept = 4 x – intercept = 3

y – intercept = 2 y – intercept = –3

Use the origin as a test point to determine shading.

$x + 2y \leq 4$

$y \geq x - 3$

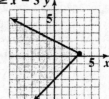

The solution set is the intersection of the shaded half-planes.

31. $x \leq 2$

$y \geq -1$

Graph the vertical line, $x = 2$, and the horizontal line, $y = -1$. Use the origin as a test point to determine shading.

$x \leq 2$

$y \geq -1$

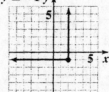

The solution set is the intersection of the shaded half-planes.

33. $-2 \le x < 5$

Since x lies between -2 and 5, graph the two vertical lines, $x = -2$ and $x = 5$. Since x lies between -2 and 5, shade between the two vertical lines.

$-2 \le x < 5$

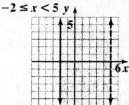

The solution is the shaded region.

35. $x - y \le 1$

  $x \ge 2$

Graph the equations.

$x - y = 1$ $x = 2$

$x - \text{intercept} = 1$ $x - \text{intercept} = 2$

$y - \text{intercept} = -1$ vertical line

Use the origin as a test point to determine shading.

$x - y \le 1$

   $x \ge 2$

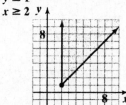

The solution set is the intersection of the shaded half-planes.

37. $x + y > 4$

  $x + y < -1$

Graph the equations using the intercepts.

$x + y = 4$ $x + y = -1$

$x - \text{intercept} = 4$ $x - \text{intercept} = -1$

$y - \text{intercept} = 4$ $y - \text{intercept} = -1$

Use the origin as a test point to determine shading.

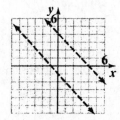

The solution set is the intersection of the shaded half-planes. Since the shaded half-planes do not intersect, there is no solution. The solution set is $\varnothing$ or $\{\ \}$.

39. $x + y > 4$

  $x + y > -1$

Graph the equations using the intercepts.

$x + y = 4$ $x + y = -1$

$x - \text{intercept} = 4$ $x - \text{intercept} = -1$

$y - \text{intercept} = 4$ $y - \text{intercept} = -1$

Use the origin as a test point to determine shading.

$x + y > 4$

$x + y > -1$

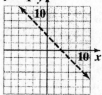

The solution set is the intersection of the shaded half-planes.

41. $x - y \le 2$

  $x \ge -2$

  $y \le 3$

Graph the equations using the intercepts.

$x - y = 2$ $y = 3$

$x - \text{intercept} = 2$ $y - \text{intercept} = 3$

$y - \text{intercept} = -2$ horizontal line

$x = -2$

$x - \text{intercept} = -2$

vertical line

Use the origin as a test point to determine shading.

$x - y \le 2$

  $x \ge -2$

  $y \le 3$

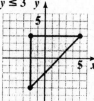

The solution set is the intersection of the shaded half-planes.

43. $x \geq 0$

$y \geq 0$

$2x + 5y \leq 10$

$3x + 4y \leq 12$

Since both x and y are greater than 0, we are concerned only with the first quadrant. Graph the other equations using the intercepts.

$2x + 5y = 10$ $3x + 4y = 12$

x – intercept = 5 x – intercept = 4

y – intercept = 2 y – intercept = 3

Use the origin as a test point to determine shading.

$x \geq 0$

$y \geq 0$

$2x + 5y \leq 10$

$3x + 4y \leq 12$

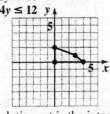

The solution set is the intersection of the shaded half-planes.

45. $3x + y \leq 6$

$2x - y \leq -1$

$x \geq -2$

$y \leq 4$

Graph the equations using the intercepts.

$z = 25x + 55y$

$\quad = 25(30) + 55(10)$

$\quad = 750 + 550 = 1300$

$z = 25x + 55y$

$\quad = 25(70) + 55(10)$

$\quad = 1750 + 550 = 2300$

Use the origin as a test point to determine shading.

$3x + y \leq 6$

$2x - y \leq -1$

$\quad\quad x \geq -2$

$\quad\quad y \leq 4$

The solution set is the intersection of the shaded half-planes. Because all inequalities are greater than or equal to or less than or equal to, the boundaries of the shaded half-planes are also included in the solution set.

47. $y \geq -2x + 4$

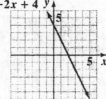

49. $x + y \leq 4$

$3x + y \leq 6$

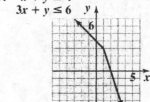

51. $-2 \leq x \leq 2$

$-3 \leq y \leq 3$

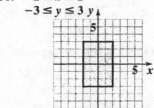

53. Find the union of solutions of $y > \dfrac{3}{2}x - 2$ and

$y < 4$.

$y > \dfrac{3}{2}x - 2$ or $y < 4$

55. The system $\begin{array}{c} 3x + 3y < 9 \\ 3x + 3y > 9 \end{array}$ has no solution. The number $3x + 3y$ cannot both be less than 9 and greater than 9 at the same time.

57. The system $\begin{array}{c} 3x + y \leq 9 \\ 3x + y \geq 9 \end{array}$ has infinitely many solutions. The solutions are all points on the line $3x + y = 9$.

59. a. The coordinates of point A are $(20,150)$. This means that a 20 year-old person with a heart rate of 150 beats per minute falls within the target zone.

b. $10 \le a \le 70$

$10 \le 20 \le 70$, true

$H \ge 0.7(220 - a)$

$150 \ge 0.7(220 - 20)$

$150 \ge 140$, true

$H \le 0.8(220 - a)$

$150 \le 0.8(220 - 20)$

$150 \le 160$, true

Since point A makes all three inequalities true, it is a solution of the system.

61. $10 \le a \le 70$

$H \ge 0.6(220 - a)$

$H \le 0.7(220 - a)$

63. a.
$$y \ge 0$$
$$x + y \ge 5$$
$$x \ge 1$$
$$200x + 100y \le 700$$

b. $y \ge 0$
$x + y \ge 5$
$x \ge 1$
$200x + 100y \le 700$

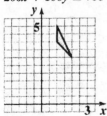

c. 2 nights

65. – 73. Answers will vary.

75. $y \le 4x + 4$

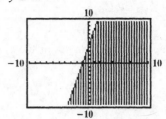

77. $2x + y \le 6$

$y \le -2x + 6$

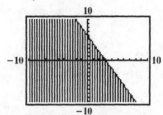

79. Answers will vary.

81. Answers will vary. For example, verify Exercise 23.

$3x + 6y \le 6$

$2x + y \le 8$

First solve both inequalities for y.

$3x + 6y \le 6$ $2x + y \le 8$

$6y \le -3x + 6$ $y \le -2x + 8$

$y \le -\dfrac{1}{2}x + 1$

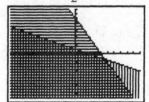

83. makes sense

85. makes sense

87. false; Changes to make the statement true will vary. A sample change is: The graph of $y \ge -x + 1$ has a solid line that falls from left to right.

89. true

91. The shading is above a dashed line of $y = -1$ and to the right of a solid line of $x = -2$. Thus, the system of inequalities is $x \ge -2$

$$y > -1.$$

93. The slope of the linear equation is
$$m = \frac{y_2 - y_1}{x_2 - x_1} = \frac{6-(-8)}{4-(-3)} = \frac{14}{7} = 2.$$
Use the point slope form to find the of the linear equation.
$$y - y_1 = m(x - x_1)$$
$$y - 6 = 2(x - 4)$$
$$\text{or}$$
$$y - 6 = 2x - 8$$
$$y = 2x - 2$$
The point $(1, 1)$ is above the line so the inequality is $y \geq 2x - 2$.

95. $y \geq nx + b$
$y \leq mx + b$

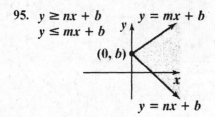

96. $3x - y = 8$
$x - 5y = -2$

Multiply the second equation by -3 and then add the two equations.
$$\begin{array}{r} 3x - y = 8 \\ -3x + 15y = 6 \\ \hline 14y = 14 \\ y = 1 \end{array}$$
Use back-substitution to find x.
$$x - 5(1) = -2$$
$$x - 5 = -2$$
$$x = 3$$
The solution set is $\{(3,1)\}$.

97. $\begin{array}{l} y = 3x - 2 \\ y = -2x + 8 \end{array}$

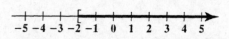

The solution is $(2,4)$.

98. $2x^6 + 20x^5y + 50x^4y^2 = x^4(2x^2 + 20xy + 50y^2)$
$$= x^4(2x^2 + 20xy + 50y^2)$$
$$= x^4(x + 5y)^2$$

99. $f(x) = \sqrt{3x + 12}$
$$f(-1) = \sqrt{3(-1) + 12}$$
$$= \sqrt{9}$$
$$= 3$$

100. $f(x) = \sqrt{3x + 12}$
$$f(-1) = \sqrt{3(8) + 12}$$
$$= \sqrt{36}$$
$$= 6$$

101. Domain: $[-4, \infty)$
Range: $[0, \infty)$

Chapter 9 Review Exerecises

1. $-6x + 3 \leq 15$
$$-6x \leq 12$$
$$\frac{-6x}{-6} \geq \frac{12}{-6}$$
$$x \geq -2$$

The solution set is $\{x | x \geq -2\}$ or $[2, \infty)$.

2. $6x - 9 \geq -4x - 3$

$10x - 9 \geq -3$

$10x \geq 6$

$x \geq \dfrac{6}{10}$

$x \geq \dfrac{3}{5}$

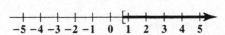

The solution set is $\left\{x \middle| x \geq \dfrac{3}{5}\right\}$ or $\left[\dfrac{3}{5}, \infty\right)$.

3. $\dfrac{x}{3} - \dfrac{3}{4} - 1 > \dfrac{x}{2}$

$12\left(\dfrac{x}{3}\right) - 12\left(\dfrac{3}{4}\right) - 12(1) > 12\left(\dfrac{x}{2}\right)$

$4x - 3(3) - 12 > 6x$

$4x - 9 - 12 > 6x$

$4x - 21 > 6x$

$-2x - 21 > 0$

$-2x > 21$

$x < -\dfrac{21}{2}$

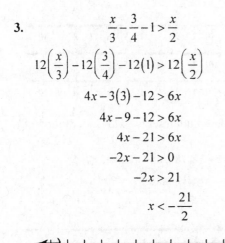

The solution set is $\left\{x \middle| x < -\dfrac{21}{2}\right\}$ or $\left(-\infty, -\dfrac{21}{2}\right)$.

4. $6x + 5 > -2(x - 3) - 25$

$6x + 5 > -2x + 6 - 25$

$6x + 5 > -2x - 19$

$8x + 5 > -19$

$8x > -24$

$x > -3$

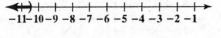

The solution set is $\left\{x \middle| x > -3\right\}$ or $(-3, \infty)$.

5. $3(2x - 1) - 2(x - 4) \geq 7 + 2(3 + 4x)$

$6x - 3 - 2x + 8 \geq 7 + 6 + 8x$

$4x + 5 \geq 13 + 8x$

$-4x + 5 \geq 13$

$-4x \geq 8$

$x \leq -2$

The solution set is $\left\{x \middle| x \leq -2\right\}$ or $(-\infty, -2]$.

6. a. $P(x) = R(x) - C(x)$

$= 125x - (40x + 357,000)$

$= 125x - 40x - 357,000$

$= 85x - 357,000$

b. $P(x) > 0$

$85x - 357,000 > 0$

$85x > 357,000$

$x > 4200$

More than 4200 toaster ovens need to be produced and sold to make a profit.

7. cost = fixed costs + variable cost

$C(x) = 360,000 + 850x$

8. revenue = price $\cdot$ quantity

$R(x) = 1150x$

9. profit = revenue − cost

$P(x) = R(x) - C(x)$

$= 1150x - (360,000 + 850x)$

$= 1150x - 360,000 - 850x$

$= 300x - 360,000$

10. $P(x) > 0$

$300x - 360,000 > 0$

$300x > 360,000$

$x > 1200$

More than 1200 computers need to be produced and sold to make a profit.

11. Let x = the number of checks written per month. The cost using the first method is $c_1 = 11 + 0.06x$.

The cost using the second method is $c_2 = 4 + 0.20x$.

The first method is a better deal if it costs less than the second method.

$$c_1 < c_2$$
$$11 + 0.06x < 4 + 0.20x$$
$$11 - 0.14x < 4$$
$$-0.14x < -7$$
$$\frac{-0.14x}{-0.14} > \frac{-7}{-0.14}$$
$$x > 50$$

The first method is a better deal when more than 50 checks per month are written.

12. Let x = the amount of sales per month in dollars. The salesperson's commission is $c = 500 + 0.20x$. We are looking for the amount of sales, x, the salesman must make to receive more than \$3200 in income.

$$c > 3200$$
$$500 + 0.20x > 3200$$
$$0.20x > 2700$$
$$x > 13500$$

The salesman must sell more than \$13,500 to receive a total income that exceeds \$3200 per month.

13. $A \cap B = \{a, c\}$

14. $A \cap C = \{a\}$

15. $A \cup B = \{a, b, c, d, e\}$

16. $A \cup C = \{a, b, c, d, f, g\}$

17. $x \le 3$ and $x < 6$

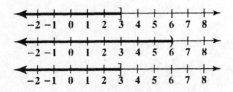

The solution set is $\{x \mid x \le 3\}$ or $(-\infty, 3]$.

18. $x \le 3$ or $x < 6$

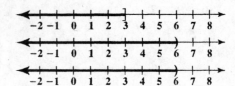

The solution set is $\{x \mid x < 6\}$ or $(-\infty, 6)$.

19. $-2x < -12$ and $x - 3 < 5$
$$\frac{-2x}{-2} > \frac{-12}{-2} \qquad x < 8$$
$$x > 6$$

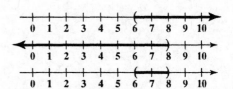

The solution set is $\{x \mid 6 < x < 8\}$ or $(6, 8)$.

20. $5x + 3 \le 18$ and $2x - 7 \le -5$
$$5x \le 15 \qquad\qquad 2x \le 2$$
$$x \le 3 \qquad\qquad x \le 1$$

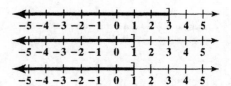

The solution set is $\{x \mid x \le 1\}$ or $(-\infty, 1]$.

21. $2x - 5 > -1$ and $3x < 3$
$$2x > 4 \qquad\qquad x < 1$$
$$x > 2$$

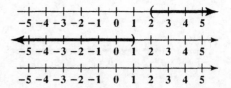

Since the two sets do not intersect, the solution set is $\varnothing$ or $\{\ \}$.

22. $2x-5>-1$　or　$3x<3$

$\qquad 2x>4 \qquad\qquad x<1$

$\qquad\quad x>2$

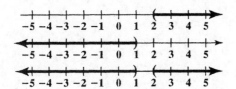

The solution set is $\{x|x<1 \text{ or } x>2\}$ or

$(-\infty,1)\cup(2,\infty)$.

23. $x+1\le-3$　or　$-4x+3<-5$

$\qquad x\le-4 \qquad\qquad -4x<-8$

$\qquad\qquad\qquad\qquad\qquad x>2$

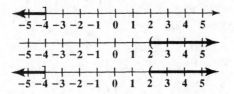

The solution set is $\{x|x\le-4 \text{ or } x>2\}$ or

$(-\infty,-4]\cup(2,\infty)$.

24. $5x-2\le-22$　or　$-3x-2>4$

$\qquad 5x\le-20 \qquad\qquad -3x>6$

$\qquad\quad x\le-4 \qquad\qquad\quad x<-2$

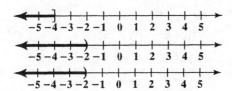

The solution set is $\{x|x<-2\}$ or $(-\infty,-2)$.

25. $5x+4\ge-11$　or　$1-4x\ge9$

$\qquad 5x\ge-15 \qquad\qquad -4x\ge8$

$\qquad\quad x\ge-3 \qquad\qquad\quad x\le-2$

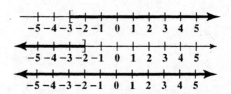

The solution set is $\mathbb{R}$, $(-\infty,\infty)$ or

$\{x|x \text{ is a real number}\}$.

26.　　$-3<x+2\le4$

$\qquad -3-2<x+2-2\le4-2$

$\qquad\qquad -5<x\le2$

$$\begin{array}{c}\text{\textbar}\;(\;\text{\textbar}\;\text{\textbar}\;\text{\textbar}\;\text{\textbar}\;\text{\textbar}\;\text{\textbar}\;\text{\textbar}\;\text{\textbar}\;]\;\text{\textbar}\;\text{\textbar}\;\text{\textbar}\\ -6\;-5\;-4\;-3\;-2\;-1\;\;0\;\;1\;\;2\;\;3\;\;4\end{array}$$

The solution set is $\{x|-5<x\le2\}$ or $(-5,2]$.

27.　　$-1\le4x+2\le6$

$\qquad -1-2\le4x+2-2\le6-2$

$\qquad\qquad -3\le4x\le4$

$\qquad\qquad -\dfrac{3}{4}\le\dfrac{4x}{4}\le\dfrac{4}{4}$

$\qquad\qquad -\dfrac{3}{4}\le x\le1$

$$\begin{array}{c}\text{\textbar}\;\text{\textbar}\;\text{\textbar}\;\text{\textbar}\;\text{\textbar}\;[\;\text{\textbar}\;]\;\text{\textbar}\;\text{\textbar}\;\text{\textbar}\;\text{\textbar}\\ -5\;-4\;-3\;-2\;-1\;\;0\;\;1\;\;2\;\;3\;\;4\;\;5\end{array}$$

The solution set is $\left\{x\left|-\dfrac{3}{4}\le x\le1\right.\right\}$ or $\left[-\dfrac{3}{4},1\right]$.

28. Let $x=$ the grade on the fifth exam.

$$80\le\dfrac{95+79+91+86+x}{5}<90$$

$$80\le\dfrac{351+x}{5}<90$$

$$5(80)\le5\left(\dfrac{351+x}{5}\right)<5(90)$$

$$400\le351+x<450$$

$$400-351\le351-351+x<450-351$$

$$49\le x<99$$

You need to score at least 49% and less than 99% on the exam to receive a B. In interval notation, the range is $[49\%,99\%)$.

29. $|2x+1| = 7$

$2x+1 = 7$ or $2x+1 = -7$

$2x = 6$ $\qquad$ $2x = -8$

$x = 3$ $\qquad$ $x = -4$

The solution set is $\{-4, 3\}$.

30. $|3x+2| = -5$

There are no values of x for which the absolute value of $3x + 2$ is a negative number. By definition, absolute values are always positive. The solution set is $\varnothing$ or $\{\ \}$.

31. $2|x-3| - 7 = 10$

$2|x-3| = 17$

$|x-3| = 8.5$

$x - 3 = 8.5$ or $x - 3 = -8.5$

$x = 11.5$ $\qquad$ $x = -5.5$

The solution set is $\{-5.5, 11.5\}$.

32. $|4x-3| = |7x+9|$

$4x-3 = 7x+9$ or $4x-3 = -7x-9$

$-3x-3 = 9$ $\qquad$ $11x-3 = -9$

$-3x = 12$ $\qquad$ $11x = -6$

$x = -4$ $\qquad$ $x = -\dfrac{6}{11}$

The solution set is $\left\{-4, -\dfrac{6}{11}\right\}$.

33. $\qquad |2x+3| \le 15$

$-15 \le 2x+3 \le 15$

$-15-3 \le 2x+3-3 \le 15-3$

$-18 \le 2x \le 12$

$-\dfrac{18}{2} \le \dfrac{2x}{2} \le \dfrac{12}{2}$

$-9 \le x \le 6$

The solution set is $\{x | -9 \le x \le 6\}$ or $[-9, 6]$.

34. $\left|\dfrac{2x+6}{3}\right| > 2$

$\dfrac{2x+6}{3} < -2$ or $\dfrac{2x+6}{3} > 2$

$2x+6 < -6$ $\qquad$ $2x+6 > 6$

$2x < -12$ $\qquad$ $2x > 0$

$x < -6$ $\qquad$ $x > 0$

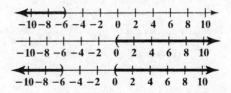

The solution set is $\{x | x < -6 \text{ or } x > 0\}$ or $(-\infty, -6) \cup (0, \infty)$.

35. $\qquad |2x+5| - 7 < -6$

$|2x+5| < 1$

$-1 < 2x+5 < 1$

$-1-5 < 2x+5-5 < 1-5$

$-6 < 2x < -4$

$-3 < x < -2$

The solution set is $\{x | -3 < x < -2\}$ or $(-3, -2)$.

36. $-4|x+2| + 5 \le -7$

$-4|x+2| \le -12$

$\dfrac{-4|x+2|}{-4} \ge \dfrac{-12}{-4}$

$|x+2| \ge 3$

$x+2 \ge 3$ or $x+2 \le -3$

$x \ge 1$ $\qquad$ $x \le -5$

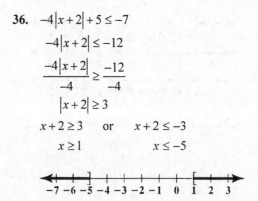

The solution set is $\{x | x \le -5 \text{ or } x \ge 1\}$ or $(-\infty, -5] \cup [1, \infty)$.

37. $|2x-3|+4 \le -10$

$\qquad |2x-3| \le -14$

There are no values of x for which the absolute value of $2x-3$ is a negative number. By definition, absolute values are always positive. The solution set is $\varnothing$ or $\{\ \}$.

38. $|h-6.5| \le 1$

$\qquad -1 \le h-6.5 \le 1$

$\qquad 5.5 \le h \le 7.5$

Approximately 90% of the population sleeps between 5.5 hours and 7.5 hours daily, inclusive.

39. $3x-4y > 12$

First, find the intercepts to the equation $3x-4y = 12$.

Find the x–intercept by setting $y = 0$.

$\qquad 3x-4y = 12$

$\qquad 3x-4(0) = 12$

$\qquad\quad 3x = 12$

$\qquad\qquad x = 4$

Find the y–intercept by setting $x = 0$.

$\qquad 3x-4y = 12$

$\qquad 3(0)-4y = 12$

$\qquad\quad -4y = 12$

$\qquad\qquad y = -3$

Next, use the origin as a test point.

$\qquad 3x-4y > 12$

$\qquad 3(0) - 4(0) > 12$

$\qquad\qquad 0 > 12$

This is a false statement. This means that the point, $(0,0)$, will not fall in the shaded half-plane.

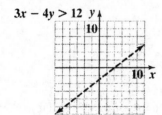

40. $x-3y \le 6$

First, find the intercepts to the equation $x-3y = 6$. Find the x–intercept by setting $y = 0$, find the y–intercept by setting $x = 0$.

$\qquad x-3y = 6 \qquad\qquad x-3y = 6$

$\qquad x-3(0) = 6 \qquad\quad 0-3y = 6$

$\qquad\qquad x = 6 \qquad\qquad -3y = 6$

$\qquad\qquad\qquad\qquad\qquad\quad y = -2$

Next, use the origin as a test point.

$\qquad 0 - 3(0) \le 6$

$\qquad\qquad 0 \le 6$

This is a true statement. This means that the point, $(0,0)$, will fall in the shaded half-plane.

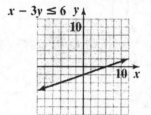

41. $y \le -\dfrac{1}{2}x+2$

Replacing the inequality symbol with an equal sign, we have $y = -\dfrac{1}{2}x+2$. Since the equation is in slope-intercept form, use the slope and the intercept to graph the equation. The y–intercept is 2 and the slope is $-\dfrac{1}{2}$.

Next, use the origin as a test point.

$\qquad y \le -\dfrac{1}{2}x+2$

$\qquad 0 \le -\dfrac{1}{2}(0) +2$

$\qquad\qquad 0 \le 2$

This is a true statement. This means that the point $(0,0)$ will fall in the shaded half-plane.

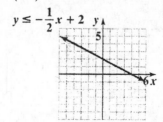

42. $y > \dfrac{3}{5}x$

Replacing the inequality symbol with an equal sign, we have $y = \dfrac{3}{5}x$. Since the equation is in slope-intercept form, use the slope and the intercept to graph the equation. The y–intercept is 0 and the slope is $\dfrac{3}{5}$.

Next, we need to find a test point. We cannot use the origin this time, because it lies on the line. Use $(1,1)$ as a test point.

$1 > \dfrac{3}{5}(1)$

$1 > \dfrac{3}{5}$

This is a true statement, so we know the point $(1,1)$ lies in the shaded half-plane.

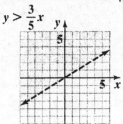

43. $x \le 2$

Replacing the inequality symbol with an equal sign, we have $x = 2$. We know that equations of the form $x = a$ are vertical lines with x–intercept = a. Next, use the origin as a test point.

$x \le 2$

$0 \le 2$

This is a true statement, so we know the point $(0,0)$ lies in the shaded half-plane.

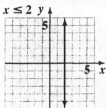

44. $y > -3$

Replacing the inequality symbol with an equal sign, we have $y = -3$. We know that equations of the form $y = b$ are horizontal lines with y–intercept = b. Next, use the origin as a test point.

$y > -3$

$0 > -3$

This is a true statement, so we know the point $(0,0)$ lies in the shaded half-plane.

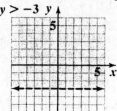

45. $2x - y \le 4$

  $x + y \ge 5$

First consider $2x - y \le 4$. If we solve for y in $2x - y = 4$, we can graph the line using the slope and the y-intercept.

$2x - y = 4$

  $-y = -2x + 4$

  $y = 2x - 4$

y-intercept = -4

slope = 2

Now, use the origin as a test point.

$2(0) - 0 \le 4$

  $0 \le 4$

This is a true statement. This means that the point $(0,0)$ will fall in the shaded half-plane.

Next consider $x + y \ge 5$. If we solve for y in $x + y = 5$, we can graph using the slope and the y-intercept.

$x + y = 5$

  $y = -x + 5$

y-intercept = 5

slope = -1

Now, use the origin as a test point.

$0 + 0 \ge 5$

  $0 \ge 5$

This is a false statement. This means that the point $(0,0)$ will not fall in the shaded half-plane.

Next, graph each of the inequalities. The solution to the system is the intersection of the shaded half-planes.

$2x - y \le 4$
$x + y \ge 5$

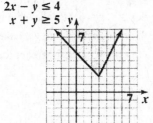

46. $y < -x + 4$

$y > x - 4$

First consider $y < -x + 4$. Change the inequality symbol to an equal sign. The line $y = -x + 4$ is in slope-intercept form and can be graphed using the slope and the y–intercept.

y–intercept = 4
slope = -1

Now, use the origin as a test point.

$0 < -\!\!\!\not 0 + 4$

$0 < 4$

This is a true statement. This means that the point $(0, 0)$ will fall in the shaded half-plane.

Next consider $y > x - 4$. Change the inequality symbol to an equal sign. The line $y = x - 4$ is in slope-intercept form and can be graphed using the slope and the y–intercept.

y–intercept = -4
slope = 1

Now, use the origin as a test point.

$0 > 0 - 4$

$0 > -4$

This is a true statement. This means that the point $(0, 0)$ will fall in the shaded half-plane.

Next, graph each of the inequalities. The solution to the system is the intersection of the shaded half-planes.

$y < -x + 4$
$y > x - 4$

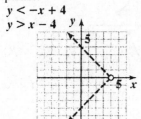

47. $-3 \le x < 5$

Rewrite the three part inequality as two separate inequalities. We have $-3 \le x$ and $x < 5$. We replace the inequality symbols with equal signs and obtain $-3 = x$ and $x = 5$. Equations of the form $x = a$ are vertical lines with x–intercept = a.

We know the shading in the graph will be between $x = -3$ and $x = 5$ because in the original inequality we see that x lies between -3 and 5.

$-3 \le x < 5$

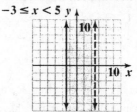

48. $-2 < y \le 6$

Rewrite the three part inequality as two separate inequalities. We have $-2 < y$ and $y \le 6$. We replace the inequality symbols with equal signs and obtain $-2 = y$ and $y = 6$. Equations of the form $y = b$ are vertical lines with y–intercept = b.

We know the shading in the graph will be between $y = -2$ and $y = 6$ because in the original inequality we see that y lies between -2 and 6.

$-2 < y \le 6$

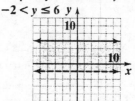

49. $x \ge 3$

$y \le 0$

First consider $x \ge 3$. Change the inequality symbol to an equal sign and we obtain the vertical line $x = 3$. Because we have $x \ge 3$, we know the shading is to the right of the line $x = 3$.

Next consider $y \le 0$. Change the inequality symbol to an equal sign and we obtain the horizontal line $y = 0$. (Recall that this is the equation of the x–axis.) Because we have $y \le 0$, we know that the shading will be below the x–axis.

Next, graph each of the inequalities. The solution to the system is the intersection of the shaded half-planes.

$x \geq 3$
$y \leq 0$

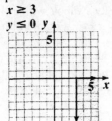

50. $2x - y > -4$
$\qquad x \geq 0$

First consider $2x - y > -4$. Replace the inequality symbol with an equal sign and we have $2x - y = -4$. Solve for y to obtain slope-intercept form.

$2x - y = -4$
$\quad -y = -2x - 4$
$\quad\quad y = 2x + 4$

y–intercept = 4
slope = 2

Now, use the origin as a test point.

$2x - y > -4$
$2(0) - 0 > -4$
$\quad\quad 0 > -4$

This is a true statement. This means that the point $(0,0)$ will fall in the shaded half-plane.

Next consider $x \geq 0$. Change the inequality symbol to an equal sign and we obtain the vertical line $x = 0$. (Recall that this is the equation of the y–axis.) Because we have $x \geq 0$, we know that the shading will be to the right of the y–axis.

Next, graph each of the inequalities. The solution to the system is the intersection of the shaded half-planes.

$2x - y > -4$
$\quad x \geq 0$

51. $x + y \leq 6$
$\qquad y \geq 2x - 3$

First consider $x + y \leq 6$. Replace the inequality symbol with an equal sign and we have $x + y = 6$. Solve for y to obtain slope-intercept form.

$x + y = 6$
$\quad\quad y = -x + 6$

y-intercept = 6
slope = -1

Now, use the origin as a test point.

$0 + 0 \leq 6$
$\quad\quad 0 \leq 6$

This is a true statement. This means that the point $(0,0)$ will fall in the shaded half-plane.

Next consider $y \geq 2x - 3$. Replace the inequality symbol with an equal sign and we have $y = 2x - 3$. The equation is in slope-intercept form, so we can use the slope and the y-intercept to graph the line.

y-intercept = -3
slope = 2

Now, use the origin as a test point.

$y \geq 2x - 3$
$0 \geq 2(0) - 3$
$0 \geq -3$

This is a true statement. This means that the point $(0,0)$ will fall in the shaded half-plane.

Next, graph each of the inequalities. The solution to the system is the intersection of the shaded half-planes.

$x + y \leq 6$
$y \geq 2x - 3$

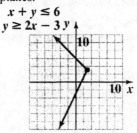

486

52. $3x + 2y \geq 4$

$x - y \leq 3$

$x \geq 0, \ y \geq 0$

First consider $3x + 2y \geq 4$. Replace the inequality symbol with an equal sign and we have $3x + 2y = 4$. Solve for y to obtain slope-intercept form.

$3x + 2y = 4$

$2y = -3x + 4$

$y = -\dfrac{3}{2}x + 2$

y–intercept $= 2$　　　slope $= -\dfrac{3}{2}$

Now, use the origin as a test point.

$3x + 2y \geq 4$

$3\cancel{(0)} + 2\cancel{(0)} \geq 4$

$0 \geq 4$

This is a false statement. This means that the point $(0,0)$ will not fall in the shaded half-plane.

Now consider $x - y \leq 3$. Replace the inequality symbol with an equal sign and we have $x - y = 3$. Solve for y to obtain slope-intercept form.

$x - y = 3$

$-y = -x + 3$

$y = x - 3$

y–intercept $= -3$　　　slope $= 1$

Now, use the origin as a test point.

$x - y \leq 3$

$0 - 0 \leq 3$

$0 \leq 3$

This is a true statement. This means that the point $(0,0)$ will fall in the shaded half-plane.

Now consider the inequalities $x \geq 0$ and $y \geq 0$. The inequalities mean that both x and y will be positive. This means that we only need to consider quadrant I.

Next, graph each of the inequalities. The solution to the system is the intersection of the shaded half-planes.

53. $2x - y > 2$

$2x - y < -2$

First consider $2x - y > 2$. Replace the inequality symbol with an equal sign and we have $2x - y = 2$. Solve for y to obtain slope-intercept form.

$2x - y = 2$

$-y = -2x + 2$

$y = 2x - 2$

y–intercept $= -2$　　　slope $= 2$

Now, use the origin as a test point.

$2x - y > 2$

$2\cancel{(0)} - 0 > 2$

$0 > 2$

This is a false statement. This means that the point $(0,0)$ will not fall in the shaded half-plane.

Now consider $2x - y < -2$. Replace the inequality symbol with an equal sign and we have $2x - y = -2$. Solve for y to obtain slope-intercept form.

$2x - y = -2$

$-y = -2x - 2$

$y = 2x + 2$

y–intercept $= 2$　　　slope $= 2$

Now, use the origin as a test point.

$2x - y < -2$

$2\cancel{(0)} - 0 < -2$

$0 < -2$

This is a false statement. This means that the point $(0,0)$ will not fall in the shaded half-plane.

Next, graph each of the inequalities. The solution to the system is the intersection of the shaded half-planes.

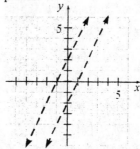

The graphs of the inequalities do not intersect, so there is no solution. The solution set is $\varnothing$ or $\{\ \}$.

$3x + 2y \geq 4$

$\ x - y \leq 3$

$x \geq 0, y \geq 0$

Chapter 9 Test

1. $3(x+4) \ge 5x-12$

$3x+12 \ge 5x-12$

$-2x+12 \ge -12$

$-2x \ge -24$

$\dfrac{-2x}{-2} \le \dfrac{-24}{-2}$

$x \le 12$

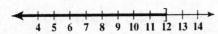

The solution set is $\{x \mid x \le 12\}$ or $(-\infty, 12]$.

2. $\dfrac{x}{6} + \dfrac{1}{8} \le \dfrac{x}{2} - \dfrac{3}{4}$

$24\left(\dfrac{x}{6}\right) + 24\left(\dfrac{1}{8}\right) \le 24\left(\dfrac{x}{2}\right) - 24\left(\dfrac{3}{4}\right)$

$4x+3 \le 12x - 6(3)$

$4x+3 \le 12x - 18$

$-8x+3 \le -18$

$-8x \le -21$

$\dfrac{-8x}{-8} \ge \dfrac{-21}{-8}$

$x \ge \dfrac{21}{8}$

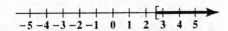

The solution set is $\left\{x \mid x \ge \dfrac{21}{8}\right\}$ or $\left[\dfrac{21}{8}, \infty\right)$.

3. a. cost = fixed costs + variable cost

$C(x) = 60,000 + 200x$

b. revenue = price $\cdot$ quantity

$R(x) = 450x$

c. profit = revenue − cost

$P(x) = R(x) - C(x)$

$= 450x - (60,000 + 200x)$

$= 450x - 60,000 - 200x$

$= 250x - 60,000$

d. $P(x) > 0$

$250x - 60,000 > 0$

$250x > 60,000$

$x > 240$

More than 240 computer desks need to be produced and sold to make a profit.

4. $\{2,4,6,8,10\} \cap \{4,6,12,14\} = \{4,6\}$

5. $\{2,4,6,8,10\} \cup \{4,6,12,14\}$

$= \{2,4,6,8,10,12,14\}$

6. $2x+4 < 2$ and $x-3 > -5$

$\qquad 2x < -2 \qquad\qquad x > -2$

$\qquad x < -1$

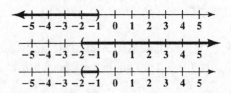

The solution set is $\{x \mid -2 < x < -1\}$ or $(-2, -1)$.

7. $x+6 \ge 4$ and $2x+3 \ge -2$

$\qquad x \ge -2 \qquad\qquad 2x \ge -5$

$\qquad\qquad\qquad\qquad\quad x \ge -\dfrac{5}{2}$

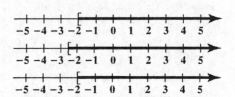

The solution set is $\{x \mid x \ge -2\}$ or $[-2, \infty)$.

8. $2x - 3 < 5$　or　$3x - 6 \leq 4$

$\quad\quad 2x < 8 \quad\quad\quad\quad 3x \leq 10$

$\quad\quad x < 4 \quad\quad\quad\quad\quad x \leq \dfrac{10}{3}$

The solution set is $\left\{x \mid x < 4\right\}$ or $(-\infty, 4)$.

9. $x + 3 \leq -1$　or　$-4x + 3 < -5$

$\quad\quad x \leq -4 \quad\quad\quad\quad -4x < -8$

$\quad\quad\quad\quad\quad\quad\quad\quad\quad\quad x > 2$

The solution set is $\left\{x \mid x \leq -4 \text{ or } x > 2\right\}$ or $(-\infty, -4] \cup (2, \infty)$.

10. $\quad\quad -3 \leq \dfrac{2x + 5}{3} < 6$

$3(-3) \leq 3\left(\dfrac{2x + 5}{3}\right) < 3(6)$

$\quad\quad -9 \leq 2x + 5 < 18$

$-9 - 5 \leq 2x + 5 - 5 < 18 - 5$

$\quad\quad -14 \leq 2x < 13$

$\quad\quad\quad -7 \leq x < \dfrac{13}{2}$

The solution set is $\left\{x \mid -7 \leq x < \dfrac{13}{2}\right\}$ or $\left[-7, \dfrac{13}{2}\right)$.

11. $|5x + 3| = 7$

$\quad 5x + 3 = 7$　or　$5x + 3 = -7$

$\quad\quad 5x = 4 \quad\quad\quad\quad 5x = -10$

$\quad\quad\quad x = \dfrac{4}{5} \quad\quad\quad\quad x = -2$

The solutions are -2 and $\dfrac{4}{5}$ and the solution set is

$\left\{-2, \dfrac{4}{5}\right\}$.

12. $|6x + 1| = |4x + 15|$

$\quad 6x + 1 = 4x + 15$

$\quad 2x + 1 = 15$

$\quad\quad 2x = 14$

$\quad\quad\quad x = 7$

or

$\quad\quad 6x + 1 = -(4x + 15)$

$\quad\quad 6x + 1 = -4x - 15$

$\quad\quad 10x + 1 = -15$

$\quad\quad\quad 10x = -16$

$\quad\quad\quad\quad x = -\dfrac{16}{10} = -\dfrac{8}{5}$

The solutions are $-\dfrac{8}{5}$ and 7 and the solution set is

$\left\{-\dfrac{8}{5}, 7\right\}$.

13. $\quad\quad\quad |2x - 1| < 7$

$\quad\quad\quad -7 < 2x - 1 < 7$

$\quad -7 + 1 < 2x - 1 + 1 < 7 + 1$

$\quad\quad\quad -6 < 2x < 8$

$\quad\quad\quad\quad -3 < x < 4$

The solution set is $\left\{x \mid -3 < x < 4\right\}$ or $(-3, 4)$.

14. $|2x - 3| \geq 5$

$$2x - 3 \leq -5 \quad \text{or} \quad 2x - 3 \geq 5$$
$$2x \leq -2 \qquad\qquad 2x \geq 8$$
$$x \leq -1 \qquad\qquad x \geq 4$$

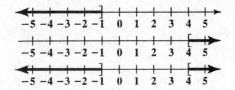

The solution set is $\left\{ x \middle| x \leq -1 \text{ or } x \geq 4 \right\}$ or $(-\infty, -1] \cup [4, \infty)$.

15. $|b - 98.6| > 8$

$$b - 98.6 > -8 \quad \text{or} \quad b - 98.6 > 8$$
$$b > 90.6 \qquad\qquad b > 106.6$$

Hyperthermia occurs when the body temperature is over $106.6°F$ and hypothermia occurs when the body temperature is below $90.6°F$.

16. $3x - 2y < 6$

First, find the intercepts to the equation $3x - 2y = 6$.

Find the *x*–intercept by setting $y = 0$.
$$3x - 2y = 6$$
$$3x - 2(0) = 6$$
$$3x = 6$$
$$x = 2$$

Find the *y*–intercept by setting $x = 0$.
$$3x - 2y = 6$$
$$3(0) - 2y = 6$$
$$-2y = 6$$
$$y = -3$$

Next, use the origin as a test point.
$$3x - 2y < 6$$
$$3(0) - 2(0) < 6$$
$$0 < 6$$

This is a true statement. This means that the point will fall in the shaded half-plane.

$3x - 2y < 6$

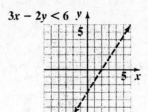

17. $y \geq \dfrac{1}{2}x - 1$

Replacing the inequality symbol with an equal sign, we have $y = \dfrac{1}{2}x - 1$. The equation is in slope-intercept form, so graph the line using the slope and the *y*-intercept.

y-intercept $= -1$ \qquad slope $= \dfrac{1}{2}$

Now, use the origin, $(0, 0)$, as a test point.

$$y \geq \frac{1}{2}x - 1$$
$$0 \geq \frac{1}{2}(0) - 1$$
$$0 \geq -1$$

This is a true statement. This means that the point will fall in the shaded half-plane.

$y \geq \dfrac{1}{2}x - 1$

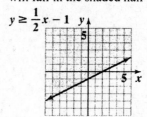

18. $y \le -1$

Replacing the inequality symbol with an equal sign, we have $y = -1$. Equations of the form $y = b$ are horizontal lines with y–intercept = b, so this is a horizontal line at $y = -1$.

Next, use the origin as a test point.
$y \le -1$
$0 \le -1$

This is a false statement, so we know the point $(0,0)$ does not lie in the shaded half-plane.

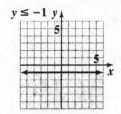

19. $x + y \ge 2$
$x - y \ge 4$

First consider $x + y \ge 2$. If we solve for y in $x + y = 2$, we can graph the line using the slope and the y–intercept.
$x + y = 2$
$y = -x + 2$
y-intercept = 2   slope = -1

Now, use the origin as a test point.
$x + y \ge 2$
$0 + 0 \ge 2$
$0 \ge 2$

This is a false statement. This means that the point will not fall in the shaded half-plane.

Next consider $x - y \ge 4$. If we solve for y in $x - y = 4$, we can graph using the slope and the y-intercept.
$x - y = 4$
$-y = -x + 4$
$y = x - 4$

y-intercept = -4 slope = 1

Now, use the origin as a test point.

$x - y \ge 4$
$0 - 0 \ge 4$
$0 \ge 4$

This is a false statement. This means that the point will not fall in the shaded half-plane.

Next, graph each of the inequalities. The solution to the system is the intersection of the shaded half-planes.
$x + y \ge 2$
$x - y \ge 4$

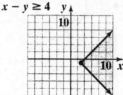

20. $3x + y \le 9$
$2x + 3y \ge 6$
$x \ge 0, \ y \ge 0$

First consider $3x + y \le 9$. If we solve for y in $3x + y = 9$, we can graph the line using the slope and the y–intercept.
$3x + y = 9$
$y = -3x + 9$
y-intercept = 9   slope = -3
Now, use the origin as a test point.
$3x + y \le 9$
$3(0) + 0 \le 9$
$0 \le 9$

This is a true statement. This means that the point will fall in the shaded half-plane.
Next consider $2x + 3y \ge 6$. If we solve for y in $2x + 3y = 6$, we can graph using the slope and the y-intercept.
$2x + 3y = 6$
$3y = -2x + 6$
$y = -\frac{2}{3}x + 2$

y-intercept = 2   slope = $-\frac{2}{3}$

Now, use the origin as a test point.
$2x + 3y \ge 6$
$2(0) + 3(0) \ge 6$
$0 \ge 6$
This is a false statement. This means that the point

will not fall in the shaded half-plane.

Next consider the inequalities $x \geq 0$ and $y \geq 0$.

When x and y are both positive, we are only concerned with the first quadrant of the coordinate system.

Graph each of the inequalities. The solution to the system is the intersection of the shaded half-planes.

$$3x + y \leq 9$$
$$2x + 3y \geq 6$$
$$x \geq 0$$
$$y \geq 0$$

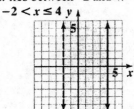

21. $-2 < x \leq 4$

Rewrite the three part inequality as two separate inequalities. We have $-2 < x$ and $x \leq 4$. We replace the inequality symbols with equal signs and obtain $-2 = x$ and $x = 4$. Equations of the form $x = a$ are vertical lines with x–intercept $= a$.

We know the shading will be between $x = -2$ and $x = 4$ because in the original inequality we see that x lies between -2 and 4.

$-2 < x \leq 4$

Cumulative Review Exercises (Chapters 1 – 9)

1.
$$5(x+1) + 2 = x - 3(2x+1)$$
$$5x + 5 + 2 = x - 6x - 3$$
$$5x + 7 = -5x - 3$$
$$10x + 7 = -3$$
$$10x = -10$$
$$x = -1$$

The solution set is $\{-1\}$.

2.
$$\frac{2(x+6)}{3} = 1 + \frac{4x-7}{3}$$
$$3\left(\frac{2(x+6)}{3}\right) = 3(1) + 3\left(\frac{4x-7}{3}\right)$$
$$2(x+6) = 3 + 4x - 7$$
$$2x + 12 = 4x - 4$$
$$-2x + 12 = -4$$
$$-2x = -16$$
$$x = 8$$

The solution set is $\{8\}$.

3.
$$\frac{-10x^2 y^4}{15x^7 y^{-3}} = \frac{-10}{15} x^{2-7} y^{4-(-3)}$$
$$= -\frac{2}{3} x^{-5} y^7 = -\frac{2y^7}{3x^5}$$

4.
$$f(x) = x^2 - 3x + 4$$
$$f(-3) = (-3)^2 - 3(-3) + 4$$
$$= 9 + 9 + 4 = 22$$
$$f(2a) = (2a)^2 - 3(2a) + 4$$
$$= 4a^2 - 6a + 4$$

5. $f(x) = 3x^2 - 4x + 1$

$g(x) = x^2 - 5x - 1$

$(f - g)(x) = f(x) - g(x)$

$\qquad = \left(3x^2 - 4x + 1\right) - \left(x^2 - 5x - 1\right)$

$\qquad = 3x^2 - 4x + 1 - x^2 + 5x + 1$

$\qquad = 2x^2 + x + 2$

$(f - g)(2) = 2(2)^2 + 2 + 2$

$\qquad = 2(4) + 2 + 2$

$\qquad = 8 + 2 + 2 = 12$

6. Since the line we are concerned with is perpendicular to the line, $y = 2x - 3$, we know the slopes are negative reciprocals. The slope of the line will be the negative reciprocal of 2 which is $-\dfrac{1}{2}$. Using the slope and the point, $(2,3)$, write the equation of the line in point-slope form.

$y - y_1 = m(x - x_1)$

$y - 3 = -\dfrac{1}{2}(x - 2)$

Solve for y to write the equation in function notation.

$y - 3 = -\dfrac{1}{2}(x - 2)$

$y - 3 = -\dfrac{1}{2}x + 1$

$y = -\dfrac{1}{2}x + 4$

$f(x) = -\dfrac{1}{2}x + 4$

7. $f(x) = 2x + 1$

$\qquad y = 2x + 1$

Find the x–intercept by setting $y = 0$, and the y–intercept by setting $x = 0$.

$\begin{array}{ll} y = 2x + 1 & y = 2x + 1 \\ 0 = 2x + 1 & y = 2(0) + 1 \\ -1 = 2x & y = 1 \\ -\dfrac{1}{2} = x & \end{array}$

$f(x) = 2x + 1$

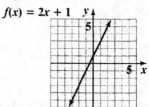

8. $y > 2x$

Consider the line $y = 2x$. Since the line is in slope-intercept form, we know that the slope is 2 and the y-intercept is 0. Use this information to graph the line.

Since the origin, $(0, 0)$, lies on the line, we cannot use it as a test point. Instead, use the point $(1, 1)$.

$y > 2x$

$1 > 2(1)$

$1 > 2$

This is a false statement. This means that the point $(1, 1)$ does not lie in the shaded region.

$y > 2x$

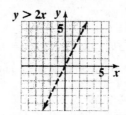

9. $2x - y \geq 6$

Graph the equation using the intercepts.

$2x - y = 6$

x – intercept = 3

y – intercept = -6

Use the origin as a test point to determine shading.

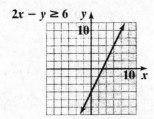

$2x - y \geq 6$

10. $f(x) = -1$

$y = -1$

Equations of the form $y = b$ are horizontal lines with y–intercept = b. This is the horizontal line at $y = -1$.

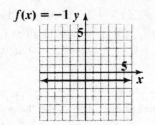

$f(x) = -1$

11. $3x - y + z = -15$

$x + 2y - z = 1$

$2x + 3y - 2z = 0$

Add the first two equations to eliminate z.

$3x - y + z = -15$

$\underline{x + 2y - z = 1}$

$4x + y = -14$

Multiply the first equation by 2 and add to the third equation.

$6x - 2y + 2z = -30$

$\underline{2x + 3y - 2z = 0}$

$8x + y = -30$

The system of two equations in two variables becomes as follows.

$4x + y = -14$

$8x + y = -30$

Multiply the first equation by -1 and add to the second equation.

$-4x - y = 14$

$\underline{8x + y = -30}$

$4x = -16$

$x = -4$

Back-substitute -4 for x to find y.

$4(-4) + y = -14$

$-16 + y = -14$

$y = 2$

Back-substitute 2 for y and -4 for x to find z.

$3x - y + z = -15$

$3(-4) - 2 + z = -15$

$-12 - 2 + z = -15$

$-14 + z = -15$

$z = -1$

The solution is $(-4, 2, -1)$ and the solution set is $\{(-4, 2, -1)\}$.

12. $-3 \leq \dfrac{2x+5}{3} < 6$

$3(-3) \leq 3\left(\dfrac{2x+5}{3}\right) < 3(6)$

$-9 \leq 2x + 5 < 18$

$-9 - 5 \leq 2x + 5 - 5 < 18 - 5$

$-14 \leq 2x < 13$

$-7 \leq x < \dfrac{13}{2}$

The solution set is $\left\{ x \middle| -7 \leq x < \dfrac{13}{2} \right\}$ or $\left[-7, \dfrac{13}{2} \right)$.

13. $|5x + 3| = 7$

$5x + 3 = 7 \quad$ or $\quad 5x + 3 = -7$

$5x = 4 \qquad\qquad 5x = -10$

$x = \dfrac{4}{5} \qquad\qquad x = -2$

The solutions are -2 and $\dfrac{4}{5}$ and the solution set is $\left\{ -2, \dfrac{4}{5} \right\}$.

14. Let x = the number of rooms with a kitchen.
Let y = the number of rooms without a kitchen.
$$x + y = 60$$
$$90x + 80y = 5260$$
Solve the first equation for y.
$$x + y = 60$$
$$y = 60 - x$$
Substitute $60 - x$ for y to find x.
$$90x + 80y = 5260$$
$$90x + 80(60 - x) = 5260$$
$$90x + 4800 - 80x = 5260$$
$$10x + 4800 = 5260$$
$$10x = 460$$
$$x = 46$$
Back-substitute 46 for x to find y.
$$y = 60 - x = 60 - 46 = 14$$
There are 46 rooms with kitchens and 14 rooms without kitchens.

15. Using the vertical line test, we see that graphs a. and b. are functions.

16.
$$\frac{x}{4} - \frac{3}{4} - 1 \le \frac{x}{2}$$
$$4\left(\frac{x}{4}\right) - 4\left(\frac{3}{4}\right) - 4(1) \le 4\left(\frac{x}{2}\right)$$
$$x - 3 - 4 \le 2x$$
$$x - 7 \le 2x$$
$$x \le 2x + 7$$
$$-x \le 7$$
$$x \ge -7$$

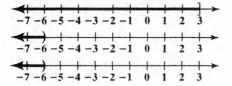

The solution set is $\{x \mid x \ge -7\}$ or $[-7, \infty)$.

17. $2x + 5 \le 11$　and　$-3x > 18$
　　　$2x \le 6$　　　　　$x < -6$
　　　$x \le 3$

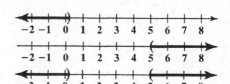

The solution set is $\{x \mid x < -6\}$ or $(-\infty, -6)$.

18. $x - 4 \ge 1$　or　$-3x + 1 \ge -5 - x$
　　$x \ge 5$　　　　$-2x + 1 \ge -5$
　　　　　　　　　$-2x \ge -6$
　　　　　　　　　$x \le 3$

The solution set is $\{x \mid x \le 3 \text{ or } x \ge 5\}$ or $(-\infty, 3] \cup [5, \infty)$.

19.　$|2x + 3| \le 17$
　$-17 \le 2x + 3 \le 17$
　$-20 \le 2x \le 14$
　$-10 \le x \le 7$

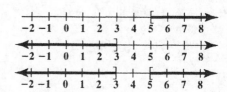

The solution set is $\{x \mid -10 \le x \le 7\}$ or $[-10, 7]$.

20. $|3x - 8| > 7$
　$3x - 8 < -7$　or　$3x - 8 > 7$
　　$3x < 1$　　　　$3x > 15$
　　$x < \dfrac{1}{3}$　　　　$x > 5$

The solution set is $\left\{x \mid x < \dfrac{1}{3} \text{ or } x > 5\right\}$ or $\left(-\infty, \dfrac{1}{3}\right) \cup (5, \infty)$.

Chapter 10
Radicals, Radical Functions, and Rational Exponents

10.1 Check Points

1. **a.** $\sqrt{64} = 8$ because $8^2 = 64$

   b. $-\sqrt{49} = -7$ because $(-7)^2 = 49$

   c. $\sqrt{\dfrac{16}{25}} = \dfrac{4}{5}$ because $\left(\dfrac{4}{5}\right)^2 = \dfrac{16}{25}$

   d. $\sqrt{0.0081} = 0.09$ because $0.09^2 = 0.0081$

   e. $\sqrt{9+16} = \sqrt{25} = 5$

   f. $\sqrt{9} + \sqrt{16} = 3 + 4 = 7$

2. **a.** $f(x) = \sqrt{12x - 20}$
   $$f(3) = \sqrt{12(3) - 20}$$
   $$= \sqrt{36 - 20}$$
   $$= \sqrt{16}$$
   $$= 4$$

   b. $g(x) = -\sqrt{9 - 3x}$
   $$g(-5) = -\sqrt{9 - 3(-5)}$$
   $$= -\sqrt{9 + 15}$$
   $$= -\sqrt{24}$$
   $$\approx -4.90$$

3. $f(x) = \sqrt{9x - 27}$
   $$9x - 27 \geq 0$$
   $$9x \geq 27$$
   $$x \geq 3$$
   Domain of f is $\{x \mid x \geq 3\}$ or $[3, \infty)$.

4. $M(x) = 0.7\sqrt{x} + 12.5$
   Because 2010 is 14 years after 1996, substitute 14 for x.
   $$M(14) = 0.7\sqrt{14} + 12.5 \approx 15.1$$
   The model indicates that there will be about 15.1 non-program minutes in an hour of prime-time cable in 2010.

5. **a.** $\sqrt{(-7)^2} = \sqrt{49} = 7$

   b. $\sqrt{(x+8)^2} = |x + 8|$

   c. $\sqrt{49x^{10}} = |7x^5|$

   d. $\sqrt{x^2 - 6x + 9} = \sqrt{(x-3)^2} = |x - 3|$

6. **a.** $f(x) = \sqrt[3]{x - 6}$
   $$f(33) = \sqrt[3]{33 - 6}$$
   $$= \sqrt[3]{27}$$
   $$= 3$$

   b. $g(x) = \sqrt[3]{2x + 2}$
   $$g(-5) = \sqrt[3]{2(-5) + 2}$$
   $$= \sqrt[3]{-8}$$
   $$= -2$$

7. $\sqrt[3]{-27x^3} = \sqrt[3]{(-3x)^3} = -3x$

8. **a.** $\sqrt[4]{16} = 2$

   b. $-\sqrt[4]{16} = -2$

   c. $\sqrt[4]{-16}$ is not a real number.

   d. $\sqrt[5]{-1} = -1$

9. **a.** $\sqrt[4]{(x+6)^4} = |x + 6|$

   b. $\sqrt[5]{(3x-2)^5} = 3x - 2$

   c. $\sqrt[6]{(-8)^6} = 8$

10.1 Exercise Set

1. $\sqrt{36} = 6$ because $6^2 = 36$

3. $-\sqrt{36} = -6$ because $(-6)^2 = 36$

5. $\sqrt{-36}$
   not a real number

7. $\sqrt{\dfrac{1}{25}} = \dfrac{1}{5}$ because $\left(\dfrac{1}{5}\right)^2 = \dfrac{1}{25}$

9. $-\sqrt{\dfrac{9}{16}} = -\dfrac{3}{4}$ because $\left(\dfrac{3}{4}\right)^2 = \dfrac{9}{16}$

11. $\sqrt{0.81} = 0.9$ because $(0.9)^2 = 0.81$

13. $-\sqrt{0.04} = -0.2$ because $(0.2)^2 = 0.04$

15. $\sqrt{25-16} = \sqrt{9} = 3$

17. $\sqrt{25} - \sqrt{16} = 5 - 4 = 1$

19. $\sqrt{16-25} = \sqrt{-9}$
    not a real number

21. $f(x) = \sqrt{x-2}$
    $f(18) = \sqrt{18-2} = \sqrt{16} = 4$
    $f(3) = \sqrt{3-2} = \sqrt{1} = 1$
    $f(2) = \sqrt{2-2} = \sqrt{0} = 0$
    $f(-2) = \sqrt{-2-2} = \sqrt{-4}$
    not a real number

23. $g(x) = -\sqrt{2x+3}$
    $g(11) = -\sqrt{2(11)+3}$
    $\quad = -\sqrt{22+3}$
    $\quad = -\sqrt{25} = -5$
    $g(1) = -\sqrt{2(1)+3}$
    $\quad = -\sqrt{2+3}$
    $\quad = -\sqrt{5} \approx -2.24$
    $g(-1) = -\sqrt{2(-1)+3}$
    $\quad = -\sqrt{-2+3}$
    $\quad = -\sqrt{1} = -1$
    $g(-2) = -\sqrt{2(-2)+3}$
    $\quad = -\sqrt{-4+3} = -\sqrt{-1}$
    not a real number

25. $h(x) = \sqrt{(x-1)^2}$
    $h(5) = \sqrt{(5-1)^2} = \sqrt{(4)^2} = |4| = 4$
    $h(3) = \sqrt{(3-1)^2} = \sqrt{(2)^2} = |2| = 2$
    $h(0) = \sqrt{(0-1)^2} = \sqrt{(-1)^2} = |-1| = 1$
    $h(-5) = \sqrt{(-5-1)^2} = \sqrt{(-6)^2}$
    $\quad = |-6| = 6$

27. To find the domain, set the radicand greater than or equal to zero and solve.
    $x - 3 \geq 0$
    $\quad x \geq 3$
    The domain of f is $\{x \mid x \geq 3\}$ or $[3, \infty)$. This corresponds to graph (c).

29. To find the domain, set the radicand greater than or equal to zero and solve.
    $3x + 15 \geq 0$
    $\quad 3x \geq -15$
    $\quad x \geq -5$
    The domain of f is $\{x \mid x \geq -5\}$ or $[-5, \infty)$. This corresponds to graph (d).

31. To find the domain, set the radicand greater than or equal to zero and solve.

$6 - 2x \geq 0$

$-2x \geq -6 \mid$

$x \leq 3$

The domain of f is $\{x \mid x \leq 3\}$ or $(-\infty, 3]$. This corresponds to graph (e).

33. $\sqrt{5^2} = |5| = 5$

35. $\sqrt{(-4)^2} = |-4| = 4$

37. $\sqrt{(x-1)^2} = |x-1|$

39. $\sqrt{36x^4} = \sqrt{(6x^2)^2} = |6x^2| = 6x^2$

41. $-\sqrt{100x^6} = -\sqrt{(10x^3)^2}$

$= -|10x^3| = -10|x^3|$

43. $\sqrt{x^2 + 12x + 36} = \sqrt{(x+6)^2} = |x+6|$

45. $-\sqrt{x^2 - 8x + 16} = -\sqrt{(x-4)^2}$

$= -|x-4|$

47. $\sqrt[3]{27} = 3$ because $3^3 = 27$

49. $\sqrt[3]{-27} = -3$ because $(-3)^3 = -27$

51. $\sqrt[3]{\dfrac{1}{125}} = \dfrac{1}{5}$ because $\left(\dfrac{1}{5}\right)^3 = \dfrac{1}{125}$

53. $\sqrt[3]{\dfrac{-27}{1000}} = -\dfrac{3}{10}$ because $\left(-\dfrac{3}{10}\right)^3 = \dfrac{-27}{1000}$

55. $f(x) = \sqrt[3]{x-1}$

$f(28) = \sqrt[3]{28-1} = \sqrt[3]{27} = 3$

$f(9) = \sqrt[3]{9-1} = \sqrt[3]{8} = 2$

$f(0) = \sqrt[3]{0-1} = \sqrt[3]{-1} = -1$

$f(-63) = \sqrt[3]{-63-1} = \sqrt[3]{-64} = -4$

57. $g(x) = -\sqrt[3]{8x-8}$

$g(2) = -\sqrt[3]{8(2)-8} = -\sqrt[3]{16-8}$

$= -\sqrt[3]{8} = -2$

$g(1) = -\sqrt[3]{8(1)-8} = -\sqrt[3]{8-8}$

$= -\sqrt[3]{0} = -0 = 0$

$g(0) = -\sqrt[3]{8(0)-8} = -\sqrt[3]{-8}$

$= -(-2) = 2$

59. $\sqrt[4]{1} = 1$ because $1^4 = 1$

61. $\sqrt[4]{16} = 2$ because $2^4 = 16$

63. $-\sqrt[4]{16} = -2$ because $2^4 = 16$

65. $\sqrt[4]{-16}$

not a real number

67. $\sqrt[5]{-1} = -1$ because $(-1)^5 = -1$

69. $\sqrt[6]{-1}$

not a real number

71. $-\sqrt[4]{256} = -4$ because $4^4 = 256$

73. $\sqrt[6]{64} = 2$ because $2^6 = 64$

75. $-\sqrt[5]{32} = -2$ because $2^5 = 32$

77. $\sqrt[3]{x^3} = x$

79. $\sqrt[4]{y^4} = |y|$

81. $\sqrt[3]{-8x^3} = -2x$

83. $\sqrt[3]{(-5)^3} = -5$

85. $\sqrt[4]{(-5)^4} = |-5| = 5$

87. $\sqrt[4]{(x+3)^4} = |x+3|$

89. $\sqrt[5]{-32(x-1)^5} = -2(x-1)$

91.

x	$f(x) = \sqrt{x} + 3$
0	$f(0) = \sqrt{0} + 3 = 0 + 3 = 3$
1	$f(1) = \sqrt{1} + 3 = 1 + 3 = 4$
4	$f(4) = \sqrt{4} + 3 = 2 + 3 = 5$
9	$f(9) = \sqrt{9} + 3 = 3 + 3 = 6$

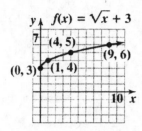

Domain: $\{x \mid x \geq 0\}$ or $[0, \infty)$

Range: $\{y \mid y \geq 3\}$ or $[3, \infty)$

93.

x	$f(x) = \sqrt{x-3}$
3	$f(3) = \sqrt{3-3} = \sqrt{0} = 0$
4	$f(4) = \sqrt{4-3} = \sqrt{1} = 1$
7	$f(7) = \sqrt{7-3} = \sqrt{4} = 2$
12	$f(12) = \sqrt{12-3} = \sqrt{9} = 3$

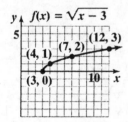

Domain: $\{x \mid x \geq 3\}$ or $[3, \infty)$

Range: $\{y \mid y \geq 0\}$ or $[0, \infty)$

95. The domain of the cube root function is all real numbers, so we only need to worry about the square root in the denominator. We need the radicand of the square root to be ≥ 0, but we also cannot divide by 0. Therefore, we have

$$30 - 2x > 0$$
$$-2x > -30$$
$$x < 15$$

The domain of f is $\{x \mid x < 15\}$ or $(-\infty, 15)$.

97. From the numerator, we need $x - 1 \geq 0$. From the denominator, we need $3 - x > 0$. We need to solve the two inequalities. The domain of the function is the overlap of the two solution sets.

$$x - 1 \geq 0 \quad \text{and} \quad 3 - x > 0$$
$$x \geq 1 \qquad\qquad -x > -3$$
$$x < 3$$

We need $x \geq 1$ and $x < 3$. Therefore, the domain of f is $\{x \mid 1 \leq x < 3\}$ or $[1, 3)$.

99. $\sqrt[3]{\sqrt[4]{16} + \sqrt{625}} = \sqrt[3]{2 + 25} = \sqrt[3]{27} = 3$

101. **a.** $f(x) = 2.9\sqrt{x} + 20.1$

$f(48) = 2.9\sqrt{48} + 20.1 \approx 40.2$

The model estimates the median height of boys who are 48 months to be 40.2 inches. This underestimates the actual median height by 0.6 inches.

b. Find $f(0)$ and $f(10)$.

$$f(x) = 2.9\sqrt{x} + 20.1$$
$$f(0) = 2.9\sqrt{0} + 20.1 = 20.1$$
$$f(10) = 2.9\sqrt{10} + 20.1 \approx 29.3$$

Average rate of change is

$$m = \frac{f(10) - f(0)}{10 - 0} = \frac{29.3 - 20.1}{10} \approx 0.9 \text{ inches per}$$

month.

c. Find $f(50)$ and $f(60)$.

$$f(x) = 2.9\sqrt{x} + 20.1$$
$$f(50) = 2.9\sqrt{50} + 20.1 \approx 40.6$$
$$f(60) = 2.9\sqrt{60} + 20.1 \approx 42.6$$

Average rate of change is

$$m = \frac{f(60) - f(50)}{60 - 50} = \frac{42.6 - 40.6}{10} \approx 0.2 \text{ inches}$$

per month.

This is a much smaller rate of change.

This is shown on the graph because the graph is not as steep between 50 and 60 as it is between 0 and 10.

103. $f(245) = \sqrt{20(245)} = \sqrt{4900} = 70$

The officer should not believe the motorist. The model predicts that the motorist's speed was 70 miles per hour. This is well above the 50 miles per hour speed limit.

105. – 113. Answers will vary.

115. $y_1 = \sqrt{x}$ $y_2 = \sqrt{x+4}$
$y_3 = \sqrt{x-3}$

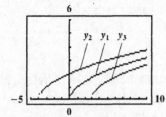

The graphs have the same shape, but differ in their orientation along the x–axis. The graphs are shifted left or right from $x = 0$.

117. $f(x) = \sqrt{x}$ $g(x) = -\sqrt{x}$
$h(x) = \sqrt{-x}$ $k(x) = -\sqrt{-x}$

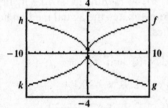

Set Builder Notation:

Function	Domain	Range
$f(x) = \sqrt{x}$	$\{x \mid x \geq 0\}$	$\{y \mid y \geq 0\}$
$g(x) = -\sqrt{x}$	$\{x \mid x \geq 0\}$	$\{y \mid y \leq 0\}$
$h(x) = \sqrt{-x}$	$\{x \mid x \leq 0\}$	$\{y \mid y \geq 0\}$
$k(x) = -\sqrt{-x}$	$\{x \mid x \leq 0\}$	$\{y \mid y \leq 0\}$

Interval Notation:

Function	Domain	Range
$f(x) = \sqrt{x}$	$[0, \infty)$	$[0, \infty)$
$g(x) = -\sqrt{x}$	$[0, \infty)$	$(-\infty, 0]$
$h(x) = \sqrt{-x}$	$(-\infty, 0]$	$[0, \infty)$
$k(x) = -\sqrt{-x}$	$(-\infty, 0]$	$(-\infty, 0]$

119. does not make sense; Explanations will vary. Sample explanation: Because the negative is raised to an even power *first*, this expression will simplify to positive 8.

121. make sense

123. false; Changes to make the statement true will vary. A sample change is: Because the expression is a cube root, the radicand is not required to be greater than zero.

125. false; Changes to make the statement true will vary.
A sample change is: $\sqrt{x^6} = \left| x^3 \right|$
$$\sqrt{(-2)^6} = \left| (-2)^3 \right|$$
$$\sqrt{64} = |-8|$$
$$8 = 8$$

127. Answers will vary. One example is $f(x) = \sqrt{5-x}$.

129. $\sqrt{(2x+3)^{10}} = \sqrt{\left((2x+3)^5 \right)^2}$
$$= \left| (2x+3)^5 \right|$$

131. $h(x) = \sqrt{x+3}$

x	$h(x) = \sqrt{x+3}$
-3	$h(-3) = \sqrt{-3+3} = \sqrt{0} = 0$
-2	$h(-2) = \sqrt{-2+3} = \sqrt{1} = 1$
1	$h(1) = \sqrt{1+3} = \sqrt{4} = 2$
6	$h(6) = \sqrt{6+3} = \sqrt{9} = 3$

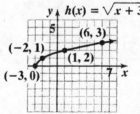

The graph of h is the graph of f shifted three units to the left.

132. $3x - 2\left[x - 3(x+5) \right] = 3x - 2\left[x - 3x - 15 \right]$
$$= 3x - 2\left[-2x - 15 \right]$$
$$= 3x + 4x + 30$$
$$= 7x + 30$$

133. $\left(-3x^{-4}y^3\right)^{-2} = (-3)^{-2}\left(x^{-4}\right)^{-2}\left(y^3\right)^{-2}$

$$= \frac{1}{(-3)^2}x^8y^{-6}$$

$$= \frac{x^8}{(-3)^2\,y^6} = \frac{x^8}{9y^6}$$

134. $|3x-4| > 11$

$3x-4 < -11$ or $3x-4 > 11$

$\quad\; 3x < -7 \qquad\qquad 3x > 15$

$\qquad x < -\dfrac{7}{3} \qquad\qquad x > 5$

The solution set is

$\left\{ x \mid x < -\dfrac{7}{3} \ \text{ or } \ x > 5 \right\}$

or $\left(-\infty, -\dfrac{7}{3}\right) \cup (5, \infty)$.

135. $(2^3x^5)(2^4x^{-6}) = 2^{3+4}x^{5+(-6)} = 2^7x^{-1} = \dfrac{2^7}{x} = \dfrac{128}{x}$

136. $\dfrac{32x^2}{16x^5} = 2x^{2-5} = 2x^{-3} = \dfrac{2}{x^3}$

137. $(x^{-2}y^3)^4 = x^{-2\cdot4}y^{3\cdot4} = x^{-8}y^{12} = \dfrac{y^{12}}{x^8}$

10.2 Check Points

1. a. $25^{1/2} = \sqrt{25} = 5$

   b. $(-8)^{1/3} = \sqrt[3]{-8} = -2$

   c. $\left(5xy^2\right)^{1/4} = \sqrt[4]{5xy^2}$

2. a. $\sqrt[4]{5xy} = (5xy)^{1/4}$

   b. $\sqrt[5]{\dfrac{a^3b}{2}} = \left(\dfrac{a^3b}{2}\right)^{1/5}$

3. a. $8^{4/3} = \left(\sqrt[3]{8}\right)^4 = 2^4 = 16$

   b. $25^{3/2} = \left(\sqrt{25}\right)^3 = 5^3 = 125$

   c. $-81^{3/4} = -\left(\sqrt[4]{81}\right)^3 = -3^3 = -27$

4. a. $\sqrt[3]{6^4} = 6^{4/3}$

   b. $\left(\sqrt[5]{2xy}\right)^7 = (2xy)^{7/5}$

5. a. $100^{-1/2} = \dfrac{1}{100^{1/2}} = \dfrac{1}{\sqrt{100}} = \dfrac{1}{10}$

   b. $8^{-1/3} = \dfrac{1}{8^{1/3}} = \dfrac{1}{\sqrt[3]{8}} = \dfrac{1}{2}$

   c. $32^{-3/5} = \dfrac{1}{32^{3/5}} = \dfrac{1}{\left(\sqrt[5]{32}\right)^3} = \dfrac{1}{2^3} = \dfrac{1}{8}$

   d. $(3xy)^{-5/9} = \dfrac{1}{(3xy)^{5/9}}$

6. a. $7^{\frac{1}{2}} \cdot 7^{\frac{1}{3}} = 7^{\frac{1}{2}+\frac{1}{3}} = 7^{\frac{5}{6}}$

   b. $\dfrac{50x^{\frac{1}{3}}}{10x^{\frac{4}{3}}} = 5 \cdot x^{\frac{1}{3}-\frac{4}{3}} = 5x^{-1} = \dfrac{5}{x}$

   c. $\left(9.1^{\frac{2}{5}}\right)^{\frac{3}{4}} = 9.1^{\frac{2}{5}\cdot\frac{3}{4}} = 9.1^{\frac{3}{10}}$

   d. $\left(x^{-\frac{3}{5}}y^{\frac{1}{4}}\right)^{\frac{1}{3}} = x^{-\frac{3}{5}\cdot\frac{1}{3}}y^{\frac{1}{4}\cdot\frac{1}{3}} = x^{-\frac{1}{5}}y^{\frac{1}{12}} = \dfrac{y^{\frac{1}{12}}}{x^{\frac{1}{5}}}$

7. a. $\sqrt[6]{x^3} = x^{\frac{3}{6}} = x^{\frac{1}{2}} = \sqrt{x}$

   b. $\sqrt[3]{8a^{12}} = \left(8a^{12}\right)^{\frac{1}{3}} = 8^{\frac{1}{3}}a^{12\cdot\frac{1}{3}} = 2a^4$

c. $\sqrt[8]{x^4 y^2} = \left(x^4 y^2\right)^{\frac{1}{8}} = \left(x^4\right)^{\frac{1}{8}} \left(y^2\right)^{\frac{1}{8}}$

$\qquad = x^{\frac{2}{4}} y^{\frac{1}{4}} = \left(x^2 y\right)^{\frac{1}{4}} = \sqrt[4]{x^2 y}$

d. $\dfrac{\sqrt{x}}{\sqrt[3]{x}} = \dfrac{x^{\frac{1}{2}}}{x^{\frac{1}{3}}} = x^{\frac{1}{2} - \frac{1}{3}} = x^{\frac{3}{6} - \frac{2}{6}} = x^{\frac{1}{6}} = \sqrt[6]{x}$

e. $\sqrt{\sqrt[3]{x}} = \left(x^{\frac{1}{3}}\right)^{\frac{1}{2}} = x^{\frac{1}{3} \cdot \frac{1}{2}} = x^{\frac{1}{6}} = \sqrt[6]{x}$

10.2 Exercise Set

1. $49^{1/2} = \sqrt{49} = 7$

3. $(-27)^{1/3} = \sqrt[3]{-27} = -3$

5. $-16^{1/4} = -\sqrt[4]{16} = -2$

7. $(xy)^{1/3} = \sqrt[3]{xy}$

9. $\left(2xy^3\right)^{1/5} = \sqrt[5]{2xy^3}$

11. $81^{3/2} = \left(\sqrt{81}\right)^3 = 9^3 = 729$

13. $125^{2/3} = \left(\sqrt[3]{125}\right)^2 = 5^2 = 25$

15. $(-32)^{3/5} = \left(\sqrt[5]{-32}\right)^3 = (-2)^3 = -8$

17. $27^{2/3} + 16^{3/4} = \left(\sqrt[3]{27}\right)^2 + \left(\sqrt[4]{16}\right)^3$

$\qquad = 3^2 + 2^3$

$\qquad = 9 + 8 = 17$

19. $(xy)^{4/7} = \left(\sqrt[7]{xy}\right)^4$ or $\sqrt[7]{(xy)^4}$

21. $\sqrt{7} = 7^{1/2}$

23. $\sqrt[3]{5} = 5^{1/3}$

25. $\sqrt[5]{11x} = (11x)^{1/5}$

27. $\sqrt{x^3} = x^{3/2}$

29. $\sqrt[5]{x^3} = x^{3/5}$

31. $\sqrt[5]{x^2 y} = \left(x^2 y\right)^{1/5}$

33. $\left(\sqrt{19xy}\right)^3 = (19xy)^{3/2}$

35. $\left(\sqrt[6]{7xy^2}\right)^5 = \left(7xy^2\right)^{5/6}$

37. $2x\sqrt[3]{y^2} = 2xy^{2/3}$

39. $49^{-1/2} = \dfrac{1}{49^{1/2}} = \dfrac{1}{\sqrt{49}} = \dfrac{1}{7}$

41. $27^{-1/3} = \dfrac{1}{27^{1/3}} = \dfrac{1}{\sqrt[3]{27}} = \dfrac{1}{3}$

43. $16^{-3/4} = \dfrac{1}{16^{3/4}} = \dfrac{1}{\left(\sqrt[4]{16}\right)^3} = \dfrac{1}{2^3} = \dfrac{1}{8}$

45. $8^{-2/3} = \dfrac{1}{8^{2/3}} = \dfrac{1}{\left(\sqrt[3]{8}\right)^2} = \dfrac{1}{2^2} = \dfrac{1}{4}$

47. $\left(\dfrac{8}{27}\right)^{-1/3} = \left(\dfrac{27}{8}\right)^{1/3} = \sqrt[3]{\dfrac{27}{8}} = \dfrac{3}{2}$

49. $(-64)^{-2/3} = \dfrac{1}{(-64)^{2/3}} = \dfrac{1}{\left(\sqrt[3]{-64}\right)^2}$

$\qquad = \dfrac{1}{(-4)^2} = \dfrac{1}{16}$

51. $(2xy)^{-7/10} = \dfrac{1}{(2xy)^{7/10}}$

$\qquad = \dfrac{1}{\sqrt[10]{(2xy)^7}}$ or $\dfrac{1}{\left(\sqrt[10]{2xy}\right)^7}$

53. $5xz^{-1/3} = \dfrac{5xz^{-1/3}}{1} = \dfrac{5x}{z^{1/3}}$

55. $3^{3/4} \cdot 3^{1/4} = 3^{(3/4)+(1/4)}$

$= 3^{4/4} = 3^1 = 3$

57. $\dfrac{16^{3/4}}{16^{1/4}} = 16^{(3/4)-(1/4)} = 16^{2/4}$

$= 16^{1/2} = \sqrt{16} = 4$

59. $x^{1/2} \cdot x^{1/3} = x^{(1/2)+(1/3)}$

$= x^{(3/6)+(2/6)} = x^{5/6}$

61. $\dfrac{x^{4/5}}{x^{1/5}} = x^{(4/5)-(1/5)} = x^{3/5}$

63. $\dfrac{x^{1/3}}{x^{3/4}} = x^{(1/3)-(3/4)} = x^{(4/12)-(9/12)}$

$= x^{-5/12} = \dfrac{1}{x^{5/12}}$

65. $\left(5^{\frac{2}{3}}\right)^3 = 5^{\frac{2}{3}\cdot 3} = 5^2 = 25$

67. $\left(y^{-2/3}\right)^{1/4} = y^{(-2/3)\cdot(1/4)} = y^{-2/12}$

$= y^{-1/6} = \dfrac{1}{y^{1/6}}$

69. $\left(2x^{1/5}\right)^5 = 2^5 x^{(1/5)\cdot5} = 32x^1 = 32x$

71. $\left(25x^4 y^6\right)^{1/2} = 25^{1/2}\left(x^4\right)^{1/2}\left(y^6\right)^{1/2}$

$= \sqrt{25}\,x^{4(1/2)} y^{6(1/2)}$

$= 5x^2 y^3$

73. $\left(x^{1/2} y^{-3/5}\right)^{1/2} = \left(\dfrac{x^{1/2} y^{-3/5}}{1}\right)^{1/2}$

$= \left(\dfrac{x^{1/2}}{y^{3/5}}\right)^{1/2}$

$= \dfrac{x^{(1/2)\cdot(1/2)}}{y^{(3/5)\cdot(1/2)}} = \dfrac{x^{1/4}}{y^{3/10}}$

75. $\dfrac{3^{1/2} \cdot 3^{3/4}}{3^{1/4}} = 3^{(1/2)+(3/4)-(1/4)}$

$= 3^{(2/4)+(3/4)-(1/4)}$

$= 3^{4/4} = 3^1 = 3$

77. $\dfrac{\left(3y^{1/4}\right)^3}{y^{1/12}} = \dfrac{3^3 y^{(1/4)\cdot3}}{y^{1/12}} = \dfrac{27y^{3/4}}{y^{1/12}}$

$= 27y^{(3/4)-(1/12)}$

$= 27y^{(9/12)-(1/12)}$

$= 27y^{8/12} = 27y^{2/3}$

79. $\sqrt[8]{x^2} = x^{2/8} = x^{1/4} = \sqrt[4]{x}$

81. $\sqrt[3]{8a^6} = 8^{1/3} a^{6/3} = 2a^2$

83. $\sqrt[5]{x^{10} y^{15}} = x^{10/5} y^{15/5} = x^2 y^3$

85. $\left(\sqrt[3]{xy}\right)^{18} = (xy)^{18/3} = (xy)^6 = x^6 y^6$

87. $\sqrt[10]{(3y)^2} = (3y)^{2/10} = (3y)^{1/5} = \sqrt[5]{3y}$

89. $\left(\sqrt[6]{2a}\right)^4 = (2a)^{4/6} = (2a)^{2/3}$

$= \left(4a^2\right)^{1/3} = \sqrt[3]{4a^2}$

91. $\sqrt[9]{x^6 y^3} = x^{6/9} y^{3/9}$

$= x^{2/3} y^{1/3} = \sqrt[3]{x^2 y}$

93. $\sqrt{2} \cdot \sqrt[3]{2} = 2^{1/2} \cdot 2^{1/3} = 2^{(1/2)+(1/3)}$

$= 2^{(3/6)+(2/6)} = 2^{5/6}$

$= \sqrt[6]{2^5} \text{ or } \sqrt[6]{32}$

95. $\sqrt[5]{x^2} \cdot \sqrt{x} = \left(x^2\right)^{1/5} \cdot x^{1/2}$

$= x^{2/5} \cdot x^{1/2} = x^{(2/5)+(1/2)}$

$= x^{(4/10)+(5/10)} = x^{9/10}$

$= \sqrt[10]{x^9}$

97. $\sqrt[4]{a^2 b} \cdot \sqrt[3]{ab} = \left(a^2 b\right)^{1/4} \cdot (ab)^{1/3}$

$\qquad = a^{1/2} b^{1/4} \cdot a^{1/3} b^{1/3}$

$\qquad = a^{(1/2)+(1/3)} b^{(1/4)+(1/3)}$

$\qquad = a^{(6/12)+(4/12)} b^{(3/12)+(4/12)}$

$\qquad = a^{10/12} b^{7/12}$

$\qquad = \sqrt[12]{a^{10} b^7}$

99. $\dfrac{\sqrt[4]{x}}{\sqrt[5]{x}} = \dfrac{x^{1/4}}{x^{1/5}} = x^{(1/4)-(1/5)}$

$\qquad = x^{(5/20)-(4/20)}$

$\qquad = x^{1/20} = \sqrt[20]{x}$

101. $\dfrac{\sqrt[3]{y^2}}{\sqrt[6]{y}} = \dfrac{y^{2/3}}{y^{1/6}} = y^{(2/3)-(1/6)}$

$\qquad = y^{(4/6)-(1/6)} = y^{3/6}$

$\qquad = y^{1/2} = \sqrt{y}$

103. $\sqrt[4]{\sqrt{x}} = \sqrt[4]{x^{1/2}} = \left(x^{1/2}\right)^{1/4}$

$\qquad = x^{(1/2)\cdot(1/4)} = x^{1/8}$

$\qquad = \sqrt[8]{x}$

105. $\sqrt{\sqrt{x^2 y}} = \sqrt{\left(x^2 y\right)^{1/2}} = \left(\left(x^2 y\right)^{1/2}\right)^{1/2}$

$\qquad = \left(x^2 y\right)^{(1/2)\cdot(1/2)} = \left(x^2 y\right)^{1/4}$

$\qquad = \sqrt[4]{x^2 y}$

107. $\sqrt[4]{\sqrt[3]{2x}} = \sqrt[4]{(2x)^{1/3}} = \left((2x)^{1/3}\right)^{1/4}$

$\qquad = (2x)^{(1/3)\cdot(1/4)} = (2x)^{1/12}$

$\qquad = \sqrt[12]{2x}$

109. $\left(\sqrt[4]{x^3 y^5}\right)^{12} = \left(\left(x^3 y^5\right)^{1/4}\right)^{12}$

$\qquad = \left(x^3 y^5\right)^{(1/4)\cdot 12}$

$\qquad = \left(x^3 y^5\right)^{12/4} = \left(x^3 y^5\right)^3$

$\qquad = x^{3\cdot 3} y^{5\cdot 3} = x^9 y^{15}$

111. $\dfrac{\sqrt[4]{a^5 b^5}}{\sqrt{ab}} = \dfrac{a^{5/4} b^{5/4}}{a^{1/2} b^{1/2}}$

$\qquad = a^{(5/4)-(1/2)} b^{(5/4)-(1/2)}$

$\qquad = a^{(5/4)-(2/4)} b^{(5/4)-(2/4)}$

$\qquad = a^{3/4} b^{3/4} = \left(a^3 b^3\right)^{1/4}$

$\qquad = \sqrt[4]{a^3 b^3}$

113. $x^{1/3}\left(x^{1/3} - x^{2/3}\right) = x^{1/3} \cdot x^{1/3} - x^{1/3} \cdot x^{2/3}$

$\qquad = x^{(1/3)+(1/3)} - x^{(1/3)+(2/3)}$

$\qquad = x^{2/3} - x^{3/3}$

$\qquad = x^{2/3} - x$

115. $\left(x^{1/2} - 3\right)\left(x^{1/2} + 5\right)$

$\qquad = x^{1/2} \cdot x^{1/2} + x^{1/2} \cdot 5 - 3 \cdot x^{1/2} - 3 \cdot 5$

$\qquad = x^{(1/2)+(1/2)} + 5x^{1/2} - 3x^{1/2} - 15$

$\qquad = x^{2/2} + 2x^{1/2} - 15$

$\qquad = x + 2x^{1/2} - 15$

117. $6x^{1/2} + 2x^{3/2} = 3 \cdot 2x^{1/2} + 2x^{(1/2)+(2/2)}$

$\qquad = 3 \cdot 2x^{1/2} + 2x^{1/2} \cdot x^{2/2}$

$\qquad = 3 \cdot 2x^{1/2} + 2x^{1/2} \cdot x$

$\qquad = 2x^{1/2}(3+x)$

119. $15x^{1/3} - 60x = 15x^{1/3} - 60x^{3/3}$

$\qquad = 15x^{1/3} - 60x^{(1/3)+(2/3)}$

$\qquad = 15x^{1/3} - 60x^{1/3} x^{2/3}$

$\qquad = 15x^{1/3} \cdot 1 - 15x^{1/3} \cdot 4x^{2/3}$

$\qquad = 15x^{1/3}\left(1 - 4x^{2/3}\right)$

121. $\left(49x^{-2}y^4\right)^{-1/2}\left(xy^{1/2}\right)$

$= (49)^{-1/2}\left(x^{-2}\right)^{-1/2}\left(y^4\right)^{-1/2}\left(xy^{1/2}\right)$

$= \dfrac{1}{49^{1/2}}x^{(-2)(-1/2)}y^{(4)(-1/2)}\left(xy^{1/2}\right)$

$= \dfrac{1}{7}x^1y^{-2}\cdot xy^{1/2} = \dfrac{1}{7}x^{1+1}y^{-2+(1/2)}$

$= \dfrac{1}{7}x^2y^{-3/2} = \dfrac{x^2}{7y^{3/2}}$

123. $\left(\dfrac{x^{-5/4}y^{1/3}}{x^{-3/4}}\right)^{-6} = \left(x^{(-5/4)-(-3/4)}y^{1/3}\right)^{-6}$

$= \left(x^{-2/4}y^{1/3}\right)^{-6} = x^{(-2/4)(-6)}y^{(1/3)(-6)}$

$= x^3y^{-2} = \dfrac{x^3}{y^2}$

125. $f(8) = 29(8)^{1/3}$

$\qquad = 29\sqrt[3]{8}$

$\qquad = 29(2) = 58$

There are 58 plant species on an 8 square mile island.

127. $f(x) = 70x^{3/4}$

$f(80) = 70(80)^{3/4} \approx 1872$

A person who weighs 80 kilograms needs about 1872 calories per day to maintain life.

129. a. $C = 35.74 + 0.6215t - 35.74v^{4/25} + 0.4275tv^{4/25}$

b. $C = 35.74 + 0.6215(25) - 35.74(30)^{4/25} + 0.4275(25)(30)^{4/25} \approx 8°F$

131. $C = 35.74 + 0.6215t - 35.74v^{4/25} + 0.4275t\cdot v^{4/25}$

a. For $t = 0$, we get

$C(v) = 35.74 - 35.74v^{4/25}$

b. $C(25) = 35.74 - 35.74(25)^{4/25} \approx -24$

When the air temperature is $0°F$ and the wind speed is 25 miles per hour, the windchill temperature is $-24°F$.

c. The solution to part (b) is represented by the point $(25, -24)$ on the graph.

133. $L + 1.25\sqrt{S} - 9.8\sqrt[3]{D} \le 16.296$

a. $L + 1.25S^{1/2} - 9.8D^{1/3} \le 16.296$

b.
$$L + 1.25S^{1/2} - 9.8D^{1/3} \leq 16.296$$
$$20.85 + 1.25(276.4)^{1/2} - 9.8(18.55)^{1/3} \leq 16.296$$
$$20.85 + 1.25\sqrt{276.4} - 9.8\sqrt[3]{18.55} \leq 16.296$$
$$20.85 + 1.25(16.625) - 9.8(2.647) \leq 16.296$$
$$20.85 + 20.781 - 25.941 \leq 16.296$$
$$15.69 \leq 16.296$$
The yacht is eligible to enter the America's Cup.

135. – 143. Answers will vary.

145.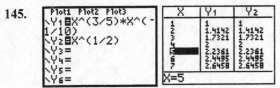

The simplification is correct.

147.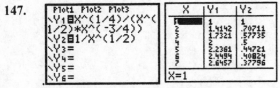

The simplification is not correct.

$$\frac{x^{1/4}}{x^{1/2} \cdot x^{-3/4}} = x^{(1/4) - \left[(1/2) + (-3/4)\right]}$$
$$= x^{(1/4) - (2/4) + (3/4)}$$
$$= x^{2/4}$$
$$= x^{1/2}$$

The new simplification is correct.

149. does not make sense; Explanations will vary. Sample explanation: It is often easier to find the *n*th root before raising the expression to the *m*th power.

151. does not make sense; Explanations will vary. Sample explanation: In the top line, $5 \cdot 5^{1/2}$ was incorrectly simplified to $25^{1/2}$. The order of operations does not allow the multiplication to occur before the exponent is evaluated.

153. false; Changes to make the statement true will vary. A sample change is: $(a+b)^{1/n} \neq a^{1/n} + b^{1/n}$. Do not confuse $(a+b)^{1/n}$ with $(ab)^{1/n}$. We can simplify $(ab)^{1/n}$ as $a^{1/n}b^{1/n}$.

155. true

157.
$$\frac{8^{-4/3}+2^{-2}}{16^{-3/4}+2^{-1}}=\frac{\dfrac{1}{8^{4/3}}+\dfrac{1}{2^{2}}}{\dfrac{1}{16^{3/4}}+\dfrac{1}{2^{1}}}=\frac{\dfrac{1}{\left(\sqrt[3]{8}\right)^{4}}+\dfrac{1}{4}}{\dfrac{1}{\left(\sqrt[4]{16}\right)^{3}}+\dfrac{1}{2}}$$

$$=\frac{\dfrac{1}{\left(2\right)^{4}}+\dfrac{1}{4}}{\dfrac{1}{\left(2\right)^{3}}+\dfrac{1}{2}}=\frac{\dfrac{1}{16}+\dfrac{1}{4}}{\dfrac{1}{8}+\dfrac{1}{2}}=\frac{\dfrac{1}{16}+\dfrac{4}{16}}{\dfrac{1}{8}+\dfrac{4}{8}}$$

$$=\frac{\dfrac{5}{16}}{\dfrac{5}{8}}=\frac{5}{16}\div\frac{5}{8}=\frac{1\cancel{5}}{2\cancel{16}}\cdot\frac{1\cancel{8}}{1\cancel{5}}=\frac{1}{2}$$

The boy is allowed to eat half of the cake. The professor will eat half of what's left over or $\dfrac{1}{2}\cdot\dfrac{1}{2}=\dfrac{1}{4}$ of the cake.

159. First simplify.
$$f(x)=(x-3)^{1/2}(x+4)^{-1/2}$$
$$=\frac{(x-3)^{1/2}(x+4)^{-1/2}}{1}$$
$$=\frac{(x-3)^{1/2}}{(x+4)^{1/2}}=\frac{\sqrt{x-3}}{\sqrt{x+4}}$$

To find the domain, we need to consider the radicals and the denominator of the fraction. Since radicands have to be greater than or equal to zero, we can write inequalities and solve. But since the denominator cannot be zero, that radicand must be strictly greater than zero.

$x-3\geq0$ and $x+4>0$

$\quad x\geq3 \qquad\qquad x>-4$

The domain of the function is the intersection of the sets, $\{x\,|\,x\geq3\}$ or $[3,\infty)$.

160. $m=\dfrac{y_{2}-y_{1}}{x_{2}-x_{1}}=\dfrac{3-1}{4-5}=\dfrac{2}{-1}=-2$

$y-3=-2(x-4)$

Solve for y to write the equation in slope–intercept form.

$y-3=-2x+8$

$\quad y=-2x+11$

$\qquad$ or

$f(x)=-2x+11$

161. $y \le -\dfrac{3}{2}x + 3$

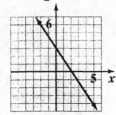

162. $f(x) = 3x^2 - 5x + 4$

$f(a+h) = 3(a+h)^2 - 5(a+h) + 4$

$\qquad = 3(a^2 + 2ah + h^2) - 5(a+h) + 4$

$\qquad = 3a^2 + 6ah + 3h^2 - 5a - 5h + 4$

163. a. $\sqrt{16} \cdot \sqrt{4} = 4 \cdot 2 = 8$

   b. $\sqrt{16 \cdot 4} = \sqrt{64} = 8$

   c. $\sqrt{16} \cdot \sqrt{4} = \sqrt{16 \cdot 4}$

164. a. $\sqrt{300} \approx 17.32$

   b. $10\sqrt{3} \approx 17.32$

   c. $\sqrt{300} = 10\sqrt{3}$

165. a. $\sqrt[3]{x^{21}} = x^{21/3} = x^7$

   b. $\sqrt[6]{y^{24}} = y^{24/6} = y^4$

10.3 Check Points

1. a. $\sqrt{5} \cdot \sqrt{11} = \sqrt{5 \cdot 11} = \sqrt{55}$

   b. $\sqrt{x+4} \cdot \sqrt{x-4} = \sqrt{(x+4)(x-4)} = \sqrt{x^2 - 16}$

   c. $\sqrt[3]{6} \cdot \sqrt[3]{10} = \sqrt[3]{6 \cdot 10} = \sqrt[3]{60}$

   d. $\sqrt[7]{2x} \cdot \sqrt[7]{6x^3} = \sqrt[7]{12x^4}$

2. a. $\sqrt{80} = \sqrt{16 \cdot 5} = \sqrt{16} \cdot \sqrt{5} = 4\sqrt{5}$

   b. $\sqrt[3]{40} = \sqrt[3]{8 \cdot 5} = \sqrt[3]{8} \cdot \sqrt[3]{5} = 2\sqrt[3]{5}$

   c. $\sqrt[4]{32} = \sqrt[4]{16 \cdot 2} = \sqrt[4]{16} \cdot \sqrt[4]{2} = 2\sqrt[4]{2}$

d. $\sqrt{200x^2 y} = \sqrt{100x^2} \cdot \sqrt{2y} = 10|x|\sqrt{2y}$

3. $f(x) = \sqrt{3x^2 - 12x + 12}$

$= \sqrt{3(x^2 - 4x + 4)}$

$= \sqrt{3(x-2)^2}$

$= \sqrt{3} \cdot \sqrt{(x-2)^2}$

$= \sqrt{3} \cdot |x-2|$

4. $\sqrt{x^9 y^{11} z^3} = \sqrt{x^8 y^{10} z^2} \cdot \sqrt{xyz} = x^4 y^5 z\sqrt{xyz}$

5. $\sqrt[3]{40x^{10}y^{14}} = \sqrt[3]{8 \cdot 5 \cdot x^9 \cdot x \cdot y^{12} \cdot y^2} = \sqrt[3]{8x^9 y^{12}}\sqrt[3]{5xy^2} = 2x^3 y^4 \sqrt[3]{5xy^2}$

6. $\sqrt[5]{32x^{12}y^2 z^8} = \sqrt[5]{32 \cdot x^{10} \cdot x^2 \cdot y^2 \cdot z^5 \cdot z^3} = \sqrt[5]{32x^{10}z^5}\sqrt[5]{x^2 y^2 z^3} = 2x^2 z\sqrt[5]{x^2 y^2 z^3}$

7. a. $\sqrt{6} \cdot \sqrt{2} = \sqrt{12} = \sqrt{4 \cdot 3} = \sqrt{4} \cdot \sqrt{3} = 2\sqrt{3}$

b. $10\sqrt[3]{16} \cdot 5\sqrt[3]{2} = 50\sqrt[3]{16 \cdot 2} = 50\sqrt[3]{32} = 50\sqrt[3]{8 \cdot 4} = 50\sqrt[3]{8} \cdot \sqrt[3]{4} = 50 \cdot 2 \cdot \sqrt[3]{4} = 100\sqrt[3]{4}$

c. $\sqrt[4]{4x^2 y} \cdot \sqrt[4]{8x^6 y^3} = \sqrt[4]{32x^8 y^4} = \sqrt[4]{16x^8 y^4} \cdot \sqrt[4]{2} = 2x^2 y\sqrt[4]{2}$

10.3 Exercise Set

1. $\sqrt{3} \cdot \sqrt{5} = \sqrt{3 \cdot 5} = \sqrt{15}$

3. $\sqrt[3]{2} \cdot \sqrt[3]{9} = \sqrt[3]{2 \cdot 9} = \sqrt[3]{18}$

5. $\sqrt[4]{11} \cdot \sqrt[4]{3} = \sqrt[4]{11 \cdot 3} = \sqrt[4]{33}$

7. $\sqrt{3x} \cdot \sqrt{11y} = \sqrt{3x \cdot 11y} = \sqrt{33xy}$

9. $\sqrt[5]{6x^3} \cdot \sqrt[5]{4x} = \sqrt[5]{6x^3 \cdot 4x} = \sqrt[5]{24x^4}$

11. $\sqrt{x+3} \cdot \sqrt{x-3} = \sqrt{(x+3)(x-3)} = \sqrt{x^2 - 9}$

13. $\sqrt[6]{x-4} \cdot \sqrt[6]{(x-4)^4} = \sqrt[6]{(x-4)(x-4)^4}$

$= \sqrt[6]{(x-4)^5}$

15. $\sqrt{\dfrac{2x}{3}} \cdot \sqrt{\dfrac{3}{2}} = \sqrt{\dfrac{2x}{3} \cdot \dfrac{3}{2}} = \sqrt{\dfrac{\cancel{12}x}{\cancel{18}} \cdot \dfrac{1\cancel{3}}{1\cancel{2}}} = \sqrt{x}$

17. $\sqrt[4]{\dfrac{x}{7}} \cdot \sqrt[4]{\dfrac{3}{y}} = \sqrt[4]{\dfrac{x}{7} \cdot \dfrac{3}{y}} = \sqrt[4]{\dfrac{3x}{7y}}$

19. $\sqrt[7]{7x^2y} \cdot \sqrt[7]{11x^3y^2} = \sqrt[7]{7x^2y \cdot 11x^3y^2}$
    $= \sqrt[7]{7 \cdot 11x^2x^3yy^2}$
    $= \sqrt[7]{77x^5y^3}$

21. $\sqrt{50} = \sqrt{25 \cdot 2} = \sqrt{25} \cdot \sqrt{2} = 5\sqrt{2}$

23. $\sqrt{45} = \sqrt{9 \cdot 5} = \sqrt{9} \cdot \sqrt{5} = 3\sqrt{5}$

25. $\sqrt{75x} = \sqrt{25 \cdot 3x} = \sqrt{25} \cdot \sqrt{3x} = 5\sqrt{3x}$

27. $\sqrt[3]{16} = \sqrt[3]{8 \cdot 2} = \sqrt[3]{8} \cdot \sqrt[3]{2} = 2\sqrt[3]{2}$

29. $\sqrt[3]{27x^3} = \sqrt[3]{27 \cdot x^3} = \sqrt[3]{27} \cdot \sqrt[3]{x^3} = 3x$

31. $\sqrt[3]{-16x^2y^3} = \sqrt[3]{-8 \cdot 2x^2y^3}$
    $= \sqrt[3]{-8y^3} \cdot \sqrt[3]{2x^2}$
    $= -2y\sqrt[3]{2x^2}$

33. $f(x) = \sqrt{36(x+2)^2} = 6|x+2|$

35. $f(x) = \sqrt[3]{32(x+2)^3}$
    $= \sqrt[3]{8 \cdot 4(x+2)^3}$
    $= \sqrt[3]{8(x+2)^3} \cdot \sqrt[3]{4}$
    $= 2(x+2)\sqrt[3]{4}$

37. $f(x) = \sqrt{3x^2 - 6x + 3}$
    $= \sqrt{3(x^2 - 2x + 1)}$
    $= \sqrt{3(x-1)^2}$
    $= |x-1|\sqrt{3}$

39. $\sqrt{x^7} = \sqrt{x^6 \cdot x} = \sqrt{x^6} \cdot \sqrt{x} = x^3\sqrt{x}$

41. $\sqrt{x^8y^9} = \sqrt{x^8y^8y} = \sqrt{x^8y^8}\sqrt{y}$
    $= x^4y^4\sqrt{y}$

43. $\sqrt{48x^3} = \sqrt{16 \cdot 3x^2x} = \sqrt{16x^2} \cdot \sqrt{3x}$
    $= 4x\sqrt{3x}$

45. $\sqrt[3]{y^8} = \sqrt[3]{y^6 \cdot y^2} = \sqrt[3]{y^6} \cdot \sqrt[3]{y^2}$
    $= y^2\sqrt[3]{y^2}$

47. $\sqrt[3]{x^{14}y^3z} = \sqrt[3]{x^{12}x^2y^3z} = \sqrt[3]{x^{12}y^3} \cdot \sqrt[3]{x^2z}$
    $= x^4y\sqrt[3]{x^2z}$

49. $\sqrt[3]{81x^8y^6} = \sqrt[3]{27 \cdot 3x^6x^2y^6}$
    $= \sqrt[3]{27x^6y^6} \cdot \sqrt[3]{3x^2}$
    $= 3x^2y^2\sqrt[3]{3x^2}$

51. $\sqrt[3]{(x+y)^5} = \sqrt[3]{(x+y)^3 \cdot (x+y)^2}$
    $= \sqrt[3]{(x+y)^3} \cdot \sqrt[3]{(x+y)^2}$
    $= (x+y)\sqrt[3]{(x+y)^2}$

53. $\sqrt[5]{y^{17}} = \sqrt[5]{y^{15} \cdot y^2} = \sqrt[5]{y^{15}} \cdot \sqrt[5]{y^2}$
    $= y^3\sqrt[5]{y^2}$

55. $\sqrt[5]{64x^6y^{17}} = \sqrt[5]{32 \cdot 2x^5xy^{15}y^2}$
    $= \sqrt[5]{32x^5y^{15}} \cdot \sqrt[5]{2xy^2}$
    $= 2xy^3\sqrt[5]{2xy^2}$

57. $\sqrt[4]{80x^{10}} = \sqrt[4]{16 \cdot 5x^8x^2}$
    $= \sqrt[4]{16x^8} \cdot \sqrt[4]{5x^2}$
    $= 2x^2\sqrt[4]{5x^2}$

59. $\sqrt[4]{(x-3)^{10}} = \sqrt[4]{(x-3)^8(x-3)^2}$
    $= \sqrt[4]{(x-3)^8} \cdot \sqrt[4]{(x-3)^2}$
    $= (x-3)^2\sqrt[4]{(x-3)^2}$
    or
    $= (x-3)^2\sqrt{x-3}$

61. $\sqrt{12} \cdot \sqrt{2} = \sqrt{12 \cdot 2} = \sqrt{24}$
$$= \sqrt{4 \cdot 6} = \sqrt{4} \cdot \sqrt{6}$$
$$= 2\sqrt{6}$$

63. $\sqrt{5x} \cdot \sqrt{10y} = \sqrt{5x \cdot 10y} = \sqrt{50xy}$
$$= \sqrt{25 \cdot 2xy} = 5\sqrt{2xy}$$

65. $\sqrt{12x} \cdot \sqrt{3x} = \sqrt{12x \cdot 3x}$
$$= \sqrt{36x^2} = 6x$$

67. $\sqrt{50xy} \cdot \sqrt{4xy^2} = \sqrt{50xy \cdot 4xy^2}$
$$= \sqrt{200x^2 y^3}$$
$$= \sqrt{100 \cdot 2x^2 y^2 y}$$
$$= \sqrt{100x^2 y^2} \cdot \sqrt{2y}$$
$$= 10xy\sqrt{2y}$$

69. $2\sqrt{5} \cdot 3\sqrt{40} = 2 \cdot 3\sqrt{5 \cdot 40} = 6\sqrt{200}$
$$= 6\sqrt{100 \cdot 2} = 6\sqrt{100} \cdot \sqrt{2}$$
$$= 6 \cdot 10\sqrt{2} = 60\sqrt{2}$$

71. $\sqrt[3]{12} \cdot \sqrt[3]{4} = \sqrt[3]{12 \cdot 4} = \sqrt[3]{48} = \sqrt[3]{8 \cdot 6}$
$$= \sqrt[3]{8} \cdot \sqrt[3]{6} = 2\sqrt[3]{6}$$

73. $\sqrt{5x^3} \cdot \sqrt{8x^2} = \sqrt{5x^3 \cdot 8x^2} = \sqrt{40x^5}$
$$= \sqrt{4 \cdot 10x^4 x}$$
$$= \sqrt{4x^4} \cdot \sqrt{10x}$$
$$= 2x^2 \sqrt{10x}$$

75. $\sqrt[3]{25x^4 y^2} \cdot \sqrt[3]{5xy^{12}}$
$$= \sqrt[3]{25x^4 y^2 \cdot 5xy^{12}}$$
$$= \sqrt[3]{125x^5 y^{14}}$$
$$= \sqrt[3]{125x^3 x^2 y^{12} y^2}$$
$$= \sqrt[3]{125x^3 y^{12}} \cdot \sqrt[3]{x^2 y^2}$$
$$= 5xy^4 \sqrt[3]{x^2 y^2}$$

77. $\sqrt[4]{8x^2 y^3 z^6} \cdot \sqrt[4]{2x^4 yz}$
$$= \sqrt[4]{8x^2 y^3 z^6 \cdot 2x^4 yz}$$
$$= \sqrt[4]{16x^6 y^4 z^7}$$
$$= \sqrt[4]{16x^4 x^2 y^4 z^4 z^3}$$
$$= \sqrt[4]{16x^4 y^4 z^4} \cdot \sqrt[4]{x^2 z^3}$$
$$= 2xyz\sqrt[4]{x^2 z^3}$$

79. $\sqrt[5]{8x^4 y^6 z^2} \cdot \sqrt[5]{8xy^7 z^4}$
$$= \sqrt[5]{8x^4 y^6 z^2 \cdot 8xy^7 z^4}$$
$$= \sqrt[5]{64x^5 y^{13} z^6}$$
$$= \sqrt[5]{32 \cdot 2x^5 y^{10} z^5 z}$$
$$= \sqrt[5]{32x^5 y^{10} z^5} \cdot \sqrt[5]{2y^3 z}$$
$$= 2xy^2 z\sqrt[5]{2y^3 z}$$

81. $\sqrt[3]{x-y} \cdot \sqrt[3]{(x-y)^7}$
$$= \sqrt[3]{(x-y) \cdot (x-y)^7}$$
$$= \sqrt[3]{(x-y)^8}$$
$$= \sqrt[3]{(x-y)^6 (x-y)^2}$$
$$= \sqrt[3]{(x-y)^6} \cdot \sqrt[3]{(x-y)^2}$$
$$= (x-y)^2 \sqrt[3]{(x-y)^2}$$

83. $-2x^2 y \left(\sqrt[3]{54x^3 y^7 z^2} \right)$
$$= -2x^2 y\sqrt[3]{27 \cdot 2x^3 y^6 yz^2}$$
$$= -2x^2 y\sqrt[3]{27x^3 y^6} \cdot \sqrt[3]{2yz^2}$$
$$= -2x^2 y \cdot 3xy^2 \cdot \sqrt[3]{2yz^2}$$
$$= -6x^3 y^3 \sqrt[3]{2yz^2}$$

85. $-3y \left(\sqrt[5]{64x^3 y^6} \right) = -3y\sqrt[5]{32 \cdot 2x^3 y^5 y}$
$$= -3y\sqrt[5]{32y^5} \cdot \sqrt[5]{2x^3 y}$$
$$= -3y \cdot 2y\sqrt[5]{2x^3 y}$$
$$= -6y^2 \sqrt[5]{2x^3 y}$$

87. $\left(-2xy^2\sqrt{3x}\right)\left(xy\sqrt{6x}\right) = -2x^2y^3\sqrt{3x\cdot 6x}$

$\qquad\qquad = -2x^2y^3\sqrt{18x^2}$

$\qquad\qquad = -2x^2y^3\sqrt{9x^2\cdot 2}$

$\qquad\qquad = -2x^2y^3\sqrt{9x^2}\cdot\sqrt{2}$

$\qquad\qquad = -2x^2y^3\left(3x\right)\sqrt{2}$

$\qquad\qquad = -6x^3y^3\sqrt{2}$

89. $\left(2x^2y\sqrt[4]{8xy}\right)\left(-3xy^2\sqrt[4]{2x^2y^3}\right)$

$\qquad = -6x^3y^3\sqrt[4]{8xy\cdot 2x^2y^3}$

$\qquad = -6x^3y^3\sqrt[4]{16x^3y^4}$

$\qquad = -6x^3y^3\sqrt[4]{16y^4\cdot x^3}$

$\qquad = -6x^3y^3\left(2y\right)\sqrt[4]{x^3}$

$\qquad = -12x^3y^4\sqrt[4]{x^3}$

91. $d(x) = \sqrt{\dfrac{3x}{2}}$

$\qquad d(72) = \sqrt{\dfrac{3(72)}{2}}$

$\qquad\qquad = \sqrt{3(36)}$

$\qquad\qquad = \sqrt{3}\cdot\sqrt{36}$

$\qquad\qquad = 6\sqrt{3} \approx 10.4$ miles

A passenger on the pool deck can see roughly 10.4 miles.

93. $W(x) = 4\sqrt{2x}$

$\qquad W(6) = 4\sqrt{2(6)} = 4\sqrt{12}$

$\qquad\qquad = 4\sqrt{4\cdot 3} = 4\sqrt{4}\cdot\sqrt{3}$

$\qquad\qquad = 4\cdot 2\sqrt{3}$

$\qquad\qquad = 8\sqrt{3} \approx 14$ feet per second

A dinosaur with a leg length of 6 feet has a walking speed of about 14 feet per second.

95. a. $C(32) = \dfrac{7.644}{\sqrt[4]{32}} = \dfrac{7.644}{\sqrt[4]{16\cdot 2}}$

$\qquad\qquad = \dfrac{7.644}{2\sqrt[4]{2}} = \dfrac{3.822}{\sqrt[4]{2}}$

The cardiac index of a 32-year-old is $\dfrac{3.822}{\sqrt[4]{2}}$.

b. $\dfrac{3.822}{\sqrt[4]{2}} = \dfrac{3.822}{1.189} = \dfrac{3.822}{1.189} \approx 3.21$

The cardiac index of a 32-year-old is 3.21 liters per minute per square meter. This is shown on the graph as the point (32, 3.21).

97. – 101. Answers will vary.

103. $\sqrt{x^4} = x^2$

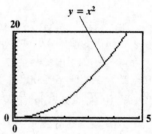

The graphs coincide, so the simplification is correct.

105. $\sqrt{3x^2 - 6x + 3} = (x-1)\sqrt{3}$

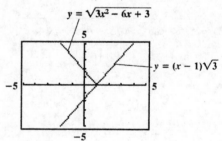

The graphs do not coincide. Correct the simplification.

$\sqrt{3x^2 - 6x + 3} = \sqrt{3\left(x^2 - 2x + 1\right)}$

$\qquad\qquad = \sqrt{3(x-1)^2} = |x-1|\sqrt{3}$

107. makes sense

109. makes sense

111. false; Changes to make the statement true will vary. A sample change is:

$2\sqrt{5}\cdot 6\sqrt{5} = 12\sqrt{5\cdot 5} = 12\cdot 5 = 60$

113. false; Changes to make the statement true will vary.
A sample change is: $\sqrt{12} = \sqrt{4 \cdot 3} = 2\sqrt{3}$

115. If a number is tripled, its square root is multiplied
by $\sqrt{3}$. For example, consider the number 4 and its
triple, 12.
$\sqrt{4} = 2$ $\sqrt{12} = \sqrt{4 \cdot 3} = 2\sqrt{3}$
Thus, if a number is tripled, its square root is
multiplied by $\sqrt{3}$.

117. $(fg)(x) = f(x) \cdot g(x)$
$$2x = \sqrt[3]{2x} \cdot g(x)$$
$$\frac{2x}{\sqrt[3]{2x}} = g(x)$$
$$\frac{(2x)^1}{(2x)^{\frac{1}{3}}} = g(x)$$
$$(2x)^{1-\frac{1}{3}} = g(x)$$
$$(2x)^{\frac{2}{3}} = g(x)$$
$$\sqrt[3]{(2x)^2} = g(x)$$
$$g(x) = \sqrt[3]{4x^2}$$

119. $2x - 1 \le 21$ and $2x + 2 \ge 12$
     $2x \le 22$ $2x \ge 10$
       $x \le 11$ $x \ge 5$
$\{x | 5 \le x \le 11\}$ or $[5, 11]$

120. Multiply the first equation by -3, the second
equation by 2 and solve by addition.
$$-15x - 6y = -6$$
$$\underline{8x + 6y = -8}$$
$$-7x = -14$$
$$x = 2$$

Back-substitute 2 for x to find y.
$$5x + 2y = 2$$
$$5(2) + 2y = 2$$
$$10 + 2y = 2$$
$$2y = -8$$
$$y = -4$$

The solution is $(2, -4)$.
The solution set is $\{(2, -4)\}$.

121. $64x^3 - 27 = (4x - 3)(16x^2 + 12x + 9)$

122. a. $21x + 10x = 31x$

    b. $21\sqrt{2} + 10\sqrt{2} = 31\sqrt{2}$

123. a. $4x - 12x = -8x$

    b. $4\sqrt[3]{2} - 12\sqrt[3]{2} = -8\sqrt[3]{2}$

124. $\dfrac{\sqrt[4]{7y^5}}{\sqrt[4]{x^{12}}} = \dfrac{\sqrt[4]{7y^5}}{x^3} = \dfrac{\sqrt[4]{y^4} \cdot \sqrt[4]{7y}}{x^3} = \dfrac{y\sqrt[4]{7y}}{x^3}$

10.4 Check Points

1. a. $8\sqrt{13} + 2\sqrt{13} = 10\sqrt{13}$

    b. $9\sqrt[3]{7} - 6x\sqrt[3]{7} + 12\sqrt[3]{7} = (9 - 6x + 12)\sqrt[3]{7}$
$$= (21 - 6x)\sqrt[3]{7}$$

    c. $7\sqrt[4]{3x} - 2\sqrt[4]{3x} + 2\sqrt[3]{3x} = (7 - 2)\sqrt[4]{3x} + 2\sqrt[3]{3x}$
$$= 5\sqrt[4]{3x} + 2\sqrt[3]{3x}$$

2. a. $3\sqrt{20} + 5\sqrt{45} = 3\sqrt{4 \cdot 5} + 5\sqrt{9 \cdot 5}$
$$= 3 \cdot 2\sqrt{5} + 5 \cdot 3\sqrt{5}$$
$$= 6\sqrt{5} + 15\sqrt{5}$$
$$= 21\sqrt{5}$$

    b. $3\sqrt{12x} - 6\sqrt{27x} = 3\sqrt{4 \cdot 3x} - 6\sqrt{9 \cdot 3x}$
$$= 3 \cdot 2\sqrt{3x} - 6 \cdot 3\sqrt{3x}$$
$$= 6\sqrt{3x} - 18\sqrt{3x}$$
$$= -12\sqrt{3x}$$

    c. $8\sqrt{5} - 6\sqrt{2}$ cannot be simplified.

3. a. $3\sqrt[3]{24} - 5\sqrt[3]{81} = 3\sqrt[3]{8 \cdot 3} - 5\sqrt[3]{27 \cdot 3}$

$$= 3 \cdot 2\sqrt[3]{3} - 5 \cdot 3\sqrt[3]{3}$$

$$= 6\sqrt[3]{3} - 15\sqrt[3]{3}$$

$$= -9\sqrt[3]{3}$$

b. $5\sqrt[3]{x^2 y} + \sqrt[3]{27x^5 y^4} = 5\sqrt[3]{x^2 y} + \sqrt[3]{27x^3 y^3 x^2 y}$

$$= 5\sqrt[3]{x^2 y} + \sqrt[3]{27x^3 y^3} \sqrt[3]{x^2 y}$$

$$= 5\sqrt[3]{x^2 y} + 3xy\sqrt[3]{x^2 y}$$

$$= (5 + 3xy)\sqrt[3]{x^2 y}$$

4. a. $\sqrt[3]{\dfrac{24}{125}} = \dfrac{\sqrt[3]{24}}{\sqrt[3]{125}} = \dfrac{\sqrt[3]{8 \cdot 3}}{\sqrt[3]{125}} = \dfrac{\sqrt[3]{8} \cdot \sqrt[3]{3}}{\sqrt[3]{125}} = \dfrac{2\sqrt[3]{3}}{5}$

b. $\sqrt{\dfrac{9x^3}{y^{10}}} = \dfrac{\sqrt{9x^3}}{\sqrt{y^{10}}} = \dfrac{\sqrt{9x^2 \cdot x}}{\sqrt{y^{10}}} = \dfrac{\sqrt{9x^2} \sqrt{x}}{\sqrt{y^{10}}} = \dfrac{3x\sqrt{x}}{y^5}$

c. $\sqrt[3]{\dfrac{8y^7}{x^{12}}} = \dfrac{\sqrt[3]{8y^7}}{\sqrt[3]{x^{12}}} = \dfrac{\sqrt[3]{8y^6 \cdot y}}{\sqrt[3]{x^{12}}} = \dfrac{2y^2 \sqrt[3]{y}}{x^4}$

5. a. $\dfrac{\sqrt{40x^5}}{\sqrt{2x}} = \sqrt{\dfrac{40x^5}{2x}} = \sqrt{20x^4} = \sqrt{4x^4 \cdot 5} = \sqrt{4x^4} \sqrt{5} = 2x^2 \sqrt{5}$

b. $\dfrac{\sqrt{50xy}}{2\sqrt{2}} = \dfrac{1}{2} \cdot \sqrt{\dfrac{50xy}{2}} = \dfrac{1}{2} \cdot \sqrt{25xy} = \dfrac{1}{2} \cdot 5\sqrt{xy} = \dfrac{5\sqrt{xy}}{2}$

c. $\dfrac{\sqrt[3]{48x^7 y}}{\sqrt[3]{6xy^{-2}}} = \sqrt[3]{\dfrac{48x^7 y}{6xy^{-2}}} = \sqrt[3]{\dfrac{48}{6} \cdot x^{7-1} \cdot y^{1-(-2)}} = \sqrt[3]{8x^6 \cdot y^3} = 2x^2 y$

10.4 Exercise Set

1. $8\sqrt{5} + 3\sqrt{5} = (8 + 3)\sqrt{5} = 11\sqrt{5}$

3. $9\sqrt[3]{6} - 2\sqrt[3]{6} = (9 - 2)\sqrt[3]{6} = 7\sqrt[3]{6}$

5. $4\sqrt[5]{2} + 3\sqrt[5]{2} - 5\sqrt[5]{2} = (4 + 3 - 5)\sqrt[5]{2}$

$$= 2\sqrt[5]{2}$$

7. $3\sqrt{13} - 2\sqrt{5} - 2\sqrt{13} + 4\sqrt{5}$

$\quad = 3\sqrt{13} - 2\sqrt{13} - 2\sqrt{5} + 4\sqrt{5}$

$\quad = (3-2)\sqrt{13} + (-2+4)\sqrt{5}$

$\quad = \sqrt{13} + 2\sqrt{5}$

9. $3\sqrt{5} - \sqrt[3]{x} + 4\sqrt{5} + 3\sqrt[3]{x}$

$\quad = 3\sqrt{5} + 4\sqrt{5} - \sqrt[3]{x} + 3\sqrt[3]{x}$

$\quad = (3+4)\sqrt{5} + (-1+3)\sqrt[3]{x}$

$\quad = 7\sqrt{5} + 2\sqrt[3]{x}$

11. $\sqrt{3} + \sqrt{27} = \sqrt{3} + \sqrt{9 \cdot 3} = \sqrt{3} + 3\sqrt{3}$

$\quad\quad\quad\quad\quad = (1+3)\sqrt{3} = 4\sqrt{3}$

13. $7\sqrt{12} + \sqrt{75} = 7\sqrt{4 \cdot 3} + \sqrt{25 \cdot 3}$

$\quad = 7 \cdot 2\sqrt{3} + 5\sqrt{3}$

$\quad = 14\sqrt{3} + 5\sqrt{3}$

$\quad = (14+5)\sqrt{3} = 19\sqrt{3}$

15. $3\sqrt{32x} - 2\sqrt{18x}$

$\quad = 3\sqrt{16 \cdot 2x} - 2\sqrt{9 \cdot 2x}$

$\quad = 3 \cdot 4\sqrt{2x} - 2 \cdot 3\sqrt{2x}$

$\quad = 12\sqrt{2x} - 6\sqrt{2x} = 6\sqrt{2x}$

17. $5\sqrt[3]{16} + \sqrt[3]{54} = 5\sqrt[3]{8 \cdot 2} + \sqrt[3]{27 \cdot 2}$

$\quad\quad\quad\quad\quad = 5 \cdot 2\sqrt[3]{2} + 3\sqrt[3]{2}$

$\quad\quad\quad\quad\quad = 10\sqrt[3]{2} + 3\sqrt[3]{2}$

$\quad\quad\quad\quad\quad = (10+3)\sqrt[3]{2} = 13\sqrt[3]{2}$

19. $3\sqrt{45x^3} + \sqrt{5x} = 3\sqrt{9 \cdot 5x^2 x} + \sqrt{5x}$

$\quad\quad\quad\quad\quad = 3 \cdot 3x\sqrt{5x} + \sqrt{5x}$

$\quad\quad\quad\quad\quad = 9x\sqrt{5x} + \sqrt{5x}$

$\quad\quad\quad\quad\quad = (9x+1)\sqrt{5x}$

21. $\sqrt[3]{54xy^3} + y\sqrt[3]{128x}$

$\quad = \sqrt[3]{27 \cdot 2xy^3} + y\sqrt[3]{64 \cdot 2x}$

$\quad = 3y\sqrt[3]{2x} + 4y\sqrt[3]{2x}$

$\quad = (3y+4y)\sqrt[3]{2x} = 7y\sqrt[3]{2x}$

23. $\sqrt[3]{54x^4} - \sqrt[3]{16x} = \sqrt[3]{27 \cdot 2x^3 x} - \sqrt[3]{8 \cdot 2x}$

$\quad\quad\quad\quad\quad = 3x\sqrt[3]{2x} - 2\sqrt[3]{2x}$

$\quad\quad\quad\quad\quad = (3x-2)\sqrt[3]{2x}$

25. $\sqrt{9x-18} + \sqrt{x-2}$

$\quad = \sqrt{9(x-2)} + \sqrt{x-2}$

$\quad = 3\sqrt{x-2} + \sqrt{x-2}$

$\quad = (3+1)\sqrt{x-2}$

$\quad = 4\sqrt{x-2}$

27. $2\sqrt[3]{x^4 y^2} + 3x\sqrt[3]{xy^2}$

$\quad = 2\sqrt[3]{x^3 xy^2} + 3x\sqrt[3]{xy^2}$

$\quad = 2x\sqrt[3]{xy^2} + 3x\sqrt[3]{xy^2}$

$\quad = (2x+3x)\sqrt[3]{xy^2}$

$\quad = 5x\sqrt[3]{xy^2}$

29. $\sqrt{\dfrac{11}{4}} = \dfrac{\sqrt{11}}{\sqrt{4}} = \dfrac{\sqrt{11}}{2}$

31. $\sqrt[3]{\dfrac{19}{27}} = \dfrac{\sqrt[3]{19}}{\sqrt[3]{27}} = \dfrac{\sqrt[3]{19}}{3}$

33. $\sqrt{\dfrac{x^2}{36y^8}} = \dfrac{\sqrt{x^2}}{\sqrt{36y^8}} = \dfrac{x}{6y^4}$

35. $\sqrt{\dfrac{8x^3}{25y^6}} = \dfrac{\sqrt{8x^3}}{\sqrt{25y^6}} = \dfrac{\sqrt{4 \cdot 2x^2 x}}{5y^3}$

$\quad\quad\quad\quad = \dfrac{2x\sqrt{2x}}{5y^3}$

37. $\sqrt[3]{\dfrac{x^4}{8y^3}} = \dfrac{\sqrt[3]{x^4}}{\sqrt[3]{8y^3}} = \dfrac{\sqrt[3]{x^3 x}}{2y} = \dfrac{x\sqrt[3]{x}}{2y}$

39. $\sqrt[3]{\dfrac{50x^8}{27y^{12}}} = \dfrac{\sqrt[3]{50x^8}}{\sqrt[3]{27y^{12}}} = \dfrac{\sqrt[3]{50x^6 x^2}}{3y^4}$

$\quad\quad\quad\quad = \dfrac{x^2\sqrt[3]{50x^2}}{3y^4}$

41. $\sqrt[4]{\dfrac{9y^6}{x^8}} = \dfrac{\sqrt[4]{9y^6}}{\sqrt[4]{x^8}} = \dfrac{\sqrt[4]{9y^4 y^2}}{x^2} = \dfrac{y\sqrt[4]{9y^2}}{x^2}$

43. $\sqrt[5]{\dfrac{64x^{13}}{y^{20}}} = \dfrac{\sqrt[5]{64x^{13}}}{\sqrt[5]{y^{20}}} = \dfrac{\sqrt[5]{32 \cdot 2x^{10} x^3}}{y^4}$

$= \dfrac{2x^2 \sqrt[5]{2x^3}}{y^4}$

45. $\dfrac{\sqrt{40}}{\sqrt{5}} = \sqrt{\dfrac{40}{5}} = \sqrt{8} = \sqrt{4 \cdot 2} = 2\sqrt{2}$

47. $\dfrac{\sqrt[3]{48}}{\sqrt[3]{6}} = \sqrt[3]{\dfrac{48}{6}} = \sqrt[3]{8} = 2$

49. $\dfrac{\sqrt{54x^3}}{\sqrt{6x}} = \sqrt{\dfrac{54x^3}{6x}} = \sqrt{9x^2} = 3x$

51. $\dfrac{\sqrt{x^5 y^3}}{\sqrt{xy}} = \sqrt{\dfrac{x^5 y^3}{xy}} = \sqrt{x^4 y^2} = x^2 y$

53. $\dfrac{\sqrt{200x^3}}{\sqrt{10x^{-1}}} = \sqrt{\dfrac{200x^3}{10x^{-1}}} = \sqrt{20x^{3-(-1)}}$

$= \sqrt{20x^4} = \sqrt{4 \cdot 5x^4} = 2x^2 \sqrt{5}$

55. $\dfrac{\sqrt{48a^8 b^7}}{\sqrt{3a^{-2} b^{-3}}} = \sqrt{\dfrac{48a^8 b^7}{3a^{-2} b^{-3}}}$

$= \sqrt{16a^{10} b^{10}}$

$= 4a^5 b^5$

57. $\dfrac{\sqrt{72xy}}{2\sqrt{2}} = \dfrac{1}{2}\sqrt{\dfrac{72xy}{2}} = \dfrac{1}{2}\sqrt{36xy}$

$= \dfrac{1}{2} \cdot 6\sqrt{xy} = 3\sqrt{xy}$

59. $\dfrac{\sqrt[3]{24x^3 y^5}}{\sqrt[3]{3y^2}} = \sqrt[3]{\dfrac{24x^3 y^5}{3y^2}} = \sqrt[3]{8x^3 y^3} = 2xy$

61. $\dfrac{\sqrt[4]{32x^{10} y^8}}{\sqrt[4]{2x^2 y^{-2}}} = \sqrt[4]{\dfrac{32x^{10} y^8}{2x^2 y^{-2}}} = \sqrt[4]{16x^8 y^{8-(-2)}}$

$= \sqrt[4]{16x^8 y^{10}} = \sqrt[4]{16x^8 y^8 y^2}$

$= 2x^2 y^2 \sqrt[4]{y^2}$ or $2x^2 y^2 \sqrt{y}$

63. $\dfrac{\sqrt[3]{x^2 + 5x + 6}}{\sqrt[3]{x+2}} = \sqrt[3]{\dfrac{x^2 + 5x + 6}{x+2}}$

$= \sqrt[3]{\dfrac{(x+2)(x+3)}{x+2}}$

$= \sqrt[3]{x+3}$

65. $\dfrac{\sqrt[3]{a^3 + b^3}}{\sqrt[3]{a+b}} = \sqrt[3]{\dfrac{a^3 + b^3}{a+b}}$

$= \sqrt[3]{\dfrac{(a+b)(a^2 - ab + b^2)}{a+b}}$

$= \sqrt[3]{a^2 - ab + b^2}$

67. $\dfrac{\sqrt{32}}{5} + \dfrac{\sqrt{18}}{7} = \dfrac{\sqrt{16 \cdot 2}}{5} + \dfrac{\sqrt{9 \cdot 2}}{7}$

$= \dfrac{\sqrt{16} \cdot \sqrt{2}}{5} + \dfrac{\sqrt{9} \cdot \sqrt{2}}{7}$

$= \dfrac{4\sqrt{2}}{5} + \dfrac{3\sqrt{2}}{7}$

$= \dfrac{28\sqrt{2}}{35} + \dfrac{15\sqrt{2}}{35}$

$= \dfrac{28\sqrt{2} + 15\sqrt{2}}{35}$

$= \dfrac{(28 + 15)\sqrt{2}}{35}$

$= \dfrac{43\sqrt{2}}{35}$

69. $3x\sqrt{8xy^2} - 5y\sqrt{32x^3} + \sqrt{18x^3 y^2}$

$= 3x\sqrt{4y^2 \cdot 2x} - 5y\sqrt{16x^2 \cdot 2x} + \sqrt{9x^2 y^2 \cdot 2x}$

$= 3x(2y)\sqrt{2x} - 5y(4x)\sqrt{2x} + 3xy\sqrt{2x}$

$= 6xy\sqrt{2x} - 20xy\sqrt{2x} + 3xy\sqrt{2x}$

$= (6 - 20 + 3)xy\sqrt{2x}$

$= -11xy\sqrt{2x}$

71. $5\sqrt{2x^3} + \dfrac{30x^3\sqrt{24x^2}}{3x^2\sqrt{3x}}$

$= 5\sqrt{2x^3} + 10x\sqrt{\dfrac{24x^2}{3x}}$

$= 5\sqrt{2x^3} + 10x\sqrt{8x}$

$= 5\sqrt{x^2 \cdot 2x} + 10x\sqrt{4 \cdot 2x}$

$= 5x\sqrt{2x} + 10x(2)\sqrt{2x}$

$= 5x\sqrt{2x} + 20x\sqrt{2x}$

$= (5+20)x\sqrt{2x} = 25x\sqrt{2x}$

73. $2x\sqrt{75xy} - \dfrac{\sqrt{81xy^2}}{\sqrt{3x^{-2}y}}$

$= 2x\sqrt{75xy} - \sqrt{\dfrac{81xy^2}{3x^{-2}y}}$

$= 2x\sqrt{75xy} - \sqrt{27x^3 y}$

$= 2x\sqrt{25 \cdot 3xy} - \sqrt{9x^2 \cdot 3xy}$

$= 2x(5)\sqrt{3xy} - 3x\sqrt{3xy}$

$= 10x\sqrt{3xy} - 3x\sqrt{3xy}$

$= (10-3)x\sqrt{3xy}$

$= 7x\sqrt{3xy}$

75. $\dfrac{15x^4\sqrt[3]{80x^3 y^2}}{5x^3\sqrt[3]{2x^2 y}} - \dfrac{75\sqrt[3]{5x^3 y}}{25\sqrt[3]{x^{-1}}}$

$= 3x\sqrt[3]{\dfrac{80x^3 y^2}{2x^2 y}} - 3\sqrt[3]{\dfrac{5x^3 y}{x^{-1}}}$

$= 3x\sqrt[3]{40xy} - 3\sqrt[3]{5x^4 y}$

$= 3x\sqrt[3]{8 \cdot 5xy} - 3\sqrt[3]{x^3 \cdot 5xy}$

$= 3x(2)\sqrt[3]{5xy} - 3x\sqrt[3]{5xy}$

$= 6x\sqrt[3]{5xy} - 3x\sqrt[3]{5xy}$

$= (6-3)x\sqrt[3]{5xy} = 3x\sqrt[3]{5xy}$

77. $\left(\dfrac{f}{g}\right)(x) = \dfrac{\sqrt{48x^5}}{\sqrt{3x^2}} = \sqrt{\dfrac{48x^5}{3x^2}}$

$= \sqrt{16x^3} = \sqrt{16x^2 \cdot x} = 4x\sqrt{x}$

To get the domain, we need $x \geq 0$ and $3x^2 > 0$.
Combining these restrictions gives us $x > 0$.
Domain: $\{x \mid x > 0\}$ or $(0, \infty)$

79. $\left(\dfrac{f}{g}\right)(x) = \dfrac{\sqrt[3]{32x^6}}{\sqrt[3]{2x^2}} = \sqrt[3]{\dfrac{32x^6}{2x^2}} = \sqrt[3]{16x^4}$

$= \sqrt[3]{8x^3 \cdot 2x} = 2x\sqrt[3]{2x}$

Our only restriction here is that we cannot divide by 0. Thus, we need $x \neq 0$.

Domain: $\{x \mid x \neq 0\}$ or $(-\infty, 0) \cup (0, \infty)$

81. Perimeter:
$P = 2l + 2w$

$= 2 \cdot \sqrt{125} + 2 \cdot 2\sqrt{20}$

$= 2 \cdot \sqrt{25 \cdot 5} + 4\sqrt{4 \cdot 5}$

$= 2 \cdot 5\sqrt{5} + 4 \cdot 2\sqrt{5}$

$= 10\sqrt{5} + 8\sqrt{5}$

$= 18\sqrt{5}$ feet

Area:
$A = lw$

$= \sqrt{125} \cdot 2\sqrt{20}$

$= 2\sqrt{125 \cdot 20}$

$= 2\sqrt{2500}$

$= 2 \cdot 50$

$= 100$ square feet

83. Perimeter:
$P = a + b + c$

$= \sqrt{45} + \sqrt{80} + \sqrt{125}$

$= \sqrt{9 \cdot 5} + \sqrt{16 \cdot 5} + \sqrt{25 \cdot 5}$

$= 3\sqrt{5} + 4\sqrt{5} + 5\sqrt{5}$

$= 12\sqrt{5}$ meters

85. a. $f(x) = 5\sqrt{x} + 34.1$

$$f(40) - f(10) = 5\sqrt{40} + 34.1 - \left(5\sqrt{10} + 34.1\right)$$

$$= 5\sqrt{4 \cdot 10} + 34.1 - 5\sqrt{10} - 34.1$$

$$= 5 \cdot 2\sqrt{10} - 5\sqrt{10}$$

$$= 10\sqrt{10} - 5\sqrt{10}$$

$$= 5\sqrt{10}$$

The projected increase in the number of Americans ages 65 – 84, in millions, from 2020 to 2050 is $5\sqrt{10}$.

b. $5\sqrt{10} \approx 15.8$

This value underestimates the difference in the projected data shown in the bar graph by 2.7 million.

87. – 93. Answers will vary.

95. $\sqrt{16x} - \sqrt{9x} = \sqrt{7x}$

The graphs do not coincide. Correct the simplification.

$$\sqrt{16x} - \sqrt{9x} = 4\sqrt{x} - 3\sqrt{x}$$

$$= (4 - 3)\sqrt{x} = \sqrt{x}$$

97. makes sense

99. does not make sense; Explanations will vary. Sample explanation:

$$3\sqrt[3]{81} + 2\sqrt[3]{54} = 3\sqrt[3]{27 \cdot 3} + 2\sqrt[3]{27 \cdot 2}$$

$$= 3 \cdot 3\sqrt[3]{3} + 2 \cdot 3\sqrt[3]{2}$$

$$= 9\sqrt[3]{3} + 6\sqrt[3]{2}$$

101. false; Changes to make the statement true will vary. A sample change is: $\sqrt{5} + \sqrt{5} = 2\sqrt{5}$

103. false; Changes to make the statement true will vary. A sample change is: In order for two radical expressions to be combined, both the index and the radicand must be the same. Just because two radical expressions are completely simplified, that does not guarantee that the index and radicands match.

105. Let x = the irrational number.

$$x - \left(2\sqrt{18} - \sqrt{50}\right) = \sqrt{2}$$

$$x - 2\sqrt{18} + \sqrt{50} = \sqrt{2}$$

$$x = \sqrt{2} + 2\sqrt{18} - \sqrt{50}$$

$$x = \sqrt{2} + 2\sqrt{9 \cdot 2} - \sqrt{25 \cdot 2}$$

$$x = \sqrt{2} + 2 \cdot 3\sqrt{2} - 5\sqrt{2}$$

$$x = \sqrt{2} + 6\sqrt{2} - 5\sqrt{2}$$

$$x = 2\sqrt{2}$$

The irrational number is $2\sqrt{2}$.

107.

$$\frac{6\sqrt{49xy}\sqrt{ab^2}}{7\sqrt{36x^{-3}y^{-5}}\sqrt{a^{-9}b^{-1}}} = \frac{6\sqrt{49xyab^2}}{7\sqrt{36x^{-3}y^{-5}a^{-9}b^{-1}}}$$

$$= \frac{6}{7}\sqrt{\frac{49x^4y^6a^{10}b^3}{36}} = \frac{6}{7}\sqrt{\frac{49x^4y^6a^{10}b^2b}{36}}$$

$$= \frac{\cancel{6} \cdot 7x^2y^3a^5b}{7 \cdot \cancel{6}}\sqrt{b} = x^2y^3a^5b\sqrt{b}$$

108. $2(3x - 1) - 4 = 2x - (6 - x)$

$$6x - 2 - 4 = 2x - 6 + x$$

$$6x - 6 = 3x - 6$$

$$3x = 0$$

$$x = 0$$

The solution set is $\{0\}$.

109. $x^2 - 8xy + 12y^2 = (x - 6y)(x - 2y)$

110.

$$\frac{2}{x^2 + 5x + 6} + \frac{3x}{x^2 + 6x + 9}$$

$$= \frac{2}{(x + 3)(x + 2)} + \frac{3x}{(x + 3)^2}$$

$$= \frac{2}{(x + 3)(x + 2)} \cdot \frac{(x + 3)}{(x + 3)} + \frac{3x}{(x + 3)^2} \cdot \frac{(x + 2)}{(x + 2)}$$

$$= \frac{2(x + 3) + 3x(x + 2)}{(x + 3)^2 (x + 2)}$$

$$= \frac{2x + 6 + 3x^2 + 6x}{(x + 3)^2 (x + 2)}$$

$$= \frac{3x^2 + 8x + 6}{(x + 3)^2 (x + 2)}$$

111. a. $7(x+5)=7x+35$

b. $\sqrt{7}\left(x+\sqrt{5}\right)=x\sqrt{7}+\sqrt{35}$

112. a. $(x+5)(6x+3)=6x^2+33x+15$

b. $\left(\sqrt{2}+5\right)\left(6\sqrt{2}+3\right)=6\sqrt{2}^2+33\sqrt{2}+15$

$$=6\cdot2+33\sqrt{2}+15$$
$$=12+33\sqrt{2}+15$$
$$=27+33\sqrt{2}$$

113. $\dfrac{10y}{\sqrt[5]{4x^3y}}\cdot\dfrac{\sqrt[5]{8x^2y^4}}{\sqrt[5]{8x^2y^4}}=\dfrac{10y\sqrt[5]{8x^2y^4}}{\sqrt[5]{4x^3y}\cdot\sqrt[5]{8x^2y^4}}$

$$=\dfrac{10y\sqrt[5]{8x^2y^4}}{\sqrt[5]{32x^5y^5}}$$

$$=\dfrac{10y\sqrt[5]{8x^2y^4}}{2xy}$$

$$=\dfrac{5\sqrt[5]{8x^2y^4}}{x}$$

Chapter 10 Mid-Chapter Check Point

1. $\sqrt{100}-\sqrt[3]{-27}=10-(-3)=10+3=13$

2. $\sqrt{8x^5y^7}=\sqrt{4x^4y^6\cdot2xy}=2x^2y^3\sqrt{2xy}$

3. $3\sqrt[3]{4x^2}+2\sqrt[3]{4x^2}=(3+2)\sqrt[3]{4x^2}=5\sqrt[3]{4x^2}$

4. $\left(3\sqrt[3]{4x^2}\right)\left(2\sqrt[3]{4x^2}\right)=6\sqrt[3]{4x^2\cdot4x^2}$

$$=6\sqrt[3]{16x^4}$$
$$=6\sqrt[3]{8x^3\cdot2x}$$
$$=6(2x)\sqrt[3]{2x}$$
$$=12x\sqrt[3]{2x}$$

5. $27^{2/3}+(-32)^{3/5}=\left(\sqrt[3]{27}\right)^2+\left(\sqrt[5]{-32}\right)^3$

$$=(3)^2+(-2)^3$$
$$=9+(-8)$$
$$=1$$

6. $\left(64x^3y^{1/4}\right)^{1/3}=(64)^{1/3}\left(x^3\right)^{1/3}\left(y^{1/4}\right)^{1/3}$

$$=\sqrt[3]{64}\cdot x^{3(1/3)}\cdot y^{(1/4)(1/3)}$$
$$=4xy^{1/12}$$

7. $5\sqrt{27}-4\sqrt{48}=5\sqrt{9\cdot3}-4\sqrt{16\cdot3}$

$$=5(3)\sqrt{3}-4(4)\sqrt{3}$$
$$=15\sqrt{3}-16\sqrt{3}$$
$$=(15-16)\sqrt{3}$$
$$=-\sqrt{3}$$

8. $\sqrt{\dfrac{500x^3}{4y^4}}=\dfrac{\sqrt{500x^3}}{\sqrt{4y^4}}=\dfrac{\sqrt{100x^2\cdot5x}}{\sqrt{4y^4}}$

$$=\dfrac{10x\sqrt{5x}}{2y^2}=\dfrac{5x\sqrt{5x}}{y^2}$$

9. $\dfrac{x}{\sqrt[4]{x}}=\dfrac{x}{x^{1/4}}=x^{1-(1/4)}$

$$=x^{(4/4)-(1/4)}=x^{3/4}$$
$$=\sqrt[4]{x^3}$$

10. $\sqrt[3]{54x^5}=\sqrt[3]{27x^3\cdot2x^2}=3x\sqrt[3]{2x^2}$

11. $\dfrac{\sqrt[3]{160}}{\sqrt[3]{2}}=\sqrt[3]{\dfrac{160}{2}}=\sqrt[3]{80}=\sqrt[3]{8\cdot10}=2\sqrt[3]{10}$

12. $\sqrt[5]{\dfrac{x^{10}}{y^{20}}}=\left(\dfrac{x^{10}}{y^{20}}\right)^{1/5}$

$$=\dfrac{\left(x^{10}\right)^{1/5}}{\left(y^{20}\right)^{1/5}}$$

$$=\dfrac{x^{10(1/5)}}{y^{20(1/5)}}=\dfrac{x^2}{y^4}$$

519

13. $\dfrac{\left(x^{2/3}\right)^2}{\left(x^{1/4}\right)^3} = \dfrac{x^{(2/3)\cdot 2}}{x^{(1/4)\cdot 3}} = \dfrac{x^{4/3}}{x^{3/4}}$

$= x^{(4/3)-(3/4)}$

$= x^{(16/12)-(9/12)}$

$= x^{7/12}$

14. $\sqrt[6]{x^6 y^4} = \left(x^6 y^4\right)^{1/6} = \left(x^6\right)^{1/6}\left(y^4\right)^{1/6}$

$= x^{6(1/6)} y^{4(1/6)} = x^1 y^{2/3}$

$= x\sqrt[3]{y^2}$

15. $\sqrt[7]{(x-2)^3} \cdot \sqrt[7]{(x-2)^6}$

$= \sqrt[7]{(x-2)^3 \cdot (x-2)^6}$

$= \sqrt[7]{(x-2)^9}$

$= \sqrt[7]{(x-2)^7 \cdot (x-2)^2}$

$= (x-2)\sqrt[7]{(x-2)^2}$

16. $\sqrt[4]{32x^{11} y^{17}} = \sqrt[4]{16x^8 y^{16} \cdot 2x^3 y}$

$= \sqrt[4]{16x^8 y^{16}} \cdot \sqrt[4]{2x^3 y}$

$= 2x^2 y^4 \sqrt[4]{2x^3 y}$

17. $4\sqrt[3]{16} + 2\sqrt[3]{54} = 4\sqrt[3]{8\cdot 2} + 2\sqrt[3]{27\cdot 2}$

$= 4\sqrt[3]{8}\cdot\sqrt[3]{2} + 2\sqrt[3]{27}\cdot\sqrt[3]{2}$

$= 4(2)\sqrt[3]{2} + 2(3)\sqrt[3]{2}$

$= 8\sqrt[3]{2} + 6\sqrt[3]{2}$

$= (8+6)\sqrt[3]{2}$

$= 14\sqrt[3]{2}$

18. $\dfrac{\sqrt[7]{x^4 y^9}}{\sqrt[7]{x^{-5} y^7}} = \sqrt[7]{\dfrac{x^4 y^9}{x^{-5} y^7}} = \sqrt[7]{x^9 y^2}$

$= \sqrt[7]{x^7 \cdot x^2 y^2}$

$= x\sqrt[7]{x^2 y^2}$

19. $(-125)^{-2/3} = \dfrac{1}{(-125)^{2/3}} = \dfrac{1}{\left(\sqrt[3]{-125}\right)^2}$

$= \dfrac{1}{(-5)^2} = \dfrac{1}{25}$

20. $\sqrt{2}\cdot\sqrt[3]{2} = 2^{1/2}\cdot 2^{1/3} = 2^{(1/2)+(1/3)}$

$= 2^{(3/6)+(2/6)} = 2^{5/6}$

$= \sqrt[6]{2^5} = \sqrt[6]{32}$

21. $\sqrt[3]{\dfrac{32x}{y^4}} \cdot \sqrt[3]{\dfrac{2x^2}{y^2}} = \sqrt[3]{\dfrac{32x}{y^4}\cdot\dfrac{2x^2}{y^2}}$

$= \sqrt[3]{\dfrac{64x^3}{y^6}} = \dfrac{\sqrt[3]{64x^3}}{\sqrt[3]{y^6}}$

$= \dfrac{4x}{y^2}$

22. $\sqrt{32xy^2}\cdot\sqrt{2x^3 y^5} = \sqrt{32xy^2 \cdot 2x^3 y^5}$

$= \sqrt{64x^4 y^7}$

$= \sqrt{64x^4 y^6 \cdot y}$

$= 8x^2 y^3 \sqrt{y}$

23. $4x\sqrt{6x^4 y^3} - 7y\sqrt{24x^6 y}$

$= 4x\sqrt{x^4 y^2 \cdot 6y} - 7y\sqrt{4x^6 \cdot 6y}$

$= 4x\left(x^2 y\right)\sqrt{6y} - 7y\left(2x^3\right)\sqrt{6y}$

$= 4x^3 y\sqrt{6y} - 14x^3 y\sqrt{6y}$

$= (4-14)x^3 y\sqrt{6y}$

$= -10x^3 y\sqrt{6y}$

24. $f(x) = \sqrt{30-5x}$

To find the domain, we set the radicand greater than or equal to 0.

$30 - 5x \ge 0$

$-5x \ge -30$

$x \le 6$

Domain: $\{x \mid x \le 6\}$ or $(-\infty, 6]$

25. $g(x) = \sqrt[3]{3x-15}$

The domain of a cube root is all real numbers. Since there are no other restrictions, we have

Domain: $\{x \mid x \text{ is a real number}\}$ or $(-\infty, \infty)$

10.5 Check Points

1. a. $\sqrt{6}\left(x + \sqrt{10}\right) = \sqrt{6} \cdot x + \sqrt{6}\sqrt{10}$

$\qquad = x\sqrt{6} + \sqrt{60}$

$\qquad = x\sqrt{6} + \sqrt{4 \cdot 15}$

$\qquad = x\sqrt{6} + 2\sqrt{15}$

b. $\sqrt[3]{y}\left(\sqrt[3]{y^2} - \sqrt[3]{7}\right) = \sqrt[3]{y} \cdot \sqrt[3]{y^2} - \sqrt[3]{y} \cdot \sqrt[3]{7}$

$\qquad = \sqrt[3]{y^3} - \sqrt[3]{7y}$

$\qquad = y - \sqrt[3]{7y}$

c.

$\left(6\sqrt{5} + 3\sqrt{2}\right)\left(2\sqrt{5} - 4\sqrt{2}\right) = \left(6\sqrt{5}\right)\left(2\sqrt{5}\right) + \left(6\sqrt{5}\right)\left(-4\sqrt{2}\right) + \left(3\sqrt{2}\right)\left(2\sqrt{5}\right) + \left(3\sqrt{2}\right)\left(-4\sqrt{2}\right)$

$\qquad = 12 \cdot 5 - 24\sqrt{10} + 6\sqrt{10} - 12 \cdot 2$

$\qquad = 60 - 24\sqrt{10} + 6\sqrt{10} - 24$

$\qquad = 36 - 18\sqrt{10}$

2. a. $\left(\sqrt{5} + \sqrt{6}\right)^2 = \left(\sqrt{5}\right)^2 + 2 \cdot \sqrt{5} \cdot \sqrt{6} + \left(\sqrt{6}\right)^2$

$\qquad = 5 + 2\sqrt{30} + 6$

$\qquad = 11 + 2\sqrt{30}$

b. $\left(\sqrt{6} + \sqrt{5}\right)\left(\sqrt{6} - \sqrt{5}\right) = \left(\sqrt{6}\right)^2 - \left(\sqrt{5}\right)^2$

$\qquad = 6 - 5$

$\qquad = 1$

c. $\left(\sqrt{a} - \sqrt{7}\right)\left(\sqrt{a} + \sqrt{7}\right) = \left(\sqrt{a}\right)^2 - \left(\sqrt{7}\right)^2$

$\qquad = a - 7$

3. a. $\dfrac{\sqrt{3}}{\sqrt{7}} = \dfrac{\sqrt{3}}{\sqrt{7}} \cdot \dfrac{\sqrt{7}}{\sqrt{7}} = \dfrac{\sqrt{21}}{\sqrt{49}} = \dfrac{\sqrt{21}}{7}$

b. $\sqrt[3]{\dfrac{2}{9}} = \dfrac{\sqrt[3]{2}}{\sqrt[3]{9}} = \dfrac{\sqrt[3]{2}}{\sqrt[3]{3^2}} = \dfrac{\sqrt[3]{2}}{\sqrt[3]{3^2}} \cdot \dfrac{\sqrt[3]{3}}{\sqrt[3]{3}} = \dfrac{\sqrt[3]{6}}{\sqrt[3]{3^3}} = \dfrac{\sqrt[3]{6}}{3}$

4. a. $\sqrt{\dfrac{2x}{7y}} = \dfrac{\sqrt{2x}}{\sqrt{7y}} = \dfrac{\sqrt{2x}}{\sqrt{7y}} \cdot \dfrac{\sqrt{7y}}{\sqrt{7y}} = \dfrac{\sqrt{14xy}}{7y}$

b.

$\dfrac{\sqrt[3]{x}}{\sqrt[3]{9y}} = \dfrac{\sqrt[3]{x}}{\sqrt[3]{3^2 y}} = \dfrac{\sqrt[3]{x}}{\sqrt[3]{3^2 y}} \cdot \dfrac{\sqrt[3]{3y^2}}{\sqrt[3]{3y^2}} = \dfrac{\sqrt[3]{3xy^2}}{\sqrt[3]{3^3 y^3}} = \dfrac{\sqrt[3]{3xy^2}}{3y}$

c. $\dfrac{6x}{\sqrt[5]{8x^2 y^4}} = \dfrac{6x}{\sqrt[5]{2^3 x^2 y^4}}$

$\qquad = \dfrac{6x}{\sqrt[5]{2^3 x^2 y^4}} \cdot \dfrac{\sqrt[5]{2^2 x^3 y}}{\sqrt[5]{2^2 x^3 y}}$

$\qquad = \dfrac{6x\sqrt[5]{2^2 x^3 y}}{\sqrt[5]{2^5 x^5 y^5}}$

$\qquad = \dfrac{6x\sqrt[5]{2^2 x^3 y}}{2xy}$

$\qquad = \dfrac{3\sqrt[5]{4x^3 y}}{y}$

5. $\dfrac{18}{2\sqrt{3} + 3} = \dfrac{18}{2\sqrt{3} + 3} \cdot \dfrac{2\sqrt{3} - 3}{2\sqrt{3} - 3}$

$\qquad = \dfrac{36\sqrt{3} - 54}{2^2 \cdot 3 - 3^2}$

$\qquad = \dfrac{36\sqrt{3} - 54}{12 - 9}$

$\qquad = \dfrac{36\sqrt{3} - 54}{3}$

$\qquad = \dfrac{3(12\sqrt{3} - 18)}{3}$

$\qquad = 12\sqrt{3} - 18$

6. $\dfrac{3 + \sqrt{7}}{\sqrt{5} - \sqrt{2}} = \dfrac{3 + \sqrt{7}}{\sqrt{5} - \sqrt{2}} \cdot \dfrac{\sqrt{5} + \sqrt{2}}{\sqrt{5} + \sqrt{2}}$

$\qquad = \dfrac{3\sqrt{5} + 3\sqrt{2} + \sqrt{35} + \sqrt{14}}{5 - 2}$

$\qquad = \dfrac{3\sqrt{5} + 3\sqrt{2} + \sqrt{35} + \sqrt{14}}{3}$

7. $\dfrac{\sqrt{x+3}-\sqrt{x}}{3} = \dfrac{\sqrt{x+3}-\sqrt{x}}{3} \cdot \dfrac{\sqrt{x+3}+\sqrt{x}}{\sqrt{x+3}+\sqrt{x}}$

$= \dfrac{\left(\sqrt{x+3}\right)^2 - \left(\sqrt{x}\right)^2}{3\sqrt{x+3}+3\sqrt{x}}$

$= \dfrac{x+3-x}{3\sqrt{x+3}+3\sqrt{x}}$

$= \dfrac{3}{3\left(\sqrt{x+3}+\sqrt{x}\right)}$

$= \dfrac{1}{\sqrt{x+3}+\sqrt{x}}$

10.5 Exercise Set

1. $\sqrt{2}\left(x+\sqrt{7}\right) = \sqrt{2}\cdot x + \sqrt{2}\sqrt{7}$

$= x\sqrt{2} + \sqrt{14}$

3. $\sqrt{6}\left(7-\sqrt{6}\right) = \sqrt{6}\cdot 7 - \sqrt{6}\sqrt{6}$

$= 7\sqrt{6} - \sqrt{36} = 7\sqrt{6} - 6$

5. $\sqrt{3}\left(4\sqrt{6} - 2\sqrt{3}\right)$

$= \sqrt{3}\cdot 4\sqrt{6} - \sqrt{3}\cdot 2\sqrt{3}$

$= 4\sqrt{18} - 2\sqrt{9}$

$= 4\sqrt{9\cdot 2} - 2\cdot 3$

$= 4\cdot 3\sqrt{2} - 6 = 12\sqrt{2} - 6$

7. $\sqrt[3]{2}\left(\sqrt[3]{6} + 4\sqrt[3]{5}\right) = \sqrt[3]{2}\cdot\sqrt[3]{6} + \sqrt[3]{2}\cdot 4\sqrt[3]{5}$

$= \sqrt[3]{12} + 4\sqrt[3]{10}$

9. $\sqrt[3]{x}\left(\sqrt[3]{16x^2} - \sqrt[3]{x}\right)$

$= \sqrt[3]{x}\cdot\sqrt[3]{16x^2} - \sqrt[3]{x}\cdot\sqrt[3]{x}$

$= \sqrt[3]{x}\cdot\sqrt[3]{8\cdot 2x^2} - \sqrt[3]{x^2}$

$= \sqrt[3]{8\cdot 2x^3} - \sqrt[3]{x^2}$

$= 2x\sqrt[3]{2} - \sqrt[3]{x^2}$

11. $\left(5+\sqrt{2}\right)\left(6+\sqrt{2}\right)$

$= 5\cdot 6 + 5\sqrt{2} + 6\sqrt{2} + \sqrt{2}\sqrt{2}$

$= 30 + (5+6)\sqrt{2} + 2$

$= 32 + 11\sqrt{2}$

13. $\left(6+\sqrt{5}\right)\left(9-4\sqrt{5}\right)$

$= 6\cdot 9 - 6\cdot 4\sqrt{5} + 9\sqrt{5} - 4\sqrt{5}\sqrt{5}$

$= 54 - 24\sqrt{5} + 9\sqrt{5} - 4\cdot 5$

$= 54 + (-24+9)\sqrt{5} - 20$

$= 34 + (-15)\sqrt{5}$

$= 34 - 15\sqrt{5}$

15. $\left(6-3\sqrt{7}\right)\left(2-5\sqrt{7}\right)$

$= 6\cdot 2 - 6\cdot 5\sqrt{7} - 2\cdot 3\sqrt{7} + 3\sqrt{7}\cdot 5\sqrt{7}$

$= 12 - 30\sqrt{7} - 6\sqrt{7} + 15\cdot 7$

$= 12 + (-30-6)\sqrt{7} + 105$

$= 117 + (-36)\sqrt{7}$

$= 117 - 36\sqrt{7}$

17. $\left(\sqrt{2}+\sqrt{7}\right)\left(\sqrt{3}+\sqrt{5}\right)$

$= \sqrt{2}\sqrt{3} + \sqrt{2}\sqrt{5} + \sqrt{7}\sqrt{3} + \sqrt{7}\sqrt{5}$

$= \sqrt{6} + \sqrt{10} + \sqrt{21} + \sqrt{35}$

19. $\left(\sqrt{2}-\sqrt{7}\right)\left(\sqrt{3}-\sqrt{5}\right)$

$= \sqrt{2}\sqrt{3} - \sqrt{2}\sqrt{5} - \sqrt{7}\sqrt{3} + \sqrt{7}\sqrt{5}$

$= \sqrt{6} - \sqrt{10} - \sqrt{21} + \sqrt{35}$

21. $\left(3\sqrt{2}-4\sqrt{3}\right)\left(2\sqrt{2}+5\sqrt{3}\right)$

$= 3\sqrt{2}\left(2\sqrt{2}\right) + 3\sqrt{2}\left(5\sqrt{3}\right)$

$\quad -4\sqrt{3}\left(2\sqrt{2}\right) - 4\sqrt{3}\left(5\sqrt{3}\right)$

$= 6\cdot 2 + 15\sqrt{6} - 8\sqrt{6} - 20\cdot 3$

$= 12 + 7\sqrt{6} - 60$

$= 7\sqrt{6} - 48 \ \text{ or } \ -48 + 7\sqrt{6}$

23. $\left(\sqrt{3}+\sqrt{5}\right)^2 = \left(\sqrt{3}\right)^2 + 2\sqrt{3}\sqrt{5} + \left(\sqrt{5}\right)^2$

$= 3 + 2\sqrt{15} + 5$

$= 8 + 2\sqrt{15}$

25. $\left(\sqrt{3x}-\sqrt{y}\right)^2$

$= \left(\sqrt{3x}\right)^2 - 2\sqrt{3x}\sqrt{y} + \left(\sqrt{y}\right)^2$

$= 3x - 2\sqrt{3xy} + y$

27. $\left(\sqrt{5}+7\right)\left(\sqrt{5}-7\right)$

$= \sqrt{5}\sqrt{5} - 7\sqrt{5} + 7\sqrt{5} - 7\cdot 7$

$= 5 - 7\sqrt{5} + 7\sqrt{5} - 49$

$= 5 - 49$

$= -44$

29. $\left(2-5\sqrt{3}\right)\left(2+5\sqrt{3}\right)$

$= 2\cdot 2 + 2\cdot 5\sqrt{3} - 2\cdot 5\sqrt{3} - 5\sqrt{3}\cdot 5\sqrt{3}$

$= 4 + 10\sqrt{3} - 10\sqrt{3} - 25\cdot 3$

$= 4 - 75$

$= -71$

31. $\left(3\sqrt{2}+2\sqrt{3}\right)\left(3\sqrt{2}-2\sqrt{3}\right)$

$= 3\sqrt{2}\cdot 3\sqrt{2} - 3\sqrt{2}\cdot 2\sqrt{3}$

$\quad + 3\sqrt{2}\cdot 2\sqrt{3} - 2\sqrt{3}\cdot 2\sqrt{3}$

$= 9\cdot 2 - 6\sqrt{6} + 6\sqrt{6} - 4\cdot 3$

$= 18 - 12$

$= 6$

33. $\left(3-\sqrt{x}\right)\left(2-\sqrt{x}\right)$

$= 3\cdot 2 - 3\sqrt{x} - 2\sqrt{x} + \sqrt{x}\sqrt{x}$

$= 6 + \left(-3-2\right)\sqrt{x} + x$

$= 6 + \left(-5\right)\sqrt{x} + x$

$= 6 - 5\sqrt{x} + x$

35. $\left(\sqrt[3]{x}-4\right)\left(\sqrt[3]{x}+5\right)$

$= \sqrt[3]{x}\sqrt[3]{x} + 5\sqrt[3]{x} - 4\sqrt[3]{x} - 4\cdot 5$

$= \sqrt[3]{x^2} + \left(5-4\right)\sqrt[3]{x} - 20$

$= \sqrt[3]{x^2} + \sqrt[3]{x} - 20$

37. $\left(x+\sqrt[3]{y^2}\right)\left(2x-\sqrt[3]{y^2}\right)$

$= x\cdot 2x - x\sqrt[3]{y^2} + 2x\sqrt[3]{y^2} - \sqrt[3]{y^2}\sqrt[3]{y^2}$

$= 2x^2 + \left(-x+2x\right)\sqrt[3]{y^2} - \sqrt[3]{y^4}$

$= 2x^2 + x\sqrt[3]{y^2} - \sqrt[3]{y^3 y}$

$= 2x^2 + x\sqrt[3]{y^2} - y\sqrt[3]{y}$

39. $\dfrac{\sqrt{2}}{\sqrt{5}} = \dfrac{\sqrt{2}}{\sqrt{5}}\cdot\dfrac{\sqrt{5}}{\sqrt{5}} = \dfrac{\sqrt{2\cdot 5}}{\sqrt{5\cdot 5}} = \dfrac{\sqrt{10}}{5}$

41. $\sqrt{\dfrac{11}{x}} = \dfrac{\sqrt{11}}{\sqrt{x}} = \dfrac{\sqrt{11}}{\sqrt{x}}\cdot\dfrac{\sqrt{x}}{\sqrt{x}}$

$= \dfrac{\sqrt{11x}}{\sqrt{x^2}} = \dfrac{\sqrt{11x}}{x}$

43. $\dfrac{9}{\sqrt{3y}} = \dfrac{9}{\sqrt{3y}}\cdot\dfrac{\sqrt{3y}}{\sqrt{3y}} = \dfrac{9\sqrt{3y}}{\sqrt{3y\cdot 3y}}$

$= \dfrac{\overset{3}{\cancel{9}}\sqrt{3y}}{\underset{1}{\cancel{3}}\,y} = \dfrac{3\sqrt{3y}}{y}$

45. $\dfrac{1}{\sqrt[3]{2}} = \dfrac{1}{\sqrt[3]{2}}\cdot\dfrac{\sqrt[3]{2^2}}{\sqrt[3]{2^2}} = \dfrac{\sqrt[3]{2^2}}{\sqrt[3]{2^3}} = \dfrac{\sqrt[3]{4}}{2}$

47. $\dfrac{6}{\sqrt[3]{4}} = \dfrac{6}{\sqrt[3]{2^2}}\cdot\dfrac{\sqrt[3]{2}}{\sqrt[3]{2}} = \dfrac{6\sqrt[3]{2}}{\sqrt[3]{2^2}\sqrt[3]{2}}$

$= \dfrac{6\sqrt[3]{2}}{\sqrt[3]{2^3}} = \dfrac{6\sqrt[3]{2}}{2}$

$= 3\sqrt[3]{2}$

49. $\sqrt[3]{\dfrac{2}{3}} = \dfrac{\sqrt[3]{2}}{\sqrt[3]{3}} = \dfrac{\sqrt[3]{2}}{\sqrt[3]{3}}\cdot\dfrac{\sqrt[3]{3^2}}{\sqrt[3]{3^2}} = \dfrac{\sqrt[3]{2\cdot 3^2}}{\sqrt[3]{3^3}}$

$= \dfrac{\sqrt[3]{2\cdot 9}}{3} = \dfrac{\sqrt[3]{18}}{3}$

51. $\dfrac{4}{\sqrt[3]{x}} = \dfrac{4}{\sqrt[3]{x}}\cdot\dfrac{\sqrt[3]{x^2}}{\sqrt[3]{x^2}} = \dfrac{4\sqrt[3]{x^2}}{\sqrt[3]{x}\sqrt[3]{x^2}}$

$= \dfrac{4\sqrt[3]{x^2}}{\sqrt[3]{x^3}} = \dfrac{4\sqrt[3]{x^2}}{x}$

53. $\sqrt[3]{\dfrac{2}{y^2}} = \dfrac{\sqrt[3]{2}}{\sqrt[3]{y^2}} = \dfrac{\sqrt[3]{2}}{\sqrt[3]{y^2}}\cdot\dfrac{\sqrt[3]{y}}{\sqrt[3]{y}}$

$= \dfrac{\sqrt[3]{2y}}{\sqrt[3]{y^3}} = \dfrac{\sqrt[3]{2y}}{y}$

55. $\dfrac{7}{\sqrt[3]{2x^2}} = \dfrac{7}{\sqrt[3]{2x^2}} \cdot \dfrac{\sqrt[3]{2^2 x}}{\sqrt[3]{2^2 x}} = \dfrac{7\sqrt[3]{2^2 x}}{\sqrt[3]{2x^2}\,\sqrt[3]{2^2 x}}$

$\qquad = \dfrac{7\sqrt[3]{4x}}{\sqrt[3]{2^3 x^3}} = \dfrac{7\sqrt[3]{4x}}{2x}$

57. $\sqrt[3]{\dfrac{2}{xy^2}} = \dfrac{\sqrt[3]{2}}{\sqrt[3]{xy^2}} = \dfrac{\sqrt[3]{2}}{\sqrt[3]{xy^2}} \cdot \dfrac{\sqrt[3]{x^2 y}}{\sqrt[3]{x^2 y}}$

$\qquad = \dfrac{\sqrt[3]{2}\,\sqrt[3]{x^2 y}}{\sqrt[3]{xy^2}\,\sqrt[3]{x^2 y}} = \dfrac{\sqrt[3]{2x^2 y}}{\sqrt[3]{x^3 y^3}}$

$\qquad = \dfrac{\sqrt[3]{2x^2 y}}{xy}$

59. $\dfrac{3}{\sqrt[4]{x}} = \dfrac{3}{\sqrt[4]{x}} \cdot \dfrac{\sqrt[4]{x^3}}{\sqrt[4]{x^3}} = \dfrac{3\sqrt[4]{x^3}}{\sqrt[4]{xx^3}}$

$\qquad = \dfrac{3\sqrt[4]{x^3}}{\sqrt[4]{x^4}} = \dfrac{3\sqrt[4]{x^3}}{x}$

61. $\dfrac{6}{\sqrt[5]{8x^3}} = \dfrac{6}{\sqrt[5]{2^3 x^3}} \cdot \dfrac{\sqrt[5]{2^2 x^2}}{\sqrt[5]{2^2 x^2}} = \dfrac{6\sqrt[5]{4x^2}}{\sqrt[5]{2^5 x^5}}$

$\qquad = \dfrac{6\sqrt[5]{4x^2}}{2x} = \dfrac{3\sqrt[5]{4x^2}}{x}$

63. $\dfrac{2x^2 y}{\sqrt[5]{4x^2 y^4}} = \dfrac{2x^2 y}{\sqrt[5]{2^2 x^2 y^4}} \cdot \dfrac{\sqrt[5]{2^3 x^3 y}}{\sqrt[5]{2^3 x^3 y}}$

$\qquad = \dfrac{2x^2 y\sqrt[5]{8x^3 y}}{\sqrt[5]{2^5 x^5 y^5}}$

$\qquad = \dfrac{\cancel{2}x^{\cancel{2}}\,\cancel{y}\sqrt[5]{8x^3 y}}{\cancel{2}\cancel{x}\cancel{y}}$

$\qquad = x\sqrt[5]{8x^3 y}$

65. $\dfrac{9}{\sqrt{3x^2 y}} = \dfrac{9}{\sqrt{x^2 \cdot 3y}} = \dfrac{9}{x\sqrt{3y}}$

$\qquad = \dfrac{9}{x\sqrt{3y}} \cdot \dfrac{\sqrt{3y}}{\sqrt{3y}}$

$\qquad = \dfrac{9\sqrt{3y}}{x\sqrt{(3y)^2}} = \dfrac{9\sqrt{3y}}{x(3y)}$

$\qquad = \dfrac{\cancel{9} \cdot 3\sqrt{3y}}{\cancel{3}xy} = \dfrac{3\sqrt{3y}}{xy}$

67. $-\sqrt{\dfrac{75a^5}{b^3}} = -\dfrac{\sqrt{75a^5}}{\sqrt{b^3}} = -\dfrac{\sqrt{25a^4 \cdot 3a}}{\sqrt{b^2 \cdot b}}$

$\qquad = -\dfrac{5a^2\sqrt{3a}}{b\sqrt{b}} = -\dfrac{5a^2\sqrt{3a}}{b\sqrt{b}} \cdot \dfrac{\sqrt{b}}{\sqrt{b}}$

$\qquad = -\dfrac{5a^2\sqrt{3ab}}{b\sqrt{b^2}} = -\dfrac{5a^2\sqrt{3ab}}{b(b)}$

$\qquad = -\dfrac{5a^2\sqrt{3ab}}{b^2}$

69. $\sqrt{\dfrac{7m^2 n^3}{14m^3 n^2}} = \sqrt{\dfrac{n}{2m}} = \dfrac{\sqrt{n}}{\sqrt{2m}}$

$\qquad = \dfrac{\sqrt{n}}{\sqrt{2m}} \cdot \dfrac{\sqrt{2m}}{\sqrt{2m}} = \dfrac{\sqrt{2mn}}{\sqrt{(2m)^2}}$

$\qquad = \dfrac{\sqrt{2mn}}{2m}$

71. $\dfrac{3}{\sqrt[4]{x^5 y^3}} = \dfrac{3}{\sqrt[4]{x^4 \cdot xy^3}} = \dfrac{3}{x\sqrt[4]{xy^3}}$

$\qquad = \dfrac{3}{x\sqrt[4]{xy^3}} \cdot \dfrac{\sqrt[4]{x^3 y}}{\sqrt[4]{x^3 y}}$

$\qquad = \dfrac{3\sqrt[4]{x^3 y}}{x\sqrt[4]{x^4 y^4}} = \dfrac{3\sqrt[4]{x^3 y}}{x(xy)}$

$\qquad = \dfrac{3\sqrt[4]{x^3 y}}{x^2 y}$

73.
$$\frac{12}{\sqrt[3]{-8x^5y^8}} = \frac{12}{\sqrt[3]{-8x^3y^6 \cdot x^2y^2}}$$

$$= \frac{12}{-2xy^2\sqrt[3]{x^2y^2}}$$

$$= \frac{12}{-2xy^2\sqrt[3]{x^2y^2}} \cdot \frac{\sqrt[3]{xy}}{\sqrt[3]{xy}}$$

$$= \frac{12\sqrt[3]{xy}}{-2xy^2\sqrt[3]{x^3y^3}} = \frac{12\sqrt[3]{xy}}{-2xy^2(xy)}$$

$$= \frac{12\sqrt[3]{xy}}{-2x^2y^3}$$

$$= -\frac{6\sqrt[3]{xy}}{x^2y^3}$$

75.
$$\frac{8}{\sqrt{5}+2} = \frac{8}{\sqrt{5}+2} \cdot \frac{\sqrt{5}-2}{\sqrt{5}-2}$$

$$= \frac{8\sqrt{5}-8\cdot2}{\sqrt{5}\sqrt{5}-2\sqrt{5}+2\sqrt{5}-2\cdot2}$$

$$= \frac{8\sqrt{5}-16}{5-2\sqrt{5}+2\sqrt{5}-4}$$

$$= \frac{8\sqrt{5}-16}{5-4} = \frac{8\sqrt{5}-16}{1}$$

$$= 8\sqrt{5}-16$$

77.
$$\frac{13}{\sqrt{11}-3} = \frac{13}{\sqrt{11}-3} \cdot \frac{\sqrt{11}+3}{\sqrt{11}+3}$$

$$= \frac{13\left(\sqrt{11}+3\right)}{\left(\sqrt{11}-3\right)\left(\sqrt{11}+3\right)}$$

$$= \frac{13\sqrt{11}+13\cdot3}{\sqrt{11}\cdot\sqrt{11}+3\sqrt{11}-3\sqrt{11}-3\cdot3}$$

$$= \frac{13\sqrt{11}+39}{11+3\sqrt{11}-3\sqrt{11}-9}$$

$$= \frac{13\sqrt{11}+39}{11-9}$$

$$= \frac{13\sqrt{11}+39}{2}$$

79.
$$\frac{6}{\sqrt{5}+\sqrt{3}}$$

$$= \frac{6}{\sqrt{5}+\sqrt{3}} \cdot \frac{\sqrt{5}-\sqrt{3}}{\sqrt{5}-\sqrt{3}}$$

$$= \frac{6\left(\sqrt{5}-\sqrt{3}\right)}{\left(\sqrt{5}+\sqrt{3}\right)\left(\sqrt{5}-\sqrt{3}\right)}$$

$$= \frac{6\sqrt{5}-6\sqrt{3}}{\sqrt{5}\sqrt{5}-\sqrt{3}\sqrt{5}+\sqrt{3}\sqrt{5}-\sqrt{3}\sqrt{3}}$$

$$= \frac{6\sqrt{5}-6\sqrt{3}}{5-\sqrt{15}+\sqrt{15}-3}$$

$$= \frac{6\sqrt{5}-6\sqrt{3}}{5-3}$$

$$= \frac{6\sqrt{5}-6\sqrt{3}}{2}$$

$$= \frac{2\left(3\sqrt{5}-3\sqrt{3}\right)}{2}$$

$$= 3\sqrt{5}-3\sqrt{3}$$

81.
$$\frac{\sqrt{a}}{\sqrt{a}-\sqrt{b}} = \frac{\sqrt{a}}{\sqrt{a}-\sqrt{b}} \cdot \frac{\sqrt{a}+\sqrt{b}}{\sqrt{a}+\sqrt{b}}$$

$$= \frac{\sqrt{a}\left(\sqrt{a}+\sqrt{b}\right)}{\left(\sqrt{a}-\sqrt{b}\right)\left(\sqrt{a}+\sqrt{b}\right)}$$

$$= \frac{\sqrt{a}\sqrt{a}+\sqrt{a}\sqrt{b}}{\sqrt{a}\sqrt{a}+\sqrt{a}\sqrt{b}-\sqrt{a}\sqrt{b}-\sqrt{b}\sqrt{b}}$$

$$= \frac{a+\sqrt{ab}}{a-b}$$

83. $\dfrac{25}{5\sqrt{2}-3\sqrt{5}} = \dfrac{25}{5\sqrt{2}-3\sqrt{5}} \cdot \dfrac{5\sqrt{2}+3\sqrt{5}}{5\sqrt{2}+3\sqrt{5}} = \dfrac{25\left(5\sqrt{2}+3\sqrt{5}\right)}{\left(5\sqrt{2}-3\sqrt{5}\right)\left(5\sqrt{2}+3\sqrt{5}\right)}$

$= \dfrac{125\sqrt{2}+75\sqrt{5}}{5\cdot 5\sqrt{2\cdot 2}+5\cdot 3\sqrt{2\cdot 5}-5\cdot 3\sqrt{2\cdot 5}-3\cdot 3\sqrt{5\cdot 5}} = \dfrac{125\sqrt{2}+75\sqrt{5}}{25\cdot 2-9\cdot 5}$

$= \dfrac{125\sqrt{2}+75\sqrt{5}}{50-45} = \dfrac{125\sqrt{2}+75\sqrt{5}}{5} = \dfrac{5\left(25\sqrt{2}+15\sqrt{5}\right)}{5} = 25\sqrt{2}+15\sqrt{5}$

85. $\dfrac{\sqrt{5}+\sqrt{3}}{\sqrt{5}-\sqrt{3}} = \dfrac{\sqrt{5}+\sqrt{3}}{\sqrt{5}-\sqrt{3}} \cdot \dfrac{\sqrt{5}+\sqrt{3}}{\sqrt{5}+\sqrt{3}} = \dfrac{\left(\sqrt{5}+\sqrt{3}\right)^2}{\left(\sqrt{5}-\sqrt{3}\right)\left(\sqrt{5}+\sqrt{3}\right)}$

$= \dfrac{\left(\sqrt{5}\right)^2+2\sqrt{5}\sqrt{3}+\left(\sqrt{3}\right)^2}{\sqrt{5}\cdot\sqrt{5}+\sqrt{5}\cdot\sqrt{3}-\sqrt{5}\cdot\sqrt{3}-\sqrt{3}\sqrt{3}} = \dfrac{5+2\sqrt{15}+3}{5+\sqrt{15}-\sqrt{15}-3}$

$= \dfrac{8+2\sqrt{15}}{5-3} = \dfrac{2\left(4+\sqrt{15}\right)}{2} = 4+\sqrt{15}$

87. $\dfrac{\sqrt{x}+1}{\sqrt{x}+3} = \dfrac{\sqrt{x}+1}{\sqrt{x}+3} \cdot \dfrac{\sqrt{x}-3}{\sqrt{x}-3} = \dfrac{\sqrt{x}\cdot\sqrt{x}-3\sqrt{x}+1\sqrt{x}-3\cdot 1}{\sqrt{x}\cdot\sqrt{x}-3\sqrt{x}+3\sqrt{x}-3\cdot 3}$

$= \dfrac{\sqrt{x^2}+(-3+1)\sqrt{x}-3}{\sqrt{x^2}-9} = \dfrac{x+(-2)\sqrt{x}-3}{x-9} = \dfrac{x-2\sqrt{x}-3}{x-9}$

89. $\dfrac{5\sqrt{3}-3\sqrt{2}}{3\sqrt{2}-2\sqrt{3}} = \dfrac{5\sqrt{3}-3\sqrt{2}}{3\sqrt{2}-2\sqrt{3}} \cdot \dfrac{3\sqrt{2}+2\sqrt{3}}{3\sqrt{2}+2\sqrt{3}} = \dfrac{5\sqrt{3}\cdot 3\sqrt{2}+5\sqrt{3}\cdot 2\sqrt{3}-3\sqrt{2}\cdot 3\sqrt{2}-3\sqrt{2}\cdot 2\sqrt{3}}{3\sqrt{2}\cdot 3\sqrt{2}+3\sqrt{2}\cdot 2\sqrt{3}-3\sqrt{2}\cdot 2\sqrt{3}-2\sqrt{3}\cdot 2\sqrt{3}}$

$= \dfrac{15\sqrt{6}+10\cdot 3-9\cdot 2-6\sqrt{6}}{9\cdot 2+6\sqrt{6}-6\sqrt{6}-4\cdot 3} = \dfrac{15\sqrt{6}+30-18-6\sqrt{6}}{18-12}$

$= \dfrac{9\sqrt{6}+12}{6} = \dfrac{3\left(3\sqrt{6}+4\right)}{3\cdot 2} = \dfrac{3\sqrt{6}+4}{2}$

91. $\dfrac{2\sqrt{x}+\sqrt{y}}{\sqrt{y}-2\sqrt{x}} = \dfrac{2\sqrt{x}+\sqrt{y}}{\sqrt{y}-2\sqrt{x}} \cdot \dfrac{\sqrt{y}+2\sqrt{x}}{\sqrt{y}+2\sqrt{x}} = \dfrac{2\sqrt{x}\sqrt{y}+2\sqrt{x}\cdot 2\sqrt{x}+\sqrt{y}\sqrt{y}+2\sqrt{x}\sqrt{y}}{\sqrt{y}\sqrt{y}+2\sqrt{x}\sqrt{y}-2\sqrt{x}\sqrt{y}-2\sqrt{x}\cdot 2\sqrt{x}}$

$= \dfrac{2\sqrt{xy}+4\sqrt{x^2}+\sqrt{y^2}+2\sqrt{xy}}{\sqrt{y^2}+2\sqrt{xy}-2\sqrt{xy}-4\sqrt{x^2}} = \dfrac{2\sqrt{xy}+4x+y+2\sqrt{xy}}{y-4x} = \dfrac{4\sqrt{xy}+4x+y}{y-4x}$

93. $\sqrt{\dfrac{3}{2}} = \dfrac{\sqrt{3}}{\sqrt{2}} \cdot \dfrac{\sqrt{3}}{\sqrt{3}} = \dfrac{\sqrt{3}\sqrt{3}}{\sqrt{2}\sqrt{3}} = \dfrac{3}{\sqrt{6}}$

95. $\dfrac{\sqrt[3]{4x}}{\sqrt[3]{y}} = \dfrac{\sqrt[3]{2^2 x}}{\sqrt[3]{y}} \cdot \dfrac{\sqrt[3]{2x^2}}{\sqrt[3]{2x^2}} = \dfrac{\sqrt[3]{2^3 x^3}}{\sqrt[3]{2x^2 y}} = \dfrac{2x}{\sqrt[3]{2x^2 y}}$

97. $\dfrac{\sqrt{x}+3}{\sqrt{x}} = \dfrac{\sqrt{x}+3}{\sqrt{x}} \cdot \dfrac{\sqrt{x}-3}{\sqrt{x}-3} = \dfrac{\sqrt{x}\cdot\sqrt{x}-3\sqrt{x}+3\sqrt{x}-3\cdot3}{\sqrt{x}\cdot\sqrt{x}-3\sqrt{x}} = \dfrac{\sqrt{x^2}-9}{\sqrt{x^2}-3\sqrt{x}} = \dfrac{x-9}{x-3\sqrt{x}}$

99. $\dfrac{\sqrt{a}+\sqrt{b}}{\sqrt{a}-\sqrt{b}} = \dfrac{\sqrt{a}+\sqrt{b}}{\sqrt{a}-\sqrt{b}} \cdot \dfrac{\sqrt{a}-\sqrt{b}}{\sqrt{a}-\sqrt{b}} = \dfrac{\sqrt{a}\cdot\sqrt{a}-\sqrt{a}\sqrt{b}+\sqrt{a}\sqrt{b}-\sqrt{b}\sqrt{b}}{\sqrt{a}\cdot\sqrt{a}-\sqrt{a}\sqrt{b}-\sqrt{a}\sqrt{b}+\sqrt{b}\sqrt{b}}$

$$= \dfrac{\sqrt{a^2}-\sqrt{ab}+\sqrt{ab}-\sqrt{b^2}}{\sqrt{a^2}-\sqrt{ab}-\sqrt{ab}+\sqrt{b^2}} = \dfrac{a-b}{a-2\sqrt{ab}+b}$$

101. $\dfrac{\sqrt{x+5}-\sqrt{x}}{5} = \dfrac{\sqrt{x+5}-\sqrt{x}}{5} \cdot \dfrac{\sqrt{x+5}+\sqrt{x}}{\sqrt{x+5}+\sqrt{x}} = \dfrac{\left(\sqrt{x+5}\right)^2+\sqrt{x+5}\cdot\sqrt{x}-\sqrt{x+5}\cdot\sqrt{x}-\left(\sqrt{x}\right)^2}{5\left(\sqrt{x+5}+\sqrt{x}\right)}$

$$= \dfrac{x+5+\sqrt{x(x+5)}-\sqrt{x(x+5)}-x}{5\left(\sqrt{x+5}+\sqrt{x}\right)} = \dfrac{5}{5\left(\sqrt{x+5}+\sqrt{x}\right)} = \dfrac{1}{\sqrt{x+5}+\sqrt{x}}$$

103. $\dfrac{\sqrt{x}+\sqrt{y}}{x^2-y^2} = \dfrac{\sqrt{x}+\sqrt{y}}{x^2-y^2} \cdot \dfrac{\sqrt{x}-\sqrt{y}}{\sqrt{x}-\sqrt{y}} = \dfrac{\left(\sqrt{x}\right)^2-\sqrt{xy}+\sqrt{xy}-\left(\sqrt{y}\right)^2}{x^2\sqrt{x}-x^2\sqrt{y}-y^2\sqrt{x}+y^2\sqrt{y}}$

$$= \dfrac{x-y}{x^2\left(\sqrt{x}-\sqrt{y}\right)-y^2\left(\sqrt{x}-\sqrt{y}\right)} = \dfrac{x-y}{\left(\sqrt{x}-\sqrt{y}\right)\left(x^2-y^2\right)}$$

$$= \dfrac{x-y}{\left(\sqrt{x}-\sqrt{y}\right)(x+y)(x-y)} = \dfrac{1}{\left(\sqrt{x}-\sqrt{y}\right)(x+y)}$$

105. $\sqrt{2} + \dfrac{1}{\sqrt{2}} = \sqrt{2} + \dfrac{1}{\sqrt{2}} \cdot \dfrac{\sqrt{2}}{\sqrt{2}}$

$$= \sqrt{2} + \dfrac{\sqrt{2}}{2} = \dfrac{2\sqrt{2}}{2} + \dfrac{\sqrt{2}}{2}$$

$$= \dfrac{2\sqrt{2}+\sqrt{2}}{2} = \dfrac{3\sqrt{2}}{2}$$

107. $\sqrt[3]{25} - \dfrac{15}{\sqrt[3]{5}} = \sqrt[3]{25} - \dfrac{15}{\sqrt[3]{5}} \cdot \dfrac{\sqrt[3]{5^2}}{\sqrt[3]{5^2}}$

$$= \sqrt[3]{25} - \dfrac{15\sqrt[3]{25}}{5}$$

$$= \sqrt[3]{25} - 3\sqrt[3]{25} = -2\sqrt[3]{25}$$

109. $\sqrt{6} - \sqrt{\dfrac{1}{6}} + \sqrt{\dfrac{2}{3}}$

$= \sqrt{6} - \dfrac{\sqrt{1}}{\sqrt{6}} + \dfrac{\sqrt{2}}{\sqrt{3}}$

$= \sqrt{6} - \dfrac{1}{\sqrt{6}} \cdot \dfrac{\sqrt{6}}{\sqrt{6}} + \dfrac{\sqrt{2}}{\sqrt{3}} \cdot \dfrac{\sqrt{3}}{\sqrt{3}}$

$= \sqrt{6} - \dfrac{\sqrt{6}}{6} + \dfrac{\sqrt{6}}{3}$

$= \dfrac{6\sqrt{6}}{6} - \dfrac{\sqrt{6}}{6} + \dfrac{2\sqrt{6}}{6}$

$= \dfrac{6\sqrt{6} - \sqrt{6} + 2\sqrt{6}}{6} = \dfrac{7\sqrt{6}}{6}$

111. $\dfrac{2}{\sqrt{2} + \sqrt{3}} + \sqrt{75} - \sqrt{50}$

$= \dfrac{2}{\sqrt{2} + \sqrt{3}} \cdot \dfrac{\sqrt{2} - \sqrt{3}}{\sqrt{2} - \sqrt{3}} + \sqrt{25 \cdot 3} - \sqrt{25 \cdot 2}$

$= \dfrac{2\sqrt{2} - 2\sqrt{3}}{\left(\sqrt{2}\right)^2 - \left(\sqrt{3}\right)^2} + 5\sqrt{3} - 5\sqrt{2}$

$= \dfrac{2\sqrt{2} - 2\sqrt{3}}{2 - 3} + 5\sqrt{3} - 5\sqrt{2}$

$= \dfrac{2\sqrt{2} - 2\sqrt{3}}{-1} + 5\sqrt{3} - 5\sqrt{2}$

$= 2\sqrt{3} - 2\sqrt{2} + 5\sqrt{3} - 5\sqrt{2}$

$= 7\sqrt{3} - 7\sqrt{2}$

113. $f(x) = x^2 - 6x - 4$

$f\left(3 - \sqrt{13}\right) = \left(3 - \sqrt{13}\right)^2 - 6\left(3 - \sqrt{13}\right) - 4$

$\qquad = 9 - 6\sqrt{13} + 13 - 18 + 6\sqrt{13} - 4$

$\qquad = 0$

115. $f(x) = \sqrt{9 + x}$

$f\left(3\sqrt{5}\right) \cdot f\left(-3\sqrt{5}\right) = \sqrt{9 + 3\sqrt{5}} \cdot \sqrt{9 - 3\sqrt{5}}$

$\qquad = \sqrt{\left(9 + 3\sqrt{5}\right)\left(9 - 3\sqrt{5}\right)}$

$\qquad = \sqrt{9^2 - \left(3\sqrt{5}\right)^2}$

$\qquad = \sqrt{81 - 9 \cdot 5}$

$\qquad = \sqrt{81 - 45}$

$\qquad = \sqrt{36}$

$\qquad = 6$

117. $\dfrac{w}{h} = \dfrac{2}{\sqrt{5}-1}$

$\qquad = \dfrac{2}{\sqrt{5}-1} \cdot \dfrac{\sqrt{5}+1}{\sqrt{5}+1}$

$\qquad = \dfrac{2\left(\sqrt{5}+1\right)}{5-1}$

$\qquad = \dfrac{2\left(\sqrt{5}+1\right)}{4}$

$\qquad = \dfrac{\sqrt{5}+1}{2}$

$\qquad \approx 1.62$

The ratio is approximately 1.62 to 1.

119. $\text{Perimeter} = 2l + 2w$

$\qquad = 2\left(\sqrt{8}+1\right) + 2\left(\sqrt{8}-1\right)$

$\qquad = 2\sqrt{8} + \cancel{2} + 2\sqrt{8} - \cancel{2}$

$\qquad = (2+2)\sqrt{8} = 4\sqrt{8}$

$\qquad = 4\sqrt{4\cdot 2} = 4 \cdot 2\sqrt{2} = 8\sqrt{2}$

The perimeter is $8\sqrt{2}$ inches.

$\text{Area} = lw = \left(\sqrt{8}+1\right)\left(\sqrt{8}-1\right)$

$\qquad = \left(\sqrt{8}\right)^2 - \cancel{\sqrt{8}} + \cancel{\sqrt{8}} - 1$

$\qquad = 8 - 1 = 7$

The area is 7 square inches.

121. $c^2 = a^2 + b^2$

$c^2 = \left(\sqrt{10}+\sqrt{2}\right)^2 + \left(\sqrt{10}-\sqrt{2}\right)^2$

$c^2 = \left(\sqrt{10}\right)^2 + 2\sqrt{10}\sqrt{2} + \left(\sqrt{2}\right)^2 + \left(\sqrt{10}\right)^2 - 2\sqrt{10}\sqrt{2} + \left(\sqrt{2}\right)^2$

$c^2 = 10 + 2\sqrt{20} + 2 + 10 - 2\sqrt{20} + 2$

$c^2 = 24$

$c = \sqrt{24}$

$c = 2\sqrt{6}$ inches

123. – 129. Answers will vary.

131. $\left(\sqrt{x}-1\right)\left(\sqrt{x}-1\right)=x+1$

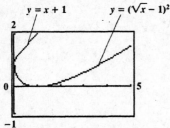

The graphs do not coincide. Correct the simplification.

$$\left(\sqrt{x}-1\right)\left(\sqrt{x}-1\right)$$
$$=\left(\sqrt{x}\right)^2-\sqrt{x}-\sqrt{x}+1$$
$$=x-2\sqrt{x}+1$$

133. $\left(\sqrt{x}+1\right)^2=x+1$

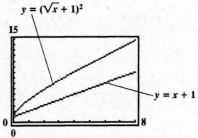

The graphs do not coincide. Correct the simplification.

$$\left(\sqrt{x}+1\right)^2=\left(\sqrt{x}\right)^2+2\sqrt{x}\cdot 1+1^2$$
$$=x+2\sqrt{x}+1$$

135. makes sense

137. does not make sense; Explanations will vary. Sample explanation: Multiplying a radical expression and its conjugate will clear the expression of any radicals.

139. false; Changes to make the statement true will vary. A sample change is: $\dfrac{\sqrt{3}+7}{\sqrt{3}-2}=\dfrac{\sqrt{3}+7}{\sqrt{3}-2}\cdot\dfrac{\sqrt{3}+2}{\sqrt{3}+2}$

$$=\dfrac{\left(\sqrt{3}\right)^2+2\sqrt{3}+7\sqrt{3}+14}{\left(\sqrt{3}\right)^2-2^2}$$

$$=\dfrac{3+9\sqrt{3}+14}{3-4}=\dfrac{17+9\sqrt{3}}{-1}$$

$$=-\left(17+9\sqrt{3}\right)=-17-9\sqrt{3}$$

141. true

143. $7\left[(2x-5)-(x+1)\right]=\left(\sqrt{7}+2\right)\left(\sqrt{7}-2\right)$

$$7\left[2x-5-x-1\right]=\left(\sqrt{7}\right)^2-(2)^2$$

$$7\left(x-6\right)=7-4$$

$$7x-42=3$$

$$7x=45$$

$$x=\frac{45}{7}$$

The solution is $\frac{45}{7}$ and the solution set is $\left\{\frac{45}{7}\right\}$.

145.

$$\frac{1}{\sqrt{2}+\sqrt{3}+\sqrt{4}}$$

$$=\frac{1}{\left(\sqrt{2}+\sqrt{3}\right)+2}\cdot\frac{\left(\sqrt{2}+\sqrt{3}\right)-2}{\left(\sqrt{2}+\sqrt{3}\right)-2}$$

$$=\frac{\sqrt{2}+\sqrt{3}-2}{\left(\sqrt{2}+\sqrt{3}\right)^2-2^2}=\frac{\sqrt{2}+\sqrt{3}-2}{2+2\sqrt{6}+3-4}$$

$$=\frac{\sqrt{2}+\sqrt{3}-2}{2\sqrt{6}+1}\cdot\frac{2\sqrt{6}-1}{2\sqrt{6}-1}$$

$$=\frac{2\sqrt{12}+2\sqrt{18}-4\sqrt{6}-\sqrt{2}-\sqrt{3}+2}{\left(2\sqrt{6}\right)^2-1^2}$$

$$=\frac{4\sqrt{3}+6\sqrt{2}-4\sqrt{6}-\sqrt{2}-\sqrt{3}+2}{4\cdot6-1}$$

$$=\frac{3\sqrt{3}+5\sqrt{2}-4\sqrt{6}+2}{24+1}$$

$$=\frac{3\sqrt{3}+5\sqrt{2}-4\sqrt{6}+2}{24-1}$$

$$=\frac{3\sqrt{3}+5\sqrt{2}-4\sqrt{6}+2}{23}$$

146.

$$\frac{2}{x-2}+\frac{3}{x^2-4}$$

$$=\frac{2}{x-2}+\frac{3}{(x+2)(x-2)}$$

$$=\frac{2(x+2)}{(x-2)(x+2)}+\frac{3}{(x+2)(x-2)}$$

$$=\frac{2(x+2)+3}{(x-2)(x+2)}=\frac{2x+4+3}{(x-2)(x+2)}$$

$$=\frac{2x+7}{(x-2)(x+2)}\text{ or }\frac{2x+7}{x^2-4}$$

147. $3x-4\le2$ and $4x+5\ge5$

$$3x\le6 \qquad\qquad 4x\ge0$$

$$x\le2 \qquad\qquad x\ge0$$

$$0\le x\le2$$

The solution interval is $\left[0,2\right]$.

148. **a.** The relation is a function.

   b. The relation is not a function.

149. $\left(\sqrt{x+4}+1\right)^2=\left(\sqrt{x+4}\right)^2+2\sqrt{x+4}+1^2$

$$=x+4+2\sqrt{x+4}+1$$

$$=x+5+2\sqrt{x+4}$$

150. $4x^2-16x+16=4(x+4)$

$$4x^2-16x+16=4x+16$$

$$4x^2-20x=0$$

$$4x(x-5)=0$$

$$x=0 \quad\text{or}\quad x-5=0$$

$$x=5$$

The solution set is $\left\{0,5\right\}$.

151. $26-11x=16-8x+x^2$

$$0=-10+3x+x^2$$

$$0=x^2+3x-10$$

$$0=(x+5)(x-2)$$

$$x+5=0 \quad\text{or}\quad x-2=0$$

$$x=-5 \qquad\qquad x=2$$

The solution set is $\left\{-5,2\right\}$.

10.6 Check Points

1. $\sqrt{3x+4} = 8$

$\left(\sqrt{3x+4}\right)^2 = 8^2$

$3x+4 = 64$

$3x = 60$

$x = 20$

Check:

$\sqrt{3x+4} = 8$

$\sqrt{3 \cdot 20 + 4} = 8$

$\sqrt{60+4} = 8$

$\sqrt{64} = 8$

$8 = 8$

The solution set is $\{20\}$.

2. $\sqrt{x-1} + 7 = 2$

$\sqrt{x-1} = -5$

A principal square root cannot be negative. Thus, this equation has no solution.
The solution set is $\{\ \}$.

3. $\sqrt{6x+7} - x = 2$

$\sqrt{6x+7} = x+2$

$\left(\sqrt{6x+7}\right)^2 = (x+2)^2$

$6x+7 = x^2 + 4x + 4$

$0 = x^2 - 2x - 3$

$0 = (x+1)(x-3)$

$x+1 = 0 \quad$ or $\quad x-3 = 0$

$x = -1 \qquad\qquad x = 3$

Check:

$\sqrt{6x+7} - x = 2$

$\sqrt{6(-1)+7} - (-1) = 2$

$\sqrt{1} + 1 = 2$

$1 + 1 = 2$

$2 = 2$

$\sqrt{6x+7} - x = 2$

$\sqrt{6(3)+7} - (3) = 2$

$\sqrt{25} - 3 = 2$

$5 - 3 = 2$

$2 = 2$

The solution set is $\{-1, 3\}$.

4. $\sqrt{x+5} - \sqrt{x-3} = 2$

$\sqrt{x+5} = \sqrt{x-3} + 2$

$\left(\sqrt{x+5}\right)^2 = \left(\sqrt{x-3} + 2\right)^2$

$x+5 = \left(\sqrt{x-3}\right)^2 + 2 \cdot 2\sqrt{x-3} + 2^2$

$x+5 = x-3 + 4\sqrt{x-3} + 4$

$x+5 = x+1 + 4\sqrt{x-3}$

$4 = 4\sqrt{x-3}$

$\dfrac{4}{4} = \dfrac{4\sqrt{x-3}}{4}$

$1 = \sqrt{x-3}$

$1^2 = \left(\sqrt{x-3}\right)^2$

$1 = x-3$

$4 = x$

Check:

$\sqrt{x+5} - \sqrt{x-3} = 2$

$\sqrt{4+5} - \sqrt{4-3} = 2$

$\sqrt{9} - \sqrt{1} = 2$

$3 - 1 = 2$

$2 = 2$

The solution set is $\{4\}$.

5. $(2x-3)^{\frac{1}{3}} + 3 = 0$

$(2x-3)^{\frac{1}{3}} = -3$

$\left((2x-3)^{\frac{1}{3}}\right)^3 = (-3)^3$

$2x-3 = -27$

$2x = -24$

$x = -12$

Check:

$(2x-3)^{\frac{1}{3}} + 3 = 0$

$(2(-12)-3)^{\frac{1}{3}} + 3 = 0$

$(-27)^{\frac{1}{3}} + 3 = 0$

$\sqrt[3]{-27} + 3 = 0$

$-3 + 3 = 0$

$0 = 0$

The solution set is $\{-12\}$.

6. $f(x) = 3.5\sqrt{x} + 38$

$73 = 3.5\sqrt{x} + 38$

$35 = 3.5\sqrt{x}$

$\dfrac{35}{3.5} = \dfrac{3.5\sqrt{x}}{3.5}$

$10 = \sqrt{x}$

$10^2 = \left(\sqrt{x}\right)^2$

$100 = x$

The model projects that 73% of U.S. women will participate in the work force 100 years after 1960, or 2060.

10.6 Exercise Set

1. $\sqrt{3x-2} = 4$

$\left(\sqrt{3x-2}\right)^2 = 4^2$

$3x - 2 = 16$

$3x = 18$

$x = 6$

Check:

$\sqrt{3(6)-2} = 4$

$\sqrt{18-2} = 4$

$\sqrt{16} = 4$

$4 = 4$

The solution set is $\{6\}$.

3. $\sqrt{5x-4} - 9 = 0$

$\sqrt{5x-4} = 9$

$\left(\sqrt{5x-4}\right)^2 = 9^2$

$5x - 4 = 81$

$5x = 85$

$x = 17$

Check:

$\sqrt{5(17)-4} - 9 = 0$

$\sqrt{85-4} - 9 = 0$

$\sqrt{81} - 9 = 0$

$9 - 9 = 0$

$0 = 0$

The solution set is $\{17\}$.

5. $\sqrt{3x+7} + 10 = 4$

$\sqrt{3x+7} = -6$

Since the square root of a number is always positive, the solution set is $\{\ \}$ or $\varnothing$.

7. $x = \sqrt{7x+8}$

$x^2 = \left(\sqrt{7x+8}\right)^2$

$x^2 = 7x + 8$

$x^2 - 7x - 8 = 0$

$(x-8)(x+1) = 0$

Apply the zero product principle.

$x - 8 = 0$ or $x + 1 = 0$

$x = 8$ $x = -1$

Check:

$8 = \sqrt{7(8)+8}$ $-1 = \sqrt{7(-1)+8}$

$8 = \sqrt{56+8}$

$8 = \sqrt{64}$

$8 = 8$

We disregard -1 because square roots are always positive. The solution set is $\{8\}$.

9. $\sqrt{5x+1} = x + 1$

$\left(\sqrt{5x+1}\right)^2 = (x+1)^2$

$5x + 1 = x^2 + 2x + 1$

$0 = x^2 - 3x$

$0 = x(x-3)$

Apply the zero product principle.

$x = 0$ or $x - 3 = 0$

$x = 3$

Both solutions check. The solution set is $\{0, 3\}$.

11.
$$x = \sqrt{2x-2} + 1$$
$$x - 1 = \sqrt{2x-2}$$
$$(x-1)^2 = \left(\sqrt{2x-2}\right)^2$$
$$x^2 - 2x + 1 = 2x - 2$$
$$x^2 - 4x + 3 = 0$$
$$(x-3)(x-1) = 0$$
Apply the zero product principle.
$$x - 3 = 0 \quad \text{or} \quad x - 1 = 0$$
$$x = 3 \qquad\qquad x = 1$$

Both solutions check. The solution set is $\{1, 3\}$.

13.
$$x - 2\sqrt{x-3} = 3$$
$$x - 3 = 2\sqrt{x-3}$$
$$(x-3)^2 = \left(2\sqrt{x-3}\right)^2$$
$$x^2 - 6x + 9 = 4(x-3)$$
$$x^2 - 6x + 9 = 4x - 12$$
$$x^2 - 10x + 21 = 0$$
$$(x-7)(x-3) = 0$$
Apply the zero product principle.
$$x - 7 = 0 \quad \text{or} \quad x - 3 = 0$$
$$x = 7 \qquad\qquad x = 3$$

Both solutions check. The solution set is $\{3, 7\}$.

15.
$$\sqrt{2x-5} = \sqrt{x+4}$$
$$\left(\sqrt{2x-5}\right)^2 = \left(\sqrt{x+4}\right)^2$$
$$2x - 5 = x + 4$$
$$x - 5 = 4$$
$$x = 9$$
The solution checks. The solution set is $\{9\}$.

17.
$$\sqrt[3]{2x+11} = 3$$
$$\left(\sqrt[3]{2x+11}\right)^3 = 3^3$$
$$2x + 11 = 27$$
$$2x = 16$$
$$x = 8$$
The solution checks. The solution set is $\{8\}$.

19.
$$\sqrt[3]{2x-6} - 4 = 0$$
$$\sqrt[3]{2x-6} = 4$$
$$\left(\sqrt[3]{2x-6}\right)^3 = 4^3$$
$$2x - 6 = 64$$
$$2x = 70$$
$$x = 35$$
The solution checks. The solution set is $\{35\}$.

21.
$$\sqrt{x-7} = 7 - \sqrt{x}$$
$$\left(\sqrt{x-7}\right)^2 = \left(7 - \sqrt{x}\right)^2$$
$$\cancel{x} - 7 = 49 - 14\sqrt{x} + \cancel{x}$$
$$-7 = 49 - 14\sqrt{x}$$
$$-56 = -14\sqrt{x}$$
$$\frac{-56}{-14} = \frac{-14\sqrt{x}}{-14}$$
$$4 = \sqrt{x}$$
$$4^2 = \left(\sqrt{x}\right)^2$$
$$16 = x$$
The solution checks. The solution set is $\{16\}$.

23.
$$\sqrt{x+2} + \sqrt{x-1} = 3$$
$$\sqrt{x+2} = 3 - \sqrt{x-1}$$
$$\left(\sqrt{x+2}\right)^2 = \left(3 - \sqrt{x-1}\right)^2$$
$$\cancel{x} + 2 = 9 - 6\sqrt{x-1} + \cancel{x} - 1$$
$$2 = 8 - 6\sqrt{x-1}$$
$$-6 = -6\sqrt{x-1}$$
$$\frac{-6}{-6} = \frac{-6\sqrt{x-1}}{-6}$$
$$1 = \sqrt{x-1}$$
$$1^2 = \left(\sqrt{x-1}\right)^2$$
$$1 = x - 1$$
$$2 = x$$
The solution checks. The solution set is $\{2\}$.

25. $2\sqrt{4x+1}-9=x-5$

$2\sqrt{4x+1}=x+4$

$\left(2\sqrt{4x+1}\right)^2=(x+4)^2$

$2^2\left(\sqrt{4x+1}\right)^2=x^2+8x+16$

$4(4x+1)=x^2+8x+16$

$16x+4=x^2+8x+16$

$0=x^2-8x+12$

$0=(x-6)(x-2)$

$x-6=0$ or $x-2=0$

$x=6$ $x=2$

Check $x=6$:

$2\sqrt{4(6)+1}-9=6-5$

$2\sqrt{25}-9=1$

$2(5)-9=1$

$10-9=1$

$1=1$

Check $x=2$:

$2\sqrt{4(2)+1}-9=2-5$

$2\sqrt{8+1}-9=-3$

$2\sqrt{9}-9=-3$

$2(3)-9=-3$

$6-9=-3$

$-3=-3$

Both solutions check. The solution set is $\{2,6\}$.

27. $(2x+3)^{1/3}+4=6$

$(2x+3)^{1/3}=2$

$\left((2x+3)^{1/3}\right)^3=2^3$

$2x+3=8$

$2x=5$

$x=\dfrac{5}{2}$

The solution checks. The solution set is $\left\{\dfrac{5}{2}\right\}$.

29. $(3x+1)^{1/4}+7=9$

$(3x+1)^{1/4}=2$

$\left((3x+1)^{1/4}\right)^4=2^4$

$3x+1=16$

$3x=15$

$x=5$

The solution checks. The solution set is $\{5\}$.

31. $(x+2)^{1/2}+8=4$

$(x+2)^{1/2}=-4$

$\sqrt{x+2}=-4$

The square root of a number must be positive. The solution set is $\varnothing$.

33. $\sqrt{2x-3}-\sqrt{x-2}=1$

$\sqrt{2x-3}=\sqrt{x-2}+1$

$\left(\sqrt{2x-3}\right)^2=\left(\sqrt{x-2}+1\right)^2$

$2x-3=x-2+2\sqrt{x-2}+1$

$2x-3=x-1+2\sqrt{x-2}$

$x-2=2\sqrt{x-2}$

$(x-2)^2=\left(2\sqrt{x-2}\right)^2$

$x^2-4x+4=4(x-2)$

$x^2-4x+4=4x-8$

$x^2-8x+12=0$

$(x-6)(x-2)=0$

$x-6=0$ $x-2=0$

$x=6$ $x=2$

Both solutions check. The solution set is $\{2,6\}$.

35. $3x^{1/3} = \left(x^2 + 17x\right)^{1/3}$

$$\left(3x^{1/3}\right)^3 = \left(\left(x^2 + 17x\right)^{1/3}\right)^3$$

$$3^3 x = x^2 + 17x$$

$$27x = x^2 + 17x$$

$$0 = x^2 - 10x$$

$$0 = x(x - 10)$$

$$x = 0 \qquad x - 10 = 0$$

$$x = 10$$

Both solutions check. The solution set is $\{0, 10\}$.

37. $(x + 8)^{1/4} = (2x)^{1/4}$

$$\left((x + 8)^{1/4}\right)^4 = \left((2x)^{1/4}\right)^4$$

$$x + 8 = 2x$$

$$8 = x$$

The solution checks. The solution set is $\{8\}$.

39. $f(x) = x + \sqrt{x + 5}$

$$7 = x + \sqrt{x + 5}$$

$$7 - x = \sqrt{x + 5}$$

$$(7 - x)^2 = \left(\sqrt{x + 5}\right)^2$$

$$49 - 14x + x^2 = x + 5$$

$$x^2 - 15x + 44 = 0$$

$$(x - 11)(x - 4) = 0$$

$$x - 11 = 0 \quad \text{or} \quad x - 4 = 0$$

$$x = 11 \qquad\qquad x = 4$$

Check $x = 11$: $11 + \sqrt{11 + 5} = 11 + \sqrt{16}$
$$= 15 \neq 7$$

Check $x = 4$: $4 + \sqrt{4 + 5} = 4 + \sqrt{9}$
$$= 7$$

Discard 11. The solution is 4.

41. $f(x) = (5x + 16)^{1/3}$; $g(x) = (x - 12)^{1/3}$

$$(5x + 16)^{1/3} = (x - 12)^{1/3}$$

$$\left[(5x + 16)^{1/3}\right]^3 = \left[(x - 12)^{1/3}\right]^3$$

$$5x + 16 = x - 12$$

$$4x = -28$$

$$x = -7$$

The solution is -7.

43. $r = \sqrt{\dfrac{3V}{\pi h}}$

$$r^2 = \left(\sqrt{\frac{3V}{\pi h}}\right)^2$$

$$r^2 = \frac{3V}{\pi h}$$

$$\pi r^2 h = 3V$$

$$V = \frac{\pi r^2 h}{3} \quad \text{or} \quad V = \frac{1}{3}\pi r^2 h$$

45. $t = 2\pi\sqrt{\dfrac{l}{32}}$

$$\frac{t}{2\pi} = \sqrt{\frac{l}{32}}$$

$$\left(\frac{t}{2\pi}\right)^2 = \left(\sqrt{\frac{l}{32}}\right)^2$$

$$\frac{t^2}{4\pi^2} = \frac{l}{32}$$

$$\frac{32t^2}{4\pi^2} = l$$

$$\frac{8t^2}{\pi^2} = l \quad \text{or} \quad l = \frac{8t^2}{\pi^2}$$

47. Let x = the number.

$$\sqrt{5x-4} = x-2$$
$$\left(\sqrt{5x-4}\right)^2 = (x-2)^2$$
$$5x-4 = x^2 - 4x + 4$$
$$0 = x^2 - 9x + 8$$
$$0 = (x-8)(x-1)$$
$$x-8 = 0 \quad \text{or} \quad x-1 = 0$$
$$x = 8 \qquad\qquad x = 1$$

Check $x = 8$:　$\sqrt{5(8)-4} = 8-2$
$$\sqrt{40-4} = 6$$
$$\sqrt{36} = 6$$
$$6 = 6$$

Check $x = 1$:　$\sqrt{5(1)-4} = 1-2$
$$\sqrt{5-4} = -1$$
$$\sqrt{1} = -1$$
$$1 \neq -1$$

Discard $x = 1$. The number is 8.

49. $f(x) = \sqrt{x+16} - \sqrt{x} - 2$

To find the x-intercepts, set the function equal to 0 and solve for x.

$$0 = \sqrt{x+16} - \sqrt{x} - 2$$
$$\sqrt{x} + 2 = \sqrt{x+16}$$
$$\left(\sqrt{x}+2\right)^2 = \left(\sqrt{x+16}\right)^2$$
$$x + 4\sqrt{x} + 4 = x + 16$$
$$4\sqrt{x} = 12$$
$$\sqrt{x} = 3$$
$$\left(\sqrt{x}\right)^2 = 3^2$$
$$x = 9$$

Check $x = 9$:

$$\sqrt{9+16} - \sqrt{9} - 2$$
$$= \sqrt{25} - \sqrt{9} - 2$$
$$= 5 - 3 - 2$$
$$= 0$$

The only x-intercept is 9.

51.
$$t = \frac{\sqrt{d}}{2}$$
$$1.16 = \frac{\sqrt{d}}{2}$$
$$2.32 = \sqrt{d}$$
$$2.32^2 = \left(\sqrt{d}\right)^2$$
$$d \approx 5.4$$

The vertical distance was about 5.4 feet.

53. It is represented by the point $(5.4, 1.16)$.

55. a. $f(x) = -4.4\sqrt{x} + 38$
$$f(25) = -4.4\sqrt{25} + 38 = 16$$

16% of Americans earning $25 thousand annually report fair or poor health. This underestimates the percent displayed in the graph by 1%.

b. $f(x) = -4.4\sqrt{x} + 38$
$$14 = -4.4\sqrt{x} + 38$$
$$-24 = -4.4\sqrt{x}$$
$$\frac{-24}{-4.4} = \frac{-4.4\sqrt{x}}{-4.4}$$
$$\frac{24}{4.4} = \sqrt{x}$$
$$\left(\frac{24}{4.4}\right)^2 = \left(\sqrt{x}\right)^2$$
$$x \approx 30$$

$30 thousand is the annual income that corresponds to 14% reporting fair or poor health.

57.
$$87 = 29x^{1/3}$$
$$\frac{87}{29} = \frac{29x^{1/3}}{29}$$
$$3 = x^{1/3}$$
$$3^3 = \left(x^{1/3}\right)^3$$
$$27 = x$$

A Galapagos island with an area of 27 square miles will have 87 plant species.

59.
$$365 = 0.2x^{3/2}$$
$$\frac{365}{0.2} = \frac{0.2x^{3/2}}{0.2}$$
$$1825 = x^{3/2}$$
$$1825^2 = \left(x^{3/2}\right)^2$$
$$3,330,625 = x^3$$
$$\sqrt[3]{3,330,625} = \sqrt[3]{x^3}$$
$$149.34 \approx x$$

The average distance of the Earth from the sun is approximately 149 million kilometers.

61. – 67. Answers will vary.

69. $\sqrt{x} + 3 = 5$

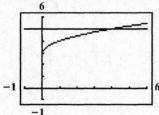

The solution set is $\{4\}$.

71. $4\sqrt{x} = x + 3$

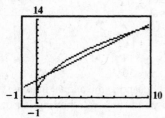

The solution set is $\{1, 9\}$.

73. does not make sense; Explanations will vary. Sample explanation: You should always substitute into the original equation. Later equations in the solution of the problem may be after extraneous roots were introduced. Substituting into such an equation will not allow you to rule out extraneous roots.

75. does not make sense; Explanations will vary. Sample explanation: Raising both sides to the first power does not introduce extraneous solutions.

77. false; Changes to make the statement true will vary. A sample change is: The first step is to square both sides, obtaining $x + 6 = x^2 + 4x + 4$.

79. true

81.
$$\left(\sqrt{x-7}\right)^2 + \left(\sqrt{x}\right)^2 = \left(1 + \sqrt{x}\right)^2$$
$$\cancel{x} - 7 + x = 1 + 2\sqrt{x} + \cancel{x}$$
$$-7 + x = 1 + 2\sqrt{x}$$
$$-8 + x = 2\sqrt{x}$$
$$(-8 + x)^2 = \left(2\sqrt{x}\right)^2$$
$$64 - 16x + x^2 = 4x$$
$$x^2 - 16x + 64 = 4x$$
$$x^2 - 20x + 64 = 0$$
$$(x - 16)(x - 4) = 0$$
$$x - 16 = 0 \quad \text{or} \quad x - 4 = 0$$
$$x = 16 \qquad \qquad \cancel{x = 4}$$

We disregard 4. If $x = 4$, one of the legs becomes $\sqrt{4-7} = \sqrt{-3}$.

The legs of the triangle are:
$$\sqrt{x-7} = \sqrt{16-7} = \sqrt{9} = 3$$
$$\sqrt{x} = \sqrt{16} = 4, \text{ and}$$
$$1 + \sqrt{x} = 1 + \sqrt{16} = 1 + 4 = 5.$$

83. $\sqrt{\sqrt{x}+\sqrt{x+9}} = 3$

$\left(\sqrt{\sqrt{x}+\sqrt{x+9}}\right)^2 = 3^2$

$\sqrt{x}+\sqrt{x+9} = 9$

$\sqrt{x+9} = 9-\sqrt{x}$

$\left(\sqrt{x+9}\right)^2 = \left(9-\sqrt{x}\right)^2$

$\cancel{x}+9 = 81-18\sqrt{x}+\cancel{x}$

$9 = 81-18\sqrt{x}$

$-72 = -18\sqrt{x}$

$4 = \sqrt{x}$

$4^2 = \left(\sqrt{x}\right)^2$

$16 = x$

Check:

$\sqrt{\sqrt{x}+\sqrt{x+9}} = 3$

$\sqrt{\sqrt{16}+\sqrt{16+9}} = 3$

$\sqrt{4+\sqrt{25}} = 3$

$\sqrt{4+5} = 3$

$\sqrt{9} = 3$

$3 = 3$

The solution checks. The solution set is $\{16\}$.

85.

$$\begin{array}{r|rrrr} -3 & 4 & -3 & 2 & -1 & -1 \\ & & -12 & 45 & -141 & 426 \\ \hline & 4 & -15 & 47 & -142 & 425 \end{array}$$

$\dfrac{4x^4 -3x^3 +2x^2 -x-1}{x+3}$

$= 4x^3 -15x^2 +47x -142 +\dfrac{425}{x+3}$

86. $\dfrac{3x^2 -12}{x^2 +2x-8} \div \dfrac{6x+18}{x+4} = \dfrac{3x^2 -12}{x^2 +2x-8} \cdot \dfrac{x+4}{6x+18}$

$= \dfrac{3\left(x^2 -4\right)}{\cancel{(x+4)}(x-2)} \cdot \dfrac{\cancel{x+4}}{6(x+3)}$

$= \dfrac{3(x+2)\cancel{(x-2)}}{\cancel{(x-2)}} \cdot \dfrac{1}{6(x+3)}$

$= \dfrac{3(x+2)}{1} \cdot \dfrac{1}{6(x+3)}$

$= \dfrac{3(x+2)}{6(x+3)} = \dfrac{x+2}{2(x+3)}$

87. $2x^2 + 12xy + 18y^2 = 2(x^2 + 6xy + 9y^2)$
$$= 2(x + 3y)^2$$

88. $(-5 + 7x) - (-11 - 6x) = -5 + 7x + 11 + 6x$
$$= 6 + 13x$$

89. $(7 - 3x)(-2 - 5x) = (7)(-2) + (7)(-5x) + (-3x)(-2) + (-3x)(-5x)$
$$= -14 - 35x + 6x + 15x^2$$
$$= 15x^2 - 29x - 14$$

90. $\dfrac{7 + 4\sqrt{2}}{2 - 5\sqrt{2}} = \dfrac{7 + 4\sqrt{2}}{2 - 5\sqrt{2}} \cdot \dfrac{2 + 5\sqrt{2}}{2 + 5\sqrt{2}}$

$$= \frac{\left(7 + 4\sqrt{2}\right)\left(2 + 5\sqrt{2}\right)}{\left(2 - 5\sqrt{2}\right)\left(2 + 5\sqrt{2}\right)}$$

$$= \frac{7 \cdot 2 + 7 \cdot 5\sqrt{2} + 4\sqrt{2} \cdot 2 + 4\sqrt{2} \cdot 5\sqrt{2}}{(2)^2 - \left(5\sqrt{2}\right)^2}$$

$$= \frac{14 + 35\sqrt{2} + 8\sqrt{2} + 20\sqrt{4}}{(2)^2 - \left(5\sqrt{2}\right)^2}$$

$$= \frac{14 + 43\sqrt{2} + 20 \cdot 2}{4 - (5)^2 \left(\sqrt{2}\right)^2}$$

$$= \frac{14 + 43\sqrt{2} + 40}{4 - 25 \cdot 2}$$

$$= \frac{54 + 43\sqrt{2}}{-46}$$

$$= -\frac{54 + 43\sqrt{2}}{46}$$

10.7 Check Points

1. a. $\sqrt{-64} = \sqrt{64 \cdot -1} = \sqrt{64} \cdot \sqrt{-1} = 8i$

b. $\sqrt{-11} = \sqrt{11 \cdot -1} = \sqrt{11} \cdot \sqrt{-1} = i\sqrt{11}$

c. $\sqrt{-48} = \sqrt{48 \cdot -1}$
$$= \sqrt{48} \cdot \sqrt{-1}$$
$$= \sqrt{16} \cdot \sqrt{3} \cdot \sqrt{-1}$$
$$= 4 \cdot \sqrt{3} \cdot i$$
$$= 4i\sqrt{3}$$

2. a. $(5-2i)+(3+3i)=5-2i+3+3i$

$$=8+i$$

b. $(2+6i)-(12-4i)=2+6i-12+4i$

$$=-10+10i$$

3. a. $7i(2-9i)=7i\cdot2-7i\cdot9i$

$$=14i-63i^2$$

$$=14i-63(-1)$$

$$=63+14i$$

b. $(5+4i)(6-7i)=30-35i+24i-28i^2$

$$=30-35i+24i-28(-1)$$

$$=30+28-35i+24i$$

$$=58-11i$$

4. $\sqrt{-5}\cdot\sqrt{-7}=\sqrt{5}\sqrt{-1}\cdot\sqrt{7}\sqrt{-1}$

$$=i\sqrt{5}\cdot i\sqrt{7}$$

$$=i^2\sqrt{35}$$

$$=(-1)\sqrt{35}$$

$$=-\sqrt{35}$$

5. $\dfrac{6+2i}{4-3i}=\dfrac{6+2i}{4-3i}\cdot\dfrac{4+3i}{4+3i}$

$$=\frac{24+18i+8i+6i^2}{4^2-(3i)^2}$$

$$=\frac{24+18i+8i+6(-1)}{16-9i^2}$$

$$=\frac{18+26i}{16-9(-1)}$$

$$=\frac{18+26i}{25}$$

$$=\frac{18}{25}+\frac{26}{25}i$$

6. $\dfrac{3-2i}{4i}=\dfrac{3-2i}{4i}\cdot\dfrac{-4i}{-4i}$

$$=\frac{-12i+8i^2}{-16i^2}$$

$$=\frac{-12i+8(-1)}{-16(-1)}$$

$$=\frac{-8-12i}{16}$$

$$=-\frac{8}{16}-\frac{12}{16}i$$

$$=-\frac{1}{2}-\frac{3}{4}i$$

7. a. $i^{16}=(i^2)^8=(-1)^8=1$

b. $i^{25}=i^{24}i=(i^2)^{12}i=(-1)^{12}i=1\cdot i=i$

c. $i^{35}=i^{34}i=(i^2)^{17}i=(-1)^{17}i=(-1)i=-i$

10.7 Exercise Set

1. $\sqrt{-100}=\sqrt{100\cdot-1}=\sqrt{100}\cdot\sqrt{-1}=10i$

3. $\sqrt{-23}=\sqrt{23\cdot-1}=\sqrt{23}\cdot\sqrt{-1}=i\sqrt{23}$

5. $\sqrt{-18}=\sqrt{9\cdot2\cdot-1}$

$$=\sqrt{9}\cdot\sqrt{2}\cdot\sqrt{-1}$$

$$=3i\sqrt{2}$$

7. $\sqrt{-63}=\sqrt{9\cdot7\cdot-1}$

$$=\sqrt{9}\cdot\sqrt{7}\cdot\sqrt{-1}$$

$$=3i\sqrt{7}$$

9. $-\sqrt{-108}=-\sqrt{36\cdot3\cdot-1}$

$$=-\sqrt{36}\cdot\sqrt{3}\cdot\sqrt{-1}$$

$$=-6i\sqrt{3}$$

11. $5+\sqrt{-36}=5+\sqrt{36\cdot-1}$

$$=5+\sqrt{36}\cdot\sqrt{-1}$$

$$=5+6i$$

13. $15+\sqrt{-3}=15+\sqrt{3\cdot-1}$

$$=15+\sqrt{3}\cdot\sqrt{-1}$$

$$=15+i\sqrt{3}$$

15. $-2 - \sqrt{-18} = -2 - \sqrt{9 \cdot 2 \cdot -1}$
$= -2 - \sqrt{9} \cdot \sqrt{2} \cdot \sqrt{-1}$
$= -2 - 3i\sqrt{2}$

17. $(3 + 2i) + (5 + i)$
$= 3 + 2i + 5 + i = 3 + 5 + 2i + i$
$= (3 + 5) + (2 + 1)i = 8 + 3i$

19. $(7 + 2i) + (1 - 4i)$
$= 7 + 2i + 1 - 4i = 7 + 1 + 2i - 4i$
$= (7 + 1) + (2 - 4)i = 8 - 2i$

21. $(10 + 7i) - (5 + 4i)$
$= 10 + 7i - 5 - 4i = 10 - 5 + 7i - 4i$
$= (10 - 5) + (7 - 4)i = 5 + 3i$

23. $(9 - 4i) - (10 + 3i)$
$= 9 - 4i - 10 - 3i = 9 - 10 - 4i - 3i$
$= (9 - 10) + (-4 - 3)i$
$= -1 + (-7)i = -1 - 7i$

25. $(3 + 2i) - (5 - 7i)$
$= 3 + 2i - 5 + 7i = 3 - 5 + 2i + 7i$
$= (3 - 5) + (2 + 7)i = -2 + 9i$

27. $(-5 + 4i) - (-13 - 11i)$
$= -5 + 4i + 13 + 11i$
$= -5 + 13 + 4i + 11i$
$= (-5 + 13) + (4 + 11)i = 8 + 15i$

29. $8i - (14 - 9i)$
$= 8i - 14 + 9i = -14 + 8i + 9i$
$= -14 + (8 + 9)i = -14 + 17i$

31. $\left(2 + i\sqrt{3}\right) + \left(7 + 4i\sqrt{3}\right)$
$= 2 + i\sqrt{3} + 7 + 4i\sqrt{3}$
$= 2 + 7 + i\sqrt{3} + 4i\sqrt{3}$
$= (2 + 7) + \left(\sqrt{3} + 4\sqrt{3}\right)i = 9 + 5i\sqrt{3}$

33. $2i(5 + 3i)$
$= 2i \cdot 5 + 2i \cdot 3i = 10i + 6i^2$
$= 10i + 6(-1) = -6 + 10i$

35. $3i(7i - 5)$
$= 3i \cdot 7i - 3i \cdot 5 = 21i^2 - 15i$
$= 21(-1) - 15i = -21 - 15i$

37. $-7i(2 - 5i)$
$= -7i \cdot 2 - (-7i)5i = -14i + 35i^2$
$= -14i + 35(-1) = -35 - 14i$

39. $(3 + i)(4 + 5i) = 12 + 15i + 4i + 5i^2$
$= 12 + 15i + 4i + 5(-1)$
$= 12 - 5 + 15i + 4i$
$= 7 + 19i$

41. $(7 - 5i)(2 - 3i)$
$= 14 - 21i - 10i + 15i^2$
$= 14 - 21i - 10i + 15(-1)$
$= 14 - 15 - 21i - 10i = -1 - 31i$

43. $(6 - 3i)(-2 + 5i)$
$= -12 + 30i + 6i - 15i^2$
$= -12 + 30i + 6i - 15(-1)$
$= -12 + 15 + 30i + 6i = 3 + 36i$

45. $(3 + 5i)(3 - 5i)$
$= 9 - \cancel{15i} + \cancel{15i} - 25i^2$
$= 9 - 25(-1) = 9 + 25$
$= 34 = 34 + 0i$

47. $(-5 + 3i)(-5 - 3i)$
$= 25 + \cancel{15i} - \cancel{15i} - 9i^2$
$= 25 - 9(-1) = 25 + 9$
$= 34 = 34 + 0i$

49. $\left(3 - i\sqrt{2}\right)\left(3 + i\sqrt{2}\right)$
$= 9 + \cancel{3i\sqrt{2}} - \cancel{3i\sqrt{2}} - 2i^2$
$= 9 - 2(-1) = 9 + 2$
$= 11 = 11 + 0i$

51. $(2 + 3i)^2$
$= 4 + 2 \cdot 6i + 9i^2 = 4 + 12i + 9(-1)$
$= 4 - 9 + 12i = -5 + 12i$

53. $(5-2i)^2 = 25 - 2 \cdot 10i + 4i^2$

$$= 25 - 20i + 4(-1)$$

$$= 25 - 4 - 20i = 21 - 20i$$

55. $\sqrt{-7} \cdot \sqrt{-2} = \sqrt{7}\sqrt{-1} \cdot \sqrt{2}\sqrt{-1}$

$$= \sqrt{7}\ i \cdot \sqrt{2}\ i = \sqrt{14}\ i^2$$

$$= \sqrt{14}(-1) = -\sqrt{14}$$

$$= -\sqrt{14} + 0i$$

57. $\sqrt{-9} \cdot \sqrt{-4}$

$$= \sqrt{9}\sqrt{-1} \cdot \sqrt{4}\sqrt{-1} = 3i \cdot 2i = 6i^2$$

$$= 6(-1) = -6 = -6 + 0i$$

59. $\sqrt{-7} \cdot \sqrt{-25} = \sqrt{7}\sqrt{-1} \cdot \sqrt{25}\sqrt{-1}$

$$= \sqrt{7}\ i \cdot 5i = 5\sqrt{7}\ i^2$$

$$= 5\sqrt{7}(-1) = -5\sqrt{7}$$

$$= -5\sqrt{7} + 0i$$

61. $\sqrt{-8} \cdot \sqrt{-3} = \sqrt{4 \cdot 2}\sqrt{-1} \cdot \sqrt{3}\sqrt{-1}$

$$= 2\sqrt{2}\ i \cdot \sqrt{3}\ i = 2\sqrt{6}\ i^2$$

$$= 2\sqrt{6}(-1) = -2\sqrt{6}$$

$$= -2\sqrt{6} + 0i$$

63. $\dfrac{2}{3+i} = \dfrac{2}{3+i} \cdot \dfrac{3-i}{3-i} = \dfrac{6-2i}{3^2 - i^2}$

$$= \dfrac{6-2i}{9-(-1)} = \dfrac{6-2i}{9+1}$$

$$= \dfrac{6-2i}{10} = \dfrac{6}{10} - \dfrac{2i}{10}$$

$$= \dfrac{3}{5} - \dfrac{1}{5}i$$

65. $\dfrac{2i}{1+i} = \dfrac{2i}{1+i} \cdot \dfrac{1-i}{1-i} = \dfrac{2i - 2i^2}{1^2 - i^2}$

$$= \dfrac{2i - 2(-1)}{1-(-1)} = \dfrac{2+2i}{1+1}$$

$$= \dfrac{2+2i}{2} = \dfrac{2}{2} + \dfrac{2i}{2} = 1 + i$$

67. $\dfrac{7}{4-3i} = \dfrac{7}{4-3i} \cdot \dfrac{4+3i}{4+3i} = \dfrac{28+21i}{4^2 - (3i)^2}$

$$= \dfrac{28+21i}{16-9i^2} = \dfrac{28+21i}{16-9(-1)}$$

$$= \dfrac{28+21i}{16+9} = \dfrac{28+21i}{25}$$

$$= \dfrac{28}{25} + \dfrac{21}{25}i$$

69. $\dfrac{6i}{3-2i} = \dfrac{6i}{3-2i} \cdot \dfrac{3+2i}{3+2i} = \dfrac{18i + 12i^2}{3^2 - (2i)^2}$

$$= \dfrac{18i + 12(-1)}{9 - 4i^2} = \dfrac{-12+18i}{9-4(-1)}$$

$$= \dfrac{-12+18i}{9+4} = \dfrac{-12+18i}{13}$$

$$= -\dfrac{12}{13} + \dfrac{18}{13}i$$

71. $\dfrac{1+i}{1-i} = \dfrac{1+i}{1-i} \cdot \dfrac{1+i}{1+i} = \dfrac{1+2i+i^2}{1^2 - i^2}$

$$= \dfrac{1+2i+(-1)}{1-(-1)} = \dfrac{2i}{2}$$

$$= i \ \text{ or } \ 0 + i$$

73. $\dfrac{2-3i}{3+i} = \dfrac{2-3i}{3+i} \cdot \dfrac{3-i}{3-i}$

$$= \dfrac{6 - 2i - 9i + 3i^2}{3^2 - i^2}$$

$$= \dfrac{6 - 11i + 3(-1)}{9 - (-1)}$$

$$= \dfrac{6 - 3 - 11i}{9 + 1}$$

$$= \dfrac{3 - 11i}{10} = \dfrac{3}{10} - \dfrac{11}{10}i$$

75. $\dfrac{5-2i}{3+2i} = \dfrac{5-2i}{3+2i} \cdot \dfrac{3-2i}{3-2i}$

$= \dfrac{15-10i-6i+4i^2}{3^2-(2i)^2}$

$= \dfrac{15-16i+4(-1)}{9-4i^2}$

$= \dfrac{15-4-16i}{9-4(-1)}$

$= \dfrac{11-16i}{9+4}$

$= \dfrac{11-16i}{13} = \dfrac{11}{13} - \dfrac{16}{13}i$

77. $\dfrac{4+5i}{3-7i} = \dfrac{4+5i}{3-7i} \cdot \dfrac{3+7i}{3+7i}$

$= \dfrac{12+28i+15i+35i^2}{3^2-(7i)^2}$

$= \dfrac{12+43i+35(-1)}{9-49i^2}$

$= \dfrac{12-35+43i}{9-49(-1)}$

$= \dfrac{-23+43i}{9+49} = \dfrac{-23+43i}{58}$

$= -\dfrac{23}{58} + \dfrac{43}{58}i$

79. $\dfrac{7}{3i} = \dfrac{7}{3i} \cdot \dfrac{-3i}{-3i} = \dfrac{-21i}{-9i^2} = \dfrac{-21i}{-9(-1)}$

$= \dfrac{-21i}{9} = -\dfrac{7}{3}i$ or $0 - \dfrac{7}{3}i$

81. $\dfrac{8-5i}{2i} = \dfrac{8-5i}{2i} \cdot \dfrac{-2i}{-2i} = \dfrac{-16i+10i^2}{-4i^2}$

$= \dfrac{-16i+10(-1)}{-4(-1)} = \dfrac{-10-16i}{4}$

$= -\dfrac{10}{4} - \dfrac{16}{4}i = -\dfrac{5}{2} - 4i$

83. $\dfrac{4+7i}{-3i} = \dfrac{4+7i}{-3i} \cdot \dfrac{3i}{3i} = \dfrac{12i+21i^2}{-9i^2}$

$= \dfrac{12i+21(-1)}{-9(-1)} = \dfrac{-21+12i}{9}$

$= -\dfrac{21}{9} + \dfrac{12}{9}i = -\dfrac{7}{3} + \dfrac{4}{3}i$

85. $i^{10} = \left(i^2\right)^5 = (-1)^5 = -1$

87. $i^{11} = \left(i^2\right)^5 i = (-1)^5 i = -i$

89. $i^{22} = \left(i^2\right)^{11} = (-1)^{11} = -1$

91. $i^{200} = \left(i^2\right)^{100} = (-1)^{100} = 1$

93. $i^{17} = \left(i^2\right)^8 i = (-1)^8 i = i$

95. $(-i)^4 = (-1)^4 i^4 = i^4 = \left(i^2\right)^2$

$\qquad = (-1)^2 = 1$

97. $(-i)^9 = (-1)^9 i^9 = (-1)\left(i^2\right)^4 i$

$\qquad = (-1)(-1)^4 i = (-1)i$

$\qquad = -i$

99. $i^{24} + i^2 = \left(i^2\right)^{12} + (-1)$

$\qquad = (-1)^{12} + (-1)$

$\qquad = 1 + (-1) = 0$

101. $(2-3i)(1-i) - (3-i)(3+i)$

$\qquad = \left(2-2i-3i+3i^2\right) - \left(3^2-i^2\right)$

$\qquad = 2-5i+3i^2 - 9+i^2$

$\qquad = -7-5i+4i^2$

$\qquad = -7-5i+4(-1)$

$\qquad = -11-5i$

103. $(2+i)^2 - (3-i)^2$

$= \left(4 + 4i + i^2\right) - \left(9 - 6i + i^2\right)$

$= 4 + 4i + i^2 - 9 + 6i - i^2$

$= -5 + 10i$

105. $5\sqrt{-16} + 3\sqrt{-81}$

$= 5\sqrt{16}\sqrt{-1} + 3\sqrt{81}\sqrt{-1}$

$= 5 \cdot 4i + 3 \cdot 9i$

$= 20i + 27i$

$= 47i$ or $0 + 47i$

107. $\dfrac{i^4 + i^{12}}{i^8 - i^7} = \dfrac{i^4 + \left(i^4\right)^3}{\left(i^4\right)^2 - \left(i^2\right)^3 i}$

$= \dfrac{1 + 1^3}{1^2 - (-1)^3 i} = \dfrac{1+1}{1+i}$

$= \dfrac{2}{1+i} = \dfrac{2}{1+i} \cdot \dfrac{1-i}{1-i}$

$= \dfrac{2-2i}{1^2 - i^2} = \dfrac{2-2i}{1+1}$

$= \dfrac{2-2i}{2} = 1 - i$

109. $f(x) = x^2 - 2x + 2$

$f(1+i) = (1+i)^2 - 2(1+i) + 2$

$\quad = 1 + 2i + i^2 - 2 - 2i + 2$

$\quad = 1 + i^2$

$\quad = 1 - 1$

$\quad = 0$

111. $f(x) = x - 3i$; $g(x) = 4x + 2i$

$f(-1) = -1 - 3i$

$g(-1) = -4 + 2i$

$(fg)(-1) = (-1 - 3i)(-4 + 2i)$

$\quad = 4 - 2i + 12i - 6i^2$

$\quad = 4 + 10i - 6(-1)$

$\quad = 10 + 10i$

113. $f(x) = \dfrac{x^2 + 19}{2 - x}$

$f(3i) = \dfrac{(3i)^2 + 19}{2 - 3i} = \dfrac{9i^2 + 19}{2 - 3i}$

$= \dfrac{9(-1) + 19}{2 - 3i} = \dfrac{10}{2 - 3i}$

$= \dfrac{10}{2 - 3i} \cdot \dfrac{2 + 3i}{2 + 3i}$

$= \dfrac{20 + 30i}{2^2 - (3i)^2} = \dfrac{20 + 30i}{4 - 9i^2}$

$= \dfrac{20 + 30i}{4 - 9(-1)} = \dfrac{20 + 30i}{13}$

$= \dfrac{20}{13} + \dfrac{30}{13}i$

115. $E = IR = (4 - 5i)(3 + 7i)$

$= 12 + 28i - 15i - 35i^2$

$= 12 + 13i - 35(-1)$

$= 12 + 35 + 13i = 47 + 13i$

The voltage of the circuit is $(47 + 13i)$ volts.

117. Sum:

$\left(5 + i\sqrt{15}\right) + \left(5 - i\sqrt{15}\right)$

$= 5 + i\sqrt{15} + 5 - i\sqrt{15}$

$= 5 + 5 = 10$

Product:

$\left(5 + i\sqrt{15}\right)\left(5 - i\sqrt{15}\right)$

$= 25 - 5i\sqrt{15} + 5i\sqrt{15} - 15i^2$

$= 25 - 15(-1) = 25 + 15 = 40$

119. – 129. Answers will vary.

131. $\sqrt{-9} + \sqrt{-16} = i\sqrt{9} + i\sqrt{16}$

$= 3i + 4i = 7i$

133. does not make sense; Explanations will vary. Sample explanation: The average of complex real numbers is never a complex imaginary number.

135. does not make sense; Explanations will vary. Sample explanation: The i in $5i$ is not a variable. It is the imaginary unit $\sqrt{-1}$.

137. false; Changes to make the statement true will vary. A sample change is: All irrational numbers are complex numbers.

139. false; Changes to make the statement true will vary. A sample change is:

$$\frac{7+3i}{5+3i} = \frac{7+3i}{5+3i} \cdot \frac{5-3i}{5-3i}$$

$$= \frac{35-21i+15i-9i^2}{5^2-(3i)^2}$$

$$= \frac{35-21i+15i-9(-1)}{25-9i^2}$$

$$= \frac{35-6i+9}{25-9(-1)}$$

$$= \frac{44-6i}{34}$$

$$= \frac{44}{34} - \frac{6}{34}i$$

$$= \frac{22}{17} - \frac{3}{17}i$$

141.

$$\frac{4}{(2+i)(3-i)} = \frac{4}{6-2i+3i-i^2}$$

$$= \frac{4}{6+i-(-1)} = \frac{4}{6+1+i}$$

$$= \frac{4}{7+i} \cdot \frac{7-i}{7-i} = \frac{28-4i}{7^2-i^2}$$

$$= \frac{28-4i}{49-(-1)} = \frac{28-4i}{50}$$

$$= \frac{28}{50} - \frac{4}{50}i = \frac{14}{25} - \frac{2}{25}i$$

143.

$$\frac{8}{1+\frac{2}{i}} = \frac{8}{\frac{i}{i}+\frac{2}{i}} = \frac{8}{\frac{2+i}{i}}$$

$$= \frac{8i}{2+i} \cdot \frac{2-i}{2-i} = \frac{8i(2-i)}{(2+i)(2-i)}$$

$$= \frac{16i-8i^2}{4-i^2} = \frac{16i-8(-1)}{4-(-1)}$$

$$= \frac{16i+8}{4+1} = \frac{8+16i}{5} = \frac{8}{5} + \frac{16i}{5}$$

144.

$$\frac{\frac{x}{y^2}+\frac{1}{y}}{\frac{y}{x^2}+\frac{1}{x}} = \frac{\frac{x}{y^2}+\frac{1}{y}}{\frac{y}{x^2}+\frac{1}{x}} \cdot \frac{x^2 y^2}{x^2 y^2}$$

$$= \frac{\frac{x}{y^2}\cdot x^2 y^2 + \frac{1}{y}\cdot x^2 y^2}{\frac{y}{x^2}\cdot x^2 y^2 + \frac{1}{x}\cdot x^2 y^2}$$

$$= \frac{x^3+x^2 y}{y^3+xy^2} = \frac{x^2(x+y)}{y^2(y+x)} = \frac{x^2}{y^2}$$

145.

$$\frac{1}{x}+\frac{1}{y} = \frac{1}{z}$$

$$\frac{1}{x}\cdot xyz + \frac{1}{y}\cdot xyz = \frac{1}{z}\cdot xyz$$

$$yz+xz = xy$$

$$yz = xy-xz$$

$$yz = x(y-z)$$

$$x = \frac{yz}{y-z}$$

146.

$$(g-f)(x) = g(x)-f(x)$$
$$(g-f)(3) = g(3)-f(3)$$
$$= [3-6]-[2(3)^2-3]$$
$$= (-3)-[18-3]$$
$$= -3-15$$
$$= -18$$

147.

$$2x^2+7x-4 = 0$$
$$(x+4)(2x-1) = 0$$
$$x+4=0 \quad \text{or} \quad 2x-1=0$$
$$x=-4 \qquad\qquad 2x=1$$
$$\qquad\qquad\qquad x=\frac{1}{2}$$

The solution set is $\left\{-4, \frac{1}{2}\right\}$.

148.

$$x^2 = 9$$
$$x^2-9 = 0$$
$$(x+3)(x-3) = 0$$
$$x+3=0 \quad \text{or} \quad x-3=0$$
$$x=-3 \qquad\qquad x=3$$

The solution set is $\{-3,3\}$.

149.

$$3x^2 = 18$$
$$3(-\sqrt{6})^2 = 18$$
$$3(-1)^2(\sqrt{6})^2 = 18$$
$$3 \cdot 1 \cdot 6 = 18$$
$$18 = 18, \text{ true}$$

$-\sqrt{6}$ is a solution.

Chapter 10 Review Exercises

1. $\sqrt{81} = 9$ because $9^2 = 81$

2. $-\sqrt{\dfrac{1}{100}} = -\dfrac{1}{10}$ because $\left(-\dfrac{1}{10}\right)^2 = \dfrac{1}{100}$

3. $\sqrt[3]{-27} = -3$ because $(-3)^3 = -27$

4. $\sqrt[4]{-16}$

not a real number
The index is even and the radicand is negative.

5. $\sqrt[5]{-32} = -2$ because $(-2)^5 = -32$

6. $f(15) = \sqrt{2(15)-5} = \sqrt{30-5}$
$\qquad = \sqrt{25} = 5$
$f(4) = \sqrt{2(4)-5} = \sqrt{8-5} = \sqrt{3} \approx 1.73$
$f\left(\dfrac{5}{2}\right) = \sqrt{2\left(\dfrac{5}{2}\right)-5} = \sqrt{5-5}$
$\qquad = \sqrt{0} = 0$
$f(1) = \sqrt{2(1)-5} = \sqrt{2-5} = \sqrt{-3}$

not a real number

7. $g(4) = \sqrt[3]{4(4)-8} = \sqrt[3]{16-8} = \sqrt[3]{8} = 2$

$g(0) = \sqrt[3]{4(0)-8} = \sqrt[3]{-8} = -2$

$g(-14) = \sqrt[3]{4(-14)-8} = \sqrt[3]{-56-8}$
$\qquad = \sqrt[3]{-64} = -4$

8. To find the domain, set the radicand greater than or equal to zero and solve the resulting inequality.
$$x - 2 \ge 0$$
$$x \ge 2$$
The domain of f is $\{x | x \ge 2\}$ or $[2, \infty)$.

9. To find the domain, set the radicand greater than or equal to zero and solve the resulting inequality.
$$100 - 4x \ge 0$$
$$-4x \ge -100$$
$$\dfrac{-4x}{-4} \le \dfrac{-100}{-4}$$
$$x \le 25$$
The domain of g is $\{x | x \le 25\}$ or $(-\infty, 25]$.

10. $\sqrt{25x^2} = 5|x|$

11. $\sqrt{(x+14)^2} = |x+14|$

12. $\sqrt{x^2 - 8x + 16} = \sqrt{(x-4)^2} = |x-4|$

13. $\sqrt[3]{64x^3} = 4x$

14. $\sqrt[4]{16x^4} = 2|x|$

15. $\sqrt[5]{-32(x+7)^5} = -2(x+7)$

16. $(5xy)^{\frac{1}{3}} = \sqrt[3]{5xy}$

17. $16^{\frac{3}{2}} = \left(\sqrt{16}\right)^3 = (4)^3 = 64$

18. $32^{\frac{4}{5}} = \left(\sqrt[5]{32}\right)^4 = (2)^4 = 16$

19. $\sqrt{7x} = (7x)^{\frac{1}{2}}$

20. $\left(\sqrt[3]{19xy}\right)^5 = (19xy)^{\frac{5}{3}}$

21. $8^{-\frac{2}{3}} = \dfrac{1}{8^{\frac{2}{3}}} = \dfrac{1}{\left(\sqrt[3]{8}\right)^2} = \dfrac{1}{(2)^2} = \dfrac{1}{4}$

22. $3x(ab)^{-\frac{4}{5}} = \dfrac{3x}{(ab)^{\frac{4}{5}}}$

$\qquad = \dfrac{3x}{\left(\sqrt[5]{ab}\right)^4}$

$\qquad = \dfrac{3x}{(ab)^{\frac{4}{5}}}$

$\qquad = \dfrac{3x}{a^{\frac{4}{5}}b^{\frac{4}{5}}}$

23. $x^{\frac{1}{3}} \cdot x^{\frac{1}{4}} = x^{\frac{1}{3}+\frac{1}{4}} = x^{\frac{4}{12}+\frac{3}{12}} = x^{\frac{7}{12}}$

24. $\dfrac{5^{\frac{1}{2}}}{5^{\frac{1}{3}}} = 5^{\frac{1}{2}-\frac{1}{3}} = 5^{\frac{3}{6}-\frac{2}{6}} = 5^{\frac{1}{6}}$

25. $\left(8x^6 y^3\right)^{\frac{1}{3}} = 8^{\frac{1}{3}} x^{6 \cdot \frac{1}{3}} y^{3 \cdot \frac{1}{3}} = 2x^2 y$

26. $\left(x^{-\frac{2}{3}} y^{\frac{1}{4}}\right)^{\frac{1}{2}} = x^{-\frac{2}{3} \cdot \frac{1}{2}} y^{\frac{1}{4} \cdot \frac{1}{2}}$

$\qquad = x^{-\frac{1}{3}} y^{\frac{1}{8}} = \dfrac{y^{\frac{1}{8}}}{x^{\frac{1}{3}}}$

27. $\sqrt[3]{x^9 y^{12}} = \left(x^9 y^{12}\right)^{\frac{1}{3}}$

$\qquad = x^{9 \cdot \frac{1}{3}} y^{12 \cdot \frac{1}{3}} = x^3 y^4$

28. $\sqrt[9]{x^3 y^9} = \left(x^3 y^9\right)^{\frac{1}{9}} = x^{3 \cdot \frac{1}{9}} y^{9 \cdot \frac{1}{9}}$

$\qquad = x^{\frac{1}{3}} y = y\sqrt[3]{x}$

29. $\sqrt{x} \cdot \sqrt[3]{x} = x^{\frac{1}{2}} x^{\frac{1}{3}} = x^{\frac{1}{2}+\frac{1}{3}} = x^{\frac{3}{6}+\frac{2}{6}}$

$\qquad = x^{\frac{5}{6}} = \sqrt[6]{x^5}$

30. $\dfrac{\sqrt[3]{x^2}}{\sqrt[4]{x^2}} = \dfrac{x^{\frac{2}{3}}}{x^{\frac{2}{4}}} = x^{\frac{2}{3}-\frac{1}{2}}$

$\qquad = x^{\frac{4}{6}-\frac{3}{6}} = x^{\frac{1}{6}} = \sqrt[6]{x}$

31. $\sqrt[5]{\sqrt[3]{x}} = \sqrt[5]{x^{\frac{1}{3}}} = \left(x^{\frac{1}{3}}\right)^{\frac{1}{5}} = x^{\frac{1}{3} \cdot \frac{1}{5}}$

$\qquad = x^{\frac{1}{15}} = \sqrt[15]{x}$

32. Since 2012 is 27 years after 1985, find $f(27)$.

$$f(27) = 350(27)^{\frac{2}{3}} = 350\left(\sqrt[3]{27}\right)^2$$

$$= 350(3)^2 = 350(9) = 3150$$

Expenditures will be $3150 million or $3,150,000,000 in the year 2012.

33. $\sqrt{3x} \cdot \sqrt{7y} = \sqrt{21xy}$

34. $\sqrt[5]{7x^2} \cdot \sqrt[5]{11x} = \sqrt[5]{77x^3}$

35. $\sqrt[6]{x-5} \cdot \sqrt[6]{(x-5)^4} = \sqrt[6]{(x-5)^5}$

36. $f(x) = \sqrt{7x^2 - 14x + 7}$

$\qquad = \sqrt{7\left(x^2 - 2x + 1\right)}$

$\qquad = \sqrt{7(x-1)^2} = \sqrt{7}\,|x-1|$

37. $\sqrt{20x^3} = \sqrt{4 \cdot 5 \cdot x^2 \cdot x} = \sqrt{4x^2 \cdot 5x}$

$\qquad = 2x\sqrt{5x}$

38. $\sqrt[3]{54x^8 y^6} = \sqrt[3]{27 \cdot 2 \cdot x^6 \cdot x^2 y^6}$

$\qquad = \sqrt[3]{27x^6 y^6 \cdot 2x^2}$

$\qquad = 3x^2 y^2 \sqrt[3]{2x^2}$

39. $\sqrt[4]{32x^3 y^{11} z^5} = \sqrt[4]{16 \cdot 2 \cdot x^3 y^8 \cdot y^3 \cdot z^5 \cdot z}$

$\qquad = \sqrt[4]{16 y^8 z^4 \cdot 2x^3 y^3 z}$

$\qquad = 2y^2 z \sqrt[4]{2x^3 y^3 z}$

40. $\sqrt{6x^3} \cdot \sqrt{4x^2} = \sqrt{24x^5} = \sqrt{4 \cdot 6 \cdot x^4 \cdot x}$
$$= \sqrt{4x^4 \cdot 6x} = 2x^2\sqrt{6x}$$

41. $\sqrt[3]{4x^2y} \cdot \sqrt[3]{4xy^4} = \sqrt[3]{16x^3y^5}$
$$= \sqrt[3]{8 \cdot 2 \cdot x^3 \cdot y^3 \cdot y^2}$$
$$= \sqrt[3]{8x^3y^3 \cdot 2y^2}$$
$$= 2xy\sqrt[3]{2y^2}$$

42. $\sqrt[5]{2x^4y^3z^4} \cdot \sqrt[5]{8xy^6z^7}$
$$= \sqrt[5]{16x^5y^9z^{11}}$$
$$= \sqrt[5]{16 \cdot x^5 \cdot y^5 \cdot y^4 \cdot z^{10} \cdot z}$$
$$= \sqrt[5]{x^5y^5z^{10} \cdot 16y^4z}$$
$$= xyz^2\sqrt[5]{16y^4z}$$

43. $\sqrt{x+1} \cdot \sqrt{x-1} = \sqrt{(x+1)(x-1)}$
$$= \sqrt{x^2 - 1}$$

44. $6\sqrt[3]{3} + 2\sqrt[3]{3} = (6+2)\sqrt[3]{3} = 8\sqrt[3]{3}$

45. $5\sqrt{18} - 3\sqrt{8} = 5\sqrt{9 \cdot 2} - 3\sqrt{4 \cdot 2}$
$$= 5 \cdot 3\sqrt{2} - 3 \cdot 2\sqrt{2}$$
$$= 15\sqrt{2} - 6\sqrt{2}$$
$$= (15-6)\sqrt{2} = 9\sqrt{2}$$

46. $\sqrt[3]{27x^4} + \sqrt[3]{xy^6}$
$$= \sqrt[3]{27x^3x} + \sqrt[3]{xy^6}$$
$$= 3x\sqrt[3]{x} + y^2\sqrt[3]{x}$$
$$= (3x + y^2)\sqrt[3]{x}$$

47. $2\sqrt[3]{6} - 5\sqrt[3]{48} = 2\sqrt[3]{6} - 5\sqrt[3]{8 \cdot 6}$
$$= 2\sqrt[3]{6} - 5 \cdot 2\sqrt[3]{6}$$
$$= 2\sqrt[3]{6} - 10\sqrt[3]{6}$$
$$= (2-10)\sqrt[3]{6} = -8\sqrt[3]{6}$$

48. $\sqrt[3]{\dfrac{16}{125}} = \sqrt[3]{\dfrac{8 \cdot 2}{125}} = \dfrac{2}{5}\sqrt[3]{2}$

49. $\sqrt{\dfrac{x^3}{100y^4}} = \sqrt{\dfrac{x^2 \cdot x}{100y^4}}$
$$= \dfrac{x}{10y^2}\sqrt{x} \text{ or } \dfrac{x\sqrt{x}}{10y^2}$$

50. $\sqrt[4]{\dfrac{3y^5}{16x^{20}}} = \sqrt[4]{\dfrac{y^4 \cdot 3y}{16x^{20}}}$
$$= \dfrac{y}{2x^5}\sqrt[4]{3y} \text{ or } \dfrac{y\sqrt[4]{3y}}{2x^5}$$

51. $\dfrac{\sqrt{48}}{\sqrt{2}} = \sqrt{\dfrac{48}{2}} = \sqrt{24} = \sqrt{4 \cdot 6} = 2\sqrt{6}$

52. $\dfrac{\sqrt[3]{32}}{\sqrt[3]{2}} = \sqrt[3]{\dfrac{32}{2}} = \sqrt[3]{16} = \sqrt[3]{8 \cdot 2} = 2\sqrt[3]{2}$

53. $\dfrac{\sqrt[4]{64x^7}}{\sqrt[4]{2x^2}} = \sqrt[4]{\dfrac{64x^7}{2x^2}} = \sqrt[4]{32x^5}$
$$= \sqrt[4]{16 \cdot 2 \cdot x^4 \cdot x}$$
$$= \sqrt[4]{16x^4 \cdot 2x} = 2x\sqrt[4]{2x}$$

54. $\dfrac{\sqrt{200x^3y^2}}{\sqrt{2x^{-2}y}} = \sqrt{\dfrac{200x^3y^2}{2x^{-2}y}} = \sqrt{100x^5y}$
$$= \sqrt{100x^4xy} = 10x^2\sqrt{xy}$$

55. $\sqrt{3}\left(2\sqrt{6} + 4\sqrt{15}\right) = 2\sqrt{18} + 4\sqrt{45}$
$$= 2\sqrt{9 \cdot 2} + 4\sqrt{9 \cdot 5}$$
$$= 2 \cdot 3\sqrt{2} + 4 \cdot 3\sqrt{5}$$
$$= 6\sqrt{2} + 12\sqrt{5}$$

56. $\sqrt[3]{5}\left(\sqrt[3]{50} - \sqrt[3]{2}\right) = \sqrt[3]{250} - \sqrt[3]{10}$
$$= \sqrt[3]{125 \cdot 2} - \sqrt[3]{10}$$
$$= 5\sqrt[3]{2} - \sqrt[3]{10}$$

57. $\left(\sqrt{7} - 3\sqrt{5}\right)\left(\sqrt{7} + 6\sqrt{5}\right)$
$$= 7 + 6\sqrt{35} - 3\sqrt{35} - 18 \cdot 5$$
$$= 7 + 3\sqrt{35} - 90$$
$$= 3\sqrt{35} - 83 \text{ or } -83 + 3\sqrt{35}$$

58. $\left(\sqrt{x}-\sqrt{11}\right)\left(\sqrt{y}-\sqrt{11}\right)$

$\quad = \sqrt{xy}-\sqrt{11x}-\sqrt{11y}+11$

59. $\left(\sqrt{5}+\sqrt{8}\right)^2 = 5+2\cdot\sqrt{5}\cdot\sqrt{8}+8$

$\quad\quad\quad\quad = 13+2\sqrt{40}$

$\quad\quad\quad\quad = 13+2\sqrt{4\cdot 10}$

$\quad\quad\quad\quad = 13+2\cdot 2\sqrt{10}$

$\quad\quad\quad\quad = 13+4\sqrt{10}$

60. $\left(2\sqrt{3}-\sqrt{10}\right)^2$

$\quad = 4\cdot 3-2\cdot 2\sqrt{3}\cdot\sqrt{10}+10$

$\quad = 12-4\sqrt{30}+10 = 22-4\sqrt{30}$

61. $\left(\sqrt{7}+\sqrt{13}\right)\left(\sqrt{7}-\sqrt{13}\right)$

$\quad = \left(\sqrt{7}\right)^2-\left(\sqrt{13}\right)^2 = 7-13 = -6$

62. $\left(7-3\sqrt{5}\right)\left(7+3\sqrt{5}\right) = 7^2-\left(3\sqrt{5}\right)^2$

$\quad\quad\quad\quad\quad\quad = 49-9\cdot 5$

$\quad\quad\quad\quad\quad\quad = 49-45 = 4$

63. $\dfrac{4}{\sqrt{6}} = \dfrac{4}{\sqrt{6}}\cdot\dfrac{\sqrt{6}}{\sqrt{6}} = \dfrac{4\sqrt{6}}{6} = \dfrac{2\sqrt{6}}{3}$

64. $\sqrt{\dfrac{2}{7}} = \dfrac{\sqrt{2}}{\sqrt{7}} = \dfrac{\sqrt{2}}{\sqrt{7}}\cdot\dfrac{\sqrt{7}}{\sqrt{7}} = \dfrac{\sqrt{14}}{7}$

65. $\dfrac{12}{\sqrt[3]{9}} = \dfrac{12}{\sqrt[3]{3^2}}\cdot\dfrac{\sqrt[3]{3}}{\sqrt[3]{3}} = \dfrac{12\sqrt[3]{3}}{\sqrt[3]{3^3}}$

$\quad\quad = \dfrac{12\sqrt[3]{3}}{3} = 4\sqrt[3]{3}$

66. $\sqrt{\dfrac{2x}{5y}} = \dfrac{\sqrt{2x}}{\sqrt{5y}}\cdot\dfrac{\sqrt{5y}}{\sqrt{5y}} = \dfrac{\sqrt{10xy}}{\sqrt{5^2 y^2}} = \dfrac{\sqrt{10xy}}{5y}$

67. $\dfrac{14}{\sqrt[3]{2x^2}} = \dfrac{14}{\sqrt[3]{2x^2}}\cdot\dfrac{\sqrt[3]{2^2 x}}{\sqrt[3]{2^2 x}} = \dfrac{14\sqrt[3]{2^2 x}}{\sqrt[3]{2^3 x^3}}$

$\quad\quad = \dfrac{14\sqrt[3]{4x}}{2x} = \dfrac{7\sqrt[3]{4x}}{x}$

68. $\sqrt[4]{\dfrac{7}{3x}} = \dfrac{\sqrt[4]{7}}{\sqrt[4]{3x}} = \dfrac{\sqrt[4]{7}}{\sqrt[4]{3x}}\cdot\dfrac{\sqrt[4]{3^3 x^3}}{\sqrt[4]{3^3 x^3}}$

$\quad\quad = \dfrac{\sqrt[4]{7\cdot 3^3 x^3}}{\sqrt[4]{3^4 x^4}} = \dfrac{\sqrt[4]{7\cdot 27 x^3}}{3x}$

$\quad\quad = \dfrac{\sqrt[4]{189 x^3}}{3x}$

69. $\dfrac{5}{\sqrt[5]{32x^4 y}} = \dfrac{5}{\sqrt[5]{2^5 x^4 y}}\cdot\dfrac{\sqrt[5]{xy^4}}{\sqrt[5]{xy^4}}$

$\quad\quad = \dfrac{5\sqrt[5]{xy^4}}{\sqrt[5]{2^5 x^5 y^5}} = \dfrac{5\sqrt[5]{xy^4}}{2xy}$

70. $\dfrac{6}{\sqrt{3}-1} = \dfrac{6}{\sqrt{3}-1}\cdot\dfrac{\sqrt{3}+1}{\sqrt{3}+1}$

$\quad\quad = \dfrac{6\left(\sqrt{3}+1\right)}{\left(\sqrt{3}\right)^2-1^2} = \dfrac{6\left(\sqrt{3}+1\right)}{3-1}$

$\quad\quad = \dfrac{6\left(\sqrt{3}+1\right)}{2} = 3\left(\sqrt{3}+1\right)$

$\quad\quad = 3\sqrt{3}+3$

71. $\dfrac{\sqrt{7}}{\sqrt{5}+\sqrt{3}} = \dfrac{\sqrt{7}}{\sqrt{5}+\sqrt{3}}\cdot\dfrac{\sqrt{5}-\sqrt{3}}{\sqrt{5}-\sqrt{3}}$

$\quad\quad = \dfrac{\sqrt{35}-\sqrt{21}}{\left(\sqrt{5}\right)^2-\left(\sqrt{3}\right)^2}$

$\quad\quad = \dfrac{\sqrt{35}-\sqrt{21}}{5-3} = \dfrac{\sqrt{35}-\sqrt{21}}{2}$

72. $\dfrac{10}{2\sqrt{5}-3\sqrt{2}}$

$\quad = \dfrac{10}{2\sqrt{5}-3\sqrt{2}}\cdot\dfrac{2\sqrt{5}+3\sqrt{2}}{2\sqrt{5}+3\sqrt{2}}$

$\quad = \dfrac{10\left(2\sqrt{5}+3\sqrt{2}\right)}{\left(2\sqrt{5}\right)^2-\left(3\sqrt{2}\right)^2} = \dfrac{10\left(2\sqrt{5}+3\sqrt{2}\right)}{4\cdot 5-9\cdot 2}$

$\quad = \dfrac{10\left(2\sqrt{5}+3\sqrt{2}\right)}{20-18} = \dfrac{10\left(2\sqrt{5}+3\sqrt{2}\right)}{2}$

$\quad = 5\left(2\sqrt{5}+3\sqrt{2}\right) = 10\sqrt{5}+15\sqrt{2}$

73. $\dfrac{\sqrt{x}+5}{\sqrt{x}-3} = \dfrac{\sqrt{x}+5}{\sqrt{x}-3} \cdot \dfrac{\sqrt{x}+3}{\sqrt{x}+3}$

$= \dfrac{x+3\sqrt{x}+5\sqrt{x}+15}{\left(\sqrt{x}\right)^2 - 3^2}$

$= \dfrac{x+8\sqrt{x}+15}{x-9}$

74. $\dfrac{\sqrt{7}+\sqrt{3}}{\sqrt{7}-\sqrt{3}} = \dfrac{\sqrt{7}+\sqrt{3}}{\sqrt{7}-\sqrt{3}} \cdot \dfrac{\sqrt{7}+\sqrt{3}}{\sqrt{7}+\sqrt{3}}$

$= \dfrac{7+2\cdot\sqrt{7}\cdot\sqrt{3}+3}{\left(\sqrt{7}\right)^2 - \left(\sqrt{3}\right)^2}$

$= \dfrac{10+2\sqrt{21}}{7-3} = \dfrac{10+2\sqrt{21}}{4}$

$= \dfrac{2\left(5+\sqrt{21}\right)}{4} = \dfrac{5+\sqrt{21}}{2}$

75. $\dfrac{2\sqrt{3}+\sqrt{6}}{2\sqrt{6}+\sqrt{3}} = \dfrac{2\sqrt{3}+\sqrt{6}}{2\sqrt{6}+\sqrt{3}} \cdot \dfrac{2\sqrt{6}-\sqrt{3}}{2\sqrt{6}-\sqrt{3}}$

$= \dfrac{4\sqrt{18}-2\cdot3+2\cdot6-\sqrt{18}}{\left(2\sqrt{6}\right)^2 - \left(\sqrt{3}\right)^2}$

$= \dfrac{3\sqrt{18}-6+12}{4\cdot6-3} = \dfrac{3\sqrt{9\cdot2}+6}{24-3}$

$= \dfrac{3\cdot3\sqrt{2}+6}{21} = \dfrac{9\sqrt{2}+6}{21}$

$= \dfrac{3\left(3\sqrt{2}+2\right)}{21} = \dfrac{3\sqrt{2}+2}{7}$

76. $\sqrt{\dfrac{2}{7}} = \dfrac{\sqrt{2}}{\sqrt{7}} = \dfrac{\sqrt{2}}{\sqrt{7}} \cdot \dfrac{\sqrt{2}}{\sqrt{2}} = \dfrac{2}{\sqrt{14}}$

77. $\dfrac{\sqrt[3]{3x}}{\sqrt[3]{y}} = \dfrac{\sqrt[3]{3x}}{\sqrt[3]{y}} \cdot \dfrac{\sqrt[3]{3^2 x^2}}{\sqrt[3]{3^2 x^2}}$

$= \dfrac{\sqrt[3]{3^3 x^3}}{\sqrt[3]{3^2 x^2 y}} = \dfrac{3x}{\sqrt[3]{9x^2 y}}$

78. $\dfrac{\sqrt{7}}{\sqrt{5}+\sqrt{3}} = \dfrac{\sqrt{7}}{\sqrt{5}+\sqrt{3}} \cdot \dfrac{\sqrt{7}}{\sqrt{7}}$

$= \dfrac{7}{\sqrt{35}+\sqrt{21}}$

79. $\dfrac{\sqrt{7}+\sqrt{3}}{\sqrt{7}-\sqrt{3}}$

$= \dfrac{\sqrt{7}+\sqrt{3}}{\sqrt{7}-\sqrt{3}} \cdot \dfrac{\sqrt{7}-\sqrt{3}}{\sqrt{7}-\sqrt{3}}$

$= \dfrac{\left(\sqrt{7}\right)^2 - \left(\sqrt{3}\right)^2}{7-2\sqrt{7}\sqrt{3}+3} = \dfrac{7-3}{10-2\sqrt{21}}$

$= \dfrac{4}{10-2\sqrt{21}} = \dfrac{4}{2\left(5-\sqrt{21}\right)}$

$= \dfrac{2}{5-\sqrt{21}}$

80. $\sqrt{2x+4} = 6$

$\left(\sqrt{2x+4}\right)^2 = 6^2$

$2x+4 = 36$

$2x = 32$

$x = 16$

The solution checks. The solution set is $\{16\}$.

81. $\sqrt{x-5}+9 = 4$

$\sqrt{x-5} = -5$

The square root of a number is always nonnegative. The solution set is $\varnothing$ or $\{\ \}$.

82. $\sqrt{2x-3}+x = 3$

$\sqrt{2x-3} = 3-x$

$\left(\sqrt{2x-3}\right)^2 = (3-x)^2$

$2x-3 = 9-6x+x^2$

$0 = 12-8x+x^2$

$0 = x^2-8x+12$

$0 = (x-6)(x-2)$

$x-6 = 0 \qquad x-2 = 0$

$x = 6 \qquad\quad x = 2$

6 is an extraneous solution. The solution set is $\{2\}$.

83. $\sqrt{x-4} + \sqrt{x+1} = 5$

$\sqrt{x-4} = 5 - \sqrt{x+1}$

$\left(\sqrt{x-4}\right)^2 = \left(5 - \sqrt{x+1}\right)^2$

$x - 4 = 25 - 10\sqrt{x+1} + x + 1$

$-30 = -10\sqrt{x+1}$

$\dfrac{-30}{-10} = \dfrac{-10\sqrt{x+1}}{-10}$

$3 = \sqrt{x+1}$

$3^2 = \left(\sqrt{x+1}\right)^2$

$9 = x + 1$

$8 = x$

The solution checks. The solution set is $\{8\}$.

84. $\left(x^2 + 6x\right)^{\frac{1}{3}} + 2 = 0$

$\left(x^2 + 6x\right)^{\frac{1}{3}} = -2$

$\sqrt[3]{x^2 + 6x} = -2$

$\left(\sqrt[3]{x^2 + 6x}\right)^3 = (-2)^3$

$x^2 + 6x = -8$

$x^2 + 6x + 8 = 0$

$(x+4)(x+2) = 0$

$\begin{array}{ll} x + 4 = 0 & x + 2 = 0 \\ \quad x = -4 & \quad x = -2 \end{array}$

Both solutions check. The solution set is $\{-4, -2\}$.

85. a. $f(x) = 0.23\sqrt{x} + 0.43$

$f(6) = 0.23\sqrt{6} + 0.43 \approx 0.99$

The average state cigarette tax per pack was approximately $0.99 6 years after 2001, or in 2007. This underestimates the tax displayed in the graph by $0.01.

b. $f(x) = 0.23\sqrt{x} + 0.43$

$1.12 = 0.23\sqrt{x} + 0.43$

$0.69 = 0.23\sqrt{x}$

$\dfrac{0.69}{0.23} = \dfrac{0.23\sqrt{x}}{0.23}$

$3 = \sqrt{x}$

$3^2 = \left(\sqrt{x}\right)^2$

$9 = x$

The average state cigarette tax will be $1.12 per pack 9 years after 2001, or 2010.

86. $20,000 = 5000\sqrt{100 - x}$

$\dfrac{20,000}{5000} = \dfrac{5000\sqrt{100 - x}}{5000}$

$4 = \sqrt{100 - x}$

$4^2 = \left(\sqrt{100 - x}\right)^2$

$16 = 100 - x$

$-84 = -x$

$84 = x$

20,000 people in the group will survive to 84 years old.

87. $\sqrt{-81} = \sqrt{81 \cdot -1} = \sqrt{81}\sqrt{-1} = 9i$

88. $\sqrt{-63} = \sqrt{9 \cdot 7 \cdot -1}$

$= \sqrt{9}\sqrt{7}\sqrt{-1} = 3i\sqrt{7}$

89. $-\sqrt{-8} = -\sqrt{4 \cdot 2 \cdot -1}$

$= -\sqrt{4}\sqrt{2}\sqrt{-1} = -2i\sqrt{2}$

90. $(7 + 12i) + (5 - 10i)$

$= 7 + 12i + 5 - 10i = 12 + 2i$

91. $(8 - 3i) - (17 - 7i) = 8 - 3i - 17 + 7i$

$= -9 + 4i$

92. $4i(3i - 2) = 4i \cdot 3i - 4i \cdot 2$

$= 12i^2 - 8i$

$= 12(-1) - 8i$

$= -12 - 8i$

93. $(7-5i)(2+3i) = 14+21i-10i-15i^2$
$$= 14+11i-15(-1)$$
$$= 14+11i+15$$
$$= 29+11i$$

94. $(3-4i)^2 = 3^2 - 2\cdot3\cdot4i + (4i)^2$
$$= 9-24i+16i^2$$
$$= 9-24i+16(-1)$$
$$= 9-24i-16$$
$$= -7-24i$$

95. $(7+8i)(7-8i) = 7^2 - (8i)^2$
$$= 49-64i^2$$
$$= 49-64(-1)$$
$$= 49+64$$
$$= 113 \text{ or } 113+0i$$

96. $\sqrt{-8}\cdot\sqrt{-3} = \sqrt{4\cdot2\cdot-1}\cdot\sqrt{3\cdot-1}$
$$= 2\sqrt{2}i\cdot\sqrt{3}i = 2\sqrt{6}i^2$$
$$= 2\sqrt{6}(-1) = -2\sqrt{6}$$
$$= -2\sqrt{6} \text{ or } -2\sqrt{6}+0i$$

97. $\dfrac{6}{5+i} = \dfrac{6}{5+i}\cdot\dfrac{5-i}{5-i} = \dfrac{30-6i}{25-i^2}$
$$= \dfrac{30-6i}{25-(-1)} = \dfrac{30-6i}{25+1}$$
$$= \dfrac{30-6i}{26} = \dfrac{30}{26}-\dfrac{6}{26}i$$
$$= \dfrac{15}{13}-\dfrac{3}{13}i$$

98. $\dfrac{3+4i}{4-2i} = \dfrac{3+4i}{4-2i}\cdot\dfrac{4+2i}{4+2i}$
$$= \dfrac{12+6i+16i+8i^2}{16-4i^2}$$
$$= \dfrac{12+22i+8(-1)}{16-4(-1)}$$
$$= \dfrac{12+22i-8}{16+4} = \dfrac{4+22i}{20}$$
$$= \dfrac{4}{20}+\dfrac{22}{20}i = \dfrac{1}{5}+\dfrac{11}{10}i$$

99. $\dfrac{5+i}{3i} = \dfrac{5+i}{3i}\cdot\dfrac{i}{i} = \dfrac{5i+i^2}{3i^2}$
$$= \dfrac{5i+(-1)}{3(-1)} = \dfrac{5i-1}{-3}$$
$$= \dfrac{-1}{-3}+\dfrac{5}{-3}i = \dfrac{1}{3}-\dfrac{5}{3}i$$

100. $i^{16} = (i^2)^8 = (-1)^8 = 1$

101. $i^{23} = i^{22}\cdot i = (i^2)^{11}i = (-1)^{11}i = (-1)i = -i$

Chapter 10 Test

1. a. $f(-14) = \sqrt{8-2(-14)}$
$$= \sqrt{8+28} = \sqrt{36} = 6$$

b. To find the domain, set the radicand greater than or equal to zero and solve the resulting inequality.
$$8-2x \ge 0$$
$$-2x \ge -8$$
$$x \le 4$$
The domain of f is $\{x|x \le 4\}$ or $(-\infty, 4]$.

2. $27^{-\frac{4}{3}} = \dfrac{1}{27^{\frac{4}{3}}} = \dfrac{1}{(\sqrt[3]{27})^4} = \dfrac{1}{(3)^4} = \dfrac{1}{81}$

3. $\left(25x^{-\frac{1}{2}}y^{\frac{1}{4}}\right)^{\frac{1}{2}} = 25^{\frac{1}{2}}x^{-\frac{1}{4}}y^{\frac{1}{8}}$
$$= 5x^{-\frac{1}{4}}y^{\frac{1}{8}}$$
$$= \dfrac{5y^{\frac{1}{8}}}{x^{\frac{1}{4}}}$$

4. $\sqrt[8]{x^4} = (x^4)^{\frac{1}{8}} = x^{4\cdot\frac{1}{8}} = x^{\frac{1}{2}} = \sqrt{x}$

5. $\sqrt[4]{x}\cdot\sqrt[5]{x} = x^{\frac{1}{4}}\cdot x^{\frac{1}{5}} = x^{\frac{1}{4}+\frac{1}{5}} = x^{\frac{5}{20}+\frac{4}{20}}$
$$= x^{\frac{9}{20}} = \sqrt[20]{x^9}$$

6. $\sqrt{75x^2} = \sqrt{25 \cdot 3x^2} = 5|x|\sqrt{3}$

7. $\sqrt{x^2 - 10x + 25} = \sqrt{(x-5)^2}$
$$= |x - 5|$$

8. $\sqrt[3]{16x^4 y^8} = \sqrt[3]{8 \cdot 2 \cdot x^3 \cdot x \cdot y^6 \cdot y^2}$
$$= \sqrt[3]{8x^3 y^6 \cdot 2xy^2}$$
$$= 2xy^2 \sqrt[3]{2xy^2}$$

9. $\sqrt[5]{-\dfrac{32}{x^{10}}} = \sqrt[5]{-\dfrac{2^5}{\left(x^2\right)^5}} = -\dfrac{2}{x^2}$

10. $\sqrt[3]{5x^2} \cdot \sqrt[3]{10y} = \sqrt[3]{50x^2 y}$

11. $\sqrt[4]{8x^3 y} \cdot \sqrt[4]{4xy^2} = \sqrt[4]{32x^4 y^3}$
$$= \sqrt[4]{16 \cdot 2 \cdot x^4 \cdot y^3}$$
$$= \sqrt[4]{16x^4 \cdot 2y^3}$$
$$= 2x\sqrt[4]{2y^3}$$

12. $3\sqrt{18} - 4\sqrt{32} = 3\sqrt{9 \cdot 2} - 4\sqrt{16 \cdot 2}$
$$= 3 \cdot 3\sqrt{2} - 4 \cdot 4\sqrt{2}$$
$$= 9\sqrt{2} - 16\sqrt{2} = -7\sqrt{2}$$

13. $\sqrt[3]{8x^4} + \sqrt[3]{xy^6} = \sqrt[3]{8x^3 \cdot x} + \sqrt[3]{xy^6}$
$$= 2x\sqrt[3]{x} + y^2\sqrt[3]{x}$$
$$= \left(2x + y^2\right)\sqrt[3]{x}$$

14. $\dfrac{\sqrt[3]{16x^8}}{\sqrt[3]{2x^4}} = \sqrt[3]{\dfrac{16x^8}{2x^4}} = \sqrt[3]{8x^4}$
$$= \sqrt[3]{8x^3 \cdot x} = 2x\sqrt[3]{x}$$

15. $\sqrt{3}\left(4\sqrt{6} - \sqrt{5}\right) = \sqrt{3} \cdot 4\sqrt{6} - \sqrt{3} \cdot \sqrt{5}$
$$= 4\sqrt{18} - \sqrt{15}$$
$$= 4\sqrt{9 \cdot 2} - \sqrt{15}$$
$$= 4 \cdot 3\sqrt{2} - \sqrt{15}$$
$$= 12\sqrt{2} - \sqrt{15}$$

16. $\left(5\sqrt{6} - 2\sqrt{2}\right)\left(\sqrt{6} + \sqrt{2}\right)$
$$= 5 \cdot 6 + 5\sqrt{12} - 2\sqrt{12} - 2 \cdot 2$$
$$= 30 + 3\sqrt{12} - 4 = 26 + 3\sqrt{4 \cdot 3}$$
$$= 26 + 3 \cdot 2\sqrt{3} = 26 + 6\sqrt{3}$$

17. $\left(7 - \sqrt{3}\right)^2 = 49 - 2 \cdot 7 \cdot \sqrt{3} + 3$
$$= 52 - 14\sqrt{3}$$

18. $\sqrt{\dfrac{5}{x}} = \dfrac{\sqrt{5}}{\sqrt{x}} \cdot \dfrac{\sqrt{x}}{\sqrt{x}} = \dfrac{\sqrt{5x}}{x}$

19. $\dfrac{5}{\sqrt[3]{5x^2}} = \dfrac{5}{\sqrt[3]{5x^2}} \cdot \dfrac{\sqrt[3]{5^2 x}}{\sqrt[3]{5^2 x}} = \dfrac{5\sqrt[3]{5^2 x}}{\sqrt[3]{5^3 x^3}}$
$$= \dfrac{5\sqrt[3]{25x}}{5x} = \dfrac{\sqrt[3]{25x}}{x}$$

20. $\dfrac{\sqrt{2} - \sqrt{3}}{\sqrt{2} + \sqrt{3}} = \dfrac{\sqrt{2} - \sqrt{3}}{\sqrt{2} + \sqrt{3}} \cdot \dfrac{\sqrt{2} - \sqrt{3}}{\sqrt{2} - \sqrt{3}}$
$$= \dfrac{2 - 2\sqrt{2}\sqrt{3} + 3}{2 - 3}$$
$$= \dfrac{5 - 2\sqrt{6}}{-1} = -5 + 2\sqrt{6}$$

21. $3 + \sqrt{2x-3} = x$

$\sqrt{2x-3} = x - 3$

$\left(\sqrt{2x-3}\right)^2 = (x-3)^2$

$2x - 3 = x^2 - 6x + 9$

$0 = x^2 - 8x + 12$

$0 = (x-6)(x-2)$

$x - 6 = 0 \qquad x - 2 = 0$

$x = 6 \qquad\quad x = 2$

2 is an extraneous solution. The solution set is $\{6\}$.

22. $\sqrt{x+9} - \sqrt{x-7} = 2$

$\sqrt{x+9} = 2 + \sqrt{x-7}$

$\left(\sqrt{x+9}\right)^2 = \left(2 + \sqrt{x-7}\right)^2$

$x + 9 = 4 + 2 \cdot 2 \cdot \sqrt{x-7} + x - 7$

$x + 9 = 4\sqrt{x-7} + x - 3$

$12 = 4\sqrt{x-7}$

$3 = \sqrt{x-7}$

$3^2 = \left(\sqrt{x-7}\right)^2$

$9 = x - 7$

$16 = x$

The solution set is $\{16\}$.

23. $(11x+6)^{\frac{1}{3}} + 3 = 0$

$(11x+6)^{\frac{1}{3}} = -3$

$\sqrt[3]{11x+6} = -3$

$\left(\sqrt[3]{11x+6}\right)^3 = (-3)^3$

$11x + 6 = -27$

$11x = -33$

$x = -3$

The solution set is $\{-3\}$.

24. $40.4 = 2.9\sqrt{x} + 20.1$

$20.3 = 2.9\sqrt{x}$

$7 = \sqrt{x}$

$7^2 = \left(\sqrt{x}\right)^2$

$49 = x$

Boys who are 49 months of age have an average height of 40.4 inches.

25. $\sqrt{-75} = \sqrt{25 \cdot 3 \cdot -1}$

$= \sqrt{25} \cdot \sqrt{3} \cdot \sqrt{-1} = 5i\sqrt{3}$

26. $(5-3i) - (6-9i) = 5 - 3i - 6 + 9i$

$= 5 - 6 - 3i + 9i$

$= -1 + 6i$

27. $(3-4i)(2+5i) = 6 + 15i - 8i - 20i^2$

$= 6 + 7i - 20(-1)$

$= 6 + 7i + 20$

$= 26 + 7i$

28. $\sqrt{-9} \cdot \sqrt{-4} = \sqrt{9 \cdot -1} \cdot \sqrt{4 \cdot -1}$

$= \sqrt{9} \cdot \sqrt{-1} \cdot \sqrt{4} \cdot \sqrt{-1}$

$= 3 \cdot i \cdot 2 \cdot i = 6i^2 = 6(-1)$

$= -6 \ \text{ or } \ -6 + 0i$

29. $\dfrac{3+i}{1-2i} = \dfrac{3+i}{1-2i} \cdot \dfrac{1+2i}{1+2i}$

$= \dfrac{3 + 6i + i + 2i^2}{1 - 4i^2}$

$= \dfrac{3 + 7i + 2(-1)}{1 - 4(-1)}$

$= \dfrac{3 + 7i - 2}{1 + 4}$

$= \dfrac{1 + 7i}{5} = \dfrac{1}{5} + \dfrac{7}{5}i$

30. $i^{35} = i^{34} \cdot i = \left(i^2\right)^{17} \cdot i$

$= (-1)^{17} \cdot i$

$= (-1)i = -i$

Cumulative Review Exercises (Chapters 1 – 10)

1. $2x - y + z = -5$
$x - 2y - 3z = 6$
$x + y - 2z = 1$

Add the first and third equations to eliminate y.

$2x - y + z = -5$
$\underline{x + y - 2z = 1}$
$3x - z = -4$

Multiply the third equation by 2 and add to the second equation.

$x - 2y - 3z = 6$
$\underline{2x + 2y - 4z = 2}$
$3x - 7z = 8$

We now have a system of two equations in two variables.

$3x - z = -4$
$3x - 7z = 8$

Multiply the first equation by -1 and add to the second equation.

$-3x + z = 4$
$\underline{3x - 7z = 8}$
$-6z = 12$
$z = -2$

Back-substitute -2 for z to find x.

$-3x + z = 4$
$-3x - 2 = 4$
$-3x = 6$
$x = -2$

Back-substitute -2 for x and z in one of the original equations to find y.

$2x - y + z = -5$
$2(-2) - y - 2 = -5$
$-4 - y - 2 = -5$
$-y - 6 = -5$
$-y = 1$
$y = -1$

The solution is $(-2, -1, -2)$ or the solution set is $\{(-2, -1, -2)\}$.

2. $3x^2 - 11x = 4$
$3x^2 - 11x - 4 = 0$
$(3x + 1)(x - 4) = 0$

Apply the zero product principle.

$3x + 1 = 0 \qquad x - 4 = 0$
$3x = -1 \qquad x = 4$
$x = -\dfrac{1}{3}$

The solution set is $\left\{-\dfrac{1}{3}, 4\right\}$.

3. $2(x + 4) < 5x + 3(x + 2)$
$2x + 8 < 5x + 3x + 6$
$2x + 8 < 8x + 6$
$-6x + 8 < 6$
$-6x < -2$
$\dfrac{-6x}{-6} > \dfrac{-2}{-6}$
$x > \dfrac{1}{3}$

The solution set is $\left\{x \,\middle|\, x > \dfrac{1}{3}\right\}$ or $\left(\dfrac{1}{3}, \infty\right)$.

4.
$$\frac{1}{x+2}+\frac{15}{x^2-4}=\frac{5}{x-2}$$

$$\frac{1}{x+2}+\frac{15}{(x+2)(x-2)}=\frac{5}{x-2}$$

So that denominators will not equal zero, x cannot equal 2 or -2. To eliminate fractions, multiply by the LCD, $(x+2)(x-2)$.

$$(x+2)(x-2)\left(\frac{1}{x+2}+\frac{15}{(x+2)(x-2)}\right)=(x+2)(x-2)\left(\frac{5}{x-2}\right)$$

$$(x+2)(x-2)\left(\frac{1}{x+2}\right)+(x+2)(x-2)\left(\frac{15}{(x+2)(x-2)}\right)=(x+2)(5)$$

$$x-2+15=5x+10$$
$$x+13=5x+10$$
$$-4x+13=10$$
$$-4x=-3$$
$$x=\frac{3}{4}$$

The solution set is $\left\{\frac{3}{4}\right\}$.

5.
$$\sqrt{x+2}-\sqrt{x+1}=1$$
$$\sqrt{x+2}=1+\sqrt{x+1}$$
$$\left(\sqrt{x+2}\right)^2=\left(1+\sqrt{x+1}\right)^2$$
$$x+2=1+2\sqrt{x+1}+x+1$$
$$x+2=2+2\sqrt{x+1}+x$$

$$0^2=\left(2\sqrt{x+1}\right)^2$$
$$0=4(x+1)$$
$$0=4x+4$$
$$-4=4x$$
$$-1=x$$

The solution checks. The solution set is $\{-1\}$.

6. $x + 2y < 2$

$2y - x > 4$

First consider $x + 2y < 2$. Replace the inequality symbol with an equal sign and we have $x + 2y = 2$. Solve for y to put the equation in slope-intercept form.

$x + 2y = 2$

$2y = -x + 2$

$y = -\dfrac{1}{2}x + 1$

slope $= -\dfrac{1}{2}$ y–intercept $= 1$

Now, use the origin as a test point.

$0 + 2(0) < 2$

$\quad 0 < 2$

This is a true statement. This means that the point $(0,0)$ will fall in the shaded half-plane.

Next consider $2y - x > 4$. Replace the inequality symbol with an equal sign and we have $2y - x = 4$. Solve for y to put the equation in slope-intercept form.

$2y - x = 4$

$2y = x + 4$

$y = \dfrac{1}{2}x + 2$

slope $= \dfrac{1}{2}$ y–intercept $= 2$

Now, use the origin as a test point.

$2(0) - 0 > 4$

$\quad 0 > 4$

This is a false statement. This means that the point $(0,0)$ will not fall in the shaded half-plane.

Next, graph each of the inequalities. The solution to the system is the intersection of the shaded half-planes.

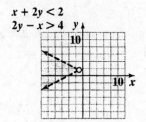

$x + 2y < 2$
$2y - x > 4$

7. $\dfrac{8x^2}{3x^2 - 12} \div \dfrac{40}{x - 2} \quad \underset{x \to \infty}{\lim}$

$= \dfrac{8x^2}{3x^2 - 12} \cdot \dfrac{x - 2}{40}$

$= \dfrac{8x^2}{3(x^2 - 4)} \cdot \dfrac{x - 2}{40}$

$= \dfrac{\overset{1}{\cancel{8}} x^2}{3(x+2)(\cancel{x-2})} \cdot \dfrac{\cancel{x-2}}{\underset{5}{\cancel{40}}}$

$= \dfrac{x^2}{3 \cdot 5(x+2)} = \dfrac{x^2}{15(x+2)}$

8. $\dfrac{x + \dfrac{1}{y}}{y + \dfrac{1}{x}} = \dfrac{x + \dfrac{1}{y}}{y + \dfrac{1}{x}} \cdot \dfrac{xy}{xy} = \dfrac{xy \cdot x + xy \cdot \dfrac{1}{y}}{xy \cdot y + xy \cdot \dfrac{1}{x}}$

$= \dfrac{x^2 y + x}{xy^2 + y} = \dfrac{x(xy + 1)}{y(xy + 1)} = \dfrac{x}{y}$

9. $(2x - 3)(4x^2 - 5x - 2)$

$= 2x \cdot 4x^2 - 2x \cdot 5x - 2x \cdot 2 - 3 \cdot 4x^2$

$\quad + 3 \cdot 5x + 3 \cdot 2$

$= 8x^3 - 10x^2 - 4x - 12x^2 + 15x + 6$

$= 8x^3 - 22x^2 + 11x + 6$

10. $\dfrac{7x}{x^2-2x-15}-\dfrac{2}{x-5}$

$=\dfrac{7x}{(x-5)(x+3)}-\dfrac{2}{x-5}$

$=\dfrac{7x}{(x-5)(x+3)}-\dfrac{2(x+3)}{(x-5)(x+3)}$

$=\dfrac{7x-2(x+3)}{(x-5)(x+3)}=\dfrac{7x-2x-6}{(x-5)(x+3)}$

$=\dfrac{5x-6}{(x-5)(x+3)}$

11. $7(8-10)^3-7+3\div(-3)$

$=7(-2)^3-7+3\div(-3)$

$=7(-8)-7+3\div(-3)$

$=-56-7+(-1)=-64$

12. $\sqrt{80x}-5\sqrt{20x}+2\sqrt{45x}$

$=\sqrt{16\cdot5x}-5\sqrt{4\cdot5x}+2\sqrt{9\cdot5x}$

$=4\sqrt{5x}-5\cdot2\sqrt{5x}+2\cdot3\sqrt{5x}$

$=4\sqrt{5x}-10\sqrt{5x}+6\sqrt{5x}=0$

13. $\dfrac{\sqrt{3}-2}{2\sqrt{3}+5}=\dfrac{\sqrt{3}-2}{2\sqrt{3}+5}\cdot\dfrac{2\sqrt{3}-5}{2\sqrt{3}-5}$

$=\dfrac{2\cdot3-5\sqrt{3}-4\sqrt{3}+10}{4\cdot3-25}$

$=\dfrac{6-9\sqrt{3}+10}{12-25}$

$=\dfrac{16-9\sqrt{3}}{-13}$

$=-\dfrac{16-9\sqrt{3}}{13}$

14.

$$
\begin{array}{r}
2x^2+\ x+5 \\
x-2\overline{)2x^3-3x^2+3x-4} \\
\underline{2x^3-4x^2} \\
x^2+3x \\
\underline{x^2-2x} \\
5x-\ 4 \\
\underline{5x-10} \\
6
\end{array}
$$

$\dfrac{2x^3-3x^2+3x-4}{x-2}=2x^2+x+5+\dfrac{6}{x-2}$

15. $(2\sqrt{3}+5\sqrt{2})(\sqrt{3}-4\sqrt{2})$

$=2\cdot3-8\sqrt{6}+5\sqrt{6}-20\cdot2$

$=6-3\sqrt{6}-40=-34-3\sqrt{6}$

16. $24x^2+10x-4=2\left(12x^2+5x-2\right)$

$\qquad\qquad\qquad=2\left(3x+2\right)\left(4x-1\right)$

17. $16x^4-1=\left(4x^2+1\right)\left(4x^2-1\right)$

$\qquad\qquad\ =\left(4x^2+1\right)\left(2x+1\right)\left(2x-1\right)$

18. Since light varies inversely as the square of the distance, we have $l=\dfrac{k}{d^2}$.

Use the given values to find k.

$l=\dfrac{k}{d^2}$

$120=\dfrac{k}{10^2}$

$120=\dfrac{k}{100}$

$12,000=k$

The equation becomes $l=\dfrac{12,000}{d^2}$. When $d=15$,

$l=\dfrac{12,000}{15^2}\approx53.3$.

At a distance of 15 feet, approximately 53 lumens are provided.

19. Let $x =$ the amount invested at 7%.
    Let $y =$ the amount invested at 9%.

    $$x + \quad y = 6000$$
    $$0.07x + 0.09y = \ 510$$

    Solve the first equation for y.
    $$x + y = 6000$$
    $$y = 6000 - x$$

    Substitute and solve.
    $$0.07x + 0.09(6000 - x) = 510$$
    $$0.07x + 540 - 0.09x = 510$$
    $$540 - 0.02x = 510$$
    $$-0.02x = -30$$
    $$x = 1500$$

    Back-substitute 1500 for x to find y.
    $$y = 6000 - x$$
    $$y = 6000 - 1500 = 4500$$

    $1500 was invested at 7% and $4500 was invested at 9%.

20. Let $x =$ the number of students enrolled last year.
    $$x - 0.12x = 2332$$
    $$0.88x = 2332$$
    $$x = 2650$$

    2650 students were enrolled last year.

Chapter 11
Quadratic Equations and Functions

11.1 Check Points

1. $4x^2 = 28$

   $x^2 = 7$

   Apply the square root property.

   $x = \pm\sqrt{7}$

   The solution set is $\left\{\pm\sqrt{7}\right\}$.

2. $3x^2 - 11 = 0$

   $3x^2 = 11$

   $x^2 = \dfrac{11}{3}$

   $x = \pm\sqrt{\dfrac{11}{3}}$

   $x = \pm\dfrac{\sqrt{11}}{\sqrt{3}} \cdot \dfrac{\sqrt{3}}{\sqrt{3}}$

   $x = \pm\dfrac{\sqrt{33}}{3}$

   The solution set is $\left\{\pm\dfrac{\sqrt{33}}{3}\right\}$.

3. $4x^2 + 9 = 0$

   $4x^2 = -9$

   $x^2 = -\dfrac{9}{4}$

   $x = \pm\sqrt{-\dfrac{9}{4}}$

   $x = \pm\dfrac{\sqrt{-9}}{\sqrt{4}}$

   $x = \pm\dfrac{3}{2}i$

   The solution set is $\left\{\pm\dfrac{3}{2}i\right\}$.

4. $(x-3)^2 = 10$

   $x - 3 = \sqrt{10}$ or $x - 3 = -\sqrt{10}$

   $x = 3 + \sqrt{10}$ $\qquad x = 3 - \sqrt{10}$

   The solution set is $\left\{3 \pm \sqrt{10}\right\}$.

5. **a.** $x^2 + 10x$

   The coefficient of the x-term is 10.

   Half of 10 is 5, and 5^2 is 25, which should be added to the binomial.

   The result is a perfect square trinomial.

   $x^2 + 10x + 25 = (x+5)^2$

   b. $x^2 - 3x$

   The coefficient of the x-term is -3.

   Half of -3 is $-\dfrac{3}{2}$, and $\left(-\dfrac{3}{2}\right)^2$ is $\dfrac{9}{4}$ which

   should be added to the binomial.

   The result is a perfect square trinomial.

   $x^2 - 3x + \dfrac{9}{4} = \left(x - \dfrac{3}{2}\right)^2$

   c. $x^2 + \dfrac{3}{4}x$

   The coefficient of the x-term is $\dfrac{3}{4}$.

   Half of $\dfrac{3}{4}$ is $\dfrac{3}{8}$, and $\left(\dfrac{3}{8}\right)^2$ is $\dfrac{9}{64}$ which should

   be added to the binomial.

   The result is a perfect square trinomial.

   $x^2 + \dfrac{3}{4}x + \dfrac{9}{64} = \left(x + \dfrac{3}{8}\right)^2$

6. $x^2 + 4x - 1 = 0$

   $x^2 + 4x \quad = 1$

   Half of 4 is 2, and 2^2 is 4, which should be added to both sides.

   $x^2 + 4x + 4 = 1 + 4$

   $x^2 + 4x + 4 = 5$

   $(x+2)^2 = 5$

   $x + 2 = \sqrt{5}$ or $x + 2 = -\sqrt{5}$

   $x = -2 + \sqrt{5}$ $\qquad x = -2 - \sqrt{5}$

   The solution set is $\left\{-2 \pm \sqrt{5}\right\}$.

7. $2x^2 + 3x - 4 = 0$

$$\frac{2x^2}{2} + \frac{3x}{2} - \frac{4}{2} = \frac{0}{2}$$

$$x^2 + \frac{3}{2}x - 2 = 0$$

$$x^2 + \frac{3}{2}x \quad = 2$$

Half of $\frac{3}{2}$ is $\frac{3}{4}$, and $\left(\frac{3}{4}\right)^2$ is $\frac{9}{16}$, which should be added to both sides.

$$x^2 + \frac{3}{2}x + \frac{9}{16} = 2 + \frac{9}{16}$$

$$\left(x + \frac{3}{4}\right)^2 = \frac{41}{16}$$

$$x + \frac{3}{4} = \sqrt{\frac{41}{16}} \quad \text{or} \quad x + \frac{3}{4} = -\sqrt{\frac{41}{16}}$$

$$x = -\frac{3}{4} + \sqrt{\frac{41}{16}} \qquad x = -\frac{3}{4} - \sqrt{\frac{41}{16}}$$

$$x = -\frac{3}{4} + \frac{\sqrt{41}}{4} \qquad x = -\frac{3}{4} + \frac{\sqrt{41}}{4}$$

$$x = \frac{-3 + \sqrt{41}}{4} \qquad x = \frac{-3 + \sqrt{41}}{4}$$

The solution set is $\left\{\dfrac{-3 \pm \sqrt{41}}{4}\right\}$.

8. $A = P(1+r)^t$

$$4320 = 3000(1+r)^2$$

$$1.44 = (1+r)^2$$

$1 + r = \sqrt{1.44} \quad \text{or} \quad 1 + r = -\sqrt{1.44}$

$1 + r = 1.2 \qquad\qquad 1 + r = -1.2$

$\quad r = 0.2 \qquad\qquad\quad r = -2.2$

Reject -2.2 because we cannot have a negative interest rate. The solution is 0.2; the annual interest rate is 20%.

9. $x^2 + 20^2 = 50^2$

$$x^2 + 400 = 2500$$

$$x^2 = 2100$$

$$x = \pm\sqrt{2100}$$

$$x = \pm 10\sqrt{21}$$

$$x \approx \pm 45.8$$

The wire is attached $10\sqrt{21}$ feet, or about 45.8 feet, up the antenna.

10. $d = \sqrt{(2-(-1))^2 + (3-(-3))^2}$

$$= \sqrt{3^2 + 6^2} = \sqrt{9+36}$$

$$= \sqrt{45} = 6.71 \text{ units}$$

11. Midpoint $= \left(\dfrac{1+7}{2}, \dfrac{2+(-3)}{2}\right)$

$$= \left(\frac{8}{2}, \frac{-1}{2}\right) = \left(4, -\frac{1}{2}\right)$$

11.1 Exercise Set

1. $3x^2 = 75$

$$x^2 = 25$$

Apply the square root property.

$$x = \pm\sqrt{25}$$

$$x = \pm 5$$

The solution set is $\{\pm 5\}$.

3. $7x^2 = 42$

$$x^2 = 6$$

Apply the square root property.

$$x = \pm\sqrt{6}$$

The solution set is $\{\pm\sqrt{6}\}$.

5. $16x^2 = 25$

$$x^2 = \frac{25}{16}$$

Apply the square root property.

$$x = \pm\sqrt{\frac{25}{16}}$$

$$x = \pm\frac{5}{4}$$

The solution set is $\left\{\pm\frac{5}{4}\right\}$.

7. $3x^2 - 2 = 0$

$$3x^2 = 2$$

$$x^2 = \frac{2}{3}$$

Apply the square root property.

$$x = \pm\sqrt{\frac{2}{3}}$$

Because the proposed solutions are opposites, rationalize both denominators at once.

$$x = \pm\sqrt{\frac{2}{3}} = \pm\frac{\sqrt{2}}{\sqrt{3}} \cdot \frac{\sqrt{3}}{\sqrt{3}} = \pm\frac{\sqrt{6}}{3}$$

The solution set is $\left\{\pm\frac{\sqrt{6}}{3}\right\}$.

9. $25x^2 + 16 = 0$

$$25x^2 = -16$$

$$x^2 = -\frac{16}{25}$$

Apply the square root property.

$$x = \pm\sqrt{-\frac{16}{25}}$$

$$x = \pm\sqrt{\frac{16}{25}}\sqrt{-1}$$

$$x = \pm\frac{4}{5}i$$

$$x = 0 \pm \frac{4}{5}i = \pm\frac{4}{5}i$$

The solution set is $\left\{\pm\frac{4}{5}i\right\}$.

11. $(x+7)^2 = 9$

Apply the square root property.

$$x + 7 = \sqrt{9} \quad \text{or} \quad x + 7 = -\sqrt{9}$$

$$x + 7 = 3 \qquad\qquad x + 7 = -3$$

$$x = -4 \qquad\qquad x = -10$$

The solution set is $\{-10, -4\}$.

13. $(x-3)^2 = 5$

Apply the square root property.

$$x - 3 = \pm\sqrt{5}$$

$$x = 3 \pm \sqrt{5}$$

The solution set is $\left\{3 \pm \sqrt{5}\right\}$.

15. $(x+2)^2 = 8$

Apply the square root property.

$$x + 2 = \pm\sqrt{8}$$

$$x + 2 = \pm\sqrt{4 \cdot 2}$$

$$x + 2 = \pm 2\sqrt{2}$$

$$x = -2 \pm 2\sqrt{2}$$

The solution set is $\left\{-2 \pm 2\sqrt{2}\right\}$.

17. $(x-5)^2 = -9$

Apply the square root property.

$$x - 5 = \pm\sqrt{-9}$$

$$x - 5 = \pm 3i$$

$$x = 5 \pm 3i$$

The solution set is $\{5 \pm 3i\}$.

19. $\left(x + \frac{3}{4}\right)^2 = \frac{11}{16}$

Apply the square root property.

$$x + \frac{3}{4} = \pm\sqrt{\frac{11}{16}}$$

$$x + \frac{3}{4} = \pm\frac{\sqrt{11}}{4}$$

$$x = -\frac{3}{4} \pm \frac{\sqrt{11}}{4} = \frac{-3 \pm \sqrt{11}}{4}$$

The solution set is $\left\{\dfrac{-3 \pm \sqrt{11}}{4}\right\}$.

21. $x^2 - 6x + 9 = 36$

$(x-3)^2 = 36$

Apply the square root property.

$x - 3 = \sqrt{36}$ or $x - 3 = -\sqrt{36}$

$x - 3 = 6$ $\quad\quad$ $x - 3 = -6$

$x = 9$ $\quad\quad\quad$ $x = -3$

The solutions are 9 and -3 and the solution set is $\{-3, 9\}$.

23. $x^2 + 2x +$ ____

Since $b = 2$, add $\left(\dfrac{b}{2}\right)^2 = \left(\dfrac{2}{2}\right)^2 = (1)^2 = 1$.

$x^2 + 2x + 1 = (x+1)^2$

25. $x^2 - 14x +$ ____

Since $b = -14$, add $\left(\dfrac{b}{2}\right)^2 = \left(\dfrac{-14}{2}\right)^2 = (-7)^2 = 49$.

$x^2 - 14x + 49 = (x-7)^2$

27. $x^2 + 7x +$ ____

Since $b = 7$, add $\left(\dfrac{b}{2}\right)^2 = \left(\dfrac{7}{2}\right)^2 = \dfrac{49}{4}$.

$x^2 + 7x + \dfrac{49}{4} = \left(x + \dfrac{7}{2}\right)^2$

29. $x^2 - \dfrac{1}{2}x +$ ____

Since $b = -\dfrac{1}{2}$, add

$\left(\dfrac{b}{2}\right)^2 = \left(\dfrac{-1}{2} \div 2\right)^2 = \left(\dfrac{-1}{2} \cdot \dfrac{1}{2}\right)^2 = \left(\dfrac{-1}{4}\right)^2 = \dfrac{1}{16}$.

$x^2 - \dfrac{1}{2}x + \dfrac{1}{16} = \left(x - \dfrac{1}{4}\right)^2$

31. $x^2 + \dfrac{4}{3}x +$ ____

Since $b = \dfrac{4}{3}$, add

$\left(\dfrac{b}{2}\right)^2 = \left(\dfrac{4}{3} \div 2\right)^2 = \left(\dfrac{4}{3} \cdot \dfrac{1}{2}\right)^2 = \left(\dfrac{2}{3}\right)^2 = \dfrac{4}{9}$.

$x^2 + \dfrac{4}{3}x + \dfrac{4}{9} = \left(x + \dfrac{2}{3}\right)^2$

33. $x^2 - \dfrac{9}{4}x +$ ____

Since $b = -\dfrac{9}{4}$, add

$\left(\dfrac{b}{2}\right)^2 = \left(-\dfrac{9}{4} \div 2\right)^2$

$\quad\quad = \left(-\dfrac{9}{4} \cdot \dfrac{1}{2}\right)^2 = \left(-\dfrac{9}{8}\right)^2 = \dfrac{81}{64}$.

$x^2 - \dfrac{9}{4}x + \dfrac{81}{64} = \left(x - \dfrac{9}{8}\right)^2$

35. $x^2 + 4x = 32$

$x^2 + 4x \quad = 32$

Since $b = 4$, add $\left(\dfrac{b}{2}\right)^2 = \left(\dfrac{4}{2}\right)^2 = (2)^2 = 4$.

$x^2 + 4x + 4 = 32 + 4$

$(x+2)^2 = 36$

Apply the square root property.

$x + 2 = \sqrt{36}$ $\quad\quad$ $x + 2 = -\sqrt{36}$

$\quad\quad\quad\quad\quad\quad$ or

$x + 2 = 6$ $\quad\quad\quad$ $x + 2 = -6$

$x = 4$ $\quad\quad\quad\quad$ $x = -8$

The solution set is $\{-8, 4\}$.

37. $\quad x^2 + 6x = -2$

$x^2 + 6x \qquad = -2$

Since $b = 6$, add $\left(\dfrac{b}{2}\right)^2 = \left(\dfrac{6}{2}\right)^2 = (3)^2 = 9$.

$x^2 + 6x + 9 = -2 + 9$

$(x+3)^2 = 7$

Apply the square root property.

$x + 3 = \pm\sqrt{7}$

$x = -3 \pm \sqrt{7}$

The solution set is $\left\{-3 \pm \sqrt{7}\right\}$.

39. $\quad x^2 - 8x + 1 = 0$

$x^2 - 8x \qquad = -1$

Since $b = -8$, add $\left(\dfrac{b}{2}\right)^2 = \left(\dfrac{-8}{2}\right)^2 = (-4)^2 = 16$.

$x^2 - 8x + 16 = -1 + 16$

$(x-4)^2 = 15$

Apply the square root property.

$x - 4 = \pm\sqrt{15}$

$x = 4 \pm \sqrt{15}$

The solution set is $\left\{4 \pm \sqrt{15}\right\}$.

41. $\quad x^2 + 2x + 2 = 0$

$x^2 + 2x \qquad = -2$

Since $b = 2$, add $\left(\dfrac{b}{2}\right)^2 = \left(\dfrac{2}{2}\right)^2 = (1)^2 = 1$.

$x^2 + 2x + 1 = -2 + 1$

$(x+1)^2 = -1$

Apply the square root property.

$x + 1 = \pm\sqrt{-1} = \pm i$

$x = -1 \pm i$

The solution set is $\left\{-1 \pm i\right\}$.

43. $\quad x^2 + 3x - 1 = 0$

$x^2 + 3x \qquad = 1$

Since $b = 3$, add $\left(\dfrac{b}{2}\right)^2 = \left(\dfrac{3}{2}\right)^2 = \dfrac{9}{4}$.

$x^2 + 3x + \dfrac{9}{4} = 1 + \dfrac{9}{4}$

$\left(x + \dfrac{3}{2}\right)^2 = \dfrac{13}{4}$

Apply the square root property.

$x + \dfrac{3}{2} = \pm\sqrt{\dfrac{13}{4}} = \pm\dfrac{\sqrt{13}}{2}$

$x = -\dfrac{3}{2} \pm \dfrac{\sqrt{13}}{2} = \dfrac{-3 \pm \sqrt{13}}{2}$

The solution set is $\left\{\dfrac{-3 \pm \sqrt{13}}{2}\right\}$.

45. $\quad x^2 + \dfrac{4}{7}x + \dfrac{3}{49} = 0$

$x^2 + \dfrac{4}{7}x \qquad = -\dfrac{3}{49}$

Since $b = \dfrac{4}{7}$, add $\left(\dfrac{1}{2}b\right)^2 = \left(\dfrac{1}{2} \cdot \dfrac{4}{7}\right)^2 \left(\dfrac{2}{7}\right)^2 = \dfrac{4}{49}$.

$x^2 + \dfrac{4}{7}x + \dfrac{4}{49} = -\dfrac{3}{49} + \dfrac{4}{49}$

$\left(x + \dfrac{2}{7}\right)^2 = \dfrac{1}{49}$

Apply the square root property.

$x + \dfrac{2}{7} = \pm\sqrt{\dfrac{1}{49}}$

$x + \dfrac{2}{7} = \pm\dfrac{1}{7}$

$x = -\dfrac{2}{7} \pm \dfrac{1}{7}$

$x = -\dfrac{2}{7} + \dfrac{1}{7} = -\dfrac{1}{7}$ or $-\dfrac{2}{7} - \dfrac{1}{7} = -\dfrac{3}{7}$

The solution set is $\left\{-\dfrac{3}{7}, -\dfrac{1}{7}\right\}$.

47. $x^2 + x - 1 = 0$

$x^2 + x \quad = 1$

Since $b = 1$, add $\left(\dfrac{b}{2}\right)^2 = \left(\dfrac{1}{2}\right)^2 = \dfrac{1}{4}$.

$x^2 + x + \dfrac{1}{4} = 1 + \dfrac{1}{4}$

$\left(x + \dfrac{1}{2}\right)^2 = \dfrac{5}{4}$

Apply the square root property.

$x + \dfrac{1}{2} = \pm\sqrt{\dfrac{5}{4}}$

$x + \dfrac{1}{2} = \pm\dfrac{\sqrt{5}}{2}$

$x = -\dfrac{1}{2} \pm \dfrac{\sqrt{5}}{2} = \dfrac{-1 \pm \sqrt{5}}{2}$

The solution set is $\left\{\dfrac{-1 \pm \sqrt{5}}{2}\right\}$.

49. $2x^2 + 3x - 5 = 0$

$x^2 + \dfrac{3}{2}x - \dfrac{5}{2} = 0$

$x^2 + \dfrac{3}{2}x \quad = \dfrac{5}{2}$

Since $b = \dfrac{3}{2}$, add

$\left(\dfrac{1}{2}b\right)^2 = \left(\dfrac{1}{2} \cdot \dfrac{3}{2}\right)^2 = \left(\dfrac{3}{4}\right)^2 = \dfrac{9}{16}$.

$x^2 + \dfrac{3}{2}x + \dfrac{9}{16} = \dfrac{5}{2} + \dfrac{9}{16}$

$\left(x + \dfrac{3}{4}\right)^2 = \dfrac{40}{16} + \dfrac{9}{16} = \dfrac{49}{16}$

Apply the square root property.

$x + \dfrac{3}{4} = \pm\sqrt{\dfrac{49}{16}} = \pm\dfrac{7}{4}$

$x = -\dfrac{3}{4} \pm \dfrac{7}{4}$

$x = -\dfrac{3}{4} + \dfrac{7}{4} \quad$ or $\quad x = -\dfrac{3}{4} - \dfrac{7}{4}$

$x = \dfrac{4}{4} \qquad\qquad x = -\dfrac{10}{4}$

$x = 1 \qquad\qquad\quad x = -\dfrac{5}{2}$

The solution set is $\left\{-\dfrac{5}{2},\, 1\right\}$.

51. $3x^2 + 6x + 1 = 0$

$x^2 + 2x + \dfrac{1}{3} = 0$

$x^2 + 2x \quad = -\dfrac{1}{3}$

Since $b = 2$, add

$\left(\dfrac{b}{2}\right)^2 = \left(\dfrac{2}{2}\right)^2 = 1^2 = 1$.

$x^2 + 2x + 1 = -\dfrac{1}{3} + 1$

$(x + 1)^2 = -\dfrac{1}{3} + \dfrac{3}{3} = \dfrac{2}{3}$

Apply the square root property.

$x + 1 = \pm\sqrt{\dfrac{2}{3}}$

$x + 1 = \pm\dfrac{\sqrt{2}}{\sqrt{3}} \cdot \dfrac{\sqrt{3}}{\sqrt{3}} = \pm\dfrac{\sqrt{6}}{3}$

$x = -1 \pm \dfrac{\sqrt{6}}{3} = \dfrac{-3 \pm \sqrt{6}}{3}$

The solution set is $\left\{\dfrac{-3 \pm \sqrt{6}}{3}\right\}$.

53. $3x^2 - 8x + 1 = 0$

$x^2 - \dfrac{8}{3}x + \dfrac{1}{3} = 0$

$x^2 - \dfrac{8}{3}x \quad = -\dfrac{1}{3}$

Since $b = -\dfrac{8}{3}$, add

$\left(\dfrac{1}{2}b\right)^2 = \left[\dfrac{1}{2}\left(-\dfrac{8}{3}\right)\right]^2 = \left(-\dfrac{4}{3}\right)^2 = \dfrac{16}{9}$.

$x^2 - \dfrac{8}{3}x + \dfrac{16}{9} = -\dfrac{1}{3} + \dfrac{16}{9}$

$\left(x - \dfrac{4}{3}\right)^2 = -\dfrac{3}{9} + \dfrac{16}{9} = \dfrac{13}{9}$

Apply the square root property.

$$x - \frac{4}{3} = \pm\sqrt{\frac{13}{9}}$$

$$x - \frac{4}{3} = \pm\frac{\sqrt{13}}{3}$$

$$x = \frac{4}{3} \pm \frac{\sqrt{13}}{3} = \frac{4 \pm \sqrt{13}}{3}$$

The solution set is $\left\{\dfrac{4 \pm \sqrt{13}}{3}\right\}$.

55. $8x^2 - 4x + 1 = 0$

$$x^2 - \frac{1}{2}x + \frac{1}{8} = 0$$

$$x^2 - \frac{1}{2}x \quad = -\frac{1}{8}$$

Since $b = -\dfrac{1}{2}$, add

$$\left(\frac{1}{2}b\right)^2 = \left[\frac{1}{2}\left(-\frac{1}{2}\right)\right]^2 = \left(-\frac{1}{4}\right)^2 = \frac{1}{16}.$$

$$x^2 - \frac{1}{2}x + \frac{1}{16} = -\frac{1}{8} + \frac{1}{16}$$

$$\left(x - \frac{1}{4}\right)^2 = -\frac{2}{16} + \frac{1}{16} = -\frac{1}{16}$$

Apply the square root property.

$$x - \frac{1}{4} = \pm\sqrt{-\frac{1}{16}}$$

$$x - \frac{1}{4} = \pm\frac{1}{4}i$$

$$x = \frac{1}{4} \pm \frac{1}{4}i$$

The solution set is $\left\{\dfrac{1}{4} \pm \dfrac{1}{4}i\right\}$.

57. $f(x) = 36$

$$(x-1)^2 = 36$$

Apply the square root property.

$$x - 1 = \pm\sqrt{36}$$

$$x - 1 = \pm 6$$

$$x = 1 \pm 6$$

$$x = 1 + 6 = 7 \text{ or } 1 - 6 = -5$$

The values are -5 and 7.

59. $g(x) = \dfrac{9}{25}$

$$\left(x - \frac{2}{5}\right)^2 = \frac{9}{25}$$

Apply the square root property.

$$x - \frac{2}{5} = \pm\sqrt{\frac{9}{25}} = \pm\frac{3}{5}$$

$$x = \frac{2}{5} \pm \frac{3}{5}$$

$$x = \frac{2}{5} + \frac{3}{5} \text{ or } x = \frac{2}{5} - \frac{3}{5}$$

$$x = \frac{5}{5} = 1 \text{ or } x = -\frac{1}{5}$$

The values are $-\dfrac{1}{5}$ and 1.

61. $h(x) = -125$

$$5(x+2)^2 = -125$$

$$(x+2)^2 = -25$$

Apply the square root property.

$$x + 2 = \pm\sqrt{-25}$$

$$x + 2 = \pm 5i$$

$$x = -2 \pm 5i$$

The values are $-2 \pm 5i$.

63. $d = \sqrt{(14-2)^2 + (8-3)^2}$

$$= \sqrt{12^2 + 5^2} = \sqrt{144 + 25}$$

$$= \sqrt{169} = 13$$

65. $d = \sqrt{(6-4)^2 + (3-1)^2}$

$$= \sqrt{2^2 + 2^2} = \sqrt{4+4}$$

$$= \sqrt{8} = \sqrt{4 \cdot 2} = 2\sqrt{2} \approx 2.83$$

67. $d = \sqrt{(-3-0)^2 + (4-0)^2}$

$= \sqrt{(-3)^2 + 4^2} = \sqrt{9+16}$

$= \sqrt{25} = 5$

69. $d = \sqrt{(3-(-2))^2 + (-4-(-6))^2}$

$= \sqrt{5^2 + 2^2} = \sqrt{25+4}$

$= \sqrt{29} \approx 5.39$

71. $d = \sqrt{(4-0)^2 + (1-(-3))^2}$

$= \sqrt{4^2 + 4^2} = \sqrt{16+16}$

$= \sqrt{32} = \sqrt{16 \cdot 2} = 4\sqrt{2} \approx 5.66$

73. $d = \sqrt{(3.5-(-0.5))^2 + (8.2-6.2)^2}$

$= \sqrt{4^2 + 2^2} = \sqrt{16+4}$

$= \sqrt{20} = \sqrt{4 \cdot 5} = 2\sqrt{5} \approx 4.47$

75. $d = \sqrt{(\sqrt{5}-0)^2 + (0-(-\sqrt{3}))^2}$

$= \sqrt{(\sqrt{5})^2 + (\sqrt{3})^2} = \sqrt{5+3}$

$= \sqrt{8} = \sqrt{4 \cdot 2} = 2\sqrt{2} \approx 2.83$

77. $d = \sqrt{(3\sqrt{3}-(-\sqrt{3}))^2 + (\sqrt{5}-4\sqrt{5})^2}$

$= \sqrt{(4\sqrt{3})^2 + (-3\sqrt{5})^2}$

$= \sqrt{16 \cdot 3 + 9 \cdot 5} = \sqrt{48+45}$

$= \sqrt{93} \approx 9.64$

79. $d = \sqrt{\left(\dfrac{7}{3}-\dfrac{1}{3}\right)^2 + \left(\dfrac{1}{5}-\dfrac{6}{5}\right)^2}$

$= \sqrt{\left(\dfrac{6}{3}\right)^2 + \left(-\dfrac{5}{5}\right)^2}$

$= \sqrt{2^2 + (-1)^2} = \sqrt{4+1}$

$= \sqrt{5} \approx 2.24$

81. Midpoint $= \left(\dfrac{6+2}{2}, \dfrac{8+4}{2}\right)$

$= \left(\dfrac{8}{2}, \dfrac{12}{2}\right) = (4, 6)$

83. Midpoint $= \left(\dfrac{-2+(-6)}{2}, \dfrac{-8+(-2)}{2}\right)$

$= \left(\dfrac{-8}{2}, \dfrac{-10}{2}\right) = (-4, -5)$

85. Midpoint $= \left(\dfrac{-3+6}{2}, \dfrac{-4+(-8)}{2}\right)$

$= \left(\dfrac{3}{2}, \dfrac{-12}{2}\right) = \left(\dfrac{3}{2}, -6\right)$

87. Midpoint $= \left(\dfrac{-\dfrac{7}{2}+\left(-\dfrac{5}{2}\right)}{2}, \dfrac{\dfrac{3}{2}+\left(-\dfrac{11}{2}\right)}{2}\right)$

$= \left(\dfrac{-\dfrac{12}{2}}{2}, \dfrac{-\dfrac{8}{2}}{2}\right)$

$= \left(-\dfrac{12}{2} \cdot \dfrac{1}{2}, -\dfrac{8}{2} \cdot \dfrac{1}{2}\right)$

$= \left(-\dfrac{12}{4}, -\dfrac{8}{4}\right) = (-3, -2)$

89. Midpoint $= \left(\dfrac{8+(-6)}{2}, \dfrac{3\sqrt{5}+7\sqrt{5}}{2}\right)$

$= \left(\dfrac{2}{2}, \dfrac{10\sqrt{5}}{2}\right) = (1, 5\sqrt{5})$

91. Midpoint $= \left(\dfrac{\sqrt{18}+\sqrt{2}}{2}, \dfrac{-4+4}{2}\right)$

$= \left(\dfrac{\sqrt{9 \cdot 2}+\sqrt{2}}{2}, \dfrac{0}{2}\right)$

$= \left(\dfrac{3\sqrt{2}+\sqrt{2}}{2}, 0\right)$

$= \left(\dfrac{4\sqrt{2}}{2}, 0\right) = (2\sqrt{2}, 0)$

93. Let x = the number.

$$3(x-2)^2 = -12$$

$$(x-2)^2 = -4$$

Apply the square root property.

$$x - 2 = \pm\sqrt{-4} = \pm 2i$$
$$x = 2 \pm 2i$$

The values are $2 + 2i$ and $2 - 2i$.

95. $h = \dfrac{v^2}{2g}$

$$2gh = v^2$$

Apply the square root property and keep only the principal square root.

$$v = \sqrt{2gh}$$

97. $A = P(1+r)^2$

$$\frac{A}{P} = (1+r)^2$$

Apply the square root property and keep only the principal square root.

$$1 + r = \sqrt{\frac{A}{P}}$$

$$1 + r = \frac{\sqrt{A}}{\sqrt{P}} \cdot \frac{\sqrt{P}}{\sqrt{P}} = \frac{\sqrt{AP}}{P}$$

$$r = \frac{\sqrt{AP}}{P} - 1$$

99. $\dfrac{x^2}{3} + \dfrac{x}{9} - \dfrac{1}{6} = 0$

$$3\left(\frac{x^2}{3} + \frac{x}{9} - \frac{1}{6}\right) = 3(0)$$

$$x^2 + \frac{1}{3}x - \frac{1}{2} = 0$$

$$x^2 + \frac{1}{3}x = \frac{1}{2}$$

Since $b = \dfrac{1}{3}$, add

$$\left(\frac{1}{2}b\right)^2 = \left(\frac{1}{2} \cdot \frac{1}{3}\right)^2 = \left(\frac{1}{6}\right)^2 = \frac{1}{36}.$$

$$x^2 + \frac{1}{3}x + \frac{1}{36} = \frac{1}{2} + \frac{1}{36}$$

$$\left(x + \frac{1}{6}\right)^2 = \frac{18}{36} + \frac{1}{36} = \frac{19}{36}$$

Apply the square root property.

$$x + \frac{1}{6} = \pm\sqrt{\frac{19}{36}} = \pm\frac{\sqrt{19}}{6}$$

$$x = -\frac{1}{6} \pm \frac{\sqrt{19}}{6} = \frac{-1 \pm \sqrt{19}}{6}$$

The solution set is $\left\{\dfrac{-1 \pm \sqrt{19}}{6}\right\}$.

101.　　$x^2 - bx = 2b^2$

$$x^2 - bx \quad = 2b^2$$

Since $-b$ is the linear coefficient, add

$$\left(\frac{-b}{2}\right)^2 = \frac{b^2}{4}.$$

$$x^2 - bx + \frac{b^2}{4} = 2b^2 + \frac{b^2}{4}$$

$$\left(x - \frac{b}{2}\right)^2 = \frac{8b^2}{4} + \frac{b^2}{4} = \frac{9b^2}{4}$$

Apply the square root property.

$$x - \frac{b}{2} = \pm\sqrt{\frac{9b^2}{4}} = \pm\frac{3b}{2}$$

$$x = \frac{b}{2} \pm \frac{3b}{2}$$

$$x = \frac{b}{2} + \frac{3b}{2} \quad \text{or} \quad x = \frac{b}{2} - \frac{3b}{2}$$

$$x = \frac{4b}{2} = 2b \quad \text{or} \quad x = -\frac{2b}{2} = -b$$

The solution set is $\{-b,\ 2b\}$.

103. a. $x^2 + 8x$

b. 16

c. $x^2 + 8x + 16$

d. $(x+4)^2$

105. $2880 = 2000(1+r)^2$

$$\frac{2880}{2000} = (1+r)^2$$

$$1.44 = (1+r)^2$$

Apply the square root property.

$1+r = \pm\sqrt{1.44}$
$1+r = \pm 1.2$
$r = -1 \pm 1.2$
$r = -1 + 1.2 \text{ or } -1 - 1.2$
$r = 0.2 \text{ or } -2.2$

Reject –2.2 because we cannot have a negative interest rate. The solution is 0.2 and the annual interest rate is 20%.

107. $1445 = 1280(1+r)^2$

$$\frac{1445}{1280} = (1+r)^2$$

$1.12890625 = (1+r)^2$

Apply the square root property.
$1+r = \pm\sqrt{1.12890625}$
$1+r = \pm 1.0625$
$r = -1 \pm 1.0625$
$r = -1 + 1.0625 \text{ or } -1 - 1.0625$
$r = 0.0625 \text{ or } -2.0625$

Reject –2.0625 because we cannot have a negative interest rate. The solution is 0.0625 and the annual interest rate is 6.25%.

109. $32,000 = 62.2x^2 + 7000$

$25,000 = 62.2x^2$

$$\frac{25,000}{62.2} = x^2$$

Apply the square root property.

$$x = \pm\sqrt{\frac{25,000}{62.2}} \approx \pm 20$$

Disregard –20 because we can't have a negative number of years. The solution is 20 and we conclude that there were about 32,000 multinational corporations in approximately $1970 + 20 = 1990$. The function models the actual number quite well.

111. $4800 = 16t^2$

$$\frac{4800}{16} = t^2$$

$300 = t^2$

Apply the square root property.

$t = \pm\sqrt{300}$
$t = \pm 10\sqrt{3} \approx \pm 17.3$

Disregard –17.3 because we can't have a negative time measurement. The solution is 17.3 and we conclude that the sky diver was in a free fall for $10\sqrt{3}$ or approximately 17.3 seconds.

113.

$x^2 = 6^2 + 3^2 = 36 + 9 = 45$

Apply the square root property.
$x = \pm\sqrt{45} = \pm\sqrt{9 \cdot 5} = \pm 3\sqrt{5}$

Disregard $-3\sqrt{5}$ because we can't have a negative length measurement. The solution is $3\sqrt{5}$ and we conclude that the pedestrian route is $3\sqrt{5}$ or approximately 6.7 miles long.

115.
$$x^2 + 10^2 = 30^2$$
$$x^2 + 100 = 900$$
$$x^2 = 800$$

Apply the square root property.

$$x = \pm\sqrt{800} = \pm\sqrt{400 \cdot 2} = \pm 20\sqrt{2}$$

Disregard $-20\sqrt{2}$ because we can't have a negative length measurement. The solution is $20\sqrt{2}$. We conclude that the ladder reaches $20\sqrt{2}$ feet, or approximately 28.3 feet, up the house.

117.

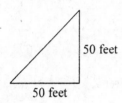

50 feet

50 feet

$$50^2 + 50^2 = x^2$$
$$2500 + 2500 = x^2$$
$$5000 = x^2$$

Apply the square root property.
$$x = \pm\sqrt{5000}$$
$$x = \pm\sqrt{2500 \cdot 2} = \pm 50\sqrt{2} \approx \pm 70.7$$

Disregard $-50\sqrt{2}$ because we cannot have a negative length measurement. The solution is $50\sqrt{2}$. We conclude that a supporting wire of $50\sqrt{2}$ feet, or approximately 70.7 feet, is required.

119.
$$A = lw$$
$$196 = (x+2+2)(x+2+2)$$
$$196 = (x+4)(x+4)$$
$$196 = (x+4)^2$$

Apply the square root property.

$$x+4 = \pm\sqrt{196} = \pm 14$$
$$x = -4 \pm 14$$
$$x = -4+14 = 10 \text{ or } x = -4-14 = -18$$

Disregard -18 because we can't have a negative length measurement. The solution is 10. We conclude that the length of the original square is 10 meters.

121. First find the distance from Bangkok to Phnom Penh.

$$d = \sqrt{\left(65 - (-115)\right)^2 + (70 - 170)^2}$$
$$= \sqrt{180^2 + (-100)^2}$$
$$= \sqrt{32400 + 10000}$$
$$= \sqrt{42400} \approx 205.9$$

The distance is approximately 205.9 miles.
$$t = \frac{d}{r} = \frac{205.9}{400} \approx 0.5$$

It will take approximately 0.5 hours or 30 minutes to make the flight.

123. – 129. Answers will vary.

131. $(x-1)^2 - 9 = 0$

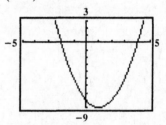

The solution set is $\{-2, 4\}$.
Check:

$x = -2$	$x = 4$
$(-2-1)^2 - 9 = 0$	$(4-1)^2 - 9 = 0$
$(-3)^2 - 9 = 0$	$(3)^2 - 9 = 0$
$9 - 9 = 0$	$9 - 9 = 0$
$0 = 0, \text{ true}$	$0 = 0, \text{ true}$

133. does not make sense; Explanations will vary. Sample explanation: After you take half of the b-term you must square that value. Once squared, the value to be added to both sides will always be positive.

135. makes sense

137. true

139. false; Changes to make the statement true will vary. A sample change is: Divide both sides of the equation by 2 before completing the square. Add $\left(\dfrac{b}{2}\right)^2 = \left(\dfrac{-3}{2}\right)^2 = \dfrac{9}{4}$.

141. $\dfrac{x^2}{a^2} + \dfrac{y^2}{b^2} = 1$

$$\dfrac{y^2}{b^2} = 1 - \dfrac{x^2}{a^2}$$

$$y^2 = b^2\left(1 - \dfrac{x^2}{a^2}\right)$$

$$\sqrt{y^2} = \pm\sqrt{b^2\left(1 - \dfrac{x^2}{a^2}\right)}$$

$$y = \pm b\sqrt{1 - \dfrac{x^2}{a^2}}$$

$$y = \pm b\sqrt{\dfrac{a^2}{a^2} - \dfrac{x^2}{a^2}}$$

$$y = \pm b\sqrt{\dfrac{a^2 - x^2}{a^2}} = \pm\dfrac{b\sqrt{a^2 - x^2}}{a}$$

143. $x^2 + bx + c = 0$

$$x^2 + bx = -c$$

$$x^2 + bx + \dfrac{b^2}{4} = -c + \dfrac{b^2}{4}$$

$$\left(x + \dfrac{b}{2}\right)^2 = -c + \dfrac{b^2}{4}$$

$$x + \dfrac{b}{2} = \pm\sqrt{-c + \dfrac{b^2}{4}}$$

$$x = -\dfrac{b}{2} \pm \sqrt{-\dfrac{4c}{4} + \dfrac{b^2}{4}}$$

$$x = -\dfrac{b}{2} \pm \sqrt{-\dfrac{4c + b^2}{4}}$$

$$x = -\dfrac{b}{2} \pm \dfrac{\sqrt{-4c + b^2}}{2}$$

$$x = \dfrac{-b \pm \sqrt{b^2 - 4c}}{2}$$

145. $4x - 2 - 3\big[4 - 2(3 - x)\big]$

$= 4x - 2 - 3\big[4 - 6 + 2x\big] = 4x - 2 - 3\big[-2 + 2x\big]$

$= 4x - 2 + 6 - 6x = 4 - 2x$

146. $1 - 8x^3 = 1^3 - (2x)^3$

$\qquad\qquad = (1 - 2x)\left(1 + 2x + 4x^2\right)$

147.

$$\begin{array}{r|rrrr}
3 & 1 & -5 & 2 & 0 & -6 \\
& & 3 & -6 & -12 & -36 \\
\hline
& 1 & -2 & -4 & -12 & -42
\end{array}$$

$$x^3 - 2x^2 - 4x - 12 - \dfrac{42}{x - 3}$$

148. a. $\quad 8x^2 + 2x - 1 = 0$

$(2x + 1)(4x - 1) = 0$

$2x + 1 = 0 \qquad$ or $\qquad 4x - 1 = 0$

$2x = -1 \qquad\qquad\qquad 4x = 1$

$x = -\dfrac{1}{2} \qquad\qquad\qquad x = \dfrac{1}{4}$

The solution set is $\left\{-\dfrac{1}{2}, \dfrac{1}{4}\right\}$.

b. $b^2 - 4ac = 2^2 - 4(8)(-1) = 36$

Yes, $b^2 - 4ac$ is a perfect square.

149. a. $9x^2 - 6x + 1 = 0$

$(3x - 1)^2 = 0$

$3x - 1 = 0$

$3x = 1$

$x = \dfrac{1}{3}$

The solution set is $\left\{\dfrac{1}{3}\right\}$.

b. $b^2 - 4ac = (-6)^2 - 4(9)(1) = 36 - 36 = 0$

150. a. $\qquad 3 + \dfrac{4}{x} = -\dfrac{2}{x^2}$

$$x^2\left(3 + \dfrac{4}{x}\right) = x^2\left(-\dfrac{2}{x^2}\right)$$

$$3x^2 + 4x = -2$$

$$3x^2 + 4x + 2 = 0$$

b. $b^2 - 4ac = 4^2 - 4(3)(2) = 16 - 24 = -8$

11.2 Check Points

1. $2x^2 + 9x - 5 = 0$

$a = 2 \quad b = 9 \quad c = -5$

$x = \dfrac{-b \pm \sqrt{b^2 - 4ac}}{2a}$

$x = \dfrac{-9 \pm \sqrt{9^2 - 4(2)(-5)}}{2(2)}$

$= \dfrac{-9 \pm \sqrt{81 + 40}}{4}$

$= \dfrac{-9 \pm \sqrt{121}}{4}$

$= \dfrac{-9 \pm 11}{4}$

Evaluate the expression to obtain two solutions.

$x = \dfrac{-9 + 11}{4} = \dfrac{1}{2}$ or $\quad x = \dfrac{-9 - 11}{4} = -5$

The solution set is $\left\{ -5, \dfrac{1}{2} \right\}$.

2. $\qquad 2x^2 = 6x - 1$

$2x^2 - 6x + 1 = 0$

$a = 2 \quad b = -6 \quad c = 1$

$x = \dfrac{-b \pm \sqrt{b^2 - 4ac}}{2a}$

$x = \dfrac{-(-6) \pm \sqrt{(-6)^2 - 4(2)(1)}}{2(2)}$

$= \dfrac{6 \pm \sqrt{36 - 8}}{4}$

$= \dfrac{6 \pm \sqrt{28}}{4}$

$= \dfrac{6 \pm 2\sqrt{7}}{4}$

$= \dfrac{3 \pm \sqrt{7}}{2}$

The solution set is $\left\{ \dfrac{3 \pm \sqrt{7}}{2} \right\}$.

3. $\qquad 3x^2 + 5 = -6x$

$3x^2 + 6x + 5 = 0$

$a = 3 \quad b = 6 \quad c = 5$

$x = \dfrac{-b \pm \sqrt{b^2 - 4ac}}{2a}$

$x = \dfrac{-6 \pm \sqrt{6^2 - 4(3)(5)}}{2(3)}$

$= \dfrac{-6 \pm \sqrt{36 - 60}}{6}$

$= \dfrac{-6 \pm \sqrt{-24}}{6}$

$= \dfrac{6 \pm 2i\sqrt{6}}{6}$

$= \dfrac{6}{6} \pm \dfrac{2i\sqrt{6}}{6}$

$= 1 \pm i\dfrac{\sqrt{6}}{3}$

The solution set is $\left\{ 1 \pm i\dfrac{\sqrt{6}}{3} \right\}$.

4. a. $b^2 - 4ac = 6^2 - 4(1)(9) = 0$

Since the discriminant is zero, there is one real rational solution.

b. $b^2 - 4ac = (-7)^2 - 4(2)(-4) = 81$

Since the discriminant is positive and a perfect square, there are two real rational solutions.

c. $b^2 - 4ac = (-2)^2 - 4(3)(4) = -44$

Since the discriminant is negative, there is no real solution. There are imaginary solutions that are complex conjugates.

5. a. Because the solution set is $\left\{ -\dfrac{3}{5}, \dfrac{1}{4} \right\}$, we have

$x = -\dfrac{3}{5}$ or $\qquad x = \dfrac{1}{4}$

$5x = -3 \qquad\qquad 4x = 1$

$5x + 3 = 0 \qquad\quad 4x - 1 = 0.$

Use the zero-product principle in reverse.

$(5x + 3)(4x - 1) = 0$

$20x^2 - 5x + 12x - 3 = 0$

$20x^2 + 7x - 3 = 0$

b. Because the solution set is $\{-7i, 7i\}$, we have

$$x = -7i \quad \text{or} \quad x = 7i$$
$$x + 7i = 0 \qquad x - 7i = 0$$

Use the zero-product principle in reverse.

$$(x + 7i)(x - 7i) = 0$$
$$x^2 - 7ix + 7ix - 49i^2 = 0$$
$$x^2 - 49(-1) = 0$$
$$x^2 + 49 = 0$$

6. $P(A) = 0.01A^2 + 0.05A + 107$

$$115 = 0.01A^2 + 0.05A + 107$$
$$0 = 0.01A^2 + 0.05A - 8$$
$$a = 0.01 \quad b = 0.05 \quad c = -8$$

$$x = \frac{-b \pm \sqrt{b^2 - 4ac}}{2a}$$

$$x = \frac{-0.05 \pm \sqrt{0.05^2 - 4(0.01)(-8)}}{2(0.01)}$$

$$= \frac{-0.05 \pm \sqrt{0.3225}}{0.02}$$

$$x = \frac{-0.05 + \sqrt{0.3225}}{0.02} \quad \text{or} \quad x = \frac{-0.05 - \sqrt{0.3225}}{0.02}$$

$$x \approx 26 \qquad\qquad x \approx -31$$

A woman's normal systolic blood pressure is 115 mm at about 26 years of age.
This is represented by the point (26,115) on the blue graph.

11.2 Exercise Set

1. $x^2 + 8x + 12 = 0$

$$a = 1 \quad b = 8 \quad c = 12$$

$$x = \frac{-8 \pm \sqrt{8^2 - 4(1)(12)}}{2(1)}$$

$$= \frac{-8 \pm \sqrt{64 - 48}}{2}$$

$$= \frac{-8 \pm \sqrt{16}}{2} = \frac{-8 \pm 4}{2}$$

Evaluate the expression to obtain two solutions.

$$x = \frac{-8 - 4}{2} = -6 \quad \text{or} \quad x = \frac{-8 + 4}{2} = -2$$

The solution set is $\{-6, -2\}$.

3. $2x^2 - 7x = -5$

$$2x^2 - 7x + 5 = 0$$
$$a = 2 \quad b = -7 \quad c = 5$$

$$x = \frac{-(-7) \pm \sqrt{(-7)^2 - 4(2)(5)}}{2(2)}$$

$$= \frac{7 \pm \sqrt{49 - 40}}{4} = \frac{7 \pm \sqrt{9}}{4} = \frac{7 \pm 3}{4}$$

Evaluate the expression to obtain two solutions.

$$x = \frac{7 - 3}{4} = 1 \quad \text{or} \quad x = \frac{7 + 3}{4} = \frac{5}{2}$$

The solution set is $\left\{1, \dfrac{5}{2}\right\}$.

5. $x^2 + 3x - 20 = 0$

$$a = 1 \quad b = 3 \quad c = -20$$

$$x = \frac{-3 \pm \sqrt{3^2 - 4(1)(-20)}}{2(1)}$$

$$= \frac{-3 \pm \sqrt{9 - (-80)}}{2} = \frac{-3 \pm \sqrt{89}}{2}$$

The solution set is $\left\{\dfrac{-3 \pm \sqrt{89}}{2}\right\}$.

7. $3x^2 - 7x = 3$

$$3x^2 - 7x - 3 = 0$$
$$a = 3 \quad b = -7 \quad c = -3$$

$$x = \frac{-(-7) \pm \sqrt{(-7)^2 - 4(3)(-3)}}{2(3)}$$

$$= \frac{7 \pm \sqrt{49 - (-36)}}{6} = \frac{7 \pm \sqrt{85}}{6}$$

The solution set is $\left\{\dfrac{7 \pm \sqrt{85}}{6}\right\}$.

9. $6x^2 = 2x + 1$

$6x^2 - 2x - 1 = 0$

$a = 6 \quad b = -2 \quad c = -1$

$x = \dfrac{-(-2) \pm \sqrt{(-2)^2 - 4(6)(-1)}}{2(6)}$

$= \dfrac{2 \pm \sqrt{4 + 24}}{12}$

$= \dfrac{2 \pm \sqrt{28}}{12}$

$= \dfrac{2 \pm 2\sqrt{7}}{12} = \dfrac{2(1 \pm \sqrt{7})}{12} = \dfrac{1 \pm \sqrt{7}}{6}$

The solution set is $\left\{ \dfrac{1 \pm \sqrt{7}}{6} \right\}$.

11. $4x^2 - 3x = -6$

$4x^2 - 3x + 6 = 0$

$a = 4 \quad b = -3 \quad c = 6$

$x = \dfrac{-(-3) \pm \sqrt{(-3)^2 - 4(4)(6)}}{2(4)}$

$= \dfrac{3 \pm \sqrt{9 - 96}}{8}$

$= \dfrac{3 \pm \sqrt{-87}}{8}$

$= \dfrac{3 \pm \sqrt{87(-1)}}{8}$

$= \dfrac{3 \pm i\sqrt{87}}{8} = \dfrac{3}{8} \pm i\dfrac{\sqrt{87}}{8}$

The solution set is $\left\{ \dfrac{3}{8} \pm i\dfrac{\sqrt{87}}{8} \right\}$.

13. $x^2 - 4x + 8 = 0$

$a = 1 \quad b = -4 \quad c = 8$

$x = \dfrac{-(-4) \pm \sqrt{(-4)^2 - 4(1)(8)}}{2(1)}$

$= \dfrac{4 \pm \sqrt{16 - 32}}{2}$

$= \dfrac{4 \pm \sqrt{-16}}{2}$

$= \dfrac{4 \pm 4i}{2} = \dfrac{4}{2} \pm \dfrac{4}{2}i = 2 \pm 2i$

The solution set is $\{ 2 \pm 2i \}$.

15. $3x^2 = 8x - 7$

$3x^2 - 8x + 7 = 0$

$a = 3 \quad b = -8 \quad c = 7$

$x = \dfrac{-(-8) \pm \sqrt{(-8)^2 - 4(3)(7)}}{2(3)}$

$= \dfrac{8 \pm \sqrt{64 - 84}}{6}$

$= \dfrac{8 \pm \sqrt{-20}}{6}$

$= \dfrac{8 \pm \sqrt{4 \cdot 5(-1)}}{6}$

$= \dfrac{8 \pm 2i\sqrt{5}}{6} = \dfrac{8}{6} \pm \dfrac{2}{6}i\sqrt{5} = \dfrac{4}{3} \pm i\dfrac{\sqrt{5}}{3}$

The solution set is $\left\{ \dfrac{4}{3} \pm i\dfrac{\sqrt{5}}{3} \right\}$.

17.
$$2x(x-2) = x+12$$
$$2x^2 - 4x = x+12$$
$$2x^2 - 5x - 12 = 0$$
$$a = 2 \quad b = -5 \quad c = -12$$

$$x = \frac{-(-5) \pm \sqrt{(-5)^2 - 4(2)(-12)}}{2(2)}$$

$$= \frac{5 \pm \sqrt{25+96}}{4} = \frac{5 \pm \sqrt{121}}{4} = \frac{5 \pm 11}{4}$$

Evaluate the expression to obtain two solutions.

$$x = \frac{5-11}{4} = -\frac{3}{2} \quad \text{or} \quad x = \frac{5+11}{4} = 4$$

The solution set is $\left\{ -\frac{3}{2}, \ 4 \right\}$.

19.
$$x^2 + 8x + 3 = 0$$
$$a = 1 \quad b = 8 \quad c = 3$$

$$b^2 - 4ac = 8^2 - 4(1)(3) = 64 - 12 = 52$$

Since the discriminant is positive and not a perfect square, there are two real irrational solutions.

21.
$$x^2 + 6x + 8 = 0$$
$$a = 1 \quad b = 6 \quad c = 8$$

$$b^2 - 4ac = (6)^2 - 4(1)(8)$$
$$= 36 - 32 = 4$$

Since the discriminant is greater than zero, there are two unequal real solutions. Also, since the discriminant is a perfect square, the solutions are real rational.

23.
$$2x^2 + x + 3 = 0$$
$$a = 2 \quad b = 1 \quad c = 3$$

$$b^2 - 4ac = 1^2 - 4(2)(3) = 1 - 24 = -23$$

Since the discriminant is negative, there are no real solutions. There are two imaginary solutions that are complex conjugates.

25.
$$2x^2 + 6x = 0$$
$$a = 2 \quad b = 6 \quad c = 0$$

$$b^2 - 4ac = (6)^2 - 4(1)(0)$$
$$= 36 - 0 = 36$$

Since the discriminant is greater than zero, there are two unequal real solutions. Also, since the discriminant is a perfect square, the solutions are real rational.

27.
$$5x^2 + 3 = 0$$
$$a = 5 \quad b = 0 \quad c = 3$$

$$b^2 - 4ac = 0^2 - 4(5)(3) = 0 - 60 = -60$$

Since the discriminant is negative, there are no real solutions. There are two imaginary solutions that are complex conjugates.

29.
$$9x^2 = 12x - 4$$
$$9x^2 - 12x + 4 = 0$$
$$a = 9 \quad b = -12 \quad c = 4$$

$$b^2 - 4ac = (-12)^2 - 4(9)(4)$$
$$= 144 - 144 = 0$$

Since the discriminant is zero, there is one repeated real rational solution.

31.
$$3x^2 - 4x = 4$$
$$3x^2 - 4x - 4 = 0$$
$$(3x+2)(x-2) = 0$$
$$3x + 2 = 0 \quad \text{or} \quad x - 2 = 0$$
$$3x = -2 \qquad\qquad x = 2$$
$$x = -\frac{2}{3}$$

The solution set is $\left\{ -\frac{2}{3}, 2 \right\}$.

33. $x^2 - 2x = 1$

Since $b = -2$, add $\left(\dfrac{b}{2}\right)^2 = \left(\dfrac{-2}{2}\right)^2 = (-1)^2 = 1$.

$x^2 - 2x + 1 = 1 + 1$

$(x-1)^2 = 2$

Apply the square root principle.

$x - 1 = \pm\sqrt{2}$

$x = 1 \pm \sqrt{2}$

The solution set is $\left\{1 \pm \sqrt{2}\right\}$.

35. $(2x-5)(x+1) = 2$

$2x^2 + 2x - 5x - 5 = 2$

$2x^2 - 3x - 7 = 0$

Apply the quadratic formula.

$a = 2 \quad b = -3 \quad c = -7$

$x = \dfrac{-(-3) \pm \sqrt{(-3)^2 - 4(2)(-7)}}{2(2)}$

$= \dfrac{3 \pm \sqrt{9 - (-56)}}{4} = \dfrac{3 \pm \sqrt{65}}{4}$

The solution set is $\left\{\dfrac{3 \pm \sqrt{65}}{4}\right\}$.

37. $(3x-4)^2 = 16$

Apply the square root property.

$3x - 4 = \sqrt{16}$ or $3x - 4 = -\sqrt{16}$

$3x - 4 = 4 \qquad\qquad 3x - 4 = -4$

$3x = 8 \qquad\qquad\quad 3x = 0$

$x = \dfrac{8}{3} \qquad\qquad\quad x = 0$

The solution set is $\left\{0, \dfrac{8}{3}\right\}$.

39. $\dfrac{x^2}{2} + 2x + \dfrac{2}{3} = 0$

Multiply both sides of the equation by 6 to clear fractions.

$3x^2 + 12x + 4 = 0$

Apply the quadratic formula.

$a = 3 \quad b = 12 \quad c = 4$

$x = \dfrac{-12 \pm \sqrt{12^2 - 4(3)(4)}}{2(3)}$

$= \dfrac{-12 \pm \sqrt{144 - 48}}{6}$

$= \dfrac{-12 \pm \sqrt{96}}{6}$

$= \dfrac{-12 \pm \sqrt{16 \cdot 6}}{6}$

$= \dfrac{-12 \pm 4\sqrt{6}}{6}$

$= \dfrac{2\left(-6 \pm 2\sqrt{6}\right)}{6} = \dfrac{-6 \pm 2\sqrt{6}}{3}$

The solution set is $\left\{\dfrac{-6 \pm 2\sqrt{6}}{3}\right\}$.

41. $(3x-2)^2 = 10$

Apply the square root property.

$3x - 2 = \pm\sqrt{10}$

$3x = 2 \pm \sqrt{10}$

$x = \dfrac{2 \pm \sqrt{10}}{3}$

The solution set is $\left\{\dfrac{2 \pm \sqrt{10}}{3}\right\}$.

43. $\dfrac{1}{x} + \dfrac{1}{x+2} = \dfrac{1}{3}$

The LCD is $3x(x+2)$.

$$3x(x+2)\left(\dfrac{1}{x} + \dfrac{1}{x+2}\right) = 3x(x+2)\left(\dfrac{1}{3}\right)$$

$$3(x+2) + 3x = x(x+2)$$

$$3x + 6 + 3x = x^2 + 2x$$

$$0 = x^2 - 4x - 6$$

Apply the quadratic formula.

$a = 1 \quad b = -4 \quad c = -6$

$$x = \dfrac{-(-4) \pm \sqrt{(-4)^2 - 4(1)(-6)}}{2(1)}$$

$$= \dfrac{4 \pm \sqrt{16 - (-24)}}{2}$$

$$= \dfrac{4 \pm \sqrt{40}}{2} = \dfrac{4 \pm 2\sqrt{10}}{2} = 2 \pm \sqrt{10}$$

The solution set is $\left\{2 \pm \sqrt{10}\right\}$.

45. $(2x-6)(x+2) = 5(x-1) - 12$

$$2x^2 + 4x - 6x - 12 = 5x - 5 - 12$$

$$2x^2 - 2x - 12 = 5x - 17$$

$$2x^2 - 7x + 5 = 0$$

$$(2x-5)(x-1) = 0$$

Apply the zero product principle.

$2x - 5 = 0 \quad$ or $\quad x - 1 = 0$

$2x = 5 \qquad\qquad x = 1$

$x = \dfrac{5}{2}$

The solution set is $\left\{1, \dfrac{5}{2}\right\}$.

47. Because the solution set is $\{-3, 5\}$, we have

$x = -3 \quad$ or $\qquad x = 5$

$x + 3 = 0 \qquad\qquad x - 5 = 0.$

Use the zero-product principle in reverse.

$$(x+3)(x-5) = 0$$

$$x^2 - 5x + 3x - 15 = 0$$

$$x^2 - 2x - 15 = 0$$

49. Because the solution set is $\left\{-\dfrac{2}{3}, \dfrac{1}{4}\right\}$, we have

$x = -\dfrac{2}{3} \quad$ or $\qquad x = \dfrac{1}{4}$

$3x = -2 \qquad\qquad 4x = 1$

$3x + 2 = 0 \qquad\qquad 4x - 1 = 0.$

Use the zero-product principle in reverse.

$$(3x+2)(4x-1) = 0$$

$$12x^2 - 3x + 8x - 2 = 0$$

$$12x^2 + 5x - 2 = 0$$

51. Because the solution set is $\{-6i, 6i\}$, we have

$x = 6i \quad$ or $\qquad x = -6i$

$x - 6i = 0 \qquad\qquad x + 6i = 0.$

Use the zero-product principle in reverse.

$$(x-6i)(x+6i) = 0$$

$$x^2 + 6i - 6i - 36i^2 = 0$$

$$x^2 - 36(-1) = 0$$

$$x^2 + 36 = 0$$

53. Because the solution set is $\left\{-\sqrt{2}, \sqrt{2}\right\}$, we have

$x = \sqrt{2} \quad$ or $\qquad x = -\sqrt{2}$

$x - \sqrt{2} = 0 \qquad\qquad x + \sqrt{2} = 0.$

Use the zero-product principle in reverse.

$$\left(x - \sqrt{2}\right)\left(x + \sqrt{2}\right) = 0$$

$$x^2 + x\sqrt{2} - x\sqrt{2} - 2 = 0$$

$$x^2 - 2 = 0$$

55. Because the solution set is $\left\{-2\sqrt{5},\ 2\sqrt{5}\right\}$ we have

$$x = 2\sqrt{5} \quad \text{or} \quad x = -2\sqrt{5}$$
$$x - 2\sqrt{5} = 0 \qquad x + 2\sqrt{5} = 0.$$

Use the zero-product principle in reverse.

$$\left(x - 2\sqrt{5}\right)\left(x + 2\sqrt{5}\right) = 0$$
$$x^2 + 2x\sqrt{5} - 2x\sqrt{5} - 4 \cdot 5 = 0$$
$$x^2 - 20 = 0$$

57. Because the solution set is $\left\{1+i,\ 1-i\right\}$, we have

$$x = 1 + i \qquad \text{or} \quad x = 1 - i$$
$$x - (1+i) = 0 \qquad x - (1-i) = 0.$$

Use the zero-product principle in reverse.

$$\left[x - (1+i)\right]\left[x - (1-i)\right] = 0$$
$$x^2 - x(1-i) - x(1+i) + (1+i)(1-i) = 0$$
$$x^2 - x + xi - x - xi + 1 - i^2 = 0$$
$$x^2 - x - x + 1 - (-1) = 0$$
$$x^2 - 2x + 2 = 0$$

59. Because the solution set is $\left\{1+\sqrt{2}, 1-\sqrt{2}\right\}$, we have

$$x = 1 + \sqrt{2} \qquad \text{or} \quad x = 1 - \sqrt{2}$$
$$x - \left(1+\sqrt{2}\right) = 0 \qquad x - \left(1-\sqrt{2}\right) = 0.$$

Use the zero-product principle in reverse.

$$\left(x - \left(1+\sqrt{2}\right)\right)\left(x - \left(1-\sqrt{2}\right)\right) = 0$$
$$x^2 - x\left(1-\sqrt{2}\right) - x\left(1+\sqrt{2}\right)$$
$$+ \left(1+\sqrt{2}\right)\left(1-\sqrt{2}\right) = 0$$
$$x^2 - x + x\sqrt{2} - x - x\sqrt{2} + 1 - 2 = 0$$
$$x^2 - 2x - 1 = 0$$

61. b. If the solutions are imaginary numbers, then the graph will not cross the *x*-axis.

63. a. The equation has two non-integer solutions $3 \pm \sqrt{2}$, so the graph crosses the *x*-axis at $3 - \sqrt{2}$ and $3 + \sqrt{2}$.

65. Let $x =$ the number.

$$x^2 - (6 + 2x) = 0$$
$$x^2 - 2x - 6 = 0$$

Apply the quadratic formula.
$$a = 1 \quad b = -2 \quad c = -6$$

$$x = \frac{-(-2) \pm \sqrt{(-2)^2 - 4(1)(-6)}}{2(1)}$$
$$= \frac{2 \pm \sqrt{4 - (-24)}}{2}$$
$$= \frac{2 \pm \sqrt{28}}{2}$$
$$= \frac{2 \pm \sqrt{4 \cdot 7}}{2} = \frac{2 \pm 2\sqrt{7}}{2} = 1 \pm \sqrt{7}$$

Disregard $1 - \sqrt{7}$ because it is negative, and we are looking for a positive number. Thus, the number is $1 + \sqrt{7}$.

67.
$$\frac{1}{x^2 - 3x + 2} = \frac{1}{x+2} + \frac{5}{x^2 - 4}$$
$$\frac{1}{(x-1)(x-2)} = \frac{1}{x+2} + \frac{5}{(x+2)(x-2)}$$

Multiply both sides of the equation by the least common denominator, $(x-1)(x-2)(x+2)$. This results in the following:

$$x + 2 = (x-1)(x-2) + 5(x-1)$$
$$x + 2 = x^2 - 2x - x + 2 + 5x - 5$$
$$x + 2 = x^2 + 2x - 3$$
$$0 = x^2 + x - 5$$

Apply the quadratic formula.
$$a = 1 \quad b = 1 \quad c = -5.$$

$$x = \frac{-1 \pm \sqrt{1^2 - 4(1)(-5)}}{2(1)} = \frac{-1 \pm \sqrt{1 - (-20)}}{2}$$
$$= \frac{-1 \pm \sqrt{21}}{2}$$

The solution set is $\left\{\dfrac{-1 \pm \sqrt{21}}{2}\right\}$.

69. $\sqrt{2}x^2 + 3x - 2\sqrt{2} = 0$

Apply the quadratic formula.

$$a = \sqrt{2} \quad b = 3 \quad c = -2\sqrt{2}$$

$$x = \frac{-3 \pm \sqrt{3^2 - 4\left(\sqrt{2}\right)\left(-2\sqrt{2}\right)}}{2\left(\sqrt{2}\right)}$$

$$= \frac{-3 \pm \sqrt{9 - (-16)}}{2\sqrt{2}}$$

$$= \frac{-3 \pm \sqrt{25}}{2\sqrt{2}} = \frac{-3 \pm 5}{2\sqrt{2}}$$

Evaluate the expression to obtain two solutions.

$$x = \frac{-3-5}{2\sqrt{2}} \quad \text{or} \quad x = \frac{-3+5}{2\sqrt{2}}$$

$$= \frac{-8}{2\sqrt{2}} \cdot \frac{\sqrt{2}}{\sqrt{2}} \qquad = \frac{2}{2\sqrt{2}} \cdot \frac{\sqrt{2}}{\sqrt{2}}$$

$$= \frac{-8\sqrt{2}}{4} \qquad = \frac{2\sqrt{2}}{4}$$

$$= -2\sqrt{2} \qquad = \frac{\sqrt{2}}{2}$$

The solution set is $\left\{-2\sqrt{2}, \ \dfrac{\sqrt{2}}{2}\right\}$.

71. $\left|x^2 + 2x\right| = 3$

$$x^2 + 2x = -3 \quad \text{or} \quad x^2 + 2x = 3$$
$$x^2 + 2x + 3 = 0 \qquad x^2 + 2x - 3 = 0$$

Apply the quadratic formula to solve
$x^2 + 2x + 3 = 0$.
$a = 1 \quad b = 2 \quad c = 3$.

$$x = \frac{-2 \pm \sqrt{2^2 - 4(1)(3)}}{2(1)}$$

$$= \frac{-2 \pm \sqrt{4 - 12}}{2}$$

$$= \frac{-2 \pm \sqrt{-8}}{2}$$

$$= \frac{-2 \pm \sqrt{2 \cdot 4 \cdot (-1)}}{2}$$

$$= \frac{-2 \pm 2i\sqrt{2}}{2} = -1 \pm i\sqrt{2}$$

Apply the zero product principle to solve
$x^2 + 2x - 3 = 0$.
$(x+3)(x-1) = 0$
$x + 3 = 0 \quad \text{or} \quad x - 1 = 0$
$\quad x = -3 \quad \text{or} \qquad x = 1$

The solution set is $\left\{-3, \ 1, \ -1 \pm i\sqrt{2}\right\}$.

73. $f(x) = 0.013x^2 - 1.19x + 28.24$

$$3 = 0.013x^2 - 1.19x + 28.24$$

$$0 = 0.013x^2 - 1.19x + 25.24$$

Apply the quadratic formula.
$a = 0.013 \quad b = -1.19 \quad c = 25.24$

$$x = \frac{-(-1.19) \pm \sqrt{(-1.19)^2 - 4(0.013)(25.24)}}{2(0.013)}$$

$$= \frac{1.19 \pm \sqrt{1.4161 - 1.31248}}{0.026}$$

$$= \frac{1.19 \pm \sqrt{0.10362}}{0.026}$$

$$\approx \frac{1.19 \pm 0.32190}{0.026}$$

$$\approx 58.15 \ \text{or} \ 33.39$$

The solutions are approximately 33.39 and 58.15. Thus, 33 year olds and 58 year olds are expected to be in 3 fatal crashes per 100 million miles driven. The function models the actual data well.

75. Use the quadratic formula to solve
$f(x) = -0.01x^2 + 0.7x + 6.1 = 0$.
$a = -0.01, b = 0.7, c = 6.1$

$$x = \frac{-0.7 \pm \sqrt{0.7^2 - 4(-0.01)(6.1)}}{2(-0.01)}$$

$$= \frac{-0.7 \pm \sqrt{0.49 + 0.244}}{-0.02}$$

$$= \frac{-0.7 \pm \sqrt{0.734}}{-0.02} \approx -7.8 \ \text{or} \ 77.8$$

Disregard -7.8 because the distance must be positive. Thus, the maximum distance is approximately 77.8 feet. Graph (b) shows the shot's path.

77. Let x = the width of the rectangle.
Let $x + 4$ = the length of the rectangle.

$A = lw$

$8 = x(x+4)$

$0 = x^2 + 4x - 8$

Apply the quadratic formula.
$a = 1 \quad b = 4 \quad c = -8$

$x = \dfrac{-4 \pm \sqrt{4^2 - 4(1)(-8)}}{2(1)}$

$= \dfrac{-4 \pm \sqrt{16 - (-32)}}{2}$

$= \dfrac{-4 \pm \sqrt{48}}{2} = \dfrac{-4 \pm 4\sqrt{3}}{2}$

$= -2 \pm 2\sqrt{3} \approx 1.5 \text{ or } -5.5$

Disregard -5.5 because the width of a rectangle cannot be negative. Thus, the solution is 1.5, and the rectangle's dimensions are 1.5 meters by $1.5 + 4 = 5.5$ meters.

79. Let x = the length of the longer leg.
Let $x - 1$ = the length of the shorter leg.
Let $x + 7$ = the length of the hypotenuse.

$x^2 + (x-1)^2 = (x+7)^2$

$x^2 + x^2 - 2x + 1 = x^2 + 14x + 49$

$2x^2 - 2x + 1 = x^2 + 14x + 49$

$x^2 - 16x - 48 = 0$

Apply the quadratic formula.
$a = 1 \quad b = -16 \quad c = -48$

$x = \dfrac{-(-16) \pm \sqrt{(-16)^2 - 4(1)(-48)}}{2(1)}$

$= \dfrac{16 \pm \sqrt{256 - (-192)}}{2}$

$= \dfrac{16 \pm \sqrt{448}}{2}$

$= \dfrac{16 \pm 8\sqrt{7}}{2}$

$= 8 \pm 4\sqrt{7} \approx 18.6 \text{ or } -2.6$

Disregard -2.6 because the length of a leg cannot be negative. The solution is 18.6, and the lengths of the triangle's legs are approximately 18.6 inches and $18.6 - 1 = 17.6$ inches.

81. $x(20 - 2x) = 13$

$20x - 2x^2 = 13$

$0 = 2x^2 - 20x + 13$

Apply the quadratic formula.
$a = 2 \quad b = -20 \quad c = 13$

$x = \dfrac{-(-20) \pm \sqrt{(-20)^2 - 4(2)(13)}}{2(2)}$

$= \dfrac{20 \pm \sqrt{400 - 104}}{4}$

$= \dfrac{20 \pm \sqrt{296}}{4}$

$= \dfrac{20 \pm 2\sqrt{74}}{4} = \dfrac{10 \pm \sqrt{74}}{2} \approx 9.3 \text{ or } 0.7$

A gutter with depth 9.3 or 0.7 inches will have a cross-sectional area of 13 square inches.

83. Let x = the time for the first person to mow the yard alone.
Let $x + 1$ = the time for the second person to mow the yard alone.

	Fractional part of job completed in 1 hour	Time working together	Fractional part of job completed in 4 hour
1st person	$\dfrac{1}{x}$	4	$\dfrac{4}{x}$
2nd person	$\dfrac{1}{x+1}$	4	$\dfrac{1}{x+1}$

$\dfrac{4}{x} + \dfrac{4}{x+1} = 1$

$x(x+1)\left(\dfrac{4}{x} + \dfrac{4}{x+1}\right) = x(x+1)1$

$4(x+1) + 4x = x^2 + x$

$4x + 4 + 4x = x^2 + x$

$0 = x^2 - 7x - 4$

Apply the quadratic formula.

$a = 1 \quad b = -7 \quad c = -4$

$$x = \frac{-(-7) \pm \sqrt{(-7)^2 - 4(1)(-4)}}{2(1)}$$

$$= \frac{7 \pm \sqrt{49 - (-16)}}{2}$$

$$= \frac{7 \pm \sqrt{65}}{2} \approx 7.5 \text{ or } -0.5$$

Disregard −0.5 because time cannot be negative. Thus, the solution is 7.5, and we conclude that the first person can mow the lawn alone in 7.5 hours, and the second mow the yard alone in 7.5 + 1 = 8.5 hours.

85. – 91. Answers will vary.

93. $f(x) = x(20 - 2x)$

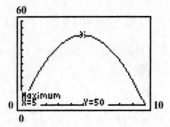

The depth of the gutter that will maximize water flow is 5 inches. This will allow for a water flow of 50 square inches. The situation described does not take full advantage of the sheets of aluminum.

95. does not make sense; Explanations will vary. Sample explanation: That is not a correct simplification because 2 is not a factor of both terms in the numerator.

97. does not make sense; Explanations will vary. Sample explanation: There are two imaginary solutions. They are not classified as irrational.

99. false; Changes to make the statement true will vary. A sample change is: The quadratic formula is developed by completing the square and the square root property.

101. true

103. The dimensions of the pool are 12 meters by 8 meters. With the tile, the dimensions will be 12 + 2*x* meters by 8 + 2*x* meters. If we take the area of the pool with the tile and subtract the area of the pool without the tile, we are left with the area of the tile only.

$$(12 + 2x)(8 + 2x) - 12(8) = 120$$

$$\cancel{96} + 24x + 16x + 4x^2 - \cancel{96} = 120$$

$$4x^2 + 40x - 120 = 0$$

$$x^2 + 10x - 30 = 0$$

$$a = 1 \quad b = 10 \quad c = -30$$

$$x = \frac{-10 \pm \sqrt{10^2 - 4(1)(-30)}}{2(1)}$$

$$= \frac{-10 \pm \sqrt{100 + 120}}{2}$$

$$= \frac{-10 \pm \sqrt{220}}{2} \approx \frac{-10 \pm 14.8}{2}$$

Evaluate the expression to obtain two solutions.

$$x = \frac{-10 + 14.8}{2} \quad \text{or} \quad x = \frac{-10 - 14.8}{2}$$

$$x = \frac{4.8}{2} \qquad\qquad x = \frac{-24.8}{2}$$

$$x = 2.4 \qquad\qquad x = -12.4$$

Disregard −12.4 because we can't have a negative width measurement. The solution is 2.4 and we conclude that the width of the uniform tile border is 2.4 meters. This is more than the 2-meter requirement, so the tile meets the zoning laws.

105. $|5x + 2| = |4 - 3x|$

$5x + 2 = 4 - 3x \quad \text{or} \quad 5x + 2 = -(4 - 3x)$

$8x + 2 = 4 \qquad\qquad 5x + 2 = -4 + 3x$

$8x = 2 \qquad\qquad\quad 2x + 2 = -4$

$x = \dfrac{1}{4} \qquad\qquad\quad 2x = -6$

$\qquad\qquad\qquad\qquad x = -3$

The solution set is $\left\{-3, \dfrac{1}{4}\right\}$.

106. $\sqrt{2x-5} - \sqrt{x-3} = 1$

$$\sqrt{2x-5} = \sqrt{x-3} + 1$$

$$\left(\sqrt{2x-5}\right)^2 = \left(\sqrt{x-3}+1\right)^2$$

$$2x-5 = x-3+2\sqrt{x-3}+1$$

$$2x-5 = x-2+2\sqrt{x-3}$$

$$x-3 = 2\sqrt{x-3}$$

$$\left(x-3\right)^2 = \left(2\sqrt{x-3}\right)^2$$

$$x^2-6x+9 = 4\left(x-3\right)$$

$$x^2-6x+9 = 4x-12$$

$$x^2-10x+21 = 0$$

$$\left(x-7\right)\left(x-3\right) = 0$$

$$x-7 = 0 \quad \text{or} \quad x-3 = 0$$
$$x = 7 \qquad\qquad x = 3$$

The solution set is $\{3,7\}$.

107. $\dfrac{5}{\sqrt{3}+x} = \dfrac{5}{\sqrt{3}+x} \cdot \dfrac{\sqrt{3}-x}{\sqrt{3}-x} = \dfrac{5\sqrt{3}-5x}{3-x^2}$

108.

x	$f(x) = x^2$	$g(x) = x^2 + 2$
-3	9	11
-2	4	6
-1	1	3
0	0	2
1	1	3
2	4	6
3	9	11

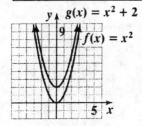

109.

x	$f(x) = x^2$	$g(x) = (x+2)^2$
-3	9	1
-2	4	0
-1	1	1
0	0	4
1	1	9
2	4	16
3	9	25

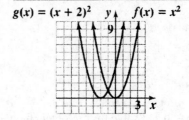

110. $f(x) = -2(x-3)^2 + 8$

$$0 = -2(x-3)^2 + 8$$

$$2(x-3)^2 = 8$$

$$(x-3)^2 = 4$$

$$\sqrt{(x-3)^2} = \pm\sqrt{4}$$

$$x-3 = \pm 2$$

$$x = 3 \pm 2$$

$$x = 3+2 \quad \text{or} \quad x = 3-2$$
$$x = 5 \qquad\qquad x = 1$$

The x-intercepts are 1 and 5.

11.3 Check Points

1. $f(x) = -(x-1)^2 + 4$

Since $a = -1$ is negative, the parabola opens downward. The vertex of the parabola is $(h,k) = (1,4)$. Replace $f(x)$ with 0 to find x–intercepts.

$$0 = -(x-1)^2 + 4$$

$$(x-1)^2 = 4$$

$$x - 1 = \pm\sqrt{4}$$

$$x - 1 = \pm 2$$

$$x - 1 = 2 \quad \text{or} \quad x - 1 = -2$$

$$x = 3 \qquad\qquad x = -1$$

The x–intercepts are -1 and 3.

Set $x = 0$ and solve for y to obtain the y–intercept.

$$y = -(0-1)^2 + 4 = 3$$

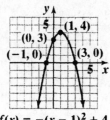

$$f(x) = -(x-1)^2 + 4$$

2. $f(x) = (x-2)^2 + 1$

Since $a = 1$ is positive, the parabola opens upward. The vertex of the parabola is $(h,k) = (2,1)$.

Replace $f(x)$ with 0 to find x–intercepts.

$$0 = (x-2)^2 + 1$$

$$(x-2)^2 = -1$$

$$x - 2 = \pm\sqrt{-1}$$

$$x = 2 \pm i$$

Because this equation has no real solutions, the parabola has no x-intercepts.

Set $x = 0$ and solve for y to obtain the y–intercept.

$$y = (0-2)^2 + 1 = 5$$

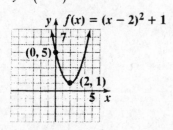

3. $f(x) = 2x^2 + 8x - 1$

The x–coordinate of the vertex of the parabola is

$$-\frac{b}{2a} = -\frac{8}{2(2)} = -\frac{8}{4} = -2,$$ and the y–coordinate of

the vertex of the parabola is

$$f\left(-\frac{b}{2a}\right) = f(-2) = 2(-2)^2 + 8(-2) - 1 = -9.$$

The vertex is $(-2,-9)$.

4. $f(x) = -x^2 + 4x + 1$

Since $a = -1$ is negative, the parabola opens downward. The x–coordinate of the vertex of the

parabola is $-\dfrac{b}{2a} = -\dfrac{4}{2(-1)} = -\dfrac{4}{-2} = 2$ and the y–

coordinate of the vertex of the parabola is

$$f\left(-\frac{b}{2a}\right) = f(2) = -(2)^2 + 4(2) + 1 = 5.$$

The vertex is $(2,5)$.

Replace $f(x)$ with 0 to find x–intercepts.

$$0 = -x^2 + 4x + 1$$

$$x = \frac{-b \pm \sqrt{b^2 - 4ac}}{2a}$$

$$x = \frac{-4 \pm \sqrt{4^2 - 4(-1)(1)}}{2(-1)}$$

$$x = 2 \pm \sqrt{5}$$

$$x \approx -0.2 \quad \text{or} \quad x \approx 4.2$$

The x–intercepts are -0.2 and 4.2.

Set $x = 0$ and solve for y to obtain the y–intercept.

$$y = -0^2 + 4 \cdot 0 + 1 = 1$$

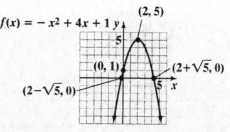

Domain: $\{x \mid x \text{ is a real number}\}$ or $(-\infty, \infty)$.

Range: $\{y \mid y \le 5\}$ or $(-\infty, 5]$.

5. $f(x) = 4x^2 - 16x + 1000$

a. Because $a > 0$, the function has a minimum value.

b. The minimum value occurs at

$$-\frac{b}{2a} = -\frac{-16}{2(4)} = -\frac{-16}{8} = 2$$

The minimum of $f(x)$ is

$$f(2) = 4 \cdot 2^2 - 16 \cdot 2 + 1000 = 984.$$

c. Like all quadratic functions, the domain is $\{x \mid x$ is a real number$\}$ or $(-\infty, \infty)$.

Because the minimum is 984, the range includes all real numbers at or above 984.

The range is $\{y \mid y \geq 984\}$ or $[984, \infty)$.

6. $g(x) = -0.04x^2 + 2.1x + 6.1$

Because $a < 0$, the function has a maximum value that occurs at

$$-\frac{b}{2a} = -\frac{2.1}{2(-0.04)} = -\frac{2.1}{-0.08} = 26.25.$$

The maximum height is given by

$$g(26.25) = -0.04(26.25)^2 + 2.1(26.25) + 6.1 \approx 33.7$$

The maximum height of about 33.7 feet occurs at a horizontal distance of 26.25 feet.

7. Let the two numbers be represented by x and y, and let the product be represented by P.
We must minimize $P = xy$.

Because the difference of the two numbers is 8, then $x - y = 8$.

Solve for y in terms of x.
$$x - y = 8$$
$$-y = -x + 8$$
$$y = x - 8$$

Write P as a function of x.
$$P = xy$$
$$P(x) = x(x - 8)$$
$$P(x) = x^2 - 8x$$

Because $a > 0$, the function has a minimum value that occurs at $x = -\dfrac{b}{2a} = -\dfrac{-8}{2(1)} = 4.$

Substitute to find the other number.
$$y = x - 8$$
$$y = 4 - 8$$
$$= -4$$

The two numbers are 4 and -4. The minimum product is $P = xy = (4)(-4) = -16$.

8. Let the two dimensions be represented by x and y, and let the area be represented by A.
We must maximize $A = xy$.

Because the perimeter, P, is 120, then
$$P = 2x + 2y$$
$$120 = 2x + 2y$$

Solve for y in terms of x.
$$120 = 2x + 2y$$
$$-2y = 2x - 120$$
$$y = -x + 60$$

Write A as a function of x.
$$A = xy$$
$$A(x) = x(-x + 60)$$
$$A(x) = -x^2 + 60x$$

Because $a < 0$, the function has a maximum value that occurs at $x = -\dfrac{b}{2a} = -\dfrac{60}{2(-1)} = 30.$

Substitute to find the other dimension.
$$y = -x + 60$$
$$y = -30 + 60$$
$$= 30$$

The two dimensions are 30 feet by 30 feet. The maximum area is
$$A = xy = (30)(30) = 900 \text{ square feet.}$$

11.3 Exercise Set

1. The vertex of the graph is the point $(1, 1)$. This means that the equation is $h(x) = (x-1)^2 + 1$.

3. The vertex of the graph is the point $(1, -1)$. This means that the equation is $j(x) = (x-1)^2 - 1$.

5. The vertex of the graph is the point $(0, -1)$. This means that the equation is
$$h(x) = (x - 0)^2 - 1 = x^2 - 1.$$

7. The vertex of the graph is the point $(1, 0)$. This means that the equation is
$$f(x) = (x - 1)^2 + 0$$
$$= (x - 1)^2 = x^2 - 2x + 1$$

9. $f(x) = 2(x - 3)^2 + 1$
The vertex is $(3, 1)$.

11. $f(x) = -2(x+1)^2 + 5$

The vertex is $(-1, 5)$.

13. $f(x) = 2x^2 - 8x + 3$

The x–coordinate of the vertex of the parabola is

$-\dfrac{b}{2a} = -\dfrac{-8}{2(2)} = -\dfrac{-8}{4} = 2$, and the y–coordinate of

the vertex of the parabola is

$f\left(-\dfrac{b}{2a}\right) = f(2) = 2(2)^2 - 8(2) + 3$

$= 2(4) - 16 + 3$

$= 8 - 16 + 3 = -5.$

The vertex is $(2, -5)$.

15. $f(x) = -x^2 - 2x + 8$

The x–coordinate of the vertex of the parabola is

$-\dfrac{b}{2a} = -\dfrac{-2}{2(-1)} = -\dfrac{-2}{-2} = -1,$

and the y–coordinate of the vertex of the parabola is

$f\left(-\dfrac{b}{2a}\right) = f(-1)$

$= -(-1)^2 - 2(-1) + 8$

$= -1 + 2 + 8 = 9.$

The vertex is $(-1, 9)$.

17. $f(x) = (x-4)^2 - 1$

Since $a = 1$ is positive, the parabola opens upward.
The vertex of the parabola is $(h, k) = (4, -1)$.

Replace $f(x)$ with 0 to find x–intercepts.

$0 = (x-4)^2 - 1$

$1 = (x-4)^2$

Apply the square root property.

$x - 4 = \pm\sqrt{1} = \pm 1$

$x = 4 \pm 1 = 5 \text{ or } 3$

The x–intercepts are 5 and 3.
Set $x = 0$ and solve for y to obtain the y–intercept.

$y = (0-4)^2 - 1 = 15$

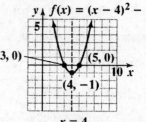

Axis of symmetry: $x = 4$.
Range: $\{y \mid y \geq -1\}$ or $[-1, \infty)$.

19. $f(x) = (x-1)^2 + 2$

Since $a = 1$ is positive, the parabola opens upward.
The vertex of the parabola is $(h, k) = (1, 2)$.

Replace $f(x)$ with 0 to find x–intercepts.

$0 = (x-1)^2 + 2$

$-2 = (x-1)^2$

Because the solutions to the equation are imaginary, there are no x–intercepts. Set $x = 0$ and solve for y to obtain the y–intercept.

$y = (0-1)^2 + 2 = (-1)^2 + 2 = 1 + 2 = 3$

The y–intercept is 3.

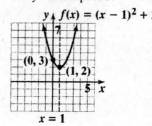

Axis of symmetry: $x = 1$.
Range: $\{y \mid y \geq 2\}$ or $[2, \infty)$.

21. $y - 1 = (x - 3)^2$

$$y = (x - 3)^2 + 1$$

$$f(x) = (x - 3)^2 + 1$$

Since $a = 1$ is positive, the parabola opens upward. The vertex of the parabola is $(h, k) = (3, 1)$.

Replace $f(x)$ with 0 to find x-intercepts.

$$0 = (x - 3)^2 + 1$$

$$-1 = (x - 3)^2$$

Because the solutions to the equation are imaginary, there are no x-intercepts. Set $x = 0$ and solve for y to obtain the y-intercept.

$$y = (0 - 3)^2 + 1 = (-3)^2 + 1 = 9 + 1 = 10$$

The y-intercept is 10.

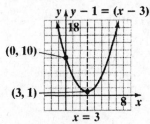

Axis of symmetry: $x = 3$.
Range: $\{y \mid y \ge 1\}$ or $[1, \infty)$.

23. $f(x) = 2(x + 2)^2 - 1$

Since $a = 2$ is positive, the parabola opens upward. The vertex of the parabola is $(h, k) = (-2, -1)$.

Replace $f(x)$ with 0 to find x-intercepts.

$$0 = 2(x + 2)^2 - 1$$

$$1 = 2(x + 2)^2$$

$$\frac{1}{2} = (x + 2)^2$$

Apply the square root property.

$$x + 2 = \pm\sqrt{\frac{1}{2}}$$

$$x = -2 \pm \sqrt{\frac{1}{2}}$$

$$x \approx -2 - \sqrt{\frac{1}{2}} \quad \text{or} \quad -2 + \sqrt{\frac{1}{2}}$$

$$x \approx -2.7 \quad \text{or} \quad -1.3$$

The x-intercepts are -1.3 and -2.7.
Set $x = 0$ and solve for y to obtain the y-intercept.

$$y = 2(0 + 2)^2 - 1$$

$$= 2(2)^2 - 1 = 2(4) - 1 = 8 - 1 = 7$$

The y-intercept is 7.

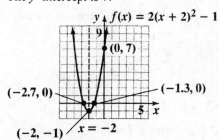

Axis of symmetry: $x = -2$.
Range: $\{y \mid y \ge -1\}$ or $[-1, \infty)$.

25. $f(x) = 4 - (x - 1)^2$

$$f(x) = -(x - 1)^2 + 4$$

Since $a = -1$ is negative, the parabola opens downward. The vertex of the parabola is $(h, k) = (1, 4)$. Replace $f(x)$ with 0 to find x-intercepts.

$$0 = -(x - 1)^2 + 4$$

$$-4 = -(x - 1)^2$$

$$4 = (x - 1)^2$$

Apply the square root property.

$$\sqrt{4} = x - 1 \quad \text{or} \quad -\sqrt{4} = x - 1$$

$$2 = x - 1 \qquad\qquad -2 = x - 1$$

$$3 = x \qquad\qquad\quad -1 = x$$

The x-intercepts are -1 and 3.
Set $x = 0$ and solve for y to obtain the y-intercept.

$$y = -(0 - 1)^2 + 4$$

$$= -(-1)^2 + 4 = -1 + 4 = 3$$

The y-intercept is 3.

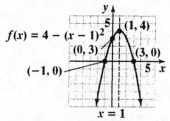

Axis of symmetry: $x = 1$.
Range: $\{y \mid y \le 4\}$ or $(-\infty, 4]$.

27. $f(x) = x^2 + 2x - 3$

Since $a = 1$ is positive, the parabola opens upward. The x–coordinate of the vertex of the parabola is

$$-\frac{b}{2a} = -\frac{2}{2(1)} = -\frac{2}{2} = -1 \text{ and the}$$

y–coordinate of the vertex of the parabola is

$$f\left(-\frac{b}{2a}\right) = f(-1)$$

$$= (-1)^2 + 2(-1) - 3$$

$$= 1 - 2 - 3 = -4.$$

The vertex is $(-1, -4)$. Replace $f(x)$ with 0 to find x–intercepts.

$$0 = x^2 + 2x - 3$$

$$0 = (x + 3)(x - 1)$$

Apply the zero product principle.

$x + 3 = 0 \quad$ or $\quad x - 1 = 0$

$\quad x = -3 \qquad\qquad x = 1$

The x–intercepts are -3 and 1. Set $x = 0$ and solve for y to obtain the y–intercept.

$$y = (0)^2 + 2(0) - 3 = -3$$

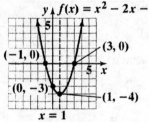

Axis of symmetry: $x = -1$.

Range: $\{y \mid y \geq -4\}$ or $[-4, \infty)$.

29. $f(x) = x^2 + 3x - 10$

Since $a = 1$ is positive, the parabola opens upward. The x–coordinate of the vertex of the parabola is

$$-\frac{b}{2a} = -\frac{3}{2(1)} = -\frac{3}{2} \text{ and the}$$

y–coordinate of the vertex of the parabola is

$$f\left(-\frac{b}{2a}\right) = f\left(-\frac{3}{2}\right)$$

$$= \left(-\frac{3}{2}\right)^2 + 3\left(-\frac{3}{2}\right) - 10$$

$$= \frac{9}{4} - \frac{9}{2} - 10 = -\frac{49}{4}.$$

The vertex is $\left(-\frac{3}{2}, -\frac{49}{4}\right)$.

Replace $f(x)$ with 0 to find the x–intercepts.

$$0 = x^2 + 3x - 10$$

$$0 = (x + 5)(x - 2)$$

Apply the zero product principle.

$x + 5 = 0 \quad$ or $\quad x - 2 = 0$

$\quad x = -5 \qquad\qquad x = 2$

The x–intercepts are -5 and 2.
Set $x = 0$ and solve for y to obtain the y–intercept.

$$y = 0^2 + 3(0) - 10 = -10$$

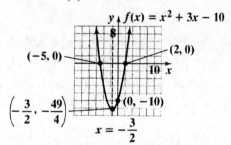

Axis of symmetry: $x = -\dfrac{3}{2}$.

Range: $\left\{y \mid y \geq -\dfrac{49}{4}\right\}$ or $\left[-\dfrac{49}{4}, \infty\right)$.

31. $f(x) = 2x - x^2 + 3$

$f(x) = -x^2 + 2x + 3$

Since $a = -1$ is negative, the parabola opens downward. The
x–coordinate of the vertex of the parabola is

$-\dfrac{b}{2a} = -\dfrac{2}{2(-1)} = -\dfrac{2}{-2} = 1$ and the y–coordinate of

the vertex of the parabola is

$f\left(-\dfrac{b}{2a}\right) = f(1) = -(1)^2 + 2(1) + 3$

$\qquad = -1 + 2 + 3 = 4.$

The vertex is $(1, 4)$. Replace $f(x)$ with 0 to find
x–intercepts.

$0 = -x^2 + 2x + 3$

$0 = x^2 - 2x - 3$

$0 = (x - 3)(x + 1)$

Apply the zero product principle.

$x - 3 = 0 \quad \text{or} \quad x + 1 = 0$

$\qquad x = 3 \qquad\qquad x = -1$

The x–intercepts are 3 and –1. Set
$x = 0$ and solve for y to obtain the

y–intercept. $y = -(0)^2 + 2(0) + 3 = 3$

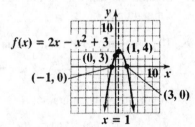

Axis of symmetry: $x = 1$.
Range: $\{y \mid y \le 4\}$ or $(-\infty, 4]$.

33. $f(x) = x^2 + 6x + 3$

$-\dfrac{b}{2a} = -\dfrac{6}{2(1)} = -3$

$f\left(-\dfrac{b}{2a}\right) = f(-3) = (-3)^2 + 6(-3) + 3 = -6$

The vertex is $(-3, -6)$.
To find x–intercepts let $y = 0$.

$0 = x^2 + 6x + 3$

$x = \dfrac{-b \pm \sqrt{b^2 - 4ac}}{2a}$

$x = \dfrac{-6 \pm \sqrt{6^2 - 4(1)(3)}}{2(1)}$

$x = -3 \pm \sqrt{6}$

$x \approx -5.4 \quad \text{or} \quad x \approx -0.6$

The x–intercepts are –5.4 and –0.6. The y–intercept
is 3.

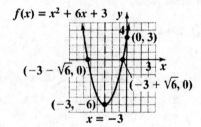

Range: $\{y \mid y \ge -6\}$ or $[-6, \infty)$.

35. $f(x) = 2x^2 + 4x - 3$

$$-\frac{b}{2a} = -\frac{4}{2(2)} = -1$$

$$f\left(-\frac{b}{2a}\right) = f(-1) = 2(-1)^2 + 4(-1) - 3 = -5$$

The vertex is $(-1, -5)$.
To find x–intercepts let $y = 0$.

$$f(x) = 2x^2 + 4x - 3$$

$$x = \frac{-b \pm \sqrt{b^2 - 4ac}}{2a}$$

$$x = \frac{-4 \pm \sqrt{4^2 - 4(2)(-3)}}{2(2)}$$

$$x = -1 \pm \frac{\sqrt{10}}{2}$$

$$x \approx -2.6 \text{ or } x \approx 0.6$$

The x–intercepts are -2.6 and 0.6. The y–intercept is -3.

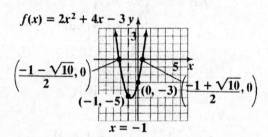

Range: $\{y \mid y \geq -5\}$ or $[-5, \infty)$.

37. $f(x) = 2x - x^2 - 2$

$$f(x) = -x^2 + 2x - 2$$

Since $a = -1$ is negative, the parabola opens downward. The x–coordinate of the vertex is

$$-\frac{b}{2a} = -\frac{2}{2(-1)} = -\frac{2}{-2} = 1 \text{ and the } y\text{–coordinate of}$$

the vertex is

$$f\left(-\frac{b}{2a}\right) = f(1) = -(1)^2 + 2(1) - 2$$

$$= -1 + 2 - 2 = -1.$$

The vertex is $(1, -1)$. Replace $f(x)$ with 0 to find x–intercepts.

$$0 = -x^2 + 2x - 2$$

$$x^2 - 2x = -2$$

Since $b = -2$, add $\left(\frac{b}{2}\right)^2 = \left(\frac{-2}{2}\right)^2 = (-1)^2 = 1$

$$x^2 - 2x + 1 = -2 + 1$$

$$(x - 1)^2 = -1$$

Because the solutions to the equation are imaginary, there are no x–intercepts. Set $x = 0$ and solve for y to obtain the y–intercept. $y = 2(0) - 0^2 - 2 = -2$

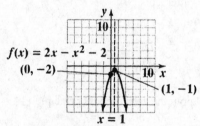

Axis of symmetry: $x = 1$.
Range: $\{y \mid y \leq -1\}$ or $(-\infty, -1]$.

39. $f(x) = 3x^2 - 12x - 1$

a. Since $a > 0$, the parabola opens upward and has a minimum.

b. The x–coordinate of the minimum is

$$-\frac{b}{2a} = -\frac{-12}{2(3)} = -\frac{-12}{6} = 2 \text{ and the } y\text{–}$$

coordinate of the minimum is

$$f\left(-\frac{b}{2a}\right) = f(2) = 3(2)^2 - 12(2) - 1 = 12 - 24 - 1 = -1$$

c. Domain: $\{x \mid x \text{ is a real number}\}$ or $(-\infty, \infty)$.
Range: $\{y \mid y \geq -13\}$ or $[-13, \infty)$.

41. $f(x) = -4x^2 + 8x - 3$

    a. Since $a < 0$, the parabola opens downward and has a maximum.

    b. The x–coordinate of the maximum is

$$-\frac{b}{2a} = -\frac{8}{2(-4)} = -\frac{8}{-8} = 1 \text{ and the } y–$$

coordinate of the maximum is

$$f\left(-\frac{b}{2a}\right) = f(1) = -4(1)^2 + 8(1) - 3 = -4 + 8 - 3 = 1.$$

    c. Domain: $\{x \mid x \text{ is a real number}\}$ or $(-\infty, \infty)$.

       Range: $\{y \mid y \leq 1\}$ or $(-\infty, 1]$.

43. $f(x) = 5x^2 - 5x$

    a. Since $a > 0$, the parabola opens upward and has a minimum.

    b. The x–coordinate of the minimum is

$$-\frac{b}{2a} = -\frac{-5}{2(5)} = -\frac{-5}{10} = \frac{1}{2} \text{ and the } y–\text{coordinate}$$

of the minimum is

$$f\left(-\frac{b}{2a}\right) = f\left(\frac{1}{2}\right) = 5\left(\frac{1}{2}\right)^2 - 5\left(\frac{1}{2}\right)$$

$$= 5\left(\frac{1}{4}\right) - \frac{5}{2} = \frac{5}{4} - \frac{10}{4} = -\frac{5}{4}.$$

    c. Domain: $\{x \mid x \text{ is a real number}\}$ or $(-\infty, \infty)$.

       Range: $\left\{y \mid y \geq -\frac{5}{4}\right\}$ or $\left[-\frac{5}{4}, \infty\right)$.

45. Since the parabola opens up, the vertex $(-1, -2)$ is a minimum point.
The domain is $\{x \mid x \text{ is a real number}\}$ or $(-\infty, \infty)$.
The range is $\{y \mid y \geq -2\}$ or $[-2, \infty)$.

47. Since the parabola has a maximum point, it opens down. The domain is $\{x \mid x \text{ is a real number}\}$ or $(-\infty, \infty)$. The range is $\{y \mid y \leq -6\}$ or $(-\infty, -6]$.

49. $(h, k) = (5, 3)$

$$f(x) = 2(x - h)^2 + k = 2(x - 5)^2 + 3$$

51. $(h, k) = (-10, -5)$

$$f(x) = 2(x - h)^2 + k$$

$$= 2[x - (-10)]^2 + (-5)$$

$$= 2(x + 10)^2 - 5$$

53. Since the vertex is a maximum, the parabola opens down and $a = -3$.

$$(h, k) = (-2, 4)$$

$$f(x) = -3(x - h)^2 + k$$

$$= -3[x - (-2)]^2 + 4$$

$$= -3(x + 2)^2 + 4$$

55. Since the vertex is a minimum, the parabola opens up and $a = 3$.

$$(h, k) = (11, 0)$$

$$f(x) = 3(x - h)^2 + k$$

$$= 3(x - 11)^2 + 0$$

$$= 3(x - 11)^2$$

57. $f(x) = 0.004x^2 - 0.094x + 2.6$

    a. $f(25) = 0.004(25)^2 - 0.094(25) + 2.6$

$$= 2.75$$

       According to the function, U.S. adult wine consumption in 2005 was 2.75 gallons per person. This underestimates the graph's value by 0.05 gallon.

    b. $\text{year} = -\dfrac{b}{2a} = -\dfrac{-0.094}{2(0.004)} \approx 12$

       Wine consumption was at a minimum about 12 years after 1980, or 1992.

$$f(12) = 0.004(12)^2 - 0.094(12) + 2.6 \approx 2.048$$

       Wine consumption was about 2.048 gallons per U.S. adult in 1992.
This seems reasonable as compared to the values in the graph.

59. $s(t) = -16t^2 + 64t + 160$

  a. The t-coordinate of the minimum is

  $$t = -\frac{b}{2a} = -\frac{64}{2(-16)} = -\frac{64}{-32} = 2.$$

  The s-coordinate of the minimum is

  $$s(2) = -16(2)^2 + 64(2) + 160$$
  $$= -16(4) + 128 + 160$$
  $$= -64 + 128 + 160 = 224$$

  The ball reaches a maximum height of 224 feet 2 seconds after it is thrown.

  b. $0 = -16t^2 + 64t + 160$

  $$0 = t^2 - 4t - 10$$

  $$a = 1 \quad b = -4 \quad c = -10$$

  $$t = \frac{-(-4) \pm \sqrt{(-4)^2 - 4(1)(-10)}}{2(1)}$$

  $$= \frac{4 \pm \sqrt{16 + 40}}{2}$$

  $$= \frac{4 \pm \sqrt{56}}{2} \approx \frac{4 \pm 7.48}{2}$$

  Evaluate the expression to obtain two solutions.

  $$x = \frac{4 + 7.48}{2} \quad \text{or} \quad x = \frac{4 - 7.48}{2}$$

  $$x = \frac{11.48}{2} \qquad\qquad x = \frac{-3.48}{2}$$

  $$x = 5.74 \qquad\qquad\quad x = -1.74$$

  Disregard -1.74 because we can't have a negative time
  measurement. The solution is 5.74 and we conclude that the ball will hit the ground in approximately 5.7 seconds.

  c. $s(0) = -16(0)^2 + 64(0) + 160$

  $$= -16(0) + 0 + 160 = 160$$

  At $t = 0$, the ball has not yet been thrown and is at a height of 160 feet. This is the height of the building.

  d.

  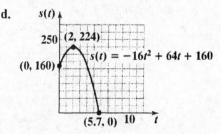

61. Let x = one of the numbers.
  Let $16 - x$ = the other number.

  The product is $f(x) = x(16 - x)$

  $$= 16x - x^2 = -x^2 + 16x$$

  The x-coordinate of the maximum is

  $$x = -\frac{b}{2a} = -\frac{16}{2(-1)} = -\frac{16}{-2} = 8.$$

  $$f(8) = -8^2 + 16(8) = -64 + 128 = 64$$

  The vertex is $(8, 64)$. The maximum product is 64. This occurs when the two number are 8 and $16 - 8 = 8$.

63. Let x = one of the numbers.
  Let $x - 16$ = the other number.

  The product is $f(x) = x(x - 16) = x^2 - 16x$.

  The x-coordinate of the minimum is

  $$x = -\frac{b}{2a} = -\frac{16}{2(1)} = -\frac{16}{2} = -8.$$

  $$f(-8) = (-8)^2 + 16(-8)$$

  $$= 64 - 128 = -64$$

  The vertex is $(-8, -64)$. The minimum product is -64. This occurs when the two number are -8 and $-8 + 16 = 8$.

65. Maximize the area of a rectangle constructed along a river with 600 feet of fencing.

  Let x = the width of the rectangle.
  Let $600 - 2x$ = the length of the rectangle.
  We need to maximize.
  $$A(x) = x(600 - 2x)$$

  $$= 600x - 2x^2 = -2x^2 + 600x$$

  Since $a = -2$ is negative, the function opens downward and has a maximum at

  $$x = -\frac{b}{2a} = -\frac{600}{2(-2)} = -\frac{600}{-4} = 150.$$

  When the width is $x = 150$ feet, the length is
  $600 - 2(150) = 600 - 300 = 300$ feet.

  The dimensions of the rectangular plot with maximum area are 150 feet by 300 feet. This gives an area of $150 \cdot 300 = 45,000$ square feet.

67. Maximize the area of a rectangle constructed with 50 yards of fencing.

Let x = the length of the rectangle. Let y = the width of the rectangle.
Since we need an equation in one variable, use the perimeter to express y in terms of x.
$2x + 2y = 50$

$\quad 2y = 50 - 2x$

$\qquad y = \dfrac{50 - 2x}{2} = 25 - x$

We need to maximize $A = xy = x(25 - x)$. Rewrite A as a function of x.

$A(x) = x(25 - x) = -x^2 + 25x$

Since $a = -1$ is negative, the function opens downward and has a maximum at

$x = -\dfrac{b}{2a} = -\dfrac{25}{2(-1)} = -\dfrac{25}{-2} = 12.5.$

When the length x is 12.5, the width y is
$y = 25 - x = 25 - 12.5 = 12.5.$

The dimensions of the rectangular region with maximum area are 12.5 yards by 12.5 yards. This gives an area of $12.5 \cdot 12.5 = 156.25$ square yards.

69. Maximize the cross-sectional area of the gutter:
$A(x) = x(20 - 2x)$

$\qquad = 20x - 2x^2 = -2x^2 + 20x.$

Since $a = -2$ is negative, the function opens downward and has a maximum at

$x = -\dfrac{b}{2a} = -\dfrac{20}{2(-2)} = -\dfrac{20}{-4} = 5.$

When the height x is 5, the width is
$20 - 2x = 20 - 2(5) = 20 - 10 = 10.$

$A(5) = -2(5)^2 + 20(5)$

$\qquad = -2(25) + 100 = -50 + 100 = 50$

The maximum cross-sectional area is 50 square inches. This occurs when the gutter is 5 inches deep and 10 inches wide.

71. a. $C(x) = 525 + 0.55x$

b. $P(x) = R(x) - C(x)$

$\quad = \left(-0.001x^2 + 3x\right) - \left(525 + 0.55x\right)$

$\quad = -0.001x^2 + 3x - 525 - 0.55x$

$\quad = -0.001x^2 + 2.45x - 525$

c. Since $a = -0.001$ is negative, the function opens down and has a maximum at

$x = -\dfrac{b}{2a}$

$\quad = -\dfrac{2.45}{2(-0.001)} = -\dfrac{2.45}{-0.002} = 1225.$

When the number of units x is 1225, the profit is
$P(1225)$

$\quad = -0.001(1225)^2 + 2.45(1225) - 525$

$\quad = -0.001(1500625) + 3001.25 - 525$

$\quad = -1500.625 + 3001.25 - 525$

$\quad = 975.625$

The store maximizes its weekly profit when 1225 roast beef sandwiches are made and sold, resulting in a profit of \$975.63.

73. – 77. Answers will vary.

79. a. $y = 2x^2 - 82x + 720$

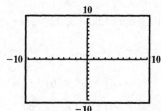

The function has no values that fall within the window.

b. $y = 2x^2 - 82x + 720$

The x–coordinate of the vertex of the parabola is

$-\dfrac{b}{2a} = -\dfrac{-82}{2(2)} = -\dfrac{-82}{4} = 20.5$ and the y–

coordinate of the vertex of the parabola is

$f\left(-\dfrac{b}{2a}\right) = f(20.5)$

$= 2(20.5)^2 - 82(20.5) + 720$

$= 2(420.25) - 1681 + 720$

$= 840.5 - 1681 + 720 = -120.5.$

The vertex is $(20.5, -120.5)$.

c. Using a Ymax of 50, we have the following.

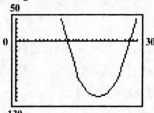

d. Answers will vary.

81. $y = -4x^2 + 20x + 160$

The x–coordinate of the vertex of the parabola is

$-\dfrac{b}{2a} = -\dfrac{20}{2(-4)} = -\dfrac{20}{-8} = 2.5$

and the y–coordinate of the vertex of the parabola is

$f\left(-\dfrac{b}{2a}\right) = f(2.5)$

$= -4(2.5)^2 + 20(2.5) + 160$

$= -4(6.25) + 50 + 160$

$= -25 + 50 + 160$

$= 185.$

The vertex is $(2.5, 185)$.

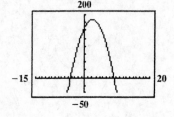

83. $y = 0.01x^2 + 0.6x + 100$

The x–coordinate of the vertex of the parabola is

$-\dfrac{b}{2a} = -\dfrac{0.6}{2(0.01)} = -\dfrac{0.6}{0.02} = -30$

and the y–coordinate of the vertex of the parabola is

$f\left(-\dfrac{b}{2a}\right) = f(-30)$

$= 0.01(-30)^2 + 0.6(-30) + 100$

$= 0.01(900) - 18 + 100$

$= 9 - 18 + 100 = 91.$

The vertex is $(-30, 91)$.

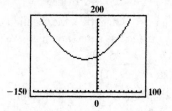

85. makes sense

87. makes sense

89. true

91. false; Changes to make the statement true will vary. A sample change is: The graph has no x–intercepts. To find x–intercepts, set $y = 0$ and solve for x.

$0 = -2(x+4)^2 - 8$

$2(x+4)^2 = -8$

$(x+4)^2 = -4$

Because the solutions to the equation are imaginary, there are no x–intercepts.

93. $f(x) = 3(x+2)^2 - 5$

Since the vertex is $(-2, -5)$, the axis of symmetry is the line $x = -2$. The point $(-1, -2)$ is on the parabola and lies one unit to the right of the axis of symmetry. This means that the point $(-3, -2)$ will also lie on the parabola since it lies one unit to the left of the axis of symmetry.

95. Start with the form $f(x) = a(x-h)^2 + k$.

Since the vertex is $(h,k) = (-3,-4)$, we have $f(x) = a(x+3)^2 - 4$. We also know that the graph passes through the point $(1,4)$, which allows us to solve for a.

$$4 = a(1+3)^2 - 4$$
$$8 = a(4)^2$$
$$8 = 16a$$
$$\frac{1}{2} = a$$

Therefore, the function is

$$f(x) = \frac{1}{2}(x+3)^2 - 4.$$

97.
$$P = 3x + 4y$$
$$1000 = 3x + 4y$$
$$1000 - 3x = 4y$$
$$\frac{1000 - 3x}{4} = y$$

$$A(x) = x\left(\frac{1000 - 3x}{4}\right)$$
$$= 250x - \frac{3}{4}x^2 = -\frac{3}{4}x^2 + 250x.$$

$$-\frac{b}{2a} = -\frac{250}{2\left(-\frac{3}{4}\right)} = -\frac{250}{-1.5} = \frac{500}{3} \approx 166.7$$

$$y = \frac{1000 - 3x}{4} = \frac{1000 - 3(500/3)}{4}$$
$$= \frac{1000 - 500}{4} = \frac{500}{4} = 125 \text{ feet}$$

The enclosed area is $A = xy = \frac{500}{3}(125) \approx 20,833$.

The dimensions of the enclosed area are approximately 166.7 feet by 125 feet with an enclosed area of 20,833 square feet.

99.

$$\frac{2}{x+5}+\frac{1}{x-5}=\frac{16}{x^2-25}$$

$$\frac{2}{x+5}+\frac{1}{x-5}=\frac{16}{(x+5)(x-5)}$$

$$(x+5)(x-5)\left(\frac{2}{x+5}+\frac{1}{x-5}\right)=(x+5)(x-5)\left(\frac{16}{(x+5)(x-5)}\right)$$

$$(x-5)(2)+(x+5)(1)=16$$

$$2x-10+x+5=16$$

$$3x-5=16$$

$$3x=21$$

$$x=7$$

The solution set is $\{7\}$.

100.

$$\frac{1+\dfrac{2}{x}}{1-\dfrac{4}{x^2}}=\frac{x^2}{x^2}\cdot\frac{1+\dfrac{2}{x}}{1-\dfrac{4}{x^2}}=\frac{x^2+x\cdot2}{x^2-4}$$

$$=\frac{x^2+2x}{(x+2)(x-2)}$$

$$=\frac{x\cancel{(x+2)}}{\cancel{(x+2)}(x-2)}=\frac{x}{x-2}$$

101. $2x+3y=6$

$x-4y=14$

Solve the second equation for x in terms of y.

$x-4y=14$

$x=4y+14$

Substitute $4y+14$ for x in the first equation.

$$2x+3y=6$$

$$2(4y+14)+3y=6$$

$$8y+28+3y=6$$

$$11y+28=6$$

$$11y=-22$$

$$y=-2$$

Back-substitute to find x.

$x=4y+14$

$x=4(-2)+14$

$x=6$

The solution set is $\{(6,-2)\}$.

102. $u^2 - 8u - 9 = 0$

$(u+1)(u-9) = 0$

$u + 1 = 0$ or $u - 9 = 0$

$u = -1$ $u = 9$

The solution set is $\{-1, 9\}$.

103. $2u^2 - u - 10 = 0$

$(u+2)(2u-5) = 0$

$u + 2 = 0$ or $2u - 5 = 0$

$u = -2$ $2u = 5$

$u = \dfrac{5}{2}$

The solution set is $\left\{-2, \dfrac{5}{2}\right\}$.

104. $5x^{\frac{2}{3}} + 11x^{\frac{1}{3}} + 2 = 0$

$5\left(x^{\frac{1}{3}}\right)^2 + 11x^{\frac{1}{3}} + 2 = 0$

$5u^2 + 11u + 2 = 0$

Chapter 11 Mid-Chapter Check Point

1. $(3x-5)^2 = 36$

Apply the square root principle.

$3x - 5 = \pm\sqrt{36} = \pm 6$

$3x = 5 \pm 6$

$x = \dfrac{5 \pm 6}{3} = \dfrac{11}{3}$ or $-\dfrac{1}{3}$

The solution set is $\left\{-\dfrac{1}{3}, \dfrac{11}{3}\right\}$.

2. $5x^2 - 2x = 7$

$5x^2 - 2x - 7 = 0$

$(5x-7)(x+1) = 0$

Apply the zero-product principle.

$5x - 7 = 0$ or $x + 1 = 0$

$5x = 7$ $x = -1$

$x = \dfrac{7}{5}$

The solution set is $\left\{-1, \dfrac{7}{5}\right\}$.

3. $3x^2 - 6x - 2 = 0$

Apply the quadratic formula.

$a = 3$ $b = -6$ $c = -2$

$x = \dfrac{-(-6) \pm \sqrt{(-6)^2 - 4(3)(-2)}}{2(3)}$

$= \dfrac{6 \pm \sqrt{36 - (-24)}}{6}$

$= \dfrac{6 \pm \sqrt{60}}{6}$

$= \dfrac{6 \pm \sqrt{4 \cdot 15}}{6} = \dfrac{6 \pm 2\sqrt{15}}{6} = \dfrac{3 \pm \sqrt{15}}{3}$

The solution set is $\left\{\dfrac{3 \pm \sqrt{15}}{3}\right\}$.

4. $x^2 + 6x = -2$

$x^2 + 6x + 2 = 0$

Apply the quadratic formula.

$a = 1$ $b = 6$ $c = 2$

$x = \dfrac{-6 \pm \sqrt{6^2 - 4(1)(2)}}{2(1)}$

$= \dfrac{-6 \pm \sqrt{36 - 8}}{2}$

$= \dfrac{-6 \pm \sqrt{28}}{2}$

$= \dfrac{-6 \pm \sqrt{4 \cdot 7}}{2} = \dfrac{-6 \pm 2\sqrt{7}}{2} = -3 \pm \sqrt{7}$

The solution set is $\left\{-3 \pm \sqrt{7}\right\}$.

5. $5x^2 + 1 = 37$

$5x^2 = 36$

$x^2 = \dfrac{36}{5}$

Apply the square root principle.

$x = \pm\sqrt{\dfrac{36}{5}} = \pm\dfrac{6}{\sqrt{5}} \cdot \dfrac{\sqrt{5}}{\sqrt{5}} = \pm\dfrac{6\sqrt{5}}{5}$

The solution set is $\left\{\pm\dfrac{6\sqrt{5}}{5}\right\}$.

6. $x^2 - 5x + 8 = 0$

Apply the quadratic formula.

$a = 1 \quad b = -5 \quad c = 8$

$x = \dfrac{-(-5) \pm \sqrt{(-5)^2 - 4(1)(8)}}{2(1)}$

$= \dfrac{5 \pm \sqrt{25 - 32}}{2}$

$= \dfrac{5 \pm \sqrt{-7}}{2}$

$= \dfrac{5 \pm \sqrt{7 \cdot (-1)}}{2}$

$= \dfrac{5 \pm i\sqrt{7}}{2}$

$= \dfrac{5}{2} \pm i\dfrac{\sqrt{7}}{2}$

The solution set is $\left\{ \dfrac{5}{2} \pm \dfrac{\sqrt{7}}{2}i \right\}$.

7. $2x^2 + 26 = 0$

$2x^2 = -26$

$x^2 = -13$

Apply the square root principle.

$x = \pm\sqrt{-13} = \pm\sqrt{13(-1)} = \pm i\sqrt{13}$.

The solution set is $\left\{ \pm i\sqrt{13} \right\}$.

8. $(2x + 3)(x + 2) = 10$

$2x^2 + 4x + 3x + 6 = 10$

$2x^2 + 7x - 4 = 0$

$(2x - 1)(x + 4) = 0$

Apply the zero-product principle.

$2x - 1 = 0 \quad$ or $\quad x + 4 = 0$

$2x = 1 \qquad\qquad x = -4$

$x = \dfrac{1}{2}$

The solution set is $\left\{ -4, \dfrac{1}{2} \right\}$.

9. $(x + 3)^2 = 24$

Apply the square root principle.

$x + 3 = \pm\sqrt{24}$

$x + 3 = \pm\sqrt{4 \cdot 6} = \pm 2\sqrt{6}$

$x = -3 \pm 2\sqrt{6}$

The solution set is $\left\{ -3 \pm 2\sqrt{6} \right\}$.

10. $\dfrac{1}{x^2} - \dfrac{4}{x} + 1 = 0$

Multiply both sides of the equation by the least common denominator x^2.

$x^2 \left(\dfrac{1}{x^2} - \dfrac{4}{x} + 1 \right) = x^2 (0)$

$1 - 4x + x^2 = 0$

$x^2 - 4x + 1 = 0$

Apply the quadratic formula.

$a = 1 \quad b = -4 \quad c = 1$

$x = \dfrac{-(-4) \pm \sqrt{(-4)^2 - 4(1)(1)}}{2(1)}$

$= \dfrac{4 \pm \sqrt{16 - 4}}{2}$

$= \dfrac{4 \pm \sqrt{12}}{2}$

$= \dfrac{4 \pm \sqrt{4 \cdot 3}}{2} = \dfrac{4 \pm 2\sqrt{3}}{2} = 2 \pm \sqrt{3}$

The solution set is $\left\{ 2 \pm \sqrt{3} \right\}$.

11. $x(2x - 3) = -4$

$2x^2 - 3x = -4$

$2x^2 - 3x + 4 = 0$

Apply the quadratic formula.

$a = 2 \quad b = -3 \quad c = 4$

$x = \dfrac{-(-3) \pm \sqrt{(-3)^2 - 4(2)(4)}}{2(2)}$

$= \dfrac{3 \pm \sqrt{9 - 32}}{4}$

$= \dfrac{3 \pm \sqrt{-23}}{4}$

$= \dfrac{3 \pm \sqrt{23(-1)}}{4} = \dfrac{3 \pm i\sqrt{23}}{4} = \dfrac{3}{4} \pm i\dfrac{\sqrt{23}}{4}$

The solution set is $\left\{ \dfrac{3}{4} \pm i\dfrac{\sqrt{23}}{4} \right\}$.

12. $\dfrac{x^2}{3} + \dfrac{x}{2} = \dfrac{2}{3}$

Multiply both sides of the equation by the least common denominator 6.

$$6\left(\dfrac{x^2}{3} + \dfrac{x}{2}\right) = 6\left(\dfrac{2}{3}\right)$$

$$2x^2 + 3x = 4$$

$$2x^2 + 3x - 4 = 0$$

Apply the quadratic formula.

$$a = 3 \quad b = 3 \quad c = -4$$

$$x = \dfrac{-3 \pm \sqrt{3^2 - 4(2)(-4)}}{2(2)}$$

$$= \dfrac{-3 \pm \sqrt{9 - (-32)}}{4} = \dfrac{-3 \pm \sqrt{41}}{4}$$

The solution set is $\left\{\dfrac{-3 \pm \sqrt{41}}{4}\right\}$.

13. $\dfrac{2x}{x^2 + 6x + 8} = \dfrac{x}{x+4} - \dfrac{2}{x+2}$

$$\dfrac{2x}{(x+4)(x+2)} = \dfrac{x}{x+4} - \dfrac{2}{x+2}$$

Multiply both sides of the equation by the least common denominator $(x+4)(x+2)$.

$$(x+4)(x+2)\left[\dfrac{2x}{(x+4)(x+2)}\right] = (x+4)(x+2)\left[\dfrac{x}{x+4} - \dfrac{2}{x+2}\right]$$

$$2x = x(x+2) - 2(x+4)$$

$$2x = x^2 + 2x - 2x - 8$$

$$0 = x^2 - 2x - 8$$

$$0 = (x-4)(x+2)$$

Apply the zero-product principle.

$$x - 4 = 0 \quad \text{or} \quad x + 2 = 0$$

$$x = 4 \qquad\qquad x = -2$$

Disregard $x = -2$ since it makes the denominator zero. Thus, the solution is 4, and the solution set is $\{4\}$.

14. $x^2 + 10x - 3 = 0$

$$x^2 + 10x = 3$$

Since $b = 10$, add $\left(\dfrac{10}{2}\right)^2 = 5^2 = 25$.

$$x^2 + 10x + 25 = 3 + 25$$

$$(x+5)^2 = 28$$

Apply the square root principle.

$$x + 5 = \pm\sqrt{28}$$

$$x + 5 = \pm\sqrt{4 \cdot 7} = \pm 2\sqrt{7}$$

$$x = -5 \pm 2\sqrt{7}$$

The solution set is $\left\{-5 \pm 2\sqrt{7}\right\}$.

15. $d = \sqrt{(-2-2)^2 + (2-(-2))^2}$

$$= \sqrt{(-4)^2 + (4)^2}$$

$$= \sqrt{16 + 16} = \sqrt{32}$$

$$= \sqrt{16 \cdot 2}$$

$$= 4\sqrt{2} \approx 5.66 \text{ units}$$

$$\left(\dfrac{2 + (-2)}{2}, \dfrac{-2 + 2}{2}\right) = \left(\dfrac{0}{2}, \dfrac{0}{2}\right) = (0, 0)$$

The length of the segment is $4\sqrt{2} \approx 5.66$ units and the midpoint is the origin, $(0, 0)$.

16. $d = \sqrt{(-10-(-5))^2 + (14-8)^2}$

$$= \sqrt{(-5)^2 + (6)^2}$$

$$= \sqrt{25 + 36}$$

$$= \sqrt{61} \approx 7.81 \text{ units}$$

$$\left(\dfrac{-5 + (-10)}{2}, \dfrac{8 + 14}{2}\right) = \left(\dfrac{-15}{2}, \dfrac{22}{2}\right)$$

$$= \left(-\dfrac{15}{2}, 11\right)$$

The length of the segment is $\sqrt{61} \approx 7.81$ units and the midpoint is $\left(-\dfrac{15}{2}, 11\right)$.

17. $f(x) = (x-3)^2 - 4$

Since $a = 1$ is positive, the parabola opens upward.

The vertex of the parabola is $(h, k) = (3, -4)$.

Replace $f(x)$ with 0 to find x–intercepts.

$0 = (x-3)^2 - 4$

$4 = (x-3)^2$

Apply the square root property.

$x - 3 = \pm\sqrt{4} = \pm 2$

$\quad x = 3 \pm 2 = 5 \text{ or } 1$

The x–intercepts are 5 and 1.

Set $x = 0$ and solve for y to obtain the y–intercept.

$y = (0-3)^2 - 4 = 5$

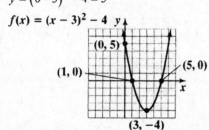

Domain: $\{x \mid x \text{ is a real number}\}$ or $(-\infty, \infty)$

Range: $\{y \mid y \geq -4\}$ or $[-4, \infty)$.

18. $g(x) = 5 - (x+2)^2$

$g(x) = -\left[x - (-2)\right]^2 + 5$

Since $a = -1$ is negative, the parabola opens downward. The vertex of the parabola is $(h, k) = (-2, 5)$. Replace $g(x)$ with 0 to find x–intercepts.

$0 = -(x+2)^2 + 5$

$-5 = -(x+2)^2$

$5 = (x+2)^2$

Apply the square root property.

$\sqrt{5} = x + 2 \quad \text{or} \quad -\sqrt{5} = x + 2$

$-2 + \sqrt{5} = x \qquad\quad -2 - \sqrt{5} = x$

$\quad 0.24 \approx x \qquad\qquad -4.24 \approx x$

The x–intercepts are $-2 + \sqrt{5} \approx 0.24$ and $-2 - \sqrt{5} \approx -4.24$.

Set $x = 0$ and solve for y to obtain the y–intercept.

$y = 5 - (0+2)^2 = 5 - 4 = 1$

The y–intercept is 1.

$g(x) = 5 - (x+2)^2$

Domain: $\{x \mid x \text{ is a real number}\}$ or $(-\infty, \infty)$

Range: $\{y \mid y \leq 5\}$ or $(-\infty, 5]$.

19. $h(x) = -x^2 - 4x + 5$

Since $a = -1$ is negative, the parabola opens downward. The x–coordinate of the vertex of the parabola is $-\dfrac{b}{2a} = -\dfrac{-4}{2(-1)} = -\dfrac{-4}{-2} = -2$ and the y–coordinate of the vertex of the parabola is

$h\left(-\dfrac{b}{2a}\right) = h(-2)$

$\qquad = -(-2)^2 - 4(-2) + 5$

$\qquad = -4 + 8 + 5 = 9$

The vertex is $(-2, 9)$. Replace $h(x)$ with 0 to find the x–intercepts.

$0 = -x^2 - 4x + 5$

$0 = x^2 + 4x - 5$

$0 = (x+5)(x-1)$

Apply the zero product principle.

$x + 5 = 0 \quad \text{or} \quad x - 1 = 0$

$\quad x = -5 \qquad\qquad x = 1$

The x–intercepts are -5 and 1. Set $x = 0$ and solve for y to obtain the y–intercept. $y = -(0)^2 - 4(0) + 5 = 5$

$h(x) = -x^2 - 4x + 5$

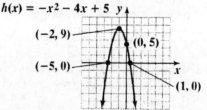

Domain: $\{x \mid x \text{ is a real number}\}$ or $(-\infty, \infty)$

Range: $\{y \mid y \leq 9\}$ or $(-\infty, 9]$.

20. $f(x) = 3x^2 - 6x + 1$

Since $a = 3$, the parabola opens upward. The x–coordinate of the vertex of the parabola is

$-\dfrac{b}{2a} = -\dfrac{-6}{2(3)} = -\dfrac{-6}{6} = 1$ and the y–coordinate of

the minimum is

$$f\left(-\frac{b}{2a}\right) = f(1)$$
$$= 3(1)^2 - 6(1) + 1$$
$$= 3 - 6 + 1 = -2$$

The vertex $(1, 1)$. Replace $f(x)$ with 0 to find the x-intercepts.

$0 = 3x^2 - 6x + 1$

Apply the quadratic formula.

$a = 3 \quad b = -6 \quad c = 1$

$$x = \frac{-(-6) \pm \sqrt{(-6)^2 - 4(3)(1)}}{2(3)}$$
$$= \frac{6 \pm \sqrt{36 - 12}}{6}$$
$$= \frac{6 \pm \sqrt{24}}{6}$$
$$= \frac{6 \pm 2\sqrt{6}}{6} = \frac{3 \pm \sqrt{6}}{3} \approx 0.18 \text{ or } 1.82$$

Set $x = 0$ and solve for y to obtain the y-intercept:

$y = 3(0)^2 - 6(0) + 1 = 1$

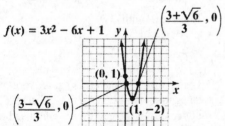

$f(x) = 3x^2 - 6x + 1$

$\left(\dfrac{3+\sqrt{6}}{3}, 0\right)$

$(0, 1)$

$\left(\dfrac{3-\sqrt{6}}{3}, 0\right)$

$(1, -2)$

Domain: $\{x \mid x$ is a real number$\}$ or $(-\infty, \infty)$

Range: $\{y \mid y \geq -2\}$ or $[-2, \infty)$.

21. $2x^2 + 5x + 4 = 0$

$a = 2 \quad b = 5 \quad c = 4$

$b^2 - 4ac = 5^2 - 4(2)(4)$

$= 25 - 32 = -7$

Since the discriminant is negative, there are no real solutions. There are two imaginary solutions that are complex conjugates.

22.
$$10x(x + 4) = 15x - 15$$
$$10x^2 + 40x = 15x - 15$$
$$10x^2 - 25x + 15 = 0$$
$$a = 10 \quad b = -25 \quad c = 15$$

$$b^2 - 4ac = (-25)^2 - 4(10)(15)$$
$$= 625 - 600 = 25$$

Since the discriminant is positive and a perfect square, there are two real rational solutions.

23. Because the solution set is $\left\{-\dfrac{1}{2}, \dfrac{3}{4}\right\}$, we have

$$x = -\frac{1}{2} \quad \text{or} \quad x = \frac{3}{4}$$
$$2x = -1 \qquad\qquad 4x = 3$$
$$2x + 1 = 0 \qquad\quad 4x - 3 = 0$$

Use the zero-product principle in reverse.

$$(2x + 1)(4x - 3) = 0$$
$$8x^2 - 6x + 4x - 3 = 0$$
$$8x^2 - 2x - 3 = 0$$

24. Because the solution set is $\left\{-2\sqrt{3}, 2\sqrt{3}\right\}$ we have

$$x = -2\sqrt{3} \quad \text{or} \qquad x = 2\sqrt{3}$$
$$x + 2\sqrt{3} = 0 \qquad\quad x - 2\sqrt{3} = 0$$

Use the zero-product principle in reverse.

$$(x + 2\sqrt{3})(x - 2\sqrt{3}) = 0$$
$$x^2 + 2x\sqrt{3} - 2x\sqrt{3} - 4 \cdot 3 = 0$$
$$x^2 - 12 = 0$$

25. $P(x) = -x^2 + 150x - 4425$

Since $a = -1$ is negative, the function opens down and has a maximum at

$$x = -\frac{b}{2a} = -\frac{150}{2(-1)} = -\frac{150}{-2} = 75 \, .$$

$$P(75) = -75^2 + 150(75) - 4425$$
$$= -5625 + 11,250 - 4425 = 1200$$

The company will maximize its profit by manufacturing and selling 75 cabinets per day. The maximum daily profit is $1200.

26. Let x = one of the numbers.
Let $-18-x$ = the other number.

The product is $f(x) = x(-18-x) = -x^2 -18x$
The x-coordinate of the maximum is
$$x = -\frac{b}{2a} = -\frac{-18}{2(-1)} = -\frac{-18}{-2} = -9.$$
$$f(-9) = -9\left[-18-(-9)\right]$$
$$= -9(-18+9) = -9(-9) = 81$$

The vertex is $(-9, 81)$. The maximum product is 81. This occurs when the two number are -9 and $-18-(-9) = -9$.

27. Let x = the measure of the height.
Let $40-2x$ = the measure of the base.
$$A = \frac{1}{2}bh$$
$$A(x) = \frac{1}{2}(40-2x)x$$
$$A(x) = -x^2 + 20x$$

Since $a = -1$ is negative, the function opens down and has a maximum at
$$x = -\frac{b}{2a} = -\frac{20}{2(-1)} = -\frac{20}{-2} = 10.$$
$$A(10) = -10^2 + 20(10)$$
$$= -100 + 200 = 100$$
A height of 10 inches will maximize the area of the triangle. The maximum area will be 100 square inches.

11.4 Check Points

1. Let $u = x^2$.
$$x^4 - 5x^2 + 6 = 0$$
$$\left(x^2\right)^2 - 5x^2 + 6 = 0$$
$$u^2 - 5u + 6 = 0$$
$$(u-3)(u-2) = 0$$
Apply the zero product principle.
$$u-3 = 0 \quad \text{or} \quad u-2 = 0$$
$$u = 3 \qquad\qquad u = 2$$
Replace u with x^2.
$$x^2 = 3 \quad \text{or} \quad x^2 = 2$$
$$x = \pm\sqrt{3} \qquad x = \pm\sqrt{2}$$
The solution set is $\left\{\pm\sqrt{2}, \pm\sqrt{3}\right\}$.

2. Let $u = \sqrt{x}$.
$$x - 2\sqrt{x} - 8 = 0$$
$$\left(\sqrt{x}\right)^2 - 2\sqrt{x} - 8 = 0$$
$$u^2 - 2u - 8 = 0$$
$$(u+2)(u-4) = 0$$
Apply the zero product principle.
$$u+2 = 0 \quad \text{or} \quad u-4 = 0$$
$$u = -2 \qquad\qquad u = 4$$
Replace u with $\sqrt{x}$.
$$\cancel{\sqrt{x} = -2} \quad \text{or} \quad \sqrt{x} = 4$$
$$x = 16$$

Disregard -2 because the square root of x cannot be a negative number. We must check 16, because both sides of the equation were raised to an even power.
Check:
$$x - 2\sqrt{x} - 8 = 0$$
$$16 - 2\sqrt{16} - 8 = 0$$
$$16 - 2\cdot 4 - 8 = 0$$
$$16 - 8 - 8 = 0$$
$$0 = 0$$
The solution set is $\{16\}$.

3. Let $u = x^2 - 4$.

$$(x^2 - 4)^2 + (x^2 - 4) - 6 = 0$$

$$u^2 + u - 6 = 0$$

$$(u + 3)(u - 2) = 0$$

Apply the zero product principle and replace u.

$$u + 3 = 0 \quad \text{or} \quad u - 2 = 0$$

$$x^2 - 4 + 3 = 0 \qquad x^2 - 4 - 2 = 0$$

$$x^2 - 1 = 0 \qquad x^2 - 6 = 0$$

$$x^2 = 1 \qquad x^2 = 6$$

$$x = \pm 1 \qquad x = \pm\sqrt{6}$$

The solution set is $\left\{ \pm\sqrt{6}, \pm 1 \right\}$.

4. Let $u = x^{-1}$.

$$2x^{-2} + x^{-1} - 1 = 0$$

$$2\left(x^{-1}\right)^2 + x^{-1} - 1 = 0$$

$$2u^2 + u - 1 = 0$$

$$(u + 1)(2u - 1) = 0$$

Apply the zero product principle.

$$u + 1 = 0 \quad \text{or} \quad 2u - 1 = 0$$

$$u = -1 \qquad 2u = 1$$

$$u = \frac{1}{2}$$

Replace u with x^{-1}.

$$u = -1 \quad \text{or} \quad u = \frac{1}{2}$$

$$x^{-1} = -1 \qquad x^{-1} = \frac{1}{2}$$

$$\left(x^{-1}\right)^{-1} = (-1)^{-1}$$

$$x = -1 \qquad \left(x^{-1}\right)^{-1} = \left(\frac{1}{2}\right)^{-1}$$

$$x = 2$$

The solution set is $\{-1, 2\}$.

5. Let $u = x^{\frac{1}{3}}$.

$$3x^{\frac{2}{3}} - 11x^{\frac{1}{3}} - 4 = 0$$

$$3\left(x^{\frac{1}{3}}\right)^2 - 11x^{\frac{1}{3}} - 4 = 0$$

$$3u^2 - 11u - 4 = 0$$

$$(3u + 1)(u - 4) = 0$$

Apply the zero product principle.

$$3u + 1 = 0 \quad \text{or} \quad u - 4 = 0$$

$$u = -\frac{1}{3} \qquad u = 4$$

Replace u with $x^{\frac{1}{3}}$.

$$u = -\frac{1}{3} \quad \text{or} \quad u = 4$$

$$x^{\frac{1}{3}} = -\frac{1}{3} \qquad x^{\frac{1}{3}} = 4$$

$$\left(x^{\frac{1}{3}}\right)^3 = \left(-\frac{1}{3}\right)^3 \qquad \left(x^{\frac{1}{3}}\right)^3 = (4)^3$$

$$x = 64$$

$$x = -\frac{1}{27}$$

The solution set is $\left\{ -\frac{1}{27}, 64 \right\}$.

11.4 Exercise Set

1. Let $u = x^2$.

$$x^4 - 5x^2 + 4 = 0$$

$$\left(x^2\right)^2 - 5x^2 + 4 = 0$$

$$u^2 - 5u + 4 = 0$$

$$(u - 4)(u - 1) = 0$$

Apply the zero product principle.

$$u - 4 = 0 \quad \text{or} \quad u - 1 = 0$$

$$u = 4 \qquad u = 1$$

Replace u with x^2.

$$x^2 = 4 \quad \text{or} \quad x^2 = 1$$

$$x = \pm 2 \qquad x = \pm 1$$

The solution set is $\{-2, -1, 1, 2\}$.

3. Let $u = x^2$.

$$x^4 - 11x^2 + 18 = 0$$

$$\left(x^2\right)^2 - 11x^2 + 18 = 0$$

$$u^2 - 11u + 18 = 0$$

$$(u-9)(u-2) = 0$$

Apply the zero product principle.

$$u - 9 = 0 \quad \text{or} \quad u - 2 = 0$$

$$u = 9 \qquad\qquad u = 2$$

Replace u with x^2.

$$x^2 = 9 \quad \text{or} \quad x^2 = 2$$

$$x = \pm 3 \qquad x = \pm\sqrt{2}$$

The solution set is $\left\{-3, -\sqrt{2}, \sqrt{2}, 3\right\}$.

5. Let $u = x^2$.

$$x^4 + 2x^2 = 8$$

$$x^4 + 2x^2 - 8 = 0$$

$$\left(x^2\right)^2 + 2x^2 - 8 = 0$$

$$u^2 + 2u - 8 = 0$$

$$(u+4)(u-2) = 0$$

Apply the zero product principle.

$$u + 4 = 0 \quad \text{or} \quad u - 2 = 0$$

$$u = -4 \qquad\qquad u = 2$$

Replace u with x^2.

$$x^2 = -4 \quad \text{or} \quad x^2 = 2$$

$$x = \pm\sqrt{-4} \qquad x = \pm\sqrt{2}$$

$$x = \pm 2i$$

The solution set is $\left\{-2i, 2i, -\sqrt{2}, \sqrt{2}\right\}$.

7. Let $u = \sqrt{x}$.

$$x + \sqrt{x} - 2 = 0$$

$$\left(\sqrt{x}\right)^2 + \sqrt{x} - 2 = 0$$

$$u^2 + u - 2 = 0$$

$$(u+2)(u-1) = 0$$

Apply the zero product principle.

$$u + 2 = 0 \quad \text{or} \quad u - 1 = 0$$

$$u = -2 \qquad\qquad u = 1$$

Replace u with $\sqrt{x}$.

$$\cancel{\sqrt{x} = -2} \quad \text{or} \quad \sqrt{x} = 1$$

$$x = 1$$

Disregard -2 because the square root of x cannot be a negative number. We must check 1, because both sides of the equation were raised to an even power.

Check:

$$1 + \sqrt{1} - 2 = 0$$

$$1 + 1 - 2 = 0$$

$$2 - 2 = 0$$

$$0 = 0$$

The solution set is $\left\{1\right\}$.

9. Let $u = x^{\frac{1}{2}}$.

$$x - 4x^{\frac{1}{2}} - 21 = 0$$

$$\left(x^{\frac{1}{2}}\right)^2 - 4x^{\frac{1}{2}} - 21 = 0$$

$$u^2 - 4u - 21 = 0$$

$$(u-7)(u+3) = 0$$

Apply the zero product principle.

$$u - 7 = 0 \quad \text{or} \quad u + 3 = 0$$

$$u = 7 \qquad\qquad u = -3$$

Replace u with $x^{\frac{1}{2}}$.

$$x^{\frac{1}{2}} = 7 \quad \text{or} \quad x^{\frac{1}{2}} = -3$$

$$\sqrt{x} = 7 \qquad \cancel{\sqrt{x} = -3}$$

$$x = 49$$

Disregard –3 because the square root of x cannot be a negative number. We must check 49, because both sides of the equation were raised to an even power.

Check:

$$49 - 4(49)^{\frac{1}{2}} - 21 = 0$$
$$49 - 4(7) - 21 = 0$$
$$49 - 28 - 21 = 0$$
$$49 - 49 = 0$$
$$0 = 0$$

The solution set is $\{49\}$.

11. Let $u = \sqrt{x}$.

$$x - 13\sqrt{x} + 40 = 0$$
$$\left(\sqrt{x}\right)^2 - 13\sqrt{x} + 40 = 0$$
$$u^2 - 13u + 40 = 0$$
$$(u - 5)(u - 8) = 0$$

Apply the zero product principle.
$$u - 5 = 0 \quad \text{or} \quad u - 8 = 0$$
$$u = 5 \qquad\qquad u = 8$$

Replace u with $\sqrt{x}$.
$$\sqrt{x} = 5 \quad \text{or} \quad \sqrt{x} = 8$$
$$x = 25 \qquad\qquad x = 64$$

Both solutions must be checked since both sides of the equation were raised to an even power.

$$x = 25$$
$$25 - 13\sqrt{25} + 40 = 0$$
$$25 - 13(5) + 40 = 0$$
$$25 - 65 + 40 = 0$$
$$65 - 65 = 0$$
$$0 = 0$$

$$x = 64$$
$$64 - 13\sqrt{64} + 40 = 0$$
$$64 - 13(8) + 40 = 0$$
$$64 - 104 + 40 = 0$$
$$0 = 0$$

Both solutions check. The solution set is $\{25, 64\}$.

13. Let $u = x - 5$.

$$(x - 5)^2 - 4(x - 5) - 21 = 0$$
$$u^2 - 4u - 21 = 0$$
$$(u - 7)(u + 3) = 0$$

Apply the zero product principle.
$$u - 7 = 0 \quad \text{or} \quad u + 3 = 0$$
$$u = 7 \qquad\qquad u = -3$$

Replace u with $x - 5$.
$$x - 5 = 7 \quad \text{or} \quad x - 5 = -3$$
$$x = 12 \qquad\qquad x = 2$$

The solution set is $\{2, 12\}$.

15. Let $u = x^2 - 1$.

$$\left(x^2 - 1\right)^2 - \left(x^2 - 1\right) = 2$$
$$\left(x^2 - 1\right)^2 - \left(x^2 - 1\right) - 2 = 0$$
$$u^2 - u - 2 = 0$$
$$(u - 2)(u + 1) = 0$$

Apply the zero product principle.
$$u - 2 = 0 \quad \text{or} \quad u + 1 = 0$$
$$u = 2 \qquad\qquad u = -1$$

Replace u with $x^2 - 1$.
$$x^2 - 1 = 2 \quad \text{or} \quad x^2 - 1 = -1$$
$$x^2 = 3 \qquad\qquad x^2 = 0$$
$$x = \pm\sqrt{3} \qquad\qquad x = 0$$

The solution set is $\left\{-\sqrt{3}, 0, \sqrt{3}\right\}$.

17. Let $u = x^2 + 3x$.

$$\left(x^2 + 3x\right)^2 - 8\left(x^2 + 3x\right) - 20 = 0$$
$$u^2 - 8u - 20 = 0$$
$$(u - 10)(u + 2) = 0$$

Apply the zero product principle.
$$u - 10 = 0 \quad \text{or} \quad u + 2 = 0$$
$$u = 10 \qquad\qquad u = -2$$

Replace u with $x^2 + 3x$.
First, consider $u = 10$.
$$x^2 + 3x = 10$$
$$x^2 + 3x - 10 = 0$$
$$(x + 5)(x - 2) = 0$$
Apply the zero product principle.
$$x + 5 = 0 \quad \text{or} \quad x - 2 = 0$$
$$x = -5 \qquad\qquad x = 2$$

Next, consider $u = -2$.
$$x^2 + 3x = -2$$
$$x^2 + 3x + 2 = 0$$
$$(x + 2)(x + 1) = 0$$
Apply the zero product principle.

$$x + 2 = 0 \quad \text{or} \quad x + 1 = 0$$
$$x = -2 \qquad\qquad x = -1$$

The solution set is $\{-5, -2, -1, 2\}$.

19. Let $u = x^{-1}$.

$$x^{-2} - x^{-1} - 20 = 0$$
$$\left(x^{-1}\right)^2 - x^{-1} - 20 = 0$$
$$u^2 - u - 20 = 0$$
$$(u - 5)(u + 4) = 0$$

Apply the zero product principle.
$$u - 5 = 0 \quad \text{or} \quad u + 4 = 0$$
$$u = 5 \qquad\qquad u = -4$$

Replace u with x^{-1}.
$$x^{-1} = 5 \quad \text{or} \quad x^{-1} = -4$$
$$\frac{1}{x} = 5 \qquad\qquad \frac{1}{x} = -4$$
$$5x = 1 \qquad\qquad -4x = 1$$
$$x = \frac{1}{5} \qquad\qquad x = -\frac{1}{4}$$

The solution set is $\left\{-\frac{1}{4}, \frac{1}{5}\right\}$.

21. Let $u = x^{-1}$.

$$2x^{-2} - 7x^{-1} + 3 = 0$$
$$2\left(x^{-1}\right)^2 - 7x^{-1} + 3 = 0$$
$$2u^2 - 7u + 3 = 0$$
$$(2u - 1)(u - 3) = 0$$

Apply the zero product principle.
$$2u - 1 = 0 \quad \text{or} \quad u - 3 = 0$$
$$2u = 1 \qquad\qquad u = 3$$
$$u = \frac{1}{2}$$

Replace u with x^{-1}.
$$x^{-1} = \frac{1}{2} \quad \text{or} \quad x^{-1} = 3$$
$$\frac{1}{x} = \frac{1}{2} \qquad\qquad \frac{1}{x} = 3$$
$$x = 2 \qquad\qquad 3x = 1$$
$$x = \frac{1}{3}$$

The solution set is $\left\{\frac{1}{3}, 2\right\}$.

23. Let $u = x^{-1}$.

$$x^{-2} - 4x^{-1} = 3$$
$$x^{-2} - 4x^{-1} - 3 = 0$$
$$\left(x^{-1}\right)^2 - 4x^{-1} - 3 = 0$$
$$u^2 - 4u - 3 = 0$$
$$a = 1 \quad b = -4 \quad c = -3$$

Use the quadratic formula.

$$u = \frac{-(-4) \pm \sqrt{(-4)^2 - 4(1)(-3)}}{2(1)}$$

$$= \frac{4 \pm \sqrt{16 + 12}}{2} = \frac{4 \pm \sqrt{28}}{2}$$

$$= \frac{4 \pm 2\sqrt{7}}{2} = \frac{2\left(2 \pm \sqrt{7}\right)}{2} = 2 \pm \sqrt{7}$$

Replace u with x^{-1}.

$$x^{-1} = 2 \pm \sqrt{7}$$
$$\frac{1}{x} = 2 \pm \sqrt{7}$$
$$\left(2 \pm \sqrt{7}\right)x = 1$$
$$x = \frac{1}{2 \pm \sqrt{7}}$$

Rationalize the denominator.

$$x = \frac{1}{2 \pm \sqrt{7}} \cdot \frac{2 \mp \sqrt{7}}{2 \mp \sqrt{7}} = \frac{2 \mp \sqrt{7}}{2^2 - \left(\sqrt{7}\right)^2}$$

$$= \frac{2 \mp \sqrt{7}}{4 - 7} = \frac{2 \mp \sqrt{7}}{-3} = \frac{-2 \pm \sqrt{7}}{3}$$

The solution set is $\left\{\dfrac{-2 \pm \sqrt{7}}{3}\right\}$.

25. Let $u = x^{\frac{1}{3}}$.

$$x^{\frac{2}{3}} - x^{\frac{1}{3}} - 6 = 0$$

$$\left(x^{\frac{1}{3}}\right)^2 - x^{\frac{1}{3}} - 6 = 0$$
$$u^2 - u - 6 = 0$$
$$(u - 3)(u + 2) = 0$$

Apply the zero product principle.

$$u - 3 = 0 \quad \text{or} \quad u + 2 = 0$$
$$u = 3 \qquad\qquad u = -2$$

Replace u with $x^{\frac{1}{3}}$.

$$x^{\frac{1}{3}} = 3 \qquad\qquad x^{\frac{1}{3}} = -2$$
$$\text{or}$$
$$\left(x^{\frac{1}{3}}\right)^3 = 3^3 \qquad \left(x^{\frac{1}{3}}\right)^3 = (-2)^3$$
$$x = 27 \qquad\qquad x = -8$$

The solution set is $\{-8, 27\}$.

27. Let $u = x^{\frac{1}{5}}$.

$$x^{\frac{2}{5}} + x^{\frac{1}{5}} - 6 = 0$$
$$\left(x^{\frac{1}{5}}\right)^2 + x^{\frac{1}{5}} - 6 = 0$$
$$u^2 + u - 6 = 0$$
$$(u + 3)(u - 2) = 0$$

Apply the zero product principle.

$$u + 3 = 0 \quad \text{or} \quad u - 2 = 0$$
$$u = -3 \qquad\qquad u = 2$$

Replace u with $x^{\frac{1}{5}}$.

$$x^{\frac{1}{5}} = -3 \quad \text{or} \quad x^{\frac{1}{5}} = 2$$
$$\left(x^{\frac{1}{5}}\right)^5 = (-3)^5 \qquad \left(x^{\frac{1}{5}}\right)^5 = (2)^5$$
$$x = -243 \qquad\qquad x = 32$$

The solution set is $\{-243, 32\}$.

29. Let $u = x^{\frac{1}{4}}$.

$$2x^{\frac{1}{2}} - x^{\frac{1}{4}} = 1$$

$$2\left(x^{\frac{1}{4}}\right)^2 - x^{\frac{1}{4}} - 1 = 0$$

$$2u^2 - u - 1 = 0$$

$$(2u + 1)(u - 1) = 0$$

$$2u + 1 = 0 \quad \text{or} \quad u - 1 = 0$$

$$2u = -1 \qquad\qquad u = 1$$

$$u = -\frac{1}{2}$$

Replace u with $x^{\frac{1}{4}}$.

$$x^{\frac{1}{4}} = -\frac{1}{2} \quad \text{or} \quad x^{\frac{1}{4}} = 1$$

$$\left(x^{\frac{1}{4}}\right)^4 = \left(-\frac{1}{2}\right)^4 \qquad \left(x^{\frac{1}{4}}\right)^4 = 1^4$$

$$\qquad\qquad\qquad\qquad x = 1$$

$$x = \frac{1}{16}$$

Since both sides of the equations were raised to an even power, the solutions must be checked.

First, check $x = \frac{1}{16}$.

$$2\left(\frac{1}{16}\right)^{\frac{1}{2}} - \left(\frac{1}{16}\right)^{\frac{1}{4}} = 1$$

$$2\left(\frac{1}{4}\right) - \frac{1}{2} = 1$$

$$\frac{1}{2} - \frac{1}{2} = 1$$

$$0 \neq 1$$

The solution does not check, so disregard $x = \frac{1}{16}$.

Next, check $x = 1$.

$$2(1)^{\frac{1}{2}} - (1)^{\frac{1}{4}} = 1$$

$$2(1) - 1 = 1$$

$$1 = 1$$

The solution checks. The solution set is $\{1\}$.

31. Let $u = x - \frac{8}{x}$.

$$\left(x - \frac{8}{x}\right)^2 + 5\left(x - \frac{8}{x}\right) - 14 = 0$$

$$u^2 + 5u - 14 = 0$$

$$(u + 7)(u - 2) = 0$$

$$u + 7 = 0 \quad \text{or} \quad u - 2 = 0$$

$$u = -7 \qquad\qquad u = 2$$

Replace u with $x - \frac{8}{x}$.

First, consider $u = -7$.

$$x - \frac{8}{x} = -7$$

$$x\left(x - \frac{8}{x}\right) = x(-7)$$

$$x^2 - 8 = -7x$$

$$x^2 + 7x - 8 = 0$$

$$(x + 8)(x - 1) = 0$$

$$x + 8 = 0 \quad \text{or} \quad x - 1 = 0$$

$$x = -8 \qquad\qquad x = 1$$

Next, consider $u = 2$.

$$x - \frac{8}{x} = 2$$

$$x\left(x - \frac{8}{x}\right) = x(2)$$

$$x^2 - 8 = 2x$$

$$x^2 - 2x - 8 = 0$$

$$(x - 4)(x + 2) = 0$$

$$x - 4 = 0 \quad \text{or} \quad x + 2 = 0$$

$$x = 4 \qquad\qquad x = -2$$

The solution set is $\{-8, -2, 1, 4\}$.

33. $f(x) = x^4 - 5x^2 + 4$

$$y = x^4 - 5x^2 + 4$$

Set $y = 0$ to find the x–intercept(s).

$$0 = x^4 - 5x^2 + 4$$

Let $u = x^2$.

$$x^4 - 5x^2 + 4 = 0$$

$$\left(x^2\right)^2 - 5x^2 + 4 = 0$$

$$u^2 - 5u + 4 = 0$$

$$(u - 4)(u - 1) = 0$$

$$u - 1 = 0 \quad \text{or} \quad u - 4 = 0$$

$$u = 1 \qquad\qquad u = 4$$

Substitute x^2 for u.

$$x^2 = 1 \quad \text{or} \quad x^2 = 4$$

$$x = \pm 1 \qquad\quad x = \pm 2$$

The intercepts are ± 1 and ± 2. The corresponding graph is graph c.

35. $f(x) = x^{\frac{1}{3}} + 2x^{\frac{1}{6}} - 3$

$$y = x^{\frac{1}{3}} + 2x^{\frac{1}{6}} - 3$$

Set $y = 0$ to find the x–intercept(s).

$$0 = x^{\frac{1}{3}} + 2x^{\frac{1}{6}} - 3$$

Let $u = x^{\frac{1}{6}}$.

$$x^{\frac{1}{3}} + 2x^{\frac{1}{6}} - 3 = 0$$

$$\left(x^{\frac{1}{6}}\right)^2 + 2x^{\frac{1}{6}} - 3 = 0$$

$$u^2 + 2u - 3 = 0$$

$$(u + 3)(u - 1) = 0$$

$$u + 3 = 0 \quad \text{or} \quad u - 1 = 0$$

$$u = -3 \qquad\qquad u = 1$$

Substitute $x^{\frac{1}{6}}$ for u.

$$x^{\frac{1}{6}} = -3 \qquad\qquad x^{\frac{1}{6}} = 1$$

$$\text{or}$$

$$\left(x^{\frac{1}{6}}\right)^6 = (-3)^6 \qquad \left(x^{\frac{1}{6}}\right)^6 = (1)^6$$

$$x = 729 \qquad\qquad\qquad x = 1$$

Since both sides of the equations were raised to an even power, the solutions must be checked.

First check $x = 729$.

$$(729)^{\frac{1}{3}} + 2(729)^{\frac{1}{6}} - 3 = 0$$

$$9 + 2(3) - 3 = 0$$

$$9 + 6 - 3 = 0$$

$$12 \neq 0$$

Next check $x = 1$.

$$(1)^{\frac{1}{3}} + 2(1)^{\frac{1}{6}} - 3 = 0$$

$$1 + 2(1) - 3 = 0$$

$$1 + 2 - 3 = 0$$

$$0 = 0$$

Since 729 does not check, disregard it. The intercept is 1. The corresponding graph is graph e.

37. $f(x) = (x + 2)^2 - 9(x + 2) + 20$

$$y = (x + 2)^2 - 9(x + 2) + 20$$

Set $y = 0$ to find the x–intercept(s).

$$(x + 2)^2 - 9(x + 2) + 20 = 0$$

Let $u = x + 2$.

$$(x + 2)^2 - 9(x + 2) + 20 = 0$$

$$u^2 - 9u + 20 = 0$$

$$(u - 5)(u - 4) = 0$$

Apply the zero product principle.

$$u - 5 = 0 \quad \text{or} \quad u - 4 = 0$$

$$u = 5 \qquad\qquad u = 4$$

Substitute $x + 2$ for u.

$$x + 2 = 5 \quad \text{or} \quad x + 2 = 4$$

$$x = 3 \qquad\qquad x = 2$$

The intercepts are 2 and 3. The corresponding graph is graph f.

39. Let $u = x^2 + 3x - 2$

$$f(x) = -16$$

$$\left(x^2 + 3x - 2\right)^2 - 10\left(x^2 + 3x - 2\right) = -16$$

$$u^2 - 10u = -16$$

$$u^2 - 10u + 16 = 0$$

$$(u - 8)(u - 2) = 0$$

Apply the zero product principle.

$u - 8 = 0$ or $u - 2 = 0$

$u = 8$ $u = 2$

Replace u with $x^2 + 3x - 2$.

First, consider $u = 8$.

$$x^2 + 3x - 2 = 8$$

$$x^2 + 3x - 10 = 0$$

$$(x + 5)(x - 2) = 0$$

Apply the zero product principle.

$x + 5 = 0$ or $x - 2 = 0$

$x = -5$ $x = 2$

Next, consider $u = 2$.

$$x^2 + 3x - 2 = 2$$

$$x^2 + 3x - 4 = 0$$

$$(x + 4)(x - 1) = 0$$

Apply the zero product principle.

$x + 4 = 0$ or $x - 1 = 0$

$x = -4$ $x = 1$

The solutions are -5, -4, 1, and 2.

41. Let $u = \dfrac{1}{x} + 1$.

$$f(x) = 2$$

$$3\left(\frac{1}{x} + 1\right)^2 + 5\left(\frac{1}{x} + 1\right) = 2$$

$$3u^2 + 5u = 2$$

$$3u^2 + 5u - 2 = 0$$

$$(3u - 1)(u + 2) = 0$$

$3u - 1 = 0$ or $u + 2 = 0$

$3u = 1$ $u = -2$

$$u = \frac{1}{3}$$

Replace u with $\dfrac{1}{x} + 1$.

First, consider $u = \dfrac{1}{3}$.

$$\frac{1}{x} + 1 = \frac{1}{3}$$

$$3x\left(\frac{1}{x} + 1\right) = 3x\left(\frac{1}{3}\right)$$

$$3 + 3x = x$$

$$2x = -3$$

$$x = -\frac{3}{2}$$

Next, consider $u = -2$.

$$\frac{1}{x} + 1 = -2$$

$$3x\left(\frac{1}{x} + 1\right) = 3x(-2)$$

$$3 + 3x = -6x$$

$$9x = -3$$

$$x = -\frac{3}{9} = -\frac{1}{3}$$

The solutions are $-\dfrac{3}{2}$ and $-\dfrac{1}{3}$.

43. Let $u = \sqrt{\dfrac{x}{x-4}}$.

$$f(x) = g(x)$$

$$\frac{x}{x-4} = 13\sqrt{\frac{x}{x-4}} - 36$$

$$\left(\sqrt{\frac{x}{x-4}}\right)^2 = 13\sqrt{\frac{x}{x-4}} - 36$$

$$u^2 = 13u - 36$$

$$u^2 - 13u + 36 = 0$$

$$(u-9)(u-4) = 0$$

$$u - 9 = 0 \quad \text{or} \quad u - 4 = 0$$

$$u = 9 \qquad\qquad u = 4$$

Replace u with $\sqrt{\dfrac{x}{x-4}}$.

First, consider $u = 9$.

$$\sqrt{\frac{x}{x-4}} = 9$$

$$\left(\sqrt{\frac{x}{x-4}}\right)^2 = 9^2$$

$$\frac{x}{x-4} = 81$$

$$81(x-4) = x$$

$$81x - 324 = x$$

$$80x = 324$$

$$x = \frac{324}{80} = \frac{81}{20}$$

Next, consider $u = 4$.

$$\sqrt{\frac{x}{x-4}} = 4$$

$$\left(\sqrt{\frac{x}{x-4}}\right)^2 = 4^2$$

$$\frac{x}{x-4} = 16$$

$$16(x-4) = x$$

$$16x - 64 = x$$

$$15x = 64$$

$$x = \frac{64}{15}$$

Since both sides of the equations were raised to an even power, the solutions must be checked. In this case, both check, so the solutions are $\dfrac{81}{20}$ and $\dfrac{64}{15}$.

45. Let $u = (x-4)^{-1}$.

$$f(x) = g(x) + 12$$

$$3(x-4)^{-2} = 16(x-4)^{-1} + 12$$

$$3\left[(x-4)^{-1}\right]^2 = 16(x-4)^{-1} + 12$$

$$3u^2 = 16u + 12$$

$$3u^2 - 16u - 12 = 0$$

$$(3u+2)(u-6) = 0$$

$$3u + 2 = 0 \quad \text{or} \quad u - 6 = 0$$

$$3u = -2 \qquad\qquad u = 6$$

$$u = -\frac{2}{3}$$

Replace u with $(x-4)^{-1}$.

First, consider $u = -\dfrac{2}{3}$.

$$(x-4)^{-1} = -\frac{2}{3}$$

$$\frac{1}{x-4} = -\frac{2}{3}$$

$$-2(x-4) = 1(3)$$

$$-2x + 8 = 3$$

$$-2x = -5$$

$$x = \frac{-5}{-2} = \frac{5}{2}$$

Next, consider $u = 6$.

$$(x-4)^{-1} = 6$$

$$\frac{1}{x-4} = 6$$

$$6(x-4) = 1$$

$$6x - 24 = 1$$

$$6x = 25$$

$$x = \frac{25}{6}$$

The solutions are $\dfrac{5}{2}$ and $\dfrac{25}{6}$.

47. $P(x) = 0.04(x+40)^2 - 3(x+40) + 104$

$60 = 0.04(x+40)^2 - 3(x+40) + 104$

$0 = 0.04(x+40)^2 - 3(x+40) + 44$

Let $u = x + 40$.

$0.04(x+40)^2 - 3(x+40) + 44 = 0$

$0.04u^2 - 3u + 44 = 0$

Solve using the quadratic formula.

$a = 0.04 \quad b = -3 \quad c = 44$

$u = \dfrac{-(-3) \pm \sqrt{(-3)^2 - 4(0.04)(44)}}{2(0.04)}$

$= \dfrac{3 \pm \sqrt{9 - 7.04}}{0.08}$

$= \dfrac{3 \pm \sqrt{1.96}}{0.08} = \dfrac{3 \pm 1.4}{0.08} = 55 \text{ or } 20$

Since x represents the number of years a person's age is above or below 40, $u = x + 40$ is the age we are looking for. The ages at which 60% of us feel that having a clean house is very important are 20 and 55. From the graph, we see that at 20, 58%, and at 55, 52% feel that a clean house if very important. The function models the data fairly well.

49. – 51. Answers will vary.

53. $3(x-2)^{-2} - 4(x-2)^{-1} + 1 = 0$

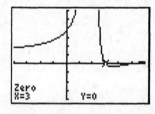

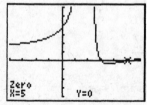

The solutions are 3 and 5. Both solutions check.

The solution set is $\{3, 5\}$.

55. $2x + 6\sqrt{x} = 8$

$2x + 6\sqrt{x} - 8 = 0$

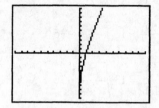

The solution is 1. The solution checks.

The solution set is $\{1\}$.

57. $(x^2 - 3x)^2 + 2(x^2 - 3x) - 24 = 0$

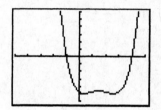

The solutions are -1 and 4.

The solution set is $\{-1, 4\}$.

59. $x^{\frac{2}{3}} - 3x^{\frac{1}{3}} + 2 = 0$

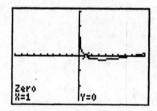

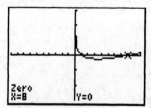

The solution set is $\{1, 8\}$.

61. makes sense

63. does not make sense; Explanations will vary. Sample explanation: Changing the order of the terms does not change the classification of this equation.

65. true

67. false; Changes to make the statement true will vary. A sample change is: To solve the equation, let $u = \sqrt{x}$.

69.
$$5x^6 + x^3 = 18$$
$$5x^6 + x^3 - 18 = 0$$

Let $u = x^3$.
$$5x^6 + x^3 - 18 = 0$$
$$5\left(x^3\right)^2 + x^3 - 18 = 0$$
$$5u^2 + u - 18 = 0$$
$$(5u - 9)(u + 2) = 0$$

$5u - 9 = 0 \qquad$ or $\qquad u + 2 = 0$
$5u = 9 \qquad\qquad\qquad u = -2$
$u = \dfrac{9}{5}$

Substitute x^3 for u.

$x^3 = \dfrac{9}{5} \qquad$ or $\qquad x^3 = -2$

$x = \sqrt[3]{\dfrac{9}{5}} \qquad\qquad x = \sqrt[3]{-2}$

Rationalize the denominator.
$$\sqrt[3]{\frac{9}{5}} = \frac{\sqrt[3]{9}}{\sqrt[3]{5}} \cdot \frac{\sqrt[3]{5^2}}{\sqrt[3]{5^2}} = \frac{\sqrt[3]{9 \cdot 5^2}}{\sqrt[3]{5^3}} = \frac{\sqrt[3]{225}}{5}$$

The solution set is $\left\{ \sqrt[3]{-2}, \dfrac{\sqrt[3]{225}}{5} \right\}$.

71. $\dfrac{2x^2}{10x^3 - 2x^2} = \dfrac{2x^2}{2x^2(5x - 1)} = \dfrac{1}{5x - 1}$

72.
$$\frac{2+i}{1-i} = \frac{2+i}{1-i} \cdot \frac{1+i}{1+i} = \frac{2 + 2i + i + i^2}{1^2 - i^2}$$
$$= \frac{2 + 3i - 1}{1 - (-1)} = \frac{1 + 3i}{2} = \frac{1}{2} + \frac{3}{2}i$$

73.
$$f(x) = \sqrt{x+1}$$
$$f(3) - f(24) = \sqrt{3+1} - \sqrt{24+1}$$
$$= \sqrt{4} - \sqrt{25}$$
$$= 2 - 5$$
$$= -3$$

74.
$$2x^2 + x = 15$$
$$2x^2 + x - 15 = 0$$
$$(x + 3)(2x - 5) = 0$$

$x + 3 = 0 \qquad$ or $\qquad 2x - 5 = 0$
$x = -3 \qquad\qquad\qquad 2x = 5$
$\qquad\qquad\qquad\qquad\qquad x = \dfrac{5}{2}$

The solution set is $\left\{ -3, \dfrac{5}{2} \right\}$.

75. $x^3 + x^2 - 4x - 4 = x^2(x+1) - 4(x+1)$
$$= (x+1)(x^2 - 4)$$
$$= (x+1)(x+2)(x-2)$$

76.
$$\frac{x+1}{x+3} - 2 = \frac{x+1}{x+3} - \frac{2(x+3)}{x+3}$$
$$= \frac{x+1}{x+3} - \frac{2x+6}{x+3}$$
$$= \frac{x+1 - 2x - 6}{x+3}$$
$$= \frac{-x - 5}{x+3}$$

11.5 Check Points

1. $x^2 - x > 20$

    $x^2 - x - 20 > 0$

    Solve the related quadratic equation to find the boundary points.

    $x^2 - x - 20 = 0$

    $(x+4)(x-5) = 0$

    $x + 4 = 0$ or $x - 5 = 0$

      $x = -4$ $x = 5$

    The boundary points are -4 and 5.

Interval	Test Value	Test	Conclusion
$(-\infty, -4)$	-5	$x^2 - x > 20$ $(-5)^2 - (-5) > 20$ $30 > 20$, true	$(-\infty, -4)$ belongs to the solution set.
$(-4, 5)$	0	$x^2 - x > 20$ $(0)^2 - (0) > 20$ $0 > 20$, false	$(-4, 5)$ does not belong to the solution set.
$(5, \infty)$	10	$x^2 - x > 20$ $(10)^2 - (10) > 20$ $90 > 20$, true	$(5, \infty)$ belongs to the solution set.

The solution set is $(-\infty, -4) \cup (5, \infty)$ or $\{x \mid x < -4 \text{ or } x > 5\}$.

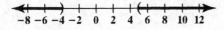

2. $x^3 + 3x^2 \le x + 3$

    $x^3 + 3x^2 - x - 3 \le 0$

    Solve the related quadratic equation to find the boundary points.

    $x^3 + 3x^2 - x - 3 = 0$

    $x^2(x+3) - 1(x+3) = 0$

      $(x+3)(x^2-1) = 0$

    $(x+3)(x-1)(x+1) = 0$

    $x + 3 = 0$ or $x - 1 = 0$ or $x + 1 = 0$

      $x = -3$ $x = 1$ $x = -1$

Interval	Test Value	Test	Conclusion
$(-\infty, -3)$	-5	$(-5)^3 + 3(-5)^2 \le (-5) + 3$ $-50 \le -2$, true	$(-\infty, -3)$ is part of the solution set
$(-3, -1)$	-2	$(-2)^3 + 3(-2)^2 \le (-2) + 3$ $4 \le 1$, false	$(-3, -1)$ is not part of the solution set
$(-1, 1)$	0	$(0)^3 + 3(0)^2 \le (0) + 3$ $0 \le 3$, true	$(-1, 1)$ is part of the solution set
$(1, \infty)$	2	$(2)^3 + 3(2)^2 \le (2) + 3$ $20 \le 5$, false	$(1, \infty)$ is not part of the solution set

The solution set is $(-\infty, -3] \cup [-1, 1]$ or $\{x \mid x \le -3 \text{ or } -1 \le x \le 1\}$.

3. $\dfrac{x-5}{x+2} < 0$

Find the values of x that make the numerator and denominator zero.

$x - 5 = 0 \qquad x + 2 = 0$

$\quad x = 5 \qquad\quad x = -2$

The boundary points are -2 and 5.

Interval	Test Value	Test	Conclusion
$(-\infty, -2)$	-3	$\dfrac{-3-5}{-3+2} < 0$ $8 < 0$, false	$(-\infty, -2)$ does not belong to the solution set.
$(-2, 5)$	0	$\dfrac{0-5}{0+2} < 0$ $-\dfrac{5}{2} < 0$, true	$(-2, 5)$ belongs to the solution set.
$(5, \infty)$	6	$\dfrac{6-5}{6+2} < 0$ $\dfrac{1}{8} < 0$, false	$(5, \infty)$ does not belong to the solution set.

The solution set is $(-2, 5)$ or $\{x \mid -2 < x < 5\}$.

615

4.

$$\frac{2x}{x+1} \geq 1$$

$$\frac{2x}{x+1} - 1 \geq 0$$

$$\frac{2x}{x+1} - \frac{x+1}{x+1} \geq 0$$

$$\frac{2x - x - 1}{x+1} \geq 0$$

$$\frac{x-1}{x+1} \geq 0$$

Find the values of x that make the numerator and denominator zero.

$x - 1 = 0$ and $x + 1 = 0$

$x = 1$ $\qquad\qquad x = -1$

The boundary points are -1 and 1.

Interval	Test Value	Test	Conclusion
$(-\infty, -1)$	-2	$\frac{2(-2)}{-2+1} \geq 1$ $4 \geq 1$, true	$(-\infty, -1)$ belongs to the solution set.
$(-1, 1)$	0	$\frac{2(0)}{0+1} \geq 1$ $0 \geq 1$, false	$(-1, 1)$ does not belong to the solution set.
$(1, \infty)$	2	$\frac{2(2)}{2+1} \geq 1$ $\frac{4}{3} \geq 1$, true	$(1, \infty)$ belongs to the solution set.

Exclude -1 from the solution set because it would make the denominator zero. The solution set is $(-\infty, -1) \cup [1, \infty)$ or $\{x \mid x < -1 \text{ or } x \geq 1\}$.

5. $s(t) = -16t^2 + 80t$

To find when the object will be more than 64 feet above the ground, solve the inequality $-16t^2 + 80t > 64$.

Solve the related quadratic equation.

$$-16t^2 + 80t = 64$$

$$-16t^2 + 80t - 64 = 0$$

$$t^2 - 5t + 4 = 0$$

$$(t-4)(t-1) = 0$$

$t - 4 = 0$ or $t - 1 = 0$

$t = 4$ $\qquad\qquad t = 1$

The boundary points are 1 and 4.

Interval	Test Value	Test	Conclusion
$(0,1)$	0.5	$-16(0.5)^2 + 80(0.5) > 64$ $36 > 64$, false	$(0,1)$ does not belong to the solution set.
$(1,4)$	2	$-16(2)^2 + 80(2) > 64$ $96 > 64$, true	$(1,4)$ belongs to the solution set.
$(4,\infty)$	5	$-16(5)^2 + 80(5) > 64$ $0 > 64$, false	$(4,\infty)$ does not belong to the solution set.

The solution set is $(1,4)$. This means that the object will be more than 64 feet above the ground between 1 and 4 seconds excluding $t = 1$ and $t = 4$.

11.5 Exercise Set

1. $(x-4)(x+2) > 0$

Solve the related quadratic equation.
$(x-4)(x+2) = 0$
$x - 4 = 0$ or $x + 2 = 0$
$\quad x = 4 \qquad\qquad x = -2$

The boundary points are –2 and 4.

Interval	Test Value	Test	Conclusion
$(-\infty,-2)$	–3	$(-3-4)(-3+2) > 0$ $7 > 0$, true	$(-\infty,-2)$ belongs to the solution set.
$(-2,4)$	0	$(0-4)(0+2) > 0$ $-8 > 0$, false	$(-2,4)$ does not belong to the solution set.
$(4,\infty)$	5	$(5-4)(5+2) > 0$ $7 > 0$, true	$(4,\infty)$ belongs to the solution set.

The solution set is $(-\infty,-2) \cup (4,\infty)$ or $\{x \mid x < -2 \text{ or } x > 4\}$.

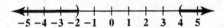

3. $(x-7)(x+3) \leq 0$

Solve the related quadratic equation.

$(x-7)(x+3) = 0$

$x - 7 = 0$ or $x + 3 = 0$

$\quad x = 7 \qquad\quad x = -3$

The boundary points are -3 and 7.

Interval	Test Value	Test	Conclusion
$(-\infty, -3)$	-4	$(-4-7)(-4+3) \leq 0$ $11 \leq 0$, false	$(-\infty, -3)$ does not belong to the solution set.
$(-3, 7)$	0	$(0-7)(0+3) \leq 0$ $-21 \leq 0$, true	$(-3, 7)$ belongs to the solution set.
$(7, \infty)$	8	$(8-7)(8+3) \leq 0$ $11 \leq 0$, false	$(7, \infty)$ does not belong to the solution set.

The solution set is $[-3, 7]$ or $\{x \mid -3 \leq x \leq 7\}$.

5. $x^2 - 5x + 4 > 0$

Solve the related quadratic equation.

$\quad x^2 - 5x + 4 = 0$

$(x-4)(x-1) = 0$

$x - 4 = 0$ or $x - 1 = 0$

$\quad x = 4 \qquad\quad x = 1$

The boundary points are 1 and 4.

Interval	Test Value	Test	Conclusion
$(-\infty, 1)$	0	$0^2 - 5(0) + 4 > 0$ $4 > 0$, true	$(-\infty, 1)$ belongs to the solution set.
$(1, 4)$	2	$2^2 - 5(2) + 4 > 0$ $-2 > 0$, false	$(1, 4)$ does not belong to the solution set.
$(4, \infty)$	5	$5^2 - 5(5) + 4 > 0$ $4 > 0$, true	$(4, \infty)$ belongs to the solution set.

The solution set is $(-\infty, 1) \cup (4, \infty)$ or $\{x \mid x < 1 \text{ or } x > 4\}$.

7. $x^2 + 5x + 4 > 0$

Solve the related quadratic equation.

$x^2 + 5x + 4 = 0$

$(x + 4)(x + 1) = 0$

$x + 4 = 0$ or $x + 1 = 0$

$x = -4$ $x = -1$

The boundary points are −1 and −4.

Interval	Test Value	Test	Conclusion
$(-\infty, -4)$	−5	$(-5)^2 + 5(-5) + 4 > 0$ $4 > 0$, true	$(-\infty, -4)$ belongs to the solution set.
$(-4, -1)$	-2	$(-2)^2 + 5(-2) + 4 > 0$ $-2 > 0$, false	$(-4, -1)$ does not belong to the solution set.
$(-1, \infty)$	0	$0^2 + 5(0) + 4 > 0$ $4 > 0$, true	$(-1, \infty)$ belongs to the solution set.

The solution set is $(-\infty, -4) \cup (-1, \infty)$ or $\{x \mid x < -4 \text{ or } x > -1\}$.

9. $x^2 - 6x + 8 \leq 0$

Solve the related quadratic equation.

$x^2 - 6x + 8 = 0$

$(x - 4)(x - 2) = 0$

$x - 4 = 0$ or $x - 2 = 0$

$x = 4$ $x = 2$

The boundary points are 2 and 4.

Interval	Test Value	Test	Conclusion
$(-\infty, 2)$	0	$0^2 - 6(0) + 8 \leq 0$ $8 \leq 0$, false	$(-\infty, 2)$ does not belong to the solution set.
$(2, 4)$	3	$3^2 - 6(3) + 8 \leq 0$ $-1 \leq 0$, true	$(2, 4)$ belongs to the solution set.
$(4, \infty)$	5	$5^2 - 6(5) + 8 \leq 0$ $3 \leq 0$, false	$(4, \infty)$ does not belong to the solution set.

The solution set is $[2, 4]$ or $\{x \mid 2 \leq x \leq 4\}$.

11. $3x^2 + 10x - 8 \leq 0$

Solve the related quadratic equation.

$$3x^2 + 10x - 8 = 0$$
$$(3x-2)(x+4) = 0$$
$$3x - 2 = 0 \quad \text{or} \quad x + 4 = 0$$
$$3x = 2 \qquad \qquad x = -4$$
$$x = \frac{2}{3}$$

The boundary points are -4 and $\frac{2}{3}$.

Interval	Test Value	Test	Conclusion
$(-\infty, -4)$	-5	$3(-5)^2 + 10(-5) - 8 \leq 0$ $17 \leq 0$, false	$(-\infty, -4)$ does not belong to the solution set.
$\left(-4, \frac{2}{3}\right)$	0	$3(0)^2 + 10(0) - 8 \leq 0$ $-8 \leq 0$, true	$\left(-4, \frac{2}{3}\right)$ belongs to the solution set.
$\left(\frac{2}{3}, \infty\right)$	1	$3(1)^2 + 10(1) - 8 \leq 0$ $5 \leq 0$, false	$\left(\frac{2}{3}, \infty\right)$ does not belong to the solution set.

The solution set is $\left[-4, \frac{2}{3}\right]$ or $\left\{x \middle| -4 \leq x \leq \frac{2}{3}\right\}$.

13. $\qquad 2x^2 + x < 15$

$$2x^2 + x - 15 < 0$$

Solve the related quadratic equation.

$$2x^2 + x - 15 = 0$$
$$(2x-5)(x+3) = 0$$
$$2x - 5 = 0 \quad \text{or} \quad x + 3 = 0$$
$$2x = 5 \qquad \qquad x = -3$$
$$x = \frac{5}{2}$$

The boundary points are -3 and $\frac{5}{2}$.

Interval	Test Value	Test	Conclusion
$(-\infty, -3)$	-4	$2(-4)^2 + (-4) < 15$ $28 < 15,$ false	$(-\infty, -3)$ does not belong to the solution set.
$\left(-3, \dfrac{5}{2}\right)$	0	$2(0)^2 + 0 < 15$ $0 < 15,$ true	$\left(-3, \dfrac{5}{2}\right)$ belongs to the solution set.
$\left(\dfrac{5}{2}, \infty\right)$	3	$2(3)^2 + 3 < 15$ $21 < 15,$ false	$\left(\dfrac{5}{2}, \infty\right)$ does not belong to the solution set.

The solution set is $\left(-3, \dfrac{5}{2}\right)$ or $\left\{ x \middle| -3 < x < \dfrac{5}{2} \right\}$.

15. $4x^2 + 7x < -3$

$4x^2 + 7x + 3 < 0$

Solve the related quadratic equation.

$4x^2 + 7x + 3 = 0$

$(4x + 3)(x + 1) = 0$

$4x + 3 = 0 \quad$ or $\quad x + 1 = 0$

$\quad 4x = -3 \qquad\qquad x = -1$

$\quad\quad x = -\dfrac{3}{4}$

The boundary points are -1 and $-\dfrac{3}{4}$.

Interval	Test Value	Test	Conclusion
$(-\infty, -1)$	-2	$4(-2)^2 + 7(-2) < -3$ $2 < -3,$ false	$(-\infty, -1)$ does not belong to the solution set.
$\left(-1, -\dfrac{3}{4}\right)$	$-\dfrac{7}{8}$	$4\left(-\dfrac{7}{8}\right)^2 + 7\left(-\dfrac{7}{8}\right) < -3$ $-3\dfrac{1}{16} < -3,$ true	$\left(-1, -\dfrac{3}{4}\right)$ belongs to the solution set.
$\left(-\dfrac{3}{4}, \infty\right)$	0	$4(0)^2 + 7(0) < -3$ $0 < -3,$ false	$\left(-\dfrac{3}{4}, \infty\right)$ does not belong to the solution set.

The solution set is $\left(-1, -\dfrac{3}{4}\right)$ or $\left\{ x \middle| -1 < x < -\dfrac{3}{4} \right\}$.

17. $x^2 - 4x \geq 0$

Solve the related quadratic equation.

$x^2 - 4x = 0$

$x(x-4) = 0$

$x = 0$ or $x - 4 = 0$

$\qquad\qquad x = 4$

The boundary points are 0 and 4.

Interval	Test Value	Test	Conclusion
$(-\infty, 0)$	-1	$(-1)^2 - 4(-1) \geq 0$ $5 \geq 0$, true	$(-\infty, 0)$ belongs to the solution set.
$(0, 4)$	1	$(1)^2 - 4(1) \geq 0$ $-3 \geq 0$, false	$(0, 4)$ does not belong to the solution set.
$(4, \infty)$	5	$(5)^2 - 4(5) \geq 0$ $5 \geq 0$, true	$(4, \infty)$ belongs to the solution set.

The solution set is $(-\infty, 0] \cup [4, \infty)$ or $\{x | x \leq 0 \text{ or } x \geq 4\}$.

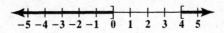

19. $2x^2 + 3x > 0$

Solve the related quadratic equation.

$2x^2 + 3x = 0$

$x(2x + 3) = 0$

$x = 0$ or $2x + 3 = 0$

$\qquad\qquad 2x = -3$

$\qquad\qquad x = -\dfrac{3}{2}$

The boundary points are $-\dfrac{3}{2}$ and 0.

Interval	Test Value	Test	Conclusion
$\left(-\infty, -\dfrac{3}{2}\right)$	-2	$2(-2)^2 + 3(-2) > 0$ $2 > 0$, true	$\left(-\infty, -\dfrac{3}{2}\right)$ belongs to the solution set.
$\left(-\dfrac{3}{2}, 0\right)$	-1	$2(-1)^2 + 3(-1) > 0$ $-1 > 0$, false	$\left(-\dfrac{3}{2}, 0\right)$ does not belong to the solution set.
$(0, \infty)$	1	$2(1)^2 + 3(1) > 0$ $5 > 0$, true	$(0, \infty)$ belongs to the solution set.

The solution set is $\left(-\infty, -\dfrac{3}{2}\right) \cup (0, \infty)$ or $\left\{ x \mid x < -\dfrac{3}{2} \text{ or } x > 0 \right\}$.

21. $-x^2 + x \geq 0$

Solve the related quadratic equation.

$-x^2 + x = 0$

$-x(x - 1) = 0$

$-x = 0 \quad \text{or} \quad x - 1 = 0$

$x = 0 \qquad\qquad x = 1$

The boundary points are 0 and 1.

Interval	Test Value	Test	Conclusion
$(-\infty, 0)$	-1	$-(-1)^2 + (-1) \geq 0$ $-2 \geq 0$, false	$(-\infty, 0)$ does not belong to the solution set.
$(0, 1)$	$\dfrac{1}{2}$	$-\left(\dfrac{1}{2}\right)^2 + \dfrac{1}{2} \geq 0$ $\dfrac{1}{4} \geq 0$, true	$(0, 1)$ belongs to the solution set.
$(1, \infty)$	2	$-(2)^2 + 2 \geq 0$ $-2 \geq 0$, false	$(1, \infty)$ does not belong to the solution set.

The solution set is $[0, 1]$ or $\left\{ x \mid 0 \leq x \leq 1 \right\}$.

23. $\qquad x^2 \leq 4x - 2$

$x^2 - 4x + 2 \leq 0$

Solve the related quadratic equation, using the quadratic formula.

$x^2 - 4x + 2 = 0$

$a = 1 \quad b = -4 \quad c = 2$

$x = \dfrac{-(-4) \pm \sqrt{(-4)^2 - 4(1)(2)}}{2(1)} = \dfrac{4 \pm \sqrt{16 - 8}}{2} = \dfrac{4 \pm \sqrt{8}}{2} = \dfrac{4 \pm \sqrt{4 \cdot 2}}{2}$

$= \dfrac{4 \pm 2\sqrt{2}}{2} = \dfrac{2(2 \pm \sqrt{2})}{2} = 2 \pm \sqrt{2}$

The boundary points are $2 - \sqrt{2}$ and $2 + \sqrt{2}$.

Interval	Test Value	Test	Conclusion
$\left(-\infty, 2-\sqrt{2}\right)$	0	$0^2 \le 4(0)-2$ $0 \le -2$, false	$\left(-\infty, 2-\sqrt{2}\right)$ does not belong to the solution set.
$\left(2-\sqrt{2}, 2+\sqrt{2}\right)$	2	$2^2 \le 4(2)-2$ $4 \le 6$, true	$\left(2-\sqrt{2}, 2+\sqrt{2}\right)$ belongs to the solution set.
$\left(2+\sqrt{2}, \infty\right)$	4	$4^2 \le 4(4)-2$ $16 \le 14$, false	$\left(2+\sqrt{2}, \infty\right)$ does not belong to the solution set.

The solution set is $\left[2-\sqrt{2}, 2+\sqrt{2}\right]$ or $\left\{x \middle| 2-\sqrt{2} \le x \le 2+\sqrt{2}\right\}$.

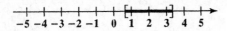

25.
$$3x^2 > 4x + 2$$
$$3x^2 - 4x - 2 > 0$$

Solve the related quadratic equation.
$$3x^2 - 4x - 2 = 0$$
$$x = \frac{-b \pm \sqrt{b^2 - 4ac}}{2a}$$
$$x = \frac{-(-4) \pm \sqrt{(-4)^2 - 4(3)(-2)}}{2(3)}$$
$$x = \frac{2 \pm \sqrt{10}}{3}$$
$$x \approx -0.39 \text{ or } 1.72$$

Interval	Test Value	Test	Conclusion
$\left(-\infty, \dfrac{2-\sqrt{10}}{3}\right)$	-1	$3(-1)^2 > 4(-1)+2$ $3 > -2$, true	$\left(-\infty, \dfrac{2-\sqrt{10}}{3}\right)$ belongs to the solution set.
$\left(\dfrac{2-\sqrt{10}}{3}, \dfrac{2+\sqrt{10}}{3}\right)$	0	$3(0)^2 > 4(0)+2$ $0 > 2$, false	$\left(\dfrac{2-\sqrt{10}}{3}, \dfrac{2+\sqrt{10}}{3}\right)$ does not belong to the solution set.
$\left(\dfrac{2+\sqrt{10}}{3}, \infty\right)$	2	$3(2)^2 > 4(2)+2$ $12 > 10$, true	$\left(\dfrac{2+\sqrt{10}}{3}, \infty\right)$ belongs to the solution set.

The solution set is $\left(-\infty, \dfrac{2-\sqrt{10}}{3}\right) \cup \left(\dfrac{2+\sqrt{10}}{3}, \infty\right)$ or $\left\{x \middle| x < \dfrac{2-\sqrt{10}}{3} \text{ or } x > \dfrac{2+\sqrt{10}}{3}\right\}$.

27. $2x^2 - 5x \geq 1$

$2x^2 - 5x - 1 \geq 0$

Solve the related quadratic equation.

$2x^2 - 5x - 1 = 0$

$x = \dfrac{-b \pm \sqrt{b^2 - 4ac}}{2a}$

$x = \dfrac{-(-5) \pm \sqrt{(-5)^2 - 4(2)(-1)}}{2(2)}$

$x = \dfrac{5 \pm \sqrt{33}}{4}$

$x \approx -0.19 \ \text{or} \ 2.69$

Interval	Test Value	Test	Conclusion
$\left(-\infty, \dfrac{5-\sqrt{33}}{4}\right)$	-1	$2(-1)^2 - 5(-1) \geq 1$ $7 \geq 1, \ \text{true}$	$\left(-\infty, \dfrac{5-\sqrt{33}}{4}\right)$ belongs to the solution set.
$\left(\dfrac{5-\sqrt{33}}{4}, \dfrac{5+\sqrt{33}}{4}\right)$	0	$2(0)^2 - 5(0) \geq 1$ $0 \geq 1, \ \text{false}$	$\left(\dfrac{5-\sqrt{33}}{4}, \dfrac{5+\sqrt{33}}{4}\right)$ does not belong to the solution set.
$\left(\dfrac{5+\sqrt{33}}{4}, \infty\right)$	3	$2(3)^2 - 5(3) \geq 1$ $3 \geq 1, \ \text{true}$	$\left(\dfrac{5+\sqrt{33}}{4}, \infty\right)$ belongs to the solution set.

The solution set is $\left(-\infty, \dfrac{5-\sqrt{33}}{4}\right] \cup \left[\dfrac{5+\sqrt{33}}{4}, \infty\right)$ or $\left\{x \ \middle| \ x \leq \dfrac{5-\sqrt{33}}{4} \ \text{or} \ x \geq \dfrac{5+\sqrt{33}}{4}\right\}$.

29. $x^2 - 6x + 9 < 0$

Solve the related quadratic equation.

$x^2 - 6x + 9 = 0$

$(x-3)^2 = 0$

$x - 3 = 0$

$x = 3$

The boundary point is 3.

Interval	Test Value	Test	Conclusion
$(-\infty, 3)$	0	$0^2 - 6(0) + 9 < 0$ $9 < 0, \ \text{False}$	$(-\infty, 3)$ does not belong to the solution set.
$(3, \infty)$	4	$4^2 - 6(4) + 9 < 0$ $1 < 0, \ \text{false}$	$(3, \infty)$ does not belong to the solution set.

There is no solution. The solution set is $\varnothing$ or $\{ \ \}$.

31. $(x-1)(x-2)(x-3) \geq 0$

Solve the related polynomial equation.
$(x-1)(x-2)(x-3) = 0$
$x-1=0$ or $x-2=0$ or $x-3=0$
$\quad x=1 \qquad\quad x=2 \qquad\quad x=3$

The boundary points are 1, 2, and 3.

Interval	Test Value	Test	Conclusion
$(-\infty,1)$	0	$(0-1)(0-2)(0-3) \geq 0$ $-6 \geq 0$, False	$(-\infty,1)$ does not belong to the solution set.
$(1,2)$	1.5	$(1.5-1)(1.5-2)(1.5-3) \geq 0$ $0.375 \geq 0$, True	$(1,2)$ belongs to the solution set.
$(2,3)$	2.5	$(2.5-1)(2.5-2)(2.5-3) \geq 0$ $-0.375 \geq 0$, False	$(2,3)$ does not belong to the solution set.
$(3,\infty)$	4	$(4-1)(4-2)(4-3) \geq 0$ $6 \geq 0$, True	$(3,\infty)$ belongs to the solution set.

The solution set is $[1,2] \cup [3,\infty)$ or $\{x | 1 \leq x \leq 2 \text{ or } x \geq 3\}$.

33. $x^3 + 2x^2 - x - 2 \geq 0$

Solve the related polynomial equation.
$\quad x^3 + 2x^2 - x - 2 = 0$
$x^2(x+2) - 1(x+2) = 0$
$\quad (x^2-1)(x+2) = 0$
$(x-1)(x+1)(x+2) = 0$
$x-1=0$ or $x+1=0$ or $x+2=0$
$\quad x=1 \qquad\quad x=-1 \qquad\quad x=-2$

The boundary points are $-2, -1,$ and 1.

Interval	Test Value	Test	Conclusion
$(-\infty,-2)$	-3	$(-3)^3 + 2(-3)^2 - (-3) - 2 \geq 0$ $-8 \geq 0$, False	$(-\infty,-2)$ does not belong to the solution set.
$(-2,-1)$	-1.5	$(-1.5)^3 + 2(-1.5)^2 - (-1.5) - 2 \geq 0$ $0.625 \geq 0$, True	$(-2,-1)$ belongs to the solution set.
$(-1,1)$	0	$0^3 + 2(0)^2 - 0 - 2 \geq 0$ $-2 \geq 0$, False	$(-1,1)$ does not belong to the solution set.
$(1,\infty)$	2	$2^3 + 2(2)^2 - 2 - 2 \geq 0$ $12 \geq 0$, True	$(1,\infty)$ belongs to the solution set.

The solution set is $[-2,-1] \cup [1,\infty)$ or $\{x \mid -2 \leq x \leq -1 \text{ or } x \geq 1\}$.

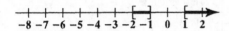

35. $x^3 - 3x^2 - 9x + 27 < 0$

Solve the related polynomial equation.

$$x^3 - 3x^2 - 9x + 27 = 0$$
$$x^2(x-3) - 9(x-3) = 0$$
$$(x^2 - 9)(x-3) = 0$$
$$(x-3)(x+3)(x-3) = 0$$
$$(x-3)^2(x+3) = 0$$
$$x - 3 = 0 \quad \text{or} \quad x + 3 = 0$$
$$x = 3 \qquad\qquad x = -3$$

The boundary points are -3 and 3.

Interval	Test Value	Test	Conclusion
$(-\infty,-3)$	-4	$(-4)^3 - 3(-4)^2 - 9(-4) + 27 < 0$ $-49 < 0$, True	$(-\infty,-3)$ belongs to the solution set.
$(-3,3)$	0	$0^3 - 3(0)^2 - 9(0) + 27 < 0$ $27 < 0$, False	$(-3,3)$ does not belong to the solution set.
$(3,\infty)$	4	$4^3 - 3(4)^2 - 9(4) + 27 < 0$ $7 < 0$, False	$(3,\infty)$ does not belong to the solution set.

The solution set is $(-\infty,-3)$ or $\{x \mid x < -3\}$.

37. $x^3 + x^2 + 4x + 4 > 0$

Solve the related polynomial equation.

$$x^3 + x^2 + 4x + 4 = 0$$
$$x^2(x+1) + 4(x+1) = 0$$
$$(x^2 + 4)(x+1) = 0$$

$$x^2 + 4 = 0 \qquad \text{or} \quad x + 1 = 0$$
$$x^2 = -4 \qquad\qquad x = -1$$
$$x = \pm\sqrt{-4}$$
$$= \pm 2i$$

The imaginary solutions will not be boundary points, so the only boundary point is -1.

Interval	Test Value	Test	Conclusion
$(-\infty, -1)$	-2	$(-2)^3 + (-2)^2 + 4(-2) + 4 > 0$ $-8 > 0$, False	$(-\infty, -1)$ does not belong to the solution set.
$(-1, \infty)$	0	$0^3 + 0^2 + 4(0) + 4 > 0$ $4 > 0$, True	$(-1, \infty)$ belongs to the solution set.

The solution set is $(-1, \infty)$ or $\{x \mid x > -1\}$.

39. $x^3 \geq 9x^2$

$$x^3 - 9x^2 \geq 0$$

Solve the related polynomial equation.

$$x^3 - 9x^2 = 0$$
$$x^2(x-9) = 0$$

$$x^2 = 0 \qquad \text{or} \quad x - 9 = 0$$
$$x = \pm\sqrt{0} = 0 \qquad\qquad x = 9$$

The boundary points are 0 and 9.

Interval	Test Value	Test	Conclusion
$(-\infty, 0)$	-1	$(-1)^3 \geq 9(-1)^2$ $-1 \geq 9$, False	$(-\infty, 0)$ does not belong to the solution set.
$(0, 9)$	1	$1^3 \geq 9(1)^2$ $1 \geq 9$, False	$(0, 9)$ does not belong to the solution set.
$(9, \infty)$	10	$10^3 \geq 9(10)^2$ $1000 \geq 900$, True	$(9, \infty)$ belongs to the solution set.

The solution set is $\{0\} \cup [9, \infty)$ or $\{x \mid x = 0 \text{ or } x \geq 9\}$.

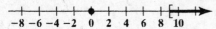

41. $\dfrac{x-4}{x+3} > 0$

Find the values of x that make the numerator and denominator zero.

$x - 4 = 0 \qquad x + 3 = 0$

$\qquad x = 4 \qquad\quad x = -3$

The boundary points are –3 and 4. Exclude –3 from the solution set, since this would make the denominator zero.

Interval	Test Value	Test	Conclusion
$(-\infty, -3)$	-4	$\dfrac{-4-4}{-4+3} > 0$ $8 > 0$, true	$(-\infty, -3)$ belongs to the solution set.
$(-3, 4)$	0	$\dfrac{0-4}{0+3} > 0$ $\dfrac{-4}{3} > 0$, false	$(-3, 4)$ does not belong to the solution set.
$(4, \infty)$	5	$\dfrac{5-4}{5+3} > 0$ $\dfrac{1}{8} > 0$, true	$(4, \infty)$ belongs to the solution set.

The solution set is $(-\infty, -3) \cup (4, \infty)$ or $\{x \mid x < -3 \text{ or } x > 4\}$.

43. $\dfrac{x+3}{x+4} < 0$

Find the values of x that make the numerator and denominator zero.

$x + 3 = 0 \qquad x + 4 = 0$

$\qquad x = -3 \qquad\quad x = -4$

The boundary points are –4 and –3.

Interval	Test Value	Test	Conclusion
$(-\infty, -4)$	-5	$\dfrac{-5+3}{-5+4} < 0$ $2 < 0$, false	$(-\infty, -4)$ does not belong to the solution set.
$(-4, -3)$	-3.5	$\dfrac{-3.5+3}{-3.5+4} < 0$ $-1 < 0$, true	$(-4, -3)$ belongs to the solution set.
$(-3, \infty)$	0	$\dfrac{0+3}{0+4} < 0$ $\dfrac{3}{4} < 0$, false	$(-3, \infty)$ does not belong to the solution set.

The solution set is $(-4, -3)$ or $\{x \mid -4 < x < -3\}$.

45. $\dfrac{-x+2}{x-4} \geq 0$

Find the values of x that make the numerator and denominator zero.

$$-x+2=0 \quad \text{and} \quad x-4=0$$
$$-x=-2 \qquad\qquad x=4$$
$$x=2$$

The boundary points are 2 and 4.

Interval	Test Value	Test	Conclusion
$(-\infty, 2)$	0	$\dfrac{-0+2}{0-4} \geq 0$ $-\dfrac{1}{2} \geq 0$, false	$(-\infty, 2)$ does not belong to the solution set.
$(2, 4)$	3	$\dfrac{-3+2}{3-4} \geq 0$ $1 \geq 0$, true	$(2, 4)$ belongs to the solution set.
$(4, \infty)$	5	$\dfrac{-5+2}{5-4} \geq 0$ $-3 \geq 0$, false	$(4, \infty)$ does not belong to the solution set.

Exclude 4 from the solution set because 4 would make the denominator zero. The solution set is $[2, 4)$ or $\{x \mid 2 \leq x < 4\}$.

47. $\dfrac{4-2x}{3x+4} \leq 0$

Find the values of x that make the numerator and denominator zero.

$$4-2x=0 \quad \text{and} \quad 3x+4=0$$
$$-2x=-4 \qquad\qquad 3x=-4$$
$$x=2 \qquad\qquad x=-\dfrac{4}{3}$$

The boundary points are $-\dfrac{4}{3}$ and 2 .

Interval	Test Value	Test	Conclusion
$\left(-\infty, -\dfrac{4}{3}\right)$	-2	$\dfrac{4-2(-2)}{3(-2)+4} \le 0$ $-4 \le 0$, true	$\left(-\infty, -\dfrac{4}{3}\right)$ belongs to the solution set.
$\left(-\dfrac{4}{3}, 2\right)$	0	$\dfrac{4-2(0)}{3(0)+4} \le 0$ $1 \le 0$, false	$\left(-\dfrac{4}{3}, 2\right)$ does not belong to the solution set.
$[2, \infty)$	3	$\dfrac{4-2(3)}{3(3)+4} \le 0$ $-\dfrac{2}{13} \le 0$, true	$[2, \infty)$ belongs to the solution set.

Exclude $-\dfrac{4}{3}$ from the solution set because $-\dfrac{4}{3}$ would make the denominator zero. The solution set is

$\left(-\infty, -\dfrac{4}{3}\right) \cup [2, \infty)$ or $\left\{x \,\middle|\, x < -\dfrac{4}{3} \text{ or } x \ge 2\right\}$.

49. $\dfrac{x}{x-3} > 0$

Find the values of x that make the numerator and denominator zero.
$x = 0$ and $x - 3 = 0$
                      $x = 3$

The boundary points are 0 and 3.

Interval	Test Value	Test	Conclusion
$(-\infty, 0)$	-1	$\dfrac{-1}{-1-3} > 0$ $\dfrac{1}{4} > 0$, true	$\left(-\infty, -\dfrac{4}{3}\right)$ belongs to the solution set.
$(0, 3)$	1	$\dfrac{1}{1-3} > 0$ $-\dfrac{1}{2} > 0$, false	$(0, 3)$ does not belong to the solution set.
$(3, \infty)$	4	$\dfrac{4}{4-3} > 0$ $4 > 0$, true	$(3, \infty)$ belongs to the solution set.

The solution set is $(-\infty, 0) \cup (3, \infty)$ or $\left\{x \,\middle|\, x < 0 \text{ or } x > 3\right\}$.

51. $\dfrac{x+1}{x+3} < 2$

Express the inequality so that one side is zero.

$$\frac{x+1}{x+3} - 2 < 0$$

$$\frac{x+1}{x+3} - \frac{2(x+3)}{x+3} < 0$$

$$\frac{x+1-2(x+3)}{x+3} < 0$$

$$\frac{x+1-2x-6}{x+3} < 0$$

$$\frac{-x-5}{x+3} < 0$$

Find the values of x that make the numerator and denominator zero.

$$-x-5=0 \qquad x+3=0$$
$$-x=5 \qquad\quad x=-3$$
$$x=-5$$

The boundary points are -5 and -3.

Interval	Test Value	Test	Conclusion
$(-\infty,-5)$	-6	$\dfrac{-6+1}{-6+3} < 2$ $\dfrac{5}{3} < 2$, true	$(-\infty,-5)$ belongs to the solution set.
$(-5,-3)$	-4	$\dfrac{-4+1}{-4+3} < 2$ $3 < 2$, false	$(-5,-3)$ does not belong to the solution set.
$(-3,\infty)$	0	$\dfrac{0+1}{0+3} < 2$ $\dfrac{1}{3} < 2$, true	$(-3,\infty)$ belongs to the solution set.

The solution set is $(-\infty,-5) \cup (-3,\infty)$ or $\{x \mid x < -5 \text{ or } x > -3\}$.

53. $\dfrac{x+4}{2x-1} \le 3$

Express the inequality so that one side is zero.

$$\frac{x+4}{2x-1} - 3 \le 0$$

$$\frac{x+4}{2x-1} - \frac{3(2x-1)}{2x-1} \le 0$$

$$\frac{x+4-3(2x-1)}{2x-1} \le 0$$

$$\frac{x+4-6x+3}{2x-1} \le 0$$

$$\frac{-5x+7}{2x-1} \le 0$$

Find the values of x that make the numerator and denominator zero.

$$-5x+7 = 0 \qquad 2x-1 = 0$$
$$-5x = -7 \qquad\quad 2x = 1$$
$$x = \frac{7}{5} \qquad\qquad x = \frac{1}{2}$$

The boundary points are $\dfrac{1}{2}$ and $\dfrac{7}{5}$.

Interval	Test Value	Test	Conclusion
$\left(-\infty, \dfrac{1}{2}\right)$	0	$\dfrac{0+4}{2(0)-1} \le 3$ $-4 \le 3$, true	$\left(-\infty, \dfrac{1}{2}\right)$ belongs to the solution set.
$\left(\dfrac{1}{2}, \dfrac{7}{5}\right)$	1	$\dfrac{1+4}{2(1)-1} \le 3$ $5 \le 3$, false	$\left(\dfrac{1}{2}, \dfrac{7}{5}\right)$ does not belong to the solution set.
$\left(\dfrac{7}{5}, \infty\right)$	2	$\dfrac{2+4}{2(2)-1} \le 3$ $2 \le 3$, true	$\left(\dfrac{7}{5}, \infty\right)$ belongs to the solution set.

Exclude $\dfrac{1}{2}$ from the solution set because $\dfrac{1}{2}$ would make the denominator zero. The solution set is $\left(-\infty, \dfrac{1}{2}\right) \cup \left[\dfrac{7}{5}, \infty\right)$ or $\left\{ x \,\middle|\, x < \dfrac{1}{2} \text{ or } x \ge \dfrac{7}{5} \right\}$.

55. $\dfrac{x-2}{x+2} \le 2$

Express the inequality so that one side is zero.

$$\frac{x-2}{x+2} - 2 \le 0$$

$$\frac{x-2}{x+2} - \frac{2(x+2)}{x+2} \le 0$$

$$\frac{x-2-2(x+2)}{x+2} \le 0$$

$$\frac{x-2-2x-4}{x+2} \le 0$$

$$\frac{-x-6}{x+2} \le 0$$

Find the values of x that make the numerator and denominator zero.

$$\begin{array}{ll} -x-6=0 & x+2=0 \\ \quad -x=6 & \quad x=-2 \\ \quad\ \ x=-6 & \end{array}$$

The boundary points are -6 and -2.

Interval	Test Value	Test	Conclusion
$(-\infty,-6)$	-7	$\dfrac{-7-2}{-7+2} \le 2$ $\dfrac{9}{5} \le 2$, true	$(-\infty,-6)$ belongs to the solution set.
$(-6,-2)$	-3	$\dfrac{-3-2}{-3+2} \le 2$ $5 \le 2$, false	$(-6,-2)$ does not belong to the solution set.
$(-2,\infty)$	0	$\dfrac{0-2}{0+2} \le 2$ $-1 \le 2$, true	$(-2,\infty)$ belongs to the solution set.

Exclude -2 from the solution set because -2 would make the denominator zero. The solution set is $(-\infty,-6]\cup(-2,\infty)$ or $\{x \mid x \le -6 \text{ or } x > -2\}$.

57. $f(x) \ge g(x)$

$2x^2 \ge 5x - 2$

$2x^2 - 5x + 2 \ge 0$

Solve the related quadratic equation.

$2x^2 - 5x + 2 = 0$

$(2x - 1)(x - 2) = 0$

Apply the zero product principle.

$2x - 1 = 0$ or $x - 2 = 0$

$2x = 1$ $x = 2$

$x = \dfrac{1}{2}$

The boundary points are $\dfrac{1}{2}$ and 2.

Interval	Test Value	Test	Conclusion
$\left(-\infty, \dfrac{1}{2}\right)$	0	$2(0)^2 \ge 5(0) - 2$ $0 \ge -2$, True	$\left(-\infty, \dfrac{1}{2}\right)$ belongs to the solution set.
$\left(\dfrac{1}{2}, 2\right)$	1	$2(1)^2 \ge 5(1) - 2$ $2 \ge 3$, False	$\left(\dfrac{1}{2}, 2\right)$ does not belong to the solution set.
$(2, \infty)$	3	$2(3)^2 \ge 5(3) - 2$ $18 \ge 13$, True	$(2, \infty)$ does not belong to the solution set.

The solution set is $\left(-\infty, \dfrac{1}{2}\right] \cup [2, \infty)$ or $\left\{ x \,\middle|\, x \le \dfrac{1}{2} \text{ or } x \ge 2 \right\}$.

59. $f(x) < g(x)$

$\dfrac{2x}{x+1} < 1$

Express the inequality so that one side is zero.

$\dfrac{2x}{x+1} - 1 < 0$

$\dfrac{2x}{x+1} - \dfrac{x+1}{x+1} < 0$

$\dfrac{2x - x - 1}{x+1} < 0$

$\dfrac{x-1}{x+1} < 0$

Find the values of x that make the numerator and denominator zero.

$x - 1 = 0$ or $x + 1 = 0$

$x = 1$ $x = -1$

The boundary points are -1 and 1.

Interval	Test Value	Test	Conclusion
$(-\infty,-1)$	-3	$\dfrac{2(-3)}{-3+1}<1$ $3<1$, false	$(-\infty,-1)$ does not belong to the solution set.
$(-1,1)$	0	$\dfrac{2(0)}{0+1}<1$ $0<1$, true	$(-1,1)$ belongs to the solution set.
$(1,\infty)$	2	$\dfrac{2(3)}{3+1}<1$ $\dfrac{3}{2}<1$, false	$(1,\infty)$ does not belong to the solution set.

The solution set is $(-1,1)$ or $\{x|-1<x<1\}$.

61. $\left|x^2+2x-36\right|>12$

Express the inequality without the absolute value symbol.

$x^2+2x-36<-12 \quad$ or $\quad x^2+2x-36>12$

$x^2+2x-24<0 \qquad\qquad x^2+2x-48>0$

Solve the related quadratic equations.

$x^2+2x-24=0 \quad$ or $\quad x^2+2x-48=0$

$(x+6)(x-4)=0 \qquad\quad (x+8)(x-6)=0$

Apply the zero product principle.

$x+6=0 \quad$ or $\quad x-4=0 \quad$ or $\quad x+8=0 \quad$ or $\quad x-6=0$

$\quad x=-6 \qquad\qquad x=4 \qquad\qquad x=-8 \qquad\qquad x=6$

The boundary points are -8, -6, 4 and 6.

Test Interval	Test Number	Test	Conclusion
$(-\infty,-8)$	-9	$\left|(-9)^2+2(-9)-36\right|>12$ $27>12$, True	$(-\infty,-8)$ belongs to the solution set.
$(-8,-6)$	-7	$\left|(-7)^2+2(-7)-36\right|>12$ $1>12$, False	$(-8,-6)$ does not belong to the solution set.
$(-6,4)$	0	$\left|0^2+2(0)-36\right|>12$ $36>12$, True	$(-6,4)$ belongs to the solution set.
$(4,6)$	5	$\left|5^2+2(5)-36\right|>12$ $1>12$, False	$(4,6)$ does not belong to the solution set.
$(6,\infty)$	7	$\left|7^2+2(7)-36\right|>12$ $27>12$, True	$(6,\infty)$ belongs to the solution set.

The solution set is $(-\infty,-8)\cup(-6,4)\cup(6,\infty)$ or $\{x|x<-8$ or $-6<x<4$ or $x>6\}$.

636

63. $\dfrac{3}{x+3} > \dfrac{3}{x-2}$

Express the inequality so that one side is zero.

$$\frac{3}{x+3} - \frac{3}{x-2} > 0$$

$$\frac{3(x-2)}{(x+3)(x-2)} - \frac{3(x+3)}{(x+3)(x-2)} > 0$$

$$\frac{3x-6-3x-9}{(x+3)(x-2)} < 0$$

$$\frac{-15}{(x+3)(x-2)} < 0$$

Find the values of x that make the denominator zero.

$x+3=0 \qquad x-2=0$

$\quad x=-3 \qquad\quad x=2$

The boundary points are -3 and 2.

Interval	Test Value	Test	Conclusion
$(-\infty,-3)$	-4	$\dfrac{3}{-4+3} > \dfrac{3}{-4-2}$ False	$(-\infty,-3)$ does not belong to the solution set.
$(-3,2)$	0	$\dfrac{3}{0+3} > \dfrac{3}{0-2}$ True	$(-3,2)$ belongs to the solution set.
$(2,\infty)$	3	$\dfrac{3}{3+3} > \dfrac{3}{3-2}$ False	$(2,\infty)$ does not belong to the solution set.

The solution set is $(-3,2)$ or $\{x\,|\,-3 < x < 2\}$.

65. $\dfrac{x^2 - x - 2}{x^2 - 4x + 3} > 0$

Find the values of x that make the numerator and denominator zero.

$x^2 - x - 2 = 0 \qquad x^2 - 4x + 3 = 0$

$(x - 2)(x + 1) = 0 \qquad (x - 3)(x - 1) = 0$

Apply the zero product principle.

$x - 2 = 0 \quad$ or $\quad x + 1 = 0 \qquad x - 3 = 0 \quad$ or $\quad x - 1 = 0$

$x = 2 \qquad\qquad x = -1 \qquad\qquad x = 3 \qquad\qquad x = 1$

The boundary points are -1, 1, 2 and 3.

Interval	Test Value	Test	Conclusion
$(-\infty, -1)$	-2	$\dfrac{(-2)^2 - (-2) - 2}{(-2)^2 - 4(-2) + 3} > 0$ $\dfrac{4}{15} > 0$, True	$(-\infty, -1)$ belongs to the solution set.
$(-1, 1)$	0	$\dfrac{0^2 - 0 - 2}{0^2 - 4(0) + 3} > 0$ $-\dfrac{2}{3} > 0$, False	$(-1, 1)$ does not belong to the solution set.
$(1, 2)$	1.5	$\dfrac{1.5^2 - 1.5 - 2}{1.5^2 - 4(1.5) + 3} > 0$ $\dfrac{5}{3} > 0$, True	$(1, 2)$ belongs to the solution set.
$(2, 3)$	2.5	$\dfrac{2.5^2 - 2.5 - 2}{2.5^2 - 4(2.5) + 3} > 0$ $-\dfrac{7}{3} > 0$, False	$(2, 3)$ does not belong to the solution set.
$(3, \infty)$	4	$\dfrac{4^2 - 4 - 2}{4^2 - 4(4) + 3} > 0$ $\dfrac{10}{3} > 0$, True	$(3, \infty)$ belongs to the solution set.

The solution set is $(-\infty, -1) \cup (1, 2) \cup (3, \infty)$ or $\{x \mid x < -1 \text{ or } 1 < x < 2 \text{ or } x > 3\}$.

67.
$$2x^3 + 11x^2 \geq 7x + 6$$
$$2x^3 + 11x^2 - 7x - 6 \geq 0$$

The graph of $f(x) = 2x^3 + 11x^2 - 7x - 6$ appears to cross the x-axis at -6, $-\frac{1}{2}$, and 1. Verify this numerically by substituting these values into the function.

$$f(-6) = 2(-6)^3 + 11(-6)^2 - 7(-6) - 6 = 2(-216) + 11(36) - (-42) - 6 = -432 + 396 + 42 - 6 = 0$$

$$f\left(-\frac{1}{2}\right) = 2\left(-\frac{1}{2}\right)^3 + 11\left(-\frac{1}{2}\right)^2 - 7\left(-\frac{1}{2}\right) - 6 = 2\left(-\frac{1}{8}\right) + 11\left(\frac{1}{4}\right) - \left(-\frac{7}{2}\right) - 6 = -\frac{1}{4} + \frac{11}{4} + \frac{7}{2} - 6 = 0$$

$$f(1) = 2(1)^3 + 11(1)^2 - 7(1) - 6 = 2(1) + 11(1) - 7 - 6 = 2 + 11 - 7 - 6 = 0$$

Thus, the boundaries are -6, $-\frac{1}{2}$, and 1. We need to find the intervals on which $f(x) \geq 0$. These intervals are

indicated on the graph where the curve is above the x-axis. Now, the curve is above the x-axis when $-6 < x < -\frac{1}{2}$ and

when $x > 1$. Thus, the solution set is $\left\{ x \mid -6 \leq x \leq -\frac{1}{2} \text{ or } x \geq 1 \right\}$ or $\left[-6, -\frac{1}{2}\right] \cup [1, \infty)$.

69.
$$\frac{1}{4(x+2)} \leq -\frac{3}{4(x-2)}$$

$$\frac{1}{4(x+2)} + \frac{3}{4(x-2)} \leq 0$$

Simplify the left side of the inequality.
$$\frac{x-2}{4(x+2)} + \frac{3(x+2)}{4(x-2)} = \frac{x-2+3x+6}{4(x+2)(x-2)} = \frac{4x+4}{4(x+2)(x-2)} = \frac{4(x+1)}{4(x+2)(x-2)} = \frac{x+1}{x^2-4}.$$

The graph of $f(x) = \frac{x+1}{x^2-4}$ crosses the x-axis at -1, and has vertical asymptotes at $x = -2$ and $x = 2$. Thus, the

boundaries are -2, -1, and 1. We need to find the intervals on which $f(x) \leq 0$. These intervals are indicated on the

graph where the curve is below the x-axis. Now, the curve is below the x-axis when $x < -2$ and when $-1 < x < 2$.

Thus, the solution set is $\left\{ x \mid x < -2 \text{ or } -1 \leq x < 2 \right\}$ or $(-\infty, -2) \cup [-1, 2)$.

71. $s(t) = -16t^2 + 48t + 160$

To find when the height exceeds the height of the building, solve the inequality $-16t^2 + 48t + 160 > 160$.

Solve the related quadratic equation.

$-16t^2 + 48t + 160 = 160$

$\qquad -16t^2 + 48t = 0$

$\qquad\quad t^2 - 3t = 0$

$\qquad\quad t(t-3) = 0$

Apply the zero product principle.

$t = 0 \quad$ or $\quad t - 3 = 0$

$\qquad\qquad\qquad t = 3$

The boundary points are 0 and 3.

Interval	Test Value	Test	Conclusion
$(0,3)$	1	$-16(1)^2 + 48(1) + 160 > 160$ $192 > 160$, true	$(0,3)$ belongs to the solution set.
$(3,\infty)$	4	$-16(4)^2 + 48(4) + 160 > 160$ $96 > 160$, false	$(3,\infty)$ does not belong to the solution set.

The solution set is $(0,3)$. This means that the ball exceeds the height of the building between 0 and 3 seconds.

73. $f(x) = 0.0875x^2 - 0.4x + 66.6$

$g(x) = 0.0875x^2 + 1.9x + 11.6$

a. $f(35) = 0.0875(35)^2 - 0.4(35) + 66.6 \approx 160$ feet

$g(35) = 0.0875(35)^2 + 1.9(35) + 11.6 \approx 185$ feet

b. Dry pavement: graph (b)
Wet pavement: graph (a)

c. The answers to part (a) model the actual stopping distances shown in the figure extremely well. The function values and the data are identical.

d. $\quad 0.0875x^2 - 0.4x + 66.6 > 540$

$0.0875x^2 - 0.4x + 473.4 > 0$

Solve the related quadratic equation.

$0.0875x^2 - 0.4x + 473.4 = 0$

$x = \dfrac{-b \pm \sqrt{b^2 - 4ac}}{2a}$

$x = \dfrac{-(-0.4) \pm \sqrt{(-0.4)^2 - 4(0.0875)(473.4)}}{2(0.0875)}$

$x \approx -71 \quad$ or $\quad 76$

Since the function's domain is $x \geq 30$, we must test the following intervals.

Interval	Test Value	Test	Conclusion
$(30, 76)$	50	$0.0875(50)^2 - 0.4(50) + 66.6 > 540$ $265.35 > 540$, False	$(30, 76)$ does not belong to the solution set.
$(76, \infty)$	100	$0.0875(100)^2 - 0.4(100) + 66.6 > 540$ $901.6 > 540$, True	$(76, \infty)$ belongs to the solution set.

On dry pavement, stopping distances will exceed 540 feet for speeds exceeding 76 miles per hour. This is represented on graph (b) to the right of point (76, 540).

75. $\overline{C}(x) = \dfrac{500,000 + 400x}{x}$

To find when the cost of producing each wheelchair does not exceed \$425, solve the inequality $\dfrac{500,000 + 400x}{x} \leq 425$.

Express the inequality so that one side is zero.

$$\frac{500,000 + 400x}{x} - 425 \leq 0$$
$$\frac{500,000 + 400x}{x} - \frac{425x}{x} \leq 0$$
$$\frac{500,000 + 400x - 425x}{x} \leq 0$$
$$\frac{500,000 - 25x}{x} \leq 0$$

Find the values of x that make the numerator and denominator zero.

$$500,000 - 25x = 0 \qquad x = 0$$
$$500,000 = 25x$$
$$20,000 = x$$

The boundary points are 0 and 20,000.

Interval	Test Value	Test	Conclusion
$(0, 20000)$	1	$\dfrac{500,000 + 400(1)}{1} \leq 425$ $500,400 \leq 425$, false	$(0, 20000)$ does not belong to the solution set.
$(20000, \infty)$	25,000	$\dfrac{500,000 + 400(25,000)}{25,000} \leq 425$ $420 \leq 425$, true	$(20000, \infty)$ belongs to the solution set.

The solution set is $[20000, \infty)$. This means that the company's production level will have to be at least 20,000 wheelchairs per week. The boundary corresponds to the point (20,000, 425) on the graph. When production is 20,000 or more per month, the average cost is \$425 or less.

77. Let x = the length of the rectangle.

Since $\text{Perimeter} = 2(\text{length}) + 2(\text{width})$, we know

$$50 = 2x + 2(\text{width})$$

$$50 - 2x = 2(\text{width})$$

$$\text{width} = \frac{50 - 2x}{2} = 25 - x$$

Now, $A = (\text{length})(\text{width})$, so we have that

$$A(x) \le 114$$

$$x(25 - x) \le 114$$

$$25x - x^2 \le 114$$

Solve the related equation

$$25x - x^2 = 114$$

$$0 = x^2 - 25x + 114$$

$$0 = (x - 19)(x - 6)$$

Apply the zero product principle.

$x - 19 = 0 \quad$ or $\quad x - 6 = 0$

$\quad x = 19 \qquad\qquad x = 6$

The boundary points are 6 and 19.

Interval	Test Value	Test	Conclusion
$(0,6)$	1	$25(1) - 1^2 \le 114$ $24 \le 114$, True	$(0,6)$ belongs to the solution set.
$(6,19)$	10	$25(10) - 10^2 \le 114$ $150 \le 114$, False	$(6,19)$ does not belong to the solution set.
$(19,\infty)$	20	$25(20) - 20^2 \le 114$ $100 \le 114$, True	$(19,\infty)$ belongs to the solution set.

If the length is 6 feet, then the width is 19 feet. If the length is less than 6 feet, then the width is greater than 19 feet. Thus, if the area of the rectangle is not to exceed 114 square feet, the length of the shorter side must be 6 feet or less.

79. – 81. Answers will vary.

83. $2x^2 + 5x - 3 \le 0$

Let $y_1 = 2x^2 + 5x - 3$.

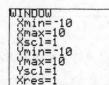 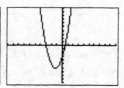

The graph is crosses the x-axis at -3 and $\frac{1}{2}$. The graph is below the x-axis when $-3 < x < \frac{1}{2}$. Thus, the solution set is

$\left\{ x \mid -3 \le x \le \frac{1}{2} \right\}$ or $\left[-3, \frac{1}{2} \right]$.

85.　$\dfrac{x+2}{x-3} \le 2$

$\dfrac{x+2}{x-3} - 2 \le 0$

Let $y_1 = \dfrac{x+2}{x-3} - 2$.

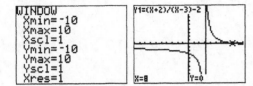

The graph is crosses the x-axis at 8. The function has a vertical asymptote at $x = 3$. The graph is below the x-axis when $x < 3$ and when $x > 8$. Thus, the solution set is $\{x \mid x < 3$ or $x \ge 8\}$ or $(-\infty, 3) \cup [8, \infty)$.

87.　$x^3 + 2x^2 - 5x - 6 > 0$

Let $y_1 = x^3 + 2x^2 - 5x - 6$

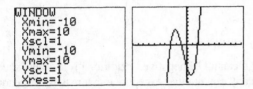

The graph is crosses the x-axis at -3, -1, and 2. The graph is above the x-axis when $-3 < x < -1$ and when $x > 2$. Thus, the solution set is $\{x \mid -3 < x < -1$ or $x > 2\}$ or $(-3, -1) \cup (2, \infty)$.

89.　a.　$f(x) = 0.1375x^2 + 0.7x + 37.8$

　b.　$0.1375x^2 + 0.7x + 37.8 > 446$

$0.1375x^2 + 0.7x + 408.2 > 0$
Solve the related quadratic equation.

$0.1375x^2 + 0.7x + 408.2 = 0$

$x = \dfrac{-b \pm \sqrt{b^2 - 4ac}}{2a}$

$x = \dfrac{-(0.7) \pm \sqrt{(0.7)^2 - 4(0.1375)(408.2)}}{2(0.1375)}$

$x \approx -57$ or 52

Since the function's domain must be $x \ge 0$, we must test the following intervals.

Interval	Test Value	Test	Conclusion
$(0, 52)$	10	$0.1375(10)^2 + 0.7(10) + 37.8 > 446$ $58.55 > 446$, False	$(0, 52)$ does not belong to the solution set.
$(52, \infty)$	100	$0.1375(100)^2 + 0.7(100) + 37.8 > 446$ $1482.8 > 446$, True	$(52, \infty)$ belongs to the solution set.

On wet pavement, stopping distances will exceed 446 feet for speeds exceeding 52 miles per hour.

91. does not make sense; Explanations will vary. Sample explanation: Polynomials are defined for all values.

93. does not make sense; Explanations will vary. Sample explanation: To solve this inequality you must first subtract 2 from both sides.

95. false; Changes to make the statement true will vary. A sample change is: The inequality cannot be solved by multiplying both sides by $x + 3$. We do not know if $x + 3$ is positive or negative. Thus, we would not know whether or not to reverse the order of the inequality.

97. true

99. Answers will vary. An example is $\dfrac{x-3}{x+4} \geq 0$.

101. $(x-2)^2 \leq 0$

Since the left hand side of the inequality is a square, we know it cannot be negative. In addition, the inequality calls for a number that is less than or equal to zero. The only possible solution is for the left hand side to equal zero. The left hand side of the inequality is zero when x is 2. Hence, the solution set is $\{2\}$.

103. $\dfrac{1}{(x-2)^2} > 0$

Since the denominator in the inequality is a square, we know it cannot be negative. Additionally, because the numerator is 1, the fraction will never be negative. As a result, x can be any real number except one that makes the denominator zero. Since 2 is the only value that makes the denominator zero, the solution set is $\{x|x$ is a real number and $x \neq 2\}$ or $(-\infty, 2) \cup (2, \infty)$.

105. The radicand must be greater than or equal to zero: $27 - 3x^2 \geq 0$
The inequality is true for values between 3 and –3. This means that the radicand is positive for values between 3 and –3, and the domain of the function is $\{x|-3 \leq x \leq 3\}$ or $[-3, 3]$.

106. $\left|\dfrac{x-5}{3}\right| < 8$

$-8 < \dfrac{x-5}{3} < 8$

$-24 < x - 5 < 24$

$-19 < x < 29$

The solution set is $\{x|-19 < x < 29\}$ or $(-19, 29)$.

107. $\dfrac{2x+6}{x^2+8x+16} \div \dfrac{x^2-9}{x^2+3x-4}$

$= \dfrac{2x+6}{x^2+8x+16} \cdot \dfrac{x^2+3x-4}{x^2-9}$

$= \dfrac{2\cancel{(x+3)}}{\cancel{(x+4)}(x+4)} \cdot \dfrac{\cancel{(x+4)}(x-1)}{\cancel{(x+3)}(x-3)}$

$= \dfrac{2(x-1)}{(x+4)(x-3)}$

108. $x^4 - 16y^4$

$= \left(x^2+4y^2\right)\left(x^2-4y^2\right)$

$= \left(x^2+4y^2\right)(x+2y)(x-2y)$

109. $f(x) = 2^x$

x	$f(x) = 2^x$	(x,y)
-3	$2^{-3} = \dfrac{1}{8}$	$\left(-3, \dfrac{1}{8}\right)$
-2	$2^{-2} = \dfrac{1}{4}$	$\left(-2, \dfrac{1}{4}\right)$
-1	$2^{-1} = \dfrac{1}{2}$	$\left(-1, \dfrac{1}{2}\right)$
0	$2^0 = 1$	$(0,1)$
1	$2^1 = 2$	$(1,2)$
2	$2^2 = 4$	$(2,4)$
3	$2^3 = 8$	$(3,8)$

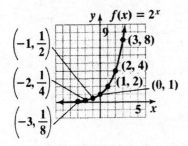

110. $f(x) = 2^{-x}$

x	$f(x) = 2^{-x}$	(x,y)
-3	$2^{-(-3)} = 8$	$(-3,8)$
-2	$2^{-(-2)} = 4$	$(-2,4)$
-1	$2^{-(-1)} = 2$	$(-1,2)$
0	$2^0 = 1$	$(0,1)$
1	$2^{-1} = \dfrac{1}{2}$	$\left(1, \dfrac{1}{2}\right)$
2	$2^{-2} = \dfrac{1}{4}$	$\left(2, \dfrac{1}{4}\right)$
3	$2^{-3} = \dfrac{1}{8}$	$\left(3, \dfrac{1}{8}\right)$

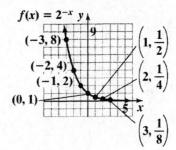

111. $f(x) = 2^x + 1$

x	$f(x) = 2^x + 1$	(x,y)
-3	$2^{-3} + 1 = 1\dfrac{1}{8}$	$\left(-3, 1\dfrac{1}{8}\right)$
-2	$2^{-2} + 1 = 1\dfrac{1}{4}$	$\left(-2, 1\dfrac{1}{4}\right)$
-1	$2^{-1} + 1 = 1\dfrac{1}{2}$	$\left(-1, 1\dfrac{1}{2}\right)$
0	$2^0 + 1 = 2$	$(0,2)$
1	$2^1 + 1 = 3$	$(1,3)$
2	$2^2 + 1 = 5$	$(2,5)$
3	$2^3 + 1 = 9$	$(3,9)$

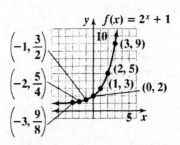

Chapter 11 Review Exercises

1. $2x^2 - 3 = 125$

$2x^2 = 128$

$x^2 = 64$

$x = \pm 8$

The solution set is $\{-8, 8\}$.

2. $3x^2 - 150 = 0$

$3x^2 = 150$

$x^2 = 50$

$x = \pm\sqrt{50}$

$x = \pm\sqrt{25 \cdot 2}$

$x = \pm 5\sqrt{2}$

The solution set is $\left\{-5\sqrt{2}, 5\sqrt{2}\right\}$.

3. $3x^2 - 2 = 0$

$3x^2 = 2$

$x^2 = \dfrac{2}{3}$

$x = \pm\sqrt{\dfrac{2}{3}}$

Rationalize the denominator.

$x = \pm\dfrac{\sqrt{2}}{\sqrt{3}} \cdot \dfrac{\sqrt{3}}{\sqrt{3}} = \pm\dfrac{\sqrt{6}}{3}$

The solution set is $\left\{-\dfrac{\sqrt{6}}{3}, \dfrac{\sqrt{6}}{3}\right\}$.

4. $(x-4)^2 = 18$

$x - 4 = \pm\sqrt{18}$

$x = 4 \pm \sqrt{9 \cdot 2}$

$x = 4 \pm 3\sqrt{2}$

The solution set is $\left\{4 - 3\sqrt{2}, 4 + 3\sqrt{2}\right\}$.

5. $(x+7)^2 = -36$

$x + 7 = \pm\sqrt{-36}$

$x = -7 \pm 6i$

The solution set is $\{-7 - 6i, -7 + 6i\}$.

6. $x^2 + 20x + \underline{\hspace{1cm}}$

Since $b = 20$, add $\left(\dfrac{b}{2}\right)^2 = \left(\dfrac{20}{2}\right)^2 = (10)^2 = 100$.

$x^2 + 20x + 100 = (x + 10)^2$

7. $x^2 - 3x + \underline{\hspace{1cm}}$

Since $b = 3$, add $\left(\dfrac{b}{2}\right)^2 = \left(\dfrac{3}{2}\right)^2 = \dfrac{9}{4}$.

$x^2 - 3x + \dfrac{9}{4} = \left(x - \dfrac{3}{2}\right)^2$

8. $x^2 - 12x + 27 = 0$

$x^2 - 12x \qquad = -27$

Since $b = -12$, add $\left(\dfrac{b}{2}\right)^2 = \left(\dfrac{-12}{2}\right)^2 = (-6)^2 = 36$.

$x^2 - 12x + 27 = 0$

$x^2 - 12x + 36 = -27 + 36$

$(x - 6)^2 = 9$

Apply the square root property.

$x - 6 = 3 \qquad x - 6 = -3$

$x = 9 \qquad\quad x = 3$

The solution set is $\{3, 9\}$.

9. $x^2 - 7x - 1 = 0$

$x^2 - 7x \qquad = 1$

Since $b = -7$, add $\left(\dfrac{b}{2}\right)^2 = \left(\dfrac{-7}{2}\right)^2 = \dfrac{49}{4}$.

$x^2 - 7x + \dfrac{49}{4} = 1 + \dfrac{49}{4}$

$\left(x - \dfrac{7}{2}\right)^2 = \dfrac{4}{4} + \dfrac{49}{4}$

$\left(x - \dfrac{7}{2}\right)^2 = \dfrac{53}{4}$

Apply the square root property.

$x - \dfrac{7}{2} = \pm\sqrt{\dfrac{53}{4}}$

$x = \dfrac{7}{2} \pm \dfrac{\sqrt{53}}{2} = \dfrac{7 \pm \sqrt{53}}{2}$

The solution set is $\left\{\dfrac{7 \pm \sqrt{53}}{2}\right\}$.

10. $2x^2 + 3x - 4 = 0$

$x^2 + \dfrac{3}{2}x - 2 = 0$

$x^2 + \dfrac{3}{2}x \quad = 2$

Since $b = \dfrac{3}{2}$, add

$\left(\dfrac{b}{2}\right)^2 = \left(\dfrac{\frac{3}{2}}{2}\right)^2 = \left(\dfrac{3}{2} \div 2\right)^2$

$= \left(\dfrac{3}{2} \cdot \dfrac{1}{2}\right)^2 = \left(\dfrac{3}{4}\right)^2 = \dfrac{9}{16}.$

$x^2 + \dfrac{3}{2}x + \dfrac{9}{16} = 2 + \dfrac{9}{16}$

$\left(x + \dfrac{3}{4}\right)^2 = \dfrac{32}{16} + \dfrac{9}{16}$

$\left(x + \dfrac{3}{4}\right)^2 = \dfrac{41}{16}$

Apply the square root property.

$x + \dfrac{3}{4} = \pm\sqrt{\dfrac{41}{16}}$

$x = -\dfrac{3}{4} \pm \dfrac{\sqrt{41}}{4}$

$x = \dfrac{-3 \pm \sqrt{41}}{4}$

The solution set is $\left\{\dfrac{-3 \pm \sqrt{41}}{4}\right\}$.

11. $A = P(1+r)^t$

$2916 = 2500(1+r)^2$

$\dfrac{2916}{2500} = (1+r)^2$

Apply the square root property.

$1 + r = \pm\sqrt{\dfrac{2916}{2500}}$

$r = -1 \pm \sqrt{1.1664}$

$r = -1 \pm 1.08$

The solutions are $-1 - 1.08 = -2.08$ and $-1 + 1.08 = 0.08$. Disregard -2.08 since we cannot have a negative interest rate. The interest rate is 0.08 or 8%.

12. $W(t) = 3t^2$

$588 = 3t^2$

$196 = t^2$

Apply the square root property.

$t^2 = 196$

$t = \pm\sqrt{196}$

$t = \pm 14$

The solutions are -14 and 14. Disregard -14, because we cannot have a negative time measurement. The fetus will weigh 588 grams after 14 weeks.

13.

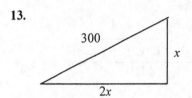

Use the Pythagorean Theorem.

$(2x)^2 + x^2 = 300^2$

$4x^2 + x^2 = 90{,}000$

$5x^2 = 90{,}000$

$x^2 = 18{,}000$

$x = \pm\sqrt{18{,}000}$

$x = \pm\sqrt{3600 \cdot 5}$

$x = \pm 60\sqrt{5}$

The solutions are $\pm 60\sqrt{5}$ meters. Disregard $-60\sqrt{5}$ meters, because we can't have a negative length measurement. Therefore, the building is $60\sqrt{5}$ meters, or approximately 134.2 meters high.

14. $d = \sqrt{\left(3 - (-2)\right)^2 + \left(9 - (-3)\right)^2}$

$= \sqrt{(3 + 2)^2 + (9 + 3)^2}$

$= \sqrt{5^2 + 12^2} = \sqrt{25 + 144}$

$= \sqrt{169} = 13$

15. $d = \sqrt{\left(-2 - (-4)\right)^2 + (5 - 3)^2}$

$= \sqrt{(-2 + 4)^2 + 2^2} = \sqrt{2^2 + 4}$

$= \sqrt{4 + 4} = \sqrt{8} = \sqrt{4 \cdot 2} = 2\sqrt{2} \approx 2.83$

16. Midpoint $= \left(\dfrac{2+(-12)}{2}, \dfrac{6+4}{2} \right)$

$= \left(\dfrac{-10}{2}, \dfrac{10}{2} \right) = (-5, 5)$

17. Midpoint $= \left(\dfrac{4+(-15)}{2}, \dfrac{-6+2}{2} \right)$

$= \left(\dfrac{-11}{2}, \dfrac{-4}{2} \right) = \left(-\dfrac{11}{2}, -2 \right)$

18. $\qquad x^2 = 2x + 4$

$x^2 - 2x - 4 = 0$

$a = 1 \quad b = -2 \quad c = -4$

$x = \dfrac{-(-2) \pm \sqrt{(-2)^2 - 4(1)(-4)}}{2(1)}$

$= \dfrac{2 \pm \sqrt{4+16}}{2}$

$= \dfrac{2 \pm \sqrt{20}}{2}$

$= \dfrac{2 \pm \sqrt{4 \cdot 5}}{2}$

$= \dfrac{2 \pm 2\sqrt{5}}{2} = \dfrac{2(1 \pm \sqrt{5})}{2} = 1 \pm \sqrt{5}$

The solution set is $\left\{ 1 \pm \sqrt{5} \right\}$.

19. $\qquad x^2 - 2x + 19 = 0$

$a = 1 \quad b = -2 \quad c = 19$

$x = \dfrac{-(-2) \pm \sqrt{(-2)^2 - 4(1)(19)}}{2(1)}$

$= \dfrac{2 \pm \sqrt{4 - 76}}{2}$

$= \dfrac{2 \pm \sqrt{-72}}{2}$

$= \dfrac{2 \pm \sqrt{-36 \cdot 2}}{2}$

$= \dfrac{2 \pm 6i\sqrt{2}}{2} = \dfrac{2(1 \pm 3i\sqrt{2})}{2} = 1 \pm 3i\sqrt{2}$

The solution set is $\left\{ 1 \pm 3i\sqrt{2} \right\}$.

20. $\qquad 2x^2 = 3 - 4x$

$2x^2 + 4x - 3 = 0$

$a = 2 \quad b = 4 \quad c = -3$

$x = \dfrac{-4 \pm \sqrt{4^2 - 4(2)(-3)}}{2(2)}$

$= \dfrac{-4 \pm \sqrt{16 + 24}}{4}$

$= \dfrac{-4 \pm \sqrt{40}}{4}$

$= \dfrac{-4 \pm \sqrt{4 \cdot 10}}{4}$

$= \dfrac{-4 \pm 2\sqrt{10}}{4}$

$= \dfrac{2(-2 \pm \sqrt{10})}{4} = \dfrac{-2 \pm \sqrt{10}}{2}$

The solution set is $\left\{ \dfrac{-2 \pm \sqrt{10}}{2} \right\}$.

21. $\qquad x^2 - 4x + 13 = 0$

$a = 1 \quad b = -4 \quad c = 13$

Find the discriminant.

$b^2 - 4ac = (-4)^2 - 4(1)(13)$

$= 16 - 52 = -36$

Since the discriminant is negative, there are two imaginary solutions which are complex conjugates.

22. $\qquad 9x^2 = 2 - 3x$

$9x^2 + 3x - 2 = 0$

$a = 9 \quad b = 3 \quad c = -2$

Find the discriminant.

$b^2 - 4ac = 3^2 - 4(9)(-2)$

$= 9 + 72 = 81$

Since the discriminant is greater than zero and a perfect square, there are two real rational solutions.

23. $\qquad 2x^2 + 4x = 3$

$2x^2 + 4x - 3 = 0$

$a = 2 \quad b = 4 \quad c = -3$

Find the discriminant.

$b^2 - 4ac = 4^2 - 4(2)(-3)$

$= 16 + 24 = 40$

Since the discriminant is greater than zero but not a perfect square, there are two real irrational solutions.

24. $3x^2 - 10x - 8 = 0$

$(3x + 2)(x - 4) = 0$

Apply the zero product principle.

$3x + 2 = 0$ and $x - 4 = 0$

$\qquad 3x = -2 \qquad\qquad x = 4$

$\qquad x = -\dfrac{2}{3}$

The solution set is $\left\{-\dfrac{2}{3}, 4\right\}$.

25. $(2x - 3)(x + 2) = x^2 - 2x + 4$

$2x^2 + 4x - 3x - 6 = x^2 - 2x + 4$

$x^2 + 3x - 10 = 0$

Use the quadratic formula.

$a = 1 \quad b = 3 \quad c = -10$

$x = \dfrac{-3 \pm \sqrt{3^2 - 4(1)(-10)}}{2(1)}$

$= \dfrac{-3 \pm \sqrt{9 - (-40)}}{2}$

$= \dfrac{-3 \pm \sqrt{49}}{2} = \dfrac{-3 \pm 7}{2} = -5 \text{ or } 2$

The solution set is $\{-5, 2\}$.

26. $5x^2 - x - 1 = 0$

Use the quadratic formula.

$a = 5 \quad b = -1 \quad c = -1$

$x = \dfrac{-(-1) \pm \sqrt{(-1)^2 - 4(5)(-1)}}{2(5)}$

$= \dfrac{1 \pm \sqrt{1 - (-20)}}{10} = \dfrac{1 \pm \sqrt{21}}{10}$

The solution set is $\left\{\dfrac{1 \pm \sqrt{21}}{10}\right\}$.

27. $x^2 - 16 = 0$

$x^2 = 16$

Apply the square root principle.

$x = \pm\sqrt{16} = \pm 4$

The solution set is $\{-4, 4\}$.

28. $(x - 3)^2 - 8 = 0$

$(x - 3)^2 = 8$

Apply the square root principle.

$x - 3 = \pm\sqrt{8}$

$\qquad x = 3 \pm \sqrt{4 \cdot 2}$

$x = 3 \pm 2\sqrt{2}$

The solution set is $\left\{3 \pm 2\sqrt{2}\right\}$.

29. $3x^2 - x + 2 = 0$

Use the quadratic formula.

$a = 3 \quad b = -1 \quad c = 2$

$x = \dfrac{-(-1) \pm \sqrt{(-1)^2 - 4(3)(2)}}{2(3)}$

$= \dfrac{1 \pm \sqrt{1 - 24}}{6}$

$= \dfrac{1 \pm \sqrt{-23}}{6} = \dfrac{1}{6} \pm i\dfrac{\sqrt{23}}{6}$

The solution set is $\left\{\dfrac{1}{6} \pm i\dfrac{\sqrt{23}}{6}\right\}$.

30. $\dfrac{5}{x+1} + \dfrac{x-1}{4} = 2$

$4(x+1)\left(\dfrac{5}{x+1} + \dfrac{x-1}{4}\right) = 4(x+1)(2)$

$20 + (x+1)(x-1) = 8x + 8$

$20 + x^2 - 1 = 8x + 8$

$x^2 - 8x + 11 = 0$

Use the quadratic formula.

$a = 1 \quad b = -8 \quad c = 11$

$x = \dfrac{-(-8) \pm \sqrt{(-8)^2 - 4(1)(11)}}{2(1)}$

$= \dfrac{8 \pm \sqrt{64 - 44}}{2}$

$= \dfrac{8 \pm \sqrt{20}}{2}$

$= \dfrac{8 \pm \sqrt{4 \cdot 5}}{2}$

$= \dfrac{8 \pm 2\sqrt{5}}{2} = \dfrac{2(4 \pm \sqrt{5})}{2} = 4 \pm \sqrt{5}$

The solution set is $\left\{4 \pm \sqrt{5}\right\}$.

31. Because the solution set is $\left\{-\dfrac{1}{3}, \dfrac{3}{5}\right\}$, we have

$$x = -\frac{1}{3} \quad \text{or} \quad x = \frac{3}{5}$$

$$3x = -1 \qquad\qquad 5x = 3$$

$$3x + 1 = 0 \qquad 5x - 3 = 0.$$

Apply the zero-product principle in reverse.

$$(3x + 1)(5x - 3) = 0$$

$$15x^2 - 9x + 5x - 3 = 0$$

$$15x^2 - 4x - 3 = 0$$

32. Because the solution set is $\{-9i,\ 9i\}$, we have

$$x = -9i \quad \text{or} \quad x = 9i$$

$$x + 9i = 0 \qquad x - 9i = 0.$$

Apply the zero-product principle in reverse.

$$(x + 9i)(x - 9i) = 0$$

$$x^2 - 9ix + 9ix - 81i^2 = 0$$

$$x^2 - 81(-1) = 0$$

$$x^2 + 81 = 0$$

33. Because the solution set is $\left\{-4\sqrt{3}, 4\sqrt{3}\right\}$, we have

$$x = -4\sqrt{3} \quad \text{or} \quad x = 4\sqrt{3}$$

$$x + 4\sqrt{3} = 0 \qquad x - 4\sqrt{3} = 0.$$

Apply the zero product principle in reverse.

$$\left(x + 4\sqrt{3}\right)\left(x - 4\sqrt{3}\right) = 0$$

$$x^2 - \left(4\sqrt{3}\right)^2 = 0$$

$$x^2 - 16 \cdot 3 = 0$$

$$x^2 - 48 = 0$$

34. a. $g(x) = 0.125x^2 + 2.3x + 27$

$$g(35) = 0.125(35)^2 + 2.3(35) + 27 \approx 261$$

On wet pavement, a motorcycle traveling at 35 miles per hour will require a stopping distance of 261 feet.
This answer overestimates the stopping distance shown in the graph by 1 foot.

b. $f(x) = 0.125x^2 - 0.8x + 99$

$$267 = 0.125x^2 - 0.8x + 99$$

$$0 = 0.125x^2 - 0.8x - 168$$

$$x = \frac{-b \pm \sqrt{b^2 - 4ac}}{2a}$$

$$x = \frac{-(-0.8) \pm \sqrt{(-0.8)^2 - 4(0.125)(-168)}}{2(0.125)}$$

$$x \approx -33.6 \text{ or } 40$$

On dry pavement, a stopping distances of 267 feet will be required for a motorcycle traveling 40 miles per hour.

35. a. $g(35) = 0.125(35)^2 + 2.3(35) + 27 \approx 261$

This value is shown in the graph by the point (35, 261).

b. $f(40) = 0.125(40)^2 - 0.8(40) + 99 = 267$

This value is shown in the graph by the point (40, 267).

36. $0 = -16t^2 + 140t + 3$

Apply the Quadratic Formula.

$$a = -16 \quad b = 140 \quad c = 3$$

$$= \frac{-140 \pm \sqrt{19{,}600 + 192}}{-32}$$

$$= \frac{-140 \pm \sqrt{19{,}792}}{-32} \approx \frac{-140 \pm 140.7}{-32}$$

$$\approx \frac{-140 - 140.7}{-32} \text{ or } \frac{-140 + 140.7}{-32}$$

$$\approx \frac{-280.7}{-32} \text{ or } \frac{0.7}{-32}$$

$$\approx 8.8 \text{ or } -0.02$$

Disregard -0.02 because we cannot have a negative time measurement. The solution is approximately 8.8. The ball will hit the ground in about 8.8 seconds.

37. $f(x) = -(x+1)^2 + 4$

Since $a = -1$ is negative, the parabola opens downward. The vertex of the parabola is $(h, k) = (-1, 4)$ and the axis of symmetry is $x = -1$.

Replace $f(x)$ with 0 to find x–intercepts.

$$0 = -(x+1)^2 + 4$$

$$(x+1)^2 = 4$$

Apply the square root property.

$x + 1 = \sqrt{4}$　or　$x + 1 = -\sqrt{4}$

$x + 1 = 2$　　　　$x + 1 = -2$

　$x = 1$　　　　　　$x = -3$

The x–intercepts are 1 and –3. Set $x = 0$ and solve for y to obtain the y–intercept.

$$y = -(0+1)^2 + 4$$

$$y = -(1)^2 + 4 = 3$$

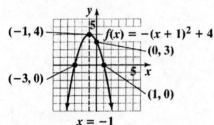

Axis of symmetry: $x = -1$.

38. $f(x) = (x+4)^2 - 2$

Since $a = 1$ is positive, the parabola opens upward. The vertex of the parabola is $(h, k) = (-4, -2)$ and the axis of symmetry is $x = -4$. Replace $f(x)$ with 0 to find x–intercepts.

$$0 = (x+4)^2 - 2$$

$$2 = (x+4)^2$$

Apply the square root property.

$x + 4 = \sqrt{2}$　　　or　　$x + 4 = -\sqrt{2}$

　$x = -4 + \sqrt{2}$　　　　$x = -4 - \sqrt{2}$

The x–intercepts are $-4 - \sqrt{2}$ and $-4 + \sqrt{2}$. Set $x = 0$ and solve for y to obtain the y–intercept.

$$y = (0+4)^2 - 2$$

$$y = 4^2 - 2$$

$$y = 16 - 2$$

$$y = 14$$

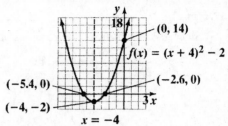

Axis of symmetry: $x = -4$.

39. $f(x) = -x^2 + 2x + 3$

Since $a = -1$ is negative, the parabola opens downward. The x–coordinate of the vertex of the parabola is $-\dfrac{b}{2a} = -\dfrac{2}{2(-1)} = -\dfrac{2}{-2} = 1$ and the y–coordinate of the vertex of the parabola is

$$f\left(-\frac{b}{2a}\right) = f(1)$$

$$= -1^2 + 2(1) + 3$$

$$= -1 + 2 + 3 = 4.$$

The vertex is (1, 4). Replace $f(x)$ with 0 to find x–intercepts.

$$0 = -x^2 + 2x + 3$$

$$0 = x^2 - 2x - 3$$

$$0 = (x-3)(x+1)$$

Apply the zero product principle.

$x - 3 = 0$　or　$x + 1 = 0$

　$x = 3$　　　　$x = -1$

The x–intercepts are –1 and 3. Set $x = 0$ and solve for y to obtain the y–intercept.

$$y = -0^2 + 2(0) + 3$$

$$y = 0 + 0 + 3$$

$$y = 3$$

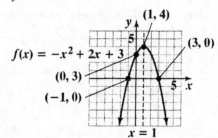

Axis of symmetry: $x = 1$.

40. $f(x) = 2x^2 - 4x - 6$

Since $a = 2$ is positive, the parabola opens upward. The x–coordinate of the vertex of the parabola is
$$-\frac{b}{2a} = -\frac{-4}{2(2)} = -\frac{-4}{4} = 1 \text{ and the}$$

y–coordinate of the vertex of the parabola is
$$f\left(-\frac{b}{2a}\right) = f(1)$$
$$= 2(1)^2 - 4(1) - 6$$
$$= 2(1) - 4 - 6$$
$$= 2 - 4 - 6 = -8.$$

The vertex is $(1, -8)$. Replace $f(x)$ with 0 to find x–intercepts.

$$0 = 2x^2 - 4x - 6$$
$$0 = x^2 - 2x - 3$$
$$0 = (x - 3)(x + 1)$$

Apply the zero product principle.

$$x - 3 = 0 \quad \text{or} \quad x + 1 = 0$$
$$x = 3 \qquad\qquad x = -1$$

The x–intercepts are -1 and 3. Set $x = 0$ and solve for y to obtain the y–intercept.

$$y = 2(0)^2 - 4(0) - 6$$
$$y = 2(0) - 0 - 6$$
$$y = 0 - 0 - 6 = -6$$

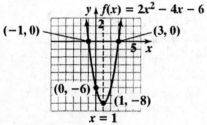

Axis of symmetry: $x = 1$.

41. $f(x) = -0.02x^2 + x + 1$

Since $a = -0.02$ is negative, the function opens downward and has a maximum at
$$x = -\frac{b}{2a} = -\frac{1}{2(-0.02)} = -\frac{1}{-0.04} = 25.$$

When 25 inches of rain falls, the maximum growth will occur.

The maximum growth is $f(25) = -0.02(25)^2 + 25 + 1$
$$= -0.02(625) + 25 + 1$$
$$= -12.5 + 25 + 1 = 13.5.$$

A maximum yearly growth of 13.5 inches occurs when 25 inches of rain falls per year.

42.　$s(t) = -16t^2 + 400t + 40$

Since $a = -16$ is negative, the function opens downward and has a maximum at $x = -\dfrac{b}{2a} = -\dfrac{400}{2(-16)} = -\dfrac{400}{-32} = 12.5$.

At 12.5 seconds, the rocket reaches its maximum height.

The maximum height is $s(12.5) = -16(12.5)^2 + 400(12.5) + 40$
$$= -16(156.25) + 5000 + 40$$
$$= -2500 + 5000 + 40 = 2540.$$

The rocket reaches a maximum height of 2540 feet in 12.5 seconds.

43.　$f(x) = 104.5x^2 - 1501.5x + 6016$

Since a is positive, the function opens upward and has a minimum at $x = -\dfrac{b}{2a} = -\dfrac{-1501.5}{2(104.5)} \approx 7.2$.

At 7.2 hours, the death rate reaches its minimum.

The minimum death rate is　$f(x) = 104.5x^2 - 1501.5x + 6016$
$$f(7.2) = 104.5(7.2)^2 - 1501.5(7.2) + 6016 \approx 622$$

U.S. men who average 7.2 hours of sleep have a death rate of about 622 per 100,000.

44.　Maximize the area using $A = lw$.

$A(x) = x(1000 - 2x)$

$A(x) = -2x^2 + 1000x$

Since $a = -2$ is negative, the function opens downward and has a maximum at $x = -\dfrac{b}{2a} = -\dfrac{1000}{2(-2)} = -\dfrac{1000}{-4} = 250$.

The maximum area is achieved when the width is 250 yards.　The maximum area is

$A(250) = 250(1000 - 2(250))$
$$= 250(1000 - 500)$$
$$= 250(500) = 125,000.$$

The area is maximized at 125,000 square yards when the width is 250 yards and the length is $1000 - 2 \cdot 250 = 500$ yards.

45.　Let x = one of the numbers.
Let $14 + x$ = the other number.
We need to minimize the function $P(x) = x(14 + x)$
$$= 14x + x^2$$
$$= x^2 + 14x.$$

The minimum is at $x = -\dfrac{b}{2a} = -\dfrac{14}{2(1)} = -\dfrac{14}{2} = -7$.

The other number is $14 + x = 14 + (-7) = 7$.

The numbers which minimize the product are 7 and -7. The minimum product is $-7 \cdot 7 = -49$.

46. Let $u = x^2$.

$$x^4 - 6x^2 + 8 = 0$$

$$\left(x^2\right)^2 - 6x^2 + 8 = 0$$

$$u^2 - 6u + 8 = 0$$

$$(u-4)(u-2) = 0$$

$$u - 4 = 0 \quad \text{or} \quad u - 2 = 0$$

$$u = 4 \qquad\qquad u = 2$$

Replace u with x^2.

$$x^2 = 4 \quad \text{or} \quad x^2 = 2$$

$$x = \pm 2 \qquad x = \pm\sqrt{2}$$

The solution set is $\left\{-2, -\sqrt{2}, \sqrt{2}, 2\right\}$.

47. Let $u = \sqrt{x}$.

$$x + 7\sqrt{x} - 8 = 0$$

$$\left(\sqrt{x}\right)^2 + 7\sqrt{x} - 8 = 0$$

$$u^2 + 7u - 8 = 0$$

$$(u+8)(u-1) = 0$$

Apply the zero product principle.

$$u + 8 = 0 \quad \text{or} \quad u - 1 = 0$$

$$u = -8 \qquad\qquad u = 1$$

Replace u with $\sqrt{x}$.

$$\cancel{\sqrt{x} = -8} \quad \text{or} \quad \sqrt{x} = 1$$

$$x = 1$$

Disregard -8 because the square root of x cannot be a negative number.
We must check 1, because both sides of the equation were raised to an even power.
Check:

$$1 + 7\sqrt{1} - 8 = 0$$

$$1 + 7(1) - 8 = 0$$

$$1 + 7 - 8 = 0$$

$$8 - 8 = 0$$

$$0 = 0$$

The solution set is $\{1\}$.

48. Let $u = x^2 + 2x$.

$$\left(x^2 + 2x\right)^2 - 14\left(x^2 + 2x\right) = 15$$

$$\left(x^2 + 2x\right)^2 - 14\left(x^2 + 2x\right) - 15 = 0$$

$$u^2 - 14u - 15 = 0$$

$$(u - 15)(u + 1) = 0$$

$$u - 15 = 0 \quad \text{or} \quad u + 1 = 0$$

$$u = 15 \qquad\qquad u = -1$$

Replace u with $x^2 + 2x$.
First, consider $u = 15$.

$$x^2 + 2x = 15$$

$$x^2 + 2x - 15 = 0$$

$$(x + 5)(x - 3) = 0$$

$$x + 5 = 0 \quad \text{or} \quad x - 3 = 0$$

$$x = -5 \qquad\qquad x = 3$$

Next, consider $u = -1$.

$$x^2 + 2x = -1$$

$$x^2 + 2x + 1 = 0$$

$$(x + 1)^2 = 0$$

$$x + 1 = 0$$

$$x = -1$$

The solution set is $\{-5, -1, 3\}$.

49. Let $u = x^{-1}$.

$$x^{-2} + x^{-1} - 56 = 0$$

$$\left(x^{-1}\right)^2 + x^{-1} - 56 = 0$$

$$u^2 + u - 56 = 0$$

$$(u + 8)(u - 7) = 0$$

$$u + 8 = 0 \quad \text{or} \quad u - 7 = 0$$

$$u = -8 \qquad\qquad u = 7$$

Replace u with x^{-1}.

$$x^{-1} = -8 \quad \text{or} \quad x^{-1} = 7$$

$$\frac{1}{x} = -8 \qquad\qquad \frac{1}{x} = 7$$

$$-8x = 1 \qquad\qquad 7x = 1$$

$$x = -\frac{1}{8} \qquad\qquad x = \frac{1}{7}$$

The solution set is $\left\{-\frac{1}{8}, \frac{1}{7}\right\}$.

50. Let $u = x^{\frac{1}{3}}$.

$$x^{\frac{2}{3}} - x^{\frac{1}{3}} - 12 = 0$$

$$\left(x^{\frac{1}{3}}\right)^2 - x^{\frac{1}{3}} - 12 = 0$$

$$u^2 - u - 12 = 0$$

$$(u-4)(u+3) = 0$$

$$u - 4 = 0 \quad \text{or} \quad u + 3 = 0$$

$$u = 4 \qquad\qquad u = -3$$

Replace u with $x^{\frac{1}{3}}$.

$$x^{\frac{1}{3}} = 4 \qquad\qquad x^{\frac{1}{3}} = -3$$

$$\text{or}$$

$$\left(x^{\frac{1}{3}}\right)^3 = 4^3 \qquad \left(x^{\frac{1}{3}}\right)^3 = (-3)^3$$

$$x = 64 \qquad\qquad x = -27$$

The solution set is $\{-27, 64\}$.

51. Let $u = x^{\frac{1}{4}}$.

$$x^{\frac{1}{2}} + 3x^{\frac{1}{4}} - 10 = 0$$

$$\left(x^{\frac{1}{4}}\right)^2 + 3x^{\frac{1}{4}} - 10 = 0$$

$$u^2 + 3u - 10 = 0$$

$$(u+5)(u-2) = 0$$

$$u + 5 = 0 \quad \text{or} \quad u - 2 = 0$$

$$u = -5 \qquad\qquad u = 2$$

Replace u with $x^{\frac{1}{4}}$.

$$x^{\frac{1}{4}} = -5 \quad \text{or} \quad x^{\frac{1}{4}} = 2$$

$$\cancel{\sqrt[4]{x} = -5} \qquad \left(x^{\frac{1}{4}}\right)^4 = 2^4$$

$$x = 16$$

Disregard -5 because the fourth root of x cannot be a negative number.
We must check 16, because both sides of the equation were raised to an even power. Check $x = 16$.

$$16^{\frac{1}{2}} + 3(16)^{\frac{1}{4}} - 10 = 0$$

$$4 + 3(2) - 10 = 0$$

$$4 + 6 - 10 = 0$$

$$10 - 10 = 0$$

$$0 = 0$$

The solution checks. The solution set is $\{16\}$.

52. $2x^2 + 5x - 3 < 0$

Solve the related quadratic equation.

$$2x^2 + 5x - 3 = 0$$
$$(2x - 1)(x + 3) = 0$$

$2x - 1 = 0$ or $x + 3 = 0$
$2x = 1$ $x = -3$
$x = \dfrac{1}{2}$

The boundary points are -3 and $\dfrac{1}{2}$.

Interval	Test Value	Test	Conclusion
$(-\infty, -3)$	-4	$2(-4)^2 + 5(-4) - 3 < 0$ false	$(-\infty, -3)$ does not belong to the solution set.
$\left(-3, \dfrac{1}{2}\right)$	0	$2(0)^2 + 5(0) - 3 < 0$ true	$\left(-3, \dfrac{1}{2}\right)$ belongs to the solution set.
$\left(\dfrac{1}{2}, \infty\right)$	1	$2(1)^2 + 5(1) - 3 < 0$ false	$\left(\dfrac{1}{2}, \infty\right)$ does not belong to the solution set.

The solution set is $\left(-3, \dfrac{1}{2}\right)$ or $\left\{x \mid -3 < x < \dfrac{1}{2}\right\}$.

53. $2x^2 + 9x + 4 \geq 0$

Solve the related quadratic equation.

$$2x^2 + 9x + 4 = 0$$
$$(2x + 1)(x + 4) = 0$$

$2x + 1 = 0$ or $x + 4 = 0$
$2x = -1$ $x = -4$
$x = -\dfrac{1}{2}$

The boundary points are -4 and $-\dfrac{1}{2}$.

Interval	Test Value	Test	Conclusion
$(-\infty, -4]$	-5	$2(-5)^2 + 9(-5) + 4 \geq 0$ true	$(-\infty, -4]$ belongs to the solution set.
$\left[-4, -\dfrac{1}{2}\right]$	-1	$2(-1)^2 + 9(-1) + 4 \geq 0$ false	$\left[-4, -\dfrac{1}{2}\right]$ does not belong to the solution set.
$\left[-\dfrac{1}{2}, \infty\right)$	0	$2(0)^2 + 9(0) + 4 \geq 0$ true	$\left[-\dfrac{1}{2}, \infty\right)$ belongs to the solution set.

The solution set is $\left(-\infty, -4\right] \cup \left[-\dfrac{1}{2}, \infty\right)$ or $\left\{x \mid x \leq -4 \text{ or } x \geq -\dfrac{1}{2}\right\}$.

54. $x^3 + 2x^2 > 3x$

Solve the related polynomial equation.

$$x^3 + 2x^2 = 3x$$

$$x^3 + 2x^2 - 3x = 0$$

$$x\left(x^2 + 2x - 3\right) = 0$$

$$x(x+3)(x-1) = 0$$

$x = 0$ or $x + 3 = 0$ or $x - 1 = 0$

$\qquad\qquad x = -3 \qquad\qquad x = 1$

The boundary points are -3, 0, and 1.

Interval	Test Value	Test	Conclusion
$(-\infty, -3)$	-4	$(-4)^3 + 2(-4)^2 > 3(-4)$ False	$(-\infty, -3)$ does not belong to the solution set.
$(-3, 0)$	-2	$(-2)^3 + 2(-2)^2 > 3(-2)$ True	$(-3, 0)$ belongs to the solution set.
$(0, 1)$	0.5	$0.5^3 + 2(0.5)^2 > 3(0.5)$ False	$(0, 1)$ does not belong to the solution set.
$(1, \infty)$	2	$2^3 + 2(2)^2 > 3(2)$ True	$(1, \infty)$ belongs to the solution set.

The solution set is $(-3, 0) \cup (1, \infty)$ or $\left\{x \mid -3 < x < 0 \text{ or } x > 1\right\}$.

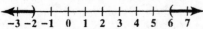

55. $\dfrac{x-6}{x+2} > 0$

Find the values of x that make the numerator and denominator zero.

$x - 6 = 0 \qquad x + 2 = 0$

$\quad x = 6 \qquad\quad x = -2$

The boundary points are -2 and 6.

Interval	Test Value	Test	Conclusion
$(-\infty, -2)$	-3	$\dfrac{-3-6}{-3+2} > 0$ true	$(-\infty, -2)$ belongs to the solution set.
$(-2, 6)$	0	$\dfrac{0-6}{0+2} > 0$ false	$(-2, 6)$ does not belong to the solution set.
$(6, \infty)$	7	$\dfrac{7-6}{7+2} > 0$ true	$(6, \infty)$ belongs to the solution set.

The solution set is $(-\infty, -2) \cup (6, \infty)$ or $\left\{x \mid x < -2 \text{ or } x > 6\right\}$.

56. $\dfrac{x+3}{x-4} \le 5$

Express the inequality so that one side is zero.

$$\frac{x+3}{x-4} - 5 \le 0$$

$$\frac{x+3}{x-4} - \frac{5(x-4)}{x-4} \le 0$$

$$\frac{x+3-5(x-4)}{x-4} \le 0$$

$$\frac{x+3-5x+20}{x-4} \le 0$$

$$\frac{-4x+23}{x-4} \le 0$$

Find the values of x that make the numerator and denominator zero.

$-4x+23 = 0 \qquad$ and $\qquad x-4 = 0$

$\qquad -4x = -23 \qquad\qquad\qquad x = 4$

$\qquad\quad x = \dfrac{23}{4}$

The boundary points are 4 and $\dfrac{23}{4}$. Exclude 4 from the solution set, since this would make the denominator zero.

Interval	Test Value	Test	Conclusion
$(-\infty, 4)$	0	$\dfrac{0+3}{0-4} \le 5$ $\dfrac{3}{-4} \le 5$, true	$(-\infty, 4)$ belongs to the solution set.
$\left(4, \dfrac{23}{4}\right]$	5	$\dfrac{5+3}{5-4} \le 5$ $8 \le 5$, false	$\left(4, \dfrac{23}{4}\right]$ does not belong to the solution set.
$\left[\dfrac{23}{4}, \infty\right)$	6	$\dfrac{6+3}{6-4} \le 5$ $\dfrac{9}{2} \le 5$, true	$\left[\dfrac{23}{4}, \infty\right)$ belongs to the solution set.

The solution set is $(-\infty, 4) \cup \left[\dfrac{23}{4}, \infty\right)$ or $\left\{x \,\middle|\, x < 4 \text{ or } x \ge \dfrac{23}{4}\right\}$.

57. $s(t) = -16t^2 + 48t$

To find when the height is more than 32 feet above the ground, solve the inequality $-16t^2 + 48t > 32$.
Solve the related quadratic equation.

$$-16t^2 + 48t = 32$$

$$-16t^2 + 48t - 32 = 0$$

$$t^2 - 3t + 2 = 0$$

$$(t-2)(t-1) = 0$$

$t - 2 = 0 \quad$ or $\quad t - 1 = 0$

$\quad t = 2 \qquad\qquad t = 1$

The boundary points are 1 and 2.

Interval	Test Value	Test	Conclusion
$(0,1)$	0.5	$-16(0.5)^2 + 48(0.5) > 32$ $20 > 32$, false	$(0,1)$ does not belong to the solution set.
$(1,2)$	1.5	$-16(1.5)^2 + 48(1.5) > 32$ $36 > 32$, true	$(1,2)$ belongs to the solution set.
$(2,\infty)$	3	$-16(3)^2 + 48(3) > 32$ $0 > 32$, false	$(2,\infty)$ does not belong to the solution set.

The solution set is $(1,2)$. This means that the ball will be more than 32 feet above the graph between 1 and 2 seconds.

58. a. $H(0) = \dfrac{15}{8}(0)^2 - 30(0) + 200 = \dfrac{15}{8}(0) - 0 + 200 = 0 - 0 + 200 = 200$

The heart rate is 200 beats per minute immediately following the workout.

b.

$$\dfrac{15}{8}x^2 - 30x + 200 > 110$$

$$\dfrac{15}{8}x^2 - 30x + 90 > 0$$

$$\dfrac{8}{15}\left(\dfrac{15}{8}x^2 - 30x + 90\right) > \dfrac{8}{15}(0)$$

$$x^2 - \dfrac{8}{15}(30x) + \dfrac{8}{15}(90) > 0$$

$$x^2 - 16x + 48 > 0$$

$$(x-12)(x-4) > 0$$

Apply the zero product principle.

$$x - 12 = 0 \quad \text{or} \quad x - 4 = 0$$
$$x = 12 \qquad\qquad x = 4$$

The boundary points are 4 and 12.

Interval	Test Value	Test	Conclusion
$(0,4)$	1	$\dfrac{15}{8}(1)^2 - 30(1) + 200 > 110$ $171\dfrac{7}{8} > 110$, true	$(0,4)$ belongs to the solution set.
$(4,12)$	5	$\dfrac{15}{8}(5)^2 - 30(5) + 200 > 110$ $96\dfrac{7}{8} > 110$, false	$(4,12)$ does not belong to the solution set.
$(12,\infty)$	13	$\dfrac{15}{8}(13)^2 - 30(13) + 200 > 110$ $126\dfrac{7}{8} > 110$, true	$(12,\infty)$ does not belong to the solution set.

The solution set is $(0,4) \cup (12,\infty)$. This means that the heart rate exceeds 110 beats per minute between 0 and 4 minutes after the workout and more than 12 minutes after the workout. Between 0 and 4 minutes provides a more realistic answer since it is unlikely that the heart rate will begin to climb again without further exertion. Model breakdown occurs for the interval $(12,\infty)$.

Chapter 11 Test

1. $2x^2 - 5 = 0$

   $2x^2 = 5$

   $x^2 = \dfrac{5}{2}$

   $x = \pm\sqrt{\dfrac{5}{2}}$

   Rationalize the denominators.

   $x = \pm\dfrac{\sqrt{5}}{\sqrt{2}} \cdot \dfrac{\sqrt{2}}{\sqrt{2}} = \pm\dfrac{\sqrt{10}}{2}$

   The solution set is $\left\{\pm\dfrac{\sqrt{10}}{2}\right\}$.

2. $(x-3)^2 = 20$

   $x - 3 = \pm\sqrt{20}$

   $x = 3 \pm\sqrt{4 \cdot 5}$

   $x = 3 \pm 2\sqrt{5}$

   The solution set is $\left\{3 \pm 2\sqrt{5}\right\}$.

3. $x^2 - 16x + \underline{\hspace{1cm}}$

   Since $b = -16$, add $\left(\dfrac{b}{2}\right)^2 = \left(\dfrac{-16}{2}\right)^2 = (-8)^2 = 64$.

   $x^2 - 16x + 64 = (x-8)^2$

4. $x^2 + \dfrac{2}{5}x + \underline{\hspace{1cm}}$

   Since $b = \dfrac{2}{5}$, add $\left(\dfrac{1}{2}b\right)^2 = \left(\dfrac{1}{2} \cdot \dfrac{2}{5}\right)^2 = \left(\dfrac{1}{5}\right)^2 = \dfrac{1}{25}$.

   $x^2 + \dfrac{2}{5}x + \dfrac{1}{25} = \left(x + \dfrac{1}{5}\right)^2$

5. $x^2 - 6x + 7 = 0$

   $x^2 - 6x \quad = -7$

   Since $b = -6$, add $\left(\dfrac{b}{2}\right)^2 = \left(\dfrac{-6}{2}\right)^2 = (-3)^2 = 9$.

   $x^2 - 6x + 9 = -7 + 9$

   $(x-3)^2 = 2$

   Apply the square root property.

   $x - 3 = \pm\sqrt{2}$

   $x = 3 \pm\sqrt{2}$

   The solution set is $\left\{3 \pm\sqrt{2}\right\}$.

6. Use the Pythagorean Theorem.

   $50^2 + 50^2 = x^2$

   $2500 + 2500 = x^2$

   $5000 = x^2$

   $\pm\sqrt{5000} = x$

   $\pm\sqrt{2500 \cdot 2} = x$

   $\pm 50\sqrt{2} = x$

   The solutions are $\pm 50\sqrt{2}$ feet. Disregard $-50\sqrt{2}$ feet because we can't have a negative length measurement. The width of the pond is $50\sqrt{2}$ feet.

7. $d = \sqrt{(2-(-1))^2 + (-3-5)^2}$

   $= \sqrt{(3)^2 + (-8)^2}$

   $= \sqrt{9 + 64} = \sqrt{73} \approx 8.54$

   The distance between the points is $\sqrt{73}$ or 8.54 units.

8. Midpoint $= \left(\dfrac{-5+12}{2}, \dfrac{-2+(-6)}{2}\right)$

   $= \left(\dfrac{7}{2}, \dfrac{-8}{2}\right) = \left(\dfrac{7}{2}, -4\right)$

   The midpoint is $\left(\dfrac{7}{2}, -4\right)$.

9. $3x^2 + 4x - 2 = 0$

$a = 3$ $b = 4$ $c = -2$

Find the discriminant.

$b^2 - 4ac = 4^2 - 4(3)(-2)$

$\qquad = 16 + 24 = 40$

Since the discriminant is greater than zero but not a perfect square, there are two real irrational solutions.

10. $x^2 = 4x - 8$

$x^2 - 4x + 8 = 0$

$a = 1$ $b = -4$ $c = 8$

Find the discriminant.

$b^2 - 4ac = (-4)^2 - 4(1)(8)$

$\qquad = 16 - 32 = -16$

Since the discriminant is negative, there are two imaginary solutions which are complex conjugates.

11. $2x^2 + 9x = 5$

$2x^2 + 9x - 5 = 0$

$(2x - 1)(x + 5) = 0$

Apply the zero product principle.

$2x - 1 = 0$ and $x + 5 = 0$

$\quad 2x = 1$ $\qquad\qquad x = -5$

$\quad\ x = \dfrac{1}{2}$

The solution set is $\left\{ -5, \dfrac{1}{2} \right\}$.

12. $x^2 + 8x + 5 = 0$

Solve using the quadratic formula.

$a = 1$ $b = 8$ $c = 5$

$x = \dfrac{-8 \pm \sqrt{8^2 - 4(1)(5)}}{2(1)}$

$\ = \dfrac{-8 \pm \sqrt{64 - 20}}{2}$

$\ = \dfrac{-8 \pm \sqrt{44}}{2}$

$\ = \dfrac{-8 \pm \sqrt{4 \cdot 11}}{2}$

$\ = \dfrac{-8 \pm 2\sqrt{11}}{2}$

$\ = \dfrac{2\left(-4 \pm \sqrt{11} \right)}{2} = -4 \pm \sqrt{11}$

The solution set is $\left\{ -4 \pm \sqrt{11} \right\}$.

13. $(x + 2)^2 + 25 = 0$

$(x + 2)^2 = -25$

Apply the square root principle.

$x + 2 = \pm\sqrt{-25}$

$\quad\ x = -2 \pm 5i$

The solution set is $\{ -2 \pm 5i \}$.

14. $2x^2 - 6x + 5 = 0$

$a = 2$ $b = -6$ $c = 5$

$x = \dfrac{-(-6) \pm \sqrt{(-6)^2 - 4(2)(5)}}{2(2)}$

$\ = \dfrac{6 \pm \sqrt{36 - 40}}{4}$

$\ = \dfrac{6 \pm \sqrt{-4}}{4}$

$\ = \dfrac{6 \pm 2i}{4} = \dfrac{6}{4} \pm \dfrac{2}{4}i = \dfrac{3}{2} \pm \dfrac{1}{2}i$

The solution set is $\left\{ \dfrac{3}{2} \pm \dfrac{1}{2}i \right\}$.

15.   Because the solution set is $\{ -3, 7 \}$, we have

$x = -3$ or $x = 7$

$x + 3 = 0$ $\qquad x - 7 = 0$

Apply the zero-product principle in reverse.

$(x + 3)(x - 7) = 0$

$x^2 - 7x + 3x - 21 = 0$

$x^2 - 4x - 21 = 0$

16.   Because the solution set is $\{ -10i, 10i \}$, we have

$x = -10i$ or $x = 10i$

$x + 10i = 0$ $\qquad x - 10i = 0$

Apply the zero-product principle in reverse.

$(x + 10i)(x - 10i) = 0$

$x^2 - 100i^2 = 0$

$x^2 - 100(-1) = 0$

$x^2 + 100 = 0$

17. a. 2009 is 4 years after 2005.

$$f(x) = -1.8x^2 + 21x + 15$$

$$f(4) = -1.8(4)^2 + 21(4) + 15$$

$$= 70.2$$

In 2009, the function predicts the percentage of new cellphones that play music is expected to be 70.2%. This overestimates the percentage shown in the graph by 0.2%.

b. $f(x) = -1.8x^2 + 21x + 15$

$$75 = -1.8x^2 + 21x + 15$$

$$0 = -1.8x^2 + 21x - 60$$

$$x = \frac{-b \pm \sqrt{b^2 - 4ac}}{2a}$$

$$x = \frac{-(21) \pm \sqrt{(21)^2 - 4(-1.8)(-60)}}{2(-1.8)}$$

$$x = 5 \text{ or } 6\frac{2}{3}$$

The percentage of new cellphones that play music will first reach 75% 5 years after 2005, or 2010.

18. $f(x) = (x + 1)^2 + 4$

Since $a = 1$ is positive, the parabola opens upward. The vertex of the parabola is $(h, k) = (-1, 4)$ and the axis of symmetry is $x = -1$. Replace $f(x)$ with 0 to find x–intercepts.

$$0 = (x + 1)^2 + 4$$

$$-4 = (x + 1)^2$$

This will be result in complex solutions. As a result, there are no x–intercepts. Set $x = 0$ and solve for y to obtain the y–intercept.

$$y = (0 + 1)^2 + 4 = 1 + 4 = 5$$

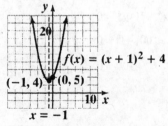

Axis of symmetry: $x = -1$.

19. $f(x) = x^2 - 2x - 3$

Since $a = 1$ is positive, the parabola opens upward. The x–coordinate of the vertex of the parabola is

$$-\frac{b}{2a} = -\frac{-2}{2(1)} = -\frac{-2}{2} = 1 \text{ and the}$$

y–coordinate of the vertex of the parabola is

$$f\left(-\frac{b}{2a}\right) = f(1) = 1^2 - 2(1) - 3$$

$$= 1 - 2 - 3 = -4.$$

The vertex is $(1, -4)$. Replace $f(x)$ with 0 to find x–intercepts.

$$0 = x^2 - 2x - 3$$

$$0 = (x - 3)(x + 1)$$

Apply the zero product principle.

$$x - 3 = 0 \quad \text{or} \quad x + 1 = 0$$

$$x = 3 \qquad\qquad x = -1$$

The x–intercepts are -1 and 3. Set $x = 0$ and solve for y to obtain the y–intercept.

$$y = 0^2 - 2(0) - 3 = -3$$

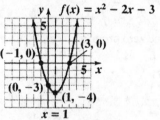

Axis of symmetry: $x = 1$.

20. $s(t) = -16t^2 + 64t + 5$

Since $a = -16$ is negative, the function opens downward and has a maximum at

$$x = -\frac{b}{2a} = -\frac{64}{2(-16)} = -\frac{64}{-32} = 2.$$

The ball reaches its maximum height in two seconds. The maximum height is

$$s(2) = -16(2)^2 + 64(2) + 5$$

$$= -16(4) + 128 + 5$$

$$= -64 + 128 + 5 = 69.$$

The baseball reaches a maximum height of 69 feet after 2 seconds.

21. $0 = -16t^2 + 64t + 5$

Solve using the quadratic formula.

$a = -16 \quad b = 64 \quad c = 5$

$x = \dfrac{-64 \pm \sqrt{64^2 - 4(-16)(5)}}{2(-16)}$

$= \dfrac{-64 \pm \sqrt{4096 + 320}}{-32}$

$= \dfrac{-64 \pm \sqrt{4416}}{-32}$

$\approx 4.1 \text{ or } -0.1$

Disregard -0.1 since we cannot have a negative time measurement. The solution is 4.1 and we conclude that the baseball hits the ground in approximately 4.1 seconds.

22. $f(x) = -x^2 + 46x - 360$

Since $a = -1$ is negative, the function opens downward and has a maximum at

$x = -\dfrac{b}{2a} = -\dfrac{46}{2(-1)} = -\dfrac{46}{-2} = 23.$

$f(23) = -23^2 + 46(23) - 360 = 169$

Profit is maximized when 23 computers are manufactured. This produces a profit of $169 hundreds or $16,900.

23. Let $u = 2x - 5$.

$(2x-5)^2 + 4(2x-5) + 3 = 0$

$u^2 + 4u + 3 = 0$

$(u+3)(u+1) = 0$

$u + 3 = 0 \quad \text{or} \quad u + 1 = 0$

$u = -3 \qquad\qquad u = -1$

Replace u with $2x - 5$.

$2x - 5 = -3 \quad \text{or} \quad 2x - 5 = -1$

$2x = 2 \qquad\qquad 2x = 4$

$x = 1 \qquad\qquad x = 2$

The solution set is $\{1, 2\}$.

24. Let $u = x^2$.

$x^4 - 13x^2 + 36 = 0$

$\left(x^2\right)^2 - 13x^2 + 36 = 0$

$u^2 - 13u + 36 = 0$

$(u-9)(u-4) = 0$

$u - 9 = 0 \quad \text{or} \quad u - 4 = 0$

$u = 9 \qquad\qquad u = 4$

Replace u with x^2.

$x^2 = 9 \quad \text{or} \quad x^2 = 4$

$x = \pm 3 \qquad\qquad x = \pm 2$

The solution set is $\{-3, -2, 2, 3\}$.

25. Let $u = x^{1/3}$.

$x^{2/3} - 9x^{1/3} + 8 = 0$

$\left(x^{1/3}\right)^2 - 9x^{1/3} + 8 = 0$

$u^2 - 9u + 8 = 0$

$(u-8)(u-1) = 0$

$u - 8 = 0 \quad \text{or} \quad u - 1 = 0$

$u = 8 \qquad\qquad u = 1$

Replace u with $x^{1/3}$.

$x^{1/3} = 8 \qquad\qquad \text{or} \qquad x^{1/3} = 1$

$x = 8^3 = 512 \qquad\qquad x = 1^3 = 1$

The solution set is $\{1, \ 512\}$.

26. $x^2 - x - 12 < 0$

Solve the related quadratic equation.

$x^2 - x - 12 = 0$

$(x-4)(x+3) = 0$

$x - 4 = 0$ or $x + 3 = 0$

$x = 4$ $x = -3$

The boundary points are -3 and 4.

Interval	Test Value	Test	Conclusion
$(-\infty, -3)$	-4	$(-4)^2 - (-4) - 12 < 0$ $8 < 0$, false	$(-\infty, -3)$ does not belong to the solution set.
$(-3, 4)$	0	$0^2 - 0 - 12 < 0$ $-12 < 0$, true	$(-3, 4)$ belongs to the solution set.
$(4, \infty)$	5	$5^2 - 5 - 12 < 0$ $8 < 0$, false	$(4, \infty)$ does not belong to the solution set.

The solution set is $(-3, 4)$ or $\{x \mid -3 < x < 4\}$.

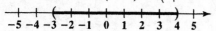

27. $\dfrac{2x+1}{x-3} \le 3$

Express the inequality so that one side is zero.

$$\frac{2x+1}{x-3} - 3 \le 0$$

$$\frac{2x+1}{x-3} - \frac{3(x-3)}{x-3} \le 0$$

$$\frac{2x+1-3(x-3)}{x-3} \le 0$$

$$\frac{2x+1-3x+9}{x-3} \le 0$$

$$\frac{-x+10}{x-3} \le 0$$

Find the values of x that make the numerator and denominator zero.

$-x + 10 = 0$ and $x - 3 = 0$

$-x = -10$ $x = 3$

$x = 10$

The boundary points are 3 and 10. Exclude 3 from the solution set, since this would make the denominator zero.

Interval	Test Value	Test	Conclusion
$(-\infty,3)$	0	$\dfrac{2(0)+1}{0-3} \le 3$ $-\dfrac{1}{3} \le 3$, true	$(-\infty,3)$ belongs to the solution set.
$(3,10]$	4	$\dfrac{2(4)+1}{4-3} \le 3$ $9 \le 3$, false	$(3,10]$ does not belong to the solution set.
$[10,\infty)$	11	$\dfrac{2(11)+1}{11-3} \le 3$ $\dfrac{23}{8} \le 3$, true	$[10,\infty)$ belongs to the solution set.

The solution set is $(-\infty,3)\cup[10,\infty)$ or $\{x \mid x < 3 \text{ or } x \ge 10\}$.

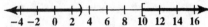

Cumulative Review Exercises (Chapters 1 – 11)

1. $9(x-1) = 1 + 3(x-2)$

    $9x - 9 = 1 + 3x - 6$

    $9x - 9 = 3x - 5$

    $6x - 9 = -5$

    $6x = 4$

    $x = \dfrac{4}{6} = \dfrac{2}{3}$

    The solution set is $\left\{\dfrac{2}{3}\right\}$.

2. $3x + 4y = -7$

    $x - 2y = -9$

    Multiply the second equation by 2 and add the result to the first equation.

    $3x + 4y = -7$

    $2x - 4y = -18$

    $5x \qquad = -25$

    $x = -5$

    Back substitute into the second equation.

    $-5 - 2y = -9$

    $-2y = -4$

    $y = 2$

    The solution set is $\{(-5,2)\}$.

3. $x - y + 3z = -9$

$2x + 3y - z = 16$

$5x + 2y - z = 15$

Multiply the second equation by 3 and add to the first equation to eliminate z.

$x - y + 3z = -9$

$\underline{6x + 9y - 3z = 48}$

$7x + 8y \quad\quad = 39$

Multiply the second equation by -1 and add to the third equation.

$-2x - 3y + z = -16$

$\underline{5x + 2y - z = 15}$

$3x - y \quad\quad = -1$

We now have a system of two equations in two variables.

$7x + 8y = 39$

$3x - y = -1$

Multiply the second equation by 8 and add to the second equation.

$7x + 8y = 39$

$\underline{24x - 8y = -8}$

$31x \quad\quad = 31$

$x = 1$

Back-substitute 1 for x to find y.

$3x - y = -1$

$3(1) - y = -1$

$3 - y = -1$

$-y = -4$

$y = 4$

Back-substitute 1 for x and 4 for y to find z.

$x - y + 3z = -9$

$1 - 4 + 3z = -9$

$-3 + 3z = -9$

$3z = -6$

$z = -2$

The solution set is $\{(1, 4, -2)\}$.

4. $7x + 18 \leq 9x - 2$

$-2x + 18 \leq -2$

$-2x \leq -20$

$\dfrac{-2x}{-2} \geq \dfrac{-20}{-2}$

$x \geq 10$

The solution set is $\{x \mid x \geq 10\}$ or $[10, \infty)$.

5. $4x-3<13$ and $-3x-4 \geq 8$

$\qquad 4x<16 \qquad\qquad -3x \geq 12$

$\qquad\quad x<4 \qquad\qquad\quad x \leq -4$

For a value to be in the solution set, it must be both less than 4 and less than or equal to -4. Now only values that are less than or equal to -4 meet both conditions. Therefore, the solution set is $\{x \mid x \leq -4\}$ or $(-\infty, -4]$.

6. $2x+4>8$ or $x-7 \geq 3$

$\qquad 2x>4 \qquad\qquad x \geq 10$

$\qquad\quad x>2$

For a value to be in the solution set, it must satisfy either of the conditions. Now, all numbers that are greater than or equal to 10 are also greater than 2. Therefore, the solution set is $\{x \mid x > 2\}$ or $(2, \infty)$.

7. $|2x-1|<5$

$-5<2x-1<5$

$-4<2x<6$

$-2<x<3$

The solution set is $\{x \mid -2<x<3\}$ or $(-2,3)$.

8. $\left| \dfrac{2}{3}x-4 \right| = 2$

$\qquad \dfrac{2}{3}x-4=-2 \qquad$ or $\qquad \dfrac{2}{3}x-4=2$

$\quad 3\left(\dfrac{2}{3}x-4 \right)=3(-2) \qquad 3\left(\dfrac{2}{3}x-4 \right)=3(2)$

$\qquad 2x-12=-6 \qquad\qquad\quad 2x-12=6$

$\qquad\quad 2x=6 \qquad\qquad\qquad\quad 2x=18$

$\qquad\quad\; x=3 \qquad\qquad\qquad\quad\; x=9$

The solution set is $\{3,9\}$.

9.

$$\dfrac{4}{x-3}-\dfrac{6}{x+3}=\dfrac{24}{x^2-9}$$

$$\dfrac{4}{x-3}-\dfrac{6}{x+3}=\dfrac{24}{(x-3)(x+3)}$$

$$(x-3)(x+3)\left[\dfrac{4}{x-3}-\dfrac{6}{x+3}\right]=(x-3)(x+3)\left[\dfrac{24}{(x-3)(x+3)}\right]$$

$$4(x+3)-6(x-3)=24$$

$$4x+12-6x+18=24$$

$$-2x+30=24$$

$$-2x=-6$$

$$x=3$$

Disregard 3 since it causes a result of 0 in the denominator of a fraction.

Thus, the equation has no solution. The solution set is $\varnothing$ or $\{\ \}$.

10. $\sqrt{x+4} - \sqrt{x-3} = 1$

$$\sqrt{x+4} = \sqrt{x-3} + 1$$

$$\left(\sqrt{x+4}\right)^2 = \left(\sqrt{x-3} + 1\right)^2$$

$$x + 4 = (x-3) + 2\sqrt{x-3} + 1$$

$$x + 4 = x - 2 + 2\sqrt{x-3}$$

$$6 = 2\sqrt{x-3}$$

$$3 = \sqrt{x-3}$$

$$(3)^2 = \left(\sqrt{x-3}\right)^2$$

$$9 = x - 3$$

$$12 = x$$

Since we square both sides of the equation, we must check to make sure 12 is not extraneous:

$$\sqrt{12+4} - \sqrt{12-3} = 1$$

$$\sqrt{16} - \sqrt{9} = 1$$

$$4 - 3 = 1$$

$$1 = 1, \text{ true}$$

Thus, the solution set is $\{12\}$.

11. $2x^2 = 5 - 4x$

$2x^2 + 4x - 5 = 0$

Apply the quadratic formula:

$a = 2 \quad b = 4 \quad c = -5$

$$x = \frac{-4 \pm \sqrt{4^2 - 4(2)(-5)}}{2(2)}$$

$$= \frac{-4 \pm \sqrt{56}}{4}$$

$$= \frac{-4 \pm \sqrt{4 \cdot 14}}{4}$$

$$= \frac{-4 \pm 2\sqrt{14}}{4} = \frac{-2 \pm \sqrt{14}}{2}$$

The solution set is $\left\{\dfrac{-2 \pm \sqrt{14}}{2}\right\}$.

12. $x^{\frac{2}{3}} - 5x^{\frac{1}{3}} + 6 = 0$

$$\left(x^{\frac{1}{3}}\right)^2 - 5x^{\frac{1}{3}} + 6 = 0$$

Let $u = x^{\frac{1}{3}}$.

$$u^2 - 5u + 6 = 0$$

$$(u-3)(u-2) = 0$$

$$u - 3 = 0 \quad \text{or} \quad u - 2 = 0$$

$$u = 3 \qquad\qquad u = 2$$

Substitute $x^{\frac{1}{3}}$ back in for u.

$$x^{\frac{1}{3}} = 3 \quad \text{or} \quad x^{\frac{1}{3}} = 2$$

$$\left(x^{\frac{1}{3}}\right)^3 = 3^3 \qquad \left(x^{\frac{1}{3}}\right)^3 = 2^3$$

$$x = 27 \qquad\qquad x = 8$$

The solution set is $\{8, 27\}$.

13. $2x^2 + x - 6 \le 0$

Solve the related quadratic equation.

$$2x^2 + x - 6 = 0$$

$$(2x - 3)(x + 2) = 0$$

$$2x - 3 = 0 \quad \text{or} \quad x + 2 = 0$$

$$2x = 3 \qquad\qquad x = -2$$

$$x = \frac{3}{2}$$

The boundary points are -2 and $\frac{3}{2}$.

Interval	Test Value	Test	Conclusion
$(-\infty, -2)$	-3	$2(-3)^2 + (-3) - 6 \le 0$ $9 \le 0$, False	$(-\infty, -2)$ does not belong to the solution set.
$\left(-2, \frac{3}{2}\right)$	0	$2(0)^2 + 0 - 6 \le 0$ $-6 \le 0$, True	$\left(-2, \frac{3}{2}\right)$ belongs to the solution set.
$\left(\frac{3}{2}, \infty\right)$	2	$2(2)^2 + 2 - 6 \le 0$ $4 \le 0$, False	$\left(\frac{3}{2}, \infty\right)$ does not belong to the solution set.

The solution set is $\left[-2, \frac{3}{2}\right]$ or $\left\{x \middle| -2 \le x \le \frac{3}{2}\right\}$.

14. $x - 3y = 6$

$$-3y = -x + 6$$

$$y = \frac{-x + 6}{-3}$$

$$y = \frac{1}{3}x - 2$$

The slope is $m = \frac{1}{3}$ and the y-intercept is $b = -2$.

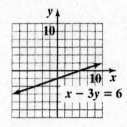

15. $f(x) = \frac{1}{2}x - 1$

This is a linear function with slope $m = \frac{1}{2}$ and y-intercept $b = -1$.

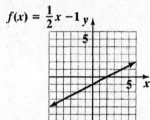

16. $3x - 2y > -6$

First, graph the equation $3x - 2y = -6$ as a dashed line.

$$3x - 2y = -6$$

$$-2y = -3x - 6$$

$$y = \frac{-3x - 6}{-2}$$

$$y = \frac{3}{2}x + 3$$

The slope is $m = \frac{3}{2}$ and the y-intercept is $b = 3$. Next, use the origin as a test point.

$$3(0) - 2(0) > -6$$

$$0 > -6$$

This is a true statement. This means that the point, $(0,0)$, will fall in the shaded half-plane.

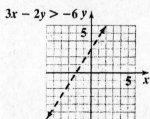

17. $f(x) = -2(x-3)^2 + 2$

Since $a = -2$ is negative, the parabola opens downward. The vertex of the parabola is $(h,k) = (3,2)$. Replace $f(x)$ with 0 to find x–intercepts.

$$0 = -2(x-3)^2 + 2$$
$$2(x-3)^2 = 2$$
$$(x-3)^2 = 1$$
$$x - 3 = \pm 1$$
$$x = 3 \pm 1$$
$$x = 3 - 1 \text{ or } 3 + 1$$
$$x = 2 \text{ or } 4$$

The x–intercepts are 2 and 4.
Set $x = 0$ to obtain the y–intercept.

$$f(0) = -2(0-3)^2 + 2$$
$$= -2(-3)^2 + 2$$
$$= -2(9) + 2 = -18 + 2 = -16$$

The y-intercept is -16.

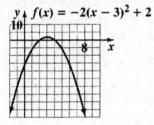

18. $4[2x - 6(x-y)] = 4(2x - 6x + 6y)$
$$= 4(-4x + 6y)$$
$$= -16x + 24y$$

19. $(-5x^3y^2)(4x^4y^{-6}) = -20x^{3+4}y^{2+(-6)}$
$$= -20x^7y^{-4}$$
$$= -\frac{20x^7}{y^4}$$

20. $(8x^2 - 9xy - 11y^2) - (7x^2 - 4xy + 5y^2)$
$$= 8x^2 - 9xy - 11y^2 - 7x^2 + 4xy - 5y^2$$
$$= x^2 - 5xy - 16y^2$$

21. $(3x-1)(2x+5) = 6x^2 + 15x - 2x - 5$
$$= 6x^2 + 13x - 5$$

22. $(3x^2 - 4y)^2$
$$= (3x^2)^2 - 2(3x^2)(4y) + (4y)^2$$
$$= 9x^4 - 24x^2y + 16y^2$$

23. $\dfrac{3x}{x+5} - \dfrac{2}{x^2 + 7x + 10}$
$$= \frac{3x}{x+5} - \frac{2}{(x+5)(x+2)}$$
$$= \frac{3x}{x+5} \cdot \frac{(x+2)}{(x+2)} - \frac{2}{(x+5)(x+2)}$$
$$= \frac{3x^2 + 6x}{(x+5)(x+2)} - \frac{2}{(x+5)(x+2)}$$
$$= \frac{3x^2 + 6x - 2}{(x+5)(x+2)}$$

24. $\dfrac{1 - \dfrac{9}{x^2}}{1 + \dfrac{3}{x}} = \dfrac{1 - \dfrac{9}{x^2}}{1 + \dfrac{3}{x}} \cdot \dfrac{x^2}{x^2}$
$$= \frac{x^2 - 9}{x^2 + 3x}$$
$$= \frac{(x-3)(x+3)}{x(x+3)} = \frac{x-3}{x}$$

25. $\dfrac{x^2 - 6x + 8}{3x + 9} \div \dfrac{x^2 - 4}{x+3}$
$$= \frac{x^2 - 6x + 8}{3x + 9} \cdot \frac{x+3}{x^2 - 4}$$
$$= \frac{(x-4)\cancel{(x-2)}}{3\cancel{(x+3)}} \cdot \frac{\cancel{x+3}}{\cancel{(x-2)}(x+2)}$$
$$= \frac{x-4}{3(x+2)} \quad \text{or} \quad \frac{x-4}{3x+6}$$

26. $\sqrt{5xy} \cdot \sqrt{10x^2y} = \sqrt{50x^3y^2}$
$$= \sqrt{25 \cdot 2 \cdot x^2 \cdot x \cdot y^2}$$
$$= 5xy\sqrt{2x}$$

27. $4\sqrt{72} - 3\sqrt{50} = 4\sqrt{36 \cdot 2} - 3\sqrt{25 \cdot 2}$
$$= 4 \cdot 6\sqrt{2} - 3 \cdot 5\sqrt{2}$$
$$= 24\sqrt{2} - 15\sqrt{2}$$
$$= 9\sqrt{2}$$

28. $(5+3i)(7-3i) = 35 - 15i + 21i - 9i^2$
$$= 35 + 6i - 9(-1)$$
$$= 35 + 6i + 9$$
$$= 44 + 6i$$

29. $81x^4 - 1 = (9x^2 + 1)(9x^2 - 1)$
$$= (9x^2 + 1)(3x + 1)(3x - 1)$$

30. $24x^3 - 22x^2 + 4x = 2x(12x^2 - 11x + 2)$
$$= 2x(4x - 1)(3x - 2)$$

31. $x^3 + 27y^3 = x^3 + (3y)^3$
$$= (x + 3y)\left[x^2 - x(3y) + (3y)^2\right]$$
$$= (x + 3y)(x^2 - 3xy + 9y^2)$$

32. $(f - g)(x) = f(x) - g(x)$
$$= (x^2 + 3x - 15) - (x - 2)$$
$$= x^2 + 3x - 15 - x + 2$$
$$= x^2 + 2x - 13$$

$(f - g)(5) = 5^2 + 2 \cdot 5 - 13$
$$= 25 + 10 - 13 = 22$$

33. $\left(\dfrac{f}{g}\right)(x) = \dfrac{f(x)}{g(x)} = \dfrac{x^2 + 3x - 15}{x - 2}$

The domain of $\dfrac{f}{g}$ is

$\{x \mid x \text{ is a real number and } x \neq 2\}$ or

$(-\infty, 2) \cup (2, \infty)$.

34. $f(x) = x^2 + 3x - 15$

$\dfrac{f(a+h) - f(a)}{h} = \dfrac{\left((a+h)^2 + 3(a+h) - 15\right) - \left(a^2 + 3a - 15\right)}{h}$

$$= \dfrac{a^2 + 2ah + h^2 + 3a + 3h - 15 - a^2 - 3a + 15}{h}$$

$$= \dfrac{2ah + h^2 + 3h}{h}$$

$$= 2a + h + 3$$

35.

$$\begin{array}{r|rrrr} -2 & 3 & -1 & 4 & 8 \\ & & -6 & 14 & -36 \\ \hline & 3 & -7 & 18 & -28 \end{array}$$

$$(3x^3 - x^2 + 4x + 8) \div (x + 2) = 3x^2 - 7x + 18 - \dfrac{28}{x + 2}$$

36. $I = \dfrac{R}{R + r}$

$$I(R + r) = R$$
$$IR + Ir = R$$
$$Ir = R - IR$$
$$Ir = (1 - I)R$$
$$\dfrac{Ir}{1 - I} = R \text{ or } R = -\dfrac{Ir}{I - 1}$$

37. $3x + y = 9$
$$y = -3x + 9$$

The line whose equation we want to find has a slope of $m = -3$, the same as that of the line above. Using this slope with the point through which the line passes, $(-2, 5)$, we first find the point-slope equation and then put it in slope-intercept form.

$$y - y_1 = m(x - x_1)$$
$$y - 5 = -3(x - (-2))$$
$$y - 5 = -3(x + 2)$$
$$y - 5 = -3x - 6$$
$$y = -3x - 1$$

The slope-intercept equation of the line through $(-2, 5)$ and parallel to $3x + y = 9$ is $y = -3x - 1$.

38. Let $x =$ the computer's original price.
$$x - 0.30x = 434$$
$$0.70x = 434$$
$$x = \dfrac{434}{0.70} = 620$$

The original price of the computer was $620.

39. Let x = the width of the rectangle.
Let $3x + 1$ = the length of the rectangle.

$$x(3x+1) = 52$$

$$3x^2 + x - 52 = 0$$

$$(3x+13)(x-4) = 0$$

$$3x + 13 = 0 \quad \text{or} \quad x - 4 = 0$$

$$x = -\frac{13}{3} \qquad\qquad x = 4$$

Disregard $-\frac{13}{3}$ because the width of a rectangle cannot be negative. If $x = 4$, then
$3x + 1 = 3(4) + 1 = 13$. Thus, the length of the rectangle is 13 yards and the width is 4 yards.

40. Let x = the amount invested at 12%.
Let $4000 - x$ = the amount invested at 14%.

$$0.12x + 0.14(4000 - x) = 508$$

$$0.12x + 560 - 0.14x = 508$$

$$-0.02x + 560 = 508$$

$$-0.02x = -52$$

$$x = 2600$$

$$4000 - x = 4000 - 2600 = 1400$$

Thus, $2600 was invested at 12% and $1400 was invested at 14%.

41. Because I varies inversely as R, we have the following for a constant k:

$$I = \frac{k}{R}$$

Use the fact that $I = 5$ when $R = 22$ to find k:

$$5 = \frac{k}{22}$$

$$k = 22 \cdot 5 = 110$$

Thus, the equation relating I and R is $I = \frac{110}{R}$.

If $R = 10$, then $I = \frac{110}{10} = 11$.

A current of 11 amperes is required when the resistance is 10 ohms.

Chapter 12
Exponential and Logarithmic Functions

12.1 Check Points

1. $f(x) = 42.2(1.56)^x$

   $f(3) = 42.2(1.56)^3 \approx 160.20876 \approx 160$

   The average amount spent after three hours at a mall is \$160. This overestimates the amount shown in the figure by \$11.

2. $f(x) = 3^x$

x	$f(x) = 3^x$	(x, y)
-3	$3^{-3} = \dfrac{1}{27}$	$\left(-3, \dfrac{1}{27}\right)$
-2	$3^{-2} = \dfrac{1}{9}$	$\left(-2, \dfrac{1}{9}\right)$
-1	$3^{-1} = \dfrac{1}{3}$	$\left(-1, \dfrac{1}{3}\right)$
0	$3^0 = 1$	$(0, 1)$
1	$3^1 = 3$	$(1, 3)$
2	$3^2 = 9$	$(2, 9)$
3	$3^3 = 27$	$(3, 27)$

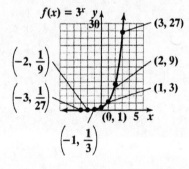

3. $f(x) = \left(\dfrac{1}{3}\right)^x = 3^{-x}$

x	$f(x) = \left(\dfrac{1}{3}\right)^x = 3^{-x}$	(x, y)
-3	$3^{-(-3)} = 27$	$(-3, 27)$
-2	$3^{-(-2)} = 9$	$(-2, 9)$
-1	$3^{-(-1)} = 3$	$(-1, 3)$
0	$3^0 = 1$	$(0, 1)$
1	$3^{-1} = \dfrac{1}{3}$	$\left(1, \dfrac{1}{3}\right)$
2	$3^{-2} = \dfrac{1}{9}$	$\left(2, \dfrac{1}{9}\right)$
3	$3^{-3} = \dfrac{1}{27}$	$\left(3, \dfrac{1}{27}\right)$

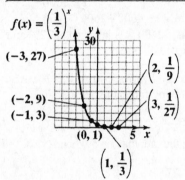

4. $f(x) = 3^x$ and $g(x) = 3^{x-1}$

x	$f(x) = 3^x$	$g(x) = 3^{x-1}$
-2	$3^{-2} = \dfrac{1}{9}$	$3^{-2-1} = 3^{-3} = \dfrac{1}{27}$
-1	$3^{-1} = \dfrac{1}{3}$	$3^{-1-1} = 3^{-2} = \dfrac{1}{9}$
0	$3^0 = 1$	$3^{0-1} = 3^{-1} = \dfrac{1}{3}$
1	$3^1 = 3$	$3^{1-1} = 3^0 = 1$
2	$3^2 = 9$	$3^{2-1} = 3^1 = 3$

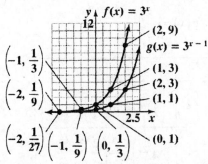

The graph of $g(x) = 3^{x-1}$ is the graph of $f(x) = 3^x$ shifted 1 unit to the right.

5. $f(x) = 2^x$ and $g(x) = 2^x + 3$

x	$f(x) = 2^x$	$g(x) = 2^x + 3$
-2	$2^{-2} = \dfrac{1}{4}$	$2^{-2} + 3 = 3\dfrac{1}{4}$
-1	$2^{-1} = \dfrac{1}{2}$	$2^{-1} + 3 = 3\dfrac{1}{2}$
0	$2^0 = 1$	$2^0 + 3 = 4$
1	$2^1 = 2$	$2^1 + 3 = 5$
2	$2^2 = 4$	$2^2 + 3 = 7$

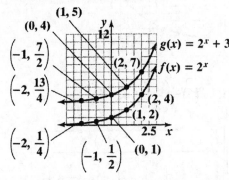

The graph of $g(x) = 2^x + 3$ is the graph of $f(x) = 2^x$ shifted up 3 units.

6. 2012 is 34 years after 1978.

$$f(x) = 1066e^{0.042x}$$

$$f(34) = 1066e^{0.042(34)} \approx 4446$$

In 2012 the gray wolf population of the Western Great Lakes is projected to be about 4446.

7. a. $A = P\left(1 + \dfrac{r}{n}\right)^{nt} = \$10,000\left(1 + \dfrac{0.08}{4}\right)^{4 \cdot 5}$

$\approx \$14,859.47$

b. $A = Pe^{rt} = \$10,000e^{0.08(5)} \approx \$14,918.25$

12.1 Exercise Set

1. $2^{3.4} \approx 10.556$

3. $3^{\sqrt{5}} \approx 11.665$

5. $4^{-1.5} = 0.125$

7. $e^{2.3} \approx 9.974$

9. $e^{-0.95} \approx 0.387$

11. $f(x) = 3^x$

x	$f(x)$
-2	$3^{-2} = \dfrac{1}{3^2} = \dfrac{1}{9}$
-1	$3^{-1} = \dfrac{1}{3^1} = \dfrac{1}{3}$
0	$3^0 = 1$
1	$3^1 = 3$
2	$3^2 = 9$

This function matches graph (d).

13. $f(x) = 3^x - 1$

x	$f(x)$
-2	$3^{-2} - 1 = \dfrac{1}{3^2} - 1 = \dfrac{1}{9} - 1 = -\dfrac{8}{9}$
-1	$3^{-1} - 1 = \dfrac{1}{3^1} - 1 = \dfrac{1}{3} - 1 = -\dfrac{2}{3}$
0	$3^0 - 1 = 1 - 1 = 0$
1	$3^1 - 1 = 3 - 1 = 2$
2	$3^2 - 1 = 9 - 1 = 8$

This function matches graph (e).

15. $f(x) = 3^{-x}$

x	$f(x)$
-2	$3^{-(-2)} = 3^2 = 9$
-1	$3^{-(-1)} = 3^1 = 3$
0	$3^{-(0)} = 3^0 = 1$
1	$3^{-(1)} = 3^{-1} = \dfrac{1}{3}$
2	$3^{-(2)} = 3^{-2} = \dfrac{1}{3^2} = \dfrac{1}{9}$

This function matches graph (f).

17. $f(x) = 4^x$

x	$f(x)$
-2	$4^{-2} = \dfrac{1}{4^2} = \dfrac{1}{16}$
-1	$4^{-1} = \dfrac{1}{4^1} = \dfrac{1}{4}$
0	$4^0 = 1$
1	$4^1 = 4$
2	$4^2 = 16$

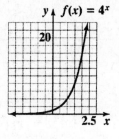

19. $g(x) = \left(\dfrac{3}{2}\right)^x$

x	$g(x)$
-2	$\left(\dfrac{3}{2}\right)^{-2} = \left(\dfrac{2}{3}\right)^2 = \dfrac{4}{9}$
-1	$\left(\dfrac{3}{2}\right)^{-1} = \left(\dfrac{2}{3}\right)^1 = \dfrac{2}{3}$
0	$\left(\dfrac{3}{2}\right)^0 = 1$
1	$\left(\dfrac{3}{2}\right)^1 = \dfrac{3}{2}$
2	$\left(\dfrac{3}{2}\right)^2 = \dfrac{9}{4}$

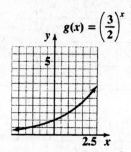

21. $h(x) = \left(\dfrac{1}{2}\right)^x$

x	$h(x)$
-2	$\left(\dfrac{1}{2}\right)^{-2} = \left(\dfrac{2}{1}\right)^2 = \dfrac{4}{1} = 4$
-1	$\left(\dfrac{1}{2}\right)^{-1} = \left(\dfrac{2}{1}\right)^1 = \dfrac{2}{1} = 2$
0	$\left(\dfrac{1}{2}\right)^0 = 1$
1	$\left(\dfrac{1}{2}\right)^1 = \dfrac{1}{2}$
2	$\left(\dfrac{1}{2}\right)^2 = \dfrac{1}{4}$

$$h(x) = \left(\frac{1}{2}\right)^x$$

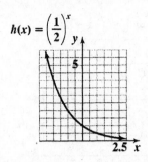

23. $f(x) = (0.6)^x = \left(\frac{6}{10}\right)^x = \left(\frac{3}{5}\right)^x$

x	$f(x)$
-2	$\left(\frac{3}{5}\right)^{-2} = \left(\frac{5}{3}\right)^2 = \frac{25}{9}$
-1	$\left(\frac{3}{5}\right)^{-1} = \left(\frac{5}{3}\right)^1 = \frac{5}{3}$
0	$\left(\frac{3}{5}\right)^0 = 1$
1	$\left(\frac{3}{5}\right)^1 = \frac{3}{5}$
2	$\left(\frac{3}{5}\right)^2 = \frac{9}{25}$

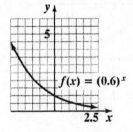

25. $f(x) = 2^x$ and $g(x) = 2^{x+1}$

x	$f(x) = 2^x$	$g(x) = 2^{x+1}$
-2	$\frac{1}{4}$	$\frac{1}{2}$
-1	$\frac{1}{2}$	1
0	1	2
1	2	4
2	4	8

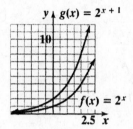

The graph of g is the graph of f shifted 1 unit to the left.

27. $f(x) = 2^x$ and $g(x) = 2^{x-2}$

x	$f(x) = 2^x$	$g(x) = 2^{x-2}$
-2	$\frac{1}{4}$	$\frac{1}{16}$
-1	$\frac{1}{2}$	$\frac{1}{8}$
0	1	$\frac{1}{2}$
1	2	1
2	4	2

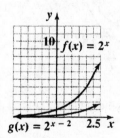

The graph of g is the graph of f shifted 2 units to the right.

29. $f(x) = 2^x$ and $g(x) = 2^x + 1$

x	$f(x) = 2^x$	$g(x) = 2^x + 1$
-2	$\dfrac{1}{4}$	$\dfrac{5}{4}$
-1	$\dfrac{1}{2}$	$\dfrac{3}{2}$
0	1	2
1	2	3
2	4	5

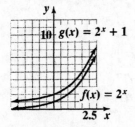

The graph of g is the graph of f shifted up 1 unit.

31. $f(x) = 2^x$ and $g(x) = 2^x - 2$

x	$f(x) = 2^x$	$g(x) = 2^x - 2$
-2	$\dfrac{1}{4}$	$-\dfrac{7}{4}$
-1	$\dfrac{1}{2}$	$-\dfrac{3}{2}$
0	1	-1
1	2	0
2	4	2

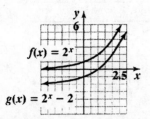

The graph of g is the graph of f shifted down 2 units.

33. $f(x) = 3^x$ and $g(x) = -3^x$

x	$f(x) = 3^x$	$g(x) = -3^x$
-2	$\dfrac{1}{9}$	$-\dfrac{1}{9}$
-1	$\dfrac{1}{3}$	$-\dfrac{1}{3}$
0	1	-1
1	3	-3
2	9	-9

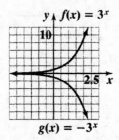

The graph of g is the graph of f reflected across the x–axis.

35. $f(x) = 2^x$ and $g(x) = 2^{x+1} - 1$

x	$f(x) = 2^x$	$g(x) = 2^{x+1} - 1$
-2	$\dfrac{1}{4}$	$-\dfrac{1}{2}$
-1	$\dfrac{1}{2}$	0
0	1	1
1	2	3
2	4	7

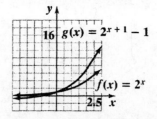

The graph of g is the graph of f shifted 1 unit down and 1 unit to the left.

37. $f(x) = 3^x$ and $g(x) = \frac{1}{3} \cdot 3^x$

x	$f(x) = 3^x$	$g(x) = \frac{1}{3} \cdot 3^x$
-2	$\frac{1}{9}$	$\frac{1}{27}$
-1	$\frac{1}{3}$	$\frac{1}{9}$
0	1	$\frac{1}{3}$
1	3	1
2	9	3

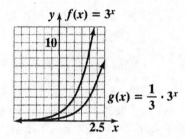

The graph of g is the graph of f compressed vertically by a factor of $\frac{1}{3}$.

39. a. $A = 10,000\left(1 + \frac{0.055}{2}\right)^{2(5)} \approx 13,116.51$

The balance in the account is \$13,116.51 after 5 years of semiannual compounding.

b. $A = 10,000\left(1 + \frac{0.055}{12}\right)^{12(5)} \approx 13,157.04$

The balance in the account is \$13,157.04 after 5 years of monthly compounding.

c. $A = Pe^{rt} = 10,000e^{0.055(5)} \approx 13,165.31$

The balance in the account is \$13,165.31 after 5 years of continuous compounding.

41. Monthly Compounding

$A = 12,000\left(1 + \frac{0.07}{12}\right)^{12(3)} \approx 14,795.11$

Continuous Compounding

$A = 12,000e^{0.0685(3)} \approx 14737.67$

Monthly compounding at 7% yields the greatest return.

43. Domain: $\{x \mid x \text{ is a real number}\}$ or $(-\infty, \infty)$
Range: $\{y \mid y > -2\}$ or $(-2, \infty)$

45. Domain: $\{x \mid x \text{ is a real number}\}$ or $(-\infty, \infty)$
Range: $\{y \mid y > 1\}$ or $(1, \infty)$

47. Domain: $\{x \mid x \text{ is a real number}\}$ or $(-\infty, \infty)$
Range: $\{y \mid y > 0\}$ or $(0, \infty)$

49.

x	$f(x) = 2^x$	$g(x) = 2^{-x}$
-2	$\frac{1}{4}$	4
-1	$\frac{1}{2}$	2
0	1	1
1	2	$\frac{1}{2}$
2	4	$\frac{1}{4}$

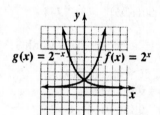

The point of intersection is $(0, 1)$.

51.

x	$y = 2^x$	y	$x = 2^y$
-2	$\frac{1}{4}$	-2	$\frac{1}{4}$
-1	$\frac{1}{2}$	-1	$\frac{1}{2}$
0	1	0	1
1	2	1	2
2	4	2	4

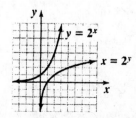

53. a. $f(0) = 574(1.026)^0$

$= 574(1) = 574$

India's population in 1974 was 574 million.

b. $f(27) = 574(1.026)^{27} \approx 1148$

India's population in 2001 will be 1148 million.

c. Since $2028 - 1974 = 54$, find

$f(54) = 574(1.026)^{54} \approx 2295$.

India's population in 2028 will be 2295 million.

d. $2055 - 1974 = 81$, find

$f(54) = 574(1.026)^{81} \approx 4590$.

India's population in 2055 will be 4590 million.

e. India's population appears to be doubling every 27 years.

55. $S = 465,000(1+0.06)^{10}$

$= 465,000(1.06)^{10} \approx 832,744$

In 10 years, the house will be worth \$832,744.

57. a. 2005 is 50 years after 1955.

$f(x) = 0.15x + 1.44$

$f(50) = 0.15(50) + 1.44 \approx 8.9$

According to the linear model, there were about 8.9 million words in the federal tax code in 2005.

b. 2005 is 50 years after 1955.

$g(x) = 1.87e^{0.0344x}$

$g(50) = 1.87e^{0.0344(50)} \approx 10.4$

According to the exponential model, there were about 10.4 million words in the federal tax code in 2005.

c. The linear model is the better model for the data in 2005.

59. a. $f(0) = 80e^{-0.5(0)} + 20$

$= 80e^0 + 20$

$= 80(1) + 20$

$= 80 + 20 = 100$

100% of information is remembered at the moment it is first learned.

b. $f(1) = 80e^{-0.5(1)} + 20$

$= 80e^{-0.5} + 20 \approx 68.522$

About 68.5% of information is remembered after one week.

c. $f(4) = 80e^{-0.5(4)} + 20$

$= 80e^{-2} + 20$

$= 10.827 + 20 = 30.827$

Approximately 30.8% of information is remembered after four weeks.

d. $f(52) = 80e^{-0.5(52)} + 20$

$= 80e^{-26} + 20$

$= (4.087 \times 10^{-10}) + 20$

≈ 20

(Note that 4.087×10^{-10} is eliminated in rounding.)

Approximately 20% of information is remembered after one year.

61. $f(30) = \dfrac{90}{1 + 270e^{-0.122(30)}}$

$= \dfrac{90}{1 + 270e^{-3.66}}$

$= \dfrac{90}{1 + 6.948} = \dfrac{90}{7.948} \approx 11.3$

Approximately 11.3% of 30-year-olds have some coronary heart disease.

63. a. $N(0) = \dfrac{30,000}{1 + 20e^{-1.5(0)}}$

$= \dfrac{30,000}{1 + 20e^0}$

$= \dfrac{30,000}{1 + 20(1)}$

$= \dfrac{30,000}{1 + 20}$

$= \dfrac{30,000}{21} \approx 1428.6$

Approximately 1429 people became ill with the flu when the epidemic began.

b. $N(3) = \dfrac{30,000}{1+20e^{-1.5(3)}}$

$= \dfrac{30,000}{1+20e^{-4.5}} \approx 24,546$

Approximately 24,546 people became ill with the flu by the end of the third week.

c. The epidemic cannot grow indefinitely because there are a limited number of people that can become ill. Because there are 30,000 people in the town, the limit is 30,000.

65. – 67. Answers will vary

69. a. $Q(t) = 10,000\left(1+\dfrac{0.05}{4}\right)^{4t}$

$M(t) = 10,000\left(1+\dfrac{0.045}{12}\right)^{12t}$

b.

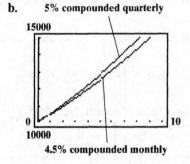

5% compounded quarterly

15000

0   10
10000
4.5% compounded monthly

The bank paying 5% compounded quarterly offers a better return.

71. does not make sense; Explanations will vary. Sample explanation: The horizontal asymptote is $y = 0$.

73. does not make sense; Explanations will vary. Sample explanation: An exponential function would be a better choice.

75. false; Changes to make the statement true will vary. A sample change is: The amount of money will not increase without bound.

77. false; Changes to make the statement true will vary. A sample change is: $f(a+b) = 2^{a+b} = 2^a \cdot 2^b$, not $f(a) + f(b) = 2^a + 2^b$.

79. Graph (a) is $y = \left(\dfrac{1}{3}\right)^x$.

Graph (b) is $y = \left(\dfrac{1}{5}\right)^x$.

Graph (c) is $y = 5^x$.

Graph (d) is $y = 3^x$.

Answers will vary. One possibility follows: A base between 0 and 1 will rise to the left and a base greater than 1 will rise to the right.

81. $D = \dfrac{ab}{a+b}$

$D(a+b) = ab$

$Da + Db = ab$

$Da = ab - Db$

$Da = (a-D)b$

$b = \dfrac{Da}{a-D}$

82. $\dfrac{2x+3}{x^2-7x+12} - \dfrac{2}{x-3} = \dfrac{2x+3}{(x-3)(x-4)} - \dfrac{2}{x-3}$

$= \dfrac{2x+3}{(x-3)(x-4)} - \dfrac{2(x-4)}{(x-3)(x-4)}$

$= \dfrac{2x+3}{(x-3)(x-4)} - \dfrac{2x-8}{(x-3)(x-4)}$

$= \dfrac{2x+3-(2x-8)}{(x-3)(x-4)}$

$= \dfrac{2x+3-2x+8}{(x-3)(x-4)}$

$= \dfrac{11}{(x-3)(x-4)}$

83.
$$x(x-3) = 10$$
$$x^2 - 3x = 10$$
$$x^2 - 3x - 10 = 0$$
$$(x-5)(x+2) = 0$$

Apply the zero product principle.
$$x - 5 = 0 \quad \text{or} \quad x + 2 = 0$$
$$x = 5 \qquad\qquad x = -2$$

The solutions are 5 and –2, and the solution set is $\{-2, 5\}$.

84. There is no method for solving $x = 2^y$ for y.

85. 25 requires a power of $\dfrac{1}{2}$ to obtain 5.
$$25^{\frac{1}{2}} = 5$$

86. $f(x) = 2x - 5$
$$y = 2x - 5$$
Interchange x and y and solve for y.
$$x = 2y - 5$$
$$x + 5 = 2y$$
$$\frac{x+5}{2} = y$$
$$f^{-1}(x) = \frac{x+5}{2}$$

12.2 Check Points

1. a. $\quad 3 = \log_7 x$
$$7^3 = x$$

b. $\quad 2 = \log_b 25$
$$b^2 = 25$$

c. $\log_4 26 = y$
$$4^y = 26$$

2. a. $2^5 = x$
$$5 = \log_2 x$$

b. $b^3 = 27$
$$3 = \log_b 27$$

c. $e^y = 33$
$$y = \log_e 33$$

3. a. $\log_{10} 100 = 2$ because $10^2 = 100$.

b. $\log_3 3 = 1$ because $3^1 = 3$.

c. $\log_{36} 6 = \dfrac{1}{2}$ because $36^{\frac{1}{2}} = \sqrt{36} = 6$.

4. a. Because $\log_b b = 1$, we conclude $\log_9 9 = 1$.

b. Because $\log_b 1 = 0$, we conclude $\log_8 1 = 0$.

5. a. Because $\log_b b^x = x$, we conclude $\log_7 7^8 = 8$.

b. Because $b^{\log_b x} = x$, we conclude $3^{\log_3 17} = 17$.

6. Set up a table of coordinates for $f(x) = 3^x$.

x	-2	-1	0	1	2	3
$f(x) = 3^x$	$\frac{1}{9}$	$\frac{1}{3}$	1	3	9	27

Reverse these coordinates to obtain the coordinates of $g(x) = \log_3 x$.

x	$\frac{1}{9}$	$\frac{1}{3}$	1	3	9	27
$g(x) = \log_3 x$	-2	-1	0	1	2	3

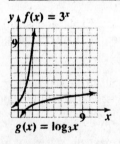

7. $x - 5 > 0$
$$x > 5$$
The domain of h is $(5, \infty)$ or $\{x \mid x > 5\}$.

8. $f(x) = 29 + 48.8\log(x+1)$

$f(10) = 29 + 48.8\log(10+1)$

$\quad = 29 + 48.8\log 11$

$\quad \approx 80$

A 10-year-old boy has attained approximately 80% of his adult height.

9. $R = \log\dfrac{I}{I_0}$

$R = \log\dfrac{10,000 I_0}{I_0}$

$\quad = \log 10,000$

$\quad = \log 10^4$

$\quad = 4$

The magnitude on the Richter Scale is 4.

10. a. The domain of f consists of all x for which $4 - x > 0$.

$4 - x > 0$

$-x > -4$

$x < 4$

The domain of f is $(-\infty, 4)$ or $\{x | x < 4\}$.

b. The domain of g consists of all x for which $x^2 > 0$. It follows that the domain is all real numbers except 0.

The domain of g is $(-\infty, 0) \cup (0, \infty)$ or $\{x | x \neq 0\}$.

11. $f(x) = 13.4\ln x - 11.6$

$f(30) = 13.4\ln 30 - 11.6$

$\quad \approx 34$

The temperature increase after 30 minutes will be 34°. The function models the actual increase shown in the figure extremely well.

12.2 Exercise Set

1. $4 = \log_2 16$

$2^4 = 16$

3. $2 = \log_3 x$

$3^2 = x$

5. $5 = \log_b 32$

$b^5 = 32$

7. $\log_6 216 = y$

$6^y = 216$

9. $2^3 = 8$

$\log_2 8 = 3$

11. $2^{-4} = \dfrac{1}{16}$

$\log_2 \dfrac{1}{16} = -4$

13. $\sqrt[3]{8} = 2$

$8^{\frac{1}{3}} = 2$

$\log_8 2 = \dfrac{1}{3}$

15. $13^2 = x$

$\log_{13} x = 2$

17. $b^3 = 1000$

$\log_b 1000 = 3$

19. $7^y = 200$

$\log_7 200 = y$

21. $\log_4 16 = y$

$4^y = 16$

$4^y = 4^2$

$y = 2$

23. $\log_2 64 = y$

$2^y = 64$

$2^y = 2^6$

$y = 6$

25. $\log_5 \dfrac{1}{5} = y$

$5^y = \dfrac{1}{5}$

$5^y = 5^{-1}$

$y = -1$

27. $\log_2 \dfrac{1}{8} = y$

$\quad 2^y = \dfrac{1}{8}$

$\quad 2^y = \dfrac{1}{2^3}$

$\quad 2^y = 2^{-3}$

$\quad\ y = -3$

29. $\log_7 \sqrt{7} = y$

$\quad 7^y = \sqrt{7}$

$\quad 7^y = 7^{\frac{1}{2}}$

$\quad\ y = \dfrac{1}{2}$

31. $\log_2 \dfrac{1}{\sqrt{2}} = y$

$\quad 2^y = \dfrac{1}{\sqrt{2}}$

$\quad 2^y = \dfrac{1}{2^{\frac{1}{2}}}$

$\quad 2^y = 2^{-\frac{1}{2}}$

$\quad\ y = -\dfrac{1}{2}$

33. $\log_{64} 8 = y$

$\quad 64^y = 8$

$\quad 64^y = 64^{\frac{1}{2}}$

$\quad\ y = \dfrac{1}{2}$

35. $\log_5 5 = y$

$\quad 5^y = 5^1$

$\quad\ y = 1$

37. $\log_4 1 = y$

$\quad 4^y = 1$

$\quad 4^y = 4^0$

$\quad\ y = 0$

39. $\log_5 5^7 = y$

$\quad 5^y = 5^7$

$\quad\ y = 7$

41. Since $b^{\log_b x} = x$, $\ 8^{\log_8 19} = 19$.

43. $f(x) = 4^x$

$\quad g(x) = \log_4 x$

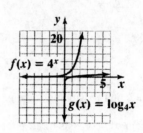

45. $f(x) = \left(\dfrac{1}{2}\right)^x$

$\quad g(x) = \log_{\frac{1}{2}} x$

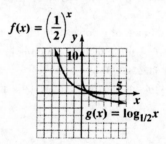

47. $f(x) = \log_5 (x+4)$

$\quad x + 4 > 0$

$\quad\quad\ x > -4$

The domain of f is $\{x \,|\, x > -4\}$ or $(-4, \infty)$.

49. $f(x) = \log_5 (2-x)$

$\quad 2 - x > 0$

$\quad\ -x > -2$

$\quad\quad\ x < 2$

The domain of f is $\{x \,|\, x < 2\}$ or $(-\infty, 2)$.

51. $f(x) = \ln(x-2)^2$

The domain of g is all real numbers for which $(x-2)^2 > 0$. The only number that must be excluded is 2. The domain of f is $\{x \mid x \neq 2\}$ or $(-\infty, 2) \cup (2, \infty)$.

53. $\log 100 = y$

$\quad 10^y = 100$

$\quad 10^y = 10^2$

$\quad\quad y = 2$

55. $\log 10^7 = y$

$\quad 10^y = 10^7$

$\quad\quad y = 7$

57. Since $10^{\log x} = x$, $10^{\log 33} = 33$.

59. $\ln 1 = y$

$\quad e^y = 1$

$\quad e^y = e^0$

$\quad\; y = 0$

61. Since $\ln e^x = x$, $\ln e^6 = 6$.

63. $\ln \dfrac{1}{e^6} = \ln e^{-6}$

Since $\ln e^x = x$, $\ln e^{-6} = -6$.

65. Since $e^{\ln x} = x$, $e^{\ln 125} = 125$.

67. Since $\ln e^x = x$, $\ln e^{9x} = 9x$.

69. Since $e^{\ln x} = x$, $e^{\ln 5x^2} = 5x^2$.

71. Since $10^{\log x} = x$, $10^{\log \sqrt{x}} = \sqrt{x}$.

73. $\log_3(x-1) = 2$

$\quad 3^2 = x - 1$

$\quad\; 9 = x - 1$

$\quad 10 = x$

The solution set is $\{10\}$.

75. $\log_4 x = -3$

$\quad 4^{-3} = x$

$\quad x = \dfrac{1}{4^3} = \dfrac{1}{64}$

The solution set is $\left\{ \dfrac{1}{64} \right\}$.

77. $\log_3(\log_7 7) = \log_3 1 = 0$

79. $\log_2(\log_3 81) = \log_2\left(\log_3 3^4\right)$

$\quad\quad\quad\quad\quad\quad = \log_2 4 = \log_2 2^2 = 2$

81. **(d)** The graph is similar to that of $y = \ln x$, but shifted left 2 units.

83. **(c)** The graph is similar to that of $y = \ln x$, but shifted up 2 units.

85. **(b)** The graph is similar to that of $y = \ln x$, but reflected across the y-axis and then shifted right 1 unit.

87. $f(13) = 62 + 35 \log(13 - 4)$

$\quad\quad\quad = 62 + 35 \log(9) \approx 95.4$

A 13-year-old girl is approximately 95.4% of her adult height.

89. a. 2004 is 35 years after 1969.

$\quad f(x) = -7.49 \ln x + 53$

$\quad f(35) = -7.49 \ln 35 + 53 \approx 26.4$

According to the function, 26.4% of first-year college men expressed antifeminist views in 2004. This underestimates the value in the graph by 1%.

b. 2010 is 41 years after 1969.

$\quad f(x) = -7.49 \ln x + 53$

$\quad f(41) = -7.49 \ln 41 + 53 \approx 25.2$

According to the function, 25.2% of first-year college men will express antifeminist views in 2010.

91. $D = 10\log\left(10^{12}\left(6.3 \times 10^{6}\right)\right)$

$\quad = 10\log\left(6.3 \times 10^{18}\right) \approx 188.0$

The decibel level of a blue whale is approximately 188 decibels. At close range, the sound could rupture the human ear drum.

93. a. The original exam was at time, $t = 0$.

$\quad f(0) = 88 - 15\ln(0+1)$

$\quad\quad = 88 - 15\ln(1) \approx 88$

The average score on the original exam was 88.

b. $f(2) = 88 - 15\ln(2+1)$

$\quad\quad = 88 - 15\ln(3) \approx 71.5$

$\quad f(4) = 88 - 15\ln(4+1)$

$\quad\quad = 88 - 15\ln(5) \approx 63.9$

$\quad f(6) = 88 - 15\ln(6+1)$

$\quad\quad = 88 - 15\ln(7) \approx 58.8$

$\quad f(8) = 88 - 15\ln(8+1)$

$\quad\quad = 88 - 15\ln(9) \approx 55.0$

$\quad f(10) = 88 - 15\ln(10+1)$

$\quad\quad = 88 - 15\ln(11) \approx 52.0$

$\quad f(12) = 88 - 15\ln(12+1)$

$\quad\quad = 88 - 15\ln(13) \approx 49.5$

The average score for the tests is as follows:

2 months: 71.5
4 months: 63.9
6 months: 58.8
8 months: 55.0
10 months: 52.0
12 months: 49.5

c.

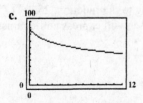

The students remembered less of the material over time.

95. – 101. Answers will vary.

103. $f(x) = \ln x \qquad g(x) = \ln x + 3$

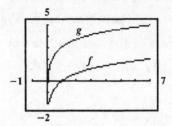

The graph of g is the graph of f shifted up 3 units.

105. $f(x) = \log x \qquad g(x) = \log(x-2) + 1$

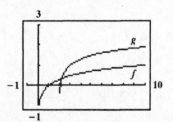

The graph of g is the graph of f shifted 2 units to the right and 1 unit up.

107. a. $f(x) = \ln(3x)$

$\quad\quad g(x) = \ln 3 + \ln x$

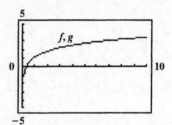

The graphs coincide.

b. $f(x) = \log\left(5x^2\right)$

$\quad\quad g(x) = \log 5 + \log x^2$

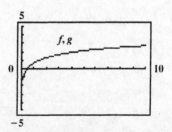

The graphs coincide.

c. $f(x) = \ln\left(2x^3\right)$

$g(x) = \ln 2 + \ln x^3$

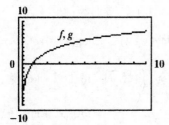

The graphs coincide.

d. In each case, the function, f, is equivalent to g. This means that $\log_b\left(MN\right) = \log_b M + \log_b N$.

e. The logarithm of a product is equal to <u>the sum of the logarithms of the factors</u>.

109. makes sense

111. makes sense

113. false; Changes to make the statement true will vary. A sample change is: $\dfrac{\log_2 8}{\log_2 4} = \dfrac{3}{2}$

115. false; Changes to make the statement true will vary. A sample change is: The domain of $f(x) = \log_2 x$ is $(0, \infty)$.

117. To evaluate $\dfrac{\log_3 81 - \log_\pi 1}{\log_{2\sqrt{2}} 8 - \log 0.001}$, consider each of the terms independently.

$\log_3 81 = y$ $\qquad$ $\log_\pi 1 = y$

$\quad 3^y = 81$ $\qquad\qquad$ $\pi^y = 1$

$\quad 3^y = 3^4$ $\qquad\qquad$ $\pi^y = \pi^0$

$\quad y = 4$ $\qquad\qquad\quad$ $y = 0$

$\log_{2\sqrt{2}} 8 = y$ $\qquad\qquad$ $\log 0.001 = y$

$\left(2\sqrt{2}\right)^y = 8$ $\qquad\qquad$ $10^y = 0.001$

$\left(2^1 2^{\frac{1}{2}}\right)^y = 2^3$ $\qquad\qquad$ $10^y = 10^{-3}$

$\left(2^{\frac{3}{2}}\right)^y = 2^3$ $\qquad\qquad\quad$ $y = -3$

$(2)^{\frac{3}{2}y} = 2^3$

$\dfrac{3}{2} y = 3$

$y = 3 \cdot \dfrac{2}{3} = 2$

$\dfrac{\log_3 81 - \log_\pi 1}{\log_{2\sqrt{2}} 8 - \log 0.001} = \dfrac{4-0}{2-(-3)} = \dfrac{4}{2+3} = \dfrac{4}{5}$

119. To determine which expression represents a greater number, rewrite the expressions in exponential notation.

$\log_4 60 = x$ $\qquad$ $\log_3 40 = y$

$\quad 4^x = 60$ $\qquad\qquad$ $3^y = 40$

First consider $4^x = 60$. We know that $4^2 = 16$ and $4^3 = 64$, so x falls between 2 and 3 and is much closer to 3. Next consider $3^y = 40$. We know that $3^3 = 27$. This means that y is greater than 3, and therefore greater than x. This means that $\log_3 40$ represents the greater number.

120. Rewrite the equations in $Ax + By = C$ form.

$2x + 5y = \quad 11$

$3x - 2y = -12$

Multiply the first equation by 2 and the second equation by 5 and solve by addition.

$\quad 4x + 10y = \quad 22$

$\underline{\quad 15x - 10y = -60\quad}$

$\qquad 19x = -38$

$\qquad\quad x = -2$

Back-substitute -2 for x to find y.

$2(-2) + 5y = 11$

$\quad -4 + 5y = 11$

$\qquad\quad 5y = 15$

$\qquad\quad\, y = 3$

The solution is $(-2, 3)$ and the solution set is $\{(-2, 3)\}$.

121. $6x^2 - 8xy + 2y^2 = 2(3x^2 - 4xy + y^2)$
$$= 2(3x - y)(x - y)$$

122. $x + 3 \leq -4$ or $2 - 7x \leq 16$
$\qquad x \leq -7 \qquad\qquad -7x \leq 14$
$$x \geq -2$$

The solution set is $\{x \mid x \leq -7 \text{ or } x \geq -2\}$ or $(-\infty, -7] \cup [-2, \infty)$.

123. a. $\log_2 32 = \log_2 2^5 = 5$

b. $\log_2 8 + \log_2 4 = \log_2 2^3 + \log_2 2^2 = 3 + 2 = 5$

c. $\log_2(8 \cdot 4) = \log_2 8 + \log_2 4$

124. a. $\log_2 16 = \log_2 2^4 = 4$

b. $\log_2 32 - \log_2 2 = \log_2 2^5 - \log_2 2 = 5 - 1 = 4$

c. $\log_2\left(\dfrac{32}{2}\right) = \log_2 32 - \log_2 2$

125. a. $\log_3 81 = \log_3 3^4 = 4$

b. $2\log_3 9 = 2\log_3 3^2 = 2 \cdot 2 = 4$

c. $\log_3 9^2 = 2\log_3 9$

12.3 Check Points

1. a. $\log_6(7 \cdot 11) = \log_6 7 + \log_6 11$

b. $\log(100x) = \log 100 + \log x$
$$= \log 10^2 + \log x$$
$$= 2 + \log x$$

2. a. $\log_8\left(\dfrac{23}{x}\right) = \log_8 23 - \log_8 x$

b. $\ln\left(\dfrac{e^5}{11}\right) = \ln e^5 - \ln 11$
$$= 5 - \ln 11$$

3. a. $\log_6 8^9 = 9\log_6 8$

b. $\ln \sqrt[3]{x} = \ln x^{\frac{1}{3}} = \dfrac{1}{3}\ln x$

c. $\log(x+4)^2 = 2\log(x+4)$

4. a. $\log_b\left(x^4 \sqrt[3]{y}\right) = \log_b x^4 + \log_b \sqrt[3]{y}$
$$= \log_b x^4 + \log_b y^{\frac{1}{3}}$$
$$= 4\log_b x + \frac{1}{3}\log_b y$$

b. $\log_5\left(\dfrac{\sqrt{x}}{25y^3}\right) = \log_5\left(\dfrac{x^{\frac{1}{2}}}{25y^3}\right)$
$$= \log_5 x^{\frac{1}{2}} - \log_5\left(25y^3\right)$$
$$= \log_5 x^{\frac{1}{2}} - \left(\log_5 25 + \log_5 y^3\right)$$
$$= \log_5 x^{\frac{1}{2}} - \log_5 25 - \log_5 y^3$$
$$= \frac{1}{2}\log_5 x - 2 - 3\log_5 y$$

5. a. $\log 25 + \log 4 = \log(25 \cdot 4)$
$$= \log 100$$
$$= 2$$

b. $\log(7x+6) - \log x = \log\left(\dfrac{7x+6}{x}\right)$

6. a. $2\ln x + \dfrac{1}{3}\ln(x+5) = \ln x^2 + \ln(x+5)^{\frac{1}{3}}$
$$= \ln x^2 + \ln \sqrt[3]{x+5}$$
$$= \ln\left(x^2 \sqrt[3]{x+5}\right)$$

b. $2\log(x-3) - \log x = \log(x-3)^2 - \log x$
$$= \log \frac{(x-3)^2}{x}$$

c. $\frac{1}{4}\log_b x - 2\log_b 5 - 10\log_b y = \log_b x^{\frac{1}{4}} - \log_b 5^2 - \log_b y^{10}$

$$= \log_b x^{\frac{1}{4}} - \left(\log_b 5^2 + \log_b y^{10}\right)$$

$$= \log_b \sqrt[4]{x} - \left(\log_b 25 + \log_b y^{10}\right)$$

$$= \log_b \sqrt[4]{x} - \left(\log_b 25 y^{10}\right)$$

$$= \log_b \left(\frac{\sqrt[4]{x}}{25 y^{10}}\right)$$

7. $\log_7 2506 = \dfrac{\log 2506}{\log 7} \approx 4.02$

8. $\log_7 2506 = \dfrac{\ln 2506}{\ln 7} \approx 4.02$

12.3 Exercise Set

1. $\log_5 (7 \cdot 3) = \log_5 7 + \log_5 3$

3. $\log_7 (7x) = \log_7 7 + \log_7 x = 1 + \log_7 x$

5. $\log(1000x) = \log 1000 + \log x$
    $$= 3 + \log x$$

7. $\log_7 \left(\dfrac{7}{x}\right) = \log_7 7 - \log_7 x = 1 - \log_7 x$

9. $\log\left(\dfrac{x}{100}\right) = \log x - \log 100 = \log x - 2$

11. $\log_4 \left(\dfrac{64}{y}\right) = \log_4 64 - \log_4 y$
     $$= 3 - \log_4 y$$

13. $\ln\left(\dfrac{e^2}{5}\right) = \ln e^2 - \ln 5 = 2 - \ln 5$

15. $\log_b x^3 = 3\log_b x$

17. $\log N^{-6} = -6\log N$

19. $\ln \sqrt[5]{x} = \ln x^{\frac{1}{5}} = \dfrac{1}{5}\ln x$

21. $\log_b x^2 y = \log_b x^2 + \log_b y$

$\qquad\qquad = 2\log_b x + \log_b y$

23. $\log_4 \left(\dfrac{\sqrt{x}}{64} \right) = \log_4 \sqrt{x} - \log_4 64$

$\qquad\qquad\qquad = \log_4 x^{\frac{1}{2}} - 3$

$\qquad\qquad\qquad = \dfrac{1}{2}\log_4 x - 3$

25. $\log_6 \left(\dfrac{36}{\sqrt{x+1}} \right) = \log_6 36 - \log_6 \sqrt{x+1}$

$\qquad\qquad\qquad = 2 - \log_6 (x+1)^{\frac{1}{2}}$

$\qquad\qquad\qquad = 2 - \dfrac{1}{2}\log_6 (x+1)$

27. $\log_b \left(\dfrac{x^2 y}{z^2} \right) = \log_b x^2 y - \log_b z^2$

$\qquad\qquad\qquad = \log_b x^2 + \log_b y - 2\log_b z$

$\qquad\qquad\qquad = 2\log_b x + \log_b y - 2\log_b z$

29. $\log \sqrt{100x} = \log (100x)^{\frac{1}{2}}$

$\qquad\qquad = \dfrac{1}{2}\log (100x)$

$\qquad\qquad = \dfrac{1}{2}\left(\log 100 + \log x \right)$

$\qquad\qquad = \dfrac{1}{2}\left(2 + \log x \right)$

$\qquad\qquad = 1 + \dfrac{1}{2}\log x$

31. $\log \sqrt[3]{\dfrac{x}{y}} = \log \left(\dfrac{x}{y} \right)^{\frac{1}{3}}$

$\qquad\qquad = \dfrac{1}{3}\log \left(\dfrac{x}{y} \right)$

$\qquad\qquad = \dfrac{1}{3}\left(\log x - \log y \right)$

$\qquad\qquad = \dfrac{1}{3}\log x - \dfrac{1}{3}\log y$

33. $\log_b\left(\dfrac{\sqrt{x}\,y^3}{z^3}\right)$

$= \log_b\left(\dfrac{x^{\frac{1}{2}}y^3}{z^3}\right)$

$= \log_b\left(x^{\frac{1}{2}}y^3\right) - \log_b z^3$

$= \log_b x^{\frac{1}{2}} + \log_b y^3 - \log_b z^3$

$= \dfrac{1}{2}\log_b x + 3\log_b y - 3\log_b z$

35. $\log_5\sqrt[3]{\dfrac{x^2 y}{25}}$

$= \log_5\left(\dfrac{x^2 y}{25}\right)^{\frac{1}{3}}$

$= \dfrac{1}{3}\log_5\left(\dfrac{x^2 y}{25}\right)$

$= \dfrac{1}{3}\left(\log_5\left[x^2 y\right] - \log_5 25\right)$

$= \dfrac{1}{3}\left(\log_5 x^2 + \log_5 y - \log_5 5^2\right)$

$= \dfrac{1}{3}\left(2\log_5 x + \log_5 y - 2\right)$

$= \dfrac{2}{3}\log_5 x + \dfrac{1}{3}\log_5 y - \dfrac{2}{3}$

37. $\log 5 + \log 2 = \log(5\cdot 2) = \log 10 = 1$

39. $\ln x + \ln 7 = \ln(x\cdot 7) = \ln(7x)$

41. $\log_2 96 - \log_2 3 = \log_2\dfrac{96}{3} = \log_2 32 = 5$

43. $\log(2x+5) - \log x = \log\left(\dfrac{2x+5}{x}\right)$

45. $\log x + 3\log y = \log x + \log y^3$
$ = \log\left(xy^3\right)$

47. $\frac{1}{2}\ln x + \ln y = \ln x^{\frac{1}{2}} + \ln y$

$$= \ln\left(x^{\frac{1}{2}} y\right) = \ln\left(y\sqrt{x}\right)$$

49. $2\log_b x + 3\log_b y = \log_b x^2 + \log_b y^3$

$$= \log_b\left(x^2 y^3\right)$$

51. $5\ln x - 2\ln y = \ln x^5 - \ln y^2$

$$= \ln\left(\frac{x^5}{y^2}\right)$$

53. $3\ln x - \frac{1}{3}\ln y = \ln x^3 - \ln y^{\frac{1}{3}}$

$$= \ln\left(\frac{x^3}{y^{\frac{1}{3}}}\right) = \ln\left(\frac{x^3}{\sqrt[3]{y}}\right)$$

55. $4\ln(x+6) - 3\ln x = \ln(x+6)^4 - \ln x^3$

$$= \ln\left[\frac{(x+6)^4}{x^3}\right]$$

57. $3\ln x + 5\ln y - 6\ln z$

$= \ln x^3 + \ln y^5 - \ln z^6$

$= \ln\left(x^3 y^5\right) - \ln z^6 = \ln\left(\frac{x^3 y^5}{z^6}\right)$

59. $\frac{1}{2}\left(\log_5 x + \log_5 y\right) - 2\log_5(x+1)$

$= \frac{1}{2}\log_5(xy) - \log_5(x+1)^2$

$= \log_5(xy)^{\frac{1}{2}} - \log_5(x+1)^2$

$= \log_5\sqrt{xy} - \log_5(x+1)^2$

$= \log_5\left[\frac{\sqrt{xy}}{(x+1)^2}\right]$

61. $\log_5 13 = \frac{\log 13}{\log 5} \approx 1.5937$

63. $\log_{14} 87.5 = \dfrac{\log 87.5}{\log 14} \approx 1.6944$

65. $\log_{0.1} 17 = \dfrac{\log 17}{\log 0.1} \approx -1.2304$

67. $\log_{\pi} 63 = \dfrac{\log 63}{\log \pi} \approx 3.6193$

69. $\log_b \dfrac{3}{2} = \log_b 3 - \log_b 2 = C - A$

71. $\log_b 8 = \log_b 2^3 = 3\log_b 2 = 3A$

73. $\log_b \sqrt{\dfrac{2}{27}} = \log_b \left(\dfrac{2}{27}\right)^{\frac{1}{2}}$

$= \dfrac{1}{2}\log_b \left(\dfrac{2}{3^3}\right)$

$= \dfrac{1}{2}\left(\log_b 2 - \log_b 3^3\right)$

$= \dfrac{1}{2}\left(\log_b 2 - 3\log_b 3\right)$

$= \dfrac{1}{2}\log_b 2 - \dfrac{3}{2}\log_b 3$

$= \dfrac{1}{2}A - \dfrac{3}{2}C$

75. false; Changes to make the statement true will vary. A sample change is: $\ln e = 1$.

77. false; Changes to make the statement true will vary. A sample change is: $\log_4 (2x)^3 = 3\log_4 (2x)$.

79. true

81. true

83. false; Changes to make the statement true will vary. A sample change is: $\log(x+3) - \log(2x) = \log\left(\dfrac{x+3}{2x}\right)$.

85. true

87. true

89. a. $\log_3 9 = 2$

    b. $\log_3 x + 4\log_3 y - 2 = \log_3 x + 4\log_3 y - \log_3 9$

$= \log_3 x + \log_3 y^4 - \log_3 9$

$= \log_3 \dfrac{xy^4}{9}$

91. a. $\log_{25} 5 = \frac{1}{2}$

b. $\log_{25} x + \log_{25}(x^2 - 1) - \log_{25}(x+1) - \frac{1}{2} = \log_{25} x + \log_{25}(x^2 - 1) - \log_{25}(x+1) - \log_{25} 5$

$$= \log_{25} \frac{x(x^2 - 1)}{5(x+1)}$$

$$= \log_{25} \frac{x(x-1)(x+1)}{5(x+1)}$$

$$= \log_{25} \frac{x(x-1)}{5}$$

93. a. $D = 10(\log I - \log I_0)$

$$= 10 \log \frac{I}{I_0}$$

b. $D = 10 \log \frac{I}{I_0}$

$$= 10 \log \frac{100 I_0}{I_0}$$

$$= 10 \log 100$$

$$= 10 \cdot 2$$

$$= 20$$

A sound that has an intensity 100 times the intensity of a softer sound, it is 20 decibels louder than the softer sound.

95. – 101. Answers will vary.

103. a. $y = \log_3 x = \frac{\log x}{\log 3}$

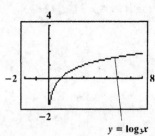

$y = \log_3 x$

b. $y = 2 + \log_3 x$

$y = \log_3 (x + 2)$

$y = -\log_3 x$

$y = \log_3 x$

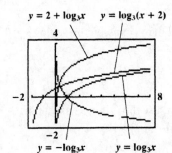

The graph of $y = 2 + \log_3 x$ is the graph of $y = \log_3 x$ shifted up two units.

The graph of $y = \log_3 (x + 2)$ is the graph of $y = \log_3 x$ shifted 2 units to the left.

The graph of $y = -\log_3 x$ is the graph of $y = \log_3 x$ reflected about the x–axis.

105. $y = \log_3 x$

$y = \log_{25} x$

$y = \log_{100} x$

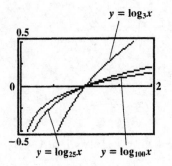

a. $y = \log_{100} x$ is on top. $y = \log_3 x$ is on the bottom.

b. $y = \log_3 x$ is on top. $y = \log_{100} x$ is on the bottom.

c. If $y = \log_b x$ is graphed for two different values of b, the graph of the one with the larger base will be on top in the interval $(0, 1)$ and the one with the smaller base will be on top in the interval $(1, \infty)$. Likewise, if $y = \log_b x$ is graphed for two different values of b, the graph of the one with the smaller base will be on the bottom in the interval $(0, 1)$ and the one with the larger base will be on the bottom in the interval $(1, \infty)$.

107. Answers will vary. One example follows. To disprove the statement $\log \frac{x}{y} = \frac{\log x}{\log y}$, let $y = 3$.

Graph $y = \log \frac{x}{3}$ and $y = \frac{\log x}{\log 3}$.

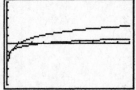

The graphs do not coincide, so the expressions are not equivalent.

109. Answers will vary. One example follows. To disprove the statement $\ln(xy) = (\ln x)(\ln y)$, let $y = 3$.
Graph $y = \ln(x \cdot 3)$ and $y = (\ln x)(\ln 3)$.

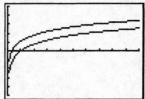

The graphs do not coincide, so the expressions are not equivalent.

111. makes sense

113. makes sense

115. true

117. false; Changes to make the statement true will vary. A sample change is: $\log_b\left(x^3 + y^3\right)$ cannot be simplified. If we were taking the logarithm of a product and not a sum, we would have been able to simplify as follows.

$$\log_b\left(x^3 y^3\right) = \log_b x^3 + \log_b y^3$$
$$= 3\log_b x + 3\log_b y$$

119. Recall that when a logarithm is written without a base, the base is 10.

$$\log e = \frac{\ln e}{\ln 10} = \frac{1}{\ln 10}$$

121. $e^{\ln 8x^5 - \ln 2x^2} = \frac{e^{\ln 8x^5}}{e^{\ln 2x^2}} = \frac{8x^5}{2x^2} = 4x^3$

122. $5x - 2y > 10$
First, find the intercepts to the equation $5x - 2y = 10$.
Find the x–intercept by setting $y = 0$.
$$5x - 2(0) = 10$$
$$5x = 10$$
$$x = 2$$
Find the y–intercept by setting $x = 0$.
$$5(0) - 2y = 10$$
$$-2y = 10$$
$$y = -5$$
Next, use the origin as a test point.
$$5(0) - 2(0) > 10$$
$$0 - 0 > 10$$
$$0 > 10$$
This is a false statement. This means that the origin will not fall in the shaded half-plane.

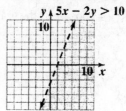

123. $x - 2(3x - 2) > 2x - 3$
$$x - 6x + 4 > 2x - 3$$
$$-5x + 4 > 2x - 3$$
$$-7x + 4 > -3$$
$$-7x > -7$$
$$x < 1$$
The solution set is $\{x | x < 1\}$ or $(-\infty, 1)$.

124. $\dfrac{\sqrt[3]{40x^2 y^6}}{\sqrt[3]{5xy}} = \sqrt[3]{\dfrac{40x^2 y^6}{5xy}}$
$$= \sqrt[3]{8xy^5} = \sqrt[3]{8xy^3 y^2} = 2y\sqrt[3]{xy^2}$$

125. $16^{\frac{3}{2}} = \left(\sqrt{16}\right)^3 = 4^3 = 64$

126. $3\ln(2x) = 3\ln\left(2 \cdot \dfrac{e^4}{2}\right)$
$$= 3\ln e^4$$
$$= 3 \cdot 4$$
$$= 12$$

127.
$$\frac{x+2}{4x+3} = \frac{1}{x}$$

$$x(4x+3)\left(\frac{x+2}{4x+3}\right) = x(4x+3)\left(\frac{1}{x}\right)$$

$$x(x+2) = 4x+3$$

$$x^2 + 2x = 4x+3$$

$$x^2 - 2x - 3 = 0$$

$$(x+1)(x-3) = 0$$

$$x+1 = 0 \quad \text{or} \quad x-3 = 0$$

$$x = -1 \qquad\qquad x = 3$$

The solution set is $\{-1, 3\}$.

Chapter 12 Mid-Chapter Check Point

1.

x	$f(x) = 2^x - 3$
-2	$-\frac{11}{4} = -2.75$
-1	$-\frac{5}{2} = -2.5$
0	-3
1	-1
2	1
3	5

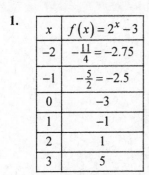

Domain: $\{x \mid x \text{ is a real number}\}$ or $(-\infty, \infty)$;

Range: $\{y \mid y > -3\}$ or $(-3, \infty)$.

2.

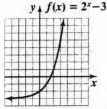

x	$f(x) = \left(\frac{1}{3}\right)^x$
-2	9
-1	3
0	1
1	$\frac{1}{3}$
2	$\frac{1}{9}$

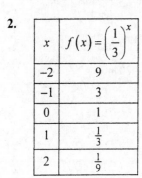

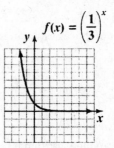

Domain: $\{x \mid x \text{ is a real number}\}$ or $(-\infty, \infty)$;

Range: $\{y \mid y > 0\}$ or $(0, \infty)$.

3.

x	$f(x) = \log_2 x$
$\frac{1}{4}$	-2
$\frac{1}{2}$	-1
1	0
2	1
4	2

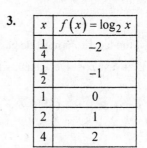

Domain: $\{x \mid x > 0\}$ or $(0, \infty)$;

Range: $\{y \mid y \text{ is a real number}\}$ or $(-\infty, \infty)$.

4.

x	$f(x) = \log_2 x + 1$
$\frac{1}{4}$	-1
$\frac{1}{2}$	0
1	1
2	2
4	3

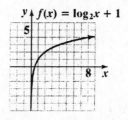

Domain: $\{x \mid x > 0\}$ or $(0, \infty)$;

Range: $\{y \mid y \text{ is a real number}\}$ or $(-\infty, \infty)$.

5. $f(x) = \log_3(x+6)$

The argument of the logarithm must be positive:
$x + 6 > 0$

$x > -6$

Domain: $\{x \mid x > -6\}$ or $(-6, \infty)$.

6. $f(x) = \log_3 x + 6$

The argument of the logarithm must be positive:
$x > 0$

Domain: $\{x \mid x > 0\}$ or $(0, \infty)$.

7. $\log_3(x+6)^2$

The argument of the logarithm must be positive.

Now $(x+6)^2$ is always positive, except when $x = -6$.

Domain: $\{x \mid x \neq -6\}$ or $(-\infty - 6) \cup (-6, \infty)$.

8. $f(x) = 3^{x+6}$

Domain: $\{x \mid x \text{ is a real number}\}$ or $(-\infty, \infty)$.

9. $\log_2 8 + \log_5 25 = \log_2 2^3 + \log_5 5^2$

$= 3 + 2 = 5$

10. $\log_3 \dfrac{1}{9} = \log_3 \dfrac{1}{3^2} = \log_3 3^{-2} = -2$

11. Let $\log_{100} 10 = y$.

$100^y = 10$

$\left(10^2\right)^y = 10^1$

$10^{2y} = 10^1$

$2y = 1$

$y = \dfrac{1}{2}$

12. $\log \sqrt[3]{10} = \log 10^{\frac{1}{3}} = \dfrac{1}{3}$

13. $\log_2(\log_3 81) = \log_2\left(\log_3 3^4\right)$

$= \log_2 4 = \log_2 2^2 = 2$

14. $\log_3\left(\log_2 \dfrac{1}{8}\right) = \log_3\left(\log_2 \dfrac{1}{2^3}\right)$

$= \log_3\left(\log_2 2^{-3}\right)$

$= \log_3(-3)$

$= \text{not possible}$

This expression is impossible to evaluate.

15. $6^{\log_6 5} = 5$

16. $\ln e^{\sqrt{7}} = \sqrt{7}$

17. $10^{\log 13} = 13$

18. $\log_{100} 0.1 = y$

$100^y = 0.1$

$\left(10^2\right)^y = \dfrac{1}{10}$

$10^{2y} = 10^{-1}$

$2y = -1$

$y = -\dfrac{1}{2}$

19. $\log_\pi \pi^{\sqrt{\pi}} = \sqrt{\pi}$

20. $\log\left(\dfrac{\sqrt{xy}}{1000}\right) = \log\left(\sqrt{xy}\right) - \log 1000$

$= \log(xy)^{\frac{1}{2}} - \log 10^3$

$= \dfrac{1}{2}\log(xy) - 3$

$= \dfrac{1}{2}(\log x + \log y) - 3$

$= \dfrac{1}{2}\log x + \dfrac{1}{2}\log y - 3$

21. $\ln\left(e^{19} x^{20}\right) = \ln e^{19} + \ln x^{20}$

$= 19 + 20 \ln x$

22. $8\log_7 x - \dfrac{1}{3}\log_7 y = \log_7 x^8 - \log_7 y^{\frac{1}{3}}$

$$= \log_7\left(\dfrac{x^8}{y^{\frac{1}{3}}}\right)$$

$$= \log_7\left(\dfrac{x^8}{\sqrt[3]{y}}\right)$$

23. $7\log_5 x + 2\log_5 x = \log_5 x^7 + \log_5 x^2$

$$= \log_5\left(x^7 \cdot x^2\right)$$

$$= \log_5 x^9$$

24. $\dfrac{1}{2}\ln x - 3\ln y - \ln(z-2)$

$$= \ln x^{\frac{1}{2}} - \ln y^3 - \ln(z-2)$$

$$= \ln\sqrt{x} - \left[\ln y^3 + \ln(z-2)\right]$$

$$= \ln\sqrt{x} - \ln\left[y^3(z-2)\right]$$

$$= \ln\left[\dfrac{\sqrt{x}}{y^3(z-2)}\right]$$

25. Continuously: $A = 8000e^{0.08(3)}$

$$\approx 10,170$$

Monthly: $A = 8000\left(1 + \dfrac{0.08}{12}\right)^{12 \cdot 3}$

$$\approx 10,162$$

$10,170 - 10,162 = 8$

Interest returned will be $8 more if compounded continuously.

12.4 Check Points

1. a. $5^{3x-6} = 125$

$$5^{3x-6} = 5^3$$

$$3x - 6 = 3$$

$$3x = 9$$

$$x = 3$$

The solution set is $\{3\}$.

b. $4^x = 32$

$$\left(2^2\right)^x = 2^5$$

$$2^{2x} = 2^5$$

$$2x = 5$$

$$x = \dfrac{5}{2}$$

The solution set is $\left\{\dfrac{5}{2}\right\}$.

2. a. Take the natural log of both sides of the equation.

$$5^x = 134$$

$$\ln 5^x = \ln 134$$

$$x\ln 5 = \ln 134$$

$$x = \dfrac{\ln 134}{\ln 5}$$

$$x \approx 3.04$$

The solution set is $\left\{\dfrac{\ln 134}{\ln 5} \approx 3.04\right\}$.

b. Take the common log of both sides of the equation.

$$10^x = 8000$$

$$\log 10^x = \log 8000$$

$$x = \log 8000$$

$$x \approx 3.90$$

The solution set is $\{\log 8000 \approx 3.90\}$.

3. Isolate the exponential expression then take the natural log of both sides of the equation.

$$7e^{2x} - 5 = 58$$

$$7e^{2x} = 63$$

$$e^{2x} = 9$$

$$\ln e^{2x} = \ln 9$$

$$2x = \ln 9$$

$$x = \frac{\ln 9}{2}$$

$$x = \frac{\ln 3^2}{2}$$

$$x = \frac{2\ln 3}{2}$$

$$x = \ln 3$$

$$x \approx 1.10$$

The solution set is $\{\ln 3 \approx 1.10\}$.

4. a. $\log_2(x-4) = 3$

$$\log_2(x-4) = 3$$

$$2^3 = x - 4$$

$$8 = x - 4$$

$$12 = x$$

12 checks. The solution set is $\{12\}$.

b. $4\ln(3x) = 8$

$$\ln(3x) = 2$$

$$e^2 = 3x$$

$$\frac{e^2}{3} = x$$

$\dfrac{e^2}{3}$ checks. The solution set is $\left\{\dfrac{e^2}{3}\right\}$.

5. $\log x + \log(x-3) = 1$

$$\log(x^2 - 3x) = 1$$

$$10^1 = x^2 - 3x$$

$$0 = x^2 - 3x - 10$$

$$0 = (x+2)(x-5)$$

$$x + 2 = 0 \quad \text{or} \quad x - 5 = 0$$

$$x = -2 \qquad\qquad x = 5$$

The number -2 does not check. The solution set is $\{5\}$.

6. $\ln(x-3) = \ln(7x-23) - \ln(x+1)$

$$\ln(x-3) = \ln\left(\frac{7x-23}{x+1}\right)$$

$$x - 3 = \frac{7x-23}{x+1}$$

$$(x+1)(x-3) = (x+1)\frac{7x-23}{x+1}$$

$$x^2 - 2x - 3 = 7x - 23$$

$$x^2 - 9x + 20 = 0$$

$$(x-4)(x-5) = 0$$

$$x - 4 = 0 \quad \text{or} \quad x - 5 = 0$$

$$x = 4 \qquad\qquad x = 5$$

Both numbers check. The solution set is $\{4,5\}$.

7. $R = 6e^{12.77x}$

$$6e^{12.77x} = 7$$

$$e^{12.77x} = \frac{7}{6}$$

$$\ln e^{12.77x} = \ln\frac{7}{6}$$

$$12.77x = \ln\frac{7}{6}$$

$$x = \frac{\ln\frac{7}{6}}{12.77}$$

$$x \approx 0.01$$

For a blood alcohol concentration of 0.01, the risk of a car accident is 7%.

8. $A = P\left(1 + \dfrac{r}{n}\right)^{nt}$

$$3600 = 1000\left(1 + \frac{0.08}{4}\right)^{4t}$$

$$3.6 = 1.02^{4t}$$

$$1.02^{4t} = 3.6$$

$$\ln 1.02^{4t} = \ln 3.6$$

$$4t \ln 1.02 = \ln 3.6$$

$$t = \frac{\ln 3.6}{4\ln 1.02}$$

$$t \approx 16.2$$

After approximately 16.2 years, the $1000 will grow to $3600.

9.
$$f(x) = 34.1\ln x + 117.7$$
$$200 = 34.1\ln x + 117.7$$
$$82.3 = 34.1\ln x$$
$$\frac{82.3}{34.1} = \ln x$$
$$x = e^{\frac{82.3}{34.1}}$$
$$x \approx 11$$

There will be 200 million Internet users in the U.S. 11 years after 1999, or 2010.

12.4 Exercise Set

1. $2^x = 64$
$$2^x = 2^6$$
$$x = 6$$

The solution set is $\{6\}$.

3. $5^x = 125$
$$5^x = 5^3$$
$$x = 3$$

The solution set is $\{3\}$.

5. $2^{2x-1} = 32$
$$2^{2x-1} = 2^5$$
$$2x - 1 = 5$$
$$2x = 6$$
$$x = 3$$

The solution set is $\{3\}$.

$$4^{2x-1} = 64$$
$$4^{2x-1} = 4^3$$
7. $2x - 1 = 3$
$$2x = 4$$
$$x = 2$$

The solution set is $\{2\}$.

9. $32^x = 8$
$$\left(2^5\right)^x = 2^3$$
$$2^{5x} = 2^3$$
$$5x = 3$$
$$x = \frac{3}{5}$$

The solution set is $\left\{\frac{3}{5}\right\}$.

11. $9^x = 27$
$$\left(3^2\right)^x = 3^3$$
$$3^{2x} = 3^3$$
$$2x = 3$$
$$x = \frac{3}{2}$$

The solution set is $\left\{\frac{3}{2}\right\}$.

13. $3^{1-x} = \frac{1}{27}$
$$3^{1-x} = \frac{1}{3^3}$$
$$3^{1-x} = 3^{-3}$$
$$1 - x = -3$$
$$-x = -4$$
$$x = 4$$

The solution set is $\{4\}$.

15. $6^{\frac{x-3}{4}} = \sqrt{6}$
$$6^{\frac{x-3}{4}} = 6^{\frac{1}{2}}$$
$$\frac{x-3}{4} = \frac{1}{2}$$
$$2(x-3) = 4(1)$$
$$2x - 6 = 4$$
$$2x = 10$$
$$x = 5$$

The solution set is $\{5\}$.

17. $4^x = \dfrac{1}{\sqrt{2}}$

$\left(2^2\right)^x = \dfrac{1}{2^{\frac{1}{2}}}$

$2^{2x} = 2^{-\frac{1}{2}}$

$2x = -\dfrac{1}{2}$

$x = \dfrac{1}{2}\left(-\dfrac{1}{2}\right) = -\dfrac{1}{4}$

The solution set is $\left\{-\dfrac{1}{4}\right\}$.

19. $e^x = 5.7$

$\ln e^x = \ln 5.7$

$x = \ln 5.7 \approx 1.74$

The solution set is $\left\{\ln 5.7 \approx 1.74\right\}$.

21. $10^x = 3.91$

$\log 10^x = \log 3.91$

$x = \log 3.91$

$x \approx 0.59$

The solution set is $\left\{\log 3.91 \approx 0.59\right\}$.

23. $5^x = 17$

$\ln 5^x = \ln 17$

$x \ln 5 = \ln 17$

$x = \dfrac{\ln 17}{\ln 5} \approx 1.76$

The solution set is $\left\{\dfrac{\ln 17}{\ln 5} \approx 1.76\right\}$.

25. $5e^x = 25$

$e^x = 5$

$\ln e^x = \ln 5$

$x = \ln 5 \approx 1.61$

The solution set is $\left\{\ln 5 \approx 1.61\right\}$.

27. $3e^{5x} = 1977$

$e^{5x} = 659$

$\ln e^{5x} = \ln 659$

$5x = \ln 659$

$x = \dfrac{\ln 659}{5} \approx 1.30$

The solution set is $\left\{\dfrac{\ln 659}{5} \approx 1.30\right\}$.

29. $e^{0.7x} = 13$

$\ln e^{0.7x} = \ln 13$

$0.7x = \ln 13$

$x = \dfrac{\ln 13}{0.7} \approx 3.66$

The solution set is $\left\{\dfrac{\ln 13}{0.7} \approx 3.66\right\}$.

31. $1250e^{0.055x} = 3750$

$e^{0.055x} = 3$

$\ln e^{0.055x} = \ln 3$

$0.055x = \ln 3$

$x = \dfrac{\ln 3}{0.055} \approx 19.97$

The solution set is $\left\{\dfrac{\ln 3}{0.055} \approx 19.97\right\}$.

33. $30 - \left(1.4\right)^x = 0$

$-1.4^x = -30$

$1.4^x = 30$

$\ln 1.4^x = \ln 30$

$x \ln 1.4 = \ln 30$

$x = \dfrac{\ln 30}{\ln 1.4} \approx 10.11$

The solution set is $\left\{\dfrac{\ln 30}{\ln 1.4} \approx 10.11\right\}$.

35. $e^{1-5x} = 793$

$\ln e^{1-5x} = \ln 793$

$1 - 5x = \ln 793$

$-5x = \ln 793 - 1$

$x = \dfrac{-(\ln 793 - 1)}{5}$

$x = \dfrac{1 - \ln 793}{5} \approx -1.14$

The solution set is $\left\{ \dfrac{1 - \ln 793}{5} \approx -1.14 \right\}$.

37. $7^{x+2} = 410$

$\ln 7^{x+2} = \ln 410$

$(x + 2) \ln 7 = \ln 410$

$x + 2 = \dfrac{\ln 410}{\ln 7}$

$x = \dfrac{\ln 410}{\ln 7} - 2 \approx 1.09$

The solution set is $\left\{ \dfrac{\ln 410}{\ln 7} - 2 \approx 1.09 \right\}$.

39. $2^{x+1} = 5^x$

$\ln 2^{x+1} = \ln 5^x$

$(x + 1) \ln 2 = x \ln 5$

$x \ln 2 + \ln 2 = x \ln 5$

$x \ln 2 = x \ln 5 - \ln 2$

$x \ln 2 - x \ln 5 = -\ln 2$

$x(\ln 2 - \ln 5) = -\ln 2$

$x = \dfrac{-\ln 2}{\ln 2 - \ln 5}$

$x = \dfrac{\ln 2}{\ln 5 - \ln 2} \approx 0.76$

The solution set is $\left\{ \dfrac{\ln 2}{\ln 5 - \ln 2} \approx 0.76 \right\}$.

41. $\log_3 x = 4$

$x = 3^4$

$x = 81$

The solution set is $\{81\}$.

43. $\log_2 x = -4$

$x = 2^{-4}$

$x = \dfrac{1}{2^4} = \dfrac{1}{16}$

The solution set is $\left\{ \dfrac{1}{16} \right\}$.

45. $\log_9 x = \dfrac{1}{2}$

$x = 9^{\frac{1}{2}}$

$x = \sqrt{9} = 3$

The solution set is $\{3\}$.

47. $\log x = 2$

$x = 10^2$

$x = 100$

The solution set is $\{100\}$.

49. $\log_4(x + 5) = 3$

$x + 5 = 4^3$

$x + 5 = 64$

$x = 59$

The solution set is $\{59\}$.

51. $\log_3(x - 4) = -3$

$x - 4 = 3^{-3}$

$x - 4 = \dfrac{1}{3^3}$

$x - 4 = \dfrac{1}{27}$

$x = \dfrac{1}{27} + 4$

$x = \dfrac{1}{27} + \dfrac{108}{27} = \dfrac{109}{27}$

The solution set is $\left\{ \dfrac{109}{27} \right\}$.

53. $\log_4(3x+2) = 3$

$$3x+2 = 4^3$$
$$3x+2 = 64$$
$$3x = 62$$
$$x = \frac{62}{3}$$

The solution set is $\left\{\dfrac{62}{3}\right\}$.

55. $\ln x = 2$

$$e^{\ln x} = e^2$$
$$x = e^2 \approx 7.39$$

The solution set is $\left\{e^2 \approx 7.39\right\}$.

57. $\ln x = -3$

$$x = e^{-3} = \frac{1}{e^3}$$

The solution set is $\left\{e^{-3} = \dfrac{1}{e^3} \approx 0.05\right\}$.

59. $5\ln(2x) = 20$

$$\ln(2x) = 4$$
$$e^{\ln(2x)} = e^4$$
$$2x = e^4$$
$$x = \frac{e^4}{2} \approx 27.30$$

The solution set is $\left\{\dfrac{e^4}{2} \approx 27.30\right\}$.

61. $6 + 2\ln x = 5$

$$2\ln x = -1$$
$$e^{\ln x} = e^{-\frac{1}{2}}$$
$$x = e^{-\frac{1}{2}} \approx 0.61$$

The solution set is $\left\{e^{-\frac{1}{2}} \approx 0.61\right\}$.

63. $\ln\sqrt{x+3} = 1$

$$\ln(x+3)^{\frac{1}{2}} = 1$$
$$\frac{1}{2}\ln(x+3) = 1$$
$$\ln(x+3) = 2$$
$$e^{\ln(x+3)} = e^2$$
$$x+3 = e^2$$
$$x = e^2 - 3 \approx 4.39$$

The solution set is $\left\{e^2 - 3 \approx 4.39\right\}$.

65. $\log_5 x + \log_5(4x-1) = 1$

$$\log_5\big(x(4x-1)\big) = 1$$
$$x(4x-1) = 5^1$$
$$4x^2 - x = 5$$
$$4x^2 - x - 5 = 0$$
$$(4x-5)(x+1) = 0$$
$$4x-5 = 0 \quad \text{and} \quad x+1 = 0$$
$$4x = 5 \qquad\qquad x = -1$$
$$x = \frac{5}{4}$$

We disregard -1 because it would result in taking the logarithm of a negative number in the original equation. The solution set is $\left\{\dfrac{5}{4}\right\}$.

67. $\log_3(x-5) + \log_3(x+3) = 2$

$$\log_3\big((x-5)(x+3)\big) = 2$$
$$(x-5)(x+3) = 3^2$$
$$x^2 - 2x - 15 = 9$$
$$x^2 - 2x - 24 = 0$$
$$(x-6)(x+4) = 0$$
$$x-6 = 0 \quad \text{and} \quad x+4 = 0$$
$$x = 6 \qquad\qquad x = -4$$

We disregard -4 because it would result in taking the logarithm of a negative number in the original equation. The solution set is $\{6\}$.

69. $\log_2(x+2) - \log_2(x-5) = 3$

$$\log_2\frac{x+2}{x-5} = 3$$

$$\frac{x+2}{x-5} = 2^3$$

$$\frac{x+2}{x-5} = 8$$

$$x+2 = 8(x-5)$$

$$x+2 = 8x-40$$

$$-7x+2 = -40$$

$$-7x = -42$$

$$x = 6$$

The solution set is $\{6\}$.

71. $\log(3x-5) - \log(5x) = 2$

$$\log\frac{3x-5}{5x} = 2$$

$$\frac{3x-5}{5x} = 10^2$$

$$\frac{3x-5}{5x} = 100$$

$$3x-5 = 500x$$

$$-5 = 497x$$

$$-\frac{5}{497} = x$$

We disregard $-\dfrac{5}{497}$ because it would result in taking the logarithm of a negative number in the original equation. Therefore, the equation has no solution. The solution set is $\varnothing$ or $\{\ \ \}$.

73. $\ln(x+1) - \ln x = 1$

$$\ln\left(\frac{x+1}{x}\right) = 1$$

$$\frac{x+1}{x} = e^1$$

$$x+1 = ex$$

$$1 = ex - x$$

$$1 = (e-1)x$$

$$x = \frac{1}{e-1} \approx 0.58$$

The solution set is $\left\{\dfrac{1}{e-1} \approx 0.58\right\}$.

75. $\log_3(x+4) = \log_3 7$

$$x+4 = 7$$

$$x = 3$$

The solution set is $\{3\}$.

77. $\log(x+4) = \log x + \log 4$

$$\log(x+4) = \log 4x$$

$$x+4 = 4x$$

$$4 = 3x$$

$$x = \frac{4}{3}$$

The solution set is $\left\{\dfrac{4}{3}\right\}$.

79. $\log(3x-3) = \log(x+1) + \log 4$

$$\log(3x-3) = \log(4x+4)$$

$$3x-3 = 4x+4$$

$$-7 = x$$

This value is rejected. The solution set is $\{\ \ \}$.

81. $2\log x = \log 25$

$$\log x^2 = \log 25$$

$$x^2 = 25$$

$$x = \pm 5$$

-5 is rejected. The solution set is $\{5\}$.

83. $\log(x+4) - \log 2 = \log(5x+1)$

$$\log\frac{x+4}{2} = \log(5x+1)$$

$$\frac{x+4}{2} = 5x+1$$

$$x+4 = 10x+2$$

$$-9x = -2$$

$$x = \frac{2}{9}$$

The solution set is $\left\{\dfrac{2}{9}\right\}$.

85. $2\log x - \log 7 = \log 112$

$\log x^2 - \log 7 = \log 112$

$\log \dfrac{x^2}{7} = \log 112$

$\dfrac{x^2}{7} = 112$

$x^2 = 784$

$x = \pm 28$

−28 is rejected. The solution set is {28}.

87. $\log x + \log(x+3) = \log 10$

$\log(x^2 + 3x) = \log 10$

$x^2 + 3x = 10$

$x^2 + 3x - 10 = 0$

$(x+5)(x-2) = 0$

$x = -5 \ \text{ or } \ x = 2$

−5 is rejected. The solution set is {2}.

89. $\ln(x-4) + \ln(x+1) = \ln(x-8)$

$\ln(x^2 - 3x - 4) = \ln(x-8)$

$x^2 - 3x - 4 = x - 8$

$x^2 - 4x + 4 = 0$

$(x-2)(x-2) = 0$

$x = 2$

2 is rejected. The solution set is { }.

91. $5^{2x} \cdot 5^{4x} = 125$

$5^{2x+4x} = 5^3$

$5^{6x} = 5^3$

$6x = 3$

$x = \dfrac{1}{2}$

The solution set is $\left\{ \dfrac{1}{2} \right\}$.

93. $3^{x^2} = 45$

$\ln 3^{x^2} = \ln 45$

$x^2 \ln 3 = \ln 45$

$x^2 = \dfrac{\ln 45}{\ln 3}$

$x = \pm \sqrt{\dfrac{\ln 45}{\ln 3}} \approx \pm 1.86$

The solution set is $\left\{ \pm \sqrt{\dfrac{\ln 45}{\ln 3}} \approx \pm 1.86 \right\}$.

95. $\log_2(x-6) + \log_2(x-4) - \log_2 x = 2$

$\log_2 \left[(x-6)(x-4) \right] - \log_2 x = 2$

$\log_2 \left[\dfrac{(x-6)(x-4)}{x} \right] = 2$

$\log_2 \left(\dfrac{x^2 - 10x + 24}{x} \right) = 2$

$\dfrac{x^2 - 10x + 24}{x} = 2^2$

$\dfrac{x^2 - 10x + 24}{x} = 4$

$x^2 - 10x + 24 = 4x$

$x^2 - 14x + 24 = 0$

$(x-12)(x-2) = 0$

Apply the zero product property:

$x - 12 = 0 \quad \text{or} \quad x - 2 = 0$

$x = 12 \qquad\qquad x = 2$

We disregard 2 because it would result in taking the logarithm of a negative number in the original equation. The solution set is {12}.

97. $5^{x^2 - 12} = 25^{2x}$

$5^{x^2 - 12} = \left(5^2 \right)^{2x}$

$5^{x^2 - 12} = 5^{4x}$

$x^2 - 12 = 4x$

$x^2 - 4x - 12 = 0$

$(x-6)(x+2) = 0$

$x - 6 = 0 \quad \text{or} \quad x + 2 = 0$

$x = 6 \qquad\qquad x = -2$

The solution set is $\{-2, \ 6\}$.

99. a. 2005 is 0 years after 2005.

$A = 36.1 e^{0.0126t}$

$A = 36.1 e^{0.0126(0)} = 36.1$

The population of California was 36.1 million in 2005.

b.
$$A = 36.1e^{0.0126t}$$
$$40 = 36.1e^{0.0126t}$$
$$\frac{40}{36.1} = e^{0.0126t}$$
$$\ln\frac{40}{36.1} = \ln e^{0.0126t}$$
$$0.0126t = \ln\frac{40}{36.1}$$
$$t = \frac{\ln\dfrac{40}{36.1}}{0.0126} \approx 8$$

The population of California will reach 40 million about 8 years after 2005, or 2013.

101.
$$f(x) = 20(0.975)^x$$
$$1 = 20(0.975)^x$$
$$\frac{1}{20} = 0.975^x$$
$$\ln\frac{1}{20} = \ln 0.975^x$$
$$\ln\frac{1}{20} = x\ln 0.975$$
$$x = \frac{\ln\dfrac{1}{20}}{\ln 0.975}$$
$$x \approx 118$$

There is 1% of surface sunlight at 118 feet. This is represented by the point (118,1).

103.
$$20000 = 12500\left(1+\frac{0.0575}{4}\right)^{4t}$$
$$20000 = 12500\left(1+0.014375\right)^{4t}$$
$$20000 = 12500\left(1.014375\right)^{4t}$$
$$\frac{20000}{12500} = \left(1.014375\right)^{4t}$$
$$1.6 = \left(1.014375\right)^{4t}$$
$$\ln 1.6 = \ln\left(1.014375\right)^{4t}$$
$$\ln 1.6 = 4t\ln 1.014375$$
$$\frac{4t\ln 1.014375}{4\ln 1.014375} = \frac{\ln 1.6}{4\ln 1.014375}$$
$$t = \frac{\ln 1.6}{4\ln 1.014375} \approx 8.2$$

It will take approximately 8.2 years.

105.
$$1400 = 1000\left(1+\frac{r}{360}\right)^{360(2)}$$
$$\frac{1400}{1000} = \left(1+\frac{r}{360}\right)^{720}$$
$$1.4 = \left(1+\frac{r}{360}\right)^{720}$$
$$\ln 1.4 = \ln\left(1+\frac{r}{360}\right)^{720}$$
$$\ln 1.4 = 720\ln\left(1+\frac{r}{360}\right)$$
$$\frac{\ln 1.4}{720} = \ln\left(1+\frac{r}{360}\right)$$
$$e^{\frac{\ln 1.4}{720}} = e^{\ln\left(1+\frac{r}{360}\right)}$$
$$e^{\frac{\ln 1.4}{720}} = 1+\frac{r}{360}$$
$$1+\frac{r}{360} = e^{\frac{\ln 1.4}{720}}$$
$$\frac{r}{360} = e^{\frac{\ln 1.4}{720}} - 1$$
$$r = 360\left(e^{\frac{\ln 1.4}{720}} - 1\right) \approx 0.168$$

The annual interest rate is approximately 16.8%.

107.
$$16000 = 8000e^{0.08t}$$
$$\frac{16000}{8000} = e^{0.08t}$$
$$2 = e^{0.08t}$$
$$\ln 2 = \ln e^{0.08t}$$
$$\ln 2 = 0.08t$$
$$t = \frac{\ln 2}{0.08} \approx 8.7$$

It will take approximately 8.7 years to double the money.

109. $7050 = 2350e^{r \cdot 7}$

$$\frac{7050}{2350} = e^{7r}$$

$$3 = e^{7r}$$

$$\ln 3 = \ln e^{7r}$$

$$\ln 3 = 7r$$

$$r = \frac{\ln 3}{7} \approx 0.157$$

The annual interest rate would have to be 15.7% to triple the money.

111. a. 2007 is 5 years after 2002.

$$f(x) = 8 + 38 \ln x$$

$$f(5) = 8 + 38 \ln 5 \approx 69$$

According to the function, 69% of new cellphones will have cameras in 2007. This overestimates the value shown in the graph by 1%.

b. $f(x) \doteq 8 + 38 \ln x$

$$87 = 8 + 38 \ln x$$

$$79 = 38 \ln x$$

$$\frac{79}{38} = \ln x$$

$$x = e^{\frac{79}{38}}$$

$$x \approx 8$$

If the trend continues, 87% of new cellphones will have cameras 8 years after 2002, or 2010.

113. $50 = 95 - 30 \log_2 x$

$$-45 = -30 \log_2 x$$

$$\frac{-45}{-30} = \log_2 x$$

$$\log_2 x = \frac{3}{2}$$

$$x = 2^{\frac{3}{2}} \approx 2.8$$

After approximately 2.8 days, only half the students recall the important features of the lecture. This is represented by the point (2.8, 50).

115. a. $pH = -\log x$

$$5.6 = -\log x$$

$$-5.6 = \log x$$

$$x = 10^{-5.6}$$

The hydrogen ion concentration is $10^{-5.6}$ mole per liter.

b. $pH = -\log x$

$$2.4 = -\log x$$

$$-2.4 = \log x$$

$$x = 10^{-2.4}$$

The hydrogen ion concentration is $10^{-2.4}$ mole per liter.

c. $\dfrac{10^{-2.4}}{10^{-5.6}} = 10^{-2.4-(-5.6)} = 10^{3.2}$

The concentration of the acidic rainfall in part (b) is $10^{3.2}$ times greater than the normal rainfall in part (a).

117. –121. Answers will vary.

123. $2^{x+1} = 8$

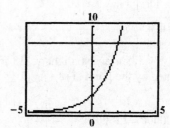

The solution is 2, and the solution set is {2}. Verify the solution algebraically:

$$2^{2+1} = 8$$

$$2^3 = 8$$

$$8 = 8$$

125. $\log_3(4x-7)=2$

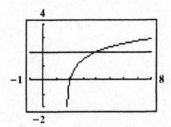

The solution is 4, and the solution set is {4}.

Verify the solution algebraically:

$\log_3(4\cdot4-7)=2$

$\log_3(16-7)=2$

$\log_3 9=2$

$\log_3 3^2=2$

$2=2$

127. $\log(x+3)+\log x=1$

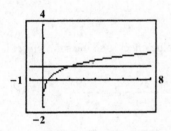

The solution is 2, and the solution set is {2}.

Verify the solution algebraically:

$\log(2+3)+\log 2=1$

$\log 5+\log 2=1$

$\log(5\cdot2)=1$

$\log 10=1$

$1=1$

129. $3^x=2x+3$

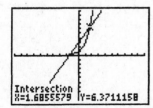

The solution set is $\{-1.39,1.69\}$.

The solutions check algebraically.

131. $f(x)=0.48\ln(x+1)+27$

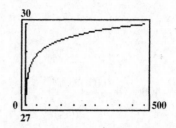

The barometric air pressure increases as the distance from the eye increases. It increases quickly at first, and the more slowly over time.

133. $P(t)=145e^{-0.092t}$

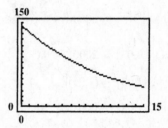

The runner's pulse will be 70 beats per minute after approximately 7.9 minutes.

Verifying algebraically:

$P(7.9)=145e^{-0.092(7.9)}$

$=145e^{-07268}\approx 70$

135. does not make sense; Explanations will vary.

Sample explanation: $2^x=15$ requires logarithms. $2^x=16$ can be solved by rewriting 16 as 2^4.

$2^x=15$

$\ln 2^x=\ln 15$

$x\ln 2=\ln 15$

$x=\dfrac{\ln 15}{\ln 2}$

$2^x=16$

$2^x=2^4$

$x=4$

137. makes sense

139. false; Changes to make the statement true will vary. A sample change is: If $\log(x+3) = 2$, then
$10^2 = x + 3$.

141. true

143. $A_{4000} = 4000\left(1 + \dfrac{0.03}{1}\right)^{1t}$

$A_{2000} = 2000\left(1 + \dfrac{0.05}{1}\right)^{1t}$

Set the right hand sides of the equations equal and solve for t.

$$4000\left(1 + \frac{0.03}{1}\right)^{1t} = 2000\left(1 + \frac{0.05}{1}\right)^{1t}$$

$$4000(1+0.03)^t = 2000(1+0.05)^t$$

$$2(1.03)^t = (1.05)^t$$

$$\ln 2(1.03)^t = \ln(1.05)^t$$

$$\ln 2 + \ln(1.03)^t = t\ln(1.05)$$

$$\ln 2 + t\ln(1.03) = t\ln(1.05)$$

$$\ln 2 = t\ln(1.05) - t\ln(1.03)$$

$$\ln 2 = t\left(\ln(1.05) - \ln(1.03)\right)$$

$$t = \frac{\ln 2}{\ln(1.05) - \ln(1.03)}$$

$$t \approx 36.0$$

In approximately 36 years, the two accounts will have the same balance.

145. $(\log x)(2\log x + 1) = 6$

Let $u = \log x$.

$$(u)(2u+1) = 6$$

$$2u^2 + u = 6$$

$$2u^2 + u - 6 = 0$$

$$(2u - 3)(u + 2) = 0$$

$$2u - 3 = 0 \quad \text{or} \quad u + 2 = 0$$

$$2u = 3 \qquad\qquad u = -2$$

$$u = \frac{3}{2}$$

Substitute $\log x$ for u.

$$u = \frac{3}{2} \qquad \text{or} \qquad u = -2$$

$$\log x = \frac{3}{2} \qquad\qquad \log x = -2$$

$$\qquad\qquad\qquad x = 10^{-2}$$

$$x = 10^{\frac{3}{2}}$$

The solution set is $\left\{10^{-2}, 10^{\frac{3}{2}}\right\}$.

147. $\sqrt{x+4} - \sqrt{x-1} = 1$

$$\sqrt{x+4} = 1 + \sqrt{x-1}$$

$$\left(\sqrt{x+4}\right)^2 = \left(1 + \sqrt{x-1}\right)^2$$

$$x + 4 = 1 + 2\sqrt{x-1} + x - 1$$

$$4 = 2\sqrt{x-1}$$

$$2 = \sqrt{x-1}$$

$$2^2 = \left(\sqrt{x-1}\right)^2$$

$$4 = x - 1$$

$$5 = x$$

This value checks, so the solution set is $\{5\}$.

148.

$$\frac{3}{x+1} - \frac{5}{x} = \frac{19}{x^2 + x}$$

$$\frac{3}{x+1} - \frac{5}{x} = \frac{19}{x(x+1)}$$

$$x(x+1)\left(\frac{3}{x+1} - \frac{5}{x}\right) = x(x+1)\left(\frac{19}{x(x+1)}\right)$$

$$x(3) - 5(x+1) = 19$$

$$3x - 5x - 5 = 19$$

$$-2x - 5 = 19$$

$$-2x = 24$$

$$x = -12$$

The solution set is $\{-12\}$.

149. $\left(-2x^3 y^{-2}\right)^{-4} = \left(-\dfrac{2x^3}{y^2}\right)^{-4} = \left(-\dfrac{y^2}{2x^3}\right)^{4} = \dfrac{y^8}{16x^{12}}$

150. $A = 10e^{-0.003t}$

a. 2006: $A = 10e^{-0.003(0)} = 10$ million

2007: $A = 10e^{-0.003(1)} \approx 9.97$ million

2008: $A = 10e^{-0.003(2)} \approx 9.94$ million

2009: $A = 10e^{-0.003(3)} \approx 9.91$ million

b. The population is decreasing.

151. An exponential function is the best choice.

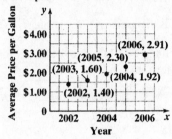

152. a. $e^{\ln 3} = 3$

b. $e^{\ln 3} = 3$

$\left(e^{\ln 3}\right)^x = 3^x$

$e^{(\ln 3)x} = 3^x$

12.5 Check Points

1. a. $A_0 = 643$. Since 2006 is 16 years after 1990, when $t = 16$, $A = 906$.

$$A = A_0 e^{kt}$$

$$906 = 643e^{k(16)}$$

$$\frac{906}{643} = e^{16k}$$

$$\ln\left(\frac{906}{643}\right) = \ln e^{16k}$$

$$\ln\left(\frac{906}{643}\right) = 16k$$

$$k = \frac{\ln\left(\frac{906}{643}\right)}{16} \approx 0.021$$

Thus, the growth function is $A = 643e^{0.021t}$.

b.

$$A = 643e^{0.021t}$$

$$2000 = 643e^{0.021t}$$

$$\frac{2000}{643} = e^{0.021t}$$

$$\ln\left(\frac{2000}{643}\right) = \ln e^{0.021t}$$

$$\ln\left(\frac{2000}{643}\right) = 0.021t$$

$$t = \frac{\ln\left(\frac{2000}{643}\right)}{0.021} \approx 54$$

Africa's population will reach 2000 million approximately 54 years after 1990, or 2044.

2. a.

$$A = A_0 e^{kt}$$

$$\frac{A_0}{2} = A_0 e^{k\,28}$$

$$\frac{1}{2} = e^{28k}$$

$$\ln\left(\frac{1}{2}\right) = \ln e^{28k}$$

$$\ln\left(\frac{1}{2}\right) = 28k$$

$$k = \frac{\ln\left(\frac{1}{2}\right)}{28} \approx -0.0248$$

Thus, the decay model is $A = A_0 e^{-0.0248t}$.

b.

$$A = A_0 e^{-0.0248t}$$

$$10 = 60e^{-0.0248t}$$

$$\frac{10}{60} = e^{-0.0248t}$$

$$\ln\left(\frac{10}{60}\right) = \ln e^{-0.0248t}$$

$$\ln\left(\frac{10}{60}\right) = -0.0248t$$

$$t = \frac{\ln\left(\frac{10}{60}\right)}{-0.0248} \approx 72.2$$

It will take about 72.2 years to decay to a level of 10 grams.

3. A logarithmic function would be a good choice for modeling the data.

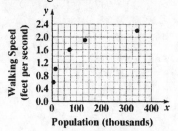

4. An exponential function would be a good choice for modeling the data although model choices may vary.

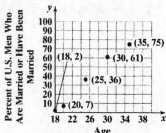

5. a. 1970 is 21 years after 1949.

$$f(x) = 0.073x + 2.316$$

$$f(21) = 0.073(21) + 2.316$$

$$\approx 3.8$$

$$g(x) = 2.569(1.017)^x$$

$$g(21) = 2.569(1.017)^{21}$$

$$\approx 3.7$$

The exponential function g serves as a better model for 1970.

b. 2012 is 63 years after 1949.

$$f(x) = 0.073x + 2.316$$

$$f(63) = 0.073(63) + 2.316$$

$$\approx 6.9$$

This value models the projected value of 7 billion fairly well.

$$g(x) = 2.569(1.017)^x$$

$$g(63) = 2.569(1.017)^{63}$$

$$\approx 7.4$$

The exponential function does not model the projected value very well.

6. $y = 4(7.8)^x$

$$= 4e^{(\ln 7.8)x}$$

Rounded to three decimal places:

$$y = 4e^{(\ln 7.8)x}$$

$$= 4e^{2.054x}$$

12.5 Exercise Set

1. Since 2006 is 0 years after 2006, find A when $t = 0$:

$$A = 127.5e^{0.001t}$$

$$A = 127.5e^{0.001(0)}$$

$$A = 127.5e^0$$

$$A = 127.5(1)$$

$$A = 127.5$$

In 2006, the population of Japan was 127.5 million.

3. Iraq has the greatest growth rate at 2.7% per year.

5. Substitute $A = 1238$ into the model for India and solve for t:

$$1238 = 1095.4e^{0.014t}$$

$$\frac{1238}{1095.4} = e^{0.014t}$$

$$\ln \frac{1238}{1095.4} = \ln e^{0.014t}$$

$$\ln \frac{1238}{1095.4} = 0.014t$$

$$t = \frac{\ln \dfrac{1238}{1095.4}}{0.014} \approx 9$$

The population of India will be 1238 million approximately 9 years after 2006, or 2015.

7. a. $A_0 = 6.04$. Since 2050 is 50 years after 2000, when $t = 50$, $A = 10$.

$$A = A_0 e^{kt}$$

$$10 = 6.04e^{k(50)}$$

$$\frac{10}{6.04} = e^{50k}$$

$$\ln \left(\frac{10}{6.04} \right) = \ln e^{50k}$$

$$\ln \left(\frac{10}{6.04} \right) = 50k$$

$$k = \frac{\ln \left(\dfrac{10}{6.04} \right)}{50} \approx 0.01$$

Thus, the growth function is $A = 6.04e^{0.01t}$.

b. $9 = 6.04e^{0.01t}$

$$\frac{9}{6.04} = e^{0.01t}$$

$$\ln\left(\frac{9}{6.04}\right) = \ln e^{0.01t}$$

$$\ln\left(\frac{9}{6.04}\right) = 0.01t$$

$$t = \frac{\ln\left(\dfrac{9}{6.04}\right)}{0.01} \approx 40$$

Now, $2000 + 40 = 2040$, so the population will be 9 million is approximately the year 2040.

9. $A = 16e^{-0.000121t}$

$A = 16e^{-0.000121(5715)}$

$A = 16e^{-0.691515}$

$A \approx 8.01$

Approximately **8 grams** of carbon-14 will be present in 5715 years.

11. After 10 seconds, there will be $16 \cdot \dfrac{1}{2} = 8$ grams present. After 20 seconds, there will be $8 \cdot \dfrac{1}{2} = 4$ grams present. After 30 seconds, there will be $4 \cdot \dfrac{1}{2} = 2$ grams present. After 40 seconds, there will be $2 \cdot \dfrac{1}{2} = 1$ grams present. After 50 seconds, there will be $1 \cdot \dfrac{1}{2} = \dfrac{1}{2}$ gram present.

13. $A = A_0 e^{-0.000121t}$

$15 = 100e^{-0.000121t}$

$$\frac{15}{100} = e^{-0.000121t}$$

$\ln 0.15 = \ln e^{-0.000121t}$

$\ln 0.15 = -0.000121t$

$$t = \frac{\ln 0.15}{-0.000121} \approx 15,679$$

The paintings are approximately 15,679 years old.

15. a. $\dfrac{1}{2} = 1e^{k1.31}$

$$\ln\frac{1}{2} = \ln e^{1.31k}$$

$$\ln\frac{1}{2} = 1.31k$$

$$k = \frac{\ln\dfrac{1}{2}}{1.31} \approx -0.52912$$

The exponential model is given by
$A = A_0 e^{-0.52912t}$.

b. $A = A_0 e^{-0.52912t}$

$0.945A_0 = A_0 e^{-0.52912t}$

$0.945 = e^{-0.52912t}$

$\ln 0.945 = \ln e^{-0.52912t}$

$\ln 0.945 = -0.52912t$

$$t = \frac{\ln 0.945}{-0.52912} \approx 0.1069$$

The age of the dinosaur bones is approximately 0.1069 billion or 106,900,000 years old.

17. $2A_0 = A_0 e^{kt}$

$2 = e^{kt}$

$\ln 2 = \ln e^{kt}$

$\ln 2 = kt$

$$t = \frac{\ln 2}{k}$$

The population will double in $t = \dfrac{\ln 2}{k}$ years.

19. $A = 4.1e^{0.01t}$

a. $k = 0.01$, so New Zealand's growth rate is 1%.

b. $A = 4.1e^{0.01t}$

$2 \cdot 4.1 = 4.1e^{0.01t}$

$2 = e^{0.01t}$

$\ln 2 = \ln e^{0.01t}$

$\ln 2 = 0.01t$

$$t = \frac{\ln 2}{0.01} \approx 69$$

New Zealand's population will double in approximately 69 years.

21. a.

Woman's Age

b. An exponential function appears to be the best choice for modeling the data.

23. a.

Intensity
(watts per meter^2)

b. A logarithmic function appears to be the best choice for modeling the data.

25. a.

Year

b. A linear function appears to be the best choice for modeling the data.

27. $y = 100(4.6)^x$

$y = 100e^{(\ln 4.6)x}$

$y = 100e^{1.526x}$

29. $y = 2.5(0.7)^x$

$y = 2.5e^{(\ln 0.7)x}$

$y = 2.5e^{-0.357x}$

31. – 37. Answers will vary.

39. a. The exponential model is $y = 200.9(1.011)^x$. Since $r \approx 0.999$ is very close to 1, the model fits the data well.

b. $y = 200.9(1.011)^x$

$y = 200.9e^{(\ln 1.011)x}$

$y = 200.9e^{0.0109x}$

Since $k = .0109$, the population of the United States is increasing by about 1% each year.

41. The linear model is $y = 2.654x + 198.015$. Since $r \approx 0.997$ is close to 1, the model fits the data very well.

43. Using r, the model of best fit is the exponential model $y = 200.9(1.011)^x$.

The model of second best fit is the linear model $y = 2.654x + 198.015$.

Using the exponential model:

$$315 = 200.9(1.011)^x$$

$$\frac{315}{200.9} = (1.011)^x$$

$$\ln\left(\frac{315}{200.9}\right) = \ln(1.011)^x$$

$$\ln\left(\frac{315}{200.9}\right) = x\ln(1.011)$$

$$x = \frac{\ln\left(\dfrac{315}{200.9}\right)}{\ln(1.011)} \approx 41$$

$$1969 + 41 = 2010$$

Using the linear model:

$$y = 2.654x + 198.015$$

$$315 = 2.654x + 198.015$$

$$116.985 = 2.654x$$

$$x = \frac{116.985}{2.654} \approx 44$$

$$1969 + 44 = 2013$$

According to the exponential model, the U.S. population will reach 315 million around the year 2010. According to the linear model, the U.S. population will reach 315 million around the year 2013. Both results are reasonably close to the result found in Example 1 (2010).

Explanations will vary.

45. Models and predictions will vary. Sample models are provided

Exercise 21: $y = 1.402(1.078)^x$

Exercise 22: $y = 2896.7(1.056)^x$

Exercise 23: $y = 120 + 4.343 \ln x$

Exercise 24: $y = -11.629 + 13.424 \ln x$

Exercise 25: $y = 1.849x + 16.947$ (where x is the number of years after 2000)

Exercise 26: $y = 5.3x + 9.5$ (where x is the number of years after 2003)

47. does not make sense; Explanations will vary. Sample explanation: This is not necessarily so. Growth rate measures how fast a population is growing relative to that population. It does not indicate how the size of a population compares to the size of another population.

49. makes sense

51. true

53. true

55. $\dfrac{x^2 - 9}{2x^2 + 7x + 3} \div \dfrac{x^2 - 3x}{2x^2 + 11x + 5}$

$= \dfrac{x^2 - 9}{2x^2 + 7x + 3} \cdot \dfrac{2x^2 + 11x + 5}{x^2 - 3x}$

$= \dfrac{\cancel{(x+3)}(x-3)}{\cancel{(2x+1)}\cancel{(x+3)}} \cdot \dfrac{\cancel{(2x+1)}(x+5)}{x\cancel{(x-3)}}$

$= \dfrac{x+5}{x}$

56. $x^{\frac{2}{3}} + 2x^{\frac{1}{3}} - 3 = 0$

Let $t = x^{\frac{1}{3}}$.

$\left(x^{\frac{1}{3}}\right)^2 + 2x^{\frac{1}{3}} - 3 = 0$

$t^2 + 2t - 3 = 0$

$(t+3)(t-1) = 0$

$t + 3 = 0$　　or　　$t - 1 = 0$

$t = -3$　　　　　　　$t = 1$

Substitute $x^{\frac{1}{3}}$ for t.

$x^{\frac{1}{3}} = -3$　　or　　$x^{\frac{1}{3}} = 1$

$\left(x^{\frac{1}{3}}\right)^3 = (-3)^3$　　$\left(x^{\frac{1}{3}}\right)^3 = (1)^3$

$x = -27$　　　　　　$x = 1$

The solution set is $\{-27, 1\}$.

57. $6\sqrt{2} - 2\sqrt{50} + 3\sqrt{98}$

$= 6\sqrt{2} - 2\sqrt{25 \cdot 2} + 3\sqrt{49 \cdot 2}$

$= 6\sqrt{2} - 2 \cdot 5\sqrt{2} + 3 \cdot 7\sqrt{2}$

$= 6\sqrt{2} - 10\sqrt{2} + 21\sqrt{2} = 17\sqrt{2}$

58. $x^2 + 4x + \underline{\quad}$

$x^2 + 4x + 4 = (x+2)^2$

59. $y^2 - 6y + \underline{\quad}$

$y^2 - 6y + 9 = (y-3)^2$

60. Graph of Circle:

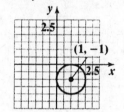

Chapter 12 Review Exercises

1. $f(x) = 4^x$

x	$f(x)$
-2	$4^{-2} = \dfrac{1}{4^2} = \dfrac{1}{16}$
-1	$4^{-1} = \dfrac{1}{4^1} = \dfrac{1}{4}$
0	$4^0 = 1$
1	$4^1 = 4$
2	$4^2 = 16$

The coordinates match graph **d**.

2. $f(x) = 4^{-x}$

x	$f(x)$
-2	$4^{-(-2)} = 4^2 = 16$
-1	$4^{-(-1)} = 4^1 = 4$
0	$4^{-0} = 4^0 = 1$
1	$4^{-1} = \dfrac{1}{4^1} = \dfrac{1}{4}$
2	$4^{-2} = \dfrac{1}{4^2} = \dfrac{1}{16}$

The coordinates match graph **a.**

3. $f(x) = -4^{-x}$

x	$f(x)$
-2	$-4^{-(-2)} = -4^2 = -16$
-1	$-4^{-(-1)} = -4^1 = -4$
0	$-4^{-0} = -4^0 = -1$
1	$-4^{-1} = -\dfrac{1}{4^1} = -\dfrac{1}{4}$
2	$-4^{-2} = -\dfrac{1}{4^2} = -\dfrac{1}{16}$

The coordinates match graph **b.**

4. $f(x) = -4^{-x} + 3$

x	$f(x)$
-2	$-4^{-(-2)} + 3 = -4^2 + 3 = -16 + 3 = -13$
-1	$-4^{-(-1)} + 3 = -4^1 + 3 = -4 + 3 = -1$
0	$-4^{-0} + 3 = -4^0 + 3 = -1 + 3 = 2$
1	$-4^{-1} + 3 = -\dfrac{1}{4^1} + 3 = -\dfrac{1}{4} + 3 = \dfrac{11}{4}$
2	$-4^{-2} + 3 = -\dfrac{1}{4^2} + 3 = -\dfrac{1}{16} + 3 = \dfrac{47}{16}$

The coordinates match graph **c.**

5. $f(x) = 2^x$ and $g(x) = 2^{x-1}$

x	$f(x)$	$g(x)$
-2	$\dfrac{1}{4}$	$\dfrac{1}{8}$
-1	$\dfrac{1}{2}$	$\dfrac{1}{4}$
0	1	$\dfrac{1}{2}$
1	2	1
2	4	2

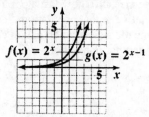

The graph of g is the graph of f shifted 1 unit to the right.

6. $f(x) = 2^x$ and $g(x) = \left(\dfrac{1}{2}\right)^x$

x	$f(x)$	$g(x)$
-2	$\dfrac{1}{4}$	4
-1	$\dfrac{1}{2}$	2
0	1	1
1	2	$\dfrac{1}{2}$
2	4	$\dfrac{1}{4}$

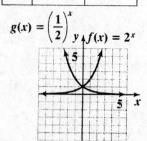

The graph of g is the graph of f reflected across the y–axis.

7. $f(x) = 3^x$ and $g(x) = 3^x - 1$

x	$f(x)$	$g(x)$
-2	$\dfrac{1}{9}$	$-\dfrac{8}{9}$
-1	$\dfrac{1}{3}$	$-\dfrac{2}{3}$
0	1	0
1	3	2
2	9	8

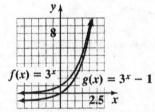

The graph of g is the graph of f shifted down 1 unit.

8. $f(x) = 3^x$ and $g(x) = -3^x$

x	$f(x)$	$g(x)$
-2	$\dfrac{1}{9}$	$-\dfrac{1}{9}$
-1	$\dfrac{1}{3}$	$-\dfrac{1}{3}$
0	1	-1
1	3	-3
2	9	-9

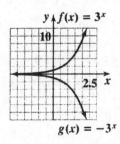

The graph of g is the graph of f reflected across the x–axis.

9. 5.5% Compounded Semiannually:

$$A = 5000\left(1 + \frac{0.055}{2}\right)^{2 \cdot 5}$$
$$= 5000(1 + 0.0275)^{10}$$
$$= 5000(1.0275)^{10} \approx 6558.26$$

5.25% Compounded Monthly:

$$A = 5000\left(1 + \frac{0.0525}{12}\right)^{12 \cdot 5}$$
$$= 5000(1 + 0.004375)^{60}$$
$$= 5000(1.004375)^{60} \approx 6497.16$$

5.5% compounded semiannually yields the greater return.

10. 7.0% Compounded Monthly:

$$A = 14000\left(1 + \frac{0.07}{12}\right)^{12 \cdot 10}$$
$$= 14000\left(1 + \frac{7}{1200}\right)^{120}$$
$$= 14000\left(\frac{1207}{1200}\right)^{120} \approx 28135.26$$

6.85% Compounded Continuously:
$$A = 14000e^{0.0685 \cdot 10}$$
$$= 14000e^{0.685} \approx 27772.81$$

7.0% compounded monthly yields the greater return.

11. a. The coffee was $200°F$ when it was first taken out of the microwave.

b. After 20 minutes, the temperature is approximately $119°F$.

c. The coffee will cool to a low of $70°F$. This means that the temperature of the room is $70°F$.

12. $\dfrac{1}{2} = \log_{49} 7$

$49^{\frac{1}{2}} = 7$

13. $3 = \log_4 x$

$4^3 = x$

14. $\log_3 81 = y$

$\qquad 3^y = 81$

15. $\qquad 6^3 = 216$

$\log_6 216 = 3$

16. $\qquad b^4 = 625$

$\log_b 625 = 4$

17. $\qquad 13^y = 874$

$\log_{13} 874 = y$

18. $\log_4 64 = \log_4 4^3 = 3$ because $\log_b b^x = x$.

19. $\log_5 \dfrac{1}{25} = \log_5 \dfrac{1}{5^2} = \log_5 5^{-2} = -2$ because

$\log_b b^x = x$.

20. $\log_3(-9)$

This logarithm cannot be evaluated because -9 is not in the domain of $y = \log_3 x$.

21. $\log_{16} 4 = y$

$\qquad 16^y = 4$

$\qquad \left(4^2\right)^y = 4$

$\qquad 4^{2y} = 4^1$

$\qquad 2y = 1$

$\qquad y = \dfrac{1}{2}$

22. $\log_{17} 17 = 1$ because $17^1 = 17$.

23. $\log_3 3^8 = 8$ because $\log_b b^x = x$.

24. Because $\ln e^x = x$, we conclude that $\ln e^5 = 5$.

25. $\log_3 \dfrac{1}{\sqrt{3}} = \log_3 \dfrac{1}{3^{\frac{1}{2}}} = \log_3 3^{-\frac{1}{2}} = -\dfrac{1}{2}$ because

$\log_b b^x = x$.

26. $\ln \dfrac{1}{e^2} = \ln e^{-2} = -2$ because $\log_b b^x = x$.

27. $\log \dfrac{1}{1000} = \log \dfrac{1}{10^3} = \log 10^{-3} = -3$ because

$\log_b b^x = x$.

28. Recall that $\log_b b = 1$ and $\log_b 1 = 0$ for all $b > 0$, $b \neq 1$. Therefore, $\log_3 \left(\log_8 8\right) = \log_3 1 = 0$.

29. $f(x) = 2^x$; $g(x) = \log_2 x$

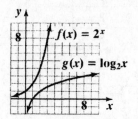

Domain of f: $\{x \mid x \text{ is a real number}\}$ or $(-\infty, \infty)$.

Range of f: $\{y \mid y > 0\}$ or $(0, \infty)$

Domain of g: $\{x \mid x > 0\}$ or $(0, \infty)$

Range of g: $\{y \mid y \text{ is a real number}\}$ or $(-\infty, \infty)$.

30. $f(x) = \left(\dfrac{1}{3}\right)^x$; $g(x) = \log_{\frac{1}{3}} x$

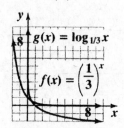

Domain of f: $\{x \mid x \text{ is a real number}\}$ or $(-\infty, \infty)$.

Range of f: $\{y \mid y > 0\}$ or $(0, \infty)$

Domain of g: $\{x \mid x > 0\}$ or $(0, \infty)$

Range of g: $\{y \mid y \text{ is a real number}\}$ or $(-\infty, \infty)$.

31. $f(x) = \log_8(x + 5)$

$x + 5 > 0$

$\qquad x > -5$

The domain of f is $\{x \mid x > -5\}$ or $(-5, \infty)$.

32. $f(x) = \log(3 - x)$

$3 - x > 0$

$-x > -3$

$x < 3$

The domain of f is $\{x \mid x < 3\}$ or $(-\infty, 3)$.

33. $f(x) = \ln(x - 1)^2$

The domain of g is all real numbers for which $(x - 1)^2 > 0$. The only number that must be excluded is 1. The domain of f is $\{x \mid x \neq 1\}$ or $(-\infty, 1) \cup (1, \infty)$.

34. Since $\ln e^x = x$, $\ln e^{6x} = 6x$.

35. Since $e^{\ln x} = x$, $e^{\ln \sqrt{x}} = \sqrt{x}$.

36. Since $10^{\log x} = x$, $10^{\log 4x^2} = 4x^2$.

37. $R = \log \dfrac{I}{I_0}$

$R = \log \dfrac{1000 I_0}{I_0}$

$R = \log 1000$

$10^R = 1000$

$10^R = 10^3$

$R = 3$

The magnitude on the Richter scale is 3.

38. a. $f(0) = 76 - 18 \log(0 + 1)$

$= 76 - 18 \log 1 = 76 - 18(0) = 76$

The average score when the exam was first given was 76.

b. $f(2) = 76 - 18 \log(2 + 1)$

$= 76 - 18 \log(3) \approx 67.4$

$f(4) = 76 - 18 \log(4 + 1)$

$= 76 - 18 \log(5) \approx 63.4$

$f(6) = 76 - 18 \log(6 + 1)$

$= 76 - 18 \log(7) \approx 60.8$

$f(8) = 76 - 18 \log(8 + 1)$

$= 76 - 18 \log(9) \approx 58.8$

$f(12) = 76 - 18 \log(12 + 1)$

$= 76 - 18 \log(13) \approx 55.9$

The average scores were as follows:

2 months	67.4
4 months	63.4
6 months	60.8
8 months	58.8
12 months	55.9.

c.

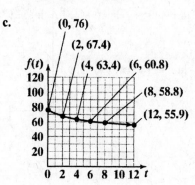

The students retain less material over time.

39. $t = \dfrac{1}{0.06} \ln\left(\dfrac{12}{12 - 5}\right)$

$= \dfrac{1}{0.06} \ln\left(\dfrac{12}{7}\right) \approx 9.0$

It will take approximately 9 weeks for the man to run 5 miles per hour.

40. $\log_6\left(36x^3\right) = \log_6 36 + \log_6 x^3 = 2 + 3 \log_6 x$

41. $\log_4 \dfrac{\sqrt{x}}{64} = \log_4 \sqrt{x} - \log_4 64$

$= \log_4 x^{\frac{1}{2}} - 3 = \dfrac{1}{2} \log_4 x - 3$

42. $\log_2\left(\dfrac{xy^2}{64}\right) = \log_2 xy^2 - \log_2 64$

$= \log_2 x + \log_2 y^2 - 6$

$= \log_2 x + 2 \log_2 y - 6$

43. $\ln \sqrt[3]{\dfrac{x}{e}} = \ln\left(\dfrac{x}{e}\right)^{\frac{1}{3}} = \dfrac{1}{3}\ln\left(\dfrac{x}{e}\right)$

$\qquad = \dfrac{1}{3}\left(\ln x - \ln e\right)$

$\qquad = \dfrac{1}{3}\left(\ln x - 1\right) = \dfrac{1}{3}\ln x - \dfrac{1}{3}$

44. $\log_b 7 + \log_b 3 = \log_b\left(7 \cdot 3\right) = \log_b 21$

45. $\log 3 - 3\log x = \log 3 - \log x^3 = \log\dfrac{3}{x^3}$

46. $3\ln x + 4\ln y = \ln x^3 + \ln y^4 = \ln\left(x^3 y^4\right)$

47. $\dfrac{1}{2}\ln x - \ln y = \ln x^{\frac{1}{2}} - \ln y$

$\qquad = \ln\sqrt{x} - \ln y = \ln\left(\dfrac{\sqrt{x}}{y}\right)$

48. $\log_6 72{,}348 = \dfrac{\log 72{,}348}{\log 6} \approx 6.2448$

49. $\log_4 0.863 = \dfrac{\log 0.863}{\log 4} \approx -0.1063$

50. true

51. false; Changes to make the statement true will vary. A sample change is:

$\log(x+9) - \log(x+1) = \log\left(\dfrac{x+9}{x+1}\right)$.

52. false; Changes to make the statement true will vary. A sample change is: $\log_2 x^4 = 4\log_2 x$.

53. true

54. $2^{4x-2} = 64$

$2^{4x-2} = 2^6$

$4x - 2 = 6$

$\qquad 4x = 8$

$\qquad x = 2$

The solution set is $\{2\}$.

55. $125^x = 25$

$\left(5^3\right)^x = 5^2$

$5^{3x} = 5^2$

$3x = 2$

$x = \dfrac{2}{3}$

The solution set is $\left\{\dfrac{2}{3}\right\}$.

56. $9^x = \dfrac{1}{27}$

$\left(3^2\right)^x = 3^{-3}$

$3^{2x} = 3^{-3}$

$2x = -3$

$x = -\dfrac{3}{2}$

The solution set is $\left\{-\dfrac{3}{2}\right\}$.

57. $8^x = 12{,}143$

$\ln 8^x = \ln 12{,}143$

$x \ln 8 = \ln 12{,}143$

$x = \dfrac{\ln 12{,}143}{\ln 8} \approx 4.52$

The solution set is $\left\{\dfrac{\ln 12{,}143}{\ln 8} \approx 4.52\right\}$.

58. $9e^{5x} = 1269$

$e^{5x} = \dfrac{1269}{9}$

$\ln e^{5x} = \ln 141$

$5x = \ln 141$

$x = \dfrac{\ln 141}{5} \approx 0.99$

The solution set is $\left\{\dfrac{\ln 141}{5} \approx 0.99\right\}$.

59. $30e^{0.045x} = 90$

$\qquad e^{0.045x} = \dfrac{90}{30}$

$\qquad \ln e^{0.045x} = \ln 3$

$\qquad 0.045x = \ln 3$

$\qquad x = \dfrac{\ln 3}{0.045} \approx 24.41$

The solution set is $\left\{ \dfrac{\ln 3}{0.045} \approx 24.41 \right\}$.

60. $\log_5 x = -3$

$\qquad x = 5^{-3}$

$\qquad x = \dfrac{1}{125}$

The solution set is $\left\{ \dfrac{1}{125} \right\}$.

61. $\log x = 2$

$\qquad x = 10^2$

$\qquad x = 100$

The solution set is $\{100\}$.

62. $\log_4(3x - 5) = 3$

$\qquad 3x - 5 = 4^3$

$\qquad 3x - 5 = 64$

$\qquad 3x = 69$

$\qquad x = 23$

The solution set is $\{23\}$.

63. $\ln x = -1$

$\qquad x = e^{-1}$

$\qquad x = \dfrac{1}{e}$

The solution set is $\left\{ \dfrac{1}{e} \right\}$.

64. $3 + 4\ln(2x) = 15$

$\qquad 4\ln(2x) = 12$

$\qquad \ln(2x) = 3$

$\qquad 2x = e^3$

$\qquad x = \dfrac{e^3}{2}$

The solution set is $\left\{ \dfrac{e^3}{2} \right\}$.

65. $\log_2(x + 3) + \log_2(x - 3) = 4$

$\qquad \log_2\big((x+3)(x-3)\big) = 4$

$\qquad \log_2\big(x^2 - 9\big) = 4$

$\qquad x^2 - 9 = 2^4$

$\qquad x^2 - 9 = 16$

$\qquad x^2 = 25$

$\qquad x = \pm 5$

We disregard -5 because it would result in taking the logarithm of a negative number in the original equation. The solution set is $\{5\}$.

66. $\log_3(x - 1) - \log_3(x + 2) = 2$

$\qquad \log_3 \dfrac{x-1}{x+2} = 2$

$\qquad \dfrac{x-1}{x+2} = 3^2$

$\qquad \dfrac{x-1}{x+2} = 9$

$\qquad x - 1 = 9(x + 2)$

$\qquad x - 1 = 9x + 18$

$\qquad -8x - 1 = 18$

$\qquad -8x = 19$

$\qquad x = -\dfrac{19}{8}$

We disregard $-\dfrac{19}{8}$ because it would result in taking the logarithm of a negative number in the original equation. There is no solution. The solution set is $\varnothing$ or $\{\ \ \}$.

67. $\log_4(3x-5) = \log_4 3$

$$3x-5 = 3$$
$$3x = 8$$
$$x = \frac{8}{3}$$

The solution set is $\left\{\frac{8}{3}\right\}$.

68. $\ln(x+4) - \ln(x+1) = \ln x$

$$\ln\frac{x+4}{x+1} = \ln x$$
$$\frac{x+4}{x+1} = x$$
$$(x+1)\frac{x+4}{x+1} = x(x+1)$$
$$x+4 = x^2 + x$$
$$x^2 = 4$$
$$x = \pm 2$$

We disregard –2 because it would result in taking the logarithm of a negative number in the original equation. The solution set is $\{2\}$.

69. $\log_6(2x+1) = \log_6(x-3) + \log_6(x+5)$

$$\log_6(2x+1) = \log_6(x^2+2x-15)$$
$$2x+1 = x^2 + 2x - 15$$
$$x^2 = 16$$
$$x = \pm 4$$

We disregard –4 because it would result in taking the logarithm of a negative number in the original equation. The solution set is $\{4\}$.

70. $P(x) = 14.7e^{-0.21x}$

$$4.6 = 14.7e^{-0.21x}$$
$$\frac{4.6}{14.7} = e^{-0.21x}$$
$$\ln\frac{4.6}{14.7} = \ln e^{-0.21x}$$
$$\ln\frac{4.6}{14.7} = -0.21x$$
$$t = \frac{\ln\frac{4.6}{14.7}}{-0.21} \approx 5.5$$

The peak of Mt. Everest is about 5.5 miles above sea level.

71. $f(t) = 364(1.005)^t$

$$560 = 364(1.005)^t$$
$$\frac{560}{364} = (1.005)^t$$
$$\ln\frac{560}{364} = \ln(1.005)^t$$
$$\ln\frac{560}{364} = t\ln 1.005$$
$$t = \frac{\ln\frac{560}{364}}{\ln 1.005} \approx 86.4$$

The carbon dioxide concentration will be double the pre-industrial level approximately 86 years after the year 2000 in the year 2086.

72. $W(x) = 0.37\ln x + 0.05$

$$3.38 = 0.37\ln x + 0.05$$
$$3.33 = 0.37\ln x$$
$$9 = \ln x$$
$$e^9 = e^{\ln x}$$
$$x = e^9 \approx 8103$$

The population of New Your City is approximately 8103 thousand, or 8,103,000.

73. $20,000 = 12,500\left(1+\frac{0.065}{4}\right)^{4t}$

$$20,000 = 12,500(1+0.01625)^{4t}$$
$$20,000 = 12,500(1.01625)^{4t}$$
$$1.6 = (1.01625)^{4t}$$
$$\ln 1.6 = \ln(1.01625)^{4t}$$
$$\ln 1.6 = 4t\ln 1.01625$$
$$\frac{\ln 1.6}{4\ln 1.01625} = \frac{4t\ln 1.01625}{4\ln 1.01625}$$
$$t = \frac{\ln 1.6}{4\ln 1.01625} \approx 7.3$$

It will take approximately 7.3 years.

74. $3(50,000) = 50,000e^{0.075t}$

$$\frac{3(50,000)}{50,000} = e^{0.075t}$$

$$3 = e^{0.075t}$$

$$\ln 3 = \ln e^{0.075t}$$

$$\ln 3 = 0.075t$$

$$t = \frac{\ln 3}{0.075} \approx 14.6$$

The money will triple in approximately 14.6 years.

75. $3 = e^{r5}$

$$\ln 3 = \ln e^{5r}$$

$$\ln 3 = 5r$$

$$r = \frac{\ln 3}{5} \approx 0.220$$

The money will triple in 5 years if the interest rate is approximately 22%.

76. a. $t = 2005 - 1990 = 15$

$$A = 22.4e^{kt}$$

$$41.9 = 22.4e^{k(15)}$$

$$\frac{41.9}{22.4} = e^{15k}$$

$$\ln \frac{41.9}{22.4} = \ln e^{15k}$$

$$\ln \frac{41.9}{22.4} = 15k$$

$$k = \frac{\ln \dfrac{41.9}{22.4}}{15} \approx 0.042$$

b. Note that 2010 is 20 years after 1990, find A for $t = 20$.

$$A = 22.4e^{0.042t}$$

$$= 22.4e^{0.042(20)} \approx 51.9$$

The population will be about 51.9 million the year 2010.

c. $60 = 22.4e^{0.042t}$

$$\frac{60}{22.4} = e^{0.042t}$$

$$\ln \frac{60}{22.4} = \ln e^{0.042t}$$

$$\ln \frac{60}{22.4} = 0.042t$$

$$t = \frac{\ln \dfrac{60}{22.4}}{0.042} \approx 23$$

The Hispanic resident population will reach 60 million approximately 23 years after 1990, or 2013.

77. Find k:

$$A = A_0 e^{kt}$$

$$\frac{A_0}{2} = A_0 e^{k \cdot 140}$$

$$\frac{1}{2} = e^{140k}$$

$$\ln \frac{1}{2} = \ln e^{140k}$$

$$\ln \frac{1}{2} = 140k$$

$$k = \frac{\ln \dfrac{1}{2}}{140} \approx -0.00495$$

Thus, $A = A_0 e^{-0.00495t}$.

$$0.20 A_0 = A_0 e^{-0.00495t}$$

$$0.20 = e^{-0.00495t}$$

$$\ln 0.20 = \ln e^{-0.00495t}$$

$$\ln 0.20 = -0.00495t$$

$$t = \frac{\ln 0.20}{-0.00495} \approx 325$$

It will take 325 days for polonium-210 to decay to 20% of its original amount.

78. a.

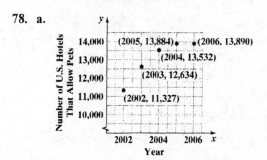

b. A logarithmic function appears to be the best choice for modeling the data.

79. a.

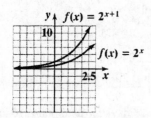

b. An exponential function appears to be the best choice for modeling the data although model choice may vary.

80. $y = 73(2.6)^x$

$y = 73e^{(\ln 2.6)x}$

$y = 73e^{0.956x}$

81. $y = 6.5(0.43)^x$

$y = 6.5e^{(\ln 0.43)x}$

$y = 6.5e^{-0.844x}$

Chapter 12 Test

1. $f(x) = 2^x$

$g(x) = 2^{x+1}$

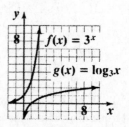

2. Semiannual Compounding:

$$A = 3000\left(1 + \frac{0.065}{2}\right)^{2(10)}$$

$$= 3000(1.0325)^{20} \approx 5687.51$$

Continuous Compounding:

$A = 3000e^{0.06(10)} = 3000e^{0.6} \approx 5466.36$

Semiannual compounding at 6.5% yields a greater return. The difference in the yields is $221.

3. $\log_5 125 = 3$

$5^3 = 125$

4. $\sqrt{36} = 6$

$36^{\frac{1}{2}} = 6$

$\log_{36} 6 = \frac{1}{2}$

5. $f(x) = 3^x$

$g(x) = \log_3 x$

Domain of f: $\{x \mid x \text{ is a real number}\}$ or $(-\infty, \infty)$.

Range of f: $\{y \mid y > 0\}$ or $(0, \infty)$

Domain of g: $\{x \mid x > 0\}$ or $(0, \infty)$

Range of g: $\{y \mid y \text{ is a real number}\}$ or $(-\infty, \infty)$.

6. Since $\ln e^x = x$, $\ln e^{5x} = 5x$.

7. $\log_b b = 1$ because $b^1 = b$.

8. $\log_6 1 = 0$ because $6^0 = 1$.

9. $f(x) = \log_5(x - 7)$

$x - 7 > 0$

$x > 7$

The domain of f is $\{x \mid x > 7\}$ or $(7, \infty)$.

10. $D = 10\log\dfrac{I}{I_0}$

$D = 10\log\dfrac{10^{12} I_0}{I_0}$

$= 10\log 10^{12} = 10(12) = 120$

The sound has a loudness of 120 decibels.

11. $\log_4\left(64x^5\right) = \log_4 64 + \log_4 x^5$

$\qquad\qquad = 3 + 5\log_4 x$

12. $\log_3 \dfrac{\sqrt[3]{x}}{81} = \log_3 \sqrt[3]{x} - \log_3 81$

$\qquad\qquad = \log_3 x^{\frac{1}{3}} - 4 = \dfrac{1}{3}\log_3 x - 4$

13. $6\log x + 2\log y = \log x^6 + \log y^2$

$\qquad\qquad\qquad = \log(x^6 y^2)$

14. $\ln 7 - 3\ln x = \ln 7 - \ln x^3 = \ln\left(\dfrac{7}{x^3}\right)$

15. $\log_{15} 71 = \dfrac{\ln 71}{\ln 15} \approx 1.5741$

16. $3^{x-2} = 81$

$3^{x-2} = 3^4$

$x - 2 = 4$

$\quad x = 6$

The solution set is $\{6\}$.

17. $\quad 5^x = 1.4$

$\ln 5^x = \ln 1.4$

$x\ln 5 = \ln 1.4$

$\quad x = \dfrac{\ln 1.4}{\ln 5} \approx 0.21$

The solution set is $\left\{\dfrac{\ln 1.4}{\ln 5} \approx 0.21\right\}$.

18. $400e^{0.005x} = 1600$

$e^{0.005x} = \dfrac{1600}{400}$

$\ln e^{0.005x} = \ln 4$

$0.005x = \ln 4$

$\quad x = \dfrac{\ln 4}{0.005} \approx 277.26$

The solution set is $\left\{\dfrac{\ln 4}{0.005} \approx 277.26\right\}$.

19. $\log_{25} x = \dfrac{1}{2}$

$x = 25^{\frac{1}{2}} = \sqrt{25} = 5$

The solution set is $\{5\}$.

20. $\log_6\left(4x - 1\right) = 3$

$4x - 1 = 6^3$

$4x - 1 = 216$

$4x = 217$

$x = \dfrac{217}{4}$

The solution set is $\left\{\dfrac{217}{4}\right\}$.

21. $2\ln(3x) = 8$

$\ln(3x) = \dfrac{8}{2}$

$e^{\ln(3x)} = e^4$

$3x = e^4$

$x = \dfrac{e^4}{3}$

The solution set is $\left\{\dfrac{e^4}{3}\right\}$.

22. $\log x + \log\left(x + 15\right) = 2$

$\log\left(x(x + 15)\right) = 2$

$x(x + 15) = 10^2$

$x^2 + 15 = 100$

$x^2 + 15 - 100 = 0$

$(x + 20)(x - 5) = 0$

$x + 20 = 0 \qquad \text{or} \qquad x - 5 = 0$

$x = -20 \qquad\qquad\qquad x = 5$

We disregard -20 because it would result in taking the logarithm of a negative number in the original equation. The solution set is $\{5\}$

23. $\ln(x-4) - \ln(x+1) = \ln 6$

$$\ln\left(\frac{x-4}{x+1}\right) = \ln 6$$

$$\frac{x-4}{x+1} = 6$$

$$(x+1)\frac{x-4}{x+1} = 6(x+1)$$

$$x - 4 = 6x + 6$$

$$-5x = 10$$

$$x = -2$$

We disregard -2 because it would result in taking the logarithm of a negative number in the original equation. The solution set is { }.

24. a. $P(0) = 82.4e^{-0.002(0)}$

$$= 82.4e^0 = 82.4(1) = 82.4$$

In 2006, the population of Germany was 82.4 million.

b. The population of Germany is decreasing. The growth rate, $k = -0.002$, is negative.

c. $81.4 = 82.4e^{-0.002t}$

$$\frac{81.4}{82.4} = e^{-0.002t}$$

$$\ln\frac{81.4}{82.4} = \ln e^{-0.002t}$$

$$\ln\frac{81.4}{82.4} = -0.002t$$

$$t = \frac{\ln\dfrac{81.4}{82.4}}{-0.002} \approx 6$$

The population of Germany will be 81.5 million approximately 6 years after 2006, or 2012.

25.

$$8000 = 4000\left(1 + \frac{0.05}{4}\right)^{4t}$$

$$\frac{8000}{4000} = (1 + 0.0125)^{4t}$$

$$2 = (1.0125)^{4t}$$

$$\ln 2 = \ln(1.0125)^{4t}$$

$$\ln 2 = 4t\ln(1.0125)$$

$$\frac{\ln 2}{4\ln(1.0125)} = \frac{4t\ln(1.0125)}{4\ln(1.0125)}$$

$$t = \frac{\ln 2}{4\ln(1.0125)} \approx 13.9$$

It will take approximately 13.9 years for the money to grow to $8000.

26. $2 = 1e^{r10}$

$$2 = e^{10r}$$

$$\ln 2 = \ln e^{10r}$$

$$\ln 2 = 10r$$

$$r = \frac{\ln 2}{10} \approx 0.069$$

The money will double in 10 years with an interest rate of approximately 6.9%.

27. Substitute $A_0 = 509$, $A = 729$, and $t = 2000 - 1990 = 10$ into the general growth function to determine the growth rate k:

$$A = A_0 e^{kt}$$

$$729 = 509e^{k(10)}$$

$$\frac{729}{509} = e^{10k}$$

$$\ln\frac{729}{509} = \ln e^{10k}$$

$$\ln\frac{729}{509} = 10k$$

$$k = \frac{\ln\dfrac{729}{509}}{10} \approx 0.036$$

The exponential growth function is $A = 509e^{0.036t}$.

28. $A = A_0 e^{-0.000121t}$

$5 = 100 e^{-0.000121t}$

$\dfrac{5}{100} = e^{-0.000121t}$

$\ln 0.05 = \ln e^{-0.000121t}$

$\ln 0.05 = -0.000121t$

$t = \dfrac{\ln 0.05}{-0.000121} \approx 24758$

The man died approximately 24,758 years ago.

29. Plot the ordered pairs.

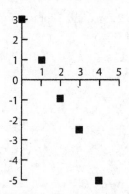

The values appear to belong to a linear function.

30. Plot the ordered pairs.

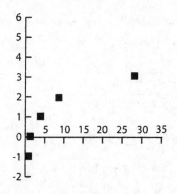

The values appear to belong to a logarithmic function.

31. Plot the ordered pairs.

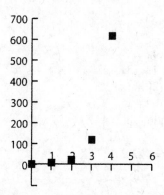

The values appear to belong to an exponential function.

32. Plot the ordered pairs.

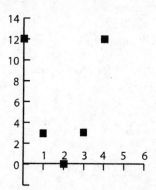

The values appear to belong to a quadratic function.

33. $y = 96(0.38)^x$

$y = 96 e^{(\ln 0.38)x}$

$y = 96 e^{-0.968x}$

Cumulative Review Exercises (Chapters 1 – 12)

1. $8-(4x-5)=x-7$
   $8-4x+5=x-7$
   $13-4x=x-7$
   $13=5x-7$
   $20=5x$
   $4=x$

   The solution set is $\{4\}$.

2. $5x+4y=22$
   $3x-8y=-18$

   Multiply the first equation by 2 and solve by addition.
   $10x+8y=44$
   $\underline{3x-8y=-18}$
   $13x=26$
   $x=2$

   Back-substitute 2 for x to find y.
   $5(2)+4y=22$
   $10+4y=22$
   $4y=12$
   $y=3$

   The solution set is $\{(2,3)\}$.

3. $-3x+2y+4z=6$
   $7x-y+3z=23$
   $2x+3y+z=7$

   Multiply the second equation by 2 and add to the first equation to eliminate y.
   $-3x+2y+4z=6$
   $\underline{14x-2y+6z=46}$
   $11x+10z=52$

   Multiply the second equation by 3 and add to the third equation to eliminate y.
   $21x-3y+9z=69$
   $\underline{2x+3y+z=7}$
   $23x+10z=76$

   The system of two variables in two equations is:
   $11x+10z=52$
   $23x+10z=76$

Multiply the first equation by -1 and add to the second equation.
$-11x-10z=-52$
$\underline{23x+10z=76}$
$12x=24$
$x=2$

Back-substitute 2 for x to find z.
$11(2)+10z=52$
$22+10z=52$
$10z=30$
$z=3$

Back-substitute 2 for x and 3 for z to find y.
$-3(2)+2y+4(3)=6$
$-6+2y+12=6$
$2y=0$
$y=0$

The solution is $\{(2, 0, 3)\}$.

4. $|x-1|>3$
   $x-1<-3$ or $x-1>3$
   $x<-2$ $\qquad$ $x>4$

   The solution set is $\{x|x<-2 \text{ or } x>4\}$ or $(-\infty,-2)\cup(4,\infty)$.

5. $\sqrt{x+4}-\sqrt{x-4}=2$
   $\sqrt{x+4}=2+\sqrt{x-4}$
   $(\sqrt{x+4})^2=(2+\sqrt{x-4})^2$
   $x+4=4+4\sqrt{x-4}+x-4$
   $\cancel{x}+4=4\sqrt{x-4}+\cancel{x}$
   $4=4\sqrt{x-4}$
   $1=\sqrt{x-4}$
   $1^2=(\sqrt{x-4})^2$
   $1=x-4$
   $5=x$

   The solution set is $\{5\}$.

6. $x - 4 \geq 0$ and $-3x \leq -6$

$x \geq 4 x \geq 2$

For a value to be in the solution set, it must satisfy both of the conditions $x \geq 4$ and $x \geq 2$. Now any value that is 4 or larger is also larger than 2. But values between 2 and 4 do not satisfy both conditions. Therefore, only values that are 4 or larger will be in the solution set. Thus, the solution set is $\{x | x \geq 4\}$ or $[4, \infty)$.

7. $ 2x^2 = 3x - 2$

$ 2x^2 - 3x + 2 = 0$

Solve using the quadratic formula.

$a = 2 b = -3 c = 2$

$$x = \frac{-(-3) \pm \sqrt{(-3)^2 - 4(2)(2)}}{2(2)}$$

$$= \frac{3 \pm \sqrt{9 - 16}}{4}$$

$$= \frac{3 \pm \sqrt{-7}}{4} = \frac{3 \pm i\sqrt{7}}{4} = \frac{3}{4} \pm \frac{\sqrt{7}}{4} i$$

The solutions are $\frac{3}{4} \pm \frac{\sqrt{7}}{4} i$, and the solution set is

$\left\{ \frac{3}{4} - \frac{\sqrt{7}}{4} i, \frac{3}{4} + \frac{\sqrt{7}}{4} i \right\}$.

8. $3x = 15 + 5y$

Find the x–intercept by setting $y = 0$ and solving.

$3x = 15 + 5(0)$

$3x = 15$

$x = 5$

Find the y–intercept by setting $x = 0$ and solving.

$3(0) = 15 + 5y$

$0 = 15 + 5y$

$-15 = 5y$

$-3 = y$

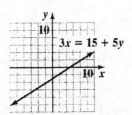

9. $2x - 3y > 6$

First, find the intercepts to the equation $2x - 3y = 6$.

Find the x–intercept by setting $y = 0$ and solving.

$2x - 3(0) = 6$

$2x = 6$

$x = 3$

Find the y–intercept by setting $x = 0$ and solving.

$2(0) - 3y = 6$

$-3y = 6$

$y = -2$

Next, use the origin as a test point.

$2(0) - 3(0) > 6$

$0 > 6$

This is a false statement. This means that the origin will not fall in the shaded half-plane.

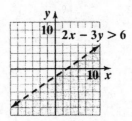

10. $f(x) = -\frac{1}{2}x + 1$

$m = -\frac{1}{2}; \; y - \text{intercept} = 1$

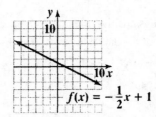

11. $f(x) = x^2 + 6x + 8$

Since $a = 1$ is positive, the parabola opens upward. The x–coordinate of the vertex of the parabola is $-\dfrac{b}{2a} = -\dfrac{6}{2(1)} = -\dfrac{6}{2} = -3$ and the y–coordinate of the vertex of the parabola is

$$f\left(-\dfrac{b}{2a}\right) = f(-3) = (-3)^2 + 6(-3) + 8$$
$$= 9 - 18 + 8 = -1.$$

The vertex is $(-3, -1)$. Replace $f(x)$ with 0 to find x–intercepts.

$$0 = x^2 + 6x + 8$$
$$0 = (x+4)(x+2)$$
$$x + 4 = 0 \quad \text{or} \quad x + 2 = 0$$
$$x = -4 \qquad\qquad x = -2$$

The x–intercepts are -4 and -2. Set $x = 0$ and solve for y to obtain the y–intercept.

$$y = 0^2 + 6(0) + 8$$
$$y = 0 + 0 + 8$$
$$y = 8$$

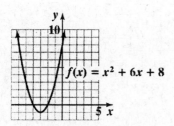

12. $f(x) = (x-3)^2 - 4$

Since $a = 1$ is positive, the parabola opens upward. The vertex of the parabola is $(h, k) = (3, -4)$ and the axis of symmetry is $x = 3$. Replace $f(x)$ with 0 to find x–intercepts.

$$0 = (x-3)^2 - 4$$
$$4 = (x-3)^2$$

Apply the square root property.
$$x - 3 = -2 \quad \text{or} \quad x - 3 = 2$$
$$x = 1 \qquad\qquad x = 5$$

The x–intercepts are 1 and 5.

Set $x = 0$ and solve for y to obtain the y–intercept.
$$y = (0-3)^2 - 4$$
$$y = (-3)^2 - 4$$
$$y = 9 - 4$$
$$y = 5$$

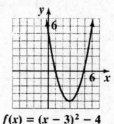

$f(x) = (x - 3)^2 - 4$

13.
$$A = \dfrac{cd}{c+d}$$
$$A(c+d) = cd$$
$$Ac + Ad = cd$$
$$Ac - cd = -Ad$$
$$c(A-d) = -Ad$$
$$c = -\dfrac{Ad}{A-d} \quad \text{or} \quad \dfrac{Ad}{d-A}$$

14.
$$f(g(x)) = \big[g(x)\big]^2 + 3\big[g(x)\big] - 15$$
$$= (x-2)^2 + 3(x-2) - 15$$
$$= x^2 - 4x + 4 + 3x - 6 - 15$$
$$= x^2 - x - 17$$

15.
$$g(f(x)) = f(x) - 2$$
$$= \left(x^2 + 3x - 15\right) - 2$$
$$= x^2 + 3x - 17$$

16.
$$g(a+h) - g(a) = ((a+h) - 2) - (a - 2)$$
$$= a + h - 2 - a + 2$$
$$= h$$

17. $f(x) = 7x - 3$

$$y = 7x - 3$$

Interchange x and y, and solve for y.

$$x = 7y - 3$$

$$x + 3 = 7y$$

$$\frac{x+3}{7} = y$$

Thus, $f^{-1}(x) = \frac{x+3}{7}$.

18. $f(x) = \frac{x-2}{x^2 - 3x + 2}$

Since a denominator cannot equal zero, exclude from the domain all values which make $x^2 - 3x + 2 = 0$.

$$x^2 - 3x + 2 = 0$$

$$(x-2)(x-1) = 0$$

$$x - 2 = 0 \quad \text{or} \quad x - 1 = 0$$

$$x = 2 \qquad\qquad x = 1$$

The domain of f is $\{x \mid x$ is a real number and $x \neq 1$ and $x \neq 2\}$ or $(-\infty, 1) \cup (1, 2) \cup (2, \infty)$.

19. $f(x) = \ln(2x - 8)$

To find the domain, find all values of x for which $2x - 8$ is greater than zero.

$$2x - 8 > 0$$

$$2x > 8$$

$$x > 4$$

The domain of f is $\{x \mid x > 4\}$ or $(4, \infty)$.

20. First, solve for y to obtain the slope of the line whose equation is $2x + y = 10$.

$$2x + y = 10$$

$$y = -2x + 10$$

The slope is –2. The line we want to find is perpendicular to this line, so we know the slope will be $\frac{1}{2}$.

Using the point, $(-2, 4)$, and the slope, $\frac{1}{2}$, we can write the equation in point-slope form.

$$y - y_1 = m(x - x_1)$$

$$y - 4 = \frac{1}{2}(x - (-2))$$

$$y - 4 = \frac{1}{2}(x + 2)$$

Solve for y to obtain slope-intercept form.

$$y - 4 = \frac{1}{2}(x + 2)$$

$$y - 4 = \frac{1}{2}x + 1$$

$$y = \frac{1}{2}x + 5$$

$$f(x) = \frac{1}{2}x + 5$$

21. $\dfrac{-5x^3 y^7}{15x^4 y^{-2}} = \dfrac{-y^7 y^2}{3x} = \dfrac{-y^9}{3x} = -\dfrac{y^9}{3x}$

22. $\left(4x^2 - 5y\right)^2$

$$= \left(4x^2\right)^2 + 2\left(4x^2\right)(-5y) + (-5y)^2$$

$$= 16x^4 - 40x^2 y + 25y^2$$

23.

$$5x+1 \overline{)5x^3-24x^2+0x+9}$$

quotient: x^2-5x+1

$$\underline{5x^3+x^2}$$
$$-25x^2+0x$$
$$\underline{-25x^2-5x}$$
$$5x+9$$
$$\underline{5x+1}$$
$$8$$

$$\frac{5x^3-24x^2+9}{5x+1}=x^2-5x+1+\frac{8}{5x+1}$$

24.

$$\frac{\sqrt[3]{32xy^{10}}}{\sqrt[3]{2xy^2}}=\sqrt[3]{\frac{32xy^{10}}{2xy^2}}$$

$$=\sqrt[3]{16y^8}$$

$$=\sqrt[3]{8\cdot 2y^6 y^2}=2y^2\sqrt[3]{2y^2}$$

25.

$$\frac{x+2}{x^2-6x+8}+\frac{3x-8}{x^2-5x+6}$$

$$=\frac{x+2}{(x-4)(x-2)}+\frac{3x-8}{(x-2)(x-3)}$$

$$=\frac{(x+2)(x-3)}{(x-4)(x-2)(x-3)}+\frac{(3x-8)(x-4)}{(x-4)(x-2)(x-3)}$$

$$=\frac{x^2-3x+2x-6+3x^2-12x-8x+32}{(x-4)(x-2)(x-3)}$$

$$=\frac{4x^2-21x+26}{(x-4)(x-2)(x-3)}$$

$$=\frac{(4x-13)\cancel{(x-2)}}{(x-4)\cancel{(x-2)}(x-3)}=\frac{4x-13}{(x-4)(x-3)}$$

26. $x^4-4x^3+8x-32$

$$=x^3(x-4)+8(x-4)$$

$$=(x-4)(x^3+8)$$

$$=(x-4)(x+2)(x^2-2x+4)$$

27. $2x^2+12xy+18y^2$

$$=2(x^2+6xy+9y^2)=2(x+3y)^2$$

28. $2\ln x-\dfrac{1}{2}\ln y=\ln x^2-\ln y^{\frac{1}{2}}$

$$=\ln\left(\frac{x^2}{y^{\frac{1}{2}}}\right)$$

$$=\ln\left(\frac{x^2}{\sqrt{y}}\right)$$

29. Let x = the width of the carpet.
Let $2x + 4$ = the length of the carpet.

$$x(2x+4)=48$$

$$2x^2+4x=48$$

$$2x^2+4x-48=0$$

$$x^2+2x-24=0$$

$$(x+6)(x-4)=0$$

$$x+6=0 \quad \text{and} \quad x-4=0$$

$$x=-6 \qquad\qquad x=4$$

We disregard –6 because we can't have a negative length measurement. The width of the carpet is 4 feet and the length of the carpet is

$$2x+4=2(4)+4=8+4=12 \text{ feet.}$$

30. Let x = time it takes when working together.

	Part done in 1 hour	Time Working Together	Part done in x hours
You	$\dfrac{1}{2}$	x	$\dfrac{x}{2}$
Your Sister	$\dfrac{1}{3}$	x	$\dfrac{x}{3}$

$$\frac{x}{2}+\frac{x}{3}=1$$

$$6\left(\frac{x}{2}+\frac{x}{3}\right)=6(1)$$

$$6\left(\frac{x}{2}\right)+6\left(\frac{x}{3}\right)=6$$

$$3x+2x=6$$

$$5x=6$$

$$x=\frac{6}{5}$$

If you and your sister work together, it will take $\dfrac{6}{5}$ hours, or 1 hour and 12 minutes, to clean the house.

31. Let $x =$ the rate of the current.

	distance d	rate r	time $t = \dfrac{d}{r}$
with the current	20	$15 + x$	$\dfrac{20}{15+x}$
against the current	10	$15 - x$	$\dfrac{10}{15-x}$

$$\frac{20}{15+x} = \frac{10}{15-x}$$
$$20(15-x) = 10(15+x)$$
$$300 - 20x = 150 + 10x$$
$$300 = 150 + 30x$$
$$150 = 30x$$
$$5 = x$$

The rate of the current is 5 miles per hour.

32.
$$A = Pe^{rt}$$
$$18,000 = 6000e^{r(10)}$$
$$3 = e^{10r}$$
$$\ln 3 = \ln e^{10r}$$
$$\ln 3 = 10r$$
$$r = \frac{\ln 3}{10} \approx 0.11$$

An interest rate of approximately 11% compounded continuously would be required for $6000 to grow to $18,000 in 10 years.

Chapter 13
Conic Sections and Systems of Nonlinear Equations

13.1 Check Points

1. $(x-h)^2 + (y-k)^2 = r^2$

   $(x-0)^2 + (y-0)^2 = 4^2$

   $x^2 + y^2 = 16$

2. $(x-h)^2 + (y-k)^2 = r^2$

   $(x-5)^2 + (y-(-6))^2 = 10^2$

   $(x-5)^2 + (y+6)^2 = 100$

3. $(x+3)^2 + (y-1)^2 = 4$

   $(x-(-3))^2 + (y-1)^2 = 2^2$

   The center is $(-3,1)$ and the radius is 2 units.

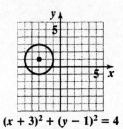

$(x + 3)^2 + (y - 1)^2 = 4$

4. $x^2 + y^2 + 4x - 4y - 1 = 0$

   $(x^2 + 4x \quad) + (y^2 - 4y \quad) = 1$

   Complete the squares.

   $\left(\frac{b}{2}\right)^2 = \left(\frac{4}{2}\right)^2 = (2)^2 = 4$

   $\left(\frac{b}{2}\right)^2 = \left(\frac{-4}{2}\right)^2 = (-2)^2 = 4$

   $(x^2 + 4x + 4) + (y^2 - 4y + 4) = 1 + 4 + 4$

   $(x+2)^2 + (y-2)^2 = 9$

   $(x-(-2))^2 + (y-2)^2 = 3^2$

The center is $(-2,2)$ and the radius is 3 units.

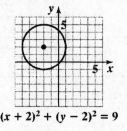

$(x + 2)^2 + (y - 2)^2 = 9$

13.1 Exercise Set

1. $(x-h)^2 + (y-k)^2 = r^2$

   $(x-0)^2 + (y-0)^2 = 7^2$

   $x^2 + y^2 = 49$

3. $(x-h)^2 + (y-k)^2 = r^2$

   $(x-3)^2 + (y-2)^2 = 5^2$

   $(x-3)^2 + (y-2)^2 = 25$

5. $(x-h)^2 + (y-k)^2 = r^2$

   $(x-(-1))^2 + (y-4)^2 = 2^2$

   $(x+1)^2 + (y-4)^2 = 4$

7. $(x-h)^2 + (y-k)^2 = r^2$

   $(x-(-3))^2 + (y-(-1))^2 = (\sqrt{3})^2$

   $(x+3)^2 + (y+1)^2 = 3$

9. $(x-h)^2 + (y-k)^2 = r^2$

   $(x-(-4))^2 + (y-0)^2 = 10^2$

   $(x+4)^2 + y^2 = 100$

11.　　　　$x^2 + y^2 = 16$

$(x-0)^2 + (y-0)^2 = 4^2$

The center is $(0,0)$ and the radius is 4 units.

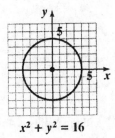

$x^2 + y^2 = 16$

13. $(x-3)^2 + (y-1)^2 = 36$

$(x-3)^2 + (y-1)^2 = 6^2$

The center is $(3,1)$ and the radius is 6 units.

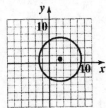

$(x - 3)^2 + (y - 1)^2 = 36$

15.　　$(x+3)^2 + (y-2)^2 = 4$

$(x-(-3))^2 + (y-2)^2 = 2^2$

The center is $(-3,2)$ and the radius is 2 units.

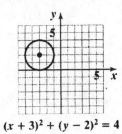

$(x + 3)^2 + (y - 2)^2 = 4$

17.　　　$(x+2)^2 + (y+2)^2 = 4$

$(x-(-2))^2 + (y-(-2))^2 = 2^2$

The center is $(-2,-2)$ and the radius is 2 units.

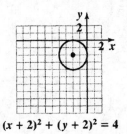

$(x + 2)^2 + (y + 2)^2 = 4$

19.　　　$x^2 + y^2 + 6x + 2y + 6 = 0$

$(x^2 + 6x \quad) + (y^2 + 2y \quad) = -6$

Complete the squares.

$\left(\dfrac{b}{2}\right)^2 = \left(\dfrac{6}{2}\right)^2 = (3)^2 = 9$

$\left(\dfrac{b}{2}\right)^2 = \left(\dfrac{2}{2}\right)^2 = (1)^2 = 1$

$(x^2 + 6x + 9) + (y^2 + 2y + 1) = -6 + 9 + 1$

$(x+3)^2 + (y+1)^2 = 4$

The center is $(-3,-1)$ and the radius is 2 units.

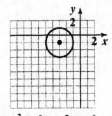

$x^2 + y^2 + 6x + 2y + 6 = 0$

21.
$$x^2 + y^2 - 10x - 6y - 30 = 0$$
$$\left(x^2 - 10x \quad\right) + \left(y^2 - 6y \quad\right) = 30$$

Complete the squares.

$$\left(\frac{b}{2}\right)^2 = \left(\frac{-10}{2}\right)^2 = (-5)^2 = 25$$

$$\left(\frac{b}{2}\right)^2 = \left(\frac{-6}{2}\right)^2 = (-3)^2 = 9$$

$$\left(x^2 - 10x + 25\right) + \left(y^2 - 6y + 9\right) = 30 + 25 + 9$$
$$(x - 5)^2 + (y - 3)^2 = 64$$

The center is $(5, 3)$ and the radius is 8 units.

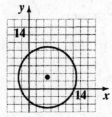

$$x^2 + y^2 - 10x - 6y - 30 = 0$$

23.
$$x^2 + y^2 + 8x - 2y - 8 = 0$$
$$\left(x^2 + 8x \quad\right) + \left(y^2 - 2y \quad\right) = 8$$

Complete the squares.

$$\left(\frac{b}{2}\right)^2 = \left(\frac{8}{2}\right)^2 = (4)^2 = 16$$

$$\left(\frac{b}{2}\right)^2 = \left(\frac{-2}{2}\right)^2 = (-1)^2 = 1$$

$$\left(x^2 + 8x + 16\right) + \left(y^2 - 2y + 1\right) = 8 + 16 + 1$$
$$(x + 4)^2 + (y - 1)^2 = 25$$

The center is $(-4, 1)$ and the radius is 5 units.

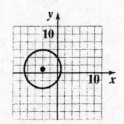

$$x^2 + y^2 + 8x - 2y - 8 = 0$$

25.
$$x^2 - 2x + y^2 - 15 = 0$$
$$\left(x^2 - 2x \quad\right) + y^2 = 15$$

Complete the square.

$$\left(\frac{b}{2}\right)^2 = \left(\frac{-2}{2}\right)^2 = (-1)^2 = 1$$

$$\left(x^2 - 2x + 1\right) + y^2 = 15 + 1$$
$$(x - 1)^2 + y^2 = 16$$

The center is $(1, 0)$ and the radius is 4 units.

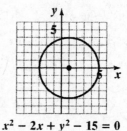

$$x^2 - 2x + y^2 - 15 = 0$$

27.
$$x^2 + y^2 = 16$$
$$x - y = 4$$

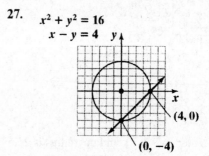

Intersection points: $(0, -4)$ and $(4, 0)$

Check $(0, -4)$:

$$0^2 + (-4)^2 = 16 \qquad 0 - (-4) = 4$$
$$16 = 16 \text{ true} \qquad 4 = 4 \text{ true}$$

Check $(4, 0)$:

$$4^2 + 0^2 = 16 \qquad 4 - 0 = 4$$
$$16 = 16 \text{ true} \qquad 4 = 4 \text{ true}$$

The solution set is $\{(0, -4), (4, 0)\}$.

29. $(x-2)^2 + (y+3)^2 = 4$

$y = x - 3$

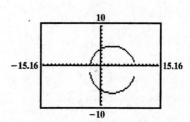

Intersection points: $(0,-3)$ and $(2,-1)$

Check $(0,-3)$:

$(0-2)^2 + (-3+3)^2 = 9$　　$-3 = 0 - 3$

$(-2)^2 + 0^2 = 4$　　　　　$-3 = -3$ true

$4 = 4$

true

Check $(2,-1)$:

$(2-2)^2 + (-1+3)^2 = 4$　　$-1 = 2 - 3$

$0^2 + 2^2 = 4$　　　　　　$-1 = -1$ true

$4 = 4$

true

The solution set is $\{(0,-3),(2,-1)\}$.

31. From the graph we can see that the center of the circle is at $(2,-1)$ and the radius is 2 units. Therefore, the equation is

$(x-2)^2 + (y-(-1))^2 = 2^2$

$(x-2)^2 + (y+1)^2 = 4$

33. From the graph we can see that the center of the circle is at $(-3,-2)$ and the radius is 1 unit. Therefore, the equation is

$(x-(-3))^2 + (y-(-2))^2 = 1^2$

$(x+3)^2 + (y+2)^2 = 1$

35. a. Since the line segment passes through the center, the center is the midpoint of the segment.

$M = \left(\dfrac{x_1 + x_2}{2}, \dfrac{y_1 + y_2}{2}\right)$

$= \left(\dfrac{3+7}{2}, \dfrac{9+11}{2}\right) = \left(\dfrac{10}{2}, \dfrac{20}{2}\right)$

$= (5,10)$

The center is $(5,10)$.

b. The radius is the distance from the center to one of the points on the circle. Using the point $(3,9)$, we get:

$d = \sqrt{(5-3)^2 + (10-9)^2}$

$= \sqrt{2^2 + 1^2} = \sqrt{4+1}$

$= \sqrt{5}$

The radius is $\sqrt{5}$ units.

c. $(x-5)^2 + (y-10)^2 = \left(\sqrt{5}\right)^2$

$(x-5)^2 + (y-10)^2 = 5$

37. If we place L.A. at the origin, then we want the equation of a circle with center at $(-2.4, -2.7)$ and radius 30.

$\left(x-(-2.4)\right)^2 + \left(y-(-2.7)\right)^2 = 30^2$

$(x+2.4)^2 + (y+2.7)^2 = 900$

39. – 43. Answers will vary.

45. $(y+1)^2 = 36 - (x-3)^2$

$y+1 = \pm\sqrt{36 - (x-3)^2}$

$y = -1 \pm \sqrt{36 - (x-3)^2}$

47. does not make sense; Explanations will vary. Sample explanation: The radius of a circle cannot be 0.

49. does not make sense; Explanations will vary. Sample explanation: Because of the negative on the constant 4, this is not the equation of a circle.

51. false; Changes to make the statement true will vary. A sample change is: The circle has a radius of 4.

53. false; Changes to make the statement true will vary. A sample change is: Because the variables are not squared, this is a linear equation, not a circle.

55. The graph of $(x-2)^2 + (y+3)^2 = 25$ is circle with radius 5 and center at $(2,-3)$. The graph of $(x-2)^2 + (y+3)^2 = 36$ is a circle of radius 6 and center at $(2,-3)$.

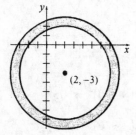

To find the area of the doughnut-shaped region, we find the area of the larger circle and subtract the area of the smaller circle.

$$A_d = A_L - A_S$$
$$= \pi(6)^2 - \pi(5)^2$$
$$= \pi(36-25)$$
$$= 11\pi \approx 35$$

The area of the doughnut-shaped region is $11\pi \approx 35$ square units.

57. $f(g(x)) = f(3x+4) = (3x+4)^2 - 2$
$$= 9x^2 + 24x + 16 - 2$$
$$= 9x^2 + 24x + 14$$

$g(f(x)) = g(x^2-2) = 3(x^2-2) + 4$
$$= 3x^2 - 6 + 4 = 3x^2 - 2$$

58.
$$2x = \sqrt{7x-3} + 3$$
$$2x - 3 = \sqrt{7x-3}$$
$$(2x-3)^2 = 7x-3$$
$$4x^2 - 12x + 9 = 7x - 3$$
$$4x^2 - 19x + 12 = 0$$
$$(4x-3)(x-4) = 0$$
$$4x - 3 = 0 \quad \text{or} \quad x - 4 = 0$$
$$4x = 3 \qquad\qquad x = 4$$
$$x = \frac{3}{4}$$

The solution $\frac{3}{4}$ does not check. The solution set is $\{4\}$.

59.
$$|2x-5| < 10$$
$$-10 < 2x - 5 < 10$$
$$-10 + 5 < 2x - 5 + 5 < 10 + 5$$
$$-5 < 2x < 15$$
$$-\frac{5}{2} < x < \frac{15}{2}$$

The solution set is $\left\{ x \middle| -\frac{5}{2} < x < \frac{15}{2} \right\}$ or $\left(-\frac{5}{2}, \frac{15}{2} \right)$.

60. $\dfrac{x^2}{9} + \dfrac{y^2}{4} = 1$

$$\frac{x^2}{9} + \frac{0^2}{4} = 1$$
$$\frac{x^2}{9} = 1$$
$$x^2 = 9$$
$$x = \pm 3$$

The x-intercepts are -3 and 3.

61. $\dfrac{x^2}{9} + \dfrac{y^2}{4} = 1$

$$\frac{0^2}{9} + \frac{y^2}{4} = 1$$
$$\frac{y^2}{4} = 1$$
$$y^2 = 4$$
$$x = \pm 2$$

The y-intercepts are -2 and 2.

62. $25x^2 + 16y^2 = 400$

$$\frac{25x^2}{400} + \frac{16y^2}{400} = \frac{400}{400}$$
$$\frac{x^2}{16} + \frac{y^2}{25} = 1$$

13.2 Check Points

1. $\dfrac{x^2}{36} + \dfrac{y^2}{9} = 1$

Because the denominator of the x^2 – term is greater than the denominator of the y^2 – term, the major axis is horizontal. Since $a^2 = 36$, $a = 6$ and the vertices are $(-6,0)$ and $(6,0)$. Since $b^2 = 9$, $b = 3$ and endpoints of the minor axis are $(0,-3)$ and $(0,3)$.

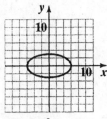

$\dfrac{x^2}{36} + \dfrac{y^2}{9} = 1$

2. $16x^2 + 9y^2 = 144$

$\dfrac{16x^2}{144} + \dfrac{9y^2}{144} = \dfrac{144}{144}$

$\dfrac{x^2}{9} + \dfrac{y^2}{16} = 1$

Because the denominator of the y^2 – term is greater than the denominator of the x^2 – term, the major axis is vertical. Since $a^2 = 16$, $a = 4$ and the vertices are $(0,-4)$ and $(0,4)$. Since $b^2 = 9$, $b = 3$ and endpoints of the minor axis are $(-3,0)$ and $(3,0)$.

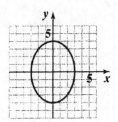

$16x^2 + 9y^2 = 144$

3. $\dfrac{(x+1)^2}{9} + \dfrac{(y-2)^2}{4} = 1$

The center of the ellipse is $(-1,2)$. Because the denominator of the x^2 – term is greater than the denominator of the y^2 – term, the major axis is horizontal. Since $a^2 = 9$, $a = 3$ and the vertices lie 3 units to the right and left of the center. Since $b^2 = 4$, $b = 2$ and endpoints of the minor axis lie 2 units above and below the center.

Center	Vertices	Endpoints Minor Axis
$(-1,2)$	$(-1-3,2)$ $= (-4,2)$	$(-1,2-2)$ $= (-1,0)$
	$(-1+3,2)$ $= (2,2)$	$(-1,2+2)$ $= (-1,4)$

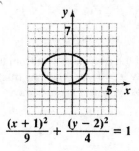

$\dfrac{(x+1)^2}{9} + \dfrac{(y-2)^2}{4} = 1$

4. Using the equation $\dfrac{x^2}{a^2} + \dfrac{y^2}{b^2} = 1$ the archway can be

expressed as $\dfrac{x^2}{20^2} + \dfrac{y^2}{10^2} = 1$ or $\dfrac{x^2}{400} + \dfrac{y^2}{100} = 1$.

Since the truck is 12 feet wide, we need to

determine the height of the archway at $\dfrac{12}{2} = 6$ feet

from the center.
Substitute 6 for x to find the height y.

$$\frac{6^2}{400} + \frac{y^2}{100} = 1$$

$$\frac{36}{400} + \frac{y^2}{100} = 1$$

$$400\left(\frac{36}{400} + \frac{y^2}{100}\right) = 400(1)$$

$$36 + 4y^2 = 400$$

$$4y^2 = 364$$

$$y^2 = 91$$

$$y = \sqrt{91} \approx 9.54$$

The height of the archway 6 feet from the center is approximately 9.54 feet. Since the truck is 9 feet high, the truck will clear the archway.

13.2 Exercise Set

1. $\dfrac{x^2}{16} + \dfrac{y^2}{4} = 1$

Because the denominator of the x^2 – term is greater than the denominator of the y^2 – term, the major axis is horizontal. Since $a^2 = 16$, $a = 4$ and the vertices are $(-4, 0)$ and $(4, 0)$. Since $b^2 = 4$, $b = 2$ and endpoints of the minor axis are $(0, -2)$ and $(0, 2)$.

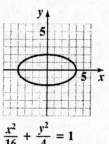

$$\frac{x^2}{16} + \frac{y^2}{4} = 1$$

3. $\dfrac{x^2}{9} + \dfrac{y^2}{36} = 1$

Because the denominator of the y^2 – term is greater than the denominator of the x^2 – term, the major axis is vertical. Since $a^2 = 36$, $a = 6$ and the vertices are $(0, -6)$ and $(0, 6)$. Since $b^2 = 9$, $b = 3$ and endpoints of the minor axis are $(-3, 0)$ and $(3, 0)$.

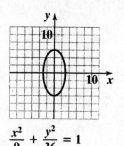

$$\frac{x^2}{9} + \frac{y^2}{36} = 1$$

5. $\dfrac{x^2}{25} + \dfrac{y^2}{64} = 1$

Because the denominator of the y^2 – term is greater than the denominator of the x^2 – term, the major axis is vertical. Since $a^2 = 64$, $a = 8$ and the vertices are $(0, -8)$ and $(0, 8)$. Since $b^2 = 25$, $b = 5$ and endpoints of the minor axis are $(-5, 0)$ and $(5, 0)$.

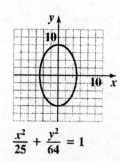

$$\frac{x^2}{25} + \frac{y^2}{64} = 1$$

7. $\dfrac{x^2}{49} + \dfrac{y^2}{81} = 1$

Because the denominator of the y^2 – term is greater than the denominator of the x^2 – term, the major axis is vertical. Since $a^2 = 81$, $a = 9$ and the vertices are $(0,-9)$ and $(0,9)$. Since $b^2 = 49$, $b = 7$ and endpoints of the minor axis are $(-7,0)$ and $(7,0)$.

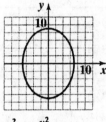

$$\dfrac{x^2}{49} + \dfrac{y^2}{81} = 1$$

9. $25x^2 + 4y^2 = 100$

$\dfrac{25x^2}{100} + \dfrac{4y^2}{100} = \dfrac{100}{100}$

$\dfrac{x^2}{4} + \dfrac{y^2}{25} = 1$

Because the denominator of the y^2 – term is greater than the denominator of the x^2 – term, the major axis is vertical. Since $a^2 = 25$, $a = 5$ and the vertices are $(0,-5)$ and $(0,5)$. Since $b^2 = 4$, $b = 2$ and endpoints of the minor axis are $(-2,0)$ and $(2,0)$.

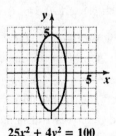

$25x^2 + 4y^2 = 100$

11. $4x^2 + 16y^2 = 64$

$\dfrac{4x^2}{64} + \dfrac{16y^2}{64} = \dfrac{64}{64}$

$\dfrac{x^2}{16} + \dfrac{y^2}{4} = 1$

Because the denominator of the x^2 – term is greater than the denominator of the y^2 – term, the major axis is horizontal. Since $a^2 = 16$, $a = 4$ and the vertices are $(-4,0)$ and $(4,0)$. Since $b^2 = 4$, $b = 2$ and endpoints of the minor axis are $(0,-2)$ and $(0,2)$.

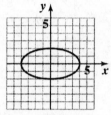

$4x^2 + 16y^2 = 64$

13. $25x^2 + 9y^2 = 225$

$\dfrac{25x^2}{225} + \dfrac{9y^2}{225} = \dfrac{225}{225}$

$\dfrac{x^2}{9} + \dfrac{y^2}{25} = 1$

Because the denominator of the y^2 – term is greater than the denominator of the x^2 – term, the major axis is vertical. Since $a^2 = 25$, $a = 5$ and the vertices are $(0,-5)$ and $(0,5)$. Since $b^2 = 9$, $b = 3$ and endpoints of the minor axis are $(-3,0)$ and $(3,0)$.

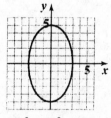

$25x^2 + 9y^2 = 225$

15. $x^2 + 2y^2 = 8$

$$\frac{x^2}{8} + \frac{2y^2}{8} = \frac{8}{8}$$

$$\frac{x^2}{8} + \frac{y^2}{4} = 1$$

Because the denominator of the x^2 – term is greater than the denominator of the y^2 – term, the major axis is horizontal. Since $a^2 = 8$, $a = \sqrt{8} = 2\sqrt{2}$ and the vertices are $\left(-2\sqrt{2}, 0\right)$ and $\left(2\sqrt{2}, 0\right)$. Since $b^2 = 4$, $b = 2$ and endpoints of the minor axis are $(0, -2)$ and $(0, 2)$.

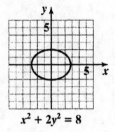

$x^2 + 2y^2 = 8$

17. From the graph, we see that the center of the ellipse is the origin, the major axis is horizontal with $a = 2$, and $b = 1$.

$$\frac{x^2}{2^2} + \frac{y^2}{1^2} = 1$$

$$\frac{x^2}{4} + \frac{y^2}{1} = 1$$

19. From the graph, we see that the center of the ellipse is the origin, the major axis is vertical with $a = 2$, and $b = 1$.

$$\frac{x^2}{1^2} + \frac{y^2}{2^2} = 1$$

$$\frac{x^2}{1} + \frac{y^2}{4} = 1$$

21. $\dfrac{(x-2)^2}{9} + \dfrac{(y-1)^2}{4} = 1$

The center of the ellipse is $(2,1)$. Because the denominator of the x^2 – term is greater than the denominator of the y^2 – term, the major axis is horizontal. Since $a^2 = 9$, $a = 3$ and the vertices lie 3 units to the left and right of the center. Since $b^2 = 4$, $b = 2$ and endpoints of the minor axis lie two units above and below the center.

Center	Vertices	Endpoints of Minor Axis
$(2,1)$	$(2-3,1)$ $= (-1,1)$	$(2,1-2)$ $= (2,-1)$
	$(2+3,1)$ $= (5,1)$	$(2,1+2)$ $= (2,3)$

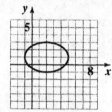

$$\frac{(x-2)^2}{9} + \frac{(y-1)^2}{4} = 1$$

23. $(x+3)^2 + 4(y-2)^2 = 16$

$$\frac{(x+3)^2}{16} + \frac{4(y-2)^2}{16} = \frac{16}{16}$$

$$\frac{(x+3)^2}{16} + \frac{(y-2)^2}{4} = 1$$

The center of the ellipse is $(-3, 2)$. Because the denominator of the x^2 – term is greater than the denominator of the y^2 – term, the major axis is horizontal. Since $a^2 = 16$, $a = 4$ and the vertices lie 4 units to the left and right of the center. Since $b^2 = 4$, $b = 2$ and endpoints of the minor axis lie two units above and below the center.

Center	Vertices	Endpoints of Minor Axis
$(-3,2)$	$(-3-4,2)$ $=(-7,2)$	$(-3,2-2)$ $=(-3,0)$
	$(-3+4,2)$ $=(1,2)$	$(-3,2+2)$ $=(-3,4)$

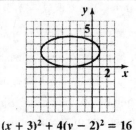

$$(x + 3)^2 + 4(y - 2)^2 = 16$$

25. $\dfrac{(x-4)^2}{9} + \dfrac{(y+2)^2}{25} = 1$

The center of the ellipse is $(4,-2)$. Because the denominator of the y^2 – term is greater than the denominator of the x^2 – term, the major axis is vertical. Since $a^2 = 25$, $a = 5$ and the vertices lie 5 units above and below the center. Since $b^2 = 9$, $b = 3$ and endpoints of the minor axis lie 3 units to the right and left of the center.

Center	Vertices	Endpoints Minor Axis
$(4,-2)$	$(4,-2-5)$ $=(4,-7)$	$(4-3,-2)$ $=(1,-2)$
	$(4,-2+5)$ $=(4,3)$	$(4+3,-2)$ $=(7,-2)$

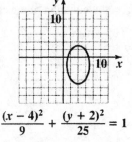

$$\dfrac{(x - 4)^2}{9} + \dfrac{(y + 2)^2}{25} = 1$$

27. $\dfrac{x^2}{25} + \dfrac{(y-2)^2}{36} = 1$

The center of the ellipse is $(0,2)$. Because the denominator of the y^2 – term is greater than the denominator of the x^2 – term, the major axis is vertical. Since $a^2 = 36$, $a = 6$ and the vertices lie 6 units above and below the center. Since $b^2 = 25$, $b = 5$ and endpoints of the minor axis lie 5 units to the left and right of the center.

Center	Vertices	Endpoint Minor Axis
$(0,2)$	$(0,2-6)$ $=(0,-4)$	$(0-5,2)$ $=(-5,2)$
	$(0,2+6)$ $=(0,8)$	$(0+5,2)$ $=(5,2)$

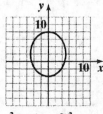

$$\dfrac{x^2}{25} + \dfrac{(y - 2)^2}{36} = 1$$

743

29. $\dfrac{(x+3)^2}{9} + (y-2)^2 = 1$

$\dfrac{(x+3)^2}{9} + \dfrac{(y-2)^2}{1} = 1$

The center of the ellipse is $(-3,2)$. Because the denominator of the x^2 – term is greater than the denominator of the y^2 – term, the major axis is horizontal. Since $a^2 = 9$, $a = 3$ and the vertices lie 3 units to the left and right of the center. Since $b^2 = 1$, $b = 1$ and endpoints of the minor axis lie two units above and below the center.

Center	Vertices	Endpoints of Minor Axis
$(-3,2)$	$(-3+3,2)$ $=(0,2)$	$(-3,2-1)$ $=(-3,1)$
	$(-3-3,2)$ $=(-6,2)$	$(-3,2+1)$ $=(-3,3)$

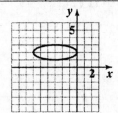

$\dfrac{(x+3)^2}{9} + (y-2)^2 = 1$

31. $9(x-1)^2 + 4(y+3)^2 = 36$

$\dfrac{9(x-1)^2}{36} + \dfrac{4(y+3)^2}{36} = \dfrac{36}{36}$

$\dfrac{(x-1)^2}{4} + \dfrac{(y+3)^2}{9} = 1$

The center of the ellipse is $(1,-3)$. Because the denominator of the y^2 – term is greater than the denominator of the x^2 – term, the major axis is vertical. Since $a^2 = 9$, $a = 3$ and the vertices lie 3 units above and below the center. Since $b^2 = 4$, $b = 2$ and endpoints of the minor axis lie 2 units to the right and left of the center.

Center	Vertices	Endpoints of Minor Axis
$(1,-3)$	$(1,-3-3)$ $=(1,-6)$	$(1-2,-3)$ $=(-1,-3)$
	$(1,-3+3)$ $=(1,0)$	$(1+2,-3)$ $=(3,-3)$

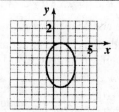

$9(x-1)^2 + 4(y+3)^2 = 36$

33. From the graph we see that the center of the ellipse is $(h,k) = (-1,1)$. We also see that the major axis is horizontal. The length of the major axis is 4 units, so $a = 2$. The length of the minor axis is 2 units so $b = 1$. Therefore, the equation of the ellipse is

$\dfrac{\left(x-(-1)\right)^2}{2^2} + \dfrac{(y-1)^2}{1^2} = 1$

$\dfrac{(x+1)^2}{4} + \dfrac{(y-1)^2}{1} = 1$

35. $x^2 + y^2 = 1$ and $x^2 + 9y^2 = 9$

$$\frac{x^2}{9} + \frac{9y^2}{9} = \frac{9}{9}$$

$$\frac{x^2}{9} + \frac{y^2}{1} = 1$$

The first equation is that of a circle with center at the origin and $r = 1$. The second equation is that of an ellipse with center at the origin, horizontal major axis of length 6 units ($a^2 = 9$, so $a = 3$), and vertical minor axis of length 2 units ($b^2 = 1$, so $b = 1$).

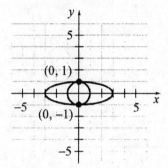

Check each intersection point.

The solutions set is $\{(0,-1),(0,1)\}$.

37. $\dfrac{x^2}{25} + \dfrac{y^2}{9} = 1$ and $y = 3$

The first equation is for an ellipse centered at the origin with horizontal major axis of length 10 units ($a^2 = 25$, so $a = 5$) and vertical minor axis of length 6 units ($b^2 = 9$, so $b = 3$). The second equation is for a horizontal line with a y-intercept of 3.

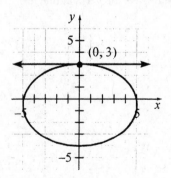

Check the intersection point.

The solution set is $\{(0,3)\}$.

39. $4x^2 + y^2 = 4$ and $2x - y = 2$

$$\frac{4x^2}{4} + \frac{y^2}{4} = \frac{4}{4} \qquad \begin{aligned} -y &= -2x + 2 \\ y &= 2x - 2 \end{aligned}$$

$$\frac{x^2}{1} + \frac{y^2}{4} = 1$$

The first equation is for an ellipse centered at the origin with vertical major axis of length 4 units ($a^2 = 4$, so $a = 2$) and horizontal minor axis of length 2 units ($b^2 = 1$, so $b = 1$). The second equation is for a line with slope 2 and y-intercept -2.

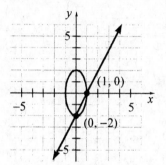

Check the intersection points.

The solution set is $\{(0,-2),(1,0)\}$.

41.
$$y = -\sqrt{16 - 4x^2}$$

$$y^2 = \left(-\sqrt{16 - 4x^2}\right)^2$$

$$y^2 = 16 - 4x^2$$

$$4x^2 + y^2 = 16$$

$$\frac{x^2}{4} + \frac{y^2}{16} = 1$$

We want to graph the bottom half of an ellipse centered at the origin with a vertical major axis of length 8 units ($a^2 = 16$, so $a = 4$) and horizontal minor axis of length 4 units ($b^2 = 4$, so $b = 2$).

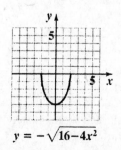

$y = -\sqrt{16-4x^2}$

43. From the figure, we see that the major axis is horizontal with $a = 15$, and $b = 10$.

$$\frac{x^2}{15^2} + \frac{y^2}{10^2} = 1$$

$$\frac{x^2}{225} + \frac{y^2}{100} = 1$$

Since the truck is 8 feet wide, we need to determine the height of the archway at $\frac{8}{2} = 4$ feet from the center.

$$\frac{4^2}{225} + \frac{y^2}{100} = 1$$

$$\frac{16}{225} + \frac{y^2}{100} = 1$$

$$900\left(\frac{16}{225} + \frac{y^2}{100}\right) = 900(1)$$

$$4(16) + 9y^2 = 900$$

$$64 + 9y^2 = 900$$

$$9y^2 = 836$$

$$y^2 = \frac{836}{9}$$

$$y = \sqrt{\frac{836}{9}} \approx 9.64$$

The height of the archway 4 feet from the center is approximately 9.64 feet. Since the truck is 7 feet high, the truck will clear the archway.

45. a. $\dfrac{x^2}{48^2} + \dfrac{y^2}{23^2} = 1$

$$\frac{x^2}{2304} + \frac{y^2}{529} = 1$$

b. $c^2 = a^2 - b^2$

$$c^2 = 48^2 - 23^2$$

$$c^2 = 2304 - 529$$

$$c^2 = 1775$$

$$c = \sqrt{1775} \approx 42.1$$

The desk was situated approximately 42 feet from the center of the ellipse.

47. – 51. Answers will vary.

53. Answers will vary. For example, consider Exercise 21.

$$\frac{(x-2)^2}{9} + \frac{(y-1)^2}{4} = 1$$

$$\frac{(y-1)^2}{4} = 1 - \frac{(x-2)^2}{9}$$

$$(y-1)^2 = 4\left(1 - \frac{(x-2)^2}{9}\right)$$

$$(y-1)^2 = 4 - \frac{4(x-2)^2}{9}$$

$$y - 1 = \pm\sqrt{4 - \frac{4(x-2)^2}{9}}$$

$$y = 1 \pm \sqrt{4 - \frac{4(x-2)^2}{9}}$$

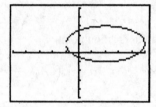

55. does not make sense; Explanations will vary. Sample explanation: An ellipse is symmetrical about both axes.

57. makes sense

59. false; Changes to make the statement true will vary. A sample change is: Ellipses are not functions. They do not pass the vertical line test.

61. true

63.　　$9x^2 + 25y^2 - 36x + 50y - 164 = 0$

$\left(9x^2 - 36x \quad\right) + \left(25y^2 + 50y \quad\right) = 164$

$9\left(x^2 - 4x \quad\right) + 25\left(y^2 + 2y \quad\right) = 164$

Complete the squares.

$\left(\dfrac{b}{2}\right)^2 = \left(\dfrac{-4}{2}\right)^2 = (-2)^2 = 4$

$\left(\dfrac{b}{2}\right)^2 = \left(\dfrac{2}{2}\right)^2 = (1)^2 = 1$

$9\left(x^2 - 4x + 4\right) + 25\left(y^2 + 2y + 1\right) = 164 + 9(4) + 25(1)$

$9(x-2)^2 + 25(y+1)^2 = 164 + 36 + 25$

$9(x-2)^2 + 25(y+1)^2 = 225$

$\dfrac{9(x-2)^2}{225} + \dfrac{25(y+1)^2}{225} = \dfrac{225}{225}$

$\dfrac{(x-2)^2}{25} + \dfrac{(y+1)^2}{9} = 1$

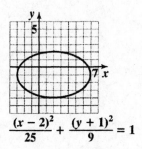

$\dfrac{(x-2)^2}{25} + \dfrac{(y+1)^2}{9} = 1$

65.　a. The perigee is $5000 - 16 - 4000 = 984$ miles above the Earth.

**　　b.** The apogee is $5000 + 16 - 4000 = 1016$ miles above the Earth.

67.　$x^3 + 2x^2 - 4x - 8 = x^2(x+2) - 4(x+2) = (x+2)(x^2 - 4)$

$= (x+2)(x+2)(x-2) = (x+2)^2(x-2)$

68.　$\sqrt[3]{40x^4y^7} = \sqrt[3]{8 \cdot 5x^3xy^6y} = 2xy^2\sqrt[3]{5xy}$

69.
$$\frac{2}{x+2} + \frac{4}{x-2} = \frac{x-1}{x^2-4}$$

$$\frac{2}{x+2} + \frac{4}{x-2} = \frac{x-1}{(x+2)(x-2)}$$

$$(x+2)(x-2)\left(\frac{2}{x+2} + \frac{4}{x-2}\right) = (x+2)(x-2)\left(\frac{x-1}{(x+2)(x-2)}\right)$$

$$2(x-2) + 4(x+2) = x-1$$

$$2x - 4 + 4x + 8 = x - 1$$

$$6x + 4 = x - 1$$

$$5x = -5$$

$$x = -1$$

The solution set is $\{-1\}$.

70. $4x^2 - 9y^2 = 36$

$$\frac{4x^2}{36} - \frac{9y^2}{36} = \frac{36}{36}$$

$$\frac{x^2}{9} - \frac{y^2}{4} = 1$$

The terms are separated by subtraction rather than by addition.

71. $\dfrac{x^2}{16} - \dfrac{y^2}{9} = 1$

a. Substitute 0 for y.

$$\frac{x^2}{16} - \frac{0^2}{9} = 1$$

$$\frac{x^2}{16} = 1$$

$$x^2 = 16$$

$$x = \pm 4$$

The x-intercepts are -4 and 4.

b.
$$\frac{0^2}{16} - \frac{y^2}{9} = 1$$

$$-\frac{y^2}{9} = 1$$

$$y^2 = -9$$

The equation $y^2 = -9$ has no real solutions.

748

72. $\dfrac{y^2}{9} - \dfrac{x^2}{16} = 1$

   a. Substitute 0 for x.

$$\frac{y^2}{9} - \frac{0^2}{16} = 1$$

$$\frac{y^2}{9} = 1$$

$$y^2 = 9$$

$$y = \pm 3$$

The y-intercepts are -3 and 3.

   b. $\dfrac{0^2}{9} - \dfrac{x^2}{16} = 1$

$$-\frac{x^2}{16} = 1$$

$$x^2 = -16$$

The equation $x^2 = -16$ has no real solutions.

13.3 Check Points

1. a. Since the x^2 – term is positive, the transverse axis lies along the x–axis. Also, since $a^2 = 25$ and $a = 5$, the vertices are $(-5, 0)$ and $(5, 0)$.

   b. Since the y^2 – term is positive, the transverse axis lies along the y–axis. Also, since $a^2 = 25$ and $a = 5$, the vertices are $(0, -5)$ and $(0, 5)$.

2. $\dfrac{x^2}{36} - \dfrac{y^2}{9} = 1$

Since the x^2 – term is positive, the transverse axis lies along the x–axis. Also, since $a^2 = 36$ and $a = 6$, the vertices are $(-6, 0)$ and $(6, 0)$. Construct a rectangle using -6 and 6 on the x–axis, and -3 and 3 on the y–axis. Draw extended diagonals to obtain the asymptotes. Draw the two branches of the hyperbola by starting at each vertex and approaching the asymptotes.

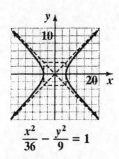

$$\frac{x^2}{36} - \frac{y^2}{9} = 1$$

3. $y^2 - 4x^2 = 4$

$$\frac{y^2}{4} - \frac{4x^2}{4} = \frac{4}{4}$$

$$\frac{y^2}{4} - \frac{x^2}{1} = 1$$

The equation is in the form $\dfrac{y^2}{a^2} - \dfrac{x^2}{b^2} = 1$ with

$a^2 = 4$ and $b^2 = 1$. We know the transverse axis lies on the y-axis and the vertices are

$(0, -2)$ and $(0, 2)$. Because $a^2 = 4$ and $b^2 = 1$,

$a = 2$ and $b = 1$. Construct a rectangle using -2 and 2 on the y–axis, and -1 and 1 on the x–axis. Draw extended diagonals to obtain the asymptotes. Draw the two branches of the hyperbola by starting at each vertex and approaching the asymptotes.

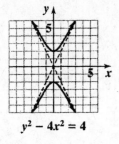

$$y^2 - 4x^2 = 4$$

13.3 Exercise Set

1. Since the x^2 – term is positive, the transverse axis lies along the x–axis. Also, since $a^2 = 4$ and $a = 2$, the vertices are $(-2, 0)$ and $(2, 0)$. This corresponds to graph (b).

3. Since the y^2 – term is positive, the transverse axis lies along the y–axis. Also, since $a^2 = 4$ and $a = 2$, the vertices are $(0, -2)$ and $(0, 2)$. This corresponds to graph (a).

5. $\dfrac{x^2}{9} - \dfrac{y^2}{25} = 1$

The equation is in the form $\dfrac{x^2}{a^2} - \dfrac{y^2}{b^2} = 1$ with

$a^2 = 9$, and $b^2 = 25$. We know the transverse axis lies on the x-axis and the vertices are

$(-3, 0)$ and $(3, 0)$. Because $a^2 = 9$ and $b^2 = 25$,

$a = 3$ and $b = 5$. Construct a rectangle using -3 and 3 on the x-axis, and -5 and 5 on the y–axis. Draw extended diagonals to obtain the asymptotes. Graph the hyperbola.

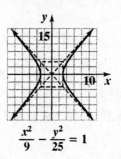

$$\frac{x^2}{9} - \frac{y^2}{25} = 1$$

7. $\dfrac{x^2}{100} - \dfrac{y^2}{64} = 1$

The equation is in the form $\dfrac{x^2}{a^2} - \dfrac{y^2}{b^2} = 1$ with

$a^2 = 100$, and $b^2 = 64$. We know the transverse axis lies on the x-axis and the vertices are

$(-10, 0)$ and $(10, 0)$. Because

$a^2 = 100$ and $b^2 = 64$, $a = 10$ and $b = 8$. Construct a rectangle using -10 and 10 on the x-axis, and -8 and 8 on the y–axis. Draw extended diagonals to obtain the asymptotes. Graph the hyperbola.

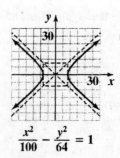

$$\frac{x^2}{100} - \frac{y^2}{64} = 1$$

7. $\dfrac{x^2}{100} - \dfrac{y^2}{64} = 1$

The equation is in the form $\dfrac{x^2}{a^2} - \dfrac{y^2}{b^2} = 1$ with

$a^2 = 100$, and $b^2 = 64$. We know the transverse axis lies on the *x*-axis and the vertices are $(-10,0)$ and $(10,0)$. Because

$a^2 = 100$ and $b^2 = 64$, $a = 10$ and $b = 8$. Construct a rectangle using -10 and 10 on the *x*–axis, and -8 and 8 on the *y*–axis. Draw extended diagonals to obtain the asymptotes. Graph the hyperbola.

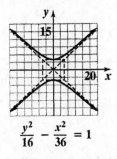

$\dfrac{y^2}{16} - \dfrac{x^2}{36} = 1$

9. $\dfrac{y^2}{25} - \dfrac{x^2}{64} = 1$

The transverse axis lies on the *y*-axis and the vertices are $(0,-5)$ and $(0,5)$. Construct a rectangle using -8 and 8 on the *x*–axis, and -5 and 5 on the *y*–axis. Draw extended diagonals to obtain the asymptotes.

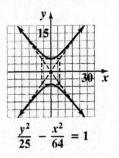

$\dfrac{y^2}{25} - \dfrac{x^2}{64} = 1$

11. $\dfrac{y^2}{36} - \dfrac{x^2}{25} = 1$

The equation is in the form $\dfrac{y^2}{a^2} - \dfrac{x^2}{b^2} = 1$ with

$a^2 = 36$, and $b^2 = 25$. We know the transverse axis lies on the *y*-axis and the vertices are $(0,-6)$ and $(0,6)$. Because $a^2 = 36$ and $b^2 = 25$, $a = 6$ and $b = 5$. Construct a rectangle using -5 and 5 on the *x*–axis, and -6 and 6 on the *y*–axis. Draw extended diagonals to obtain the asymptotes. Graph the hyperbola.

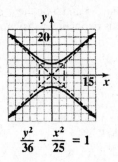

$\dfrac{y^2}{36} - \dfrac{x^2}{25} = 1$

13. $9x^2 - 4y^2 = 36$

$\dfrac{9x^2}{36} - \dfrac{4y^2}{36} = \dfrac{36}{36}$

$\dfrac{x^2}{4} - \dfrac{y^2}{9} = 1$

The equation is in the form $\dfrac{x^2}{a^2} - \dfrac{y^2}{b^2} = 1$ with

$a^2 = 4$ and $b^2 = 9$. We know the transverse axis lies on the *x*-axis and the vertices are $(-2,0)$ and $(2,0)$. Because $a^2 = 4$ and $b^2 = 9$, $a = 2$ and $b = 3$. Construct a rectangle using -2 and 2 on the *x*–axis, and -3 and 3 on the *y*–axis. Draw extended diagonals to obtain the asymptotes. Graph the hyperbola.

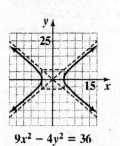

$9x^2 - 4y^2 = 36$

15. $9y^2 - 25x^2 = 225$

$$\frac{9y^2}{225} - \frac{25x^2}{225} = \frac{225}{225}$$

$$\frac{y^2}{25} - \frac{x^2}{9} = 1$$

The equation is in the form $\dfrac{y^2}{a^2} - \dfrac{x^2}{b^2} = 1$ with

$a^2 = 25$ and $b^2 = 9$. We know the transverse axis lies on the y-axis and the vertices are

$(0, -5)$ and $(0, 5)$. Because $a^2 = 25$ and $b^2 = 9$,

$a = 5$ and $b = 3$. Construct a rectangle using -3 and 3 on the x-axis, and -5 and 5 on the y-axis. Draw extended diagonals to obtain the asymptotes. Graph the hyperbola.

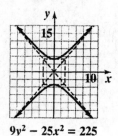

$9y^2 - 25x^2 = 225$

17. $4x^2 = 4 + y^2$

$$4x^2 - y^2 = 4$$

$$\frac{4x^2}{4} - \frac{y^2}{4} = \frac{4}{4}$$

$$\frac{x^2}{1} - \frac{y^2}{4} = 1$$

The equation is in the form $\dfrac{x^2}{a^2} - \dfrac{y^2}{b^2} = 1$ with

$a^2 = 1$ and $b^2 = 4$. We know the transverse axis lies on the x-axis and the vertices are

$(-1, 0)$ and $(1, 0)$. Because $a^2 = 1$ and $b^2 = 4$,

$a = 1$ and $b = 2$. Construct a rectangle using -1 and 1 on the x-axis, and -2 and 2 on the y-axis. Draw extended diagonals to obtain the asymptotes.

Graph the hyperbola.

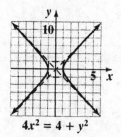

$4x^2 = 4 + y^2$

19. The graph shows that the transverse axis lies along the x-axis and the vertices are $(-3, 0)$ and $(3, 0)$. This means that $a = 3$. We also see that $b = 5$.

$$\frac{x^2}{a^2} - \frac{y^2}{b^2} = 1$$

$$\frac{x^2}{3^2} - \frac{y^2}{5^2} = 1$$

$$\frac{x^2}{9} - \frac{y^2}{25} = 1$$

21. The graph shows that the transverse axis lies along the y-axis and the vertices are $(0, -2)$ and $(0, 2)$.

This means that $a = 2$. We also see that $b = 3$.

$$\frac{y^2}{a^2} - \frac{x^2}{b^2} = 1$$

$$\frac{y^2}{2^2} - \frac{x^2}{3^2} = 1$$

$$\frac{y^2}{4} - \frac{x^2}{9} = 1$$

23. $\dfrac{x^2}{9} - \dfrac{y^2}{16} = 1$

The equation is for a hyperbola in standard form with the transverse axis on the x-axis. We have $a^2 = 9$ and $b^2 = 16$, so $a = 3$ and $b = 4$. Therefore, the vertices are at $(\pm a, 0)$ or $(\pm 3, 0)$.

Using a dashed line, we construct a rectangle using the ± 3 on the x-axis and ± 4 on the y-axis. Then use dashed lines to draw extended diagonals for the rectangle. These represent the asymptotes of the graph.

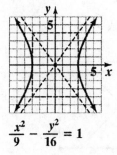

$$\dfrac{x^2}{9} - \dfrac{y^2}{16} = 1$$

From the graph we determine the following:
Domain: $\{x \mid x \le -3 \ \text{or} \ x \ge 3\}$ or $(-\infty, -3] \cup [3, \infty)$
Range: $\{y \mid y \ \text{is a real number}\}$ or $(-\infty, \infty)$

25. $\dfrac{x^2}{9} + \dfrac{y^2}{16} = 1$

The equation is for an ellipse in standard form with major axis along the y-axis. We have $a^2 = 16$ and $b^2 = 9$, so $a = 4$ and $b = 3$. Therefore, the vertices are $(0, \pm a)$ or $(0, \pm 4)$. The endpoints of the minor axis are $(\pm b, 0)$ or $(\pm 3, 0)$.

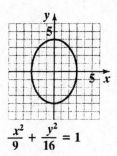

$$\dfrac{x^2}{9} + \dfrac{y^2}{16} = 1$$

From the graph we determine the following:
Domain: $\{x \mid -3 \le x \le 3\}$ or $[-3, 3]$
Range: $\{y \mid -4 \le y \le 4\}$ or $[-4, 4]$.

27. $\dfrac{y^2}{16} - \dfrac{x^2}{9} = 1$

The equation is in standard form with the transverse axis on the y-axis. We have $a^2 = 16$ and $b^2 = 9$, so $a = 4$ and $b = 3$. Therefore, the vertices are at $(0, \pm a)$ or $(0, \pm 4)$. Using a dashed line, we construct a rectangle using the ± 4 on the y-axis and ± 3 on the x-axis. Then use dashed lines to draw extended diagonals for the rectangle. These represent the asymptotes of the graph.

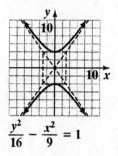

$$\dfrac{y^2}{16} - \dfrac{x^2}{9} = 1$$

From the graph we determine the following:
Domain: $\{x \mid x \ \text{is a real number}\}$ or $(-\infty, \infty)$
Range: $\{y \mid y \le -4 \ \text{or} \ y \ge 4\}$ or $(-\infty, -4] \cup [4, \infty)$

29. $x^2 - y^2 = 4$
$x^2 + y^2 = 4$

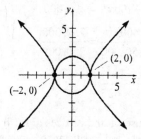

Check $(-2, 0)$:

$(-2)^2 - 0^2 = 4 \qquad (-2)^2 + 0^2 = 4$
$\qquad 4 - 0 = 4 \qquad\qquad 4 + 0 = 4$
$\qquad\qquad 4 = 4 \ \text{true} \qquad\qquad 4 = 4 \ \text{true}$

Check $(2, 0)$:

$(2)^2 - 0^2 = 4 \qquad (2)^2 + 0^2 = 4$
$\qquad 4 - 0 = 4 \qquad\qquad 4 + 0 = 4$
$\qquad\qquad 4 = 4 \ \text{true} \qquad\qquad 4 = 4 \ \text{true}$

The solution set is $\{(-2, 0), (2, 0)\}$.

31. $9x^2 + y^2 = 9$ or $\dfrac{x^2}{1} + \dfrac{y^2}{9} = 1$

$y^2 - 9x^2 = 9$ $\qquad \dfrac{y^2}{9} - \dfrac{x^2}{1} = 1$

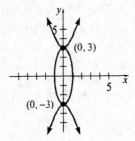

Check $(0, -3)$:

$9(0)^2 + (-3)^2 = 9 \qquad (-3)^2 - 9(0)^2 = 9$

$\qquad 0 + 9 = 9 \qquad\qquad 9 - 0 = 9$

$\qquad\qquad 9 = 9$ true $\qquad\qquad 9 = 9$ true

Check $(0, 3)$:

$9(0)^2 + (3)^2 = 9 \qquad (3)^2 - 9(0)^2 = 9$

$\qquad 0 + 9 = 9 \qquad\qquad 9 - 0 = 9$

$\qquad\qquad 9 = 9$ true $\qquad\qquad 9 = 9$ true

The solution set is $\{(0, -3), (0, 3)\}$.

33. $\qquad 625y^2 - 400x^2 = 250,000$

$\dfrac{625y^2}{250,000} - \dfrac{400x^2}{250,000} = \dfrac{250,000}{250,000}$

$\dfrac{y^2}{400} - \dfrac{x^2}{625} = 1$

Since the houses at the vertices of the hyperbola will be closest, find the distance between the vertices. Since $a^2 = 400$, $a = 20$. The houses are $20 + 20 = 40$ yards apart.

35. – 39. Answers will vary.

41. $\dfrac{x^2}{4} - \dfrac{y^2}{9} = 0$

Solve the equation for y.

$\dfrac{x^2}{4} = \dfrac{y^2}{9}$

$9x^2 = 4y^2$

$\dfrac{9}{4}x^2 = y^2$

$\pm\sqrt{\dfrac{9}{4}x^2} = y$

$\pm\dfrac{3}{2}x = y$

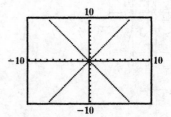

The graph is not a hyperbola. The graph is two lines.

43. does not make sense; Explanations will vary. Sample explanation: This would change the ellipse to a hyperbola.

45. makes sense

47. false; Changes to make the statement true will vary. A sample change is: If a hyperbola has a transverse axis along the x–axis and one of the branches is removed, the remaining branch does not define a function of x.

49. true

51. $\dfrac{(x-2)^2}{16} - \dfrac{(y-3)^2}{9} = 1$

This is the equation of a hyperbola with center $(2,3)$. The transverse axis is horizontal and the vertices lie 4 units to the right and left of $(2,3)$ at $(2-4,3) = (-2,3)$ and $(2+4,3) = (6,3)$.

Construct two sides of a rectangle using –2 and 6 on the x–axis. The remaining two sides of the rectangle are constructed 3 units above and 3 units below the center at $3-3 = 0$ and $3+3 = 6$. Draw extended diagonals to obtain the asymptotes.

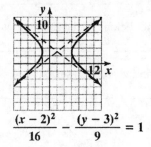

$$\dfrac{(x - 2)^2}{16} - \dfrac{(y - 3)^2}{9} = 1$$

53. $(x-3)^2 - 4(y+3)^2 = 4$

$\dfrac{(x-3)^2}{4} - \dfrac{4(y+3)^2}{4} = \dfrac{4}{4}$

$\dfrac{(x-3)^2}{4} - \dfrac{(y+3)^2}{1} = 1$

This is the equation of a hyperbola with center $(3,-3)$. The transverse axis is horizontal and the vertices lie 2 units to the right and left of $(3,-3)$ at $(3-2,-3) = (1,-3)$ and $(3+2,-3) = (5,-3)$.

Construct two sides of a rectangle using 1 and 5 on the x–axis. The remaining two sides of the rectangle are constructed 1 unit above and below center at $-3-1 = -4$ and $-3+1 = -2$. Draw extended diagonals to obtain the asymptotes.

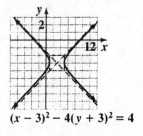

$(x - 3)^2 - 4(y + 3)^2 = 4$

55. Since the vertices are $(6,0)$ and $(-6,0)$, we know that the transverse axis lies along the x–axis and $a = 6$. Use the equation of the asymptote, $y = 4x$, to find b.

$y = 4x = 4(6) = 24$

This means that $b = \pm 24$.

$\dfrac{x^2}{6^2} - \dfrac{y^2}{24^2} = 1$

$\dfrac{x^2}{36} - \dfrac{y^2}{576} = 1$

57. $y = -x^2 - 4x + 5$

The x–coordinate of the vertex is

$-\dfrac{b}{2a} = -\dfrac{-4}{2(-1)} = -\dfrac{-4}{-2} = -2$ and the y–coordinate of the vertex is

$f\left(-\dfrac{b}{2a}\right) = f(-2) = -(-2)^2 - 4(-2) + 5$

$\qquad\qquad = -4 + 8 + 5 = 9.$

The vertex is at (–2, 9). The x–intercepts are –5 and 1. The y–intercept is 5.

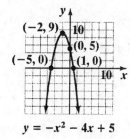

$y = -x^2 - 4x + 5$

58. $3x^2 - 11x - 4 \geq 0$

$$3x^2 - 11x - 4 = 0$$
$$(3x+1)(x-4) = 0$$

$3x+1 = 0 \quad$ or $\quad x - 4 = 0$

$\quad 3x = -1 \qquad\qquad x = 4$

$\quad x = -\dfrac{1}{3}$

Interval	Test Value	Substitution	Conclusion
$\left(-\infty, -\dfrac{1}{3}\right]$	-1	$3(-1)^2 - 11(-1) - 4 \geq 0$ $10 \geq 0$, true	$\left(-\infty, -\dfrac{1}{3}\right]$ belongs in the solution set
$\left[-\dfrac{1}{3}, 4\right]$	0	$3(0)^2 - 11(0) - 4 \geq 0$ $-4 \geq 0$, false	$\left[-\dfrac{1}{3}, 4\right]$ does not belong in the solution set.
$[4, \infty)$	5	$3(5)^2 - 11(5) - 4 \geq 0$ $16 \geq 0$, true	$[4, \infty)$ belongs in the solution set.

The solution set is $\left(-\infty, -\dfrac{1}{3}\right] \cup [4, \infty)$ or $\left\{ x \middle| x \leq -\dfrac{1}{3} \text{ or } x \geq 4 \right\}$.

59. $\log_4(3x+1) = 3$

$$3x + 1 = 4^3$$
$$3x + 1 = 64$$
$$3x = 63$$
$$x = 21$$

The solution set is $\{21\}$.

60. $y = x^2 + 4x - 5$

Since $a = 1$ is positive, the parabola opens upward. The x-coordinate of the vertex is $x = -\dfrac{b}{2a} = -\dfrac{4}{2(1)} = -2$. The y-coordinate of the vertex is $y = (-2)^2 + 4(-2) - 5 = -9$.

Vertex: $(-2, -9)$.

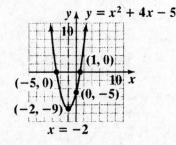

61. $y = -3(x-1)^2 + 2$

Since $a = -3$ is negative, the parabola opens downward. The vertex of the parabola is $(h,k) = (1,2)$.
The y-intercept is -1.

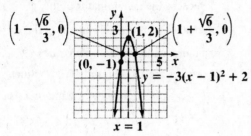

62. Find the y-intercepts.

$$x = -3(y-1)^2 + 2$$
$$0 = -3(y-1)^2 + 2$$
$$3(y-1)^2 = 2$$
$$(y-1)^2 = \frac{2}{3}$$
$$y - 1 = \pm\sqrt{\frac{2}{3}}$$
$$y = 1 \pm \sqrt{\frac{2}{3}}$$
$$y \approx 0.2 \text{ or } 1.8$$

The y-intercepts are approximately 0.2 and 1.8.

Chapter 13 Mid-Chapter Check Point

1. $x^2 + y^2 = 9$

Center: $(0,0)$

Radius: $r = \sqrt{9} = 3$

We plot points that are 3 units to the left, right, above, and below the center. These points are $(-3,0)$, $(3,0)$, $(0,3)$ and $(0,-3)$.

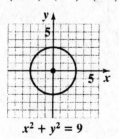

2. $(x-3)^2 + (y+2)^2 = 25$

Center: $(3,-2)$

Radius: $r = \sqrt{25} = 5$

We plot the points that are 5 units to the left, right, above and below the center.
These points are $(-2,-2)$, $(8,-2)$, $(3,3)$, and $(3,-7)$.

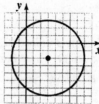

3. $x^2 + (y-1)^2 = 4$

Center: $(0,1)$

Radius: $r = \sqrt{4} = 2$

We plot the points that are 2 units to the left, right, above, and below the center. These points are $(-2,1)$, $(2,1)$, $(0,3)$, and $(0,-1)$.

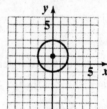

4. $x^2 + y^2 - 4x - 2y - 4 = 0$

Complete the square in both x and y to get the equation in standard form.

$$\left(x^2 - 4x\right) + \left(y^2 - 2y\right) = 4$$

$$\left(x^2 - 4x + 4\right) + \left(y^2 - 2y + 1\right) = 4 + 4 + 1$$

$$(x-2)^2 + (y-1)^2 = 9$$

Center: $(2,1)$

Radius: $r = \sqrt{9} = 3$

We plot the points that are 3 units to the left, right, above, and below the center. These points are $(-1,1)$, $(5,1)$, $(2,4)$, and $(2,-2)$.

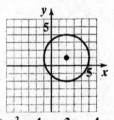

$x^2 + y^2 - 4x - 2y - 4 = 0$

5. $\dfrac{x^2}{25} + \dfrac{y^2}{4} = 1$

Center: $(0,0)$

Because the denominator of the x^2-term is greater than the denominator of the y^2-term, the major axis is horizontal. Since $a^2 = 25$, $a = 5$ and the vertices are $(-5,0)$ and $(5,0)$. Since $b^2 = 4$, $b = 2$ and endpoints of the minor axis are $(0,-2)$ and $(0,2)$.

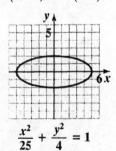

$\dfrac{x^2}{25} + \dfrac{y^2}{4} = 1$

6. $9x^2 + 4y^2 = 36$

Divide both sides by 36 to get the standard form:

$$\frac{x^2}{4} + \frac{y^2}{9} = 1$$

Center: $(0,0)$

Because the denominator of the y^2-term is greater than the denominator of the x^2-term, the major axis is vertical. Since $a^2 = 9$, $a = 3$ and the vertices are $(0,-3)$ and $(0,3)$. Since $b^2 = 4$, $b = 2$ and endpoints of the minor axis are $(-2,0)$ and $(2,0)$.

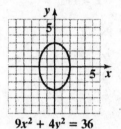

$9x^2 + 4y^2 = 36$

7. $\dfrac{(x-2)^2}{16} + \dfrac{(y+1)^2}{25} = 1$

Center: $(2,-1)$

Because the denominator of the y^2-term is greater than the denominator of the x^2-term, the major axis is vertical. We have $a^2 = 25$ and $b^2 = 16$, so $a = 5$ and $b = 4$. The vertices lie 5 units above and below the center. The endpoints of the minor axis lie 4 units to the left and right of the center.

Vertices: $(2,4)$ and $(2,-6)$

Minor endpoints: $(-2,-1)$ and $(6,-1)$

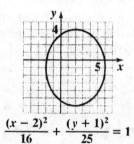

$\dfrac{(x-2)^2}{16} + \dfrac{(y+1)^2}{25} = 1$

8. $\dfrac{(x+2)^2}{25} + \dfrac{(y-1)^2}{16} = 1$

Center: $(-2,1)$

Because the denominator of the x^2 – term is greater than the denominator of the y^2 – term, the major axis is horizontal. We have $a^2 = 25$ and $b^2 = 16$, so $a = 5$ and $b = 4$. The vertices lie 5 units to the left and right of the center. The endpoints of the minor axis lie 4 units above and below the center.

Vertices: $(-7,1)$ and $(3,1)$

Minor endpoints: $(-2,5)$ and $(-2,-3)$

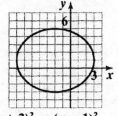

$$\dfrac{(x+2)^2}{25} + \dfrac{(y-1)^2}{16} = 1$$

9. $\dfrac{x^2}{9} - y^2 = 1$

The equation is for a hyperbola in standard form with the transverse axis on the x-axis. We have $a^2 = 9$ and $b^2 = 1$, so $a = 3$ and $b = 1$. Therefore, the vertices are at $(\pm a, 0)$ or $(\pm 3, 0)$. Using a dashed line, we construct a rectangle using the ± 3 on the x-axis and ± 1 on the y-axis. Then use dashed lines to draw extended diagonals for the rectangle. These represent the asymptotes of the graph. Graph the hyperbola.

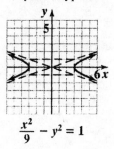

$$\dfrac{x^2}{9} - y^2 = 1$$

10. $\dfrac{y^2}{9} - x^2 = 1$

The equation is in the form $\dfrac{y^2}{a^2} - \dfrac{x^2}{b^2} = 1$ with $a^2 = 9$, and $b^2 = 1$. We know the transverse axis lies on the y-axis and the vertices are $(0,-3)$ and $(0,3)$. Because $a^2 = 9$ and $b^2 = 1$, $a = 3$ and $b = 1$. Construct a rectangle using -1 and 1 on the x-axis, and -3 and 3 on the y-axis. Draw extended diagonals to obtain the asymptotes. Graph the hyperbola.

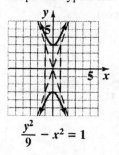

$$\dfrac{y^2}{9} - x^2 = 1$$

11. $y^2 - 4x^2 = 16$

$$\dfrac{y^2}{16} - \dfrac{x^2}{4} = 1$$

The equation is in the form $\dfrac{y^2}{a^2} - \dfrac{x^2}{b^2} = 1$ with $a^2 = 16$, and $b^2 = 4$. We know the transverse axis lies on the y-axis and the vertices are $(0,-4)$ and $(0,4)$. Because $a^2 = 16$ and $b^2 = 4$, $a = 4$ and $b = 2$. Construct a rectangle using -2 and 2 on the x-axis, and -4 and 4 on the y-axis. Draw extended diagonals to obtain the asymptotes. Graph the hyperbola.

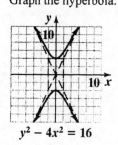

$$y^2 - 4x^2 = 16$$

12. $4x^2 - 49y^2 = 196$

$$\frac{x^2}{49} - \frac{y^2}{4} = 1$$

The equation is for a hyperbola in standard form with the transverse axis on the x-axis. We have $a^2 = 49$ and $b^2 = 4$, so $a = 7$ and $b = 2$. Therefore, the vertices are at $(\pm a, 0)$ or $(\pm 7, 0)$.

Using a dashed line, we construct a rectangle using the ± 7 on the x-axis and ± 2 on the y-axis. Then use dashed lines to draw extended diagonals for the rectangle. These represent the asymptotes of the graph.
Graph the hyperbola.

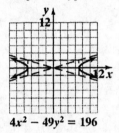

$4x^2 - 49y^2 = 196$

13. $x^2 + y^2 = 4$

This is the equation of a circle centered at the origin with radius $r = \sqrt{4} = 2$.
We can plot points that are 2 units to the left, right, above, and below the origin and then graph the circle. The points are $(-2, 0)$, $(2, 0)$, $(0, 2)$, and $(0, -2)$.

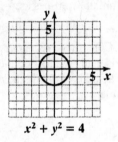

$x^2 + y^2 = 4$

14. $x + y = 4$

$y = -x + 4$

This is the equation of a line with slope $m = -1$ and a y-intercept of 4. We can plot the point $(0, 4)$, use the slope to get an additional point, connect the points with a straight line and then extend the line to represent the graph of the equation.

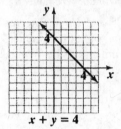

$x + y = 4$

15. $x^2 - y^2 = 4$

$$\frac{x^2}{4} - \frac{y^2}{4} = 1$$

The equation is for a hyperbola in standard form with the transverse axis on the x-axis. We have $a^2 = 4$ and $b^2 = 4$, so $a = 2$ and $b = 2$. Therefore, the vertices are at $(\pm a, 0)$ or $(\pm 2, 0)$.

Using a dashed line, we construct a rectangle using the ± 2 on the x-axis and ± 2 on the y-axis. Then use dashed lines to draw extended diagonals for the rectangle. These represent the asymptotes of the graph.
Graph the hyperbola.

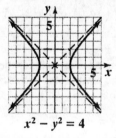

$x^2 - y^2 = 4$

16. $x^2 + 4y^2 = 4$

$\dfrac{x^2}{4} + \dfrac{y^2}{1} = 1$

Center: $(0,0)$

Because the denominator of the x^2 – term is greater than the denominator of the y^2 – term, the major axis is horizontal. We have $a^2 = 4$ and $b^2 = 1$, so $a = 2$ and $b = 1$. The vertices lie 2 units to the left and right of the center. The endpoints of the minor axis lie 1 unit above and below the center.

Vertices: $(-2,0)$ and $(2,0)$

Minor endpoints: $(0,-1)$ and $(0,1)$

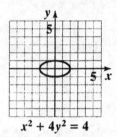

$x^2 + 4y^2 = 4$

17. $(x+1)^2 + (y-1)^2 = 4$

Center: $(-1,1)$

Radius: $r = \sqrt{4} = 2$

We plot the points that are 2 units to the left, right, above and below the center.

These points are $(-3,1)$, $(1,1)$, $(-1,3)$, and $(-1,-1)$.

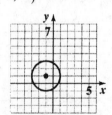

$(x + 1)^2 + (y - 1)^2 = 4$

18. $x^2 + 4(y-1)^2 = 4$

$\dfrac{x^2}{4} + \dfrac{(y-1)^2}{1} = 1$

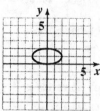

$x^2 + 4(y - 1)^2 = 4$

13.4 Check Points

1. $x = -(y-2)^2 + 1$

This is a parabola of the form $x = a(y-k)^2 + h$. Since $a = -1$ is negative, the parabola opens to the left. The vertex of the parabola is $(1,2)$. The axis of symmetry is $y = 2$.
Replace y with 0 to find the x–intercept.

$x = -(y-2)^2 + 1$

$ = -(0-2)^2 + 1$

$ = -4 + 1$

$ = -3$

Replace x with 0 to find the y–intercepts.

$x = -(y-2)^2 + 1$

$0 = -(y-2)^2 + 1$

$0 = -(y^2 - 4y + 4) + 1$

$0 = -y^2 + 4y - 4 + 1$

$0 = -y^2 + 4y - 3$

$0 = y^2 - 4y + 3$

$0 = (y-1)(y-3)$

$y - 1 = 0 \quad$ or $\quad y - 3 = 0$

$y = 1 \qquad\qquad\quad y = 3$

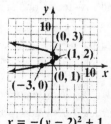

$x = -(y - 2)^2 + 1$

2. $x = y^2 + 8y + 7$

This is a parabola of the form $x = ay^2 + by + c$.
Since $a = 1$ is positive, the parabola opens to the right.
The y–coordinate of the vertex is

$$y = -\frac{b}{2a} = -\frac{8}{2(1)} = -\frac{8}{2} = -4.$$

The x–coordinate of the vertex is

$$x = y^2 + 8y + 7 = (-4)^2 + 8(-4) + 7 = -9.$$

The vertex of the parabola is $(-9, -4)$. The axis of symmetry is $y = -4$.

Replace y with 0 to find the x–intercept.

$$x = y^2 + 8y + 7 = 0^2 + 8(0) + 7 = 7$$

Replace x with 0 to find the y–intercepts.

$$x = y^2 + 8y + 7$$
$$0 = y^2 + 8y + 7$$
$$0 = (y + 7)(y + 1)$$
$$y + 7 = 0 \quad \text{or} \quad y + 1 = 0$$
$$y = -7 \qquad\qquad y = -1$$

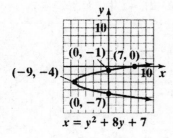

$x = y^2 + 8y + 7$

3. a. Collect x^2– and y^2–terms on the same side of the equation.

$$x^2 = 4y^2 + 16$$
$$x^2 - 4y^2 = 16$$

Because x^2 and y^2 have opposite signs, the equation's graph is a hyperbola.

b. Collect x^2– and y^2–terms on the same side of the equation.

$$x^2 = 16 - 4y^2$$
$$x^2 + 4y^2 = 16$$

Because x^2 and y^2 have different positive coefficients, the equation's graph is an ellipse.

c. Collect x^2– and y^2–terms on the same side of the equation.

$$4x^2 = 16 - 4y^2$$
$$4x^2 + 4y^2 = 16$$
$$\text{or}$$
$$x^2 + y^2 = 4$$

Because x^2 and y^2 have equal positive coefficients, the equation's graph is a circle.

d. $x = -4y^2 + 16y$

Since only one variable is squared, the graph of the equation is a parabola. Furthermore, this parabola is in the form $x = ay^2 + by + c$, with a negative coefficient of the y^2–term. Thus this horizontal parabola opens to the left.

13.4 Exercise Set

1. Since $a = 1$, the parabola opens to the right. The vertex of the parabola is $(-1, 2)$.
The equation's graph is b.

3. Since $a = 1$, the parabola opens to the right. The vertex of the parabola is $(1, -2)$.
The equation's graph is f.

5. Since $a = -1$, the parabola opens to the left. The vertex of the parabola is $(1, 2)$.
The equation's graph is a.

7. $x = 2y^2$

$$x = 2(y - 0)^2 + 0$$

The vertex is the point $(0, 0)$.

9. $x = (y - 2)^2 + 3$

The vertex is the point $(3, 2)$.

11. $x = -4(y + 2)^2 - 1$

The vertex is the point $(-1, -2)$.

13. $x = 2(y - 6)^2$

$$x = 2(y - 6)^2 + 0$$

The vertex is the point $(0, 6)$.

15. $x = y^2 - 6y + 6$

The y–coordinate of the vertex is

$$-\frac{b}{2a} = -\frac{-6}{2(1)} = -\frac{-6}{2} = 3.$$

The x–coordinate of the vertex is

$$f(3) = 3^2 - 6(3) + 6 = 9 - 18 + 6 = -3.$$

The vertex is the point $(-3, 3)$.

17. $x = 3y^2 + 6y + 7$

The y–coordinate of the vertex is

$$-\frac{b}{2a} = -\frac{6}{2(3)} = -\frac{6}{6} = -1.$$

The x–coordinate of the vertex is

$$f(-1) = 3(-1)^2 + 6(-1) + 7$$
$$= 3(1) - 6 + 7 = 3 - 6 + 7 = 4.$$

The vertex is the point $(4, -1)$.

19. $x = (y - 2)^2 - 4$

This is a parabola of the form $x = a(y - k)^2 + h$. Since $a = 1$ is positive, the parabola opens to the right. The vertex of the parabola is $(-4, 2)$. The axis of symmetry is $y = 2$. Replace y with 0 to find the x–intercept.

$$x = (0 - 2)^2 - 4 = 4 - 4 = 0$$

The x–intercept is 0. Replace x with 0 to find the y–intercepts.

$$0 = (y - 2)^2 - 4$$
$$0 = y^2 - 4y + 4 - 4$$
$$0 = y^2 - 4y$$
$$0 = y(y - 4)$$

Apply the zero product principle.
$$y = 0 \quad \text{and} \quad y - 4 = 0$$
$$y = 4$$

The y–intercepts are 0 and 4.

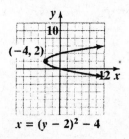

$x = (y - 2)^2 - 4$

21. $x = (y - 3)^2 - 5$

This is a parabola of the form $x = a(y - k)^2 + h$. Since $a = 1$ is positive, the parabola opens to the right. The vertex of the parabola is $(-5, 3)$. The axis of symmetry is $y = 3$. Replace y with 0 to find the x–intercept.

$$x = (0 - 3)^2 - 5 = (-3)^2 - 5 = 9 - 5 = 4$$

The x–intercept is 0. Replace x with 0 to find the y–intercepts.

$$0 = (y - 3)^2 - 5$$
$$0 = y^2 - 6y + 9 - 5$$
$$0 = y^2 - 6y + 4$$

Solve using the quadratic formula.

$$x = \frac{-b \pm \sqrt{b^2 - 4ac}}{2a}$$

$$= \frac{-(-6) \pm \sqrt{(-6)^2 - 4(1)4}}{2(1)}$$

$$= \frac{6 \pm \sqrt{36 - 16}}{2} = \frac{6 \pm \sqrt{20}}{2}$$

$$= \frac{6 \pm 2\sqrt{5}}{2} = 3 \pm \sqrt{5}$$

The y–intercepts are $3 - \sqrt{5}$ and $3 + \sqrt{5}$.

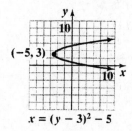

$x = (y - 3)^2 - 5$

23. $x = -(y-5)^2 + 4$

This is a parabola of the form $x = a(y-k)^2 + h$. Since $a = -1$ is negative, the parabola opens to the left. The vertex of the parabola is $(4,5)$. The axis of symmetry is $y = 5$. Replace y with 0 to find the x–intercept.

$$x = -(0-5)^2 + 4 = -(-5)^2 + 4$$
$$= -25 + 4 = -21$$

The x–intercept is 0. Replace x with 0 to find the y–intercepts.

$$0 = -(y-5)^2 + 4$$
$$0 = -(y^2 - 10y + 25) + 4$$
$$0 = -y^2 + 10y - 25 + 4$$
$$0 = -y^2 + 10y - 21$$
$$0 = y^2 - 10y + 21$$
$$0 = (y-7)(y-3)$$
$$y - 7 = 0 \quad \text{or} \quad y - 3 = 0$$
$$y = 7 \qquad\qquad y = 3$$

The y–intercepts are 3 and 7.

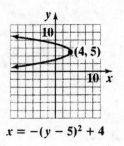

$x = -(y - 5)^2 + 4$

25. $x = (y-4)^2 + 1$

This is a parabola of the form $x = a(y-k)^2 + h$. Since $a = 1$ is positive, the parabola opens to the right. The vertex of the parabola is $(1,4)$. The axis of symmetry is $y = 4$. Replace y with 0 to find the x–intercept.

$$x = (0-4)^2 + 1$$
$$= (-4)^2 + 1$$
$$= 16 + 1$$
$$= 17$$

The x–intercept is 0. Replace x with 0 to find the y–intercepts.

$$0 = (y-4)^2 + 1$$
$$0 = y^2 - 8y + 16 + 1$$
$$0 = y^2 - 8y + 17$$

Solve using the quadratic formula.

$$y = \frac{-b \pm \sqrt{b^2 - 4ac}}{2a}$$
$$= \frac{-(-8) \pm \sqrt{(-8)^2 - 4(1)(17)}}{2(1)}$$
$$= \frac{8 \pm \sqrt{64 - 68}}{2}$$
$$= \frac{8 \pm \sqrt{-4}}{2}$$
$$= \frac{8 \pm 2i}{2}$$
$$= 4 \pm i$$

The solutions are complex, so there are no y–intercepts.

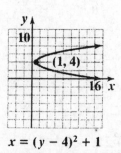

$x = (y - 4)^2 + 1$

27. $x = -3(y-5)^2 + 3$

This is a parabola of the form $x = a(y-k)^2 + h$. Since $a = -3$ is negative, the parabola opens to the left. The vertex of the parabola is $(3,5)$. The axis of symmetry is $y = 5$. Replace y with 0 to find the x–intercept.

$$x = -3(0-5)^2 + 3 = -3(-5)^2 + 3$$
$$= -3(25) + 3 = -75 + 3 = -72$$

The x–intercept is 0. Replace x with 0 to find the y–intercepts.

$$0 = -3(y-5)^2 + 3$$
$$0 = -3(y^2 - 10y + 25) + 3$$
$$0 = -3y^2 + 30y - 75 + 3$$
$$0 = -3y^2 + 30y - 72$$
$$0 = y^2 - 10y + 24$$
$$0 = (y-6)(y-4)$$
$$y - 6 = 0 \quad \text{or} \quad y - 4 = 0$$
$$y = 6 \qquad\qquad y = 4$$

The y–intercepts are 4 and 6.

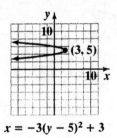

$x = -3(y - 5)^2 + 3$

29. $x = -2(y+3)^2 - 1$

This is a parabola of the form $x = a(y-k)^2 + h$. Since $a = -2$ is negative, the parabola opens to the left. The vertex of the parabola is $(-1,-3)$. The axis of symmetry is $y = -3$. Replace y with 0 to find the x–intercept.

$$x = -2(0+3)^2 - 1 = -2(3)^2 - 1$$
$$= -2(9) - 1 = -18 - 1 = -19$$

The x–intercept is 0. Replace x with 0 to find the y–intercepts.

$$0 = -2(y+3)^2 - 1$$
$$0 = -2(y^2 + 6x + 9) - 1$$
$$0 = -2y^2 - 12x - 18 - 1$$
$$0 = -2y^2 - 12x - 19$$
$$0 = 2y^2 + 12x + 19$$

Solve using the quadratic formula.

$$y = \frac{-12 \pm \sqrt{12^2 - 4(2)(19)}}{2(2)}$$
$$= \frac{-12 \pm \sqrt{144 - 152}}{4}$$
$$= \frac{-12 \pm \sqrt{-8}}{4}$$

Since the solutions will be complex, there are no y–intercepts.

$x = -2(y + 3)^2 - 1$

31. $x = \frac{1}{2}(y+2)^2 + 1$

This is a parabola of the form $x = a(y-k)^2 + h$.

Since $a = \frac{1}{2}$ is positive, the parabola opens to the

right. The vertex of the parabola is $(1,-2)$. The axis of symmetry is $y = -2$. Replace y with 0 to find the x–intercept.

$$x = \frac{1}{2}(0+2)^2 + 1$$

$$= \frac{1}{2}(4) + 1 = 2 + 1 = 3$$

The x–intercept is 3. Replace x with 0 to find the y–intercepts.

$$0 = \frac{1}{2}(y+2)^2 + 1$$

$$0 = \frac{1}{2}(y^2 + 4y + 4) + 1$$

$$0 = \frac{1}{2}y^2 + 2y + 2 + 1$$

$$0 = \frac{1}{2}y^2 + 2y + 3$$

$$0 = y^2 + 4y + 6$$

Solve using the quadratic formula.

$$y = \frac{-4 \pm \sqrt{4^2 - 4(1)(6)}}{2(1)}$$

$$= \frac{-4 \pm \sqrt{16 - 24}}{2}$$

$$= \frac{-4 \pm \sqrt{-8}}{2}$$

Since the solutions will be complex, there are no y–intercepts.

$(1, -2)$

$x = \frac{1}{2}(y + 2)^2 + 1$

33. $x = y^2 + 2y - 3$

This is a parabola of the form $x = ay^2 + by + c$.

Since $a = 1$ is positive, the parabola opens to the right. The y–coordinate of the vertex is

$-\frac{b}{2a} = -\frac{2}{2(1)} = -\frac{2}{2} = -1$. The x–coordinate of the

vertex is $x = (-1)^2 + 2(-1) - 3 = 1 - 2 - 3 = -4$.

The vertex of the parabola is $(-4, -1)$. The axis of symmetry is $y = -1$. Replace y with 0 to find the x–intercept.

$$x = 0^2 + 2(0) - 3 = 0 + 0 - 3 = -3$$

The x–intercept is -3. Replace x with 0 to find the y–intercepts.

$$0 = y^2 + 2y - 3$$

$$0 = (y+3)(y-1)$$

$$y + 3 = 0 \quad \text{or} \quad y - 1 = 0$$

$$y = -3 \qquad\qquad y = 1$$

The y–intercepts are -3 and 1.

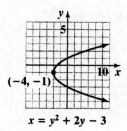

$(-4, -1)$

$x = y^2 + 2y - 3$

35. $x = -y^2 - 4y + 5$

This is a parabola of the form $x = ay^2 + by + c$. Since $a = -1$ is negative, the parabola opens to the left. The y–coordinate of the vertex is

$$-\frac{b}{2a} = -\frac{-4}{2(-1)} = -\frac{-4}{-2} = -2. \text{ The } x\text{–coordinate of}$$

the vertex is

$$x = -(-2)^2 - 4(-2) + 5 = -4 + 8 + 5 = 9.$$

The vertex of the parabola is $(9, -2)$. The axis of symmetry is $y = -2$. Replace y with 0 to find the x–intercept.

$$x = -0^2 - 4(0) + 5 = 0 - 0 + 5 = 5$$

The x–intercept is 5. Replace x with 0 to find the y–intercepts.

$$0 = -y^2 - 4y + 5$$
$$0 = y^2 + 4y - 5$$
$$0 = (y + 5)(y - 1)$$
$$y + 5 = 0 \quad \text{or} \quad y - 1 = 0$$
$$y = -5 \qquad\qquad y = 1$$

The y–intercepts are -5 and 1.

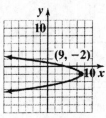

$x = -y^2 - 4y + 5$

37. $x = y^2 + 6y$

This is a parabola of the form $x = ay^2 + by + c$. Since $a = 1$ is positive, the parabola opens to the right. The y–coordinate of the vertex is

$$-\frac{b}{2a} = -\frac{6}{2(1)} = -\frac{6}{2} = -3.$$

The x–coordinate of the vertex is

$$x = (-3)^2 + 6(-3) = 9 - 18 = -9.$$

The vertex of the parabola is $(-9, -3)$. The axis of symmetry is $y = -3$. Replace y with 0 to find the x–intercept.

$$x = 0^2 + 6(0) = 0$$

The x–intercept is 0. Replace x with 0 to find the y–intercepts.

$$0 = y^2 + 6y$$
$$0 = y(y + 6)$$
$$y = 0 \quad \text{or} \quad y + 6 = 0$$
$$\qquad\qquad y = -6$$

The y–intercepts are -6 and 0.

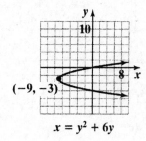

$x = y^2 + 6y$

39. $x = -2y^2 - 4y$

This is a parabola of the form $x = ay^2 + by + c$. Since $a = -2$ is negative, the parabola opens to the left. The y–coordinate of the vertex is
$$-\frac{b}{2a} = -\frac{-4}{2(-2)} = -\frac{-4}{-4} = -1.$$
The x–coordinate of the vertex is
$$x = -2(-1)^2 - 4(-1) = -2(1) + 4$$
$$= -2 + 4 = 2.$$

The vertex of the parabola is $(2, -1)$. The axis of symmetry is $y = -1$. Replace y with 0 to find the x–intercept.
$$x = -2(0)^2 - 4(0) = -2(0) - 0 = 0$$

The x–intercept is 0. Replace x with 0 to find the y–intercepts.
$$0 = -2y^2 - 4y$$
$$0 = y^2 + 2y$$
$$0 = y(y + 2)$$
$$y = 0 \quad \text{or} \quad y + 2 = 0$$
$$y = -2$$

The y–intercepts are –2 and 0.

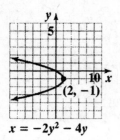

$$x = -2y^2 - 4y$$

41. $x = -2y^2 - 4y + 1$

This is a parabola of the form $x = ay^2 + by + c$. Since $a = -2$ is negative, the parabola opens to the left. The y–coordinate of the vertex is
$$-\frac{b}{2a} = -\frac{-4}{2(-2)} = -\frac{-4}{-4} = -1.$$

The x–coordinate of the vertex is
$$x = -2(-1)^2 - 4(-1) + 1 = -2(1) + 4 + 1$$
$$= -2 + 4 + 1 = 3$$

The vertex of the parabola is $(3, -1)$. The axis of symmetry is $y = -1$. Replace y with 0 to find the x–intercept.
$$x = -2(0)^2 - 4(0) + 1 = -2(0) - 0 + 1$$
$$= 0 - 0 + 1 = 1$$

The x–intercept is 1. Replace x with 0 to find the y–intercepts.
$$0 = -2y^2 - 4y + 1$$

Solve using the quadratic formula.
$$y = \frac{-b \pm \sqrt{b^2 - 4ac}}{2a}$$
$$= \frac{-(-4) \pm \sqrt{(-4)^2 - 4(-2)(1)}}{2(-2)}$$
$$= \frac{4 \pm \sqrt{16 + 8}}{-4} = \frac{4 \pm \sqrt{24}}{-4} = \frac{4 \pm 2\sqrt{6}}{-4}$$
$$= \frac{2(2 \pm \sqrt{6})}{-4} = \frac{2 \pm \sqrt{6}}{-2} = \frac{-(2 \pm \sqrt{6})}{2}$$
$$= \frac{-2 \pm \sqrt{6}}{2}$$

The y–intercepts are $\dfrac{-2 \pm \sqrt{6}}{2}$.

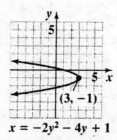

$$x = -2y^2 - 4y + 1$$

43. a. Since the squared term is y, the parabola is horizontal.

　　b. Since $a = 2$ is positive, the parabola opens to the right.

　　c. The vertex is the point $(2, 1)$.

45. a. Since the squared term is x, the parabola is vertical.

　　b. Since $a = 2$ is positive, the parabola opens up.

　　c. The vertex is the point $(1, 2)$.

47. a. Since the squared term is x, the parabola is vertical.

　　b. Since $a = -1$ is negative, the parabola opens down.

　　c. The vertex is the point $(-3, 4)$.

49. a. Since the squared term is y, the parabola is horizontal.

　　b. Since $a = -1$ is negative, the parabola opens to the left.

　　c. The vertex is the point $(4, -3)$.

51. a. Since the squared term is x, the parabola is vertical.

　　b. Since $a = 1$ is positive, the parabola opens up.

　　c. The x–coordinate of the vertex is
$$-\frac{b}{2a} = -\frac{-4}{2(1)} = -\frac{-4}{2} = 2.$$

　　The y–coordinate of the vertex is
$$f(2) = 2^2 - 4(2) - 1$$
$$= 4 - 8 - 1 = -5.$$

　　The vertex is the point $(2, -5)$.

53. a. Since the squared term is y, the parabola is horizontal.

　　b. Since $a = -1$ is negative, the parabola opens to the left.

　　c. The y–coordinate of the vertex is
$$-\frac{b}{2a} = -\frac{4}{2(-1)} = -\frac{4}{-2} = 2.$$

　　The x–coordinate of the vertex is
$$f(2) = -(2)^2 + 4(2) + 1$$
$$= -4 + 8 + 1 = 5.$$

　　The vertex is the point $(5, 2)$.

55. $x - 7 - 8y = y^2$

　Since only one variable is squared, the graph of the equation is a parabola.

57. $\quad 4x^2 = 36 - y^2$
$$4x^2 + y^2 = 36$$

　Because x^2 and y^2 have different positive coefficients, the equation's graph is an ellipse.

59. $\quad x^2 = 36 + 4y^2$
$$x^2 - 4y^2 = 36$$

　Because x^2 and y^2 have opposite signs, the equation's graph is a hyperbola.

61. $\quad 3x^2 = 12 - 3y^2$
$$3x^2 + 3y^2 = 12$$

　Because x^2 and y^2 have the same positive coefficient, the equation's graph is a circle.

63. $\quad 3x^2 = 12 + 3y^2$
$$3x^2 - 3y^2 = 12$$

　Because x^2 and y^2 have opposite signs, the equation's graph is a hyperbola.

65. $x^2 - 4y^2 = 16$

Because x^2 and y^2 have opposite signs, the equation's graph is a hyperbola.

$$\frac{x^2}{16} - \frac{4y^2}{16} = \frac{16}{16}$$

$$\frac{x^2}{16} - \frac{y^2}{4} = 1$$

The equation is in the form $\frac{x^2}{a^2} - \frac{y^2}{b^2} = 1$ with

$a^2 = 16$, and $b^2 = 4$. We know the transverse axis lies on the x-axis and the vertices are $(-4, 0)$ and

$(4, 0)$. Because $a^2 = 16$ and $b^2 = 4$,

$a = 4$ and $b = 2$. Construct a rectangle using –4 and 4 on the x–axis, and –2 and 2 on the y–axis. Draw extended diagonals to obtain the asymptotes. Graph the hyperbola.

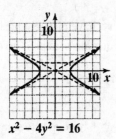

$x^2 - 4y^2 = 16$

67. $4x^2 + 4y^2 = 16$

Because x^2 and y^2 have the same positive coefficient, the equation's graph is a circle.

$$\frac{4x^2}{4} + \frac{4y^2}{4} = \frac{16}{4}$$

$$x^2 + y^2 = 4$$

The center is $(0, 0)$ and the radius is 2 units.

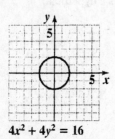

$4x^2 + 4y^2 = 16$

69. $x^2 + 4y^2 = 16$

Because x^2 and y^2 have different positive coefficients, the equation's graph is an ellipse.

$$\frac{x^2}{16} + \frac{4y^2}{16} = \frac{16}{16}$$

$$\frac{x^2}{16} + \frac{y^2}{4} = 1$$

The vertices are $(-4, 0)$ and $(4, 0)$. The endpoints of the minor axis are $(0, -2)$ and $(0, 2)$.

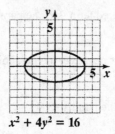

$x^2 + 4y^2 = 16$

71. $x = (y - 1)^2 - 4$

Since only one variable is squared, the graph of the equation is a parabola.

This is a parabola of the form $x = a(y - k)^2 + h$.
Since $a = 1$ is positive, the parabola opens to the right. The vertex of the parabola is $(-4, 1)$. The axis of symmetry is $y = 1$. Replace y with 0 to find the x–intercept.

$$x = (0 - 1)^2 - 4 = (-1)^2 - 4 = 1 - 4 = -3$$

The x–intercept is -3. Replace x with 0 to find the y–intercepts.

$$0 = (y - 1)^2 - 4$$
$$0 = y^2 - 2y + 1 - 4$$
$$0 = y^2 - 2y - 3$$
$$0 = (y - 3)(y + 1)$$
$$y - 3 = 0 \quad \text{or} \quad y + 1 = 0$$
$$y = 3 \qquad\qquad y = -1$$

The *y*–intercepts are –1 and 3.

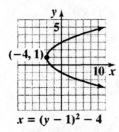

$x = (y - 1)^2 - 4$

73. $(x-2)^2 + (y+1)^2 = 16$

Because x^2 and y^2 have the same positive coefficient, the equation's graph is a circle.

The center is $(2,-1)$ and the radius is 4 units.

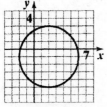

$(x - 2)^2 + (y + 1)^2 = 16$

75. The *y*-coordinate of the vertex is

$$y = -\frac{b}{2a} = -\frac{6}{2(1)} = -3$$

The *x*-coordinate of the vertex is

$$x = (-3)^2 + 6(-3) + 5 = -4$$

The vertex is $(-4,-3)$.

Since the squared term is *y* and $a > 0$, the graph opens to the right.

Domain: $\{x \mid x \geq -4\}$ or $[-4, \infty)$
Range: $\{y \mid y \text{ is a real number}\}$ or $(-\infty, \infty)$

The relation is not a function.

77. The *x*-coordinate of the vertex is

$$x = -\frac{b}{2a} = -\frac{(4)}{2(-1)} = 2$$

The *y*-coordinate of the vertex is

$$y = -(2)^2 + 4(2) - 3 = 1$$

The vertex is $(2,1)$.

Since the squared term is *x* and $a < 0$, the graph opens down.

Domain: $\{x \mid x \text{ is a real number}\}$ or $(-\infty, \infty)$
Range: $\{y \mid y \leq 1\}$ or $(-\infty, 1]$

The relation is a function.

79. The equation is in the form $x = a(y - k)^2 + h$
From the equation, we can see that the vertex is $(3,1)$.

Since the squared term is *y* and $a < 0$, the graph opens to the left.

Domain: $\{x \mid x \leq 3\}$ or $(-\infty, 3]$
Range: $\{y \mid y \text{ is a real number}\}$ or $(-\infty, \infty)$

The relation is not a function.

81. $x = (y-2)^2 - 4$

$$y = -\frac{1}{2}x$$

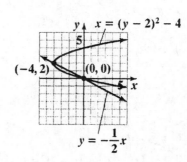

Check $(-4, 2)$:

$-4 = (2-2)^2 - 4$ $2 = -\frac{1}{2}(-4)$

$-4 = 0 - 4$ $2 = 2$ true

$-4 = -4$ true

Check $(0, 0)$:

$0 = (0-2)^2 - 4$ $0 = -\frac{1}{2}(0)$

$0 = 4 - 4$ $0 = 0$ true

$0 = 0$ true

The solution set is $\{(-4, 2), (0, 0)\}$.

83. $x = y^2 - 3$

$x = y^2 - 3y$

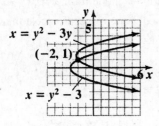

Check $(-2, 1)$:

$-2 = (1)^2 - 3$ $-2 = (1)^2 - 3(1)$

$-2 = 1 - 3$ $-2 = 1 - 3$

$-2 = -2$ true $-2 = -2$ true

The solution set is $\{(-2, 1)\}$.

85. $x = (y+2)^2 - 1$

$(x-2)^2 + (y+2)^2 = 1$

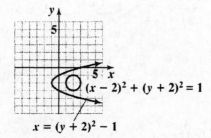

The two graphs do not cross. Therefore, the solution set is the empty set, $\{\ \}$ or $\varnothing$.

87. a. $y = ax^2$

$316 = a(1750)^2$

$316 = a(3062500)$

$\frac{316}{3062500} = a$

$0.0001032 = a$

The equation is $y = 0.0001032x^2$.

b. To find the height of the cable 1000 feet from the tower, find y when $x = 1750 - 1000 = 750$.

$y = 0.0001032(750)^2$

$= 0.0001032(562,500) = 58.05$

The height of the cable is about 58 feet.

89. a. $y = ax^2$

$2 = a(6)^2$

$2 = a(36)$

$\frac{2}{36} = a$

$\frac{1}{18} = a$

The equation is $y = \frac{1}{18}x^2$.

b. $a = \frac{1}{4p}$

$\frac{1}{18} = \frac{1}{4p}$

$4p = 18$

$p = \frac{18}{4} = 4.5$

The receiver should be placed 4.5 feet from the base of the dish.

91. a. ellipse

b. $x^2 + 4y^2 = 4$

93. – 99. Answers will vary.

101.
$$y^2 + 2y - 6x + 13 = 0$$
$$y^2 + 2y + (-6x + 13) = 0$$
$$a = 1 \qquad b = 2 \qquad c = -6x + 13$$

$$y = \frac{-2 \pm \sqrt{2^2 - 4(1)(-6x + 13)}}{2(1)}$$

$$= \frac{-2 \pm \sqrt{4 - 4(-6x + 13)}}{2}$$

$$= \frac{-2 \pm \sqrt{4 + 24x - 52}}{2}$$

$$= \frac{-2 \pm \sqrt{24x - 48}}{2} = \frac{-2 \pm \sqrt{4(6x - 12)}}{2}$$

$$= \frac{-2 \pm 2\sqrt{6x - 12}}{2}$$

$$= -1 \pm \sqrt{6x - 12}$$

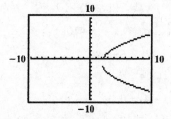

103. Answers will vary. For example, consider Exercise 19.

$$x = (y - 2)^2 - 4$$
$$x + 4 = (y - 2)^2$$
$$\pm\sqrt{x + 4} = y - 2$$
$$2 \pm \sqrt{x + 4} = y$$

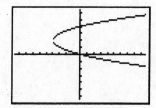

105. makes sense

107. makes sense

109. true

111. false; Changes to make the statement true will vary. A sample change is: $x = a(y - k) + h$ is not a parabola. There is no squared variable.

113.
$$y = ax^2$$
$$-50 = a(100)^2$$
$$-50 = a(10,000)$$
$$a = -\frac{50}{10,000} = -\frac{1}{200}$$

The equation of the parabola is $y = -\frac{1}{200}x^2$.

$$y = -\frac{1}{200}(30)^2 = -\frac{1}{200}(900) = -4.5$$

The height of the arch is 50 feet. The archway comes down 4.5 feet, so the height of the arch 30 feet from the center is $50 - 4.5 = 45.5$ feet.

114. $f(x) = 2^{1-x}$

x	$f(x)$
-2	8
-1	4
0	2
1	1
2	$\frac{1}{2}$

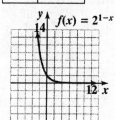

115. $f(x) = \frac{1}{3}x - 5$

$$y = \frac{1}{3}x - 5$$

Interchange x and y and solve for y.

$$x = \frac{1}{3}y - 5$$
$$x + 5 = \frac{1}{3}y$$
$$3x + 15 = y$$
$$f^{-1}(x) = 3x + 15$$

116.
$$(x+1)^2 + (x+3)^2 = 4$$
$$x^2 + 2x + 1 + x^2 + 6x + 9 = 4$$
$$2x^2 + 8x + 10 = 4$$
$$2x^2 + 8x + 6 = 0$$
$$x^2 + 4x + 3 = 0$$
$$(x+1)(x+3) = 0$$
$$x+1 = 0 \quad \text{or} \quad x+3 = 0$$
$$x = -1 \qquad x = -3$$
The solution set is $\{-3, -1\}$.

117.
$$4x + 3y = 4$$
$$4x + 3\overbrace{(2x-7)}^{y} = 4$$
$$4x + 6x - 21 = 4$$
$$10x - 21 = 4$$
$$10x = 25$$
$$x = 2.5$$
Substitute to find y.
$$y = 2x - 7$$
$$y = 2(2.5) - 7 = -2$$
The solution is $(2.5, -2)$.
The solution set is $\{(2.5, -2)\}$.

118.
$$2x + 4y = -4$$
$$3x + 5y = -3$$
Multiply the first equation by 3 and the second equation by -2.
$$6x + 12y = -12$$
$$\underline{-6x - 10y = 6}$$
$$2y = -6$$
$$y = -3$$
Back-substitute to find x.
$$2x + 4y = -4$$
$$2x + 4(-3) = -4$$
$$2x - 12 = -4$$
$$2x = 8$$
$$x = 4$$
The solution is $(4, -3)$.
The solution set is $\{(4, -3)\}$.

119.
$$x^2 = 2(3x - 9) + 10$$
$$x^2 = 6x - 18 + 10$$
$$x^2 = 6x - 8$$
$$x^2 - 6x + 8 = 0$$
$$(x-2)(x-4) = 0$$
$$x - 2 = 0 \quad \text{or} \quad x - 4 = 0$$
$$x = 2 \qquad x = 4$$
The solution set is $\{2, 4\}$.

13.5 Check Points

1. $x^2 = y - 1$

$4x - y = -1$

Solve the first equation for y.

$y = x^2 + 1$

Substitute the expression $x^2 + 1$ for y in the second equation and solve for x.

$4x - y = -1$

$$4x - \overbrace{(x^2 + 1)}^{y} = -1$$

$4x - x^2 - 1 = -1$

$x^2 - 4x = 0$

$x(x - 4) = 0$

$x = 0$ or $x - 4 = 0$

$x = 4$

If $x = 0$, $y = (0)^2 + 1 = 1$.

If $x = 4$, $y = (4)^2 + 1 = 17$.

The solution set is $\{(0, 1), (4, 17)\}$.

2. $x + 2y = 0$

$(x - 1)^2 + (y - 1)^2 = 5$

Solve the first equation for x.

$x = -2y$

Substitute the expression $-2y$ for x in the second equation and solve for y.

$(x - 1)^2 + (y - 1)^2 = 5$

$$(\overbrace{-2y}^{x} - 1)^2 + (y - 1)^2 = 5$$

$4y^2 + 4y + 1 + y^2 - 2y + 1 = 5$

$5y^2 + 2y - 3 = 0$

$(5y - 3)(y + 1) = 0$

$5y - 3 = 0$ or $y + 1 = 0$

$y = \dfrac{3}{5}$ or $y = -1$

If $y = \dfrac{3}{5}$, $x = -2\left(\dfrac{3}{5}\right) = -\dfrac{6}{5}$.

If $y = -1$, $x = -2(-1) = 2$.

The solution set is $\left\{\left(-\dfrac{6}{5}, \dfrac{3}{5}\right), (2, -1)\right\}$.

3. $3x^2 + 2y^2 = 35$

$4x^2 + 3y^2 = 48$

Eliminate the y^2-term by multiplying the first equation by -3 and the second equation by 2. Add the resulting equations.

$-9x^2 - 6y^2 = -105$

$\underline{8x^2 + 6y^2 = 96}$

$-x^2 = -9$

$x^2 = 9$

$x = \pm 3$

If $x = 3$,

$3(3)^2 + 2y^2 = 35$

$y^2 = 4$

$y = \pm 2$

If $x = -3$,

$3(-3)^2 + 2y^2 = 35$

$y^2 = 4$

$y = \pm 2$

The solution set is $\{(3, 2), (3, -2), (-3, 2), (-3, -2)\}$.

4. $y = x^2 + 5$

$x^2 + y^2 = 25$

Arrange the first equation so that variable terms appear on the left, and constants appear on the right.

Add the resulting equations to eliminate the x^2-terms and solve for y.

$-x^2 + y = 5$

$\underline{x^2 + y^2 = 25}$

$y^2 + y = 30$

$y^2 + y - 30 = 0$

$(y + 6)(y - 5) = 0$

$y + 6 = 0$ or $y - 5 = 0$

$y = -6$ or $y = 5$

If $y = -6$,

$x^2 + (-6)^2 = 25$

$x^2 = -11$

When $y = -6$ there is no real solution.

If $y = 5$,

$$x^2 + (5)^2 = 25$$
$$x^2 = 0$$
$$x = 0$$

The solution set is $\{(0, 5)\}$.

5. $2x + 2y = 20$

$$xy = 21$$

Solve the second equation for x.

$$x = \frac{21}{y}$$

Substitute the expression $\frac{21}{y}$ for x in the first

equation and solve for y.

$$2x + 2y = 20$$
$$2\left(\frac{21}{y}\right) + 2y = 20$$
$$\frac{42}{y} + 2y = 20$$
$$y^2 - 10y + 21 = 0$$
$$(y - 7)(y - 3) = 0$$
$$y - 7 = 0 \quad \text{or} \quad y - 3 = 0$$
$$y = 7 \quad \text{or} \quad y = 3$$

If $y = 7$, $x = \frac{21}{7} = 3$.

If $y = 3$, $x = \frac{21}{3} = 7$.

The dimensions are 7 feet by 3 feet.

13.5 Exercise Set

1. $x + y = 2$

$$y = x^2 - 4$$

Substitute $x^2 + 4$ for y in the first equation and
solve for x.

$$x + \left(x^2 - 4\right) = 2$$
$$x + x^2 - 4 = 2$$
$$x^2 + x - 6 = 0$$
$$(x + 3)(x - 2) = 0$$
$$x + 3 = 0 \quad \text{or} \quad x - 2 = 0$$
$$x = -3 \qquad x = 2$$

Substitute -3 and 2 for x in the second equation to
find y.

$$x = -3 \qquad \text{or} \quad x = 2$$
$$y = (-3)^2 - 4 \qquad y = 2^2 - 4$$
$$y = 9 - 4 \qquad y = 4 - 4$$
$$y = 5 \qquad y = 0$$

The solution set is $\left\{(-3, 5), \ (2, 0)\right\}$.

3. $x + y = 2$

$$y = x^2 - 4x + 4$$

Substitute $x^2 - 4x + 4$ for y in the first equation and
solve for x.

$$x + x^2 - 4x + 4 = 2$$
$$x^2 - 3x + 4 = 2$$
$$x^2 - 3x + 2 = 0$$
$$(x - 2)(x - 1) = 0$$
$$x - 2 = 0 \quad \text{or} \quad x - 1 = 0$$
$$x = 2 \qquad x = 1$$

Substitute 1 and 2 for x to find y.

$$x = 2 \quad \text{or} \qquad x = 1$$
$$x + y = 2 \qquad x + y = 2$$
$$2 + y = 2 \qquad 1 + y = 2$$
$$y = 0 \qquad y = 1$$

The solution set is $\left\{(1, 1), (2, 0)\right\}$.

5. $y = x^2 - 4x - 10$

$y = -x^2 - 2x + 14$

Substitute $-x^2 - 2x + 14$ for y in the first equation and solve for x.

$-x^2 - 2x + 14 = x^2 - 4x - 10$

$0 = 2x^2 - 2x - 24$

$0 = x^2 - x - 12$

$0 = (x-4)(x+3)$

$x - 4 = 0$ or $x + 3 = 0$

$x = 4$ $\qquad$ $x = -3$

Substitute -3 and 4 for x to find y.

$x = 4$

$y = 4^2 - 4(4) - 10$

$= 16 - 16 - 10 = -10$

$x = -3$

$y = (-3)^2 - 4(-3) - 10$

$= 9 + 12 - 10 = 11$

The solution set is $\{(-3,11),(4,-10)\}$.

7. $x^2 + y^2 = 25$

$x - y = 1$

Solve the second equation for x.

$x - y = 1$

$x = y + 1$

Substitute $y + 1$ for x to find y.

$x^2 + y^2 = 25$

$(y+1)^2 + y^2 = 25$

$y^2 + 2y + 1 + y^2 = 25$

$2y^2 + 2y + 1 = 25$

$2y^2 + 2y - 24 = 0$

$y^2 + y - 12 = 0$

$(y+4)(y-3) = 0$

$y + 4 = 0$ or $y - 3 = 0$

$y = -4$ $\qquad$ $y = 3$

Substitute -4 and 3 for y to find x.

$y = -4$ $\qquad$ $y = 3$

$x = -4 + 1$ $\qquad$ $x = 3 + 1$

$x = -3$ $\qquad$ $x = 4$

The solution set is $\{(-3,-4),\ (4,3)\}$.

9. $xy = 6$

$2x - y = 1$

Solve the first equation for y.

$xy = 6$

$y = \dfrac{6}{x}$

Substitute $\dfrac{6}{x}$ for y in the second equation and solve for x.

$$2x - \frac{6}{x} = 1$$

$$x\left(2x - \frac{6}{x}\right) = x(1)$$

$$2x^2 - 6 = x$$

$$2x^2 - x - 6 = 0$$

$$(2x+3)(x-2) = 0$$

$x - 2 = 0$ or $2x + 3 = 0$

$x = 2$ $\qquad$ $2x = -3$

$\qquad\qquad\qquad x = -\frac{3}{2}$

Substitute 2 and $-\frac{3}{2}$ for x to find y.

$x = 2$ $\qquad$ or $\qquad$ $x = -\dfrac{3}{2}$

$2y = 6$ $\qquad\qquad\qquad$ $-\dfrac{3}{2}y = 6$

$y = 3$ $\qquad\qquad\qquad$ $-\dfrac{2}{3}\left(-\dfrac{3}{2}\right)y = \left(-\dfrac{2}{3}\right)6$

$\qquad\qquad\qquad\qquad\qquad y = -4$

The solution set is $\left\{(2,3),\left(-\frac{3}{2},-4\right)\right\}$.

11. $y^2 = x^2 - 9$

$2y = x - 3$

Solve the second equation for x.

$2y = x - 3$

$2y + 3 = x$

Substitute $2y + 3$ for x to find y.

$y^2 = (2y + 3)^2 - 9$

$y^2 = 4y^2 + 12y + 9 - 9$

$y^2 = 4y^2 + 12y$

$0 = 3y^2 + 12y$

$0 = 3y(y + 4)$

$3y = 0$ or $y + 4 = 0$

$y = 0$ $\qquad y = -4$

Substitute -4 and 0 for y to find x.

$\qquad y = 0$ or $\qquad y = -4$

$2(0) + 3 = x$ $\qquad 2(-4) + 3 = x$

$\qquad 3 = x$ $\qquad -8 + 3 = x$

$\qquad\qquad -5 = x$

The solution set is $\{(-5, -4),\ (3, 0)\}$.

13. $xy = 3$

$x^2 + y^2 = 10$

Solve the first equation for y.

$xy = 3$

$y = \dfrac{3}{x}$

Substitute $\dfrac{3}{x}$ for y to find x.

$x^2 + \left(\dfrac{3}{x}\right)^2 = 10$

$x^2 + \dfrac{9}{x^2} = 10$

$x^2\left(x^2 + \dfrac{9}{x^2}\right) = x^2(10)$

$x^4 + 9 = 10x^2$

$x^4 - 10x^2 + 9 = 0$

$(x^2 - 9)(x^2 - 1) = 0$

$(x + 3)(x - 3)(x + 1)(x - 1) = 0$

Apply the zero product principle.

$x + 3 = 0$ $\qquad x - 3 = 0$

$x = -3$ $\qquad x = 3$

$x + 1 = 0$ $\qquad x - 1 = 0$

$x = -1$ $\qquad x = 1$

Substitute ± 1 and ± 3 for x to find y.

$x = -3$ $\qquad x = 3$

$y = \dfrac{3}{-3}$ $\qquad y = \dfrac{3}{3}$

$y = -1$ $\qquad y = 1$

$x = -1$ $\qquad x = 1$

$y = \dfrac{3}{-1}$ $\qquad y = \dfrac{3}{1}$

$y = -3$ $\qquad y = 3$

The solution set is $\{(-3, -1), (-1, -3), (1, 3), (3, 1)\}$.

15. $\qquad x + y = 1$

$x^2 + xy - y^2 = -5$

Solve the first equation for y.

$x + y = 1$

$y = -x + 1$

Substitute $-x + 1$ for y and solve for x.

$x^2 + x(-x + 1) - (-x + 1)^2 = -5$

$x^2 - x^2 + x - (x^2 - 2x + 1) = -5$

$x^2 - x^2 + x - x^2 + 2x - 1 = -5$

$-x^2 + 3x - 1 = -5$

$-x^2 + 3x + 4 = 0$

$x^2 - 3x - 4 = 0$

$(x - 4)(x + 1) = 0$

$x - 4 = 0$ or $x + 1 = 0$

$x = 4$ $\qquad x = -1$

Substitute -1 and 4 for x to find y.

$x = 4$ $\qquad$ or $\quad x = -1$

$y = -4 + 1$ $\qquad y = -(-1) + 1$

$y = -3$ $\qquad\quad y = 1 + 1$

$\qquad\qquad\qquad y = 2$

The solution set is $\{(4, -3),\ (-1, 2)\}$.

17.
$$x + y = 1$$
$$(x-1)^2 + (y+2)^2 = 10$$

Solve the first equation for y.
$$x + y = 1$$
$$y = -x + 1$$

Substitute $-x+1$ for y to find x.
$$(x-1)^2 + ((-x+1)+2)^2 = 10$$
$$(x-1)^2 + (-x+1+2)^2 = 10$$
$$(x-1)^2 + (-x+3)^2 = 10$$
$$x^2 - 2x + 1 + x^2 - 6x + 9 = 10$$
$$2x^2 - 8x + 10 = 10$$
$$2x^2 - 8x = 0$$
$$2x(x-4) = 0$$
$$2x = 0 \quad \text{or} \quad x - 4 = 0$$
$$x = 0 \qquad\qquad x = 4$$

Substitute 0 and 4 for x to find y.
$$x = 0 \quad \text{or} \quad x = 4$$
$$y = -0 + 1 \qquad y = -4 + 1$$
$$y = 1 \qquad\qquad y = -3$$

The solution set is $\{(0,1),\ (4,-3)\}$.

19. Solve the system by addition.
$$x^2 + y^2 = 13$$
$$\underline{x^2 - y^2 = 5}$$
$$2x^2 = 18$$
$$x^2 = 9$$
$$x = \pm 3$$

Substitute ± 3 for x to find y.
$$x = \pm 3$$
$$(\pm 3)^2 + y^2 = 13$$
$$9 + y^2 = 13$$
$$y^2 = 4$$
$$y = \pm 2$$

The solution set is $\{(-3,-2),(-3,2),(3,-2),(3,2)\}$.

21.
$$x^2 - 4y^2 = -7$$
$$3x^2 + y^2 = 31$$

Multiply the first equation by -3 and add to the second equation.
$$-3x^2 + 12y^2 = 21$$
$$\underline{3x^2 + y^2 = 31}$$
$$13y^2 = 52$$
$$y^2 = 4$$
$$y = \pm 2$$

Substitute -2 and 2 for y to find x.
$$y = \pm 2$$
$$3x^2 + (\pm 2)^2 = 31$$
$$3x^2 + 4 = 31$$
$$3x^2 = 27$$
$$x^2 = 9$$
$$x = \pm 3$$

The solution set is $\{(-3,-2),(-3,2),(3,-2),(3,2)\}$.

23.
$$3x^2 + 4y^2 - 16 = 0$$
$$2x^2 - 3y^2 - 5 = 0$$

Multiply the first equation by 3 and the second equation by 4 and solve by addition.
$$9x^2 + 12y^2 - 48 = 0$$
$$\underline{8x^2 - 12y^2 - 20 = 0}$$
$$17x^2 - 68 = 0$$
$$17x^2 = 68$$
$$x^2 = 4$$
$$x = \pm 2$$

Substitute ± 2 for x to find y.
$$2(\pm 2)^2 - 3y^2 - 5 = 0$$
$$2(4) - 3y^2 - 5 = 0$$
$$8 - 3y^2 - 5 = 0$$
$$3 - 3y^2 = 0$$
$$3 = 3y^2$$
$$1 = y^2$$
$$\pm 1 = y$$

The solution set is $\{(-2,-1),(-2,1),(2,-1),(2,1)\}$.

25.
$$x^2 + y^2 = 25$$
$$(x-8)^2 + y^2 = 41$$

Multiply the first equation by -1 and solve by addition.

$$-x^2 \qquad -y^2 = -25$$
$$\underline{(x-8)^2 + y^2 = \ 41}$$
$$-x^2 + (x-8)^2 = 16$$
$$-x^2 + x^2 - 16x + 64 = 16$$
$$-16x + 64 = 16$$
$$-16x = -48$$
$$x = 3$$

Substitute 3 for x to find y.
$$x = 3$$
$$3^2 + y^2 = 25$$
$$9 + y^2 = 25$$
$$y^2 = 16$$
$$y = \pm 4$$

The solution set is $\{(3,-4), \ (3,4)\}$.

27.
$$y^2 - x = 4$$
$$x^2 + y^2 \quad = 4$$

Multiply the first equation by -1 and solve by addition.

$$-y^2 + x = -4$$
$$\underline{x^2 + y^2 \quad = \ 4}$$
$$x^2 + x = 0$$
$$x(x+1) = 0$$

Apply the zero product principle.
$$x = 0 \quad \text{or} \quad x + 1 = 0$$
$$x = -1$$

Substitute -1 and 0 for x to find y.

$$
\begin{array}{ll}
x = 0 & x = -1 \\
\quad\quad\text{or} & \\
y^2 - 0 = 4 & y^2 - (-1) = 4 \\
y^2 = 4 & y^2 + 1 = 4 \\
y = \pm 2 & y^2 = 3 \\
& y = \pm\sqrt{3}
\end{array}
$$

The solution set is
$$\left\{\left(-1,-\sqrt{3}\right),\left(-1,\sqrt{3}\right),(0,-2),(0,2)\right\}.$$

29.
$$3x^2 + 4y^2 = 16$$
$$2x^2 - 3y^2 = 5$$

Multiply the first equation by -2 and the second equation by 3 and solve by addition.

$$-6x^2 - 8y^2 = -32$$
$$\underline{6x^2 - 9y^2 = \ 15}$$
$$-17y^2 = -17$$
$$y^2 = 1$$
$$y = \pm 1$$

Substitute ± 1 for y to find x.
$$y = \pm 1$$
$$3x^2 + 4(\pm 1)^2 = 16$$
$$3x^2 + 4(1) = 16$$
$$3x^2 + 4 = 16$$
$$3x^2 = 12$$
$$x^2 = 4$$
$$x = \pm 2$$

The solution set is $\{(-2,-1),(-2,1),(2,-1),$
$(2,1)\}$.

31. $2x^2 + y^2 = 18$
$$xy = 4$$

Solve the second equation for y.
$$xy = 4 \;\; \rightarrow \;\; y = \frac{4}{x}$$

Substitute $\dfrac{4}{x}$ for y in the second equation and solve for x.

$$2x^2 + \left(\frac{4}{x}\right)^2 = 18$$

$$2x^2 + \frac{16}{x^2} = 18$$

$$x^2\left(2x^2 + \frac{16}{x^2}\right) = x^2(18)$$

$$2x^4 + 16 = 18x^2$$

$$2x^4 - 18x^2 + 16 = 0$$

$$x^4 - 9x^2 + 8 = 0$$

$$\left(x^2 - 8\right)\left(x^2 - 1\right) = 0$$

$$\left(x^2 - 8\right)(x+1)(x-1) = 0$$

Apply the zero product principle.

$x^2 - 8 = 0$	$x + 1 = 0$	$x - 1 = 0$
$x^2 = 8$	$x = -1$	$x = 1$
$x = \pm\sqrt{8}$		
$x = \pm 2\sqrt{2}$		

Substitute $\pm 2\sqrt{2}$ and ± 1 for x to find y.

$x = 1$	$x = -1$
$y = \dfrac{4}{1}$	$y = \dfrac{4}{-1}$
$y = 4$	$y = -4$

$x = 2\sqrt{2}$	$x = -2\sqrt{2}$
$y = \dfrac{4}{2\sqrt{2}}$	$y = \dfrac{4}{-2\sqrt{2}}$
$y = \dfrac{2}{\sqrt{2}} \cdot \dfrac{\sqrt{2}}{\sqrt{2}}$	$y = -\dfrac{2}{\sqrt{2}} \cdot \dfrac{\sqrt{2}}{\sqrt{2}}$
$y = \dfrac{2\sqrt{2}}{2}$	$y = -\dfrac{2\sqrt{2}}{2}$
$y = \sqrt{2}$	$y = -\sqrt{2}$

The solution set is $\left\{\left(-2\sqrt{2}, -\sqrt{2}\right),\right.$ $\left.(-1, -4), (1, 4), \left(2\sqrt{2}, \sqrt{2}\right)\right\}$.

33. $x^2 + 4y^2 = 20$
$$x + 2y = 6$$

Solve the second equation for x.
$$x + 2y = 6$$
$$x = 6 - 2y$$

Substitute $6 - 2y$ for x to find y.

$$(6 - 2y)^2 + 4y^2 = 20$$

$$36 - 24y + 4y^2 + 4y^2 = 20$$

$$36 - 24y + 8y^2 = 20$$

$$8y^2 - 24y + 16 = 0$$

$$y^2 - 3y + 2 = 0$$

$$(y - 2)(y - 1) = 0$$

$$y - 2 = 0 \quad \text{or} \quad y - 1 = 0$$

$$y = 2 \qquad\qquad y = 1$$

Substitute 1 and 2 for y to find x.

$y = 2$	or	$y = 1$
$x = 6 - 2(2)$		$x = 6 - 2(1)$
$x = 6 - 4$		$x = 6 - 2$
$x = 2$		$x = 4$

The solution set is $\left\{(2, 2), (4, 1)\right\}$.

35. Eliminate y by adding the two equations.

$$x^3 + y = 0$$
$$\underline{x^2 - y = 0}$$
$$x^3 + x^2 = 0$$
$$x^2(x+1) = 0$$
$$x = 0 \quad \text{or} \quad x = -1$$

Substitute -1 and 0 for x to find y.

$$x = 0 \quad \text{or} \qquad\qquad x = -1$$
$$0^2 - y = 0 \qquad\quad (-1)^2 - y = 0$$
$$-y = 0 \qquad\qquad\quad 1 - y = 0$$
$$y = 0 \qquad\qquad\qquad -y = -1$$
$$\qquad\qquad\qquad\qquad y = 1$$

The solution set is $\{(-1,1),\ (0,0)\}$.

37. $x^2 + (y-2)^2 = 4$
$$x^2 - 2y = 0$$

Solve the second equation for x^2.
$$x^2 - 2y = 0$$
$$x^2 = 2y$$

Substitute $2y$ for x^2 in the first equation and solve for y.

$$2y + (y-2)^2 = 4$$
$$2y + y^2 - 4y + 4 = 4$$
$$y^2 - 2y + 4 = 4$$
$$y^2 - 2y = 0$$
$$y(y-2) = 0$$
$$y = 0 \quad \text{or} \quad y - 2 = 0$$
$$y = 2$$

Substitute 0 and 2 for y to find x.

$$y = 0 \qquad \text{or} \quad y = 2$$
$$x^2 = 2(0) \qquad x^2 = 2(2)$$
$$x^2 = 0 \qquad\quad x^2 = 4$$
$$x = 0 \qquad\quad x = \pm 2$$

The solution set is $\{(0,0),\ (-2,2),(2,2)\}$.

39. $\quad y = (x+3)^2 \displaystyle\lim_{x\to\infty}$
$$x + 2y = -2$$

Substitute $(x+3)^2$ for y in the second equation.

$$x + 2(x+3)^2 = -2$$
$$x + 2(x^2 + 6x + 9) = -2$$
$$x + 2x^2 + 12x + 18 = -2$$
$$2x^2 + 13x + 18 = -2$$
$$2x^2 + 13x + 20 = 0$$
$$(2x+5)(x+4) = 0$$
$$2x + 5 = 0 \quad \text{or} \quad x + 4 = 0$$
$$x = -\frac{5}{2} \qquad\qquad x = -4$$

Substitute $-\frac{5}{2}$ and -4 for x to find y.

$$x = -\frac{5}{2} \quad \text{or} \qquad x = -4$$
$$\qquad\qquad\qquad\qquad -4 + 2y = -2$$
$$-\frac{5}{2} + 2y = -2 \qquad 2y = 2$$
$$-5 + 4y = -4 \qquad\quad y = 1$$
$$4y = 1$$
$$y = \frac{1}{4}$$

The solution set is $\left\{(-4,1),\left(-\frac{5}{2},\frac{1}{4}\right)\right\}$.

41. $x^2 + y^2 + 3y = 22$
$$2x + y = -1$$

Solve the second equation for y.
$$2x + y = -1$$
$$y = -2x - 1$$

Substitute $-2x - 1$ for y to find x.

$$x^2 + (-2x-1)^2 + 3(-2x-1) = 22$$
$$x^2 + 4x^2 + 4x + 1 - 6x - 3 = 22$$
$$5x^2 - 2x - 2 = 22$$
$$5x^2 - 2x - 24 = 0$$
$$(5x-12)(x+2) = 0$$
$$5x - 12 = 0 \quad \text{or} \quad x + 2 = 0$$
$$5x = 12 \qquad\qquad\quad x = -2$$
$$x = \frac{12}{5}$$

Substitute -2 and $\dfrac{12}{5}$ for x to find y.

$x = \dfrac{12}{5}$ or $x = -2$

$y = -2\left(\dfrac{12}{5}\right) - 1$ $y = -2(-2) - 1$

$y = -\dfrac{24}{5} - \dfrac{5}{5}$ $y = 4 - 1$

         $y = 3$

$y = -\dfrac{29}{5}$

The solution set is $\left\{\left(\dfrac{12}{5}, -\dfrac{29}{5}\right), (-2, 3)\right\}$.

43. Let x = one of the numbers.
Let y = the other number.
$x + y = 10$
$xy = 24$

Solve the second equation for y.
$xy = 24$

$y = \dfrac{24}{x}$

Substitute $\dfrac{24}{x}$ for y in the first equation and solve
for x.

$x + \dfrac{24}{x} = 10$

$x\left(x + \dfrac{24}{x}\right) = x(10)$

$x^2 + 24 = 10x$

$x^2 - 10x + 24 = 0$

$(x - 6)(x - 4) = 0$

$x - 6 = 0$ or $x - 4 = 0$

$x = 6$ $x = 4$

Substitute 6 and 4 for x to find y.

$x = 6$ $x = 4$

$y = \dfrac{24}{6}$ or $y = \dfrac{24}{4}$

$y = 4$ $y = 6$

The numbers are 4 and 6.

45. Let x = one of the numbers.
Let y = the other number.
$x^2 - y^2 = 3$
$\underline{2x^2 + y^2 = 9}$
$\quad\quad 3x^2 = 12$
$\quad\quad\ \ x^2 = 4$
$\quad\quad\ \ \ x = \pm 2$

Substitute ± 2 for x to find y.
$\quad x = \pm 2$
$(\pm 2)^2 - y^2 = 3$
$\quad 4 - y^2 = 3$
$\quad\ \ -y^2 = -1$
$\quad\quad y^2 = 1$
$\quad\quad\ y = \pm 1$

The numbers are either 2 and -1, 2 and 1, -2 and -1, or -2 and 1.

47. $2x^2 + xy = 6$

$x^2 + 2xy = 0$

Multiply the first equation by -2 and add the two equations.
$-4x^2 - 2xy = -12$
$\underline{\ \ x^2 + 2xy = 0}$
$\quad -3x^2 = -12$
$\quad\ \ x^2 = 4$
$\quad\ \ \ x = \pm 2$

Back-substitute these values for x in the second equation and solve for y.

For $x = -2$: $(-2)^2 + 2(-2)y = 0$
$\quad\quad\quad\quad\quad 4 - 4y = 0$
$\quad\quad\quad\quad\quad\ -4y = -4$
$\quad\quad\quad\quad\quad\quad\ y = 1$

For $x = 2$: $(2)^2 + 2(2)y = 0$
$\quad\quad\quad\quad\ 4 + 4y = 0$
$\quad\quad\quad\quad\quad 4y = -4$
$\quad\quad\quad\quad\quad\ y = -1$

The solution set is $\{(-2, 1), (2, -1)\}$.

49. $-4x + y = 12$

$y = x^3 + 3x^2$

Substitute $x^3 + 3x^2$ for y in the first equation and solve for x.

$$-4x + \left(x^3 + 3x^2\right) = 12$$

$$x^3 + 3x^2 - 4x - 12 = 0$$

$$x^2(x+3) - 4(x+3) = 0$$

$$(x+3)\left(x^2 - 4\right) = 0$$

$$(x+3)(x-2)(x+2) = 0$$

$x = -3$, $x = 2$, or $x = -2$

Substitute these values for x in the second equation and solve for y.

For $x = -3$: $\begin{aligned} y &= (-3)^3 + 3(-3)^2 \\ &= -27 + 27 \\ &= 0 \end{aligned}$

For $x = 2$: $\begin{aligned} y &= (2)^3 + 3(2)^2 \\ &= 8 + 12 \\ &= 20 \end{aligned}$

For $x = -2$: $\begin{aligned} y &= (-2)^3 + 3(-2)^2 \\ &= -8 + 12 \\ &= 4 \end{aligned}$

The solution set is $\left\{(-3,0),(2,20),(-2,4)\right\}$.

51. $\dfrac{3}{x^2} + \dfrac{1}{y^2} = 7$

$\dfrac{5}{x^2} - \dfrac{2}{y^2} = -3$

Multiply the first equation by 2 and add the equations.

$$\dfrac{6}{x^2} + \dfrac{2}{y^2} = 14$$

$$\dfrac{5}{x^2} - \dfrac{2}{y^2} = -3$$

$$\overline{\qquad\qquad\qquad}$$

$$\dfrac{11}{x^2} = 11$$

$$x^2 = 1$$

$$x = \pm 1$$

Back-substitute these values for x in the first equation and solve for y.

For $x = -1$:

$$\dfrac{3}{(-1)^2} + \dfrac{1}{y^2} = 7$$

$$3 + \dfrac{1}{y^2} = 7$$

$$\dfrac{1}{y^2} = 4$$

$$y^2 = \dfrac{1}{4}$$

$$y = \pm\dfrac{1}{2}$$

For $x = 1$:

$$\dfrac{3}{(1)^2} + \dfrac{1}{y^2} = 7$$

$$3 + \dfrac{1}{y^2} = 7$$

$$\dfrac{1}{y^2} = 4$$

$$y^2 = \dfrac{1}{4}$$

$$y = \pm\dfrac{1}{2}$$

The solution set is

$$\left\{\left(-1,-\dfrac{1}{2}\right),\left(-1,\dfrac{1}{2}\right),\left(1,-\dfrac{1}{2}\right),\left(1,\dfrac{1}{2}\right)\right\}.$$

53. Answers will vary. One example:

Circle: $x^2 + y^2 = 9$

Ellipse: $\dfrac{x^2}{9} + \dfrac{y^2}{49} = 1$

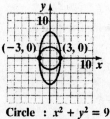

Circle : $x^2 + y^2 = 9$

Ellipse : $\dfrac{x^2}{9} + \dfrac{y^2}{49} = 1$

Solutions: $(-3,0)$ and $(3,0)$.

55. $16x^2 + 4y^2 = 64$

$$y = x^2 - 4$$

Solve the second equation for x^2.

$$y = x^2 - 4$$

$$y + 4 = x^2$$

Substitute $y + 4$ for x^2 in the first equation and solve for y.

$$16(y+4) + 4y^2 = 64$$

$$16y + 64 + 4y^2 = 64$$

$$16y + 4y^2 = 0$$

$$4y(4+y) = 0$$

$4y = 0$ or $4 + y = 0$

$y = 0$ $y = -4$

Substitute 0 and 4 for y to find x.

$y = 0$ or $y = -4$

$0 = x^2 - 4$ $-4 = x^2 - 4$

$4 = x^2$ $0 = x^2$

$\pm 2 = x$ $0 = x$

The comet intersects the planet's orbit at the points $(2,0), (-2,0)$ and $(0,-4)$.

57. Let x = the length of the rectangle.
Let y = the width of the rectangle.
Perimeter: $2x + 2y = 36$
Area: $xy = 77$

Solve the second equation for y.
$xy = 77$

$$y = \frac{77}{x}$$

Substitute $\dfrac{77}{x}$ for y in the first equation and solve for x.

$$2x + 2\left(\frac{77}{x}\right) = 36$$

$$2x + \frac{154}{x} = 36$$

$$x\left(2x + \frac{154}{x}\right) = x(36)$$

$$2x^2 + 154 = 36x$$

$$2x^2 - 36x + 154 = 0$$

$$x^2 - 18x + 77 = 0$$

$$(x-7)(x-11) = 0$$

$x - 7 = 0$ or $x - 11 = 0$

$x = 7$ $x = 11$

Substitute 7 and 11 for x to find y.
$x = 7$ or $x = 11$

$$y = \frac{77}{7} \qquad y = \frac{77}{11}$$

$y = 11$ $y = 7$

The dimensions of the rectangle are 7 feet by 11 feet.

59. Let x = the length of the screen.
Let y = the width of the screen.
$$x^2 + y^2 = 10^2$$
$$xy = 48$$

Solve the second equation for y.
$xy = 48$

$$y = \frac{48}{x}$$

Substitute $\dfrac{48}{x}$ for y to find x.

$$x^2 + \left(\frac{48}{x}\right)^2 = 10^2$$

$$x^2 + \frac{2304}{x^2} = 100$$

$$x^2\left(x^2 + \frac{2304}{x^2}\right) = x^2(100)$$

$$x^4 + 2304 = 100x^2$$

$$x^4 - 100x^2 + 2304 = 0$$

$$\left(x^2 - 64\right)\left(x^2 - 36\right) = 0$$

$$(x+8)(x-8)(x+6)(x-6) = 0$$

Apply the zero product principle.

$$x + 8 = 0 \qquad x - 8 = 0$$
$$x = -8 \qquad x = 8$$

$$x + 6 = 0 \qquad x + 6 = 0$$
$$x = -6 \qquad x = -6$$

Disregard –8 and –6 because we cannot have a negative length. Substitute 8 and 6 for x to find y.

$$x = 8 \quad \text{or} \quad x = 6$$
$$y = \frac{48}{8} \qquad y = \frac{48}{6}$$
$$y = 6 \qquad y = 8$$

The dimensions of the screen are 8 inches by 6 inches.

61. $\quad x^2 - y^2 = 21$
$\qquad 4x + 2y = 24$

Solve for y in the second equation.
$$4x + 2y = 24$$
$$2y = 24 - 4x$$
$$y = 12 - 2x$$

Substitute $12 - 2x$ for y and solve for x.

$$x^2 - \left(12 - 2x\right)^2 = 21$$
$$x^2 - \left(144 - 48x + 4x^2\right) = 21$$
$$x^2 - 144 + 48x - 4x^2 = 21$$
$$-3x^2 + 48x - 144 = 21$$
$$-3x^2 + 48x - 165 = 0$$
$$x^2 - 16x + 55 = 0$$
$$\left(x - 5\right)\left(x - 11\right) = 0$$
$$x - 5 = 0 \quad \text{or} \quad x - 11 = 0$$
$$x = 5 \qquad x = 11$$

Substitute 5 and 11 for x to find y.
$$x = 5 \qquad \text{or} \quad x = 11$$
$$y = 12 - 2(5) \qquad y = 12 - 2(11)$$
$$y = 12 - 10 \qquad y = 12 - 22$$
$$y = 2 \qquad y = -10$$

Disregard –10 because we can't have a negative length measurement. The larger square is 5 meters by 5 meters and the smaller square to be cut out is 2 meters by 2 meters.

63. a. It appears from the graphs that the percentage of white-collar workers was the same as blue-collar workers between the 1940s and 1960s.

b. $\quad 0.5x - y = -18$
$\qquad y = -0.004x^2 + 0.23x + 41$

Substitute $-0.004x^2 + 0.23x + 41$ in the first equation and solve for x.

$$0.5x - y = -18$$
$$0.5x - \overbrace{\left(-0.004x^2 + 0.23x + 41\right)}^{y} = -18$$
$$0.5x + 0.004x^2 - 0.23x - 41 = -18$$
$$0.004x^2 + 0.27x - 23 = 0$$

Use the quadratic formula.

$$x = \frac{-b \pm \sqrt{b^2 - 4ac}}{2a}$$

$$x = \frac{-(0.27) \pm \sqrt{0.27^2 - 4(0.004)(-23)}}{2(0.004)}$$

$$x \approx 49 \text{ or } -117$$

The percentage of white-collar workers was the same as blue-collar workers 49 years after 1900, or 1949.

Let $x = 49$ and solve for y in the white-collar model.
$$0.5x - y = -18$$
$$0.5(49) - y = -18$$
$$24.5 - y = -18$$
$$-y = -42.5$$
$$y \approx 43$$

The percentage of white-collar workers in 1949 was about 43%

Let $x = 49$ and solve for y in the blue-collar model.
$$y = -0.004(49)^2 + 0.23(49) + 41 \approx 43\%$$

The percentage of blue-collar workers in 1949 was about 43%.

c. According to the graph, the percentage of white-collar workers was the same as farmers in 1920. The percentages of white-collar workers and farmers in 1920 were both 28%.

d. $0.5x - y = -18$

$\dfrac{0.4x + y = 35}{0.9x \qquad = 17}$

$x = \dfrac{17}{0.9}$

$x \approx 19$

According to the models, the percentage of white-collar workers was the same as farmers 19 years after 1900, or 1919.

Let $x = 19$ and solve for y in the white-collar model.

$0.5x - y = -18$

$0.5(19) - y = -18$

$9.5 - y = -18$

$-y = -27.5$

$y = 27.5$

The percentage of white-collar workers in 1919 was about 27.5%

Let $x = 19$ and solve for y in the farming model.

$0.4x + y = 35$

$0.4(19) + y = 35$

$0.4(19) + y = 35$

$7.6 + y = 35$

$y = 27.4$

The percentage of farm workers in 1919 was about 27.4%

These answers model the actual data from part c (the graph) fairly well.

65. – 67. Answers will vary.

69. Answers will vary. For example, consider the following system.

$y = x$

$y = x^2 + 5$

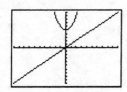

Since the graphs do not intersect, there are no ordered pairs that are real numbers in the solution set.

71. does not make sense; Explanations will vary. Sample explanation: Since the orbits of earth and Mars do not intersect, their system of equations will have no solution.

73. makes sense

75. true

77. false; Changes to make the statement true will vary. A sample change is: It is possible that a system of two equations in two variables whose graphs represent a parabola and a circle to have one real ordered-pair solution. This will occur if the graphs intersect in a single point.

79. $\log_y x = 3$

$\log_y (4x) = 5$

Rewrite the equations.

$y^3 = x$

$y^5 = 4x$

Substitute y^3 for x in the second equation and solve for y.

$$y^5 = 4y^3$$

$$y^5 - 4y^3 = 0$$

$$y^3 \left(y^2 - 4 \right) = 0$$

$$y^3 (y + 2)(y - 2) = 0$$

Apply the zero product principle.

$y^3 = 0 \qquad y + 2 = 0 \qquad y - 2 = 0$

$y = 0 \qquad\quad y = -2 \qquad\quad y = 2$

Disregard 0 and –2 because the base of a logarithm must be greater than zero.

Substitute 2 for y to find x.

$y^3 = x$

$2^3 = x$

$8 = x$

The solution is $(8, 2)$ and the solution set is $\{(8, 2)\}$.

81. $3x - 2y \le 6$

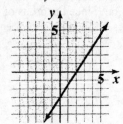

82. $m = \dfrac{5 - (-3)}{1 - (-2)} = \dfrac{5 + 3}{1 + 2} = \dfrac{8}{3}$

The slope is $\dfrac{8}{3}$.

83. $2x^2 - 4x + 3$

$\underline{\ 3x - 2}$

$6x^3 - 12x^2 + 9x$

$\underline{ -4x^2 + 8x - 6}$

$6x^3 - 16x^2 + 17x - 6$

84. For $n = 1$; $\dfrac{(-1)^n}{3^n - 1} = \dfrac{(-1)^1}{3^1 - 1} = \dfrac{-1}{3 - 1} = -\dfrac{1}{2}$

For $n = 2$; $\dfrac{(-1)^n}{3^n - 1} = \dfrac{(-1)^2}{3^2 - 1} = \dfrac{1}{9 - 1} = \dfrac{1}{8}$

For $n = 3$; $\dfrac{(-1)^n}{3^n - 1} = \dfrac{(-1)^3}{3^3 - 1} = \dfrac{-1}{27 - 1} = -\dfrac{1}{26}$

For $n = 4$; $\dfrac{(-1)^n}{3^n - 1} = \dfrac{(-1)^4}{3^4 - 1} = \dfrac{1}{81 - 1} = \dfrac{1}{80}$

85. $5 \cdot 4 \cdot 3 \cdot 2 \cdot 1 = 120$

86. For $n = 1$; $n^2 + 1 = 1^2 + 1 = 1 + 1 = 2$

For $n = 2$; $n^2 + 1 = 2^2 + 1 = 4 + 1 = 5$

For $n = 3$; $n^2 + 1 = 3^2 + 1 = 9 + 1 = 10$

For $n = 4$; $n^2 + 1 = 4^2 + 1 = 16 + 1 = 17$

For $n = 5$; $n^2 + 1 = 5^2 + 1 = 25 + 1 = 26$

For $n = 6$; $n^2 + 1 = 6^2 + 1 = 36 + 1 = 37$

$2 + 5 + 10 + 17 + 26 + 37 = 97$

Chapter 13 Review Exercises

1. $(x - 0)^2 + (y - 0)^2 = 3^2$

$ x^2 + y^2 = 9$

2. $(x - (-2))^2 + (y - 4)^2 = 6^2$

$ (x + 2)^2 + (y - 4)^2 = 36$

3. $ x^2 + y^2 = 1$

$(x - 0)^2 + (y - 0)^2 = 1^2$

The center is $(0,0)$ and the radius is 1 units.

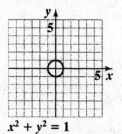

$x^2 + y^2 = 1$

4. $ (x + 2)^2 + (y - 3)^2 = 9$

$(x - (-2))^2 + (y - 3)^2 = 3^2$

The center is $(-2,3)$ and the radius is 3 units.

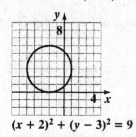

$(x + 2)^2 + (y - 3)^2 = 9$

5.
$$x^2 + y^2 - 4x + 2y - 4 = 0$$
$$\left(x^2 - 4x \quad\right) + \left(y^2 + 2y \quad\right) = 4$$

Complete the squares.

$$\left(\frac{b}{2}\right)^2 = \left(\frac{-4}{2}\right)^2 = (-2)^2 = 4$$

$$\left(\frac{b}{2}\right)^2 = \left(\frac{2}{2}\right)^2 = 1^2 = 1$$

$$\left(x^2 - 4x + 4\right) + \left(y^2 + 2y + 1\right) = 4 + 4 + 1$$

$$(x - 2)^2 + (y + 1)^2 = 9$$

$$(x - 2)^2 + \left(y - (-1)\right)^2 = 3^2$$

The center is $(2, -1)$ and the radius is 3 units.

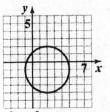

$x^2 + y^2 - 4x + 2y - 4 = 0$

6.
$$x^2 + y^2 - 4y = 0$$
$$x^2 + \left(y^2 - 4y \quad\right) = 0$$

Complete the square.

$$\left(\frac{b}{2}\right)^2 = \left(\frac{-4}{2}\right)^2 = (-2)^2 = 4$$

$$x^2 + \left(y^2 - 4y + 4\right) = 0 + 4$$

$$(x - 0)^2 + (y - 2)^2 = 4$$

$$(x - 0)^2 + (y - 2)^2 = 2^2$$

The center is $(0, 2)$ and the radius is 2 units.

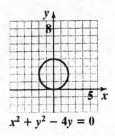

$x^2 + y^2 - 4y = 0$

7. $\dfrac{x^2}{36} + \dfrac{y^2}{25} = 1$

Because the denominator of the x^2 – term is greater than the denominator of the y^2 – term, the major axis is horizontal. Since $a^2 = 36$, $a = 6$ and the vertices are $(-6, 0)$ and $(6, 0)$. Since $b^2 = 25$, $b = 5$ and endpoints of the minor axis are $(0, -5)$ and $(0, 5)$.

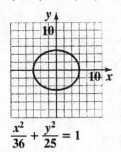

$\dfrac{x^2}{36} + \dfrac{y^2}{25} = 1$

8. $\dfrac{x^2}{25} + \dfrac{y^2}{16} = 1$

Because the denominator of the x^2 – term is greater than the denominator of the y^2 – term, the major axis is horizontal. Since $a^2 = 25$, $a = 5$ and the vertices are $(-5, 0)$ and $(5, 0)$. Since $b^2 = 16$, $b = 4$ and endpoints of the minor axis are $(0, -4)$ and $(0, 4)$.

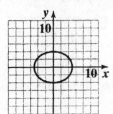

$\dfrac{x^2}{25} + \dfrac{y^2}{16} = 1$

9. $4x^2 + y^2 = 16$

$$\frac{4x^2}{16} + \frac{y^2}{16} = \frac{16}{16}$$

$$\frac{x^2}{4} + \frac{y^2}{16} = 1$$

Because the denominator of the y^2 – term is greater than the denominator of the x^2 – term, the major axis is vertical. Since $a^2 = 16$, $a = 4$ and the vertices are $(0,-4)$ and $(0,4)$. Since $b^2 = 4$, $b = 2$ and endpoints of the minor axis are $(-2,0)$ and $(2,0)$.

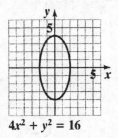

$4x^2 + y^2 = 16$

10. $4x^2 + 9y^2 = 36$

$$\frac{4x^2}{36} + \frac{9y^2}{36} = \frac{36}{36}$$

$$\frac{x^2}{9} + \frac{y^2}{4} = 1$$

Because the denominator of the x^2 – term is greater than the denominator of the y^2 – term, the major axis is horizontal. Since $a^2 = 9$, $a = 3$ and the vertices are $(-3,0)$ and $(3,0)$. Since $b^2 = 4$, $b = 2$ and endpoints of the minor axis are $(0,-2)$ and $(0,2)$.

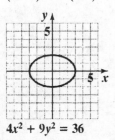

$4x^2 + 9y^2 = 36$

11. $\dfrac{(x-1)^2}{16} + \dfrac{(y+2)^2}{9} = 1$

The center of the ellipse is $(1,-2)$. Because the denominator of the x^2 – term is greater than the denominator of the y^2 – term, the major axis is horizontal. Since $a^2 = 16$, $a = 4$ and the vertices lie 4 units to the left and right of the center. Since $b^2 = 9$, $b = 3$ and endpoints of the minor axis lie 3 units above and below the center.

Center	Vertices	Endpoints of Minor Axis
$(1,-2)$	$(1-4,-2)$ $= (-3,-2)$	$(1,-2-3)$ $= (1,-5)$
	$(1+4,-2)$ $= (5,-2)$	$(1,-2+3)$ $= (1,1)$

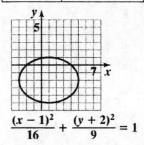

$\dfrac{(x-1)^2}{16} + \dfrac{(y+2)^2}{9} = 1$

12. $\dfrac{(x+1)^2}{9} + \dfrac{(y-2)^2}{16} = 1$

The center of the ellipse is $(-1,2)$. Because the denominator of the y^2 – term is greater than the denominator of the x^2 – term the major axis is vertical. Since $a^2 = 16$, $a = 4$ and the vertices lie 4 units above and below the center. Since $b^2 = 9$, $b = 3$ and endpoints of the minor axis lie 3 units to the left and right of the center.

Center	Vertices	Endpoints of Minor Axis
$(-1,2)$	$(-1,2-4)$ $= (-1,-2)$	$(-1-3,2)$ $= (-4,2)$
	$(-1,2+4)$ $= (-1,6)$	$(-1+3,2)$ $= (2,2)$

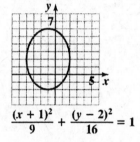

$\dfrac{(x+1)^2}{9} + \dfrac{(y-2)^2}{16} = 1$

13. From the figure, we see that the major axis is horizontal with $a = 25$, and $b = 15$.

$$\frac{x^2}{25^2} + \frac{y^2}{15^2} = 1$$

$$\frac{x^2}{625} + \frac{y^2}{225} = 1$$

Since the truck is 14 feet wide, determine the height of the archway at 14 feet to the right of center.

$$\frac{14^2}{625} + \frac{y^2}{225} = 1$$

$$\frac{196}{625} + \frac{y^2}{225} = 1$$

$$5625\left(\frac{196}{625} + \frac{y^2}{225}\right) = 5625(1)$$

$$9(196) + 25y^2 = 5625$$

$$1764 + 25y^2 = 5625$$

$$25y^2 = 3861$$

$$y^2 = \frac{3861}{25}$$

$$y = \sqrt{\frac{3861}{25}} \approx 12.43$$

The height of the archway 14 feet from the center is approximately 12.43 feet. Since the truck is 12 feet high, the truck will clear the archway.

14. $\dfrac{x^2}{16} - y^2 = 1$

$$\frac{x^2}{16} - \frac{y^2}{1} = 1$$

The equation is in the form $\dfrac{x^2}{a^2} - \dfrac{y^2}{b^2} = 1$ with $a^2 = 16$, and $b^2 = 1$. We know the transverse axis lies on the x-axis and the vertices are $(-4,0)$ and $(4,0)$. Because $a^2 = 16$ and $b^2 = 1$, $a = 4$ and $b = 1$. Construct a rectangle using –4 and 4 on the x-axis, and –1 and 1 on the y–axis. Draw extended diagonals to obtain the asymptotes. Graph the hyperbola.

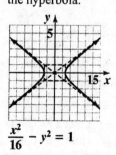

$\dfrac{x^2}{16} - y^2 = 1$

15. $\dfrac{y^2}{16} - x^2 = 1$

$\dfrac{y^2}{16} - \dfrac{x^2}{1} = 1$

The equation is in the form $\dfrac{y^2}{a^2} - \dfrac{x^2}{b^2} = 1$ with

$a^2 = 16$, and $b^2 = 1$. We know the transverse axis lies on the y-axis and the vertices are

$(0,-4)$ and $(0,4)$. Because $a^2 = 16$ and $b^2 = 1$,

$a = 4$ and $b = 1$. Construct a rectangle using -1 and 1 on the x–axis, and -4 and 4 on the y–axis. Draw extended diagonals to obtain the asymptotes. Graph the hyperbola.

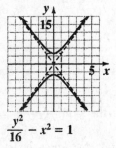

$\dfrac{y^2}{16} - x^2 = 1$

16. $9x^2 - 16y^2 = 144$

$\dfrac{9x^2}{144} - \dfrac{16y^2}{144} = \dfrac{144}{144}$

$\dfrac{x^2}{16} - \dfrac{y^2}{9} = 1$

The equation is in the form $\dfrac{x^2}{a^2} - \dfrac{y^2}{b^2} = 1$ with

$a^2 = 16$, and $b^2 = 9$. We know the transverse axis lies on the x-axis and the vertices are

$(-4,0)$ and $(4,0)$. Because $a^2 = 16$ and $b^2 = 9$,

$a = 4$ and $b = 3$. Construct a rectangle using -4 and 4 on the x–axis, and -3 and 3 on the y–axis. Draw extended diagonals to obtain the asymptotes. Graph the hyperbola.

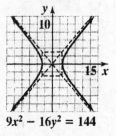

$9x^2 - 16y^2 = 144$

17. $4y^2 - x^2 = 16$

$\dfrac{4y^2}{16} - \dfrac{x^2}{16} = \dfrac{16}{16}$

$\dfrac{y^2}{4} - \dfrac{x^2}{16} = 1$

The equation is in the form $\dfrac{y^2}{a^2} - \dfrac{x^2}{b^2} = 1$ with

$a^2 = 4$, and $b^2 = 16$. We know the transverse axis lies on the y-axis and the vertices are

$(0,-2)$ and $(0,2)$. Because $a^2 = 4$ and $b^2 = 16$,

$a = 2$ and $b = 4$. Construct a rectangle using -2 and 2 on the y–axis, and -4 and 4 on the x–axis. Draw extended diagonals to obtain the asymptotes. Graph the hyperbola.

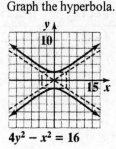

$4y^2 - x^2 = 16$

18. $x = (y-3)^2 - 4$

This is a parabola of the form $x = a(y-k)^2 + h$.
Since $a = 1$ is positive, the parabola opens to the right. The vertex of the parabola is $(-4, 3)$. The axis of symmetry is $y = 3$. Replace y with 0 to find the x–intercept.

$x = (0-3)^2 - 4 = (-3)^2 - 4 = 9 - 4 = 5$

The x–intercept is 5. Replace x with 0 to find the y–intercepts.

$0 = (y-3)^2 - 4$

$0 = y^2 - 6y + 9 - 4$

$0 = y^2 - 6y + 5$

$0 = (y-5)(y-1)$

$y - 5 = 0$　　or　　$y - 1 = 0$

　$y = 5$　　　　　　$y = 1$

The y–intercepts are 1 and 5.

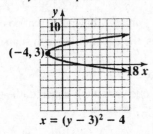

$x = (y - 3)^2 - 4$

19. $x = -2(y+3)^2 + 2$

This is a parabola of the form $x = a(y-k)^2 + h$.
Since $a = -2$ is negative, the parabola opens to the left. The vertex of the parabola is $(2, -3)$. The axis of symmetry is $y = -3$. Replace y with 0 to find the x–intercept.

$x = -2(0+3)^2 + 2 = -2(3)^2 + 2$

　　$= -2(9) + 2 = -18 + 2 = -16$

The x–intercept is -16. Replace x with 0 to find the y–intercepts.

$0 = -2(y+3)^2 + 2$

$0 = -2\left(y^2 + 6y + 9\right) + 2$

$0 = -2y^2 - 12y - 18 + 2$

$0 = -2y^2 - 12y - 16$

$0 = y^2 + 6y + 8$

$0 = (y+4)(y+2)$

$y + 4 = 0$　　or　　$y + 2 = 0$

　$y = -4$　　　　　　$y = -2$

The y–intercepts are -4 and -2.

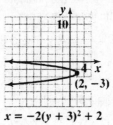

$x = -2(y + 3)^2 + 2$

20. $x = y^2 - 8y + 12$

This is a parabola of the form $x = ay^2 + by + c$.
Since $a = 1$ is positive, the parabola opens to the right. The y–coordinate of the vertex is

$-\dfrac{b}{2a} = -\dfrac{-8}{2(1)} = -\dfrac{-8}{2} = 4$. The x–coordinate of the

vertex is $x = 4^2 - 8(4) + 12 = 16 - 32 + 12$

　　　　　　　$= 16 - 32 + 12 = -4$.

The vertex of the parabola is $(-4, 4)$. The axis of symmetry is $y = 4$.

Replace y with 0 to find the x–intercept.

$x = 0^2 - 8(0) + 12 = 12$

The x–intercept is 12. Replace x with 0 to find the y–intercepts.

$0 = y^2 - 8y + 12$

$0 = (y-6)(y-2)$

$y - 6 = 0$　　or　　$y - 2 = 0$

　$y = 6$　　　　　　$y = 2$

The y–intercepts are 2 and 6.

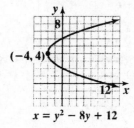

$x = y^2 - 8y + 12$

21. $x = -y^2 - 4y + 6$

This is a parabola of the form $x = ay^2 + by + c$.
Since $a = -1$ is negative, the parabola opens to the left. The y–coordinate of the vertex is
$-\dfrac{b}{2a} = -\dfrac{-4}{2(-1)} = -\dfrac{-4}{-2} = -2$. The x–coordinate of

the vertex is $x = -(-2)^2 - 4(-2) + 6$
$$= -4 + 8 + 6 = 10.$$
The vertex of the parabola is $(10, -2)$. The axis of symmetry is $y = -2$.
Replace y with 0 to find the x–intercept.
$x = -0^2 - 4(0) + 6 = 0^2 - 0 + 6 = 6$

The x–intercept is 6. Replace x with 0 to find the y–intercepts.
$0 = -y^2 - 4y + 6$

Solve using the quadratic formula.
$$y = \frac{-b \pm \sqrt{b^2 - 4ac}}{2a}$$

$$= \frac{-(-4) \pm \sqrt{(-4)^2 - 4(-1)(6)}}{2(-1)} = \frac{4 \pm \sqrt{16 + 24}}{-2}$$

$$= \frac{4 \pm \sqrt{40}}{-2} = \frac{4 \pm 2\sqrt{10}}{-2} = -2 \pm \sqrt{10}$$

The y–intercepts are $-2 \pm \sqrt{10}$.

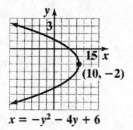

$x = -y^2 - 4y + 6$

22. $x + 8y = y^2 + 10$

Since only one variable is squared, the graph of the equation is a parabola.

23. $16x^2 = 32 - y^2$

$16x^2 + y^2 = 32$

Because x^2 and y^2 have different positive coefficients, the equation's graph is an ellipse.

24. $x^2 = 25 + 25y^2$

$x^2 - 25y^2 = 25$

Because x^2 and y^2 have opposite signs, the equation's graph is a hyperbola.

25. $x^2 = 4 - y^2$

$x^2 + y^2 = 4$

Because x^2 and y^2 have the same positive coefficient, the equation's graph is a circle.

26. $36y^2 = 576 + 16x^2$

$36y^2 - 16x^2 = 576$

Because x^2 and y^2 have opposite signs, the equation's graph is a hyperbola.

27. $\dfrac{(x+3)^2}{9} + \dfrac{(y-4)^2}{25} = 1$

Because x^2 and y^2 have different positive coefficients, the equation's graph is an ellipse.

28. $y = x^2 + 6x + 9$

Since only one variable is squared, the graph of the equation is a parabola.

29. $5x^2 + 5y^2 = 180$

Because x^2 and y^2 have the same positive coefficient, the equation's graph is a circle. Divide both sides of the equation by 5.
$x^2 + y^2 = 36$

The center is (0, 0) and the radius is 6 units.

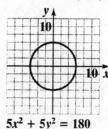

$5x^2 + 5y^2 = 180$

30. $4x^2 + 9y^2 = 36$

Because x^2 and y^2 have different positive coefficients, the equation's graph is an ellipse.

$$\frac{4x^2}{36} + \frac{9y^2}{36} = \frac{36}{36}$$

$$\frac{x^2}{9} + \frac{y^2}{4} = 1$$

Because the denominator of the x^2–term is greater than the denominator of the y^2–term, the major axis is horizontal. Since $a^2 = 9$, $a = 3$ and the vertices are $(-3,0)$ and $(3,0)$. Since $b^2 = 4$, $b = 2$ and endpoints of the minor axis are $(0,-2)$ and $(0,2)$.

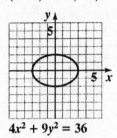

$4x^2 + 9y^2 = 36$

31. $4x^2 - 9y^2 = 36$

Because x^2 and y^2 have opposite signs, the equation's graph is a hyperbola.

$$\frac{4x^2}{36} - \frac{9y^2}{36} = \frac{36}{36}$$

$$\frac{x^2}{9} - \frac{y^2}{4} = 1$$

The equation is in the form $\dfrac{x^2}{a^2} - \dfrac{y^2}{b^2} = 1$ with

$a^2 = 9$, and $b^2 = 4$. We know the transverse axis lies on the x-axis and the vertices are $(-3,0)$ and $(3,0)$. Because $a^2 = 9$ and $b^2 = 4$, $a = 3$ and $b = 2$. Construct a rectangle using -3 and 3 on the x-axis, and -2 and 2 on the y-axis. Draw extended diagonals to obtain the asymptotes. Graph the hyperbola.

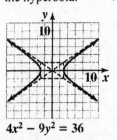

$4x^2 - 9y^2 = 36$

32. $\dfrac{x^2}{25} + \dfrac{y^2}{1} = 1$

Because x^2 and y^2 have different positive coefficients, the equation's graph is an ellipse. Because the denominator of the x^2–term is greater than the denominator of the y^2–term, the major axis is horizontal. Since $a^2 = 25$, $a = 5$ and the vertices are $(-5,0)$ and $(5,0)$. Since $b^2 = 1$, $b = 1$ and endpoints of the minor axis are $(0,-1)$ and $(0,1)$.

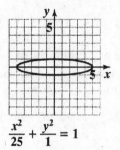

$$\frac{x^2}{25} + \frac{y^2}{1} = 1$$

33. $x + 3 = -y^2 + 2y$

$$x = -y^2 + 2y - 3$$

Since only one variable is squared, the graph of the equation is a parabola.

This is a parabola of the form $x = ay^2 + by + c$.

Since $a = -1$ is negative, the parabola opens to the left. The y–coordinate of the vertex is

$-\dfrac{b}{2a} = -\dfrac{2}{2(-1)} = -\dfrac{2}{-2} = 1$. The x–coordinate of

the vertex is $x = -1^2 + 2(1) - 3 = -1 + 2 - 3 = -2$.

The vertex of the parabola is $(-2,1)$.

Replace y with 0 to find the x–intercept.

$x = -0^2 + 2(0) - 3 = 0 + 0 - 3 = -3$

The x–intercept is -3. Replace x with 0 to find the y–intercepts.

$0 = -y^2 + 2y - 3$

Solve using the quadratic formula.

$$y = \frac{-2 \pm \sqrt{2^2 - 4(-1)(-3)}}{2(-1)}$$

$$= \frac{-2 \pm \sqrt{4 - 12}}{-2} = \frac{-2 \pm \sqrt{-8}}{-2}$$

We do not need to simplify further. The solutions are complex and there are no y–intercepts.

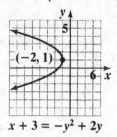

$$x + 3 = -y^2 + 2y$$

34. $y - 3 = x^2 - 2x$

$\quad\quad y = x^2 - 2x + 3$

Since only one variable is squared, the graph of the equation is a parabola.

This is a parabola of the form $y = ax^2 + bx + c$.

Since $a = 1$ is positive, the parabola opens to the right. The x–coordinate of the vertex is

$-\dfrac{b}{2a} = -\dfrac{-2}{2(1)} = -\dfrac{-2}{2} = 1$. The y–coordinate of the

vertex is $y = 1^2 - 2(1) + 3 = 1 - 2 + 3 = 2$.

The vertex of the parabola is $(1, 2)$.

Replace x with 0 to find the y–intercept.

$y = 0^2 - 2(0) + 3 = 0 - 0 + 3 = 3$

The y–intercept is 3. Replace y with 0 to find the x–intercepts.

$0 = x^2 - 2x + 3$

Solve using the quadratic formula.

$$x = \frac{-2 \pm \sqrt{2^2 - 4(-1)(-3)}}{2(-1)}$$

$$= \frac{-2 \pm \sqrt{4 - 12}}{-2} = \frac{-2 \pm \sqrt{-8}}{-2}$$

We do not need to simplify further. The solutions are complex and there are no x–intercepts.

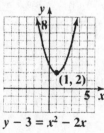

$$y - 3 = x^2 - 2x$$

35. $\dfrac{(x+2)^2}{16} + \dfrac{(y-5)^2}{4} = 1$

Because x^2 and y^2 have different positive coefficients, the equation's graph is an ellipse. The center of the ellipse is $(-2, 5)$. Because the denominator of the x^2 – term is greater than the denominator of the y^2 – term, the major axis is horizontal. Since $a^2 = 16$, $a = 4$ and the vertices lie 4 units to the left and right of the center. Since $b^2 = 4$, $b = 2$ and endpoints of the minor axis lie two units above and below the center.

Center	Vertices	Endpoints of Minor Axis
$(-2,5)$	$(-2-4,5)$ $=(-6,5)$	$(-2,5-2)$ $=(-2,3)$
	$(-2+4,5)$ $=(2,5)$	$(-2,5+2)$ $=(-2,7)$

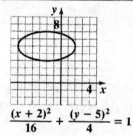

$$\frac{(x+2)^2}{16} + \frac{(y-5)^2}{4} = 1$$

36. $(x-3)^2 + (y+2)^2 = 4$

Because x^2 and y^2 have the same positive coefficient, the equation's graph is a circle.

$(x-3)^2 + (y+2)^2 = 4$

The center is $(3, -2)$ and the radius is 2 units.

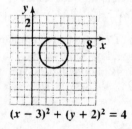

$$(x - 3)^2 + (y + 2)^2 = 4$$

37. $x^2 + y^2 + 6x - 2y + 6 = 0$

Because x^2 and y^2 have the same positive coefficient, the equation's graph is a circle.

$$x^2 + y^2 + 6x - 2y + 6 = 0$$

$$\left(x^2 + 6x \quad\right) + \left(y^2 - 2y \quad\right) = -6$$

Complete the squares.

$$\left(\frac{b}{2}\right)^2 = \left(\frac{6}{2}\right)^2 = (3)^2 = 9$$

$$\left(\frac{b}{2}\right)^2 = \left(\frac{-2}{2}\right)^2 = (-1)^2 = 1$$

$$\left(x^2 + 6x + 9\right) + \left(y^2 - 2y + 1\right) = -6 + 9 + 1$$

$$(x+3)^2 + (y-1)^2 = 4$$

The center is $(-3, 1)$ and the radius is 2 units.

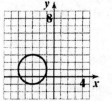

$x^2 + y^2 + 6x - 2y + 6 = 0$

38. a. Using the point (6, 3), substitute for x and y to find a in $y = ax^2$.

$$3 = a(6)^2$$

$$3 = a(36)$$

$$a = \frac{3}{36} = \frac{1}{12}$$

The equation for the parabola is $y = \frac{1}{12}x^2$.

b. $a = \frac{1}{4p}$

$$\frac{1}{12} = \frac{1}{4p}$$

$$4p = 12$$

$$p = 3$$

The light source should be placed at the point (0, 3). This is the point 3 inches above the vertex.

39. $5y = x^2 - 1$

$x - y = 1$

Solve the second equation for y.

$x - y = 1$

$-y = -x + 1$

$y = x - 1$

Substitute $x - 1$ for y in the first equation.

$$5(x-1) = x^2 - 1$$

$$5x - 5 = x^2 - 1$$

$$0 = x^2 - 5x + 4$$

$$0 = (x-4)(x-1)$$

$x - 4 = 0$ or $x - 1 = 0$

$x = 4$ $\qquad$ $x = 1$

Back-substitute 1 and 4 for x to find y.

$x = 4$	or	$x = 1$
$y = x - 1$		$y = x - 1$
$y = 4 - 1$		$y = 1 - 1$
$y = 3$		$y = 0$

The solution set is $\{(1, 0), (4, 3)\}$.

40. $y = x^2 + 2x + 1$

$x + y = 1$

Solve the second equation for y.

$x + y = 1$

$y = -x + 1$

Substitute $-x + 1$ for y in the first equation.

$$-x + 1 = x^2 + 2x + 1$$

$$0 = x^2 + 3x$$

$$0 = x(x + 3)$$

$x = 0$ or $x + 3 = 0$

$\qquad\qquad\quad x = -3$

Back-substitute -3 and 0 for x to find y.

$x = 0$	or	$x = -3$
$y = -x + 1$		$y = -x + 1$
$y = -0 + 1$		$y = -(-3) + 1$
$y = 1$		$y = 3 + 1$
		$y = 4$

The solution set is $\{(-3, 4), (0, 1)\}$.

41. $x^2 + y^2 = 2$

$x + y = 0$

Solve the second equation for y.

$x + y = 0$

$y = -x$

Substitute $-x$ for y in the first equation.

$x^2 + (-x)^2 = 2$

$x^2 + x^2 = 2$

$2x^2 = 2$

$x^2 = 1$

$x = \pm 1$

Back-substitute -1 and 1 for x to find y.

$x = -1$	or	$x = 1$
$y = -x$		$y = -x$
$y = -(-1)$		$y = -1$
$y = 1$		

The solution set is $\{(-1,1),\ (1,-1)\}$.

42. $2x^2 + y^2 = 24$

$x^2 + y^2 = 15$

Multiple the second equation by -1 and add to the first equation.

$2x^2 + y^2 = 24$

$\underline{-x^2 - y^2 = -15}$

$x^2 = 9$

$x = \pm 3$

Back-substitute -3 and 3 for x to find y.

$x = \pm 3$

$(\pm 3)^2 + y^2 = 15$

$9 + y^2 = 15$

$y^2 = 6$

$y = \pm\sqrt{6}$

The solution set is $\{(-3, -\sqrt{6}),$

$(-3, \sqrt{6}), (3, -\sqrt{6}),\ (3, \sqrt{6})\}$.

43. $xy - 4 = 0$

$y - x = 0$

Solve the second equation for y.

$y - x = 0$

$y = x$

Substitute x for y in the first equation and solve for x.

$x(x) - 4 = 0$

$x^2 - 4 = 0$

$(x + 2)(x - 2) = 0$

$x + 2 = 0$	or	$x - 2 = 0$
$x = -2$		$x = 2$

Back-substitute -2 and 2 for x to find y.

$x = -2$	or	$x = 2$
$y = x$		$y = x$
$y = -2$		$y = 2$

The solution set is $\{(-2,-2),\ (2,2)\}$.

44. $y^2 = 4x$

$x - 2y + 3 = 0$

Solve the second equation for x.

$x - 2y + 3 = 0$

$x = 2y - 3$

Substitute $2y - 3$ for x in the first equation and solve for y.

$y^2 = 4(2y - 3)$

$y^2 = 8y - 12$

$y^2 - 8y + 12 = 0$

$(y - 6)(y - 2) = 0$

$y - 6 = 0$	or	$y - 2 = 0$
$y = 6$		$y = 2$

Back-substitute 2 and 6 for y to find x.

$y = 6$	or	$y = 2$
$x = 2y - 3$		$x = 2y - 3$
$x = 2(6) - 3$		$x = 2(2) - 3$
$x = 12 - 3$		$x = 4 - 3$
$x = 9$		$x = 1$

The solution set is $\{(1,2),(9,6)\}$.

45. $x^2 + y^2 = 10$

$y = x + 2$

Substitute $x + 2$ for y in the first equation and solve for x.

$x^2 + (x+2)^2 = 10$

$x^2 + x^2 + 4x + 4 = 10$

$2x^2 + 4x + 4 = 10$

$2x^2 + 4x - 6 = 0$

$x^2 + 2x - 3 = 0$

$(x+3)(x-1) = 0$

$x + 3 = 0$ or $x - 1 = 0$

$x = -3$ $x = 1$

Back-substitute -3 and 1 for x to find y.

$x = -3$ or $x = 1$

$y = x + 2$ $y = x + 2$

$y = -3 + 2$ $y = 1 + 2$

$y = -1$ $y = 3$

The solution set is $\{(-3, -1),\ (1, 3)\}$.

46. $xy = 1$

$y = 2x + 1$

Substitute $2x + 1$ for y in the first equation and solve for x.

$x(2x+1) = 1$

$2x^2 + x = 1$

$2x^2 + x - 1 = 0$

$(2x-1)(x+1) = 0$

$2x - 1 = 0$ or $x + 1 = 0$

$2x = 1$ $x = -1$

$x = \dfrac{1}{2}$

Back-substitute -1 and $\dfrac{1}{2}$ for x to find y.

$x = -1$ or $x = \dfrac{1}{2}$

$y = 2x + 1$ $y = 2x + 1$

$y = 2(-1) + 1$ $y = 2\left(\dfrac{1}{2}\right) + 1$

$y = -2 + 1$ $y = 1 + 1$

$y = -1$ $y = 2$

The solution set is $\left\{(-1, -1), \left(\dfrac{1}{2}, 2\right)\right\}$.

47. $x + y + 1 = 0$

$x^2 + y^2 + 6y - x = -5$

Solve for y in the first equation.

$x + y + 1 = 0$

$y = -x - 1$

Substitute $-x - 1$ for y in the second equation and solve for x.

$x^2 + (-x-1)^2 + 6(-x-1) - x = -5$

$x^2 + x^2 + 2x + 1 - 6x - 6 - x = -5$

$2x^2 - 5x - 5 = -5$

$2x^2 - 5x = 0$

$x(2x - 5) = 0$

$x = 0$ or $2x - 5 = 0$

$2x = 5$

$x = \dfrac{5}{2}$

Back-substitute 0 and $\dfrac{5}{2}$ for x to find y.

$x = 0$ or $x = \dfrac{5}{2}$

$y = -x - 1$ $y = -x - 1$

$y = -0 - 1$ $y = -\dfrac{5}{2} - 1$

$y = -1$ $y = -\dfrac{7}{2}$

The solution set is $\left\{(0, -1), \left(\dfrac{5}{2}, -\dfrac{7}{2}\right)\right\}$.

48. $x^2 + y^2 = 13$

$x^2 - y = 7$

Solve for x^2 in the second equation.

$x^2 - y = 7$

$x^2 = y + 7$

Substitute $y + 7$ for x^2 in the first equation and solve for y.

$(y+7) + y^2 = 13$

$y^2 + y + 7 = 13$

$y^2 + y - 6 = 0$

$(y+3)(y-2) = 0$

$y + 3 = 0$ or $y - 2 = 0$

$y = -3$ $y = 2$

Back-substitute -3 and 2 for y to find x.

$y = -3$ or $y = 2$

$x^2 = y + 7$ $x^2 = y + 7$

$x^2 = -3 + 7$ $x^2 = 2 + 7$

$x^2 = 4$ $x^2 = 9$

$x = \pm 2$ $x = \pm 3$

The solution set is $\{(-3,2),(-2,-3),(2,-3),(3,2)\}$.

49. $2x^2 + 3y^2 = 21$

$3x^2 - 4y^2 = 23$

Multiply the first equation by 4 and the second equation by 3.

$8x^2 + 12y^2 = 84$

$\underline{9x^2 - 12y^2 = 69}$

$17x^2 = 153$

$x^2 = 9$

$x = \pm 3$

Back-substitute ± 3 for x to find y.

$x = \pm 3$

$2(\pm 3)^2 + 3y^2 = 21$

$2(9) + 3y^2 = 21$

$18 + 3y^2 = 21$

$3y^2 = 3$

$y^2 = 1$

$y = \pm 1$

The solution set is $\{(-3,-1),(-3,1),(3,-1),(3,1)\}$.

50. Let x = the length of the rectangle.
Let y = the width of the rectangle.

$2x + 2y = 26$

$xy = 40$

Solve the first equation for y.

$2x + 2y = 26$

$x + y = 13$

$y = 13 - x$

Substitute $13 - x$ for y in the second equation.

$x(13 - x) = 40$

$13x - x^2 = 40$

$0 = x^2 - 13x + 40$

$0 = (x-8)(x-5)$

$x - 8 = 0$ or $x - 5 = 0$

$x = 8$ $x = 5$

Back-substitute 5 and 8 for x to find y.

$x = 8$ or $x = 5$

$y = 13 - 8$ $y = 13 - 5$

$y = 5$ $y = 8$

The solutions are the same. The dimensions are 8 meters by 5 meters.

51. $2x + y = 8$

$xy = 6$

Solve the first equation for y.

$2x + y = 8$

$y = -2x + 8$

Substitute $-2x + 8$ for y in the second equation.

$x(-2x + 8) = 6$

$-2x^2 + 8x = 6$

$-2x^2 + 8x - 6 = 0$

$x^2 - 4x + 3 = 0$

$(x-3)(x-1) = 0$

$x - 3 = 0$ or $x - 1 = 0$

$x = 3$ $x = 1$

Back-substitute 1 and 3 for x to find y.

$x = 3$ or $x = 1$

$y = -2x + 8$ $y = -2x + 8$

$y = -2(3) + 8$ $y = -2(1) + 8$

$y = -6 + 8$ $y = -2 + 8$

$y = 2$ $y = 6$

The solutions are the points $(1,6)$ and $(3,2)$.

52. Using the formula for the area, we have

$x^2 + y^2 = 2900$.   Since there are 240 feet of fencing available, we have

$x + (x + y) + y + y + (x - y) + x = 240$

$\quad x + x + y + y + y + x - y + x = 240$

$\qquad\qquad\qquad\qquad\quad 4x + 2y = 240.$

The system of two variables in two equations is as follows.

$x^2 + y^2 = 2900$

$4x + 2y = 240$

Solve the second equation for y.

$4x + 2y = 240$

$\quad 2y = -4x + 240$

$\quad\ y = -2x + 120$

Substitute $-2x + 120$ for y to find x.

$x^2 + (-2x + 120)^2 = 2900$

$x^2 + 4x^2 - 480x + 14400 = 2900$

$\quad 5x^2 - 480x + 11500 = 0$

$\qquad x^2 - 96x + 2300 = 0$

$\qquad (x - 50)(x - 46) = 0$

$x - 50 = 0 \quad$ or $\quad x - 46 = 0$

$\quad x = 50 \qquad\qquad\quad x = 46$

Back-substitute 46 and 50 for x to find y.

$x = 50 \qquad\qquad x = 46$

$y = -2x + 120 \qquad y = -2x + 120$

$y = -2(50) + 120 \quad y = -2(46) + 120$

$y = -100 + 120 \qquad y = -92 + 120$

$y = 20 \qquad\qquad y = 28$

The solutions are $x = 50$ feet and $y = 20$ feet or $x = 46$ feet and $y = 28$ feet.

Chapter 13 Test

1. $(x - 3)^2 + (y - (-2))^2 = 5^2$

$\quad (x - 3)^2 + (y + 2)^2 = 25$

2. $\quad (x - 5)^2 + (y + 3)^2 = 49$

$(x - 5)^2 + (y - (-3))^2 = 7^2$

The center is $(5, -3)$ and the radius is 7 units.

3. $\qquad x^2 + y^2 + 4x - 6y - 3 = 0$

$\left(x^2 + 4x \quad\right) + \left(y^2 - 6y \quad\right) = 3$

Complete the squares.

$\left(\dfrac{b}{2}\right)^2 = \left(\dfrac{4}{2}\right)^2 = (2)^2 = 4$

$\left(\dfrac{b}{2}\right)^2 = \left(\dfrac{-6}{2}\right)^2 = (-3)^2 = 9$

$\left(x^2 + 4x + 4\right) + \left(y^2 - 6y + 9\right) = 3 + 4 + 9$

$\qquad\qquad (x + 2)^2 + (y - 3)^2 = 16$

$\qquad\quad (x - (-2))^2 + (y - 3)^2 = 4^2$

The center is $(-2, 3)$ and the radius is 4 units.

4. $x = -2(y + 3)^2 + 7$

$x = -2(y - (-3))^2 + 7$

The vertex of the parabola is $(7, -3)$.

5. $x = y^2 + 10y + 23$

The y–coordinate of the vertex is

$-\dfrac{b}{2a} = -\dfrac{10}{2(1)} = -\dfrac{10}{2} = -5.$

The x–coordinate of the vertex is

$x = (-5)^2 + 10(-5) + 23$

$\quad = 25 - 50 + 23$

$\quad = 25 - 50 + 23 = -2.$

The vertex of the parabola is $(-2, -5)$.

6. $\dfrac{x^2}{4} - \dfrac{y^2}{9} = 1$

Because x^2 and y^2 have opposite signs, the equation's graph is a hyperbola.

The equation is in the form $\dfrac{x^2}{a^2} - \dfrac{y^2}{b^2} = 1$ with

$a^2 = 4$, and $b^2 = 9$. We know the transverse axis lies on the x-axis and the vertices are $(-2, 0)$ and $(2, 0)$. Because $a^2 = 4$ and $b^2 = 9$, $a = 2$ and $b = 3$. Construct a rectangle using –2 and 2 on the x–axis, and –3 and 3 on the y–axis. Draw extended diagonals to obtain the asymptotes. Graph the hyperbola.

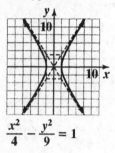

$\dfrac{x^2}{4} - \dfrac{y^2}{9} = 1$

7. $4x^2 + 9y^2 = 36$

Because x^2 and y^2 have different positive coefficients, the equation's graph is an ellipse.

$$\dfrac{4x^2}{36} + \dfrac{9y^2}{36} = \dfrac{36}{36}$$

$$\dfrac{x^2}{9} + \dfrac{y^2}{4} = 1$$

Because the denominator of the x^2 – term is greater than the denominator of the y^2 – term, the major axis is horizontal.

Since $a^2 = 9$, $a = 3$ and the vertices are $(-3, 0)$ and $(3, 0)$. Since $b^2 = 4$, $b = 2$ and endpoints of the minor axis are $(0, -2)$ and $(0, 2)$.

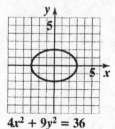

$4x^2 + 9y^2 = 36$

8. $x = (y + 1)^2 - 4$

Since only one variable is squared, the graph of the equation is a parabola.

This is a parabola of the form $x = a(y - k)^2 + h$.

Since $a = 1$ is positive, the parabola opens to the right. The vertex of the parabola is $(-4, -1)$.

Replace y with 0 to find the x–intercept.

$x = (0 + 1)^2 - 4 = (1)^2 - 4 = 1 - 4 = -3$.

The x–intercept is –3. Replace x with 0 to find the y–intercepts.

$$0 = (y + 1)^2 - 4$$

$$0 = y^2 + 2y + 1 - 4$$

$$0 = y^2 + 2y - 3$$

$$0 = (y + 3)(y - 1)$$

$$y + 3 = 0 \qquad \text{or} \qquad y - 1 = 0$$

$$y = -3 \qquad\qquad\qquad y = 1$$

The y–intercepts are –3 and 1.

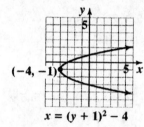

$x = (y + 1)^2 - 4$

9. $16x^2 + y^2 = 16$

Because x^2 and y^2 have different positive coefficients, the equation's graph is an ellipse.

$$\frac{16x^2}{16} + \frac{y^2}{16} = \frac{16}{16}$$

$$\frac{x^2}{1} + \frac{y^2}{16} = 1$$

Because the denominator of the y^2–term is greater than the denominator of the x^2–term, the major axis is vertical. Since $a^2 = 16$, $a = 4$ and the vertices are $(0, -4)$ and $(0, 4)$. Since $b^2 = 1$, $b = 1$ and endpoints of the minor axis are $(-1, 0)$ and $(1, 0)$.

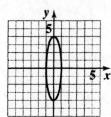

$16x^2 + y^2 = 16$

10. $25y^2 = 9x^2 + 225$

$25y^2 - 9x^2 = 225$

Because x^2 and y^2 have opposite signs, the equation's graph is a hyperbola.

$$\frac{25y^2}{225} - \frac{9x^2}{225} = \frac{225}{225}$$

$$\frac{y^2}{9} - \frac{x^2}{25} = 1$$

The equation is in the form $\dfrac{y^2}{a^2} - \dfrac{x^2}{b^2} = 1$ with

$a^2 = 9$, and $b^2 = 25$. We know the transverse axis lies on the y-axis and the vertices are $(0, -3)$ and $(0, 3)$. Because $a^2 = 9$ and $b^2 = 25$, $a = 3$ and $b = 5$. Construct a rectangle using -5 and 5 on the x–axis, and -3 and 3 on the y–axis. Draw extended diagonals to obtain the asymptotes.

Graph the hyperbola.

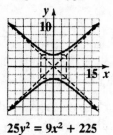

$25y^2 = 9x^2 + 225$

11. $x = -y^2 + 6y$

Since only one variable is squared, the graph of the equation is a parabola.

This is a parabola of the form $x = ay^2 + by + c$.

Since $a = -1$ is negative, the parabola opens to the left. The y–coordinate of the vertex is

$-\dfrac{b}{2a} = -\dfrac{6}{2(-1)} = -\dfrac{6}{-2} = 3$. The x–coordinate of

the vertex is $x = -3^2 + 6(3) = -9 + 18 = 9$.

The vertex of the parabola is $(9, 3)$.

Replace y with 0 to find the x–intercept

$x = -0^2 + 6(0) = 0 + 0 = 0$

The x–intercept is 0. Replace x with 0 to find the y–intercepts.

$0 = -y^2 + 6y$

$0 = -y(y - 6)$

$-y = 0$ or $y - 6 = 0$

$y = 0$ $y = 6$

The y–intercepts are 0 and 6.

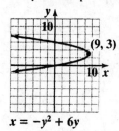

$x = -y^2 + 6y$

12. $\dfrac{(x-2)^2}{16}+\dfrac{(y+3)^2}{9}=1$

Because x^2 and y^2 have different positive coefficients, the equation's graph is an ellipse. The center of the ellipse is $(2,-3)$. Because the denominator of the x^2-term is greater than the denominator of the y^2-term, the major axis is horizontal. Since $a^2=16$, $a=4$ and the vertices lie 4 units to the left and right of the center. Since $b^2=9$, $b=3$ and endpoints of the minor axis lie 3 units above and below the center.

Center	Vertices	Endpoints of Minor Axis
$(2,-3)$	$(2-4,-3)$ $=(-2,-3)$	$(2,-3-3)$ $=(2,-6)$
	$(2+4,-3)$ $=(6,-3)$	$(2,-3+3)$ $=(2,0)$

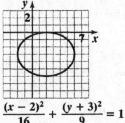

$\dfrac{(x-2)^2}{16}+\dfrac{(y+3)^2}{9}=1$

13. $(x+1)^2+(y+2)^2=9$

Because x^2 and y^2 have the same positive coefficient, the equation's graph is a circle. The center of the circle is $(-1,-2)$ and the radius is 3.

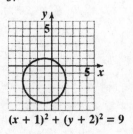

$(x+1)^2+(y+2)^2=9$

14. $\dfrac{x^2}{4}+\dfrac{y^2}{4}=1$

$$4\left(\dfrac{x^2}{4}+\dfrac{y^2}{4}\right)=4(1)$$

$$x^2+y^2=4$$

Because x^2 and y^2 have the same positive coefficient, the equation's graph is a circle. The circle has center $(0,0)$ and radius 2.

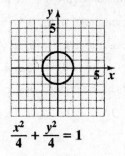

$\dfrac{x^2}{4}+\dfrac{y^2}{4}=1$

15. $x^2+y^2=25$

$\qquad x+y=1$

Solve the second equation for y.

$x+y=1$

$\qquad y=-x+1$

Substitute $-x+1$ for y to find x.

$$x^2+(-x+1)^2=25$$

$$x^2+x^2-2x+1=25$$

$$2x^2-2x+1=25$$

$$2x^2-2x-24=0$$

$$x^2-x-12=0$$

$$(x-4)(x+3)=0$$

$x-4=0 \quad$ or $\quad x+3=0$

$\quad x=4 \qquad\qquad\quad x=-3$

Back-substitute -3 and 4 for x to find y.

$x=4 \qquad$ or $\qquad x=-3$

$y=-x+1 \qquad\qquad y=-x+1$

$y=-4+1 \qquad\qquad y=-(-3)+1$

$y=-3 \qquad\qquad\quad y=3+1$

$\qquad\qquad\qquad\qquad\quad y=4$

The solution set is $\{(-3,4),\ (4,-3)\}$.

16. $2x^2 - 5y^2 = -2$

$3x^2 + 2y^2 = 35$

Multiply the first equation by 2 and the second equation by 5.

$4x^2 - 10y^2 = -4$

$\underline{15x^2 + 10y^2 = 175}$

$19x^2 = 171$

$x^2 = 9$

$x = \pm 3$

In this case, we can back-substitute 9 for x^2 to find y.

$x^2 = 9$

$2x^2 - 5y^2 = -2$

$2(9) - 5y^2 = -2$

$18 - 5y^2 = -2$

$-5y^2 = -20$

$y^2 = 4$

$y = \pm 2$

We have $x = \pm 3$ and $y = \pm 2$, the solution set is

$\{(-3,-2),(-3,2),(3,-2),(3,2)\}$.

17. $2x + y = 39$

$xy = 180$

Solve the first equation for y.

$2x + y = 39$

$y = 39 - 2x$

Substitute $39 - 2x$ for y to find x.

$x(39 - 2x) = 180$

$39x - 2x^2 = 180$

$0 = 2x^2 - 39x + 180$

$0 = (2x - 15)(x - 12)$

$2x - 15 = 0 \quad$ or $\quad x - 12 = 0$

$2x = 15 \qquad\qquad x = 12$

$x = \dfrac{15}{2}$

Back-substitute $\dfrac{15}{2}$ and 12 for x to find y.

$x = \dfrac{15}{2} \qquad$ or $\qquad x = 12$

$y = 39 - 2x \qquad\qquad y = 39 - 2x$

$y = 39 - 2\left(\dfrac{15}{2}\right) \qquad y = 39 - 2(12)$

$y = 39 - 15 \qquad\qquad y = 39 - 24$

$y = 24 \qquad\qquad\quad y = 15$

The dimensions are 15 feet by 12 feet or 24 feet by $\dfrac{15}{2}$ or 7.5 feet.

18. Let x = the length of the rectangle

Let y = the width of the rectangle

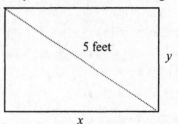

Using the Pythagorean Theorem, we obtain

$x^2 + y^2 = 5^2$. Since the perimeter is 14 feet, we have $2x + 2y = 14$.

The system of two equations in two variables is as follows.

$x^2 + y^2 = 25$

$2x + 2y = 14$

Solve the second equation for y.

$2x + 2y = 14$

$2y = 14 - 2x$

$y = 7 - x$

Substitute $7 - x$ for y to find x.

$x^2 + (7 - x)^2 = 25$

$x^2 + 49 - 14x + x^2 = 25$

$2x^2 - 14x + 49 = 25$

$2x^2 - 14x + 24 = 0$

$x^2 - 7x + 12 = 0$

$(x - 4)(x - 3) = 0$

$x - 4 = 0 \quad$ or $\quad x - 3 = 0$

$x = 4 \qquad\qquad x = 3$

Back-substitute 3 and 4 for x to find y.

$$x = 4 \quad \text{or} \quad x = 3$$
$$y = 7 - x \qquad y = 7 - x$$
$$y = 7 - 4 \qquad y = 7 - 3$$
$$y = 3 \qquad\quad y = 4$$

The solutions are the same. The dimensions are 4 feet by 3 feet.

Cumulative Review Exercises (Chapters 1-13)

1. $3x + 7 > 4 \quad$ or $\quad 6 - x < 1$
$$\qquad 3x > -3 \qquad\quad -x < -5$$
$$\qquad\quad x > -1 \qquad\qquad x > 5$$

The solution set is $\{x \mid x > -1\}$ or $(-1, \infty)$.

2. $\qquad x(2x - 7) = 4$
$$2x^2 - 7x = 4$$
$$2x^2 - 7x - 4 = 0$$
$$(2x + 1)(x - 4) = 0$$
$$2x + 1 = 0 \quad \text{or} \quad x - 4 = 0$$
$$2x = -1 \qquad\qquad x = 4$$
$$x = -\frac{1}{2}$$

The solution set is $\left\{ -\frac{1}{2}, 4 \right\}$.

3. $\dfrac{5}{x-3} = 1 + \dfrac{30}{x^2 - 9}$

$\dfrac{5}{x-3} = 1 + \dfrac{30}{(x+3)(x-3)}$

Multiply both sides of the equation by the LCD, $(x+3)(x-3)$.

$$(x+3)(x-3)\left(\frac{5}{x-3}\right) = (x+3)(x-3)\left(1 + \frac{30}{(x+3)(x-3)}\right)$$
$$(x+3)(5) = (x+3)(x-3) + 30$$
$$5x + 15 = x^2 - 9 + 30$$
$$15 = x^2 - 5x + 21$$
$$0 = x^2 - 5x + 6$$
$$0 = (x-3)(x-2)$$
$$x - 3 = 0 \quad \text{or} \quad x - 2 = 0$$
$$x = 3 \qquad\qquad x = 2$$

Disregard 3 because it would make the denominator zero. The solution set is $\{2\}$.

4. $3x^2 + 8x + 5 < 0$

Solve the related quadratic equation.

$3x^2 + 8x + 5 = 0$

$(3x + 5)(x + 1) = 0$

$3x + 5 = 0 \quad$ or $\quad x + 1 = 0$

$3x = -5 \qquad\qquad x = -1$

$x = -\dfrac{5}{3}$

The boundary points are $-\dfrac{5}{3}$ and -1.

Interval	Test Value	Test	Conclusion
$\left(-\infty, -\dfrac{5}{3}\right)$	-2	$3(-2)^2 + 8(-2) + 5 < 0$ $1 < 0$, false	$\left(-\infty, -\dfrac{5}{3}\right)$ does not belong to the solution set.
$\left(-\dfrac{5}{3}, -1\right)$	$-\dfrac{4}{3}$	$3\left(-\dfrac{4}{3}\right)^2 + 8\left(-\dfrac{4}{3}\right) + 5 < 0$ $-\dfrac{1}{3} < 0$, true	$\left(-\dfrac{5}{3}, -1\right)$ belongs to the solution set.
$(-1, \infty)$	0	$3(0)^2 + 8(0) + 5 < 0$ $5 < 0$, false	$(-1, \infty)$ does not belong to the solution set.

The solution set is $\left(-\dfrac{5}{3}, -1\right)$ or $\left\{x \,\middle|\, -\dfrac{5}{3} < x < -1\right\}$.

5. $3^{2x-1} = 81$

$3^{2x-1} = 3^4$

$2x - 1 = 4$

$2x = 5$

$x = \dfrac{5}{2}$

The solution set is $\left\{\dfrac{5}{2}\right\}$.

6. $30e^{0.7x} = 240$

$e^{0.7x} = 80$

$\ln e^{0.7x} = \ln 8$

$0.7x = \ln 8$

$x = \dfrac{\ln 8}{0.7} = \dfrac{2.08}{0.7} \approx 2.97$

The solution set is $\left\{\dfrac{\ln 8}{0.7} \approx 2.97\right\}$.

7. $3x^2 + 4y^2 = 39$

    $5x^2 - 2y^2 = -13$

    Multiply the second equation by 2 and add to the first equation.

    $3x^2 + 4y^2 = 39$

    $\underline{10x^2 - 4y^2 = -26}$

    $\qquad 13x^2 = 13$

    $\qquad x^2 = 1$

    $\qquad x = \pm 1$

    We can back-substitute 1 for x^2 to find y.

    $x^2 = 1$

    $3x^2 + 4y^2 = 39$

    $3(1) + 4y^2 = 39$

    $3 + 4y^2 = 39$

    $4y^2 = 36$

    $y^2 = 9$

    $y = \pm 3$

    We have $x = \pm 1$ and $y = \pm 3$, the solution set is $\{(-1, -3), (-1, 3), (1, -3), (1, 3)\}$.

8. $f(x) = -\dfrac{2}{3}x + 4$

    $y = -\dfrac{2}{3}x + 4$

    The y–intercept is 4 and the slope is $-\dfrac{2}{3}$. We can write the slope as $m = \dfrac{-2}{3} = \dfrac{\text{rise}}{\text{run}}$ and use the intercept and the slope to graph the function.

    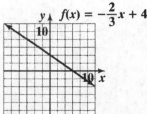

9. $3x - y > 6$

    First, find the intercepts to the equation $3x - y = 6$.

    Find the x–intercept by setting y equal to zero.

    $3x - 0 = 6$

    $3x = 6$

    $x = 2$

    Find the y–intercept by setting x equal to zero.

    $3(0) - y = 6$

    $-y = 6$

    $y = -6$

    Next, use the origin as a test point.

    $3(0) - 0 > 6$

    $0 - 0 > 6$

    $0 > 6$

    This is a false statement. This means that the origin will not fall in the shaded half-plane.

    $3x - y > 6$

    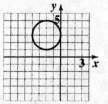

10. $x^2 + y^2 + 4x - 6y + 9 = 0$

    Because x^2 and y^2 have the same positive coefficient, the equation's graph is a circle.

    $\left(x^2 + 4x \quad\right) + \left(y^2 - 6y \quad\right) = -9$

    Complete the squares.

    $\left(\dfrac{b}{2}\right)^2 = \left(\dfrac{4}{2}\right)^2 = (2)^2 = 4$

    $\left(\dfrac{b}{2}\right)^2 = \left(\dfrac{-6}{2}\right)^2 = (-3)^2 = 9$

    $\left(x^2 + 4x + 4\right) + \left(y^2 - 6y + 9\right) = -9 + 4 + 9$

    $(x + 2)^2 + (y - 3)^2 = 4$

    The circle has center $(-2, 3)$ and radius 2.

    $x^2 + y^2 + 4x - 6y + 9 = 0$

11. $9x^2 - 4y^2 = 36$

Because x^2 and y^2 have opposite signs, the equation's graph is a hyperbola.

$$\frac{9x^2}{36} - \frac{4y^2}{36} = \frac{36}{36}$$

$$\frac{x^2}{4} - \frac{y^2}{9} = 1$$

The equation is in the form $\frac{x^2}{a^2} - \frac{y^2}{b^2} = 1$ with

$a^2 = 4$, and $b^2 = 9$. We know the transverse axis lies on the x-axis and the vertices are

$(-2, 0)$ and $(2, 0)$. Because $a^2 = 4$ and $b^2 = 9$,

$a = 2$ and $b = 3$. Construct a rectangle using -2 and 2 on the x–axis, and -3 and 3 on the y–axis. Draw extended diagonals to obtain the asymptotes. Graph the hyperbola.

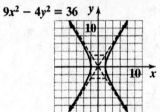

12. $-2\left(3^2 - 12\right)^3 - 45 \div 9 - 3$

$= -2\left(9 - 12\right)^3 - 45 \div 9 - 3$

$= -2\left(-3\right)^3 - 45 \div 9 - 3$

$= -2\left(-27\right) - 45 \div 9 - 3$

$= 54 - 5 - 3 = 46$

13. $\left(3x^3 - 19x^2 + 17x + 4\right) \div \left(3x - 4\right)$

Rewrite the polynomials in descending order and divide.

$$\require{enclose}
\begin{array}{r}
x^2 - 5x - 1 \\
3x - 4 \enclose{longdiv}{3x^3 - 19x^2 + 17x + 4} \\
\underline{3x^3 - 4x^2} \\
-15x^2 + 17x \\
\underline{-15x^2 + 20x} \\
-3x + 4 \\
\underline{-3x + 4} \\
0
\end{array}$$

$$\frac{3x^3 - 19x^2 + 17x + 4}{3x - 4} = x^2 - 5x - 1$$

14. $\sqrt[3]{4x^2 y^5} \cdot \sqrt[3]{4xy^2}$

$= \sqrt[3]{4x^2 y^5 \, 4xy^2} = \sqrt[3]{16x^3 y^7}$

$= \sqrt[3]{8 \cdot 2x^3 y^6 y} = 2xy^2 \sqrt[3]{2y}$

15. $(2 + 3i)(4 - i)$

$= 8 - 2i + 12i - 3i^2 = 8 + 10i - 3(-1)$

$= 8 + 10i + 3 = 11 + 10i$

16. $12x^3 - 36x^2 + 27x = 3x\left(4x^2 - 12x + 9\right)$

$= 3x\left(2x - 3\right)^2$

17. $x^3 - 2x^2 - 9x + 18$

$= x^2(x - 2) - 9(x - 2)$

$= (x - 2)\left(x^2 - 9\right)$

$= (x - 2)(x + 3)(x - 3)$

18. Since the radicand must be positive, the domain will exclude all values of x which make the radicand less than zero.

$6 - 3x \geq 0$

$-3x \geq -6$

$x \leq 2$

The domain of $f = \left\{x \,|\, x \leq 2\right\}$ or $(-\infty, 2]$.

19. $\dfrac{1-\sqrt{x}}{1+\sqrt{x}} = \dfrac{1-\sqrt{x}}{1+\sqrt{x}} \cdot \dfrac{1-\sqrt{x}}{1-\sqrt{x}}$

$\qquad = \dfrac{\left(1-\sqrt{x}\right)^2}{1^2 - \left(\sqrt{x}\right)^2}$

$\qquad = \dfrac{\left(1-\sqrt{x}\right)^2}{1-x}$ or $\dfrac{1-2\sqrt{x}+x}{1-x}$

20. $\dfrac{1}{3}\ln x + 7\ln y = \ln x^{\frac{1}{3}} + \ln y^7 = \ln\left(x^{\frac{1}{3}}y^7\right)$

21. $\left(3x^3 - 5x^2 + 2x - 1\right) \div (x-2)$

$$
\begin{array}{r|rrrr}
2 & 3 & -5 & 2 & -1 \\
  &   & 6 & 2 & 8 \\
\hline
  & 3 & 1 & 4 & 7
\end{array}
$$

$\left(3x^3 - 5x^2 + 2x - 1\right) \div (x-2)$

$= 3x^2 + x + 4 + \dfrac{7}{x-2}$

22. $x = -2\sqrt{3}$ or $x = 2\sqrt{3}$

$\quad x + 2\sqrt{3} = 0 \qquad x - 2\sqrt{3} = 0$

Multiply the factors to obtain the polynomial.

$\left(x + 2\sqrt{3}\right)\left(x - 2\sqrt{3}\right) = 0$

$x^2 - \left(2\sqrt{3}\right)^2 = 0$

$x^2 - 4 \cdot 3 = 0$

$x^2 - 12 = 0$

23. Let x = the rate of the slower car

	r	t	d
Fast	$x + 10$	2	$2(x + 10)$
Slow	x	2	$2x$

$2(x+10) + 2x = 180$

$2x + 20 + 2x = 180$

$4x + 20 = 180$

$4x = 160$

$x = 40$

The rate of the slower car is 40 miles per hour and the rate of the faster car is $40 + 10 = 50$ miles per hour.

24. Let x = the number of miles driven in a day.

$C_R = 39 + 0.16x$

$C_A = 25 + 0.24x$

Set the costs equal.

$39 + 0.16x = 25 + 0.24x$

$\qquad 39 = 25 + 0.08x$

$\qquad 14 = 0.08x$

$\qquad \dfrac{14}{0.08} = x$

$\qquad x = 175$

The cost is the same when renting from either company when 175 miles are driven in a day.

$C_R = 39 + 0.16(175) = 39 + 28 = 67$

When 175 miles are driven, the cost is \$67.

25. Let x = the number of apples.

Let y = the number of bananas.

$3x + 2y = 354$

$2x + 3y = 381$

Multiply the first equation by -3 and the second equation by 2 and solve by addition.

$-9x - 6y = -1062$

$\underline{4x + 6y = 762}$

$\qquad -5x = -300$

$\qquad\quad x = 60$

Back-substitute 60 for x to find y.

$3(60) + 2y = 354$

$\quad 180 + 2y = 354$

$\qquad\quad 2y = 174$

$\qquad\quad y = 87$

There are 60 calories in an apple and 87 calories in a banana.

Chapter 14
Sequences, Series, and the Binomial Theorem

14.1 Check Points

1. a. $a_n = 2n + 5$

$a_1 = 2(1) + 5 = 7$

$a_2 = 2(2) + 5 = 9$

$a_3 = 2(3) + 5 = 11$

$a_4 = 2(4) + 5 = 13$

The first four terms are 7, 9, 11, and 13.

b. $a_n = \dfrac{(-1)^n}{2^n + 1}$

$a_1 = \dfrac{(-1)^1}{2^1 + 1} == \dfrac{-1}{3} - \dfrac{1}{3}$

$a_2 = \dfrac{(-1)^2}{2^2 + 1} = \dfrac{1}{5}$

$a_3 = \dfrac{(-1)^3}{2^3 + 1} = \dfrac{-1}{9} = -\dfrac{1}{9}$

$a_4 = \dfrac{(-1)^4}{2^4 + 1} = \dfrac{1}{17}$

The first four terms are $-\frac{1}{3}, \frac{1}{5}, -\frac{1}{9},$ and $\frac{1}{17}$.

2. $a_n = \dfrac{20}{(n+1)!}$

$a_1 = \dfrac{20}{(1+1)!} = \dfrac{20}{2!} = 10$

$a_2 = \dfrac{20}{(2+1)!} = \dfrac{20}{3!} = \dfrac{20}{6} = \dfrac{10}{3}$

$a_3 = \dfrac{20}{(3+1)!} = \dfrac{20}{4!} = \dfrac{20}{24} = \dfrac{5}{6}$

$a_4 = \dfrac{20}{(4+1)!} = \dfrac{20}{5!} = \dfrac{20}{120} = \dfrac{1}{6}$

The first four terms are $10, \frac{10}{3}, \frac{5}{6},$ and $\frac{1}{6}$.

3. a. $\displaystyle\sum_{i=1}^{6} 2i^2$

$= 2(1)^2 + 2(2)^2 + 2(3)^2$

$\quad + 2(4)^2 + 2(5)^2 + 2(6)^2$

$= 2 + 8 + 18 + 32 + 50 + 72$

$= 182$

b. $\displaystyle\sum_{k=3}^{5} \left(2^k - 3\right)$

$= \left(2^3 - 3\right) + \left(2^4 - 3\right) + \left(2^5 - 3\right)$

$= (8 - 3) + (16 - 3) + (32 - 3)$

$= 5 + 13 + 29$

$= 47$

c. $\displaystyle\sum_{i=1}^{5} 4 = 4 + 4 + 4 + 4 + 4 = 20$

4. a. The sum has nine terms, each of the form i^2, starting at $i = 1$ and ending at $i = 9$.

$$1^2 + 2^2 + 3^2 + \cdots + 9^2 = \sum_{i=1}^{9} i^2$$

b. The sum has n terms, each of the form $\dfrac{1}{2^{i-1}}$, starting at $i = 1$ and ending at $i = n$.

$$1 + \frac{1}{2} + \frac{1}{4} + \frac{1}{8} + \cdots + \frac{1}{2^{n-1}} = \sum_{i=1}^{n} \frac{1}{2^{i-1}}$$

14.1 Exercise Set

1. $a_n = 3n + 2$

$a_1 = 3(1) + 2 = 3 + 2 = 5$

$a_2 = 3(2) + 2 = 6 + 2 = 8$

$a_3 = 3(3) + 2 = 9 + 2 = 11$

$a_4 = 3(4) + 2 = 12 + 2 = 14$

The first four terms are 5, 8, 11, 14.

3. $a_n = 3^n$

$a_1 = 3^1 = 3$

$a_2 = 3^2 = 9$

$a_3 = 3^3 = 27$

$a_4 = 3^4 = 81$

The first four terms are 3, 9, 27, 81.

5. $a_n = (-3)^n$

$a_1 = (-3)^1 = -3$

$a_2 = (-3)^2 = 9$

$a_3 = (-3)^3 = -27$

$a_4 = (-3)^4 = 81$

The first four terms are $-3, 9, -27, 81$.

7. $a_n = (-1)^n (n+3)$

$a_1 = (-1)^1 (1+3) = -1(4) = -4$

$a_2 = (-1)^2 (2+3) = 1(5) = 5$

$a_3 = (-1)^3 (3+3) = -1(6) = -6$

$a_4 = (-1)^4 (4+3) = 1(7) = 7$

The first four terms are $-4, 5, -6, 7$.

9. $a_n = \dfrac{2n}{n+4}$

$a_1 = \dfrac{2(1)}{1+4} = \dfrac{2}{5}$

$a_2 = \dfrac{2(2)}{2+4} = \dfrac{4}{6} = \dfrac{2}{3}$

$a_3 = \dfrac{2(3)}{3+4} = \dfrac{6}{7}$

$a_4 = \dfrac{2(4)}{4+4} = \dfrac{8}{8} = 1$

The first four terms are $\dfrac{2}{5}, \dfrac{2}{3}, \dfrac{6}{7}, 1$.

11. $a_n = \dfrac{(-1)^{n+1}}{2^n - 1}$

$a_1 = \dfrac{(-1)^{1+1}}{2^1 - 1} = \dfrac{(-1)^2}{2-1} = \dfrac{1}{1} = 1$

$a_2 = \dfrac{(-1)^{2+1}}{2^2 - 1} = \dfrac{(-1)^3}{4-1} = \dfrac{-1}{3} = -\dfrac{1}{3}$

$a_3 = \dfrac{(-1)^{3+1}}{2^3 - 1} = \dfrac{(-1)^4}{8-1} = \dfrac{1}{7}$

$a_4 = \dfrac{(-1)^{4+1}}{2^4 - 1} = \dfrac{(-1)^5}{16-1} = \dfrac{-1}{15} = -\dfrac{1}{15}$

The first four terms are $1, -\dfrac{1}{3}, \dfrac{1}{7}, -\dfrac{1}{15}$.

13. $a_n = \dfrac{n^2}{n!}$

$a_1 = \dfrac{1^2}{1!} = \dfrac{1}{1} = 1$

$a_2 = \dfrac{2^2}{2!} = \dfrac{4}{2 \cdot 1} = \dfrac{4}{2} = 2$

$a_3 = \dfrac{3^2}{3!} = \dfrac{9}{3 \cdot 2 \cdot 1} = \dfrac{3}{2}$

$a_4 = \dfrac{4^2}{4!} = \dfrac{16}{4 \cdot 3 \cdot 2 \cdot 1} = \dfrac{2}{3}$

The first four terms are $1, 2, \dfrac{3}{2}, \dfrac{2}{3}$.

15. $a_n = 2(n+1)!$

$a_1 = 2(1+1)! = 2 \cdot 2! = 2 \cdot 2 \cdot 1 = 4$

$a_2 = 2(2+1)! = 2 \cdot 3! = 2 \cdot 3 \cdot 2 \cdot 1 = 12$

$a_3 = 2(3+1)! = 2 \cdot 4! = 2 \cdot 4 \cdot 3 \cdot 2 \cdot 1$

$\quad = 48$

$a_4 = 2(4+1)! = 2 \cdot 5! = 2 \cdot 5 \cdot 4 \cdot 3 \cdot 2 \cdot 1$

$\quad = 240$

17. $\displaystyle\sum_{i=1}^{6} 5i = 5(1) + 5(2) + 5(3) + 5(4) + 5(5) + 5(6) = 5 + 10 + 15 + 20 + 25 + 30 = 105$

19. $\displaystyle\sum_{i=1}^{4} 2i^2 = 2(1)^2 + 2(2)^2 + 2(3)^2 + 2(4)^2 = 2(1) + 2(4) + 2(9) + 2(16)$

$\quad = 2 + 8 + 18 + 32 = 60$

21. $\displaystyle\sum_{k=1}^{5} k(k+4) = 1(1+4) + 2(2+4) + 3(3+4) + 4(4+4) + 5(5+4)$

$\quad = 1(5) + 2(6) + 3(7) + 4(8) + 5(9) = 5 + 12 + 21 + 32 + 45 = 115$

23. $\displaystyle\sum_{i=1}^{4} \left(-\frac{1}{2}\right)^i = \left(-\frac{1}{2}\right)^1 + \left(-\frac{1}{2}\right)^2 + \left(-\frac{1}{2}\right)^3 + \left(-\frac{1}{2}\right)^4 = -\frac{1}{2} + \frac{1}{4} + \left(-\frac{1}{8}\right) + \frac{1}{16}$

$\quad = -\frac{1}{2} \cdot \frac{8}{8} + \frac{1}{4} \cdot \frac{4}{4} + \left(-\frac{1}{8}\right)\frac{2}{2} + \frac{1}{16} = -\frac{8}{16} + \frac{4}{16} - \frac{2}{16} + \frac{1}{16}$

$\quad = \frac{-8+4-2+1}{16} = -\frac{5}{16}$

25. $\displaystyle\sum_{i=5}^{9} 11 = 11 + 11 + 11 + 11 + 11 = 55$

27. $\displaystyle\sum_{i=0}^{4} \frac{(-1)^i}{i!} = \frac{(-1)^0}{0!} + \frac{(-1)^1}{1!} + \frac{(-1)^2}{2!} + \frac{(-1)^3}{3!} + \frac{(-1)^4}{4!} = \frac{1}{1} + \frac{-1}{1} + \frac{1}{2 \cdot 1} + \frac{-1}{3 \cdot 2 \cdot 1} + \frac{1}{4 \cdot 3 \cdot 2 \cdot 1}$

$\quad = 1 - 1 + \frac{1}{2} - \frac{1}{6} + \frac{1}{24} = \frac{1}{2} \cdot \frac{12}{12} - \frac{1}{6} \cdot \frac{4}{4} + \frac{1}{24} = \frac{12}{24} - \frac{4}{24} + \frac{1}{24} = \frac{12-4+1}{24} = \frac{9}{24} = \frac{3}{8}$

29. $\displaystyle\sum_{i=1}^{5} \frac{i!}{(i-1)!} = \frac{1!}{(1-1)!} + \frac{2!}{(2-1)!} + \frac{3!}{(3-1)!} + \frac{4!}{(4-1)!} + \frac{5!}{(5-1)!} = \frac{1!}{0!} + \frac{2!}{1!} + \frac{3!}{2!} + \frac{4!}{3!} + \frac{5!}{4!}$

$\quad = \frac{1}{1} + \frac{2 \cdot \cancel{1!}}{\cancel{1!}} + \frac{3 \cdot \cancel{2!}}{\cancel{2!}} + \frac{4 \cdot \cancel{3!}}{\cancel{3!}} + \frac{5 \cdot \cancel{4!}}{\cancel{4!}} = 1 + 2 + 3 + 4 + 5 = 15$

31. $1^2 + 2^2 + 3^2 + \ldots + 15^2 = \displaystyle\sum_{i=1}^{15} i^2$

33. $2 + 2^2 + 2^3 + \ldots + 2^{11} = \displaystyle\sum_{i=1}^{11} 2^i$

35. $1 + 2 + 3 + \ldots + 30 = \displaystyle\sum_{i=1}^{30} i$

37. $\dfrac{1}{2} + \dfrac{2}{3} + \dfrac{3}{4} + \ldots + \dfrac{14}{14+1} = \displaystyle\sum_{i=1}^{14} \dfrac{i}{i+1}$

39. $4 + \dfrac{4^2}{2} + \dfrac{4^3}{3} + \ldots + \dfrac{4^n}{n} = \displaystyle\sum_{i=1}^{n} \dfrac{4^i}{i}$

41. $1 + 3 + 5 + \ldots + (2n-1) = \displaystyle\sum_{i=1}^{n} (2i-1)$

43. $5 + 7 + 9 + 11 + \ldots + 31 = \displaystyle\sum_{k=2}^{15} (2k+1)$ or

$\qquad\qquad = \displaystyle\sum_{k=1}^{14} (2k+3)$

45. $a + ar + ar^2 + \ldots + ar^{12} = \displaystyle\sum_{k=0}^{12} ar^k$

47. $a + (a+d) + (a+2d) + \ldots + a(a+nd)$

$\qquad = \displaystyle\sum_{k=0}^{n} (a+kd)$

49. $\displaystyle\sum_{i=1}^{5} (a_i^2 + 1) = \left((-4)^2 + 1\right) + \left((-2)^2 + 1\right) + \left((0)^2 + 1\right) + \left((2)^2 + 1\right) + \left((4)^2 + 1\right)$

$\qquad\qquad = 17 + 5 + 1 + 5 + 17$

$\qquad\qquad = 45$

51. $\displaystyle\sum_{i=1}^{5} (2a_i + b_i) = \left(2(-4)+4\right) + \left(2(-2)+2\right) + \left(2(0)+0\right) + \left(2(2)+(-2)\right) + \left(2(4)+(-4)\right)$

$\qquad\qquad = -4 + (-2) + 0 + 2 + 4 = 0$

53. $\displaystyle\sum_{i=4}^{5} \left(\dfrac{a_i}{b_i}\right)^2 = \left(\dfrac{2}{-2}\right)^2 + \left(\dfrac{4}{-4}\right)^2 = (-1)^2 + (-1)^2 = 1 + 1 = 2$

55. $\displaystyle\sum_{i=1}^{5} a_i^2 + \sum_{i=1}^{5} b_i^2 = \left((-4)^2 + (-2)^2 + 0^2 + 2^2 + 4^2\right) + \left(4^2 + 2^2 + 0^2 + (-2)^2 + (-4)^2\right)$

$\qquad\qquad = (16 + 4 + 0 + 4 + 16) + (16 + 4 + 0 + 4 + 16)$

$\qquad\qquad = 80$

57. a. $\sum_{i=1}^{6} a_i = 113 + 417 + 225 + 243 + 175 + 181 = 1354$

From 2001 through 2006, the total number of books published about the September 11 attacks was 1354

b. $\dfrac{\sum_{i=1}^{6} a_i}{6} = \dfrac{1354}{6} \approx 226$

From 2001 through 2006, the average number of books published each year about the September 11 attacks was approximately 226.

59. a. $\dfrac{1}{7}\sum_{i=1}^{7} a_i = \dfrac{1}{7}(8.1 + 7.2 + 6.1 + 8.1 + 10.0 + 13.1 + 16.7) = \dfrac{1}{7}(69.3) = 9.9$

From 2000 through 2006, Online ad spending averaged $9.9 billion per year.

b. $a_n = 0.5n^2 - 1.5n + 8$ $a_4 = 0.5(4)^2 - 1.5(4) + 8 = 10$

$a_1 = 0.5(1)^2 - 1.5(2) + 8 = 7$ $a_5 = 0.5(5)^2 - 1.5(5) + 8 = 13$

$a_2 = 0.5(2)^2 - 1.5(2) + 8 = 7$ $a_6 = 0.5(6)^2 - 1.5(6) + 8 = 17$

$a_3 = 0.5(3)^2 - 1.5(3) + 8 = 8$ $a_7 = 0.5(7)^2 - 1.5(7) + 8 = 22$

$\dfrac{1}{7}\sum_{i=1}^{7} a_i = \dfrac{1}{7}(7 + 7 + 8 + 10 + 13 + 17 + 22) = \dfrac{1}{7}(84) = 12$

This overestimates the actual sum by $2.1 billion.

61. $a_{20} = 6000\left(1 + \dfrac{0.06}{4}\right)^{20} = 6000(1 + 0.015)^{20} = 6000(1.015)^{20} = 8081.13$

The balance in the account after 5 years if $8081.13.

63. – 67. Answers will vary.

69. Answers will vary.

71. $a_n = \dfrac{n}{n+1}$;

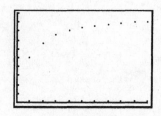

As n gets larger, the terms get closer to 1.

73. $a_n = \dfrac{2n^2 + 5n - 7}{n^3}$

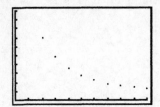

As n gets larger, the terms get closer to 0.

75. does not make sense; Explanations will vary. Sample explanation: There is nothing that implies that there is a negative number of sheep.

77. makes sense

79. false; Changes to make the statement true will vary. A sample change is:

$$\sum_{i=1}^{2} (-1)^i 2^i = (-1)^1 \, 2^1 + (-1)^2 \, 2^2 = -1(2) + 1(4) = -2 + 4 = 2$$

81. true

83. $a_n = \dfrac{1}{n}$

85. $a_n = (-1)^n$

87. $a_n = \dfrac{n+2}{n+1}$

89. $a_n = \dfrac{(n+1)^2}{n}$

91. $\dfrac{600!}{599!} = \dfrac{600 \cdot \cancel{599!}}{\cancel{599!}} = 600$

93. $\dfrac{n!}{(n-3)!} = \dfrac{n(n-1)(n-2)\cancel{(n-3)!}}{\cancel{(n-3)!}}$

$\qquad = n(n-1)(n-2)$

$\qquad = n(n^2 - 3n + 2)$

$\qquad = n^3 - 3n^2 + 2n$

95. $\displaystyle\sum_{i=2}^{4} 2i \log x = 2(2)\log x + 2(3)\log x + 2(4)\log x$

$\qquad\qquad = 4\log x + 6\log x + 8\log x$

$\qquad\qquad = \log x^4 + \log x^6 + \log x^8$

$\qquad\qquad = \log\left(x^4 \cdot x^6 \cdot x^8\right) = \log x^{18}$

97. $\sqrt[3]{40x^4y^7} = \sqrt[3]{8 \cdot 5x^3xy^6y} = 2xy^2\sqrt[3]{5xy}$

98. $27x^3 - 8 = (3x-2)(9x^2 + 6x + 4)$

99.
$$\frac{6}{x} + \frac{6}{x+2} = \frac{5}{2}$$
$$2x(x+2)\left(\frac{6}{x} + \frac{6}{x+2}\right) = 2x(x+2)\left(\frac{5}{2}\right)$$
$$2(x+2)(6) + 2x(6) = x(x+2)(5)$$
$$12(x+2) + 12x = 5x(x+2)$$
$$12x + 24 + 12x = 5x^2 + 10x$$
$$24x + 24 = 5x^2 + 10x$$
$$0 = 5x^2 - 14x - 24$$
$$0 = (5x+6)(x-4)$$

Apply the zero product principle.
$$5x + 6 = 0 \quad \text{or} \quad x - 4 = 0$$
$$5x = -6 \qquad\qquad x = 4$$
$$x = -\frac{6}{5}$$

The solution set is $\left\{-\dfrac{6}{5}, 4\right\}$.

100. $a_2 - a_1 = 3 - 8 = -5$
$a_3 - a_2 = -2 - 3 = -5$
$a_4 - a_3 = -7 - (-2) = -5$
$a_5 - a_4 = -12 - (-7) = -5$
The difference between consecutive terms is always -5.

101. $a_2 - a_1 = (4(2)-3) - (4(1)-3) = 4$
$a_3 - a_2 = (4(3)-3) - (4(2)-3) = 4$
$a_4 - a_3 = (4(4)-3) - (4(3)-3) = 4$
$a_5 - a_4 = (4(5)-3) - (4(4)-3) = 4$
The difference between consecutive terms is always 4.

102. $a_n = 4 + (n-1)(-7)$
$a_8 = 4 + (8-1)(-7) = 4 + (7)(-7) = 4 - 49 = -45$

14.2 Check Points

1. $a_1 = 100$

    $a_2 = 100 + (-30) = 70$

    $a_3 = 70 + (-30) = 40$

    $a_4 = 40 + (-30) = 10$

    $a_5 = 10 + (-30) = -20$

    $a_6 = -20 + (-30) = -50$

2. $a_1 = 6, \ d = -5$

    To find the ninth term, a_9, replace n in the formula with 9, a_1 with 6, and d with -5.

    $a_n = a_1 + (n-1)d$

    $a_9 = 6 + (9-1)(-5)$

    $\quad = 6 + 8(-5)$

    $\quad = 6 + (-40)$

    $\quad = -34$

3. **a.** $a_n = a_1 + (n-1)d$

    $\quad = 32 + (n-1)0.7$

    $\quad = 0.7n + 31.3$

    b. $a_n = 0.7n + 31.3$

    $a_{11} = 0.7(11) + 31.3 = 39$

    In 2014 Americans will average 39 car meals.

4. $3, 6, 9, 12, \ldots$

    To find the sum of the first 15 terms, S_{15}, replace n in the formula with 15.

    $S_n = \dfrac{n}{2}(a_1 + a_n)$

    $S_{15} = \dfrac{15}{2}(a_1 + a_{15})$

    Use the formula for the general term of a sequence to find a_{15}. The common difference, d, is 3, and the first term, a_1, is 3.

    $a_n = a_1 + (n-1)d$

    $a_{15} = 3 + (15-1)(3)$

    $\quad = 3 + 14(3)$

    $\quad = 3 + 42$

    $\quad = 45$

    Thus, $S_{15} = \dfrac{15}{2}(3 + 45) = \dfrac{15}{2}(48) = 360$.

5. $\sum\limits_{i=1}^{30}(6i-11)=(6\cdot1-11)+(6\cdot2-11)+(6\cdot3-11)+...+(6\cdot30-11)=-5+1+7+...+169$

The first term, a_1, is -5; the common difference, d, is $1-(-5)=6$; the last term, a_{30}, is 169. Substitute

$n=30$, $a_1=-5$, and $a_{30}=169$ in the formula $S_n=\frac{n}{2}(a_1+a_n)$.

$S_{30}=\frac{30}{2}(-5+169)=15(164)=2460$

Thus, $\sum\limits_{i=1}^{30}(6i-11)=2460$

6. $a_n=1800n+64,130$

$a_1=1800(1)+64,130=65,930$

$a_{10}=1800(10)+64,130=82,130$

$S_n=\dfrac{n}{2}\left(a_1+a_n\right)$

$S_{10}=\dfrac{10}{2}\left(a_1+a_{10}\right)$

$\qquad=5\left(65,930+82,130\right)$

$\qquad=5\left(148,060\right)$

$\qquad=\$740,300$

It would cost $7400 for the ten-year period beginning in 2009

14.2 Exercise Set

1. Since $6-2=4$, $d=4$.

3. Since $-2-(-7)=5$, $d=5$.

5. Since $711-714=-3$, $d=-3$.

7. $a_1=200$

$a_2=200+20=220$

$a_3=220+20=240$

$a_4=240+20=260$

$a_5=260+20=280$

$a_6=280+20=300$

9. $a_1=-7$

$a_2=-7+4=-3$

$a_3=-3+4=1$

$a_4=1+4=5$

$a_5=5+4=9$

$a_6=9+4=13$

11. $a_1 = 300$

$a_2 = 300 - 90 = 210$

$a_3 = 210 - 90 = 120$

$a_4 = 120 - 90 = 30$

$a_5 = 30 - 90 = -60$

$a_6 = -60 - 90 = -150$

13. $a_1 = \dfrac{5}{2}$

$a_2 = \dfrac{5}{2} - \dfrac{1}{2} = \dfrac{4}{2} = 2$

$a_3 = \dfrac{4}{2} - \dfrac{1}{2} = \dfrac{3}{2}$

$a_4 = \dfrac{3}{2} - \dfrac{1}{2} = \dfrac{2}{2} = 1$

$a_5 = 1 - \dfrac{1}{2} = \dfrac{1}{2}$

$a_6 = \dfrac{1}{2} - \dfrac{1}{2} = 0$

15. $a_1 = -0.4$

$a_2 = -0.4 - 1.6 = -2$

$a_3 = -2 - 1.6 = -3.6$

$a_4 = -3.6 - 1.6 = -5.2$

$a_5 = -5.2 - 1.6 = -6.8$

$a_6 = -6.8 - 1.6 = -8.4$

17. $a_6 = 13 + (6-1)4 = 13 + (5)4$

$\qquad = 13 + 20 = 33$

19. $a_{50} = 7 + (50-1)5 = 7 + (49)5$

$\qquad = 7 + 245 = 252$

21. $a_{200} = -40 + (200-1)5 = -40 + (199)5$

$\qquad = -40 + 995 = 955$

23. $a_{60} = 35 + (60-1)(-3) = 35 + (59)(-3)$

$\qquad = 35 + (-177) = -142$

25. $a_n = a_1 + (n-1)d = 1 + (n-1)4$

$\qquad = 1 + 4n - 4 = 4n - 3$

$a_{20} = 4(20) - 3 = 80 - 3 = 77$

27. $a_n = a_1 + (n-1)d = 7 + (n-1)(-4)$
$$= 7 - 4n + 4 = 11 - 4n$$

$$a_{20} = 11 - 4(20) = 11 - 80 = -69$$

29. $a_n = a_1 + (n-1)d = -20 + (n-1)(-4)$
$$= -20 - 4n + 4 = -4n - 16$$

$$a_{20} = -4(20) - 16 = -80 - 16 = -96$$

31. $a_n = a_1 + (n-1)d = -\dfrac{1}{3} + (n-1)\left(\dfrac{1}{3}\right)$

$$= -\dfrac{1}{3} + \dfrac{1}{3}n - \dfrac{1}{3} = \dfrac{1}{3}n - \dfrac{2}{3}$$

$$a_{20} = \dfrac{1}{3}(20) - \dfrac{2}{3} = \dfrac{20}{3} - \dfrac{2}{3} = \dfrac{18}{3} = 6$$

33. $a_n = a_1 + (n-1)d = 4 + (n-1)(-0.3)$
$$= 4 - 0.3n + 0.3 = 4.3 - 0.3n$$

$$a_{20} = 4.3 - 0.3(20) = 4.3 - 6 = -1.7$$

35. First find a_{20}.

$$a_{20} = 4 + (20-1)6 = 4 + (19)6$$
$$= 4 + 114 = 118$$

$$S_{20} = \dfrac{20}{2}(4 + 118) = 10(122) = 1220$$

37. First find a_{50}.

$$a_{50} = -10 + (50-1)4 = -10 + (49)4$$
$$= -10 + 196 = 186$$

$$S_{50} = \dfrac{50}{2}(-10 + 186) = 25(176) = 4400$$

39. First find a_{100}.

$$a_{100} = 1 + (100-1)1 = 1 + (99)1$$
$$= 1 + 99 = 100$$

$$S_{100} = \dfrac{100}{2}(1 + 100) = 50(101) = 5050$$

41. First find a_{60}.

$$a_{60} = 2 + (60-1)2 = 2 + (59)2$$
$$= 2 + 118 = 120$$

$$S_{60} = \frac{60}{2}(2 + 120) = 30(122) = 3660$$

43. The even integers between 21 and 45 start with 22 and end with 44.

$$44 = 22 + (n-1)2$$
$$22 = 2(n-1)$$
$$11 = n-1$$
$$12 = n$$

$$S_{12} = \frac{12}{2}(22 + 44) = 6(66) = 396$$

45. $\sum\limits_{i=1}^{17}(5i+3) = (5(1)+3) + (5(2)+3) + (5(3)+3) + ... + (5(17)+3)$

$$= (5+3) + (10+3) + (15+3) + ... + (85+3) = 8 + 13 + 18 + ... + 88$$

$$S_{17} = \frac{17}{2}(8+88) = \frac{17}{2}(96) = 17(48) = 816$$

$$S_{20} = \frac{20}{2}(2 + 116) = 10(118) = 1180$$

47. $\sum\limits_{i=1}^{30}(-3i+5) = (-3(1)+5) + (-3(2)+5) + (-3(3)+5) + ... + (-3(30)+5)$

$$= (-3+5) + (-6+5) + (-9+5) + ... + (-90+5) = 2 + (-1) + (-4) + ... + (-85)$$

$$S_{30} = \frac{30}{2}(2 + (-85)) = 15(-83) = -1245$$

49. $\sum\limits_{i=1}^{100} 4i = 4(1) + 4(2) + 4(3) + ... + 4(100) = 4 + 8 + 12 + ... + 400$

$$S_{100} = \frac{100}{2}(4 + 400) = 50(404) = 20,200$$

51. First find a_{14} and b_{12}:

$$a_{14} = a_1 + (n-1)d$$
$$= 1 + (14-1)(-3-1) = -51$$

$$b_{12} = b_1 + (n-1)d$$
$$= 3 + (12-1)(8-3) = 58$$

So, $a_{14} + b_{12} = -51 + 58 = 7$.

53. $a_n = a_1 + (n-1)d$

$-83 = 1 + (n-1)(-3-1)$

$-83 = 1 + -4(n-1)$

$-84 = -4n + 4$

$-88 = -4n$

$n = 22$

There are 22 terms.

55. $S_n = \dfrac{n}{2}(a_1 + a_n)$

For $\{a_n\}$: $S_{14} = \dfrac{14}{2}(a_1 + a_{14}) = 7(1 + (-51)) = -350$

For $\{b_n\}$: $S_{14} = \dfrac{14}{2}(b_1 + b_{14}) = 7(3 + 68) = 497$

So $\displaystyle\sum_{n=1}^{14} b_n - \sum_{n=1}^{14} a_n = 497 - (-350) = 847$

57. Two points on the graph are $(1, 1)$ and $(2, -3)$. Finding the slope of the line; $m = \dfrac{y_2 - y_1}{x_2 - x_1} = \dfrac{-3-1}{2-1} = \dfrac{-4}{1} = -4$

Using the point-slope form of an equation of a line;

$y - y_2 = m(x - x_2)$

$y - 1 = -4(x - 1)$

$y - 1 = -4x + 4$

$y = -4x + 5$

Thus, $f(x) = -4x + 5$.

59. Using $a_n = a_1 + (n-1)d$ and $a_2 = 4$:

$a_2 = a_1 + (2-1)d$

$4 = a_1 + d$

And since $a_6 = 16$:

$a_6 = a_1 + (6-1)d$

$16 = a_1 + 5d$

The system of equations is

$4 = a_1 + d$

$16 = a_1 + 5d$

Solving the first equation for a_1:

$a_1 = 4 - d$

Substituting the value into the second equation and solving for d:

$16 = (4-d) + 5d$

$16 = 4 + 4d$

$12 = 4d$

$3 = d$

Then $a_n = a_1 + (n-1)d$

$a_n = 1 + (n-1)3$

$a_n = 1 + 3n - 3$

$a_n = 3n - 2$

61. a. $a_n = a_1 + (n-1)d$

$a_n = 10 + (n-1)(0.77)$

$a_n = 10 + 0.77n - 0.77$

$a_n = 0.77n + 9.23$

b. 2011 is 27 years after 1984.

$a_{27} = 0.77(27) + 9.23 \approx 30.0$

If trends continue, in 2011 the percentage of Americans with no close friends will be 30.0%

63. Company A

$a_n = 24000 + (n-1)1600$

$\quad = 24000 + 1600n - 1600$

$\quad = 1600n + 22400$

$a_{10} = 1600(10) + 22400$

$\quad = 16000 + 22400 = 38400$

Company B

$a_n = 28000 + (n-1)1000$

$\quad = 28000 + 1000n - 1000$

$\quad = 1000n + 27000$

$a_{10} = 1000(10) + 27000$

$\quad = 10000 + 27000 = 37000$

Company A will pay \$1400 more in year 10.

65. a. Total cost: $\$4694 + \$5132 + \$5491 + \$5836 = \$21,153$

b. $a_1 = 379(1) + 4342 = 4721$

$a_4 = 379(4) + 4342 = 5858$

$S_n = \dfrac{n}{2}(a_1 + a_n)$

$S_4 = \dfrac{4}{2}(4721 + 5858) = 2(10,579) = \$21,158$

The model overestimates the actual sum by \$5.

67. Answers will vary.

69. Company A

$$a_n = 19,000 + (n-1)2600$$
$$= 19,000 + 2600n - 2600$$
$$= 2600n + 16,400$$

$$a_{10} = 2600(10) + 16400$$
$$= 26,000 + 16,400 = 42,400$$

$$S_n = \frac{n}{2}(a_1 + a_{10})$$
$$S_{10} = \frac{10}{2}(19,000 + 42,400)$$
$$= 5(61,400) = \$307,000$$

Company B

$$a_n = 27,000 + (n-1)1200$$
$$= 27,000 + 1200n - 1200$$
$$= 1200n + 25,800$$

$$a_{10} = 1200(10) + 25,800$$
$$= 12,000 + 25,800$$
$$= 37,800$$

$$S_n = \frac{n}{2}(a_1 + a_{10})$$
$$S_{10} = \frac{10}{2}(27,000 + 37,800)$$
$$= 5(64,800) = \$324,000$$

Company B pays the greater total amount.

71. $a_{38} = a_1 + (n-1)d$
$$= 20 + (38-1)(3)$$
$$= 20 + 37(3) = 131$$

$$S_{38} = \frac{n}{2}(a_1 + a_{38})$$
$$= \frac{38}{2}(20 + 131)$$
$$= 19(151)$$
$$= 2869$$

There are 2869 seats in this section of the stadium.

73. Answers will vary.

77. Answers will vary. For example, consider Exercise 45.

$$\sum_{i=1}^{17}(5i+3)$$

```
sum(seq(5I+3,I,1
,17))
              816
```

79. makes sense

81. makes sense

83. false; Changes to make the statement true will vary. A sample change is: The common difference is −2.

85. true

87. From the sequence, we see that $a_1 = 21700$ and $d = 23172 - 21700 = 1472$.

We know that $a_n = a_1 + (n-1)d$. We can substitute what we know to find n.

$$314,628 = 21,700 + (n-1)1472$$
$$292,928 = (n-1)1472$$
$$\frac{292,928}{1472} = \frac{(n-1)1472}{1472}$$
$$199 = n-1$$
$$200 = n$$

314,628 is the 200th term of the sequence.

89. $1+3+5+\ldots+(2n-1)$

$$S_n = \frac{n}{2}(a_1 + a_n) = \frac{n}{2}\big(1+(2n-1)\big)$$
$$= \frac{n}{2}(1+2n-1) = \frac{n}{2}(2n)$$
$$= n(n) = n^2$$

90. $\log(x^2 - 25) - \log(x + 5) = 3$

$$\log\left(\frac{x^2 - 25}{x + 5}\right) = 3$$

$$\log\left(\frac{(x-5)(\cancel{x+5})}{\cancel{x+5}}\right) = 3$$

$$\log(x - 5) = 3$$

$$x - 5 = 10^3$$

$$x - 5 = 1000$$

$$x = 1005$$

The solution set is $\{1005\}$.

91. $x^2 + 3x \leq 10$

Solve the related quadratic equation.

$$x^2 + 3x - 10 = 0$$

$$(x + 5)(x - 2) = 0$$

Apply the zero product principle.

$x + 5 = 0$ or $x - 2 = 0$

$\quad x = -5 \qquad\quad x = 2$

The boundary points are -5 and 2.

Interval	Test Value	Test	Conclusion
$(-\infty, -5]$	-6	$(-6)^2 + 3(-6) \leq 10$ $18 \leq 10$, false	$(-\infty, -5]$ does not belong to the solution set.
$[-5, 2]$	0	$0^2 + 3(0) \leq 10$ $0 \leq 10$, true	$[-5, 2]$ belongs to the solution set.
$[2, \infty)$	3	$3^2 + 3(3) \leq 10$ $18 \leq 10$, false	$[2, \infty)$ does not belong to the solution set.

The solution set is $[-5, 2]$ or $\{x \mid -5 \leq x \leq 2\}$.

92.
$$A = \frac{Pt}{P + t}$$

$$A(P + t) = Pt$$

$$AP + At = Pt$$

$$AP - Pt = -At$$

$$P(A - t) = -At$$

$$P = -\frac{At}{A - t} \text{ or } \frac{At}{t - A}$$

93. $\dfrac{a_2}{a_1} = \dfrac{-2}{1} = -2$

$\dfrac{a_3}{a_2} = \dfrac{4}{-2} = -2$

$\dfrac{a_4}{a_3} = \dfrac{-8}{4} = -2$

$\dfrac{a_5}{a_4} = \dfrac{16}{-8} = -2$

The ratio of a term to the term that directly precedes it is always -2.

94. $\dfrac{a_2}{a_1} = \dfrac{3 \cdot 5^2}{3 \cdot 5^1} = 5$

$\dfrac{a_3}{a_2} = \dfrac{3 \cdot 5^3}{3 \cdot 5^2} = 5$

$\dfrac{a_4}{a_3} = \dfrac{3 \cdot 5^4}{3 \cdot 5^3} = 5$

$\dfrac{a_5}{a_4} = \dfrac{3 \cdot 5^5}{3 \cdot 5^4} = 5$

The ratio of a term to the term that directly precedes it is always 5.

95. $a_n = a_1 3^{n-1}$

$a_7 = 11 \cdot 3^{7-1} = 11 \cdot 3^6 = 11 \cdot 729 = 8019$

14.3 Check Points

1. $a_1 = 12, \; r = \dfrac{1}{2}$

$a_2 = 12\left(\dfrac{1}{2}\right)^1 = 6$

$a_3 = 12\left(\dfrac{1}{2}\right)^2 = \dfrac{12}{4} = 3$

$a_4 = 12\left(\dfrac{1}{2}\right)^3 = \dfrac{12}{8} = \dfrac{3}{2}$

$a_5 = 12\left(\dfrac{1}{2}\right)^4 = \dfrac{12}{16} = \dfrac{3}{4}$

$a_6 = 12\left(\dfrac{1}{2}\right)^5 = \dfrac{12}{32} = \dfrac{3}{8}$

The first six terms are $12, 6, 3, \dfrac{3}{2}, \dfrac{3}{4},$ and $\dfrac{3}{8}$.

2. $a_1 = 5, \; r = -3$

$a_n = a_1 r^{n-1}$

$a_7 = 5(-3)^{7-1} = 5(-3)^6 = 5(729) = 3645$

The seventh term is 3645.

3. $3, 6, 12, 24, 48, \ldots$

$r = \dfrac{6}{3} = 2, \; a_1 = 3$

$a_n = 3(2)^{n-1}$

$a_8 = 3(2)^{8-1} = 3(2)^7 = 3(128) = 384$

The eighth term is 384.

4. $a_1 = 2, \; r = \dfrac{-6}{2} = -3$

$S_n = \dfrac{a_1(1 - r^n)}{1 - r}$

$S_9 = \dfrac{2\left(1 - (-3)^9\right)}{1 - (-3)} = \dfrac{2(19,684)}{4} = 9842$

The sum of the first nine terms is 9842.

5. $\displaystyle\sum_{i=1}^{8} 2 \cdot 3^i$

$a_1 = 2 \cdot (3)^1 = 6, \; r = 3$

$S_n = \dfrac{a_1(1 - r^n)}{1 - r}$

$S_8 = \dfrac{6\left(1 - 3^8\right)}{1 - 3} = \dfrac{6(-6560)}{-2} = 19,680$

Thus, $\displaystyle\sum_{i=1}^{8} 2 \cdot 3^i = 19,680$.

6. $a_1 = 30,000, \; r = 1.06$

$S_n = \dfrac{a_1(1 - r^n)}{1 - r}$

$S_{30} = \dfrac{30,000\left(1 - (1.06)^{30}\right)}{1 - 1.06} \approx 2,371,746$

The total lifetime salary is $2,371,746.

7. a. $A = \dfrac{P\left[\left(1+\frac{r}{n}\right)^{nt}-1\right]}{\frac{r}{n}}$

$P = 100,\ r = 0.095,\ n = 12,\ t = 35$

$A = \dfrac{100\left[\left(1+\frac{0.095}{12}\right)^{12\cdot 35}-1\right]}{\frac{0.095}{12}} \approx 333{,}946$

The value of the IRA will be \$333,946.

b. Interest = Value of IRA − Total deposits

$\approx \$333{,}946 - \$100 \cdot 12 \cdot 35$

$\approx \$333{,}946 - \$42{,}000$

$\approx \$291{,}946$

8. $3 + 2 + \dfrac{4}{3} + \dfrac{8}{9} + \cdots$

$a_1 = 3,\ r = \dfrac{2}{3}$

$S = \dfrac{a_1}{1-r}$

$S = \dfrac{3}{1-\frac{2}{3}} = \dfrac{3}{\frac{1}{3}} = 9$

The sum of this infinite geometric series is 9.

9. $0.\overline{9} = 0.9999\cdots = \dfrac{9}{10} + \dfrac{9}{100} + \dfrac{9}{1000} + \cdots$

$a_1 = \dfrac{9}{10},\ r = \dfrac{1}{10}$

$S = \dfrac{\frac{9}{10}}{1-\frac{1}{10}} = \dfrac{\frac{9}{10}}{\frac{9}{10}} = 1$

An equivalent fraction for $0.\overline{9}$ is 1.

10. $a_1 = 1000(0.8) = 800,\ r = 0.8$

$S = \dfrac{800}{1-0.8} = 4000$

The total amount spent is \$4000.

14.3 Exercise Set

1. $r = \dfrac{a_2}{a_1} = \dfrac{15}{5} = 3$

3. $r = \dfrac{a_2}{a_1} = \dfrac{30}{-15} = -2$

5. $r = \dfrac{a_2}{a_1} = \dfrac{\frac{9}{2}}{3} = \dfrac{9}{2} \cdot \dfrac{1}{3} = \dfrac{3}{2}$

7. $r = \dfrac{a_2}{a_1} = \dfrac{0.04}{-0.4} = -0.1$

9. The first term is 2.
The second term is $2 \cdot 3 = 6$.
The third term is $6 \cdot 3 = 18$.
The fourth term is $18 \cdot 3 = 54$.
The fifth term is $54 \cdot 3 = 162$.

11. The first term is 20.
The second term is $20 \cdot \dfrac{1}{2} = 10$.
The third term is $10 \cdot \dfrac{1}{2} = 5$.
The fourth term is $5 \cdot \dfrac{1}{2} = \dfrac{5}{2}$.
The fifth term is $\dfrac{5}{2} \cdot \dfrac{1}{2} = \dfrac{5}{4}$.

13. The first term is –4.
The second term is $-4(-10) = 40$.
The third term is $40(-10) = -400$.
The fourth term is $-400(-10) = 4000$.
The fifth term is $4000(-10) = -40{,}000$.

15. The first term is $-\dfrac{1}{4}$.
The second term is $-\dfrac{1}{4}(-2) = \dfrac{1}{2}$.
The third term is $\dfrac{1}{2}(-2) = -1$.
The fourth term is $-1(-2) = 2$.
The fifth term is $2(-2) = -4$.

17. $a_8 = 6(2)^{8-1} = 6(2)^7 = 6(128) = 768$

19. $a_{12} = 5(-2)^{12-1} = 5(-2)^{11}$

$\qquad = 5(-2048) = -10,240$

21. $a_6 = 6400\left(-\dfrac{1}{2}\right)^{6-1} = 6400\left(-\dfrac{1}{2}\right)^5$

$\qquad = -200$

23. $a_8 = 1,000,000(0.1)^{8-1}$

$\qquad = 1,000,000(0.1)^7$

$\qquad = 1,000,000(0.0000001) = 0.1$

25. $r = \dfrac{a_2}{a_1} = \dfrac{12}{3} = 4$

$\qquad a_n = a_1 r^{n-1} = 3(4)^{n-1}$

$\qquad a_7 = 3(4)^{7-1} = 3(4)^6$

$\qquad = 3(4096) = 12,288$

27. $r = \dfrac{a_2}{a_1} = \dfrac{6}{18} = \dfrac{1}{3}$

$\qquad a_n = a_1 r^{n-1} = 18\left(\dfrac{1}{3}\right)^{n-1}$

$\qquad a_7 = 18\left(\dfrac{1}{3}\right)^{7-1} = 18\left(\dfrac{1}{3}\right)^6$

$\qquad = 18\left(\dfrac{1}{729}\right) = \dfrac{18}{729} = \dfrac{2}{81}$

$\qquad a_7 = 12\left(\dfrac{1}{2}\right)^{7-1} = 12\left(\dfrac{1}{2}\right)^6 = 12\left(\dfrac{1}{64}\right) = \dfrac{3}{16}$

29. $r = \dfrac{a_2}{a_1} = \dfrac{-3}{1.5} = -2$

$\qquad a_n = a_1 r^{n-1} = 1.5(-2)^{n-1}$

$\qquad a_7 = 1.5(-2)^{7-1} = 1.5(-2)^6$

$\qquad = 1.5(64) = 96$

31. $r = \dfrac{a_2}{a_1} = \dfrac{-0.004}{0.0004} = -10$

$\qquad a_n = a_1 r^{n-1} = 0.0004(-10)^{n-1}$

$\qquad a_7 = 0.0004(-10)^{7-1} = 0.0004(-10)^6$

$\qquad = 0.0004(1000000) = 400$

33. $r = \dfrac{a_2}{a_1} = \dfrac{6}{2} = 3$

$\qquad S_{12} = \dfrac{2(1-3^{12})}{1-3} = \dfrac{2(1-531,441)}{-2}$

$\qquad = \dfrac{2(-531,440)}{-2} = \dfrac{-1,062,880}{-2}$

$\qquad = 531,440$

35. $r = \dfrac{a_2}{a_1} = \dfrac{-6}{3} = -2$

$\qquad S_{11} = \dfrac{a_1(1-r^n)}{1-r} = \dfrac{3(1-(-2)^{11})}{1-(-2)}$

$\qquad = \dfrac{\cancel{3}(1-(-2048))}{\cancel{3}} = 2049$

37. $r = \dfrac{a_2}{a_1} = \dfrac{3}{-\dfrac{3}{2}} = 3 \div \left(-\dfrac{3}{2}\right)$

$\qquad = 3 \cdot \left(-\dfrac{2}{3}\right) = -2$

$\qquad S_{14} = \dfrac{a_1(1-r^n)}{1-r} = \dfrac{-\dfrac{3}{2}\left(1-(-2)^{14}\right)}{1-(-2)}$

$\qquad = \dfrac{-\dfrac{3}{2}(1-(16,384))}{3} = \dfrac{-\dfrac{3}{2}(-16,383)}{3}$

$\qquad = -\dfrac{3}{2}(-16,383) \div 3 = \dfrac{49,149}{2} \cdot \dfrac{1}{3}$

$\qquad = \dfrac{16,383}{2}$

39. $\displaystyle\sum_{i=1}^{8} 3^i = \dfrac{3(1-3^8)}{1-3} = \dfrac{3(1-6561)}{-2}$

$\qquad = \dfrac{3(-6560)}{-2} = \dfrac{-19,680}{-2} = 9840$

41. $\displaystyle\sum_{i=1}^{10} 5 \cdot 2^i = \dfrac{10(1-2^{10})}{1-2} = \dfrac{10(1-1024)}{-1}$

$\qquad = \dfrac{10(-1023)}{-1} = 10,230$

43. $\displaystyle\sum_{i=1}^{6}\left(\frac{1}{2}\right)^{i+1} = \frac{\frac{1}{4}\left(1-\left(\frac{1}{2}\right)^{6}\right)}{1-\frac{1}{2}} = \frac{\frac{1}{4}\left(1-\frac{1}{64}\right)}{\frac{1}{2}}$

$= \dfrac{\frac{1}{4}\left(\frac{64}{64}-\frac{1}{64}\right)}{\frac{1}{2}} = \dfrac{\frac{1}{4}\left(\frac{63}{64}\right)}{\frac{1}{2}}$

$= \frac{1}{4}\left(\frac{63}{64}\right)\div\frac{1}{2} = \frac{1}{\cancel{4}_{2}}\left(\frac{63}{64}\right)\cdot\frac{\cancel{2}}{1}$

$= \dfrac{63}{128}$

45. $r = \dfrac{a_2}{a_1} = \dfrac{\frac{1}{3}}{1} = \dfrac{1}{3}$

$S = \dfrac{a_1}{1-r} = \dfrac{1}{1-\frac{1}{3}} = \dfrac{1}{\frac{2}{3}} = 1\div\frac{2}{3} = 1\cdot\frac{3}{2} = \frac{3}{2}$

47. $r = \dfrac{a_2}{a_1} = \dfrac{\frac{3}{4}}{3} = \frac{3}{4}\div 3 = \frac{3}{4}\cdot\frac{1}{3} = \frac{1}{4}$

$S = \dfrac{a_1}{1-r} = \dfrac{3}{1-\frac{1}{4}} = \dfrac{3}{\frac{3}{4}} = 3\div\frac{3}{4}$

$= 3\cdot\frac{4}{3} = \frac{12}{3} = 4$

49. $r = \dfrac{a_2}{a_1} = \dfrac{-\frac{1}{2}}{1} = -\frac{1}{2}$

$S = \dfrac{a_1}{1-r} = \dfrac{1}{1-\left(-\frac{1}{2}\right)} = \dfrac{1}{\frac{3}{2}} = 1\div\frac{3}{2}$

$= 1\cdot\frac{2}{3} = \frac{2}{3}$

51. $r = -0.3$

$a_1 = 26(-0.3)^{1-1} = 26(-0.3)^{0}$
$\quad = 26(1) = 26$

$S = \dfrac{26}{1-(-0.3)} = \dfrac{26}{1.3} = 20$

53. $0.\overline{5} = \dfrac{a_1}{1-r} = \dfrac{\frac{5}{10}}{1-\frac{1}{10}} = \dfrac{\frac{5}{10}}{\frac{9}{10}} = \frac{5}{10}\div\frac{9}{10}$

$= \frac{5}{10}\cdot\frac{10}{9} = \frac{5}{9}$

55. $0.\overline{47} = \dfrac{a_1}{1-r} = \dfrac{\frac{47}{100}}{1-\frac{1}{100}} = \dfrac{\frac{47}{100}}{\frac{99}{100}}$

$= \frac{47}{100}\div\frac{99}{100} = \frac{47}{100}\cdot\frac{100}{99} = \frac{47}{99}$

57. $0.\overline{257} = \dfrac{a_1}{1-r} = \dfrac{\frac{257}{1000}}{1-\frac{1}{1000}} = \dfrac{\frac{257}{1000}}{\frac{999}{1000}}$

$= \frac{257}{1000}\div\frac{999}{1000} = \frac{257}{1000}\cdot\frac{1000}{999}$

$= \dfrac{257}{999}$

59. The sequence is arithmetic with common difference $d = 1$.

61. The sequence is geometric with common ratio $r = 2$.

63. The sequence is neither arithmetic nor geometric.

65. First find a_{10} and b_{10}:

$a_{10} = a_1 r^{n-1}$

$\quad = (-5)\left(\dfrac{10}{-5}\right)^{10-1} = (-5)(-2)^9 = 2560$

$b_{10} = b_1 + (n-1)d$

$\quad = 10 + (10-1)(-5-10)$

$\quad = 10 + (9)(-15) = -125$

So, $a_{10} + b_{10} = 2560 + (-125) = 2435$.

67. From Exercise 65, $a_{10} = 2560$ and $b_{10} = -125$.

For $\{a_n\}$,

$$r = \frac{10}{-5} = -2$$

$$S_{10} = \frac{a_1(1-r^n)}{1-r} = \frac{(-5)\left(1-(-2)^{10}\right)}{1-(-2)}$$

$$= \frac{(-5)(-1023)}{3} = 1705$$

For $\{b_n\}$,

$$S_n = \frac{n}{2}(b_1 + b_n) = \frac{10}{2}(10 + (-125))$$

$$= 5(-115) = -575$$

So, $\displaystyle\sum_{n=1}^{10} a_n - \sum_{n=1}^{10} b_n = 1705 - (-575) = 2280$

69. For $\{a_n\}$,

$$S_6 = \frac{a_1(1-r^n)}{1-r} = \frac{(-5)\left(1-(-2)^6\right)}{1-(-2)}$$

$$= \frac{(-5)(-63)}{3} = 105$$

For $\{c_n\}$,

$$S = \frac{a_1}{1-r} = \frac{-2}{1-\frac{1}{-2}} = \frac{-2}{\frac{3}{2}} = -\frac{4}{3}$$

So, $S_6 \cdot S = 105\left(-\dfrac{4}{3}\right) = -140$

71. It is given that $a_4 = 27$. Using the formula $a_n = a_1 r^{n-1}$ when $n = 4$ we have:

$$27 = 8r^{4-1}$$

$$\frac{27}{8} = r^3$$

$$r = \sqrt[3]{\frac{27}{8}} = \frac{3}{2}$$

Then

$$a_n = a_1 r^{n-1}$$

$$a_2 = 8\left(\frac{3}{2}\right)^{2-1} = 8\left(\frac{3}{2}\right) = 12$$

$$a_3 = 8\left(\frac{3}{2}\right)^{3-1} = 8\left(\frac{3}{2}\right)^2 = 8\left(\frac{9}{4}\right) = 18$$

73. Find the total value of the lump-sum investment.

$$A = P(1+r)^t = 30,000(1+0.05)^{20} \approx 79,599$$

Find the total value of the annuity.

$$A = \frac{P\left[\left(1+\dfrac{r}{n}\right)^{nt} - 1\right]}{\dfrac{r}{n}} = \frac{1500\left[\left(1+\dfrac{0.05}{1}\right)^{20} - 1\right]}{\dfrac{0.05}{1}} \approx 49,599$$

$$\$79,599 - \$49,599 = \$30,000$$

You will have $30,000 more from the lump-sum investment.

75. $\quad r = \dfrac{a_2}{a_1} = \dfrac{2}{1} = 2$

$$a_{15} = 1(2)^{15-1} = (2)^{14} = 16,384$$

On the fifteenth day, you will put aside $16,384 for savings.

77. $\quad r = 1.04$

$$a_7 = 3,000,000(1.04)^{7-1}$$

$$= 3,000,000(1.04)^6$$

$$= 3,000,000(1.265319)$$

$$= 3,795,957$$

The athlete's salary for year 7 will be $3,795,957.

79. a. $\quad r_{2003 \text{ to } 2004} = \dfrac{35.89}{35.48} \approx 1.01$

$$r_{2004 \text{ to } 2005} = \dfrac{36.13}{35.89} \approx 1.01$$

$$r_{2005 \text{ to } 2006} = \dfrac{36.46}{36.13} \approx 1.01$$

r is approximately 1.01 for each division.

b. $\quad a_n = a_1 r^{n-1}$

$$a_n = 35.48(1.01)^{n-1}$$

c. Since year 2010 is the 8th term, find a_8.

$$a_n = 35.48(1.01)^{n-1}$$

$$a_8 = 35.48(1.01)^{8-1} \approx 38.04$$

The population of California will be approximately 38.04 million in 2010.

81. $r = \dfrac{a_2}{a_1} = \dfrac{2}{1} = 2$

$$S_{15} = \frac{a_1\left(1-r^n\right)}{1-r} = \frac{1\left(1-(2)^{15}\right)}{1-2}$$

$$= \frac{(1-32,768)}{-1} = \frac{(-32,767)}{-1} = 32,767$$

Your savings will be \$32,767 over the 15 days.

83. $r = 1.05$

$$S_{20} = \frac{a_1\left(1-r^n\right)}{1-r} = \frac{24,000\left(1-(1.05)^{20}\right)}{1-1.05}$$

$$= \frac{24,000\left(1-2.6533\right)}{-0.05}$$

$$= \frac{24,000\left(-1.6533\right)}{-0.05} = 793,583$$

The total lifetime salary over the 20 years is \$793,583.

85. $r = 0.9$

$$S_{10} = \frac{a_1\left(1-r^n\right)}{1-r} = \frac{20\left(1-(0.9)^{10}\right)}{1-0.9}$$

$$= \frac{20\left(1-0.348678\right)}{0.1}$$

$$= \frac{20\left(0.651322\right)}{0.1} = 130.264$$

After 10 swings, the pendulum covers a distance of approximately 130.26 inches.

87. a. $A = \dfrac{P\left[\left(1+\dfrac{r}{n}\right)^{nt}-1\right]}{\dfrac{r}{n}} = \dfrac{2000\left[\left(1+\dfrac{0.075}{1}\right)^{5}-1\right]}{\dfrac{0.075}{1}} \approx \$11,617$

b. $\$11,617 - 5 \times \$2000 = \$1617$

89. a. $A = \dfrac{P\left[\left(1+\dfrac{r}{n}\right)^{nt}-1\right]}{\dfrac{r}{n}} = \dfrac{50\left[\left(1+\dfrac{0.055}{12}\right)^{12\times40}-1\right]}{\dfrac{0.055}{12}} \approx \$87,052$

b. $\$87,052 - \$50 \cdot 12 \cdot 40 = \$63,052$

91. a. $A = \dfrac{P\left[\left(1+\frac{r}{n}\right)^{nt}-1\right]}{\frac{r}{n}} = \dfrac{10,000\left[\left(1+\frac{0.105}{4}\right)^{4\times10}-1\right]}{\frac{0.105}{4}} \approx \$693,031$

b. $\$693,031 - \$10,000 \cdot 4 \cdot 10 = \$293,031$

93. $r = 60\% = 0.6$

$a_1 = 6(.6) = 3.6$

$S = \dfrac{3.6}{1-0.6} = \dfrac{3.6}{0.4} = 9$

The total economic impact of the factory will be \$9 million per year.

95. $r = \dfrac{1}{4}$

$S = \dfrac{\frac{1}{4}}{1-\frac{1}{4}} = \dfrac{\frac{1}{4}}{\frac{3}{4}} = \dfrac{1}{4} \div \dfrac{3}{4} = \dfrac{1}{4} \cdot \dfrac{4}{3} = \dfrac{1}{3}$

Eventually $\dfrac{1}{3}$ of the largest square will be shaded.

97. – 103. Answers will vary.

105. Answers will vary. For example, consider Exercise 25.

$a_n = 3(4)^{n-1}$

```
seq(3(4)^(N-1),N
,7,7)
          {12288}
```

107. $f(x) = \dfrac{2\left[1 - \left(\dfrac{1}{3}\right)^{x}\right]}{1 - \dfrac{1}{3}}$

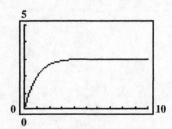

$$S = \frac{2}{1 - \dfrac{1}{3}} = \frac{2}{\dfrac{2}{3}} = 2 \div \frac{2}{3} = 2 \cdot \frac{3}{2} = 3$$

The sum of the series and the asymptote of the function are both 3.

109. makes sense

111. makes sense

113. false; Changes to make the statement true will vary. A sample change is: The sequence is not geometric. There is not a common ratio.

115. false; Changes to make the statement true will vary. A sample change is: The sum of the sequence is $\dfrac{10}{1 - \left(-\dfrac{1}{2}\right)}$.

117. $\qquad S = \dfrac{a_1}{1 - r}$

$20,000 = \dfrac{x}{1 - 0.9}$

$20,000 = \dfrac{x}{0.1}$

$20,000(0.1) = x$

$\qquad 2000 = x$

To keep 20,000 flies in the population, 2000 flies should be released each day.

119. $\sqrt{28} - 3\sqrt{7} + \sqrt{63} = \sqrt{4 \cdot 7} - 3\sqrt{7} + \sqrt{9 \cdot 7}$
$\qquad\qquad\qquad\qquad\quad = 2\sqrt{7} - 3\sqrt{7} + 3\sqrt{7}$
$\qquad\qquad\qquad\qquad\quad = 2\sqrt{7}$

120. $2x^2 = 4 - x$

$2x^2 + x - 4 = 0$

$a = 2 \quad b = 1 \quad c = -4$

Solve using the quadratic formula.

$$x = \frac{-1 \pm \sqrt{1^2 - 4(2)(-4)}}{2(2)}$$

$$= \frac{-1 \pm \sqrt{1 + 32}}{4} = \frac{-1 \pm \sqrt{33}}{4}$$

The solution set is $\left\{ \dfrac{-1 \pm \sqrt{33}}{4} \right\}$.

121. $\dfrac{6}{\sqrt{3} - \sqrt{5}} = \dfrac{6}{\sqrt{3} - \sqrt{5}} \cdot \dfrac{\sqrt{3} + \sqrt{5}}{\sqrt{3} + \sqrt{5}}$

$$= \frac{6\left(\sqrt{3} + \sqrt{5}\right)}{3 - 5}$$

$$= \frac{6\left(\sqrt{3} + \sqrt{5}\right)}{-2}$$

$$= -3\left(\sqrt{3} + \sqrt{5}\right)$$

122. The exponents begin with the exponent on $a + b$ and decrease by 1 in each successive term.

123. The exponents begin with 0, increase by 1 in each successive term, and end with the exponent on $a + b$.

124. The sum of the exponents is the exponent on $a + b$.

Chapter 14 Mid-Chapter Check Point

1. $a_n = (-1)^{n+1} \dfrac{n}{(n-1)!}$

$a_1 = (-1)^{1+1} \dfrac{1}{(1-1)!} = (-1)^2 \dfrac{1}{0!} = 1 \cdot 1 = 1$

$a_2 = (-1)^{2+1} \dfrac{2}{(2-1)!} = (-1)^3 \dfrac{2}{1!} = (-1)(2) = -2$

$a_3 = (-1)^{3+1} \dfrac{3}{(3-1)!} = (-1)^4 \dfrac{3}{2!} = 1 \cdot \dfrac{3}{2} = \dfrac{3}{2}$

$a_4 = (-1)^{4+1} \dfrac{4}{(4-1)!} = (-1)^5 \dfrac{4}{3!} = (-1)\dfrac{4}{6} = -\dfrac{2}{3}$

$a_5 = (-1)^{5+1} \dfrac{5}{(5-1)!} = (-1)^6 \dfrac{5}{4!} = 1 \cdot \dfrac{5}{24} = \dfrac{5}{24}$

2. Using $a_n = a_1 + (n-1)d$;

$a_1 = 5$

$a_2 = 5 + (2-1)(-3) = 5 + 1(-3) = 5 - 3 = 2$

$a_3 = 5 + (3-1)(-3) = 5 + 2(-3) = 5 - 6 = -1$

$a_4 = 5 + (4-1)(-3) = 5 + 3(-3) = 5 - 9 = -4$

$a_5 = 5 + (5-1)(-3) = 5 + 4(-3) = 5 - 12 = -7$

3. Using $a_n = a_1 r^{n-1}$;

$a_1 = 5$

$a_2 = 5(-3)^{2-1} = 5(-3)^1 = 5(-3) = -15$

$a_3 = 5(-3)^{3-1} = 5(-3)^2 = 5(9) = 45$

$a_4 = 5(-3)^{4-1} = 5(-3)^3 = 5(-27) = -135$

$a_5 = 5(-3)^{5-1} = 5(-3)^4 = 5(81) = 405$

4. $d = a_2 - a_1 = 6 - 2 = 4$

$\begin{aligned} a_n &= a_1 + (n-1)d \\ &= 2 + (n-1)4 \\ &= 2 + 4n - 4 \\ &= 4n - 2 \end{aligned}$

$a_{20} = 4(20) - 2 = 78$

5. $r = \dfrac{a_2}{a_1} = \dfrac{6}{3} = 2$

$a_n = a_1 r^{n-1} = 3(2)^{n-1}$

$a_{10} = 3(2)^{10-1} = 3(2)^9 = 1536$

6. $d = a_2 - a_1 = 1 - \dfrac{3}{2} = -\dfrac{1}{2}$

$a_n = a_1 + (n-1)d = \dfrac{3}{2} + (n-1)\left(-\dfrac{1}{2}\right)$

$\quad = \dfrac{3}{2} - \dfrac{1}{2}n + \dfrac{1}{2} = -\dfrac{1}{2}n + 2$

$a_{30} = -\dfrac{1}{2}(30) + 2 = -15 + 2 = -13$

7. First find r;

$r = \dfrac{a_2}{a_1} = \dfrac{10}{5} = 2$

$S_{10} = \dfrac{a_1(1 - r^n)}{1 - r} = \dfrac{5(1 - (2)^{10})}{1 - (2)}$

$\quad = \dfrac{5(1 - 1024)}{-1} = \dfrac{5(-1023)}{-1} = 5115$

8. First find a_{10};

$$d = a_2 - a_1 = 0 - (-2) = 2$$

$$a_{50} = a_1 + (n-1)d$$
$$= -2 + (50-1)(2) = -2 + 49(2) = 96$$

$$S_{50} = \frac{n}{2}(a_1 + a_n) = \frac{50}{2}(-2 + 96) = 25(94)$$
$$= 2350$$

9. First find r;

$$r = \frac{a_2}{a_1} = \frac{40}{-20} = -2$$

$$S_{10} = \frac{a_1(1 - r^n)}{1 - r} = \frac{-20(1 - (-2)^{10})}{1 - (-2)}$$

$$= \frac{-20(-1023)}{3} = \frac{20460}{3} = 6820$$

10. First find a_{100};

$$d = a_2 - a_1 = -2 - 4 = -6$$

$$a_{100} = a_1 + (n-1)d = 4 + (100-1)(-6)$$
$$= 4 + 99(-6) = -590$$

$$S_{100} = \frac{n}{2}(a_1 + a_n) = \frac{100}{2}(4 - 590)$$
$$= 50(-586)$$
$$= -29,300$$

11. $\displaystyle\sum_{i=1}^{4}(i+4)(i-1) = (1+4)(1-1) + (2+4)(2-1) + (3+4)(3-1) + (4+4)(4-1)$

$$= 5(0) + 6(1) + 7(2) + 8(3) = 0 + 6 + 14 + 24 = 44$$

12. $\displaystyle\sum_{i=1}^{50}(3i-2) = (3\cdot 1 - 2) + (3\cdot 2 - 2) + (3\cdot 3 - 2) + \ldots + (3\cdot 50 - 2)$

$$= (3-2) + (6-2) + (9-2) + \ldots + (150-2)$$
$$= 1 + 4 + 7 + \ldots + 148$$

The sum of this arithmetic sequence is given by $S_n = \dfrac{n}{2}(a_1 + a_n)$;

$$S_{50} = \frac{50}{2}(1 + 148) = 25(149) = 3725$$

13. $\displaystyle\sum_{i=1}^{6}\left(\frac{3}{2}\right)^i$

$$= \left(\frac{3}{2}\right)^1 + \left(\frac{3}{2}\right)^2 + \left(\frac{3}{2}\right)^3 + \left(\frac{3}{2}\right)^4 + \left(\frac{3}{2}\right)^5 + \left(\frac{3}{2}\right)^6$$

$$= \frac{3}{2} + \frac{9}{4} + \frac{27}{8} + \frac{81}{16} + \frac{243}{32} + \frac{729}{64} = \frac{1995}{64}$$

14. $\displaystyle\sum_{i=1}^{\infty}\left(-\frac{2}{5}\right)^{i-1}$

$=\left(-\dfrac{2}{5}\right)^{1-1}+\left(-\dfrac{2}{5}\right)^{2-1}+\left(-\dfrac{2}{5}\right)^{3-1}+\dots$

$=\left(-\dfrac{2}{5}\right)^{0}+\left(-\dfrac{2}{5}\right)^{1}+\left(-\dfrac{2}{5}\right)^{2}+\dots$

$=1+\left(-\dfrac{2}{5}\right)+\dfrac{4}{25}+\dots$

This is an infinite geometric sequence with $r=\dfrac{a_2}{a_1}=\dfrac{-\frac{2}{5}}{1}=-\dfrac{2}{5}$.

Using $S=\dfrac{a_1}{1-r}=\dfrac{1}{1-\left(-\frac{2}{5}\right)}=\dfrac{1}{\frac{7}{5}}=\dfrac{5}{7}$

15. $0.\overline{45}=\dfrac{a_1}{1-r}=\dfrac{\dfrac{45}{100}}{1-\dfrac{1}{100}}=\dfrac{\dfrac{45}{100}}{\dfrac{99}{100}}$

$=\dfrac{45}{100}\div\dfrac{99}{100}=\dfrac{45}{100}\cdot\dfrac{100}{99}=\dfrac{45}{99}=\dfrac{5}{11}$

16. Answers will vary. An example is $\displaystyle\sum_{i=1}^{18}\dfrac{i}{i+2}$.

17. The arithmetic sequence is 16, 48, 80, 112, ...

First find a_{15} where $d=a_2-a_1=48-16=32$.

$a_{15}=a_1+(n-1)d=16+(15-1)(32)$
$\quad\;\;=16+14(32)$
$\quad\;\;=16+448=464$

The distance the skydiver falls during the 15th second is 464 feet.

$S_{15}=\dfrac{n}{2}(a_1+a_n)=\dfrac{15}{2}(16+464)$
$\quad\;\;=7.5(480)=3600$

The total distance the skydiver falls in 15 seconds is 3600 feet.

18. $A=P(1+r)^t$

$\quad=500,000(1+0.10)^8$

$\quad\approx 1,071,794$

The value of the house after 8 years will be \$1,071,794.

14.4 Check Points

1. a. $\dbinom{6}{3} = \dfrac{6!}{3!(6-3)!} = \dfrac{6!}{3!3!} = \dfrac{5 \cdot 4}{1} = 20$

b. $\dbinom{6}{0} = \dfrac{6!}{0!(6-0)!} = \dfrac{6!}{6!} = 1$

c. $\dbinom{8}{2} = \dfrac{8!}{2!(8-2)!} = \dfrac{8!}{2!6!} = \dfrac{8 \cdot 7}{2} = 28$

d. $\dbinom{3}{3} = \dfrac{3!}{3!(3-3)!} = \dfrac{3!}{3!0!} = \dfrac{3!}{3!} = 1$

2. $(x+1)^4 = \dbinom{4}{0}x^4 + \dbinom{4}{1}x^3 + \dbinom{4}{2}x^2 + \dbinom{4}{3}x + \dbinom{4}{4} = x^4 + 4x^3 + 6x^2 + 4x + 1$

3. $(x-2y)^5 = \dbinom{5}{0}x^5(-2y)^0 + \dbinom{5}{1}x^4(-2y)^1 + \dbinom{5}{2}x^3(-2y)^2 + \dbinom{5}{3}x^2(-2y)^3 + \dbinom{5}{4}x(-2y)^4 + \dbinom{5}{5}x^0(-2y)^5$

$$= \quad x^5 \quad\quad -5x^4(2y) \quad +10x^3(4y^2) \quad -10x^2(8y^3) \quad +5x(16y^4) \quad\quad -32y^5$$

$$= x^5 - 10x^4y + 40x^3y^2 - 80x^2y^3 + 80xy^4 - 32y^5$$

4. $(2x+y)^9$

fifth term $= \dbinom{9}{4}(2x)^5 y^4 = \dfrac{9!}{4!5!}(32x^5)y^4 = 4032x^5y^4$

14.4 Exercise Set

1. $\dbinom{8}{3} = \dfrac{8!}{3!(8-3)!} = \dfrac{8!}{3!5!} = \dfrac{8 \cdot 7 \cdot \cancel{6} \cdot \cancel{5!}}{\cancel{3} \cancel{2} \cdot 1 \cdot \cancel{5!}} = 56$

3. $\dbinom{12}{1} = \dfrac{12!}{1!(12-1)!} = \dfrac{12 \cdot \cancel{11!}}{1 \cdot \cancel{11!}} = 12$

5. $\dbinom{6}{6} = \dfrac{\cancel{6!}}{\cancel{6!}\,(6-6)!} = \dfrac{1}{0!} = \dfrac{1}{1} = 1$

7. $\dbinom{100}{2} = \dfrac{100!}{2!(100-2)!} = \dfrac{100 \cdot 99 \cdot \cancel{98!}}{2 \cdot 1 \cdot \cancel{98!}}$

$\quad\quad = 4950$

9. Applying the Binomial Theorem to $(x+2)^3$, we have $a=x$, $b=2$, and $n=3$.

$$(x+2)^3 = \binom{3}{0}x^3 + \binom{3}{1}x^2(2) + \binom{3}{2}x(2)^2 + \binom{3}{3}2^3$$

$$= \frac{3!}{0!(3-0)!}x^3 + \frac{3!}{1!(3-1)!}2x^2 + \frac{3!}{2!(3-2)!}4x + \frac{3!}{3!(3-3)!}8$$

$$= \frac{3!}{1\cdot 3!}x^3 + \frac{3\cdot 2!}{1\cdot 2!}2x^2 + \frac{3\cdot 2!}{2!1!}4x + \frac{3!}{3!0!}8 = x^3 + 3(2x^2) + 3(4x) + 1(8)$$

$$= x^3 + 6x^2 + 12x + 8$$

11. Applying the Binomial Theorem to $(3x+y)^3$, we have $a=3x$, $b=y$, and $n=3$.

$$(3x+y)^3 = \binom{3}{0}(3x)^3 + \binom{3}{1}(3x)^2 y + \binom{3}{2}(3x)y^2 + \binom{3}{3}y^3$$

$$= \frac{3!}{0!(3-0)!}27x^3 + \frac{3!}{1!(3-1)!}9x^2 y + \frac{3!}{2!(3-2)!}3xy^2 + \frac{3!}{3!(3-3)!}y^3$$

$$= \frac{3!}{1\cdot 3!}27x^3 + \frac{3\cdot 2!}{1\cdot 2!}9x^2 y + \frac{3\cdot 2!}{2!1!}3xy^2 + \frac{3!}{3!0!}y^3 = 27x^3 + 3(9x^2 y) + 3(3xy^2) + 1(y^3)$$

$$= 27x^3 + 27x^2 y + 9xy^2 + y^3$$

13. Applying the Binomial Theorem to $(5x-1)^3$, we have $a=5x$, $b=-1$, and $n=3$.

$$(5x-1)^3 = \binom{3}{0}(5x)^3 + \binom{3}{1}(5x)^2(-1) + \binom{3}{2}(5x)(-1)^2 + \binom{3}{3}(-1)^3$$

$$= \frac{3!}{0!(3-0)!}125x^3 - \frac{3!}{1!(3-1)!}25x^2 + \frac{3!}{2!(3-2)!}5x(1) - \frac{3!}{3!(3-3)!}$$

$$= \frac{3!}{1\cdot 3!}125x^3 - \frac{3\cdot 2!}{1\cdot 2!}25x^2 + \frac{3\cdot 2!}{2!1!}5x - \frac{3!}{3!0!} = 125x^3 - 3(25x^2) + 3(5x) - 1$$

$$= 125x^3 - 75x^2 + 15x - 1$$

15. Applying the Binomial Theorem to $(2x+1)^4$, we have $a=2x$, $b=1$, and $n=4$.

$$(2x+1)^4 = \binom{4}{0}(2x)^4 + \binom{4}{1}(2x)^3 + \binom{4}{2}(2x)^2 + \binom{4}{3}2x + \binom{4}{4}$$

$$= \frac{4!}{0!(4-0)!}16x^4 + \frac{4!}{1!(4-1)!}8x^3\cdot 1 + \frac{4!}{2!(4-2)!}4x^2\cdot 1^2 + \frac{4!}{3!(4-3)!}2x\cdot 1^3 + \frac{4!}{4!(4-4)!}\cdot 1^4$$

$$= \frac{4!}{0!\,4!}16x^4 + \frac{4!}{1!3!}8x^3\cdot 1 + \frac{4!}{2!2!}4x^2\cdot 1 + \frac{4!}{3!1!}2x\cdot 1 + \frac{4!}{4!0!}\cdot 1$$

$$= 1(16x^4) + \frac{4\cdot 3!}{1\cdot 3!}8x^3 + \frac{4\cdot 3\cdot 2!}{2\cdot 1\cdot 2!}4x^2 + \frac{4\cdot 3!}{3!\cdot 1}2x + 1 = 16x^4 + 4(8x^3) + 6(4x^2) + 4(2x) + 1$$

$$= 16x^4 + 32x^3 + 24x^2 + 8x + 1$$

17. Applying the Binomial Theorem to $\left(x^2+2y\right)^4$, we have $a=x^2$, $b=2y$, and $n=4$.

$$\left(x^2+2y\right)^4=\binom{4}{0}\left(x^2\right)^4+\binom{4}{1}\left(x^2\right)^3(2y)+\binom{4}{2}\left(x^2\right)^2(2y)^2+\binom{4}{3}x^2(2y)^3+\binom{4}{4}(2y)^4$$

$$=\frac{4!}{0!(4-0)!}x^8+\frac{4!}{1!(4-1)!}2x^6y+\frac{4!}{2!(4-2)!}4x^4y^2+\frac{4!}{3!(4-3)!}8x^2y^3+\frac{4!}{4!(4-4)!}16y^4$$

$$=\frac{\cancel{4!}}{0!\,\cancel{4!}}x^8+\frac{4!}{1!3!}2x^6y+\frac{4!}{2!2!}4x^4y^2+\frac{4!}{3!1!}8x^2y^3+\frac{\cancel{4!}}{\cancel{4!}0!}16y^4$$

$$=1\left(x^8\right)+\frac{4\cdot\cancel{3!}}{1\cdot\cancel{3!}}2x^6y+\frac{4\cdot3\cdot\cancel{2!}}{2\cdot1\cdot\cancel{2!}}4x^4y^2+\frac{4\cdot\cancel{3!}}{\cancel{3!}\cdot1}8x^2y^3+16y^4$$

$$=x^8+4\left(2x^6y\right)+6\left(4x^4y^2\right)+4\left(8x^2y^3\right)+16y^4$$

$$=x^8+8x^6y+24x^4y^2+32x^2y^3+16y^4$$

19. Applying the Binomial Theorem to $(y-3)^4$, we have $a=y$, $b=-3$, and $n=4$.

$$(y-3)^4=\binom{4}{0}y^4+\binom{4}{1}y^3(-3)+\binom{4}{2}y^2(-3)^2+\binom{4}{3}y(-3)^3+\binom{4}{4}(-3)^4$$

$$=\frac{4!}{0!(4-0)!}y^4-\frac{4!}{1!(4-1)!}3y^3+\frac{4!}{2!(4-2)!}9y^2-\frac{4!}{3!(4-3)!}27y+\frac{4!}{4!(4-4)!}81$$

$$=\frac{\cancel{4!}}{0!\,\cancel{4!}}y^4-\frac{4!}{1!3!}3y^3+\frac{4!}{2!2!}9y^2-\frac{4!}{3!1!}27y+\frac{\cancel{4!}}{\cancel{4!}0!}81$$

$$=1\left(y^4\right)-\frac{4\cdot\cancel{3!}}{1\cdot\cancel{3!}}3y^3+\frac{4\cdot3\cdot\cancel{2!}}{2\cdot1\cdot\cancel{2!}}9y^2-\frac{4\cdot\cancel{3!}}{\cancel{3!}\cdot1}27y+81$$

$$=y^4-4\left(3y^3\right)+6\left(9y^2\right)-4(27y)+81=y^4-12y^3+54y^2-108y+81$$

21. Applying the Binomial Theorem to $\left(2x^3-1\right)^4$, we have $a=2x^3$, $b=-1$, and $n=4$.

$$\left(2x^3-1\right)^4=\binom{4}{0}\left(2x^3\right)^4+\binom{4}{1}\left(2x^3\right)^3(-1)+\binom{4}{2}\left(2x^3\right)^2(-1)^2+\binom{4}{3}\left(2x^3\right)(-1)^3+\binom{4}{4}(-1)^4$$

$$=\frac{4!}{0!(4-0)!}16x^{12}-\frac{4!}{1!(4-1)!}8x^9+\frac{4!}{2!(4-2)!}4x^6-\frac{4!}{3!(4-3)!}2x^3+\frac{4!}{4!(4-4)!}$$

$$=\frac{\cancel{4!}}{0!\,\cancel{4!}}16x^{12}-\frac{4!}{1!3!}8x^9+\frac{4!}{2!2!}4x^6-\frac{4!}{3!1!}2x^3+\frac{\cancel{4!}}{\cancel{4!}0!}$$

$$=1\left(16x^{12}\right)-\frac{4\cdot\cancel{3!}}{1\cdot\cancel{3!}}8x^9+\frac{4\cdot3\cdot\cancel{2!}}{2\cdot1\cdot\cancel{2!}}4x^6-\frac{4\cdot\cancel{3!}}{\cancel{3!}\cdot1}2x^3+1$$

$$=16x^{12}-4\left(8x^9\right)+6\left(4x^6\right)-4\left(2x^3\right)+1=16x^{12}-32x^9+24x^6-8x^3+1$$

23. Applying the Binomial Theorem to $(c+2)^5$, we have $a = c$, $b = 2$, and $n = 5$.

$$(c+2)^5 = \binom{5}{0}c^5 + \binom{5}{1}c^4(2) + \binom{5}{2}c^3(2)^2 + \binom{5}{3}c^2(2)^3 + \binom{5}{4}c(2)^4 + \binom{5}{5}2^5$$

$$= \frac{5!}{0!\,5!}c^5 + \frac{5!}{1!(5-1)!}2c^4 + \frac{5!}{2!(5-2)!}4c^3 + \frac{5!}{3!(5-3)!}8c^2 + \frac{5!}{4!(5-4)!}16c + \frac{5!}{5!(5-5)!}32$$

$$= 1c^5 + \frac{5\cdot 4!}{1\cdot 4!}2c^4 + \frac{5\cdot 4\cdot 3!}{2\cdot 1\cdot 3!}4c^3 + \frac{5\cdot 4\cdot 3!}{3!2\cdot 1}8c^2 + \frac{5\cdot 4!}{4!\cdot 1}16c + \frac{5!}{5!0!}32$$

$$= c^5 + 5\left(2c^4\right) + 10\left(4c^3\right) + 10\left(8c^2\right) + 5\left(16c\right) + 1(32) = c^5 + 10c^4 + 40c^3 + 80c^2 + 80c + 32$$

25. Applying the Binomial Theorem to $(x-1)^5$, we have $a = x$, $b = -1$, and $n = 5$.

$$(x-1)^5 = \binom{5}{0}x^5 + \binom{5}{1}x^4(-1) + \binom{5}{2}x^3(-1)^2 + \binom{5}{3}x^2(-1)^3 + \binom{5}{4}x(-1)^4 + \binom{5}{5}(-1)^5$$

$$= \frac{5!}{0!\,5!}x^5 - \frac{5!}{1!(5-1)!}x^4 + \frac{5!}{2!(5-2)!}x^3 - \frac{5!}{3!(5-3)!}x^2 + \frac{5!}{4!(5-4)!}x - \frac{5!}{5!(5-5)!}$$

$$= 1x^5 - \frac{5\cdot 4!}{1\cdot 4!}x^4 + \frac{5\cdot 4\cdot 3!}{2\cdot 1\cdot 3!}x^3 - \frac{5\cdot 4\cdot 3!}{3!2\cdot 1}x^2 + \frac{5\cdot 4!}{4!\cdot 1}x - \frac{5!}{5!0!}$$

$$= x^5 - 5x^4 + 10x^3 - 10x^2 + 5x - 1$$

27. Applying the Binomial Theorem to $(3x-y)^5$, we have $a = 3x$, $b = -y$, and $n = 5$.

$$(3x-y)^5$$

$$= \binom{5}{0}(3x)^5 + \binom{5}{1}(3x)^4(-y) + \binom{5}{2}(3x)^3(-y)^2 + \binom{5}{3}(3x)^2(-y)^3 + \binom{5}{4}3x(-y)^4 + \binom{5}{5}(-y)^5$$

$$= \frac{5!}{0!\,5!}243x^5 - \frac{5!}{1!(5-1)!}81x^4y + \frac{5!}{2!(5-2)!}27x^3y^2$$

$$\qquad\qquad - \frac{5!}{3!(5-3)!}9x^2y^3 + \frac{5!}{4!(5-4)!}3xy^4 - \frac{5!}{5!(5-5)!}y^5$$

$$= 243x^5 - \frac{5\cdot 4!}{1\cdot 4!}81x^4y + \frac{5\cdot 4\cdot 3!}{2\cdot 1\cdot 3!}27x^3y^2 - \frac{5\cdot 4\cdot 3!}{3!2\cdot 1}9x^2y^3 + \frac{5\cdot 4!}{4!\cdot 1}3xy^4 - \frac{5!}{5!0!}y^5$$

$$= 243x^5 - 5\left(81x^4y\right) + 10\left(27x^3y^2\right) - 10\left(9x^2y^3\right) + 5\left(3xy^4\right) - y^5$$

$$= 243x^5 - 405x^4y + 270x^3y^2 - 90x^2y^3 + 15xy^4 - y^5$$

29. Applying the Binomial Theorem to $(2a+b)^6$, we have $a=2a$, $b=b$, and $n=6$.

$(2a+b)^6$

$= \binom{6}{0}(2a)^6 + \binom{6}{1}(2a)^5 b + \binom{6}{2}(2a)^4 b^2 + \binom{6}{3}(2a)^3 b^3 + \binom{6}{4}(2a)^2 b^4 + \binom{6}{5}2ab^5 + \binom{6}{6}b^6$

$= \dfrac{6!}{0!(6-0)!}64a^6 + \dfrac{6!}{1!(6-1)!}32a^5 b + \dfrac{6!}{2!(6-2)!}16a^4 b^2 + \dfrac{6!}{3!(6-3)!}8a^3 b^3$

$\qquad + \dfrac{6!}{4!(6-4)!}4a^2 b^4 + \dfrac{6!}{5!(6-5)!}2ab^5 + \dfrac{6!}{6!(6-6)!}b^6$

$= \dfrac{\cancel{6!}}{1\cancel{6!}}64a^6 + \dfrac{6\cdot\cancel{5!}}{1\cdot\cancel{5!}}32a^5 b + \dfrac{6\cdot 5\cdot\cancel{4!}}{2\cdot 1\cdot\cancel{4!}}16a^4 b^2 + \dfrac{\cancel{6}\cdot 5\cdot 4\cdot\cancel{3!}}{3\cancel{2}\cdot 1\cdot\cancel{3!}}8a^3 b^3$

$\qquad + \dfrac{6\cdot 5\cdot\cancel{4!}}{\cancel{4!}2\cdot 1}4a^2 b^4 + \dfrac{6\cdot\cancel{5!}}{\cancel{5!}1}2ab^5 + \dfrac{\cancel{6!}}{\cancel{6!}\cdot 1}b^6$

$= 64a^6 + 6(32a^5 b) + 15(16a^4 b^2) + 20(8a^3 b^3) + 15(4a^2 b^4) + 6(2ab^5) + 1b^6$

$= 64a^6 + 192a^5 b + 240a^4 b^2 + 160a^3 b^3 + 60a^2 b^4 + 12ab^5 + b^6$

31. $(x+2)^8$

First Term $(r=0)$: $\binom{n}{r}a^{n-r}b^r = \binom{8}{0}x^{8-0}2^0 = \dfrac{8!}{0!(8-0)!}x^8 \cdot 1 = \dfrac{\cancel{8!}}{0!\cancel{8!}}x^8 = x^8$

Second Term $(r=1)$: $\binom{n}{r}a^{n-r}b^r = \binom{8}{1}x^{8-1}2^1 = \dfrac{8!}{1!(8-1)!}2x^7 = \dfrac{8\cdot\cancel{7!}}{1\cdot\cancel{7!}}2x^7 = 8\cdot 2x^7 = 16x^7$

Third Term $(r=2)$: $\binom{n}{r}a^{n-r}b^r = \binom{8}{2}x^{8-2}2^2 = \dfrac{8!}{2!(8-2)!}4x^6 = \dfrac{8\cdot 7\cdot\cancel{6!}}{2\cdot 1\cdot\cancel{6!}}4x^6 = 28\cdot 4x^6 = 112x^6$

33. $(x-2y)^{10}$

First Term $(r=0)$: $\binom{n}{r}a^{n-r}b^r = \binom{10}{0}x^{10-0}(-2y)^0 = \dfrac{10!}{0!(10-0)!}x^{10}\cdot 1 = \dfrac{\cancel{10!}}{0!\cancel{10!}}x^{10} = x^{10}$

Second Term $(r=1)$: $\binom{n}{r}a^{n-r}b^r = \binom{10}{1}x^{10-1}(-2y)^1 = -\dfrac{10!}{1!(10-1)!}2x^9 y = -\dfrac{10\cdot\cancel{9!}}{1\cdot\cancel{9!}}2x^9 y = -10\cdot 2x^9 y = -20x^9 y$

Third Term $(r=2)$: $\binom{n}{r}a^{n-r}b^r = \binom{10}{2}x^{10-2}(-2y)^2 = \dfrac{10!}{2!(10-2)!}4x^8 y^2 = \dfrac{10\cdot 9\cdot\cancel{8!}}{2\cdot 1\cdot\cancel{8!}}4x^8 y^2 = 45\cdot 4x^8 y^2 = 180x^8 y^2$

35. $\left(x^2+1\right)^{16}$

First Term $(r=0)$: $\binom{n}{r}a^{n-r}b^r = \binom{16}{0}\left(x^2\right)^{16-0}1^0 = \dfrac{16!}{0!(16-0)!}x^{32}\cdot 1 = \dfrac{16!}{0!\,16!}x^{32} = x^{32}$

Second Term $(r=1)$: $\binom{n}{r}a^{n-r}b^r = \binom{16}{1}\left(x^2\right)^{16-1}1^1 = \dfrac{16!}{1!(16-1)!}x^{30}\cdot 1 = \dfrac{16\cdot 15!}{1\cdot 15!}x^{30} = 16x^{30}$

Third Term $(r=2)$: $\binom{n}{r}a^{n-r}b^r = \binom{16}{2}\left(x^2\right)^{16-2}1^2 = \dfrac{16!}{2!(16-2)!}x^{28}\cdot 1 = \dfrac{16\cdot 15\cdot 14!}{2\cdot 1\cdot 14!}x^{28} = 120x^{28}$

37. $\left(y^3-1\right)^{20}$

First Term $(r=0)$: $\binom{n}{r}a^{n-r}b^r = \binom{20}{0}\left(y^3\right)^{20-0}(-1)^0 = \dfrac{20!}{0!(20-0)!}y^{60}\cdot 1 = \dfrac{20!}{0!\,20!}y^{60} = y^{60}$

Second Term $(r=1)$: $\binom{n}{r}a^{n-r}b^r = \binom{20}{1}\left(y^3\right)^{20-1}(-1)^1 = \dfrac{20!}{1!(20-1)!}y^{57}\cdot(-1) = -\dfrac{20\cdot 19!}{1\cdot 19!}y^{57} = -20y^{57}$

Third Term $(r=2)$: $\binom{n}{r}a^{n-r}b^r = \binom{20}{2}\left(y^3\right)^{20-2}(-1)^2 = \dfrac{20!}{2!(20-2)!}y^{54}\cdot 1 = \dfrac{20\cdot 19\cdot 18!}{2\cdot 1\cdot 18!}y^{54} = 190y^{54}$

39. $\left(2x+y\right)^6$

Third Term $(r=2)$: $\binom{n}{r}a^{n-r}b^r = \binom{6}{2}(2x)^{6-2}y^2 = \dfrac{6!}{2!(6-2)!}16x^4y^2 = \dfrac{6\cdot 5\cdot 4!}{2\cdot 1\cdot 4!}16x^4y^2 = 15\left(16x^4y^2\right) = 240x^4y^2$

41. $\left(x-1\right)^9$

Fifth Term $(r=4)$: $\binom{n}{r}a^{n-r}b^r = \binom{9}{4}x^{9-4}(-1)^4 = \dfrac{9!}{4!(9-4)!}x^5\cdot 1 = \dfrac{9\cdot 8\cdot 7\cdot 6\cdot 5!}{4\cdot 3\cdot 2\cdot 1\cdot 5!}x^5 = 126x^5$

43. $\left(x^2+y^3\right)^8$

Sixth Term $(r=5)$: $\binom{n}{r}a^{n-r}b^r = \binom{8}{5}\left(x^2\right)^{8-5}\left(y^3\right)^5 = \dfrac{8!}{5!(8-5)!}x^6y^{15} = \dfrac{8\cdot 7\cdot 6\cdot 5!}{5!\cdot 3\cdot 2\cdot 1}x^6y^{15} = 56x^6y^{15}$

45. $\left(x-\dfrac{1}{2}\right)^9$

Fourth Term $(r=3)$: $\binom{n}{r}a^{n-r}b^r = \binom{9}{3}x^{9-3}\left(-\dfrac{1}{2}\right)^3 = -\dfrac{9!}{3!(9-3)!}\cdot\dfrac{1}{8}x^6 = -\dfrac{9\cdot 8\cdot 7\cdot 6!}{3\cdot 2\cdot 1\cdot 6!}\cdot\dfrac{1}{8}x^6 = -\dfrac{21}{2}x^6$

47. $\left(x^2+y\right)^{22}$

y^{14} will occur in the fifteenth term

Fifteenth Term $(r=14)$: $\dbinom{n}{r}a^{n-r}b^r = \dbinom{22}{14}\left(x^2\right)^{22-14}(y)^{14} = \dfrac{22!}{14!(22-14)!}\left(x^2\right)^{8}y^{14} = 319,770x^{16}y^{14}$

49. $\left(x^3+x^{-2}\right)^4 = \dbinom{4}{0}\left(x^3\right)^4 + \dbinom{4}{1}\left(x^3\right)^3\left(x^{-2}\right) + \dbinom{4}{2}\left(x^3\right)^2\left(x^{-2}\right)^2 + \dbinom{4}{3}\left(x^3\right)^1\left(x^{-2}\right)^3 + \dbinom{4}{4}\left(x^{-2}\right)^4$

$= \dfrac{4!}{0!(4-0)!}x^{12} + \dfrac{4!}{1!(4-1)!}x^9x^{-2} + \dfrac{4!}{2!(4-2)!}x^6x^{-4} + \dfrac{4!}{3!(4-3)!}x^3x^{-6} + \dfrac{4!}{4!(4-4)!}x^{-8}$

$= \dfrac{\cancel{4!}}{0!\cdot\cancel{4!}}x^{12} + \dfrac{4\cdot\cancel{3!}}{1!\cdot\cancel{3!}}x^7 + \dfrac{4\cdot3\cdot\cancel{2!}}{2\cdot1\cdot\cancel{2!}}x^2 + \dfrac{4\cdot\cancel{3!}}{\cancel{3!}\cdot1!}x^{-3} + \dfrac{\cancel{4!}}{\cancel{4!}\cdot0!}x^{-8}$

$= x^{12} + 4x^7 + 6x^2 + \dfrac{4}{x^3} + \dfrac{1}{x^8}$

51. $\left(x^{\frac{1}{3}} - x^{-\frac{1}{3}}\right)^3 = \left(x^{\frac{1}{3}} + \left(-x^{-\frac{1}{3}}\right)\right)^3$

$= \dbinom{3}{0}\left(x^{\frac{1}{3}}\right)^3 + \dbinom{3}{1}\left(x^{\frac{1}{3}}\right)^2\left(-x^{-\frac{1}{3}}\right) + \dbinom{3}{2}\left(x^{\frac{1}{3}}\right)^1\left(-x^{-\frac{1}{3}}\right)^2 + \dbinom{3}{3}\left(-x^{-\frac{1}{3}}\right)^3$

$= \dfrac{3!}{0!(3-0)!}x + \dfrac{3!}{1!(3-1)!}x^{\frac{2}{3}}\left(-x^{-\frac{1}{3}}\right) + \dfrac{3!}{2!(3-2)!}x^{\frac{1}{3}}x^{-\frac{2}{3}} + \dfrac{3!}{3!(3-3)!}\left(-x^{-1}\right)$

$= x - 3x^{\frac{1}{3}} + 3x^{-\frac{1}{3}} - x^{-1}$

$= x - 3x^{\frac{1}{3}} + \dfrac{3}{x^{\frac{1}{3}}} - \dfrac{1}{x}$

53. $\left(-1+i\sqrt3\right)^3 = \dbinom{3}{0}(-1)^3 + \dbinom{3}{1}(-1)^2\left(i\sqrt3\right) + \dbinom{3}{2}(-1)^1\left(i\sqrt3\right)^2 + \dbinom{3}{3}\left(i\sqrt3\right)^3$

$= \dfrac{3!}{0!(3-0)!}\cdot-1 + \dfrac{3!}{1!(3-1)!}\cdot1\cdot i\sqrt3 + \dfrac{3!}{2!(3-2)!}\cdot-1\cdot-3 + \dfrac{3!}{3!(3-3)!}\cdot-3i\sqrt3$

$= \dfrac{\cancel{3!}}{0!\cancel{3!}}(-1) + \dfrac{3\cdot\cancel{2!}}{1!\cancel{2!}}i\sqrt3 + \dfrac{3\cdot\cancel{2!}}{\cancel{2!}1!}\cdot3 + \dfrac{\cancel{3!}}{\cancel{3!}0!}\cdot-3i\sqrt3$

$= -1 + 3i\sqrt3 + 9 - 3i\sqrt3$

$= -1 + 9 + 3i\sqrt3 - 3i\sqrt3$

$= 8$

55. $f(x) = x^4 + 7$

$$\frac{f(x+h) - f(x)}{h} = \frac{(x+h)^4 + 7 - \left(x^4 + 7\right)}{h} = \frac{\binom{4}{0}x^4 + \binom{4}{1}x^3h + \binom{4}{2}x^2h^2 + \binom{4}{3}xh^3 + \binom{4}{4}h^4 + 7 - x^4 - 7}{h}$$

$$= \frac{\dfrac{4!}{0!(4-0)!}x^4 + \dfrac{4!}{1!(4-1)!}x^3h + \dfrac{4!}{2!(4-2)!}x^2h^2 + \dfrac{4!}{3!(4-3)!}xh^3 + \dfrac{4!}{4!(4-4)!}h^4 - x^4}{h}$$

$$= \frac{\dfrac{\cancel{4!}}{0!\cancel{4!}}x^4 + \dfrac{4 \cdot \cancel{3!}}{1!\cancel{3!}}x^3h + \dfrac{4 \cdot 3 \cdot \cancel{2!}}{\cancel{2!} \cdot 2 \cdot 1}x^2h^2 + \dfrac{4 \cdot \cancel{3!}}{\cancel{3!}1!}xh^3 + \dfrac{\cancel{4!}}{\cancel{4!}0!}h^4 - x^4}{h}$$

$$= \frac{\cancel{x^4} + 4x^3h + 6x^2h^2 + 4xh^3 + h^4 - \cancel{x^4}}{h} = \frac{h(4x^3 + 6x^2h + 4xh^2 + h^3)}{h}$$

$$= 4x^3 + 6x^2h + 4xh^2 + h^3$$

57. We want to find the $(5+1) = 6^{th}$ term.

$$\binom{n}{r}a^{n-r}b^r = \binom{10}{5}\left(\frac{3}{x}\right)^{10-5}\left(\frac{x}{3}\right)^5 = \frac{10!}{5!(10-5)!}\left(\frac{3}{x}\right)^5\left(\frac{x}{3}\right)^5$$

$$= \frac{10 \cdot 9 \cdot 8 \cdot 7 \cdot 6 \cdot \cancel{5!}}{\cancel{5!} \cdot \cancel{5} \cdot 4 \cdot 3 \cdot \cancel{2} \cdot 1}\left(\frac{3}{x}\right)^5\left(\frac{x}{3}\right)^5 = 252 \cdot \frac{3^5}{x^5} \cdot \frac{x^5}{3^5} = 252$$

59. $(0.28 + 0.72)^5$

Third Term $(r = 2)$: $\binom{n}{r}a^{n-r}b^r = \binom{5}{2}0.28^{5-2}0.72^2 = \frac{5!}{2!(5-2)!}0.28^{5-2}0.72^2 = \frac{5!}{2!3!}0.28^30.72^2 \approx 0.1138$

61. – 67. Answers will vary.

69. $f_1(x) = (x+2)^3$

$f_2(x) = x^3$

$f_3(x) = x^3 + 6x^2$

$f_4(x) = x^3 + 6x^2 + 12x$

$f_5(x) = x^3 + 6x^2 + 12x + 8$

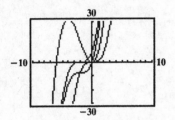

Graphs f_1 and f_5 are the same. This means that the functions are equivalent. Graphs f_2 through f_4 are increasingly similar to the graphs of f_1 and f_5.

71. Applying the Binomial Theorem to $(x-1)^3$, we have $a = x$, $b = -1$, and $n = 3$.

$$(x-1)^3 = \binom{3}{0}x^3 + \binom{3}{1}x^2(-1) + \binom{3}{2}x(-1)^2 + \binom{3}{3}(-1)^3$$

$$= \frac{3!}{0!(3-0)!}x^3 - \frac{3!}{1!(3-1)!}x^2 + \frac{3!}{2!(3-2)!}x(1) - \frac{3!}{3!(3-3)!}$$

$$= \frac{3!}{1 \cdot 3!}x^3 - \frac{3 \cdot 2!}{1 \cdot 2!}x^2 + \frac{3 \cdot 2!}{2!1!}x - \frac{3!}{3!0!} = x^3 - 3x^2 + 3x - 1$$

Graph using the method from Exercises 69 and 70.

$f_1(x) = (x-1)^3$ $f_2(x) = x^3$

$f_3(x) = x^3 + 3x^2$ $f_4(x) = x^3 + 3x^2 + 3x$

$f_5(x) = x^3 - 3x^2 + 3x - 1$

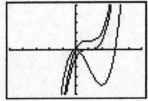

Graphs f_1 and f_5 are the same. This means that the functions are equivalent. Graphs f_2 through f_4 are increasingly similar to the graphs of f_1 and f_5.

73. Applying the Binomial Theorem to $(x+2)^6$, we have $a = x$, $b = 2$, and $n = 6$.

$$(x+2)^6 = \binom{6}{0}x^6 + \binom{6}{1}x^5 2 + \binom{6}{2}x^4 2^2 + \binom{6}{3}x^3 2^3 + \binom{6}{4}x^2 2^4 + \binom{6}{5}x 2^5 + \binom{6}{6}2^6$$

$$= \frac{6!}{0!(6-0)!}x^6 + \frac{6!}{1!(6-1)!}2x^5 + \frac{6!}{2!(6-2)!}4x^4 + \frac{6!}{3!(6-3)!}8x^3 + \frac{6!}{4!(6-4)!}16x^2$$

$$+ \frac{6!}{5!(6-5)!}32x + \frac{6!}{6!(6-6)!}64$$

$$= \frac{6!}{1 6!}x^6 + \frac{6 \cdot 5!}{1 \cdot 5!}2x^5 + \frac{6 \cdot 5 \cdot 4!}{2 \cdot 1 \cdot 4!}4x^4 + \frac{6 \cdot 5 \cdot 4 \cdot 3!}{3 \cdot 2 \cdot 1 \cdot 3!}8x^3 + \frac{6 \cdot 5 \cdot 4!}{4! 2 \cdot 1}16x^2 + \frac{6 \cdot 5!}{5! 1}32x + \frac{6!}{6! \cdot 1}64$$

$$= x^6 + 6\left(2x^5\right) + 15\left(4x^4\right) + 20\left(8x^3\right) + 15\left(16x^2\right) + 6(32x) + 1(64)$$

$$= x^6 + 12x^5 + 60x^4 + 160x^3 + 240x^2 + 192x + 64$$

Graph using the method from Exercises 69 and 70.

$$f_1(x) = (x+2)^6$$

$$f_2(x) = x^6$$

$$f_3(x) = x^6 + 12x^5$$

$$f_4(x) = x^6 + 12x^5 + 60x^4$$

$$f_5(x) = x^6 + 12x^5 + 60x^4 + 160x^3$$

$$f_6(x) = x^6 + 12x^5 + 60x^4 + 160x^3 + 240x^2$$

$$f_7(x) = x^6 + 12x^5 + 60x^4 + 160x^3 + 240x^2 + 192x$$

$$f_8(x) = x^6 + 12x^5 + 60x^4 + 160x^3 + 240x^2 + 192x + 64$$

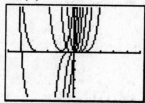

Graphs f_1 and f_8 are the same. This means that the functions are equivalent. Graphs f_2 through f_7 are increasingly similar to the graphs of f_1 and f_8.

75. makes sense

77. does not make sense; Explanations will vary. Sample explanation: $7 \neq 2 + 4$

79. true

81. false; Changes to make the statement true will vary. A sample change is: There are values of a and b for which $(a+b)^4 = a^4 + b^4$. Consider $a = 0$ and $b = 1$. $(0+1)^4 = 0^4 + 1^4$

$$(1)^4 = 0 + 1$$

$$1 = 1$$

83. In $\left(x^2 + y^2\right)^5$, the term containing x^4 is the term in which $a = x^2$ is squared. Applying the Binomial Theorem, the following pattern results. In the first term, x^2 is taken to the fifth power. In the second term, x^2 is taken to the fourth power. In the third term x^2 is taken to the third power. In the fourth term, x^2 is taken to the second power. This is the term we are looking for. Applying the Binomial Theorem to $\left(x^2 + y^2\right)^5$, we have $a = x^2$, $b = y^2$, and $n = 5$. We are looking for the 4th term where $r + 1 = 4$ so $r = 3$.

$$\binom{n}{r} a^{n-r} b^r = \binom{5}{3}\left(x^2\right)^2 \left(y^2\right)^3$$

$$= \frac{5!}{3!(5-3)!} x^4 y^6$$

$$= \frac{5!}{3!2!} x^4 y^6$$

$$= \frac{5 \cdot 4 \cdot \cancel{3!}}{\cancel{3!}2 \cdot 1} x^4 y^6$$

$$= 10x^4 y^6$$

84. $f(a+1) = (a+1)^2 + 2(a+1) + 3$

$\qquad = a^2 + 2a + 1 + 2a + 2 + 3$

$\qquad = a^2 + 4a + 6$

85. $f(x) = x^2 + 5x \qquad g(x) = 2x - 3$

$f(g(x)) = f(2x - 3)$

$\qquad = (2x - 3)^2 + 5(2x - 3)$

$\qquad = 4x^2 - 12x + 9 + 10x - 15$

$\qquad = 4x^2 - 2x - 6$

$g(f(x)) = g\left(x^2 + 5x\right)$

$\qquad = 2\left(x^2 + 5x\right) - 3$

$\qquad = 2x^2 + 10x - 3$

86.

$$\frac{x}{x+3}-\frac{x+1}{2x^2-2x-24}$$

$$=\frac{x}{x+3}-\frac{x+1}{2\left(x^2-x-12\right)}$$

$$=\frac{x}{x+3}-\frac{x+1}{2(x-4)(x+3)}$$

$$=\frac{x}{x+3}\cdot\frac{2(x-4)}{2(x-4)}-\frac{x+1}{2(x-4)(x+3)}$$

$$=\frac{x}{x+3}\cdot\frac{2x-8}{2(x-4)}-\frac{x+1}{2(x-4)(x+3)}$$

$$=\frac{2x^2-8x}{2(x-4)(x+3)}-\frac{x+1}{2(x-4)(x+3)}$$

$$=\frac{2x^2-8x-(x+1)}{2(x-4)(x+3)}$$

$$=\frac{2x^2-8x-x-1}{2(x-4)(x+3)}$$

$$=\frac{2x^2-9x-1}{2(x-4)(x+3)}$$

Chapter 14 Review Exercises

1. $a_n=7n-4$

$a_1=7(1)-4=7-4=3$

$a_2=7(2)-4=14-4=10$

$a_3=7(3)-4=21-4=17$

$a_4=7(4)-4=28-4=24$

The first four terms are 3, 10, 17, 24.

2. $a_n=(-1)^n\dfrac{n+2}{n+1}$

$a_1=(-1)^1\dfrac{1+2}{1+1}=-\dfrac{3}{2}$

$a_2=(-1)^2\dfrac{2+2}{2+1}=\dfrac{4}{3}$

$a_3=(-1)^3\dfrac{3+2}{3+1}=-\dfrac{5}{4}$

$a_4=(-1)^4\dfrac{4+2}{4+1}=\dfrac{6}{5}$

The first four terms are $-\dfrac{3}{2},\dfrac{4}{3},-\dfrac{5}{4},\dfrac{6}{5}$.

3. $a_n = \dfrac{1}{(n-1)!}$

$a_1 = \dfrac{1}{(1-1)!} = \dfrac{1}{0!} = \dfrac{1}{1} = 1$

$a_2 = \dfrac{1}{(2-1)!} = \dfrac{1}{1!} = \dfrac{1}{1} = 1$

$a_3 = \dfrac{1}{(3-1)!} = \dfrac{1}{2!} = \dfrac{1}{2\cdot 1} = \dfrac{1}{2}$

$a_4 = \dfrac{1}{(4-1)!} = \dfrac{1}{3!} = \dfrac{1}{3\cdot 2\cdot 1} = \dfrac{1}{6}$

The first four terms are $1, 1, \dfrac{1}{2}, \dfrac{1}{6}$.

4. $a_n = \dfrac{(-1)^{n+1}}{2^n}$

$a_1 = \dfrac{(-1)^{1+1}}{2^1} = \dfrac{(-1)^2}{2} = \dfrac{1}{2}$

$a_2 = \dfrac{(-1)^{2+1}}{2^2} = \dfrac{(-1)^3}{4} = -\dfrac{1}{4}$

$a_3 = \dfrac{(-1)^{3+1}}{2^3} = \dfrac{(-1)^4}{8} = \dfrac{1}{8}$

$a_4 = \dfrac{(-1)^{4+1}}{2^4} = \dfrac{(-1)^5}{16} = -\dfrac{1}{16}$

The first four terms are $\dfrac{1}{2}, -\dfrac{1}{4}, \dfrac{1}{8}, -\dfrac{1}{16}$.

5. $\displaystyle\sum_{i=1}^{5}\left(2i^2 - 3\right) = \left(2(1)^2 - 3\right) + \left(2(2)^2 - 3\right) + \left(2(3)^2 - 3\right) + \left(2(4)^2 - 3\right) + \left(2(5)^2 - 3\right) + \left(2(6)^2 - 3\right)$

$= \left(2(1) - 3\right) + \left(2(4) - 3\right) + \left(2(9) - 3\right) + \left(2(16) - 3\right) + \left(2(25) - 3\right)$

$= (2-3) + (8-3) + (18-3) + (32-3) + (50-3) = -1 + 5 + 15 + 29 + 47 = 95$

6. $\displaystyle\sum_{i=0}^{4}(-1)^{i+1}\, i! = (-1)^{0+1}\, 0! + (-1)^{1+1}\, 1! + (-1)^{2+1}\, 2! + (-1)^{3+1}\, 3! + (-1)^{4+1}\, 4!$

$= (-1)^1\, 1 + (-1)^2\, 1 + (-1)^3\, 2\cdot 1 + (-1)^4\, 3\cdot 2\cdot 1 + (-1)^5\, 4\cdot 3\cdot 2\cdot 1$

$= -1 + 1 - 2 + 6 - 24 = -20$

7. $\dfrac{1}{3} + \dfrac{2}{4} + \dfrac{3}{5} + \ldots + \dfrac{15}{17} = \displaystyle\sum_{i=1}^{15}\dfrac{i}{i+2}$

8. $4^3 + 5^3 + 6^3 + \ldots + 13^3 = \sum_{i=4}^{13} i^3$ or $\sum_{i=1}^{10} (i+3)^3$

9. $a_1 = 7$

$a_2 = 7 + 4 = 11$

$a_3 = 11 + 4 = 15$

$a_4 = 15 + 4 = 19$

$a_5 = 19 + 4 = 23$

$a_6 = 23 + 4 = 27$

The first six terms are $7, 11, 15, 19, 23, 27$

10. $a_1 = -4$

$a_2 = -4 - 5 = -9$

$a_3 = -9 - 5 = -14$

$a_4 = -14 - 5 = -19$

$a_5 = -19 - 5 = -24$

$a_6 = -24 - 5 = -29$

The first six terms are $-4, -9, -14, -19, -24, -29$.

11. $a_1 = \dfrac{3}{2}$

$a_2 = \dfrac{3}{2} - \dfrac{1}{2} = \dfrac{2}{2} = 1$

$a_3 = 1 - \dfrac{1}{2} = \dfrac{1}{2}$

$a_4 = \dfrac{1}{2} - \dfrac{1}{2} = 0$

$a_5 = 0 - \dfrac{1}{2} = -\dfrac{1}{2}$

$a_6 = -\dfrac{1}{2} - \dfrac{1}{2} = -\dfrac{2}{2} = -1$

The first six terms are $\dfrac{3}{2}, 1, \dfrac{1}{2}, 0, -\dfrac{1}{2}, -1$.

12. $a_6 = 5 + (6 - 1)3 = 5 + (5)3 = 5 + 15 = 20$

13. $a_{12} = -8 + (12 - 1)(-2) = -8 + 11(-2) = -8 + (-22) = -30$

14. $a_{14} = 14 + (14 - 1)(-4) = 14 + 13(-4) = 14 + (-52) = -38$

15. $d = -3 - (-7) = 4$

$a_n = -7 + (n-1)4 = -7 + 4n - 4$

$ = 4n - 11$

$a_{20} = 4(20) - 11 = 80 - 11 = 69$

16. $a_n = 200 + (n-1)(-20)$

$ = 200 - 20n + 20$

$ = 220 - 20n$

$a_{20} = 220 - 20(20)$

$\phantom{a_{20}} = 220 - 400 = -180$

17. $a_n = -12 + (n-1)\left(-\dfrac{1}{2}\right)$

$ = -12 - \dfrac{1}{2}n + \dfrac{1}{2}$

$ = -\dfrac{24}{2} - \dfrac{1}{2}n + \dfrac{1}{2}$

$ = -\dfrac{1}{2}n - \dfrac{23}{2}$

$a_{20} = -\dfrac{1}{2}(20) - \dfrac{23}{2}$

$\phantom{a_{20}} = -\dfrac{20}{2} - \dfrac{23}{2} = -\dfrac{43}{2}$

18. $d = 8 - 15 = -7$

$a_n = 15 + (n-1)(-7) = 15 - 7n + 7$

$ = 22 - 7n$

$a_{20} = 22 - 7(20) = 22 - 140 = -118$

19. First, find d.

$d = 12 - 5 = 7$

Next, find a_{22}.

$a_{22} = 5 + (22 - 1)7 = 5 + (21)7$

$\phantom{a_{22}} = 5 + 147 = 152$

Now, find the sum.

$S_{22} = \dfrac{22}{2}(5 + 152) = 11(157) = 1727$

20. First, find d.

$$d = -3 - (-6) = 3$$

Next, find a_{15}.

$$a_{15} = -6 + (15 - 1)3 = -6 + (14)3$$
$$= -6 + 42 = 36$$

Now, find the sum.

$$S_{15} = \frac{15}{2}(-6 + 36) = \frac{15}{2}(30) = 225$$

21. We are given that $a_{100} = 300$, $a_1 = 3$, and $n = 100$.

$$S_{100} = \frac{100}{2}(3 + 300)$$
$$= 50(303)$$
$$= 15,150$$

22. $\displaystyle\sum_{i=1}^{16}(3i + 2) = (3(1) + 2) + (3(2) + 2) + (3(3) + 2) + \ldots + (3(16) + 2)$

$$= (3 + 2) + (6 + 2) + (9 + 2) + \ldots + (48 + 2)$$
$$= 5 + 8 + 11 + \ldots + 50$$

$$S_{16} = \frac{16}{2}(5 + 50) = 8(55) = 440$$

23. $\displaystyle\sum_{i=1}^{25}(-2i + 6) = (-2(1) + 6) + (-2(2) + 6) + (-2(3) + 6) + \ldots + (-2(25) + 6)$

$$= (-2 + 6) + (-4 + 6) + (-6 + 6) + \ldots + (-50 + 6)$$
$$= 4 + 2 + 0 + \ldots + (-44)$$

$$S_{25} = \frac{25}{2}(4 + (-44)) = \frac{25}{2}(-40) = -500$$

24. $\displaystyle\sum_{i=1}^{30}(-5i) = (-5(1)) + (-5(2)) + (-5(3)) + \ldots + (-5(30))$

$$= -5 + (-10) + (-15) + \ldots + (-150)$$

$$S_{30} = \frac{30}{2}(-5 + (-150)) = 15(-155) = -2325$$

25. a. $a_n = 39 + (n-1)(4.75)$

$\qquad = 39 + 4.75n - 4.75$

$\qquad = 4.75n + 34.25$

b. $a_n = 4.75n + 34.25$

$a_{12} = 4.75(13) + 34.25$

$\qquad = 96$

The percentage of students ages $12 - 18$ who will report seeing security cameras at school in the year 2013 will be approximately 96%.

26. $a_{10} = 31500 + (10-1)2300$

$\qquad = 31500 + (9)2300$

$\qquad = 31500 + 20700 = 52200$

$S_{10} = \dfrac{10}{2}(31500 + 52200)$

$\qquad = 5(83700) = 418500$

The total salary over a ten-year period is \$418,500.

27. $a_{35} = 25 + (35-1)1 = 25 + (34)1$

$\qquad = 25 + 34 = 59$

$S_{35} = \dfrac{35}{2}(25 + 59) = \dfrac{35}{2}(84) = 1470$

There are 1470 seats in the theater.

28. The first term is 3.

The second term is $3 \cdot 2 = 6$.

The third term is $6 \cdot 2 = 12$.

The fourth term is $12 \cdot 2 = 24$.

The fifth term is $24 \cdot 2 = 48$.

29. The first term is $\dfrac{1}{2}$.

The second term is $\dfrac{1}{2} \cdot \dfrac{1}{2} = \dfrac{1}{4}$.

The third term is $\dfrac{1}{4} \cdot \dfrac{1}{2} = \dfrac{1}{8}$.

The fourth term is $\dfrac{1}{8} \cdot \dfrac{1}{2} = \dfrac{1}{16}$.

The fifth term is $\dfrac{1}{16} \cdot \dfrac{1}{2} = \dfrac{1}{32}$.

30. The first term is 16.

The second term is $16 \cdot -\dfrac{1}{4} = -4$.

The third term is $-4 \cdot -\dfrac{1}{4} = 1$.

The fourth term is $1 \cdot -\dfrac{1}{4} = -\dfrac{1}{4}$.

The fifth term is $-\dfrac{1}{4} \cdot -\dfrac{1}{4} = \dfrac{1}{16}$.

31. The first term is –5.

The second term is $-5 \cdot -1 = 5$.

The third term is $5 \cdot -1 = -5$.

The fourth term is $-5 \cdot -1 = 5$.

The fifth term is $5 \cdot -1 = -5$.

32. $a_7 = 2(3)^{7-1} = 2(3)^6 = 2(729) = 1458$

33. $a_6 = 16\left(\dfrac{1}{2}\right)^{6-1} = 16\left(\dfrac{1}{2}\right)^5 = 16\left(\dfrac{1}{32}\right) = \dfrac{1}{2}$

34. $a_5 = -3(2)^{5-1} = -3(2)^4 = -3(16) = -48$

35. $a_n = a_1 r^{n-1} = 1(2)^{n-1}$

$a_8 = 1(2)^{8-1} = 1(2)^7 = 1(128) = 128$

36. $a_n = a_1 r^{n-1} = 100\left(\dfrac{1}{10}\right)^{n-1}$

$a_8 = 100\left(\dfrac{1}{10}\right)^{8-1} = 100\left(\dfrac{1}{10}\right)^7$

$\qquad = 100\left(\dfrac{1}{10,000,000}\right) = \dfrac{1}{100,000}$

37. $d = \dfrac{-4}{12} = -\dfrac{1}{3}$

$a_n = a_1 r^{n-1} = 12\left(-\dfrac{1}{3}\right)^{n-1}$

$a_8 = 12\left(-\dfrac{1}{3}\right)^{8-1} = 12\left(-\dfrac{1}{3}\right)^7$

$\qquad = 12\left(-\dfrac{1}{2187}\right) = -\dfrac{12}{2187} = -\dfrac{4}{729}$

38. $r = \dfrac{a_2}{a_1} = \dfrac{-15}{5} = -3$

$$S_{15} = \frac{5\left(1-(-3)^{15}\right)}{1-(-3)} = \frac{5\left(1-(-14348907)\right)}{4}$$

$$= \frac{5(14348908)}{4} = \frac{71744540}{4}$$

$$= 17,936,135$$

39. $r = \dfrac{a_2}{a_1} = \dfrac{4}{8} = \dfrac{1}{2}$

$$S_7 = \frac{8\left(1-\left(\frac{1}{2}\right)^7\right)}{1-\frac{1}{2}} = \frac{8\left(1-\frac{1}{128}\right)}{\frac{1}{2}}$$

$$= \frac{8\left(\frac{128}{128}-\frac{1}{128}\right)}{\frac{1}{2}} = \frac{8\left(\frac{127}{128}\right)}{\frac{1}{2}}$$

$$= \frac{8}{1}\left(\frac{127}{128}\right) \div \frac{1}{2} = \frac{8}{1}\left(\frac{127}{128}\right) \cdot \frac{2}{1}$$

$$= \frac{2032}{128} = \frac{127}{8} = 15.875$$

40. $\displaystyle\sum_{i=1}^{6} 5^i = \frac{5\left(1-5^6\right)}{1-5} = \frac{5(1-15625)}{-4}$

$$= \frac{5(-15624)}{-4} = 5(3906)$$

$$= 19,530$$

41. $\displaystyle\sum_{i=1}^{7} 3(-2)^i = \frac{-6\left(1-(-2)^7\right)}{1-(-2)} = \frac{-6\left(1-(-128)\right)}{3}$

$$= \frac{-6(129)}{3} = -2(129) = -258$$

42. $\displaystyle\sum_{i=1}^{5} 2\left(\frac{1}{4}\right)^{i-1} = \frac{2\left(1-\left(\frac{1}{4}\right)^5\right)}{1-\left(\frac{1}{4}\right)} = \frac{2\left(1-\frac{1}{1024}\right)}{\frac{3}{4}}$

$$= \frac{2 \cdot \frac{1023}{1024}}{\frac{3}{4}} = \frac{1023}{512} \cdot \frac{4}{3} = \frac{341}{128}$$

43. $r = \dfrac{a_2}{a_1} = \dfrac{3}{9} = \dfrac{1}{3}$

$$S = \frac{9}{1-\frac{1}{3}} = \frac{9}{\frac{2}{3}} = 9 \div \frac{2}{3} = 9 \cdot \frac{3}{2} = \frac{27}{2}$$

44. $r = \dfrac{a_2}{a_1} = \dfrac{-1}{2} = -\dfrac{1}{2}$

$$S = \frac{2}{1-\left(-\frac{1}{2}\right)} = \frac{2}{\frac{3}{2}} = 2 \div \frac{3}{2} = 2 \cdot \frac{2}{3} = \frac{4}{3}$$

45. $r = \dfrac{a_2}{a_1} = \dfrac{4}{-6} = -\dfrac{2}{3}$

$$S = \frac{-6}{1-\left(-\frac{2}{3}\right)} = \frac{-6}{\frac{5}{3}} = -6 \div \frac{5}{3}$$

$$= -6 \cdot \frac{3}{5} = -\frac{18}{5}$$

46. $\displaystyle\sum_{i=1}^{\infty} 5(0.8)^i = \frac{4}{1-0.8} = \frac{4}{0.2} = 20$

47. $0.\overline{6} = \dfrac{a_1}{1-r} = \dfrac{\frac{6}{10}}{1-\frac{1}{10}} = \dfrac{\frac{6}{10}}{\frac{9}{10}} = \dfrac{6}{10} \div \dfrac{9}{10}$

$$= \frac{6}{10} \cdot \frac{10}{9} = \frac{2}{3}$$

48. $0.\overline{47} = \dfrac{a_1}{1-r} = \dfrac{\frac{47}{100}}{1-\frac{1}{100}} = \dfrac{\frac{47}{100}}{\frac{99}{100}}$

$$= \frac{47}{100} \div \frac{99}{100} = \frac{47}{100} \cdot \frac{100}{99} = \frac{47}{99}$$

49. a. Divide each value by the previous value:

$$r_{2000-2010} \approx \frac{a_2}{a_1} = \frac{5.9}{4.2} \approx 1.4$$

$$r_{2010-2020} \approx \frac{a_3}{a_2} = \frac{8.3}{5.9} \approx 1.4$$

$$r_{2020-2030} \approx \frac{a_4}{a_3} = \frac{11.6}{8.3} \approx 1.4$$

$$r_{2030-2040} \approx \frac{a_5}{a_4} = \frac{16.2}{11.6} \approx 1.4$$

$$r_{2040-2050} \approx \frac{a_6}{a_5} = \frac{22.7}{16.2} \approx 1.4$$

r is approximately 1.4 for each division.

b. $a_n = 4.2(1.4)^{n-1}$

c. 2080 is 9 decades after 1990 so $n = 9$.

$$a_n = 4.2(1.4)^{n-1}$$

$$a_9 = 4.2(1.4)^{9-1} \approx 62.0$$

In 2080, the model predicts the U.S. population, ages 85 and older, will be 62.0 million.

50. $r = 1.06$

$$a_n = a_1 r^{n-1} = 32000(1.06)^{n-1}$$

$$a_6 = \$32,000(1.06)^{6-1} \approx \$42,823$$

The salary in the sixth year is approximately \$42,823.

$$S_n = \frac{a_1\left(1 - r^n\right)}{1 - r}$$

$$S_6 = \frac{32000\left(1 - (1.06)^6\right)}{1 - 1.06} \approx 223,210$$

The total salary over the six years is approximately \$223,210.

51. a. $A = \dfrac{P\left[\left(1 + \frac{r}{n}\right)^{nt} - 1\right]}{\frac{r}{n}}$

$P = \$520, \ r = 0.06, \ n = 1, \ t = 20$

$$A = \frac{\$520\left[\left(1 + \dfrac{0.06}{1}\right)^{1 \cdot 20} - 1\right]}{\dfrac{0.06}{1}} = \frac{\$520\left[(1.06)^{20} - 1\right]}{0.06} \approx \$19,129$$

The value of the annuity will be \$19,129.

b. Interest = Value of annuity − Total deposits

$$\approx \ \$19,129 \ - \ \$520 \cdot 20$$

$$\approx \ \$8729$$

52. **a.** $A = \dfrac{P\left[\left(1+\frac{r}{n}\right)^{nt}-1\right]}{\frac{r}{n}}$

$P = 100, \; r = 0.055, \; n = 12, \; t = 30$

$A = \dfrac{\$100\left[\left(1+\dfrac{0.055}{12}\right)^{12\cdot30}-1\right]}{\dfrac{0.055}{12}} \approx \$91,361$

The value of the IRA will be \$91,361.

b. Interest = Value of IRA − Total deposits

$\approx \quad \$91,361 \; - \; \$100\cdot12\cdot30$

$\approx \quad \$55,361$

53. $r = 70\% = 0.7$

$a_1 = 4(.7) = 2.8$

$S = \dfrac{2.8}{1-0.7} = \dfrac{2.8}{0.3} = 9\dfrac{1}{3}$

The total spending in the town will be approximately $\$9\dfrac{1}{3}$ million each year.

54. $\dbinom{11}{8} = \dfrac{11!}{8!(11-8)!} = \dfrac{11!}{8!3!} = \dfrac{11\cdot10\cdot9\cdot\cancel{8!}}{\cancel{8!}3!} = 165$

55. $\dbinom{90}{2} = \dfrac{90!}{2!(90-2)!} = \dfrac{90\cdot89\cdot\cancel{88!}}{2\cdot1\cdot\cancel{88!}} = 4005$

56. Applying the Binomial Theorem to $(2x+1)^3$, we have $a = 2x, \; b = 1, \;$ and $n = 3$.

$(2x+1)^3 = \dbinom{3}{0}(2x)^3 + \dbinom{3}{1}(2x)^2 \cdot 1 + \dbinom{3}{2}(2x)\cdot1^2 + \dbinom{3}{3}1^3$

$\qquad = \dfrac{3!}{0!(3-0)!}8x^3 + \dfrac{3!}{1!(3-1)!}4x^2\cdot1 + \dfrac{3!}{2!(3-2)!}2x\cdot1 + \dfrac{3!}{3!(3-3)!}1$

$\qquad = \dfrac{\cancel{3!}}{1\cdot\cancel{3!}}8x^3 + \dfrac{3\cdot\cancel{2!}}{1\cdot\cancel{2!}}4x^2 + \dfrac{3\cdot\cancel{2!}}{\cancel{2!}1!}2x + \dfrac{\cancel{3!}}{\cancel{3!}0!}$

$\qquad = 8x^3 + 3\left(4x^2\right) + 3(2x) + 1$

$\qquad = 8x^3 + 12x^2 + 6x + 1$

57. Applying the Binomial Theorem to $\left(x^2-1\right)^4$, we have $a=x^2$, $b=-1$, and $n=4$.

$$\left(x^2-1\right)^4 = \binom{4}{0}\left(x^2\right)^4 + \binom{4}{1}\left(x^2\right)^3(-1) + \binom{4}{2}\left(x^2\right)^2(-1)^2 + \binom{4}{3}x^2(-1)^3 + \binom{4}{4}(-1)^4$$

$$= \frac{4!}{0!(4-0)!}x^8 - \frac{4!}{1!(4-1)!}x^6 + \frac{4!}{2!(4-2)!}x^4 - \frac{4!}{3!(4-3)!}x^2 + \frac{4!}{4!(4-4)!}1$$

$$= \frac{\cancel{4!}}{0!\,\cancel{4!}}x^8 - \frac{4!}{1!3!}x^6 + \frac{4!}{2!2!}x^4 - \frac{4!}{3!1!}x^2 + \frac{\cancel{4!}}{\cancel{4!}0!}$$

$$= 1\left(x^8\right) - \frac{4\cdot\cancel{3!}}{1\cdot\cancel{3!}}x^6 + \frac{4\cdot3\cdot\cancel{2!}}{2\cdot1\cdot\cancel{2!}}x^4 - \frac{4\cdot\cancel{3!}}{\cancel{3!}\cdot1}x^2 + 1$$

$$= x^8 - 4x^6 + 6x^4 - 4x^2 + 1$$

58. Applying the Binomial Theorem to $(x+2y)^5$, we have $a=x$, $b=2y$, and $n=5$.

$$(x+2y)^5 = \binom{5}{0}x^5 + \binom{5}{1}x^4(2y) + \binom{5}{2}x^3(2y)^2 + \binom{5}{3}x^2(2y)^3 + \binom{5}{4}x(2y)^4 + \binom{5}{5}(2y)^5$$

$$= \frac{\cancel{5!}}{0!\,\cancel{5!}}x^5 + \frac{5!}{1!(5-1)!}2x^4y + \frac{5!}{2!(5-2)!}4x^3y^2 + \frac{5!}{3!(5-3)!}8x^2y^3 + \frac{5!}{4!(5-4)!}16xy^4 + \frac{5!}{5!(5-5)!}32y^5$$

$$= 1x^5 + \frac{5\cdot\cancel{4!}}{1\cdot\cancel{4!}}2x^4y + \frac{5\cdot4\cdot\cancel{3!}}{2\cdot1\cdot\cancel{3!}}4x^3y^2 + \frac{5\cdot4\cdot\cancel{3!}}{\cancel{3!}2\cdot1}8x^2y^3 + \frac{5\cdot\cancel{4!}}{\cancel{4!}\cdot1}16xy^4 + \frac{\cancel{5!}}{\cancel{5!}0!}32y^5$$

$$= x^5 + 5\left(2x^4y\right) + 10\left(4x^3y^2\right) + 10\left(8x^2y^3\right) + 5\left(16xy^4\right) + 1\left(32y^5\right)$$

$$= x^5 + 10x^4y + 40x^3y^2 + 80x^2y^3 + 80xy^4 + 32y^5$$

59. Applying the Binomial Theorem to $(x-2)^6$, we have $a=x$, $b=-2$, and $n=6$.

$$(x-2)^6 = \binom{6}{0}x^6 + \binom{6}{1}x^5(-2) + \binom{6}{2}x^4(-2)^2 + \binom{6}{3}x^3(-2)^3 + \binom{6}{4}x^2(-2)^4 + \binom{6}{5}x(-2)^5 + \binom{6}{6}(-2)^6$$

$$= \frac{6!}{0!(6-0)!}x^6 + \frac{6!}{1!(6-1)!}x^5(-2) + \frac{6!}{2!(6-2)!}x^4(-2)^2 + \frac{6!}{3!(6-3)!}x^3(-2)^3$$

$$+ \frac{6!}{4!(6-4)!}x^2(-2)^4 + \frac{6!}{5!(6-5)!}x(-2)^5 + \frac{6!}{6!(6-6)!}(-2)^6$$

$$= \frac{\cancel{6!}}{1\cancel{6!}}x^6 - \frac{6\cdot\cancel{5!}}{1\cdot\cancel{5!}}2x^5 + \frac{6\cdot5\cdot\cancel{4!}}{2\cdot1\cdot\cancel{4!}}4x^4 - \frac{\cancel{6}\cdot5\cdot4\cdot\cancel{3!}}{\cancel{3}\cancel{2}\cdot1\cdot\cancel{3!}}8x^3$$

$$+ \frac{6\cdot5\cdot\cancel{4!}}{\cancel{4!}2\cdot1}16x^2 - \frac{6\cdot\cancel{5!}}{\cancel{5!}1}32x + \frac{\cancel{6!}}{\cancel{6!}\cdot1}64$$

$$= x^6 - 6\left(2x^5\right) + 15\left(4x^4\right) - 20\left(8x^3\right) + 15\left(16x^2\right) - 6(32x) + 1\cdot64$$

$$= x^6 - 12x^5 + 60x^4 - 160x^3 + 240x^2 - 192x + 64$$

60. $\left(x^2 + 3\right)^8$

First Term $(r = 0)$: $\dbinom{n}{r} a^{n-r} b^r = \dbinom{8}{0}\left(x^2\right)^{8-0} 3^0 = \dfrac{8!}{0!(8-0)!} x^{16} \cdot 1 = \dfrac{\cancel{8!}}{0! \cancel{8!}} x^{16} = x^{16}$

Second Term $(r = 1)$: $\dbinom{n}{r} a^{n-r} b^r = \dbinom{8}{1}\left(x^2\right)^{8-1} 3^1 = \dfrac{8!}{1!(8-1)!} 3x^{14} = \dfrac{8 \cdot \cancel{7!}}{1 \cdot \cancel{7!}} 3x^{14} = 8 \cdot 3x^{14} = 24x^{14}$

Third Term $(r = 2)$: $\dbinom{n}{r} a^{n-r} b^r = \dbinom{8}{2}\left(x^2\right)^{8-2} 3^2 = \dfrac{8!}{2!(8-2)!} 9x^{12} = \dfrac{8 \cdot 7 \cdot \cancel{6!}}{2 \cdot 1 \cdot \cancel{6!}} 9x^{12} = 28 \cdot 9x^{12} = 252x^{12}$

61. $(x - 3)^9$

First Term $(r = 0)$: $\dbinom{n}{r} a^{n-r} b^r = \dbinom{9}{0} x^{9-0} (-3)^0 = \dfrac{9!}{0!(9-0)!} x^9 \cdot 1 = \dfrac{\cancel{9!}}{0! \cancel{9!}} x^9 = x^9$

Second Term $(r = 1)$: $\dbinom{n}{r} a^{n-r} b^r = \dbinom{9}{1} x^{9-1} (-3)^1 = -\dfrac{9!}{1!(9-1)!} 3x^8 = -\dfrac{9 \cdot \cancel{8!}}{1 \cdot \cancel{8!}} 3x^8 = -9 \cdot 3x^8 = -27x^8$

Third Term $(r = 2)$: $\dbinom{n}{r} a^{n-r} b^r = \dbinom{9}{2} x^{9-2} (-3)^2 = \dfrac{9!}{2!(9-2)!} 9x^7 = \dfrac{9 \cdot 8 \cdot \cancel{7!}}{2 \cdot 1 \cdot \cancel{7!}} 9x^7 = 36 \cdot 9x^7 = 324x^7$

62. $(x + 2)^5$

Fourth Term $(r = 3)$: $\dbinom{n}{r} a^{n-r} b^r = \dbinom{5}{3} x^{5-3} (2)^3 = \dfrac{5!}{3!(5-3)!} 8x^2 = \dfrac{5!}{3!2!} 8x^2 = \dfrac{5 \cdot 4 \cdot \cancel{3!}}{\cancel{3!} \cdot 2 \cdot 1} 8x^2 = (10)8x^2 = 80x^2$

63. $(2x - 3)^6$

Fifth Term $(r = 4)$: $\dbinom{n}{r} a^{n-r} b^r = \dbinom{6}{4} (2x)^{6-4} (-3)^4 = \dfrac{6!}{4!(6-4)!} 4x^2 (81)$

$= \dfrac{6!}{4!2!} 324x^2 = \dfrac{6 \cdot 5 \cdot \cancel{4!}}{\cancel{4!} \cdot 2 \cdot 1} 324x^2 = (15)324x^2 = 4860x^2$

Chapter 14 Test

1. $a_n = \dfrac{(-1)^{n+1}}{n^2}$

   $a_1 = \dfrac{(-1)^{1+1}}{1^2} = \dfrac{(-1)^2}{1} = \dfrac{1}{1} = 1$

   $a_2 = \dfrac{(-1)^{2+1}}{2^2} = \dfrac{(-1)^3}{4} = \dfrac{-1}{4} = -\dfrac{1}{4}$

   $a_3 = \dfrac{(-1)^{3+1}}{3^2} = \dfrac{(-1)^4}{9} = \dfrac{1}{9}$

   $a_4 = \dfrac{(-1)^{4+1}}{4^2} = \dfrac{(-1)^5}{16} = \dfrac{-1}{16} = -\dfrac{1}{16}$

   $a_5 = \dfrac{(-1)^{5+1}}{5^2} = \dfrac{(-1)^6}{25} = \dfrac{1}{25}$

2. $\displaystyle\sum_{i=1}^{5}\left(i^2+10\right) = \left(1^2+10\right)+\left(2^2+10\right)+\left(3^2+10\right)+\left(4^2+10\right)+\left(5^2+10\right)$

   $= (1+10)+(4+10)+(9+10)+(16+10)+(25+10)$

   $= 11+14+19+26+35 = 105$

3. $\dfrac{2}{3}+\dfrac{3}{4}+\dfrac{4}{5}+\ldots+\dfrac{21}{22} = \displaystyle\sum_{i=2}^{21}\dfrac{i}{i+1}$ or $\displaystyle\sum_{i=1}^{20}\dfrac{i+1}{i+2}$

4. $d = 9-4 = 5$

   $a_n = 4+(n-1)5 = 4+5n-5 = 5n-1$

   $a_{12} = 5(12)-1 = 60-1 = 59$

5. $d = \dfrac{a_2}{a_1} = \dfrac{4}{16} = \dfrac{1}{4}$

   $a_n = a_1 r^{n-1} = 16\left(\dfrac{1}{4}\right)^{n-1}$

   $a_{12} = 16\left(\dfrac{1}{4}\right)^{12-1} = 16\left(\dfrac{1}{4}\right)^{11} = 16\left(\dfrac{1}{4,194,304}\right) = \dfrac{16}{4,194,304} = \dfrac{1}{262,144}$

6. First, find d.
   $d = -14-(-7) = -7$

   Next, find a_{10}.
   $a_{10} = -7+(10-1)(-7) = -7-63 = -70$

   Now, find the sum.
   $S_{10} = \dfrac{10}{2}\left(-7+(-70)\right) = 5(-77) = -385$

7. $\displaystyle\sum_{i=1}^{20}(3i-4) = \big(3(1)-4\big)+\big(3(2)-4\big)+\big(3(3)-4\big)+\ldots+\big(3(20)-4\big)$

$$= (3-4)+(6-4)+(9-4)+\ldots+(60-4)$$

$$= -1+2+5+\ldots+56$$

$$S_{20} = \frac{20}{2}(-1+56) = 10(55) = 550$$

8. $r = \dfrac{a_2}{a_1} = \dfrac{-14}{7} = -2$

$$S_{10} = \frac{7\big(1-(-2)^{10}\big)}{1-(-2)} = \frac{7(1-1024)}{3} = \frac{7(-1023)}{3} = -2387$$

9. $\displaystyle\sum_{i=1}^{15}(-2)^i = \frac{-2\big(1-(-2)^{15}\big)}{1-(-2)} = \frac{-2\big(1-(-32,768)\big)}{3} = \frac{-2(32,769)}{3} = -21,846$

10. $r = \dfrac{1}{2}$

$$S = \frac{4}{1-\dfrac{1}{2}} = \frac{4}{\dfrac{1}{2}} = 4 \div \frac{1}{2} = 4 \cdot \frac{2}{1} = 8$$

11. $0.\overline{73} = \dfrac{a_1}{1-r} = \dfrac{\dfrac{73}{100}}{1-\dfrac{1}{100}} = \dfrac{\dfrac{73}{100}}{\dfrac{99}{100}} = \dfrac{73}{100} \div \dfrac{99}{100} = \dfrac{73}{100} \cdot \dfrac{100}{99} = \dfrac{73}{99}$

12. $r = 1.04$

$$S_8 = \frac{a_1\big(1-r^n\big)}{1-r} = \frac{30,000\big(1-(1.04)^8\big)}{1-1.04} \approx 276,427$$

The total salary over the eight years is approximately \$276,427.

13. $\dbinom{9}{2} = \dfrac{9!}{2!(9-2)!} = \dfrac{9!}{2!7!} = \dfrac{9 \cdot 8 \cdot \cancel{7!}}{2 \cdot 1 \cdot \cancel{7!}} = 36$

14. Applying the Binomial Theorem to $\left(x^2-1\right)^5$, we have $a=x^2$, $b=-1$, and $n=5$.

$$\left(x^2-1\right)^5=\binom{5}{0}\left(x^2\right)^5+\binom{5}{1}\left(x^2\right)^4(-1)+\binom{5}{2}\left(x^2\right)^3(-1)^2+\binom{5}{3}\left(x^2\right)^2(-1)^3+\binom{5}{4}\left(x^2\right)(-1)^4+\binom{5}{5}(-1)^5$$

$$=\frac{5!}{0!\,5!}x^{10}-\frac{5!}{1!(5-1)!}x^8+\frac{5!}{2!(5-2)!}x^6-\frac{5!}{3!(5-3)!}x^4+\frac{5!}{4!(5-4)!}x^2-\frac{5!}{5!(5-5)!}$$

$$=1x^{10}-\frac{5\cdot4!}{1\cdot4!}x^8+\frac{5\cdot4\cdot3!}{2\cdot1\cdot3!}x^6-\frac{5\cdot4\cdot3!}{3!2\cdot1}x^4+\frac{5\cdot4!}{4!\cdot1}x^2-\frac{5!}{5!0!}$$

$$=x^{10}-5x^8+10x^6-10x^4+5x^2-1$$

15. $\left(x+y^2\right)^8$

First Term $(r=0)$: $\binom{n}{r}a^{n-r}b^r=\binom{8}{0}x^{8-0}\left(y^2\right)^0=\frac{8!}{0!(8-0)!}x^8\cdot1=\frac{8!}{0!\,8!}x^8=x^8$

Second Term $(r=1)$: $\binom{n}{r}a^{n-r}b^r=\binom{8}{1}x^{8-1}\left(y^2\right)^1=\frac{8!}{1!(8-1)!}x^7y^2=\frac{8\cdot7!}{1\cdot7!}x^7y^2=8x^7y^2$

Third Term $(r=2)$: $\binom{n}{r}a^{n-r}b^r=\binom{8}{2}x^{8-2}\left(y^2\right)^2=\frac{8!}{2!(8-2)!}x^6y^4=\frac{8\cdot7\cdot6!}{2\cdot1\cdot6!}x^6y^4=28x^6y^4$

Cumulative Review Exercises (Chapters 1 – 14)

1.
$$\sqrt{2x+5}-\sqrt{x+3}=2$$
$$-\sqrt{x+3}=2-\sqrt{2x+5}$$
$$x+3=\left(2-\sqrt{2x+5}\right)^2$$
$$x+3=4-4\sqrt{2x+5}+2x+5$$
$$x+3=-4\sqrt{2x+5}+2x+9$$
$$-x-6=-4\sqrt{2x+5}$$
$$(-x-6)^2=16(2x+5)$$
$$x^2+12x+36=32x+80$$
$$x^2-20x-44=0$$
$$(x-22)(x+2)=0$$
$$x=22 \text{ or } x=-2$$

The solution set is $\{22\}$.

2.
$$(x-5)^2=-49$$
$$x-5=\pm\sqrt{-49}$$
$$x=5\pm\sqrt{-49}$$
$$x=5\pm7i$$

3. $x^2 + x > 6$

Solve the related quadratic equation.
$$x^2 + x = 6$$
$$x^2 + x - 6 = 0$$
$$(x+3)(x-2) = 0$$
$$x + 3 = 0 \quad \text{or} \quad x - 2 = 0$$
$$x = -3 \qquad\qquad x = 2$$

The boundary points are -3 and 2.

Interval	Test Value	Test	Conclusion
$(-\infty, -3)$	-4	$(-4)^2 + (-4) > 6$ $12 > 6$, true	$(-\infty, -3)$ does belong to the solution set.
$(-3, 2)$	0	$(0)^2 + 0 > 6$ $0 > 6$, false	$(-3, 2)$ does not belong to the solution set.
$(2, \infty)$	3	$(3)^2 + 3 > 6$ $12 > 6$, true	$(2, \infty)$ does belong to the solution set.

The solution set is $(-\infty, -3) \cup (2, \infty)$ or $\{x \mid x < -3 \text{ or } x > 2\}$.

4. $6x - 3(5x + 2) = 4(1 - x)$
$$6x - 15x - 6 = 4 - 4x$$
$$-9x - 6 = 4 - 4x$$
$$-5x = 10$$
$$x = -2$$
The solution set is $\{-2\}$.

5.
$$\frac{2}{x-3} - \frac{3}{x+3} = \frac{12}{x^2 - 9}$$

$$\frac{2}{x-3} - \frac{3}{x+3} = \frac{12}{(x-3)(x+3)}$$

$$(x-3)(x+3)\left(\frac{2}{x-3} - \frac{3}{x+3}\right) = (x-3)(x+3)\left(\frac{12}{(x-3)(x+3)}\right)$$

$$2(x+3) - 3(x-3) = 12$$
$$2x + 6 - 3x + 9 = 12$$
$$-x + 15 = 12$$
$$-x = -3$$
$$x = 3$$

Since 3 would make one or more of the denominators in the original equation zero, we disregard it and conclude that there is no solution. The solution set is $\varnothing$ or $\{\ \}$.

6. $3x+2 < 4$ and $4-x > 1$

$\quad\quad 3x < 2 \quad\quad\quad -x > -3$

$\quad\quad x < \dfrac{2}{3} \quad\quad\quad x < 3$

The solution set is $\left\{ x \mid x < \dfrac{2}{3} \right\}$ or $\left(-\infty, \dfrac{2}{3} \right)$.

7. $3x - 2y + z = 7$

$\quad 2x + 3y - z = 13$

$\quad x - y + 2z = -6$

Multiply the second equation by 2 and add to the third equation.

$\quad 4x + 6y - 2z = 26$

$\quad \underline{x - y + 2z = -6}$

$\quad\quad 5x + 5y = 20$

Add the first equation to the second equation.

$\quad 3x - 2y + z = 7$

$\quad \underline{2x + 3y - z = 13}$

$\quad\quad 5x + y = 20$

We now have a system of two equations in two variables.

$5x + 5y = 20$

$5x + y = 20$

Multiply the second equation by -1 and add to the first equation.

$\quad 5x + 5y = 20$

$\quad \underline{-5x - y = -20}$

$\quad\quad 4y = 0$

$\quad\quad y = 0$

Back-substitute 0 for y to find x.

$5x + y = 20$

$5x + 0 = 20$

$\quad 5x = 20$

$\quad x = 4$

Back-substitute 4 for x and 0 for y to find z.

$\quad 3x - 2y + z = 7$

$3(4) - 2(0) + z = 7$

$\quad 12 - 0 + z = 7$

$\quad\quad 12 + z = 7$

$\quad\quad\quad z = -5$

The solution set is $\{(4, 0, -5)\}$.

8. $\log_9 x + \log_9 (x - 8) = 1$

$\quad \log_9 (x(x - 8)) = 1$

$\quad\quad x(x - 8) = 9^1$

$\quad\quad x^2 - 8x = 9$

$\quad\quad x^2 - 8x - 9 = 0$

$\quad\quad (x - 9)(x + 1) = 0$

$x - 9 = 0$ or $x + 1 = 0$

$\quad x = 9 \quad\quad\quad x = -1$

Since we cannot take a log of a negative number, we disregard -1 and conclude that the solution set is $\{9\}$.

9. $2x^2 - 3y^2 = 5$

$\quad 3x^2 + 4y^2 = 16$

Multiply the first equation by -3 and the second equation by 2 and solve by addition.

$\quad -6x^2 + 9y^2 = -15$

$\quad \underline{6x^2 + 8y^2 = 32}$

$\quad\quad 17y^2 = 17$

$\quad\quad y^2 = 1$

$\quad\quad y = \pm 1$

Back-substitute ± 1 for y to find x.

$\quad\quad y = \pm 1$

$2x^2 - 3(\pm 1)^2 = 5$

$\quad 2x^2 - 3(1) = 5$

$\quad\quad 2x^2 - 3 = 5$

$\quad\quad 2x^2 = 8$

$\quad\quad x^2 = 4$

$\quad\quad x = \pm 2$

The solutions are $(-2, -1), (-2, 1),$ $(2, -1)$ and $(2, 1)$ and the solution set is $\{(-2, -1), (-2, 1), (2, -1), (2, 1)\}$.

10. $2x^2 - y^2 = -8$
   $x - y = 6$

Solve the second equation for x.
$x - y = 6$
   $x = y + 6$

Substitute $y + 6$ for x.

$$2(y+6)^2 - y^2 = -8$$

$$2(y^2 + 12x + 36) - y^2 = -8$$

$$2y^2 + 24x + 72 - y^2 = -8$$

$$y^2 + 24x + 72 = -8$$

$$y^2 + 24x + 80 = 0$$

$$(y+20)(y+4) = 0$$

$y + 20 = 0$ or $y + 4 = 0$
   $y = -20$ $y = -4$

Back-substitute -4 and -20 for y to find x.

$$\begin{array}{ccc} & y = -20 & \text{or} \quad y = -4 \\ f(x) = (x+2)^2 - 4 & x = -20 + 6 & x = -4 + 6 \\ & x = -14 & x = 2 \end{array}$$

The solution set is $\{(-14, -20), (2, -4)\}$.

11.

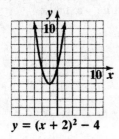

$y = (x + 2)^2 - 4$

12. $y < -3x + 5$

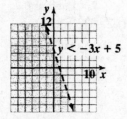

13. $f(x) = 3^{x-2}$

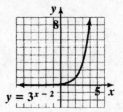

$y = 3^{x-2}$

14. $\dfrac{x^2}{16} + \dfrac{y^2}{4} = 1$

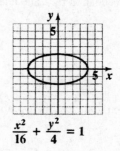

$\dfrac{x^2}{16} + \dfrac{y^2}{4} = 1$

15. $x^2 - y^2 = 9$

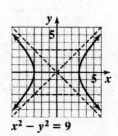

$x^2 - y^2 = 9$

16. $\dfrac{2x+1}{x-5} - \dfrac{4}{x^2 - 3x - 10}$

$$= \dfrac{2x+1}{x-5} - \dfrac{4}{(x-5)(x+2)}$$

$$= \dfrac{2x+1}{x-5} \cdot \dfrac{x+2}{x+2} - \dfrac{4}{(x-5)(x+2)}$$

$$= \dfrac{(2x+1)(x+2) - 4}{(x-5)(x+2)}$$

$$= \dfrac{2x^2 + 5x + 2 - 4}{(x-5)(x+2)}$$

$$= \dfrac{2x^2 + 5x - 2}{(x-5)(x+2)}$$

17. $\dfrac{\dfrac{1}{x-1}+1}{\dfrac{1}{x+1}-1} = \dfrac{\dfrac{1}{x-1}+\dfrac{x-1}{x-1}}{\dfrac{1}{x+1}-\dfrac{x+1}{x+1}}$

$= \dfrac{\dfrac{1+(x-1)}{x-1}}{\dfrac{1-(x+1)}{x+1}}$

$= \dfrac{\dfrac{x}{x-1}}{\dfrac{-x}{x+1}} =$

$\dfrac{x}{x-1}\cdot\dfrac{x+1}{-x}$

$= -\dfrac{x+1}{x-1}$

18. $\dfrac{6}{\sqrt{5}-\sqrt{2}}\cdot\dfrac{\sqrt{5}+\sqrt{2}}{\sqrt{5}+\sqrt{2}} = \dfrac{6\left(\sqrt{5}+\sqrt{2}\right)}{5-2}$

$= \dfrac{6\left(\sqrt{5}+\sqrt{2}\right)}{3}$

$= 2\left(\sqrt{5}+\sqrt{2}\right)$

$= 2\sqrt{5}+2\sqrt{2}$

19. $8\sqrt{45}+2\sqrt{5}-7\sqrt{20}$

$= 8\sqrt{9\cdot5}+2\sqrt{5}-7\sqrt{4\cdot5}$

$= 8\cdot3\sqrt{5}+2\sqrt{5}-7\cdot2\sqrt{5}$

$= 24\sqrt{5}+2\sqrt{5}-14\sqrt{5}$

$= 12\sqrt{5}$

20. $\dfrac{5}{\sqrt[3]{2x^2y}} = \dfrac{5}{\sqrt[3]{2x^2y}}\cdot\dfrac{\sqrt[3]{2^2xy^2}}{\sqrt[3]{2^2xy^2}}$

$= \dfrac{5\sqrt[3]{4xy^2}}{\sqrt[3]{2^3x^3y^3}} = \dfrac{5\sqrt[3]{4xy^2}}{2xy}$

21. $5ax+5ay-4bx-4by$

$= 5a(x+y)-4b(x+y)$

$= (x+y)(5a-4b)$

22. $5\log x-\dfrac{1}{2}\log y = \log x^5-\log y^{\frac{1}{2}}$

$= \log\left(\dfrac{x^5}{y^{\frac{1}{2}}}\right) = \log\left(\dfrac{x^5}{\sqrt{y}}\right)$

23. $\dfrac{1}{p}+\dfrac{1}{q}=\dfrac{1}{f}$;

$\dfrac{1}{p}=\dfrac{1}{f}-\dfrac{1}{q}$

$\dfrac{1}{p}=\dfrac{1}{f}\cdot\dfrac{q}{q}-\dfrac{1}{q}\cdot\dfrac{f}{f}$

$\dfrac{1}{p}=\dfrac{q}{qf}-\dfrac{f}{qf}$

$\dfrac{1}{p}=\dfrac{q-f}{qf}$

$p(q-f)=qf$

$p=\dfrac{qf}{q-f}$

24. $d=\sqrt{\left(6-(-3)\right)^2+\left(-1-(-4)\right)^2}$

$= \sqrt{9^2+3^2} = \sqrt{81+9} = \sqrt{90}$

$= \sqrt{9\cdot10} = 3\sqrt{10} \approx 9.49$

The distance is $3\sqrt{10}$, or about 9.49, units.

25. $\displaystyle\sum_{i=2}^{5}\left(i^3-4\right)$

$= \left(2^3-4\right)+\left(3^3-4\right)+\left(4^3-4\right)+\left(5^3-4\right)$

$= (8-4)+(27-4)+(64-4)+(125-4)$

$= 4+23+60+121 = 208$

26. First, find d.
$d = 6-2 = 4$

Next, find a_{30}.

$a_{30} = 2+(30-1)4 = 2+(29)4$

$= 2+116 = 118$

Now, find the sum.

$S_{30} = \dfrac{30}{2}(2+118) = 15(120) = 1800$

27. $0.\overline{3} = \dfrac{a_1}{1-r} = \dfrac{\dfrac{3}{10}}{1-\dfrac{1}{10}} = \dfrac{\dfrac{3}{10}}{\dfrac{9}{10}} = \dfrac{3}{10} \div \dfrac{9}{10}$

$\qquad = \dfrac{3}{10} \cdot \dfrac{10}{9} = \dfrac{1}{3}$

28. Applying the Binomial Theorem to $\left(2x - y^3\right)^4$, we have $a = 2x$, $b = -y^3$, and $n = 4$.

$\left(2x - y^3\right)^4$

$= \dbinom{4}{0}(2x)^4 + \dbinom{4}{1}(2x)^3\left(-y^3\right) + \dbinom{4}{2}(2x)^2\left(-y^3\right)^2 + \dbinom{4}{3}2x\left(-y^3\right)^3 + \dbinom{4}{4}\left(-y^3\right)^4$

$= \dfrac{4!}{0!(4-0)!}16x^4 - \dfrac{4!}{1!(4-1)!}8x^3 y^3 + \dfrac{4!}{2!(4-2)!}4x^2 y^6 - \dfrac{4!}{3!(4-3)!}2xy^9 + \dfrac{4!}{4!(4-4)!}y^{12}$

$= \dfrac{\cancel{4!}}{0!\,\cancel{4!}}16x^4 - \dfrac{4!}{1!3!}8x^3 y^3 + \dfrac{4!}{2!2!}4x^2 y^6 - \dfrac{4!}{3!1!}2xy^9 + \dfrac{\cancel{4!}}{\cancel{4!}0!}y^{12}$

$= 1\left(16x^4\right) - \dfrac{4 \cdot \cancel{3!}}{1 \cdot \cancel{3!}}8x^3 y^3 + \dfrac{4 \cdot 3 \cdot \cancel{2!}}{2 \cdot 1 \cdot \cancel{2!}}4x^2 y^6 - \dfrac{4 \cdot \cancel{3!}}{\cancel{3!} \cdot 1}2xy^9 + y^{12}$

$= 16x^4 - 4\left(8x^3 y^3\right) + 6\left(4x^2 y^6\right) - 4\left(2xy^9\right) + y^{12}$

$= 16x^4 - 32x^3 y^3 + 24x^2 y^6 - 8xy^9 + y^{12}$

29. $f(x) = \dfrac{2}{x^2 + 2x - 15}$

Set the denominator equal to 0 to find the domain:

$x^2 + 2x - 15 = 0$

$(x + 5)(x - 3) = 0$

$x = -5,\ x = 3$

So, $\{x \mid x$ is a real number

$\qquad$ and $x \neq -5$ and $x \neq 3\}$

or $(-\infty, -5) \cup (-5, 3) \cup (3, \infty)$.

30. $f(x) = \sqrt{2x - 6}$;

We can not take the square root of a negative number.

$2x - 6 \geq 0$

$\quad 2x \geq 6$

$\quad\ x \geq 3$

So, $\left\{x \mid x \geq 3\right\}$ or $[3, \infty)$.

31. $f(x) = \ln(1 - x)$

We can only take the natural logarithm of positive numbers.
$$1 - x > 0$$
$$-x > -1$$
$$x < 1$$

So, $\{x \mid x < 1\}$ or $(-\infty, 1)$.

32. Let $w = $ width of the rectangle. Then $l = 2w + 2$.

The perimeter of a rectangle is given by $P = 2w + 2l$.

$$P = 2w + 2l$$
$$22 = 2w + 2(2w + 2)$$
$$22 = 2w + 4w + 4$$
$$22 = 6w + 4$$
$$18 = 6w$$
$$w = 3$$

Thus, $l = 2w + 2 = 2(3) + 2 = 8$

The dimension of the rectangle is 8 feet by 3 feet.

33. $A = P(1 + r)^t$
$$19610 = P(1 + 0.06)^1$$
$$19610 = 1.06P$$
$$P = 18500$$
Your salary before the raise is \$18,500.

34. $F(t) = 1 - k \ln(t + 1)$

$$\frac{1}{2} = 1 - k \ln(3 + 1)$$

$$\frac{1}{2} = 1 - k \ln 4$$

$$k \ln 4 = \frac{1}{2}$$

$$k = \frac{1}{2 \ln 4} \approx 0.3607$$

$$F(t) \approx 1 - 0.3607 \ln(t + 1)$$
$$F(6) \approx 1 - 0.3607 \ln(6 + 1)$$
$$\approx 1 - 0.3607 \ln(7)$$
$$\approx 1 - 0.7019$$
$$\approx 0.298 \text{ or } \frac{298}{1000}$$

Appendices

Check Points Appendix A

1. **a.** $\dfrac{10+20+30+40+50}{5} = \dfrac{150}{5} = 30$

   b. $\dfrac{3+10+10+10+117}{5} = \dfrac{150}{5} = 30$

2. **a.** First arrange the data items from smallest to largest: 25, 28, <u>35</u>, 40, 42.
      The number of data items is odd, so the median is the middle number. The median is 35.

   b. First arrange the data items from smallest to largest: 61, 72, <u>79</u>, <u>85</u>, 87, 93.
      The number of data items is even, so the median is the mean of the two middle data items.
      The median is $\dfrac{79+85}{2} = \dfrac{164}{2} = 82$.

3. **a.** Mean $= \dfrac{417.4+46.9+37.1+35.0+32.8+27.2+20.8+19.2+19.1+13.9}{10} = \$66.94$ billion

   b. Position of mean: $\dfrac{n+1}{2} = \dfrac{10+1}{2} = 5.5$ position
      The median is the mean of the data items in positions 5 and 6.
      Thus, the median is $\dfrac{27.2+32.8}{2} = \$30$ billion.

   c. The mean is so much greater than the median because one data item ($417.4) is much greater than the others.

4. **a.** 3, 8, 5, 8, 9, 10
      8 occurs most often. The mode is 8.

   b. 3, 8, 5, 8, 9, 3
      Both 3 and 8 occur most often. The modes are 3 and 8.

   c. 3, 8, 5, 6, 9, 10
      Each data item occurs the same number of times. There is no mode.

5. *Mean*: Mean $= \dfrac{107+136+138+138+172+173+190+191}{8} = \dfrac{1245}{8} \approx 155.6$
   Median: The data items are arranged in order: 107, 136, 138, <u>138</u>, <u>172</u>, 173, 190, 191
   The number of data items is even, so the median is the mean of the two middle data items.
   The median is $\dfrac{138+172}{2} = \dfrac{310}{2} = 155$.
   Mode: The number 138 occurs more often than any other. The mode is 138.

Exercise Set Appendix A

1. $\dfrac{7+4+3+2+8+5+1+3}{8} = \dfrac{33}{8} = 4.125$

3. $\dfrac{91+95+99+97+93+95}{6} = \dfrac{570}{6} = 95$

5. $\dfrac{100+40+70+40+60}{5} = \dfrac{310}{5} = 62$

7. $\dfrac{1.6+3.8+5.0+2.7+4.2+4.2+3.2+4.7+3.6+2.5+2.5}{11} = \dfrac{38}{11} \approx 3.45$

9. First arrange the data items from smallest to largest: 1, 2, 3, 3, 4, 5, 7, 8
   The number of data items is even, so the median is the mean of the two middle data items. The median is 3.5.

11. First arrange the data items from smallest to largest: 91, 93, 95, 95, 97, 99
    The number of data items is even, so the median is the mean of the two middle data items.
    Median $= \dfrac{95+95}{2} = 95$

13. First arrange the data items from smallest to largest: 40, 40, 60, 70, 100
    The number of data items is odd, so the median is the middle number. The median is 60.

15. First arrange the data items from smallest to largest: 1.6, 2.5, 2.5, 2.7, 3.2, 3.6, 3.8, 4.2, 4.2, 4.7, 5.0
    The number of data items is odd, so the median is the middle number. The median is 3.6.

17. The mode is 3.

19. The mode is 95.

21. The mode is 40.

23. The modes are 2.5 and 4.2.

25. The data items are 52, 56, 59, 68, 69, 71, 71, 74, 87.
    Mean: Mean $= \dfrac{52+56+59+68+69+71+71+74+87}{9} = \dfrac{607}{9} \approx 67.4$

    Median: The data items are arranged in order. The number of data items is odd, so the median is the middle number. The median is 69.
    Mode: The number 71 occurs more often than any other. The mode is 71.

27. The data items are $17,500, $19,000, $22,000, $27,500, $98,500
    Mean: Mean $= \dfrac{\$17,500+\$19,000+\$22,000+\$27,500+\$98,500}{5} = \dfrac{\$184,500}{5} \approx \$36,900$

    Median: The data items are arranged in order. The number of data items is odd, so the median is the middle number. The median is $22,000.
    The median is more representative. Explanations will vary.

Check Points Appendix B

1. a. $\begin{bmatrix} 4 & 12 & -20 & | & 8 \\ 1 & 6 & -3 & | & 7 \\ -3 & -2 & 1 & | & -9 \end{bmatrix} R_1 \leftrightarrow R_2 = \begin{bmatrix} 1 & 6 & -3 & | & 7 \\ 4 & 12 & -20 & | & 8 \\ -3 & -2 & 1 & | & -9 \end{bmatrix}$

b. $\begin{bmatrix} 4 & 12 & -20 & | & 8 \\ 1 & 6 & -3 & | & 7 \\ -3 & -2 & 1 & | & -9 \end{bmatrix} \frac{1}{4} R_1 = \begin{bmatrix} 1 & 3 & -5 & | & 2 \\ 1 & 6 & -3 & | & 7 \\ -3 & -2 & 1 & | & -9 \end{bmatrix}$

c. $\begin{bmatrix} 4 & 12 & -20 & | & 8 \\ 1 & 6 & -3 & | & 7 \\ -3 & -2 & 1 & | & -9 \end{bmatrix} 3R_2 + R_3 = \begin{bmatrix} 4 & 12 & -20 & | & 8 \\ 1 & 6 & -3 & | & 7 \\ 0 & 16 & -8 & | & 12 \end{bmatrix}$

2. $2x - y = -4$

$x + 3y = 5$

Write the augmented matrix for the system.

$\begin{bmatrix} 2 & -1 & | & -4 \\ 1 & 3 & | & 5 \end{bmatrix}$

We want a 1 in the upper left position. One way to do this is to interchange row 1 and row 2.

$\begin{bmatrix} 2 & -1 & | & -4 \\ 1 & 3 & | & 5 \end{bmatrix} R_1 \leftrightarrow R_2 = \begin{bmatrix} 1 & 3 & | & 5 \\ 2 & -1 & | & -4 \end{bmatrix}$

Now we want a 0 below the 1 in the first column.

$\begin{bmatrix} 1 & 3 & | & 5 \\ 2 & -1 & | & -4 \end{bmatrix} -2R_1 + R_2 = \begin{bmatrix} 1 & 3 & | & 5 \\ 0 & -7 & | & -14 \end{bmatrix}$

Next we want a 1 in the second row, second column.

$\begin{bmatrix} 1 & 3 & | & 5 \\ 0 & -7 & | & -14 \end{bmatrix} \frac{1}{-7} R_2 = \begin{bmatrix} 1 & 3 & | & 5 \\ 0 & 1 & | & 2 \end{bmatrix}$

The resulting system is:

$x + 3y = 5$

$y = 2$

Back-substitute 2 for y in the first equation.

$x + 3y = 5$

$x + 3(2) = 5$

$x + 6 = 5$

$x = -1$

$(-1, 2)$ satisfies both equations.

The solution set is $\{(-1, 2)\}$.

3. $2x + y + 2z = 18$

$x - y + 2z = 9$

$x + 2y - z = 6$

Write the augmented matrix for the system.

$$\begin{bmatrix} 2 & 1 & 2 & | & 18 \\ 1 & -1 & 2 & | & 9 \\ 1 & 2 & -1 & | & 6 \end{bmatrix}$$

We want a 1 in the upper left position. One way to do this is to interchange row 1 and row 2.

$$\begin{bmatrix} 2 & 1 & 2 & | & 18 \\ 1 & -1 & 2 & | & 9 \\ 1 & 2 & -1 & | & 6 \end{bmatrix} \quad R_1 \leftrightarrow R_2 = \begin{bmatrix} 1 & -1 & 2 & | & 9 \\ 2 & 1 & 2 & | & 18 \\ 1 & 2 & -1 & | & 6 \end{bmatrix}$$

Now we want zeros below the 1 in the first column.

$$\begin{bmatrix} 1 & -1 & 2 & | & 9 \\ 2 & 1 & 2 & | & 18 \\ 1 & 2 & -1 & | & 6 \end{bmatrix} \quad -2R_1 + R_2 = \begin{bmatrix} 1 & -1 & 2 & | & 9 \\ 0 & 3 & -2 & | & 0 \\ 1 & 2 & -1 & | & 6 \end{bmatrix}$$

$$\begin{bmatrix} 1 & -1 & 2 & | & 9 \\ 0 & 3 & -2 & | & 0 \\ 1 & 2 & -1 & | & 6 \end{bmatrix} \quad -R_1 + R_3 = \begin{bmatrix} 1 & -1 & 2 & | & 9 \\ 0 & 3 & -2 & | & 0 \\ 0 & 3 & -3 & | & -3 \end{bmatrix}$$

Next we want a 1 in the second row, second column.

$$\begin{bmatrix} 1 & -1 & 2 & | & 9 \\ 0 & 3 & -2 & | & 0 \\ 0 & 3 & -3 & | & -3 \end{bmatrix} \quad \frac{1}{3}R_2 = \begin{bmatrix} 1 & -1 & 2 & | & 9 \\ 0 & 1 & -\frac{2}{3} & | & 0 \\ 0 & 3 & -3 & | & -3 \end{bmatrix}$$

Now we want a zero below the 1 in the second row, second column.

$$\begin{bmatrix} 1 & -1 & 2 & | & 9 \\ 0 & 1 & -\frac{2}{3} & | & 0 \\ 0 & 3 & -3 & | & -3 \end{bmatrix} \quad -3R_2 + R_3 = \begin{bmatrix} 1 & -1 & 2 & | & 9 \\ 0 & 1 & -\frac{2}{3} & | & 0 \\ 0 & 0 & -1 & | & -3 \end{bmatrix}$$

Next we want a 1 in the third row, third column.

$$\begin{bmatrix} 1 & -1 & 2 & | & 9 \\ 0 & 1 & -\frac{2}{3} & | & 0 \\ 0 & 0 & -1 & | & -3 \end{bmatrix} \quad -R_3 = \begin{bmatrix} 1 & -1 & 2 & | & 9 \\ 0 & 1 & -\frac{2}{3} & | & 0 \\ 0 & 0 & 1 & | & 3 \end{bmatrix}$$

The resulting system is:

$x - y + 2z = 9$

$y - \frac{2}{3}z = 0$

$z = 3$

Back-substitute 3 for z in the second equation.

$y - \frac{2}{3}(3) = 0$

$y - 2 = 0$

$y = 2$

Back-substitute 2 for y and 3 for z in the first equation.

$$x - y + 2z = 9$$
$$x - (2) + 2(3) = 9$$
$$x - 2 + 6 = 9$$
$$x + 4 = 9$$
$$x = 5$$

$(5, 2, 3)$ satisfies both equations.

The solution set is $\{(5, 2, 3)\}$.

Exercise Set Appendix B

1. $\begin{bmatrix} 2 & 2 & | & 5 \\ 1 & -\dfrac{3}{2} & | & 5 \end{bmatrix} \; R_1 \leftrightarrow R_2 = \begin{bmatrix} 1 & -\dfrac{3}{2} & | & 5 \\ 2 & 2 & | & 5 \end{bmatrix}$

3. $\begin{bmatrix} -6 & 8 & | & -12 \\ 3 & 5 & | & -2 \end{bmatrix} \; -\dfrac{1}{6} R_1 = \begin{bmatrix} 1 & -\dfrac{4}{3} & | & 2 \\ 3 & 5 & | & -2 \end{bmatrix}$

5. $\begin{bmatrix} 1 & -3 & | & 5 \\ 2 & 6 & | & 4 \end{bmatrix} \; -2R_1 + R_2 = \begin{bmatrix} 1 & -3 & | & 5 \\ 0 & 12 & | & -6 \end{bmatrix}$

7. $\begin{bmatrix} 1 & -\dfrac{3}{2} & | & \dfrac{7}{2} \\ 3 & 4 & | & 2 \end{bmatrix} \; -3R_1 + R_2 = \begin{bmatrix} 1 & -\dfrac{3}{2} & | & \dfrac{7}{2} \\ 0 & \dfrac{17}{2} & | & -\dfrac{17}{2} \end{bmatrix}$

9. $\begin{bmatrix} 2 & -6 & 4 & | & 10 \\ 1 & 5 & -5 & | & 0 \\ 3 & 0 & 4 & | & 7 \end{bmatrix} \; \dfrac{1}{2} R_1 = \begin{bmatrix} 1 & -3 & 2 & | & 5 \\ 1 & 5 & -5 & | & 0 \\ 3 & 0 & 4 & | & 7 \end{bmatrix}$

11. $\begin{bmatrix} 1 & -3 & 2 & | & 0 \\ 3 & 1 & -1 & | & 7 \\ 2 & -2 & 1 & | & 3 \end{bmatrix} \; -3R_1 + R_2 = \begin{bmatrix} 1 & -3 & 2 & | & 0 \\ 0 & 10 & -7 & | & 7 \\ 2 & -2 & 1 & | & 3 \end{bmatrix}$

13. $\begin{bmatrix} 1 & 1 & -1 & | & 6 \\ 2 & -1 & 1 & | & -3 \\ 3 & -1 & -1 & | & 4 \end{bmatrix} \; \begin{array}{l} -2R_1 + R_2 \\ -3R_1 + R_3 \end{array} = \begin{bmatrix} 1 & 1 & -1 & | & 6 \\ 0 & -3 & 3 & | & -15 \\ 0 & -4 & 2 & | & -14 \end{bmatrix}$

15. $\begin{bmatrix} 1 & 1 & | & 6 \\ 1 & -1 & | & 2 \end{bmatrix} -R_1 + R_2 = \begin{bmatrix} 1 & 1 & | & 6 \\ 0 & -2 & | & -4 \end{bmatrix} -\dfrac{1}{2}R_2$

$\qquad\qquad = \begin{bmatrix} 1 & 1 & | & 6 \\ 0 & 1 & | & 2 \end{bmatrix}$

The resulting system is:

$x + y = 6$

$\qquad y = 2$

Back-substitute 2 for y in the first equation.

$x + 2 = 6$

$\qquad x = 4$

The solution set is $\{(4, 2)\}$.

17. $\begin{bmatrix} 2 & 1 & | & 3 \\ 1 & -3 & | & 12 \end{bmatrix} R_1 \leftrightarrow R_2$

$= \begin{bmatrix} 1 & -3 & | & 12 \\ 2 & 1 & | & 3 \end{bmatrix} -2R_1 + R_2$

$= \begin{bmatrix} 1 & -3 & | & 12 \\ 0 & 7 & | & -21 \end{bmatrix} \dfrac{1}{7}R_2$

$= \begin{bmatrix} 1 & -3 & | & 12 \\ 0 & 1 & | & -3 \end{bmatrix}$

The system is:

$x - 3y = 12$

$\qquad y = -3$

Back-substitute -3 for y in the first equation.

$\qquad x - 3y = 12$

$x - 3(-3) = 12$

$\qquad x + 9 = 12$

$\qquad\qquad x = 3$

The solution set is $\{(3, -3)\}$.

19. $\begin{bmatrix} 5 & 7 & | & -25 \\ 11 & 6 & | & -8 \end{bmatrix} \quad \frac{1}{5}R_1$

$= \begin{bmatrix} 1 & \dfrac{7}{5} & | & -5 \\ 11 & 6 & | & -8 \end{bmatrix} \quad -11R_1 + R_2$

$= \begin{bmatrix} 1 & \dfrac{7}{5} & | & -5 \\ 0 & -\dfrac{47}{5} & | & 47 \end{bmatrix} \quad -\dfrac{5}{47}R_2$

$= \begin{bmatrix} 1 & \dfrac{7}{5} & | & -5 \\ 0 & 1 & | & -5 \end{bmatrix}$

The resulting system is:

$x + \dfrac{7}{5}y = -5$

$y = -5$

Back-substitute -5 for y in the first equation.

$x + \dfrac{7}{5}y = -5$

$x + \dfrac{7}{5}(-5) = -5$

$x - 7 = -5$

$x = 2$

The solution set is $\{(2, -5)\}$.

21. $\begin{bmatrix} 4 & -2 & | & 5 \\ -2 & 1 & | & 6 \end{bmatrix} \quad \frac{1}{4}R_1$

$= \begin{bmatrix} 1 & -\dfrac{1}{2} & | & \dfrac{5}{4} \\ -2 & 1 & | & 6 \end{bmatrix} \quad 2R_1 + R_2$

$= \begin{bmatrix} 1 & -\dfrac{1}{2} & | & \dfrac{5}{4} \\ 0 & 0 & | & \dfrac{17}{2} \end{bmatrix}$

The resulting system is:

$x - \dfrac{1}{2}y = \dfrac{5}{4}$

$0x + 0y = \dfrac{17}{2}$

This is a contradiction. The system is inconsistent. There is no solution.

23. $\begin{bmatrix} 1 & -2 & | & 1 \\ -2 & 4 & | & -2 \end{bmatrix} \quad 2R_1 + R_2$

$= \begin{bmatrix} 1 & -2 & | & 1 \\ 0 & 0 & | & 0 \end{bmatrix}$

The resulting system is:

$x - 2y = 1$

$0x + 0y = 0$

The system is dependent. There are infinitely many solutions.

25. $\begin{bmatrix} 1 & 1 & -1 & | & -2 \\ 2 & -1 & 1 & | & 5 \\ -1 & 2 & 2 & | & 1 \end{bmatrix} \quad -2R_1 + R_2$

$= \begin{bmatrix} 1 & 1 & -1 & | & -2 \\ 0 & -3 & 3 & | & 9 \\ -1 & 2 & 2 & | & 1 \end{bmatrix} \quad R_1 + R_3$

$= \begin{bmatrix} 1 & 1 & -1 & | & -2 \\ 0 & -3 & 3 & | & 9 \\ 0 & 3 & 1 & | & -1 \end{bmatrix} \quad R_2 + R_3$

$= \begin{bmatrix} 1 & 1 & -1 & | & -2 \\ 0 & -3 & 3 & | & 9 \\ 0 & 0 & 4 & | & 8 \end{bmatrix} \quad \begin{matrix} -\dfrac{1}{3}R_2 \\ \dfrac{1}{4}R_3 \end{matrix}$

$= \begin{bmatrix} 1 & 1 & -1 & | & -2 \\ 0 & 1 & -1 & | & -3 \\ 0 & 0 & 1 & | & 2 \end{bmatrix}$

The resulting system is:

$x + y - z = -2$

$y - z = -3$

$z = 2$

Back-substitute 2 for z to find y.

$y - z = -3$

$y - 2 = -3$

$y = -1$

Back-substitute 2 for z and -1 for y to find x.

$x + y - z = -2$

$x - 1 - 2 = -2$

$x - 3 = -2$

$x = 1$

The solution set is $\{(1, -1, 2)\}$.

27. $\begin{bmatrix} 1 & 3 & 0 & | & 0 \\ 1 & 1 & 1 & | & 1 \\ 3 & -1 & -1 & | & 11 \end{bmatrix}$ $\quad -R_1 + R_2$

$= \begin{bmatrix} 1 & 3 & 0 & | & 0 \\ 0 & -2 & 1 & | & 1 \\ 3 & -1 & -1 & | & 11 \end{bmatrix}$ $\quad -3R_1 + R_3$

$= \begin{bmatrix} 1 & 3 & 0 & | & 0 \\ 0 & -2 & 1 & | & 1 \\ 0 & -10 & -1 & | & 11 \end{bmatrix}$ $\quad -\dfrac{1}{2}R_2$

$= \begin{bmatrix} 1 & 3 & 0 & | & 0 \\ 0 & 1 & -\dfrac{1}{2} & | & -\dfrac{1}{2} \\ 0 & -10 & -1 & | & 11 \end{bmatrix}$ $\quad -\dfrac{1}{10}R_3$

$= \begin{bmatrix} 1 & 3 & 0 & | & 0 \\ 0 & 1 & -\dfrac{1}{2} & | & -\dfrac{1}{2} \\ 0 & 1 & \dfrac{1}{10} & | & -\dfrac{11}{10} \end{bmatrix}$ $\quad -R_2 + R_3$

$= \begin{bmatrix} 1 & 3 & 0 & | & 0 \\ 0 & 1 & -\dfrac{1}{2} & | & -\dfrac{1}{2} \\ 0 & 0 & \dfrac{3}{5} & | & -\dfrac{3}{5} \end{bmatrix}$ $\quad \dfrac{5}{3}R_3$

$= \begin{bmatrix} 1 & 3 & 0 & | & 0 \\ 0 & 1 & -\dfrac{1}{2} & | & -\dfrac{1}{2} \\ 0 & 0 & 1 & | & -1 \end{bmatrix}$

The resulting system is:

$x + 3y \qquad = 0$

$y - \dfrac{1}{2}z = -\dfrac{1}{2}$

$z = -1$

Back-substitute -1 for z and solve for y.

$y - \dfrac{1}{2}z = -\dfrac{1}{2}$

$y - \dfrac{1}{2}(-1) = -\dfrac{1}{2}$

$y + \dfrac{1}{2} = -\dfrac{1}{2}$

$y = -1$

Back-substitute -1 for y to find x.

$x + 3y = 0$

$x + 3(-1) = 0$

$x - 3 = 0$

$x = 3$

The solution set is $\{(3, -1, -1)\}$.

29. $\begin{bmatrix} 2 & 2 & 7 & | & -1 \\ 2 & 1 & 2 & | & 2 \\ 4 & 6 & 1 & | & 15 \end{bmatrix}$ $\quad \dfrac{1}{2}R_1$

$= \begin{bmatrix} 1 & 1 & \dfrac{7}{2} & | & -\dfrac{1}{2} \\ 2 & 1 & 2 & | & 2 \\ 4 & 6 & 1 & | & 15 \end{bmatrix}$ $\quad -2R_1 + R_2$

$= \begin{bmatrix} 1 & 1 & \dfrac{7}{2} & | & -\dfrac{1}{2} \\ 0 & -1 & -5 & | & 3 \\ 4 & 6 & 1 & | & 15 \end{bmatrix}$ $\quad -R_2$

$= \begin{bmatrix} 1 & 1 & \dfrac{7}{2} & | & -\dfrac{1}{2} \\ 0 & 1 & 5 & | & -3 \\ 4 & 6 & 1 & | & 15 \end{bmatrix}$ $\quad -4R_1 + R_3$

$= \begin{bmatrix} 1 & 1 & \dfrac{7}{2} & | & -\dfrac{1}{2} \\ 0 & 1 & 5 & | & -3 \\ 0 & 2 & -13 & | & 17 \end{bmatrix}$ $\quad -2R_2 + R_3$

$= \begin{bmatrix} 1 & 1 & \dfrac{7}{2} & | & -\dfrac{1}{2} \\ 0 & 1 & 5 & | & -3 \\ 0 & 0 & -23 & | & 23 \end{bmatrix}$ $\quad -\dfrac{1}{23}R_3$

$= \begin{bmatrix} 1 & 1 & \dfrac{7}{2} & | & -\dfrac{1}{2} \\ 0 & 1 & 5 & | & -3 \\ 0 & 0 & 1 & | & -1 \end{bmatrix}$

The resulting system is:

$x + y + \dfrac{7}{2}z = -\dfrac{1}{2}$

$y + 5z = -3$

$z = -1$

Back-substitute -1 for z to find y.

$y + 5z = -3$

$y + 5(-1) = -3$

$y - 5 = -3$

$y = 2$

Back-substitute -1 for z and 2 for y to find x.

$$x + y + \frac{7}{2}z = -\frac{1}{2}$$

$$x + 2 + \frac{7}{2}(-1) = -\frac{1}{2}$$

$$x + 2 - \frac{7}{2} = -\frac{1}{2}$$

$$x - \frac{3}{2} = -\frac{1}{2}$$

$$x = 1$$

The solution set is $\{(1, 2, -1)\}$.

31. $\begin{bmatrix} 1 & 1 & 1 & | & 6 \\ 1 & 0 & -1 & | & -2 \\ 0 & 1 & 3 & | & 11 \end{bmatrix}$ $R_2 \leftrightarrow R_3$

$= \begin{bmatrix} 1 & 1 & 1 & | & 6 \\ 0 & 1 & 3 & | & 11 \\ 1 & 0 & -1 & | & -2 \end{bmatrix}$ $-R_1 + R_3$

$= \begin{bmatrix} 1 & 1 & 1 & | & 6 \\ 0 & 1 & 3 & | & 11 \\ 0 & -1 & -2 & | & -8 \end{bmatrix}$ $R_2 + R_3$

$= \begin{bmatrix} 1 & 1 & 1 & | & 6 \\ 0 & 1 & 3 & | & 11 \\ 0 & 0 & 1 & | & 3 \end{bmatrix}$

The resulting system is:

$$x + y + z = 6$$
$$y + 3z = 11$$
$$z = 3$$

Back-substitute 3 for z to find y.

$$y + 3z = 11$$
$$y + 3(3) = 11$$
$$y + 9 = 11$$
$$y = 2$$

Back-substitute 3 for z and 2 for y to find x.

$$x + y + z = 6$$
$$x + 2 + 3 = 6$$
$$x + 5 = 6$$
$$x = 1$$

The solution set is $\{(1, 2, 3)\}$.

33. $\begin{bmatrix} 1 & -1 & 3 & | & 4 \\ 2 & -2 & 6 & | & 7 \\ 3 & -1 & 5 & | & 14 \end{bmatrix}$ $\begin{matrix} -2R_1 + R_2 \\ \text{and} \\ -3R_1 + R_3 \end{matrix}$

$= \begin{bmatrix} 1 & -1 & 3 & | & 4 \\ 0 & 0 & 0 & | & -1 \\ 0 & 2 & -4 & | & 2 \end{bmatrix}$

The resulting system is:

$$x - y + 3z = 4$$
$$0x + 0y + 0z = -1$$
$$2y - 4z = 2$$

The second row is a contradiction, since $0x + 0y + 0z$ cannot equal -1. We conclude that the system is inconsistent and there is no solution.

35. $\begin{bmatrix} 1 & -2 & 1 & | & 4 \\ 5 & -10 & 5 & | & 20 \\ -2 & 4 & -2 & | & -8 \end{bmatrix}$ $\frac{1}{5}R_2$

$= \begin{bmatrix} 1 & -2 & 1 & | & 4 \\ 1 & -2 & 1 & | & 4 \\ -2 & 4 & -2 & | & -8 \end{bmatrix}$

R_1 and R_2 are the same. The system is dependent and there are infinitely many solutions.

37. $\begin{bmatrix} 1 & 1 & 0 & | & 1 \\ 0 & 1 & 2 & | & -2 \\ 2 & 0 & -1 & | & 0 \end{bmatrix}$ $-2R_1 + R_3 \multimap$

$= \begin{bmatrix} 1 & 1 & 0 & | & 1 \\ 0 & 1 & 2 & | & -2 \\ 0 & -2 & -1 & | & -2 \end{bmatrix}$ $2R_2 + R_3$

$= \begin{bmatrix} 1 & 1 & 0 & | & 1 \\ 0 & 1 & 2 & | & -2 \\ 0 & 0 & 3 & | & -6 \end{bmatrix}$ $\frac{1}{3}R_3$

$= \begin{bmatrix} 1 & 1 & 0 & | & 1 \\ 0 & 1 & 2 & | & -2 \\ 0 & 0 & 1 & | & -2 \end{bmatrix}$

The resulting system is:

$$x + y = 1$$
$$y + 2z = -2$$
$$z = -2$$

Back-substitute -2 for z to find y.

$$y + 2z = -2$$
$$y + 2(-2) = -2$$
$$y - 4 = -2$$
$$y = 2$$

Back-substitute 2 for y to find x.

$$x + y = 1$$
$$x + 2 = 1$$
$$x = -1$$

The solution set is $\{(-1, 2, -2)\}$.

39. The system is

$$w - x + y + z = 3$$
$$x - 2y - z = 0$$
$$y + 6z = 17$$
$$z = 3$$

Back-substitute $z = 3$ to solve for y.

$$y + 6(3) = 17$$
$$y + 18 = 17$$
$$y = -1$$

Back-substitute $z = 3$ and $y = -1$ to solve for x.

$$x - 2(-1) - (3) = 0$$
$$x + 2 - 3 = 0$$
$$x = 1$$

Back-substitute $x = 1$, $y = -1$ and $z = 3$ to solve for w.

$$w - (1) + (-1) + (3) = 3$$
$$w - 1 - 1 + 3 = 3$$
$$w = 2$$

The solution set is $\{(2, 1, -1, 3)\}$.

41.

$$\begin{bmatrix} 1 & -1 & 1 & 1 & | & 3 \\ 0 & 1 & -2 & -1 & | & 0 \\ 2 & 0 & 3 & 4 & | & 11 \\ 5 & 1 & 2 & 4 & | & 6 \end{bmatrix} \begin{matrix} \\ \\ -2R_1 + R_3 \\ -5R_1 + R_4 \end{matrix}$$

$$= \begin{bmatrix} 1 & -1 & 1 & 1 & | & 3 \\ 0 & 1 & -2 & -1 & | & 0 \\ 0 & 2 & 1 & 2 & | & 5 \\ 0 & 6 & -3 & -1 & | & -9 \end{bmatrix}$$

43.

$$\begin{bmatrix} 1 & 1 & 1 & 1 & | & 4 \\ 2 & 1 & -2 & -1 & | & 0 \\ 1 & -2 & -1 & -2 & | & -2 \\ 3 & 2 & 1 & 3 & | & 4 \end{bmatrix} \begin{matrix} \\ -2R_1 + R_2 \\ -1R_1 + R_3 \\ -3R_1 + R_4 \end{matrix}$$

$$= \begin{bmatrix} 1 & 1 & 1 & 1 & | & 4 \\ 0 & -1 & -4 & -3 & | & -8 \\ 0 & -3 & -2 & -3 & | & -6 \\ 0 & -1 & -2 & 0 & | & -8 \end{bmatrix} \begin{matrix} \\ -1R_2 \\ \\ \end{matrix}$$

$$= \begin{bmatrix} 1 & 1 & 1 & 1 & | & 4 \\ 0 & 1 & 4 & 3 & | & 8 \\ 0 & -3 & -2 & -3 & | & -6 \\ 0 & -1 & -2 & 0 & | & -8 \end{bmatrix} \begin{matrix} \\ \\ 3R_2 + R_3 \\ R_2 + R_4 \end{matrix}$$

$$= \begin{bmatrix} 1 & 1 & 1 & 1 & | & 4 \\ 0 & 1 & 4 & 3 & | & 8 \\ 0 & 0 & 10 & 6 & | & 18 \\ 0 & 0 & 2 & 3 & | & 0 \end{bmatrix} \begin{matrix} \\ \\ \frac{1}{2}R_4 \\ R_3 \end{matrix}$$

$$= \begin{bmatrix} 1 & 1 & 1 & 1 & | & 4 \\ 0 & 1 & 4 & 3 & | & 8 \\ 0 & 0 & 1 & \frac{3}{2} & | & 0 \\ 0 & 0 & 10 & 6 & | & 18 \end{bmatrix} \begin{matrix} \\ \\ \\ -10R_3 + R_4 \end{matrix}$$

$$= \begin{bmatrix} 1 & 1 & 1 & 1 & | & 4 \\ 0 & 1 & 4 & 3 & | & 8 \\ 0 & 0 & 1 & \frac{3}{2} & | & 0 \\ 0 & 0 & 0 & -9 & | & 18 \end{bmatrix} \begin{matrix} \\ \\ \\ -\frac{1}{9}R_4 \end{matrix}$$

$$= \begin{bmatrix} 1 & 1 & 1 & 1 & | & 4 \\ 0 & 1 & 4 & 3 & | & 8 \\ 0 & 0 & 1 & \frac{3}{2} & | & 0 \\ 0 & 0 & 0 & 1 & | & -2 \end{bmatrix}$$

The resulting system is

$$w + x + y + z = 4$$
$$x + 4y + 3z = 8$$
$$y + \frac{3}{2}z = 0$$
$$z = -2$$

Back-substitute $z = -2$ to solve for y.

$$y + \frac{3}{2}(-2) = 0$$
$$y - 3 = 0$$
$$y = 3$$

Back-substitute $y = 3$ and $z = -2$ to solve for x.

$$x + 4(3) + 3(-2) = 8$$
$$x + 12 - 6 = 8$$
$$x = 2$$

Back-substitute $x = 2$, $y = 3$, and $z = -2$ to solve for w.

$$w + (2) + (3) + (-2) = 4$$
$$w + 5 - 2 = 4$$
$$w = 1$$

The solution set is $\{(1, 2, 3, -2)\}$.

Check Points Appendix C

1. a. $\begin{vmatrix} 10 & 9 \\ 6 & 5 \end{vmatrix} = 10(5) - 6(9) = 50 - 54 = -4$

b. $\begin{vmatrix} 4 & 3 \\ -5 & -8 \end{vmatrix} = 4(-8) - (-5)(3) = -32 + 15 = -17$

2. $D = \begin{vmatrix} 5 & 4 \\ 3 & -6 \end{vmatrix} = 5(-6) - 3(4) = -30 - 12 = -42$

$D_x = \begin{vmatrix} 12 & 4 \\ 24 & -6 \end{vmatrix} = 12(-6) - 24(4) = -72 - 96 = -168$

$D_y = \begin{vmatrix} 5 & 12 \\ 3 & 24 \end{vmatrix} = 5(24) - 3(12) = 120 - 36 = 84$

$x = \dfrac{D_x}{D} = \dfrac{-168}{-42} = 4$

$y = \dfrac{D_y}{D} = \dfrac{84}{-42} = -2$

The solution set is $\{(4, -2)\}$.

3. $\begin{vmatrix} 2 & 1 & 7 \\ -5 & 6 & 0 \\ -4 & 3 & 1 \end{vmatrix} = 2\begin{vmatrix} 6 & 0 \\ 3 & 1 \end{vmatrix} - (-5)\begin{vmatrix} 1 & 7 \\ 3 & 1 \end{vmatrix} - 4\begin{vmatrix} 1 & 7 \\ 6 & 0 \end{vmatrix}$

$$= 2(6(1) - 3(0)) + 5(1(1) - 3(7)) - 4(1(0) - 6(7))$$
$$= 2(6) + 5(-20) - 4(-42)$$
$$= 12 - 100 + 168$$
$$= 80$$

4. $3x - 2y + z = 16$

$2x + 3y - z = -9$

$x + 4y + 3z = 2$

$$D = \begin{vmatrix} 3 & -2 & 1 \\ 2 & 3 & -1 \\ 1 & 4 & 3 \end{vmatrix} = 3\begin{vmatrix} 3 & -1 \\ 4 & 3 \end{vmatrix} - 2\begin{vmatrix} -2 & 1 \\ 4 & 3 \end{vmatrix} + 1\begin{vmatrix} -2 & 1 \\ 3 & -1 \end{vmatrix} = 58$$

$$D_x = \begin{vmatrix} 16 & -2 & 1 \\ -9 & 3 & -1 \\ 2 & 4 & 3 \end{vmatrix} = 16\begin{vmatrix} 3 & -1 \\ 4 & 3 \end{vmatrix} - (-9)\begin{vmatrix} -2 & 1 \\ 4 & 3 \end{vmatrix} + 2\begin{vmatrix} -2 & 1 \\ 3 & -1 \end{vmatrix} = 116$$

$$D_y = \begin{vmatrix} 3 & 16 & 1 \\ 2 & -9 & -1 \\ 1 & 2 & 3 \end{vmatrix} = 3\begin{vmatrix} -9 & -1 \\ 2 & 3 \end{vmatrix} - 2\begin{vmatrix} 16 & 1 \\ 2 & 3 \end{vmatrix} + 1\begin{vmatrix} 16 & 1 \\ -9 & -1 \end{vmatrix} = -174$$

$$D_z = \begin{vmatrix} 3 & -2 & 16 \\ 2 & 3 & -9 \\ 1 & 4 & 2 \end{vmatrix} = 3\begin{vmatrix} 3 & -9 \\ 4 & 2 \end{vmatrix} - 2\begin{vmatrix} -2 & 16 \\ 4 & 2 \end{vmatrix} + 1\begin{vmatrix} -2 & 16 \\ 3 & -9 \end{vmatrix} = 232$$

$$x = \frac{D_x}{D} = \frac{116}{58} = 2 \qquad y = \frac{D_y}{D} = \frac{-174}{58} = -3 \qquad z = \frac{D_z}{D} = \frac{232}{58} = 4$$

The solution set is $\{(2, -3, 4)\}$.

Exercise Set Appendix C

1. $\begin{vmatrix} 5 & 7 \\ 2 & 3 \end{vmatrix} = 5(3) - 2(7) = 15 - 14 = 1$

3. $\begin{vmatrix} -4 & 1 \\ 5 & 6 \end{vmatrix} = -4(6) - 5(1) = -24 - 5 = -29$

5. $\begin{vmatrix} -7 & 14 \\ 2 & -4 \end{vmatrix} = -7(-4) - 2(14) = 28 - 28 = 0$

7. $\begin{vmatrix} -5 & -1 \\ -2 & -7 \end{vmatrix} = -5(-7) - (-2)(-1) = 35 - 2 = 33$

9. $\begin{vmatrix} \dfrac{1}{2} & \dfrac{1}{2} \\ \dfrac{1}{8} & -\dfrac{3}{4} \end{vmatrix} = \dfrac{1}{2}\left(-\dfrac{3}{4}\right) - \dfrac{1}{8}\left(\dfrac{1}{2}\right) = -\dfrac{3}{8} - \dfrac{1}{16} = -\dfrac{6}{16} - \dfrac{1}{16} = -\dfrac{7}{16}$

11. $D = \begin{vmatrix} 1 & 1 \\ 1 & -1 \end{vmatrix} = 1(-1) - 1(1) = -1 - 1 = -2$

$D_x = \begin{vmatrix} 7 & 1 \\ 3 & -1 \end{vmatrix} = 7(-1) - 3(1) = -7 - 3 = -10$

$D_y = \begin{vmatrix} 1 & 7 \\ 1 & 3 \end{vmatrix} = 1(3) - 1(7) = 3 - 7 = -4$

$x = \dfrac{D_x}{D} = \dfrac{-10}{-2} = 5$; $\quad y = \dfrac{D_y}{D} = \dfrac{-4}{-2} = 2$

The solution set is $\{(5,2)\}$.

13. $D = \begin{vmatrix} 12 & 3 \\ 2 & -3 \end{vmatrix} = 12(-3) - 2(3) = -36 - 6 = -42$

$D_x = \begin{vmatrix} 15 & 3 \\ 13 & -3 \end{vmatrix} = 15(-3) - 13(3) = -45 - 39 = -84$

$D_y = \begin{vmatrix} 12 & 15 \\ 2 & 13 \end{vmatrix} = 12(13) - 2(15) = 156 - 30 = 126$

$x = \dfrac{D_x}{D} = \dfrac{-84}{-42} = 2$; $\quad y = \dfrac{D_y}{D} = \dfrac{126}{-42} = -3$

The solution set is $\{(2,-3)\}$.

15. $D = \begin{vmatrix} 4 & -5 \\ 2 & 3 \end{vmatrix} = 4(3) - 2(-5) = 12 + 10 = 22$

$D_x = \begin{vmatrix} 17 & -5 \\ 3 & 3 \end{vmatrix} = 17(3) - 3(-5) = 51 + 15 = 66$

$D_y = \begin{vmatrix} 4 & 17 \\ 2 & 3 \end{vmatrix} = 4(3) - 2(17) = 12 - 34 = -22$

$x = \dfrac{D_x}{D} = \dfrac{66}{22} = 3$; $\quad y = \dfrac{D_y}{D} = \dfrac{-22}{22} = -1$

The solution set is $\{(3,-1)\}$.

17. $D = \begin{vmatrix} 1 & -3 \\ 3 & -4 \end{vmatrix} = 1(-4) - 3(-3) = -4 + 9 = 5$

$D_x = \begin{vmatrix} 4 & -3 \\ 12 & -4 \end{vmatrix} = 4(-4) - 12(-3) = -16 + 36 = 20$

$D_y = \begin{vmatrix} 1 & 4 \\ 3 & 12 \end{vmatrix} = 1(12) - 3(4) = 12 - 12 = 0$

$x = \dfrac{D_x}{D} = \dfrac{20}{5} = 4$; $\quad y = \dfrac{D_y}{D} = \dfrac{0}{5} = 0$

The solution set is $\{(4,0)\}$.

19. $D = \begin{vmatrix} 3 & -4 \\ 2 & 2 \end{vmatrix} = 3(2) - 2(-4) = 6 + 8 = 14$

$D_x = \begin{vmatrix} 4 & -4 \\ 12 & 2 \end{vmatrix} = 4(2) - 12(-4) = 8 + 48 = 56$

$D_y = \begin{vmatrix} 3 & 4 \\ 2 & 12 \end{vmatrix} = 3(12) - 2(4) = 36 - 8 = 28$

$x = \dfrac{D_x}{D} = \dfrac{56}{14} = 4 ; \quad y = \dfrac{D_y}{D} = \dfrac{28}{14} = 2$

The solution set is $\{(4,2)\}$.

21. First, rewrite the system in standard form.

$2x - 3y = 2$

$5x + 4y = 51$

$D = \begin{vmatrix} 2 & -3 \\ 5 & 4 \end{vmatrix} = 2(4) - 5(-3) = 8 + 15 = 23$

$D_x = \begin{vmatrix} 2 & -3 \\ 51 & 4 \end{vmatrix} = 2(4) - 51(-3) = 8 + 153 = 161$

$D_y = \begin{vmatrix} 2 & 2 \\ 5 & 51 \end{vmatrix} = 2(51) - 5(2) = 102 - 10 = 92$

$x = \dfrac{D_x}{D} = \dfrac{161}{23} = 7 ; \quad y = \dfrac{D_y}{D} = \dfrac{92}{23} = 4$

The solution set is $\{(7,4)\}$.

23. First, rewrite the system in standard form.

$3x + 3y = 2$.

$2x + 2y = 3$

$D = \begin{vmatrix} 3 & 3 \\ 2 & 2 \end{vmatrix} = 3(2) - 2(3) = 6 - 6 = 0$

$D_x = \begin{vmatrix} 2 & 3 \\ 3 & 2 \end{vmatrix} = 2(2) - 3(3) = 4 - 9 = -5$

$D_y = \begin{vmatrix} 3 & 2 \\ 2 & 3 \end{vmatrix} = 3(3) - 2(2) = 9 - 4 = 5$

Because $D = 0$ and at least one of the determinants for the numerators is not zero, the system is inconsistent and the solution set is $\varnothing$. Using matrices, we would get:

$\begin{bmatrix} 3 & 3 & | & 2 \\ 2 & 2 & | & 3 \end{bmatrix} \quad \dfrac{1}{3}R_1$

$= \begin{bmatrix} 1 & 1 & | & 2/3 \\ 2 & 2 & | & 3 \end{bmatrix} \quad -2R_1 + R_2$

$= \begin{bmatrix} 1 & 1 & | & 2/3 \\ 0 & 0 & | & 5/3 \end{bmatrix}$

This is a contradiction. There are no values for x and y for which $0 = 5/3$. The solution set is $\varnothing$ and the system is inconsistent.

25. First, rewrite the system in standard form.

$3x + 4y = 16$

$6x + 8y = 32$

$$D = \begin{vmatrix} 3 & 4 \\ 6 & 8 \end{vmatrix} = 3(8) - 6(4) = 24 - 24 = 0$$

$$D_x = \begin{vmatrix} 16 & 4 \\ 32 & 8 \end{vmatrix} = 16(8) - 32(4) = 128 - 128 = 0$$

$$D_y = \begin{vmatrix} 3 & 16 \\ 6 & 32 \end{vmatrix} = 3(32) - 6(16) = 96 - 96 = 0$$

Since $D = 0$ and all determinants in the numerators are 0, the equations in the system are dependent and there are infinitely many solutions.

27. $\begin{vmatrix} 3 & 0 & 0 \\ 2 & 1 & -5 \\ 2 & 5 & -1 \end{vmatrix} = 3 \begin{vmatrix} 1 & -5 \\ 5 & -1 \end{vmatrix} - 2 \begin{vmatrix} 0 & 0 \\ 5 & -1 \end{vmatrix} + 2 \begin{vmatrix} 0 & 0 \\ 1 & -5 \end{vmatrix}$

$= 3(1(-1) - 5(-5)) - 2(0(-1) - 5(0)) + 2(0(-5) - 1(0)) = 3(24) - 2(0) + 2(0) = 72$

29. $\begin{vmatrix} 3 & 1 & 0 \\ -3 & 4 & 0 \\ -1 & 3 & -5 \end{vmatrix} = 3 \begin{vmatrix} 4 & 0 \\ 3 & -5 \end{vmatrix} - (-3) \begin{vmatrix} 1 & 0 \\ 3 & -5 \end{vmatrix} + (-1) \begin{vmatrix} 1 & 0 \\ 4 & 0 \end{vmatrix}$

$= 3(4(-5) - 3(0)) + 3(1(-5) - 3(0)) - 1(1(0) - 4(0)) = 3(-20) + 3(-5) - 1(0) = -75$

31. $\begin{vmatrix} 1 & 1 & 1 \\ 2 & 2 & 2 \\ -3 & 4 & -5 \end{vmatrix} = 1 \begin{vmatrix} 2 & 2 \\ 4 & -5 \end{vmatrix} - 2 \begin{vmatrix} 1 & 1 \\ 4 & -5 \end{vmatrix} + (-3) \begin{vmatrix} 1 & 1 \\ 2 & 2 \end{vmatrix}$

$= 1(2(-5) - 4(2)) - 2(1(-5) - 4(1)) - 3(1(2) - 2(1)) = 1(-18) - 2(-9) - 3(0) = 0$

33. $x + y + z = 0$
$2x - y + z = -1$
$-x + 3y - z = -8$

$$D = \begin{vmatrix} 1 & 1 & 1 \\ 2 & -1 & 1 \\ -1 & 3 & -1 \end{vmatrix} = 1\begin{vmatrix} -1 & 1 \\ 3 & -1 \end{vmatrix} - 2\begin{vmatrix} 1 & 1 \\ 3 & -1 \end{vmatrix} - 1\begin{vmatrix} 1 & 1 \\ -1 & 1 \end{vmatrix} = 4$$

$$D_x = \begin{vmatrix} 0 & 1 & 1 \\ -1 & -1 & 1 \\ -8 & 3 & -1 \end{vmatrix} = 0\begin{vmatrix} -1 & 1 \\ 3 & -1 \end{vmatrix} - (-1)\begin{vmatrix} 1 & 1 \\ 3 & -1 \end{vmatrix} - 8\begin{vmatrix} 1 & 1 \\ -1 & 1 \end{vmatrix} = -20$$

$$D_y = \begin{vmatrix} 1 & 0 & 1 \\ 2 & -1 & 1 \\ -1 & -8 & -1 \end{vmatrix} = 1\begin{vmatrix} -1 & 1 \\ -8 & -1 \end{vmatrix} - 2\begin{vmatrix} 0 & 1 \\ -8 & -1 \end{vmatrix} - 1\begin{vmatrix} 0 & 1 \\ -1 & 1 \end{vmatrix} = -8$$

$$D_z = \begin{vmatrix} 1 & 1 & 0 \\ 2 & -1 & -1 \\ -1 & 3 & -8 \end{vmatrix} = 1\begin{vmatrix} -1 & -1 \\ 3 & -8 \end{vmatrix} - 2\begin{vmatrix} 1 & 0 \\ 3 & -8 \end{vmatrix} - 1\begin{vmatrix} 1 & 0 \\ -1 & -1 \end{vmatrix} = 28$$

$$x = \frac{D_x}{D} = \frac{-20}{4} = -5 \qquad y = \frac{D_y}{D} = \frac{-8}{4} = -2 \qquad z = \frac{D_z}{D} = \frac{28}{4} = 7$$

The solution set is $\{(-5, -2, 7)\}$.

35. $4x - 5y - 6z = -1$
$x - 2y - 5z = -12$
$2x - y \quad\;\; = 7$

$$D = \begin{vmatrix} 4 & -5 & -6 \\ 1 & -2 & -5 \\ 2 & -1 & 0 \end{vmatrix} = 4\begin{vmatrix} -2 & -5 \\ -1 & 0 \end{vmatrix} - 1\begin{vmatrix} -5 & -6 \\ -1 & 0 \end{vmatrix} + 2\begin{vmatrix} -5 & -6 \\ -2 & -5 \end{vmatrix} = 12$$

$$D_x = \begin{vmatrix} -1 & -5 & -6 \\ -12 & -2 & -5 \\ 7 & -1 & 0 \end{vmatrix} = -1\begin{vmatrix} -2 & -5 \\ -1 & 0 \end{vmatrix} - (-12)\begin{vmatrix} -5 & -6 \\ -1 & 0 \end{vmatrix} + 7\begin{vmatrix} -5 & -6 \\ -2 & -5 \end{vmatrix} = 24$$

$$D_y = \begin{vmatrix} 4 & -1 & -6 \\ 1 & -12 & -5 \\ 2 & 7 & 0 \end{vmatrix} = 4\begin{vmatrix} -12 & -5 \\ 7 & 0 \end{vmatrix} - 1\begin{vmatrix} -1 & -6 \\ 7 & 0 \end{vmatrix} + 2\begin{vmatrix} -1 & -6 \\ -12 & -5 \end{vmatrix} = -36$$

$$D_z = \begin{vmatrix} 4 & -5 & -1 \\ 1 & -2 & -12 \\ 2 & -1 & 7 \end{vmatrix} = 4\begin{vmatrix} -2 & -12 \\ -1 & 7 \end{vmatrix} - 1\begin{vmatrix} -5 & -1 \\ -1 & 7 \end{vmatrix} + 2\begin{vmatrix} -5 & -1 \\ -2 & -12 \end{vmatrix} = 48$$

$$x = \frac{D_x}{D} = \frac{24}{12} = 2 \qquad y = \frac{D_y}{D} = \frac{-36}{12} = -3 \qquad z = \frac{D_z}{D} = \frac{48}{12} = 4$$

The solution set is $\{(2, -3, 4)\}$.

37. $x + y + z = 4$
$x - 2y + z = 7$
$x + 3y + 2z = 4$

$$D = \begin{vmatrix} 1 & 1 & 1 \\ 1 & -2 & 1 \\ 1 & 3 & 2 \end{vmatrix} = 1\begin{vmatrix} -2 & 1 \\ 3 & 2 \end{vmatrix} - 1\begin{vmatrix} 1 & 1 \\ 3 & 2 \end{vmatrix} + 1\begin{vmatrix} 1 & 1 \\ -2 & 1 \end{vmatrix} = -3$$

$$D_x = \begin{vmatrix} 4 & 1 & 1 \\ 7 & -2 & 1 \\ 4 & 3 & 2 \end{vmatrix} = 4\begin{vmatrix} -2 & 1 \\ 3 & 2 \end{vmatrix} - 7\begin{vmatrix} 1 & 1 \\ 3 & 2 \end{vmatrix} + 4\begin{vmatrix} 1 & 1 \\ -2 & 1 \end{vmatrix} = -9$$

$$D_y = \begin{vmatrix} 1 & 4 & 1 \\ 1 & 7 & 1 \\ 1 & 4 & 2 \end{vmatrix} = 1\begin{vmatrix} 7 & 1 \\ 4 & 2 \end{vmatrix} - 1\begin{vmatrix} 4 & 1 \\ 4 & 2 \end{vmatrix} + 1\begin{vmatrix} 4 & 1 \\ 7 & 1 \end{vmatrix} = 3$$

$$D_z = \begin{vmatrix} 1 & 1 & 4 \\ 1 & -2 & 7 \\ 1 & 3 & 4 \end{vmatrix} = 1\begin{vmatrix} -2 & 7 \\ 3 & 4 \end{vmatrix} - 1\begin{vmatrix} 1 & 4 \\ 3 & 4 \end{vmatrix} + 1\begin{vmatrix} 1 & 4 \\ -2 & 7 \end{vmatrix} = -6$$

$$x = \frac{D_x}{D} = \frac{-9}{-3} = 3 \qquad y = \frac{D_y}{D} = \frac{3}{-3} = -1 \qquad z = \frac{D_z}{D} = \frac{-6}{-3} = 2$$

The solution set is $\{(3, -1, 2)\}$.

39. $x \qquad + 2z = 4$
$2y - z = 5$
$2x + 3y \qquad = 13$

$$D = \begin{vmatrix} 1 & 0 & 2 \\ 0 & 2 & -1 \\ 2 & 3 & 0 \end{vmatrix} = 1\begin{vmatrix} 2 & -1 \\ 3 & 0 \end{vmatrix} - 0\begin{vmatrix} 0 & 2 \\ 3 & 0 \end{vmatrix} + 2\begin{vmatrix} 0 & 2 \\ 2 & -1 \end{vmatrix} = -5$$

$$D_x = \begin{vmatrix} 4 & 0 & 2 \\ 5 & 2 & -1 \\ 13 & 3 & 0 \end{vmatrix} = 4\begin{vmatrix} 2 & -1 \\ 3 & 0 \end{vmatrix} - 5\begin{vmatrix} 0 & 2 \\ 3 & 0 \end{vmatrix} + 13\begin{vmatrix} 0 & 2 \\ 2 & -1 \end{vmatrix} = -10$$

$$D_y = \begin{vmatrix} 1 & 4 & 2 \\ 0 & 5 & -1 \\ 2 & 13 & 0 \end{vmatrix} = 1\begin{vmatrix} 5 & -1 \\ 13 & 0 \end{vmatrix} - 0\begin{vmatrix} 4 & 2 \\ 13 & 0 \end{vmatrix} + 2\begin{vmatrix} 4 & 2 \\ 5 & -1 \end{vmatrix} = -15$$

$$D_z = \begin{vmatrix} 1 & 0 & 4 \\ 0 & 2 & 5 \\ 2 & 3 & 13 \end{vmatrix} = 1\begin{vmatrix} 2 & 5 \\ 3 & 13 \end{vmatrix} - 0\begin{vmatrix} 0 & 4 \\ 3 & 13 \end{vmatrix} + 2\begin{vmatrix} 0 & 4 \\ 2 & 5 \end{vmatrix} = -5$$

$$x = \frac{D_x}{D} = \frac{-10}{-5} = 2 \qquad y = \frac{D_y}{D} = \frac{-15}{-5} = 3 \qquad z = \frac{D_z}{D} = \frac{-5}{-5} = 1$$

The solution set is $\{(2, 3, 1)\}$.

41. $$\begin{vmatrix} \begin{vmatrix} 3 & 1 \\ -2 & 3 \end{vmatrix} & \begin{vmatrix} 7 & 0 \\ 1 & 5 \end{vmatrix} \\ \begin{vmatrix} 3 & 0 \\ 0 & 7 \end{vmatrix} & \begin{vmatrix} 9 & -6 \\ 3 & 5 \end{vmatrix} \end{vmatrix} = \begin{vmatrix} 3(3) - (-2)(1) & 7(5) - 1(0) \\ 3(7) - 0(0) & 9(5) - 3(-6) \end{vmatrix} = \begin{vmatrix} 11 & 35 \\ 21 & 63 \end{vmatrix} = -42$$

43. From $D = \begin{vmatrix} 2 & -4 \\ 3 & 5 \end{vmatrix}$ we obtain the coefficients of the variables in our equations:

$2x - 4y = c_1$

$3x + 5y = c_2$

From $D_x = \begin{vmatrix} 8 & -4 \\ -10 & 5 \end{vmatrix}$ we obtain the constant coefficients: 8 and -10

$2x - 4y = 8$

$3x + 5y = -10$

45.
$$\begin{vmatrix} -2 & x \\ 4 & 6 \end{vmatrix} = 32$$

$$-2(6) - 4(x) = 32$$

$$-12 - 4x = 32$$

$$-4x = 44$$

$$x = -11$$

The solution set is $\{-11\}$.

47.
$$\begin{vmatrix} 1 & x & -2 \\ 3 & 1 & 1 \\ 0 & -2 & 2 \end{vmatrix} = -8$$

$$1\begin{vmatrix} 1 & 1 \\ -2 & 2 \end{vmatrix} - 3\begin{vmatrix} x & -2 \\ -2 & 2 \end{vmatrix} + 0\begin{vmatrix} x & -2 \\ 1 & 1 \end{vmatrix} = -8$$

$$1\left[1(2) - (-2)(1)\right] - 3\left[2x - (-2)(-2)\right] = -8$$

$$(2 + 2) - 3(2x - 4) = -8$$

$$-6x + 16 = -8$$

$$-6x = -24$$

$$x = 4$$

The solution set is $\{4\}$.

49. Area $= \pm\dfrac{1}{2}\begin{vmatrix} 3 & -5 & 1 \\ 2 & 6 & 1 \\ -3 & 5 & 1 \end{vmatrix} = \pm\dfrac{1}{2}\left[3\begin{vmatrix} 6 & 1 \\ 5 & 1 \end{vmatrix} - 2\begin{vmatrix} -5 & 1 \\ 5 & 1 \end{vmatrix} - 3\begin{vmatrix} -5 & 1 \\ 6 & 1 \end{vmatrix} \right]$

$$= \pm\frac{1}{2}\left[3(6(1) - 5(1)) - 2(-5(1) - 5(1)) - 3(-5(1) - 6(1)) \right]$$

$$= \pm\frac{1}{2}\left[3(6 - 5) - 2(-5 - 5) - 3(-5 - 6) \right]$$

$$= \pm\frac{1}{2}\left[3(1) - 2(-10) - 3(-11) \right]$$

$$\pm\frac{1}{2}[56]$$

$$= \pm 28$$

The area is 28 square units.

51. $\begin{vmatrix} 3 & -1 & 1 \\ 0 & -3 & 1 \\ 12 & 5 & 1 \end{vmatrix} = 3\begin{vmatrix} -3 & 1 \\ 5 & 1 \end{vmatrix} - 0\begin{vmatrix} -1 & 1 \\ 5 & 1 \end{vmatrix} + 12\begin{vmatrix} -1 & 1 \\ -3 & 1 \end{vmatrix}$

$$= 3\big(-3(1)-5(1)\big)+12\big(-1(1)-(-3)1\big)$$
$$= 3(-3-5)+12(-1+3)$$
$$= -24+24$$
$$= 0$$

Because the determinant is equal to zero, the points are collinear.

53. $\begin{vmatrix} x & y & 1 \\ 3 & -5 & 1 \\ -2 & 6 & 1 \end{vmatrix} = x\begin{vmatrix} -5 & 1 \\ 6 & 1 \end{vmatrix} - 3\begin{vmatrix} y & 1 \\ 6 & 1 \end{vmatrix} - 2\begin{vmatrix} y & 1 \\ -5 & 1 \end{vmatrix}$

$$= x\big(-5(1)-6(1)\big)-3\big(y(1)-6(1)\big)-2\big(y(1)-(-5)1\big)$$
$$= x(-5-6)-3(y-6)-2(y+5)$$
$$= x(-11)-3y+18-2y-10$$
$$= -11x-5y+8$$

To find the equation of the line, set the determinant equal to zero.
$-11x-5y+8=0$

Solve for y to obtain slope-intercept form.
$-11x-5y+8=0$
$$-5y=11x-8$$
$$y=-\frac{11}{5}x+\frac{8}{5}$$